T0111941

Encyclopedia of Algorithms

Ming-Yang Kao
Editor

Encyclopedia of Algorithms

Second Edition

Volume 2

G–P

With 379 Figures and 51 Tables

Editor
Ming-Yang Kao
Department of Electrical Engineering
and Computer Science
Northwestern University
Evanston, IL, USA

ISBN 978-1-4939-2863-7 ISBN 978-1-4939-2864-4 (eBook)
ISBN 978-1-4939-2865-1 (print and electronic bundle)
DOI 10.1007/ 978-1-4939-2864-4

Library of Congress Control Number: 2015958521

Printed on acid-free paper

This Springer imprint is published by SpringerNature
The registered company is Springer Science+Business Media LLC New York

Preface

The Encyclopedia of Algorithms provides researchers, students, and practitioners of algorithmic research with a mechanism to efficiently and accurately find the names, definitions, and key results of important algorithmic problems. It also provides further readings on those problems.

This *encyclopedia* covers a broad range of algorithmic areas; each area is summarized by a collection of entries. The entries are written in a clear and concise structure so that they can be readily absorbed by the readers and easily updated by the authors. A typical encyclopedia entry is an in-depth mini-survey of an algorithmic problem written by an expert in the field. The entries for an algorithmic area are compiled by area editors to survey the representative results in that area and can form the core materials of a course in the area.

This 2nd edition of the encyclopedia contains a wide array of important new research results. Highlights include works in tile self-assembly (nanotechnology), bioinformatics, game theory, Internet algorithms, and social networks. Overall, more than 70 % of the entries in this edition and new entries are updated.

This reference work will continue to be updated on a regular basis via a live site to allow timely updates and fast search. Knowledge accumulation is an ongoing community project. Please take ownership of this body of work. If you have feedback regarding a particular entry, please feel free to communicate directly with the author or the area editor of that entry. If you are interested in authoring a future entry, please contact a suitable area editor. If you have suggestions on how to improve the Encyclopedia as a whole, please contact me at kao@northwestern.edu. The credit of this Encyclopedia goes to the area editors, the entry authors, the entry reviewers, and the project editors at Springer, including Melissa Fearon, Michael Hermann, and Sylvia Blago.

About the Editor

Ming-Yang Kao is a Professor of Computer Science in the Department of Electrical Engineering and Computer Science at Northwestern University. He has published extensively in the design, analysis, and applications of algorithms. His current interests include discrete optimization, bioinformatics, computational economics, computational finance, and nanotechnology. He serves as the Editor-in-Chief of Algorithmica.

He obtained a B.S. in Mathematics from National Taiwan University in 1978 and a Ph.D. in Computer Science from Yale University in 1986. He previously taught at Indiana University at Bloomington, Duke University, Yale University, and Tufts University. At Northwestern University, he has served as the Department Chair of Computer Science. He has also cofounded the Program in Computational Biology and Bioinformatics and served as its Director. He currently serves as the Head of the EECS Division of Computing, Algorithms, and Applications and is a Member of the Theoretical Computer Science Group.

For more information, please see www.cs.northwestern.edu/~kao

Area Editors

Algorithm Engineering

Giuseppe F. Italiano* Department of Computer and Systems Science, University of Rome, Rome, Italy

Department of Information and Computer Systems, University of Rome, Rome, Italy

Rajeev Raman* Department of Computer Science, University of Leicester, Leicester, UK

Algorithms for Modern Computers

Alejandro López-Ortiz David R. Cheriton School of Computer Science, University of Waterloo, Waterloo, ON, Canada

Algorithmic Aspects of Distributed Sensor Networks

Sotiris Nikoletseas Computer Engineering and Informatics Department, University of Patras, Patras, Greece

Computer Technology Institute and Press "Diophantus", Patras, Greece

Approximation Algorithms

Susanne Albers* Technical University of Munich, Munich, Germany

Chandra Chekuri* Department of Computer Science, University of Illinois, Urbana-Champaign, Urbana, IL, USA

Department of Mathematics and Computer Science, The Open University of Israel, Raanana, Israel

Ming-Yang Kao Department of Electrical Engineering and Computer Science, Northwestern University, Evanston, IL, USA

Sanjeev Khanna* University of Pennsylvania, Philadelphia, PA, USA

Samir Khuller* Computer Science Department, University of Maryland, College Park, MD, USA

---

*Acknowledgment for first edition contribution

Average Case Analysis

Paul (Pavlos) Spirakis* Computer Engineering and Informatics, Research and Academic Computer Technology Institute, Patras University, Patras, Greece

Computer Science, University of Liverpool, Liverpool, UK

Computer Technology Institute (CTI), Patras, Greece

Bin Packing

Leah Epstein Department of Mathematics, University of Haifa, Haifa, Israel

Bioinformatics

Miklós Csürös Department of Computer Science, University of Montréal, Montréal, QC, Canada

Certified Reconstruction and Mesh Generation

Siu-Wing Cheng Department of Computer Science and Engineering, Hong Kong University of Science and Technology, Hong Kong, China

Tamal Krishna Dey Department of Computer Science and Engineering, The Ohio State University, Columbus, OH, USA

Coding Algorithms

Venkatesan Guruswami* Department of Computer Science and Engineering, University of Washington, Seattle, WA, USA

Combinatorial Group Testing

Ding-Zhu Du Computer Science, University of Minnesota, Minneapolis, MN, USA

Department of Computer Science, The University of Texas at Dallas, Richardson, TX, USA

Combinatorial Optimization

Samir Khuller* Computer Science Department, University of Maryland, College Park, MD, USA

Compressed Text Indexing

Tak-Wah Lam Department of Computer Science, University of Hong Kong, Hong Kong, China

Compression of Text and Data Structures

Gonzalo Navarro Department of Computer Science, University of Chile, Santiago, Chile

Computational Biology

Bhaskar DasGupta Department of Computer Science, University of Illinois, Chicago, IL, USA

Tak-Wah Lam Department of Computer Science, University of Hong Kong, Hong Kong, China

Computational Counting

Xi Chen Computer Science Department, Columbia University, New York, NY, USA

Computer Science and Technology, Tsinghua University, Beijing, China

Computational Economics

Xiaotie Deng AIMS Laboratory (Algorithms-Agents-Data on Internet, Market, and Social Networks), Department of Computer Science and Engineering, Shanghai Jiao Tong University, Shanghai, China

Department of Computer Science, City University of Hong Kong, Hong Kong, China

Computational Geometry

Sándor Fekete Department of Computer Science, Technical University Braunschweig, Braunschweig, Germany

Computational Learning Theory

Rocco A. Servedio Computer Science, Columbia University, New York, NY, USA

Data Compression

Paolo Ferragina* Department of Computer Science, University of Pisa, Pisa, Italy

Differential Privacy

Aaron Roth Department of Computer and Information Sciences, University of Pennsylvania, Levine Hall, PA, USA

Distributed Algorithms

Sergio Rajsbaum Instituto de Matemáticas, Universidad Nacional Autónoma de México (UNAM) México City, México

Dynamic Graph Algorithms

Giuseppe F. Italiano* Department of Computer and Systems Science, University of Rome, Rome, Italy

Department of Information and Computer Systems, University of Rome, Rome, Italy

Enumeration Algorithms

Takeaki Uno National Institute of Informatics, Chiyoda, Tokyo, Japan

Exact Exponential Algorithms

Fedor V. Fomin Department of Informatics, University of Bergen, Bergen, Norway

External Memory Algorithms

Herman Haverkort Department of Computer Science, Eindhoven University of Technology, Eindhoven, The Netherlands

Game Theory

Mohammad Taghi Hajiaghayi Department of Computer Science, University of Maryland, College Park, MD, USA

Geometric Networks

Andrzej Lingas Department of Computer Science, Lund University, Lund, Sweden

Graph Algorithms

Samir Khuller* Computer Science Department, University of Maryland, College Park, MD, USA

Seth Pettie Electrical Engineering and Computer Science (EECS) Department, University of Michigan, Ann Arbor, MI, USA

Vijaya Ramachandran* Computer Science, University of Texas, Austin, TX, USA

Liam Roditty Department of Computer Science, Bar-Ilan University, Ramat-Gan, Israel

Dimitrios Thilikos AlGCo Project-Team, CNRS, LIRMM, France

Department of Mathematics, National and Kapodistrian University of Athens, Athens, Greece

Graph Drawing

Seokhee Hong School of Information Technologies, University of Sydney, Sydney, NSW, Australia

Internet Algorithms

Edith Cohen Tel Aviv University, Tel Aviv, Israel

Stanford University, Stanford, CA, USA

I/O-Efficient Algorithms

Herman Haverkort Department of Computer Science, Eindhoven University of Technology, Eindhoven, The Netherlands

Kernels and Compressions

Gregory Gutin Department of Computer Science, Royal Holloway, University of London, Egham, UK

Massive Data Algorithms

Herman Haverkort Department of Computer Science, Eindhoven University of Technology, Eindhoven, The Netherlands

Mathematical Optimization

Ding-Zhu Du Computer Science, University of Minnesota, Minneapolis, MN, USA

Department of Computer Science, The University of Texas at Dallas, Richardson, TX, USA

Mechanism Design

Yossi Azar* Tel-Aviv University, Tel Aviv, Israel

Mobile Computing

Xiang-Yang Li* Department of Computer Science, Illinois Institute of Technology, Chicago, IL, USA

Modern Learning Theory

Maria-Florina Balcan Department of Machine Learning, Carnegie Mellon University, Pittsburgh, PA, USA

Online Algorithms

Susanne Albers* Technical University of Munich, Munich, Germany

Yossi Azar* Tel-Aviv University, Tel Aviv, Israel

Marek Chrobak Computer Science, University of California, Riverside, CA, USA

Alejandro López-Ortiz David R. Cheriton School of Computer Science, University of Waterloo, Waterloo, ON, Canada

Parameterized Algorithms

Dimitrios Thilikos AlGCo Project-Team, CNRS, LIRMM, France

Department of Mathematics, National and Kapodistrian University of Athens, Athens, Greece

Parameterized Algorithms and Complexity

Saket Saurabh Institute of Mathematical Sciences, Chennai, India

University of Bergen, Bergen, Norway

Parameterized and Exact Algorithms

Rolf Niedermeier* Department of Mathematics and Computer Science, University of Jena, Jena, Germany

Institut für Softwaretechnik und Theoretische Informatik, Technische Universität Berlin, Berlin, Germany

Price of Anarchy

Yossi Azar* Tel-Aviv University, Tel Aviv, Israel

Probabilistic Algorithms

Sotiris Nikoletseas Computer Engineering and Informatics Department, University of Patras, Patras, Greece

Computer Technology Institute and Press “Diophantus”, Patras, Greece

Paul (Pavlos) Spirakis* Computer Engineering and Informatics, Research and Academic Computer Technology Institute, Patras University, Patras, Greece

Computer Science, University of Liverpool, Liverpool, UK

Computer Technology Institute (CTI), Patras, Greece

Quantum Computing

Andris Ambainis Faculty of Computing, University of Latvia, Riga, Latvia

Radio Networks

Marek Chrobak Computer Science, University of California, Riverside, CA, USA

Scheduling

Leah Epstein Department of Mathematics, University of Haifa, Haifa, Israel

Scheduling Algorithms

Viswanath Nagarajan University of Michigan, Ann Arbor, MI, USA

Kirk Pruhs* Department of Computer Science, University of Pittsburgh, Pittsburgh, PA, USA

Social Networks

Mohammad Taghi Hajiaghayi Department of Computer Science, University of Maryland, College Park, MD, USA

Grant Schoenebeck Computer Science and Engineering, University of Michigan, Ann Arbor, MI, USA

Stable Marriage Problems, k-SAT Algorithms

Kazuo Iwama Computer Engineering, Kyoto University, Sakyo, Kyoto, Japan

School of Informatics, Kyoto University, Sakyo, Kyoto, Japan

String Algorithms and Data Structures

Paolo Ferragina* Department of Computer Science, University of Pisa, Pisa, Italy

Gonzalo Navarro Department of Computer Science, University of Chile, Santiago, Chile

Steiner Tree Algorithms

Ding-Zhu Du Computer Science, University of Minnesota, Minneapolis, MN, USA

Department of Computer Science, The University of Texas at Dallas, Richardson, TX, USA

Sublinear Algorithms

Andrew McGregor School of Computer Science, University of Massachusetts, Amherst, MA, USA

Sofya Raskhodnikova Computer Science and Engineering Department, Pennsylvania State University, University Park, State College, PA, USA

Tile Self-Assembly

Robert Schweller Department of Computer Science, University of Texas Rio Grande Valley, Edinburg, TX, USA

VLSI CAD Algorithms

Hai Zhou Electrical Engineering and Computer Science (EECS) Department, Northwestern University, Evanston, IL, USA

Contributors

Karen Aardal Centrum Wiskunde & Informatica (CWI), Amsterdam, The Netherlands

Department of Mathematics and Computer Science, Eindhoven University of Technology, Eindhoven, The Netherlands

Ittai Abraham Microsoft Research, Silicon Valley, Palo Alto, CA, USA

Adi Akavia Department of Electrical Engineering and Computer Science, MIT, Cambridge, MA, USA

Réka Albert Department of Biology and Department of Physics, Pennsylvania State University, University Park, PA, USA

Mansoor Alicherry Bell Laboratories, Alcatel-Lucent, Murray Hill, NJ, USA

Noga Alon Department of Mathematics and Computer Science, Tel-Aviv University, Tel-Aviv, Israel

Srinivas Aluru Department of Electrical and Computer Engineering, Iowa State University, Ames, IA, USA

Andris Ambainis Faculty of Computing, University of Latvia, Riga, Latvia

Christoph Ambühl Department of Computer Science, University of Liverpool, Liverpool, UK

Nina Amenta Department of Computer Science, University of California, Davis, CA, USA

Amihood Amir Department of Computer Science, Bar-Ilan University, Ramat-Gan, Israel

Department of Computer Science, Johns Hopkins University, Baltimore, MD, USA

Spyros Angelopoulos Sorbonne Universités, L'Université Pierre et Marie Curie (UPMC), Université Paris 06, Paris, France

Anurag Anshu Center for Quantum Technologies, National University of Singapore, Singapore, Singapore

Alberto Apostolico College of Computing, Georgia Institute of Technology, Atlanta, GA, USA

Vera Asodi Center for the Mathematics of Information, California Institute of Technology, Pasadena, CA, USA

Peter Auer Chair for Information Technology, Montanuniversitaet Leoben, Leoben, Austria

Pranjal Awasthi Department of Computer Science, Princeton University, Princeton, NJ, USA

Department of Electrical Engineering, Indian Institute of Technology Madras, Chennai, Tamilnadu, India

Adnan Aziz Department of Electrical and Computer Engineering, University of Texas, Austin, TX, USA

Moshe Babaioff Microsoft Research, Herzliya, Israel

David A. Bader College of Computing, Georgia Institute of Technology, Atlanta, GA, USA

Michael Bader Department of Informatics, Technical University of Munich, Garching, Germany

Maria-Florina Balcan Department of Machine Learning, Carnegie Mellon University, Pittsburgh, PA, USA

Hideo Bannai Department of Informatics, Kyushu University, Fukuoka, Japan

Nikhil Bansal Eindhoven University of Technology, Eindhoven, The Netherlands

Jérémy Barbay Department of Computer Science (DCC), University of Chile, Santiago, Chile

Sanjoy K. Baruah Department of Computer Science, The University of North Carolina, Chapel Hill, NC, USA

Surender Baswana Department of Computer Science and Engineering, Indian Institute of Technology (IIT), Kanpur, Kanpur, India

MohammadHossein Bateni Google Inc., New York, NY, USA

Luca Becchetti Department of Information and Computer Systems, University of Rome, Rome, Italy

Xiaohui Bei Division of Mathematical Sciences, School of Physical and Mathematical Sciences, Nanyang Technological University, Singapore, Singapore

József Békési Department of Computer Science, Juhász Gyula Teachers Training College, Szeged, Hungary

Djamal Belazzougui Department of Computer Science, Helsinki Institute for Information Technology (HIIT), University of Helsinki, Helsinki, Finland

Aleksandrs Belovs Computer Science and Artificial Intelligence Laboratory, MIT, Cambridge, MA, USA

Aaron Bernstein Department of Computer Science, Columbia University, New York, NY, USA

Vincent Berry Institut de Biologie Computationnelle, Montpellier, France

Randeep Bhatia Bell Laboratories, Alcatel-Lucent, Murray Hill, NJ, USA

Andreas Björklund Department of Computer Science, Lund University, Lund, Sweden

Eric Blais University of Waterloo, Waterloo, ON, Canada

Mathieu Blanchette Department of Computer Science, McGill University, Montreal, QC, Canada

Markus Bläser Department of Computer Science, Saarland University, Saarbrücken, Germany

Avrim Blum School of Computer Science, Carnegie Mellon University, Pittsburgh, PA, USA

Hans L. Bodlaender Department of Computer Science, Utrecht University, Utrecht, The Netherlands

Sergio Boixo Quantum A.I. Laboratory, Google, Venice, CA, USA

Paolo Boldi Dipartimento di Informatica, Università degli Studi di Milano, Milano, Italy

Glencora Borradaile Department of Computer Science, Brown University, Providence, RI, USA

School of Electrical Engineering and Computer Science, Oregon State University, Corvallis, OR, USA

Ulrik Brandes Department of Computer and Information Science, University of Konstanz, Konstanz, Germany

Andreas Brandstädt Computer Science Department, University of Rostock, Rostock, Germany

Department of Informatics, University of Rostock, Rostock, Germany

Gilles Brassard Université de Montréal, Montréal, QC, Canada

Vladimir Braverman Department of Computer Science, Johns Hopkins University, Baltimore, MD, USA

Tian-Ming Bu Software Engineering Institute, East China Normal University, Shanghai, China

Adam L. Buchsbaum Madison, NJ, USA

Costas Busch Department of Computer Science, Lousiana State University, Baton Rouge, LA, USA
Jaroslaw Byrka Centrum Wiskunde & Informatica (CWI), Amsterdam, The Netherlands

Department of Mathematics and Computer Science, Eindhoven University of Technology, Eindhoven, The Netherlands

Jin-Yi Cai Beijing University, Beijing, China

Computer Sciences Department, University of Wisconsin–Madison, Madison, WI, USA

Mao-cheng Cai Chinese Academy of Sciences, Institute of Systems Science, Beijing, China

Yang Cai Computer Science, McGill University, Montreal, QC, Canada

Gruia Calinescu Department of Computer Science, Illinois Institute of Technology, Chicago, IL, USA

Colin Campbell Department of Physics, Pennsylvania State University, University Park, PA, USA

Luca Castelli Aleardi Laboratoire d'Informatique (LIX), École Polytechnique, Bâtiment Alan Turing, Palaiseau, France

Katarína Cechlárová Faculty of Science, Institute of Mathematics, P. J. Šafárik University, Košice, Slovakia

Nicolò Cesa-Bianchi Dipartimento di Informatica, Università degli Studi di Milano, Milano, Italy

Amit Chakrabarti Department of Computer Science, Dartmouth College, Hanover, NH, USA

Deeparnab Chakrabarty Microsoft Research, Bangalore, Karnataka, India

Erin W. Chambers Department of Computer Science and Mathematics, Saint Louis University, St. Louis, MO, USA

Chee Yong Chan National University of Singapore, Singapore, Singapore

Mee Yee Chan Department of Computer Science, University of Hong Kong, Hong Kong, China

Wun-Tat Chan College of International Education, Hong Kong Baptist University, Hong Kong, China

Tushar Deepak Chandra IBM Watson Research Center, Yorktown Heights, NY, USA

Kun-Mao Chao Department of Computer Science and Information Engineering, National Taiwan University, Taipei, Taiwan

Bernadette Charron-Bost Laboratory for Informatics, The Polytechnic School, Palaiseau, France

Ioannis Chatzigiannakis Department of Computer Engineering and Informatics, University of Patras and Computer Technology Institute, Patras, Greece

Shuchi Chawla Department of Computer Science, University of Wisconsin–Madison, Madison, WI, USA

Shiri Chechik Department of Computer Science, Tel Aviv University, Tel Aviv, Israel

Chandra Chekuri Department of Computer Science, University of Illinois, Urbana-Champaign, Urbana, IL, USA

Department of Mathematics and Computer Science, The Open University of Israel, Raanana, Israel

Danny Z. Chen Department of Computer Science and Engineering, University of Notre Dame, Notre Dame, IN, USA

Ho-Lin Chen Department of Electrical Engineering, National Taiwan University, Taipei, Taiwan

Jianer Chen Department of Computer Science, Texas A&M University, College Station, TX, USA

Ning Chen Division of Mathematical Sciences, School of Physical and Mathematical Sciences, Nanyang Technological University, Singapore, Singapore

Xi Chen Computer Science Department, Columbia University, New York, NY, USA

Computer Science and Technology, Tsinghua University, Beijing, China

Siu-Wing Cheng Department of Computer Science and Engineering, Hong Kong University of Science and Technology, Hong Kong, China

Xiuzhen Cheng Department of Computer Science, George Washington University, Washington, DC, USA

Huang Chien-Chung Chalmers University of Technology and University of Gothenburg, Gothenburg, Sweden

Markus Chimani Faculty of Mathematics/Computer, Theoretical Computer Science, Osnabrück University, Osnabrück, Germany

Francis Y.L. Chin Department of Computer Science, University of Hong Kong, Hong Kong, China

Rajesh Chitnis Department of Computer Science, University of Maryland, College Park, MD, USA

Minsik Cho IBM T. J. Watson Research Center, Yorktown Heights, NY, USA

Rezaul A. Chowdhury Department of Computer Sciences, University of Texas, Austin, TX, USA

Stony Brook University (SUNY), Stony Brook, NY, USA

George Christodoulou University of Liverpool, Liverpool, UK

Marek Chrobak Computer Science, University of California, Riverside, CA, USA

Chris Chu Department of Electrical and Computer Engineering, Iowa State University, Ames, IA, USA

Xiaowen Chu Department of Computer Science, Hong Kong Baptist University, Hong Kong, China

Julia Chuzhoy Toyota Technological Institute, Chicago, IL, USA

Edith Cohen Tel Aviv University, Tel Aviv, Israel

Stanford University, Stanford, CA, USA

Jason Cong Department of Computer Science, UCLA, Los Angeles, CA, USA

Graham Cormode Department of Computer Science, University of Warwick, Coventry, UK

Derek G. Corneil Department of Computer Science, University of Toronto, Toronto, ON, Canada

Bruno Courcelle Laboratoire Bordelais de Recherche en Informatique (LaBRI), CNRS, Bordeaux University, Talence, France

Lenore J. Cowen Department of Computer Science, Tufts University, Medford, MA, USA

Nello Cristianini Department of Engineering Mathematics, and Computer Science, University of Bristol, Bristol, UK

Maxime Crochemore Department of Computer Science, King's College London, London, UK

Laboratory of Computer Science, University of Paris-East, Paris, France

Université de Marne-la-Vallée, Champs-sur-Marne, France

Miklós Csürös Department of Computer Science, University of Montréal, Montréal, QC, Canada

Fabio Cunial Department of Computer Science, Helsinki Institute for Information Technology (HIIT), University of Helsinki, Helsinki, Finland

Marek Cygan Institute of Informatics, University of Warsaw, Warsaw, Poland

Artur Czumaj Department of Computer Science, Centre for Discrete Mathematics and Its Applications, University of Warwick, Coventry, UK

Bhaskar DasGupta Department of Computer Science, University of Illinois, Chicago, IL, USA

Constantinos Daskalakis EECS, Massachusetts Institute of Technology, Cambridge, MA, USA

Mark de Berg Department of Mathematics and Computer Science, TU Eindhoven, Eindhoven, The Netherlands

Xavier Défago School of Information Science, Japan Advanced Institute of Science and Technology (JAIST), Ishikawa, Japan

Daniel Delling Microsoft, Silicon Valley, CA, USA

Erik D. Demaine MIT Computer Science and Artificial Intelligence Laboratory, Cambridge, MA, USA

Camil Demetrescu Department of Computer and Systems Science, University of Rome, Rome, Italy

Department of Information and Computer Systems, University of Rome, Rome, Italy

Ping Deng Department of Computer Science, The University of Texas at Dallas, Richardson, TX, USA

Xiaotie Deng AIMS Laboratory (Algorithms-Agents-Data on Internet, Market, and Social Networks), Department of Computer Science and Engineering, Shanghai Jiao Tong University, Shanghai, China

Department of Computer Science, City University of Hong Kong, Hong Kong, China

Vamsi Krishna Devabathini Center for Quantum Technologies, National University of Singapore, Singapore, Singapore

Olivier Devillers Inria Nancy – Grand-Est, Villers-lès-Nancy, France

Tamal Krishna Dey Department of Computer Science and Engineering, The Ohio State University, Columbus, OH, USA

Robert P. Dick Department of Electrical Engineering and Computer Science, University of Michigan, Ann Arbor, MI, USA

Walter Didimo Department of Engineering, University of Perugia, Perugia, Italy

Ling Ding Institute of Technology, University of Washington Tacoma, Tacoma, WA, USA

Yuzheng Ding Xilinx Inc., Longmont, CO, USA

Michael Dom Department of Mathematics and Computer Science, University of Jena, Jena, Germany

Riccardo Dondi Università degli Studi di Bergamo, Bergamo, Italy

Gyorgy Dosa University of Pannonia, Veszprém, Hungary

David Doty Computing and Mathematical Sciences, California Institute of Technology, Pasadena, CA, USA

Ding-Zhu Du Computer Science, University of Minnesota, Minneapolis, MN, USA

Department of Computer Science, The University of Texas at Dallas, Richardson, TX, USA

Hongwei Du Department of Computer Science and Technology, Shenzhen Graduate School, Harbin Institute of Technology, Shenzhen, China

Ran Duan Institute for Interdisciplinary Information Sciences, Tsinghua University, Beijing, China

Devdatt Dubhashi Department of Computer Science, Chalmers University of Technology, Gothenburg, Sweden

Gothenburg University, Gothenburg, Sweden

Adrian Dumitrescu Computer Science, University of Wisconsin–Milwaukee, Milwaukee, WI, USA

Iréne Durand Laboratoire Bordelais de Recherche en Informatique (LaBRI), CNRS, Bordeaux University, Talence, France

Stephane Durocher University of Manitoba, Winnipeg, MB, Canada

Pavlos Efraimidis Department of Electrical and Computer Engineering, Democritus University of Thrace, Xanthi, Greece

Charilaos Efthymiou Department of Computer Engineering and Informatics, University of Patras, Patras, Greece

Michael Elkin Department of Computer Science, Ben-Gurion University, Beer-Sheva, Israel

Matthias Englert Department of Computer Science, University of Warwick, Coventry, UK

David Eppstein Donald Bren School of Information and Computer Sciences, Computer Science Department, University of California, Irvine, CA, USA

Leah Epstein Department of Mathematics, University of Haifa, Haifa, Israel

Jeff Erickson Department of Computer Science, University of Illinois, Urbana, IL, USA

Constantine G. Evans Division of Biology and Bioengineering, California Institute of Technology, Pasadena, CA, USA

Eyal Even-Dar Google, New York, NY, USA

Rolf Fagerberg Department of Mathematics and Computer Science, University of Southern Denmark, Odense, Denmark

Jittat Fakcharoenphol Department of Computer Engineering, Kasetsart University, Bangkok, Thailand

Piotr Faliszewski AGH University of Science and Technology, Krakow, Poland

Lidan Fan Department of Computer Science, The University of Texas, Tyler, TX, USA

Qizhi Fang School of Mathematical Sciences, Ocean University of China, Qingdao, Shandong Province, China

Martín Farach-Colton Department of Computer Science, Rutgers University, Piscataway, NJ, USA

Panagiota Fatourou Department of Computer Science, University of Ioannina, Ioannina, Greece

Jonathan Feldman Google, Inc., New York, NY, USA

Vitaly Feldman IBM Research – Almaden, San Jose, CA, USA

Henning Fernau Fachbereich 4, Abteilung Informatikwissenschaften, Universität Trier, Trier, Germany

Institute for Computer Science, University of Trier, Trier, Germany

Paolo Ferragina Department of Computer Science, University of Pisa, Pisa, Italy

Johannes Fischer Technical University Dortmund, Dortmund, Germany

Nathan Fisher Department of Computer Science, Wayne State University, Detroit, MI, USA

Abraham Flaxman Theory Group, Microsoft Research, Redmond, WA, USA

Paola Flocchini School of Electrical Engineering and Computer Science, University of Ottawa, Ottawa, ON, Canada

Fedor V. Fomin Department of Informatics, University of Bergen, Bergen, Norway

Dimitris Fotakis Department of Information and Communication Systems Engineering, University of the Aegean, Samos, Greece

Kyle Fox Institute for Computational and Experimental Research in Mathematics, Brown University, Providence, RI, USA

Pierre Fraigniaud Laboratoire d'Informatique Algorithmique: Fondements et Applications, CNRS and University Paris Diderot, Paris, France

Fabrizio Frati School of Information Technologies, The University of Sydney, Sydney, NSW, Australia

Engineering Department, Roma Tre University, Rome, Italy

Ophir Frieder Department of Computer Science, Illinois Institute of Technology, Chicago, IL, USA

Hiroshi Fujiwara Shinshu University, Nagano, Japan

Stanley P.Y. Fung Department of Computer Science, University of Leicester, Leicester, UK

Stefan Funke Department of Computer Science, Universität Stuttgart, Stuttgart, Germany

Martin Fürer Department of Computer Science and Engineering, The Pennsylvania State University, University Park, PA, USA

Travis Gagie Department of Computer Science, University of Eastern Piedmont, Alessandria, Italy

Department of Computer Science, University of Helsinki, Helsinki, Finland

Gábor Galambos Department of Computer Science, Juhász Gyula Teachers Training College, Szeged, Hungary

Jianjiong Gao Computational Biology Center, Memorial Sloan-Kettering Cancer Center, New York, NY, USA

Jie Gao Department of Computer Science, Stony Brook University, Stony Brook, NY, USA

Xiaofeng Gao Department of Computer Science, Shanghai Jiao Tong University, Shanghai, China

Juan Garay Bell Laboratories, Murray Hill, NJ, USA

Minos Garofalakis Technical University of Crete, Chania, Greece

Olivier Gascuel Institut de Biologie Computationnelle, Laboratoire d'Informatique, de Robotique et de Microélectronique de Montpellier (LIRMM), CNRS and Université de Montpellier, Montpellier cedex 5, France

Leszek Gąsieniec University of Liverpool, Liverpool, UK

Serge Gaspers Optimisation Research Group, National ICT Australia (NICTA), Sydney, NSW, Australia

School of Computer Science and Engineering, University of New SouthWales (UNSW), Sydney, NSW, Australia

Maciej Gazda Department of Mathematics and Computer Science, Eindhoven University of Technology, Eindhoven, The Netherlands

Raffaele Giancarlo Department of Mathematics and Applications, University of Palermo, Palermo, Italy

Gagan Goel Google Inc., New York, NY, USA

Andrew V. Goldberg Microsoft Research – Silicon Valley, Mountain View, CA, USA

Oded Goldreich Department of Computer Science, Weizmann Institute of Science, Rehovot, Israel

Jens Gramm WSI Institute of Theoretical Computer Science, Tübingen University, Tübingen, Germany

Fabrizio Grandoni IDSIA, USI-SUPSI, University of Lugano, Lugano, Switzerland

Roberto Grossi Dipartimento di Informatica, Università di Pisa, Pisa, Italy

Lov K. Grover Bell Laboratories, Alcatel-Lucent, Murray Hill, NJ, USA

Xianfeng David Gu Department of Computer Science, Stony Brook University, Stony Brook, NY, USA

Joachim Gudmundsson DMiST, National ICT Australia Ltd, Alexandria, Australia

School of Information Technologies, University of Sydney, Sydney, NSW, Australia

Rachid Guerraoui School of Computer and Communication Sciences, EPFL, Lausanne, Switzerland

Heng Guo Computer Sciences Department, University of Wisconsin–Madison, Madison, WI, USA

Jiong Guo Department of Mathematics and Computer Science, University of Jena, Jena, Germany

Manoj Gupta Indian Institute of Technology (IIT) Delhi, Hauz Khas, New Delhi, India

Venkatesan Guruswami Department of Computer Science and Engineering, University of Washington, Seattle, WA, USA

Gregory Gutin Department of Computer Science, Royal Holloway, University of London, Egham, UK

Michel Habib LIAFA, Université Paris Diderot, Paris Cedex 13, France

Mohammad Taghi Hajiaghayi Department of Computer Science, University of Maryland, College Park, MD, USA

Sean Hallgren Department of Computer Science and Engineering, The Pennsylvania State University, University Park, State College, PA, USA

Dan Halperin School of Computer Science, Tel-Aviv University, Tel Aviv, Israel

Moritz Hardt IBM Research – Almaden, San Jose, CA, USA

Ramesh Hariharan Strand Life Sciences, Bangalore, India

Aram W. Harrow Department of Physics, Massachusetts Institute of Technology, Cambridge, MA, USA

Prahladh Harsha Tata Institute of Fundamental Research, Mumbai, Maharashtra, India

Herman Haverkort Department of Computer Science, Eindhoven University of Technology, Eindhoven, The Netherlands

Meng He School of Computer Science, University of Waterloo, Waterloo, ON, Canada

Xin He Department of Computer Science and Engineering, The State University of New York, Buffalo, NY, USA

Lisa Hellerstein Department of Computer Science and Engineering, NYU Polytechnic School of Engineering, Brooklyn, NY, USA

Michael Hemmer Department of Computer Science, TU Braunschweig, Braunschweig, Germany

Danny Hendler Department of Computer Science, Ben-Gurion University of the Negev, Beer-Sheva, Israel

Monika Henzinger University of Vienna, Vienna, Austria

Maurice Herlihy Department of Computer Science, Brown University, Providence, RI, USA

Ted Herman Department of Computer Science, University of Iowa, Iowa City, IA, USA

John Hershberger Mentor Graphics Corporation, Wilsonville, OR, USA

Timon Hertli Department of Computer Science, ETH Zürich, Zürich, Switzerland

Edward A. Hirsch Laboratory of Mathematical Logic, Steklov Institute of Mathematics, St. Petersburg, Russia

Wing-Kai Hon Department of Computer Science, National Tsing Hua University, Hsin Chu, Taiwan

Seokhee Hong School of Information Technologies, University of Sydney, Sydney, NSW, Australia

Paul G. Howard Akamai Technologies, Cambridge, MA, USA

Peter Høyer University of Calgary, Calgary, AB, Canada

Li-Sha Huang Department of Computer Science and Technology, Tsinghua University, Beijing, China

Yaocun Huang Department of Computer Science, The University of Texas at Dallas, Richardson, TX, USA

Zhiyi Huang Department of Computer Science, The University of Hong Kong, Hong Kong, Hong Kong

Falk Hüffner Department of Math and Computer Science, University of Jena, Jena, Germany

Thore Husfeldt Department of Computer Science, Lund University, Lund, Sweden

Lucian Ilie Department of Computer Science, University of Western Ontario, London, ON, Canada

Sungjin Im Electrical Engineering and Computer Sciences (EECS), University of California, Merced, CA, USA

Csanad Imreh Institute of Informatics, University of Szeged, Szeged, Hungary

Robert W. Irving School of Computing Science, University of Glasgow, Glasgow, UK

Alon Itai Technion, Haifa, Israel

Giuseppe F. Italiano Department of Computer and Systems Science, University of Rome, Rome, Italy

Department of Information and Computer Systems, University of Rome, Rome, Italy

Kazuo Iwama Computer Engineering, Kyoto University, Sakyo, Kyoto, Japan

School of Informatics, Kyoto University, Sakyo, Kyoto, Japan

Jeffrey C. Jackson Department of Mathematics and Computer Science, Duquesne University, Pittsburgh, PA, USA

Ronald Jackups Department of Pediatrics, Washington University, St. Louis, MO, USA

Riko Jacob Institute of Computer Science, Technical University of Munich, Munich, Germany

IT University of Copenhagen, Copenhagen, Denmark

Rahul Jain Department of Computer Science, Center for Quantum Technologies, National University of Singapore, Singapore, Singapore

Klaus Jansen Department of Computer Science, University of Kiel, Kiel, Germany

Jesper Jansson Laboratory of Mathematical Bioinformatics, Institute for Chemical Research, Kyoto University, Gokasho, Uji, Kyoto, Japan

Stacey Jeffery David R. Cheriton School of Computer Science, University of Waterloo, Waterloo, ON, Canada

Madhav Jha Sandia National Laboratories, Livermore, CA, USA

Zenefits, San Francisco, CA, USA

David S. Johnson Department of Computer Science, Columbia University, New York, NY, USA

AT&T Laboratories, Algorithms and Optimization Research Department, Florham Park, NJ, USA

Mark Jones Department of Computer Science, Royal Holloway, University of London, Egham, UK

Tomasz Jurdziński Institute of Computer Science, University of Wrocław, Wrocław, Poland

Yoji Kajitani Department of Information and Media Sciences, The University of Kitakyushu, Kitakyushu, Japan

Shahin Kamali David R. Cheriton School of Computer Science, University of Waterloo, Waterloo, ON, Canada

Andrew Kane David R. Cheriton School of Computer Science, University of Waterloo, Waterloo, ON, Canada

Mamadou Moustapha Kanté Clermont-Université, Université Blaise Pascal, LIMOS, CNRS, Aubière, France

Ming-Yang Kao Department of Electrical Engineering and Computer Science, Northwestern University, Evanston, IL, USA

Alexis Kaporis Department of Information and Communication Systems Engineering, University of the Aegean, Karlovasi, Samos, Greece

George Karakostas Department of Computing and Software, McMaster University, Hamilton, ON, Canada

Juha Kärkkäinen Department of Computer Science, University of Helsinki, Helsinki, Finland

Petteri Kaski Department of Computer Science, School of Science, Aalto University, Helsinki, Finland

Helsinki Institute for Information Technology (HIIT), Helsinki, Finland

Hans Kellerer Department of Statistics and Operations Research, University of Graz, Graz, Austria

Andrew A. Kennings Department of Electrical and Computer Engineering, University of Waterloo, Waterloo, ON, Canada

Kurt Keutzer Department of Electrical Engineering and Computer Science, University of California, Berkeley, CA, USA

Mohammad Reza Khani University of Maryland, College Park, MD, USA

Samir Khuller Computer Science Department, University of Maryland, College Park, MD, USA

Donghyun Kim Department of Mathematics and Physics, North Carolina Central University, Durham, NC, USA

Jin Wook Kim HM Research, Seoul, Korea

Yoo-Ah Kim Computer Science and Engineering Department, University of Connecticut, Storrs, CT, USA

Valerie King Department of Computer Science, University of Victoria, Victoria, BC, Canada

Zoltán Király Department of Computer Science, Eötvös Loránd University, Budapest, Hungary

Egerváry Research Group (MTA-ELTE), Eötvös Loránd University, Budapest, Hungary

Lefteris Kirousis Department of Computer Engineering and Informatics, University of Patras, Patras, Greece

Jyrki Kivinen Department of Computer Science, University of Helsinki, Helsinki, Finland

Masashi Kiyomi International College of Arts and Sciences, Yokohama City University, Yokohama, Kanagawa, Japan

Kim-Manuel Klein University Kiel, Kiel, Germany

Rolf Klein Institute for Computer Science, University of Bonn, Bonn, Germany

Adam Klivans Department of Computer Science, University of Texas, Austin, TX, USA

Koji M. Kobayashi National Institute of Informatics, Chiyoda-ku, Tokyo, Japan

Stephen Kobourov Department of Computer Science, University of Arizona, Tucson, AZ, USA

Kirill Kogan IMDEA Networks, Madrid, Spain

Christian Komusiewicz Institute of Software Engineering and Theoretical Computer Science, Technical University of Berlin, Berlin, Germany

Goran Konjevod Department of Computer Science and Engineering, Arizona State University, Tempe, AZ, USA

Spyros Kontogiannis Department of Computer Science, University of Ioannina, Ioannina, Greece

Matias Korman Graduate School of Information Sciences, Tohoku University, Miyagi, Japan

Guy Kortsarz Department of Computer Science, Rutgers University, Camden, NJ, USA

Nitish Korula Google Research, New York, NY, USA

Robin Kothari Center for Theoretical Physics, Massachusetts Institute of Technology, Cambridge, MA, USA

David R. Cheriton School of Computer Science, Institute for Quantum Computing, University of Waterloo, Waterloo, ON, Canada

Ioannis Koutis Computer Science Department, University of Puerto Rico-Rio Piedras, San Juan, PR, USA

Dariusz R. Kowalski Department of Computer Science, University of Liverpool, Liverpool, UK

Evangelos Kranakis Department of Computer Science, Carleton, Ottawa, ON, Canada

Dieter Kratsch UFM MIM – LITA, Université de Lorraine, Metz, France

Stefan Kratsch Department of Software Engineering and Theoretical Computer Science, Technical University Berlin, Berlin, Germany

Robert Krauthgamer Weizmann Institute of Science, Rehovot, Israel

IBM Almaden Research Center, San Jose, CA, USA

Stephan Kreutzer Chair for Logic and Semantics, Technical University, Berlin, Germany

Sebastian Krinninger Faculty of Computer Science, University of Vienna, Vienna, Austria

Ravishankar Krishnaswamy Computer Science Department, Princeton University, Princeton, NJ, USA

Danny Krizanc Department of Computer Science, Wesleyan University, Middletown, CT, USA

Piotr Krysta Department of Computer Science, University of Liverpool, Liverpool, UK

Gregory Kucherov CNRS/LIGM, Université Paris-Est, Marne-la-Vallée, France

Fabian Kuhn Department of Computer Science, ETH Zurich, Zurich, Switzerland

V.S. Anil Kumar Virginia Bioinformatics Institute, Virginia Tech, Blacksburg, VA, USA

Tak-Wah Lam Department of Computer Science, University of Hong Kong, Hong Kong, China

Giuseppe Lancia Department of Mathematics and Computer Science, University of Udine, Udine, Italy

Gad M. Landau Department of Computer Science, University of Haifa, Haifa, Israel

Zeph Landau Department of Computer Science, University of California, Berkelely, CA, USA

Michael Langberg Department of Electrical Engineering, The State University of New York, Buffalo, NY, USA

Department of Mathematics and Computer Science, The Open University of Israel, Raanana, Israel

Elmar Langetepe Department of Computer Science, University of Bonn, Bonn, Germany

Ron Lavi Faculty of Industrial Engineering and Management, Technion, Haifa, Israel

Thierry Lecroq Computer Science Department and LITIS Faculty of Science, Université de Rouen, Rouen, France

James R. Lee Department of Computer Science and Engineering, University of Washington, Seattle, WA, USA

Stefano Leonardi Department of Information and Computer Systems, University of Rome, Rome, Italy

Pierre Leone Informatics Department, University of Geneva, Geneva, Switzerland

Henry Leung Department of Computer Science, The University of Hong Kong, Hong Kong, China

Christos Levcopoulos Department of Computer Science, Lund University, Lund, Sweden

Asaf Levin Faculty of Industrial Engineering and Management, The Technion, Haifa, Israel

Moshe Lewenstein Department of Computer Science, Bar-Ilan University, Ramat-Gan, Israel

Li (Erran) Li Bell Laboratories, Alcatel-Lucent, Murray Hill, NJ, USA

Mengling Li Division of Mathematical Sciences, Nanyang Technological University, Singapore, Singapore

Ming Li David R. Cheriton School of Computer Science, University of Waterloo, Waterloo, ON, Canada

Ming Min Li Computer Science and Technology, Tsinghua University, Beijing, China

Xiang-Yang Li Department of Computer Science, Illinois Institute of Technology, Chicago, IL, USA

Vahid Liaghat Department of Computer Science, University of Maryland, College Park, MD, USA

Jie Liang Department of Bioengineering, University of Illinois, Chicago, IL, USA

Andrzej Lingas Department of Computer Science, Lund University, Lund, Sweden

Maarten Löffler Department of Information and Computing Sciences, Utrecht University, Utrecht, The Netherlands

Daniel Lokshtanov Department of Informatics, University of Bergen, Bergen, Norway

Alejandro López-Ortiz David R. Cheriton School of Computer Science, University of Waterloo, Waterloo, ON, Canada

Chin Lung Lu Institute of Bioinformatics and Department of Biological Science and Technology, National Chiao Tung University, Hsinchu, Taiwan

Pinyan Lu Microsoft Research Asia, Shanghai, China

Zaixin Lu Department of Mathematics and Computer Science, Marywood University, Scranton, PA, USA

Feng Luo Department of Mathematics, Rutgers University, Piscataway, NJ, USA

Haiming Luo Department of Computer Science and Technology, Shenzhen Graduate School, Harbin Institute of Technology, Shenzhen, China

Rune B. Lyngsø Department of Statistics, Oxford University, Oxford, UK

Winton Capital Management, Oxford, UK

Bin Ma David R. Cheriton School of Computer Science, University of Waterloo, Waterloo, ON, Canada

Department of Computer Science, University of Western Ontario, London, ON, Canada

Mohammad Mahdian Yahoo! Research, Santa Clara, CA, USA

Hamid Mahini Department of Computer Science, University of Maryland, College Park, MD, USA

Veli Mäkinen Department of Computer Science, Helsinki Institute for Information Technology (HIIT), University of Helsinki, Helsinki, Finland

Dahlia Malkhi Microsoft, Silicon Valley Campus, Mountain View, CA, USA

Mark S. Manasse Microsoft Research, Mountain View, CA, USA

David F. Manlove School of Computing Science, University of Glasgow, Glasgow, UK

Giovanni Manzini Department of Computer Science, University of Eastern Piedmont, Alessandria, Italy

Department of Science and Technological Innovation, University of Piemonte Orientale, Alessandria, Italy

Madha V. Marathe IBM T.J. Watson Research Center, Hawthorne, NY, USA

Alberto Marchetti-Spaccamela Department of Information and Computer Systems, University of Rome, Rome, Italy

Igor L. Markov Department of Electrical Engineering and Computer Science, University of Michigan, Ann Arbor, MI, USA

Alexander Matveev Computer Science and Artificial Intelligence Laboratory, MIT, Cambridge, MA, USA

Eric McDermid Cedar Park, TX, USA

Catherine C. McGeoch Department of Mathematics and Computer Science, Amherst College, Amherst, MA, USA

Lyle A. McGeoch Department of Mathematics and Computer Science, Amherst College, Amherst, MA, USA

Andrew McGregor School of Computer Science, University of Massachusetts, Amherst, MA, USA

Brendan D. McKay Department of Computer Science, Australian National University, Canberra, ACT, Australia

Nicole Megow Institut für Mathematik, Technische Universität Berlin, Berlin, Germany

Manor Mendel Department of Mathematics and Computer Science, The Open University of Israel, Raanana, Israel

George B. Mertzios School of Engineering and Computing Sciences, Durham University, Durham, UK

Julián Mestre Department of Computer Science, University of Maryland, College Park, MD, USA

School of Information Technologies, The University of Sydney, Sydney, NSW, Australia

Pierre-Étienne Meunier Le Laboratoire d'Informatique Fondamentale de Marseille (LIF), Aix-Marseille Université, Marseille, France

Ulrich Meyer Department of Computer Science, Goethe University Fankfurt am Main, Frankfurt, Germany

Daniele Micciancio Department of Computer Science, University of California, San Diego, La Jolla, CA, USA

István Miklós Department of Plant Taxonomy and Ecology, Eötvös Loránd University, Budapest, Hungary

Shin-ichi Minato Graduate School of Information Science and Technology, Hokkaido University, Sapporo, Japan

Vahab S. Mirrokni Theory Group, Microsoft Research, Redmond, WA, USA

Neeldhara Misra Department of Computer Science and Automation, Indian Institute of Science, Bangalore, India

Joseph S.B. Mitchell Department of Applied Mathematics and Statistics, Stony Brook University, Stony Brook, NY, USA

Shuichi Miyazaki Academic Center for Computing and Media Studies, Kyoto University, Kyoto, Japan

Alistair Moffat Department of Computing and Information Systems, The University of Melbourne, Melbourne, VIC, Australia

Mark Moir Sun Microsystems Laboratories, Burlington, MA, USA

Ashley Montanaro Department of Computer Science, University of Bristol, Bristol, UK

Tal Mor Department of Computer Science, Technion – Israel Institute of Technology, Haifa, Israel

Michele Mosca Canadian Institute for Advanced Research, Toronto, ON, Canada

Combinatorics and Optimization/Institute for Quantum Computing, University of Waterloo, Waterloo, ON, Canada

Perimeter Institute for Theoretical Physics, Waterloo, ON, Canada

Thomas Moscibroda Systems and Networking Research Group, Microsoft Research, Redmond, WA, USA

Yoram Moses Department of Electrical Engineering, Technion – Israel Institute of Technology, Haifa, Israel

Shay Mozes Efi Arazi School of Computer Science, The Interdisciplinary Center (IDC), Herzliya, Israel

Marcin Mucha Faculty of Mathematics, Informatics and Mechanics, Institute of Informatics, Warsaw, Poland

Priyanka Mukhopadhyay Center for Quantum Technologies, National University of Singapore, Singapore, Singapore

Kamesh Munagala Levine Science Research Center, Duke University, Durham, NC, USA

J. Ian Munro David R. Cheriton School of Computer Science, University of Waterloo, Waterloo, ON, Canada

Joong Chae Na Department of Computer Science and Engineering, Sejong University, Seoul, Korea

Viswanath Nagarajan University of Michigan, Ann Arbor, MI, USA

Shin-ichi Nakano Department of Computer Science, Gunma University, Kiryu, Japan

Danupon Nanongkai School of Computer Science and Communication, KTH Royal Institute of Technology, Stockholm, Sweden

Giri Narasimhan Department of Computer Science, Florida International University, Miami, FL, USA

School of Computing and Information Sciences, Florida International University, Miami, FL, USA

Gonzalo Navarro Department of Computer Science, University of Chile, Santiago, Chile

Ashwin Nayak Department of Combinatorics and Optimization, and Institute for Quantum Computing, University of Waterloo, Waterloo, ON, Canada

Amir Nayyeri Department of Electrical Engineering and Computer Science, Oregon State University, Corvallis, OR, USA

Jesper Nederlof Technical University of Eindhoven, Eindhoven, The Netherlands

Ofer Neiman Department of Computer Science, Ben-Gurion University of the Negev, Beer Sheva, Israel

Yakov Nekrich David R. Cheriton School of Computer Science, University of Waterloo, Waterloo, ON, Canada

Jelani Nelson Harvard John A. Paulson School of Engineering and Applied Sciences, Cambridge, MA, USA

Ragnar Nevries Computer Science Department, University of Rostock, Rostock, Germany

Alantha Newman CNRS-Université Grenoble Alpes and G-SCOP, Grenoble, France

Hung Q. Ngo Computer Science and Engineering, The State University of New York, Buffalo, NY, USA

Patrick K. Nicholson Department D1: Algorithms and Complexity, Max Planck Institut für Informatik, Saarbrücken, Germany

Rolf Niedermeier Department of Mathematics and Computer Science, University of Jena, Jena, Germany

Institut für Softwaretechnik und Theoretische Informatik, Technische Universität Berlin, Berlin, Germany

Sergey I. Nikolenko Laboratory of Mathematical Logic, Steklov Institute of Mathematics, St. Petersburg, Russia

Sotiris Nikoletseas Computer Engineering and Informatics Department, University of Patras, Patras, Greece

Computer Technology Institute and Press "Diophantus", Patras, Greece

Aleksandar Nikolov Department of Computer Science, Rutgers University, Piscataway, NJ, USA

Nikola S. Nikolov Department of Computer Science and Information Systems, University of Limerick, Limerick, Republic of Ireland

Kobbi Nisim Department of Computer Science, Ben-Gurion University, Beer Sheva, Israel

Lhouari Nourine Clermont-Université, Université Blaise Pascal, LIMOS, CNRS, Aubière, France

Yoshio Okamoto Department of Information and Computer Sciences, Toyohashi University of Technology, Toyohashi, Japan

Michael Okun Weizmann Institute of Science, Rehovot, Israel

Rasmus Pagh Theoretical Computer Science, IT University of Copenhagen, Copenhagen, Denmark

David Z. Pan Department of Electrical and Computer Engineering, University of Texas, Austin, TX, USA

Peichen Pan Xilinx, Inc., San Jose, CA, USA

Debmalya Panigrahi Department of Computer Science, Duke University, Durham, NC, USA

Fahad Panolan Institute of Mathematical Sciences, Chennai, India

Vicky Papadopoulou Department of Computer Science, University of Cyprus, Nicosia, Cyprus

Fabio Pardi Institut de Biologie Computationnelle, Laboratoire d'Informatique, de Robotique et de Microélectronique de Montpellier (LIRMM), CNRS and Université de Montpellier, Montpellier cedex 5, France

Kunsoo Park School of Computer Science and Engineering, Seoul National University, Seoul, Korea

Srinivasan Parthasarathy IBM T.J. Watson Research Center, Hawthorne, NY, USA

Apoorva D. Patel Centre for High Energy Physics, Indian Institute of Science, Bangalore, India

Matthew J. Patitz Department of Computer Science and Computer Engineering, University of Arkansas, Fayetteville, AR, USA

Mihai Pătraşcu Computer Science and Artificial Intelligence Laboratory (CSAIL), Massachusetts Institute of Technology (MIT), Cambridge, MA, USA

Maurizio Patrignani Engineering Department, Roma Tre University, Rome, Italy

Boaz Patt-Shamir Department of Electrical Engineering, Tel-Aviv University, Tel-Aviv, Israel

Ramamohan Paturi Department of Computer Science and Engineering, University of California at San Diego, San Diego, CA, USA

Christophe Paul CNRS, Laboratoire d'Informatique Robotique et Microélectronique de Montpellier, Université Montpellier 2, Montpellier, France

Andrzej Pelc Department of Computer Science, University of Québec-Ottawa, Gatineau, QC, Canada

Jean-Marc Petit Université de Lyon, CNRS, INSA Lyon, LIRIS, Lyon, France

Seth Pettie Electrical Engineering and Computer Science (EECS) Department, University of Michigan, Ann Arbor, MI, USA

Marcin Pilipczuk Institute of Informatics, University of Bergen, Bergen, Norway

Institute of Informatics, University of Warsaw, Warsaw, Poland

Michał Pilipczuk Institute of Informatics, University of Warsaw, Warsaw, Poland

Institute of Informatics, University of Bergen, Bergen, Norway

Yuri Pirola Università degli Studi di Milano-Bicocca, Milan, Italy

Olivier Powell Informatics Department, University of Geneva, Geneva, Switzerland

Amit Prakash Microsoft, MSN, Redmond, WA, USA

Eric Price Department of Computer Science, The University of Texas, Austin, TX, USA

Kirk Pruhs Department of Computer Science, University of Pittsburgh, Pittsburgh, PA, USA

Teresa M. Przytycka Computational Biology Branch, NCBI, NIH, Bethesda, MD, USA

Pavel Pudlák Academy of Science of the Czech Republic, Mathematical Institute, Prague, Czech Republic

Simon J. Puglisi Department of Computer Science, University of Helsinki, Helsinki, Finland

Balaji Raghavachari Computer Science Department, The University of Texas at Dallas, Richardson, TX, USA

Md. Saidur Rahman Department of Computer Science and Engineering, Bangladesh University of Engineering and Technology, Dhaka, Bangladesh

Naila Rahman University of Hertfordshire, Hertfordshire, UK

Rajmohan Rajaraman Department of Computer Science, Northeastern University, Boston, MA, USA

Sergio Rajsbaum Instituto de Matemáticas, Universidad Nacional Autónoma de México (UNAM), México City, México

Vijaya Ramachandran Computer Science, University of Texas, Austin, TX, USA

Rajeev Raman Department of Computer Science, University of Leicester, Leicester, UK

M.S. Ramanujan Department of Informatics, University of Bergen, Bergen, Norway

Edgar Ramos School of Mathematics, National University of Colombia, Medellín, Colombia

Satish Rao Department of Computer Science, University of California, Berkeley, CA, USA

Christoforos L. Raptopoulos Computer Science Department, University of Geneva, Geneva, Switzerland

Computer Technology Institute and Press "Diophantus", Patras, Greece

Research Academic Computer Technology Institute, Greece and Computer Engineering and Informatics Department, University of Patras, Patras, Greece

Sofya Raskhodnikova Computer Science and Engineering Department, Pennsylvania State University, University Park, PA, USA

Rajeev Rastogi Amazon, Seattle, WA, USA

Joel Ratsaby Department of Electrical and Electronics Engineering, Ariel University of Samaria, Ariel, Israel

Kaushik Ravindran National Instruments, Berkeley, CA, USA

Michel Raynal Institut Universitaire de France and IRISA, Université de Rennes, Rennes, France

Ben W. Reichardt Electrical Engineering Department, University of Southern California (USC), Los Angeles, CA, USA

Renato Renner Institute for Theoretical Physics, Zurich, Switzerland

Elisa Ricci Department of Electronic and Information Engineering, University of Perugia, Perugia, Italy

Andréa W. Richa School of Computing, Informatics, and Decision Systems Engineering, Ira A. Fulton Schools of Engineering, Arizona State University, Tempe, AZ, USA

Peter C. Richter Department of Combinatorics and Optimization, and Institute for Quantum Computing, University of Waterloo, Waterloo, ON, Canada

Department of Computer Science, Rutgers, The State University of New Jersey, New Brunswick, NJ, USA

Liam Roditty Department of Computer Science, Bar-Ilan University, Ramat-Gan, Israel

Marcel Roeloffzen Graduate School of Information Sciences, Tohoku University, Sendai, Japan

Martin Roetteler Microsoft Research, Redmond, WA, USA

Heiko Röglin Department of Computer Science, University of Bonn, Bonn, Germany

José Rolim Informatics Department, University of Geneva, Geneva, Switzerland

Dana Ron School of Electrical Engineering, Tel-Aviv University, Ramat-Aviv, Israel

Frances Rosamond Parameterized Complexity Research Unit, University of Newcastle, Callaghan, NSW, Australia

Jarek Rossignac Georgia Institute of Technology, Atlanta, GA, USA

Matthieu Roy Laboratory of Analysis and Architecture of Systems (LAAS), Centre National de la Recherche Scientifique (CNRS), Université Toulouse, Toulouse, France

Ronitt Rubinfeld Massachusetts Institute of Technology (MIT), Cambridge, MA, USA

Tel Aviv University, Tel Aviv-Yafo, Israel

Atri Rudra Department of Computer Science and Engineering, State University of New York, Buffalo, NY, USA

Eric Ruppert Department of Computer Science and Engineering, York University, Toronto, ON, Canada

Frank Ruskey Department of Computer Science, University of Victoria, Victoria, BC, Canada

Luís M.S. Russo Departamento de Informática, Instituto Superior Técnico, Universidade de Lisboa, Lisboa, Portugal

INESC-ID, Lisboa, Portugal

Wojciech Rytter Institute of Informatics, Warsaw University, Warsaw, Poland

Kunihiko Sadakane Graduate School of Information Science and Technology, The University of Tokyo, Tokyo, Japan

S. Cenk Sahinalp Laboratory for Computational Biology, Simon Fraser University, Burnaby, BC, USA

Michael Saks Department of Mathematics, Rutgers, State University of New Jersey, Piscataway, NJ, USA

Alejandro Salinger Department of Computer Science, Saarland University, Saarbücken, Germany

Sachin S. Sapatnekar Department of Electrical and Computer Engineering, University of Minnesota, Minneapolis, MN, USA

Shubhangi Saraf Department of Mathematics and Department of Computer Science, Rutgers University, Piscataway, NJ, USA

Srinivasa Rao Satti Department of Computer Science and Engineering, Seoul National University, Seoul, South Korea

Saket Saurabh Institute of Mathematical Sciences, Chennai, India

University of Bergen, Bergen, Norway

Guido Schäfer Institute for Mathematics and Computer Science, Technical University of Berlin, Berlin, Germany

Dominik Scheder Institute for Interdisciplinary Information Sciences, Tsinghua University, Beijing, China

Institute for Computer Science, Shanghai Jiaotong University, Shanghai, China

Christian Scheideler Department of Computer Science, University of Paderborn, Paderborn, Germany

André Schiper EPFL, Lausanne, Switzerland

Christiane Schmidt The Selim and Rachel Benin School of Computer Science and Engineering, The Hebrew University of Jerusalem, Jerusalem, Israel

Markus Schmidt Institute for Computer Science, University of Freiburg, Freiburg, Germany

Dominik Schultes Institute for Computer Science, University of Karlsruhe, Karlsruhe, Germany

Robert Schweller Department of Computer Science, University of Texas Rio Grande Valley, Edinburg, TX, USA

Shinnosuke Seki Department of Computer Science, Helsinki Institute for Information Technology (HIIT), Aalto University, Aalto, Finland

Pranab Sen School of Technology and Computer Science, Tata Institute of Fundamental Research, Mumbai, India

Sandeep Sen Indian Institute of Technology (IIT) Delhi, Hauz Khas, New Delhi, India

Maria Serna Department of Language and System Information, Technical University of Catalonia, Barcelona, Spain

Rocco A. Servedio Computer Science, Columbia University, New York, NY, USA

Comandur Seshadhri Sandia National Laboratories, Livermore, CA, USA

Department of Computer Science, University of California, Santa Cruz, CA, USA

Jay Sethuraman Industrial Engineering and Operations Research, Columbia University, New York, NY, USA

Jiří Sgall Computer Science Institute, Charles University, Prague, Czech Republic

Rahul Shah Department of Computer Science, Louisiana State University, Baton Rouge, LA, USA

Shai Shalev-Shwartz School of Computer Science and Engineering, The Hebrew University, Jerusalem, Israel

Vikram Sharma Department of Computer Science, New York University, New York, NY, USA

Nir Shavit Computer Science and Artificial Intelligence Laboratory, MIT, Cambridge, MA, USA

School of Computer Science, Tel-Aviv University, Tel-Aviv, Israel

Yaoyun Shi Department of Electrical Engineering and Computer Science, University of Michigan, Ann Arbor, MI, USA

Ayumi Shinohara Graduate School of Information Sciences, Tohoku University, Sendai, Japan

Eugene Shragowitz Department of Computer Science and Engineering, University of Minnesota, Minneapolis, MN, USA

René A. Sitters Department of Econometrics and Operations Research, VU University, Amsterdam, The Netherlands

Balasubramanian Sivan Microsoft Research, Redmond, WA, USA

Daniel Sleator Department of Computer Science, Carnegie Mellon University, Pittsburgh, PA, USA

Michiel Smid School of Computer Science, Carleton University, Ottawa, ON, Canada

Adam Smith Computer Science and Engineering Department, Pennsylvania State University, University Park, State College, PA, USA

Dina Sokol Department of Computer and Information Science, Brooklyn College of CUNY, Brooklyn, NY, USA

Rolando D. Somma Theoretical Division, Los Alamos National Laboratory, Los Alamos, NM, USA

Wen-Zhan Song School of Engineering and Computer Science, Washington State University, Vancouver, WA, USA

Bettina Speckmann Department of Mathematics and Computer Science, Technical University of Eindhoven, Eindhoven, The Netherlands

Paul (Pavlos) Spirakis Computer Engineering and Informatics, Research and Academic Computer Technology Institute, Patras University, Patras, Greece

Computer Science, University of Liverpool, Liverpool, UK

Computer Technology Institute (CTI), Patras, Greece

Aravind Srinivasan Department of Computer Science, University of Maryland, College Park, MD, USA

Venkatesh Srinivasan Department of Computer Science, University of Victoria, Victoria, BC, Canada

Gerth Stølting Department of Computer Science, University of Aarhus, Århus, Denmark

Jens Stoye Faculty of Technology, Genome Informatics, Bielefeld University, Bielefeld, Germany

Scott M. Summers Department of Computer Science, University of Wisconsin – Oshkosh, Oshkosh, WI, USA

Aries Wei Sun Department of Computer Science, City University of Hong Kong, Hong Kong, China

Vijay Sundararajan Broadcom Corp, Fremont, CA, USA

Wing-Kin Sung Department of Computer Science, National University of Singapore, Singapore, Singapore

Mario Szegedy Department of Combinatorics and Optimization, and Institute for Quantum Computing, University of Waterloo, Waterloo, ON, Canada

Stefan Szeider Department of Computer Science, Durham University, Durham, UK

Tadao Takaoka Department of Computer Science and Software Engineering, University of Canterbury, Christchurch, New Zealand

Masayuki Takeda Department of Informatics, Kyushu University, Fukuoka, Japan

Kunal Talwar Microsoft Research, Silicon Valley Campus, Mountain View, CA, USA

Christino Tamon Department of Computer Science, Clarkson University, Potsdam, NY, USA

Akihisa Tamura Department of Mathematics, Keio University, Yokohama, Japan

Tiow-Seng Tan School of Computing, National University of Singapore, Singapore, Singapore

Shin-ichi Tanigawa Research Institute for Mathematical Sciences (RIMS), Kyoto University, Kyoto, Japan

Eric Tannier LBBE Biometry and Evolutionary Biology, INRIA Grenoble Rhône-Alpes, University of Lyon, Lyon, France

Alain Tapp Université de Montréal, Montréal, QC, Canada

Stephen R. Tate Department of Computer Science, University of North Carolina, Greensboro, NC, USA

Gadi Taubenfeld Department of Computer Science, Interdiciplinary Center Herzlia, Herzliya, Israel

Kavitha Telikepalli CSA Department, Indian Institute of Science, Bangalore, India

Barbara M. Terhal JARA Institute for Quantum Information, RWTH Aachen University, Aachen, Germany

Alexandre Termier IRISA, University of Rennes, 1, Rennes, France

My T. Thai Department of Computer and Information Science and Engineering, University of Florida, Gainesville, FL, USA

Abhradeep Thakurta Department of Computer Science, Stanford University, Stanford, CA, USA

Microsoft Research, CA, USA

Justin Thaler Yahoo! Labs, New York, NY, USA

Sharma V. Thankachan School of CSE, Georgia Institute of Technology, Atlanta, USA

Dimitrios Thilikos AlGCo Project-Team, CNRS, LIRMM, France

Department of Mathematics, National and Kapodistrian University of Athens, Athens, Greece

Haitong Tian Department of Electrical and Computer Engineering, University of Illinois at Urbana-Champaign, Urbana, IL, USA

Ioan Todinca INSA Centre Val de Loire, Universite d'Orleans, Orléans, France

Alade O. Tokuta Department of Mathematics and Physics, North Carolina Central University, Durham, NC, USA

Laura Toma Department of Computer Science, Bowdoin College, Brunswick, ME, USA

Etsuji Tomita The Advanced Algorithms Research Laboratory, The University of Electro-Communications, Chofu, Tokyo, Japan

Csaba D. Tóth Department of Computer Science, Tufts University, Medford, MA, USA

Department of Mathematics, California State University Northridge, Los Angeles, CA, USA

Luca Trevisan Department of Computer Science, University of California, Berkeley, CA, USA

John Tromp CWI, Amsterdam, The Netherlands

Nicolas Trotignon Laboratoire de l'Informatique du Parallélisme (LIP), CNRS, ENS de Lyon, Lyon, France

Jakub Truszkowski Cancer Research UK Cambridge Institute, University of Cambridge, Cambridge, UK

European Molecular Biology Laboratory, European Bioinformatics Institute (EMBL-EBI), Wellcome Trust Genome Campus, Hinxton, Cambridge, UK

Esko Ukkonen Department of Computer Science, Helsinki Institute for Information Technology (HIIT), University of Helsinki, Helsinki, Finland

Jonathan Ullman Department of Computer Science, Columbia University, New York, NY, USA

Takeaki Uno National Institute of Informatics, Chiyoda, Tokyo, Japan

Ruth Urner Department of Machine Learning, Carnegie Mellon University, Pittsburgh, USA

Jan Vahrenhold Department of Computer Science, Westfälische Wilhelms-Universität Münster, Münster, Germany

Daniel Valenzuela Department of Computer Science, Helsinki Institute for Information Technology (HIIT), University of Helsinki, Helsinki, Finland

Marc van Kreveld Department of Information and Computing Sciences, Utrecht University, Utrecht, The Netherlands

Rob van Stee University of Leicester, Leicester, UK

Stefano Varricchio Department of Computer Science, University of Roma, Rome, Italy

José Verschae Departamento de Matemáticas and Departamento de Ingeniería Industrial y de Sistemas, Pontificia Universidad Católica de Chile, Santiago, Chile

Stéphane Vialette IGM-LabInfo, University of Paris-East, Descartes, France

Sebastiano Vigna Dipartimento di Informatica, Università degli Studi di Milano, Milano, Italy

Yngve Villanger Department of Informatics, University of Bergen, Bergen, Norway

Paul Vitányi Centrum Wiskunde & Informatica (CWI), Amsterdam, The Netherlands

Jeffrey Scott Vitter University of Kansas, Lawrence, KS, USA

Berthold Vöcking Department of Computer Science, RWTH Aachen University, Aachen, Germany

Tjark Vredeveld Department of Quantitative Economics, Maastricht University, Maastricht, The Netherlands

Magnus Wahlström Department of Computer Science, Royal Holloway, University of London, Egham, UK

Peng-Jun Wan Department of Computer Science, Illinois Institute of Technology, Chicago, IL, USA

Chengwen Chris Wang Department of Computer Science, Carnegie Mellon University, Pittsburgh, PA, USA

Feng Wang Mathematical Science and Applied Computing, Arizona State University at the West Campus, Phoenix, AZ, USA

Huijuan Wang Shandong University, Jinan, China

Joshua R. Wang Department of Computer Science, Stanford University, Stanford, CA, USA

Lusheng Wang Department of Computer Science, City University of Hong Kong, Hong Kong, Hong Kong

Wei Wang School of Mathematics and Statistics, Xi'an Jiaotong University, Xi'an, Shaanxi, China

Weizhao Wang Google Inc., Irvine, CA, USA

Yu Wang Department of Computer Science, University of North Carolina, Charlotte, NC, USA

Takashi Washio The Institute of Scientific and Industrial Research, Osaka University, Ibaraki, Osaka, Japan

Matthew Weinberg Computer Science, Princeton University, Princeton, NJ, USA

Tobias Weinzierl School of Engineering and Computing Sciences, Durham University, Durham, UK

Renato F. Werneck Microsoft Research Silicon Valley, La Avenida, CA, USA

Matthias Westermann Department of Computer Science, TU Dortmund University, Dortmund, Germany

Tim A.C. Willemse Department of Mathematics and Computer Science, Eindhoven University of Technology, Eindhoven, The Netherlands

Ryan Williams Department of Computer Science, Stanford University, Stanford, CA, USA

Tyson Williams Computer Sciences Department, University of Wisconsin–Madison, Madison, WI, USA

Andrew Winslow Department of Computer Science, Tufts University, Medford, MA, USA

Paul Wollan Department of Computer Science, University of Rome La Sapienza, Rome, Italy

Martin D.F. Wong Department of Electrical and Computer Engineering, University of Illinois at Urbana-Champaign, Urbana, IL, USA

Prudence W.H. Wong University of Liverpool, Liverpool, UK

David R. Wood School of Mathematical Sciences, Monash University, Melbourne, VIC, Australia

Damien Woods Computer Science, California Institute of Technology, Pasadena, CA, USA

Lidong Wu Department of Computer Science, The University of Texas, Tyler, TX, USA

Weili Wu College of Computer Science and Technology, Taiyuan University of Technology, Taiyuan, Shanxi Province, China

Department of Computer Science, California State University, Los Angeles, CA, USA

Department of Computer Science, The University of Texas at Dallas, Richardson, TX, USA

Christian Wulff-Nilsen Department of Computer Science, University of Copenhagen, Copenhagen, Denmark

Mingji Xia The State Key Laboratory of Computer Science, Chinese Academy of Sciences, Beijing, China

David Xiao CNRS, Université Paris 7, Paris, France

Dong Xu Bond Life Sciences Center, University of Missouri, Columbia, MO, USA

Wen Xu Department of Computer Science, The University of Texas at Dallas, Richardson, TX, USA

Katsuhisa Yamanaka Department of Electrical Engineering and Computer Science, Iwate University, Iwate, Japan

Hiroki Yanagisawa IBM Research – Tokyo, Tokyo, Japan

Honghua Hannah Yang Strategic CAD Laboratories, Intel Corporation, Hillsboro, OR, USA

Qiuming Yao University of Missouri, Columbia, MO, USA

Chee K. Yap Department of Computer Science, New York University, New York, NY, USA

Yinyu Ye Department of Management Science and Engineering, Stanford University, Stanford, CA, USA

Anders Yeo Engineering Systems and Design, Singapore University of Technology and Design, Singapore, Singapore

Department of Mathematics, University of Johannesburg, Auckland Park, South Africa

Chih-Wei Yi Department of Computer Science, National Chiao Tung University, Hsinchu City, Taiwan

Ke Yi Hong Kong University of Science and Technology, Hong Kong, China

Yitong Yin Nanjing University, Jiangsu, Nanjing, Gulou, China

S.M. Yiu Department of Computer Science, University of Hong Kong, Hong Kong, China

Makoto Yokoo Department of Information Science and Electrical Engineering, Kyushu University, Nishi-ku, Fukuoka, Japan

Evangeline F.Y. Young Department of Computer Science and Engineering, The Chinese University of Hong Kong, Hong Kong, China

Neal E. Young Department of Computer Science and Engineering, University of California, Riverside, CA, USA

Bei Yu Department of Electrical and Computer Engineering, University of Texas, Austin, TX, USA

Yaoliang Yu Machine Learning Department, Carnegie Mellon University, Pittsburgh, PA, USA

Raphael Yuster Department of Mathematics, University of Haifa, Haifa, Israel

Morteza Zadimoghaddam Google Research, New York, NY, USA

Francis Zane Lucent Technologies, Bell Laboraties, Murray Hill, NJ, USA

Christos Zaroliagis Department of Computer Engineering and Informatics, University of Patras, Patras, Greece

Norbert Zeh Faculty of Computer Science, Dalhousie University, Halifax, NS, Canada

Li Zhang Microsoft Research, Mountain View, CA, USA

Louxin Zhang Department of Mathematics, National University of Singapore, Singapore, Singapore

Shengyu Zhang The Chinese University of Hong Kong, Hong Kong, China

Zhang Zhao College of Mathematics Physics and Information Engineering, Zhejiang Normal University, Zhejiang, Jinhua, China

Hai Zhou Electrical Engineering and Computer Science (EECS) Department, Northwestern University, Evanston, IL, USA

Yuqing Zhu Department of Computer Science, California State University, Los Angeles, CA, USA

Department of Computer Science, The University of Texas at Dallas, Richardson, TX, USA

Sandra Zilles Department of Computer Science, University of Regina, Regina, SK, Canada

Aaron Zollinger Department of Electrical Engineering and Computer Science, University of California, Berkeley, CA, USA

Uri Zwick Department of Mathematics and Computer Science, Tel-Aviv University, Tel-Aviv, Israel

G

Gate Sizing

Vijay Sundararajan
Broadcom Corp, Fremont, CA, USA

Keywords

Fast and exact transistor sizing

Years and Authors of Summarized Original Work

2002; Sundararajan, Sapatnekar, Parhi

Problem Definition

For a detailed exposition of the solution approach presented in this entry, please refer to [15]. As evidenced by the successive announcement of ever-faster computer systems in the past decade, increasing the speed of VLSI systems continues to be one of the major requirements for VLSI system designers today. Faster integrated circuits are making possible newer applications that were traditionally considered difficult to implement in hardware. In this scenario of increasing circuit complexity, reduction of circuit delay in integrated circuits is an important design objective. Transistor sizing is one such task that has been employed for speeding up circuits for quite some time now [6]. Given the circuit topology, the delay of a combinational circuit can be controlled by varying the sizes of transistors in the circuit. Here, the size of a transistor is measured in terms of its channel width, since the channel lengths of MOS transistors in a digital circuit are generally uniform. In any case, what really matters is the ratio of channel width to channel length, and if channel lengths are not uniform, this ratio can be considered as the size. In coarse terms, the circuit delay can usually be reduced by increasing the sizes of certain transistors in the circuit from the minimum size. Hence, making the circuit faster usually entails the penalty of increased circuit area relative to a minimum-sized circuit, and the area-delay trade-off involved here is the problem of transistor size optimization. A related problem to transistor sizing is called gate sizing, where a logic gate in a circuit is modeled as an equivalent inverter and the sizing optimization is carried out on this modified circuit with equivalent inverters in place of more complex gates. There is, therefore, a reduction in the number of size parameters corresponding to every gate in the circuit. Needless to say, this is an easier problem to solve than the general transistor sizing problem. Note that gate sizing mentioned here is distinct from library-specific gate sizing that is a discrete optimization problem targeted to selecting appropriate gate sizes from an underlying cell library. The gate sizing problem targeted here is one of continuous gate sizing where the gate sizes are allowed to vary in a continuous manner between a minimum and a maximum size. There has been a large amount of work done on transistor sizing

M.-Y. Kao (ed.), *Encyclopedia of Algorithms*,
DOI 10.1007/978-1-4939-2864-4

[1–3, 5, 6, 9, 10, 12, 13], that underlines the importance of this optimization technique. Starting from a minimum-sized circuit, TILOS, [6], uses a greedy strategy for transistor sizing by iteratively sizing transistors in the critical path. A sensitivity factor is calculated for every transistor in the critical path to quantify the gain in circuit speed achieved by a unit upsizing of the transistor. The most sensitive transistor is then bumped up in size by a small constant factor to speed up the circuit. This process is repeated iteratively until the timing requirements are met. The technique is extremely simple to implement and has run-time behavior proportional to the size of the circuit. Its chief drawback is that it does not have guaranteed convergence properties and hence is not an exact optimization technique.

Key Results

The solution presented in the entry heretofore referred to as MINFLOTRANSIT was a novel way of solving the transistor sizing problem exactly and in an extremely fast manner. Even though the entry treats transistor sizing, in the description, the results apply as well to the less general problem of continuous gate sizing as described earlier. The proposed approach has some similarity in form to [2, 5, 8] which will be subsequently explained, but the similarity in content is minimal and the details of implementation are vastly different.

In essence, the proposed technique and the techniques in [2, 5, 8] are iterative relaxation approaches that involve a two-step optimization strategy. The first step involves a delay budgeting step where optimal delays are computed for transistors/gates. The second step involves sizing transistors optimally under this "constant delay" model to achieve these delay budgets. The two steps are iteratively alternated until the solution converges, i.e., until the delay budgets calculated in the first step are exactly satisfied by the transistor sizes determined by the second step.

The primary features of the proposed approach are:

- It is computationally fast and is comparable to TILOS in its run-time behavior.
- It can be used for true transistor sizing as well as the relaxed problem of gate sizing. Additionally, the approach can easily incorporate wire sizing [15].
- It can be adapted for more general delay models than the Elmore delay model [15].

The starting point for the proposed approach is a fast guess solution. This could be obtained, for example, from a circuit that has been optimized using TILOS to meet the given delay requirements. The proposed approach, as outlined earlier, is an iterative relaxation procedure that involves an alternating two-phase relaxed optimization sequence that is repeated iteratively until convergence is achieved. The two phases in the proposed approach are:

- The **D-phase** where transistor sizes are assumed fixed and transistor delays are regarded as variable parameters. Irrespective of the delay model employed, this phase can be formulated as the dual of a min-cost network flow problem. Using $|V|$ to denote the number of transistors and $|E|$ the number of wires in the circuit, this step in our application has worst-case complexity of $O(|V||E|\log(\log|V|))$ [7].
- The **W-phase** where transistor/gate delays are assumed fixed and their sizes are regarded as variable parameters. As long as the gate delay can be expressed as a separable function of the transistor sizes, this step can be solved as a Simple Monotonic Program (SMP) [11]. The complexity of SMP is similar to an all-pairs shortest-path algorithm in a directed graph, [4, 11], i.e., $O(V||E|)$.

The objective function for the problem is the minimization of circuit area. In the W-phase, this

Gate Sizing, Table 1 Comparison of TILOS and MINFLOTRANSIT on a Sun Ultrasparc 10 workstation for ISCAS85 and MCNC91 benchmarks for 0.13 um technology. The delay specs are with respect to a minimum-sized circuit. The optimization approach followed here was gate sizing

Circuit	# Gates	Area saved over TILOS (%)	Delay specs. (D_{min})	CPU time (TILOS) (s)	CPU time (OURS) (s)
Adder32	480	≤ 1	0.5	2.2	5
Adder256	3,840	≤ 1	0.5	262	608
Cm163a	65	2.1	0.55	0.13	0.32
Cm162a	71	10.4	0.5	0.23	0.96
Parity8	89	37	0.45	0.68	2.15
Frg1	177	1.9	0.7	0.55	1.49
Population	518	6.7	0.4	57	179
Pmult8	1,431	5	0.5	637	1476
Alu2	826	2.6	0.6	28	71
C432	160	9.4	0.4	0.5	4.8
C499	202	7.2	0.57	1.47	11.26
C880	383	4	0.4	2.7	8, 2
C1355	546	9.5	0.4	29	76
C1908	880	4.6	0.4	36	84
C2670	1,193	9.1	0.4	27	69
C3540	1,669	7.7	0.4	226	651
C5315	2,307	2	0.4	90	201
C6288	2,416	16.5	0.4	1,677	4,138
C7552	3,512	3.3	0.4	320	683

objective is addressed directly, and in the D-phase the objective is chosen to facilitate a move in the solution space in a direction that is known to lead to a reduction in the circuit area.

Applications

The primary application of the solution provided here is circuit and system optimization in automated VLSI design. The solution provided here can enable electronic design automation (EDA) tools that take a holistic approach toward transistor sizing. This will in turn enable making custom circuit design flows more realizable in practice. The mechanics of some of the elements of the solution provided here especially the **D-phase** have been used to address other circuit optimization problems [14].

Open Problems

The related problem of discrete gate sizing optimization matching gate sized to available gate sizes from a standard cell library is a provably hard optimization problem which could be aided by the development of efficient heuristics and probabilistic algorithms.

Experimental Results

A relative comparison of MINFLOTRANSIT with TILOS is provided in Table 1 for gate sizing of ISACS85 and mcnc91 benchmark circuits. As can be seen a significant performance improvement is observed with a tolerable loss in execution time.

Cross-References

- ▸ Circuit Retiming
- ▸ Wire Sizing

Recommended Reading

1. Chen CP, Chu CN, Wong DF (1998) Fast and exact simultaneous gate and wire sizing by Lagrangian relaxation. In: Proceedings of the 1998 IEEE/ACM international conference on computer-aided design, San Jose, pp 617–624
2. Chen HY, Kang SM (1991) iCOACH: a circuit optimization aid for CMOS high-performance circuits. Intergr VLSI J 10(2):185–212
3. Conn AR, Coulman PK, Haring RA, Morrill GL, Visweshwariah C, Wu CW (1998) Jiffy Tune: circuit optimization using time-domain sensitivities. IEEE Trans Comput Aided Des Intergr Circuits Syst 17(12):1292–1309
4. Cormen TH, Leiserson CE, Rivest RL (1990) Introduction to algorithms. McGraw-Hill, New York
5. Dai Z, Asada K (1989) MOSIZ: a two-step transistor sizing algorithm based on optimal timing assignment method for multi-stage complex gates. In: Proceedings of the 1989 custom integrated circuits conference, New York, pp 17.3.1–17.3.4
6. Fishburn JP, Dunlop AE (1985) TILOS: a posynomial programming approach to transistor sizing. In: Proceedings of the 1985 international conference on computer-aided design, Santa Clara, pp 326–328
7. Goldberg AV, Grigoriadis MD, Tarjan RE (1991) Use of dynamic trees in a network simplex algorithm for the maximum flow problem. Math Program 50(3):277–290
8. Grodstein J, Lehman E, Harkness H, Grundmann B, Watanabe Y (1995) A delay model for logic synthesis of continuously sized networks. In: Proceedings of the 1995 international conference on computer-aided design, San Jose, pp 458–462
9. Marple DP (1986) Performance optimization of digital VLSI circuits. Technical report CSL-TR-86-308, Stanford University
10. Marple DP (1989) Transistor size optimization in the tailor layout system. In: Proceedings of the 26th ACM/IEEE design automation conference, Las Vegas, pp 43–48
11. Papaefthymiou MC (1998) Asymptotically efficient retiming under setup and hold constraints. In: Proceedings of the IEEE/ACM international conference on computer-aided design, San Jose, pp 288–295
12. Sapatnekar SS, Rao VB, Vaidya PM, Kang SM (1993) An exact solution to the transistor sizing problem for CMOS circuits using convex optimization. IEEE Trans Comput Aided Des 12(11): 1621–1634
13. Shyu JM, Sangiovanni-Vincentelli AL, Fishburn JP, Dunlop AE (1988) Optimization-based transistor sizing. IEEE J Solid State Circuits 23(2):400–409
14. Sundararajan V, Parhi K (1999) Low power synthesis of dual threshold voltage CMOS VLSI circuits. In: Proceedings of the international symposium on low power electronics and design, San Diego, pp 139–144
15. Sundararajan V, Sapatnekar SS, Parhi KK (2002) Fast and exact transistor sizing based on iterative relaxation. Comput-Aided Design Intergr Circuits Syst IEEE Trans 21(5):568–581

General Equilibrium

Li-Sha Huang
Department of Computer Science and Technology, Tsinghua University, Beijing, China

Keywords

Competitive market equilibrium

Years and Authors of Summarized Original Work

2002; Deng, Papadimitriou, Safra

Problem Definition

This problem is concerned with the computational complexity of finding an exchange market equilibrium. The exchange market model consists of a set of agents, each with an initial endowment of commodities, interacting through a market, trying to maximize each's utility function. The equilibrium prices are determined by a clearance condition. That is, all commodities are bought, collectively, by all the utility maximizing agents, subject to their budget constraints (determined by the values of their initial endowments of commodities at the market price). The work of Deng, Papadimitriou and Safra [3] studies the complexity, approximability, inapproximability, and communication complexity of finding equilibrium prices. The work shows the NP-hardness of approximating the equilibrium in a market with indivisible goods. For markets with divisible goods and linear utility functions, it develops a pseudo-polynomial time algorithm for computing an ϵ-equilibrium. It also gives a communication complexity lower bound for computing Pareto allocations in markets with non-strictly concave utility functions.

Market Model

In a pure exchange economy, there are m traders, labeled by $i = 1, 2, ..., m$, and n types of commodities, labeled by $j = 1, 2, ..., n$. The commodities could be divisible or indivisible. Each trader i comes to the market with initial endowment of commodities, denoted by a vector $w_i \in \mathbb{R}_+^n$, whose j-th entry is the amount of commodity j held by trader i.

Associate each trader i a *consumption set* X_i to represents the set of possible commodity bundles for him. For example, when there are n_1 divisible commodities and $(n - n_1)$ indivisible commodities, X_i can be $\mathbb{R}_+^{n_1} \times \mathbb{Z}_+^{n-n_1}$. Each trader has a utility function $X_i \mapsto \mathbb{R}_+$ to present his utility for a bundle of commodities. Usually, the utility function is required to be concave and nondecreasing.

In the market, each trader acts as both a buyer and a seller to maximize his utility. At a certain price $p \in \mathbb{R}_+^n$, trader i is is solving the following optimization problem, under his budget constraint:

$$\max \ u_i(x_i) \ \ s.t. \ \ x_i \in X_i \text{ and } \langle p, x_i \rangle \le \langle p, w_i \rangle.$$

Definition 1 An equilibrium in a pure exchange economy is a price vector $\bar{p} \in \mathbb{R}_+^n$ and bundles of commodities $\{\bar{x}_i \in \mathbb{R}_+^n, i = 1, ..., m\}$, such that

$$\bar{x}_i \in \text{argmax}\{u_i(x_i) | x_i \in X_i \text{ and } \langle x_i, \bar{p} \rangle \le \langle w_i, \bar{p} \rangle\}, \quad \forall 1 \le i \le m$$

$$\sum_{i=1}^{m} \bar{x}_{ij} \le \sum_{i=1}^{m} w_{ij}, \forall 1 \le j \le n.$$

The concept of approximate equilibrium was introduced in [3]:

Definition 2 ([3)] An ϵ-approximate equilibrium in an exchange market is a price vector $\bar{p} \in \mathbb{R}_+^n$ and bundles of goods $\{\bar{x}_i \in \mathbb{R}_+^n, i = 1, ..., m\}$, such that

$$u_i(\bar{x}_i) \ge \frac{1}{1+\epsilon} \max\{u_i(x_i) | x_i \in X_i, \langle x_i, \bar{p} \rangle \le \langle w_i, \bar{p} \rangle\}, \forall i \qquad (1)$$

$$\langle \bar{x}_i, \bar{p} \rangle \le (1+\epsilon) \langle w_i, \bar{p} \rangle, \forall i \qquad (2)$$

$$\sum_{i=1}^{m} \bar{x}_{ij} \le (1+\epsilon) \sum_{i=1}^{m} w_{ij}, \forall j\,. \qquad (3)$$

Key Results

A linear market is a market in which all the agents have linear utility functions. The deficiency of a market is the smallest $\epsilon \ge 0$ for which an ϵ-approximate equilibrium exists.

Theorem 1 *The deficiency of a linear market with indivisible goods is NP-hard to compute, even if the number of agents is two. The deficiency is also NP-hard to approximate within 1/3.*

Theorem 2 *There is a polynomial-time algorithm for finding an equilibrium in linear markets with bounded number of divisible goods. Ditto for a polynomial number of agents.*

Theorem 3 *If the number of goods is bounded, there is a polynomial-time algorithm which, for any linear indivisible market for which a price equilibrium exists, and for any $\epsilon > 0$, finds an ϵ-approximate equilibrium.*

If the utility functions are strictly concave and the equilibrium prices are broadcasted to all agents, the equilibrium allocation can be computed distributely without any communication, since each agent's basket of goods is uniquely determined. However, if the utility functions are not strictly concave, e.g., linear functions, communications are needed to coordinate the agents' behaviors.

G

Theorem 4 *Any protocol with binary domains for computing Pareto allocations of m agents and n divisible commodities with concave utility functions (resp. ϵ-Pareto allocations for indivisible commodities, for any $\epsilon < 1$) must have market communication complexity $\Omega(m \log(m + n))$ bits.*

Applications

This concept of market equilibrium is the outcome of a sequence of efforts trying to fully understand the laws that govern human commercial activities, starting with the "invisible hand" of Adam Smith, and finally, the mathematical conclusion of Arrow and Debreu [1] that there exists a set of prices that bring supply and demand into equilibrium, under quite general conditions on the agent utility functions and their optimization behavior.

The work of Deng, Papadimitriou and Safra [3] explicitly called for an algorithmic complexity study of the problem, and developed interesting complexity results and approximation algorithms for several classes of utility functions. There has since been a surge of algorithmic study for the computation of the price equilibrium problem with continuous variables, discovering and rediscovering polynomial time algorithms for many classes of utility functions, see [2, 4–9].

Significant progress has been made in the above directions but only as a first step. New ideas and methods have already been invented and applied in reality. The next significant step will soon manifest itself with many active studies in microeconomic behavior analysis for E-commercial markets. Nevertheless the algorithmic analytic foundation in [3] will be an indispensable tool for further development in this reincarnated exciting field.

Open Problems

The most important open problem is what is the computational complexity for finding the equilibrium price, as guaranteed by the Arrow–Debreu theorem. To the best of the author's knowledge, only the markets whose set of equilibria is convex can be solved in polynomial time with current techniques. And approximating equilibria in some markets with disconnected set of equilibria, e.g., Leontief economies, are shown to be PPAD-hard. Is the convexity or (weakly) gross substitutability a necessary condition for a market to be polynomial-time solvable?

Second, how to handle the dynamic case is especially interesting in theory, mathematical modeling, and algorithmic complexity as bounded rationality. Great progress must be made in those directions for any theoretical work to be meaningful in practice.

Third, incentive compatible mechanism design protocols for the auction models have been most actively studied recently, especially with the rise of E-Commerce. Especially at this level, a proper approximate version of the equilibrium concept handling price dynamics should be especially important.

Cross-References

▸ Complexity of Core
▸ Leontief Economy Equilibrium
▸ Non-approximability of Bimatrix Nash Equilibria

Recommended Reading

1. Arrow KJ, Debreu G (1954) Existence of an equilibrium for a competitive economy. Econometrica 22(3):265–290
2. Codenotti B, McCune B, Varadarajan K (2005) Market equilibrium via the excess demand function. In: Proceedings STOC'05. ACM, Baltimore, pp 74–83
3. Deng X, Papadimitriou C, Safra S (2002) On the complexity of price equilibria. J Comput Syst Sci 67(2):311–324
4. Devanur NR, Papadimitriou CH, Saberi A, Vazirani VV (2002) Market equilibria via a primal-dual-type algorithm. In: Proceedings of FOCS'02. IEEE Computer Society, Vancouver, pp 389–395
5. Eaves BC (1985) Finite solution for pure trade markets with Cobb-Douglas utilities. Math Program Study 23:226–239

6. Garg R, Kapoor S (2004) Auction algorithms for market equilibrium. In: Proceedings of STOC'04. ACM, Chicago, pp 511–518
7. Jain K (2004) A polynomial time algorithm for computing the Arrow-Debreu market equilibrium for linear utilities. In: Proceeding of FOCS'04. IEEE Computer Society, Rome, pp 286–294
8. Nenakhov E, Primak M (1983) About one algorithm for finding the solution of the Arrow-Debreu model. Kibernetica 3:127–128
9. Ye Y (2008) A path to the Arrow-Debreu competitive market equilibrium. Math Program 111(1–2):315–348

Generalized Steiner Network

Julia Chuzhoy
Toyota Technological Institute, Chicago, IL, USA

Keywords

Survivable network design

Years and Authors of Summarized Original Work

2001; Jain

Problem Definition

The generalized Steiner network problem is a network design problem, where the input consists of a graph together with a collection of connectivity requirements, and the goal is to find the cheapest subgraph meeting these requirements.

Formally, the input to the generalized Steiner network problem is an undirected multigraph $G = (V, E)$, where each edge $e \in E$ has a non-negative cost $c(e)$, and for each pair of vertices $i, j \in V$, there is a connectivity requirement $r_{i,j} \in \mathbb{Z}$. A feasible solution is a subset $E' \subseteq E$ of edges, such that every pair $i, j \in V$ of vertices is connected by at least $r_{i,j}$ edge-disjoint path in graph $G' = (V, E')$. The generalized Steiner network problem asks to find a solution E' of minimum cost $\sum_{e \in E'} c(e)$.

This problem generalizes several classical network design problems. Some examples include minimum spanning tree, Steiner tree and Steiner forest. The most general special case for which a 2-approximation was previously known is the Steiner forest problem [1, 4].

Williamson et al. [8] were the first to show a non-trivial approximation algorithm for the generalized Steiner network problem, achieving a $2k$-approximation, where $k = \max_{i,j \in V}\{r_{i,j}\}$. This result was improved to $O(\log k)$-approximation by Goemans et al. [3].

Key Results

The main result of [6] is a factor-2 approximation algorithm for the generalized Steiner network problem. The techniques used in the design and the analysis of the algorithm seem to be of independent interest.

The 2-approximation is achieved for a more general problem, defined as follows. The input is a multigraph $G = (V, E)$ with costs $c(\cdot)$ on edges, and connectivity requirement function $f : 2^V \to \mathbb{Z}$. Function f is weakly submodular, i.e., it has the following properties:

1. $f(V) = 0$.
2. For all $A, B \subseteq V$, at least one of the following two conditions holds:
   - $f(A) + f(B) \leq f(A \setminus B) + f(B \setminus A)$.
   - $f(A) + f(B) \leq f(A \cap B) + f(A \cup B)$.

For any subset $S \subseteq V$ of vertices, let $\delta(S)$ denote the set of edges with exactly one endpoint in S. The goal is to find a minimum-cost subset of edges $E' \subseteq E$, such that for every subset $S \subseteq V$ of vertices, $|\delta(S) \cap E'| \geq f(S)$.

This problem can be equivalently expressed as an integer program. For each edge $e \in E$, let x_e be the indicator variable of whether e belongs to the solution.

$$\text{(IP)} \qquad \min \sum_{e \in E} c(e) x_e$$

subject to:

$$\sum_{e \in \delta(S)} x_e \geq f(S) \qquad \forall S \subseteq V \qquad (1)$$

$$x_e \in \{0, 1\} \qquad \forall e \in E \qquad (2)$$

It is easy to see that the generalized Steiner network problem is a special case of (IP), where for each $S \subseteq V$, $f(S) = \max_{i \in S, j \notin S}\{r_{i,j}\}$.

Techniques

The approximation algorithm uses the LP-rounding technique. The initial linear program (LP) is obtained from (IP) by replacing the integrality constraint (2) with:

$$0 \leq x_e \leq 1 \qquad \forall e \in E \qquad (3)$$

It is assumed that there is a separation oracle for (LP). It is easy to see that such an oracle exists if (LP) is obtained from the generalized Steiner network problem. The key result used in the design and the analysis of the algorithm is summarized in the following theorem.

Theorem 1 *In any basic solution of (LP), there is at least one edge $e \in E$ with $x_e \geq 1/2$.*

The approximation algorithm works by iterative LP-rounding. Given a basic optimal solution of (LP), let $E^* \subseteq E$ be the subset of edges e with $x_e \geq 1/2$. The edges of E^* are removed from the graph (and are eventually added to the solution), and the problem is then solved recursively on the residual graph, by solving (LP) on $G^* = (V, E \setminus E^*)$, where for each subset $S \subseteq V$, the new requirement is $f(S) - |\delta(S) \cap E^*|$. The main observation that leads to factor-2 approximation is the following: if E' is a 2-approximation for the residual problem, then $E' \cup E^*$ is a 2-approximation for the original problem.

Given any solution to (LP), set $S \subseteq V$ is called *tight* iff constraint (1) holds with equality for S. The proof of Theorem 1 involves constructing a large *laminar family* of tight sets (a family where for every pair of sets, either one set contains the other, or the two sets are disjoint). After that a clever accounting scheme that charges edges to the sets of the laminar family is used to show that there is at least one edge $e \in E$ with $x_e \geq 1/2$.

Applications

Generalized Steiner network is a very basic and natural network design problem that has many applications in different areas, including the design of communication networks, VLSI design and vehicle routing. One example is the design of survivable communication networks, which remain functional even after the failure of some network components (see [5] for more details).

Open Problems

The 2-approximation algorithm of Jain [6] for generalized Steiner network is based on LP-rounding, and it has high running time. It would be interesting to design a combinatorial approximation algorithm for this problem.

It is not known whether a better approximation is possible for generalized Steiner network. Very few hardness of approximation results are known for this type of problems. The best current hardness factor stands on 1.01063 [2], and this result is valid even for the special case of Steiner tree.

Cross-References

▸ Steiner Forest
▸ Steiner Trees

Recommended Reading

1. Agrawal A, Klein P, Ravi R (1995) When trees collide: an approximation algorithm for the generalized Steiner problem on networks. J SIAM Comput 24(3):440–456
2. Chlebik M, Chlebikova J (2002) Approximation hardness of the Steiner tree problem on graphs. In: 8th Scandinavian workshop on algorithm theory LNCS, vol 2368, pp 170–179
3. Goemans MX, Goldberg AV, Plotkin SA, Shmoys DB, Tardos É, Williamson DP (1994) Improved approximation algorithms for network design problems. In: Proceedings of the fifth annual ACM-SIAM symposium on discrete algorithms (SODA), pp 223–232

4. Goemans MX, Williamson DP (1995) A general approximation technique for constrained forest problems. SIAM J Comput 24(2):296–317
5. Grötschel M, Monma CL, Stoer M (1995) Design of survivable networks. In: Network models, handbooks in operations research and management science. North Holland Press, Amsterdam
6. Jain K (2001) A factor 2 approximation algorithm for the generalized Steiner network problem. Combinatorica 21(1):39–60
7. Vazirani VV (2001) Approximation algorithms. Springer, Berlin
8. Williamson DP, Goemans MX, Mihail M, Vazirani VV (1995) A primal-dual approximation algorithm for generalized Steiner network problems. Combinatorica 15(3):435–454

Generalized Two-Server Problem

René A. Sitters
Department of Econometrics and Operations Research, VU University, Amsterdam, The Netherlands

Keywords

CNN problem

Years and Authors of Summarized Original Work

2006; Sitters, Stougie

Problem Definition

In the *generalized two-server problem*, we are given two servers: one moving in a metric space $\mathbb{X}$ and one moving in a metric space $\mathbb{Y}$. They are to serve requests $r \in \mathbb{X} \times \mathbb{Y}$which arrive one by one. A request $r = (x, y)$ is served by moving either the $\mathbb{X}$-server to point x or the $\mathbb{Y}$-server to point y. The decision as to which server to move to the next request is irrevocable and has to be taken without any knowledge about future requests. The objective is to minimize the total distance traveled by the two servers (Fig. 1).

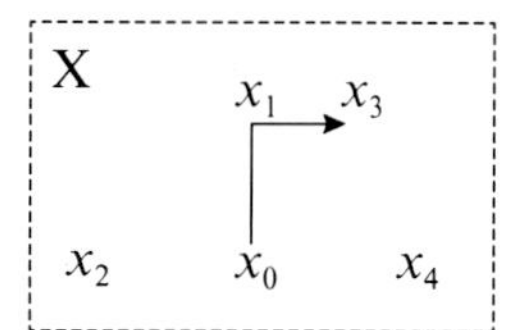

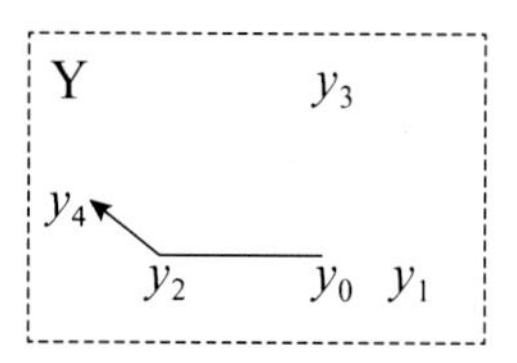

Generalized Two-Server Problem, Fig. 1 In this example, both servers move in the plane and start from the configuration (x_0, y_0). The $\mathbb{X}$-server moves through requests 1 and 3, and the $\mathbb{Y}$-server takes care of requests 2 and 4. The cost of this solution is the sum of the path-lengths

Online Routing Problems

The generalized two-server problem belongs to a class of routing problems called *metrical service systems* [4, 10]. Such a system is defined by a metric space $\mathbb{M}$ of all possible system configurations, an initial configuration $\mathcal{C}_0$, and a set $\mathcal{R}$ of possible requests, where each request $r \in \mathcal{R}$ is a subset of $\mathbb{M}$. Given a sequence, $r_1, r_2 \ldots, r_n$, of requests, a feasible solution is a sequence, $\mathcal{C}_1, \mathcal{C}_2, \ldots, \mathcal{C}_n$, of configurations such that $\mathcal{C}_i \in r_i$ for all $i \in \{1, \ldots, n\}$.

When we model the generalized two-server problem as a metrical service system we have $\mathbb{M} = \mathbb{X} \times \mathbb{Y}$ and $\mathcal{R} = \{\{x \times \mathbb{Y}\} \cup \{\mathbb{X} \times y\} | x \in \mathbb{X}, y \in \mathbb{Y}\}$. In the classical *two-server problem*, both servers move in the same space and receive the same requests, that is, $\mathbb{M} = \mathbb{X} \times \mathbb{X}$ and $\mathcal{R} = \{\{x \times \mathbb{Y}\} \cup \{\mathbb{X} \times x\} | x \in \mathbb{X}\}$.

The performance of algorithms for online optimization problems is often measured using *competitive analysis*. We say that an algorithm is *α-competitive* ($\alpha \geq 1$) for some minimization problem if for every possible instance the cost of the algorithm's solution is at most α times the cost of an optimal solution for the instance.

A standard algorithm that performs provably well for several elementary routing problems is the so-called work function algorithm [2, 5, 8]; after each request, the algorithm moves to a configuration with low cost and which is not too far from the current configuration. More precisely, if the system's configuration after serving a sequence σ is $\mathcal{C}$ and $r \subseteq \mathbb{M}$ is the next request, then the work function algorithm with parameter $\lambda \geq 1$ moves to a configuration $\mathcal{C}' \in r$ that minimizes

$$\lambda W_{\sigma,r}(\mathcal{C}') + d(\mathcal{C}, \mathcal{C}'),$$

where $d(\mathcal{C}, \mathcal{C}')$ is the distance between configurations $\mathcal{C}$ and $\mathcal{C}'$, and $W_{\sigma,r}(\mathcal{C}')$ is the cost of an optimal solution that serves all requests (in order) in σ plus request r with the restriction that it ends in configuration $\mathcal{C}'$.

Key Results

The main result in [11] is a sufficient condition for a metrical service system to have a constant-competitive algorithm. Additionally, the authors show that this condition holds for the generalized two-server problem.

For a fixed metrical service system $\mathcal{S}$ with metric space $\mathbb{M}$, denote by $A(\mathcal{C}, \sigma)$ the cost of algorithm A on input sequence σ, starting in configuration $\mathcal{C}$. Let $\mathrm{OPT}(\mathcal{C}, \sigma)$ be the cost of the corresponding optimal solution. We say that a path T in $\mathbb{M}$ *serves* a sequence σ if it visits all requests in order. Hence, a feasible path is a path that serves the sequence and starts in the initial configuration.

Paths T_1 and T_2 are said to be *independent* if they are far apart in the following way: $|T_1| + |T_2| < d(C_1^g, C_1^t) + d(C_1^g, C_1^t)$, where C_i^g and C_i^t are, respectively, the start and end point of path $T_i (i \in \{1,2\})$. Notice, for example, that two intersecting paths are not independent.

Theorem 1 *Let $\mathcal{S}$ be a metrical service system with metric space $\mathbb{M}$. Suppose there exists an algorithm A and constants $\alpha \geq 1$, $\beta \geq 0$, and $m \geq 2$ such that for any point $\mathcal{C} \in \mathbb{M}$, sequence σ and pairwise independent paths $T_1, T_2, \ldots, T_m$ that serve σ*

$$A(\mathcal{C}, \sigma) \leq \alpha \mathrm{OPT}(\mathcal{C}, \sigma) + \beta \sum_{i=1}^{m} |T_i|. \quad (1)$$

Then there exists an algorithm B that is constant competitive for $\mathcal{S}$.

The proof in [11] of the theorem above provides an explicit formulation of B. This algorithm combines algorithm A with the work function algorithm and operates in phases. In each phase, it applies algorithm A until its cost becomes too large compared to the optimal cost. Then, it makes one step of the work function algorithm and a new phase starts. In each phase, algorithm A makes a restart, that is, it takes the final configuration of the previous phase as the initial configuration, whereas the work function algorithm remembers the whole request sequence.

For the generalized two-server problem the so-called balance algorithm satisfies condition (1). This algorithm stores the cumulative costs of the two servers and with each request it moves the server that minimizes the maximum of the two new values. The balance algorithm itself is not constant competitive but Theorem 1 says that, if we combine it in a clever way with the work function algorithm, then we get an algorithm that is constant competitive.

Applications

A set of metrical service systems can be combined to get what is called in [9] the *sum system.* A request of the sum system consists of one request for each system, and to serve it we need to serve at least one of the individual requests. The generalized two-server problem should be considered as one of the simplest sum systems since the two individual problems are completely trivial: There is one server and each request consists of a single point.

Sum systems are particularly interesting to model systems for information storage and retrieval. To increase stability or efficiency, one may store copies of the same information in multiple systems (e.g., databases, hard disks). To retrieve one piece of information, we may read it from any system. However, to read information it may be necessary to change the configuration of the system. For example, if the database is stored in a binary search tree, then it is efficient to make online changes to the structure of the tree, that is, to use dynamic search trees [12].

Open Problems

A proof that the work function algorithm is competitive for the generalized two-server problem (as conjectured in [9] and [11]) is still lacking.

Also, a randomized algorithm with a smaller competitive ratio than that of [11] is not known. No results (except for a lower bound) are known for the generalized problem with more than two servers. It is not even clear if the work function algorithm may be competitive here.

There are systems for which the work function algorithm is not competitive. It would be interesting to have a nontrivial property that implies competitiveness of the work function algorithm.

Cross-References

- ▶ Algorithm DC-TREE for k-Servers on Trees
- ▶ Metrical Task Systems
- ▶ Online Paging and Caching
- ▶ Work-Function Algorithm for k-Servers

Recommended Reading

1. Borodin A, El-Yaniv R (1998) Online computation and competitive analysis. Cambridge University Press, Cambridge
2. Burley WR (1996) Traversing layered graphs using the work function algorithm. J Algorithms 20:479–511
3. Chrobak M (2003) Sigact news online algorithms column 1. ACM SIGACT News 34:68–77
4. Chrobak M, Larmore LL (1992) Metrical service systems: deterministic strategies. Tech. Rep. UCR-CS-93-1, Department of Computer Science, University of California at Riverside
5. Chrobak M, Sgall J (2004) The weighted 2-server problem. Theor Comput Sci 324:289–312
6. Chrobak M, Karloff H, Payne TH, Vishwanathan S (1991) New results on server problems. SIAM J Discret Math 4:172–181
7. Fiat A, Ricklin M (1994) Competitive algorithms for the weighted server problem. Theor Comput Sci 130:85–99
8. Koutsoupias E, Papadimitriou CH (1995) On the k-server conjecture. J ACM 42:971–983
9. Koutsoupias E, Taylor DS (2004) The CNN problem and other k-server variants. Theor Comput Sci 324:347–359
10. Manasse MS, McGeoch LA, Sleator DD (1990) Competitive algorithms for server problems. J Algorithms **11**, 208–230
11. Sitters RA, Stougie L (2006) The generalized two-server problem. J ACM 53:437–458
12. Sleator DD, Tarjan RE (1985) Self-adjusting binary search trees. J ACM 32:652–686

Generalized Vickrey Auction

Makoto Yokoo
Department of Information Science and Electrical Engineering, Kyushu University, Nishi-ku, Fukuoka, Japan

Keywords

Generalized Vickrey auction; GVA; VCG; Vickrey–Clarke–Groves mechanism

Years and Authors of Summarized Original Work

1995; Varian

Problem Definition

Auctions are used for allocating goods, tasks, resources, etc. Participants in an auction include an auctioneer (usually a seller) and bidders (usually buyers). An auction has well-defined rules that enforce an agreement between the auctioneer and the winning bidder. Auctions are often used when a seller has difficulty in estimating the value of an auctioned good for buyers.

The Generalized Vickrey Auction protocol (GVA) [5] is an auction protocol that can be used for combinatorial auctions [3] in which multiple items/goods are sold simultaneously. Although conventional auctions sell a single item at a time, combinatorial auctions sell multiple items/goods. These goods may have interdependent values, e.g., these goods are complementary/substitutable and bidders can bid on any combination of goods. In a combinatorial auction, a bidder can express complementary/substitutable preferences over multiple bids. By taking into account complementary/substitutable preferences, the participants' utilities and the revenue of the seller can be increased. The GVA is one instance of the Clarke mechanism [2, 4]. It is also called the Vickrey–Clarke–Groves mechanism (VCG). As its name suggests, it is a generalized version of the well-known Vickrey (or second-price) auction protocol [6], proposed by an American economist W. Vickrey, a 1996 Nobel Prize winner.

Assume there is a set of bidders $N = \{1, 2, \dots, n\}$ and a set of goods $M = \{1, 2, \dots, m\}$. Each bidder i has his/her preferences over a bundle, i.e., a subset of goods $B \subseteq M$. Formally, this can be modeled by supposing that bidder i privately observes a parameter, or signal, θ_i, which determines his/her preferences. The parameter θ_i is called the *type* of bidder i. A bidder is assumed to have a *quasilinear, private value* defined as follows.

Definition 1 (Utility of a Bidder) The utility of bidder i, when i obtains $B \subseteq M$ and pays p_i, is represented as $v(B, \theta_i) - p_i$.

Here, the valuation of a bidder is determined independently of other bidders' valuations. Also, the utility of a bidder is linear in terms of the payment. Thus, this model is called a quasilinear, private value model.

Definition 2 (Incentive Compatibility) An auction protocol is (dominant-strategy) *incentive compatible* (or *strategy-proof*) if declaring the true type/evaluation values is a dominant strategy for each bidder, i.e., an optimal strategy regardless of the actions of other bidders.

A combination of dominant strategies of all bidders is called a *dominant-strategy equilibrium*.

Definition 3 (Individual Rationality) An auction protocol is *individually rational* if no participant suffers any loss in a dominant-strategy equilibrium, i.e., the payment never exceeds the evaluation value of the obtained goods.

Definition 4 (Pareto Efficiency) An auction protocol is *Pareto efficient* when the sum of all participants' utilities (including that of the auctioneer), i.e., the social surplus, is maximized in a dominant-strategy equilibrium.

The goal is to design an auction protocol that is incentive compatible, individually rational, and Pareto efficient. It is clear that individual rationality and Pareto efficiency are desirable. Regarding the incentive compatibility, the *revelation principle* states that in the design of an auction protocol, it is possible to restrict attention only to incentive compatible protocols without loss of generality [4]. In other words, if a certain property (e.g., Pareto efficiency) can be achieved using some auction protocol in a dominant-strategy equilibrium, then the property can also be achieved using an incentive-compatible auction protocol.

Key Results

A *feasible* allocation is defined as a vector of n bundles $\vec{B} = \langle B_1, \dots, B_n \rangle$, where $\bigcup_{j \in N} B_j \subseteq M$ and for all $j \neq j'$, $B_j \cap B_{j'} = \emptyset$ hold.

The GVA protocol can be described as follows.

1. Each bidder i declares his/her type $\hat{\theta}_i$, which can be different from his/her true type θ_i.
2. The auctioneer chooses an optimal allocation $\vec{B}^*$ according to the declared types. More precisely, the auctioneer chooses $\vec{B}^*$ defined as follows:

$$\vec{B}^* = \arg\max_{\vec{B}} \sum_{j \in N} v\left(B_j, \hat{\theta}_j\right).$$

3. Each bidder i pays p_i, which is defined as follows ($B_j^{\sim i}$ and B_j^* are the jth element of $\vec{B}^{\sim i}$ and $\vec{B}^*$, respectively):

$$p_i = \sum_{j \in N \setminus \{i\}} v\left(B_j^{\sim i}, \hat{\theta}_j\right) - \sum_{j \in N \setminus \{i\}} v\left(B_j^*, \hat{\theta}_j\right),$$

$$\text{where } \vec{B}^{\sim i} = \arg\max_{\vec{B}} \sum_{j \in N \setminus \{i\}} v\left(B_j, \hat{\theta}_j\right). \quad (1)$$

The first term in Eq. (1) is the social surplus when bidder i does not participate. The second term is the social surplus except bidder i when i does participate. In the GVA, the payment of bidder i can be considered as the decreased amount of the other bidders' social surplus resulting from his/her participation.

A description of how this protocol works is given below.

Example 1 Assume there are two goods a and b, and three bidders, 1, 2, and 3, whose types are θ_1, θ_2, and θ_3, respectively. The evaluation value for a bundle $v(B, \theta_i)$ is determined as follows.

	$\{a\}$	$\{b\}$	$\{a,b\}$
θ_1	\$6	\$0	\$6
θ_2	\$0	\$0	\$8
θ_3	\$0	\$5	\$5

Here, bidder 1 wants good a only, and bidder 3 wants good b only. Bidder 2's utility is all-or-nothing, i.e., he/she wants both goods at the same time and having only one good is useless.

Assume each bidder i declares his/her true type θ_i. The optimal allocation is to allocate good a to bidder 1 and b to bidder 3, i.e., $\vec{B}^* = \langle\{a\}, \{\}, \{b\}\rangle$. The payment of bidder 1 is calculated as follows. If bidder 1 does not participate, the optimal allocation would have been allocating both items to bidder 2, i.e., $\vec{B}^{\sim 1} = \langle\{\}, \{a,b\}, \{\}\rangle$ and the social surplus, i.e., $\sum_{j \in N \setminus \{1\}} v\left(B_j^{\sim 1}, \hat{\theta}_j\right)$ is equal to \$8. When bidder 1 does participate, bidder 3 obtains $\{b\}$, and the social surplus except for bidder 1, i.e., $\sum_{j \in N \setminus \{1\}} v\left(B_j^*, \hat{\theta}_j\right)$, is 5. Therefore, bidder 1 pays the difference \$8 – \$5 = \$3. The obtained utility of bidder 1 is \$6 – \$3 = \$3. The payment of bidder 3 is calculated as \$8 – \$6 = \$2.

The intuitive explanation of why truth telling is the dominant strategy in the GVA is as follows. In the GVA, goods are allocated so that the social surplus is maximized. In general, the utility of society as a whole does not necessarily mean maximizing the utility of each participant. Therefore, each participant might have an incentive for lying if the group decision is made so that the social surplus is maximized.

However, the payment of each bidder in the GVA is cleverly determined so that the utility of each bidder is maximized when the social surplus is maximized. Figure 1 illustrates the relationship between the payment and utility of bidder 1 in Example 1. The payment of bidder 1 is defined as the difference between the social surplus when bidder 1 does not participate (i.e., the length of the upper shaded bar) and the social surplus except bidder 1 when bidder 1 does participate (the length of the lower black bar), i.e., \$8 – \$5 = \$3.

On the other hand, the utility of bidder 1 is the difference between the evaluation value of the obtained item and the payment, which equals \$6 – \$3 = \$3. This amount is equal to the difference between the total length of the lower bar and the upper bar. Since the length of the upper bar is determined independently of bidder 1's declaration, bidder 1 can maximize his/her utility by maximizing the length of the lower bar. However, the length of the lower bar represents the social surplus. Thus, bidder 1 can maximize his/her utility when the social surplus is maximized. Therefore, bidder 1 does not have an incentive for lying since the group decision is made so that the social surplus is maximized.

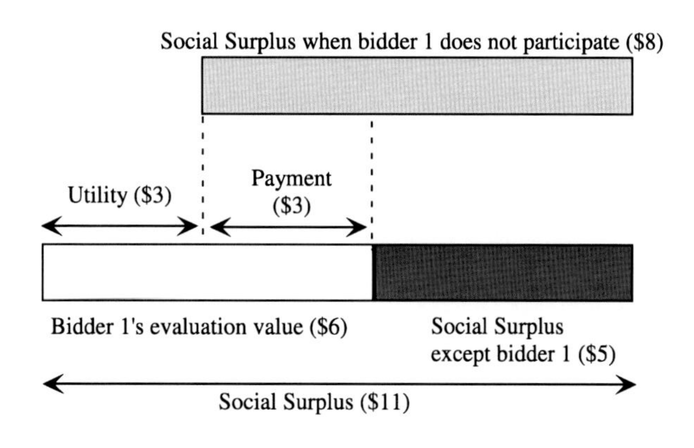

Generalized Vickrey Auction, Fig. 1 Utilities and Payments in the GVA

Theorem 1 *The GVA is incentive compatible.*

Proof Since the utility of bidder i is assumed to be quasilinear, it can be represented as

$$v(B_i, \theta_i) - p_i = v(B_i, \theta_i) - \left[\sum_{j \in N \setminus \{i\}} v\left(B_j^{\sim i}, \hat{\theta}_j\right) - \sum_{j \in N \setminus \{i\}} v\left(B_j^*, \hat{\theta}_j\right) \right]$$

$$= \left[v(B_i, \theta_i) + \sum_{j \in N \setminus \{i\}} v\left(B_j^*, \hat{\theta}_j\right) \right] - \sum_{j \in N \setminus \{i\}} v\left(B_j^{\sim i}, \hat{\theta}_j\right) \quad (2)$$

The second term in Eq. (2) is determined independently of bidder i's declaration. Thus, bidder 1 can maximize his/her utility by maximizing the first term. However, $\vec{B}^*$ is chosen so that $\sum_{j \in N} v\left(B_j, \hat{\theta}_j\right)$ is maximized. Therefore, bidder i can maximize his/her utility by declaring $\hat{\theta}_i = \theta_i$, i.e., by declaring his/her true type. □

Theorem 2 *The GVA is individually rational.*

Proof This is clear from Eq. (2), since the first term is always larger than (or at least equal to) the second term. □

Theorem 3 *The GVA is Pareto efficient.*

Proof From Theorem 1, truth telling is a dominant-strategy equilibrium. From the way of choosing the allocation, the social surplus is maximized if all bidders declare their true types. □

Applications

The GVA can be applied to combinatorial auctions, which have lately attracted considerable attention [3]. The US Federal Communications Commission has been conductingauctions for allocating spectrum rights. Clearly, there exist interdependencies among the values of spectrum rights. For example, a bidder may desire licenses for adjoining regions simultaneously, i.e., these licenses are complementary. Thus, the spectrum auctions is a promising application field of combinatorial auctions and have been a major driving force for activating the research on combinatorial auctions.

Open Problems

Although the GVA has these good characteristics (Pareto efficiency, incentive compatibility, and individual rationality), these characteristics cannot be guaranteed when bidders can submit *false-name* bids. Furthermore, [1] pointed out several other limitations such as vulnerability to the collusion of the auctioneer and/or losers.

Also, to execute the GVA, the auctioneer must solve a complicated optimization problem. Various studies have been conducted to introduce search techniques, which were developed in the artificial intelligence literature, for solving this optimization problem [3].

Cross-References

▶ False-Name-Proof Auction

Recommended Reading

1. Ausubel LM, Milgrom PR (2002) Ascending auctions with package bidding. Front Theor Econ 1(1). Article 1
2. Clarke EH (1971) Multipart pricing of public goods. Public Choice 2:19–33
3. Cramton P, Steinberg R, Shoham Y (eds) (2005) Combinatorial auctions. MIT, Cambridge
4. Mas-Colell A, Whinston MD, Green JR (1995) Microeconomic theory. Oxford University Press, Oxford
5. Varian HR (1995) Economic mechanism design for computerized agents. In: Proceedings of the 1st Usenix workshop on electronic commerce
6. Vickrey W (1961) Counter speculation, auctions, and competitive sealed tenders. J Financ 16:8–37

Geographic Routing

Aaron Zollinger
Department of Electrical Engineering and Computer Science, University of California, Berkeley, CA, USA

Keywords

Directional routing; Geometric routing; Location-based routing; Position-based routing

Years and Authors of Summarized Original Work

2003; Kuhn, Wattenhofer, Zollinger

Problem Definition

Geographic routing is a type of routing particularly well suited for dynamic ad hoc networks. Sometimes also called directional, geometric, location-based, or position-based routing, it is based on two principal assumptions. First, it is assumed that every node knows its own and its network neighbors' positions. Second, the source of a message is assumed to be informed about the position of the destination. Geographic routing is defined on a Euclidean graph, that is a graph whose nodes are embedded in the Euclidean plane. Formally, geographic ad hoc routing algorithms can be defined as follows:

Definition 1 (Geographic Ad Hoc Routing Algorithm) Let $G = (V, E)$ be a Euclidean graph. The task of a geographic ad hoc routing algorithm $\mathcal{A}$ is to transmit a message from a source $s \in V$ to a destination $t \in V$ by sending packets over the edges of G while complying with the following conditions:

- All nodes $v \in V$ know their geographic positions as well as the geographic positions of all their neighbors in G.
- The source s is informed about the position of the destination t.
- The control information which can be stored in a packet is limited by $O(\log n)$ bits, that is, only information about a constant number of nodes is allowed.
- Except for the temporary storage of packets before forwarding, a node is not allowed to maintain any information.

Geographic routing is particularly interesting, as it operates without any routing tables whatsoever. Furthermore, once the position of the destination is known, all operations are strictly local, that is, every node is required to keep track only of its direct neighbors. These two factors – absence of necessity to keep routing tables up to date and independence of remotely occurring topology changes – are among the foremost reasons why geographic routing is exceptionally suitable for operation in ad hoc networks. Furthermore, in a sense, geographic routing can be considered a lean version of source routing appropriate for dynamic networks: While in source routing the complete hop-by-hop route to be followed by the message is specified by the source, in geographic routing the source simply addresses the message with the position of the destination. As the destination can generally be expected to move slowly compared to the frequency of topology changes between the source and the destination, it makes sense to keep track of the position of the destination instead of maintaining network topology information up to date; if the destination does not move too fast, the message is delivered regardless of possible topology changes among intermediate nodes.

The cost bounds presented in this entry are achieved on *unit disk graphs*. A unit disk graph is defined as follows:

Definition 2 (Unit Disk Graph) Let $V \subset \mathbb{R}^2$ be a set of points in the 2-dimensional plane. The graph with edges between all nodes with distance at most 1 is called the unit disk graph of V.

Unit disk graphs are often employed to model wireless ad hoc networks.

The routing algorithms considered in this entry operate on planar graphs, graphs that contain no two intersecting edges. There exist strictly local algorithms constructing such planar graphs given a unit disk graph. The edges of planar graphs partition the Euclidean plane into contiguous areas, so-called faces. The algorithms cited in this entry are based on these faces.

Key Results

The first geographic routing algorithm shown to always reach the destination was Face Routing introduced in [14].

Theorem 1 *If the source and the destination are connected, Face Routing executed on an arbitrary planar graph always finds a path to the destination. It thereby takes at most $O(n)$ steps, where n is the total number of nodes in the network.*

There exists however a geographic routing algorithm whose cost is bounded not only with respect to the total number of nodes, but in relation to the *shortest path* between the source and the destination: The GOAFR$^+$ algorithm [15, 16, 18, 24] (pronounced as "gopher-plus") combines *greedy routing* – where every intermediate node relays the message to be routed to its neighbor located nearest to the destination – with face routing. Together with the locally computable *Gabriel Graph* planarization technique, the effort expended by the GOAFR$^+$ algorithm is bounded as follows:

Theorem 2 *Let c be the cost of an optimal path from s to t in a given unit disk graph. GOAFR$^+$ reaches t with cost $O(c^2)$ if s and t are connected. If s and t are not connected, GOAFR$^+$ reports so to the source.*

On the other hand it can be shown that – on certain worst-case graphs – no geographic routing algorithm operating in compliance with the above definition can perform asymptotically better than GOAFR$^+$:

Theorem 3 *There exist graphs where any deterministic (randomized) geographic ad hoc routing algorithm has (expected) cost $\Omega(c^2)$.*

This leads to the following conclusion:

Theorem 4 *The cost expended by GOAFR$^+$ to reach the destination on a unit disk graph is asymptotically optimal.*

In addition, it has been shown that the GOAFR$^+$ algorithm is not only guaranteed to have low worst-case cost but that it also performs well in average-case networks with nodes randomly placed in the plane [15, 24].

Applications

By its strictly local nature geographic routing is particularly well suited for application in potentially highly dynamic wireless ad hoc networks. However, also its employment in dynamic networks in general is conceivable.

Open Problems

A number of problems related to geographic routing remain open. This is true above all with respect to the dissemination within the network of information about the destination position and on the other hand in the context of node mobility as well as network dynamics. Various approaches to these problems have been described in [7] as well as in chapters 11 and 12 of [24]. More generally, taking geographic routing one step further towards its application in practical wireless ad hoc networks [12, 13] is a field yet largely open. A more specific open problem is finally posed by the question whether geographic routing can be adapted to networks with nodes embedded in three-dimensional space.

Experimental Results

First experiences with geographic and in particular face routing in practical networks have been made [12, 13]. More specifically, problems in connection with graph planarization that can occur in practice were observed, documented, and tackled.

Cross-References

- ▶ Local Computation in Unstructured Radio Networks
- ▶ Planar Geometric Spanners
- ▶ Routing in Geometric Networks

Recommended Reading

1. Barrière L, Fraigniaud P, Narayanan L (2001) Robust position-based routing in wireless ad hoc networks with unstable transmission ranges. In: Proceedings of the 5th international workshop on discrete algorithms and methods for mobile computing and communications (DIAL-M). ACM Press, New York, pp 19–27
2. Bose P, Brodnik A, Carlsson S, Demaine E, Fleischer R, López-Ortiz A, Morin P, Munro J (2000) Online routing in convex subdivisions. In: International symposium on algorithms and computation (ISAAC). LNCS, vol 1969. Springer, Berlin/New York, pp 47–59
3. Bose P, Morin P (1999) Online routing in triangulations. In: Proceedings of 10th internatrional symposium on algorithms and computation (ISAAC). LNCS, vol 1741. Springer, Berlin, pp 113–122
4. Bose P, Morin P, Stojmenovic I, Urrutia J (1999) Routing with guaranteed delivery in ad hoc wireless networks. In: Proceedings of the 3rd international workshop on discrete algorithms and methods for mobile computing and communications (DIAL-M), pp 48–55
5. Datta S, Stojmenovic I, Wu J (2002) Internal node and shortcut based routing with guaranteed delivery in wireless networks. In: Cluster computing, vol 5. Kluwer Academic, Dordrecht, pp 169–178
6. Finn G (1987) Routing and addressing problems in large metropolitan-scale internetworks. Technical report ISI/RR-87-180, USC/ISI
7. Flury R, Wattenhofer R (2006) MLS: an efficient location service for mobile ad hoc networks. In: Proceedings of the 7th ACM international symposium on mobile ad-hoc networking and computing (MobiHoc), Florence
8. Fonseca R, Ratnasamy S, Zhao J, Ee CT, Culler D, Shenker S, Stoica I (2005) Beacon vector routing: scalable point-to-point routing in wireless sensornets. In: 2nd symposium on networked systems design & implementation (NSDI), Boston
9. Gao J, Guibas L, Hershberger J, Zhang L, Zhu A (2001) Geometric spanner for routing in mobile networks. In: Proceedings of 2nd ACM international symposium on mobile ad-hoc networking and computing (MobiHoc), Long Beach
10. Hou T, Li V (1986) Transmission range control in multihop packet radio networks. IEEE Trans Commun 34:38–44
11. Karp B, Kung H (2000) GPSR: greedy perimeter stateless routing for wireless networks. In: Proceedings of 6th annual international conference on mobile computing and networking (MobiCom), pp 243–254
12. Kim YJ, Govindan R, Karp B, Shenker S (2005) Geographic routing made practical. In: Proceedings of the second USENIX/ACM symposium on networked system design and implementation (NSDI 2005), Boston
13. Kim YJ, Govindan R, Karp B, Shenker S (2005) On the pitfalls of geographic face routing. In: Proceedings of the ACM joint workshop on foundations of mobile computing (DIALM-POMC), Cologne
14. Kranakis E, Singh H, Urrutia J (1999) Compass routing on geometric networks. In: Proceedings of 11th Canadian conference on computational geometry, Vancouver, pp 51–54
15. Kuhn F, Wattenhofer R, Zhang Y, Zollinger A (2003) Geometric routing: of theory and practice. In: Proceedings of the 22nd ACM symposium on the principles of distributed computing (PODC)
16. Kuhn F, Wattenhofer R, Zollinger A (2002) Asymptotically optimal geometric mobile ad-hoc routing. In: Proceedings of 6th international workshop on discrete algorithms and methods for mobile computing and communications (Dial-M). ACM, New York, pp 24–33
17. Kuhn F, Wattenhofer R, Zollinger A (2003) Ad-hoc networks beyond unit disk graphs. In: 1st ACM joint workshop on foundations of mobile computing (DIALM-POMC), San Diego
18. Kuhn F, Wattenhofer R, Zollinger A (2003) Worst-case optimal and average-case efficient geometric ad-hoc routing. In: Proceedings of 4th ACM international symposium on mobile ad-hoc networking and computing (MobiHoc)
19. Leong B, Liskov B, Morris R (2006) Geographic routing without planarization. In: 3rd symposium on networked systems design & implementation (NSDI), San Jose
20. Leong B, Mitra S, Liskov B (2005) Path vector face routing: geographic routing with local face information. In: 13th IEEE international conference on network protocols (ICNP), Boston
21. Takagi H, Kleinrock L (1984) Optimal transmission ranges for randomly distributed packet radio terminals. IEEE Trans Commun 32:246–257
22. Urrutia J (2002) Routing with guaranteed delivery in geometric and wireless networks. In: Stojmenovic I (ed) Handbook of wireless networks and mobile computing, ch. 18. Wiley, Hoboken, pp. 393–406

23. Wattenhofer M, Wattenhofer R, Widmayer P (2005) Geometric routing without geometry. In: 12th colloquium on structural information and communication complexity (SIROCCO), Le Mont Saint-Michel
24. Zollinger A (2005) Networking unleashed: geographic routing and topology control in ad hoc and sensor networks. PhD thesis, ETH Zurich, Switzerland Dissertation, ETH 16025

Geometric Approaches to Answering Queries

Aleksandar Nikolov
Department of Computer Science, Rutgers University, Piscataway, NJ, USA

Keywords

Convex geometry; Convex optimization; Differential privacy; Query release

Years and Authors of Summarized Original Work

2014; Nikolov, Talwar, Zhang
2015; Dwork, Nikolov, Talwar

Problem Definition

The central problem of private data analysis is to extract meaningful information from a statistical database without revealing too much about any particular individual represented in the database. Here, by a *statistical database*, we mean a multiset $D \in \mathcal{X}^n$ of n rows from the *data universe* $\mathcal{X}$. The notation $|D| \triangleq n$ denotes the *size* of the database. Each row represents the information belonging to a single individual. The universe $\mathcal{X}$ depends on the domain. A natural example to keep in mind is $\mathcal{X} = \{0, 1\}^d$, i.e., each row of the database gives the values of d binary attributes for some individual.

Differential privacy formalizes the notion that an adversary should not learn too much about any individual as a result of a private computation. The formal definition follows.

Definition 1 ([8]) A randomized algorithm $\mathcal{A}$ satisfies (ε, δ)-differential privacy if for any two databases D and D' that differ in at most a single row (i.e., $|D \triangle D'| \leq 1$), and any measurable event S in the range of $\mathcal{A}$,

$$\Pr[\mathcal{A}(D) \in S] \leq e^{\varepsilon} \Pr[\mathcal{A}(D') \in S] + \delta.$$

Above, probabilities are taken over the internal coin tosses of $\mathcal{A}$.

Differential privacy guarantees to a data owner that allowing her data to be used for analysis does not risk much more than she would if she did not allow her data to be used.

In the sequel, we shall call databases D and D' that differ in a single row *neighboring databases*, denoted $D \sim D'$. Usually, the parameter ε is set to be a small constant so that $e^{\varepsilon} \approx 1 + \varepsilon$, and δ is set to be no bigger than n^{-2} or even $n^{-\omega(1)}$. The case of $\delta = 0$ often requires different techniques from the case $\delta > 0$; as is common in the literature, we shall call the two cases *pure differential privacy* and *approximate differential privacy*.

Query Release

In the query release problem, we are given a set $\mathcal{Q}$ of queries, where each $q \in \mathcal{Q}$ is a function $q : \mathcal{X}^n \to \mathbb{R}$. Our goal is to design a differentially private algorithm $\mathcal{A}$ which takes as input a database D and outputs a list of answers to the queries in $\mathcal{Q}$. We shall call such an algorithm a *(query answering) mechanism*. Here, we treat the important special case of query release for sets of *linear queries*. A linear query q is specified by a function $q : \mathcal{X} \to [-1, 1]$, and, slightly abusing notation, we define the value of the query as $q(D) \triangleq \sum_{\chi \in D} q(e)$. When $q : \mathcal{X} \to \{0, 1\}$ is a predicate, $q(D)$ is a *counting query*: it simply counts the number of rows of D that satisfy the predicate.

It is easy to see that a differentially private algorithm (with any reasonable choice of ε and δ) cannot answer a nontrivial set of queries exactly. For this reason, we need to have a measure of error, and here we introduce the two most commonly used ones: average and worst-case error. Assume that on an input database D and a set of linear queries $\mathcal{Q}$, the algorithm $\mathcal{A}$ gives answer $\tilde{q}^{\mathcal{A}}(D)$ for query $q \in \mathcal{Q}$. The *average error*of $\mathcal{A}$ on the query set $\mathcal{Q}$ for databases of size at most n is equal to

$$\mathrm{err}_{\mathrm{avg}}(\mathcal{A}, \mathcal{Q}, n) \triangleq \max_{D:|D|\leq n} \sqrt{\frac{1}{|\mathcal{Q}|}\mathbb{E}\left[\sum_{q\in\mathcal{Q}} |\tilde{q}^{\mathcal{A}}(D) - q(D)|^2\right]}.$$

The *worst-case error* is equal to

$$\mathrm{err}_{\mathrm{wc}}(\mathcal{A}, \mathcal{Q}, n) \triangleq \max_{D:|D|\leq n} \mathbb{E} \max_{q\in\mathcal{Q}} |\tilde{q}^{\mathcal{A}}(D) - q(D)|.$$

In both definitions above, expectations are taken over the coin throws of $\mathcal{A}$. We also define $\mathrm{err}_{\mathrm{avg}}(\mathcal{A}, \mathcal{Q}) = \sup_n \mathrm{err}_{\mathrm{avg}}(\mathcal{A}, \mathcal{Q}, n)$, and respectively $\mathrm{err}_{\mathrm{wc}}(\mathcal{A}, \mathcal{Q}) = \sup_n \mathrm{err}_{\mathrm{wc}}(\mathcal{A}, \mathcal{Q}, n)$, to be the maximum error over all database sizes. The objective in the query release problem is to *minimize error subject to privacy constraints.*

Marginal Queries

An important class of counting queries are the *marginal queries*. A k-way marginal query $\mathrm{mar}_{S,\alpha} : \{0,1\}^d \to \{0,1\}$ is specified by a subset of attributes $S \subseteq \{1, \ldots, d\}$ of size k and a vector $\alpha \in \{0,1\}^S$. The query evaluates to 1 on those rows that agree with α on all attributes in S, i.e., $\mathrm{mar}_{S,\alpha}(\chi) = \bigwedge_{i\in S} \chi_i = \alpha_i$ for any $\chi \in \{0,1\}^d$. Recall that, using the notation we introduced above, this implies that $\mathrm{mar}_{S,\alpha}(D)$ counts the number of rows in the database D that agree with α on S. Marginal queries capture contingency tables in statistics and OLAP cubes in databases. They are widely used in the sciences and are released by a number of official agencies.

Matrix Notation

It will be convenient to encode the query release problem for linear queries using matrix notation. A common and very useful representation of a database $D \in \mathcal{X}^n$ is the *histogram representation*: the histogram of D is a vector $x \in \mathbb{P}^{\mathcal{X}}$ ($\mathbb{P}$ is the set of nonnegative integers) such that for any $\chi \in \mathcal{X}$, x_χ is equal to the number of copies of χ in D. Notice that $\|x\|_1 = n$ and also that if x and x' are, respectively, the histograms of two neighboring databases D and D', then $\|x - x'\|_1 \leq 1$ (here $\|x\|_1 = \sum_\chi |x_\chi|$ is the standard ℓ_1 norm). Linear queries are a linear transformation of x. More concretely, let us define the *query matrix* $A \in [-1,1]^{\mathcal{Q}\times\mathcal{X}}$ associated with a set of linear queries $\mathcal{Q}$ by $a_{q,\chi} = q(\chi)$. Then it is easy to see that the vector Ax gives the answers to the queries $\mathcal{Q}$ on a database D with histogram x.

Key Results

A central object of study in geometric approaches to the query release problem is a convex body associated with a set of linear queries. Before introducing some of the main results and algorithms, we define this body.

The Sensitivity Polytope

Let A be the query matrix for some set of queries $\mathcal{Q}$, and let x and x' be the histograms of two neighboring databases, respectively, D and D'. Above, we observed that $D \sim D'$ implies that $\|x - x\|_1 \leq 1$. Let us use the notation $B_1^{\mathcal{X}} \triangleq \{x : \|x\|_1 \leq 1\}$ for the unit ball of the ℓ_1 norm in $\mathbb{R}^{\mathcal{X}}$. Then, $Ax - Ax' \in K_{\mathcal{Q}}$, where

$$K_{\mathcal{Q}} \triangleq \{Ax : \|x\|_1 \leq 1\} = A \cdot B_1^{\mathcal{X}}$$

is the *sensitivity polytope* associated with $\mathcal{Q}$. In other words, the sensitivity polytope is the smallest convex body such that $Ax' \in Ax + K_{\mathcal{Q}}$ for any histogram x and any histogram x' of a neighboring database. In this sense, $K_{\mathcal{Q}}$ describes how the answers to the queries $\mathcal{Q}$ can change between neighboring databases, which motivates the terminology. Informally, a differentially private algorithm must "hide" where in $Ax + K_{\mathcal{Q}}$ the true query answers are.

Another very useful property of the sensitivity polytope is that the vector Ax of query answers for any database of size at most n is contained in $n \cdot K_{\mathcal{Q}} = \{Ax : \|x\|_1 \leq n\}$.

Geometrically, $K_{\mathcal{Q}}$ is a convex polytope in $\mathbb{R}^{\mathcal{Q}}$, centrally symmetric around 0, i.e., $K_{\mathcal{Q}} = -K_{\mathcal{Q}}$. It is the convex hull of the points $\{\pm a_\chi : \chi \in \mathcal{X}\}$, where a_χ is the column of A indexed by the universe element χ, i.e., $a_\chi = (q(\chi))_{q \in \mathcal{Q}}$.

The sensitivity polytope was introduced by Hardt and Talwar [12]. The name was suggested by Li Zhang.

The Generalized Gaussian Mechanism

We mentioned informally that a differentially private mechanism must hide where in $Ax + K_{\mathcal{Q}}$ the true query answers lie. A simple formalization of this intuition is the Gaussian Mechanism, which we present here in a generalized geometric variant.

Recall that an *ellipsoid* in $\mathbb{R}^m$ is an affine transformation $F \cdot B_2^m + y$ of the unit Euclidean ball $B_2^m \triangleq \{x \in \mathbb{R}^m : \|x\|_2 \leq 1\}$ ($\|x\|_2$ is the usual Euclidean, i.e., ℓ_2 norm). In this article, we will only consider centrally symmetric ellipsoids, i.e., ellipsoids of the form $E = F \cdot B_2^m$.

Algorithm 1: Generalized Gaussian Mechanism $\mathcal{A}_E$

Input: *(Public)* Query set $\mathcal{Q}$; ellipsoid $E = F \cdot B_2^{\mathcal{Q}}$ such that $K_{\mathcal{Q}} \subseteq E$.

Input: *(Private)* Database D.

Sample a vector $g \sim N(0, c_{\varepsilon,\delta}^2)^{\mathcal{Q}}$, where $c_{\varepsilon,\delta} = \frac{0.5\sqrt{\varepsilon}+\sqrt{2\ln(1/\delta)}}{\varepsilon}$;

Compute the query matrix A and the histogram x for the database D;

Output: Vector of query answers $Ax + Fg$.

The *generalized Gaussian mechanism* $\mathcal{A}_E$ is shown as Algorithm 1. The notation $g \sim N(0, c_{\varepsilon,\delta}^2)^{\mathcal{Q}}$ means that each coordinate of g is an independent Gaussian random variable with mean 0 and variance $c_{\varepsilon,\delta}^2$. In the special case, when the ellipsoid E is just the Euclidean ball $\Delta_2 \cdot B_2^{\mathcal{Q}}$, with radius equal to the diameter $\Delta_2 \triangleq \max_{y \in K_{\mathcal{Q}}} \|y\|_2$ of $K_{\mathcal{Q}}$, $\mathcal{A}_E$ is the well-known Gaussian mechanism, whose privacy was analyzed in [6–8]. The diameter Δ_2 is also known as the ℓ_2-*sensitivity* of $\mathcal{Q}$ and for linear queries is always upper bounded by $\sqrt{|\mathcal{Q}|}$. The privacy of the generalized version is an easy corollary of the privacy of the standard Gaussian mechanism (see [16] for a proof).

Theorem 1 ([6–8, 16]) *For any ellipsoid E containing $K_{\mathcal{Q}}$, $\mathcal{A}_E$ satisfies (ε, δ)-differential privacy.*

It is not hard to analyze the error of the mechanism $\mathcal{A}_E$. Let $E = F \cdot B_2^{\mathcal{Q}}$, and recall the Hilbert-Schmidt norm $\|F\|_{HS} = \sqrt{\mathrm{tr}(FF^\intercal)}$ and the 1-to-2 norm $\|F^\intercal\|_{1\to 2}$ which is equal to the largest ℓ_2 norm of any row of F. Geometrically, $\|F\|_{HS}$ is equal to the square root of the sum of squared major axis lengths of E, and $\|F^\intercal\|_{1\to 2}$ is equal to the largest ℓ_∞ norm of any point in E. We have the error bounds

$$\mathrm{err}_{\mathrm{avg}}(\mathcal{A}, \mathcal{Q}) = O(c_{\varepsilon,\delta}) \cdot \frac{1}{\sqrt{|\mathcal{Q}|}} \|F\|_{HS};$$

$$\mathrm{err}_{\mathrm{wc}}(\mathcal{A}, \mathcal{Q}) = O(c_{\varepsilon,\delta}\sqrt{\log|\mathcal{Q}|}) \cdot \|F^\intercal\|_{1\to 2}.$$

Surprisingly, for any query set $\mathcal{Q}$, there exists an ellipsoid E such that the generalized Gaussian noise mechanism $\mathcal{A}_E$ is nearly optimal *among all differentially private mechanisms* for $\mathcal{Q}$. In order to formulate the result, let us define $\mathrm{opt}_{\mathrm{avg}}^{\varepsilon,\delta}(\mathcal{Q})$ (respectively, $\mathrm{opt}_{\mathrm{wc}}^{\varepsilon,\delta}(\mathcal{Q})$) to be the infimum of $\mathrm{err}_{\mathrm{avg}}(\mathcal{A}, \mathcal{Q})$ (respectively, $\mathrm{err}_{\mathrm{wc}}(\mathcal{A}, \mathcal{Q})$) over all (ε, δ)-differentially private mechanisms $\mathcal{A}$.

Theorem 2 ([16]) *Let $E = F \cdot B_2^{\mathcal{Q}}$ be the ellipsoid that minimizes $\|F\|_{HS}$ over all ellipsoids E containing $K_{\mathcal{Q}}$. Then*

$$\mathrm{err}_{\mathrm{avg}}(\mathcal{A}_E, \mathcal{Q}) = O(\log|\mathcal{Q}|\sqrt{\log 1/\delta}) \cdot \mathrm{opt}_{\mathrm{avg}}^{\varepsilon,\delta}(\mathcal{Q}).$$

If $E = F \cdot B_2^{\mathcal{Q}}$ minimizes $\|F^\intercal\|_{1\to 2}$ subject to $K_{\mathcal{Q}} \subseteq E$, then

$$\mathrm{err}_{\mathrm{wc}}(\mathcal{A}_E, \mathcal{Q}) = O((\log|\mathcal{Q}|)^{3/2}\sqrt{\log 1/\delta}) \cdot \mathrm{opt}_{\mathrm{wc}}^{\varepsilon,\delta}(\mathcal{Q}).$$

Minimizing $\|F\|_{HS}$ or $\|F^\intercal\|_{1\to 2}$ subject to $K_{\mathcal{Q}} \subseteq F \cdot B_2^{\mathcal{Q}}$ is a convex minimization problem.

An optimal solution can be approximated to within any prescribed degree of accuracy in time polynomial in $|\mathcal{Q}|$ and $|\mathcal{X}|$ via the ellipsoid algorithm. In fact, more efficient solutions are available: both problems can be formulated as semidefinite programs and solved via interior point methods, or one can also use the Plotkin-Shmoys-Tardos framework [1, 17]. Algorithm $\mathcal{A}_E$ also runs in time polynomial in $n, |\mathcal{Q}|, |\mathcal{X}|$, since it only needs to compute the true query answers and sample $|\mathcal{Q}|$ many Gaussian random variables. Thus, Theorem 2 gives an efficient approximation to the optimal differentially private mechanism for any set of linear queries.

The near-optimal mechanisms of Theorem 2 are closely related to the matrix mechanism [13]. The matrix mechanism, given a set of queries $\mathcal{Q}$ with query matrix A, solves an optimization problem to find a strategy matrix M, then computes answers $\tilde{y}$ to the queries Mx using the standard Gaussian mechanism, and outputs $AM^{-1}\tilde{y}$. The generalized Gaussian mechanism $\mathcal{A}_E$ instantiated with ellipsoid $E = F \cdot B_2^{\mathcal{Q}}$ is equivalent to the matrix mechanism with strategy matrix $F^{-1}A$.

The proof of optimality for the generalized Gaussian mechanism is related to a fundamental geometric fact: if all ellipsoids containing a convex body are "large," then the body itself must be "large." In particular, if the sum of squared major axis lengths of any ellipsoid containing $K_{\mathcal{Q}}$ is large, then $K_{\mathcal{Q}}$ must contain a simplex of proportionally large volume. Moreover, this simplex is the convex hull of a subset of the contact points of $K_{\mathcal{Q}}$ with the optimal ellipsoid. Since the contact points must be vertices of $K_{\mathcal{Q}}$, and all vertices of $K_{\mathcal{Q}}$ are either columns of the query matrix A or their negations, this guarantees the existence of a submatrix of A with large determinant. Determinants of submatrices in turn bound $\mathrm{opt}_{\mathrm{avg}}^{\varepsilon,\delta}(\mathcal{Q})$ from below (this is a consequence of a connection between combinatorial discrepancy and privacy [15], and the determinant lower bound on discrepancy [14]). This phenomenon is related to the Restricted Invertibility Principle of Bourgain and Tzafriri [4] and was established for the closely related minimum volume ellipsoid by Vershynin [18].

The Gaussian noise mechanism can only provide approximate privacy guarantees: when $\delta = 0$, the noise variance scaling factor $c_{\varepsilon,\delta}$ is unbounded. The case of pure privacy requires different techniques. Nevertheless, $(\varepsilon, 0)$-differentially private algorithms with efficiency and optimality guarantees analogous to these in Theorem 2 are known [2,12]. They use a more complicated noise distribution. In the important special case when $K_{\mathcal{Q}}$ is "well rounded" (technically, when $K_{\mathcal{Q}}$ is isotropic), the noise vector is sampled uniformly from $r \cdot K_{\mathcal{Q}}$, where r is a Γ-distributed random variable. Optimality is established conditional on the Hyperplane Conjecture [12] or unconditionally using Klartag's proof of an isomorphic version of the conjecture [2].

The Projection Mechanism

Despite the near-optimality guarantees, the generalized Gaussian mechanism has some drawbacks that can limit its applicability. One issue is that in some natural scenarios, the universe size $|\mathcal{X}|$ can be huge, and running time linear in $|\mathcal{X}|$ is impractical. Another is that its error is sometimes larger even than the database size, making the query answers unusable. We shall see that a simple modification of the Gaussian mechanism, based on an idea from statistics, goes a long way towards addressing these issues.

It is known that there exist sets of linear queries $\mathcal{Q}$ for which $\mathrm{opt}_{\mathrm{avg}}^{\varepsilon,\delta}(\mathcal{Q}) = \Omega(\sqrt{|\mathcal{Q}|})$ for any small enough constant ε and δ [6, 9]. However, this lower bound only holds for large databases, and algorithms with significantly better error guarantees are known when $n = o(|\mathcal{Q}|)$ [3, 10, 11]. We now know that there are (ε, δ)-differentially private algorithms that answer any set $\mathcal{Q}$ of linear queries on any database $D \in \mathcal{X}^n$ with average error at most

$$O\left(\frac{\sqrt{n}(\log|\mathcal{X}|)^{1/4}(\log 1/\delta)^{1/4}}{\sqrt{\varepsilon}}\right). \quad (1)$$

Moreover, for k-way marginal queries, this much error is necessary, up to factors logarithmic in n and $|\mathcal{Q}|$ [5]. Here, we describe a simple geometric algorithm from [16] that achieves this error bound for any $\mathcal{Q}$.

G

We know that for any set of queries $\mathcal{Q}$ and a database of size n, the true query answers are between 0 and n. Therefore, it is always safe to take the noisy answers $\tilde{y}$ output by the Gaussian mechanism and truncate them inside the interval $[0, n]$. However, we can do better by using knowledge of the query set $\mathcal{Q}$. For any database D of size n, the true query answers $y = Ax$ lie in $n \cdot K_{\mathcal{Q}}$. This suggests a regression approach: find the vector of answers $\hat{y} \in n \cdot K_{\mathcal{Q}}$ which is closest to the noisy output $\tilde{y}$ from the Gaussian mechanism. This is the main insight used in the projection mechanism (Algorithm 2).

Algorithm 2: Projection Mechanism $\mathcal{A}_{\text{proj}}$

Input: *(Public)* Query set $\mathcal{Q}$;
Input: *(Private)* Database $D \in \mathcal{X}^n$.
Compute a noisy vector of query answers $\tilde{y}$ with $\mathcal{A}_E$ for $E = \sqrt{|\mathcal{Q}|} \cdot B_2^{\mathcal{Q}}$.
Compute a projection $\hat{y}$ of $\tilde{y}$ onto $n \cdot K_{\mathcal{Q}}$:
$\hat{y} \triangleq \arg\min_{\hat{y} \in n \cdot K_{\mathcal{Q}}} \|\tilde{y} - \hat{y}\|_2$.
Output: Vector of answers $\hat{y}$.

The fact that the projection mechanism is (ε, δ)-differentially private is immediate, because its only interaction with the database is via the (ε, δ)-differentially private Gaussian mechanism, and post-processing cannot break differential privacy.

The projection step in $\mathcal{A}_{\text{proj}}$ reduces the noise significantly when $n = o(|\mathcal{Q}|/\varepsilon)$. Intuitively, in this case, $n \cdot K_{\mathcal{Q}}$ is small enough so that projection cancels a significant portion of the noise. Let us sketch the analysis. Let $y = Ax$ be the true query answers and $g = \tilde{y} - y$ the Gaussian noise vector. A simple geometric argument shows that $\|y - \hat{y}\|_2^2 \le 2|\langle y - \hat{y}, g\rangle|$: the main observation is that in the triangle formed by y, $\tilde{y}$, and $\hat{y}$, the angle at $\hat{y}$ is an obtuse or right angle; see Fig. 1. Since $\hat{y} \in n \cdot K_{\mathcal{Q}}$, there exists some histogram vector $\hat{x}$ with $\|\hat{x}\|_1 \le n$ such that $\hat{y} = A\hat{x}$. We can rewrite the inner product $\langle y - \hat{y}, g\rangle$ as $\langle x - \hat{x}, A^\intercal g\rangle$. Now, we apply Hölder's inequality and get

$$\begin{aligned} \mathbb{E}_g \|y - \hat{y}\|_2^2 &\le 2\mathbb{E}_g |\langle x - \hat{x}, A^\intercal g\rangle| \\ &\le 2\mathbb{E}_g \|x - \hat{x}\|_1 \|A^\intercal g\|_\infty \\ &\le 4n\mathbb{E}_g \|A^\intercal g\|_\infty. \end{aligned} \quad (2)$$

The term $\mathbb{E}_g \|A^\intercal g\|_\infty$ is the expected maximum of $|\mathcal{X}|$ Gaussian random variables, each with mean 0 and variance $c_{\varepsilon,\delta}^2 |\mathcal{Q}|^2$, and standard techniques give the bound $O(c_{\varepsilon,\delta}|\mathcal{Q}|\sqrt{\log|\mathcal{X}|})$. Plugging this into (2) shows that $\text{err}_{\text{avg}}(\mathcal{A}_{\text{proj}}, \mathcal{Q})$ is always bounded by (1). It is useful to note that $\|A^\intercal g\|_\infty$ is equal to $h_{K_{\mathcal{Q}}}(g) = \max_{y \in K_{\mathcal{Q}}} |\langle y, g\rangle|$, where $h_{K_{\mathcal{Q}}}$ is the *support function* of $K_{\mathcal{Q}}$. Geometrically, $h_{K_{\mathcal{Q}}}(g)$ is equal to half the width of $K_{\mathcal{Q}}$ in the direction of g, scaled by the Euclidean length of g (see Fig. 1). Thus, the average error of $\mathcal{A}_{\text{proj}}$ scales with the expected width of $n \cdot K_{\mathcal{Q}}$ in a random direction.

Running in Time Sublinear in $|\mathcal{X}|$

An important example when running time linear in $|\mathcal{X}|$ is impractical is marginal queries: the size of the universe is 2^d, which is prohibitive even for a moderate number of attributes. Notice, however, that in order to compute $\tilde{y}$ in Algorithm 2, we only need to compute the true query answers and add independent Gaussian noise to each. This can be done in time $O(n|\mathcal{Q}|)$. The computa-

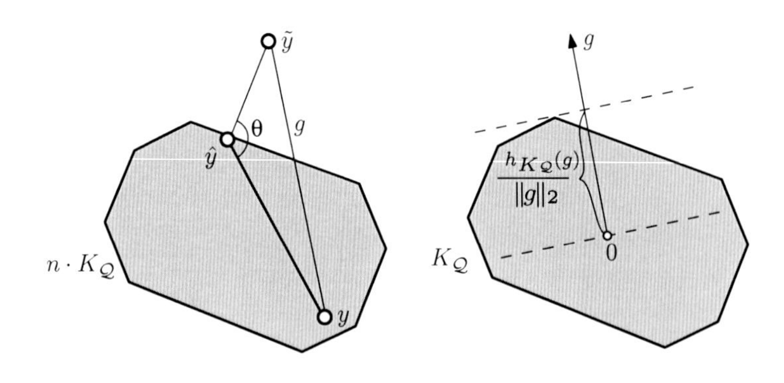

Geometric Approaches to Answering Queries, Fig. 1 The projection mechanism $\mathcal{A}_{\text{proj}}$ on the *left*: the angle θ is necessarily obtuse or right. The figure on the *right* shows the value of the support function $h_{K_{\mathcal{Q}}}(g)$ is equal to $\frac{1}{2}\|g\|_2$ times the width of $K_{\mathcal{Q}}$ in the direction of g

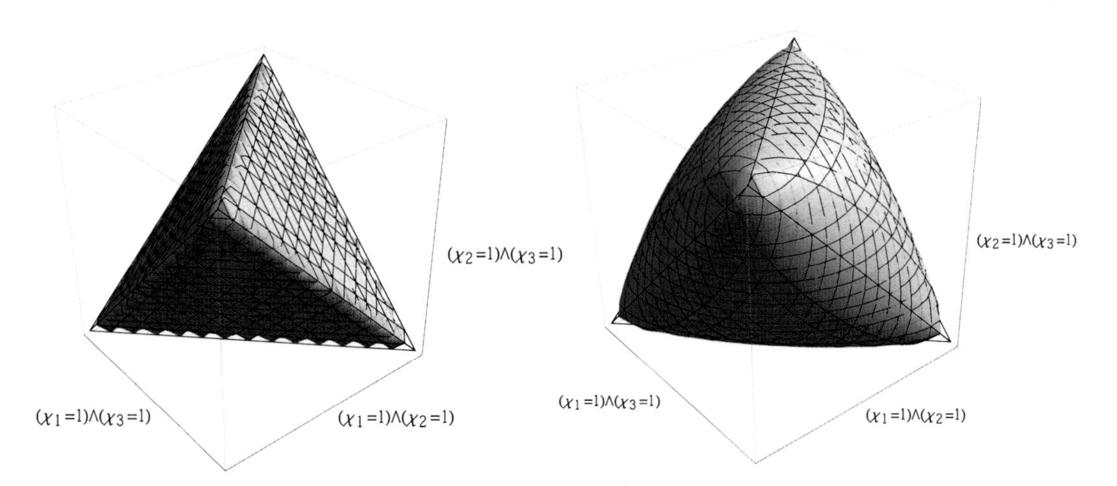

Geometric Approaches to Answering Queries, Fig. 2 The sensitivity polytope for 2-way marginals on 3 attributes (*left*) and a spectrahedral relaxation of the polytope (*right*). A projection onto 3 of the 12 queries restricted to the positive orthant is shown

tionally expensive operation then is computing the projection $\tilde{y}$. This is a convex optimization problem, and can be solved using the ellipsoid algorithm, provided we have a separation oracle for $K_{\mathcal{Q}}$. (A more practical approach is to use the Frank-Wolfe algorithm which can be implemented efficiently as long as we can solve arbitrary linear programs with feasible region $K_{\mathcal{Q}}$.) For k-way marginals, after a linear transformation that doesn't significantly affect error, $K_{\mathcal{Q}}$ can be assumed to be the convex hull of $\{\pm\chi^{\otimes k} : \chi \in \{-1,1\}^d\}$, where $\chi^{\otimes k}$ is the k-fold tensor power of χ. Unfortunately, even for $k = 2$, separation for this convex body is NP-hard. Nevertheless, a small modification of the analysis of $\mathcal{A}_{\text{proj}}$ shows that the algorithm achieves asymptotically the same error bound if we project onto a convex body L such that $K_{\mathcal{Q}} \subseteq L$ and $\mathbb{E}_g h_L(g) \leq O(1) \cdot \mathbb{E}_g h_{K_{\mathcal{Q}}}(g)$. In other words, we need a convex L that relaxes $K_{\mathcal{Q}}$ but is not too much wider than $K_{\mathcal{Q}}$ in a random direction. If we can find such an L with an efficient separation oracle, we can implement $\mathcal{A}_{\text{proj}}$ in time polynomial in $\mathcal{Q}$ and n while only increasing the error by a constant factor. For 2-way marginals, an appropriate relaxation can be derived from Grothendieck's inequality and is formulated using semidefinite programming. The sensitivity polytope $K_{\mathcal{Q}}$ and the relaxation L are shown for 2-way marginals on $\{0,1\}^3$ in Fig. 2. Finding a relaxation L for k-way marginals with efficient separation and mean width bound $\mathbb{E}_g h_L(g) \leq O(1) \cdot \mathbb{E}_g h_{K_{\mathcal{Q}}}(g)$ is an open problem for $k \geq 3$.

Optimal Error for Small Databases

We can refine the optimal error $\text{opt}_{\text{avg}}^{\varepsilon,\delta}(\mathcal{Q})$ to a curve $\text{opt}_{\text{avg}}^{\varepsilon,\delta}(\mathcal{Q}, n)$, where $\text{opt}_{\text{avg}}^{\varepsilon,\delta}(\mathcal{Q}, n)$ is the infimum of $\text{err}_{\text{avg}}(\mathcal{A}, \mathcal{Q}, n)$ over all (ε, δ)-differentially private algorithms $\mathcal{A}$. There exists an algorithm that, for any database of size at most n and any query set $\mathcal{Q}$, has an average error only a polylogarithmic (in $|\mathcal{Q}|$, $|\mathcal{X}|$, and $1/\delta$) factor larger than $\text{opt}_{\text{avg}}^{\varepsilon,\delta}(\mathcal{Q}, n)$ [16]. The algorithm is similar to $\mathcal{A}_{\text{proj}}$. However, the noise distribution used is the optimal one from Theorem 2. The post-processing step is also slightly more complicated, but the key step is again noise reduction via projection onto a convex body. The running time is polynomial in $n, |\mathcal{Q}|, |\mathcal{X}|$. Giving analogous guarantees for worst-case error remains open.

Recommended Reading

1. Arora S, Hazan E, Kale S (2012) The multiplicative weights update method: a meta-algorithm and applications. Theory Comput 8(1):121–164
2. Bhaskara A, Dadush D, Krishnaswamy R, Talwar K (2012) Unconditional differentially private mechanisms for linear queries. In: Proceedings of the 44th

symposium on theory of computing (STOC'12), New York. ACM, New York, pp 1269–1284. DOI 10.1145/2213977.2214089, http://doi.acm.org/10.1145/2213977.2214089
3. Blum A, Ligett K, Roth A (2008) A learning theory approach to non-interactive database privacy. In: Proceedings of the 40th annual ACM symposium on theory of computing (STOC'08), Victoria. ACM, New York, pp 609–618. http://doi.acm.org/10.1145/1374376.1374464
4. Bourgain J, Tzafriri L (1987) Invertibility of large submatrices with applications to the geometry of banach spaces and harmonic analysis. Isr J Math 57(2):137–224
5. Bun M, Ullman J, Vadhan S (2013) Fingerprinting codes and the price of approximate differential privacy. arXiv preprint arXiv:13113158
6. Dinur I, Nissim K (2003) Revealing information while preserving privacy. In: Proceedings of the 22nd ACM symposium on principles of database systems, San Diego, pp 202–210
7. Dwork C, Nissim K (2004) Privacy-preserving datamining on vertically partitioned databases. In: Advances in cryptology – CRYPTO'04, Santa Barbara, pp 528–544
8. Dwork C, Mcsherry F, Nissim K, Smith A (2006) Calibrating noise to sensitivity in private data analysis. In: TCC, New York. http://www.cs.bgu.ac.il/~kobbi/papers/sensitivity-tcc-final.pdf
9. Dwork C, McSherry F, Talwar K (2007) The price of privacy and the limits of LP decoding. In: Proceedings of the thirty-ninth annual ACM symposium on theory of computing (STOC'07), San Diego. ACM, New York, pp 85–94. DOI 10.1145/1250790.1250804, http://doi.acm.org/10.1145/1250790.1250804
10. Gupta A, Roth A, Ullman J (2012) Iterative constructions and private data release. In: TCC, Taormina, pp 339–356. http://dx.doi.org/10.1007/978-3-642-28914-9_19
11. Hardt M, Rothblum G (2010) A multiplicative weights mechanism for privacy-preserving data analysis. In: Proceedings of the 51st foundations of computer science (FOCS), Las Vegas. IEEE
12. Hardt M, Talwar K (2010) On the geometry of differential privacy. In: Proceedings of the 42nd ACM symposium on theory of computing (STOC'10), Cambridge. ACM, New York, pp 705–714. DOI 10.1145/1806689.1806786, http://doi.acm.org/10.1145/1806689.1806786
13. Li C, Hay M, Rastogi V, Miklau G, McGregor A (2010) Optimizing linear counting queries under differential privacy. In: Proceedings of the twenty-ninth ACM SIGMOD-SIGACT-SIGART symposium on principles of database systems (PODS'10), Indianapolis. ACM, New York, pp 123–134. http://doi.acm.org/10.1145/1807085.1807104
14. Lovász L, Spencer J, Vesztergombi K (1986) Discrepancy of set-systems and matrices. Eur J Comb 7(2):151–160
15. Muthukrishnan S, Nikolov A (2012) Optimal private halfspace counting via discrepancy. In: Proceedings of the 44th symposium on theory of computing (STOC'12), New York. ACM, New York, pp 1285–1292. DOI 10.1145/2213977.2214090, http://doi.acm.org/10.1145/2213977.2214090
16. Nikolov A, Talwar K, Zhang L (2013) The geometry of differential privacy: the sparse and approximate cases. In: Proceedings of the 45th annual ACM symposium on symposium on theory of computing (STOC'13), Palo Alto. ACM, New York, pp 351–360. DOI 10.1145/2488608.2488652, http://doi.acm.org/10.1145/2488608.2488652
17. Plotkin SA, Shmoys DB, Tardos E (1995) Fast approximation algorithms for fractional packing and covering problems. Math Oper Res 20(2):257–301. DOI 10.1287/moor.20.2.257, http://dx.doi.org/10.1287/moor.20.2.257
18. Vershynin R (2001) John's decompositions: selecting a large part. Isr J Math 122(1):253–277

Geometric Dilation of Geometric Networks

Rolf Klein
Institute for Computer Science, University of Bonn, Bonn, Germany

Keywords

Detour; Spanning ratio; Stretch factor

Years and Authors of Summarized Original Work

2006; Dumitrescu, Ebbers-Baumann, Grüne, Klein, Knauer, Rote

Problem Definition

Urban street systems can be modeled by *plane geometric networks* $G = (V, E)$ whose edges $e \in E$ are piecewise smooth curves that connect the vertices $v \in V \subset \mathbb{R}^2$. Edges do not intersect, except at common endpoints in V. Since streets are lined with houses, the quality of such a network can be measured by the length of the connections

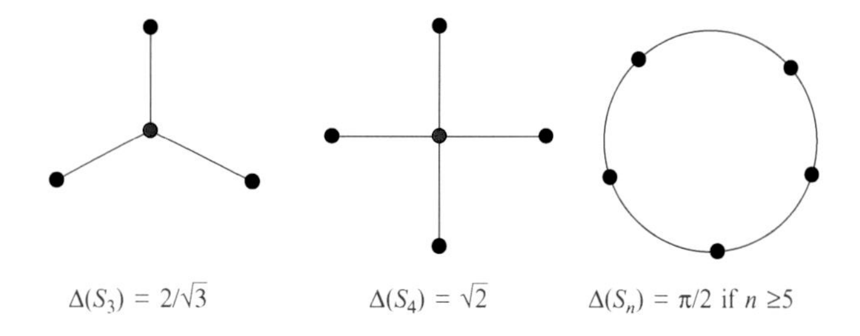

Geometric Dilation of Geometric Networks, Fig. 1 Minimum dilation embeddings of regular point sets

it provides between two arbitrary points p and q on G.

Let $\xi_G(p,q)$ denote a shortest path from p to q in G. Then

$$\delta(p,q) := \frac{|\xi_G(p,q)|}{|pq|} \quad (1)$$

is the detour one encounters when using network G, in order to get from p to q, instead of walking straight. Here, $|\,.\,|$ denotes the Euclidean length. The *geometric dilation of network G* is defined by

$$\delta(G) := \sup_{p \neq q \in G} \delta(p,q). \quad (2)$$

This definition differs from the notion of stretch factor (or spanning ratio) used in the context of spanners; see the monographs by Eppstein [6] or Narasimhan and Smid [11]. In the latter, only the paths between the vertices $p, q \in V$ are considered, whereas the geometric dilation involves all points on the edges as well. As a consequence, the stretch factor of a triangle T equals 1, but its geometric dilation is given by $\delta(T) = \sqrt{2/(1-\cos\alpha)} \geq 2$, where $\alpha \leq 60°$ is the most acute angle of T.

Presented with a finite set S of points in the plane, one would like to find a finite geometric network containing S whose geometric dilation is as small as possible. The value of

$$\Delta(S) := \inf\{\delta(G); G \text{ finite plane geometric network containing } S\}$$

is called the *geometric dilation of point set S*. The problem is in computing, or bounding, $\Delta(S)$ for a given set S.

Key Results

Theorem 1 ([4]) *Let S_n denote the set of corners of a regular n-gon. Then, $\Delta(S_3) = 2/\sqrt{3}, \Delta(S_4) = \sqrt{2}$, and $\Delta(S_n) = \pi/2$ for all $n \geq 5$.*

The networks realizing these minimum values are shown in Fig. 1. The proof of minimality uses the following two lemmata that may be interesting in their own right. Lemma 1 was independently obtained by Aronov et al. [1].

Lemma 1 *Let T be a tree containing S_n. Then $\delta(T) \geq n/\pi$.*

Lemma 2 follows from a result of Gromov's [7]. It can more easily be proven by applying Cauchy's surface area formula; see [4].

Lemma 2 *Let C denote a simple closed curve in the plane. Then $\delta(C) \geq \pi/2$.*

Clearly, Lemma 2 is tight for the circle. The next lemma implies that the circle is the only closed curve attaining the minimum geometric dilation of $\pi/2$.

Lemma 3 ([3]) *Let C be a simple closed curve of geometric dilation $<\pi/2 + \epsilon(\delta)$. Then C is contained in an annulus of width δ.*

For points in general position, computing their geometric dilation seems quite complicated. Only for sets $S = \{A, B, C\}$ of size three is the solution completely known.

Theorem 2 ([5]) *The plane geometric network of minimum geometric dilation containing three given points $\{A, B, C\}$ is either a line segment, or a Steiner tree as depicted in Fig. 1, or a simple path consisting of two line segments and one segment of an exponential spiral; see Fig. 2.*

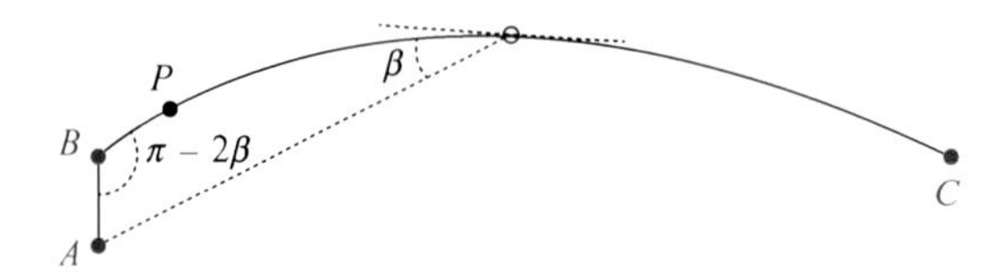

Geometric Dilation of Geometric Networks, Fig. 2 The minimum dilation embedding of points A, B, and C

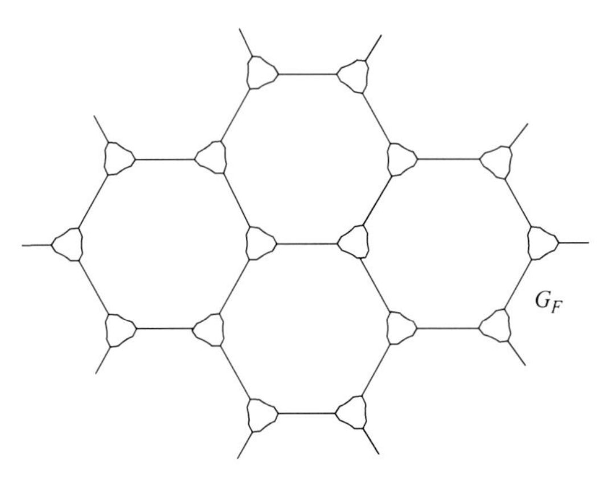

Geometric Dilation of Geometric Networks, Fig. 3 A network of geometric dilation $\approx$1,6778

The optimum path shown in Fig. 2 contains a degree two Steiner vertex, P, situated at distance $|AB|$ from B. The path runs straight between A, B and B, P. From P to C, it follows an exponential spiral centered at A.

The next results provide upper and lower bounds to $\Delta(S)$.

Theorem 3 ([4]) *For each finite point set S, the estimate $\Delta(S) < 1.678$ holds.*

To prove this general upper bound, one can replace each vertex of the hexagonal tiling of $\mathbb{R}^2$ with a certain closed Zindler curve (by definition, all point pairs bisecting the perimeter of a Zindler curve have identical distance). This results in a network G_F of geometric dilation ≈ 1.6778; see Fig. 3. Given a finite point set S, one applies a slight deformation to a scaled version of G_F, such that all points of S lie on a finite part, G, of the deformed net. By Dirichlet's result on simultaneous approximation of real numbers by rationals, a deformation small as compared to the cell size is sufficient, so that the dilation is not affected. See [8] for the history and properties of Zindler curves.

Theorem 4 ([3]) *There exists a finite point set S such that $\Delta(S) > (1 + 10^{-11})\pi/2$.*

Theorem 4 holds for the set S of 19×19 vertices of the integer grid. Roughly, if S were contained in a geometric network G of dilation close to $\pi/2$, the boundaries of the faces of G must be contained in small annuli, by Lemma 3. To the inner and outer circles of these annuli, one can now apply a result by Kuperberg et al. [9] stating that an enlargement, by a certain factor, of a packing of disks of radius ≤ 1 cannot cover a square of size 4.

Applications

The geometric dilation has applications in the theory of knots; see, e.g., Kusner and Sullivan [10] and Denne and Sullivan [2]. With respect to urban planning, the above results highlight principal dilation bounds for connecting given sites with plane geometric networks.

Open Problems

For practical applications, one would welcome upper bounds to the weight (= total edge length) of a geometric network, in addition to upper bounds on its geometric dilation. Some theoretical questions require further investigation, too. Is $\Delta(S)$ always attained by a finite network? How to compute, or approximate, $\Delta(S)$ for a given finite set S? What is the precise value of $\sup\{\Delta(S); S \text{finite}\}$?

Cross-References

▸ Dilation of Geometric Networks

Recommended Reading

1. Aronov B, de Berg M, Cheong O, Gudmundsson J, Haverkort H, Vigneron A (2008) Sparse geometric graphs with small dilation. Comput Geom Theory Appl 40(3):207–219
2. Denne E, Sullivan JM (2004) The distortion of a knotted curve. http://www.arxiv.org/abs/math.GT/0409438
3. Dumitrescu A, Ebbers-Baumann A, Grüne A, Klein R, Rote G (2006) On the geometric dilation of closed curves, graphs, and point sets. Comput Geom Theory Appl 36(1):16–38
4. Ebbers-Baumann A, Grüne A, Klein R (2006) On the geometric dilation of finite point sets. Algorithmica 44(2):137–149
5. Ebbers-Baumann A, Klein R, Knauer C, Rote G (2006) The geometric dilation of three points. Manuscript
6. Eppstein D (1999) Spanning trees and spanners. In: Sack J-R, Urrutia J (eds) Handbook of computational geometry, pp 425–461. Elsevier, Amsterdam
7. Gromov M (1981) Structures Métriques des Variétés Riemanniennes. Textes Math. CEDIX, vol 1. F. Nathan, Paris
8. Grüne A (2006) Geometric dilation and halving distance. Ph.D. thesis, Institut für Informatik I, Universität Bonn
9. Kuperberg K, Kuperberg W, Matousek J, Valtr P (1999) Almost tiling the plane with ellipses. Discrete Comput Geom 22(3):367–375
10. Kusner RB, Sullivan JM (1998) On distortion and thickness of knots. In: Whittington SG et al (eds) Topology and geometry in polymer science. IMA volumes in mathematics and its applications, vol 103. Springer, New York, pp 67–78
11. Narasimhan G, Smid M (2007) Geometric spanner networks. Cambridge University Press, Cambridge/New York

Geometric Object Enumeration

Shin-ichi Nakano
Department of Computer Science, Gunma University, Kiryu, Japan

Keywords

Enumeration; Floor plan; Generation; Listing; Plane triangulation; Reverse search

Years and Authors of Summarized Original Work

2001; Li, Nakano
2001; Nakano
2004; Nakano

Problem Definition

Enumerating objects with the given property is one of basic problems in mathematics. We review some geometric objects enumeration problems and algorithms to solve them.

A graph is *planar* if it can be embedded in the plane so that no two edges intersect geometrically except at a vertex to which they are both incident. A *plane* graph is a planar graph with a fixed planar embedding. A plane graph divides the plane into connected regions called *faces*. The unbounded face is called the outer face, and other faces are called inner faces.

A plane graph is a *floor plan* if each face (including the outer face) is a rectangle. A *based* floor plan is a floor plan with one designated line on the contour of the outer face. The designated line is called the *base* line and we always draw the base line as the lowermost horizontal line of the drawing. The 25 based floor plans having 4 inner faces are shown in Fig. 1. Given an integer f the problem of *floor plan enumeration* asks for generating all floor plans with exactly f inner faces.

A plane graph is a *plane triangulation* if each inner face has exactly three edges on its contour. A *based* plane triangulation is a plane triangulation with one designated edge on the contour of the outer face. The designated edge is called the *base* edge. Triangulations are important model for 3D modeling. A graph is biconnected if removing any vertex always results in a connected graph. A graph is triconnected if removing any two vertices always results in a connected graph. Given two integers n and r, the problem of *biconnected plane triangulation enumeration* asks for generating all biconnected

Geometric Object Enumeration, Fig. 1 The 25 based floor plans having 4 inner faces

Geometric Object Enumeration, Fig. 2 The sequence

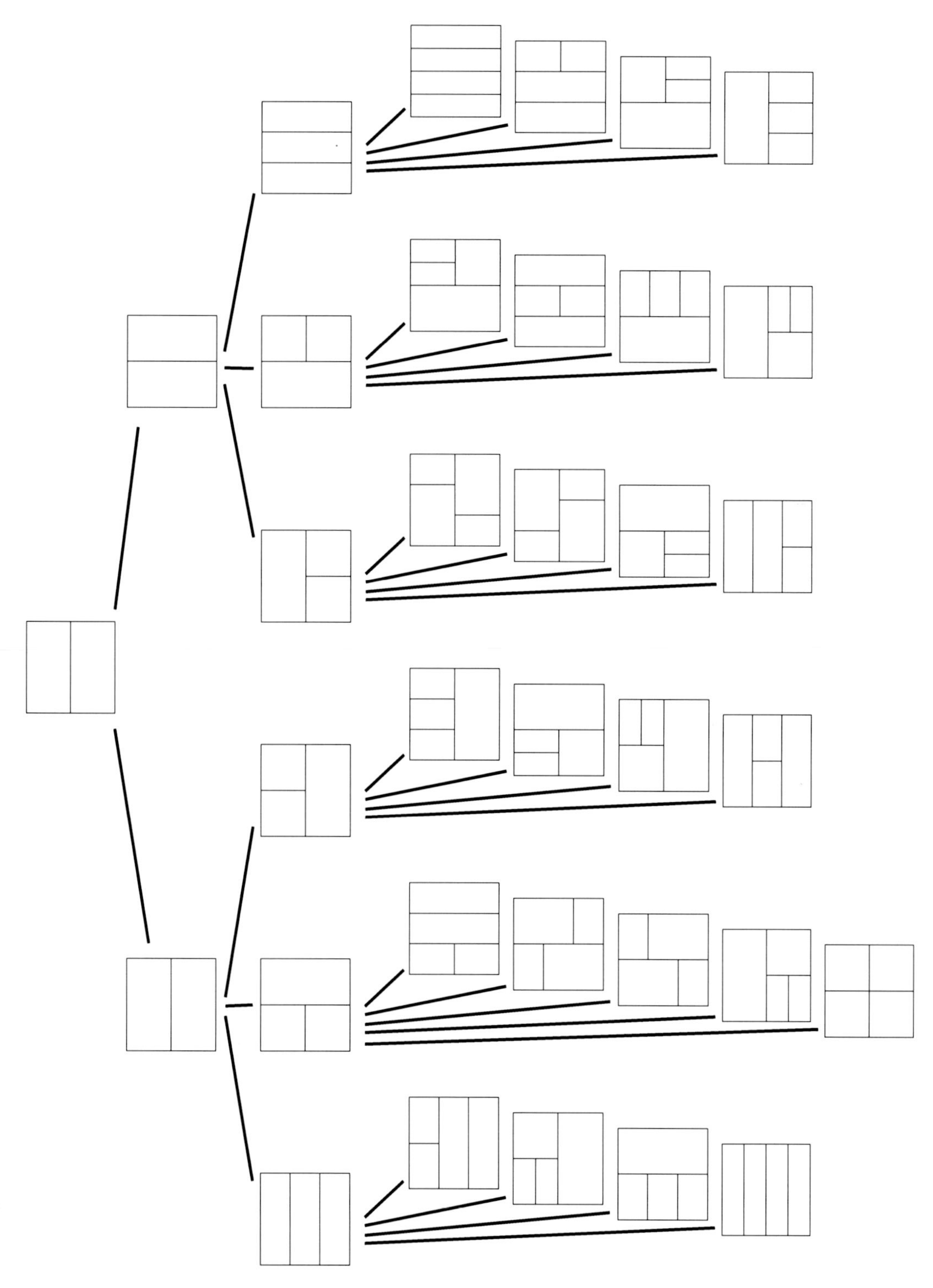

Geometric Object Enumeration, Fig. 3 The tree F_4

plane triangulations with exactly n vertices including exactly r vertices on the outer face. Given two integers n and r, the problem of *triconnected plane triangulation enumeration* asks for generating all triconnected plane triangulations with exactly n vertices including exactly r vertices on the outer face.

Key Results

Enumeration of All Floor Plans

Using *reverse search method* [1], one can enumerate all based floor plans with f inner faces in $O(1)$ time for each [3]. We sketch the method in [3].

Let S_f be the set of all based floor plans with $f > 1$ inner faces. Let R be a based floor plan in S_f and F a face of R having the upper right corner of R. We have two cases. If R has a vertical line segment with upper end at the lower left corner of F, then by continually shrinking R to the uppermost horizontal line of R with preserving the width of F and enlarging the faces below R, we can have a based floor plan with one less inner face. If R has no vertical line segment with the upper end at the lower left corner of F, then R has a horizontal line segment with the right end at the lower left corner of F, and then by continually shrinking R to the rightmost vertical line of R with preserving the height of F and enlarging the faces locating the left of R, we can have a base floor plan with one less inner face. Repeating this results in the sequence of based floor plans which always ends with the based floor plan with one inner face. See an example in Fig. 2. If we merge the sequence of all R in S_f, then we have the tree T_f in which every R in S_f appears as a leaf in T_f. See Fig. 3.

The reverse search method efficiently traverses the tree (without storing the tree in the memory) and output each based floor plan in S_f at each corresponding leaf. Thus, we can efficiently enumerate all based floor plans in S_f. The algorithm enumerates all based floor plans in S_f in $O(1)$ time for each.

Enumeration of Triangulations

Similarly, using *reverse search method* [1], given two integers n and r, one can enumerate all based biconnected triangulations having exactly n vertices including exactly r vertices on the outer face in $O(1)$ time for each [2], all based triconnected triangulations having exactly n vertices including exactly r vertices on the outer face in $O(1)$ time for each [3], and all triconnected (non-based) plane triangulation having exactly n vertices including exactly r vertices on the outer face in $O(r^2 n)$ time for each [3]. Also one can enumerate all based triangulation having exactly n vertices with exactly three vertices on the outer face in $O(1)$ time for each [3].

Cross-References

- ▸ Enumeration of Paths, Cycles, and Spanning Trees
- ▸ Reverse Search; Enumeration Algorithms
- ▸ Tree Enumeration

Recommended Reading

1. Avis D, Fukuda K (1996) Reverse search for enumeration. Discret Appl Math 65(1–3):21–46
2. Li Z, Nakano S (2001) Efficient generation of plane triangulations without repetitions. In: Proceedings of the 28th international colloquium on automata, languages and programming, (ICALP 2001), Crete. LNCS, vol 2076, pp 433–443
3. Nakano S (2004) Efficient generation of triconnected plane triangulations. Comput Geom 27:109–122
4. Nakano S (2001) Enumerating floorplans with n rooms. In: Proceedings of the 12th international symposium on algorithms and computation (ISAAC 2001), Christchurch. LNCS, vol 2223, pp 107–115

Geometric Shortest Paths in the Plane

John Hershberger
Mentor Graphics Corporation, Wilsonville, OR, USA

Keywords

Computational geometry; Continuous Dijkstra method; Shortest path; Shortest path map; Visibility graph

Years and Authors of Summarized Original Work

1996; Mitchell
1999; Hershberger, Suri

Problem Definition

Finding the shortest path between a source and a destination is a natural optimization problem with many applications. Perhaps the oldest variant of the problem is the geometric shortest path problem, in which the domain is physical space: the problem is relevant to human travelers, migrating animals, and even physical phenomena like wave propagation. The key feature that distinguishes the geometric shortest path problem from the corresponding problem in graphs or other discrete spaces is the unbounded number of paths in a multidimensional space. To solve the problem efficiently, one must use the "shortness" criterion to limit the search.

In computational geometry, physical space is modeled abstractly as the union of some number of constant-complexity primitive elements. The traditional formulation of the shortest path problem considers paths in a domain bounded by linear elements – line segments in two dimensions, triangles in three dimensions, and $(d-1)$-dimensional simplices in $d > 3$. Canny and Reif showed that the three-dimensional shortest path problem is NP-complete [2], so this article will focus on the two-dimensional problem.

We consider paths in a free space P bounded by h polygons – one outer boundary and $(h-1)$ obstacles – with a total of n vertices. The free space is closed, so paths may touch the boundary. The source and destination of the shortest path are points s and t inside or on the boundary of P. The goal of the shortest path problem is to find the shortest path from s to t inside P, denoted by $\pi(s,t)$, as efficiently as possible, where running time and memory use are expressed as functions of n and h. The length of $\pi(s,t)$ is denoted by $dist(s,t)$; in some applications, it may be desirable to compute $dist(s,t)$ without finding $\pi(s,t)$ explicitly.

Key Results

Visibility Graph Algorithms

Early approaches to the two-dimensional shortest path problem exploited the *visibility graph* to reduce the continuous shortest path problem to a discrete graph problem [1, 12, 18]. The visibility graph is a graph whose nodes are s, t, and the vertices of P and whose edges (u,v) connect vertex pairs such that the line segment $\overline{uv}$ is contained in P. It is convenient and customary to identify the edges of the abstract visibility graph with the line segments they represent. The visibility graph is important because the edges of the shortest path $\pi(s,t)$ are a subset of the visibility graph edges. This is easy to understand intuitively, because of subpath optimality – for any two points $a, b \in \pi(s,t)$, the subpath of $\pi(s,t)$ between a and b is also the shortest path between a and b. In particular, if $\overline{ab}$ is contained in P, then $\pi(s,t)$ coincides with $\overline{ab}$ between a and b, and the distance $dist(a,b)$ is equal to $|ab|$, the length of the segment $\overline{ab}$. If $\pi(s,t)$ has a bend anywhere except at a vertex of P, an infinitesimal subpath near the bend can be shortened by a straight shortcut, implying that $\pi(s,t)$ is *not* the shortest path. Hence every segment of $\pi(s,t)$ is an edge of the visibility graph.

This observation leads directly to an algorithm for computing shortest paths: compute the visibility graph of P and then run Dijkstra's algorithm to find the shortest path from s to t in the visibility graph. The visibility graph can be constructed in $O(n \log n + m)$ time, where m is the number of edges, using an algorithm of Ghosh and Mount [7]. Dijkstra's algorithm takes $O(n \log n + m)$ time on a graph with n nodes and m edges [4], so this is the running time of the straightforward visibility graph solution to the shortest path problem. This algorithm can be quadratic in the worst case, since m, the number of visibility graph edges, can be as large as $\Theta(n^2)$.

The running time can be improved somewhat by noting that only a subset of the visibility graph edges can belong to a shortest path. In particular, any shortest path must turn toward the boundary of P at any path vertex. This limits the edges to common tangents of the polygons of P. We omit

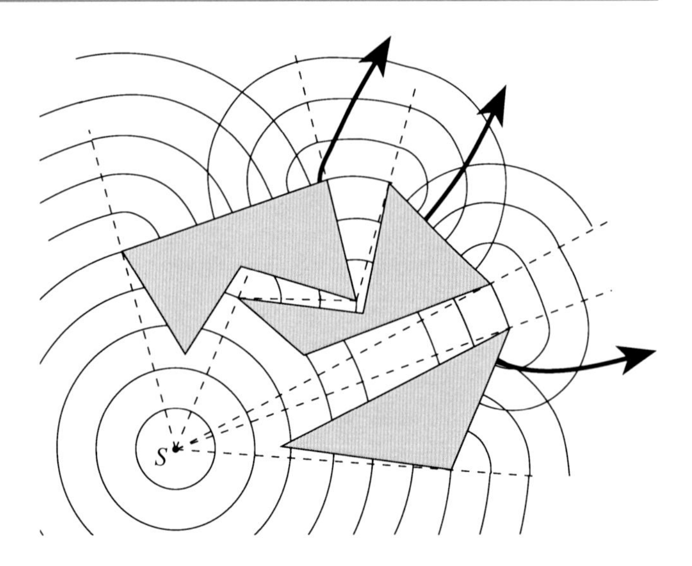

Geometric Shortest Paths in the Plane, Fig. 1 The spreading wavefront

the details, but note that if s and t are known, the common tangent restriction limits the number of visibility graph edges that may belong to $\pi(s,t)$ to $O(n + h^2)$ [11]. These useful edges can be computed in $O(n \log n + h^2)$ time [17], and so the shortest path can be computed in the same time bound by applying Dijkstra's algorithm to the subgraph [11].

Continuous Dijkstra Algorithms

Visibility graph approaches to finding the shortest path run in quadratic time in the worst case, since h may be $\Theta(n)$. This led Mitchell to propose an alternative approach called the *continuous Dijkstra method* [15]. Imagine a wavefront that spreads at unit speed inside P, starting from s. The wavefront at time τ is the set of points in P whose geodesic (shortest path) distance from s is exactly τ. Said another way, the shortest path distance from s to a point $p \in P$ is equal to the time at which the wavefront reaches p.

The wavefront at time τ is a union of paths and cycles bounding the region whose geodesic distance from s is at most τ. Each path or cycle is a sequence of circular arc *wavelets*, each centered on a vertex v that is its *root*. The radius of the wavelet is $\tau - dist(s,v)$, with $dist(s,v) < \tau$. As τ increases, the combinatorial structure of the wavefront changes at discrete event times when the wavefront hits the free space boundary, collides with itself, or eliminates wavelets squeezed between neighboring wavelets. See Fig. 1. The continuous Dijkstra method simulates the spread of this wavefront, computing the shortest path distance to every point of the free space in the process.

Mitchell used the continuous Dijkstra method to compute shortest paths under the L_1 metric in $O(n \log n)$ time [15]. He later extended the approach to compute L_2 (Euclidean) shortest paths in $O(n^{5/3+\epsilon})$ time, for ϵ arbitrarily small [16]. Hershberger and Suri gave an alternative implementation of the continuous Dijkstra scheme, using different data structures, that computes Euclidean shortest paths in $O(n \log n)$ time [9]. The next two subsections discuss these algorithms in more detail.

Continuous Dijkstra with Sector Propagation Queries

If p is a point in P, $\pi(s,p)$ is a shortest path from s to p, and the *predecessor* of p is the vertex of $\pi(s,p)$ adjacent to (immediately preceding) p in the path. If a point is reached by multiple shortest paths, it has multiple predecessors. The *shortest path map* is a linear-complexity partition of P into regions such that every point inside a region has the same predecessor. See Fig. 2. The *root* of each region is the predecessor of all points in the region. The edges of the shortest path map are polygon edges and *bisectors* (curves with two distinct predecessors, namely, the roots of the regions separated by the bisector).

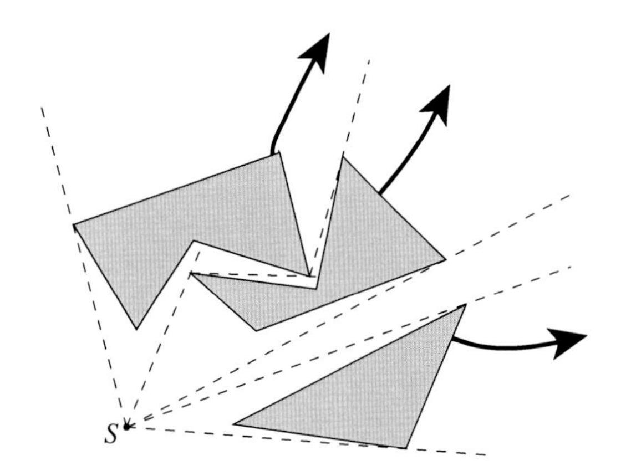

Geometric Shortest Paths in the Plane, Fig. 2 The shortest path map for the wavefront in Fig. 1

Mitchell's shortest path algorithm simulates the spread of the wavefront inside the shortest path map. This may seem a bit peculiar, since the shortest path map is not known until the shortest paths have been computed. The trick is that the algorithm builds the shortest path map as it runs, and it propagates a *pseudo-wavefront* inside its current model of the shortest path map at each step. The true wavefront is a subset of the pseudo-wavefront. This pseudo-wavefront is locally correct – each wavelet's motion is determined by its neighbors in the pseudo-wavefront and the shortest path map known so far – but it may overrun itself. When an overrun is detected, the algorithm revises its model of the shortest path map in the neighborhood of the overrun.

To be more specific, each wavelet w is a circular arc centered at a root vertex $r(w)$. The endpoints of w move along left and right *tracks* $\alpha(w)$ and $\beta(w)$. Each track is either a straight line segment (a polygon edge or an extension of a visibility edge) or a bisector determined by w and the left/right neighbor wavefront $L(w)$ or $R(w)$. For example, if $r = r(w)$ and $r' = r(L(w))$, the left bisector is the set of points x such that $dist(s,r) + |rx| = dist(s,r') + |r'x|$; consequently, the bisector is a hyperbolic arc. For every wavelet w, the algorithm computes a next event, which is the next value of τ where w reaches an endpoint of one of its tracks, the left and right tracks collide, or w hits a polygon vertex between its left and right tracks. (Collisions with polygon edges or other wavefront arcs are *not* detected.) The events for all wavelets are placed in a global priority queue and processed in order of increasing τ values.

When the algorithm processes an event, it updates the wavelets involved and their events in the priority queue. Processing wavelet collisions with a polygon vertex v is the most complicated case: To detect possible previous collisions with polygon edges, the algorithm performs a ray shooting query from $r(w)$ toward v [8]. If the ray hits an edge, the algorithm traces the edge through the current shortest path map regions and updates the corresponding wavelets. If v is reached for the first time by w, then the algorithm updates the wavefront with a new wavelet rooted at v. If vertex v was previously reached by another wavelet, then there are previously undiscovered bisectors between w and the other wavelet. The algorithm traces these bisectors through its local shortest path map model and carves off portions that are reached by a shorter path following another route. Processing other events (track vertices and track collisions) is similar.

Mitchell shows that even though vertices may be reached more than once by different wavelets, and portions of the shortest path map are carved off and discarded when they are discovered to be invalid, no vertex is reached more than $O(1)$ times, and the total shortest path map complexity, even including discarded portions, is $O(n)$. The most costly part of the algorithm is finding the first polygon vertex hit by each wavelet. All the rest of the algorithm – ray shooting, priority queue, bisector tracing, and maintaining the shortest path map structure – can be done in $O(n \log n)$ total time.

The complexity of Mitchell's algorithm is dominated by *wavelet dragging queries*, which find the first obstacle vertex hit by a wavelet w between the left and right tracks $\alpha(w)$ and $\beta(w)$. Mitchell phrases this as a ten-dimensional optimization problem dependent on the position and distance from s for the root of w and its neighbors in the wavefront, plus the start time τ. Although Euclidean distances are square roots of quadratics (by Pythagoras), Mitchell is able,

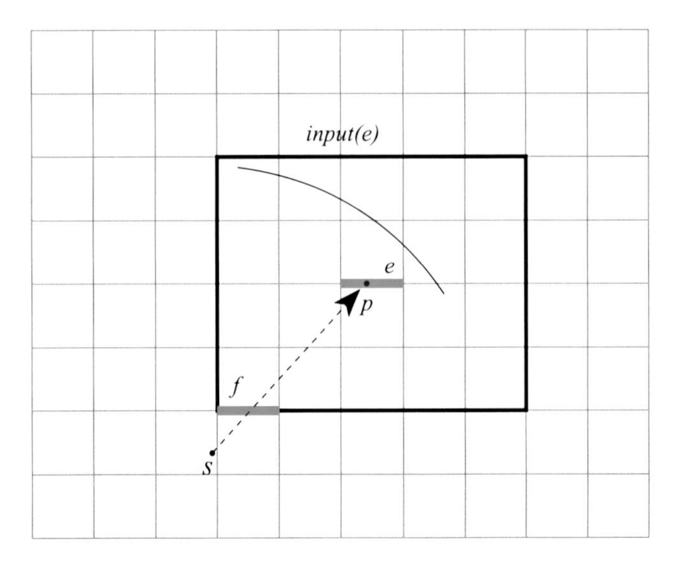

Geometric Shortest Paths in the Plane, Fig. 3 The well-covering region for edge e is bounded by $input(e)$

by squaring, substitution, and simplification, to convert the distance minimization problem into a linear optimization range query over a constant-size polyhedron in $\Re^5$. (The objects in the range query are n 5-dimensional points, images of the polygon vertices.) There are $O(n)$ such queries to be performed. Using known bounds and balancing preprocessing against query time, the $O(n)$ queries can be answered in $O(n^{5/3+\epsilon})$ time and space [3, 13, 14]. All other parts of the algorithm take near-linear time, so the total time for Mitchell's algorithm to find the Euclidean shortest path map is $O(n^{5/3+\epsilon})$ [16].

Continuous Dijkstra in a Conforming Subdivision

The challenge of implementing the continuous Dijkstra paradigm is that detecting and processing wavefront events in strict temporal order is difficult to do efficiently, but processing events out of order may lead to incorrect results or to processing too many invalid events. Mitchell addresses the challenge by detecting only one subclass of events in temporal order (wavelet contacts with polygon vertices) and repairing errors in the shortest path map structure as they are discovered. Hershberger and Suri achieve a better time bound (optimal $O(n \log n)$) by processing events in an even more relaxed order [9]. The key to their approach is a subdivision of the free space in which spatial locality is used to bound the temporal inaccuracy of wavefront event processing.

As a simple example, consider a wavefront propagating across an obstacle-free plane that has been subdivided into a grid of unit squares. Each edge e of the grid lies at the center of a 4×5 rectangle of squares. The distance from e to each of the 18 edges on the rectangle boundary is at least 2. If the wavefront source is outside the rectangle, then the first wavelet that reaches any point $p \in e$ must pass through the rectangle boundary at least two time units before it reaches p. By the triangle inequality, an edge of length δ is completely covered by the wavefront within time δ of the time the wavefront first hits it. It follows that if the shortest path to $p \in e$ passes through an edge f on the rectangle boundary, edge f is completely covered by the wavefront at least one time unit before the wavefront reaches p. See Fig. 3.

The algorithm propagates the wavefront from edge to edge in the grid. For each edge e, let $input(e)$ be the edges on the boundary of the 4×5 rectangle around e. The algorithm computes a *cover time* for e, denoted $cover(e)$, that is an upper bound on the time when the wavefront completely covers e. If $fvc(e)$ is the time at which the wavefront first contacts a vertex of e, and $|e|$ is the length of e, then $cover(e)$ is defined to be $fvc(e) + |e|$. For each edge e, $cover(e)$ is determined by a wavefront passing through $f \in$ *input*(e), and *cover*$(f) <$ *cover*(e).

The propagation algorithm processes edges in order of cover time, computing the wavefront at e by combining the wavefronts from edges $f \in$ *input*(e) with *cover*$(f) <$ *cover*(e). The combination algorithm is linear in the number of features of the shortest path map that lie inside the rectangle for e. The algorithm computes a one-dimensional representation of the intersection of the shortest path map with each edge of the grid; each bisector that has an event (an arc endpoint) within the input region of an edge e is flagged in the wavefront representation for e. To turn the one-dimensional wavefront representation at edges into a two-dimensional representation of the shortest path map, the algorithm combines the wavefronts of the edges on each cell's boundary to compute the shortest path map inside the cell. (The algorithm computes additively weighted Voronoi diagrams [6] for the wavelet roots whose bisectors have events (endpoints) in the cell, plus compact representations of the groups of bisectors that have no endpoints in the cell.)

The key feature of the grid subdivision is *well-covering property*: each edge e is surrounded by a region that is the union of $O(1)$ cells, and the distance from e to the region boundary is relatively large. In particular, if f is an edge on the boundary, *dist*$(e, f) \geq 2 \cdot \max(|e|, |f|)$. This property allows the algorithm to perform *spatial* (not temporal) wavefront propagation at discrete cover times. Hershberger and Suri show how to extend the well-covering property to a special *conforming subdivision* of free space made up of $O(n)$ constant-complexity cells. The wavefront propagation algorithm carries over from the grid to the conforming subdivision of free space with only a few changes to handle the obstacle vertices. As on the grid, the number of propagation steps and data structure changes is $O(n)$. Including the overhead of a priority queue and data structure updates (full persistence is needed [5]) increases the time and space by a factor of $O(\log n)$, so the overall algorithm runs in $O(n \log n)$ time and space.

Extensions

Hershberger and Suri's algorithm supports multiple wavefront sources, including line-segment sources. Hence the algorithm can be used to compute geodesic Voronoi diagrams, including Voronoi diagrams whose sites are points, segments, polygons, or combinations of all these.

Since the publication of Hershberger and Suri's optimal-time algorithm for shortest paths among polygonal obstacles, their result has been extended to other two-dimensional domains. Schreiber showed how to find shortest paths on the surface of a convex polyhedron in $O(n \log n)$ time [20]. His algorithm decomposes the surface into cells and then propagates wavefronts between cell edges similarly to Hershberger and Suri's algorithm. Schreiber extended his algorithm for polyhedra to work for polygonal terrains as well, assuming that the maximum gradient of the terrain is bounded by a constant [19]. More recently, Hershberger, Suri, and Yıldız [10] extended the algorithm for polygonal obstacles [9] to find shortest paths in a free space bounded by curved obstacle edges. The conforming subdivision for the free space is very similar to that for polygonal obstacles; the chief difficulty is computing the positions of bisector events (intersections). Bisectors for polygonal obstacles are hyperbolic arcs, but they are much more complicated curves for curved obstacles. The algorithm of [10] approximates the bisector events using primitive tangent-finding operations on individual obstacle curves, with the result that the algorithm's running time is $O(n \log(n/\epsilon))$, where ϵ is the relative error of the computed path length.

Cross-References

- ▶ Range Searching
- ▶ Single-Source Shortest Paths
- ▶ Voronoi Diagrams and Delaunay Triangulations

Recommended Reading

1. Asano T, Asano T, Guibas LJ, Hershberger J, Imai H (1986) Visibility of disjoint polygons. Algorithmica 1:49–63
2. Canny J, Reif JH (1987) New lower bound techniques for robot motion planning problems. In: Proceedings of the 28th annual IEEE symposium on foundations of computer Science, Washington, DC, pp 49–60
3. Chazelle B, Sharir M, Welzl E (1992) Quasi-optimal upper bounds for simplex range searching and new zone theorems. Algorithmica 8:407–429
4. Cormen TH, Leiserson CE, Rivest RL, Stein C (2001) Introduction to algorithms, 2nd edn. MIT Press, Cambridge
5. Driscoll JR, Sarnak N, Sleator DD, Tarjan RE (1989) Making data structures persistent. J Comput Syst Sci 38:86–124
6. Fortune SJ (1987) A sweepline algorithm for Voronoi diagrams. Algorithmica 2:153–174
7. Ghosh SK, Mount DM (1991) An output-sensitive algorithm for computing visibility graphs. SIAM J Comput 20:888–910
8. Hershberger J, Suri S (1995) A pedestrian approach to ray shooting: Shoot a ray, take a walk. J Algorithms 18:403–431
9. Hershberger J, Suri S (1999) An optimal algorithm for Euclidean shortest paths in the plane. SIAM J Comput 28(6):2215–2256
10. Hershberger J, Suri S, Yıldız H (2013) A near-optimal algorithm for shortest paths among curved obstacles in the plane. In: Proceedings of the 29th annual symposium on computational geometry, SoCG '13. ACM, New York, pp 359–368
11. Kapoor S, Maheshwari SN, Mitchell JSB (1997) An efficient algorithm for Euclidean shortest paths among polygonal obstacles in the plane. Discret Comput Geom 18:377–383
12. Lozano-Perez T, Wesley MA (1979) An algorithm for planning collision-free paths among polyhedral obstacles. Commun ACM 22:560–570
13. Matoušek J (1993) Range searching with efficient hierarchical cuttings. Discret Comput Geom 10(2):157–182
14. Megiddo N (1983) Applying parallel computation algorithms in the design of serial algorithms. J ACM 30(4):852–865
15. Mitchell JSB (1992) L_1 shortest paths among polygonal obstacles in the plane. Algorithmica 8:55–88
16. Mitchell JSB (1996) Shortest paths among obstacles in the plane. Int J Comput Geom Appl 6:309–332
17. Pocchiola M, Vegter G (1996) Topologically sweeping visibility complexes via pseudotriangulations. Discret Comput Geom 16(4):419–453
18. Rohnert H (1988) Time and space efficient algorithms for shortest paths between convex polygons. Inf Process Lett 27:175–179
19. Schreiber Y (2010) An optimal-time algorithm for shortest paths on realistic polyhedra. Discret Comput Geom 43(1):21–53
20. Schreiber Y, Sharir M (2008) An optimal-time algorithm for shortest paths on a convex polytope in three dimensions. Discret Comput Geom 39(1–3):500–579

Geometric Spanners

Joachim Gudmundsson[1,2], Giri Narasimhan[3,4], and Michiel Smid[5]
[1]DMiST, National ICT Australia Ltd, Alexandria, Australia
[2]School of Information Technologies, University of Sydney, Sydney, NSW, Australia
[3]Department of Computer Science, Florida International University, Miami, FL, USA
[4]School of Computing and Information Sciences, Florida International University, Miami, FL, USA
[5]School of Computer Science, Carleton University, Ottawa, ON, Canada

Keywords

Computational geometry; Dilation; Geometric networks; Spanners

Years and Authors of Summarized Original Work

2002; Gudmundsson, Levcopoulos, Narasimhan

Problem Definition

Consider a set S of n points in d-dimensional Euclidean space. A network on S can be modeled as an undirected graph G with vertex set S of size n and an edge set E where every edge (u, v) has a weight. A geometric (Euclidean) network is a network where the weight of the edge (u, v) is the Euclidean distance $|uv|$ between its end points. Given a real number $t > 1$, we say that G is a *t-spanner* for S, if for each pair of points $u, v \in S$, there exists a path in G of weight at most t times the Euclidean distance between u and v. The minimum t such that G is a t-spanner for S is called the stretch factor, or dilation, of G. For a detailed description of many constructions of t-spanners, see the book by Narasimhan and Smid [30]. The problem considered is the construction of t-spanners given a set S of n points in $\mathcal{R}^d$ and a positive real value $t > 1$, where d is a constant. The aim is to compute a good t-spanner for S with respect to the following quality measures:

size: the number of edges in the graph
degree: the maximum number of edges incident on a vertex
weight: the sum of the edge weights
spanner diameter: the smallest integer k such that for any pair of vertices u and v in S, there is a path in the graph of length at most $t \cdot |uv|$ between u and v containing at most k edges
fault tolerance: the resilience of the graph to edge, vertex, or region failures

Thus, good t-spanners require large fault tolerance and small size, degree, weight, and spanner diameter. Additionally, the time required to compute such spanners must be as small as possible.

Key Results

This section contains descriptions of several known approaches for constructing a t-spanner of a set of points in Euclidean space. We also present descriptions of the construction of fault-tolerant spanners, spanners among polygonal obstacles, and, finally, a short note on dynamic and kinetic spanners.

Spanners of Points in Euclidean Space

The most well-known classes of t-spanner networks for points in Euclidean space include Θ-graphs, WSPD graphs, and greedy spanners. In the following sections, the main idea of each of these classes is given, together with the known bounds on the quality measures.

The Θ-Graph

The Θ-graph was discovered independently by Clarkson and Keil in the late 1980s. The general idea is to process each point $p \in S$ independently as follows: partition $\mathcal{R}^d$ into k simplicial cones of angular diameter at most θ and apex at p, where $k = O(1/\theta^{d-1})$. For each nonempty cone C, an edge is added between p and the point in C whose orthogonal projection onto some fixed ray in C emanating from p is closest to p; see Fig. 1a. The resulting graph is called the Θ-graph on S. The following result is due to Arya et al. [9].

Theorem 1 *The Θ-graph is a t-spanner of S for $t = \frac{1}{\cos\theta - \sin\theta}$ with $O\left(\frac{n}{\theta^{d-1}}\right)$ edges and can be computed in $O\left(\frac{n}{\theta^{d-1}} \log^{d-1} n\right)$ time using $O\left(\frac{n}{\theta^{d-1}} + n \log^{d-2} n\right)$ space.*

The following variants of the Θ-graph also give bounds on the degree, spanner diameter, and weight.

Skip-List Spanners

The idea is to generalize skip lists and apply them to the construction of spanners. Construct a

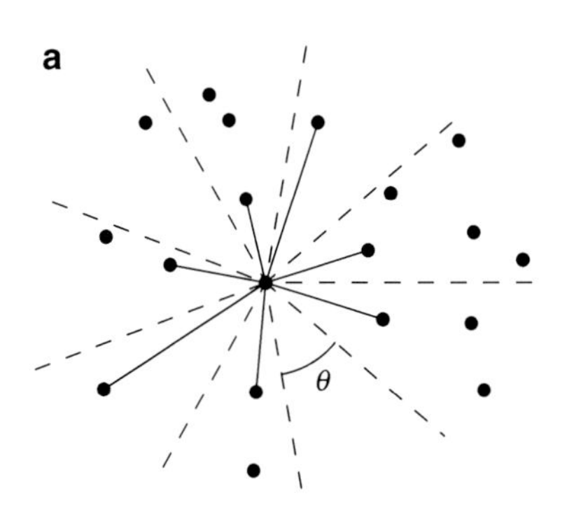

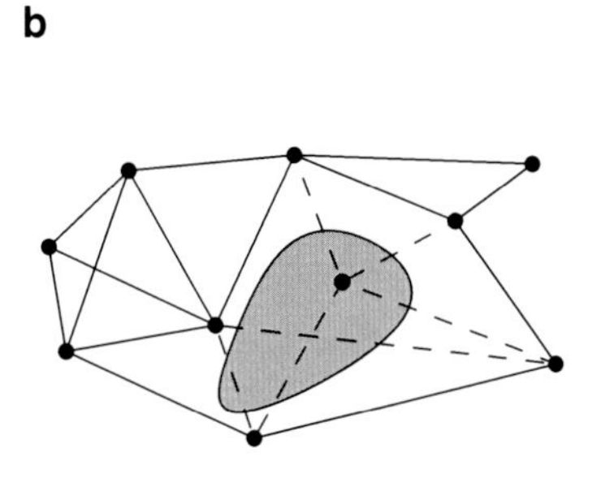

Geometric Spanners, Fig. 1 (**a**) Illustrating the Θ-graph and (**b**) a graph with a region fault

sequence of h subsets, $S_1, \ldots, S_h$, where $S_1 = S$ and S_i is constructed from S_{i-1} as follows (reminiscent of the levels in a skip list). For each point in S_{i-1}, flip a fair coin. The set S_i is the set of all points of S_{i-1} whose coin flip produced heads. The construction stops if $S_i = \emptyset$. For each subset, a Θ-graph is constructed. The union of the graphs is the skip-list spanner of S with dilation t, having $O\left(\frac{n}{\theta^{d-1}}\right)$ edges and $O(\log n)$ spanner diameter with high probability [9].

Gap Greedy

A set of directed edges is said to satisfy the *gap property* if the sources of any two distinct edges in the set are separated by a distance that is at least proportional to the length of the shorter of the two edges. Arya and Smid [6] proposed an algorithm that uses the gap property to decide whether or not an edge should be added to the t-spanner graph. Using the gap property, the constructed spanner can be shown to have degree $O(1/\theta^{d-1})$ and weight $O(\log n \cdot \text{wt}(\text{MST}(S)))$, where $\text{wt}(\text{MST}(S))$ is the weight of the minimum spanning tree of S.

The WSPD Graph

The well-separated pair decomposition (WSPD) was developed by Callahan and Kosaraju [12]. The construction of a t-spanner using the well-separated pair decomposition is done by first constructing a WSPD of S with respect to a separation constant $s = \frac{4(t+1)}{(t-1)}$. Initially set the spanner graph $G = (S, \emptyset)$ and add edges iteratively as follows. For each well-separated pair $\{A, B\}$ in the decomposition, an edge (a, b) is added to the graph, where a and b are arbitrary points in A and B, respectively. The resulting graph is called the WSPD graph on S.

Theorem 2 *The WSPD graph is a t-spanner for S with $O(s^d \cdot n)$ edges and can be constructed in time $O(s^d n + n \log n)$, where $s = 4(t+1)/(t-1)$.*

There are modifications that can be made to obtain bounded spanner diameter or bounded degree.

Bounded spanner diameter: Arya, Mount, and Smid [7] showed how to modify the construction algorithm such that the spanner diameter of the graph is bounded by $2 \log n$. Instead of selecting an arbitrary point in each well-separated set, their algorithm carefully chooses a representative point for each set.

Bounded degree: A single point v can be part of many well-separated pairs, and each of these pairs may generate an edge with an end point at v. Arya et al. [8] suggested an algorithm that retains only the shortest edge for each cone direction, thus combining the Θ-graph approach with the WSPD graph. By adding a postprocessing step that handles all high-degree vertices, a t-spanner of degree $O\left(\frac{1}{(t-1)^{2d-1}}\right)$ is obtained.

The Greedy Spanner

The greedy algorithm was first presented in 1989 by Bern, and since then, the greedy algorithm has been subject to considerable research. The graph constructed using the greedy algorithm is called a Greedy spanner, and the general idea is that the algorithm iteratively builds a graph G. The edges in the complete graph are processed in order of increasing edge length. Testing an edge (u, v) entails a shortest path query in the partial spanner graph G. If the shortest path in G between u and v is at most $t \cdot |uv|$, then the edge (u, v) is discarded; otherwise, it is added to the partial spanner graph G.

Das, Narasimhan, and Salowe [22] proved that the Greedy spanner fulfills the so-called *leapfrog property*. A set of undirected edges E is said to satisfy the t-leapfrog property, if for every $k \geq 2$, and for every possible sequence $\{(p_1, q_1), \ldots, (p_k, q_k)\}$ of pairwise distinct edges of E,

$$t \cdot |p_1 q_1| < \sum_{i=2}^{k} |p_i q_i| + t \cdot \left(\sum_{i=1}^{k-1} |q_i p_{i+1}| + |p_k q_1| \right).$$

Using the leapfrog property, it has been shown that the total edge weight of the graph is within a constant factor of the weight of a minimum spanning tree of S.

Using Dijkstra's shortest-path algorithm, the greedy spanner can be constructed in $O(n^3 \log n)$ time. Bose et al. [10] improved the time to $O(n^2 \log n)$, while using $O(n^2)$ space. Alewijnse et al. [4] improved the space bound to $O(n)$, while slightly increasing the time bound to $O(n^2 \log^2 n)$.

Das and Narasimhan [21] observed that an approximation of the greedy spanner can be constructed while maintaining the leapfrog property. This observation allowed for faster construction algorithms.

Theorem 3 ([27]) *The greedy spanner is a t-spanner of S with* $O\left(\frac{n}{(t-1)^d}\log(\frac{1}{t-1})\right)$ *edges, maximum degree* $O\left(\frac{1}{(t-1)^d}\log(\frac{1}{t-1})\right)$, *and weight* $O\left(\frac{1}{(t-1)^{2d}}\cdot \mathrm{wt}(\mathrm{MST}(S))\right)$ *and can be computed in time* $O\left(\frac{n}{(t-1)^{2d}}\log n\right)$.

The Transformation Technique

Chandra et al. [16, 17] introduced a transformation technique for general metrics that transforms an algorithm for constructing spanners with small stretch factor and size into an algorithm for constructing spanners with the same asymptotic stretch factor and size, but with the additional feature of small weight. Elkin and Solomon [24] refined their approach to develop a transformation technique that achieved the following: It takes an algorithm for constructing spanners with small stretch factor, small size, small degree, and small spanner diameter and transforms it into an algorithm for constructing spanners with a small increase in stretch factor, size, degree, and spanner diameter, but that also has small weight and running time.

Using the transformation technique allowed Elkin and Solomon to prove the following theorem.

Theorem 4 ([24]) *For any set of n points in Euclidean space of any constant dimension d, any* $\epsilon > 0$, *and any parameter* $\rho \geq 2$, *there exists a* $(1+\epsilon)$*-spanner with* $O(n)$ *edges, degree* $O(\rho)$, *spanner diameter* $O(\log_\rho n + \alpha(\rho))$, *andweight* $O(\rho \cdot \log_\rho n \cdot \mathrm{wt}(\mathrm{MST}))$, *which can be constructed in time* $O(n \log n)$.

Given the lower bounds proved by Chan and Gupta [13] and Dinitz et al. [23], these results represent optimal tradeoffs in the entire range of the parameter ρ.

Fault-Tolerant Spanners

The concept of fault-tolerant spanners was first introduced by Levcopoulos et al. [28] in 1998: After one or more vertices or edges fail, the spanner should retain its good properties. In particular, there should still be a short path between any two vertices in what remains of the spanner after the fault. Czumaj and Zhao [19] showed that a greedy approach produces a k-vertex (or k-edge) fault-tolerant geometric t-spanner with degree $O(k)$ and total weight $O(k^2 \cdot \mathrm{wt}(\mathrm{MST}(S)))$; these bounds are asymptotically optimal. Chan et al. [15] used a "standard net-tree with cross-edge framework" developed by [14, 26] to design an algorithm that produces a k-vertex (or k-edge) fault-tolerant geometric $(1+\epsilon)$-spanner with degree $O(k^2)$, diameter $O(\log n)$, and total weight $O(k^2 \log n \cdot \mathrm{wt}(\mathrm{MST}(S)))$. Such a spanner can be constructed in $O(n \log n + k^2 n)$ time.

For geometric spanners, it is natural to consider *region faults*, i.e., faults that destroy all vertices and edges intersecting some geometric fault region. For a fault region F, let $G \ominus F$ be the part of G that remains after the points from S inside F and all edges that intersect F have been removed from the graph; see Fig. 1b. Abam et al. [2] showed how to construct region-fault tolerant t-spanners of size $O(n \log n)$ that are fault tolerant to any convex region fault. If one is allowed to use Steiner points, then a linear size t-spanner can be achieved.

Spanners Among Obstacles

The visibility graph of a set of pairwise non-intersecting polygons is a graph of intervisible locations. Each polygonal vertex is a vertex in the graph and each edge represents a visible connection between them, that is, if two vertices can see each other, an edge is drawn between them. This graph is useful since it contains the shortest obstacle avoiding path between any pair of vertices.

Das [20] showed that a t-spanner of the visibility graph of a point set in the Euclidean plane can be constructed by using the Θ-graph approach followed by a pruning step. The obtained graph has linear size and constant degree.

Dynamic and Kinetic Spanners

Arya et al. [9] designed a data structure of size $O(n \log^d n)$ that maintains the skip-list spanner, described in section "The Θ-Graph," in $O(\log^d n \log\log n)$ expected amortized time per insertion and deletion in the model of random updates.

Gao et al. [26] showed how to maintain a t-spanner of size $O\left(\frac{n}{(t-1)^d}\right)$ and maximum degree $O\left(\frac{1}{(t-2)^d} \log \alpha\right)$, in time $O\left(\frac{\log \alpha}{(t-1)^d}\right)$ per insertion and deletion, where α denotes the aspect ratio of S, i.e., the ratio of the maximum pairwise distance to the minimum pairwise distance. The idea is to use an hierarchical structure T with $O(\log \alpha)$ levels, where each level contains a set of centers (subset of S). Each vertex v on level i in T is connected by an edge to all other vertices on level i within distance $O\left(\frac{2^i}{t-1}\right)$ of v. The resulting graph is a t-spanner of S and it can be maintained as stated above. The approach can be generalized to the kinetic case so that the total number of events in maintaining the spanner is $O(n^2 \log n)$ under pseudo-algebraic motion. Each event can be updated in $O\left(\frac{\log \alpha}{(t-1)^d}\right)$ time.

The problem of maintaining a spanner under insertions and deletions of points was settled by Gottlieb and Roditty [5]: For every set of n points in a metric space of bounded doubling dimension, there exists a $(1+\epsilon)$-spanner whose maximum degree is $O(1)$ and that can be maintained under insertions and deletions of points, in $O(\log n)$ time per operation.

Recently several papers have considered the kinetic version of the spanner construction problem. Abam et al. [1, 3] gave the first data structures for maintaining the Θ-graph, which was later improved by Rahmati et al. [32]. Assuming the trajectories of the points can be described by polynomials whose degrees are at most a constant s, the data structure uses $O(n \log^d n)$ space and handles $O(n^2)$ events with a total cost of $O\left(n\lambda_{2s+2}(n) \log^{d+1} n\right)$, where $\lambda_{2s+2}(n)$ is the maximum length of Davenport-Schinzel sequences of order $2s+2$ on n symbols. The kinetic data structure is compact, efficient, responsive (in an amortized sense), and local.

Applications

The construction of sparse spanners has been shown to have numerous application areas such as metric space searching [31], which includes query by content in multimedia objects, text retrieval, pattern recognition, and function approximation. Another example is broadcasting in communication networks [29]. Several well-known theoretical results also use the construction of t-spanners as a building block, for example, Rao and Smith [33] made a breakthrough by showing an optimal $O(n \log n)$-time approximation scheme for the well-known Euclidean *traveling salesperson problem*, using t-spanners (or banyans). Similarly, Czumaj and Lingas [18] showed approximation schemes for minimum-cost multi-connectivity problems in geometric networks.

Open Problems

A few open problems are mentioned below:

1. Determine if there exists a fault-tolerant t-spanner of linear size for convex region faults.
2. Can the k-vertex fault-tolerant spanner be computed in $O(n \log n + kn)$ time?

Experimental Results

The problem of constructing spanners has received considerable attention from a theoretical perspective but not much attention from a practical or experimental perspective. Navarro and Paredes [31] presented four heuristics for point sets in high-dimensional space ($d = 20$) and showed by empirical methods that the running time was $O(n^{2.24})$ and the number of edges in the produced graphs was $O(n^{1.13})$. Farshi and Gudmundsson [25] performed a thorough comparison of the construction algorithms discussed in section "Spanners of Points in Euclidean Space." The results showed that the spanner produced by the original greedy algorithm is superior compared to the graphs produced by the other approaches discussed in section "Spanners of Points in Euclidean Space" when it comes to number of edges, maximum degree, and weight. However, the greedy algorithm requires $O(n^2 \log n)$ time [10] and uses quadratic space, which restricted experiments in [25] to instances containing at most 13,000 points. Alewijnse et al. [4] showed how to reduce the space usage to linear only paying an additional $O(\log n)$ factor in the running time. In their experiments, they could handle more than a million points. In a follow-up paper, Bouts et al. [11] gave further experimental improvements.

Cross-References

- ▶ Applications of Geometric Spanner Networks
- ▶ Dilation of Geometric Networks
- ▶ Simple Algorithms for Spanners in Weighted Graphs
- ▶ Single-Source Shortest Paths
- ▶ Sparse Graph Spanners
- ▶ Well Separated Pair Decomposition

Recommended Reading

1. Abam MA, de Berg M (2011) Kinetic spanners in $\mathbb{R}^d$. Discret Comput Geom 45(4):723–736
2. Abam MA, de Berg M, Farshi M, Gudmundsson J (2009) Region-fault tolerant geometric spanners. Discret Comput Geom 41:556–582
3. Abam MA, de Berg M, Gudmundsson J (2010) A simple and efficient kinetic spanner. Comput Geom 43(3):251–256
4. Alewijnse SPA, Bouts QW, ten Brink AP, Buchin K (2013) Computing the greedy spanner in linear space. In: Bodlaender HL, Italiano GF (eds) 21st annual European symposium on algorithms. Lecture notes in computer science, vol 8125. Springer, Heidelberg, pp 37–48
5. Arikati SR, Chen DZ, Chew LP, Das G, Smid M, Zaroliagis CD (1996) Planar spanners and approximate shortest path queries among obstacles in the plane. In: Proceedings of 4th European symposium on algorithms. Lecture notes in computer science, vol 1136. Springer Berlin/Heidelberg, pp 514–528
6. Arya S, Smid M (1997) Efficient construction of a bounded-degree spanner with low weight. Algorithmica 17:33–54
7. Arya S, Mount DM, Smid M (1994) Randomized and deterministic algorithms for geometric spanners of small diameter. In: Proceedings of 35th IEEE symposium on foundations of computer science, Milwaukee, pp 703–712
8. Arya S, Das G, Mount DM, Salowe JS, Smid M (1995) Euclidean spanners: short, thin, and lanky. In: Proceedings of 27th ACM symposium on theory of computing, Las Vegas, pp 489–498
9. Arya S, Mount DM, Smid M (1999) Dynamic algorithms for geometric spanners of small diameter: randomized solutions. Comput Geom – Theory Appl 13(2):91–107
10. Bose P, Carmi P, Farshi M, Maheshwari A, Smid MHM (2010) Computing the greedy spanner in near-quadratic time. Algorithmica 58(3):711–729
11. Bouts QW, ten Brink AP, Buchin K (2014) A framework for computing the greedy spanner. In: Cheng SW, Devillers O (eds) Symposium on computational geometry, Kyoto. ACM, pp 11–20
12. Callahan PB, Kosaraju SR (1995) A decomposition of multidimensional point sets with applications to k-nearest-neighbors and n-body potential fields. J ACM 42:67–90
13. Chan HTH, Gupta A (2009) Small hop-diameter sparse spanners for doubling metrics. Discret Comput Geom 41(1):28–44
14. Chan HTH, Gupta A, Maggs B, Zhou S (2005) On hierarchical routing in doubling metrics. In: Symposium on discrete algorithms, Vancouver. ACM, pp 762–771
15. Chan HTH, Li M, Ning L, Solomon S (2013) New doubling spanners: better and simpler. In: Automata, languages, and programming. Springer, Berlin/Heidelberg, pp 315–327
16. Chandra B, Das G, Narasimhan G, Soares J (1992) New sparseness results on graph spanners. In: Proceedings of 8th annual symposium on computational geometry, Berlin, pp 192–201
17. Chandra B, Das G, Narasimhan G, Soares J (1995) New sparseness results on graph spanners. Int J Comput Geom Appl 5:124–144

18. Czumaj A, Lingas A (2000) Fast approximation schemes for Euclidean multi-connectivity problems. In: Proceedings of 27th international colloquium on automata, languages and programming. Lecture notes in computer science, vol 1853. Springer, Berlin/Heidelberg, pp 856–868
19. Czumaj A, Zhao H (2004) Fault-tolerant geometric spanners. Discret Comput Geom 32(2): 207–230
20. Das G (1997) The visibility graph contains a bounded-degree spanner. In: Proceedings of 9th Canadian conference on computational geometry, Kingston
21. Das G, Narasimhan G (1997) A fast algorithm for constructing sparse Euclidean spanners. Int J Comput Geom Appl 7:297–315
22. Das G, Narasimhan G, Salowe J (1995) A new way to weigh malnourished Euclidean graphs. In: Proceedings of 6th ACM-SIAM symposium on discrete algorithms, San Francisco, pp 215–222
23. Dinitz Y, Elkin M, Solomon S (2010) Low-light trees, and tight lower bounds for Euclidean spanners. Discret Comput Geom 43(4):736–783
24. Elkin M, Solomon S (2013) Optimal Euclidean spanners: really short, thin and lanky. In: Proceedings of the forty-fifth annual ACM symposium on theory of computing, Palo Alto, pp 645–654
25. Farshi M, Gudmundsson J (2009) Experimental study of geometric t-spanners. ACM J Exp Algorithmics 14(1):3–39
26. Gao J, Guibas LJ, Nguyen A (2004) Deformable spanners and applications. In: Proceedings of 20th ACM symposium on computational geometry, Brooklyn, pp 190–199
27. Gudmundsson J, Levcopoulos C, Narasimhan G (2002) Improved greedy algorithms for constructing sparse geometric spanners. SIAM J Comput 31(5):1479–1500
28. Levcopoulos C, Narasimhan G, Smid M (2002) Improved algorithms for constructing fault-tolerant spanners. Algorithmica 32(1):144–156
29. Li XY (2003) Applications of computational geomety in wireless ad hoc networks. In: Cheng XZ, Huang X, Du DZ (eds) Ad Hoc wireless networking. Kluwer, Dordrecht, pp 197–264
30. Narasimhan G, Smid M (2007) Geometric spanner networks. Cambridge University Press, Cambridge/New York
31. Navarro G, Paredes R (2003) Practical construction of metric t-spanners. In: Proceedings of 5th workshop on algorithm engineering and experiments, Baltimore, Maryland. SIAM Press, pp 69–81
32. Rahmati Z, Abam MA, King V, Whitesides S (2014) Kinetic data structures for the semi-Yao graph and all nearest neighbors in $\mathbb{R}^d$. In: He M, Zeh N (eds) Canadian conference on computational geometry
33. Rao S, Smith WD (1998) Approximating geometrical graphs via spanners and banyans. In: Proceedings of 30th ACM symposium on theory of computing, Dallas, pp 540–550

Global Minimum Cuts in Surface-Embedded Graphs

Erin W. Chambers[1], Jeff Erickson[2], Kyle Fox[3], and Amir Nayyeri[4]
[1]Department of Computer Science and Mathematics, Saint Louis University, St. Louis, MO, USA
[2]Department of Computer Science, University of Illinois, Urbana, IL, USA
[3]Institute for Computational and Experimental Research in Mathematics, Brown University, Providence, RI, USA
[4]Department of Electrical Engineering and Computer Science, Oregon State University, Corvallis, OR, USA

Keywords

Covering spaces; Fixed-parameter tractability; Graph embedding; Homology; Minimum cuts; Topological graph theory

Years and Authors of Summarized Original Work

2009; Chambers, Erickson, Nayyeri
2011; Erickson, Nayyeri
2012; Erickson, Fox, Nayyeri

Problem Definition

Given a graph G in which every edge has a nonnegative capacity, the goal of the minimum-cut problem is to find a subset of edges of G with minimum total capacity whose deletion disconnects G. The closely related minimum (s,t)-cut problem further requires two specific vertices s and t to be separated by the deleted edges. Minimum cuts and their generalizations play a central role in divide-and-conquer and network optimization algorithms.

The fastest algorithms known for computing minimum cuts in arbitrary graphs run in roughly $O(mn)$ time for graphs with n vertices and m edges. However, even faster algorithms are known for graphs with additional topological structure. This entry sketches algorithms to compute minimum cuts in near-linear time when the input graph can be drawn on a surface with bounded genus – informally, a sphere with a bounded number of handles.

Problem 1 (Minimum (s,t)-Cut)

INPUT: *An undirected graph* $G = (V, E)$ *embedded on an orientable surface of genus* g, *a nonnegative capacity function* $c: E \to \mathbb{R}$, *and two vertices* s *and* t. OUTPUT: *A minimum-capacity* (s,t)*-cut in* G.

Problem 2 (Global Minimum Cut)

INPUT: *An undirected graph* $G = (V, E)$ *embedded on an orientable surface of genus* g *and a nonnegative capacity function* $c: E \to \mathbb{R}$. OUTPUT: *A minimum-capacity cut in* G.

Key Results

Topological Background

A surface is a compact space in which each point has a neighborhood homeomorphic to either the plane or a closed half plane. Points with half-plane neighborhoods comprise the boundary of the surface, which is the union of disjoint simple cycles. The genus is the maximum number of disjoint simple cycles whose deletion leaves the surface connected. A surface is orientable if it does not contain a MÃbius band. An embedding is a drawing of a graph on a surface, with vertices drawn as distinct points and edges as simple interior-disjoint paths, whose complement is a collection of disjoint open disks called the faces of the embedding.

An even subgraph of G is a subgraph in which every vertex has even degree; each component of an even subgraph is Eulerian. Two even subgraphs of an embedded graph G are $\mathbb{Z}_2$-homologous, or in the same $\mathbb{Z}_2$-homology class, if their symmetric difference is the boundary of a subset of the surface. If G is embedded on a surface of genus g with $b > 0$ boundary cycles, the even subgraphs of G fall into 2^{2g+b-1} $\mathbb{Z}_2$-homology classes. An even subgraph of G is $\mathbb{Z}_2$-minimal if it has minimum total cost within its $\mathbb{Z}_2$-homology class. Each component of a $\mathbb{Z}_2$-minimal even subgraph is itself $\mathbb{Z}_2$-minimal.

Every embedded graph G has a dual graph G^*, embedded on the same surface, whose vertices correspond to faces of G and whose edges correspond to pairs of faces that share an edge in G. The cost of a dual edge in G^* is the capacity of the corresponding primal edge in G.

Duality maps cut to certain sets of cycles and vice versa. For example, the minimum-capacity (s,t)-cut in any planar graph G is dual to the minimum-cost cycle in G^* that separates the dual faces s^* and t^*. If we remove s^* and t^* from the sphere, the dual of the minimum cut is the shortest generating cycle of the resulting annulus [11, 15]. More generally, let X denote the set of edges that cross some minimum (s,t)-cut in an embedded graph G, and let X^* denote the corresponding subgraph of the dual graph G^*. Then X^* is a minimum-cost even subgraph of G^* that separates s^* and t^*. If we remove s^* and t^* from the surface, X^* becomes a $\mathbb{Z}_2$-minimal subgraph homologous with the boundary of s^*.

Crossing Sequences

Our first algorithm [6] reduces computing a minimum (s,t)-cut in a graph embedded on a genus-g surface to $g^{O(g)}$ instances of the planar minimum-cut problem.

The algorithm begins by constructing a collection A of $2g + 1$ paths in G^*, called a greedy system of arcs, with three important properties. First, the endpoints of each path are incident to the boundary faces s^* and t^*. Second, each path is composed of two shortest paths plus at most one additional edge. Finally, the complement $\Sigma \setminus A$ of the paths is a topological open disk. A greedy system of arcs can be computed in $O(gn)$ time [5, 9].

We regard each component of X^* as a closed walk, and we enumerate all possible sequences of crossings between the components of X^* with the arcs in A. The components of any $\mathbb{Z}_2$-minimal even subgraph cross any shortest path, and therefore any arc in A, at most $O(g)$ times. It follows that we need to consider at most $g^{O(g)}$ crossing sequences, each of length at most $O(g^2)$. Following Kutz [13], the shortest closed walk with a given crossing sequence is the shortest generating cycle in an annulus obtained by gluing together $O(g^2)$ copies of the disk $\Sigma \setminus A$, which can be computed in $O(g^2 n \log\log n)$ time using the planar minimum-cut algorithm of Italiano et al. [12]. The overall running time of this algorithm is $g^{O(g)} n \log\log n$.

Surprisingly, a reduction from MAXCUT implies that finding the minimum-cost even subgraph in an arbitrary $\mathbb{Z}_2$-homology class is NP-hard. Different reductions imply that it is NP-hard to find the minimum-cost *closed walk* [5] or *simple cycle* [2] in a given $\mathbb{Z}_2$-homology class.

$\mathbb{Z}_2$-Homology Cover

Our second algorithm [9] finds the minimum-cost closed walks in G^* in every $\mathbb{Z}_2$-homology class by searching a certain covering space and then assembles X^* from these closed walks via dynamic programming.

As in our first algorithm, we first compute a greedy system of arcs A. The homology class of any cycle γ is determined by the parity of the number of crossings of γ with each arc in A. Each arc $\alpha_i \in A$ appears as two paths α_i^+ and α_i^- on the boundary of the disk $D = \Sigma \setminus A$.

We then construct a new surface $\overline{\Sigma}$, called the $\mathbb{Z}_2$-*homology cover* of Σ, by gluing together several copies of D as follows. We associate each homology class $h \in \mathbb{Z}_2^{2g+1}$ with a vector of $2g+1$ bits. Let $h \wedge i$ denote the bit vector obtained from h by flipping its ith bit. For each bit vector h, we construct a copy D_h of D; let $\alpha_{i,h}^+$ and $\alpha_{i,h}^-$ denote the copies of α_i^+ and α_i^- on the boundary of D_h. Finally, we construct $\overline{\Sigma}$ by identifying the paths $\alpha_{i,h}^+$ and $\alpha_{i,h\wedge i}^-$ for each homology class h and index i.

This construction also yields a graph $\overline{G}$ embedded in $\overline{\Sigma}$, with 2^{2g+1} vertices v_h and edges e_h for each vertex v and edge e of G^*. Each edge e_h of $\overline{G}$ inherits the cost of the corresponding edge e in G^*. Any walk in $\overline{G}$ projects to a walk in G^* by dropping subscripts; in particular, any walk in $\overline{G}$ from v_0 to v_h projects to a closed walk in G^* with homology class h that starts and ends at vertex v. Conversely, the *shortest* closed walk in G^* in any homology class h is the projection of the *shortest* path from v_0 to v_h, for some vertex v.

Any cycle in any nontrivial homology class crosses some path α_i, an odd number of times, and therefore at least once. To find all such cycles for each index i, we slice $\overline{G}$ along the lifted path $\alpha_{i,0}$ to obtain a new boundary cycle, and then compute the shortest path from each vertex v_0 on this cycle to every other vertex v_h in $2^{O(g)} n \log n$ time, using an algorithm of Chambers et al. [3]. Altogether, we find the shortest closed walk in every $\mathbb{Z}_2$-homology class in $2^{O(g)} n \log n$ time. The dual minimum cut X^* can then be built from these $\mathbb{Z}_2$-minimal cycles in $2^{O(g)}$ additional time via dynamic programming.

Global Minimum Cuts

Our final result generalizes the recent $O(n \log\log n)$-time algorithm for planar graphs by Lacki and Sankowski [14], which in turn relies on the $O(n \log\log n)$-time algorithm for planar minimum (s,t)-cuts of Italiano et al. [12].

The global minimum cut X in a surface graph G is dual to the minimum-cost nonempty separating subgraph of the dual graph G^*. In particular, if G is planar, X is dual to the shortest nonempty cycle in G^*. There are two cases to consider: either X^* is a simple contractible cycle, or it isn't. We describe two algorithms, one of which is guaranteed to return the minimum-cost separating subgraph.

To handle the contractible cycle case, we first slice the surface Σ to make it planar, first along the shortest non-separating cycle α in G^*, which we compute in $g^{O(g)} n \log\log n$ time using a variant of our crossing sequence algorithm, and then along a greedy system of arcs A connecting the resulting boundary cycles. Call the resulting planar graph D; each edge of $\alpha \cup A$ appears as two edges on the boundary of D. Let e^+ and e^- be edges on the boundary of D corresponding to some edge e of α. Using the planar algorithm of Lacki and Sankowski [14], we find the shortest cycle γ^+ in $D \setminus e^+$ and the shortest cycle γ^- in $D \setminus e^-$. The shorter of these two cycles projects to a closed walk γ in the original dual graph G^*.

Results of Cabello [2] imply that if γ is a simple cycle, it is the shortest contractible simple cycle in G^*; otherwise, X^* is not a simple cycle.

Our second algorithm begins by enumerating all $2^{O(g)}$ $\mathbb{Z}_2$-minimal even subgraphs in G^* in $g^{O(g)} n \log\log n$ time, using our crossing sequence algorithm. Our algorithm marks the faces on either side of an arbitrary edge of each $\mathbb{Z}_2$-minimal even subgraphs in G^*. If X^* is not a simple contractible cycle, then some pair of marked faces must be separated by X^*. In other words, in $g^{O(g)} n \log\log n$ time, we identify a set T of $2^{O(g)}$ vertices of G, at least two of which are separated by the global minimum cut. Thus, if we fix an arbitrary source vertex s and compute the minimum (s,t)-cut for each vertex $t \in T$ in $g^{O(g)} n \log\log n$ time, the smallest such cut is the global minimum cut X.

Open Problems

Extending these algorithms to *directed* surface graphs remains an interesting open problem; currently the only effective approach known is to compute a maximum (s,t)-flow and apply the maxflow-mincut theorem. The recent algorithm of Borradaile and Klein [1] computes maximum flows in directed *planar* graphs in $O(n \log n)$ time. For higher-genus graphs, Chambers et al. [7] describes maximum-flow algorithms that run in $g^{O(g)} n^{3/2}$ time for arbitrary capacities and in $O(g^8 n \log^2 n \log^2 C)$ for integer capacities that sum to C.

Another open problem is reducing the dependencies on the genus from exponential to polynomial. Even though there are near-quadratic algorithms to compute minimum cuts, the only known approach to achieving near-linear time for bounded-genus graphs with weighted edges is to solve an NP-hard problem.

Finally, it is natural to ask whether minimum cuts can be computed quickly in other minor-closed families of graphs, for which embeddings on to bounded-genus surfaces may not exist. Such results already exist for one-crossing-minor-free families [4] and in particular, graphs of bounded treewidth [10].

Cross-References

▶ Max Cut
▶ Separators in Graphs
▶ Shortest Paths in Planar Graphs with Negative Weight Edges
▶ Sparsest Cut

Recommended Reading

1. Borradaile G, Klein P (2009) An $O(n \log n)$ algorithm for maximum st-flow in a directed planar graph. J ACM 56(2):9:1–30
2. Cabello S (2010) Finding shortest contractible and shortest separating cycles in embedded graphs. ACM Trans Algorithms 6(2):24:1–24:18
3. Cabello S, Chambers EW, Erickson J (2013) Multiple-source shortest paths in embedded graphs. SIAM J Comput 42(4):1542–1571
4. Chambers E, Eppstein D (2013) Flows in one-crossing-minor-free graphs. J Graph Algorithms Appl 17(3):201–220. doi:10.7155/jgaa.00291
5. Chambers EW, Colin de Verdière É, Erickson J, Lazarus F, Whittlesey K (2008) Splitting (complicated) surfaces is hard. Comput Geom Theory Appl 41(1–2):94–110
6. Chambers EW, Erickson J, Nayyeri A (2009) Minimum cuts and shortest homologous cycles. In: Proceedings of the 25th annual symposium computational geometry, Aarhus, pp 377–385
7. Chambers EW, Erickson J, Nayyeri A (2012) Homology flows, cohomology cuts. SIAM J Comput 41(6):1605–1634
8. Erickson J (2012) Combinatorial optimization of cycles and bases. In: Zomorodian A (ed) Advances in applied and computational topology. Invited survey for an AMS short course on computational topology at the 2011 joint mathematics meetings, New Orleans. Proceedings of symposia in applied mathematics, vol 70. American Mathematical Society, pp 195–228
9. Erickson J, Nayyeri A (2011) Minimum cuts and shortest non-separating cycles via homology covers. In: Proceedings of the 22nd annual ACM-SIAM symposium on discrete algorithms, San Francisco, pp 1166–1176
10. Hagerup T, Katajainen J, Nishimura N, Ragde P (1998) Characterizing multiterminal flow networks and computing flows in networks of small treewidth. J Comput Syst Sci 57(3):366–375
11. Itai A, Shiloach Y (1979) Maximum flow in planar networks. SIAM J Comput 8:135–150
12. Italiano GF, Nussbaum Y, Sankowski P, Wulff-Nilsen C (2011) Improved algorithms for min cut and max flow in undirected planar graphs. In: Proceedings of the 43rd annual ACM symposium theory of computing, San Jose, pp 313–322

G

13. Kutz M (2006) Computing shortest non-trivial cycles on orientable surfaces of bounded genus in almost linear time. In: Proceedings of the 22nd annual symposium on computational geometry, Sedona, pp 430–438
14. Łącki J, Sankowski P (2011) Min-cuts and shortest cycles in planar graphs in $O(n \log\log n)$ time. In: Proceedings of the 19th annual European symposium on algorithms, Saarbrücken. Lecture notes in computer science, vol 6942. Springer, pp 155–166
15. Reif J (1983) Minimum s-t cut of a planar undirected network in $O(n \log^2 n)$ time. SIAM J Comput 12:71–81

Global Routing

Minsik Cho[1] and David Z. Pan[2]
[1]IBM T. J. Watson Research Center, Yorktown Heights, NY, USA
[2]Department of Electrical and Computer Engineering, University of Texas, Austin, TX, USA

Keywords

EDA; Global routing; Graph search; Layout; Mathematical programming; Netlist; Shortest path

Years and Authors of Summarized Original Work

2006; Cho, Pan

Problem Definition

Global routing is a key step in VLSI physical design after floor planning and placement. Its main goal is to reduce the overall routing complexity and guide the detailed router by planning the approximate routing path of each net. The commonly used objectives during global routing include minimizing total wirelength, mitigating routing congestion, or meeting routing resource constraints. If timing critical paths are known, they can also be put in the design objectives during global routing, along with other metrics such as manufacturability and noise.

The global routing problem can be formulated using graph models. For a given netlist graph $G(C, N)$, vertices C represent pins on placed objects such as standard cells or IP blocks, and edges N represent nets connecting the pins. The routing resources on a chip can be modeled in another graph $G(V, E)$ by dividing the entire global routing region into a set of smaller regions, so-called global routing cells (G-cells), where $v \in V$ represents a G-cell and $e \in E$ represents the boundary between two adjacent G-cells with a given routing capacity (c_e). Figure 1 shows how the chip can be abstracted into a 2-dimensional global routing graph. Such abstraction can be easily extended to 3-dimensional global routing graph to perform layer assignment (e.g., [15]). Since all standard cells and IP blocks are placed before the global routing stage (e.g., C can be mapped into V), the goal of global routing is to find G-cell to G-cell paths for N while trying to meet certain objectives such as routability optimization and wirelength minimization.

A straightforward mathematical optimization for global routing can be formulated as a 0-1 integer linear programming (ILP). Let R_i be a set of Steiner trees on G for net $n_i \in N$, and $x_{i,j}$ be the binary variable to indicate whether $r_{i,j} \in R_i$ is selected as the routing solution. Then an example ILP formulation can be written as follows:

The above formulation minimizes the total routing capacity utilization under the maximum routing capacity constraint for each edge $e \in E$. In fact, minimizing the total routing capacity is equivalent to minimizing total wirelength, because a unit wirelength in the global routing utilizes one routing resource (i.e., crossing the boundary between two adjacent G-cells). Other objectives/constraints can include timing optimization, noise reduction [10, 14], or manufacturability (e.g., CMP) [8].

Key Results

The straightforward formulation using ILP, e.g., as in Fig. 2, is NP-complete which cannot be solved efficiently for modern VLSI designs. One common technique to solve ILP is to use linear

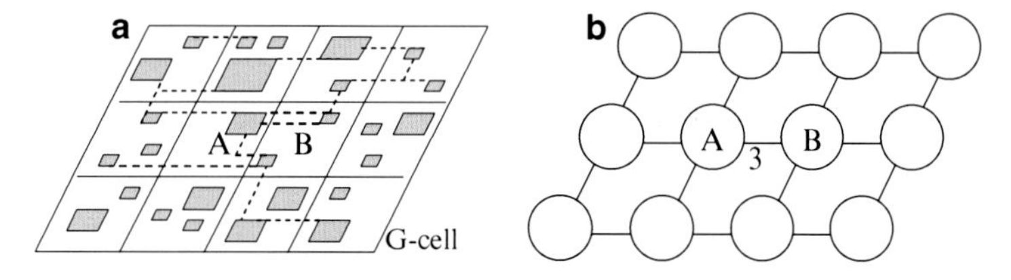

Global Routing, Fig. 1 Graph model for global routing. (**a**) Real circuit with G-cells. (**b**) Global routing graph

$$\min : \sum_{e \in E} u_e$$

$$\text{s.t} : \sum_{r_{i,j} \in R_i} x_{i,j} = 1, \forall n_i \in N$$

$$u_e = \sum_{(i,j): e \in r_{i,j}} x_{i,j}, \forall e \in E$$

$$u_e \leq c_e, \forall e \in E$$

Global Routing, Fig. 2 An example of global routing formulation using ILP

programming relaxation where the binary variables are made continuous, $x_{i,j} \in [0, 1]$, and that can be solved in polynomial time. Once a linear programming solution is obtained, rounding technique is used to find the binary solution. Another technique is a hierarchical divide-and-conquer scheme to limit the complexity, which solves many independent subproblems of similar sizes. These approaches may suffer from large amount of rounding errors or lack of interactions between subproblems, resulting in poor quality.

BoxRouter [1, 6] proposed a new approach to divide the entire routing region into a set of synergistic subregions. The key idea in BoxRouter is the progressive ILP based on the routing box expansion, which pushes congestion outward progressively from the highly congested region. Unlike conventional hierarchical divide-and-conquer approach, BoxRouter solves a sequence of ILP problems where an early problem is a subset of a later problem. BoxRouter progressively applies box expansion to build a sequence of ILP problems starting from the most congested region which is obtained through a very fast pre-routing stage. Figure 3 illustrates the concept of box expansion.

The advantage of BoxRouter over conventional approach is that each problem synergically reflects the decisions made so far by taking the previous solutions as constraints, in order to enhance congestion distribution and shorten the wirelength. In that sense, the first ILP problem has the largest flexibility which motivates the box expansion originating from the most congested region. Even though the last box can cover the whole design, the effective ILP size remains tractable in BoxRouter, as ILP is only performed on the wires between two subsequent boxes.

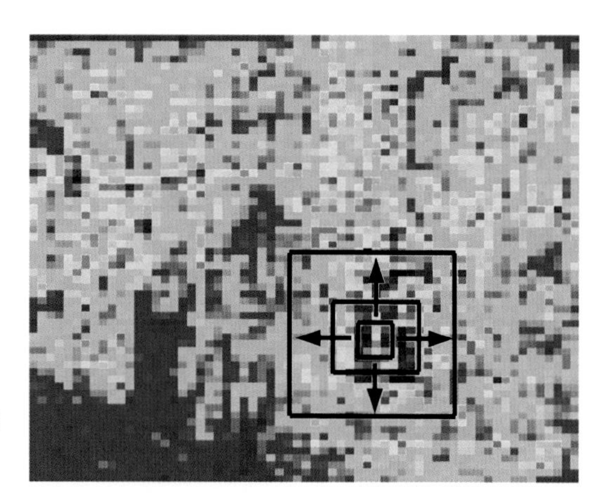

Global Routing, Fig. 3 Box expansion for Progressive ILP during BoxRouter

Compared with the formulation in Fig. 2 which directly minimizes the total wirelength, progressive ILP in BoxRouter maximizes the completion rate (e.g., minimizing unrouted nets) which can be more important and practical than minimizing total wirelength as shown in Fig. 4. Wirelength minimization in BoxRouter is indirectly achieved by allowing only the minimum rectilinear Steiner trees for each binary variable (i.e., $x_{i,j}$) and being augmented with the post-maze routing step. Such change in the objective function provides higher computation efficiency: it is found that the BoxRouter formulation can be solved significantly faster

$$\max : \sum x_{i,j}$$

$$\text{s.t} : \sum_{r_{i,j} \in R_i} x_{i,j} \leq 1, \forall n_i \in N$$

$$\sum_{(i,j):e \in r_{i,h}} x_{i,j} \leq c_e, \forall e \in E$$

Global Routing, Fig. 4 Global routing formulation in BoxRouter for minimal unrouted nets

than the traditional formulations due to its simple and well-exploited knapsack structure [7].

In case some nets remain unrouted after each ILP problem either due to insufficient routing resources inside a box or a limited number of Steiner graphs for each net, BoxRouter applies adaptive maze routing which penalizes using routing resources outside the current box, in order to reserve them for subsequence problems. Based on the new ILP techniques, BoxRouter has obtained much better results than previous state-of-the-art global routers [4, 9] and motivated many further studies in global routing (e.g., [5, 11–13, 15]) and global routing contests at ISPD 2007 and ISPD 2008 [2, 3].

Cross-References

▸ Circuit Placement
▸ Rectilinear Steiner Tree
▸ Routing

Recommended Reading

1. (2007) Minsik Cho, Kun Yuan, Katrina Lu and David Z. Pan http://www.cerc.utexas.edu/utda/download/BoxRouter.htm
2. (2007) http://archive.sigda.org/ispd2007/contest.html
3. (2008) http://archive.sigda.org/ispd2008/contests/ispd08rc.html
4. Albrecht C (2001) Global routing by new approximation algorithms for multicommodity flow. IEEE Trans Comput-Aided Des Integr Circuits Syst 20(5):622–632
5. Chang YJ, Lee YT, Wang TC (2008) NTHU-Route 2.0: a fast and stable global router. In: Proceedings of the international conference on computer aided design, San Jose, pp 338–343
6. Cho M, Pan DZ (2006) BoxRouter: a new global router based on box expansion and progressive ILP. In: Proceedings of the design automation conference, San Francisco, pp 373–378
7. Cho M, Pan DZ (2007) BoxRouter: a new global router based on box expansion and progressive ILP. IEEE Trans Comput-Aided Des Integr Circuits Syst 26(12):2130–2143
8. Cho M, Xiang H, Puri R, Pan DZ (2006) Wire density driven global routing for CMP variation and timing. In: Proceedings of the international conference on computer aided design, San Jose
9. Kastner R, Bozorgzadeh E, Sarrafzadeh M (2002) Pattern routing: use and theory for increasing predictability and avoiding coupling. IEEE Trans Comput-Aided Des Integr Circuits Syst 21(7):777–790
10. Kay R, Rutenbar RA (2000) Wire packing: a strong formulation of crosstalk-aware chip-level track/layer assignment with an efficient integer programming solution. In: Proceedings of the international symposium on physical design, San Diego
11. Moffitt MD (2008) Maizerouter: engineering an effective global router. In: Proceedings of the Asia and South Pacific design automation conference, Seoul
12. Ozdal M, Wong M (2007) Archer: a history-driven global routing algorithm. In: Proceedings of the international conference on computer aided design, San Jose, pp 488–495
13. Pan M, Chu C (2007) Fastroute 2.0: a high-quality and efficient global router. In: Proceedings of the Asia and South Pacific design automation conference, Yokohama
14. Wu D, Hu J, Mahapatra R, Zhao M (2004) Layer assignment for crosstalk risk minimization. In: Proceedings of the Asia and South Pacific design automation conference, Yokohama
15. Wu TH, Davoodi A, Linderoth JT (2009) GRIP: scalable 3D global routing using integer programming. In: Proceedings of the design automation conference, San Francisco, pp 320–325

Gomory-Hu Trees

Debmalya Panigrahi
Department of Computer Science, Duke University, Durham, NC, USA

Keywords

Cut trees; Minimum *s-t* cut; Undirected graph connectivity

Years and Authors of Summarized Original Work

2007; Bhalgat, Hariharan, Kavitha, Panigrahi

Problem Definition

Let $G = (V, E)$ be an undirected graph with $|V| = n$ and $|E| = m$. The edge connectivity of two vertices $s, t \in V$, denoted by $\lambda(s, t)$, is defined as the size of the smallest cut that separates s and t; such a cut is called a minimum $s - t$ cut. Clearly, one can represent the $\lambda(s, t)$ values for all pairs of vertices s and t in a table of size $O(n^2)$. However, for reasons of efficiency, one would like to represent all the $\lambda(s, t)$ values in a more succinct manner. *Gomory-Hu trees* (also known as *cut trees*) offer one such succinct representation of linear (i.e., $O(n)$) space and constant (i.e., $O(1)$) lookup time. It has the additional advantage that apart from representing all the $\lambda(s, t)$ values, it also contains structural information from which a minimum $s - t$ cut can be retrieved easily for any pair of vertices s and t.

Formally, a Gomory-Hu tree $T = (V, F)$ of an undirected graph $G = (V, E)$ is a weighted undirected tree defined on the vertices of the graph such that the following properties are satisfied:

- For any pair of vertices $s, t \in V, \lambda(s, t)$ is equal to the minimum weight on an edge in the unique path connecting s to t in T. Call this edge $e(s, t)$. If there are multiple edges with the minimum weight on the s to t path in T, any one of these edges is designated as $e(s, t)$.
- For any pair of vertices s and t, the bipartition of vertices into components produced by removing $e(s, t)$ (if there are multiple candidates for $e(s, t)$, this property holds for each candidate edge) from T corresponds to a minimum $s - t$ cut in the original graph G.

To understand this definition better, consider the following example. Figure 1 shows an undirected graph and a corresponding Gomory-Hu tree. Focus on a pair of vertices, for instance, 3 and 5.

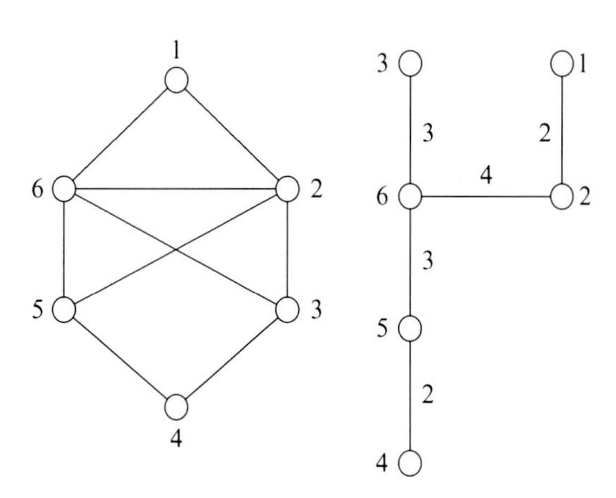

Gomory-Hu Trees, Fig. 1 An undirected graph (*left*) and a corresponding Gomory-Hu tree (*right*)

Clearly, the edge (6,5) of weight 3 is a minimum-weight edge on the 3 to 5 path in the Gomory-Hu tree. It is easy to see that $\lambda(3, 5) = 3$ in the original graph. Moreover, removing edge (6,5) in the Gomory-Hu tree produces the vertex bipartition ({1,2,3,6},{4,5}), which is a cut of size 3 in the original graph.

It is not immediate that such Gomory-Hu trees exist for all undirected graphs. In a classical result in 1961, Gomory and Hu [8] showed that not only do such trees exist for all undirected graphs but that they can also be computed using $n - 1$ minimum s-t cut (or equivalently maximum s-t flow) computations. In fact, a graph can have multiple Gomory-Hu trees.

All previous algorithms for constructing Gomory-Hu trees for undirected graphs used maximum-flow subroutines. Gomory and Hu gave an algorithm to compute a cut tree T using $n - 1$ maximum-flow computations and graph contractions. Gusfield [9] proposed an algorithm that does not use graph contractions; all $n - 1$ maximum-flow computations are performed on the input graph. Goldberg and Tsioutsiouliklis [7] did an experimental study of the algorithms due to Gomory and Hu and due to Gusfield for the cut tree problem and described efficient implementations of these algorithms. Examples were shown by Benczúr [1] that cut trees do not exist for directed graphs.

Any maximum-flow-based approach for constructing a Gomory-Hu tree would have a running time of $(n-1)$ times the time for computing a single maximum flow. Till now, faster algorithms for Gomory-Hu trees were by-products of faster algorithms for computing a maximum flow. The current fastest $\tilde{O}(m+n\lambda(s,t))$ (polylog n factors ignored in $\tilde{O}$ notation) maximum-flow algorithm, due to Karger and Levine [11], yields the current best expected running time of $\tilde{O}(n^3)$ for Gomory-Hu tree construction on simple unweighted graphs with n vertices. Bhalgat et al. [2] improved this time complexity to $\tilde{O}(mn)$. Note that both Karger and Levine's algorithm and Bhalgat et al.'s algorithm are randomized Las Vegas algorithms. The fastest deterministic algorithm for the Gomory-Hu tree construction problem is a by-product of Goldberg and Rao's maximum-flow algorithm [6] and has a running time of $\tilde{O}(nm^{1/2}\min(m,n^{3/2}))$.

Since the publication of the results of Bhalgat et al. [2], it has been observed that the maximum-flow subroutine of Karger and Levine [11] can also be used to obtain an $\tilde{O}(mn)$ time Las Vegas algorithm for constructing the Gomory-Hu tree of an unweighted graph. However, this algorithm does not yield partial Gomory-Hu trees which are defined below. For planar undirected graphs, Borradaile et al. [3] gave an $\tilde{O}(mn)$ time algorithm for constructing a Gomory-Hu tree.

It is important to note that in spite of the tremendous recent progress in approximate maximum s-t flow (or approximate minimum s-t cut) computation, this does not immediately translate to an improved algorithm for approximate Gomory-Hu tree construction. This is because of two reasons: first, the property of uncrossability of minimum s-t cuts used by Gomory and Hu in their minimum s-t cut based cut tree construction algorithm does not hold for approximate minimum s-t cuts, and second, the errors introduced in individual minimum s-t cut computation can add up to create large errors in the Gomory-Hu tree.

Key Results

Bhalgat et al. [2] considered the problem of designing an efficient algorithm for constructing a Gomory-Hu tree on unweighted undirected graphs. The main theorem shown in this entry is the following.

Theorem 1 *Let $G=(V,E)$ be a simple unweighted graph with m edges and n vertices. Then a Gomory-Hu tree for G can be built in expected time $\tilde{O}(mn)$.*

Their algorithm is always faster by a factor of $\tilde{\Omega}(n^{2/9})$ (polylog n factors ignored in $\tilde{\Omega}$ notation) compared to the previous best algorithm.

Instead of using maximum-flow subroutines, they use a Steiner connectivity algorithm. The *Steiner connectivity* of a set of vertices S (called the *Steiner set*) in an undirected graph is the minimum size of a cut which splits S into two parts; such a cut is called a *minimum Steiner cut.* Generalizing a tree-packing algorithm given by Gabow [5] for finding the edge connectivity of a graph, Cole and Hariharan [4] gave an algorithm for finding the Steiner connectivity k of a set of vertices in either undirected or directed Eulerian unweighted graphs in $\tilde{O}(mk^2)$ time. (For undirected graphs, their algorithm runs a little faster in time $\tilde{O}(m+nk^3)$.) Bhalgat et al. improved this result and gave the following theorem.

Theorem 2 *In an undirected or directed Eulerian unweighted graph, the Steiner connectivity k of a set of vertices can be determined in time $\tilde{O}(mk)$.*

The algorithm in [4] was used by Hariharan et al. [10] to design an algorithm with expected running time $\tilde{O}(m+nk^3)$ to compute a *partial* Gomory-Hu tree for representing the $\lambda(s,t)$ values for all pairs of vertices s,t that satisfied $\lambda(s,t)\leq k$. Replacing the algorithm in [4] by the new algorithm for computing Steiner connectivity yields an algorithm to compute a partial Gomory-Hu tree in expected running time $\tilde{O}(m+nk^2)$. Bhalgat et al. showed that using a more detailed analysis, this result can be improved to give the following theorem.

Theorem 3 *The partial Gomory-Hu tree of an undirected unweighted graph to represent all $\lambda(s,t)$ values not exceeding k can be constructed in expected time $\tilde{O}(mk)$.*

Since $\lambda(s,t) < n$ for all s,t vertex pairs in an unweighted (and simple) graph, setting k to n in Theorem 3 implies Theorem 1.

Applications

Gomory-Hu trees have many applications in multiterminal network flows and are an important data structure in graph connectivity literature.

Open Problems

The problem of derandomizing the algorithm due to Bhalgat et al. [2] to produce an $\tilde{O}(mn)$ time deterministic algorithm for constructing Gomory-Hu trees for unweighted undirected graphs remains open. The other main challenge is to extend the results in [2] to weighted graphs.

Experimental Results

Goldberg and Tsioutsiouliklis [7] did an extensive experimental study of the cut tree algorithms due to Gomory and Hu [8] and that due to Gusfield [9]. They showed how to efficiently implement these algorithms and also introduced and evaluated heuristics for speeding up the algorithms. Their general observation was that while Gusfield's algorithm is faster in many situations, Gomory and Hu's algorithm is more robust. For more detailed results of their experiments, refer to [7].

No experimental results are reported for the algorithm due to Bhalgat et al. [2].

Recommended Reading

1. Benczúr AA (1995) Counterexamples for directed and node capacitated cut-trees. SIAM J Comput 24(3):505–510
2. Bhalgat A, Hariharan R, Kavitha T, Panigrahi D (2007) An $\tilde{O}(mn)$ Gomory-Hu tree construction algorithm for unweighted graphs. In: Proceedings of the 39th annual ACM symposium on theory of computing, San Diego
3. Borradaile G, Sankowski P, Wulff-Nilsen C (2010) Min st-cut Oracle for planar graphs with near-linear preprocessing time. In: Proceedings of the 51th annual IEEE symposium on foundations of computer science, Las Vegas, pp 601–610
4. Cole R, Hariharan R (2003) A fast algorithm for computing steiner edge connectivity. In: Proceedings of the 35th annual ACM symposium on theory of computing, San Diego, pp 167–176
5. Gabow HN (1995) A matroid approach to finding edge connectivity and packing arborescences. J Comput Syst Sci 50:259–273
6. Goldberg AV, Rao S (1998) Beyond the flow decomposition barrier. J ACM 45(5):783–797
7. Goldberg AV, Tsioutsiouliklis K (2001) Cut tree algorithms: an experimental study. J Algorithms 38(1):51–83
8. Gomory RE, Hu TC (1961) Multi-terminal network flows. J Soc Ind Appl Math 9(4):551–570
9. Gusfield D (1990) Very simple methods for all pairs network flow analysis. SIAM J Comput 19(1):143–155
10. Hariharan R, Kavitha T, Panigrahi D (2007) Efficient algorithms for computing all low s-t edge connectivities and related problems. In: Proceedings of the 18th annual ACM-SIAM symposium on discrete algorithms, New Orleans, pp 127–136
11. Karger D, Levine M (2002) Random sampling in residual graphs. In: Proceedings of the 34th annual ACM symposium on theory of computing, Montreal, pp 63–66

Grammar Compression

Hideo Bannai
Department of Informatics, Kyushu University, Fukuoka, Japan

Keywords

Balanced binary grammar; Context-free grammar; LZ77 factorization; Smallest grammar problem; Straight line program

Years and Authors of Summarized Original Work

2000; Kieffer, Yang
2003; Rytter
2004; Sakamoto et al.
2005; Charikar et al.

Problem Definition

Given an input string S, the grammar-based compression problem is to find a small

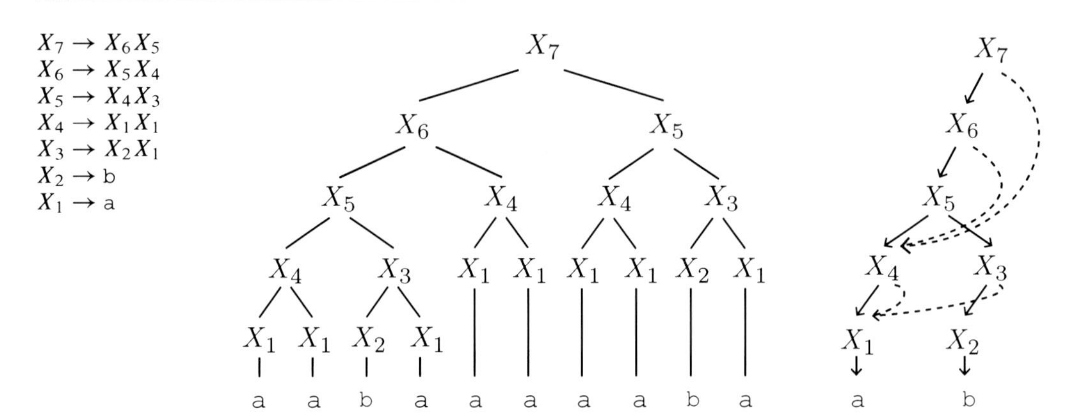

Grammar Compression, Fig. 1 (*Left*) an SLP G that represents string aabaaaaaba, where the variable X_7 is the start symbol. (*Center*) derivation tree of G. (*Right*) an ordered DAG corresponding to G, where the *solid* and *dashed edges* respectively correspond to the first and second child of each node

description of S that is based on a deterministic context-free grammar that generates a language consisting only of S. We will call such a context-free grammar, a grammar that represents S.

Generally, grammar-based compression can be divided into two phases [8], the *grammar transform* phase, where a context-free grammar G that represents string S is computed, and the *grammar encoding* phase, where an encoding for G is computed. Kieffer and Yang [8] showed that if a grammar transform is *irreducible*, namely, if the resulting grammar that represents S satisfies the following three conditions: (1) distinct variables derive different strings, (2) every variable other than the start symbol is used more than once (rule utility), and (3) all pairs of symbols have at most one nonoverlapping occurrence in the right-hand side of production rules (di-gram uniqueness); then, the grammar-based code using a zero order arithmetic code for encoding the grammar is universal.

Grammar-based compression algorithms differ mostly by how they perform the grammar transform, which can be stated as the following problem.

Problem 1 (Smallest Grammar Problem) Given an input string S of length N, output the smallest context-free grammar that represents S.

Here, the *size* of the grammar is defined as the total length of the right-hand side of the production rules in the grammar. Often, grammars are considered to be in the Chomsky normal form, in which case the grammar is called a *straight line program* (SLP) [7], i.e., the right-hand side of each production rule is either a terminal character or a pair of variables. Note that any grammar of size n can be converted into an SLP of size $O(n)$. Figure 1 shows an example of an SLP that represents string aabaaaaaba. A grammar representing a string can be considered as an ordered directed acyclic graph. Another important feature is that grammars allow for exponential compression, that is, the size of a grammar that represents a string of length N can be as small as $O(\log N)$.

Grammar-based compression is known to be especially suitable for compressing *highly repetitive strings*, for example, multiple whole genomes, where, although each individual string may not be easily compressible, the ensemble of strings is very compressible since each string is very similar to each other. Also, due to its ease of manipulation, grammar-based representation of strings is a frequently used model for *compressed string processing*, where the aim is to efficiently process compressed strings without explicit decompression. Such an approach allows for theoretical and practical speedups compared to a naive decompress-then-process approach.

Key Results

Hardness

The smallest grammar problem is known to be NP-hard [21]. The approximation ratio of a grammar-based algorithm A is defined as $\max_{S \in \Sigma^*} \frac{|G_S^A|}{|G_S^{opt}|}$, where $|G_S^A|$ is the size of the grammar that represents string S produced by A and $|G_S^{opt}|$ is the size of the smallest grammar that represents string S. Charikar et al. [3] showed that there is no polynomial-time algorithm for the smallest grammar problem with approximation ratio less than $\frac{8,569}{8,568}$, unless $P = NP$. Furthermore, they show that for a given set $\{k_1, \ldots, k_m\}$ of positive integers, the smallest grammar for string $\mathtt{a}^{k_1}\mathtt{ba}^{k_2}\mathtt{b}\cdots\mathtt{ba}^{k_m}$ is within a constant factor of the smallest number of multiplications required to compute $x^{k_1}, \ldots, x^{k_m}$, given a real number x. This is a well-studied problem known as the addition chain problem, whose best-known approximation algorithm has an approximation ratio of $O(\frac{\log N}{\log\log N})$ [23]. Thus, achieving $o(\frac{\log N}{\log\log N})$ approximation for the smallest grammar problem may be difficult.

Algorithms for Finding Small Grammars

Heuristics

Below, we give brief descriptions of several grammar-based compression algorithms based on simple greedy heuristics for which approximation ratios have been analyzed [3] (see Table 1).

- LZ78 [24] can be considered as constructing a grammar. Recall that each LZ78 factor of length at least two consists of a previous factor and a letter and can be expressed as a production rule of a grammar.
- SEQUITUR [15] processes the string in an online manner and adds a new character of the string to the right-hand side of the production rule of the start symbol, which is initially empty. For each new character, the algorithm updates the grammar, adding or removing production rules and replacing corresponding symbols in the grammar so that the di-gram uniqueness and rule utility properties are satisfied. The algorithm can be implemented to run in expected linear time. The grammar produced by SEQUITUR is not necessarily irreducible, and thus a revised version called SEQUENTIAL was proposed in [8].
- RE-PAIR [11] greedily and recursively replaces the most frequent di-gram in the string with a new symbol until no di-gram occurs more than once. Each such replacement corresponds to a new production rule in the final grammar. The algorithm can be implemented to run in linear time.
- LONGEST MATCH [8] greedily and recursively replaces the longest substring that has more than one nonoverlapping occurrence. The algorithm can be implemented to run in linear time by carefully maintaining a structure based on the suffix tree, through the course of the algorithm [10, 14].
- GREEDY [1] (originally called OFF-LINE, but coined in [3]) greedily and recursively replaces substrings that give the highest compression (with several variations in its definition). The algorithm can be implemented to run in $O(N \log N)$ time for each production rule, utilizing a data structure called minimal augmented suffix trees, which augments the suffix tree in order to consider the total number of nonoverlapping occurrences of a given substring.
- BISECTION [9] recursively partitions the string S into strings L and R of lengths 2^i and $N - 2^i$, where $i = \lceil \log N \rceil - 1$, each time forming a production rule $X_S \to X_L X_R$. A new production rule is created only for each distinct substring, and the rule is shared for identical substrings. The algorithm can be viewed as fixing the shape of the derivation tree and then computing the smallest grammar whose derivation tree is of the given shape.

Approximation Algorithms

Rytter [16] and Charikar et al. [3] independently and almost simultaneously developed linear time

Grammar Compression, Table 1 Known upper and lower bounds on approximation ratios for the simple heuristic algorithms (Taken from [3] with corrections)

Algorithm	Upper bound	Lower bound
LZ78 [24]	$O((N/\log N)^{2/3})$	$\Omega(N^{2/3}/\log N)$
RE-PAIR [11]		$\Omega(\sqrt{\log N})$
LONGEST MATCH [8]		$\Omega(\log\log N)$
GREEDY [1]		$(5\log 3)/(3\log 5) > 1.137\ldots$
SEQUENTIAL [8]	$O((N/\log N)^{3/4})$	$\Omega(N^{1/3})$
BISECTION [9]	$O((N/\log N)^{1/2})$	$\Omega(N^{1/2}/\log N)$

Grammar Compression, Table 2 Approximation algorithms for the smallest grammar problem. N is the size of the input string, and n is the size of the output grammar

Algorithm	Approximation ratio	Working space	Running time
Charikar et al. [3]	$O(\log(N/G_S^{\text{opt}}))$	$O(N)$	$O(N)$
Rytter [16]			
LEVELWISE-REPAIR [18]			
Jeż [5]			
Jeż [6]			
LCA [19]	$O((\log N)\log G_S^{\text{opt}})$	$O(n)$	$O(N)$ expected
LCA* [20]	$O((\log^* N)\log N)$	$O(n)$	$O(N\log^* N)$ expected
OLCA [12]	$O(\log^2 N)$	$O(n)$	$O(N)$ expected
FOLCA [13]	$O(\log^2 N)$	$2n\log n(1+o(1))+2n$ bits	$O(N\log N)$

algorithms which achieve the currently best approximation ratio of $O(\log(N/G_S^{\text{opt}}))$, essentially relying on the same two key ideas: the LZ77 factorization and balanced binary grammars. Below, we briefly describe the approach by Rytter to obtain an $O(\log N)$ approximation algorithm.

The string is processed from left to right, and the LZ77 factorization of the string helps to reuse, as much as possible, the grammar of previously occurring substrings. For string S, let $S = f_1 \ldots f_z$ be the LZ77 factorization of S. The algorithm sequentially processes each LZ factor f_i, maintaining a grammar G_i for $f_1 \ldots f_i$. Recall that by definition, each factor f_i of length at least 2 occurs in $f_1 \ldots f_{i-1}$. Therefore, there exists a sequence of $O(h_{i-1})$ variables of grammar G_{i-1} whose concatenation represents f_i, where h_{i-1} is the height of the derivation tree of G_{i-1}. Using this sequence of variables, a grammar for f_i is constructed, which is then subsequently appended to G_{i-1} to finally construct G_i.

A balanced binary grammar is a grammar in which the shape of the derivation tree resembles a balanced binary tree. Rytter proposed AVL (height balanced) grammars, where the height of sibling sub-trees differ by at most one. By restricting the grammar to AVL grammars, the height of the grammar is bounded by $O(\log N)$, and the above operations can be performed in $O(\log N)$ time for each LZ77 factor, by adding $O(\log N)$ new variables and using techniques resembling those of binary balanced search trees for re-balancing the tree. The resulting time complexity as well as the size of the grammar is $O(z\log N)$.

Finally, an important observation is that the size of the LZ77 factorization of a string S is a lower bound on the size of any grammar G that represents S.

Theorem 1 ([3, 16]) *For string S, let $S = f_1 \ldots f_z$ be the LZ77 factorization of S. Then, for any grammar G that represents S, $z \le |G|$.*

Thus, the total size of the grammar is $O(G_S^{\mathrm{opt}} \log N)$, achieving an $O(\log N)$ approximation ratio. Instead of AVL grammars, Charikar et al. use α balanced (length balanced) grammars, where the ratio between the lengths of sibling sub-trees is between $\frac{\alpha}{1-\alpha}$ and $\frac{1-\alpha}{\alpha}$ for some constant $0 < \alpha \leq \frac{1}{2}$, but the remaining arguments are similar.

Several other linear time algorithms that achieve $O(\log(N/G_S^{\mathrm{opt}}))$ approximation have been proposed [5, 6, 18]. These algorithms resemble RE-PAIR in that they basically replace di-grams in the string with a new symbol in a bottom-up fashion but with specific mechanisms to choose the di-grams so that a good approximation ratio is achieved.

LCA and its variants [12, 13, 19, 20] are approximation algorithms shown to be among the most scalable and practical. The approximation ratios are slightly weaker, but the algorithm can be made to run in an online manner and to use small space (see Table 2). Although seemingly proposed independently, the core idea of LCA is essentially the same as LCP [17] which constructs a grammar based on a technique called *locally consistent parsing*. The parsing is a partitioning of the string that can be computed using only local characteristics and guarantees that for any two occurrences of a given substring, the partitioning in the substring will be almost identical with exceptions in a sufficiently short prefix and suffix of the substring. This allows the production rules of the grammar to be more or less the same for repeated substrings, thus bounding the approximation ratio.

Decompression

The string that a grammar represents can be recovered in linear time by a simple depth-first left-to-right traversal on the grammar. Given an SLP G of size n that represents a string S of length N, G can be preprocessed in $O(n)$ time and space so that each variable holds the length of the string it derives. Using this information, it is possible to access $S[i]$ for any $1 \leq i \leq N$ in $O(h)$ time, where h is the height of the SLP, by simply traversing down the production rules starting from the start symbol until reaching a terminal character corresponding to $S[i]$. Balanced SLPs have height $O(\log N)$ and, therefore, allow access to any position of S in $O(\log N)$ time. For any grammar G, G can be preprocessed in $O(n)$ time and space, so that an arbitrary substring of length l of S can be obtained in $O(l + \log N)$ time [2]. Also, G can be preprocessed in $O(n)$ time and space so that the prefix or suffix of any length l for any variable in G can be obtained in $O(l)$ time [4]. On the other hand, it has been shown that using any data structure of size polynomial in n, the time for retrieving a character at an arbitrary position is at least $(\log N)^{1-\epsilon}$ for any constant $\epsilon > 0$ [22].

URLs to Code and Data Sets

Publicly available implementations of SEQUITUR:

- http://www.sequitur.info

Publicly available implementations of RE-PAIR:

- http://www.dcc.uchile.cl/~gnavarro/software/repair.tgz
- http://www.cbrc.jp/~rwan/en/restore.html, and
- https://code.google.com/p/re-pair/

Publicly available implementations of GREEDY (OFF-LINE):

- http://www.cs.ucr.edu/~stelo/Offline/.

Publicly available implementations of LCA variants:

- https://code.google.com/p/lcacomp/
- https://github.com/tb-yasu/olca-plus-plus

Cross-References

▶ Arithmetic Coding for Data Compression
▶ Lempel-Ziv Compression
▶ Pattern Matching on Compressed Text

Recommended Reading

1. Apostolico A, Lonardi S (2000) Off-line compression by greedy textual substitution. Proc IEEE 88(11):1733–1744
2. Bille P, Landau GM, Raman R, Sadakane K, Satti SR, Weimann O (2011) Random access to grammar-compressed strings. In: Proceedings of the SODA'11, San Francisco, pp 373–389
3. Charikar M, Lehman E, Liu D, Panigrahy R, Prabhakaran M, Sahai A, Shelat A (2005) The smallest grammar problem. IEEE Trans. Inf. Theory 51(7):2554–2576
4. Gąsieniec L, Kolpakov R, Potapov I, Sant P (2005) Real-time traversal in grammar-based compressed files. In: Proceedings of the DCC'05, Snowbird, p 458
5. Jeż A (2013) Approximation of grammar-based compression via recompression. In: Proceedings of the CPM'13, Bad Herrenalb, pp 165–176
6. Jeż A (2014) A *really* simple approximation of smallest grammar. In: Proceedings of the CPM'14, Moscow, pp 182–191
7. Karpinski M, Rytter W, Shinohara A (1997) An efficient pattern-matching algorithm for strings with short descriptions. Nord J Comput 4:172–186
8. Kieffer JC, Yang EH (2000) Grammar-based codes: a new class of universal lossless source codes. IEEE Trans Inf Theory 46(3):737–754
9. Kieffer J, Yang E, Nelson G, Cosman P (2000) Universal lossless compression via multilevel pattern matching. IEEE Trans Inf Theory 46(4):1227–1245
10. Lanctôt JK, Li M, Yang E (2000) Estimating DNA sequence entropy. In: Proceedings of the SODA'00, San Francisco, pp 409–418
11. Larsson NJ, Moffat A (2000) Off-line dictionary-based compression. Proc IEEE 88(11):1722–1732
12. Maruyama S, Sakamoto H, Takeda M (2012) An online algorithm for lightweight grammar-based compression. Algorithms 5(2):214–235
13. Maruyama S, Tabei Y, Sakamoto H, Sadakane K (2013) Fully-online grammar compression. In: Proceedings of the SPIRE'13, Jerusalem, pp 218–229
14. Nakamura R, Inenaga S, Bannai H, Funamoto T, Takeda M, Shinohara A (2009) Linear-time text compression by longest-first substitution. Algorithms 2(4):1429–1448
15. Nevill-Manning CG, Witten IH (1997) Identifying hierarchical structure in sequences: a linear-time algorithm. J Artif Intell Res 7(1):67–82
16. Rytter W (2003) Application of Lempel-Ziv factorization to the approximation of grammar-based compression. Theor Comput Sci 302(1–3):211–222
17. Sahinalp SC, Vishkin U (1995) Data compression using locally consistent parsing. Technical report, UMIACS Technical Report
18. Sakamoto H (2005) A fully linear-time approximation algorithm for grammar-based compression. J Discret Algorithms 3(2–4):416–430
19. Sakamoto H, Kida T, Shimozono S (2004) A space-saving linear-time algorithm for grammar-based compression. In: Proceedings of the SPIRE'04, Padova, pp 218–229
20. Sakamoto H, Maruyama S, Kida T, Shimozono S (2009) A space-saving approximation algorithm for grammar-based compression. IEICE Trans 92-D(2):158–165
21. Storer JA (1977) NP-completeness results concerning data compression. Technical report 234, Department of Electrical Engineering and Computer Science, Princeton University
22. Verbin E, Yu W (2013) Data structure lower bounds on random access to grammar-compressed strings. In: Proceedings of the CPM'13, Bad Herrenalb, pp 247–258
23. Yao ACC (1976) On the evaluation of powers. SIAM J Comput 5(1):100–1–03
24. Ziv J, Lempel A (1978) Compression of individual sequences via variable-length coding. IEEE Trans Inf Theory 24(5):530–536

Graph Bandwidth

James R. Lee
Department of Computer Science and Engineering, University of Washington, Seattle, WA, USA

Keywords

Approximation algorithms; Graph bandwidth; Metric embeddings

Years and Authors of Summarized Original Work

1998; Feige
2000; Feige

Problem Definition

The *graph bandwidth problem* concerns producing a linear ordering of the vertices of a graph $G = (V, E)$ so as to minimize the maximum "stretch" of any edge in the ordering. Formally, let $n = |V|$, and consider any one-to-one mapping $\pi : V \to \{1, 2, \ldots, n\}$. The *bandwidth* of this ordering is $\mathsf{bw}_\pi(G) = \max_{\{u,v\}\in E} |\pi(u) - \pi(v)|$. The *bandwidth of G* is given by the bandwidth of the best possible ordering: $\mathsf{bw}(G) = \min_\pi \mathsf{bw}_\pi(G)$.

The original motivation for this problem lies in the preprocessing of sparse symmetric square matrices. Let A be such an $n \times n$ matrix, and consider the problem of finding a permutation matrix P such that the non-zero entries of $P^{\mathrm{T}}AP$ all lie in as narrow a band as possible about the diagonal. This problem is equivalent to minimizing the bandwidth of the graph G whose vertex set is $\{1, 2, \ldots, n\}$ and which has an edge $\{u, v\}$ precisely when $A_{u,v} \neq 0$.

In lieu of this fact, one tries to efficiently compute a linear ordering π for which $\mathsf{bw}_\pi(G) \leq A \cdot \mathsf{bw}(G)$, with the *approximation factor* A is as small as possible. There is even evidence that achieving any value $A = O(1)$ is NP-hard [18]. Much of the difficulty of the bandwidth problem is due to the objective function being a maximum over all edges of the graph. This makes divide-and-conquer approaches ineffective for graph bandwidth, whereas they often succeed for related problems like Minimum Linear Arrangement [6] (here the objective is to minimize $\sum_{\{u,v\}\in E} |\pi(u) - \pi(v)|$). Instead, a more global algorithm is required. To this end, a good lower bound on the value of $\mathsf{bw}(G)$ has to be initially discussed.

The Local Density

For any pair of vertices $u, v \in V$, let $d(u, v)$ to be the shortest path distance between u and v in the graph G. Then, define $B(v, r) = \{u \in V : d(u, v) \leq r\}$ as the *ball of radius* r about a vertex $v \in V$. Finally, the *local density of G* is defined by $D(G) = \max_{v\in V, r\geq 1} |B(v, r)|/(2r)$. It is not difficult to see that $\mathsf{bw}(G) \geq D(G)$. Although it was conjectured that an upper bound of the form $\mathsf{bw}(G) \leq \mathrm{poly}(\log n) \cdot D(G)$ holds, it was not proven until the seminal work of Feige [7].

Key Results

Feige proved the following.

Theorem 1 *There is an efficient algorithm that, given a graph* $G = (V, E)$ *as input, produces a linear ordering* $\pi : V \to \{1, 2, \ldots, n\}$ *for which* $\mathsf{bw}_\pi(G) \leq O\left((\log n)^3 \sqrt{\log n \log\log n}\right) \cdot D(G)$. *In particular, this provides a* $\mathrm{poly}(\log n)$*-approximation algorithm for the bandwidth problem in general graphs.*

Feige's algorithmic framework can be described quite simply as follows.

1. Compute a representation $f : V \to \mathbb{R}^n$ of G in Euclidean space.
2. Let $u_1, u_2, \ldots, u_n$ be independent $N(0, 1)$. ($N(0; 1)$ denotes a standard normal random variable with mean 0 and variance 1.) random variables, and for each vertex $v \in V$, compute $h(v) = \sum_{i=1}^{n} u_i f_i(v)$, where $f_i(v)$ is the ith coordinate of the vector $f(v)$.
3. Sort the vertices by the value $h(v)$, breaking ties arbitrarily, and output the induced linear ordering.

An equivalent characterization of steps (2) and (3) is to choose a uniformly random vector $\mathbf{a} \in S^{n-1}$ from the $(n-1)$-dimensional sphere $S^{n-1} \subseteq \mathbb{R}^n$ and output the linear ordering induced by the values $h(v) = \langle \mathbf{a}, f(v) \rangle$, where $\langle \cdot, \cdot \rangle$ denotes the usual inner product on $\mathbb{R}^n$. In other words, the algorithm first computes a map $f : V \to \mathbb{R}^n$, projects the images of the vertices onto a randomly oriented line, and then outputs the induced ordering; step (2) is the standard way that such a random projection is implemented.

Volume-Respecting Embeddings

The only step left unspecified is (1); the function f has to somehow preserve the structure of the graph G in order for the algorithm to output a low-bandwidth ordering. The inspiration for the existence of such an f comes from the field of *low-distortion metric embeddings* (see, e.g., [2, 14]). Feige introduced a generalization of low-distortion embeddings to mappings called *volume respecting embeddings.* Roughly, the map f should be non-expansive, in the sense that $\|f(u) - f(v)\| \leq 1$ for every edge $\{u, v\} \in E$, and should satisfy the following property: For any set of k vertices $v_1, \ldots, v_k$, the $(k-1)$-dimensional volume of the convex hull of the points $f(v_1), \ldots, f(v_k)$ should be as large as possible. The proper value of k is chosen to optimize the performance of the algorithm. Refer to [7, 10, 11] for precise definitions on volume-respecting embeddings, and a detailed discussion of their construction. Feige showed that a modification of Bourgain's embedding [2] yields a mapping $f: V \to \mathbb{R}^n$ which is good enough to obtain the results of Theorem 1.

The requirement $\|f(u) - f(v)\| \leq 1$ for every edge $\{u, v\}$ is natural since $f(u)$ and $f(v)$ need to have similar projections onto the random direction $\mathbf{a}$; intuitively, this suggests that u and v will not be mapped too far apart in the induced linear ordering. But even if $|h(u) - h(v)|$ is small, it may be that many vertices project between $h(u)$ and $h(v)$, causing u and v to incur a large stretch. To prevent this, the images of the vertices should be sufficiently "spread out," which corresponds to the volume requirement on the convex hull of the images.

Applications

As was mentioned previously, the graph bandwidth problem has applications to preprocessing sparse symmetric matrices. Minimizing the bandwidth of matrices helps in improving the efficiency of certain linear algebraic algorithms like Gaussian elimination; see [3, 8, 17]. Follow-up work has shown that Feige's techniques can be applied to VLSI layout problems [19].

Open Problems

First, state the *bandwidth conjecture* (see, e.g., [13]).

Conjecture: For any n-node graph $G = (V, E)$, one has $\mathsf{bw}(G) = O(\log n) \cdot D(G)$.

The conjecture is interesting and unresolved even in the special case when G is a tree (see [9] for the best results for trees). The best-known bound in the general case follows from [7, 10], and is of the form $\mathsf{bw}(G) = O(\log n)^{3.5} \cdot D(G)$. It is known that the conjectured upper bound is best possible, even for trees [4]. One suspects that these combinatorial studies will lead to improved approximation algorithms.

However, the best approximation algorithms, which achieve ratio $O((\log n)^3 (\log \log n)^{1/4})$, are not based on the local density bound. Instead, they are a hybrid of a semi-definite programming approach of [1, 5] with the arguments of Feige, and the volume-respecting embeddings constructed in [12, 16]. Determining the approximability of graph bandwidth is an outstanding open problem, and likely requires improving both the upper and lower bounds.

Recommended Reading

1. Blum A, Konjevod G, Ravi R, Vempala S (2000) Semi-definite relaxations for minimum bandwidth and other vertex-ordering problems. Theor Comput Sci 235(1):25–42, Selected papers in honor of Manuel Blum (Hong Kong, 1998)
2. Bourgain J (1985) On Lipschitz embedding of finite-metric spaces in Hilbert space. Israel J Math 52(1–2):46–52
3. Chinn PZ, Chvátalová J, Dewdney AK, Gibbs NE (1982) The bandwidth problem for graphs and matrices – a survey. J Graph Theory 6(3):223–254
4. Chung FRK, Seymour PD (1989) Graphs withsmall bandwidth and cutwidth. Discret Math 75(1–3):113–119, Graph theory and combinatorics, Cambridge (1988)
5. Dunagan J, Vempala S (2001) On Euclidean embeddings and bandwidth minimization. In: Randomization, approximation, and combinatorial optimization. Springer, pp 229–240
6. Even G, Naor J, Rao S, Schieber B (2000) Divide-and-conquer approximation algorithms via spreading metrics. J ACM 47(4):585–616
7. Feige U (2000) Approximating the bandwidth via volume respecting embeddings. J Comput Syst Sci 60(3):510–539
8. George A, Liu JWH (1981) Computer solution of large sparse positive definite systems. Prentice-hall series in computational mathematics. Prentice-Hall, Englewood Cliffs
9. Gupta A (2001) Improved bandwidth approximation for trees and chordal graphs. J Algorithms 40(1):24–36

10. Krauthgamer R, Lee JR, Mendel M, Naor A (2005) Measured descent: a new embedding method for finite metrics. Geom Funct Anal 15(4):839–858
11. Krauthgamer R, Linial N, Magen A (2004) Metric embeddings – beyond one-dimensional distortion. Discret Comput Geom 31(3):339–356
12. Lee JR (2006) Volume distortion for subsets of Euclidean spaces. In: Proceedings of the 22nd annual symposium on computational geometry. ACM, Sedona, pp 207–216
13. Linial N (2002) Finite metric-spaces – combinatorics, geometry and algorithms. In: Proceedings of the international congress of mathematicians, vol. III, Beijing, 2002. Higher Ed. Press, Beijing, pp 573–586
14. Linial N, London E, Rabinovich Y (1995) The geometry of graphs and some of its algorithmic applications. Combinatorica 15(2):215–245
15. Papadimitriou CH (1976) The NP-completeness of the bandwidth minimization problem. Computing 16(3):263–270
16. Rao S (1999) Small distortion and volume preserving embeddings for planar and Euclidean metrics. In: Proceedings of the 15th annual symposium on computational geometry. ACM, New York, pp 300–306
17. Strang G (1980) Linear algebra and its applications, 2nd edn. Academic [Harcourt Brace Jovanovich Publishers], New York
18. Unger W (1998) The complexity of the approximation of the bandwidth problem. In: 39th annual symposium on foundations of computer science. IEEE, 8–11 Oct 1998, pp 82–91
19. Vempala S (1998) Random projection: a new approach to VLSI layout. In: 39th annual symposium on foundations of computer science. IEEE, 8–11 Oct 1998, pp 389–398

Graph Coloring

Michael Langberg [1,3] and Chandra Chekuri[2,3]
[1]Department of Electrical Engineering, The State University of New York, Buffalo, NY, USA
[2]Department of Computer Science, University of Illinois, Urbana-Champaign, Urbana, IL, USA
[3]Department of Mathematics and Computer Science, The Open University of Israel, Raanana, Israel

Keywords

Approximation algorithms; Clique cover; Graph coloring; Semidefinite

Years and Authors of Summarized Original Work

1994, 1998; Karger, Motwani, Sudan

Problem Definition

An independent set in an undirected graph $G = (V, E)$ is a set of vertices that induce a subgraph which does not contain any edges. The size of the maximum independent set in G is denoted by $\alpha(G)$. For an integer k, a k-coloring of G is a function $\sigma : V \rightarrow [1 \ldots k]$ which assigns colors to the vertices of G. A valid k-coloring of G is a coloring in which each color class is an independent set. The chromatic number $\chi(G)$ of G is the smallest k for which there exists a valid k-coloring of G. Finding $\chi(G)$ is a fundamental NP-hard problem. Hence, when limited to polynomial time algorithms, one turns to the question of estimating the value of $\chi(G)$ or to the closely related problem of *approximate coloring*.

Problem 1 (Approximate coloring)
INPUT: Undirected graph $G = (V, E)$.
OUTPUT: A valid coloring of G with $r \cdot \chi(G)$ colors, for some approximation ratio $r \geq 1$.
OBJECTIVE: Minimize r.

Let G be a graph of size n. The approximate coloring of G can be solved efficiently within an approximation ratio of $r = O\left(\frac{n(\log\log n)^2}{\log^3 n}\right)$ [12]. This holds also for the approximation of $\alpha(G)$ [8]. These results may seem rather weak; however, it is NP-hard to approximate $\alpha(G)$ and $\chi(G)$ within a ratio of $n^{1-\varepsilon}$ for any constant $\varepsilon > 0$ [9, 14, 23]. Under stronger complexity assumptions, there is some constant $0 < \delta < 1$ such that neither problem can be approximated within a ratio of $n/2^{\log^\delta n}$ [19,23]. This entry will concentrate on the problem of coloring graphs G for which $\chi(G)$ is *small*. As will be seen, in this case the approximation ratio achievable significantly improves.

Vector Coloring of Graphs

The algorithms achieving the best ratios for approximate coloring when $\chi(G)$ is small are all based on the idea of *vector coloring*, introduced by Karger, Motwani, and Sudan [17]. (Vector coloring as presented in [17] is closely related to the Lovász θ function [21]. This connection will be discussed shortly.)

Definition 1 A vector k-coloring of a graph is an assignment of unit vectors to its vertices, such that for every edge, the inner product of the vectors assigned to its endpoints is at most (in the sense that it can only be more negative) $-1/(k-1)$.

The *vector chromatic number* $\vec{\chi}(G)$ of G is the smallest k for which there exists a vector k-coloring of G. The vector chromatic number can be formulated as follows:

$$\begin{array}{lll} \vec{\chi}(G) & \text{Minimize } k & \\ & \text{subject to: } \langle v_i, v_j \rangle \leq -\frac{1}{k-1} & \forall (i,j) \in E \\ & \langle v_i, v_i \rangle = 1 & \forall i \in V \end{array}$$

Here, assume that $V = [1, \ldots, n]$ and that the vectors $\{v_i\}_{i=1}^n$ are in R^n. Every k-colorable graph is also vector k-colorable. This can be seen by identifying each color class with one vertex of a perfect $(k-1)$-dimensional simplex centered at the origin. Moreover, unlike the chromatic number, a vector k-coloring (when it exists) can be found in polynomial time using semidefinite programming (up to an arbitrarily small error in the inner products).

Claim 1 (Complexity of vector coloring [17]) Let $\varepsilon > 0$. If a graph G has a vector k-coloring, then a vector $(k+\varepsilon)$-coloring of the graph can be constructed in time polynomial in n and $\log(1/\varepsilon)$.

One can strengthen Definition 1 to obtain a different notion of vector coloring and the vector chromatic number:

$$\begin{array}{lll} \vec{\chi}_2(G) & \text{Minimize } k & \\ & \text{subject to: } \langle v_i, v_j \rangle = -\frac{1}{k-1} & \forall (i,j) \in E \\ & \langle v_i, v_i \rangle = 1 & \forall i \in V \end{array}$$

$$\begin{array}{lll} \vec{\chi}_3(G) & \text{Minimize } k & \\ & \text{subject to: } \langle v_i, v_j \rangle = -\frac{1}{k-1} & \forall (i,j) \in E \\ & \langle v_i, v_j \rangle \geq -\frac{1}{k-1} & \forall i,j \in V \\ & \langle v_i, v_i \rangle = 1 & \forall i \in V \end{array}$$

The function $\vec{\chi}_2(G)$ is referred to as the *strict* vector chromatic number of G and is equal to the Lovász θ function on $\bar{G}$ [17, 21], where $\bar{G}$ is the *complement* graph of G. The function $\vec{\chi}_3(G)$ is referred to as the *strong* vector chromatic number. An analog to Claim 1 holds for both $\vec{\chi}_2(G)$ and $\vec{\chi}_3(G)$. Let $\omega(G)$ denote the size of the maximum clique in G; it holds that $\omega(G) \leq \vec{\chi}(G) \leq \vec{\chi}_2(G) \leq \vec{\chi}_3(G) \leq \chi(G)$.

Key Results

In what follows, assume that G has n vertices and maximal degree Δ. The $\tilde{O}(\cdot)$ and $\tilde{\Omega}(\cdot)$ notation are used to suppress polylogarithmic factors. We now state the key result of Karger, Motwani, and Sudan [17]:

Theorem 1 ([17]) *If $\vec{\chi}(G) = k$, then G can be colored in polynomial time using $\min\{\tilde{O}(\Delta^{1-2/k}), \tilde{O}(n^{1-3/(k+1)})\}$ colors.*

As mentioned above, the use of vector coloring in the context of approximate coloring was initiated in [17]. Roughly speaking, once given a vector coloring of G, the heart of the algorithm in [17] finds a large independent set in G. In a nutshell, this independent set corresponds to a set of vectors in the vector coloring which are *close* to one another (and thus by definition cannot share an edge). Combining this with the ideas of Wigderson [22] mentioned below yields Theorem 1.

We proceed to describe related work. The first two theorems below appeared prior to the work of Karger, Motwani, and Sudan [17].

Theorem 2 ([22]) *If $\chi(G) = k$, then G can be colored in polynomial time using $O(kn^{1-1/(k-1)})$ colors.*

Theorem 3 ([1]) *If $\chi(G) = 3$, then G can be colored in polynomial time using $\tilde{O}(n^{3/8})$ colors. If $\chi(G) = k \geq 4$ then G can be colored in polynomial time using at most $\tilde{O}(n^{1-1/(k-3/2)})$ colors.*

Combining the techniques of [17] and [1], the following results were obtained for graphs G with $\chi(G) = 3, 4$ (these results were also extended for higher values of $\chi(G)$).

Theorem 4 ([2]) *If $\chi(G) = 3$, then G can be colored in polynomial time using $\tilde{O}(n^{3/14})$ colors.*

Theorem 5 ([13]) *If $\chi(G) = 4$, then G can be colored in polynomial time using $\tilde{O}(n^{7/19})$ colors.*

The currently best known result for coloring a 3-colorable graph is presented in [16]. The algorithm of [16] combines enhanced notions of vector coloring presented in [5] with the combinatorial coloring techniques of [15].

Theorem 6 ([16]) *If $\chi(G) = 3$, then G can be colored in polynomial time using $O(n^{0.19996})$ colors.*

To put the above theorems in perspective, it is NP-hard to color a 3-colorable graph G with 4 colors [11, 18] and a k-colorable graph (for sufficiently large k) with $k^{\frac{\log k}{25}}$ colors [19]. Under stronger complexity assumptions (related to the unique games conjecture [20]) for any constant k, it is hard to color a k-colorable graph with any constant number of colors [6]. The wide gap between these hardness results and the approximation ratios presented in this section has been a major initiative in the study of approximate coloring.

Finally, the limitations of vector coloring are addressed. Namely, are there graphs for which $\vec{\chi}(G)$ is a poor estimate of $\chi(G)$? One would expect the answer to be "yes" as estimating $\chi(G)$ beyond a factor of $n^{1-\varepsilon}$ is a hard problem. As will be stated below, this is indeed the case (even when $\vec{\chi}(G)$ is small). Some of the results that follow are stated in terms of the maximum independent set $\alpha(G)$ in G. As $\chi(G) \geq n/\alpha(G)$, these results imply a lower bound on $\chi(G)$. Theorem 7 (i) states that the original analysis of [17] is essentially tight. Theorem 7 (ii) presents bounds for the case of $\vec{\chi}(G) = 3$. Theorem 7 (iii) and Theorem 8 present graphs G in which there is an extremely large gap between $\chi(G)$ and the relaxations $\vec{\chi}(G)$ and $\vec{\chi}_2(G)$.

Theorem 7 ([10]) *(i) For every constant $\varepsilon > 0$ and constant $k > 2$, there are infinitely many graphs G with $\vec{\chi}(G) = k$ and $\alpha(G) \leq n/\Delta^{1-\frac{2}{k}-\varepsilon}$ (here $\Delta > n^{\delta}$ for some constant $\delta > 0$). (ii) There are infinitely many graphs G with $\vec{\chi}(G) = 3$ and $\alpha(G) \leq n^{0.843}$. (iii) For some constant c, there are infinitely many graphs G with $\vec{\chi}(G) = O(\frac{\log n}{\log\log n})$ and $\alpha(G) \leq \log^c n$.*

Theorem 8 ([7]) *For some constant c, there are infinitely many graphs G with $\vec{\chi}_2(G) \leq 2^{\sqrt{\log n}}$ and $\chi(G) \geq n/2^{c\sqrt{\log n}}$.*

Vector colorings, including the Lovász θ function and its variants, have been extensively studied in the context of approximation algorithms for problems other than Problem 1. These include approximating $\alpha(G)$, approximating the minimum vertex cover problem, and combinatorial optimization in the context of random graphs.

Applications

Besides its theoretical significance, graph coloring has several concrete applications (see, e.g., [3,4]).

Open Problems

By far the major open problem in the context of approximate coloring addresses the wide gap between what is known to be hard and what can be obtained in polynomial time. The case of constant $\chi(G)$ is especially intriguing, as the best known upper bounds (on the approximation ratio) are polynomial while the lower bounds are of constant nature. Regarding the vector coloring paradigm, a majority of the results stated in section "Key Results" use the weakest from of vector coloring $\vec{\chi}(G)$ in their proof, while stronger relaxations may also be considered. It would be very interesting to improve upon the algorithmic results stated above using stronger relaxations, as would a matching analysis of the limitations of these relaxations.

Recommended Reading

1. Blum A (1994) New approximations for graph coloring. J ACM 41(3):470–516
2. Blum A, Karger D (1997) An $\tilde{O}(n^{3/14})$-coloring for 3-colorable graphs. Inf Process Lett 61(6):49–53
3. Chaitin GJ (1982) Register allocation & spilling via graph coloring. In: Proceedings of the 1982 SIGPLAN symposium on compiler construction, pp 98–105
4. Chaitin GJ, Auslander MA, Chandra AK, Cocke J, Hopkins ME, Markstein PW (1981) Register allocation via coloring. Comput Lang 6:47–57
5. Chlamtac E (2007) Approximation algorithms using hierarchies of semidefinite programming relaxations. In: Proceedings of the 48th annual IEEE symposium on foundations of computer science, pp 691–701
6. Dinur I, Mossel E, Regev O (2009) Conditional hardness for approximate coloring. SIAM J Comput 39(3):843–873
7. Feige U (1997) Randomized graph products, chromatic numbers, and the Lovász theta function. Combinatorica 17(1):79–90
8. Feige U (2004) Approximating maximum clique by removing subgraphs. SIAM J Discret Math 18(2):219–225
9. Feige U, Kilian J (1998) Zero knowledge and the chromatic number. J Comput Syst Sci 57:187–199
10. Feige U, Langberg M, Schechtman G (2004) Graphs with tiny vector chromatic numbers and huge chromatic numbers. SIAM J Comput 33(6):1338–1368
11. Guruswami V, Khanna S (2000) On the hardness of 4-coloring a 3-colorable graph. In: Proceedings of the 15th annual IEEE conference on computational complexity, pp 188–197
12. Halldorsson M (1993) A still better performance guarantee for approximate graph coloring. Inf Process Lett 45:19–23
13. Halperin E, Nathaniel R, Zwick U (2002) Coloring k-colorable graphs using smaller palettes. J Algorithms 45:72–90
14. Håstad J (1999) Clique is hard to approximate within $n^{1-\varepsilon}$. Acta Math 182(1):105–142
15. Kawarabayashi K, Thorup M (2012) Combinatorial coloring of 3-Colorable graphs. In: Proceedings of the 53rd annual IEEE symposium on foundations of computer science, pp 68–75
16. Kawarabayashi K, Thorup M (2014) Coloring 3-colorable graphs with $o(n^{1/5})$ colors. In: Proceedings of the 31st international symposium on theoretical aspects of computer science, pp 458–469
17. Karger D, Motwani R, Sudan M (1998) Approximate graph coloring by semidefinite programming. J ACM 45(2):246–265
18. Khanna S, Linial N, Safra S (2000) On the hardness of approximating the chromatic number. Combinatorica 20:393–415
19. Khot S (2001) Improved inapproximability results for max clique, chromatic number and approximate graph coloring. In: Proceedings of the 42nd annual IEEE symposium on foundations of computer science, pp 600–609
20. Khot S (2002) On the power of unique 2-prover 1-round games. In: Proceedings of the 34th annual ACM symposium on theory of computing, pp 767–775
21. Lovász L (1979) On the Shannon capacity of a graph. IEEE Trans Inf Theory 25:2–13
22. Wigderson A (1983) Improving the performance guarantee for approximate graph coloring. J ACM 30(4):729–735
23. Zuckerman D (2006) Linear degree extractors and the inapproximability of max clique and chromatic number. In: Proceedings of the 38th annual ACM symposium on theory of computing, pp 681–690

Graph Connectivity

Samir Khuller[1] and Balaji Raghavachari[2]
[1]Computer Science Department, University of Maryland, College Park, MD, USA
[2]Computer Science Department, The University of Texas at Dallas, Richardson, TX, USA

Keywords

Highly connected subgraphs; Sparse certificates

Years and Authors of Summarized Original Work

1994; Khuller, Vishkin

Problem Definition

An undirected graph is said to be k-connected (specifically, k-vertex-connected) if the removal of any set of $k - 1$ or fewer vertices (with their incident edges) does not disconnect G. Analogously, it is k-edge-connected if the removal of any set of $k - 1$ edges does not disconnect

G. Menger's theorem states that a k-vertex-connected graph has at least k openly vertex-disjoint paths connecting every pair of vertices. For k-edge-connected graphs there are k edge-disjoint paths connecting every pair of vertices. The connectivity of a graph is the largest value of k for which it is k-connected. Finding the connectivity of a graph, and finding k disjoint paths between a given pair of vertices can be found using algorithms for maximum flow. An edge is said to be *critical* in a k-connected graph if upon its removal the graph is no longer k-connected.

The problem of finding a minimum-cardinality k-vertex-connected (k-edge-connected) subgraph that spans all vertices of a given graph is called k-VCSS (k-ECSS) and is known to be nondeterministic polynomial-time hard for $k \geq 2$. We review some results in finding approximately minimum solutions to k-VCSS and k-ECSS. We focus primarily on simple graphs. A simple approximation algorithm is one that considers the edges in some order and removes edges that are not critical. It thus outputs a k-connected subgraph in which all edges are critical and it can be shown that it is a 2-approximation algorithm (that outputs a solution with at most kn edges in an n-vertex graph, and since each vertex has to have degree at least k, we can claim that $kn/2$ edges are necessary).

Approximation algorithms that do better than the simple algorithm mentioned above can be classified into two categories: depth first search (DFS) based, and matching based.

Key Results

Lower Bounds for *k*-Connected Spanning Subgraphs

Each node of a k-connected graph has at least k edges incident to it. Therefore, the sum of the degrees of all its nodes is at least kn, where n is the number of its nodes. Since each edge is counted twice in this degree-sum, the cardinality of its edges is at least $kn/2$. This is called the *degree lower bound*. Expanding on this idea yields a stronger lower bound on the cardinality of a k-connected spanning subgraph of a given graph. Let D_k be a subgraph in which the degree of each node is at least k. Unlike a k-connected subgraph, D_k has no connectivity constraints. The counting argument above shows that any D_k has at least $kn/2$ edges. A minimum cardinality D_k can be computed in polynomial time by reducing the problem to matching, and it is called the *matching lower bound*.

DFS-Based Approaches

The following natural algorithm finds a 3/2 approximation for 2-ECSS. Root the tree at some node r and run DFS. All edges of the graph are now either tree edges or back edges. Process the DFS tree in postorder. For each subtree, if the removal of the edge from its root to its parent separates the graph into two components, then add a farthest-back edge from this subtree, whose other end is closest to r. It can be shown that the number of back edges added by the algorithm is at most half the size of *Opt*.

This algorithm has been generalized to solve the 2-VCSS problem with the same approximation ratio, by adding carefully chosen back edges that allow the deletion of tree edges. Wherever it is unable to delete a tree edge, it adds a vertex to an independent set I. In the final analysis, the number of edges used is less than $n + |I|$. Since *Opt* is at least $\max(n, 2|I|)$, it obtains a 3/2-approximation ratio.

The algorithm can also be extended to the k-ECSS problem by repeating these ideas $k/2$ times, augmenting the connectivity by 2 in each round. It has been shown that this algorithm achieves a performance of about 1.61.

Matching-Based Approaches

Several approximation algorithms for k-ECSS and k-VCSS problems have used a minimum cardinality D_k as a starting solution, which is then augmented with additional edges to satisfy the connectivity constraints. This approach yields better ratios than the DFS-based approaches.

$1 + \frac{1}{k}$ Algorithm for k-VCSS

Find a minimum cardinality D_{k-1}. Add just enough additional edges to it to make the subgraph k-connected. In this step, it is ensured that the edges added are critical. It is known by a theorem of Mader that in a k-connected graph, a cycle of critical edges contains at least one node of degree k. Since the edges added by the algorithm in the second step are all critical, there can be no cycle induced by these edges because the degree of all the nodes on such a cycle would be at least $k + 1$. Therefore, at most $n - 1$ edges are added in this step. The number of edges added in the first step, in the minimum D_{k-1} is at most $Opt - n/2$. The total number of edges in the solution thus computed is at most $(1 + 1/k)$ times the number of edges in an optimal k-VCSS.

$1 + \frac{2}{k+1}$ Algorithm for k-ECSS

Mader's theorem about cycles induced by critical edges is valid only for vertex connectivity and not edge connectivity, Therefore, a different algorithm is proposed for k-ECSS in graphs that are k-edge-connected, but not k-connected. This algorithm finds a minimum cardinality D_k and augments it with a minimal set of edges to make the subgraph k-edge-connected. The number of edges added in the last step is at most $\frac{k}{k+1}(n - 1)$. Since the number of edges added in the first step is at most Opt, the total number of edges is at most $(1 + \frac{2}{k+1})Opt$.

Better Algorithms for Small k

For $k \in \{2, 3\}$, better algorithms have been obtained by implementing the abovementioned algorithms carefully, deleting unnecessary edges, and by getting better lower bounds. For $k = 2$, a 4/3 approximation can be obtained by generating a path/cycle cover from a minimum cardinality D_2 and 2-connecting them one at a time to a "core" component. Small cycles/paths allow an edge to be deleted when they are 2-connected to the core, which allows a simple amortized analysis. This method also generalizes to the 3-ECSS problem, yielding a 4/3 ratio.

Hybrid approaches have been proposed which use the path/cycle cover to generate a specific DFS tree of the original graph and then 2-connect the tree, trying to delete edges wherever possible. The best ratios achieved using this approach are 5/4 for 2-ECSS, 9/7 for 2-VCSS, and 5/4 for 2-VCSS in 3-connected graphs.

Applications

Network design is one of the main application areas for this work. This involves the construction of low-cost highly connected networks.

Recommended Reading

For additional information on DFS, matchings and path/cycle covers, see [3]. Fast 2-approximation algorithms for k-ECSS and k-VCSS were studied by Nagamochi and Ibaraki [13]. DFS-based algorithms for 2-connectivity were introduced by Khuller and Vishkin [11]. They obtained 3/2 for 2-ECSS, 5/3 for 2-VCSS, and 2 for weighted k-ECSS. The ratio for 2-VCSS was improved to 3/2 by Garg et al. [6], 4/3 by Vempala and Vetta [14], and 9/7 by Gubbala and Raghavachari [7]. Khuller and Raghavachari [10] gave an algorithm for k-ECSS, which was later improved by Gabow [4], who showed that the algorithm obtains a ratio of about 1.61. Cheriyan et al. [2] studied the k-VCSS problem with edge weights and designed an $O(\log k)$ approximation algorithm in graphs with at least $6k^2$ vertices.

The matching-based algorithms were introduced by Cheriyan and Thurimella [1]. They proposed algorithms with ratios of $1 + \frac{1}{k}$ for k-VCSS, $1 + \frac{2}{k+1}$ for k-ECSS, $1 + \frac{1}{k}$ for k-VCSS in directed graphs, and $1 + \frac{4}{\sqrt{k}}$ for k-ECSS in directed graphs. Vempala and Vetta [14] obtained a ratio of 4/3 for 2-VCSS. The ratios were further improved by Krysta and Kumar [12], who introduced the hybrid approach, which was used to derive a 5/4 algorithm by Jothi et al. [9]. A 3/2-approximation algorithm for 3-ECSS has been proposed by Gabow [5] that works on multigraphs, whereas the earlier algorithm of Cheriyan and Thurimella gets the same ratio in simple graphs only. This ratio has been improved to 4/3 by Gubbala and Raghavachari [8].

1. Cheriyan J, Thurimella R (2000) Approximating minimum-size k-connected spanning subgraphs via matching. SIAM J Comput 30(2):528–560
2. Cheriyan J, Vempala S, Vetta A (2003) An approximation algorithm for the minimum-cost k-vertex connected subgraph. SIAM J Comput 32(4):1050–1055
3. Cook WJ, Cunningham WH, Pulleyblank WR, Schrijver A (1998) Combinatorial optimization. Wiley, New York
4. Gabow HN (2003) Better performance bounds for finding the smallest k-edge connected spanning subgraph of a multigraph. In: SODA, pp 460–469
5. Gabow HN (2004) An ear decomposition approach to approximating the smallest 3-edge connected spanning subgraph of a multigraph. SIAM J Discret Math 18(1):41–70
6. Garg N, Vempala S, Singla A (1993) Improved approximation algorithms for biconnected subgraphs via better lower bounding techniques. In: SODA, pp 103–111
7. Gubbala P, Raghavachari B (2005) Approximation algorithms for the minimum cardinality two-connected spanning subgraph problem. In: Jünger M, Kaibel V (eds) IPCO, vol 3509, Lecture notes in computer science. Springer, Berlin, pp 422–436
8. Gubbala P, Raghavachari B (2007) A 4/3-approximation algorithm for minimum 3-edge-connectivity. In: Proceedings of the workshop on algorithms and data structures (WADS) August 2007, Halifax, pp 39–51
9. Jothi R, Raghavachari B, Varadarajan S (2003) A 5/4-approximation algorithm for minimum 2-edge-connectivity. In: SODA, pp 725–734
10. Khuller S, Raghavachari B (1996) Improved approximation algorithms for uniform connectivity problems. J Algorithms 21(2):434–450
11. Khuller S, Vishkin U (1994) Biconnectivity approximations and graph carvings. J ACM 41(2): 214–235
12. Krysta P, Kumar VSA (2001) Approximation algorithms for minimum size 2-connectivity problems. In: Ferreira A, Reichel H (eds) STACS. Lecture notes in computer science, vol 2010. Springer, Berlin, pp 431–442
13. Nagamochi H, Ibaraki T (1992) A linear-time algorithm for finding a sparse k-connected spanning subgraph of a k-connected graph. Algorithmica 7(5–6):583–596
14. Vempala S, Vetta A (2000) Factor 4/3 approximations for minimum 2-connected subgraphs. In: Jansen K, Khuller S (eds) APPROX. Lecture notes in computer science, vol 1913. Springer, Berlin, pp 262–273

Graph Isomorphism

Brendan D. McKay
Department of Computer Science, Australian National University, Canberra, ACT, Australia

Keywords

Graph matching; Symmetry group

Years and Authors of Summarized Original Work

1980; McKay

Problem Definition

The problem of determining isomorphism of two combinatorial structures is a ubiquitous one, with applications in many areas. The paradigm case of concern in this chapter is isomorphism of two graphs. In this case, an isomorphism consists of a bijection between the vertex sets of the graphs which induces a bijection between the edge sets of the graphs. One can also take the second graph to be a copy of the first, so that isomorphisms map a graph onto themselves. Such isomorphisms are called *automorphisms* or, less formally, *symmetries*. The set of all automorphisms forms a group under function composition called the *automorphism group*. Computing the automorphism group is a problem rather similar to that of determining isomorphisms.

Graph isomorphism is closely related to many other types of isomorphism of combinatorial structures. In the section entitled "Applications", several examples are given.

Formal Description

A *graph* is a pair $G = (V, E)$ of finite sets, with E being a set of 2-tuples (v, w) of elements of V. The elements of V are called *vertices* (also *points*, *nodes*), while the elements of E are called

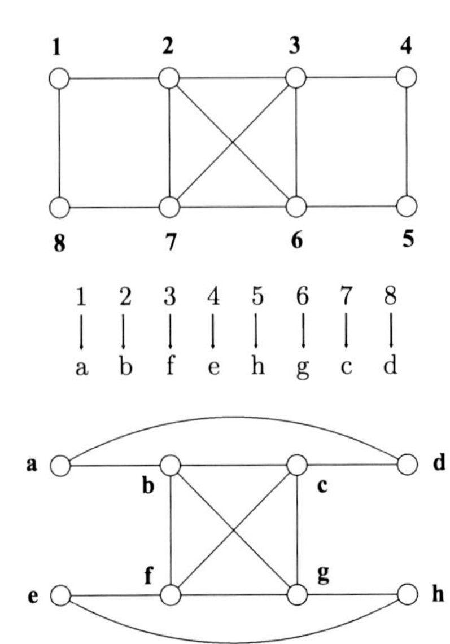

Graph Isomorphism, Fig. 1 Example of an isomorphism and an automorphism group

directed edges (also *arcs*). A complementary pair $(v, w), (w, v)$ of directed edges $(v \neq w)$ will be called an *undirected edge* and denoted $\{v, w\}$. A directed edge of the form (v, v) will also be considered an undirected edge, called a *loop* (also *self-loop*). The word "edges" without qualification will indicate undirected edges, directed edges, or both.

Given two graphs $G_1 = (V_1, E_1)$ and $G_2 = (V_2, E_2)$, an *isomorphism* from G_1 to G_2 is a bijection from V_1 to V_2 such that the induced action on E_1 is a bijection onto E_2. If $G_1 = G_2$, then the isomorphism is an *automorphism* of G_1. The set of all *automorphisms* of G_1 is a group under function composition, called the *automorphism group* of G_1, and denoted Aut (G_1).

In Fig. 1 two isomorphic graphs are shown, together with an isomorphism between them and the automorphism group of the first.

Canonical Labeling

Practical applications of graph isomorphism testing do not usually involve individual pairs of graphs. More commonly, one must decide whether a certain graph is isomorphic to any of a collection of graphs (the database lookup problem) or one has a collection of graphs and needs to identify the isomorphism classes in it (the graph sorting problem). Such applications are not well served by an algorithm that can only test graphs in pairs.

An alternative is a *canonical labeling* algorithm. The essential idea is that in each isomorphism class there is a unique, *canonical* graph which the algorithm can find, given as input any graph in the isomorphism class. The canonical graph might be, for example, the least graph in the isomorphism class according to some ordering (such as lexicographic) of the graphs in the class. Practical algorithms usually compute a canonical form designed for efficiency rather than ease of description.

Key Results

The graphisomorphism problem plays a key role in modern complexity theory. It is not known to be solvable in polynomial time, nor to be NP-complete, nor is it known to be in the class co-NP. See [3, 8] for details. Polynomial-time algorithms are known for many special classes, notably graphs with bounded genus, bounded degree, bounded tree-width, and bounded eigenvalue multiplicity. The fastest theoretical algorithm for general graphs requires $\exp(n^{1/2+o(1)})$ time [1], but it is not known to be practical.

In this entry, the focus is on the program nauty, which is generally regarded as the most successful for practical use. McKay wrote the first version of nauty in 1976 and described its method of operation in [5]. It is known [7] to have exponential worst-case time, but in practice the worst case is rarely encountered.

The input to nauty is a graph with colored vertices. Two outputs are produced. The first is a set of generators for the color-preserving automorphism group. Though it is rarely necessary, the full group can also be developed element by element. The second, optional, output is a canonical graph. The canonical graph has the following property: two input graphs with the same number of vertices of each color have the same canonical graph if and only if they are isomorphic by a color-preserving isomorphism.

Two graph data structures are supported: a packed adjacency matrix suitable for small dense graphs and a linked list suitable for large sparse graphs.

Applications

As mentioned, nauty can handle graphs with colored vertices. In this section, it is described how several other types of isomorphism problems can be solved by mapping them onto a problem for vertex-colored graphs.

Isomorphism of Edge-Colored Graphs

An isomorphism of two graphs, each with both vertices and edges colored, is defined in the obvious way. An example of such a graph appears at the left of Fig. 2.

In the center of the figure the colors are identified with the integers $1, 2, 3$. At the right of the figure an equivalent vertex-colored graph is shown. In this case there are two layers, each with its own color. Edges of color 1 are represented as an edge in the first (lowest) layer, edges of color 2 are represented as an edge in the second layer, and edges of color 3 are represented as edges in both layers. It is now easy to see that the automorphism group of the new graph (specifically, its action on the first layer) is the automorphism group of the original graph. Moreover, the order in which a canonical labeling of the new graph labels the vertices of the first layer can be taken to be a canonical labeling of the original graph.

More generally, if the edge colors are integers in $\{1, 2, \ldots, 2^d - 1\}$, there are d layers, and the binary expansion of each color number dictates which layers contain edges. The vertical threads (each corresponding to one vertex of the original graph) can be connected using either paths or cliques. If the original graph has n vertices and k colors, the new graph has $O(n \log k)$ vertices. This can be improved to $O(n \sqrt{\log k})$ vertices by also using edges that are not horizontal.

Isomorphism of Hypergraphs and Designs

A *hypergraph* is similar to an undirected graph except that the edges can be vertex sets of any size, not just of size 2. Such a structure is also called a *design*.

On the left of Fig. 3 there is a hypergraph with five vertices, two edges of size 2, and one edge of size 3. On the right is an equivalent vertex-colored graph. The vertices on the left, colored with one color, represent the hypergraph edges, while the edges on the right, colored with a

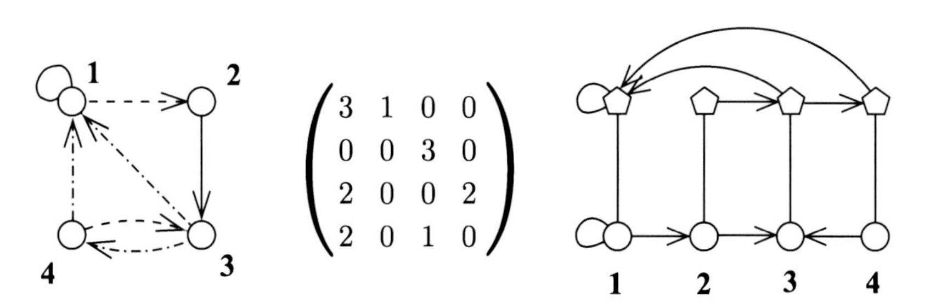

Graph Isomorphism, Fig. 2 Graph isomorphism with colored edges

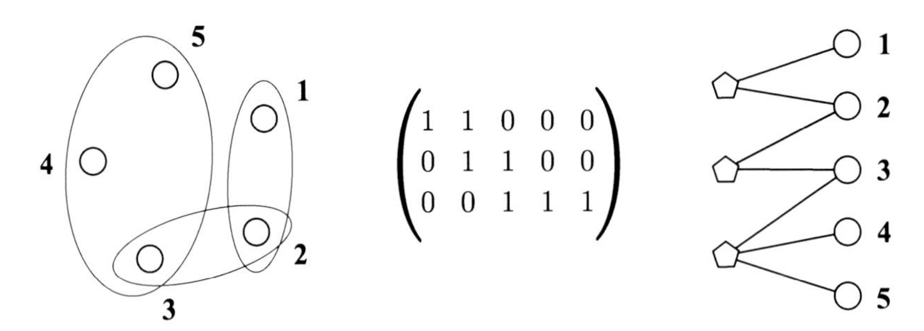

Graph Isomorphism, Fig. 3 Hypergraph/design isomorphism as graph isomorphism

different color, represent the hypergraph vertices. The edges of the graph indicate the hypergraph incidence (containment) relationship.

The edge-vertex incidence matrix appears in the center of the figure. This can be any binary matrix at all, which correctly suggests that the problem under consideration is just that of determining the 0–1 matrix equivalence under independent permutation of the rows and columns. By combining this idea with the previous construction, such an equivalence relation on the set of matrices with arbitrary entries can be handled.

Other Examples

For several applications to equivalence operations such as isotopy, important for Latin squares and quasigroups, see [6].

Another important type of equivalence relates matrices over $\{-1, +1\}$. As well as permuting rows and columns, it allows multiplication of rows and columns by -1. A method of converting this *Hadamard equivalence* problem to a graph isomorphism problem is given in [4].

Experimental Results

`Nauty` gives a choice of sparse and dense data structures, and some special code for difficult graph classes. For the following timing examples, the best of the various options are used for a single CPU of a 2.4 GHz Intel Core-duo processor.

1. Random graph with 10,000 vertices, $p = \frac{1}{2}$: 0.014 s for group only, 0.4 s for canonical labeling as well.
2. Random cubic graph with 100,000 vertices: 8 s.
3. 1-skeleton of 20-dimensional cube (1,048,576 vertices, group size 2.5×10^{24}): 92 s.
4. 3-dimensional mesh of size 50 (125,000 vertices): 0.7 s.
5. 1027-vertex strongly regular graph from random Steiner triple system: 0.6 s.

Examples of more difficult graphs can be found in the `nauty` documentation.

URL to Code

The source code of `nauty` is available at http://cs.anu.edu.au/~bdm/nauty/. Another implementation of the automorphism group portion of nauty, highly optimized for large sparse graphs, is available as `saucy` [2]. `Nauty` is also incorporated into a number of general-purpose packages, including GAP, Magma, and MuPad.

Cross-References

- ▶ Abelian Hidden Subgroup Problem
- ▶ Parameterized Algorithms for Drawing Graphs

Recommended Reading

1. Babai L, Luks E (1983) Canonical labelling of graphs. In: Proceedings of the 15th annual ACM symposium on theory of computing. ACM, New York, pp 171–183
2. Darga PT, Liffiton MH, Sakallah KA, Markov IL (2004) Exploiting structure in symmetry generation for CNF. In: Proceedings of the 41st design automation

conference, pp 530–534. Source code at http://vlsicad.eecs.umich.edu/BK/SAUCY/
3. Köbler J, Schöning U, Torán J (1993) The graph isomorphism problem: its structural complexity. Birkhäuser, Boston
4. McKay BD (1979) Hadamard equivalence via graph isomorphism. Discret Math 27:213–214
5. McKay BD (1981) Practical graph isomorphism. Congr Numer 30:45–87
6. McKay BD, Meynert A, Myrvold W (2007) Small Latin squares, quasigroups and loops. J Comb Des 15:98–119
7. Miyazaki T (1997) The complexity of McKay's canonical labelling algorithm. In: Groups and computation, II. DIMACS Series in Discrete Mathematics and Theoretical Computer Science, vol 28. American Mathematical Society, Providence, pp 239–256
8. Toran J (2004) On the hardness of graph isomorphism. SIAM J Comput 33:1093–1108

Graph Sketching

Andrew McGregor
School of Computer Science, University of Massachusetts, Amherst, MA, USA

Keywords

Connectivity; Data streams; Linear projections

Years and Authors of Summarized Original Work

2012a; Ahn, Guha, McGregor

Problem Definition

The basic problem we consider is testing whether an undirected graph G on n nodes $\{v_1, \ldots, v_n\}$ is connected. We consider this problem in the following two related models:

1. **Dynamic Graph Stream Model:** The graph G is defined by a sequence of edge insertions and deletions; the edges of G are the set of edges that have been inserted but not subsequently deleted. An algorithm for analyzing G may only read the input sequence from left to right and has limited working memory. If the available memory was $O(n^2)$ bits, then the algorithm could maintain the exact set of edges that have been inserted but not deleted. The primary objective in designing an algorithm in the stream model is to reduce the amount of memory required. Ideally, the time to process each element of the stream and the post-processing time should be small but ensuring this is typically a secondary objective.
2. **Simultaneous Communication Model:** We consider the n rows of the adjacency matrix of G to be partitioned between n players $P_1, \ldots, P_n$ where P_i receives the ith row of the matrix. This means that the existence of any edge is known by exactly two players. An additional player Q wants to evaluate a property of G, and to facilitate this, each player P_i simultaneously sends a message m_i to Q such that Q may evaluate the property given the messages $m_1, m_2, \ldots, m_n$. With n-bit messages from each player, Q could learn the entire graph and the problem would be uninteresting. The objective is to minimize the number of bits sent by each player. Note that the P_i players may not communicate to each other and that each message m_i must be constructed given only the ith row of the adjacency matrix and possibly a set of random bits that is known to all the players.

If there were no edge deletions in the data stream setting, it would be simple to determine whether G was connected using $O(n \log n)$ memory since it is possible to maintain the connected components of the graph; whenever an edge is added, we merge the connected components containing the endpoints of this edge. This algorithm is optimal in terms of space [19]. Such an approach does not extend if edges may also be deleted since it is unclear how the connected components should be updated when an edge is deleted within a connected component.

To illustrate the challenge in the simultaneous communication model, suppose G is connected but $G \setminus \{e\}$ is disconnected for some edge e. The player Q can only learn about the existence of the edge $e = \{v_i, v_j\}$ from either player P_i or player

P_j, but since both of these players have limited knowledge of the graph, neither will realize the important role this edge plays in determining the connectivity of the graph.

Linear Sketches

For both models the best known algorithms are based on random linear projections, aka *linear sketches*. If we denote the n rows of the adjacency matrix by $\mathbf{x}_1, \ldots, \mathbf{x}_n \in \{0,1\}^n$, then the linear sketches of the graph are $\mathcal{A}_1(\mathbf{x}_1), \ldots, \mathcal{A}_n(\mathbf{x}_n)$ where each $\mathcal{A}_i \in \mathbb{R}^{d \times n}$ is a random matrix chosen according to a specific distribution. Note that the matrices $\mathcal{A}_1, \ldots, \mathcal{A}_n$ need not be chosen independently.

In the simultaneous communication model, the message from player P_i is $m_i = \mathcal{A}_i(\mathbf{x}_i)$, and, assuming that the entries of $\mathcal{A}_i$ have polynomial precision, each of these messages requires $O(d \text{ polylog } n)$ bits. In the dynamic graph stream model, the algorithm constructs each $\mathcal{A}_i(\mathbf{x}_i)$ using $O(nd \text{ polylog } n)$ bits of space. Note that each $\mathcal{A}_i(\mathbf{x}_i)$ can be constructed incrementally using the following update rules:

$$\begin{aligned}
\text{on the insertion of } \{v_i, v_j\}: &\quad \mathcal{A}_i(\mathbf{x}_i) \leftarrow \mathcal{A}_i(\mathbf{x}_i) + \mathcal{A}_i(\mathbf{e}_j) \\
\text{on the deletion of } \{v_i, v_j\}: &\quad \mathcal{A}_i(\mathbf{x}_i) \leftarrow \mathcal{A}_i(\mathbf{x}_i) - \mathcal{A}_i(\mathbf{e}_j) \\
\text{on the insertion/deletion of } \{v_j, v_k\} \text{ for } i \notin \{j,k\}: &\quad \mathcal{A}_i(\mathbf{x}_i) \leftarrow \mathcal{A}_i(\mathbf{x}_i)
\end{aligned}$$

where $\mathbf{e}_j$ is the characteristic vector of the set $\{j\}$. Hence, we have transformed the problem of designing an efficient algorithm into finding the minimum d such that there exists a distribution of matrices $\mathcal{A}_1, \ldots, \mathcal{A}_n \in \mathbb{R}^{d \times n}$ such that for any graph G, we can determine (with high probability) whether G is connected given $\mathcal{A}_1(\mathbf{x}_1), \ldots, \mathcal{A}_n(\mathbf{x}_n)$.

Key Results

The algorithm for connectivity that we present in this entry, and much of the subsequent work on graph sketching, fits the following template. First, we consider a basic "non-sketch" algorithm for the graph problem in question. Second, we design sketches $\mathcal{A}_i$ such that it is possible to emulate the steps of the basic algorithm given only the projections $\mathcal{A}_i(\mathbf{x}_i) \in \mathbb{R}^d$ where $d = O(\text{polylog } n)$.

Connectivity

Basic Non-sketch Algorithm

We pick an incident edge for each node arbitrarily and collapse the resulting connected components into a set of "supernodes." In each subsequent round of the algorithm, we pick an edge from each supernode to another supernode (if one exists) and collapse the connected components into new supernodes. It can be shown that this process terminates after $O(\log n)$ rounds and that the set of edges picked during the different rounds include a spanning forest of the graph. From this we can deduce whether the graph is connected.

Designing the Sketches

There are two main steps required in constructing the sketches for the connectivity algorithm:

An Alternative Graph Representation. Rather than consider the rows of the adjacency matrix $\mathbf{x}_i$, it will be convenient to consider an alternative representation $\mathbf{a}_i \in \{-1,0,1\}^{\binom{n}{2}}$ with entries indexed by pairs

$$\mathbf{a}_i[\{j,k\}] = \begin{cases} 1 & \text{if } i = j < k \text{ and } \{v_j, v_k\} \in E \\ -1 & \text{if } j < k = i \text{ and } \{v_j, v_k\} \in E \\ 0 & \text{otherwise} \end{cases}$$

These vectors have the useful property that for any subset of nodes $\{v_i\}_{i \in S}$, the non-zero entries of $\sum_{i\in S} \mathbf{a}_i$ correspond exactly to the edges across the cut $(S, V \setminus S)$.

For example, consider the graph on nodes $\{v_1, v_2, v_3, v_4\}$ with edges $\{v_1, v_2\}$, $\{v_2, v_3\}$, $\{v_3, v_4\}$, and $\{v_1, v_4\}$. Then

$$\begin{aligned}\mathbf{a}_1 &= (\;\;1\;\;0\;\;1\;\;0\;\;0\;\;0)\\ \mathbf{a}_2 &= (-1\;\;0\;\;0\;\;1\;\;0\;\;0)\\ \mathbf{a}_3 &= (\;\;0\;\;0\;\;0\;-1\;\;0\;\;1)\\ \mathbf{a}_4 &= (\;\;0\;\;0\;-1\;\;0\;\;0\;-1)\end{aligned}$$

where the entries correspond to the pairs $\{1,2\}$, $\{1,3\}$, $\{1,4\}$, $\{2,3\}$, $\{2,4\}$, $\{3,4\}$ in that order. Note that the nonzero entries of

$$\mathbf{a}_1 + \mathbf{a}_2 = (\;0\;\;0\;\;1\;\;1\;\;0\;\;0)$$

correspond to $\{1,4\}$ and $\{2,3\}$ which are exactly the edges across the cut $(S, V \setminus S)$ for $S = \{v_1, v_2\}$.

ℓ_0-Sampling via Linear Sketches. ℓ_0-sampling is a technique that has found numerous applications in data stream processing. We appeal to a result by Jowhari et al. [12] that shows the existence of a distribution over matrices $\mathcal{M} \in \mathbb{R}^{\text{polylog}(n)\times\text{poly}(n)}$ such that for any nonzero vector $\mathbf{z} \in \mathbb{R}^{\text{poly}(n)}$, the index of some nonzero entry of $\mathbf{z}$ can be reconstructed with high probability given $\mathcal{M}(\mathbf{z}) \in \mathbb{R}^{\text{polylog}(n)}$. Note that we do not get to choose which entry is reconstructed.

Emulation Basic Algorithm via Sketches

Let $\mathcal{M}_1, \ldots, \mathcal{M}_r$ be $r = O(\log n)$ independent sketch matrices for ℓ_0-sampling. Given $\mathcal{M}_j(\mathbf{a}_i)$ for all $j \in [r]$ and $i \in [n]$, we can emulate the basic algorithm as follows:

1. Given $\mathcal{M}_1(\mathbf{a}_1), \mathcal{M}_1(\mathbf{a}_2), \ldots, \mathcal{M}_1(\mathbf{a}_n)$, we may emulate the first round of the algorithm since from each $\mathcal{M}_1(\mathbf{a}_i)$ we may reconstruct a nonzero entry of $\mathbf{a}_i$, and these nonzero entries correspond to edges incident to v_i.
2. To emulate round $j > 1$ of the algorithm, suppose S is one of the connected components already constructed. Then, given

$$\sum_{i\in S} \mathcal{M}_j(\mathbf{a}_i) = \mathcal{M}_j\Big(\sum_{i\in S} \mathbf{a}_i\Big)$$

we may reconstruct a nonzero entry of $\sum_{i\in S} \mathbf{a}_i$ which corresponds to an edge across the cut $(S, V \setminus S)$.

Extensions and Further Work

Subsequent work has extended the above results significantly. If d is increased to $O(k \,\text{polylog}\, n)$ then, it is possible to test whether every cut has at least k edges [1]. With $d = O(\epsilon^{-2} \,\text{polylog}\, n)$, it is possible to construct graph sparsifiers that can be used to estimate the size of *every* cut up to a $(1+\epsilon)$ factor [2] along with spectral properties such as the eigenvalues of the graph [14]. With $d = O(\epsilon^{-1} k \,\text{polylog}\, n)$, it is possible to distinguish graphs which are not k-vertex connected from those that are at least $(1+\epsilon)k$-vertex connected [11]. Some of the above results have also been extended to hypergraphs [11]. The algorithm presented in this entry can be implemented with $O(\text{polylog}\, n)$ update time in the dynamic graph stream model, but a connectivity query may take $\Omega(n)$ time. This was addressed in subsequent work by Kapron et al. [15].

More generally, solving graph problems via linear sketches has become a very active area of research [1–8, 10, 11, 13, 14, 16, 17]. Other problems that have been considered include approximating the densest subgraph [6, 9, 18], maximum matching [5, 7, 8, 16], vertex cover and hitting set [8], correlation clustering [4], and estimating the number of triangles [17].

Cross-References

▶ Counting Triangles in Graph Streams

Recommended Reading

1. Ahn KJ, Guha S, McGregor A (2012) Analyzing graph structure via linear measurements. In: Twenty-third annual ACM-SIAM symposium on discrete algorithms, SODA 2012, pp 459–467. http://

portal.acm.org/citation.cfm?id=2095156&CFID=63838676&CFTOKEN=79617016
2. Ahn KJ, Guha S, McGregor A (2012) Graph sketches: sparsification, spanners, and subgraphs. In: 31st ACM SIGMOD-SIGACT-SIGART symposium on principles of database systems, pp 5–14. doi:10.1145/2213556.2213560, http://doi.acm.org/10.1145/2213556.2213560
3. Ahn KJ, Guha S, McGregor A (2013) Spectral sparsification in dynamic graph streams. In: APPROX, pp 1–10. doi:10.1007/978-3-642-40328-6_1, http://dx.doi.org/10.1007/978-3-642-40328-6_1
4. Ahn KJ, Cormode G, Guha S, McGregor A, Wirth A (2015) Correlation clustering in data streams. In: ICML, Lille
5. Assadi S, Khanna S, Li Y, Yaroslavtsev G (2015) Tight bounds for linear sketches of approximate matchings. CoRR abs/1505.01467. http://arxiv.org/abs/1505.01467
6. Bhattacharya S, Henzinger M, Nanongkai D, Tsourakakis CE (2015) Space- and time-efficient algorithm for maintaining dense subgraphs on one-pass dynamic streams. In: STOC, Portland
7. Bury M, Schwiegelshohn C (2015) Sublinear estimation of weighted matchings in dynamic data streams. CoRR abs/1505.02019. http://arxiv.org/abs/1505.02019
8. Chitnis RH, Cormode G, Esfandiari H, Hajiaghayi M, McGregor A, Monemizadeh M, Vorotnikova S (2015) Kernelization via sampling with applications to dynamic graph streams. CoRR abs/1505.01731. http://arxiv.org/abs/1505.01731
9. Esfandiari H, Hajiaghayi M, Woodruff DP (2015) Applications of uniform sampling: densest subgraph and beyond. CoRR abs/1506.04505. http://arxiv.org/abs/1506.04505
10. Goel A, Kapralov M, Post I (2012) Single pass sparsification in the streaming model with edge deletions. CoRR abs/1203.4900. http://arxiv.org/abs/1203.4900
11. Guha S, McGregor A, Tench D (2015) Vertex and hypergraph connectivity in dynamic graph streams. In: PODS, Melbourne
12. Jowhari H, Saglam M, Tardos G (2011) Tight bounds for lp samplers, finding duplicates in streams, and related problems. In: PODS, Athens, pp 49–58
13. Kapralov M, Woodruff DP (2014) Spanners and sparsifiers in dynamic streams. In: ACM symposium on principles of distributed computing, PODC '14, Paris, 15–18 July 2014, pp 272–281. doi:10.1145/2611462.2611497, http://doi.acm.org/10.1145/2611462.2611497
14. Kapralov M, Lee YT, Musco C, Musco C, Sidford A (2014) Single pass spectral sparsification in dynamic streams. In: FOCS, Philadelphia
15. Kapron B, King V, Mountjoy (2013) Dynamic graph connectivity in polylogarithmic worst case time. In: SODA, New Orleans, pp 1131–1142
16. Konrad C (2015) Maximum matching in turnstile streams. CoRR abs/1505.01460. http://arxiv.org/abs/1505.01460
17. Kutzkov K, Pagh R (2014) Triangle counting in dynamic graph streams. In: Algorithm theory – SWAT 2014 – 14th Scandinavian symposium and workshops, proceedings, Copenhagen, 2–4 July 2014, pp 306–318. doi:10.1007/978-3-319-08404-6_27, http://dx.doi.org/10.1007/978-3-319-08404-6_27
18. McGregor A, Tench D, Vorotnikova S, Vu H (2015) Densest subgraph in dynamic graph streams. In: Mathematical foundations of computer science 2015 – 40th international symposium, MFCS 2014, Proceedings, Part I, Milano, 24–28 Aug 2014
19. Sun X, Woodruff D (2015) Tight bounds for graph problems in insertion streams. In: Approximation algorithms for combinatorial optimization, eighteenth international workshop, APPROX 2015, Proceedings, Princeton, 24–26 Aug 2015

Greedy Approximation Algorithms

Weili Wu[1,2,3] and Feng Wang[4]
[1]College of Computer Science and Technology, Taiyuan University of Technology, Taiyuan, Shanxi Province, China
[2]Department of Computer Science, California State University, Los Angeles, CA, USA
[3]Department of Computer Science, The University of Texas at Dallas, Richardson, TX, USA
[4]Mathematical Science and Applied Computing, Arizona State University at the West Campus, Phoenix, AZ, USA

Keywords

Technique for analysis of greedy approximation

Problem Definition

Consider a graph $G = (V, E)$. A subset C of V is called a *dominating set* if every vertex is either in C or adjacent to a vertex in C. If, furthermore, the subgraph induced by C is connected, then C is called a *connected dominating set*.

Given a connected graph G, find a connecting dominating set of minimum cardinality. This problem is denoted by MCDS and is NP-hard. Its optimal solution is called a *minimum connected dominating set*. The following is a greedy approximation with potential function f.

Greedy Algorithm A:
$C \leftarrow \emptyset$;
while $f(C) > 2$**do**
choose a vertex x to maximize $f(C) - f(C \cup \{x\})$ and
$C \leftarrow C \cup \{x\}$; output C.

Here, f is defined as $f(C) = p(C) + q(C)$ where $p(C)$ is the number of connected components of subgraph induced by C and $q(C)$ is the number of connected components of subgraph with vertex set V and edge set $\{(u, v) \in E \mid u \in C \text{ or } v \in C\}$. f has an important property that C is a connected dominating set if and only if $f(C) = 2$.

If C is a connected dominating set, then $p(C) = q(C) = 1$, and hence $f(C) = 2$. Conversely, suppose $f(C \cup \{x\}) = 2$. Since $p(C) \geq 1$ and $q(C) \geq 1$, one has $p(C) = q(C) = 1$ which implies that C is a connected dominating set. f has another property, for G with at least three vertices, that if $f(C) > 2$, then there exists $x \in V$ such that $f(C) - f(C \cup \{x\}) > 0$. In fact, for $C = \emptyset$, since G is a connected graph with at least three vertices, there must exist a vertex x with degree at least two, and for such a vertex x, $f(C \cup \{x\}) < f(C)$. For $C \neq \emptyset$, consider a connected component of the subgraph induced by C. Let B denote its vertex set which is a subset of C. For every vertex y adjacent to B, if y is adjacent to a vertex not adjacent to B and not in C, then $p(C \cup \{y\}) < p(C)$ and $q(C \cup \{y\}) \leq q(C)$; if y is adjacent to a vertex in $C - B$, then $p(C \cup \{y\}) \leq p(C)$ and $q(C \cup \{y\}) < q(C)$.

Now, look at a possible analysis for the above greedy algorithm: Let $x_1, \ldots, X_g$ be vertices chosen by the greedy algorithm in the ordering of their appearance in the algorithm. Denote $C_i = \{x_1, \ldots, x_i\}$. Let $C^* = \{y_1, \ldots, y_{opt}\}$ be a minimum connected dominating set. Since adding C^* to C_i will reduce the potential function value from $f(C_i)$ to 2, the value of f reduced by a vertex in C^* would be $(f(C_i) - 2)/opt$ in average. By the greedy rule for choosing $x_i + 1$, one has

$$f(C_i) - f(C_{i+1}) \geq \frac{f(C_i) - 2}{opt}.$$

Hence,

$$\begin{aligned} f(C_{i+1}) - 2 &\leq (f(C_i) - 2)\left(1 - \tfrac{1}{opt}\right) \\ &\leq (f(\emptyset) - 2)\left(1 - \tfrac{1}{opt}\right)^{i+1} \\ &= (n-2)\left(1 - \tfrac{1}{opt}\right)^{i+1}, \end{aligned}$$

where $n = |V|$. Note that $1 - 1/opt \leq e^{-1/opt}$. Hence,

$$f(C_i) - 2 \leq (n-2)e^{-i/opt}.$$

Choose i such that $f(C_i) \geq opt + 2 > f(C_{i+1})$. Then

$$opt \leq (n-2)\, e^{-i/opt}$$

and

$$g - i \leq opt.$$

Therefore,

$$g \leq opt + i \leq opt\left(1 + \ln \frac{n-2}{opt}\right).$$

Is this analysis correct? The answer is NO. Why? How could one give a correct analysis? This entry will answer those questions and introduce a new general technique, analysis of greedy approximation with nonsubmodular potential function.

Key Results

The Role of Submodularity

Consider a set X and a function f defined on the power set 2^X, i.e., the family of all subsets of X. f is said to be *submodular* if for any two subsets A and B in 2^X,

$$f(A) + f(B) \geq f(A \cap B) + f(A \cup B).$$

For example, consider a connected graph G. Let X be the vertex set of G. The function $-q(C)$

defined in the last section is submodular. To see this, first mention a property of submodular functions.

A submodular function f is *normalized* if $f(\emptyset) = 0$. Every submodular function f can be normalized by setting $g(A) = f(A) - f(\emptyset)$. A function f is *monotone increasing* if $f(A) \leq f(B)$ for $A \subset B$. Denote $\Delta_x f(A) = f(A \cup \{x\}) - f(A)$.

Lemma 1 *A function $f : 2^X \rightarrow R$ is submodular if and only if $\Delta_x f(A) \leq \Delta_x f(B)$ for any $x \in X - B$ and $A \subseteq B$. Moreover, f is monotone increasing if and only if $\Delta_x f(A) \leq \Delta_x f(B)$ for any $x \in B$ and $A \subseteq B$.*

Proof If f is submodular, then for $x \in X - B$ and $A \subseteq B$, one has

$$\begin{aligned} & f(A \cup \{x\}) + f(B) \\ & \quad \geq f((A \cup \{x\}) \cup B) + f(A \cup \{x\}) \cap B) \\ & \quad = f(B \cup \{x\}) + f(A), \end{aligned}$$

that is,

$$\Delta_x f(A) \geq \Delta_x f(B). \qquad (1)$$

Conversely, suppose (1) holds for any $x \in B$ and $A \subseteq B$. Let C and D be two sets and $C/D = \{x_1, \ldots, x_k\}$. Then

$$\begin{aligned} & f(C \cup D) - f(D) \\ & \quad = \sum_{i=1}^{k} \Delta_{x_i} f(D \cup \{x_1, \ldots, x_{i-1}) \\ & \quad \leq \sum_{i=1}^{k} \Delta_{x_i} f((C \cap D) \cup \{x_1, \ldots, x_{i-1}) \\ & \quad = f(C) - f(C \cap D). \end{aligned}$$

If f is monotone increasing, then for $A \subseteq B$, $f(A) \leq f(B)$. Hence, for $x \in B$,

$$\Delta)xf(A) \geq 0 = \Delta_x f(B).$$

Conversely, if $\Delta_x f(A) \geq \Delta_x f(B)$ for any $x \in B$ and $A \subseteq B$, then for any x and A, $\Delta_x f(A) \geq \Delta_x f(A \cup \{x\}) = 0$, that is, $f(A) \leq f(A \cup \{x\})$. Let $B - A = \{x_1, \ldots, x_k\}$. Then

$$\begin{aligned} & f(A) \leq f(A \cup \{x_1\}) \\ & \qquad \leq f(A \cup \{x_1, x_2\}) \leq \cdots \leq f(B). \end{aligned}$$

Next, the submodularity of $-q(A)$ is studied. □

Lemma 2 *If $A \subset B$, then $\Delta_y q(A) \geq \Delta_y q(B)$.*

Proof Note that each connected component of graph $(V, D(B))$ is constituted by one or more connected components of graph $(V, D(A))$ since $A \subset B$. Thus, the number of connected components of $(V, D(B))$ dominated by y is no more than the number of connected components of $(V, D(A))$ dominated by y. Therefore, the lemma holds.

The relationship between submodular functions and greedy algorithms has been established for a long time [2].

Let f be a normalized, monotone increasing, submodular integer function. Consider the minimization problem

$$\begin{aligned} & \min \quad c(A) \\ & \text{subject to } A \in C_f. \end{aligned}$$

where c is a nonnegative cost function defined on 2^X and $C_f = \{C \text{mid} f(C \cup \{x\}) - f(C) = 0 \text{ for all } x \in X\}$. The following is a greedy algorithm to produce approximation solution for this problem. □

Greedy Algorithm B

input submodular function f and cost function c;
$A \leftarrow \emptyset$;
while there exists $x \in E$ such that $\Delta_x f(A) > 0$
do select a vertex x that maximizes $\Delta_x f(A)/c(x)$ and set
$A \leftarrow A \cup \{x\}$;
return A.

The following two results are well known.

Theorem 1 *If f is a normalized, monotone increasing, submodular integer function, then Greedy Algorithm B produces an approximation solution within a factor of $H(\gamma)$ from optimal, where $\gamma = max_{x \in E} f(\{x\})$.*

Theorem 2 *Let f be a normalized, monotone increasing, submodular function and c a nonnegative cost function. If in Greedy Algorithm B,* *selected x always satisfies $\Delta_x f(A_{i-1})/c(x) \geq 1$, then it produces an approximation solution within a factor of $1 + \ln(f^*/opt)$ from optimal for the above minimization problem where $f^* = f(A^*)$ and $opt = c(A^*)$ for optimal solution A^*.*

Now, come back to the analysis of Greedy Algorithm A for the MCDS. It looks like that the submodularity of f is not used. Actually, the submodularity was implicitly used in the following statement:

"Since adding C^* to C_i will reduce the potential function value from $f(C_i\ )$ to 2, the value of f reduced by a vertex in C^* would be $(f(C_i - 2)/opt$ in average. By the greedy rule for choosing $x_i + 1$, one has

$$f(C_i) - f(C_{i+1}) \geq \frac{f(C_i) - 2}{opt}."$$

To see this, write this argument more carefully.

Let $C^* = \{y_1, \ldots, y_{opt}\}$ and denote $C_j^* = \{y_1, \ldots, y_j\}$. Then

$$\begin{aligned} f(C_i) - 2 &= f(C_i) - f(C_i \cup C^*) \\ &= \sum_{opt}^{j=1} [f(C_i \cup C_{j-1}^*) - f(C_i \cup C_j^*)] \end{aligned}$$

where $C_0^* = \emptyset$. By the greedy rule for choosing $x_i + 1$, one has

$$f(C_i) - f(C_{i+1}) \geq f(C_i) - f(C_i \cup \{y_j\})$$

for $j = 1, \ldots, opt$. Therefore, it needs to have

$$\begin{aligned} -\Delta_{y_j} f(C_i) &= f(C_i) - f(C_i \cup \{y_j\}) \\ &\geq f(C_i \cup C_{j-1}^*) - f(C_i \cup C_j^*) \\ &= -\Delta_{y_j} f(C_i \cup C_{j-1}^*) \end{aligned} \tag{2}$$

in order to have

$$f(C_i) - f(C_{i+1}) \geq \frac{f(C_i) - 2}{opt}.$$

Equation (2) asks the submodularity of $-f$. Unfortunately, $-f$ is not submodular. A counterexample can be found in [2]. This is why the analysis of Greedy Algorithm A in section "Problem Definition" is incorrect.

Giving Up Submodularity

Giving up submodularity is a challenge task since it is open for a long time. But, it is possible based on the following observation on (2) by Du et al. [1]: **The submodularity of $-f$ is applied to increment of a vertex y_j belonging to optimal solution C^*.**

Since the ordering of y_j's is flexible, one may arrange it to make $\Delta_{yj} f(C_i) - \Delta_{yj} f(C_i \cup C_{j-1}^*)$ under control. This is a successful idea for the MCDS.

Lemma 3 *Let y_j's be ordered in the way that for any $j = 1, \ldots, opt, \{y_1, \ldots, y_j\}$ induces a connected subgraph. Then*

$$\Delta_{y_j} f(C_i) - \Delta_{y_j} f(C_i \cup C_{j-1}^*) \leq 1.$$

Proof Since all $y_1, \ldots, y_{j-1}$ are connected, y_j can dominate at most one additional connected component in the subgraph induced by $C_{i-1} \cup C_{j-1}^*$ than in the subgraph induced by $c_i - 1$. Hence,

$$\Delta_{y_j} p(C_i) - \Delta_{y_j} f(C_i \cup C_{j-1}^*) \leq 1.$$

Moreover, since $-q$ is submodular,

$$\Delta_{y_j} q(C_i) - \Delta_{y_j} q(C_i \cup C_{j-1}^*) \leq 0.$$

Therefore,

$$\Delta_{y_j} f(C_i) - \Delta_{y_j} f(C_i \cup C_{j-1}^*) \leq 1.$$

Now, one can give a correct analysis for the greedy algorithm for the MCDS [3].

By Lemma 3,

$$f(C_i) - f(C_{i+1}) \geq \frac{f(C_i) - 2}{opt} - 1.$$

Hence,

$$\begin{aligned} f(C_{i+1}) - 2 - opt &\le (f(C_i) - 2 + opt)\left(1 - \frac{1}{opt}\right) \\ &\le (f(\emptyset) - 2 - opt)\left(1 - \frac{1}{opt}\right)^{i+1} \\ &= (n - 2 - opt)\left(1 - \frac{1}{opt}\right)^{i+1}, \end{aligned}$$

where $n = |V|$. Note that $1 - 1/opt \le e^{-1/opt}$. Hence,

$$f(C_i) - 2 - opt \le (n-2)e^{-i/opt}.$$

Choose i such that $f(C_i) \ge 2 \cdot opt + 2 > f(C_{i+1})$. Then

$$opt \le (n-2)e^{-i/opt}$$

and

$$g - i \le 2 \cdot opt.$$

Therefore,

$$g \le 2 \cdot opt + i \le opt\left(2 + \ln \frac{n-2}{opt}\right) \le opt(2 + \ln \delta)$$

where δ is the maximum degree of input graph G. □

Applications

The technique introduced in the previous section has many applications, including analysis of iterated 1-Steiner trees for minimum Steiner tree problem and analysis of greedy approximations for optimization problems in optical networks [3] and wireless networks [2].

Open Problems

Can one show the performance ratio $1 + H(\delta)$ for Greedy Algorithm B for the MCDS? The answer is unknown. More generally, it is unknown how to get a clean generalization of Theorem 1.

Cross-References

▸ Connected Dominating Set
▸ Exact Algorithms for k SAT Based on Local Search
▸ Steiner Trees

Acknowledgments Weili Wu is partially supported by NSF grant ACI-0305567.

Recommended Reading

1. Du D-Z, Graham RL, Pardalos PM, Wan P-J, Wu W, Zhao W (2008) Analysis of greedy approximations with nonsubmodular potential functions. In: ACM-SIAM Symposium on Discrete Algorithms (SODA), San Francisco
2. Nemhauser GL, Wolsey LA (1999) Integer and combinatorial optimization. Wiley, Hoboken
3. Ruan L, Du H, Jia X, Wu W, Li Y, Ko K-I (2004) A greedy approximation for minimum connected dominating set. Theor Comput Sci 329:325–330
4. Ruan L, Wu W (2005) Broadcast routing with minimum wavelength conversion in WDM optical networks. J Comb Optim 9:223–235

Greedy Set-Cover Algorithms

Neal E. Young
Department of Computer Science and Engineering, University of California, Riverside, CA, USA

Keywords

Dominating set; Greedy algorithm; Hitting set; Minimizing a linear function subject to a submodular constraint; Set cover

Years and Authors of Summarized Original Work

1974–1979; Chvátal, Johnson, Lovász, Stein

Problem Definition

Given a collection $\mathcal{S}$ of sets over a universe U, a *set cover* $C \subseteq \mathcal{S}$ is a subcollection of the sets whose union is U. The *set-cover problem* is, given $\mathcal{S}$, to find a minimum-cardinality set cover. In the *weighted set-cover problem*, for each set $s \in \mathcal{S}$, a weight $w_s \geq 0$ is also specified, and the goal is to find a set-cover C of minimum total weight $\sum_{S \in C} w_S$.

Weighted set cover is a special case of *minimizing a linear function subject to a submodular constraint*, defined as follows. Given a collection $\mathcal{S}$ of objects, for each object s a nonnegative weight w_s, and a nondecreasing submodular function $f : 2^{\mathcal{S}} \rightarrow \mathbb{R}$, the goal is to find a subcollection $C \subseteq \mathcal{S}$ such that $f(C) = f(\mathcal{S})$ minimizing $\sum_{s \in C} w_s$. (Taking $f(C) = |\cup_{s \in C} s|$ gives weighted set cover.)

Key Results

The *greedy algorithm* for weighted set cover builds a cover by repeatedly choosing a set s that minimizes the weight w_s divided by the number of elements in s not yet covered by chosen sets. It stops and returns the chosen sets when they form a cover:

Let H_k denote $\sum_{i=1}^{k} 1/i \approx \ln k$, where k is the largest set size.

greedy-set-cover(S, w)
1. Initialize $C \leftarrow \emptyset$. Define $f(C) \doteq |\cup_{s \in C} s|$.
2. Repeat until $f(C) = f(S)$:
3. Choose $s \in S$ minimizing the price per element $w_s/[f(C \cup \{s\}) - f(C)]$.
4. Let $C \leftarrow C \cup \{s\}$.
5. Return C.

Theorem 1 *The greedy algorithm returns a set cover of weight at most H_k times the minimum weight of any cover.*

Proof When the greedy algorithm chooses a set s, imagine that it charges the price per element for that iteration to each element newly covered by s. Then, the total weight of the sets chosen by the algorithm equals the total amount charged, and each element is charged once.

Consider any set $s = \{x_k, x_{k-1}, \ldots, x_1\}$ in the optimal set cover C^*. Without loss of generality, suppose that the greedy algorithm covers the elements of s in the order given: $x_k, x_{k-1}, \ldots, x_1$. At the start of the iteration in which the algorithm covers element x_i of s, at least i elements of s remain uncovered. Thus, if the greedy algorithm were to choose s in that iteration, it would pay a cost per element of at most w_s / i. Thus, in this iteration, the greedy algorithm pays at most w_s / i per element covered. Thus, it charges element x_i at most w_s / i to be covered. Summing over i, the total amount charged to elements in s is at most $w_s H_k$. Summing over $s \in C^*$ and noting that every element is in some set in C^*, the total amount charged to elements overall is at most $\sum_{s \in C^*} W_s H_k = H_k \text{OPT}$. □

The theorem was shown first for the unweighted case (each $w_s = 1$) by Johnson [5], Lovász [8], and Stein [13] and then extended to the weighted case by Chvátal [2].

Since then a few refinements and improvements have been shown, including the following:

Theorem 2 *Let S be a set system over a universe with n elements and weights $w_s \leq 1$. The total weight of the cover C returned by the greedy algorithm is at most $[1 + \ln(n/OPT)]\, OPT + 1$ (compare to [12]).*

Proof Assume without loss of generality that the algorithm covers the elements in order $x_n, x_{n-1}, \ldots, x_1$. At the start of the iteration in which the algorithm covers x_i, there are at least i elements left to cover, and all of them could be covered using multiple sets of total cost OPT. Thus, there is some set that covers not-yet-covered elements at a cost of at most OPT/i per element.

Recall the charging scheme from the previous proof. By the preceding observation, element x_i is charged at most OPT/i. Thus, the total charge to elements $x_n, \ldots, x_i$ is at most $(H_n - H_{i-1})\text{OPT}$. Using the assumption that each $w_s \leq 1$, the charge to each of the remaining elements is at most 1 per element. Thus, the total charge to all elements is at most $i - 1 + (H_n -$

H_{i-1})OPT. Taking $i = 1 + \lceil \text{OPT} \rceil$, the total charge is at most $\lceil \text{OPT} \rceil + (H_n - H_{\lceil \text{OPT} \rceil})\text{OPT} \leq 1 + \text{OPT}(1 + \ln(n/\text{OPT}))$. □

Each of the above proofs implicitly constructs a linear-programming primal-dual pair to show the approximation ratio. The same approximation ratios can be shown with respect to any fractional optimum (solution to the fractional set-cover linear program).

Other Results

The greedy algorithm has been shown to have an approximation ratio of $\ln n - \ln \ln n + O(1)$ [11]. For the special case of set systems whose duals have finite Vapnik-Chervonenkis (VC) dimension, other algorithms have substantially better approximation ratio [1]. Constant-factor approximation algorithms are known for geometric variants of the closely related k-median and facility location problems.

The greedy algorithm generalizes naturally to many problems. For example, for minimizing a linear function subject to a submodular constraint (defined above), the natural extension of the greedy algorithm gives an H_k -approximate solution, where $k = \max_{s \in \mathcal{S}} f(\{s\}) - f(\emptyset)$, assuming f is integer valued [10].

The set-cover problem generalizes to allow each element x to require an arbitrary number r_x of sets containing it to be in the cover. This generalization admits a polynomial-time $O(\log n)$-approximation algorithm [7].

The special case when each element belongs to at most r sets has a simple r-approximation algorithm ([15] § 15.2). When the sets have uniform weights ($w_s = 1$), the algorithm reduces to the following: select any maximal collection of elements, no two of which are contained in the same set; return all sets that contain a selected element.

The variant "Max k-coverage" asks for a set collection of total weight at most k covering as many of the elements as possible. This variant has a $(1 - 1/e)$-approximation algorithm ([15] Problem 2.18) (see [6] for sets with nonuniform weights).

For a general discussion of greedy methods for approximate combinatorial optimization, see ([4] Ch. 4).

Finally, under likely complexity-theoretic assumptions, the $\ln n$ approximation ratio is essentially the best possible for any polynomial-time algorithm [3, 9].

Applications

Set cover and its generalizations and variants are fundamental problems with numerous applications. Examples include:

- Selecting a small number of nodes in a network to store a file so that all nodes have a nearby copy
- Selecting a small number of sentences to be uttered to tune all features in a speech-recognition model [14]
- Selecting a small number of telescope snapshots to be taken to capture light from all galaxies in the night sky
- Finding a short string having each string in a given set as a contiguous sub-string

Recommended Reading

1. Brönnimann H, Goodrich MT (1995) Almost optimal set covers in finite VC-dimension. Discret Comput Geom 14(4):463–479
2. Chvátal V (1979) A greedy heuristic for the set-covering problem. Math Oper Res 4(3): 233–235
3. Feige U (1998) A threshold of ln n for approximating set cover. J ACM 45(4):634–652
4. Gonzalez TF (2007) Handbook of approximation algorithms and metaheuristics. Chapman & Hall/CRC computer & information science series. Chapman & Hall/CRC, Boca Raton
5. Johnson DS (1974) Approximation algorithms for combinatorial problems. J Comput Syst Sci 9:256–278
6. Khuller S, Moss A, Naor J (1999) The budgeted maximum coverage problem. Inform Process Lett 70(1):39–45
7. Kolliopoulos SG, Young NE (2001) Tight approximation results for general covering integer programs.

In: Proceedings of the forty-second annual IEEE symposium on foundations of computer science, Las Vegas, pp 522–528
8. Lovász L (1975) On the ratio of optimal integral and fractional covers. Discret Math 13:383–390
9. Lund C, Yannakakis M (1994) On the hardness of approximating minimization problems. J ACM 41(5):960–981
10. Nemhauser GL, Wolsey LA (1988) Integer and combinatorial optimization. Wiley, New York
11. Slavik P (1997) A tight analysis of the greedy algorithm for set cover. J Algorithms 25(2): 237–254
12. Srinivasan A (1995) Improved approximations of packing and covering problems. In: Proceedings of the twenty-seventh annual ACM symposium on theory of computing, Heraklion, pp 268–276
13. Stein SK (1974) Two combinatorial covering theorems. J Comb Theor A 16:391–397
14. van Santen JPH, Buchsbaum AL (1997) Methods for optimal text selection. In: Proceedings of the European conference on speech communication and technology, Rhodos, vol 2, pp 553–556
15. Vazirani VV (2001) Approximation algorithms. Springer, Berlin/Heidelberg

Hamilton Cycles in Random Intersection Graphs

Charilaos Efthymiou[1] and Paul (Pavlos) Spirakis[2,3,4]
[1]Department of Computer Engineering and Informatics, University of Patras, Patras, Greece
[2]Computer Engineering and Informatics, Research and Academic Computer Technology Institute, Patras University, Patras, Greece
[3]Computer Science, University of Liverpool, Liverpool, UK
[4]Computer Technology Institute (CTI), Patras, Greece

Keywords

Stochastic order relations between Erdös–Rényi random graph model and random intersection graphs; Threshold for appearance of Hamilton cycles in random intersection graphs

Years and Authors of Summarized Original Work

2005; Efthymiou, Spirakis

Problem Definition

E. Marczewski proved that every graph can be represented by a list of sets where each vertex corresponds to a set and the edges to nonempty intersections of sets. It is natural to ask what sort of graphs would be most likely to arise if the list of sets is generated randomly.

Consider the model of random graphs where each vertex chooses randomly from a universal set the members of its corresponding set, each independently of the others. The probability space that is created is the space of random intersection graphs, $G_{n,m,p}$, where n is the number of vertices, m is the cardinality of a universal set of elements and p is the probability for each vertex to choose an element of the universal set. The model of random intersection graphs was first introduced by M. Karoński, E. Scheinerman, and K. Singer-Cohen in [4]. A rigorous definition of the model of random intersection graphs follows:

Definition 1 Let n, m be positive integers and $0 \leq p \leq 1$. The random intersection graph $G_{n,m,p}$ is a probability space over the set of graphs on the vertex set $\{1, \ldots, n\}$ where each vertex is assigned a random subset from a fixed set of m elements. An edge arises between two vertices when their sets have at least a common element. Each random subset assigned to a vertex is determined by

$$\mathbf{Pr}\,[\text{vertex } i \text{ chooses element } j] = p$$

with these events mutually independent.

A common question for a graph is whether it has a cycle, a set of edges that form a path so that the

M.-Y. Kao (ed.), *Encyclopedia of Algorithms*,
DOI 10.1007/978-1-4939-2864-4

first and the last vertex is the same, that visits *all* the vertices of the graph exactly once. We call this kind of cycle the *Hamilton cycle* and the graph that contains such a cycle is called a *Hamiltonian graph*.

Definition 2 Consider an undirected graph $G = (V, E)$ where V is the set of vertices and E the set of edges. This graph contains a Hamilton cycle if and only if there is a simple cycle that contains each vertex in V.

Consider an instance of $G_{n,m,p}$, for specific values of its parameters n, m, and p, what is the probability of that instance to be Hamiltonian? Taking the parameter p, of the model, to be a function of n and m, in [2], a threshold function $P(n,m)$ has been found for the graph property "Contains a Hamilton cycle"; i.e., a function $P(n,m)$ is derived such that

if $p(n,m) \ll P(n,m)$

$$\lim_{n,m\to\infty} \mathbf{Pr}\left[G_{n,m,p}\text{Contains Hamilton cycle}\right] = 0$$

if $p(n,m) \gg P(n,m)$

$$\lim_{n,m\to\infty} \mathbf{Pr}\left[G_{n,m,p}\text{Contains Hamilton cycle}\right] = 1$$

When a graph property, such as "Contains a Hamilton cycle," holds with probability that tends to 1 (or 0) as n, m tend to infinity, then it is said that this property holds (does not hold), "almost surely" or "almost certainly."

If in $G_{n,m,p}$ the parameter m is very small compared to n, the model is not particularly interesting and when m is exceedingly large (compared to n) the behavior of $G_{n,m,p}$ is essentially the same as the Erdös–Rényi model of random graphs (see [3]). If someone takes $m = \lceil n^{\alpha} \rceil$, for fixed real $\alpha > 0$, then there is some deviation from the standard models, while allowing for a natural progression from sparse to dense graphs. Thus, the parameter m is assumed to be of the form $m = \lceil n^{\alpha} \rceil$ for some fixed positive real α.

The proof of existence of a Hamilton cycle in $G_{n,m,p}$ is mainly based on the establishment of a *stochastic order relation* between the model $G_{n,m,p}$ and the Erdös–Rényi random graph model $G_{n,\hat{p}}$.

Definition 3 Let n be a positive integer, $0 \le \hat{p} \le 1$. The random graph $G(n,\hat{p})$ is a probability space over the set of graphs on the vertex set $\{1,\ldots,n\}$ determined by

$$\mathbf{Pr}[i,j] = \hat{p}$$

with these events mutually independent.

The stochastic order relation between the two models of random graphs is established in the sense that if $\mathcal{A}$ is an increasing graph property, then it holds that

$$\mathbf{Pr}\left[G_{n,\hat{p}} \in \mathcal{A}\right] \le \mathbf{Pr}\left[G_{n,m,p} \in \mathcal{A}\right]$$

where $\hat{p} = f(p)$. A graph property $\mathcal{A}$ is increasing if and only if given that $\mathcal{A}$ holds for a graph $G(V,E)$ then $\mathcal{A}$ holds for any $G(V,E')$: $E' \supseteq E$.

Key Results

Theorem 1 *Let $m = \lceil n^{\alpha} \rceil$, where α is a fixed real positive, and C_1, C_2 be sufficiently large constants. If*

$$p \ge C_1 \frac{\log n}{m} \quad \text{for} \quad 0 < \alpha < 1 \quad \text{or}$$

$$p \ge C_2 \sqrt{\frac{\log n}{nm}} \quad \text{for} \quad \alpha > 1$$

then almost all $G_{n,m,p}$ are Hamiltonian. Our bounds are asymptotically tight.

Note that the theorem above says nothing when $m = n$, i.e., $\alpha = 1$.

Applications

The Erdös–Rényi model of random graphs, $G_{n,p}$, is exhaustively studied in computer science because it provides a framework for studying practical problems such as "reliable network computing" or it provides a "typical instance" of a graph and thus it is used for average case analysis of graph algorithms. However, the simplicity of $G_{n,p}$ means it is not able to capture satisfactorily many practical problems in computer science. Basically, this is because of the fact that in many problems independent edge-events are not well justified. For example, consider a graph whose vertices represent a set of objects that either are placed or move in a specific geographical region, and the edges are radio communication links. In such a graph, we expect that, any two vertices u, w are more likely to be adjacent to each other, than any other, arbitrary, pair of vertices, if both are adjecent to a third vertex v. Even epidemiological phenomena (like the spread of disease) tend to be more accurately captured by this proximity-sensitive random intersection graph model. Other applications may include oblivious resource sharing in a distributive setting, interaction of mobile agents traversing the web etc.

The model of random intersection graphs $G_{n,m,p}$ was first introduced by M. Karońsky, E. Scheinerman, and K. Singer-Cohen in [4] where they explored the evolution of random intersection graphs by studying the thresholds for the appearance and disappearance of small induced subgraphs. Also, J.A. Fill, E.R. Scheinerman, andK. Singer Cohen in [3] proved an equivalence theorem relating the evolution of $G_{n,m,p}$ and $G_{n,p}$, in particular they proved that when $m = n^{\alpha}$ where $\alpha > 6$, the total variation distance between the graph random variables has limit 0. S. Nikoletseas, C. Raptopoulos, and P. Spirakis in [8] studied the existence and the efficient algorithmic construction of close to optimal independent sets in random intersection graphs. D. Stark in [11] studied the degree of the vertices of the random intersection graphs. However, after [2], Spirakis and Raptopoulos, in [10], provide algorithms that construct Hamilton cycles in instances of $G_{n,m,p}$, for p above the Hamiltonicity threshold. Finally, Nikoletseas et al. in [7] study the mixing time and cover time as the parameter p of the model varies.

Open Problems

As in many other random structures, e.g., $G_{n,p}$ and random formulae, properties of random intersection graphs also appear to have threshold behavior. So far threshold behavior has been studied for the induced subgraph appearance and hamiltonicity.

Other fields of research for random intersection graphs may include the study of connectivity behavior, of the model i.e., the path formation, the formation of giant components. Additionally, a very interesting research question is how cover and mixing times vary with the parameter p, of the model.

Cross-References

▶ Independent Sets in Random Intersection Graphs

Recommended Reading

1. Alon N, Spencer JH (2000) The probabilistic method, 2nd edn. Wiley, New York
2. Efthymiou C, Spirakis PG (2005) On the existence of Hamilton cycles in random intersection graphs. In: Proceedings of the 32nd ICALP. LNCS, vol 3580. Springer, Berlin/Heidelberg, pp 690–701
3. Fill JA, Scheinerman ER, Singer-Cohen KB (2000) Random intersection graphs when m = ω(n): an equivalence theorem relating the evolution of the G(n, m, p) and G(n, p) models. Random Struct Algorithms 16:156–176
4. Karoński M, Scheinerman ER, Singer-Cohen K (1999) On random intersection graphs: the subgraph problem. Comb Probab Comput 8:131–159
5. Komlós J, Szemerédi E (1983) Limit distributions for the existence of Hamilton cycles in a random graph. Discret Math 43:55–63
6. Korshunov AD (1977) Solution of a problem of P. Erdös and A. Rényi on Hamilton cycles in non-oriented graphs. Metody Diskr Anal Teoriy Upr Syst Sb Trubov Novosibrirsk 31:17–56
7. Nikoletseas S, Raptopoulos C, Spirakis P (2007) Expander properties and the cover time of random intersection graphs. In: Proceedings of the 32nd MFCS. Springer, Berlin/Heidelberg, pp 44–55
8. Nikoletseas S, Raptopoulos C, Spirakis P (2004) The existence and efficient construction of large independent sets in general random intersection graphs.

In: Proceedings of the 31st ICALP. LNCS, vol 3142. Springer, Berlin/Heidelberg, pp 1029–1040
9. Singer K (1995) Random intersection graphs. PhD thesis, The Johns Hopkins University, Baltimore
10. Spirakis PG, Raptopoulos C (2005) Simple and efficient greedy algorithms for Hamilton cycles in random intersection graphs. In: Proceedings of the 16th ISAAC. LNCS, vol 3827. Springer, Berlin/Heidelberg, pp 493–504
11. Stark D (2004) The vertex degree distribution of random intersection graphs. Random Struct Algorithms 24:249–258

Haplotype Inference on Pedigrees Without Recombinations

Mee Yee Chan[1], Wun-Tat Chan[2], Francis Y.L. Chin[1], Stanley P.Y. Fung[3], and Ming-Yang Kao[4]
[1]Department of Computer Science, University of Hong Kong, Hong Kong, China
[2]College of International Education, Hong Kong Baptist University, Hong Kong, China
[3]Department of Computer Science, University of Leicester, Leicester, UK
[4]Department of Electrical Engineering and Computer Science, Northwestern University, Evanston, IL, USA

Keywords

Computational biology; Haplotype inference; Pedigree; Recombination

Years aud Authors of Summarized Original Work

2009; Chan, Chan, Chin, Fung, Kao

Problem Definition

In many diploid organisms like humans, chromosomes come in pairs. Genetic variation occurs in some "positions" along the chromosomes. These genetic variations are commonly modelled in the form of *single nucleotide polymorphisms* (SNPs) [5], which are the nucleotide sites where more than one nucleotide can occur. A *haplotype* is the sequence of linked SNP genetic markers (small segments of DNA) on a single chromosome. However, experiments often yield *genotypes*, which is a blend of the two haplotypes of the chromosome pair. It is more useful to have information on the haplotypes, thus giving rise to the computational problem of inferring haplotypes from genotypes.

The physical position of a marker on a chromosome is called a *locus* and its state is called an *allele*. SNP are often *biallelic*, i.e., the allele can take on two different states, corresponding to two different nucleotides. In the language of computer science, the allele of a biallelic SNP can be denoted by 0 and 1, and a haplotype with m loci is represented as a length-m string in $\{0,1\}^m$ and a genotype as a length-m string in $\{0,1,2\}^m$. Consider a haplotype pair $\langle h_1, h_2\rangle$ and a corresponding genotype g. For each locus, if both haplotypes show a 0, then the genotype must also be 0, and if both haplotypes show a 1, the genotype must also be 1. These loci are called *homozygous*. If however one of the haplotypes shows a 0 and the other a 1, the genotype shows a 2 and the locus is called *heterozygous*. This is called *SNP consistency*. For example, considering a single individual, the genotype $g = 012212$ has four SNP-consistent haplotype pairs: $\{\langle 01\underline{11}1\underline{1}, 01\underline{00}1\underline{0}\rangle, \langle 01\underline{11}1\underline{0}, 01\underline{00}1\underline{1}\rangle, \langle 01\underline{10}1\underline{1}, 01\underline{01}1\underline{0}\rangle, \langle 01\underline{10}1\underline{0}, 01\underline{01}1\underline{1}\rangle\}$. In general, if a genotype has s heterozygous loci, it can have 2^{s-1} SNP-consistent haplotype solutions.

Haplotypes are passed down from an individual to its descendants. *Mendelian consistency* requires that, in the absence of *recombinations* or mutations, each child inherits one haplotype from one of the two haplotypes of the father and inherits the other haplotype from the mother similarly. This gives us more information to infer haplotypes when we are given a *pedigree*. The computational problem is therefore, given a pedigree with n individuals where each individual is associated with a genotype of length m, find an assignment of a pair of haplotypes to each individual such that SNP consistency

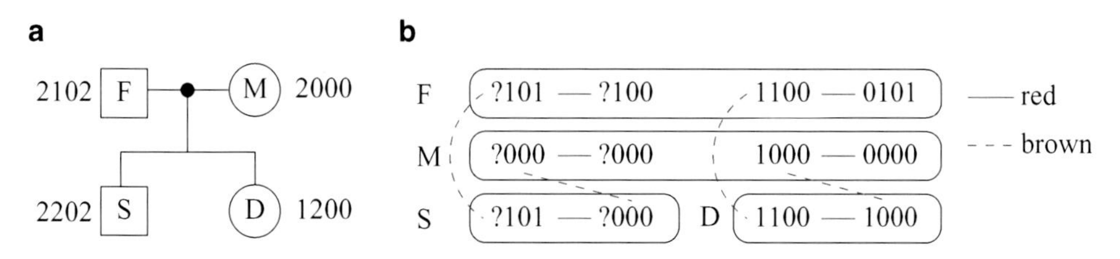

Haplotype Inference on Pedigrees Without Recombinations, Fig. 1 (**a**) Example of a pedigree with four nodes. (**b**) The graph G with 12 vertices, 6 *red edges*, and 4 *brown edges*. Each vector is a vertex in G. Vector pairs enclosed by *rounded rectangles* belong to the same individual

and Mendelian consistency are obeyed for each individual. In rare cases (especially for humans) [3], the pedigree may contain *mating loops*: a mating loop is formed when, for example, there is a marriage between descendants of a common ancestor.

As a simple example, consider the pedigree in Fig. 1a for a family of four individuals and their genotypes. Due to SNP consistency, mother M's haplotypes must be $\langle 0000, 1000\rangle$ (the order does not matter). Similarly, daughter D's haplotypes must be $\langle 1000, 1100\rangle$. Now we apply Mendelian consistency to deduce that D must obtain the 1000 haplotype from M since neither of father F's haplotypes can be 1000 (considering locus 2). Therefore, D obtains 1100 from F, and F's haplotypes must be $\langle 0101, 1100\rangle$. With F's and M's haplotypes known, the only solution for the haplotypes of son S that is consistent with his genotype 2202 is $\langle 0101, 1000\rangle$.

Key Results

While this kind of deduction might appear to be enough to resolve all haplotype values, it is not the case. As we will shortly see, there are "long-distance" constraints that need to be considered. These constraints can be represented by a system of linear equations in GF(2) and solved using Gaussian elimination. This gives a $O(m^3n^3)$ time algorithm [3]. Subsequent papers try to capture or solve the constraints more economically. The time complexity was improved in [6] to $O(mn^2 + n^3 \log^2 n \log\log n)$ by eliminating redundant equations and using low-stretch spanning trees. A different approach was used in [1], representing the constraints by the parity of edge labels of some auxiliary graphs and finding solutions of these constraints using graph traversal without (directly) solving a system of linear equations. This gives a linear $O(mn)$ time algorithm, although it only works for the case with no mating loops and only produces one particular solution even when the pedigree admits more than one solution. Later algorithms include [4] which returns the full set of solutions in optimal time (again without mating loops) and [2] which can handle mating loops and runs in $O(kmn + k^2m)$ time where k is the number of mating loops.

In the following we sketch the idea behind the linear time algorithm in [1]. Each individual only has a pair of haplotypes, but the algorithm first produces a number of vector pairs for each individual, one vector pair for each trio (a father-mother-child triplet) that this individual belongs to. Each vector pair represents the information about the two haplotypes of this individual that can be derived by considering this trio only. These vector pairs will eventually be "unified" to become a single pair.

For the pedigree in Fig. 1a, the algorithm first produces the graph G in Fig. 1b, which has two connected components for the two trios F-M-S and F-M-D. The rule for enforcing SNP consistency (Mendelian consistency) is that the unresolved loci values, i.e., the ? values, must be different (same) at opposite ends of a red (brown) edge. There is only one way to unify the vector pairs of F consistently (due to locus 4): ?101 must correspond to 0101. We add an edge between these two vectors to represent the fact that they should be identical. Then all ? values

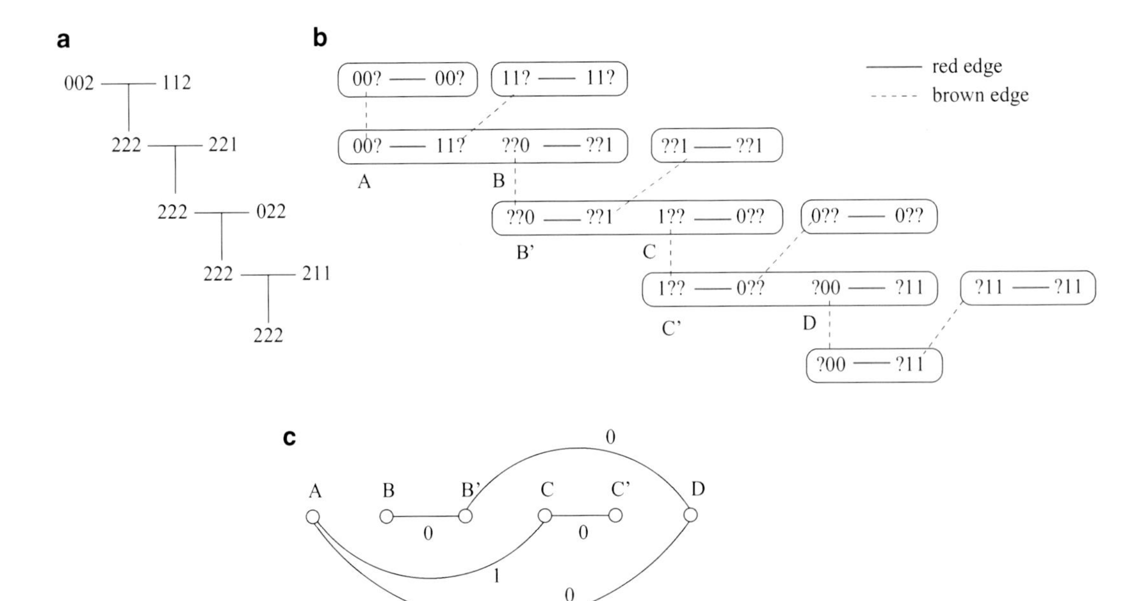

Haplotype Inference on Pedigrees Without Recombinations, Fig. 2 An example showing how constraints are represented by labeled edges in another graph. (**a**) The pedigree. (**b**) The local graph G. (**c**) The parity constraint graph J. Three constraints are added

can be resolved by traversing the now-connected graph and applying the aforementioned rules for enforcing consistency.

However, consider another pedigree in Fig. 2a. The previous steps can only produce Fig. 2b, which has four connected components and unresolved loci. We need to decide for A and B whether $A = 00?$ should connect to $B = ??0$ or its complement ??1 and similarly for B' and C, etc. Observe that a path between A and C must go through an odd number of red edges since locus 1 changes from 0 to 1. To capture this type of long-distance constraints, we construct a *parity constraint graph* J where the edge labels represent the parity constraints; see Fig. 2c. In effect, J represents a set of linear equations in GF(2); in Fig. 2c, the equations are $x_{AB} + x_{B'C} = 1$, $x_{B'C} + x_{C'D} = 0$, and $x_{AB} + x_{B'C} + x_{C'D} = 0$.

Finally, we can traverse J along the unique path between any two nodes; the parity of this path tells us how to merge the vector pairs in G. For example, the parity between A and B should be 0, indicating 00? in A should connect to ??0 in B (so both become 000), while the parity between B' and C is 1, so B' and C should be 000 and 111, respectively.

Cross-References

- ▶ Beyond Evolutionary Trees
- ▶ Musite: Tool for Predicting Protein Phosphorylation Sites
- ▶ Sequence and Spatial Motif Discovery in Short Sequence Fragments

Recommended Reading

1. Chan MY, Chan WT, Chin FYL, Fung SPY, Kao MY (2009) Linear-time haplotype inference on pedigrees without recombinations and mating loops. SIAM J Comput 38(6):2179–2197
2. Lai EY, Wang WB, Jiang T, Wu KP (2012) A linear-time algorithm for reconstructing zero-recombinant haplotype configuration on a pedigree. BMC Bioinformatics 13(S-17):S19
3. Li J, Jiang T (2003) Efficient inference of haplotypes from genotypes on a pedigree. J. Bioinformatics Comput Biol 1(1):41–69
4. Liu L and Jiang T (2010) A linear-time algorithm for reconstructing zero-recombinant haplotype configuration on pedigrees without mating loops. J Combin Optim 19:217–240
5. Russo E, Smaglik P (1999) Single nucleotide polymorphism: big pharmacy hedges its bets. The Scientist, 13(15):1, July 19, 1999

6. Xiao J, Liu L, Xia L, Jiang T (2009) Efficient algorithms for reconstructing zero-recombinant haplotypes on a pedigree based on fast elimination of redundant linear equations. SIAM J Comput 38(6):2198–2219

Hardness of Proper Learning

Vitaly Feldman
IBM Research – Almaden, San Jose, CA, USA

Keywords

DNF; function representation; NP-hardness of learning; PAC learning; Proper learning; representation-based hardness

Years and Authors of Summarized Original Work

1988; Pitt, Valiant

Problem Definition

The work of Pitt and Valiant [18] deals with learning Boolean functions in the Probably Approximately Correct (PAC) learning model introduced by Valiant [19]. A learning algorithm in Valiant's original model is given random examples of a function $f : \{0,1\}^n \rightarrow \{0,1\}$ from a representation class $\mathcal{F}$ and produces a hypothesis $h \in \mathcal{F}$ that closely approximates f. Here, a *representation class* is a set of functions and a language for describing the functions in the set. The authors give examples of natural representation classes that are NP-hard to learn in this model, whereas they can be learned if the learning algorithm is allowed to produce hypotheses from a richer representation class $\mathcal{H}$. Such an algorithm is said to learn $\mathcal{F}$ by $\mathcal{H}$; learning $\mathcal{F}$ by $\mathcal{F}$ is called *proper* learning.

The results of Pitt and Valiant were the first to demonstrate that the choice of representation of hypotheses can have a dramatic impact on the computational complexity of a learning problem. Their specific reductions from NP-hard problems are the basis of several other follow-up works on the hardness of proper learning [1, 3, 7].

Notation

Learning in the PAC model is based on the assumption that the unknown function (or *concept*) belongs to a certain class of concepts $\mathcal{C}$. In order to discuss algorithms that learn and output functions, one needs to define how these functions are represented. Informally, a representation for a concept class $\mathcal{C}$ is a way to describe concepts from $\mathcal{C}$ that defines a procedure to evaluate a concept in $\mathcal{C}$ on any input. For example, one can represent a conjunction of input variables by listing the variables in the conjunction. More formally, a representation class can be defined as follows.

Definition 1 A *representation class* $\mathcal{F}$ is a pair $(L, \mathcal{R})$ where

- L is a language over some fixed finite alphabet (e.g., $\{0,1\}$);
- $\mathcal{R}$ is an algorithm that for $\sigma \in L$, on input $(\sigma, 1^n)$ returns a Boolean circuit over $\{0,1\}^n$.

In the context of efficient learning, only efficient representations are considered, or, representations for which $\mathcal{R}$ is a polynomial-time algorithm. The concept class represented by $\mathcal{F}$ is the set of functions over $\{0,1\}^n$ defined by the circuits in $\{\mathcal{R}(\sigma, 1^n) \mid \sigma \in L\}$. For a Boolean function f, "$f \in \mathcal{F}$" means that f belongs to the concept class represented by $\mathcal{F}$ and that there is a $\sigma \in L$ whose associated Boolean circuit computes f. For most of the representations discussed in the context of learning, it is straightforward to construct a language L and the corresponding translating function $\mathcal{R}$, and therefore, they are not specified explicitly.

Associated with each representation is the complexity of describing a Boolean function using this representation. More formally, for a Boolean function $f \in \mathcal{C}$, $\mathcal{F}\texttt{-size}(f)$ is the length of the shortest way to represent f using $\mathcal{F}$, or $\min\{|\sigma| \mid \sigma \in L, \mathcal{R}(\sigma, 1^n) \equiv f\}$.

We consider Valiant's PAC model of learning [19], as generalized by Pitt and Valiant [18].

In this model, for a function f and a distribution $\mathcal{D}$ over X, an *example oracle* $\mathrm{EX}(f, \mathcal{D})$ is an oracle that, when invoked, returns an example $\langle x, f(x)\rangle$, where x is chosen randomly with respect to $\mathcal{D}$, independently of any previous examples. For $\epsilon \geq 0$, we say that function g ϵ-approximates a function f with respect to distribution $\mathcal{D}$ if $\mathbf{Pr}_{\mathcal{D}}[f(x) \neq g(x)] \leq \epsilon$.

Definition 2 A representation class $\mathcal{F}$ is *PAC learnable* by representation class $\mathcal{H}$ if there exists an algorithm that for every $\epsilon > 0$, $\delta > 0$, n, $f \in \mathcal{F}$, and distribution $\mathcal{D}$ over X, given ϵ, δ, and access to $\mathrm{EX}(f, \mathcal{D})$, runs in time polynomial in $n, s = \mathcal{F}\texttt{-size}(c)$, $1/\epsilon$ and $1/\delta$, and outputs, with probability at least $1-\delta$, a hypothesis $h \in \mathcal{H}$ that ϵ-approximates f.

A DNF expression is defined as an OR of ANDs of literals, where a *literal* is a possibly negated input variable. We refer to the ANDs of a DNF formula as its *terms*. Let $\mathsf{DNF}(k)$ denote the representation class of k-term DNF expressions. Similarly, a CNF expression is an OR of ANDs of literals. Let k-CNF denote the representation class of CNF expressions with each AND having at most k literals.

For a real-valued vector $c \in \mathbb{R}^n$ and $\theta \in \mathbb{R}$, a *linear threshold function* (also called a *halfspace*) $T_{c,\theta}(x)$ is the function that equals 1 if and only if $\sum_{i \leq n} c_i x_i \geq \theta$. The representation class of Boolean threshold functions consists of all linear threshold functions with $c \in \{0,1\}^n$ and θ an integer.

Key Results

Theorem 1 ([18]) *For every $k \geq 2$, the representation class of* $\mathsf{DNF}(k)$ *is not properly learnable unless* $\mathsf{RP} = \mathsf{NP}$.

More specifically, Pitt and Valiant show that learning $\mathsf{DNF}(k)$ by $\mathsf{DNF}(\ell)$ is at least as hard as coloring a k-colorable graph using ℓ colors. For the case $k = 2$, they obtain the result by reducing from Set Splitting (see [9] for details on the problems). Theorem 1 is in sharp contrast with the fact that $\mathsf{DNF}(k)$ is learnable by k-CNF [19].

Theorem 2 ([18]) *The representation class of Boolean threshold functions is not properly learnable unless* $\mathsf{RP} = \mathsf{NP}$.

This result is obtained via a reduction from the NP-complete Zero-One Integer Programming problem (see [9] (p.245) for details on the problem). The result is contrasted by the fact that general linear thresholds are properly learnable [4].

These results show that using a specific representation of hypotheses forces the learning algorithm to solve a combinatorial problem that can be NP-hard. In most machine learning applications it is not important which representation of hypotheses is used as long as the value of the unknown function is predicted correctly. Therefore, learning in the PAC model is now defined without any restrictions on the output hypothesis (other than it being efficiently evaluatable). Hardness results in this setting are usually based on cryptographic assumptions (cf. [15]).

Hardness results for proper learning based on assumption $\mathsf{NP} \neq \mathsf{RP}$ are now known for several other representation classes and for other variants and extensions of the PAC learning model. Blum and Rivest show that for any $k \geq 3$, unions of k halfspaces are not properly learnable [3]. Hancock et al. prove that decision trees (cf. [16] for the definition of this representation) are not learnable by decision trees of somewhat larger size [11]. This result was strengthened by Alekhnovich et al. who also proved that intersections of two halfspaces are not learnable by intersections of k halfspaces for any constant k, general DNF expressions are not learnable by unions of halfspaces (and in particular are not properly learnable) and k*-juntas* are not properly learnable [1]. Further, DNF expressions remain NP-hard to learn properly even if *membership queries*, or the ability to query the unknown function at any point, are allowed [7]. Khot and Saket show that the problem of learning intersections of two halfspaces remains NP-hard even if a hypothesis with any constant error smaller than $1/2$ is required [17]. No efficient algorithms or hardness results are known for any of the above learning problems if no restriction is placed on the representation of hypotheses.

The choice of representation is important even in powerful learning models. Feldman proved that n^c-term DNF are not properly learnable for any constant c even when the distribution of examples is assumed to be uniform and membership queries are available [7]. This contrasts with Jackson's celebrated algorithm for learning DNF in this setting [13], which is not proper.

In the *agnostic learning* model of Haussler [12] and Kearns et al. [14], even the representation classes of conjunctions, decision lists, halfspaces, and parity functions are NP-hard to learn properly (cf. [2, 6, 8, 10] and references therein). Here again the status of these problems in the representation-independent setting is largely unknown.

Applications

A large number of practical algorithms use representations for which hardness results are known (most notably decision trees, halfspaces, and neural networks). Hardness of learning $\mathcal{F}$ by $\mathcal{H}$ implies that an algorithm that uses $\mathcal{H}$ to represent its hypotheses will not be able to learn $\mathcal{F}$ in the PAC sense. Therefore such hardness results elucidate the limitations of algorithms used in practice. In particular, the reduction from an NP-hard problem used to prove the hardness of learning $\mathcal{F}$ by $\mathcal{H}$ can be used to generate hard instances of the learning problem.

Open Problems

A number of problems related to proper learning in the PAC model and its extensions are open. Almost all hardness of proper learning results are for learning with respect to unrestricted distributions. For most of the problems mentioned in section "Key Results" it is unknown whether the result is true if the distribution is restricted to belong to some natural class of distributions (e.g., product distributions). It is unknown whether decision trees are learnable properly in the PAC model or in the PAC model with membership queries. This question is open even in the PAC model restricted to the uniform distribution only. Note that decision trees are learnable (non-properly) if membership queries are available [5] and are learnable properly in time $O(n^{\log s})$, where s is the number of leaves in the decision tree [1].

An even more interesting direction of research would be to obtain hardness results for learning by richer representation classes, such as AC^0 circuits, classes of neural networks and, ultimately, unrestricted circuits.

Cross-References

► Cryptographic Hardness of Learning
► Graph Coloring
► Learning DNF Formulas
► PAC Learning

Recommended Reading

1. Alekhnovich M, Braverman M, Feldman V, Klivans A, Pitassi T (2008) The complexity of properly learning simple classes. J Comput Syst Sci 74(1):16–34
2. Ben-David S, Eiron N, Long PM (2003) On the difficulty of approximately maximizing agreements. J Comput Syst Sci 66(3):496–514
3. Blum AL, Rivest RL (1992) Training a 3-node neural network is NP-complete. Neural Netw 5(1):117–127
4. Blumer A, Ehrenfeucht A, Haussler D, Warmuth M (1989) Learnability and the Vapnik-Chervonenkis dimension. J ACM 36(4):929–965
5. Bshouty N (1995) Exact learning via the monotone theory. Inform Comput 123(1):146–153
6. Feldman V, Gopalan P, Khot S, Ponuswami A (2009) On agnostic learning of parities, monomials and halfspaces. SIAM J Comput 39(2):606–645
7. Feldman V (2009) Hardness of approximate two-level logic minimization and pac learning with membership queries. J Comput Syst Sci 75(1):13–26
8. Feldman V, Guruswami V, Raghavendra P, Wu Y (2012) Agnostic learning of monomials by halfspaces is hard. SIAM J Comput 41(6):1558–1590
9. Garey M, Johnson DS (1979) Computers and intractability. W.H. Freeman, San Francisco
10. Guruswami V, Raghavendra P (2009) Hardness of learning halfspaces with noise. SIAM J Comput 39(2):742–765
11. Hancock T, Jiang T, Li M, Tromp J (1995) Lower bounds on learning decision lists and trees. In: Mayr

EW, Puech C (eds) STACS 95: 12th Annual Symposium on Theoretical Aspects of Computer Science, Munich, pp 527–538. Springer, Berlin/Heidelberg
12. Haussler D (1992) Decision theoretic generalizations of the PAC model for neural net and other learning applications. Inform Comput 100(1):78–150
13. Jackson J (1997) An efficient membership-query algorithm for learning DNF with respect to the uniform distribution. J Comput Syst Sci 55:414–440
14. Kearns M, Schapire R, Sellie L (1994) Toward efficient agnostic learning. Mach Learn 17(2–3):115–141
15. Kearns M, Valiant L (1994) Cryptographic limitations on learning boolean formulae and finite automata. J ACM 41(1):67–95
16. Kearns M, Vazirani U (1994) An introduction to computational learning theory. MIT, Cambridge
17. Khot S, Saket R (2011) On the hardness of learning intersections of two halfspaces. J Comput Syst Sci 77(1):129–141
18. Pitt L, Valiant L (1988) Computational limitations on learning from examples. J ACM 35(4):965–984
19. Valiant LG (1984) A theory of the learnable. Commun ACM 27(11):1134–1142

Harmonic Algorithm for Online Bin Packing

Leah Epstein
Department of Mathematics, University of Haifa, Haifa, Israel

Keywords

Bin packing; Bounded space algorithms; Competitive ratio

Years and Authors of Summarized Original Work

1985; Lee, Lee

Problem Definition

One of the goals of the design of the harmonic algorithm (or class of algorithms) was to provide an online algorithm for the classic bin packing problem that performs well with respect to the asymptotic competitive ratio, which is the standard measure for online algorithms for bin packing type problems. The competitive ratio for a given input is the ratio between the costs of the algorithm and of an optimal off-line solution. The asymptotic competitive ratio is the worst-case competitive ratio of inputs for which the optimal cost is sufficiently large. In the *online* (standard) bin packing problem, items of rational sizes in $(0, 1]$ are presented one by one. The algorithm must pack each item into a bin before the following item is presented. The total size of items packed into a bin cannot exceed 1, and the goal is to use the minimum number of bins, where a bin is used if at least one item was packed into it. All items must be packed, and the supply of bins is unlimited.

When an algorithm acts on an input, it can decide to *close* some of its bins and never use them again. A bin is called *closed* in such a case, while otherwise a used bin (which already has at least one item) is called *open*. The motivation for closing bins is to obtain fast running times per item (so that the algorithm will pack it into a bin selected out of a small number of options). Simple algorithms such as First Fit (FF), Best Fit (BF), and Worst Fit (WF) have worst-case running times of $O(\log N)$ per item, where N is the number of items at the time of assignment of the new item. On the other hand, the simple algorithm Next Fit (NF), which keeps at most of open bin and closes it when a new item cannot be packed there (before it uses a new bin for the new item), has a worst-case running time of $O(1)$ per item. Algorithms that keep a constant number of open bins are called *bounded space*. In many practical applications, this property is desirable, since the number of candidate bins for a new item is small and it does not increase with the input size.

Algorithm HARM$_k$ (for an integer $k \geq 3$) was defined by Lee and Lee [7]. The fundamental and natural idea of "harmonic-based" algorithms is classify each item by size first (for online algorithms, the classification of an item must be done immediately upon arrival) and then pack it according to its class (instead of letting the exact size influence packing decisions). For the classification of items, $HARM_k$ splits the interval $(0, 1]$ into subintervals. There are $k - 1$ subintervals of the form $(\frac{1}{i+1}, \frac{1}{i}]$ for $i = 1, \ldots, k - 1$ and one final subinterval $(0, \frac{1}{k}]$. Each bin will contain only items from one subinterval (type). Every type is packed independently into its own bins using NF. Thus, there are at most $k - 1$ open bins at each time (since for items of sizes above $\frac{1}{2}$, two items cannot share a bin, and any bin can be closed once it receives an item). Moreover, for $i < k$, as the items of type i have sizes no larger than $\frac{1}{i}$ but larger than $\frac{1}{i+1}$, every closed bin of this type will have exactly i items. For type k, a closed bin will contain at least k items, but it may contain many more items. This defines a class of algorithms (containing one algorithm for any $k \geq 3$). The term *the harmonic algorithm* (or simply HARM) refers to $HARM_k$ for a sufficiently large value of k, and its asymptotic competitive ratio is the infimum value that can be achieved as the asymptotic competitive ratio of any algorithm of this class.

Key Results

It was shown in paper [7] that for k tending to infinity, the asymptotic ratio of HARM is a sum of series denoted by Π_∞ (see below), and it is equal to approximately 1.69103. Moreover, this is the best possible asymptotic competitive ratio of any online bounded space algorithm for standard bin packing.

The crucial item sizes are of the form $\frac{1}{\ell} + \varepsilon$, where $\varepsilon > 0$ is small and ℓ is an integer. These are items of type $\ell - 1$, and bins consisting of such items contain $\ell - 1$ items (except for the last bin used for this type that may contain a smaller number of items). However, a bin (of an off-line solution) that already contains an item of size $\frac{1}{2} + \varepsilon_1$ and an item of size $\frac{1}{3} + \varepsilon_2$ (for some small $\varepsilon_1, \varepsilon_2 > 0$) cannot contain also an item whose size is slightly above $\frac{1}{4}$. The largest item of this form would be slightly larger than $\frac{1}{7}$. Thus, the following sequence was defined [7]. Let $\pi_1 = 1$ and, for $j > 1$, $\pi_j = \pi_{j-1}(\pi_{j-1} + 1)$ (note that $\pi_{j'}$ is divisible by any π_j for $j < j'$). It turns out that the crucial item sizes are just above $\frac{1}{\pi_j + 1}$. The series $\sum_{j=1}^{\infty} \frac{1}{\pi_j}$ give the asymptotic competitive ratio of the HARM, Π_∞. For a long time the best lower bound on the asymptotic competitive ratio of (unbounded space) online algorithms was the one by van Vliet [8, 13], proved using this sequence (but the current best lower bound was proved using another set of inputs [1]).

In order to prove the upper bound Π_∞ on the competitive ratio, weights were used [12]. In this case weights are defined (for a specific value of k) quite easily such that all bins (except for the bins that remain open when the algorithm terminates) have total weights of at least 1. The weight of an item of type $i < k$ is $\frac{1}{i}$. The bins of type k are almost full for sufficiently large values of k (a bin can be closed only if the total size of its items exceeds $1 - \frac{1}{k}$). Assigning such an item a weight that is $\frac{k}{k-1}$ times its size will allow one to show that all bins except for a constant number of bins (at most $k-1$ bins) have total weights of at least 1. It is possible to show that the total weight of any packed bin is sufficiently close to Π_∞ for large values of k. As both $HARM_k$ and an optimal solution pack the same items, the competitive ratio is implied. To show the upper bound on the total weight of any packed bin, it is required to show that the worst-case bin contains exactly one item of size just above $\frac{1}{\pi_j + 1}$ for $\pi_j \leq k - 1$ (and the remaining space can only contain items of type k). Roughly speaking, this holds as once it was proved that the bin contains the largest such items, the largest possible additional weight can be obtained only by adding the next such item.

Proving that no better bounded space algorithms exist can be done as follows. Let j' be a fixed integer. Let N be a large integer and consider a sequence containing N items of each size $\frac{1}{\pi_j} + \delta$ for a sufficiently small $\delta > 0$, for any $j = j', j' - 1, \cdots, 1$. If δ is chosen appropriately, we have $\sum_{j=1}^{j'} \frac{1}{\pi_j + 1} + j'\delta < 1$, so the items can be packed (off-line) into N bins. However, if items are presented in this order (sorted by nondecreasing size), after all items of one size have been presented, only a constant number of bins can receive larger items, and thus the items of each size are packed almost independently.

Related Results

The space of a bounded space algorithm is the number of open bins that it can have. The space of NF is 1, while the space of harmonic algorithms increases with k. A bounded space algorithm with space 2 and the same asymptotic competitive ratio as FF and BF have been designed [3] (for comparison, $HARM_3$ has an asymptotic competitive ratio of $\frac{7}{4}$). A modification where smaller space is used to obtain the same competitive ratios of harmonic algorithms (or alternatively, smaller competitive ratios were obtained using the same space) was designed by Woeginger [15]. Thus, there exists another sequence of bounded space algorithms, with an increasing sequence of open bins, where their sequence of competitive ratios tends to Π_∞ such that the space required for every competitive ratio is much smaller than that of [7].

One drawback of the model above is that an off-line algorithm can rearrange the items and does not have to process them as a sequence. The variant where it must process them in the same order as an online algorithm was studied as well [2]. Algorithms that are based on partitioning into classes and have smaller asymptotic competitive ratios (but they are obviously not bounded space) were designed [7, 9, 11].

Generalizations have been studied too, in particular, bounded space bin packing with cardinality constraints (where an item cannot receive more than t items for a fixed integer $t \geq 2$) [5], parametric bin packing (where there is an upper bound strictly smaller than 1 on item sizes) [14], bin packing with rejection (where an item i has a rejection penalty r_i associated with it, and it can be either packed, or rejected for the cost r_i) [6], variable-sized bin packing (where bins of multiple sizes are available for packing) [10], and bin packing with resource augmentation (where the online algorithm can use bins of size $b > 1$ for a fixed rational number b, while an off-line algorithm still uses bins of size 1) [4]. In this last variant, the sequences of critical item sizes were redefined as a function of b, while variable-sized bin packing required a more careful partition into intervals.

Cross-References

▸ Current Champion for Online Bin Packing

Recommended Reading

1. Balogh J, Békési J, Galambos G (2012) New lower bounds for certain classes of bin packing algorithms. Theor Comput Sci 440–441:1–13
2. Chrobak M, Sgall J, Woeginger GJ (2011) Two-bounded-space bin packing revisited. In: Proceedings of the 19th annual European symposium on algorithms (ESA2011), Saarbrücken, Germany, pp 263–274
3. Csirik J, Johnson DS (2001) Bounded space on-line bin packing: best is better than first. Algorithmica 31:115–138
4. Csirik J, Woeginger GJ (2002) Resource augmentation for online bounded space bin packing. J Algorithms 44(2):308–320
5. Epstein L (2006) Online bin packing with cardinality constraints. SIAM J Discret Math 20(4):1015–1030
6. Epstein L (2010) Bin packing with rejection revisited. Algorithmica 56(4):505–528
7. Lee CC, Lee DT (1985) A simple online bin packing algorithm. J ACM 32(3):562–572
8. Liang FM (1980) A lower bound for on-line bin packing. Inf Process Lett 10(2):76–79
9. Ramanan P, Brown DJ, Lee CC, Lee DT (1989) Online bin packing in linear time. J Algorithms 10:305–326
10. Seiden SS (2001) An optimal online algorithm for bounded space variable-sized bin packing. SIAM J Discret Math 14(4):458–470
11. Seiden SS (2002) On the online bin packing problem. J ACM 49(5):640–671
12. Ullman JD (1971) The performance of a memory allocation algorithm. Technical report 100, Princeton University, Princeton
13. van Vliet A (1992) An improved lower bound for online bin packing algorithms. Inf Process Lett 43(5):277–284
14. van Vliet A (1996) On the asymptotic worst case behavior of Harmonic Fit. J Algorithms 20(1):113–136
15. Woeginger GJ (1993) Improved space for bounded-space online bin packing. SIAM J Discret Math 6(4):575–581

Hierarchical Self-Assembly

David Doty
Computing and Mathematical Sciences,
California Institute of Technology, Pasadena,
CA, USA

Keywords

Hierarchical assembly; Intrinsic universality; Running time; Self-assembly; Verification

Years and Authors of Summarized Original Work

2005; Aggarwal, Cheng, Goldwasser, Kao, Espanes, Schweller
2012; Chen, Doty
2013; Cannon, Demaine, Demaine, Eisenstat, Patitz, Schweller, Summers, Winslow

Problem Definition

The general idea of *hierarchical* self-assembly (a.k.a., *multiple tile* [2], *polyomino* [8, 10], *two-handed* [3, 5, 6]) is to model self-assembly of tiles in which attachment of two multi-tile assemblies is allowed, as opposed to all attachments being that of a single tile onto a larger assembly. Several problems concern comparing hierarchical self-assembly to its single-tile-attachment variant (called the "seeded" model of self-assembly), so we define both models here. The model of hierarchical self-assembly was first defined (in a slightly different form that restricted the size of assemblies that could attach) by Aggarwal, Cheng, Goldwasser, Kao, Moisset de Espanes, and Schweller [2]. Several generalizations of the model exist that incorporated staged mixing of test tubes, "dissolvable" tiles, active signaling across tiles, etc., but here we restrict attention to the model closest to the seeded model of Winfree [9], different from that model only in the absence of a seed and the ability of two large assemblies to attach.

Definitions

A *tile type* is a unit square with four sides, each consisting of a *glue label* (often represented as a finite string) and a nonnegative integer *strength*. We assume a finite set T of tile types, but an infinite number of copies of each tile type, each copy referred to as a *tile*. An *assembly* is a positioning of tiles on the integer lattice $\mathbb{Z}^2$, i.e., a partial function $\alpha : \mathbb{Z}^2 \dashrightarrow T$. We write $|\alpha|$ to denote $|\text{dom}\ \alpha|$. Write $\alpha \sqsubseteq \beta$ to denote that α is a *subassembly* of β, which means that $\text{dom}\ \alpha \subseteq \text{dom}\ \beta$ and $\alpha(p) = \beta(p)$ for all points $p \in \text{dom}\ \alpha$. We abuse notation and take a tile type t to be equivalent to the single-tile assembly containing only t (at the origin if not otherwise specified). Two adjacent tiles in an assembly *interact* if the glue labels on their abutting sides are equal and have positive strength. Each assembly induces a *binding graph*, a grid graph whose vertices are tiles, with an edge between two tiles if they interact. The assembly is τ-*stable* if every cut of its binding graph has strength at least τ, where the weight of an edge is the strength of the glue it represents. That is, the assembly is stable if at least energy τ is required to separate the assembly into two parts.

We now define both the seeded and hierarchical variants of the tile assembly model. A *seeded tile system* is a triple $\mathcal{T} = (T, \sigma, \tau)$, where T is a finite set of tile types, $\sigma : \mathbb{Z}^2 \dashrightarrow T$ is a finite, τ-stable *seed assembly*, and τ is the *temperature*. If $\mathcal{T}$ has a single seed tile $s \in T$ (i.e., $\sigma(0,0) = s$ for some $s \in T$ and is undefined elsewhere), then we write $\mathcal{T} = (T, s, \tau)$. Let $|\mathcal{T}|$ denote $|T|$. An assembly α is *producible* if either $\alpha = \sigma$ or if β is a producible assembly and α can be obtained from β by the stable binding of a single tile. In this case, write $\beta \to_1 \alpha$ (α is producible from β by the attachment of one tile), and write $\beta \to \alpha$ if $\beta \to_1^* \alpha$ (α is producible from β by the attachment of zero or more tiles). An assembly is *terminal* if no tile can be τ-stably attached to it.

A *hierarchical tile system* is a pair $\mathcal{T} = (T, \tau)$, where T is a finite set of tile types and $\tau \in \mathbb{N}$

Supported by NSF grants CCF-1219274, CCF-1162589, and 1317694.

is the temperature. An assembly is *producible* if either it is a single tile from T or it is the τ-stable result of translating two producible assemblies without overlap. Therefore, if an assembly α is producible, then it is produced via an *assembly tree*, a full binary tree whose root is labeled with α, whose $|\alpha|$ leaves are labeled with tile types, and each internal node is a producible assembly formed by the stable attachment of its two child assemblies. An assembly α is *terminal* if for every producible assembly β, α and β cannot be τ-stably attached. If α can grow into β by the attachment of zero or more assemblies, then we write $\alpha \rightarrow \beta$.

In either model, let $\mathcal{A}[\mathcal{T}]$ be the set of producible assemblies of $\mathcal{T}$, and let $\mathcal{A}_{\Box}[\mathcal{T}] \subseteq \mathcal{A}[\mathcal{T}]$ be the set of producible, terminal assemblies of $\mathcal{T}$. A TAS $\mathcal{T}$ is *directed* (a.k.a., *deterministic, confluent*) if $|\mathcal{A}_{\Box}[\mathcal{T}]| = 1$. If $\mathcal{T}$ is directed with unique producible terminal assembly α, we say that $\mathcal{T}$ *uniquely produces* α. It is easy to check that in the seeded aTAM, $\mathcal{T}$ uniquely produces α if and only if every producible assembly $\beta \sqsubseteq \alpha$. In the hierarchical model, a similar condition holds, although it is more complex since hierarchical assemblies, unlike seeded assemblies, do not have a "canonical translation" defined by the seed position. $\mathcal{T}$ uniquely produces α if and only if for every producible assembly β, there is a translation β' of β such that $\beta' \sqsubseteq \alpha$. In particular, if there is a producible assembly $\beta \neq \alpha$ such that dom α = dom β, then α is not uniquely produced. Since dom β = dom α, every nonzero translation of β has some tiled position outside of dom α, whence no such translation can be a subassembly of α, implying α is not uniquely produced.

Power of Hierarchical Assembly Compared to Seeded

One sense in which we can conclude that one model of computation M is at least as powerful as another model of computation M' is to show that any machine defined by M' can be "simulated efficiently" by a machine defined by M. In self-assembly, there is a natural definition of what it means for one tile system $\mathcal{S}$ to "simulate" another $\mathcal{T}$. We now discuss intuitively how to define such a notion. There are several intricacies to the full formal definition that are discussed in further detail in [3, 5].

First, we require that there is a constant $k \in \mathbb{Z}^+$ (the "resolution loss") such that each tile type t in $\mathcal{T}$ is "represented" by one or more $k \times k$ blocks β of tiles in $\mathcal{S}$. In this case, we write $r(\beta) = t$, where $\beta : \{1, \ldots, k\}^2 \dashrightarrow S$ and S is the tile set of $\mathcal{S}$. Then β represents a $k \times k$ block of such tiles, possibly with empty positions at points $\mathbf{x}$ where $\beta(\mathbf{x})$ is undefined. We call such a $k \times k$ block in $\mathcal{S}$ a "macrotile." We can extend r to a function R that, given an assembly $\alpha_{\mathcal{S}}$ partitioned into $k \times k$ macrotiles, outputs an assembly $\alpha_{\mathcal{T}}$ of $\mathcal{T}$ such that, for each macrotile β of $\alpha_{\mathcal{S}}$, $r(\beta) = t$, where t is the tile type at the corresponding position in $\alpha_{\mathcal{T}}$.

Given such a representation function R indicating how to interpret assemblies of $\mathcal{S}$ as representing assemblies of $\mathcal{T}$, we now define what it means to say that $\mathcal{S}$ *simulates* $\mathcal{T}$. For each producible assembly $\alpha_{\mathcal{T}}$ of $\mathcal{T}$, there is a producible assembly $\alpha_{\mathcal{S}}$ of $\mathcal{S}$ such that $R(\alpha_{\mathcal{S}}) = \mathcal{T}$, and furthermore, for every producible assembly $\alpha_{\mathcal{S}}$, if $R(\alpha_{\mathcal{S}}) = \mathcal{T}$, then $\mathcal{T}$ is producible in $\mathcal{T}$. Finally, we require that R respects the "single attachment" dynamics of $\mathcal{T}$: there is a single tile that can be attached to $\alpha_{\mathcal{T}}$ to result in $\alpha'_{\mathcal{T}}$ if and only if there is some sequence of attachments to $\alpha_{\mathcal{S}}$ that results in assembly $\alpha'_{\mathcal{S}}$ such that $R(\alpha'_{\mathcal{S}}) = \alpha'_{\mathcal{T}}$.

With such an idea in mind, we can ask, "Is the hierarchical model at least as powerful as the seeded model?"

Problem 1 For every seeded tile system $\mathcal{T}$, design a hierarchical tile system $\mathcal{S}$ that simulates $\mathcal{T}$.

Another interpretation of a solution to Problem 1 is that, to the extent that the hierarchical model is more realistic than the seeded model by incorporating the reality that tiles may aggregate even in the absence of a seed, such a solution shows how to enforce seeded growth even in such an unfriendly environment that permits non-seeded growth.

Assembly Time

We now define time complexity for hierarchical systems (this definition first appeared in [4], where it is explained in more detail). We treat each assembly as a single molecule. If two assemblies α and β can attach to create an assembly γ, then we model this as a chemical reaction $\alpha + \beta \rightarrow \gamma$, in which the rate constant is assumed to be equal for all reactions (and normalized to 1). In particular, if α and β can be attached in two different ways, this is modeled as two different reactions, even if both result in the same assembly.

At an intuitive level, the model we define can be explained as follows. We imagine dumping all tiles into solution at once, and at the same time, we grab one particular tile and dip it into the solution as well, pulling it out of the solution when it has assembled into a terminal assembly. Under the seeded model, the tile we grab will be a seed, assumed to be the only copy in solution (thus requiring that it appears only once in any terminal assembly). In the seeded model, no reactions occur other than the attachment of individual tiles to the assembly we are holding. In the hierarchical model, other reactions are allowed to occur in the background (we model this using the standard mass-action model of chemical kinetics [7]), but only those reactions with the assembly we are holding move it "closer" to completion. The other background reactions merely change concentrations of other assemblies (although these indirectly affect the time it will take our chosen assembly to complete, by changing the rate of reactions with our chosen assembly).

More formally, let $\mathcal{T} = (T, \tau)$ be a hierarchical TAS, and let $\rho : T \rightarrow [0, 1]$ be a concentrations function, giving the *initial* concentration of each tile type (we require that $\sum_{t \in T} \rho(t) = 1$, a condition known as the "finite density constraint"). Let $\mathbb{R}^+ = [0, \infty)$, and let $t \in \mathbb{R}^+$. For $\alpha \in \mathcal{A}[\mathcal{T}]$, let $[\alpha]_\rho(t)$ (abbreviated $[\alpha](t)$ when ρ is clear from context) denote the concentration of α at time t with respect to initial concentrations ρ, defined as follows. Given two assemblies α and β that can attach to form γ, we model this event as a chemical reaction $R : \alpha + \beta \rightarrow \gamma$. Say that a reaction $\alpha + \beta \rightarrow \gamma$ is *symmetric* if $\alpha = \beta$. Define the *propensity* (a.k.a., *reaction rate*) of R at time $t \in \mathbb{R}^+$ to be $\rho_R(t) = [\alpha](t) \cdot [\beta](t)$ if R is not symmetric and $\rho_R(t) = \frac{1}{2} \cdot [\alpha](t)^2$ if R is symmetric.

If α is consumed in reactions $\alpha + \beta_1 \rightarrow \gamma_1, \ldots, \alpha + \beta_n \rightarrow \gamma_n$ and produced in asymmetric reactions $\beta'_1 + \gamma'_1 \rightarrow \alpha, \ldots, \beta'_m + \gamma'_m \rightarrow \alpha$ and symmetric reactions $\beta''_1 + \beta''_1 \rightarrow \alpha, \ldots, \beta''_p + \beta''_p \rightarrow \alpha$, then the concentration $[\alpha](t)$ of α at time t is described by the differential equation:

$$\frac{d[\alpha](t)}{dt} = \sum_{i=1}^{m} [\beta'_i](t) \cdot [\gamma'_i](t) + \sum_{i=1}^{p} \frac{1}{2} \cdot [\beta''_i](t)^2 - \sum_{i=1}^{n} [\alpha](t) \cdot [\beta_i](t), \tag{1}$$

with boundary conditions $[\alpha](0) = \rho(r)$ if α is an assembly consisting of a single tile r and $[\alpha](0) = 0$ otherwise. In other words, the propensities of the various reactions involving α determine its rate of change, negatively if α is consumed and positively if α is produced.

This completes the definition of the dynamic evolution of concentrations of producible assemblies; it remains to define the time complexity of assembling a terminal assembly. Although we have distinguished between seeded and hierarchical systems, for the purpose of defining a model of time complexity in hierarchical systems and comparing them to the seeded system time complexity model of [1], it is convenient to introduce a seedlike "timekeeper tile" into the hierarchical system, in order to stochastically analyze the growth of this tile when it reacts in a solution that is itself evolving according to the continuous model described above. The seed does not have the purpose of nucleating growth but is introduced merely to focus attention on a single molecule that has not yet assembled anything, in order to ask how long it will take to assemble into a terminal assembly. The choice of which tile type to pick will be a parameter of the definition, so that a system may have different assembly times depending on the choice of timekeeper tile.

Fix a copy of a tile type s to designate as a "timekeeper seed." The assembly of s into some terminal assembly $\hat{\alpha}$ is described as a time-dependent continuous-time Markov process in which each state represents a producible assembly containing s, and the initial state is the size-1 assembly with only s. For each state α representing a producible assembly with s at the origin, and for each pair of producible assemblies β, γ such that $\alpha + \beta \to \gamma$ (with the translation assumed to happen only to β so that α stays "fixed" in position), there is a transition in the Markov process from state α to state γ with transition rate $[\beta](t)$.

We define $\mathbf{T}_{\mathcal{T},\rho,s}$ to be the random variable representing the time taken for the copy of s to assemble into a terminal assembly via some sequence of reactions as defined above. We define the time complexity of a directed hierarchical TAS $\mathcal{T}$ with concentrations ρ and timekeeper s to be $\mathsf{T}(\mathcal{T}, \rho, s) = \mathrm{E}\left[\mathbf{T}_{\mathcal{T},\rho,s}\right]$.

For a shape $S \subset \mathbb{Z}^2$ (finite and connected), define the *diameter* of S to be $\mathrm{diam}(S) = \max_{\mathbf{u},\mathbf{v} \in S} \|\mathbf{u} - \mathbf{v}\|_1$, where $\|\mathbf{w}\|_1$ is the L_1 norm of $\mathbf{w}$.

Problem 2 Design a hierarchical tile system $\mathcal{T} = (T, \tau)$ such that every producible terminal assembly $\hat{\alpha}$ has the same shape S, and for some $s \in T$ and concentrations function $\rho : T \to [0, 1]$, $\mathsf{T}(\mathcal{T}, \rho, s) = o(\mathrm{diam}(S))$.

It is provably impossible to achieve this with the seeded model [1, 4], since all assemblies in that model require expected time at least proportional to their diameter.

Key Results

Power of Hierarchical Assembly Compared to Seeded

Cannon, Demaine, Demaine, Eisenstat, Patitz, Schweller, Summers, and Winslow [3] showed a solution to Problem 1. (They also showed several other ways in which the hierarchical model is more powerful than the seeded model, but we restrict attention to simulation here.) For the most part, temperature 2 seeded systems are as powerful as those at higher temperatures, but the simulation results of [3] hold for higher temperatures as well. In particular, they showed that every seeded temperature ≥ 4 tile system $\mathcal{T}$ can be simulated by a hierarchical temperature 4 tile system (as well as showing it is possible for temperature τ hierarchical tile systems to simulate temperature τ seeded tile systems for $\tau \in \{2, 3\}$, using similar logic to the higher-temperature construction). The definition of simulation has a parameter k indicating the resolution loss of the simulation. In fact, the simulation described in [3] requires only resolution loss $k = 5$.

Figure 1 shows an example of $\mathcal{S}$ simulating $\mathcal{T}$. The construction enforces the "simulation of dynamics" constraint that if and only if a single tile can attach in $\mathcal{T}$, and then a 5×5 macrotile representing it in $\mathcal{S}$ can assemble. It is critical that each tile type in $\mathcal{T}$ is represented by *more than one* type of macrotile in $\mathcal{S}$: each different type of macrotile represents a different subset of sides that can cooperate to allow the tile to bind. To achieve this, each macrotile consists of a central "brick" (itself a 3×3 block composed of 9 unique tile types with held together with strength-4 glues) surrounded by "mortar" (forming a ring around the central brick). Figure 1 shows "mortar rectangles" but, similarly to the brick, these are just 3×1 assemblies of 3 individual tile types with strength-4 glues. The logic of the system is such that if a brick B designed for a subset of cooperating sides $C \subseteq \{\mathsf{N}, \mathsf{S}, \mathsf{E}, \mathsf{W}\}$, then only if the mortar for all sides in C is present can B attach. Its attachment is required to fill in the remaining mortar representing the other sides in $\{\mathsf{N}, \mathsf{S}, \mathsf{E}, \mathsf{W}\} \setminus C$ that may not be present. Finally, those tiles enable the assembly of mortar in *adjacent* 5×5 blocks, to be ready for possible cooperation to bind bricks in those blocks.

Assembly Time

Chen and Doty [4] showed a solution to Problem 2, by proving that for infinitely many $n \in \mathbb{N}$, there is a (non-directed) hierarchical TAS

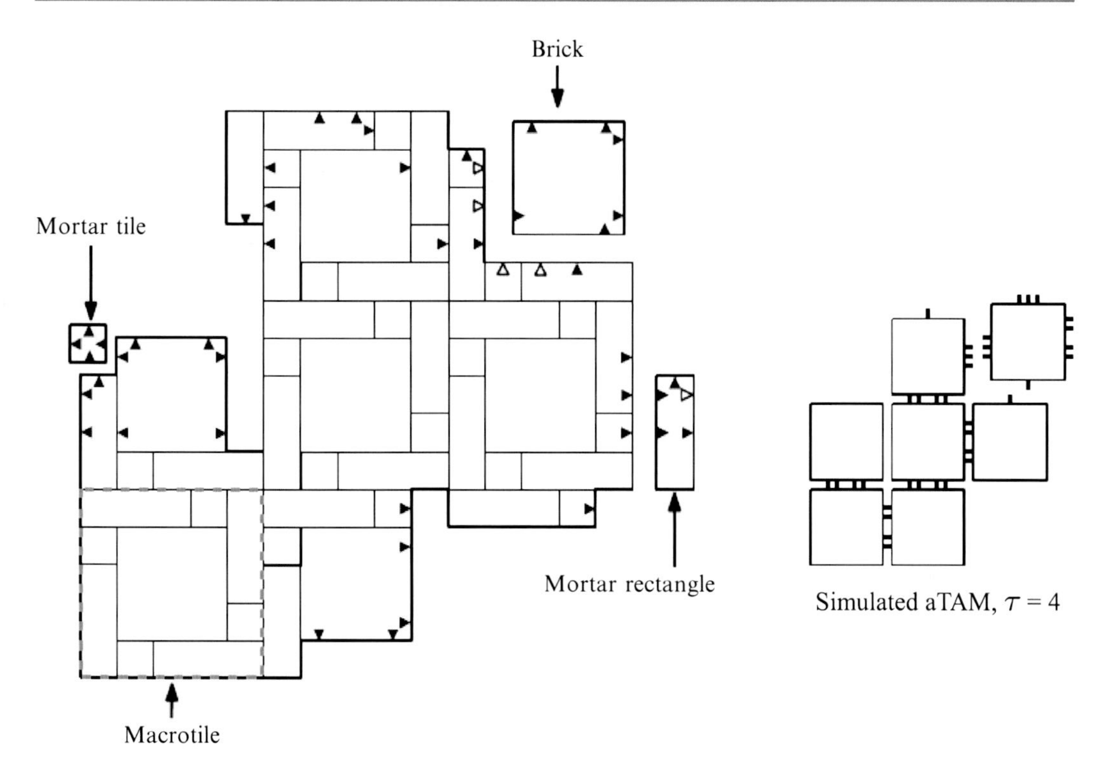

Hierarchical Self-Assembly, Fig. 1 Simulation of a seeded tile system $\mathcal{T}$ of temperature ≥ 4 by a hierarchical tile system $\mathcal{S}$ of temperature 4 (Figure taken from [3]). *Filled arrows* represent glues of strength 2, and *unfilled arrows* represent glues of strength 1. In the seeded tile system, the number of *dashes* on the side of a tile represent its strength

$\mathcal{T} = (T, 2)$ that strictly self-assembles an $n \times n'$ rectangle S, where $n' = o(n)$ (hence $\text{diam}(S) = \Theta(n)$), such that $|T| = O(\log n)$ and there is a tile type $s \in T$ and concentrations function $\rho : T \to [0, 1]$ such that $\mathsf{T}(\mathcal{T}, \rho, s) = O(n^{4/5} \log n)$.

The construction consists of $m = n^{1/5}$ stages shown in Fig. 2, where each stage consists of the attachment of two "horizontal bars" to a single "vertical bar" as shown in Fig. 3. The vertical bar of the next stage then attaches to the right of the two horizontal bars, which cooperate to allow the binding because they each have a single strength 1 glue. All vertical bars are identical when they attach, but attachment triggers the growth of some tiles (shown in orange in Figs. 2 and 3) that make the attachment sites on the right side different from their locations in the previous stage, which is how the stages "count down" from m to 1.

The bars themselves are assembled in a "standard" way that requires time linear in the diameter of the bar, which is $w = n^{4/5}$ for a horizontal bar and $mk^2 = n^{3/5}$ (where k is a parameter that we set to be $n^{1/5}$) for a vertical bar. The speedup comes from the fact that each horizontal bar can attach to one of k different binding sites on a vertical bar, so the expected time for this to happen is factor k lower than if there were only a single binding site. The vertical "arm" on the left of each horizontal bar has the purpose of preventing any other horizontal bars from binding near it. Each stage also requires filler tiles to fill in the gap regions, but the time required for this is negligible compared to the time for all vertical and horizontal bars to attach.

Note that this construction is not directed: although every producible terminal assembly has the shape of an $n \times n'$ rectangle, there are many such terminal assemblies. Chen and Doty [4] also showed that for a class of directed systems called "partial order tile systems," no solution to Problem 2 exists: provably any such tile system

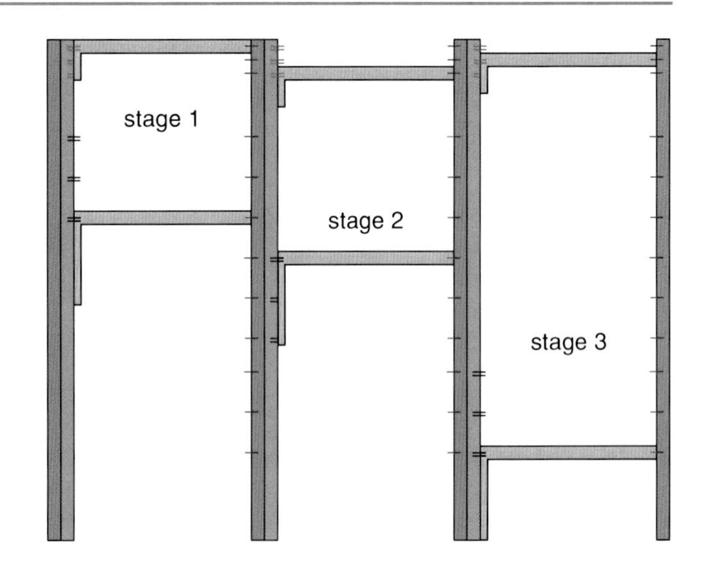

Hierarchical Self-Assembly, Fig. 2 High-level overview of interaction of "vertical bars" and "horizontal bars" to create the rectangle in the solution to Problem 2 that assembles in time sublinear in its diameter. Filler tiles fill in the empty regions. If glues overlap two regions then represent a formed bond. If glues overlap one region but not another, they are glues from the former region but are mismatched (and thus "covered and protected") by the latter region

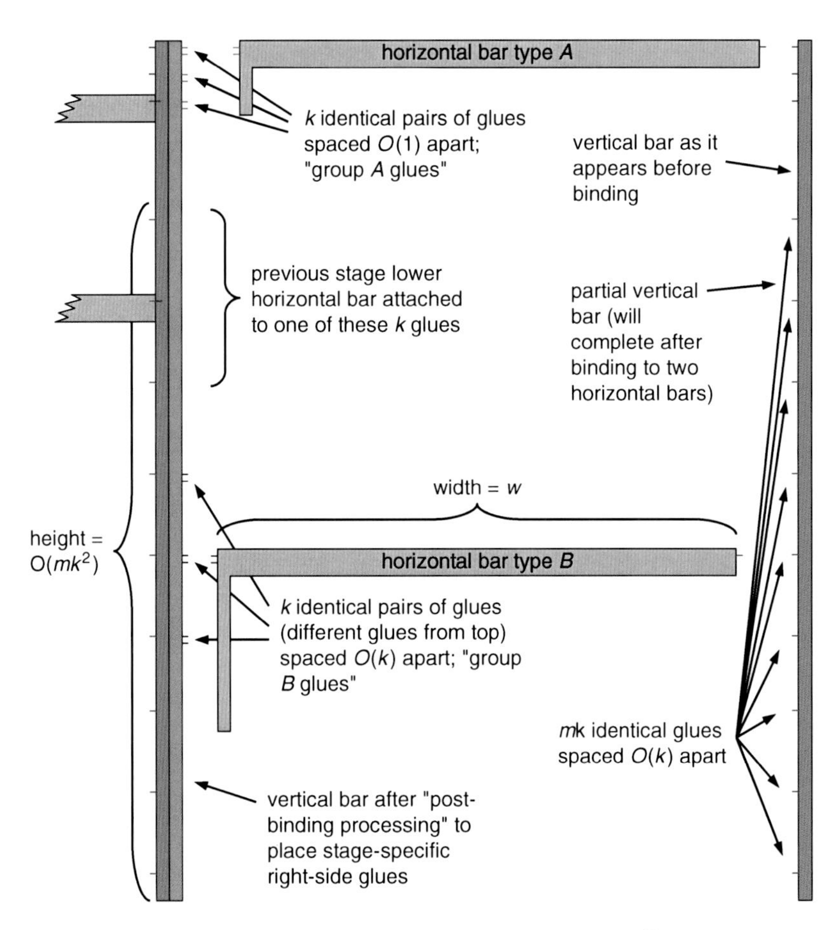

Hierarchical Self-Assembly, Fig. 3 "Vertical bars" for the construction of a fast-assembling square, and their interaction with horizontal bars, as shown for a single stage of Fig. 2. "Type B" horizontal bars have a longer vertical arm than "Type A" since the glues they must block are farther apart

assembling a shape of diameter d requires expected time $\Omega(d)$.

Open Problems

It is known [2] that the tile complexity of assembling an $n \times k$ rectangle in the seeded aTAM, if $k < \frac{\log n}{\log \log n - \log \log \log n}$, is asymptotically lower bounded by $\Omega\left(\frac{n^{1/k}}{k}\right)$ and upper bounded by $O(n^{1/k})$. For the hierarchical model, the upper bound holds as well [2], but the strongest known lower bound is the information-theoretic $\Omega\left(\frac{\log n}{\log \log n}\right)$.

Question 1 What is the tile complexity of assembling an $n \times k$ rectangle in the hierarchical model, when $k < \frac{\log n}{\log \log n - \log \log \log n}$?

Cross-References

- ▸ Experimental Implementation of Tile Assembly
- ▸ Patterned Self-Assembly Tile Set Synthesis
- ▸ Robustness in Self-Assembly
- ▸ Self-Assembly at Temperature 1
- ▸ Self-Assembly of Fractals
- ▸ Self-Assembly with General Shaped Tiles
- ▸ Staged Assembly
- ▸ Temperature Programming in Self-Assembly

Recommended Reading

1. Adleman LM, Cheng Q, Goel A, Huang M-D (2001) Running time and program size for self-assembled squares. In: STOC 2001: proceedings of the thirty-third annual ACM symposium on theory of computing, Hersonissos. ACM, pp 740–748
2. Aggarwal G, Cheng Q, Goldwasser MH, Kao M-Y, Moisset de Espanés P, Schweller RT (2005) Complexities for generalized models of self-assembly. SIAM J Comput 34:1493–1515. Preliminary version appeared in SODA 2004
3. Cannon S, Demaine ED, Demaine ML, Eisenstat S, Patitz MJ, Schweller RT, Summers SM, Winslow A (2013) Two hands are better than one (up to constant factors). In: STACS 2013: proceedings of the thirtieth international symposium on theoretical aspects of computer science, Kiel, pp 172–184
4. Chen H-L, Doty D (2012) Parallelism and time in hierarchical self-assembly. In: SODA 2012: proceedings of the 23rd annual ACM-SIAM symposium on discrete algorithms, Kyoto, pp 1163–1182
5. Demaine ED, Patitz MJ, Rogers T, Schweller RT, Summers SM, Woods D (2013) The two-handed tile assembly model is not intrinsically universal. In: ICALP 2013: proceedings of the 40th international colloquium on automata, languages and programming, Riga, July 2013
6. Doty D, Patitz MJ, Reishus D, Schweller RT, Summers SM (2010) Strong fault-tolerance for self-assembly with fuzzy temperature. In: FOCS 2010: proceedings of the 51st annual IEEE symposium on foundations of computer science, Las Vegas, pp 417–426
7. Epstein IR, Pojman JA (1998) An introduction to nonlinear chemical dynamics: oscillations, waves, patterns, and chaos. Oxford University Press, Oxford
8. Luhrs C (2010) Polyomino-safe DNA self-assembly via block replacement. Nat Comput 9(1):97–109. Preliminary version appeared in DNA 2008
9. Winfree E (1998) Algorithmic self-assembly of DNA. PhD thesis, California Institute of Technology, June 1998
10. Winfree E (2006) Self-healing tile sets. In: Chen J, Jonoska N, Rozenberg G (eds) Nanotechnology: science and computation. Natural computing series. Springer, Berlin/New York, pp 55–78

Hierarchical Space Decompositions for Low-Density Scenes

Mark de Berg
Department of Mathematics and Computer Science, TU Eindhoven, Eindhoven, The Netherlands

Keywords

Binary space partitions; Compressed quadtrees; Computational geometry; Hierarchical space decompositions; Realistic input models

Years and Authors of Summarized Original Work

2000; De Berg
2010; De Berg, Haverkort, Thite, Toma

Problem Definition

Many algorithmic problems on spatial data can be solved efficiently if a suitable decomposition of the ambient space is available. Two desirable properties of the decomposition are that its cells have a nice shape – convex and/or of constant complexity – and that each cell intersects only a few objects from the given data set. Another desirable property is that the decomposition is hierarchical, meaning that the space is partitioned in a recursive manner. Popular hierarchical space decompositions include quadtrees and binary space partitions.

When the objects in the given data set are nonpoint objects, they can be fragmented by the partitioning process. This fragmentation has a negative impact on the storage requirements of the decomposition and on the efficiency of algorithms operating on it. Hence, it is desirable to minimize fragmentation. In this chapter, we describe methods to construct linear-size compressed quadtrees and binary space partitions for so-called low-density sets. To simplify the presentation, we describe the constructions in the plane. We use S to denote the set of n objects for which we want to construct a space decomposition and assume for simplicity that the objects in S are disjoint, convex, and of nonzero area.

Binary Space Partitions

A *binary space partition* for a set S of n objects in the plane is a recursive decomposition of the plane by lines, typically such that each cell in the final decomposition intersects only a few objects from S. The tree structure modeling this decomposition is called a *binary space partition tree*, or BSP *tree* for short – see Fig. 1 for an illustration. Thus, a BSP tree $\mathcal{T}$ for S can be defined as follows.

- If a predefined stopping criterion is met – often this is when $|S|$ is sufficiently small – then $\mathcal{T}$ consists of a single leaf where the set S is stored.
- Otherwise the root node v of $\mathcal{T}$ stores a suitably chosen *splitting line* ℓ. Let ℓ^- and ℓ^+ denote the half-planes lying to the left and to the right of ℓ, respectively (or, if ℓ is horizontal, below and above ℓ).
  - The left subtree of v is a BSP tree for $S^- := \{o \cap \ell^- : o \in S\}$, the set of object fragments lying in the half-plane ℓ^-.
  - The right subtree of v is a BSP tree for $S^+ := \{o \cap \ell^+ : o \in S\}$, the set of object fragments lying in the half-plane ℓ^+.

The *size* of a BSP tree is the total number of object fragments stored in the tree.

Compressed Quadtrees

Let $U = [0, 1]^2$ be the unit square. We say that a square $\sigma \subseteq U$ is a *canonical square* if there is an integer $k \geqslant 0$ such that σ is a cell of the regular subdivision of U into $2^k \times 2^k$ squares. A *donut* is the set-theoretic difference $\sigma_{\text{out}} \setminus \sigma_{\text{in}}$ of a canonical square σ_{out} and a canonical square $\sigma_{\text{in}} \subset \sigma_{\text{out}}$. A *compressed quadtree* $\mathcal{T}$ for a set P of points inside a canonical square σ defined as follows; see also Fig. 2 (middle).

- If a predefined stopping criterion is met – usually this is when $|P|$ is sufficiently small – then $\mathcal{T}$ consists of a single leaf storing the set P.
- If the stopping criterion is not met, then $\mathcal{T}$ is defined as follows. Let σ_{NE} denote the northeast quadrant of σ and let $P_{\text{NE}} := P \cap \sigma_{\text{NE}}$. Define σ_{SE}, σ_{SW}, σ_{NW} and P_{SE}, P_{SW}, P_{NW} similarly for the other three quadrants. (Here we should make sure that points on the boundary between quadrants are assigned to quadrants in a consistent manner.) Now $\mathcal{T}$ consists of a root node v with four or two children, depending on how many of the sets P_{NE}, P_{SE}, P_{SW}, P_{NW} are nonempty:
  - If at least two of the sets P_{NE}, P_{SE}, P_{SW}, P_{NW} are nonempty, then v has four children v_{NE}, v_{SE}, v_{SW}, v_{NW}. The child v_{NE} is the root of a compressed quadtree for the set P_{NE} inside the square σ_{NW}; the other three children are defined similarly for the point sets inside the other quadrants.
  - If only one of P_{NE}, P_{SE}, P_{SW}, P_{NW} is nonempty, then v has two children v_{in} and v_{out}. The child v_{in} is the root of a

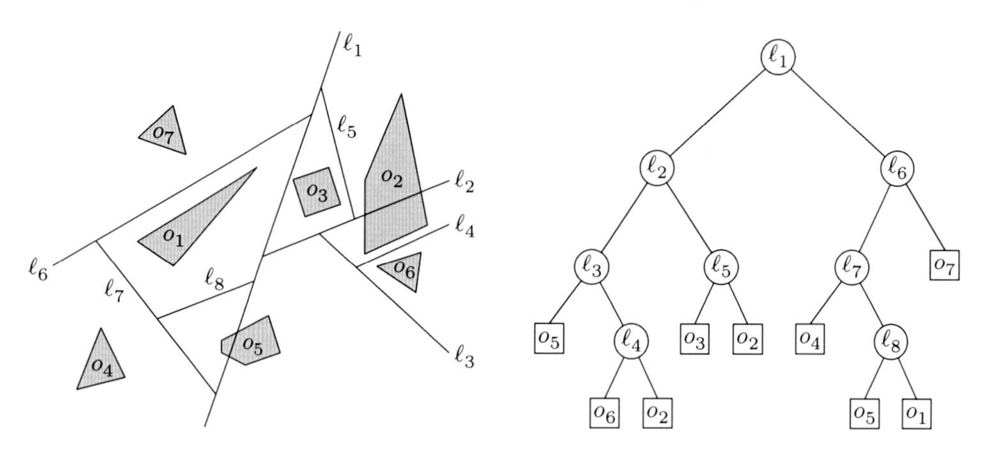

Hierarchical Space Decompositions for Low-Density Scenes, Fig. 1 A binary space partition for a set of polygons (*left*) and the corresponding BSP tree (*right*)

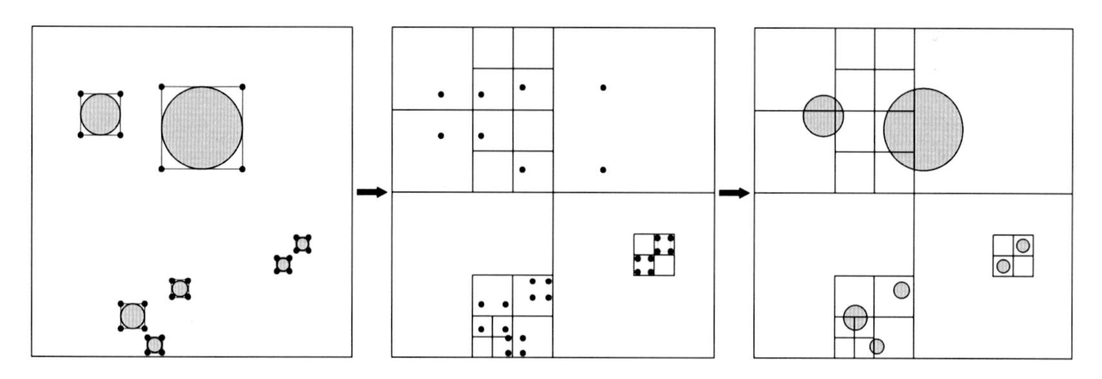

Hierarchical Space Decompositions for Low-Density Scenes, Fig. 2 Construction of a compressed quadtree for a set of disks: take the bounding-box vertices (*left*), construct a compressed quadtree for the vertices (*middle*), and put the disks back in (*right*)

compressed quadtree for P inside σ_{in}, where σ_{in} is the smallest canonical square containing all points from P. The other child is a leaf corresponding to the donut $\sigma \setminus \sigma_{\text{in}}$.

A compressed quadtree for a set of n points has size $O(n)$.

Above we defined compressed quadtrees for point sets. In this chapter, we are interested in compressed quadtrees for nonpoint objects. These are defined similarly: each internal node corresponds to a canonical square, and each leaf is a canonical square or a donut. This time donuts need not be empty, but may intersect objects (although not too many). The right picture in Fig. 2 shows a compressed quadtree for a set of disks. The *size* of a compressed quadtree for a set of nonpoint objects is defined as the total number of object fragments stored in the tree. Because nonpoint objects may be split into fragments during the subdivision process, a compressed quadtree for nonpoint objects is not guaranteed to have linear size.

Low-Density Scenes

The main question we are interested in is the following: given a set S of n objects, can we construct a compressed quadtree or BSP tree with $O(n)$ leaves such that each leaf region intersects $O(1)$ objects? In general, the answer to this question is no. For compressed quadtrees, this can be seen by considering a set S of slanted parallel segments that are very close together. A linear-size BSP tree cannot be guaranteed either: there are sets of n disjoint segments in the plane for which any BSP tree has size $\Omega(n \log n / \log \log n)$ [7]. In $\mathbb{R}^3$ the situation is even worse: there are sets of n disjoint triangles for which any BSP tree has size $\Omega(n^2)$ [5]. (Both bounds are tight: there are algorithms that guarantee a BSP tree of size $O(n \log n / \log \log n)$ in the plane [8] and of size $O(n^2)$ in $\mathbb{R}^3$ [6].) Fortunately, in practice, the objects for which we want to construct a space decomposition are often distributed nicely, which allows us to construct much smaller decompositions than for the worst-case examples mentioned above. To formalize this, we define the concept of *density* of a set of objects in $\mathbb{R}^d$.

Definition 1 The *density* of a set S of objects in $\mathbb{R}^d$, denoted density(S), is defined as the smallest number λ such that the following holds: any ball $b \subset \mathbb{R}^d$ intersects at most λ objects $o \in S$ such that diam$(o) \geq$ diam(b), where diam$(\cdot)$ denotes the diameter of an object.

As illustrated in Fig. 3(i), a set of n parallel segments can have density n if the segments are very close together. In most practical situations, however, the input objects are distributed nicely and the density will be small. For many classes of objects, one can even prove that the density is $O(1)$. For example, a set of disjoint disks in the plane has density at most 5. More generally, any set of disjoint objects that are *fat* – examples of fat objects are disks, squares, triangles whose minimum angle is lower bounded – has density $O(1)$ [3]. The main question now is: Is low density sufficient to guarantee a hierarchical space decomposition of linear size? The answer is yes, and constructing the space decomposition is surprisingly easy.

Key Results

The construction of space decompositions for low-density sets is based on the following lemma. In the lemma, the square σ is considered to be open, that is, σ does not include its boundary. Let bb(o) denote the axis-aligned bounding box of an object o.

Lemma 1 *Let S be a set of n objects in the plane and let B_S denote the set of $4n$ vertices of the bounding boxes* bb(o) *of the objects $o \in S$. Let σ be any square region in the plane. Then the number of objects in S intersecting σ is at most $k + 4\lambda$, where k is the number of bounding-box vertices inside σ and $\lambda :=$* density(S).

With Lemma 1 in hand, it is surprisingly simple to construct BSP trees or compressed quadtrees of small size for any given set S whose density is small.

Binary Space Partitions

For BSP trees we proceed as follows. Let B_S be the set of vertices of the bounding boxes of the objects in S. In a generic step in the recursive construction of $\mathcal{T}$, we are given a square σ and the set of points $B_S(\sigma) := B_S \cap \sigma$. Initially σ is a square containing all points of B_S. When $B_S = \emptyset$, then $\mathcal{T}$ consists of a single leaf and the

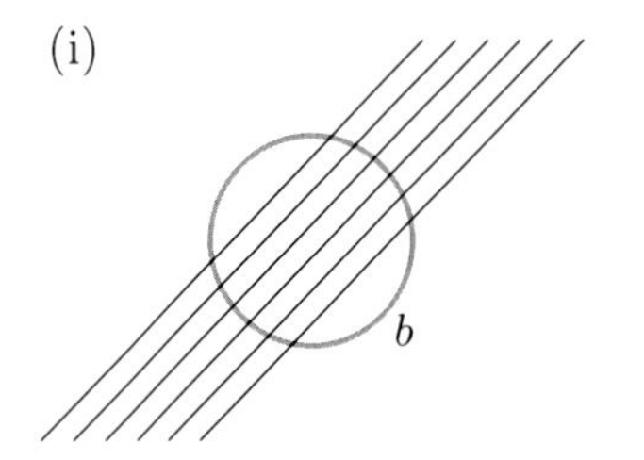

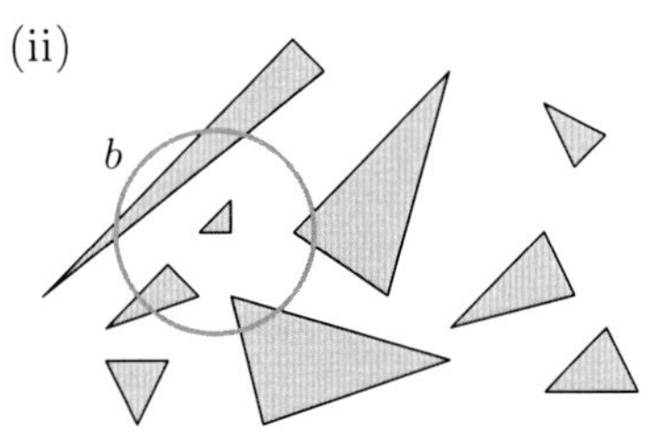

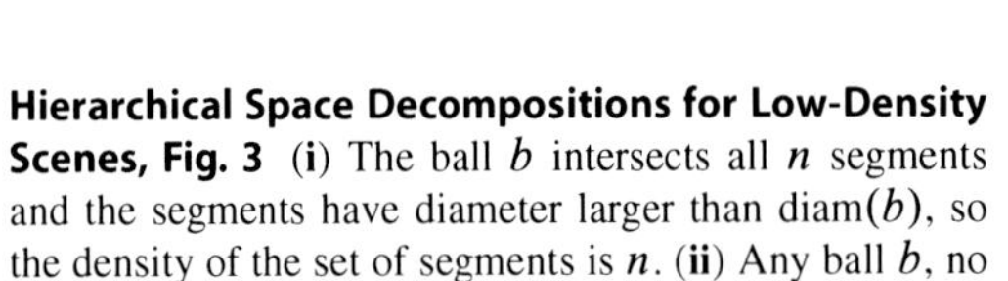

Hierarchical Space Decompositions for Low-Density Scenes, Fig. 3 (**i**) The ball b intersects all n segments and the segments have diameter larger than diam(b), so the density of the set of segments is n. (**ii**) Any ball b, no matter where it is placed or what its size is, intersects at most three triangles with diameter at least diam(b), so the density of the set of triangles is 3

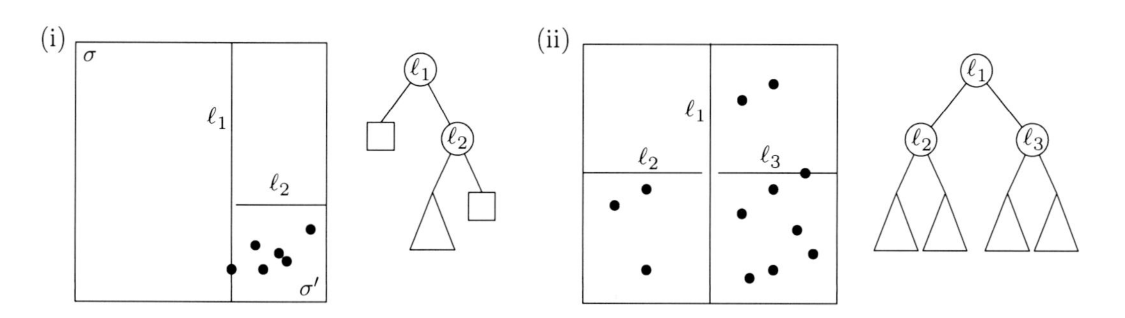

Hierarchical Space Decompositions for Low-Density Scenes, Fig. 4 Two cases in the construction of the BSP tree

recursion ends; otherwise we proceed as follows. Let σ_{NE}, σ_{SE}, σ_{SW}, and σ_{NW} denote the four quadrants of σ. We now have two cases, illustrated in Fig. 4.

Case (i): all points in $B_S(\sigma)$ lie in the same quadrant. Let σ' be the smallest square sharing a corner with σ and containing all points from $B_S(\sigma)$ in its interior or on its boundary. Split σ into three regions using a vertical and a horizontal splitting line such that σ' is one of those regions; see Fig. 4(i). Recursively construct a BSP tree for the square σ' with respect to the set $B_S(\sigma')$ of points lying in the interior of σ'.

Case (ii): not all points in $B_S(\sigma)$ lie in the same quadrant. Split σ into four quadrants using a vertical and two horizontal splitting lines; see Fig. 4(ii). Recursively construct a BSP tree for each quadrant with respect to the points lying in its interior.

The construction produces a subdivision of the initial square into $O(n)$ leaf regions, which are squares or rectangles and which do not contain points from B_S in their interior. Using Lemma 1, one can argue that each leaf region intersects $O(\lambda)$ objects.

Compressed Quadtrees

The construction of a compressed quadtree for a low-density set S is also based on the set B_S of bounding-box vertices: we construct a compressed quadtree for B_S, where we stop the recursive construction when a square contains bounding-box vertices from at most one object in S or when all bounding-box vertices inside the square coincide. Figure 2 illustrates the process. The resulting compressed quadtree has $O(n)$ leaf regions, which are canonical squares or donuts. Again using Lemma 1, one can argue that each leaf region intersects $O(\lambda)$ objects.

Improvements and Generalizations

The constructions above guarantee that each region in the space decomposition is intersected by $O(\lambda)$ objects and that the number of regions is $O(n)$. Hence, the total number of the object fragments is $O(\lambda n)$. The main idea behind the introduction of the density λ is that in practice λ is often a small constant. Nevertheless, it is (at least from a theoretical point of view) desirable to get rid of the dependency on λ in the number of fragments. This is possible by reducing the number of regions in the decomposition to (n/λ). To this end, we allow leaf regions to contain up to $O(\lambda)$ bounding-box vertices. Note that Lemma 1 implies that a square with $O(\lambda)$ bounding-box vertices inside intersects $O(\lambda)$ objects. If implemented correctly, this idea leads to decompositions with $O(n/\lambda)$ regions each of which intersects $O(\lambda)$ objects, both for binary space partitions [2, Section 12.5] and for compressed quadtrees [4]. The results can also be generalized to higher dimensions, giving the following theorem.

Theorem 1 *Let S be a set of n objects in $\mathbb{R}^d$ and let $\lambda :=$ density(S). There is a binary space partition for S consisting of $O(n/\lambda)$ leaf regions, each intersecting $O(\lambda)$ objects. Similarly, there is a compressed quadtree with $O(n/\lambda)$ leaf regions, each intersecting $O(\lambda)$ objects.*

Cross-References

▶ Binary Space Partitions
▶ Quadtrees and Morton Indexing

Recommended Reading

1. De Berg M (2000) Linear size binary space partitions for uncluttered scenes. Algorithmica 28(3):353–366
2. De Berg M, Cheong O, Van Kreveld M, Overmars M (2008) Computational geometry: algorithms and applications, 3rd edn. Springer, Berlin/Heidelberg
3. De Berg M, Katz M, Van der Stappen AF, Vleugels J (2002) Realistic input models for geometric algorithms. Algorithmica 34(1):81–97
4. De Berg M, Haverkort H, Thite S, Toma L (2010) Star-quadtrees and guard-quadtrees: I/O-efficient indexes for fat triangulations and low-density planar subdivisions. Comput Geom Theory Appl 43:493–513
5. Chazelle B (1984) Convex partitions of polyhedra: a lower bound and worst-case optimal algorithm. SIAM J Comput 13(3):488–507
6. Paterson MS and Yao FF (1990) Efficient binary space partitions for hidden-surface removal and solid modeling. Discret Comput Geom 5(5):485–503
7. Tóth CD (2003) A note on binary plane partitions. Discret Comput Geom 30(1):3–16
8. Tóth CD (2011) Binary plane partitions for disjoint line segments. Discret Comput Geom 45(4):617–646

High Performance Algorithm Engineering for Large-Scale Problems

David A. Bader
College of Computing, Georgia Institute of Technology, Atlanta, GA, USA

Keywords

Experimental algorithmics

Years and Authors of Summarized Original Work

2005; Bader

Problem Definition

Algorithm engineering refers to the process required to transform a pencil-and-paper algorithm into a robust, efficient, well tested, and easily usable implementation. Thus it encompasses a number of topics, from modeling cache behavior to the principles of good software engineering; its main focus, however, is experimentation. In that sense, it may be viewed as a recent outgrowth of *Experimental Algorithmics* [14], which is specifically devoted to the development of methods, tools, and practices for assessing and refining algorithms through experimentation. The *ACM Journal of Experimental Algorithmics (JEA)*, at URL www.jea.acm.org, is devoted to this area.

High-performance algorithm engineering [2] focuses on one of the many facets of algorithm engineering: speed. The high-performance aspect does not immediately imply parallelism; in fact, in any highly parallel task, most of the impact of high-performance algorithm engineering tends to come from refining the serial part of the code.

The term *algorithm engineering* was first used with specificity in 1997, with the organization of the first *Workshop on Algorithm Engineering (WAE 97)*. Since then, this workshop has taken place every summer in Europe. The 1998 *Workshop on Algorithms and Experiments (ALEX98)* was held in Italy and provided a discussion forum for researchers and practitioners interested in the design, analyzes and experimental testing of exact and heuristic algorithms. A sibling workshop was started in the Unites States in 1999, the *Workshop on Algorithm Engineering and Experiments (ALENEX99)*, which has taken place every winter, colocated with the *ACM/SIAM Symposium on Discrete Algorithms (SODA)*.

Key Results

Parallel computing has two closely related main uses. First, with more memory and storage resources than available on a single workstation, a parallel computer can solve correspondingly larger instances of the same problems. This increase in size can translate into running higher-fidelity simulations, handling higher volumes of information in data-intensive applications, and answering larger numbers of queries and datamining requests in corporate databases. Secondly, with more processors and larger aggregate memory subsystems than available on a single workstation, a parallel computer can often solve problems faster. This increase in speed can also translate into all of the advantages listed above, but perhaps its crucial advantage is in turnaround time. When the computation is part of a real-time system, such as weather forecasting, financial investment decision-making, or tracking and guidance systems, turnaround time is obviously the critical issue. A less obvious benefit of shortened turnaround time is higher-quality work: when a computational experiment takes less than an hour, the researcher can afford the luxury of exploration – running several different scenarios in order to gain a better understanding of the phenomena being studied.

In algorithm engineering, the aim is to present repeatable results through experiments that apply to a broader class of computers than the specific make of computer system used during the experiment. For sequential computing, empirical results are often fairly machine-independent. While machine characteristics such as word size, cache and main memory sizes, and processor and bus speeds differ, comparisons across different uniprocessor machines show the same trends. In particular, the number of memory accesses and processor operations remains fairly constant (or within a small constant factor). In high-performance algorithm engineering with parallel computers, on the other hand, this portability is usually absent: each machine and environment is its own special case. One obvious reason is major differences in hardware that affect the balance of communication and computation costs – a true shared-memory machine exhibits very different behavior from that of a cluster based on commodity networks.

Another reason is that the communication libraries and parallel programming environments (e.g., MPI [12], OpenMP [16], and High-Performance Fortran [10]), as well as the parallel algorithm packages (e.g., fast Fourier transforms using FFTW [6] or parallelized linear algebra routines in ScaLAPACK [4]), often exhibit differing performance on different types of parallel platforms. When multiple library packages exist for the same task, a user may observe different running times for each library version even on the same platform. Thus a running-time analysis should clearly separate the time spent in the user code from that spent in various library calls. Indeed, if particular library calls contribute significantly to the running time, the number of such calls and running time for each call should be recorded and used in the analysis, thereby helping library developers focus on the most cost-effective improvements. For example, in a simple message-passing program, one can characterize the work done by keeping track of sequential work, communication volume, and number of communications. A more general program using the collective communication routines of MPI could also count the number of calls to these routines. Several packages are available to instrument MPI codes in order to capture such data (e.g., MPICH's nupshot [8], Pablo [17], and Vampir [15]). The SKaMPI benchmark [18] allows running-time predictions based on such measurements even if the target machine is not available for program development. SKaMPI was designed for robustness, accuracy, portability, and efficiency; For example, SKaMPI adaptively controls how often measurements are repeated, adaptively refines message-length and step-width at "interesting" points, recovers from crashes, and automatically generates reports.

Applications

The following are several examples of algorithm engineering studies for high-performance and parallelcomputing.

1. Bader's prior publications (see [2] and http://www.cc.gatech.edu/~bader) contain many empirical studies of parallel algorithms for

combinatorial problems like sorting, selection, graph algorithms, and image processing.

2. In a recent demonstration of the power of high-performance algorithm engineering, a million-fold speed-up was achieved through a combination of a 2,000-fold speedup in the serial execution of the code and a 512-fold speedup due to parallelism (a speed-up, however, that will scale to any number of processors) [13]. (In a further demonstration of algorithm engineering, additional refinements in the search and bounding strategies have added another speedup to the serial part of about 1,000, for an overall speedup in excess of 2 billion)
3. JáJá and Helman conducted empirical studies for prefix computations, sorting, and list-ranking, on symmetric multiprocessors. The sorting research (see [9]) extends Vitter's external Parallel Disk Model to the internal memory hierarchy of SMPs and uses this new computational model to analyze a general-purpose sample sort that operates efficiently in shared-memory. The performance evaluation uses nine well-defined benchmarks. The benchmarks include input distributions commonly used for sorting benchmarks (such as keys selected uniformly and at random), but also benchmarks designed to challenge the implementation through load imbalance and memory contention and to circumvent algorithmic design choices based on specific input properties (such as data distribution, presence of duplicate keys, pre-sorted inputs, etc.).
4. In [3] Blelloch et al. compare through analysis and implementation three sorting algorithms on the Thinking Machines CM-2. Despite the use of an outdated (and no longer available) platform, this paper is a gem and should be required reading for every parallel algorithm designer. In one of the first studies of its kind, the authors estimate running times of four of the machine's primitives, then analyze the steps of the three sorting algorithms in terms of these parameters. The experimental studies of the performance are normalized to provide clear comparison of how the algorithms scale with input size on a $32K$-processor CM-2.
5. Vitter et al. provide the canonical theoretic foundation for I/O-intensive experimental algorithmics using external parallel disks (e.g., see [1, 19, 20]). Examples from sorting, FFT, permuting, and matrix transposition problems are used to demonstrate the parallel disk model.
6. Juurlink and Wijshoff [11] perform one of the first detailed experimental accounts on the preciseness of several parallel computation models on five parallel platforms. The authors discuss the predictive capabilities of the models, compare the models to find out which allows for the design of the most efficient parallel algorithms, and experimentally compare the performance of algorithms designed with the model versus those designed with machine-specific characteristics in mind. The authors derive model parameters for each platform, analyses for a variety of algorithms (matrix multiplication, bitonic sort, sample sort, all-pairs shortest path), and detailed performance comparisons.
7. The LogP model of Culler et al. [5] provides a realistic model for designing parallel algorithms for message-passing platforms. Its use is demonstrated for a number of problems, including sorting.
8. Several research groups have performed extensive algorithm engineering for high-performance numerical computing. One of the most prominent efforts is that led by Dongarra for ScaLAPACK [4], a scalable linear algebra library for parallel computers. ScaLAPACK encapsulates much of the high-performance algorithm engineering with significant impact to its users who require efficient parallel versions of matrix–matrix linear algebra routines. New approaches for automatically tuning the sequential library (e.g., LAPACK) are now available as the ATLAS package [21].

Open Problems

All of the tools and techniques developed over the last several years for algorithm engineering are applicable to high-performance algorithm engineering. However, many of these tools need further refinement. For example, cache-efficient programming is a key to performance but it is not yet well understood, mainly because of complex machine-dependent issues like limited associativity, virtual address translation, and increasingly deep hierarchies of high-performance machines. A key question is whether one can find simple models as a basis for algorithm development. For example, cache-oblivious algorithms [7] are efficient at all levels of the memory hierarchy in theory, but so far only few work well in practice. As another example, profiling a running program offers serious challenges in a serial environment (any profiling tool affects the behavior of what is being observed), but these challenges pale in comparison with those arising in a parallel or distributed environment (for instance, measuring communication bottlenecks may require hardware assistance from the network switches or at least reprogramming them, which is sure to affect their behavior). Designing efficient and portable algorithms for commodity multicore and manycore processors is an open challenge.

Cross-References

- ▸ Analyzing Cache Misses
- ▸ Cache-Oblivious B-Tree
- ▸ Cache-Oblivious Model
- ▸ Cache-Oblivious Sorting
- ▸ Engineering Algorithms for Computational Biology
- ▸ Engineering Algorithms for Large Network Applications
- ▸ Engineering Geometric Algorithms
- ▸ Experimental Methods for Algorithm Analysis
- ▸ External Sorting and Permuting
- ▸ Implementation Challenge for Shortest Paths
- ▸ Implementation Challenge for TSP Heuristics
- ▸ I/O-Model
- ▸ Visualization Techniques for Algorithm Engineering

Recommended Reading

1. Aggarwal A, Vitter J (1988) The input/output complexity of sorting and related problems. Commun ACM 31:1116–1127
2. Bader DA, Moret BME, Sanders P (2002) Algorithm engineering for parallel computation. In: Fleischer R, Meineche-Schmidt E, Moret BME (eds) Experimental algorithmics. Lecture notes in computer science, vol 2547. Springer, Berlin, pp 1–23
3. Blelloch GE, Leiserson CE, Maggs BM, Plaxton CG, Smith SJ, Zagha M (1998) An experimental analysis of parallel sorting algorithms. Theory Comput Syst 31(2):135–167
4. Choi J, Dongarra JJ, Pozo R, Walker DW (1992) ScaLAPACK: a scalable linear algebra library for distributed memory concurrent computers. In: The 4th symposium on the frontiers of massively parallel computations. McLean, pp 120–127
5. Culler DE, Karp RM, Patterson DA, Sahay A, Schauser KE, Santos E, Subramonian R, von Eicken T (1993) LogP: towards a realistic model of parallel computation. In: 4th symposium on principles and practice of parallel programming. ACM SIGPLAN, pp 1–12
6. Frigo M, Johnson SG (1998) FFTW: an adaptive software architecture for the FFT. In: Proceedings of IEEE international conference on acoustics, speech, and signal processing, Seattle, vol 3, pp 1381–1384
7. Frigo M, Leiserson CE, Prokop H, Ramachandran, S (1999) Cacheoblivious algorithms. In: Proceedings of 40th annual symposium on foundations of computer science (FOCS-99), New York. IEEE, pp 285–297
8. Gropp W, Lusk E, Doss N, Skjellum A (1996) A high-performance, portable implementation of the MPI message passing interface standard. Technical report, Argonne National Laboratory, Argonne. www.mcs.anl.gov/mpi/mpich/
9. Helman DR, JáJá J (1998) Sorting on clusters of SMP's. In: Proceedings of 12th international parallel processing symposium, Orlando, pp 1–7
10. High Performance Fortran Forum (1993) High performance Fortran language specification, 1.0 edn., May 1993
11. Juurlink BHH, Wijshoff HAG (1998) A quantitative comparison of parallel computation models. ACM Trans Comput Syst 13(3):271–318
12. Message Passing Interface Forum (1995) MPI: a message-passing interface standard. Technical report, University of Tennessee, Knoxville, June 1995. Version 1.1
13. Moret BME, Bader DA, Warnow T (2002) High-performance algorithm engineering for computational phylogenetics. J Supercomput 22:99–111, Special issue on the best papers from ICCS'01
14. Moret BME, Shapiro HD (2001) Algorithms and experiments: the new (and old) methodology. J Univ Comput Sci 7(5):434–446
15. Nagel WE, Arnold A, Weber M, Hoppe HC, Solchenbach K (1996) VAMPIR: visualization and analysis of MPI resources. Supercomputer 63 12(1):69–80
16. OpenMP Architecture Review Board (1997) OpenMP: a proposed industry standard API for shared memory programming. www.openmp.org

17. Reed DA, Aydt RA, Noe RJ, Roth PC, Shields KA, Schwartz B, Tavera LF (1993) Scalable performance analysis: the Pablo performance analysis environment. In: Skjellum A (ed) Proceedings of scalable parallel libraries conference, Mississippi State University. IEEE Computer Society Press, pp 104–113
18. Reussner R, Sanders P, Träff J (1998, accepted) SKaMPI: a comprehensive benchmark for public benchmarking of MPI. Scientific programming, 2001. Conference version with Prechelt, L., Müller, M. In: Proceedings of EuroPVM/MPI
19. Vitter JS, Shriver EAM (1994) Algorithms for parallel memory. I: two-level memories. Algorithmica 12(2/3):110–147
20. Vitter JS, Shriver EAM (1994) Algorithms for parallel memory II: hierarchical multilevel memories. Algorithmica 12(2/3):148–169
21. Whaley R, Dongarra J (1998) Automatically tuned linear algebra software (ATLAS). In: Proceedings of supercomputing 98, Orlando. www.netlib.org/utk/people/JackDongarra/PAPERS/atlas-sc98.ps

Holant Problems

Jin-Yi Cai[1,2], Heng Guo[2], and Tyson Williams[2]
[1]Beijing University, Beijing, China
[2]Computer Sciences Department, University of Wisconsin–Madison, Madison, WI, USA

Keywords

Computational complexity; Counting complexity; Holant problems; Partition functions

Years and Authors of Summarized Original Work

2011; Cai, Lu, Xia
2012; Huang, Lu
2013; Cai, Guo, Williams
2013; Guo, Lu, Valiant
2014; Cai, Guo, Williams

Problem Definition

The framework of Holant problems is intended to capture a class of sum-of-product computations in a more refined way than counting CSP problems and is inspired by Valiant's holographic algorithms [12] (also cf. entry ▶ Holographic Algorithms). A constraint function f, or *signature*, is a mapping from $[\kappa]^n$ to $\mathbb{C}$, representing a local contribution to a global sum. Here, $[\kappa]$ is a finite domain set, and n is the arity of f. The range is usually taken to be $\mathbb{C}$, but it can be replaced by any commutative semiring. A Holant problem Holant($\mathcal{F}$) is parameterized by a set of constraint functions $\mathcal{F}$. We usually focus on the Boolean domain, namely, $\kappa = 2$. For consideration of models of computation, we restrict function values to be complex algebraic numbers.

We allow multigraphs, namely, graphs with self-loops and parallel edges. A *signature grid* $\Omega = (G, \pi)$ of Holant($\mathcal{F}$) consists of a graph $G = (V, E)$, where π assigns each vertex $v \in V$ and its incident edges with some $f_v \in \mathcal{F}$ and its input variables. We say Ω is a *planar signature grid* if G is planar. The Holant problem on instance Ω is to evaluate

$$\text{Holant}(\Omega; \mathcal{F}) = \sum_{\sigma} \prod_{v \in V} f_v(\sigma \mid_{E(v)}),$$

a sum over all edge labelings $\sigma : E \to [\kappa]$, where $E(v)$ denotes the incident edges of v and $\sigma \mid_{E(v)}$ denotes the restriction of σ to $E(v)$. This is also known as the partition function in the statistical physics literature.

Formally, a set of signatures $\mathcal{F}$ defines the following Holant problem:

Name Holant($\mathcal{F}$)
Instance A *signature grid* $\Omega = (G, \pi)$
Output Holant($\Omega; \mathcal{F}$)

The problem Pl-Holant($\mathcal{F}$) is defined similarly using a planar signature grid.

A function f_v can be represented by listing its values in lexicographical order as in a truth table, which is a vector in $\mathbb{C}^{\kappa^{\deg(v)}}$ or as a tensor in $(\mathbb{C}^{\kappa})^{\otimes \deg(v)}$. Special focus has been put on *symmetric* signatures, which are functions invariant under any permutation of the input. An example is the EQUALITY signature $=_n$ of arity n. A Boolean symmetric function f of arity n can be listed as $[f_0, f_1, \ldots, f_n]$, where f_w is the function value of f when the input has Hamming weight w. Using this notation, an EQUALITY signature is $[1, 0, \dots, 0, 1]$. Another example is the EXACTONE signature $[0, 1, 0, \dots, 0]$. Clearly, the Holant problem defined by this signature counts the number of perfect matchings.

The set $\mathcal{F}$ is allowed to be an infinite set. For Holant($\mathcal{F}$) to be tractable, the problem must be computable in polynomial time even when the description of the signatures in the input Ω is included in the input size. In contrast, we say Holant($\mathcal{F}$) is #P-hard if there exists a finite subset of $\mathcal{F}$ for which the problem is #P-hard.

The Holant framework is a generalization and refinement of both counting graph homomorphisms and counting constraint satisfaction problems (see entry ► Complexity Dichotomies for Counting Graph Homomorphisms for more details and results).

Key Results

The Holant problem was introduced by Cai, Lu, and Xia [3], which also contains a dichotomy of Holant* for symmetric Boolean complex functions. The notation Holant* means that all unary functions are assumed to be available. This restriction is later weakened to only allow two constant functions that pin a variable to 0 or 1. This framework is called Holant^c. In [5], a dichotomy of Holant^c is obtained. The need to assume some freely available functions is finally avoided in [10]. In this paper, Huang and Lu proved a dichotomy for Holant but with the caveat that the functions must be real weighted. This result was later improved by Cai, Guo, and Williams [6], who proved a dichotomy for Holant parameterized by any set of symmetric Boolean complex functions.

We will give some necessary definitions and then state the dichotomy from [6]. First are several tractable families of functions over the Boolean domain.

Definition 1 A signature f of arity n is *degenerate* if there exist unary signatures $u_j \in \mathbb{C}^2$ $(1 \le j \le n)$ such that $f = u_1 \otimes \cdots \otimes u_n$. A symmetric degenerate signature has the form $u^{\otimes n}$.

Definition 2 A k-ary function $f(x_1, \dots, x_k)$ is of *affine* type if it has the form

$$\lambda \chi_{Ax=0} \cdot \sqrt{-1}^{\sum_{j=1}^{n} \langle \alpha_j, x \rangle},$$

where $\lambda \in \mathbb{C}$, $x = (x_1, x_2, \dots, x_k, 1)^{\mathrm{T}}$, A is a matrix over $\mathbb{F}_2$, α_j is a vector over $\mathbb{F}_2$, and χ is a 0–1 indicator function such that $\chi_{Ax=0}$ is 1 iff $Ax = 0$. Note that the dot product $\langle \alpha_j, x \rangle$ is calculated over $\mathbb{F}_2$, while the summation $\sum_{j=1}^{n}$ on the exponent of $i = \sqrt{-1}$ is evaluated as a sum mod 4 of 0–1 terms. We use $\mathscr{A}$ to denote the set of all affine-type functions.

An alternative but equivalent form for an affine-type function is $\lambda \chi_{Ax=0} \cdot \sqrt{-1}^{Q(x_1, x_2, \dots, x_k)}$ where $Q(\cdot)$ is a quadratic form with integer coefficients that are even for every cross term.

Definition 3 A function is of *product type* if it can be expressed as a product of unary functions, binary equality functions ($[1, 0, 1]$), and binary disequality functions ($[0, 1, 0]$), each applied to some of its variables. We use $\mathscr{P}$ to denote the set of product-type functions.

Definition 4 A function f is called *vanishing* if the value Holant($\Omega; \{f\}$) is 0 for every signature grid Ω. We use $\mathscr{V}$ to denote the set of vanishing functions.

For vanishing signatures, we need some more definitions.

Definition 5 An arity n symmetric signature of the form $f = [f_0, f_1, \dots, f_n]$ is in $\mathscr{R}_t^+$ for a nonnegative integer $t \ge 0$ if $t > n$ or for any $0 \le k \le n - t$, $f_k, \dots, f_{k+t}$ satisfy the recurrence relation

$$\binom{t}{t} i^t f_{k+t} + \binom{t}{t-1} i^{t-1} f_{k+t-1} + \cdots + \binom{t}{0} i^0 f_k = 0. \quad (1)$$

We define $\mathscr{R}_t^-$ similarly but with $-i$ in place of i in (1).

With $\mathscr{R}_t^{\pm}$, one can define the recurrence degree of a function f.

Definition 6 For a nonzero symmetric signature f of arity n, it is of *positive* (resp. *negative*) *recurrence degree* $t \leq n$, denoted by $\text{rd}^+(f) = t$ (resp. $\text{rd}^-(f) = t$), if and only if $f \in \mathscr{R}_{t+1}^+ - \mathscr{R}_t^+$ (resp. $f \in \mathscr{R}_{t+1}^- - \mathscr{R}_t^-$). If f is the all-zero signature, we define $\text{rd}^+(f) = \text{rd}^-(f) = -1$.

In [6], it is shown that $f \in \mathscr{V}$ if and only if for either $\sigma = +$ or $-$, we have $2\text{rd}^\sigma(f) < \text{arity}(f)$. Accordingly, we split the set $\mathscr{V}$ of vanishing signatures in two.

Definition 7 We define $\mathscr{V}^\sigma$ for $\sigma \in \{+, -\}$ as

$$\mathscr{V}^\sigma = \{f \mid 2\text{rd}^\sigma(f) < \text{arity}(f)\}.$$

To state the dichotomy, we also need the notion of $\mathcal{F}$-transformable. For a matrix $T \in \mathbb{C}^{2\times 2}$, and a signature set $\mathcal{F}$, define $T\mathcal{F} = \{g \mid \exists f \in \mathcal{F}$ of arity n, $g = T^{\otimes n} f\}$. Here, we view the signatures as column vectors. Let $=_2$ be the equality function of arity 2.

Definition 8 A signature set $\mathcal{F}'$ is $\mathcal{F}$-transformable if there exists a non-singular matrix $T \in \mathbb{C}^{2\times 2}$ such that $\mathcal{F}' \subseteq T\mathcal{F}$ and $(=_2)T^{\otimes 2} \in \mathcal{F}$.

If a set of functions $\mathcal{F}'$ is $\mathcal{F}$-transformable and $\mathcal{F}$ is a tractable set, then Holant($\mathcal{F}'$) is tractable as well.

The dichotomy of Holant problems over symmetric Boolean complex functions is stated as follows.

Theorem 1 ([6]) *Let $\mathcal{F}$ be any set of symmetric, complex-valued signatures in Boolean variables. Then,* Holant($\mathcal{F}$) *is #P-hard unless $\mathcal{F}$ satisfies one of the following conditions, in which case the problem is in P:*

1. *All nondegenerate signatures in $\mathcal{F}$ are of arity at most* 2;
2. *$\mathcal{F}$ is $\mathscr{A}$-transformable;*
3. *$\mathcal{F}$ is $\mathscr{P}$-transformable;*
4. *$\mathcal{F} \subseteq \mathscr{V}^\sigma \cup \{f \in \mathscr{R}_2^\sigma \mid \text{arity}(f) = 2\}$ for some $\sigma \in \{+, -\}$;*
5. *All nondegenerate signatures in $\mathcal{F}$ are in $\mathscr{R}_2^\sigma$ for some $\sigma \in \{+, -\}$.*

Theorem 1 is about Holant problems parameterized by symmetric Boolean complex functions over general graphs. Holant problems are studied in other settings as well. For planar graphs, [2] contains a dichotomy for Holant^c with real symmetric functions. There are signature sets that are #P-hard over general graphs but tractable over planar graphs. The algorithms for such sets are due to Valiant's holographic algorithms and the theory of matchgates [1, 12].

Another generalization looks at a broader range of functions. One may consider asymmetric functions as in [4], which contains a dichotomy for Holant* problems defined by asymmetric Boolean complex functions. One can also consider functions of larger domain size. For domain size 3, [7] contains a dichotomy for a single arity 3 symmetric complex function in the Holant* setting. For any constant domain size, [8] contains a dichotomy for a single arity 3 complex weighted function that satisfies a strong symmetry property.

One can consider constraint functions with a range other than $\mathbb{C}$. Replacing $\mathbb{C}$ by some finite field $\mathbb{F}_p$ for some prime p defines counting problems modulo p. The case $p = 2$ is called parity Holant problems. It is of special interest because computing the permanent modulo 2 is tractable, which implies a family of tractable matchgate functions even over general graphs. For parity Holant problems, a complete dichotomy for symmetric functions is obtained by Guo, Lu, and Valiant [9].

Open Problems

Unlike the progress in the general graph setting, the strongest known dichotomy results for planar Holant problems are rather limited. These planar dichotomies showed that newly tractable problems over planar graphs are captured by holographic algorithms with matchgates, but with restrictions like symmetric functions or regular graphs. The theory of holographic algorithms with matchgates can be applied to planar graphs and asymmetric signatures. A true test of its power would be to obtain an asymmetric complex weighted dichotomy of planar Holant problems. The situation is similarly limited for higher domain sizes, where things seem considerably more complicated. A reasonable first step in this direction would be to consider some restricted (yet still powerful) family of functions.

Despite the success for $\mathbb{F}_2$, little is known about the complexity of Holant problems over other finite fields or semirings. As Valiant showed in [11], counting problems modulo some finite modulus include some interesting and surprising phenomena. It deserves further research.

Cross-References

▸ Complexity Dichotomies for Counting Graph Homomorphisms
▸ Holographic Algorithms

Recommended Reading

1. Cai JY, Lu P (2011) Holographic algorithms: from art to science. J Comput Syst Sci 77(1):41–61
2. Cai JY, Lu P, Xia M (2010) Holographic algorithms with matchgates capture precisely tractable planar #CSP. In: FOCS, Las Vegas. IEEE Computer Society, pp 427–436
3. Cai JY, Lu P, Xia M (2011) Computational complexity of Holant problems. SIAM J Comput 40(4):1101–1132
4. Cai JY, Lu P, Xia M (2011) Dichotomy for Holant* problems of Boolean domain. In: SODA, San Francisco. SIAM, pp 1714–1728
5. Cai JY, Huang S, Lu P (2012) From Holant to #CSP and back: dichotomy for Holant^c problems. Algorithmica 64(3):511–533
6. Cai JY, Guo H, Williams T (2013) A complete dichotomy rises from the capture of vanishing signatures (extended abstract). In: STOC, Palo Alto. ACM, pp 635–644
7. Cai JY, Lu P, Xia M (2013) Dichotomy for Holant* problems with domain size 3. In: SODA, New Orleans. SIAM, pp 1278–1295
8. Cai JY, Guo H, Williams T (2014) The complexity of counting edge colorings and a dichotomy for some higher domain Holant problems. In: FOCS, Philadelphia. IEEE, pp 601–610
9. Guo H, Lu P, Valiant LG (2013) The complexity of symmetric Boolean parity Holant problems. SIAM J Comput 42(1):324–356
10. Huang S, Lu P (2012) A dichotomy for real weighted Holant problems. In: CCC, Porto. IEEE Computer Society, pp 96–106
11. Valiant LG (2006) Accidental algorithms. In: FOCS, Berkeley. IEEE, pp 509–517
12. Valiant LG (2008) Holographic algorithms. SIAM J Comput 37(5):1565–1594

Holographic Algorithms

Jin-Yi Cai[1,2], Pinyan Lu[3], and Mingji Xia[4]
[1]Beijing University, Beijing, China
[2]Computer Sciences Department, University of Wisconsin–Madison, Madison, WI, USA
[3]Microsoft Research Asia, Shanghai, China
[4]The State Key Laboratory of Computer Science, Chinese Academy of Sciences, Beijing, China

Keywords

Bases; Counting problems; Holographic algorithms; Perfect matchings; Planar graphs

Years and Authors of Summarized Original Work

2006, 2008; Valiant
2008, 2009, 2010, 2011; Cai, Lu
2009; Cai, Choudhary, Lu
2014; Cai, Gorenstein

Problem Definition

Holographic algorithm, introduced by L. Valiant [11], is an algorithm design technique rather than a single algorithm for a particular problem. In essence, these algorithms are reductions to the FKT algorithm [7–9] to count the number of perfect matchings in a planar graph in polynomial time. Computation in these algorithms is expressed and interpreted through a choice of linear basis vectors in an exponential "holographic" mix, and then it is carried out by the FKT method via the Holant Theorem. This methodology has produced polynomial time algorithms for a variety of problems ranging from restrictive versions of satisfiability, vertex cover, to other graph problems such as edge orientation and node/edge deletion. No polynomial time algorithms were known for these problems, and some minor variations are known to be NP-hard (or even #P-hard).

Let $G = (V, E, W)$ be a weighted undirected planar graph, where V, E, and W are sets of vertices, edges, and edge weights, respectively. A matchgate is a tuple (G, X) where $X \subseteq V$ is a set of external nodes on the outer face. A matchgate is considered a generator or a recognizer matchgate when the external nodes are considered output or input nodes, respectively. They differ mainly in the way they are transformed. The external nodes are ordered clockwise on the external face. Γ is called an odd (resp. even) matchgate if it has an odd (resp. even) number of nodes.

Each matchgate is assigned a *signature* tensor. A generator Γ with m output nodes is assigned a contravariant tensor $\mathbf{G} \in V_0^m$ of type $\binom{m}{0}$, where V_0^m is the tensor space spanned by the m-fold tensor products of the standard basis $\mathbf{b} = [\mathbf{b}_0, \mathbf{b}_1] = \left[\begin{pmatrix}1\\0\end{pmatrix}, \begin{pmatrix}0\\1\end{pmatrix}\right]$. The tensor $\mathbf{G}$ under the standard basis $\mathbf{b}$ has the form

$$\sum G^{i_1 i_2 \ldots i_m} \mathbf{b}_{i_1} \otimes \mathbf{b}_{i_2} \otimes \cdots \otimes \mathbf{b}_{i_m},$$

where

$$G^{i_1 i_2 \ldots i_m} = \text{PerfMatch}(G - Z).$$

Here Z is the subset of the output nodes of Γ having the characteristic sequence $\chi_Z = i_1 i_2 \ldots i_m \in \{0,1\}^m$, $\text{PerfMatch}(G - Z) = \sum_M \prod_{(i,j)\in M} w_{ij}$ is a sum over all perfect matchings M in the graph $G - Z$ obtained from G by removing Z and its incident edges, and w_{ij} is the weight of the edge (i, j). Similarly a recognizer Γ' with underlying graph G' having m input nodes is assigned a covariant tensor $\mathbf{R} \in V_m^0$ of type $\binom{0}{m}$. This tensor under the standard (dual) basis $\mathbf{b}^*$ has the form

$$\sum R_{i_1 i_2 \ldots i_m} \mathbf{b}^{i_1} \otimes \mathbf{b}^{i_2} \otimes \cdots \otimes \mathbf{b}^{i_m},$$

where

$$R_{i_1 i_2 \ldots i_m} = \text{PerfMatch}(G' - Z),$$

and Z is the subset of the input nodes of Γ' having the characteristic sequence $\chi_Z = i_1 i_2 \ldots i_m$.

As a contravariant tensor, $\mathbf{G}$ transforms as follows. Under a basis transformation $\boldsymbol{\beta}_j = \sum_i \mathbf{b}_i t_j^i$,

$$(G')^{j_1 j_2 \ldots j_m} = \sum G^{i_1 i_2 \ldots i_m} \tilde{t}_{i_1}^{j_1} \tilde{t}_{i_2}^{j_2} \ldots \tilde{t}_{i_m}^{j_m},$$

where $(\tilde{t}_i^j)$ is the inverse matrix of (t_j^i). Similarly, $\mathbf{R}$ transforms as a covariant tensor, namely,

$$(R')_{j_1 j_2 \ldots j_m} = \sum R_{i_1 i_2 \ldots i_m} t_{j_1}^{i_1} t_{j_2}^{i_2} \ldots t_{j_m}^{i_m}.$$

A signature is *symmetric* if each entry only depends on the Hamming weight of the index $i_1 i_2 \ldots i_m$. This notion is invariant under a basis transformation. A symmetric signature is denoted by $[\sigma_0, \sigma_1, \ldots, \sigma_m]$, where σ_i denotes the value of a signature entry whose Hamming weight of its index is i.

A *matchgrid* $\Omega = (A, B, C)$ is a weighted planar graph consisting of a disjoint union of: a set of g generators $A = (A_1, \ldots, A_g)$, a set of r recognizers $B = (B_1, \ldots, B_r)$, and a set of f connecting edges $C = (C_1, \ldots, C_f)$, where each C_i edge has weight 1 and joins an output node of a generator with an input node of a recognizer, so that every input and output node in every constituent matchgate has exactly one such incident connecting edge.

Let $\mathbf{G} = \bigotimes_{i=1}^g \mathbf{G}(A_i)$ be the tensor product of all the generator signatures, and let $\mathbf{R} = \bigotimes_{j=1}^r \mathbf{R}(B_j)$ be the tensor product of all the recognizer signatures. Then Holant_Ω is defined to be the contraction of the two product tensors, under some basis $\boldsymbol{\beta}$, where the corresponding indices match up according to the f connecting edges C_k:

$$\text{Holant}_{\Omega} = \langle \mathbf{R}, \mathbf{G} \rangle = \sum_{x \in \boldsymbol{\beta}^{\otimes f}} \left\{ [\Pi_{1 \le i \le g} \mathbf{G}(A_i, x|_{A_i})] \cdot [\Pi_{1 \le j \le r} \mathbf{R}(B_j, x^*|_{B_j})] \right\}. \quad (1)$$

If we write the covariant tensor **R** as a row vector of dimension 2^f, write the contravariant tensor **G** as a column vector of dimension 2^f, both indexed by some common ordering of the connecting edges, then Holant_{Ω} is just the dot product of these two vectors. Valiant's beautiful Holant Theorem is as follows:

Theorem 1 (Valiant) *For any matchgrid Ω over any basis $\boldsymbol{\beta}$, let G be its underlying weighted graph, then*

$$\text{Holant}_{\Omega} = \text{PerfMatch}(G).$$

The FKT algorithm can compute the perfect matching polynomial $\text{PerfMatch}(G)$ for a planar graph in polynomial time. This gives a polynomial time algorithm to compute Holant_{Ω}.

Key Results

To design a holographic algorithm for a given problem, the creative part is to formalize the given problem as a Holant problem. The theory of holographic algorithms is trying to answer the second question: given a Holant problem, can we find a basis transformation so that all the signatures in the Holant problem can be realized by some matchgates on that basis? More formally, we want to solve the following simultaneous realizability problem (SRP).

Definition 1 Simultaneous Realizability Problem (SRP):

Input: A set of constraint functions for generators and recognizers.

Output: A common basis under which these functions can be simultaneously realized by matchgate signatures, if any exists; "NO" if they are not simultaneously realizable.

The theory of matchgates and holographic algorithms provides a systematic understanding of which constraint functions can be realized by matchgates, the structure for the bases, and finally solve the simultaneous realizability problem.

Matchgate Identities

There is a set of algebraic identities [1, 6] which completely characterizes signatures directly realizable without basis transformation by matchgates for any number of inputs and outputs. These identities are derived from Grassmann-Plücker identities for Pfaffians.

Patterns α, β are m-bit strings, i.e., $\alpha, \beta \in \{0,1\}^m$. A position vector $P = \{p_i\}, i \in [l]$ is a subsequence of $\{1, 2, \ldots, m\}$, i.e., $p_i \in [m]$ and $p_1 < p_2 < \cdots < p_l$. We also use p to denote the m-bit string, whose $(p_1, p_2, \ldots, p_l)$-th bits are 1 and others are 0. Let $e_i \in \{0,1\}^m$ be the pattern with 1 in the i-th bit and 0 elsewhere. Let $\alpha, \beta \in \{0,1\}^m$ be any pattern, and let $P = \{p_i\} = \alpha + \beta$, $i \in [l]$ be their bit-wise XOR as a position vector. Then, we have the following identity:

$$\sum_{i=1}^{l} (-1)^i G^{\alpha + e_{p_i}} G^{\beta + e_{p_i}} = 0. \quad (2)$$

A tensor $\mathbf{G} = (G^{i_1, \ldots, i_m})$ is realizable as the signature, without basis transformation, of some planar matchgate iff it satisfies the matchgate identities (2) for all α and β.

Basis Collapse

When we consider basis transformations for holographic algorithms, we mainly focus on invertible transformations, and these are bases of dimension 2. However, in a paper called "accidental algorithm" [10], Valiant showed that a basis of dimension 4 can be used to solve in P an interesting (restrictive SAT) counting problem mod 7. In a later paper [4], we have shown, among other things, that for this particular problem, this use of bases of size 2 is unnecessary. Then, in a sequence of two papers [2, 3], we completely resolve the problem of the power of higher dimensional bases. We prove that 2-dimensional bases are universal for holographic algorithms in the Boolean domain.

Theorem 2 (Basis Collapse Theorem) *Any holographic algorithm on a basis of any dimension which employs at least one nondegenerate generator can be efficiently transformed to a holographic algorithm in a basis of dimension 2. More precisely, if generators* $G_1, G_2, \ldots, G_s$ *and recognizers* $R_1, R_2, \ldots, R_t$ *are simultaneously realizable on a basis* T *of any dimension, and not all generators are degenerate, then all the generators and recognizers are simultaneously realizable in a basis* $\hat{T}$ *of dimension 2.*

From Art to Science

Based on the characterization for matchgate signatures and basis transformations, we can solve the simultaneous realizability problem [5]. In order to investigate the realizability of signatures, it is useful to introduce a basis manifold $\mathcal{M}$, which is defined to be the set of all possible bases modulo an equivalence relation. One can characterize in terms of $\mathcal{M}$ all realizable symmetric signatures under basis transformations. This structural understanding gives: (i) a uniform account of all the previous successes of holographic algorithms using symmetric signatures [10, 11]; (ii) generalizations to solve other problems, when this is possible; and (iii) a proof when this is not possible.

Applications

In this section, we list a few problems which can be solved by holographic algorithms.

#PL-3-NAE-ICE

INPUT: A planar graph $G = (V, E)$ of maximum degree 3.

OUTPUT: The number of orientations such that no node has all incident edges directed toward it or all incident edges directed away from it.

Hence, #PL-3-NAE-ICE counts the number of no-sink-no-source orientations. A node of degree one will preclude such an orientation. We assume every node has degree 2 or 3. To solve this problem by a holographic algorithm with matchgates, we design a signature grid based on G as follows: We attach to each node of degree 3 a generator with signature $[0, 1, 1, 0]$. This represents a NOT-ALL-EQUAL or NAE gate of arity 3. For any node of degree 2, we use a generator with the binary NAE (i.e., a binary DISEQUALITY) signature $(\neq_2) = [0, 1, 0]$. For each edge in E, we use a recognizer with signature $(\neq_2)$, which stands for an orientation from one node to the other. (To express such a problem, it is completely arbitrary

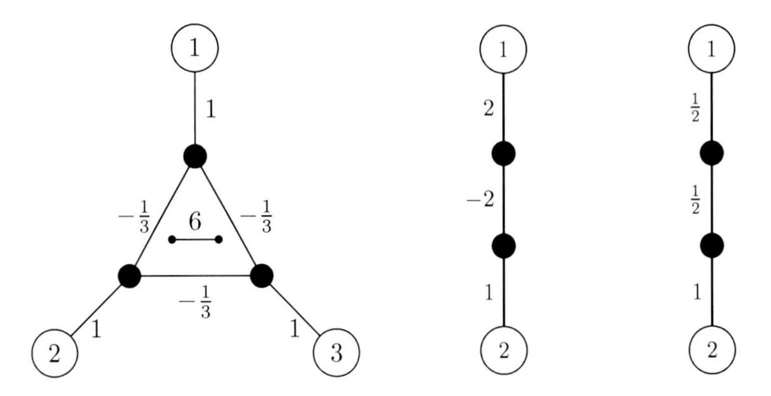

Holographic Algorithms, Fig. 1 Some matchgates used in #PL-3-NAE-ICE

to label one side as generators and the other side as recognizers.) From the given planar graph G, we obtain a signature grid Ω, where the underlying graph G' is the edge-vertex incidence graph of G. By definition, Holant_Ω is an exponential sum where each term is a product of appropriate entries of the signatures. Each term is indexed by a 0–1 assignment on all edges of G'; it has a value of 0 or 1, and it has a value of 1 iff it corresponds to an orientation of G such that at every vertex of G the local NAE constraint is satisfied. Therefore, Holant_Ω is precisely the number of valid orientations required by #PL-3-NAE-ICE.

Note that the signature $[0, 1, 1, 0]$ is not the signature of any matchgate. A simple reason for this is that a matchgate signature, being defined in terms of perfect matchings, cannot have nonzero values for inputs of both odd and even Hamming weights.

However, under a holographic transformation using $H = \begin{bmatrix} 1 & 1 \\ 1 & -1 \end{bmatrix}$,

$$H^{\otimes 3}[0, 1, 1, 0] = H^{\otimes 3}\left\{ \begin{bmatrix} 1 \\ 1 \end{bmatrix}^{\otimes 3} - \begin{bmatrix} 1 \\ 0 \end{bmatrix}^{\otimes 3} - \begin{bmatrix} 0 \\ 1 \end{bmatrix}^{\otimes 3} \right\} = [6, 0, -2, 0],$$

$$H^{\otimes 2}[0, 1, 0] = H^{\otimes 2}\left\{ \begin{bmatrix} 1 \\ 1 \end{bmatrix}^{\otimes 2} - \begin{bmatrix} 1 \\ 0 \end{bmatrix}^{\otimes 2} - \begin{bmatrix} 0 \\ 1 \end{bmatrix}^{\otimes 2} \right\} = [2, 0, -2],$$

and

$$[0, 1, 0](H^{-1})^{\otimes 2} = \frac{1}{2}[1, 0, -1].$$

These signatures are all realizable as matchgate signatures by verifying all the matchgate identities. More concretely, we can exhibit the requisite three matchgates in Fig. 1.

Hence, #PL-3-NAE-ICE is precisely the following Holant problem on planar graphs:

$$\begin{aligned} &\text{Holant}([0, 1, 0] \mid [0, 1, 0], [0, 1, 1, 0]) \\ &\equiv_T \text{Holant}(\tfrac{1}{2}[1, 0, -1] \mid [2, 0, -2], [6, 0, -2, 0]). \end{aligned}$$

Now we may replace each signature $\frac{1}{2}[1, 0, -1]$, $[2, 0, -2]$, and $[6, 0, -2, 0]$ in Ω by their corresponding matchgates, and then we can compute Holant_Ω in polynomial time by Kasteleyn's algorithm.

The next problem is a satisfiability problem.

#PL-3-NAE-SAT

INPUT: A planar formula Φ consisting of a conjunction of NAE clauses each of size 3.

OUTPUT: The number of satisfying assignments of Φ.

This is a variant of 3SAT. A Boolean formula is planar if it can be represented by a planar graph where vertices represent variables and clauses, and there is an edge iff the variable or its negation appears in that clause. The SAT problem is when the gate for each clause is the Boolean OR. When SAT is restricted to planar formulae, it is still NP-complete, and its corresponding counting problem is #P-complete. Moreover, for many connectives other than NAE (e.g., EXACTLY ONE), the unrestricted or the planar decision problems are still NP-complete, and the corresponding counting problems are #P-complete.

We design a signature grid as follows: To each NAE clause, we assign a generator with signature $[0, 1, 1, 0]$. To each Boolean variable, we assign a generator with signature $(=_k)$ where k is the number of clauses the variable appears, either negated or unnegated. Further, if a variable occurrence is negated, we have a recognizer $[0, 1, 0]$ along the edge that joins the variable generator and the NAE generator, and if the variable occurrence is unnegated, then we use a recognizer $[1, 0, 1]$ instead. Under a holographic transformation using H, $(=_k)$ is transformed to

$$H^{\otimes k}\left\{\begin{bmatrix}1\\0\end{bmatrix}^{\otimes k}+\begin{bmatrix}0\\1\end{bmatrix}^{\otimes k}\right\}=\begin{bmatrix}1\\1\end{bmatrix}^{\otimes k}+\begin{bmatrix}1\\-1\end{bmatrix}^{\otimes k}$$
$$=2[1,0,1,0,\ldots].$$

It can be verified that all the signatures used satisfy all matchgate identities and thus can be realized by matchgates under the holographic transformation.

Recommended Reading

1. Cai JY, Gorenstein A (2014) Matchgates revisited. Theory Comput 10(7):167–197
2. Cai JY, Lu P (2008) Basis collapse in holographic algorithms. Comput Complex 17(2):254–281
3. Cai JY, Lu P (2009) Holographic algorithms: the power of dimensionality resolved. Theor Comput Sci Comput Sci 410(18):1618–1628
4. Cai JY, Lu P (2010) On symmetric signatures in holographic algorithms. Theory Comput Syst 46(3):398–415
5. Cai JY, Lu P (2011) Holographic algorithms: from art to science. J Comput Syst Sci 77(1):41–61
6. Cai JY, Choudhary V, Lu P (2009) On the theory of matchgate computations. Theory Comput Syst 45(1):108–132
7. Kasteleyn PW (1961) The statistics of dimers on a lattice. Physica 27:1209–1225
8. Kasteleyn PW (1967) Graph theory and crystal physics. In: Harary F (ed) Graph theory and theoretical physics. Academic, London, pp 43–110
9. Temperley HNV, Fisher ME (1961) Dimer problem in statistical mechanics – an exact result. Philos Mag 6:1061–1063
10. Valiant LG (2006) Accidental algorthims. In: FOCS '06: proceedings of the 47th annual IEEE symposium on foundations of computer science. IEEE Computer Society, Washington, pp 509–517. doi:http://dx.doi.org/10.1109/FOCS.2006.7
11. Valiant LG (2008) Holographic algorithms. SIAM J Comput 37(5):1565–1594. doi:http://dx.doi.org/10.1137/070682575

Hospitals/Residents Problem

David F. Manlove
School of Computing Science, University of Glasgow, Glasgow, UK

Keywords

Matching; Stability

Synonyms

College admissions problem; Stable admissions problem; Stable assignment problem; Stable b-matching problem; University admissions problem

Years and Authors of Summarized Original Work

1962; Gale, Shapley

Problem Definition

An instance I of the *Hospitals/Residents problem* (HR) [6, 7, 18] involves a set $R = \{r_1, \ldots, r_n\}$ of *residents* and a set $H = \{h_1, \ldots, h_m\}$ of *hospitals*. Each hospital $h_j \in H$ has a positive integral *capacity*, denoted by c_j. Also, each resident $r_i \in R$ has a *preference list* in which he ranks in strict order a subset of H. A pair $(r_i, h_j) \in R \times H$ is said to be *acceptable* if h_j appears in r_i's preference list; in this case r_i is said to *find h_j acceptable*. Similarly each hospital $h_j \in H$ has a preference list in which it ranks in strict order those residents who find h_j acceptable. Given any three agents $x, y, z \in R \cup H$, x is said to *prefer* y to z if x finds each of y and z acceptable, and y precedes z on x's preference list. Let $C = \sum_{h_j \in H} c_j$.

Let A denote the set of acceptable pairs in I, and let $L = |A|$. An *assignment* M is a subset of A. If $(r_i, h_j) \in M$, r_i is said to be *assigned to h_j*, and h_j is *assigned r_i*. For each $q \in R \cup H$, the set of assignees of q in M is denoted by $M(q)$. If $r_i \in R$ and $M(r_i) = \emptyset$, r_i is said to be *unassigned*; otherwise r_i is *assigned*. Similarly, any hospital $h_j \in H$ is *under-subscribed*, *full*, or *over-subscribed* according as $|M(h_j)|$ is less than, equal to, or greater than c_j, respectively.

A *matching* M is an assignment such that $|M(r_i)| \leq 1$ for each $r_i \in R$ and $|M(h_j)| \leq c_j$ for each $h_j \in H$ (i.e., no resident is assigned to an unacceptable hospital, each resident is assigned to at most one hospital, and no hospital is over-subscribed). For notational convenience, given a matching M and a resident $r_i \in R$ such that $M(r_i) \neq \emptyset$, where there is no ambiguity, the notation $M(r_i)$ is also used to refer to the single member of $M(r_i)$.

A pair $(r_i, h_j) \in A \backslash M$ *blocks* a matching M or is a *blocking pair* for M, if the following conditions are satisfied relative to M:

1. r_i is unassigned or prefers h_j to $M(r_i)$;
2. h_j is under-subscribed or prefers r_i to at least one member of $M(h_j)$ (or both).

A matching M is said to be *stable* if it admits no blocking pair. Given an instance I of HR, the problem is to find a stable matching in I.

Key Results

HR was first defined by Gale and Shapley [6] under the name "College Admissions Problem." In their seminal paper, the authors' primary consideration is the classical *Stable Marriage problem* (SM; see Entries ▶ Stable Marriage and ▶ Optimal Stable Marriage), which is a special case of HR in which $n = m$, $A = R \times H$, and $c_j = 1$ for all $h_j \in H$ – in this case, the residents and hospitals are more commonly referred to as the *men* and *women*, respectively. Gale and Shapley showed that every instance I of HR admits at least one stable matching. Their proof of this result is constructive, i.e., an algorithm for finding a stable matching in I is described. This algorithm has become known as the *Gale/Shapley algorithm*.

An extended version of the Gale/Shapley algorithm for HR is shown in Fig. 1. The algorithm involves a sequence of *apply* and *delete* operations. At each iteration of the while loop, some unassigned resident r_i with a nonempty preference list applies to the first hospital h_j on his list and becomes provisionally assigned to h_j (this assignment could subsequently be broken). If h_j becomes over-subscribed as a result of this assignment, then h_j rejects its worst assigned resident r_k. Next, if h_j is full (irrespective of whether h_j was over-subscribed earlier in the same loop iteration), then for each resident r_l that h_j finds less desirable than its worst assigned resident r_k, the algorithm *deletes the pair* (r_l, h_j), which comprises deleting h_j from r_l's preference list and vice versa.

Given that the above algorithm involves residents applying to hospitals, it has become known as the *Resident-oriented* Gale/Shapley algorithm, or RGS algorithm for short [7, Section 1.6.3]. The RGS algorithm terminates with a stable matching, given an instance of HR [6] [7, Theorem 1.6.2]. Using a suitable choice of data structures (extending those described in [7, Section 1.2.3]), the RGS algorithm can be implemented to run in $O(L)$ time. This algorithm produces the unique stable matching that is simultaneously best possible for all residents [6] [7, Theorem 1.6.2]. These observations may be summarized as follows:

Theorem 1 *Given an instance of HR, the RGS algorithm constructs, in $O(L)$ time, the unique stable matching in which each assigned resident obtains the best hospital that he could obtain in any stable matching, while each unassigned resident is unassigned in every stable matching.*

A counterpart of the RGS algorithm, known as the *Hospital-oriented Gale/Shapley algorithm*, or HGS algorithm for short [7, Section 1.6.2], gives the unique stable matching that similarly satisfies an optimality property for the hospitals [7, Theorem 1.6.1].

Although there may be many stable matchings for a given instance I of HR, some key structural properties hold regarding unassigned residents and under-subscribed hospitals with respect to all stable matchings in I, as follows.

Theorem 2 *For a given instance of HR:*

- *The same residents are assigned in all stable matchings;*
- *Each hospital is assigned the same number of residents in all stable matchings;*
- *Any hospital that is under-subscribed in one stable matching is assigned exactly the same set of residents in all stable matchings.*

These results are collectively known as the "Rural Hospitals Theorem" (see [7, Section 1.6.4] for further details). Furthermore, the set of stable matchings in I forms a distributive lattice under a natural dominance relation [7, Section 1.6.5].

```
M := ∅;
while (some resident r_i is unassigned and r_i has a nonempty list) {
    h_j := first hospital on r_i's list;
    /* r_i applies to h_j */
    M := M ∪ {(r_i, h_j)};
    if (h_j is over-subscribed) {
        r_k := worst resident in M(h_j) according to h_j's list;
        M := M\{(r_k, h_j)};
    }
    if (h_j is full) {
        r_k := worst resident in M(h_j) according to h_j's list;
        for (each successor r_l of r_k on h_j's list)
            delete the pair (r_l, h_j);
    }
}
```

Hospitals/Residents Problem, Fig. 1 Gale/Shapley algorithm for HR

Applications

Practical applications of HR are widespread, most notably arising in the context of centralized automated matching schemes that assign applicants to posts (e.g., medical students to hospitals, school leavers to universities, and primary school pupils to secondary schools). Perhaps the largest and best-known example of such a scheme is the National Resident Matching Program (NRMP) in the USA [8], which annually assigns around 31,000 graduating medical students (known as residents) to their first hospital posts, taking into account the preferences of residents over hospitals and vice versa and the hospital capacities. Counterparts of the NRMP are in existence in other countries, including Canada [9] and Japan [10]. These matching schemes essentially employ extensions of the RGS algorithm for HR.

Centralized matching schemes based largely on HR also occur in other practical contexts, such as school placement in New York [1], university faculty recruitment in France [3], and university admission in Spain [16]. Further applications are described in [15, Section 1.3.7].

Indeed, the Nobel Prize in Economic Sciences was awarded in 2012 to Alvin Roth and Lloyd Shapley, partly for their theoretical work on HR and its variants [6, 18] and partly for their contribution to the widespread deployment of algorithms for HR in practical settings such as junior doctor allocation as noted above.

Extensions of HR

One key extension of HR that has considerable practical importance arises when an instance may involve a set of *couples*, each of which submits a joint preference list over pairs of hospitals (typically in order that the members of the couple can be located geographically close to one another). The extension of HR in which couples may be involved is denoted by HRC; the stability definition in HRC is a natural extension of that in HR (see [15, Section 5.3] for a formal definition of HRC). It is known that an instance of HRC need not admit a stable matching (see [4]). Moreover, the problem of deciding whether an HRC instance admits a stable matching is NP-complete [17].

HR may be regarded as a many-one generalization of SM. A further generalization of SM is to a many-many stable matching problem, in which both residents and hospitals may be multiply assigned subject to capacity constraints. In this case, residents and hospitals are more commonly referred to as *workers* and *firms*, respectively. There are two basic variations of the many-many stable matching problem according to whether workers rank (i) individual acceptable firms in order of preference and vice versa or (ii) acceptable *subsets* of firms in order of preference and vice versa. Previous work relating to both models is surveyed in [15, Section 5.4].

Other variants of HR may be obtained if preference lists include ties. This extension is again important from a practical perspective, since it may be unrealistic to expect a popular hospital to rank a large number of applicants in strict order, particularly if it is indifferent among groups of applicants. The extension of HR in which preference lists may include ties is denoted by HRT. In this context three natural stability definitions arise, the so-called *weak stability*, *strong stability*, and *super-stability* (see [15, Section 1.3.5] for formal definitions of these concepts). Given an instance I of HRT, it is known that weakly stable matchings may have different sizes, and the problem of finding a maximum cardinality weakly stable matching is NP-hard (see entry ▸ Stable Marriage with Ties and Incomplete Lists for further details). On the other hand, in contrast to the case for weak stability, a super-stable matching in I need not exist, though there is an $O(L)$ algorithm to find such a matching if one does [11]. Analogous results hold in the case of strong stability – in this case, an $O(L^2)$ algorithm [13] was improved by an $O(CL)$ algorithm [14] and extended to the many-many case [5]. Furthermore, counterparts of the Rural Hospitals Theorem hold for HRT under each of the super-stability and strong stability criteria [11, 19].

A further generalization of HR arises when each hospital may be split into several departments, where each department has a capacity, and residents rank individual departments in order of preference. This variant is modeled by the *Student-Project Allocation problem* [15, Section 5.5]. Finally, the *Hospitals/Residents problem under Social Stability* [2] is an extension of HR in which an instance is augmented by a *social network graph* G (a bipartite graph whose vertices correspond to residents and hospitals and whose edges form a subset of A) such that a blocking pair must additionally satisfy the property that it forms an edge of G. Edges in G correspond to resident–hospital pairs that are acquainted with one another and therefore more likely to block a matching in practice.

Open Problems

As noted in Section "Applications," ties in the hospitals' preference lists may arise naturally in practical applications. In an HRT instance, weak stability is the most commonly-studied stability criterion, due to the guaranteed existence of such a matching. Attempting to match as many residents as possible motivates the search for large weakly stable matchings. Several approximation algorithms for finding a maximum cardinality weakly stable matching have been formulated (see ▸ Stable Marriage with Ties and Incomplete Lists and [15, Section 3.2.6] for further details). It remains open to find tighter upper and lower bounds for the approximability of this problem.

URL to Code

Ada implementations of the RGS and HGS algorithms for HR may be found via the following URL: http://www.dcs.gla.ac.uk/research/algorithms/stable.

Cross-References

- ▸ Optimal Stable Marriage
- ▸ Ranked Matching
- ▸ Stable Marriage
- ▸ Stable Marriage and Discrete Convex Analysis
- ▸ Stable Marriage with Ties and Incomplete Lists
- ▸ Stable Partition Problem

Recommended Reading

1. Abdulkadiroğlu A, Pathak PA, Roth AE (2005) The New York city high school match. Am Econ Rev 95(2):364–367
2. Askalidis G, Immorlica N, Kwanashie A, Manlove DF, Pountourakis E (2013) Socially stable matchings in the Hospitals/Residents problem. In: Proceedings

H

of the 13th Algorithms and Data Structures Symposium (WADS' 13), London, Canada. Lecture Notes in Computer Science, vol 8037. Springer, pp 85–96
3. Baïou M, Balinski M (2004) Student admissions and faculty recruitment. Theor Comput Sci 322(2):245–265
4. Biró P, Klijn F (2013) Matching with couples: a multidisciplinary survey. Int Game Theory Rev 15(2):Article number 1340008
5. Chen N, Ghosh A (2010) Strongly stable assignment. In: Proceedings of the 18th annual European Symposium on Algorithms (ESA' 10), Liverpool, UK. Lecture Notes in Computer Science, vol 6347. Springer, pp 147–158
6. Gale D, Shapley LS (1962) College admissions and the stability of marriage. Am Math Mon 69:9–15
7. Gusfield D, Irving RW (1989) The Stable Marriage Problem: Structure and Algorithms. MIT Press, Cambridge, USA
8. http://www.nrmp.org (National Resident Matching Program website)
9. http://www.carms.ca (Canadian Resident Matching Service website)
10. http://www.jrmp.jp (Japan Resident Matching Program website)
11. Irving RW, Manlove DF, Scott S (2000) The Hospitals/Residents problem with Ties. In: Proceedings of the 7th Scandinavian Workshop on Algorithm Theory (SWAT'00), Bergen, Norway. Lecture Notes in Computer Science, vol 1851. Springer, pp 259–271
12. Irving RW, Manlove DF, Scott S (2002) Strong stability in the Hospitals/Residents problem. Technical report TR-2002-123, Department of Computing Science, University of Glasgow. Revised May 2005
13. Irving RW, Manlove DF, Sott S (2003) Strong stability in the Hospitals/Residents problem. In: Proceedings of the 20th Annual Symposium on Theoretical Aspects of Computer Science (STACS'03), Berlin, Germany. Lecture Notes in Computer Science, vol 2607. Springer, pp 439–450. Full version available as [12]
14. Kavitha T, Mehlhorn K, Michail D, Paluch KE (2007) Strongly stable matchings in time $O(nm)$ and extension to the Hospitals-Residents problem. ACM Trans Algorithms 3(2):Article number 15
15. Manlove DF (2013) Algorithmics of matching under preferences. World Scientific, Hackensack, Singapore
16. Romero-Medina A (1998) Implementation of stable solutions in a restricted matching market. Rev Econ Des 3(2):137–147
17. Ronn E (1990) NP-complete stable matching problems. J Algorithms 11:285–304
18. Roth AE, Sotomayor MAO (1990) Two-sided matching: a study in game-theoretic modeling and analysis. Econometric society monographs, vol 18. Cambridge University Press, Cambridge/New York, USA
19. Scott S (2005) A study of stable marriage problems with ties. PhD thesis, Department of Computing Science, University of Glasgow

Hospitals/Residents Problems with Quota Lower Bounds

Huang Chien-Chung
Chalmers University of Technology and University of Gothenburg, Gothenburg, Sweden

Keywords

Hospitals/Residents problem; Lower quotas; Stable matching

Years and Authors of Summarized Original Work

2010; Biró, Fleiner, Irving, Manlove;
2010; Huang;
2011; Hamada, Iwama, Miyazaki;
2012; Fleiner, Kamiyama

Problem Definition

The Hospitals/Residents (**HR**) problem is the many-to-one version of the stable marriage problem introduced by Gale and Shapley. In this problem, a bipartite graph $G = (\mathcal{R} \cup \mathcal{H}, E)$ is given. Each vertex in $\mathcal{H}$ represents a hospital and each vertex in $\mathcal{R}$ a resident. Each vertex has a preference over its neighboring vertices. Each hospital h has an upper quota $u(h)$ specifying the maximum number of residents it can take in a matching. The goal is to find a stable matching while respecting the upper quotas of the hospitals.

The original **HR** has been well studied in the past decades. A recent trend is to assume that each hospital h also comes with a lower quota $l(h)$. In this context, it is required (if possible) that a matching satisfies both the upper and the lower quotas of each hospital. The introduction of such lower quotas is to enforce some policy in hiring or to make the outcome more fair. It is well-known that hospitals in some rural areas suffer from the shortage of doctors.

With the lower quotas, the definition of stability in **HR** and the objective of the problem depend on the applications. Below we summarize three variants that have been considered in the literature.

Minimizing the Number of Blocking Pairs

In this variant, a matching M is feasible if, for each hospital h, $l(h) \leq |M(h)| \leq u(h)$. Given a feasible matching, a resident r and a hospital h form a blocking pair if the following condition holds. (i) $(r, h) \in E \backslash M$, (ii) r is unassigned in M or r prefers h to his assignment $M(r)$, and (iii) $|M(h)| < u(h)$ or h prefers r to one of its assigned residents. A matching is stable if the number of blocking pairs is 0. It is straightforward to check whether a stable matching exists. We assume that the given instance has no stable matching and the objective is to find a matching with the minimum number of blocking pairs. We call this problem **Min-BP HR**. An alternative objective is to minimize the number of residents that are part of a blocking pair in a matching. We call this problem **Min-BR HR**.

HR with the Option of Closing a Hospital

The following variation of **HR** is motivated by the higher education system in Hungary. Instead of requiring all hospitals to have enough residents to meet their lower quotas, it is allowed that a hospital be closed as long as there is not too much demand for it.

Precisely, in this variant, a matching M is feasible if, for each hospital h, $|M(h)| = 0$ or $l(h) \leq |M(h)| \leq u(h)$. In the former case, a hospital is closed; in the latter case, a hospital is opened. Given a feasible matching M, it is stable if

1. There is no opened hospital h and resident r so that (i) $(h, r) \in E \backslash M$, (ii) r is unassigned in M or r prefers h to his assignment $M(r)$, and (iii) $|M(h)| < u(h)$ or h prefers r to one of its assigned residents;
2. There is no closed hospital h and a set $R \subseteq \mathcal{R}$ of residents so that (i) $|R| \geq |l(h)|$, (ii) for each $r \in R$, $(r, h) \in E \backslash M$, and (iii) each resident $r \in R$ is either unassigned or prefers h to his assigned hospital $M(r)$.

With the above definition of stability, we refer to the question of the existence of a stable matching as **HR woCH**.

Classified HR

Motivated by the practice in academic hiring, Huang introduced a more generalized variant of **HR**. In this variant, a hospital h has a classification h_C over its neighboring residents. Each class $C \in h_C$ comes with a upper quota $u(C)$ and a lower quota $l(C)$. A matching M is feasible if, for each hospital h and for each of its classes $c \in h_C$, $l(C) \leq |M(h)| \leq u(C)$. A feasible matching M is stable if the following condition holds: there is no hospital h such that

1. There exists a resident r so that $(r, h) \in E \backslash M$, and r is either unassigned in M or r prefers h to his assignment $M(r)$;
2. For every class $C \in h_C$, $l(C) \leq |M(h) \cup \{r\}| \leq u(C)$, or there exists another resident $r' \in M(h)$ so that h prefers r to r' and for every class $C \in h_C$, $l(C) \leq |M(h) \cup \{r\} \backslash \{r'\}| \leq u(C)$.

With the above definition of stability, we refer to the question of the existence of a stable matching as **CHR**.

Key Results

For the first variant where the objective is to minimize the number of blocking pairs, Hamada et al. showed the following tight results.

Theorem 1 ([3]) *For any positive constant $\epsilon > 0$, there is no polynomial-time $(|\mathcal{R}| + |\mathcal{H}|)^{1-\epsilon}$-approximation algorithm for* ***Min-BP HR*** *unless P=NP. This holds true even if the given bipartite graph is complete and all upper quotas are 1 and all lower quotas are 0 or 1.*

Theorem 2 ([3]) *There is a polynomial-time* $(|\mathcal{R}| + |\mathcal{H}|)$*-approximation algorithm for* ***Min-BP HR***.

In the case that the objective is to minimize the number of residents involved in blocking pairs, Hamada et al. showed the following.

Theorem 3 ([3]) ***Min-BR HR*** *is NP-hard. This holds true even if the given bipartite graph is complete and all hospitals have the same preference over the residents.*

Theorem 4 ([3]) *There is a polynomial-time* $\sqrt{|\mathcal{R}|}$*-approximation algorithm for* ***Min-BR HR***.

For the second variant, where a hospital is allowed to be closed, Biró et al. showed the following.

Theorem 5 ([1]) *The problem* ***HR woCH*** *is NP-complete. This holds true even if all upper quotas are at most 3.*

For the last variant where each hospital is allowed to classify the neighboring residents and sets the upper and lower quotas for each of its classes, Huang showed that if all classifications of the hospitals are laminar families, the problem is in P. Fleiner and Kamiyama later proved the same result by a significantly simpler matroid-based technique.

Theorem 6 ([2,4]) *In* ***CHR***, *if all classifications of the hospitals are laminar families, then one find a stable matching or detect its absence in the given instance in* $O(nm)$ *time, where* $n = |\mathcal{R} \cup \mathcal{H}|$ *and* $m = |E|$.

Recommended Reading

1. Biro P, Fleiner T, Irving RW, Manlove DF (2010) The College Admissions problem with lower and common quotas. Theor Comput Sci 411(34–36):3136–3153
2. Fleiner T, Kamiyama N (2012) A matroid approach to stable matchings with lower quotas In: 23nd annual ACM-SIAM symposium on discrete algorithms, Kyoto, pp 135–142
3. Hamada K, Iwama K, Miyazaki S (2011) The Hospitals/Residents problem with quota lower bounds. In: 19th annual European symposium on algorithms, Saarbrücken, vol 411(34–36), pp 180–191
4. Huang C-C (2010) Classified stable matching. In: 21st annual ACM-SIAM symposium on discrete algorithms, Austin, pp 1235–1253

Hub Labeling (2-Hop Labeling)

Daniel Delling[1], Andrew V. Goldberg[2], and Renato F. Werneck[3]
[1]Microsoft, Silicon Valley, CA, USA
[2]Microsoft Research – Silicon Valley, Mountain View, CA, USA
[3]Microsoft Research Silicon Valley, La Avenida, CA, USA

Keywords

Distance oracles; Labeling algorithms; Shortest paths

Years and Authors of Summarized Original Work

2003; Cohen, Halperin, Kaplan, Zwick
2012; Abraham, Delling, Goldberg, Werneck
2013; Akiba, Iwata, Yoshida
2014; Delling, Goldberg, Pajor, Werneck
2014; Delling, Goldberg, Savchenko, Werneck

Problem Definition

Given a directed graph $G = (V, A)$ (with $n = |V|$ and $m = |A|$) with a length function $\ell : A \to \mathbf{R}^+$ and a pair of vertices s, t, a *distance oracle* returns the distance $\text{dist}(s, t)$ from s to t. A *labeling algorithm* [18] implements distance oracles in two stages. The *preprocessing* stage computes a *label* for each vertex of the input graph. Then, given s and t, the *query* stage computes $\text{dist}(s, t)$ using only the labels of s and t; the query does not explicitly use G and ℓ.

Hub labeling (HL) (or 2-hop labeling) is a special kind of labeling algorithm. The label $L(v)$ of a vertex v consists of two parts: the *forward label* $L_f(v)$ is a collection of vertices w with their distances $\text{dist}(v, w)$ from v, while the

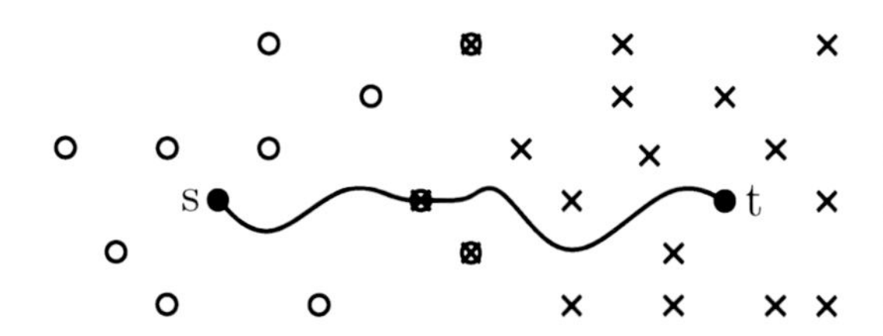

Hub Labeling (2-Hop Labeling), Fig. 1 Example of a hub labeling. The hubs of s are *circles*; the hubs of t are *crosses* (Taken from [3])

backward label $L_b(v)$ is a collection of vertices u with their distances $\text{dist}(u, v)$ to v. (If the graph is undirected, a single label per vertex suffices.) The vertices in v's label are the *hubs* of v. The labels must obey the *cover property*: for any two vertices s and t, the set $L_f(s) \cap L_b(t)$ must contain at least one hub that is on the shortest $s - t$ path. Given the labels, HL queries are straightforward: to find $\text{dist}(s, t)$, simply find the hub $x \in L_f(s) \cap L_b(t)$ that minimizes $\text{dist}(s, x) + \text{dist}(x, t)$ (see Fig. 1 for an example). If the hubs in each label are sorted by ID, queries consist of a simple linear sweep over the labels, as in mergesort.

The size of a forward (backward) label, $|L_f(v)|$ ($|L_b(v)|$), is the number of hubs it contains. The *size of a labeling* L is the sum of the average label sizes, $(L_f(v) + L_b(v))/2$, over all vertices. The memory footprint of the algorithm is proportional to the size of the labeling, while query times are determined by the maximum label size. Queries themselves are trivial; the hard part is an efficient implementation of a preprocessing algorithm that, given G and ℓ, computes a small hub labeling.

Key Results

We describe an approximation algorithm for finding labelings of size within $O(\log n)$ of the optimal [9], as well as its generalization to other objectives, including the maximum label size [6]. Although polynomial, these approximation algorithms do not scale to large networks. For more practical alternatives, we discuss *hierarchical hub labelings* (HHLS), a subclass of HL. We show that HHLs are closely related to vertex orderings and present efficient algorithms for computing the minimal HHL for a given ordering, as well as heuristics for finding vertex orderings that lead to small labels. In particular, the RXL algorithm uses sampling to efficiently approximate a greedy vertex order, leading to empirically small labels. RXL can handle large problems from several application domains. We then discuss representations of hub labels that allow various trade-offs between space and query time.

General Hub Labelings

The time and space efficiency of the distance oracles we discuss depend on the label size. If labels are big, HL is impractical. Gavoille et al. [15] show that there exist graphs for which general labelings must have size $\tilde{\Theta}(n^2)$. For planar graphs, they give an $\tilde{\Omega}(n^{4/3})$ lower and $\tilde{O}(n^{3/2})$ upper bound. They also show that graphs with k-separators have hub labelings of size $\tilde{O}(nk)$. Abraham et al. [1] show that graphs with small highway dimension (which they conjecture include road networks) have small hub labelings.

Given a particular graph, computing a labeling with the smallest size is NP-hard. Cohen et al. [9] developed an $O(\log n)$-approximation algorithm for the problem. Next we discuss this *general HL* (GHL) algorithm.

A *partial labeling* is a labeling that does not necessarily satisfy the cover property. Given a partial labeling $L = (L_f, L_b)$, we say that a vertex pair $[u, w]$ is *covered* if $L_f(u) \cap L_b(w)$ contains a vertex on a shortest path from u to w and *uncovered* otherwise. GHL maintains a partial labeling L (initially empty) and the corresponding set U of uncovered vertex pairs. Each iteration of the algorithm selects a vertex v and two subsets $X', Y' \subseteq V$, adds $(v, \text{dist}(x, v))$ to $L_f(x)$ for all $x \in X'$, and adds $(y, \text{dist}(v, y))$ to $L_b(y)$ for all $y \in Y'$. Then, GHL deletes from U the set $U(v, X', Y')$ of vertex pairs that become covered by this augmentation. Among all $v \in V$ and $X', Y' \subseteq V$, the triple (v, X', Y') picked in each iteration is one that maximizes $|U(v, X', Y')|/(|X'| + |Y'|)$, i.e., the ratio of the number of paths covered over the increase in label size.

Cohen et al.'s efficient implementation of GHL uses the notion of *center graphs*. Given a set U of vertex pairs and a vertex v, the center graph $G_v = (X, Y, A_v)$ is a bipartite graph with $X = Y = V$ such that an arc $(u, w) \in A_v$ if $[u, w] \in U$ and some shortest path from u to w in G go through v. If U is the set of uncovered vertex pairs, then, for a fixed vertex v, maximizing $|U(v, X', Y')|/(|X'| + |Y'|)$ over all $X', Y' \subseteq V$ is (by definition) the same as finding the vertex induced-subgraph of G_v with maximum *density* (defined as its number of arcs divided by its number of vertices). This *maximum density subgraph* (MDS) problem can be solved in polynomial time using parametric flows (see e.g., [14]). To maximize the ratio over all triples (v, X', Y'), GHL solves an MDS problem for center graphs G_v and picks the densest of the n resulting subgraphs. It then adds the corresponding vertex v^* to the labels of the vertices given by the sides of the MDS. Arcs corresponding to newly covered pairs are removed from center graphs between iterations.

Cohen et al. show that GHL is a special case of the greedy set cover algorithm [8] and thus gives an $O(\log n)$-optimal labeling. They also show that the same guarantee holds if one uses a constant-factor approximation to the MDS. We refer to a k approximation of MDS as a k-AMDS. Using a linear-time 2-AMDS algorithm by Kortsarz and Peleg [17], each GHL iteration is dominated by n AMDS computations on graphs with $O(n^2)$ arcs. Since each iteration increases the size of the labeling, the number of iterations is at most $O(n^2)$. The total running time of GHL is thus $O(n^5)$.

Delling et al. [11] improve the time bound for GHL to $O(n^3 \log n)$ using *eager* and *lazy* evaluation. Intuitively, eager evaluation finds an AMDS G' of G such that deleting G' reduces the MDS value of G by a constant factor. More precisely, given a graph G, an upper bound μ on the MDS value of G and a parameter $\alpha > 1$, *α-eager evaluation* attempts to find a (2α)-AMDS G' of G such that the MDS value of G with the arcs of G' deleted is at most μ/α. If the evaluation fails to find such G', the MDS value of G is at most μ/α. *Lazy evaluation* was introduced by Cohen et al. [9] to speed up their implementation of GHL and refined by Stengel et al. [20]. It is based on the observation that the MDS value of a center graph does not increase as the algorithm adds vertices to labels and removes arcs from center graphs.

The eager-lazy algorithm maintains upper bounds on the center subgraph densities μ_v computed in previous iterations. These values are computed during initialization and updated in a lazy fashion as follows. In each iteration, the algorithm picks the maximum μ_v and applies α-eager evaluation to G_v. If the evaluation succeeds, the labels are updated. Regardless of whether the evaluation succeeds or not, μ_v/α is a valid upper bound on the density of G_v at the end of the iteration. This can be used to show that each vertex is selected by $O(n \log n)$ iterations, each taking $O(n^2)$ time.

Babenko et al. [6] generalize the definition of a labeling size as follows. Suppose vertex IDs are $1, 2, \ldots, n$. Define a $(2n)$-dimensional vector $\mathcal{L}$ by $\mathcal{L}_{2i-1} = |L_f(i)|$ and $\mathcal{L}_{2i} = |L_b(i)|$. The p-norm of $\mathcal{L}$ is defined as $\|\mathcal{L}\|_p = (\sum_{i=0}^{2n-1} \mathcal{L}_i^p)^{1/p}$, where p is a natural number and $\|\mathcal{L}\|_\infty = \max \mathcal{L}_i$. Note that $\|\mathcal{L}\|_1/2$ is the total size of the labeling and $\|\mathcal{L}\|_\infty$ is the maximum label size. Babenko et al. [6] generalize the algorithm of Cohen et al. to obtain an $O(\log n)$-approximation algorithm for this more general problem in $O(n^5)$ time. Delling et al. [11] show that the eager-lazy approach yields an $O(\log n)$-approximation algorithm running in time $O(n^3 \log n \min(p, \log n))$.

Hierarchical Hub Labelings

Even with the performance improvements mentioned above, GHL requires too much time and space to work on large networks. To overcome this problem, one may use heuristics that have no known theoretical guarantees on the label size but produce small labels for large instances from a wide variety of domains. The most successful current heuristics use a restricted class of labelings called *hierarchical hub labeling* (HHL) [4]. Hierarchical labels have the cover property and implement exact distance oracles.

Given a labeling, let $v \lesssim w$ if w is a hub of $L(v)$. HL is *hierarchical* if $\lesssim$ is a partial order. (Intuitively, $v \lesssim w$ if w is "more important" than v.) We say that an HHL *respects* a given (total) order on the vertices if the partial order $\lesssim$ induced by the HHL is consistent with the order.

Consider an order defined by a permutation *rank*, with $rank(v) < rank(w)$ if v appears before (is less important than) w. The *canonical labeling* L for *rank* is defined as follows [4]. Vertex v belongs to $L_f(u)$ if and only if there exists w such that v is the highest-ranked vertex that hits $[u, w]$. Similarly, v belongs to $L_b(w)$ if and only if there exists u such that v is the highest-ranked vertex that hits $[u, w]$.

Abraham et al. [4] prove that the canonical labeling for a given vertex order *rank* is the minimum-sized labeling that respects *rank*. This suggests a two-stage approach for finding a small hierarchical hub labeling: first, find a "good" vertex order, and then compute its corresponding canonical labeling. We first discuss the latter step and then the former.

From Orderings to Labelings

We first consider how, given an order *rank*, one can compute the canonical hierarchical labeling L that respects *rank*.

The straightforward way is to just apply the definition: for every pair $[u, w]$ of vertices, find the maximum-ranked vertex on any shortest u–w path, and then add it to $L_f(u)$ and $L_b(w)$. Although polynomial, this algorithm is too slow in practice.

A faster (but still natural) algorithm is as follows [4]. Start with an empty (partial) labeling L, and process vertices from the most to least important. When processing v, for every uncovered pair $[u, w]$ that v covers, add v to $L_f(u)$ and $L_b(w)$. (In other words, add v to the labels of all end points of arcs in the center graph G_v.) Abraham et al. [4] show how to implement this in $O(mn \log n)$ time and $\Theta(n^2)$ space, which is still impractical for large instances.

When labels are not too large, a much more efficient solution is the *pruned labeling* (PL) algorithm by Akiba et al. [5]. Starting from empty labels, PL also processes vertices from the most to least important, with the iteration that processes vertex v, adding v to all relevant labels. The crucial observation is that, when processing v, one only needs to look at uncovered pairs containing v itself; if $[u, v]$ is not covered, PL adds v to $L_f(u)$; if $[v, w]$ is not covered, it adds v to $L_b(w)$. This is enough because of the subpath optimality property of the shortest paths.

To process v efficiently, PL runs two pruned Dijkstra searches [13] from v. The first search works on the forward graph (out of v) as follows. Before scanning a vertex w (with distance label $d(w)$ within the Dijkstra search), it computes a v–w distance estimate q by performing an HL query with the current partial labels. (If the labels do not intersect, set $q = \infty$.) If $q \leq d(w)$, the $[v, w]$ pair is already covered by previous hubs, so PL prunes the search (ignores w). Otherwise (if $q > d(w)$), PL adds $(v, \text{dist}(v, w))$ to $L_b(w)$ and scans w as usual. The second Dijkstra search uses the reverse graph and is pruned similarly; it adds $(v, \text{dist}(w, v))$ to $L_f(w)$ for all scanned vertices w. Note that the number of Dijkstra scans equals the size of the labeling. Since each visited vertex requires an HL query using partial labels, the running time can be quadratic in the average label size. It is easy to see that PL produces canonical labelings.

The final algorithm we discuss, due to Abraham et al. [4], computes a hierarchical hub labeling from a vertex ordering recursively. Its basic building block is the *shortcut operation* (see e.g., [16]). To shortcut a vertex v, the operation deletes v from the graph and adds arcs to ensure that the distances between the remaining vertices remain unchanged. For every pair consisting of an incoming arc (u, v) and an outgoing arc (v, w), the algorithm checks if $(u, v) \cdot (v, w)$ is the only shortest u–w path (by running a partial Dijkstra search from u or w) and, if so, adds a new arc (u, w) with length $\ell(u, w) = \ell(u, v) + \ell(v, w)$.

The recursive algorithm computes one label at a time, from the bottom up (from the least to the most important vertex). It starts by shortcutting the least important vertex v from G to get a graph

G' (same as G, but without v and its incident arcs and with the added shortcuts). It then recursively finds a labeling for G', which gives correct distances (in G) for all pairs of vertices not containing v. Then, the algorithm computes the label of v from the labels of its neighbors. We describe how to compute $L_f(v)$; $L_b(v)$ is computed similarly. The crucial observation is that any nontrivial shortest path starting at v must go through one of its neighbors. Accordingly, we initialize $L_f(v)$ with entry $(v, 0)$ (to cover the trivial path from v to itself), and then, for every neighbor w of v in G and every entry $(x, \text{dist}(w, x)) \in L_f(w)$, add $(x, \ell(v, w) + \text{dist}(w, x))$ to $L_f(v)$. If x already is a hub of v, we only keep the smallest entry for x. Finally, we prune from $L_f(v)$ the entries $(x, \ell(v, w) + \text{dist}(w, x))$ for which $\ell(v, w) + \text{dist}(w, x) > \text{dist}(v, x)$. (This can happen if the shortest path from v to x through another neighbor w' of v is shorter than the one through w.) Note that $\text{dist}(v, x)$ can be computed using the labels of v and x. In general, the shortcut operation can make the graph dense, limiting the efficiency of the bottom-up approach. On some network classes, such as road networks, the graph remains sparse and the approach scales to large problems.

Vertex Ordering Heuristics

As mentioned above, the size of the labeling is determined by the ordering. The most natural approach to capture the notion of importance is attributed to Abraham et al. [4], whose *greedy ordering algorithm* obtains good orderings on a wide class of problems. It orders vertices from the most to least important using a greedy selection rule. In each iteration, it selects as the next most important hub the vertex v that hits the most vertex pairs not covered by previously selected vertices.

When the shortest paths are unique, this can be implemented relatively efficiently. The algorithm maintains (initially full) the shortest-path trees from each vertex in the graph. The tree T_s rooted at s implicitly represents all shortest paths starting at s. The total number of descendants of a vertex v (in aggregate over all trees) is exactly the number of paths it covers. Once such a vertex v is picked as the next hub, we restore this invariant for the remaining paths by removing all of v's descendants (including v itself) from all trees. Abraham et al. [4] show how the entire greedy order can be found in $O(nm \log n)$ time. An alternative algorithm (in the same spirit) works even if the shortest paths are not unique, but takes $O(n^3)$ time [12].

The *weighted greedy ordering algorithm* is similar but selects v so as to maximize the ratio of the number of uncovered paths that v covers to the increase in the label size if v is selected next. This gives slightly better results and can be implemented in the same time bounds as the greedy ordering algorithm [4, 12]. Although faster than GHL, none of these greedy variants scale to large graphs.

To cope with this problem, Delling et al. [12] developed RXL (Robust eXact Labeling), which can be seen as a sampling version of the greedy ordering algorithm. In each iteration, RXL finds a vertex v that *approximately* maximizes the number of pairs covered. Rather than maintaining n shortest-path trees, RXL maintains shortest-path trees from a small number of roots picked uniformly at random. It estimates the coverage of v based on how many descendants it has in these trees. To reduce the bias in this estimation, the algorithm discards outliers before taking the average number of descendants. Moreover, as the original trees shrink (because some of its subtrees become covered), new subtrees (from other roots) are added. These new trees are not full, however; they are pruned from the start (using PL), ensuring the total space (and time) usage remains under control.

For certain graph classes, simpler ordering techniques can be used. Akiba et al. [5] show that ordering by degree works well on a subclass of complex networks. Abraham et al. [2,4] show that the order induced by the contraction hierarchies (CH) algorithm [16] works well on road networks and other sparse inputs. CH order vertices from the bottom up: using only local information, it determines the least important vertex, shortcuts it, and repeats the process in the remaining graph. The most relevant signals to estimate the importance of v are the arc difference (of number of arcs removed and added if v were shortcut) and how many neighbors of v have already been shortcut.

Label Representation and Queries

Given a source s and a target t, one can compute the minimum of $\text{dist}(s, v) + \text{dist}(v, t)$ over all $v \in L_f(s) \cap L_b(t)$ in $O(|L_f(s)| + |L_f(t)|)$ time. If vertex labels are represented as arrays sorted by hub IDs, one can compute $L_f(s) \cap L_b(t)$ by a coordinated sweep of the corresponding arrays, as in mergesort. This is very cache efficient and works well when the two labels have similar sizes.

In some applications, label sizes can be very different. Assuming (without loss of generality) that $|L_f(s)| \ll |L_f(t)|$, one can compute $L_f(s) \cap L_b(t)$ in time $O(|L_f(s)| + \log(|L_b(t)|))$ by performing a binary search for each hub $v \in L_f(s)$ to determine if v is in $L_b(t)$. In fact, this *set intersection problem* can be solved even faster, in $O(\min(|L_f(s)|, |L_b(t)|))$ time [19].

As each label can be stored in a contiguous memory block, HL queries are well suited for an external memory (or even distributed) implementations, including relational databases [3] or key-value stores. In such cases, query times depend on the time to fetch two blocks of data.

For in-memory implementations of HL, storage may be a bottleneck. One can trade space for time using label compression, which interprets each label as a tree and stores common subtrees only once; this reduces space consumption by an order of magnitude, but queries become much less cache efficient [10, 12]. Another technique to reduce the space consumption is to store vertices and a constant number of their neighbors as superhubs in the labels [5]; on unweighted and undirected graphs, distances from a vertex v to all elements of a superhub can be represented compactly in difference form. This works well on some social and communication networks [5].

HL has efficient extensions to problems beyond point-to-point shortest paths, including one-to-many and via-point queries. These are important for applications in road networks, such as finding the closest points of interest, ride sharing, and path prediction [3].

Experimental Results

Even for very small (constant) sample sizes, the labels produced by RXL are typically no more than about 10% bigger [12] than those produced by the full greedy hierarchical algorithms, which in turn are not much worse than those produced by GHL [11]. Scalability is much different, however. In a few hours in a modern CPU, GHL can only handle graphs with about 10,000 vertices [11]; for the greedy hierarchical algorithms, the practical limit is about 100,000 [4]. In contrast, as long as labels remain small, RXL scales to problems with millions of vertices [12] from a wide variety of graph classes, including meshes, grids, random geometric graphs (sensor networks), road networks, social networks, collaboration networks, and web graphs. For example, for a web graph with 18.5 million vertices and almost 300 million arcs, one can find labels with fewer than 300 hubs on average in about half a day [12]; queries then take less than 2 µs.

For some graph classes, other methods have faster preprocessing. For continental road networks with tens of millions of vertices, a hybrid approach combining weighted greedy (for the top few thousand vertices) with the CH order (for all other vertices) provides the best trade-off between preprocessing times and label size [2, 4]. On a benchmark data set representing Western Europe (about 18 million vertices, 42.5 million arcs), it takes roughly an hour to compute labels with about 70 hubs on average, leading to average query times of about 0.5 µs, roughly the time of ten random memory accesses. With additional improvements, one can further reduce query times (but not the label sizes) by half [2], making it the fastest algorithm for this application [7]. For some unweighted and undirected complex (social, communication, and collaboration) networks, simply sorting vertices by degree [5] produces labels that are not much bigger than those computed by a more sophisticated ordering technique.

Overall, RXL is the most robust method. For all instances tested in the literature, its preprocessing is never much slower than any other methods (and often much faster), and query times are similar. In particular, CH-based ordering is too costly for large complex networks (as contraction tends to create dense graphs), and the degree-based order leads to prohibitively large labels for road networks and web graphs.

Recommended Reading

1. Abraham I, Fiat A, Goldberg AV, Werneck RF (2010) Highway dimension, shortest paths, and provably efficient algorithms. In: Proceedings of 21st ACM-SIAM symposium on discrete algorithms, Austin, pp 782–793
2. Abraham I, Delling D, Goldberg AV, Werneck RF (2011) A hub-based labeling algorithm for shortest paths on road networks. In: Proceedings of the 10th international symposium on experimental algorithms (SEA'11), Chania. Volume 6630 of Lecture notes in computer science. Springer, pp 230–241
3. Abraham I, Delling D, Fiat A, Goldberg AV, Werneck RF (2012) HLDB: location-based services in databases. In: Proceedings of the 20th ACM SIGSPATIAL international symposium on advances in geographic information systems (GIS'12), Redondo Beach. ACM, pp 339–348
4. Abraham I, Delling D, Goldberg AV, Werneck RF (2012) Hierarchical hub labelings for shortest paths. In: Proceedings of the 20th annual European symposium on algorithms (ESA'12), Ljubljana. Volume 7501 of Lecture notes in computer science. Springer, pp 24–35
5. Akiba T, Iwata Y, Yoshida Y (2013) Fast exact shortest-path distance queries on large networks by pruned landmark labeling. In: Proceedings of the 2013 ACM SIGMOD international conference on management of data, SIGMOD'13, New York. ACM, pp 349–360
6. Babenko M, Goldberg AV, Gupta A, Nagarajan V (2013) Algorithms for hub label optimization. In: Fomin FV, Freivalds R, Kwiatkowska M, Peleg D (eds) Proceedings of 30th ICALP, Riga. Lecture notes in computer science, vol 7965. Springer, pp 69–80
7. Bast H, Delling D, Goldberg AV, Müller–Hannemann M, Pajor T, Sanders P, Wagner D, Werneck RF (2014) Route planning in transportation networks. Technical report MSR-TR-2014-4, Microsoft research
8. Chvátal V (1979) A greedy heuristic for the set-covering problem. Math Oper Res 4(3): 233–235
9. Cohen E, Halperin E, Kaplan H, Zwick U (2003) Reachability and distance queries via 2-hop labels. SIAM J Comput 32:1338–1355
10. Delling D, Goldberg AV, Werneck RF (2013) Hub label compression. In: Proceedings of the 12th international symposium on experimental algorithms (SEA'13), Rome. Volume 7933 of Lecture notes in computer science. Springer, pp 18–29
11. Delling D, Goldberg AV, Savchenko R, Werneck RF (2014) Hub labels: theory and practice. In: Proceedings of the 13th international symposium on experimental algorithms (SEA'14), Copenhagen. Lecture notes in computer science. Springer
12. Delling D, Goldberg AV, Pajor T, Werneck RF (2014, to appear) Robust distance queries on massive networks. In: Proceedings of the 22nd annual European symposium on algorithms (ESA'14), Wroclaw. Lecture notes in computer science. Springer
13. Dijkstra EW (1959) A note on two problems in connexion with graphs. Numer Math 1: 269–271
14. Gallo G, Grigoriadis MD, Tarjan RE (1989) A fast parametric maximum flow algorithm and applications. SIAM J Comput 18:30–55
15. Gavoille C, Peleg D, Pérennes S, Raz R (2004) Distance labeling in graphs. J Algorithms 53(1): 85–112
16. Geisberger R, Sanders P, Schultes D, Vetter C (2012) Exact routing in large road networks using contraction hierarchies. Transp Sci 46(3):388–404
17. Kortsarz G, Peleg D (1994) Generating sparse 2-spanners. J Algorithms 17:222–236
18. Peleg D (2000) Proximity-preserving labeling schemes. J Graph Theory 33(3):167–176
19. Sanders P, Transier F (2007) Intersection in integer inverted indices. SIAM, Philadelphia, pp 71–83
20. Schenkel R, Theobald A, Weikum G (2004) HOPI: an efficient connection index for complex XML document collections. In: Advances in database technology – EDBT 2004. Springer, Berlin/Heidelberg, pp 237–255

Huffman Coding

Alistair Moffat
Department of Computing and Information Systems, The University of Melbourne, Melbourne, VIC, Australia

Keywords

Compression; Huffman code; Minimum-redundancy code

Years and Authors of Summarized Original Work

1952; Huffman
1976; van Leeuwen
1995; Moffat, Katajainen

Problem Definition

A sequence of n positive weights or frequencies is given, $\langle w_i > 0 \mid 0 \le i < n \rangle$, together with an output radix r, with $r = 2$ in the case of binary output strings.

Objective To determine a sequence of integral codeword lengths $\langle \ell_i \mid 0 \le i < n \rangle$ such that: (a) $\sum_{i=0}^{n-1} r^{-\ell_i} \le 1$, and (b) $C = \sum_{i=0}^{n-1} \ell_i \cdot w_i$ is minimized. Any sequence of codeword lengths $\langle \ell_i \rangle$ that satisfies these two properties describes a *minimum-redundancy code* for the weights $\langle w_i \rangle$. Once a set of minimum-redundancy codeword lengths $\langle \ell_i \rangle$ has been identified, a prefix-free r-ary code in which symbol i is assigned a codeword of length ℓ_i can always be constructed.

Constraints

1. **Long messages.** In one application, each weight w_i is the frequency of symbol i in a message M of length $m = |M| = \sum_{i=0}^{n-1} w_i$, and C is the number of symbols required by a compressed representation of M. In this application it is usual to assume that $m \gg n$.
2. **Entropy-based limit.** Define $W = \sum_{i=0}^{n-1} w_i$ to be the sum of the weights and $p_i = w_i / W$ to be the corresponding probability of symbol i. Define $H_0 = -\sum_{i=0}^{n-1} (p_i \log_2 p_i)$ to be the zero-order entropy of the distribution. Then when $r = 2$, $\lceil n H_0 \rceil \le C \le n \lceil \log_2 n \rceil$.

Key Results

A minimum-redundancy code can be identified in $O(n)$ time if the weights w_i are nondecreasing and in $O(n \log n)$ time if the weights must be sorted first.

Example Weights

The $n = 10$ weights $\langle 1, 1, 1, 1, 3, 4, 4, 7, 9, 9 \rangle$ with $W = 40$ are used as an example.

Huffman's Algorithm

In 1952 David Huffman [3] described a process for calculating minimum-redundancy codes, developed in response to a term-paper challenge set the year before by his MIT class instructor, Robert Fano, a problem that Fano and his collaborator Claude Shannon had already tackled unsuccessfully [7]. In his solution Huffman created a classic algorithm that is taught to most undergraduate computing students as part of algorithms classes. Initially the sequence of input weights $\langle w_i \rangle$ is regarded as being the leaves of a tree, with no internal nodes, and each leaf the root of its own subtree. The two subtrees (whether singleton leaves or internal nodes) with the smallest root nodes are then combined by making both of them children of a new parent, with an assigned weight calculated as the sum of the two original nodes. The pool of subtrees decreases by one at each cycle of this process; after $n-1$ iterations a total of $n-1$ internal nodes has been added, and all of the original nodes must be leaves in a single tree and descendants of that tree's root node.

Figure 1 shows an example of codeword length computation, with the original weights across the top. Each iteration takes the two least-weight elements (leaf or internal) and combines them to make a new internal node; note that the internal nodes are created in nondecreasing weight order. Once the Huffman tree has been constructed, the sequence $\langle \ell_i \rangle$ can be read from it, by computing the depth of each corresponding leaf node. In Fig. 1, for example, one of the elements of weight 4 is at depth three in the tree, and one is at depth four from the root, hence $\ell_5 = 4$ and $\ell_6 = 3$. A set of codewords can be assigned at the same time as the depths are being computed; one possible assignment of codewords that satisfies the computed sequence $\langle \ell_i \rangle$ is shown in the second row in the lower box. Decoding throughput is considerably faster if codewords are assigned systematically based on codeword length in the manner shown, rather than by strictly following the edge labeling of the Huffman tree from which the codeword lengths were extracted [6].

Because ties can occur and can be broken arbitrarily, different codes are also possible. The sequence $\langle \ell_i \rangle = \langle 6, 6, 6, 6, 4, 3, 3, 3, 2, 2 \rangle$ has the same cost of $C = 117$ bits as the one shown

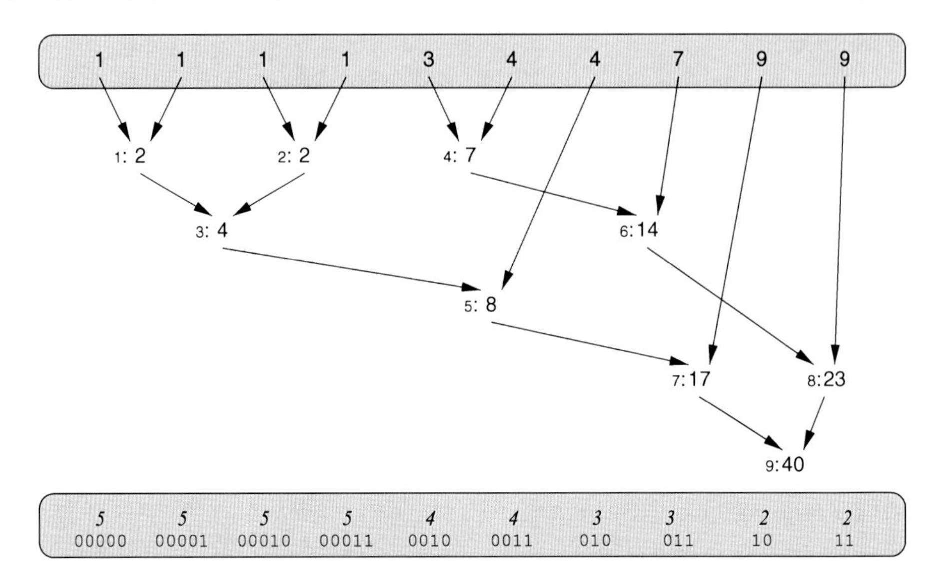

Huffman Coding, Fig. 1 Example of (binary) codeword lengths calculated using Huffman's algorithm, showing the order in which internal nodes are formed, and their weights. The input weights in the top section are used to compute the corresponding codeword lengths in the bottom box. A valid assignment of prefix-free codewords is also shown

in the figure. For the example weights, $H_0 = 2.8853$ bits per symbol, providing a lower bound of $\lceil 115.41 \rceil = 116$ bits on the total cost C for the input weights. In this case, the minimum-redundancy codes listed are just 1 bit inferior to the entropy-based lower limit.

Implementing Huffman's Algorithm

Huffman's algorithm is often used in algorithms textbooks as an example of a process that requires a dynamic priority queue. If a heap is used, for example, the n initial and $n-2$ subsequent insert operations, take a total of $O(n \log n)$ time, as do the $2(n-1)$ extract-min operations.

A simpler approach is to first sort the n weights into increasing order and then apply an $O(n)$-time algorithm due to van Leeuwen [10]. Two sorted lists are maintained: a static one of original weights, representing the leaves of the Huffman tree, and a dynamic queue of internal nodes that is initially empty, to which new internal nodes are appended as they are created. Each iteration compares front-of-list elements from the two lists and combines the two that have the least weight and then adds the new internal node at the tail of the queue. The algorithm stops when the queue contains only one node; it is the last item that was added and is the root of the Huffman tree.

If the input weights are provided in an array $w_i = A[i \mid 0 \leq i < n]$ of sorted integers, that array can be processed in situ into an output array $\ell_i = A[i]$ in $O(n)$ time by van Leeuwen's technique using an implementation described by Moffat and Katajainen [5]. Each array element takes on values that are, variously, input weight, internal node weight, parent pointer, and then, finally, codeword length. Algorithm 1 is taken from Moffat and Katajainen [5] and describes this process in detail. There are three phases of operation. In the first phase, in steps 2–10, leaf weights in $A[\mathit{leaf} \ldots n-1]$ are combined with a queue of internal node weights in $A[\mathit{root} \ldots \mathit{next}-1]$ to form a list of parent pointers in $A[0 \ldots \mathit{root}-1]$. At the end of this phase, $A[0 \ldots n-3]$ is a list of parents, $A[n-2]$ is the sum of the weights, and $A[n-1]$ is unused.

In phase 2 (steps 12–15), the set of parent pointers of internal nodes is converted to a set of internal node depths. This mapping is done by processing the tree from the root down, making the depth of each node one greater than the depth of its parent.

Algorithm 1 Compute Huffman codeword lengths

```
 0: function calc_huff_lens(A, n)                              ▷ Input: A[i − 1] ≤ A[i] for 0 < i < n
 1:    // Phase 1
 2:    set leaf ← 0 and root ← 0
 3:    for next ← 0 to n − 2 do
 4:       if leaf ≥ n or (root < next and A[root] < A[leaf]) then
 5:          set A[next] ← A[root] and A[root] ← next and root ← root + 1    ▷ Use internal node
 6:       else
 7:          set A[next] ← A[leaf] and leaf ← leaf + 1                      ▷ Use leaf node
 8:       end if
 9:       repeat steps 4–8, but adding to A[next] rather than assigning to it  ▷ Find second child
10:    end for
11:    // Phase 2
12:    set A[n − 2] ← 0
13:    for next ← n − 3 downto 0 do
14:       set A[next] ← A[A[next]] + 1                    ▷ Compute depths of internal nodes
15:    end for
16:    // Phase 3
17:    set avail ← 1 and used ← 0 and depth ← 0 and root ← n − 2 and next ← n − 1
18:    while avail > 0 do
19:       while root ≥ 0 and A[root] = depth do
                                                  ▷ Count internal nodes used at depth depth
20:          set used ← used + 1 and root ← root − 1
21:       end while
22:       while avail > used do              ▷ Assign as leaves any nodes that are not internal
23:          set A[next] ← d and next ← next − 1 and avail ← avail − 1
24:       end while
25:       set avail ← 2 · used and depth ← depth + 1 and used ← 0      ▷ Move to next depth
26:    end while
27:    return A                               ▷ Output: A[i] is the length ℓ_i of the i th codeword
28: end function
```

Phase 3 (steps 17–26) then processes those internal node depths and converts them to a list of leaf depths. At each depth, some total number *avail* of nodes exist, being twice the number of internal nodes at the previous depth. Some number *used* of those are internal nodes; the balance must thus be leaf nodes at this depth and can be assigned as codeword lengths. Initially there is one node available at $depth = 0$, representing the root of the whole Huffman tree. Table 1 shows several snapshots of the Moffat and Katajainen code construction process when applied to the example sequence of weights.

Nonbinary Output Alphabets

The example Huffman tree developed in Fig. 1 and the process shown in Algorithm 1 assume that the output alphabet is binary. Huffman noted in his original paper that for r-ary alphabets all that is required is to add additional dummy symbols of weight zero, so as to bring the total number of symbols to be one more than a multiple of $(r - 1)$. Each merging step then combines r leaf or internal nodes to form a new root node and decreases the number of items by $r - 1$.

Dynamic Huffman Coding

Another assumption made by the processes described so far is that the symbol weights are known in advance and that the code that is computed can be *static*. This assumption can be satisfied, for example, by making a first pass over the message that is to be encoded. In a *dynamic* coding system, symbols must be coded on the fly, as soon as they are received by the encoder. To achieve this, the code must be adaptive, so that it can be altered after each symbol. Vitter [11] summarizes earlier work by Gallager [2], Knuth [4], and Cormack and Horspool [1] and describes a mechanism in which the total encoding cost, including the cost of keeping the code tree up to date, is

Huffman Coding, Table 1 Sequence of values computed by Algorithm 1 for the example weights. The first row shows the initial state of the array, with $A[i] = w_i$. Values "-2-" indicate parent pointers of internal nodes that have already been merged; italic values "*7*" indicate weights of internal nodes before being merged; values "(4)" indicate depths of internal nodes; bold values "**5**" indicate depths of leaves; and values "–" are unused

	i									
	0	1	2	3	4	5	6	7	8	9
Initial arrangement, $A[i] = w_i$	1	1	1	1	3	4	4	7	9	9
Phase 1, $root = 3, next = 5, leaf = 7$	-2-	-2-	-4-	*7*	*8*	–	–	7	9	9
Phase 1, finished, $root = 8$	-2-	-2-	-4-	-5-	-6-	-7-	-8-	-8-	*40*	–
Phase 2, $next = 4$	-2-	-2-	-4-	-5-	-6-	(2)	(1)	(1)	(0)	–
Phase 2, finished	(4)	(4)	(3)	(3)	(2)	(2)	(1)	(1)	(0)	–
Phase 3, $next = 5, avail = 4$	(4)	(4)	(3)	(3)	(2)	(2)	**3**	**3**	**2**	**2**
Final arrangement, $A[i] = \ell_i$	**5**	**5**	**5**	**5**	**4**	**4**	**3**	**3**	**2**	**2**

$O(1)$ per output bit. Turpin and Moffat [9] describe an alternative approximate algorithm that reduces the time required by a constant factor, by collecting the frequency updates into batches and allowing controlled inefficiency in the length of the coded output sequence. Their "GEO" Coding method is faster than dynamic Huffman Coding and also faster than dynamic Arithmetic Coding, which is comparable in speed to dynamic Huffman Coding, but uses less space for the dynamic frequency-counting data structure.

Applications

Minimum-redundancy codes have widespread use in data compression systems. The sequences of weights are usually conditioned according to a *model*, rather than taken as plain symbol frequency counts in the source message. The use of multiple conditioning contexts, and hence multiple codes, one per context, allows improved compression when symbols are not independent in the message, as is the case in natural language data. However, when the contexts are sufficiently specific that highly biased probability distributions arise, Arithmetic Coding will yield superior compression effectiveness.

Turpin and Moffat [8] consider several ancillary components of Huffman Coding, including methods for transmitting the description of the code to the decoder.

Cross-References

▶ Arithmetic Coding for Data Compression
▶ Compressing Integer Sequences

Recommended Reading

1. Cormack GV, Horspool RN (1984) Algorithms for adaptive Huffman codes. Inf Process Lett 18(3):159–165
2. Gallager RG (1978) Variations on a theme by Huffman. IEEE Trans Inf Theory IT-24(6):668–674
3. Huffman DA (1952) A method for the construction of minimum-redundancy codes. Proc Inst Radio Eng 40(9):1098–1101
4. Knuth DE (1985) Dynamic Huffman coding. J Algorithms 6(2):163–180
5. Moffat A, Katajainen J (1995) In-place calculation of minimum-redundancy codes. In: Proceedings of the Workshop on Algorithms and Data Structures, Kingston, pp 393–402
6. Moffat A, Turpin A (1997) On the implementation of minimum-redundancy prefix codes. IEEE Trans Commun 45(10):1200–1207
7. Stix G (1991) Profile: information theorist David A. Huffman. Sci Am 265(3):54–58. Reproduced at http://www.huffmancoding.com/my-uncle/david-bio. Accessed 15 July 2014
8. Turpin A, Moffat A (2000) Housekeeping for prefix coding. IEEE Trans Commun 48(4):622–628
9. Turpin A, Moffat A (2001) On-line adaptive canonical prefix coding with bounded compression loss. IEEE Trans Inf Theory 47(1):88–98
10. van Leeuwen J (1976) On the construction of Huffman trees. In: Proceedings of the International Conference on Automata, Languages, and Programming, Edinburgh University, Edinburgh, pp 382–410
11. Vitter JS (1987) Design and analysis of dynamic Huffman codes. J ACM 34(4):825–845

I

I/O-Model

Norbert Zeh[1] and Ulrich Meyer[2]
[1]Faculty of Computer Science, Dalhousie University, Halifax, NS, Canada
[2]Department of Computer Science, Goethe University Fankfurt am Main, Fankfurt, Germany

Keywords

Disk access model (DAM); External-memory model

Years and Authors of Summarized Original Work

1988; Aggarwal, Vitter

Definition

The input/output model (I/O model) [1] views the computer as consisting of a *processor*, *internal memory* (RAM), and *external memory* (disk). See Fig. 1. The internal memory is of limited size, large enough to hold M data items. The external memory is of conceptually unlimited size and is divided into *blocks* of B consecutive data items. All computation has to happen on data in internal memory. Data is brought into internal memory and written back to external memory using *I/O operations* (I/Os), which are performed explicitly by the algorithm. Each such operation reads or writes one block of data from or to external memory. The complexity of an algorithm in this model is the number of I/Os it performs.

The *parallel disk model* (PDM) [15] is an extension of the I/O model that allows the external memory to consist of $D \geq 1$ parallel disks. See Fig. 2. In this model, a single I/O operation is capable of reading or writing up to D independent blocks, as long as each of them is stored on a different disk.

The *parallel external memory* (PEM) [5] model is a simple multiprocessor extension of the I/O model. See Fig. 3. It consists of P processing units, each having a private cache of size M. Data exchange between the processors takes place via a shared main memory of conceptually unlimited size: in a parallel I/O operation, each processor can transfer one block of size B between its private cache and the shared memory.

The relationship between the PEM model and the very popular MapReduce framework is discussed in [8]. A survey of realistic computer models can be found in [2].

Key Results

A few complexity bounds are of importance to virtually every I/O-efficient algorithm or data structure. The *searching bound* of $\Theta(\log_B n)$ I/Os, which can be achieved using a Btree [6], is

M.-Y. Kao (ed.), *Encyclopedia of Algorithms*,
DOI 10.1007/978-1-4939-2864-4

I/O-Model, Fig. 1 The I/O model

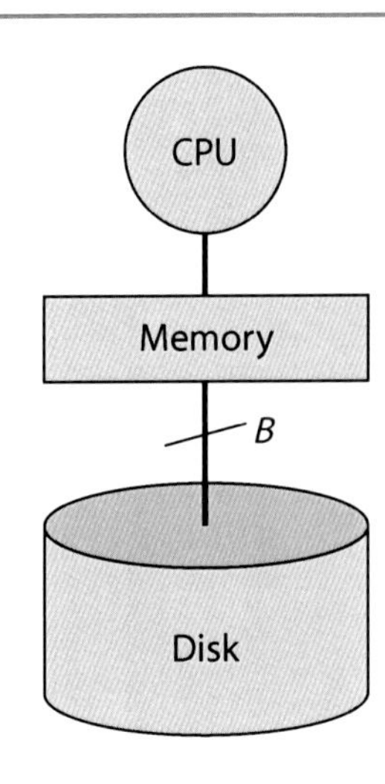

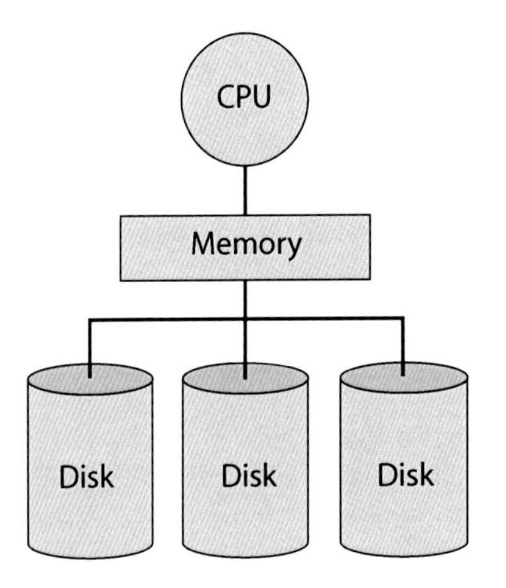

I/O-Model, Fig. 2 The parallel disk model

the cost of searching for an element in an ordered collection of n elements using comparisons only. It is thus the equivalent of the $\Theta(\log n)$ searching bound in internal memory.

Scanning a list of n consecutive data items obviously takes $\lceil n/B \rceil$ I/Os. This *scanning bound* is usually referred to as a "linear number of I/Os" because it is the equivalent of the $O(n)$ time bound required to do the same in internal memory. The respective PDM and PEM bounds are $\lceil n/DB \rceil$ and $\lceil n/PB \rceil$.

The *sorting bound* of $\text{sort}(n) = \Theta((n/B) \log_{M/B}(n/B))$ I/Os denotes the cost of sorting n elements using comparisons only. It is thus the equivalent of the $\Theta(n \log n)$ sorting bound in internal memory. In the PDM and PEM model, the sorting bound becomes $\Theta((n/DB)\log_{M/B}(n/B))$ and $\Theta((n/PB)\log_{M/B}(n/B))$, respectively. The sorting bound can be achieved using a range of sorting algorithms, including external merge sort [1, 5, 10] and distribution sort [1, 5, 9].

Arguably, the most interesting bound is the *permutation bound*, that is, the cost of rearranging n elements in a given order, which is $\Theta(\min(\text{sort}(n), n))$ [1] or, in the PDM, $\Theta(\min(\text{sort}(n), n/D))$ [15]. For all practical purposes, this is the same as the sorting bound. Note the contrast to internal memory where, up to constant factors, permuting has the same cost as a linear scan. Since almost all nontrivial algorithmic problems include a permutation problem, this implies that only exceptionally simple problems can be solved in $O(\text{scan}(n))$ I/Os; most problems have an $\Omega(\text{perm}(n))$, that is, essentially an $\Omega(\text{sort}(n))$ lower bound. Therefore, while internal-memory algorithms aiming for linear time have to carefully avoid the use of sorting as a tool, external-memory algorithms can sort without fear of significantly exceeding the lower bound. This makes the design of I/O-optimal algorithms potentially easier than the design of optimal internal-memory algorithms. It is, however, counterbalanced by the fact that, unlike in internal memory, the sorting bound is *not* equal to n times the searching bound, which implies that algorithms based on querying a tree-based search structure $O(n)$ times usually do not translate into I/O-efficient algorithms. *Buffer trees* [4] achieve an amortized search bound of $O((1/B)\log_{M/B}(N/B))$ I/Os but can be used only if the entire update and query sequence is known in advance and thus provide only a limited solution to this problem.

Apart from these fundamental results, there exist a wide range of interesting techniques, particularly for solving geometric and graph problems. For surveys, refer to [3, 14]. Also, many I/O-efficient algorithms have been derived from fast and work-efficient *parallel* algorithms; see the book entry on external-memory list ranking for a well-known example of this technique.

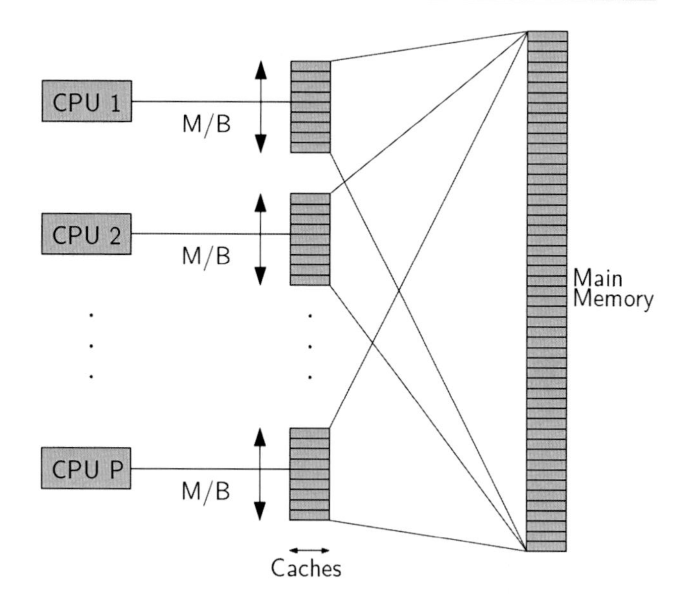

I/O-Model, Fig. 3 The PEM model

Applications

Modern computers are equipped with memory hierarchies consisting of several levels of cache memory, main memory (RAM), and disk(s). Access latencies increase with the distance from the processor, as do the sizes of the memory levels. To amortize these increasing access latencies, data are transferred between different levels of cache in blocks of consecutive data items. As a result, the cost of a memory access depends on the level in the memory hierarchy currently holding the data item – the difference in access latency between L1 cache and disk is about 10^6 – and the cost of a sequence of accesses to data items stored at the same level depends on the number of blocks over which these items are distributed.

Traditionally, algorithms were designed to minimize the number of computation steps; the access locality necessary to solve a problem using few data transfers between memory levels was largely ignored. Hence, the designed algorithms work well on data sets of moderate size but do not take noticeable advantage of cache memory and usually break down completely in out-of-core computations. Since the difference in access latencies is largest between main memory and disk, the I/O model focuses on minimizing this I/O bottleneck. This two-level view of the memory hierarchy keeps the model simple and useful for analyzing sophisticated algorithms while providing a good prediction of their practical performance. The picture is slightly more complex for flash memory-based solid state disks, which have recently become quite popular (also due to their *energy* efficiency [7]): not only do they internally use different block sizes for reading and writing, but their (reading) latency is also significantly smaller compared to traditional hard disks. Nevertheless, the latency gap of solid state disks compared to main memory remains large, and optimized device controllers or translation layers manage to hide the read/write discrepancy in most practical settings. Thus, the I/O model still provides reasonable estimates on flash memory, but extended models with different block sizes and access costs for reading and writing are more accurate.

Much effort has been made already to translate provably I/O-efficient algorithms into highly efficient implementations. Examples include TPIE [12] and STXXL [11], two libraries that aim to provide highly optimized and powerful primitives for the implementation of I/O-efficient algorithms. In particular, TPIE has been used to realize a number of geometric and GIS applications, whereas STXXL has served as a basis for the implementation of various graph algorithms.

In spite of these efforts, a significant gap between the theory and practice of I/O-efficient algorithms remains (see next section).

Open Problems

There are a substantial number of open problems in the area of I/O-efficient algorithms. The most important ones concern graph and geometric problems.

Traditional graph algorithms usually apply a well-organized graph traversal such as depth-first search or breadth-first search to gain information about the structure of the graph and then use this information to solve the problem at hand. For massive sparse graphs, no I/O-efficient depth-first search algorithm is known, and for breadth-first search and shortest paths, only limited progress has been made on undirected graphs. Some recent results concern dynamic and approximation variants or all-pairs shortest paths problems. For directed graphs, even such simple problems as deciding whether there exists a directed path between two vertices are currently still open. The main research focus in this area is therefore to either develop (or disprove the existence of) I/O-efficient general traversal algorithms or to continue the current strategy of devising graph algorithms that depart from traditional traversal-based approaches.

Techniques for solving geometric problems I/O efficiently are much better understood than is the case for graph algorithms, at least in two dimensions. Nevertheless, there are a few important frontiers that remain. Despite new results on some range reporting problems in three and higher dimensions, arguably the most important frontier is the development of I/O-efficient algorithms and data structures for higher-dimensional geometric problems. Motivated by database applications, results on specialized range searching variants (such as coloured and top-K range searching) have begun to appear in the literature. Little work has been done in the past on solving proximity problems, which pose another frontier currently being explored. Motivated by the need for such structures in a range of application areas and in particular in geographic information systems, there has been some recent focus on the development of multifunctional data structures, that is, structures that can answer different types of queries efficiently. This is in contrast to most existing structures, which are carefully tuned to efficiently support *one* particular type of query.

We also face a significant lack of external-memory lower bounds. Classic results concern permuting and sorting (see [14] for an overview), and more recent results concentrate on I/O-efficient data structure problems such as dynamic membership [13]. The optimality of many basic external-memory algorithms, however, is completely open. For instance, it is unclear whether sparse graph traversal (and hence probably a large number of advanced graph problems) will ever be solvable in an I/O-efficient manner.

For both I/O-efficient graph algorithms and computational geometry, there is still a substantial gap between the obtained theoretical results and what is known to be practical, even though quite some *algorithm engineering* work has been done during the last decade. Thus, if I/O-efficient algorithms in these areas are to have more practical impact, increased efforts are needed to bridge this gap by developing practically I/O-efficient algorithms that are still *provably* efficient.

Cross-References

For details on ▸ External Sorting and Permuting and ▸ List-Ranking, please refer to the corresponding entries. Details on one- and higher-dimensional searching are provided in the entries on ▸ B-trees and ▸ R-Trees. The reader interested in algorithms that focus on efficiency at all levels of the memory hierarchy should consult the entry on the ▸ Cache-Oblivious Model.

Recommended Reading

1. Aggarwal A, Vitter JS (1988) The input/output complexity of sorting and related problems. Commun ACM 31(9):1116–1127
2. Ajwani D, Meyerhenke H (2010) Realistic computer models. In: Müller-Hannemann M, Schirra S (eds) Algorithm engineering: bridging the gap between algorithm theory and practice. Volume 5971 of LNCS. Springer, Berlin/Heidelberg, pp 194–236
3. Arge L (2002) External memory data structures. In: Abello J, Pardalos PM, Resende MGC (eds) Handbook of massive data sets. Kluwer Academic, Dordrecht, pp 313–357
4. Arge L (2003) The buffer tree: a technique for designing batched external data structures. Algorithmica 37(1):1–24
5. Arge L, Goodrich MT, Nelson MJ, Sitchinava N (2008) Fundamental parallel algorithms for private-cache chip multiprocessors. In: Proceedings of the 20th annual ACM symposium on parallelism in algorithms and architectures, Munich, pp 197–206
6. Bayer R, McCreight E (1972) Organization of large ordered indexes. Acta Inform 1:173–189
7. Beckmann A, Meyer U, Sanders P, Singler S (2011) Energy-efficient sorting using solid state disks. Sustain Comput Inform Syst 1(2):151–163
8. Greiner G, Jacob R (2012) The efficiency of MapReduce in parallel external memory. In: Proceedings of the 10th Latin American symposium on theoretical informatic (LATIN). Volume 7256 of LNCS. Springer, Berlin/Heidelberg, pp 433–445
9. Nodine MH, Vitter JS (1993) Deterministic distribution sort in shared and distributed memory multiprocessors. In: Proceedings of the 5th annual ACM symposium on parallel algorithms and architectures, Velen, pp 120–129, June/July 1993
10. Nodine MH, Vitter JS (1995) Greed sort: an optimal sorting algorithm for multiple disks. J ACM 42(4):919–933
11. STXXL: C++ standard library for extra large data sets. http://stxxl.sourceforge.net. Accessed 23 June 2014
12. TPIE – a transparent parallel I/O-environment. http://www.madalgo.au.dk/tpie. Accessed 23 June 2014
13. Verbin E, Zhang Q (2013) The limits of buffering: a tight lower bound for dynamic membership in the external memory model. SIAM J Comput 42(1):212–229
14. Vitter JS (2006) Algorithms and data structures for external memory. Found Trends Theor Comput Sci 2(4):305–474
15. Vitter JS, Shriver EAM (1994) Algorithms for parallel memory I: two-level memories. Algorithmica 12(2–3):110–147

Implementation Challenge for Shortest Paths

Camil Demetrescu[1,2], Andrew V. Goldberg[3], and David S. Johnson[4,5]

[1]Department of Computer and Systems Science, University of Rome, Rome, Italy

[2]Department of Information and Computer Systems, University of Rome, Rome, Italy

[3]Microsoft Research – Silicon Valley, Mountain View, CA, USA

[4]Department of Computer Science, Columbia University, New York, NJ, USA

[5]AT&T Laboratories, Algorithms and Optimization Research Department, Florham Park, NJ, USA

Keywords

DIMACS; Test sets and experimental evaluation of computer programs for solving shortest path problems

Years and Authors of Summarized Original Work

2006; Demetrescu, Goldberg, Johnson

Problem Definition

DIMACS Implementation Challenges (http://dimacs.rutgers.edu/Challenges/) are scientific events devoted to assessing the practical performance of algorithms in experimental settings, fostering effective technology transfer and establishing common benchmarks for fundamental computing problems. They are organized by DIMACS, the Center for Discrete Mathematics and Theoretical Computer Science. One of the main goals of DIMACS Implementation Challenges is to address questions of determining realistic algorithm performance where worst case analysis is overly pessimistic and probabilistic models are too unrealistic: experimentation can provide guides to realistic algorithm performance where analysis fails. Experimentation also brings algorithmic questions closer to the original problems that motivated theoretical work. It also tests many assumptions about implementation methods and data structures. It provides an opportunity to develop and test problem instances, instance generators, and other methods of testing and comparing performance of algorithms. And it is a step in technology transfer by providing leading edge implementations of algorithms for others to adapt.

The first Challenge was held in 1990–1991 and was devoted to *Network flows and Matching*. Other addressed problems included: *Maximum Clique, Graph Coloring, and Satisfiability* (1992–1993), *Parallel Algorithms for Combinatorial Problems* (1993–1994), *Fragment Assembly and Genome Rearrangements* (1994–1995), *Priority Queues, Dictionaries, and Multi-Dimensional Point Sets* (1995–1996), *Near Neighbor Searches* (1998–1999), *Semidefinite and Related Optimization Problems* (1999–2000), and *The Traveling Salesman Problem* (2000–2001).

This entry addresses the goals and the results of the *9th DIMACS Implementation Challenge*, held in 2005–2006 and focused on *Shortest Path* problems.

The 9th DIMACS Implementation Challenge: The Shortest Path Problem

Shortest path problems are among the most fundamental combinatorial optimization problems with many applications, both direct and as subroutines in other combinatorial optimization algorithms. Algorithms for these problems have been studied since the 1950s and still remain an active area of research.

One goal of this Challenge was to create a reproducible picture of the state of the art in the area of shortest path algorithms, identifying a standard set of benchmark instances and generators, as well as benchmark implementations of well-known shortest path algorithms. Another goal was to enable current researchers to compare their codes with each other, in hopes of identifying the more effective of the recent algorithmic innovations that have been proposed.

Challenge participants studied the following variants of the shortest paths problem:

- *Point to point shortest paths* [4, 5, 6, 9, 10, 11, 14]: the problem consists of answering multiple online queries about the shortest paths between pairs of vertices and/or their lengths. The most efficient solutions for this problem preprocess the graph to create a data structure that facilitates answering queries quickly.
- *External-memory shortest paths* [2]: the problem consists of finding shortest paths in a graph whose size is too large to fit in internal memory. The problem actually addressed in the Challenge was single-source shortest paths in undirected graphs with unit edge weights.
- *Parallel shortest paths* [8, 12]: the problem consists of computing shortest paths using multiple processors, with the goal of achieving good speedups over traditional sequential implementations. The problem actually addressed in the Challenge was single-source shortest paths.
- *K-shortest paths* [13, 15]: the problem consists of ranking paths between a pair of vertices by non decreasing order of their length.
- *Regular-language constrained shortest paths:* [3] the problem consists of a generalization of shortest path problems where paths must satisfy certain constraints specified by a regular language. The problems studied in the context of the Challenge were single-source and point-to-point shortest paths, with applications ranging from transportation science to databases.

The Challenge culminated in a Workshop held at the DIMACS Center at Rutgers University, Piscataway, New Jersey on November 13–14, 2006. Papers presented at the conference are available at the URL: http://www.dis.uniroma1.it/~challenge9/papers.shtml. Selected contributions are expected to appear in a book published by the American Mathematical Society in the DIMACS Book Series.

Key Results

The main results of the 9th DIMACS Implementation Challenge include:

- Definition of common file formats for several variants of the shortest path problem, both static and dynamic. These include an extension of the famous DIMACS graph file format used by several algorithmic software libraries. Formats are described at the URL: http://www.dis.uniroma1.it/~challenge9/formats.shtml.
- Definition of a common set of core input instances for evaluating shortest path algorithms.
- Definition of benchmark codes for shortest path problems.
- Experimental evaluation of state-of-the-art implementations of shortest path codes on the core input families.
- A discussion of directions for further research in the area of shortest paths, identifying problems critical in real-world applications for which efficient solutions still remain unknown.

The chief information venue about the 9th DIMACS Implementation Challenge is the website http://www.dis.uniroma1.it/~challenge9.

Applications

Shortest path problems arise naturally in a remarkable number of applications. A limited list includes transportation planning, network optimization, packet routing, image segmentation, speech recognition, document formatting, robotics, compilers, traffic information systems, and dataflow analysis. It also appears as a subproblem of several other combinatorial optimization problems such as network flows. A comprehensive discussion of applications of shortest path problems appears in [1].

Open Problems

There are several open questions related to shortest path problems, both theoretical and practical. One of the most prominent discussed at the 9th DIMACS Challenge Workshop is modeling traffic fluctuations in point-to-point shortest paths. The current fastest implementations preprocess the input graph to answer point-to-point queries efficiently, and this operation may take hours on graphs arising in large-scale road map navigation systems. A change in the traffic conditions may require rescanning the whole graph several times. Currently, no efficient technique is known for updating the preprocessing information without rebuilding it from scratch. This would have a major impact on the performance of routing software.

Data Sets

The collection of benchmark inputs of the 9th DIMACS Implementation Challenge includes both synthetic and real-world data. All graphs are strongly connected. Synthetic graphs include random graphs, grids, graphs embedded on a torus, and graphs with small-world properties. Real-world inputs consist of graphs representing the road networks of Europe and USA. Europe graphs are provided by courtesy of the PTV company, Karlsruhe, Germany, subject to signing a (no-cost) license agreement. They include the road networks of 17 European countries: AUT, BEL, CHE, CZE, DEU, DNK, ESP, FIN, FRA, GBR, IRL, ITA, LUX, NDL, NOR, PRT, SWE, with a total of about 19 million nodes and 23 million edges. USA graphs are derived from the *UA Census 2000 TIGER/Line Files* produced by the Geography Division of the US Census Bureau, Washington, DC. The TIGER/Line collection is available at: http://www.census.gov/geo/www/tiger/tigerua/ua_tgr2k.html. The Challenge USA core family contains a graph representing the full USA road system with about 24 million nodes and 58 million edges, plus 11 subgraphs obtained by cutting it along different bounding boxes as shown in Table 1. Graphs in the collection include also node coordinates and are given in DIMACS format.

The benchmark input package also features query generators for the single-source and point-to-point shortest path problems. For the single-source version, sources are randomly chosen. For the point-to-point problem, both random and

Implementation Challenge for Shortest Paths, Table 1 USA road networks derived from the TIGER/Line collection

Name	Description	Nodes	Arcs	Bounding box latitude (N)	Bounding box longitude (W)
USA	Full USA	23 947 347	58 333 344	–	–
CTR	Central USA	14 081 816	34 292 496	[25.0; 50.0]	[79.0; 100.0]
W	Western USA	6 262 104	15 248 146	[27.0; 50.0]	[100.0; 130.0]
E	Eastern USA	3 598 623	8 778 114	[24.0; 50.0]	[-∞; 79.0]
LKS	Great Lakes	2 758 119	6 885 658	[41.0; 50.0]	[74.0; 93.0]
CAL	California and Nevada	1 890 815	4 657 742	[32.5; 42.0]	[114.0; 125.0]
NE	Northeast USA	1 524 453	3 897 636	[39.5, 43.0]	[-∞; 76.0]
NW	Northwest USA	1 207 945	2 840 208	[42.0; 50.0]	[116.0; 126.0]
FLA	Florida	1 070 376	2 712 798	[24.0; 31.0]	[79; 87.5]
COL	Colorado	435 666	1 057 066	[37.0; 41.0]	[102.0; 109.0]
BAY	Bay Area	321 270	800 172	[37.0; 39.0]	[121; 123]
NY	New York City	264 346	733 846	[40.3; 41.3]	[73.5; 74.5]

local queries are considered. Local queries of the form (s, t) are generated by randomly picking t among the nodes with rank in $[2^i, 2^{i+1})$ in the ordering in which nodes are scanned by Dijkstra's algorithm with source s, for any parameter i. Clearly, the smaller i is, the closer nodes s and t are in the graph. Local queries are important to test how the algorithms' performance is affected by the distance between query endpoints.

The core input families of the 9th DIMACS Implementation Challenge are available at the URL: http://www.dis.uniroma1.it/~challenge9/download.shtml.

Experimental Results

One of the main goals of the Challenge was to compare different techniques and algorithmic approaches. The most popular topic was the point-to-point shortest path problem, studied by six research groups in the context of the Challenge. For this problem, participants were additionally invited to join a competition aimed at assessing the performance and the robustness of different implementations. The competition consisted of preprocessing a version of the full USA graph of Table 1 with unit edge lengths and answering a sequence of 1,000 random distance queries. The details were announced on the first day of the workshop and the results were due on the second day. To compare experimental results by different participants on different platforms, each participant ran a Dijkstra benchmark code [7] on the USA graph to do machine calibration. The final ranking was made by considering each query time divided by the time required by the benchmark code on the same platform (benchmark ratio). Other performance measures taken into account were space usage and the average number of nodes scanned by query operations.

Six point-to-point implementations were run successfully on the USA graph defined for the competition. Among them, the fastest query time was achieved by the *HH-based transit* code [14]. Results are reported in Table 2. Codes *RE* and *REAL*(16, 1) [9] were not eligible for the competition, but used by the organizers as a proof that the problem is feasible. Some other codes were not able to deal with the size of the full USA graph, or incurred runtime errors.

Experimental results for other variants of the shortest paths problem are described in the papers presented at the Challenge Workshop.

URL to Code

Generators of problem families and benchmark solvers for shortest paths problems are avail-

Implementation Challenge for Shortest Paths, Table 2 Results of the Challenge competition on the USA graph (23.9 million nodes and 58.3 million arcs) with unit arc lengths. The benchmark ratio is the average query time divided by the time required to answer a query using the Challenge Dijkstra benchmark code on the same platform. Query times and node scans are average values per query over 1000 random queries

	Preprocessing		Query		
Code	Time (minutes)	Space (MB)	Node scans	Time (ms)	Benchmark ratio
HH-based transit [14]	104	3664	n.a.	0.019	$4.78 \cdot 10^{-6}$
TRANSIT [4]	720	n.a.	n.a.	0.052	$10.77 \cdot 10^{-6}$
HH Star [6]	32	2662	1082	1.14	$287.32 \cdot 10^{-6}$
REAL(16,1) [9]	107	2435	823	1.42	$296.30 \cdot 10^{-6}$
HH with DistTab [6]	29	2101	1671	1.61	$405.77 \cdot 10^{-6}$
RE [9]	88	861	3065	2.78	$580.08 \cdot 10^{-6}$

able at the URL: http://www.dis.uniroma1.it/~challenge9/download.shtml.

Cross-References

- ▸ Engineering Algorithms for Large Network Applications
- ▸ Experimental Methods for Algorithm Analysis
- ▸ High Performance Algorithm Engineering for Large-Scale Problems
- ▸ Implementation Challenge for TSP Heuristics
- ▸ LEDA: a Library of Efficient Algorithms

Recommended Reading

1. Ahuja R, Magnanti T, Orlin J (1993) Network flows: theory, algorithms and applications. Prentice Hall, Englewood Cliffs
2. Ajwani D, Dementiev U, Meyer R, Osipov V (2006) Breadth first search on massive graphs. In: 9th DIMACS implementation challenge workshop: shortest paths. DIMACS Center, Piscataway
3. Barrett C, Bissett K, Holzer M, Konjevod G, Marathe M, Wagner D (2006) Implementations of routing algorithms for transportation networks. In: 9th DIMACS implementation challenge workshop: shortest paths. DIMACS Center, Piscataway
4. Bast H, Funke S, Matijevic D (2006) Transit: ultrafast shortest-path queries with linear-time preprocessing. In: 9th DIMACS implementation challenge workshop: shortest paths. DIMACS Center, Piscataway
5. Delling D, Holzer M, Muller K, Schulz F, Wagner D (2006) High performance multi-level graphs. In: 9th DIMACS implementation challenge workshop: shortest paths. DIMACS Center, Piscataway
6. Delling D, Sanders P, Schultes D, Wagner D (2006) Highway hierarchies star. In: 9th DIMACS implementation challenge workshop: shortest paths. DIMACS Center, Piscataway
7. Dijkstra E (1959) A note on two problems in connexion with graphs. Numer Math 1:269–271
8. Edmonds N, Breuer A, Gregor D, Lumsdaine A (2006) Single source shortest paths with the parallel boost graph library. In: 9th DIMACS implementation challenge workshop: shortest paths. DIMACS Center, Piscataway
9. Goldberg A, Kaplan H, Werneck R (2006) Better landmarks within reach. In: 9th DIMACS implementation challenge workshop: shortest paths. DIMACS Center, Piscataway
10. Köhler E, Möhring R, Schilling H (2006) Fast point-to-point shortest path computations with arc-flags. In: 9th DIMACS implementation challenge workshop: shortest paths. DIMACS Center, Piscataway
11. Lauther U (2006) An experimental evaluation of point-to-point shortest path calculation on road networks with precalculated edge-flags. In: 9th DIMACS implementation challenge workshop: shortest paths. DIMACS Center, Piscataway
12. Madduri K, Bader D, Berry J, Crobak J (2006) Parallel shortest path algorithms for solving large-scale instances. In: 9th DIMACS implementation challenge workshop: shortest paths. DIMACS Center, Piscataway
13. Pascoal M (2006) Implementations and empirical comparison of k shortest loopless path algorithms. In: 9th DIMACS implementation challenge workshop: shortest paths. DIMACS Center, Piscataway
14. Sanders P, Schultes D (2006) Robust, almost constant time shortest-path queries in road networks. In: 9th DIMACS implementation challenge workshop: shortest paths. DIMACS Center, Piscataway
15. Santos J (2006) K shortest path algorithms. In: 9th DIMACS implementation challenge workshop: shortest paths. DIMACS Center, Piscataway

Implementation Challenge for TSP Heuristics

Lyle A. McGeoch
Department of Mathematics and Computer Science, Amherst College, Amherst, MA, USA

Keywords

Concorde; Held-Karp; Lin-Kernighan; Three-opt; TSPLIB; Two-opt

Years and Authors of Summarized Original Work

2002; Johnson, McGeoch

Problem Definition

The Eighth DIMACS Implementation Challenge, sponsored by DIMACS, the Center for Discrete Mathematics and Theoretical Computer Science, concerned heuristics for the symmetric Traveling Salesman Problem. The Challenge began in June 2000 and was organized by David S. Johnson, Lyle A. McGeoch, Fred Glover and César Rego. It explored the state-of-the-art in the area of TSP heuristics, with researchers testing a wide range of implementations on a common (and diverse) set of input instances. The Challenge remained ongoing in 2007, with new results still being accepted by the organizers and posted on the Challenge website: www.research.att.com/~dsj/chtsp. A summary of the submissions through 2002 appeared in a book chapter by Johnson and McGeoch [5].

Participants tested their heuristics on four types of instances, chosen to test the robustness and scalability of different approaches:

1. The 34 instances that have at least 1000 cities in TSPLIB, the instance library maintained by Gerd Reinelt.
2. A set of 26 instances consisting of points uniformly distributed in the unit square, with sizes ranging from 1000 to 10,000,000 cities.
3. A set of 23 randomly generated clustered instances, with sizes ranging from 1000 to 316,000 cities.
4. A set of 7 instances based on random distance matrices, with sizes ranging from 1000 to 10,000 cities.

The TSPLIB instances and generators for the random instances are available on the Challenge website. In addition, the website contains a collection of instances for the asymmetric TSP problem.

For each instance upon which a heuristic was tested, the implementers reported the machine used, the tour length produced, the user time, and (if possible) memory usage. Some heuristics could not be applied to all of the instances, either because the heuristics were inherently geometric or because the instances were too large. To help facilitate timing comparisons between heuristics tested on different machines, participants ran a benchmark heuristic (provided by the organizers) on instances of different sizes. The benchmark times could then be used to normalize, at least approximately, the observed running times of the participants' heuristics.

The quality of a tour was computed from a submitted tour length in two ways: as a ratio over the optimal tour length for the instance (if known), and as a ratio over the Held-Karp (HK) lower bound for the instance. The Concorde optimization package of Applegate et al. [1] was able to find the optimum for 58 of the instances in reasonable time. Concorde was used in a second way to compute the HK lower bound for all but the three largest instances. A third algorithm, based on Lagrangian relaxation, was used to compute an approximate HK bound, a lower bound on true HK bound, for the remaining instances. The Challenge website reports on each of these three algorithms, presenting running times and a comparison of the bounds obtained for each instance.

The Challenge website permits a variety of reports to be created:

1. For each heuristic, tables can be generated with results for each instance, including tour length, tour quality, and raw and normalized running times.

2. For each instance, a table can be produced showing the tour quality and normalized running time of each heuristic.
3. For each pair of heuristics, tables and graphs can be produced that compare tour quality and running time for instances of different type and size.

Heuristics for which results were submitted to the Challenge fell into several broad categories:

Heuristics designed for speed. These heuristics – all of which target geometric instances – have running times within a small multiple of the time needed to read the input instance. Examples include the strip and spacefilling-curve heuristics. The speed requirement affects tour quality dramatically. Two of these algorithms produced tours with 14 % of the HK lower bound for a particular TSPLIB instance, but none came within 25 % on the other 89 instances.

Tour construction heuristics. These heuristics construct tours in various ways, without seeking to find improvements once a single tour passing through all cities is found. Some are simple, such as the nearest-neighbor and greedy heuristics, while others are more complex, such as the famous Christofides heuristic. These heuristics offer a number of options in trading time for tour quality, and several produce tours within 15 % of the HK lower bound on most instances in reasonable time. The best of them, a variant of Christofides, produces tours within 8 % on uniform instances but is much more time-consuming than the other algorithms.

Simple local improvement heuristics. These include the well-known two-opt and three-opt heuristics and variants of them. These heuristics outperform tour construction heuristics in terms of tour quality on most types of instances. For example, 3-opt gets within about 3 % of the HK lower bound on most uniform instances. The submissions in this category explored various implementation choices that affect the time-quality tradeoff.

Lin-Kernighan and its variants. These heuristics extend the local search neighborhood used in 3-opt. Lin-Kernighan can produce high-quality tours (for example, within 2 % of the HK lower bound on uniform instances) in reasonable time. One variant, due to Helsgaun [3], obtains tours within 1 % on a wide variety of instances, although the running time can be substantial.

Repeated local search heuristics. These heuristics are based on repeated executions of a heuristic such as Lin-Kernighan, with random kicks applied to the tour after a local optimum is found. These algorithms can yield high-quality tours at increased running time.

Heuristics that begin with repeated local search. One example is the tour-merge heuristic [2], which runs repeated local search multiple times, builds a graph containing edges found in the best tours, and does exhaustive search within the resulting graph. This approach yields the best known tours for some of the instances in the Challenge.

The submissions to the Challenge demonstrated the remarkable effectiveness of heuristics for the traveling salesman problem. They also showed that implementation details, such a choice of data structure or whether to approximate aspects of the computation, can affect running time and/or solution quality greatly. Results for a given heuristic also varied enormously depending on the type of instance to which it is applied.

URL to Code

www.research.att.com/~dsj/chtsp

Cross-References

▸ TSP-Based Curve Reconstruction

Recommended Reading

1. Applegate D, Bixby R, Chvátal V, Cook W (1998) On the solution of traveling salesman problems. Documenta Mathematica, Extra Volume Proceedings ICM, vol III. Deutsche Mathematiker-Vereinigung, Berlin, pp 645–656

2. Applegate D, Bixby R, Chvátal V, Cook W (1999) Finding tours in the TSP. Technical report 99885. Research Institute for Discrete Mathematics, Universität Bonn
3. Helsgaun K (2000) An effective implementation of the Lin-Kernighan traveling salesman heuristic. Eur J Oper Res 126(1):106–130
4. Johnson DS, McGeoch LA (1997) The traveling salesman problem: a case study. In: Aarts E, Lenstra JK (eds) Local search in combinatorial optimization. Wiley, Chichester, pp 215–310
5. Johnson DS, McGeoch LA (2002) Experimental analysis of heuristics for the STSP. In: Gutin G, Punnen AP (eds) The traveling salesman problem and its variants. Kluwer, Dordrecht, pp 369–443

Implementing Shared Registers in Asynchronous Message-Passing Systems

Eric Ruppert
Department of Computer Science and Engineering, York University, Toronto, ON, Canada

Keywords

Emulation; Simulation

Years and Authors of Summarized Original Work

1995; Attiya, Bar-Noy, Dolev

Problem Definition

A distributed system is composed of a collection of n processes which communicate with one another. Two means of interprocess communication have been heavily studied. *Message-passing systems* model computer networks where each process can send information over message channels to other processes. In *shared-memory systems*, processes communicate less directly by accessing information in shared data structures. Distributed algorithms are often easier to design for shared-memory systems because of their similarity to single-process system architectures. However, many real distributed systems are constructed as message-passing systems. Thus, a key problem in distributed computing is the implementation of shared memory in message-passing systems. Such implementations are also called simulations or emulations of shared memory.

The most fundamental type of shared data structure to implement is a *(read-write) register*, which stores a value, taken from some domain D. It is initially assigned a value from D and can be accessed by two kinds of operations, read and write(v), where $v \in D$. A register may be either *single-writer*, meaning only one process is allowed to write it, or *multi-writer*, meaning any process may write to it. Similarly, it may be either *single-reader* or *multi-reader*. Attiya and Welch [4] give a survey of how to build multi-writer, multi-reader registers from single-writer, single-reader ones.

If reads and writes are performed one at a time, they have the following effects: a read returns the value stored in the register to the invoking process, and a write(v) changes the value stored in the register to v and returns an acknowledgment, indicating that the operation is complete. When many processes apply operations concurrently, there are several ways to specify a register's behavior [14]. A single-writer register is *regular* if each read returns either the argument of the write that completed most recently before the read began or the argument of some write operation that runs concurrently with the read. (If there is no write that completes before the read begins, the read may return either the initial value of the register or the value of a concurrent write operation.) A register is *atomic* (see ▸ Linearizability) if each operation appears to take place instantaneously. More precisely, for any concurrent execution, there is a total order of the operations such that each read returns the value written by the last write that precedes it in the order (or the initial value of the register, if there is no such write).

Moreover, this total order must be consistent with the temporal order of operations: if one operation finishes before another one begins, the former must precede the latter in the total order. Atomicity is a stronger condition than regularity, but it is possible to implement atomic registers from regular ones with some complexity overhead [12].

This article describes the problem of implementing registers in an asynchronous message-passing system in which processes may experience crash failures. Each process can send a message, containing a finite string, to any other process. To make the descriptions of algorithms more uniform, it is often assumed that processes can send messages to themselves. All messages are eventually delivered. In the algorithms described below, senders wait for an acknowledgment of each message before sending the next message, so it is not necessary to assume that the message channels are first-in-first-out. The system is totally asynchronous: there is no bound on the time required for a message to be delivered to its recipient or for a process to perform a step of local computation. A process that fails by crashing stops executing its code, but other processes cannot distinguish between a process that has crashed and one that is running very slowly. (Failures of message channels [3] and more malicious kinds of process failures [15] have also been studied.)

A *t-resilient* register implementation provides programmes to be executed by processes to simulate read and write operations. These programmes can include any standard control structures and accesses to a process's local memory, as well as instructions to send a message to another process and to read the process's buffer, where incoming messages are stored. The implementation should also specify how the processes' local variables are initialized to reflect any initial value of the implemented register. In the case of a single-writer register, only one process may execute the write programme. A process may invoke the read and write programmes repeatedly, but it must wait for one invocation to complete before starting the next one. In any such execution where at most t processes crash, each of a process's invocations of the read or write programme should eventually terminate. Each read operation returns a result from the set D, and these results should satisfy regularity or atomicity.

Relevant measures of algorithm complexity include the number of messages transmitted in the system to perform an operation, the number of bits per message, and the amount of local memory required at each process. One measure of time complexity is the time needed to perform an operation, under the optimistic assumption that the time to deliver messages is bounded by Δ and local computation is instantaneous (although algorithms must work correctly even without these assumptions).

Key Results

Implementing a Regular Register

One of the core ideas for implementing shared registers in message-passing systems is a construction that implements a regular single-writer multi-reader register. It was introduced by Attiya, Bar-Noy and Dolev [3] and made more explicit by Attiya [2]. A write(v) sends the value v to all processes and waits until a majority of the processes ($\lfloor \frac{n}{2} \rfloor + 1$, including the writer itself) return an acknowledgment. A reader sends a request to all processes for their latest values. When it has received responses from a majority of processes, it picks the most recently written value among them. If a write completes before a read begins, at least one process that answers the reader has received the write's value prior to sending its response to the reader. This is because any two sets that each contain a majority of the processes must overlap. The time required by operations when delivery times are bounded is 2Δ.

This algorithm requires the reader to determine which of the values it receives is most recent. It does this using *timestamps* attached to the values. If the writer uses increasing integers as timestamps, the messages grow without bound as the algorithm runs. Using the bounded timestamp scheme of Israeli and Li [13] instead yields the following theorem.

Theorem 1 (Attiya [2]) *There is an* $\lceil \frac{n-2}{2} \rceil$*-resilient implementation of a regular single-writer, multi-reader register in a message-passing system of n processes. The implementation uses* $\Theta(n)$ *messages per operation, with* $\Theta(n^3)$ *bits per message. The writer uses* $\Theta(n^4)$ *bits of local memory and each reader uses* $\Theta(n^3)$ *bits.*

Theorem 1 is optimal in terms of fault-tolerance. If $\lceil \frac{n}{2} \rceil$ processes can crash, the network can be partitioned into two halves of size $\lfloor \frac{n}{2} \rfloor$, with messages between the two halves delayed indefinitely. A write must terminate before any evidence of the write is propagated to the half not containing the writer, and then a read performed by a process in that half cannot return an up-to-date value. For $t \geq \lceil \frac{n}{2} \rceil$, registers can be implemented in a message-passing system only if some degree of synchrony is present in the system. The exact amount of synchrony required was studied by Delporte-Gallet et al. [6].

Theorem 1 is within a constant factor of the optimal number of messages per operation. Evidence of each write must be transmitted to at least $\lceil \frac{n}{2} \rceil - 1$ processes, requiring $\Omega(n)$ messages; otherwise this evidence could be obliterated by crashes. A write must terminate even if only $\lfloor \frac{n}{2} \rfloor + 1$ processes (including the writer) have received information about the value written, since the rest of the processes could have crashed. Thus, a read must receive information from at least $\lceil \frac{n}{2} \rceil$ processes (including itself) to ensure that it is aware of the most recent write operation.

A t-resilient implementation, for $t < \lceil \frac{n}{2} \rceil$, that uses $\Theta(t)$ messages per operation is obtained by the following adaptation. A set of $2t + 1$ processes is preselected to be data storage servers. Writes send information to the servers, and wait for $t + 1$ acknowledgments. Reads wait for responses from $t + 1$ of the servers and choose the one with the latest timestamp.

Implementing an Atomic Register

Attiya, Bar-Noy and Dolev [3] gave a construction of an atomic register in which readers forward the value they return to all processes and wait for an acknowledgment from a majority. This is done to ensure that a read does not return an older value than another read that precedes it. Using unbounded integer timestamps, this algorithm uses $\Theta(n)$ messages per operation. The time needed per operation when delivery times are bounded is 2Δ for writes and 4Δ for reads. However, their technique of bounding the timestamps increases the number of messages per operation to $\Theta(n^2)$ (and the time per operation to 12Δ). A better implementation of atomic registers with bounded message size is given by Attiya [2]. It uses the regular registers of Theorem 1 to implement atomic registers using the "handshaking" construction of Haldar and Vidyasankar [12], yielding the following result.

Theorem 2 (Attiya [2]) *There is an* $\lceil \frac{n-2}{2} \rceil$*-resilient implementation of an atomic single-writer, multi-reader register in a message-passing system of n processes. The implementation uses* $\Theta(n)$ *messages per operation, with* $\Theta(n^3)$ *bits per message. The writer uses* $\Theta(n^5)$ *bits of local memory and each reader uses* $\Theta(n^4)$ *bits.*

Since atomic registers are regular, this algorithm is optimal in terms of fault-tolerance and within a constant factor of optimal in terms of the number of messages. The time used when delivery times are bounded is at most 14Δ for writes and 18Δ for reads.

Applications

Any distributed algorithm that uses shared registers can be adapted to run in a message-passing system using the implementations described above. This approach yielded new or improved message-passing solutions for a number of problems, including randomized consensus [1], multi-writer registers [4], and snapshot objects ▶ Distributed Snapshots. The reverse simulation is also possible, using a straightforward implementation of message channels by single-writer, single-reader registers. Thus, the two asynchronous models are equivalent, in terms of the set of problems that they can solve, assuming only a minority of processes crash. However there is some complexity overhead in using the simulations.

If a shared-memory algorithm is implemented in a message-passing system using the algorithms described here, processes must continue to operate even when the algorithm terminates, to help other processes execute their reads and writes. This cannot be avoided: if each process must stop taking steps when its algorithm terminates, there are some problems solvable with shared registers that are not solvable in the message-passing model [5].

Using a majority of processes to "validate" each read and write operation is an example of a quorum system, originally introduced for replicated data by Gifford [10]. In general, a quorum system is a collection of sets of processes, called quorums, such that every two quorums intersect. Quorum systems can also be designed to implement shared registers in other models of message-passing systems, including dynamic networks and systems with malicious failures. For examples, see [7, 9, 11, 15].

Open Problems

Although the algorithms described here are optimal in terms of fault-tolerance and message complexity, it is not known if the number of bits used in messages and local memory is optimal. The exact time needed to do reads and writes when messages are delivered within time Δ is also a topic of ongoing research. (See, for example, [8].) As mentioned above, the simulation of shared registers can be used to implement shared-memory algorithms in message-passing systems. However, because the simulation introduces considerable overhead, it is possible that some of those problems could be solved more efficiently by algorithms designed specifically for message-passing systems.

Cross-References

▶ Linearizability
▶ Quorums
▶ Registers

Recommended Reading

1. Aspnes J (2003) Randomized protocols for asynchronous consensus. Distrib Comput 16(2–3):165–175
2. Attiya H (2000) Efficient and robust sharing of memory in message-passing systems. J Algorithms 34(1):109–127
3. Attiya H, Bar-Noy A, Dolev D (1995) Sharing memory robustly in message-passing systems. J ACM 42(1):124–142
4. Attiya H, Welch J (2004) Distributed computing: fundamentals, simulations and advanced topics, 2nd edn. Wiley-Interscience, Hoboken
5. Chor B, Moscovici L (1989) Solvability in asynchronous environments. In: Proceedings of the 30th symposium on foundations of computer science, pp 422–427
6. Delporte-Gallet C, Fauconnier H, Guerraoui R, Hadzilacos V, Kouznetsov P, Toueg S (2004) The weakest failure detectors to solve certain fundamental problems in distributed computing. In: Proceedings of the 23rd ACM symposium on principles of distributed computing, St. John's, 25–28 Jul 2004, pp 338–346
7. Dolev S, Gilbert S, Lynch NA, Shvartsman AA, Welch JL (2005) GeoQuorums: implementing atomic memory in mobile ad hoc networks. Distrib Comput 18(2):125–155
8. Dutta P, Guerraoui R, Levy RR, Chakraborty A (2004) How fast can a distributed atomic read be? In: Proceedings of the 23rd ACM symposium on principles of distributed computing, St. John's, 25–28 Jul 2004, pp 236–245
9. Englert B, Shvartsman AA (2000) Graceful quorum reconfiguration in a robust emulation of shared memory. In: Proceedings of the 20th IEEE international conference on distributed computing systems, Taipei, 10–13 Apr 2000, pp 454–463
10. Gifford DK (1979) Weighted voting for replicated data. In: Proceedings of the 7th ACM symposium on operating systems principles, Pacific Grove, 10–12 Dec 1979, pp 150–162
11. Gilbert S, Lynch N, Shvartsman A (2003) Rambo II: rapidly reconfigurable atomic memory for dynamic

networks. In: Proceedings of the international conference on dependable systems and networks, San Francisco, 22–25 Jun 2003, pp 259–268
12. Haldar S, Vidyasankar K (1995) Constructing 1-writer multireader multivalued atomic variables from regular variables. J ACM 42(1):186–203
13. Israeli A, Li M (1993) Bounded time-stamps. Distrib Comput 6(4):205–209
14. Lamport L (1986) On interprocess communication: part II: algorithms. Distrib Comput 1(2):86–101
15. Malkhi D, Reiter M (1998) Byzantine quorum systems. Distrib Comput 11(4):203–213

Incentive Compatible Selection

Xi Chen
Computer Science Department, Columbia University, New York, NY, USA
Computer Science and Technology, Tsinghua University, Beijing, China

Keywords

Algorithmic mechanism design; Incentive compatible ranking; Incentive compatible selection

Years and Authors of Summarized Original Work

2006; Chen, Deng, Liu

Problem Definition

Ensuring truthful evaluation of alternatives in human activities has always been an important issue throughout history. In sports, in particular, such an issue is vital and practice of the fair-play principle has been consistently put forward as a matter of foremost priority. In addition to relying on the code of ethics and professional responsibility of players and coaches, the design of game rules is an important measure in enforcing fair play.

Ranking alternatives through pairwise comparisons (or competitions) is the most common approach in sports tournaments. Its goal is to find out the "true" ordering among alternatives through complete or partial pairwise competitions [1, 3–7]. Such studies have been mainly based on the assumption that all the players play truthfully, i.e., with their maximal effort. It is, however, possible that some players form a coalition and cheat for group benefit. An interesting example can be found in [2].

Problem Description

The work of Chen, Deng, and Liu [2] considers the problem of choosing m winners out of n candidates.

Suppose a tournament is held among n players $P_n = \{p_1, \dots p_n\}$ and m winners are expected to be selected by a selection protocol. Here a protocol $f_{n,m}$ is a predefined function (which will become clear later) to choose winners through pairwise competitions, with the intention of finding m players of highest capacity. When the tournament starts, a distinct ID in $N_n = \{1, 2, \dots n\}$ is assigned to each player in P_n by a randomly picked indexing function $I : P_n \rightarrow N_n$. Then a match is played between each pair of players. The competition outcomes will form a graph G, whose vertex set is N_n and edges represent the results of all the matches. Finally, the graph will be treated as the input to $f_{n,m}$, and it will output a set of m winners. Now it should be clear that $f_{n,m}$ maps every possible tournament graph G to a subset (of cardinality m) of N_n.

Suppose there exists a group of bad players who play dishonestly, i.e., they might lose a match on purpose to gain overall benefit for the whole group, while the rest of the players always play truthfully, i.e., they try their best to win matches. The group of bad players gains benefit if they are able to have more winning positions than that according to the true ranking. Given knowledge of the selection protocol $f_{n,m}$, the indexing function I, and the true ranking of all players, the bad players try to find a cheating strategy that can fool the protocol and gain benefit.

The problem is discussed under two models in which the characterizations of bad players are different. Under the *collective incentive compatible model*, bad players are willing to sacrifice themselves to win group benefit, while the ones under the *alliance incentive compatible model* only cooperate if their individual interests are well maintained in the cheating strategy.

The goal is to find an "ideal" protocol, under which players or groups of players maximize their benefits only by strictly following the fair-play principle, i.e., always play with maximal effort.

Formal Definitions

When the tournament begins, an indexing function I is randomly picked, which assigns ID $I(p) \in N_n$ to each player $p \in P_n$. Then a match is played between each pair of players, and the results are represented as a directed graph G. Finally, G is fed into the predefined selection protocol $f_{n,m}$, to produce a set of m winners $I^{-1}(W)$, where $W = f_{n,m}(G) \subset N_n$.

Notations

An indexing function I for a tournament attended by n players $P_n = \{p_1, p_2, \ldots p_n\}$ is a one-to-one correspondence from P_n to the set of IDs: $N_n = \{1, 2, \ldots n\}$. A ranking function R is a one-to-one correspondence from P_n to $\{1, 2, \ldots n\}$. $R(p)$ represents the underlying true ranking of player p among the n players. The smaller, the stronger.

A tournament graph of size n is a directed graph $G = (N_n, E)$ such that for all $i \neq j \in N_n$, either $ij \in E$ (player with ID i beats player with ID jn) or $ji \in E_n$. Let K_n denote the set of all such graphs. A selection protocol $f_{n,m}$, which chooses m winners out of n candidates, is a function from K_n to $\{S \subset N_n \text{ and } |S| = m\}$.

A tournament T_n among players P_n is a pair $T_n = (R, B)$ where R is a ranking function from P_n to N_n and $B \subset P_n$ is the group of bad players.

Definition 1 (Benefit) Given a protocol $f_{n,m}$, a tournament $T_n = (R, B)$, an indexing function I, and a tournament graph $G \in K_n$, the benefit of the group of bad players is

$$\text{Ben}(f_{n,m}, T_n, I, G) = |\{i \in f_{n,m}(G), I^{-1}(i) \in B\}| - |\{p \in B, R(p) \leq m\}|.$$

Given knowledge of $f_{n,m}$, T_n, and I, not every $G \in K_n$ is a feasible strategy for B: the group of bad players. First, it depends on the tournament $T_n = (R, B)$, e.g., a player $p_b \in B$ cannot win a player $p_g \notin B$ if $R(p_b) > R(p_g)$. Second, it depends on the property of bad players which is specified by the model considered. Tournament graphs, which are recognized as feasible strategies, are characterized below, for each model. The key difference is that a bad player in the alliance incentive compatible model is not willing to sacrifice his/her own winning position, while a player in the other model fights for group benefit at all costs.

Definition 2 (Feasible Strategy) Given $f_{n,m}$, $T_n = (R, B)$, and I, graph $G \in K_n$ is *c-feasible* if

1. For every two players $p_i, p_j \notin B$, if $R(p_i) < R(p_j)$, then $I(p_i)I(p_j) \in E$;
2. For all $p_g \notin B$ and $p_b \in B$, if $R(p_g) < R(p_b)$, then edge $I(p_g)I(p_b) \in E$.

Graph $G \in K_n$ is *a-feasible* if it is c-feasible and also satisfies

1. For every bad player $p \in B$, if $R(p) \leq m$, then $I(p) \in f_{n,m}(G)$.

A cheating strategy is then a feasible tournament graph G that can be employed by the group of bad players to gain positive benefit.

Definition 3 (Cheating Strategy) Given $f_{n,m}$, $T_n = (R, B)$, and I, a cheating strategy for the group of bad players under the collective incentive compatible (alliance incentive compatible) model is a graph $G \in K_n$ which is c-feasible (a-feasible) and satisfies $\text{Ben}(f_{n,m}, T_n, I, G) > 0$.

The following two problems are studied in [2]: **(1)** Is there a protocol $f_{n,m}$ such that for all T_n and I no cheating strategy exists under the collective incentive compatible model? (**2**) Is there a protocol $f_{n,m}$ such that for all T_n and I, no cheating strategy exists under the alliance incentive compatible model?

Key Results

Definition 4 For all integers n and m such that $2 \leq m \leq n-2$, a tournament graph $G_{n,m} = (N_n, E) \in K_n$, which consists of three parts T_1, T_2, and T_3, is defined as follows:

1. $T_1 = \{1, 2, \ldots m-2\}$.

For all $i < j \in T_1$, edge $ij \in E$;

2. $T_2 = \{m-1, m, m+1\}$.

$(m-1)m, m(m+1), (m+1)(m-1) \in E$;

3. $T_3 = \{m+2, m+3, \ldots n\}$.

For all $i < j \in T_3$, edge $ij \in E$;

4. For all $i' \in T_i$ and $j' \in T_j$ such that $i < j$, edge $i'j' \in E$.

Theorem 1 *Under the collective incentive compatible model, for every selection protocol $f_{n,m}$ with $2 \leq m \leq n-2$, if $T_n = (R, B)$ satisfies (1) at least one bad player ranks as high as $m-1$, (2) the ones ranked $m+1$ and $m+2$ are both bad players, and (3) the one ranked m is a good player, then there always exists an indexing function I such that $G_{n,m}$ is a cheating strategy.*

Theorem 2 *Under the alliance incentive compatible model, if $n - m \geq 3$, then there exists a selection protocol $f_{n,m}$ [2] such that for every tournament T_n, indexing function I, and a-feasible strategy $G \in K_n$, $\mathrm{Ben}(f_{n,m}, T_n, I, G) \leq 0$.*

Applications

The result shows that if players are willing to sacrifice themselves, no protocol is able to prevent malicious coalitions from obtaining undeserved benefits.

The result may have potential applications in the design of output truthful mechanisms.

Open Problems

Under the collective incentive compatible model, the work of Chen, Deng, and Liu indicates that cheating strategies are available in at least 1/8 of tournaments, assuming the probability for each player to be in the bad group is 1/2. Could this bound be improved? Or could one find a good selection protocol in the sense that the number of tournaments with cheating strategies is close to this bound? On the other hand, although no ideal protocol exists in this model, does there exist any randomized protocol, under which the probability of having cheating strategies is negligible?

Cross-References

▶ Parity Games

Recommended Reading

1. Chang P, Mendonca D, Yao X, Raghavachari M (2004) An evaluation of ranking methods for multiple incomplete round-robin tournaments. In: Proceedings of the 35th annual meeting of decision sciences institute, Boston, 20–23 Nov 2004
2. Chen X, Deng X, Liu BJ (2006) On incentive compatible competitive selection protocol. In: Proceedings of the 12th annual international computing and combinatorics conference (COCOON'06), Taipei, 15–18 Aug 2006, pp 13–22
3. Harary F, Moser L (1966) The theory of round robin tournaments. Am Math Mon 73(3):231–246
4. Jech T (1983) The ranking of incomplete tournaments: a mathematician's guide to popular sports. Am Math Mon 90(4):246–266
5. Mendonca D, Raghavachari M (1999) Comparing the efficacy of ranking methods for multiple round-robin tournaments. Eur J Oper Res 123:593–605
6. Rubinstein A (1980) Ranking the participants in a tournament. SIAM J Appl Math 38(1):108–111
7. Steinhaus H (1950) Mathematical snapshots. Oxford University Press, New York

Independent Sets in Random Intersection Graphs

Sotiris Nikoletseas[1,3], Christoforos L. Raptopoulos[2,3,4], and Paul (Pavlos) Spirakis[5,6,7]
[1]Computer Engineering and Informatics Department, University of Patras, Patras, Greece
[2]Computer Science Department, University of Geneva, Geneva, Switzerland
[3]Computer Technology Institute and Press "Diophantus", Patras, Greece
[4]Research Academic Computer Technology Institute, Greece and Computer Engineering and Informatics Department, University of Patras, Patras, Greece
[5]Computer Engineering and Informatics, Research and Academic Computer Technology Institute, Patras University, Patras, Greece
[6]Computer Science, University of Liverpool, Liverpool, UK
[7]Computer Technology Institute (CTI), Patras, Greece

Keywords

Existence and efficient construction of independent sets of vertices in general random intersection graphs

Years and Authors of Summarized Original Work

2004; Nikoletseas, Raptopoulos, Spirakis

Problem Definition

This problem is concerned with the efficient construction of an independent set of vertices (i.e., a set of vertices with no edges between them) with maximum cardinality, when the input is an instance of the uniform random intersection graphs model. This model was introduced by Karoński, Sheinerman, and Singer-Cohen in [4] and Singer-Cohen in [10] and it is defined as follows

Definition 1 (Uniform random intersection graph) Consider a universe $M=\{1,2,\ldots,m\}$ of elements and a set of vertices $V=\{v_1,v_2,\ldots,v_n\}$. If one assigns independently to each vertex $v_j, j=1,2,\ldots,n$, a subset S_{v_j} of M by choosing each element independently with probability p and puts an edge between two vertices v_{j_1}, v_{j_2} if and only if $S_{v_{j_1}} \cap S_{v_{j_2}} \neq \emptyset$, then the resulting graph is an instance of the uniform random intersection graph $G_{n,m,p}$.

The universe M is sometimes called *label set* and its elements *labels*. Also, denote by L_l, for $l \in M$, the set of vertices that have chosen label l.

Because of the dependence of edges, this model can abstract more accurately (than the Bernoulli random graphs model $G_{n,p}$ that assumes independence of edges) many real-life applications. Furthermore, Fill, Sheinerman, and Singer-Cohen show in [3] that for some ranges of the parameters n, m, p $(m = n^{\alpha}, \alpha > 6)$, the spaces $G_{n,m,p}$ and $G_{n,\hat{p}}$ are equivalent in the sense that the total variation distance between the graph random variables has limit 0. The work of Nikoletseas, Raptopoulos, and Spirakis [7] introduces two new models, namely the *general random intersection graphs model* $G_{n,m,\vec{p}}, \vec{p} = [p_1, p_2, \ldots, p_m]$ and the *regular random intersection graphs model* $G_{n,m,\lambda}, \lambda > 0$ that use a different way to randomly assign labels to vertices, but the edge appearance rule remains the same. The $G_{n,m,\vec{p}}$ model is a generalization of the uniform model where each label $i \in M$ is chosen independently with probability p_i, whereas in the $G_{n,m,\lambda}$ model each vertex chooses a random subset of M with exactly λ labels.

The authors in [7] first consider the existence of independent sets of vertices of a given cardinality in general random intersection graphs and provide exact formulae for the mean and variance of the number of independent sets of vertices of cardinality k. Furthermore, they present and analyze three polynomial time (on the number of labels m and the number of vertices n) algorithms for constructing large independent sets of vertices when the input is an instance of the $G_{n,m,p}$ model. To the best knowledge of the entry authors, this work is the first to consider algorithmic issues for these models of random graphs.

Key Results

The following theorems concern the existence of independent sets of vertices of cardinality k in general random intersection graphs. The proof of Theorem 1 uses the linearity of expectation of sums of random variables.

Theorem 1 *Let $X^{(k)}$ denote the number of independent sets of size k in a random intersection graph $G(n, m, \vec{p})$, where $\vec{p} = [p_1, p_2, \ldots, p_m]$. Then*

$$E\left[X^{(k)}\right] = \binom{n}{k} \prod_{i=1}^{m} \left((1-p_i)^k + kp_i(1-p_i)^{k-1}\right).$$

Theorem 2 *Let $X^{(k)}$ denote the number of independent sets of size k in a random intersection graph $G(n, m, \vec{p})$, where $\vec{p} = [p_1, p_2, \ldots, p_m]$. Then*

$$\mathrm{Var}\left(X^{(k)}\right) = \sum_{s=1}^{k} \binom{n}{2k-s} \binom{2k-s}{s} \left(\gamma(k,s) \frac{E[X^{(k)}]}{\binom{n}{k}} - \frac{E^2[X^{(k)}]}{\binom{n}{k}^2}\right)$$

where $E\left[X^{(k)}\right]$ is the mean number of independent sets of size k and

$$\gamma(k,s) = \prod_{i=1}^{m} \left((1-p_i)^{k-s} + (k-s)p_i(1-p_i)^{k-s-1} \left(1 - \frac{sp_i}{1+(k-1)p_i}\right)\right).$$

Theorem 2 is proved by first writing the variance as the sum of covariances and then applying a vertex contraction technique that merges several vertices into one supervertex with similar probabilistic behavior in order to compute the covariances. By using the second moment method (see [1]) one can derive thresholds for the existence of independent sets of size k.

One of the three algorithms that were proposed in [7] is presented below. The algorithm starts with V (i.e., the set of vertices of the graph) as its "candidate" independent set. In every subsequent step it chooses a label and removes from the current candidate independent set all vertices having that label in their assigned label set except for one. Because of the edge appearance rule, this ensures that after doing this for every label in M, the final candidate independent set will contain only vertices that do not have edges between them and so it will be indeed an independent set.

Algorithm:
Input: A random intersection graph $G_{n,m,p}$.
Output: An independent set of vertices A_m.

1. set $A_0 := V$; set $L := M$;
2. **for** $i = 1$ **to** m **do**
3. **begin**
4. select a random label $l_i \in L$; set $L := L - \{l_i\}$;
5. set $D_i := \{v \in A_{i-1} : l_i \in S_v\}$;
6. **if** $(|D_i| \geq 1)$ **then** select a random vertex $u \in D_i$ and set $D_i := D_i - \{u\}$;
7. set $A_i := A_{i-1} - D_i$;
8. **end**
9. **output** A_m;

The following theorem concerns the cardinality of the independent set produced by the algorithm. The analysis of the algorithm uses Wald's equation (see [9]) for sums of a random number of random variables to calculate the mean value of $|A_m|$, and also Chernoff bounds (see e.g., [6]) for concentration around the mean.

Theorem 3 *For the case $mp = \alpha \log n$, for some constant $\alpha > 1$ and $m \geq n$, and for some constant $\beta > 0$, the following hold with high probability:*

1. *If $np \to \infty$ then $|A_m| \geq (1-\beta)\frac{n}{\log n}$.*
2. *If $np \to b$ where $b > 0$ is a constant then $|A_m| \geq (1-\beta)n(1-e^{-b})$.*
3. *If $np \to 0$ then $|A_m| \geq (1-\beta)n$.*

The above theorem shows that the algorithm manages to construct a quite large independent set with high probability.

Applications

First of all, note that (as proved in [5]) any graph can be transformed into an intersection graph. Thus, the random intersection graphs models can be very general. Furthermore, for some ranges of the parameters n, m, p ($m = n^{\alpha}, \alpha > 6$) the spaces $G_{n,m,p}$ and $G_{n,p}$ are equivalent (as proved by Fill, Sheinerman, and Singer-Cohen in [3], showing that in this range the total variation distance between the graph random variables has limit 0).

Second, random intersection graphs (and in particular the general intersection graphs model of [7]) may model real-life applications more accurately (compared to the $G_{n,p}$ case). In particular, such graphs can model resource allocation in networks, e.g., when network nodes (abstracted by vertices) access shared resources (abstracted by labels): the intersection graph is in fact the conflict graph of such resource allocation problems.

Other Related Work

In their work [4] Karoński et al. consider the problem of the emergence of graphs with a constant number of vertices as induced subgraphs of $G_{n,m,p}$ graphs. By observing that the $G_{n,m,p}$ model generates graphs via clique covers (for example the sets $L_l, l \in M$ constitute an obvious clique cover) they devise a natural way to use them together with the first and second moment methods in order to find thresholds for the appearance of any fixed graph H as an induced subgraph of $G_{n,m,p}$ for various values of the parameters n, m and p.

The connectivity threshold for $G_{n,m,p}$ was considered by Singer-Cohen in [10]. She studies the case $m = n^{\alpha}, \alpha > 0$ and distinguishes two cases according to the value of α. For the case $\alpha > 1$, the results look similar to the $G_{n,p}$ graphs, as the mean number of edges at the connectivity thresholds are (roughly) the same. On the other hand, for $\alpha \leq 1$ we get denser graphs in the $G_{n,m,p}$ model. Besides connectivity, [10] examines also the size of the largest clique in uniform random intersection graphs for certain values of n, m and p.

The existence of Hamilton cycles in $G_{n,m,p}$ graphs was considered by Efthymiou and Spirakis in [2]. The authors use coupling arguments to show that the threshold of appearance of Hamilton cycles is quite close to the connectivity threshold of $G_{n,m,p}$. Efficient probabilistic algorithms for finding Hamilton cycles in uniform random intersection graphs were presented by Raptopoulos and Spirakis in [8]. The analysis of those algorithms verify that they perform well w.h.p. even for values of p that are close to the connectivity threshold of $G_{n,m,p}$. Furthermore, in the same work, an expected polynomial algorithm for finding Hamilton cycles in $G_{n,m,p}$ graphs with constant p is given.

In [11] Stark gives approximations of the distribution of the degree of a fixed vertex in the $G_{n,m,p}$ model. More specifically, by applying a sieve method, the author provides an exact formula for the probability generating function of the degree of some fixed vertex and then analyzes this formula for different values of the parameters n, m and p.

Open Problems

A number of problems related to random intersection graphs remain open. Nearly all the algorithms proposed so far concerning constructing large independent sets and finding Hamilton cycles in random intersection graphs are greedy. An interesting and important line of research would be to find more sophisticated algorithms for these problems that outperform the greedy ones. Also, all these algorithms were presented and analyzed in the uniform random intersection graphs model. Very little is known about how the same algorithms would perform when their input was an instance of the general or even the regular random intersection graph models.

Of course, many classical problems concerning random graphs have not yet been studied.

One such example is the size of the minimum dominating set (i.e., a set of vertices that has the property that all vertices of the graph either belong to this set or are connected to it) in a random intersection graph. Also, what is the degree sequence of $G_{n,m,p}$ graphs? Note that this is very different from the problem addressed in [11].

Finally, notice that none of the results presented in the bibliography for general or uniform random intersection graphs carries over immediately to regular random intersection graphs. Of course, for some values of n, m, p and λ, certain graph properties shown for $G_{n,m,p}$ could also be proved for $G_{n,m,\lambda}$ by showing concentration of the number of labels chosen by any vertex via Chernoff bounds. Other than that, the fixed sizes of the sets assigned to each vertex impose more dependencies to the model.

Cross-References

▶ Hamilton Cycles in Random Intersection Graphs

Recommended Reading

1. Alon N, Spencer H (2000) The probabilistic method. Wiley
2. Efthymiou C, Spirakis P (2005) On the existence of Hamiltonian cycles in random intersection graphs. In: Proceedings of 32nd international colloquium on automata, languages and programming (ICALP). Springer, Berlin/Heidelberg, pp 690–701
3. Fill JA, Sheinerman ER, Singer-Cohen KB (2000) Random intersection graphs when m = ω(n): an equivalence theorem relating the evolution of the g(n, m, p) and g(n, p)models. Random Struct Algorithms 16(2):156–176
4. Karoński M, Scheinerman ER, Singer-Cohen KB (1999) On random intersection graphs: the subgraph problem. Adv Appl Math 8:131–159
5. Marczewski E (1945) Sur deux propriétés des classes d'ensembles. Fundam Math 33:303–307
6. Motwani R, Raghavan P (1995) Randomized algorithms. Cambridge University Press
7. Nikoletseas S, Raptopoulos C, Spirakis P (2004) The existence and efficient construction of large independent sets in general random intersection graphs. In: Proceedings of 31st international colloquium on Automata, Languages and Programming (ICALP). Springer, Berlin/Heidelberg, pp 1029–1040. Also in the Theoretical Computer Science (TCS) Journal, accepted, to appear in 2008
8. Raptopoulos C, Spirakis P (2005) Simple and efficient greedy algorithms for Hamiltonian cycles in random intersection graphs. In: Proceedings of the 16th international symposium on algorithms and computation (ISAAC). Springer, Berlin/Heidelberg, pp 493–504
9. Ross S (1995) Stochastic processes. Wiley
10. Singer-Cohen KB (1995) Random intersection graphs. Ph.D. thesis, John Hopkins University, Baltimore
11. Stark D (2004) The vertex degree distribution of random intersection graphs. Random Struct Algorithms 24:249–258

Indexed Approximate String Matching

Wing-Kin Sung
Department of Computer Science, National University of Singapore, Singapore, Singapore

Keywords

Edit distance; Hamming distance; Indexed inexact pattern matching; Indexed k-difference problem; Indexed k-mismatch problem; Suffix tree

Years and Authors of Summarized Original Work

1994; Myers
2007; Mass, Nowak
2008; Lam, Sung, Wong
2010; Chan, Lam, Sung, Tam, Wong
2010; Tsur
2011; Chan, Lam, Sung, Tam, Wong
2014; Belazzougui

Problem Definition

Consider a text $S[1 \ldots n]$ over a finite alphabet Σ. The problem is to build an index for S such that for any query pattern $P[1 \ldots m]$ and any integer $k \geq 0$, all locations in S that match P with at most k errors can be reported efficiently. If the error is measured in terms of the Hamming distance (number of character substitutions), the problem is called k-mismatch problem. If the error is measured in terms of the edit distance (number of character substitutions, insertions, or deletions), the problem is called k-difference problem. The two problems are formally defined as follows.

Problem 1 (k-mismatch problem) Consider a text $S[1 \ldots n]$ over a finite alphabet Σ. For any pattern P and threshold k, position i is an occurrence of P if the Hamming distance between P and $S[i \ldots i']$ is less than k for some i'. The k-mismatch problem asks for an index I for S such that, for any pattern P, all occurrences of P in S can be reported efficiently.

Problem 2 (k-difference problem) Consider a text $S[1 \ldots n]$ over a finite alphabet Σ. For any pattern P and threshold k, position i is an occurrence of P if the edit distance between P and $S[i \ldots i']$ is less than k for some i'. The k-difference problem asks for an index I for S such that, for any pattern P, all occurrences of P in S can be reported efficiently.

These two problems are also called indexed inexact pattern matching problem or indexed pattern searching problem based on Hamming distance or edit distance.

The major concern of these two problems is how to achieve efficient pattern searching without using a large amount of space for storing the index.

Key Results

For indexed k-mismatch or k-difference string matching, a naive solution either requires an index of size $\Omega(n^k)$ or supports the query using $\Omega(m^k)$ time. The first non-trival solution is by Cole et al. [10]. They modify suffix tree to give an $O(n \log^k n)$-word index that supports k-difference query using $O(m + \text{occ} + \frac{1}{k!}(c \log n)^k \log \log n)$ time. After that, a number of indexes are proposed that support k-mismatch/k-difference pattern query for any $k > 0$. All these indexes are created by augmenting the suffix tree and its variants.

Tables 1 and 2 summarize the related results in the literature for $k = 1$ and $k \geq 2$. Below, the current best results are briefly summarized.

Indexed Approximate String Matching, Table 1 Known results for 1-difference matching. ϵ is some positive constant smaller than 1 and occ is the number of 1-difference occurrences in the text

Space	Running time	
$O(\|\Sigma\| n \log n)$ words in avg	$O(m + \text{occ})$	[15]
$O(\|\Sigma\| n \log n)$ words	$O(m + \text{occ})$ in avg	[15]
$O(n \log^2 n)$ words	$O(m \log n \log \log n + \text{occ})$	[1]
$O(n \log n)$ words	$O(m \log \log n + \text{occ})$	[4]
	$O(m + \text{occ} + \log n \log \log n)$	[10]
$O(n)$ words	$O(\min\{n, \|\Sigma\| m^2\} + \text{occ})$	[8]
	$O(\|\Sigma\| m \log n + \text{occ})$	[13]
	$O(n^\epsilon \log n)$	[16]
	$O(n^\epsilon)$	[17]
	$O(m + \text{occ} + \|\Sigma\| \log^3 n \log \log n)$	[5]
	$O(m + \text{occ} + \log n \log \log n)$	[6]
$O(n(\log n \log \log n)^2 \log \|\Sigma\|)$ bits	$O(m + \text{occ})$	[2]
$O(n \sqrt{\log n} \log \|\Sigma\|)$ bits	$O(\|\Sigma\| m \log \log n + \text{occ})$	[14]
$O(n \log^\epsilon n \log \|\Sigma\|)$ bits	$O(\|\Sigma\| m + \text{occ})$	[3]
$O(n \log \log n \log \|\Sigma\|)$ bits	$O((\|\Sigma\| m + \text{occ}) \log \log n)$	[3]
$O(n \log \|\Sigma\|)$ bits	$O(\|\Sigma\| m \log^2 n + \text{occ} \log n)$	[13]
	$O((\|\Sigma\| m \log \log n + \text{occ}) \log^\epsilon n)$	[14]
	$O(m + (\text{occ} + \|\Sigma\| \log^4 n \log \log n) \log^\epsilon n)$	[5]

Indexed Approximate String Matching, Table 2 Known results for k-difference matching for $k \geq 2$. c and d are some positive constants and ϵ is some positive constant smaller than 1. occ is the number of k-difference occurrences in the text

Space	Running time	
$O(n^{1+\epsilon})$ words	$O(m + \log\log n + \text{occ})$	[19]
$O(\|\Sigma\|^k n \log^k n)$ words in avg	$O(m + \text{occ})$	[15]
$O(\|\Sigma\|^k n \log^k n)$ words	$O(m + \text{occ})$ in avg	[15]
$O(n \log^k n)$ words in avg	$O(3^k m^{k+1} + \text{occ})$	[9]
$O(\frac{d^k}{k!} n \log^k n)$ words	$O(m + 3^k \text{occ} + \frac{1}{k!}(c \log n)^k \log\log n)$	[10]
$O(n \log^{k-1} n)$ words	$O(m + k^3 3^k \text{occ} + \frac{1}{k!}(c \log n)^k \log\log n)$	[5]
$O(n)$ words	$O(\min\{n, \|\Sigma\|^k m^{k+2}\} + \text{occ})$	[8]
	$O((\|\Sigma\| m)^k \max(k, \log n) + \text{occ})$	[13]
	$O(m + k^3 3^k \text{occ} + (c \log n)^{k(k+1)} \log\log n)$	[5]
	$O((2\|\Sigma\|)^{k-1} m^{k-1} \log n \log\log n + \text{occ})$	[6]
$O(n\sqrt{\log n} \log\|\Sigma\|)$ bits	$O((\|\Sigma\| m)^k (k + \log\log n) + \text{occ})$	[14]
$O(n \log\|\Sigma\|)$ bits	$O((\|\Sigma\| m)^k \max(k, \log^2 n) + \text{occ} \log n)$	[13]
	$O(((\|\Sigma\| m)^k (k + \log\log n) + \text{occ}) \log^\epsilon n)$	[14]
	$O(m + (k^3 3^k \text{occ} + (c \log n)^{k^2+2k} \log\log n) \log^\epsilon n)$	[5]

Inexact Matching When *k* = 1

For 1-mismatch and 1-difference approximate matching problem, the theorems below give the current best solutions. Both algorithms try to handle long and short patterns separately. Short patterns of size polylog(n) can be handled using index of size $O(\text{polylog}(n))$ space by brute force. Long patterns can be handled with the help of some augmented suffix tree.

When the index is of size $O(n \log|\Sigma|)$ bits, the next theorem is the current best result.

Theorem 1 (Chan, Lam, Sung, Tam, and Wong [5]) *Given an index of size $O(n \log|\Sigma|)$ bits, 1-mismatch or 1-difference query can be supported in $O(m + (\text{occ} + |\Sigma| \log^4 n \log\log n) \log^\epsilon n)$ time where ϵ is any positive constant smaller than or equal to 1.*

When we allow a bit more space, Belazzougui can further reduce the query time, as shown in the following theorem.

Theorem 2 (Belazzougui [3]) *Given an index of size $O(n \log^\epsilon n \log|\Sigma|)$ bits (or $O(n \log\log n \log|\Sigma|)$ bits, respectively), 1-mismatch/1-difference lookup can be supported in $O(|\Sigma| m + \text{occ})$ (or $O((|\Sigma| m + \text{occ}) \log\log n)$, respectively) time.*

Inexact Matching When $k \geq 2$

For k-mismatch and k-difference approximate matching problem where $k \geq 2$, existing solutions are all based on the so-called k-error suffix trees and its variants (following the idea of Cole et al.).

Some current solutions create indexes whose sizes depend on k. Theorems 3–6 summarize the current best results in this direction.

Theorem 3 (Maas and Nowak [15]) *Given an index of size $O(|\Sigma|^k n \log^k n)$ words, k-mismatch/k-difference lookup can be supported in $O(m + \text{occ})$ expected time.*

Theorem 4 (Maas and Nowak [15]) *Consider a uniformly and independently generated text of length n. There exists an index of size $O(|\Sigma|^k n \log^k n)$ words on average such that an k-mismatch/k-difference lookup query can be supported in $O(m + \text{occ})$ worst-case time.*

Theorem 5 (Chan, Lam, Sung, Tam, and Wong [5]) *Given an index of size $O(n \log^{k-h+1} n)$ words where $h \le k$, k-mismatch lookup can be supported in $O(m + \text{occ} + c^{k^2} \log^{\max\{kh,k+h\}} n \log\log n)$ time where c is a positive constant. For k-difference lookup, the term* occ *becomes* $k^3 3^k \text{occ}$.

Theorem 6 (Chan, Lam, Sung, Tam, and Wong [6]) *Given an index of size $O(n \log^{k-1} n)$ words, k-mismatch/k-difference lookup can be supported in $O(m + \text{occ} + \log^k n \log\log n)$ time.*

Theorems 7–12 summarize the current best results when the index size is independent of k.

Theorem 7 (Chan, Lam, Sung, Tam, and Wong [5]) *Given an index of size $O(n)$ words, k-mismatch lookup can be supported in $O(m + \text{occ} + (c \log n)^{k(k+1)} \log\log n)$ time where c is a positive constant. For k-difference lookup, the term* occ *becomes* $k^3 3^k \text{occ}$.

Theorem 8 (Chan, Lam, Sung, Tam, and Wong [5]) *Given an index of size $O(n \log|\Sigma|)$ bits, k-mismatch lookup can be supported in $O(m + (\text{occ} + (c \log n)^{k(k+2)} \log\log n) \log^{\epsilon} n)$ time where c is a positive constant and ϵ is any positive constant smaller than or equal to 1. For k-difference lookup, the term* occ *becomes* $k^3 3^k \text{occ}$.

Theorem 9 (Lam, Sung, and Wong [14]) *Given an index of size $O(n\sqrt{\log n} \log|\Sigma|)$ bits, k-mismatch/k-difference lookup can be supported in $O((|\Sigma|m)^k (k + \log\log n) + \text{occ})$ time.*

Theorem 10 (Lam, Sung, and Wong [14]) *Given an index of size $O(n \log|\Sigma|)$ bits, k-mismatch/k-difference lookup can be supported in $O(((|\Sigma|m)^k (k + \log\log n) + \text{occ}) \log^{\epsilon} n)$ time where ϵ is any positive constant smaller than or equal to 1.*

Theorem 11 (Chan, Lam, Sung, Tam, and Wong [6]) *Given an index of size $O(n)$ words, k-mismatch/k-difference lookup can be supported in $O((2|\Sigma|)^{k-1} m^{k-1} \log n \log\log n + \text{occ})$ time.*

Theorem 12 (Tsur [19]) *Given an index of size $O(n^{1+\epsilon})$ words, k-mismatch/k-difference lookup can be supported in $O(m + \text{occ} + \log\log n)$ time.*

Practically Fast Inexact Matching

In addition, there are indexes which are efficient in practice for small k/m but give no worst-case complexity guarantees. Those methods are based on filtration. The basic idea is to partition the pattern into short segments and locate those short segments in the text allowing zero or a small number of errors. Those short segments help to identify candidate regions for the occurrences of the pattern. Finally, by verifying those candidate regions, all occurrences of the pattern are recovered. See [18] for a summary of those results. One of the best results based on filtration is stated in the following theorem.

Theorem 13 (Myers [16] and Navarro and Baeza-Yates [17]) *If $k/m < 1 - O(1/\sqrt{|\Sigma|})$, k-mismatch/k-difference search can be supported in $O(n^{\epsilon})$ expected time, where ϵ is a positive constant smaller than 1, with an index of size $O(n)$ words.*

Other methods with good performance on average include [11] and [12].

All the above approaches either try to index the strings with errors or are based on filtering. There are also solutions which use radically different approaches. For instance, there are solutions which transform approximate string searching into range queries in metric space [7].

Applications

Due to the advance in both the Internet and biological technologies, enormous text data is accumulated. For example, 60G genomic sequence data are currently available in GenBank. The data size is expected to grow exponentially.

To handle the huge data size, indexing techniques are vital to speed up the pattern matching queries. Moreover, exact pattern matching is no longer sufficient for both the Internet and biological data. For example, biological data usually contains a lot of differences due to experimental errors and due to mutation and evolution. Therefore, approximate pattern matching becomes more appropriate. This gives the motivation for developing indexing techniques that allow pattern matching with errors.

Open Problems

The complexity for indexed approximate matching is still not fully understood. A number of questions are still open. For instance, there are two open questions: (1) Given a fixed index size of $O(n)$ words, what is the best time complexity of a k-mismatch/k-difference query? (2) Fixed the k-mismatch/k-difference query time to be $O(m + \text{occ})$, what is the best space complexity of the index?

Cross-References

▸ Approximate String Matching
▸ Suffix Trees and Arrays

Recommended Reading

1. Amir A, Keselman D, Landau GM, Lewenstein M, Lewenstein N, Rodeh M (2000) Indexing and dictionary matching with one error. J Algorithms, 37(2):309–325
2. Belazzougui D (2009) Faster and space-optimal edit distance "1" dictionary. In: Proceedings of the 20th annual symposium on combinatorial pattern matching (CPM), Lille, pp 154–167
3. Belazzougui D (2014) Improved space-time tradeoffs for approximate full-text indexing with one edit error. Algorithmica. doi:10.1007/s00453-014-9873-9
4. Buchsbaum AL, Goodrich MT, Westbrook JR (2000) Range searching over tree cross products. In: Proceedings of European symposium on algorithms, Saarbrücken, pp 120–131
5. Chan H-L, Lam T-W, Sung W-K, Tam S-L, Wong S-S (2011) A linear size index for approximate pattern matching. J Discr Algorithms 9(4):358–364
6. Chan H-L, Lam T-W, Sung W-K, Tam S-L, Wong S-S (2010) Compressed indexes for approximate string matching. Algorithmica 58(2):263–281
7. Navarro G, Chávez E (2006) A metric index for approximate string matching. Theor Comput Sci 352(1–3):266–279
8. Cobbs A (1995) Fast approximate matching using suffix trees. In: Proceedings of symposium on combinatorial pattern matching, Espoo, pp 41–54
9. Coelho LP, Oliveira AL (2006) Dotted suffix trees: a structure for approximate text indexing. In: SPIRE, Glasgow, pp 329–336
10. Cole R, Gottlieb LA, Lewenstein M (2004) Dictionary matching and indexing with errors and don't cares. In: Proceedings of symposium on theory of computing, Chicago, pp 91–100
11. Epifanio C, Gabriele A, Mignosi F, Restivo A, Sciortino M (2007) Languages with mismatches. Theor Comput Sci 385(1–3):152–166
12. Gabriele A, Mignosi F, Restivo A, Sciortino M (2003) Indexing structures for approximate string matching. In: Proceedings of the 5th Italian conference on algorithms and complexity (CIAC), Rome, pp 140–151
13. Huynh TND, Hon WK, Lam TW, Sung WK (2006) Approximate string matching using compressed suffix arrays. Theor Comput Sci 352(1–3):240–249
14. Lam TW, Sung WK, Wong SS (2008) Improved approximate string matching using compressed suffix data structures. Algorithmica 51(3): 298–314
15. Maaß MG, Nowak J (2007) Text indexing with errors. J Discr Algorithms 5(4):662–681
16. Myers EG (1994) A sublinear algorithm for approximate keyword searching. Algorithmica 12: 345–374
17. Navarro G, Baeza-Yates R (2000) A hybrid indexing method for approximate string matching. J Discr Algorithms 1(1):205–209
18. Navarro G, Baeza-Yates RA, Sutinen E, Tarhio J (2001) Indexing methods for approximate string matching. IEEE Data Eng Bull 24(4):19–27
19. Tsur D (2010) Fast index for approximate string matching. J Discr Algorithms 8(4):339–345

Indexed Regular Expression Matching

Chee Yong Chan[1], Minos Garofalakis[2], and Rajeev Rastogi[3]
[1]National University of Singapore, Singapore, Singapore
[2]Technical University of Crete, Chania, Greece
[3]Amazon, Seattle, WA, USA

Keywords

Regular expression indexing; Regular expression retrieval

Years and Authors of Summarized Work

2003; Chan, Garofalakis, Rastogi

Problem Definition

Regular expressions (REs) provide an expressive and powerful formalism for capturing the structure of messages, events, and documents. Consequently, they have been used extensively in the specification of a number of languages for important application domains, including the XPath pattern language for XML documents [6] and the policy language of the *Border Gateway Protocol* (BGP) for propagating routing information between autonomous systems in the Internet [12]. Many of these applications have to manage large databases of RE specifications and need to provide an effective matching mechanism that, given an input string, quickly identifies all the REs in the database that match it. This RE retrieval problem is therefore important for a variety of software components in the middleware and networking infrastructure of the Internet.

The RE retrieval problem can be stated as follows: Given a large set S of REs over an alphabet Σ, where each RE $r \in S$ defines a regular language $L(r)$, construct a data structure on S that efficiently answers the following query: given an arbitrary input string $w \in \Sigma^*$, find the subset S_w of REs in S whose defined regular languages include the string w. More precisely, $r \in S_w$ iff $w \in L(r)$. Since S is a large, dynamic, disk-resident collection of REs, the data structure should be dynamic and provide efficient support of updates (insertions and deletions) to S. Note that this problem is the opposite of the more traditional RE search problem where $S \subseteq \Sigma^*$ is a collection of strings and the task is to efficiently find all strings in S that match an input regular expression.

Notations

An RE r over an alphabet Σ represents a subset of strings in σ^* (denoted by $L(r)$) that can be defined recursively as follows [9]: (1) the constants ϵ and $\emptyset$ are REs, where $L(\epsilon) = \{\epsilon\}$ and $L(\emptyset) = \emptyset$; (2) for any letter $a \in \sigma$, a is an RE where $L(a) = \{a\}$; (3) if r_1 and r_2 are REs, then their union, denoted by $r_1 + r_2$, is an RE where $L(r_1 + r_2) = L(r_1) \cup L(r_2)$; (4) if r_1 and r_2 are REs, then their concatenation, denoted by $r_1.r_2$, is an RE where $L(r_1.r_2) = \{s_1 s_2 \mid s_1 \in L(r_1), s_2 \in L(r_2)\}$; (5) if r is an RE, then its closure, denoted by r^*, is an RE where $L(r^*) = L(\epsilon) \cup L(r) \cup L(rr) \cup L(rrr) \cup \cdots$; and (6) if r is an RE, then a parenthesized r, denoted by (r), is an RE where $L((r)) = L(r)$. For example, if $\sigma = \{a, b, c\}$, then $(a+b).(a+b+c)^*.c$ is an RE representing the set of strings that begins with either a "a" or a "b" and ends with a "c." A string $s \in \sigma^*$ is said to match an RE r if $s \in L(r)$.

The language $L(r)$ defined by an RE r can be recognized by a *finite automaton (FA)* M that decides if an input string w is in $L(r)$ by reading each letter in w sequentially and updating its current state such that the outcome is determined by the final state reached by M after w has been processed [9]. Thus, M is an FA for r if the language accepted by M, denoted by $L(M)$, is equal to $L(r)$. An FA is classified as a *deterministic finite automaton* (DFA) if its current state is always updated to a single state; otherwise, it is a *nondeterministic finite automaton* (NFA) if its current state could refer to multiple possible states. The trade-off between a DFA and an NFA representations for an RE is that the latter is more space efficient, while the former is more time efficient for recognizing a matching string by checking a single path of state transitions. Let $|L(M)|$ denote the size of $L(M)$ and $|L_n(M)|$ denote the number of length-n strings in $L(M)$. Given a set $\mathcal{M}$ of finite automata, let $L(\mathcal{M})$ denote the language recognized by the automata in $\mathcal{M}$; i.e., $L(\mathcal{M}) = \bigcup_{M_i \in \mathcal{M}} L(M_i)$.

Key Results

The RE retrieval problem was first studied for a restricted class of REs in the context of content-based dissemination of XML documents using

XPath-based subscriptions (e.g., [1, 3, 7]), where each XPath expression is processed in terms of a collection of path expressions. While the XPath language [6] allows rich patterns with tree structure to be specified, the path expressions that it supports lack the full expressive power of REs (e.g., XPath does not permit the RE operators *, + and · to be arbitrarily nested in path expressions), and thus extending these XML-filtering techniques to handle general REs may not be straightforward. Further, all of the XPath-based methods are designed for indexing main-memory resident data. Another possible approach would be to coalesce the automata for all the REs into a single NFA and then use this structure to determine the collection of matching REs. It is unclear, however, if the performance of such an approach would be superior to a simple sequential scan over the database of REs; furthermore, it is not easy to see how such a scheme could be adapted for disk-resident RE data sets.

The first disk-based data structure that can handle the storage and retrieval of REs in their full generality is the *RE-tree* [4, 5]. Similar to the R-tree [8], an RE-tree is a dynamic, height-balanced, hierarchical index structure, where the leaf nodes contain data entries corresponding to the indexed REs, and the internal nodes contain "directory" entries that point to nodes at the next level of the index. Each leaf node entry is of the form (*id*, M), where *id* is the unique identifier of an RE r and M is a finite automaton representing r. Each internal node stores a collection of finite automata, and each node entry is of the form (M, *ptr*), where M is a finite automaton and *ptr* is a pointer to some node N (at the next level) such that the following *containment property* is satisfied: If $\mathcal{M}_N$ is the collection of automata contained in node N, then $L(\mathcal{M}_N) \subseteq L(M)$. The automaton M is referred to as the *bounding automaton* for $\mathcal{M}_N$. The containment property is key to improving the search performance of hierarchical index structures like RE-trees: if a query string w is not contained in $L(M)$, then it follows that $w \notin L(M_i)$ for all $M_i \in \mathcal{M}_N$. As a result, the entire subtree rooted at N can be pruned from the search space. Clearly, the closer $L(M)$ is to $L(\mathcal{M}_N)$, the more effective this search-space pruning will be.

In general, there are an infinite number of bounding automata for $\mathcal{M}_N$ with different degrees of precision from the least precise bounding automaton with $L(M) = \sigma^*$ to the most precise bounding automaton, referred to as the *minimal bounding automaton*, with $L(M) = L(\mathcal{M}_N)$. Since the storage space for an automaton is dependent on its complexity (in terms of the number of its states and transitions), there is a space-precision trade-off involved in the choice of a bounding automaton for each internal node entry. Thus, even though minimal bounding automata result in the best pruning due to their tightness, it may not be desirable (or even feasible) to always store minimal bounding automata in RE-trees since their space requirement can be too large (possibly exceeding the size of an index node), thus resulting in an index structure with a low fan-out. Therefore, to maintain a reasonable fan-out for RE-trees, a space constraint is imposed on the maximum number of states (denoted by α) permitted for each bounding automaton in internal RE-tree nodes. The automata stored in RE-tree nodes are, in general, NFAs with a minimum number of states. Also, for better space utilization, each individual RE-tree node is required to contain at least m entries. Thus, the RE-tree height is $O(\log_m(|S|))$.

RE-trees are conceptually similar to other hierarchical, spatial index structures, like the R-tree [8] that is designed for indexing a collection of multidimensional rectangles, where each internal entry is represented by a minimal bounding rectangle (MBR) that contains all the rectangles in the node pointed to by the entry. RE-tree search simply proceeds top-down along (possibly) multiple paths whose bounding automaton accepts the input string; RE-tree updates try to identify a "good" leaf node for insertion and can lead to node splits (or, node merges for deletions) that can propagate all the way up to the root. There is, however, a fundamental difference between the RE-tree and the R-tree in the indexed data types: regular languages typically represent *infinite* sets with no well-defined notion of spatial locality. This difference mandates the development of novel algorithmic solutions for the core RE-tree operations. To optimize for search performance, the core RE-tree operations are designed to keep each bounding automaton M in every internal node to be as "tight" as possible. Thus, if M is the bounding automaton for $\mathcal{M}_N$, then $L(M)$ should be as close to $L(\mathcal{M}_N)$ as possible.

There are three core operations that need to be addressed in the RE-tree context: (P1) selection of an optimal insertion node, (P2) computing an optimal node split, and (P3) computing an optimal bounding automaton. The goal of (P1) is to choose an insertion path for a new RE that leads to "minimal expansion" in the bounding automaton of each internal node of the insertion path. Thus, given the collection of automata $\mathcal{M}(N)$ in an internal index node N and a new automaton M, an optimal $M_i \in \mathcal{M}(N)$ needs to be chosen to insert M such that $|L(M_i) \cap L(M)|$ is maximum. The goal of (P2), which arises when splitting a set of REs during an RE-tree node-split, is to identify a partitioning that results in the minimal amount of "covered area" in terms of the languages of the resulting partitions. More formally, given the collection of automata $\mathcal{M} = \{M_1, M_2, \ldots, M_k\}$ in an overflowed index node, find the optimal partition of $\mathcal{M}$ into two disjoint subsets $\mathcal{M}_1$ and $\mathcal{M}_2$ such that $|\mathcal{M}_1| \geq m$, $|\mathcal{M}_2| \geq m$, and $|L(\mathcal{M}_1)| + |L(\mathcal{M}_2)|$ is minimum. The goal of (P3), which arises during insertions, node-splits, or node-merges, is to identify a bounding automaton for a set of REs that does not cover too much "dead space." Thus, given a collection of automata $\mathcal{M}$, the goal is to find the optimal bounding automaton M such that the number of states of M is no more than α, $L(\mathcal{M}) \subseteq L(M)$ and $|L(M)|$ is minimum.

The objective of the above three operations is to maximize the pruning during search by keeping bounding automata tight. In (P1), the optimal automaton M_i selected (within an internal node) to accommodate a newly inserted automaton M is to maximize $|L(M_i) \cap L(M)|$. The set of automata $\mathcal{M}$ are split into two tight clusters in (P2), while in (P3), the most precise automaton (with no more than α states) is computed to cover the set of automata in $\mathcal{M}$. Note that (P3) is unique to RE-trees, while both (P1) and (P2) have their equivalents in R-trees. The heuristics solutions [2, 8] proposed for (P1) and (P2) in R-trees aim to minimize the number of visits to nodes that do not lead to any qualifying data entries. Although the minimal bounding automata in RE-trees (which correspond to regular languages) are very different from the MBRs in R-trees, the intuition behind minimizing the area of MBRs (total area or overlapping area) in R-trees should be effective for RE-trees as well. The counterpart for area in an RE-tree is $|L(M)|$, the size of the regular language for M. However, since a regular language is generally an infinite set, new measures need to be developed for the size of a regular language or for comparing the sizes of two regular languages.

One approach to compare the relative sizes of two regular languages is based on the following definition: for a pair of automata M_i and M_j, $L(M_i)$ is said to be larger than $L(M_j)$ if there exists a positive integer N such that for all $k \geq N$, $\sum_{l=1}^{k} |L_l(M_i)| \geq \sum_{l=1}^{k} |L_l(M_j)|$. Based on the above intuition, three increasingly sophisticated measures are proposed to capture the size of an infinite regular language. The *max-count measure* simply counts the number of strings in the language up to a certain size λ; i.e., $|L(M)| = \sum_{i=1}^{\lambda} |L_i(M)|$. This measure is useful for applications where the maximum length of all the REs to be indexed is known and is not too large so that λ can be set to some value slightly larger than the maximum length of the REs. A second more robust measure that is less sensitive to the λ parameter value is the *rate-of-growth measure* which is based on the intuition that a larger language grows at a faster rate than a smaller language. The size of a language is approximated by computing the rate of change of its size from one "window" of lengths to the next consecutive "window" of lengths: if λ is a length parameter that denote the start of the first window and θ is a window-size parameter, then $|L(M)| = \sum_{\lambda+\theta}^{\lambda+2\theta-1} |L_i(M)| / \sum_{\lambda}^{\lambda+\theta-1} |L_i(M)|$. As in the max-count measure, the parameters λ

and θ should be chosen to be slightly greater than the number of states of M to ensure that strings involving a substantial portion of paths, cycles, and accepting states are counted in each window. However, there are cases where the rate-of-growth measure also fails to capture the "larger than" relationship between regular languages [4]. To address some of the shortcomings of the first two metrics, a third information-theoretic measure is proposed that is based on Rissanen's minimum description length (MDL) principle [11]. The intuition is that if $L(M_i)$ is larger than $L(M_j)$, then the per-symbol cost of an MDL-based encoding of a random string in $L(M_i)$ using M_i is very likely to be higher than that of a string in $L(M_j)$ using M_j, where the per-symbol cost of encoding a string $w \in L(M)$ is the ratio of the cost of an MDL-based encoding of w using M to the length of w. More specifically, if $w = w_1.w_2.\ldots.w_n \in L(M)$ and $s_0, s_1, \ldots, s_n$ is the unique sequence of states visited by w in M, then the MDL-based encoding cost of w using M is given by $\sum_{i=0}^{n-1} \lceil \log_2(n_i) \rceil$, where each n_i denotes the number of transitions out of state s_i, and $\log_2(n_i)$ is the number of bits required to specify the transition out of state s_i. Thus, a reasonable measure for the size of a regular language $L(M)$ is the expected per-symbol cost of an MDL-based encoding for a random sample of strings in $L(M)$.

To utilize the above metrics for measuring $L(M)$, one common operation needed is the computation of $|L_n(M)|$, the number of length-n strings in $L(M)$. While $|L_n(M)|$ can be efficiently computed when M is a DFA, the problem becomes #P-complete when M is an NFA [10]. Two approaches were proposed to approximate $|L_n(M)|$ when N is an NFA [10]. The first approach is an unbiased estimator for $|L_n(M)|$, which can be efficiently computed but can have a very large standard deviation. The second approach is a more accurate randomized algorithm for approximating $|L_n(M)|$ but it is not very useful in practice due to its high time complexity of $O(n^{\log(n)})$. A more practical approximation algorithm with a time complexity of $O(n^2|M|^2 \min\{|\sigma|, |M|\})$ was proposed in [4].

The RE-tree operations (P1) and (P2) require frequent computations of $|L(M_i \cap M_j)|$ and $|L(M_i \cup M_j)|$ to be performed for pairs of automata M_i, M_j. These computations can adversely affect RE-tree performance since construction of the intersection and union automaton M can be expensive. Furthermore, since the final automaton M may have many more states than the two initial automata M_i and M_j, the cost of measuring $|L(M)|$ can be high. The performance of these computations can, however, be optimized by using sampling. Specifically, if the counts and samples for each $L(M_i)$ are available, then this information can be utilized to derive approximate counts and samples for $L(M_i \cap M_j)$ and $L(M_i \cup M_j)$ without incurring the overhead of constructing the automata $M_i \cap M_j$ and $M_i \cup M_j$ and counting their sizes. The sampling techniques used are based on the following results for approximating the sizes of and generating uniform samples of unions and intersections of arbitrary sets:

Theorem 1 (Chan, Garofalakis, Rastogi [4]) *Let r_1 and r_2 be uniform random samples of sets S_1 and S_2, respectively.*

1. *$(|r_1 \cap S_2||S_1|)/|r_1|$ is an unbiased estimator of the size of $S_1 \cap S_2$.*
2. *$r_1 \cap S_2$ is a uniform random sample of $S_1 \cap S_2$ with size $|r_1 \cap S_2|$.*
3. *If the sets S_1 and S_2 are disjoint, then a uniform random sample of $S_1 \cup S_2$ can be computed in $O(|r_1| + |r_2|)$ time. If S_1 and S_2 are not disjoint, then an approximate uniform random sample of $S_1 \cup S_2$ can be computed with the same time complexity.*

Applications

The RE retrieval problem also arises in the context of both XML document classification, which identifies matching DTDs for XML documents, as well as BGP routing, which assigns appropriate priorities to BGP advertisements based on their matching routing-system sequences.

Experimental Results

Experimental results with synthetic data sets [5] clearly demonstrate that the RE-tree index is significantly more effective than performing a sequential search for matching REs and, in a number of cases, outperforms sequential search by up to an order of magnitude.

Cross-References

▶ Regular Expression Matching

Recommended Reading

1. Altinel M, Franklin M (2000) Efficient filtering of XML documents for selective dissemination of information. In: Proceedings of 26th international conference on very large data bases, Cairo. Morgan Kaufmann, Missouri, pp 53–64
2. Beckmann N, Kriegel H-P, Schneider R, Seeger B (1990) The R*-tree: an efficient and robust access method for points and rectangles. In: Proceedings of the ACM international conference on management of data, Atlantic City. ACM, New York, pp 322–331
3. Chan C-Y, Felber P, Garofalakis M, Rastogi R (2002) Efficient filtering of XML documents with XPath expressions. In: Proceedings of the 18th international conference on data engineering, San Jose. IEEE Computer Society, Piscataway, pp 235–244
4. Chan C-Y, Garofalakis M, Rastogi R (2002) RE-tree: an efficient index structure for regular expressions. In: Proceedings of 28th international conference on very large data bases, Hong Kong. Morgan Kaufmann, Missouri, pp 251–262
5. Chan C-Y, Garofalakis M, Rastogi R (2003) RE-tree: an efficient index structure for regular expressions. VLDB J 12(2):102–119
6. Clark J, DeRose S (1999) XML Path Language (XPath) Version 1.0. W3C Recommendation. http://www.w3.org./TR/xpath. Accessed Nov 1999
7. Diao Y, Fischer P, Franklin M, To R (2002) YFilter: efficient and scalable filtering of XML documents. In: Proceedings of the 18th international conference on data engineering, San Jose. IEEE Computer Society, Piscataway, pp 341–342
8. Guttman A (1984) R-trees: a dynamic index structure for spatial searching. In: Proceedings of the ACM international conference on management of data, Boston. ACM, New York, pp 47–57
9. Hopcroft J, Ullman J (1979) Introduction to automata theory, languages, and computation. Addison-Wesley, Reading
10. Kannan S, Sweedyk Z, Mahaney S (1995) Counting and random generation of strings in regular languages. In: Proceedings of the 6th ACM-SIAM symposium on discrete algorithms, San Francisco. ACM, New York, pp 551–557
11. Rissanen J (1978) Modeling by shortest data description. Automatica 14:465–471
12. Stewart JW (1998) BGP4, inter-domain routing in the Internet. Addison Wesley, Reading

Indexed Two-Dimensional String Matching

Joong Chae Na[1], Paolo Ferragina[2], Raffaele Giancarlo[3], and Kunsoo Park[4]

[1]Department of Computer Science and Engineering, Sejong University, Seoul, Korea

[2]Department of Computer Science, University of Pisa, Pisa, Italy

[3]Department of Mathematics and Applications, University of Palermo, Palermo, Italy

[4]School of Computer Science and Engineering, Seoul National University, Seoul, Korea

Keywords

Index data structures for matrices or images; Indexing for matrices or images; Two-dimensional indexing for pattern matching; Two-dimensional index data structures

Years and Authors of Summarized Original Work

2007; Na, Giancarlo, Park

2011; Kim, Na, Sim, Park

Problem Definition

This entry is concerned with designing and building indexes of a two-dimensional matrix, which is basically the generalization of indexes of a string, the *suffix tree* [12] and the *suffix array* [11], to a two-dimensional matrix. This problem was first introduced by Gonnet [7]. Informally, a two-dimensional analog of the suffix tree is a tree data structure storing all submatrices of an $n \times m$ matrix, $n \geq m$. The *submatrix tree* [2] is an incarnation of such indexes. Unfortunately, building such indexes requires $\Omega(nm^2)$ time [2]. Therefore, much of the attention paid has been restricted to square matrices and submatrices, the important special case in which much better results are available.

For square matrices, the *Lsuffix tree* and its array form, storing all *square submatrices* of an $n \times n$ matrix, have been proposed [3, 9, 10]. Moreover, the general framework for these index families is also introduced [4, 5]. Motivated by LZ1-type image compression [14], the online case, i.e., the matrix is given one row or column at a time, has been also considered. These data structures can be built in time close to n^2. Building these data structures is a nontrivial extension of the algorithms for the standard suffix tree and suffix array. Generally, a tree data structure and its array form of this type for square matrices are referred to as the *two-dimensional suffix tree* and the *two-dimensional suffix array*, which are the main concerns of this entry.

Notations

Let A be an $n \times n$ matrix with entries defined over a finite alphabet Σ. $A[i \dots k, j \dots l]$ denotes the submatrix of A with corners (i, j), (k, j), (i, l), and (k, l). When $i = k$ or $j = l$, one of the repeated indexes is omitted. For $1 \leq i, j \leq n$, the *suffix* $A(i, j)$ of A is the largest square submatrix of A that starts at position (i, j) in A. That is, $A(i, j) = A[i \dots i + k, j \dots j + k]$, where $k = n - \max(i, j)$. Let $\$_i$ be a special symbol not in Σ such that $\$_i$ is lexicographically smaller than any other character in Σ. Assume that $\$_i$ is lexicographically smaller than $\$_j$ for $i < j$. For notational convenience, assume that the last entries of the ith row and column are $\$_i$. It makes all suffixes distinct. See Fig. 1a, b for an example.

Let $L\sum = \bigcup_{i=1}^{\infty} \Sigma^{2i-1}$. The strings of $L\Sigma$ are referred to as *Lcharacters*, and each of them is considered as an atomic item. $L\Sigma$ is called the *alphabet of Lcharacters*. Two Lcharacters are equal if and only if they are equal as strings over Σ. Moreover, given two Lcharacters La and Lb of equal length, La is *lexicographically smaller than or equal to* Lb if and only if the string corresponding to La is lexicographically smaller than or equal to that corresponding to Lb. A *chunk* is the concatenation of Lcharacters with the following restriction: an Lcharacter in Σ^{2i-1} can precede only one in $\Sigma^{2(i+1)-1}$ and succeed only one in $\Sigma^{2(i-1)-1}$. An *Lstring* is a chunk such that the first Lcharacter is in Σ.

For dealing with matrices as strings, a linear representation of square matrices is needed. Given $A[1 \dots n, 1 \dots n]$, divide A into n Lshaped characters. Let $a(i)$ be the concatenation of row $A[i, 1 \dots i-1]$ and column $A[1 \dots i, i]$. Then, $a(i)$ can be regarded as an Lcharacter. The linearized string of matrix A, called the Lstring of matrix A, is the concatenation of Lcharacters $a(1), \dots, a(n)$. See Fig. 1c for an example. Slightly different linearizations have been used [9, 10, 13], but they are essentially the same in the aspect of two-dimensional functionality.

Two-Dimensional Suffix Trees

The suffix tree of matrix A is a compacted trie over the alphabet $L\Sigma$ that represents Lstrings corresponding to all suffixes of A. Formally, the *two-dimensional suffix tree* of matrix A is a rooted tree that satisfies the following conditions (see Fig. 1d for an example):

1. Each edge is labeled with a chunk.
2. There is no internal node of outdegree one.
3. Chunks assigned to sibling edges start with different Lcharacters, which are of the same length as strings in Σ^*.
4. The concatenation of the chunks labeling the edges on the path from the root to a leaf gives the Lstring of exactly one suffix of A, say $A(i, j)$. It is said that this leaf is associated with $A(i, j)$.
5. There is exactly one leaf associated with each suffix.

Conditions 4 and 5 mean that there is a one-to-one correspondence between the leaves of the tree and the suffixes of A (which are all distinct because $\$_i$ is unique).

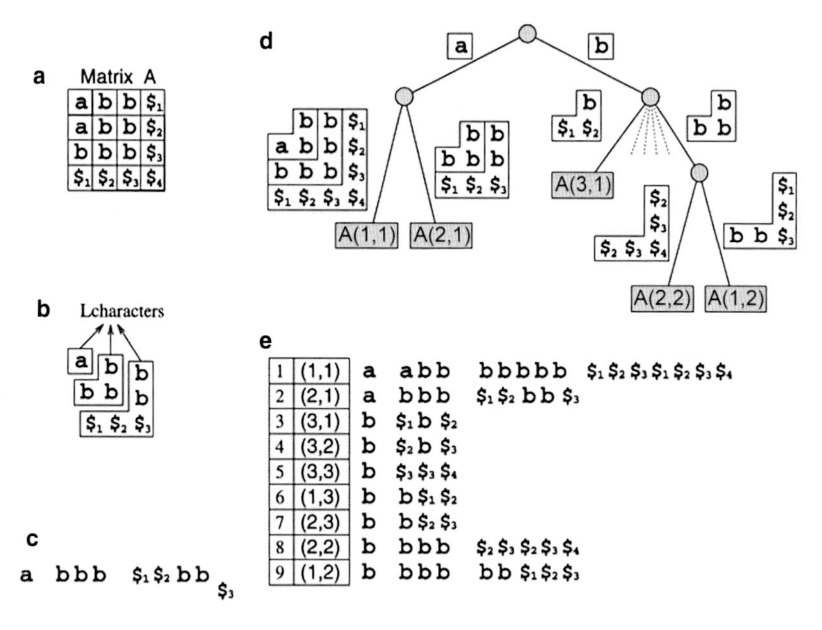

Indexed Two-Dimensional String Matching, Fig. 1 (**a**) A matrix A, (**b**) the suffix $A(2,1)$ and Lcharacters composing $A(2,1)$, (**c**) the Lstring of $A(2,1)$, (**d**) the suffix tree of A, and (**e**) the suffix array of A (omitting the suffixes started with $\$_i$)

Problem 1 (Construction of 2D suffix tree)

INPUT: An $n \times n$ matrix A.

OUTPUT: A two-dimensional suffix tree storing all square submatrices of A.

Online Suffix Trees

Assume that A is read *online* in row major order (column major order can be considered similarly). Let $A_t = A[1 \ldots t, 1 \ldots n]$ and $row_t = A[t, 1 \ldots n]$. At time $t - 1$, nothing but A_{t-1} is known about A. At time t, row_t is read and so A_t is known. After time t, the online suffix tree of A is storing all suffixes of A_t . Note that Condition 4 may not be satisfied during the online construction of the suffix tree. A leaf may be associated with more than one suffix, because the suffixes of A_t are not all distinct.

Problem 2 (Online construction of 2D suffix tree)

INPUT: A sequence of rows of $n \times n$ matrix A, $row_1, row_2, \ldots, row_n$.

OUTPUT: A two-dimensional suffix tree storing all square submatrices of A_t after reading row_t .

Two-Dimensional Suffix Arrays

The *two-dimensional suffix array* of matrix A is basically a sorted list of all Lstrings corresponding to suffixes of A. Formally, the kth element of the array has the start position (i, j) if and only if the Lstring of $A(i, j)$ is the kth smallest one among the Lstrings of all suffixes of A. See Fig. 1e for an example. The two-dimensional suffix array is also coupled with additional information tables, called *Llcp* and *Rlcp*, to enhance its performance like the standard suffix array. The two-dimensional suffix array can be constructed from the two-dimensional suffix tree in linear time.

Problem 3 (Construction of 2D suffix array)

INPUT: An $n \times n$ matrix A.

OUTPUT: The two-dimensional suffix array storing all square submatrices of A.

Submatrix Trees

The *submatrix tree* is a tree data structure storing all submatrices. This entry just gives a result on submatrix trees. See [2] for details.

Problem 4 (Construction of a submatrix tree)

INPUT: An $n \times m$ matrix B, $n \geq m$.
OUTPUT: The submatrix tree and its array form storing all submatrices of B.

Key Results

Theorem 1 (Kim et al. 2011 [10], Cole and Hariharan 2003 [1]) *Given an $n \times n$ matrix A over an integer alphabet, one can construct the two-dimensional suffix tree in $O(n^2)$ time.*

Kim and Park's result is a deterministic algorithm, while Cole and Hariharan's result is a randomized one. For an arbitrary alphabet, one needs first to sort it and then to apply the theorem above.

Theorem 2 (Na et al. 2007 [13]) *Given an $n \times n$ matrix A, one can construct online the two-dimensional suffix tree of A in $O(n^2 \log n)$ time.*

Theorem 3 (Kim et al. 2003 [9]) *Given an $n \times n$ matrix A, one can construct the two-dimensional suffix array of A in $O(n^2 \log n)$ time without constructing the two-dimensional suffix tree.*

Theorem 4 (Giancarlo 1993 [2]) *Given an $n \times m$ matrix B, one can construct the submatrix tree of B in $O(nm^2 \log(nm))$ time.*

Applications

Two-dimensional indexes can be used for many pattern-matching problems of two-dimensional applications such as low-level image processing, image compression, visual data bases, and so on [3, 6]. Given an $n \times n$ text matrix and an $m \times m$ pattern matrix over an alphabet Σ, the *two-dimensional pattern retrieval problem*, which is a basic pattern-matching problem, is to find all occurrences of the pattern in the text. The two-dimensional suffix tree and array of the text can be queried in $O(m^2 \log |\Sigma| + occ)$ time and $O(m^2 + \log n + occ)$ time, respectively, where occ is the number of occurrences of the pattern in the text. This problem can be easily extended to a set of texts. These queries have the same procedure and performance as those of indexes for strings. Online construction of the two-dimensional suffix tree can be applied to LZ-1-type image compression [6].

Open Problems

The main open problems on two-dimensional indexes are to construct indexes in optimal time. The linear-time construction algorithm for two-dimensional suffix trees is already known [10]. The online construction algorithm due to [13] is optimal for unbounded alphabets, but not for integer or constant alphabets. Another open problem is to construct two-dimensional suffix arrays directly in linear time.

Experimental Results

An experiment that compares construction algorithms of two-dimensional suffix trees and suffix arrays was presented in [8]. Giancarlo's algorithm [2] and Kim et al.'s algorithm [8] were implemented for two-dimensional suffix trees and suffix arrays, respectively. Random matrices of sizes $200 \times 200 \sim 800 \times 800$ and alphabets of sizes 2, 4, 16 were used for input data. According to experimental results, the construction of two-dimensional suffix arrays is ten times faster and five times more space efficient than that of two-dimensional suffix trees.

Cross-References

- ▸ Multidimensional String Matching
- ▸ Suffix Array Construction
- ▸ Suffix Tree Construction

Recommended Reading

1. Cole R, Hariharan R (2003) Faster suffix tree construction with missing suffix links. SIAM J Comput 33:26–42

2. Giancarlo R (1993) An index data structure for matrices, with applications to fast two-dimensional pattern matching. In: Proceedings of workshop on algorithm and data structures, Montréal. Springer Lecture notes in computer science, vol 709, pp 337–348
3. Giancarlo R (1995) A generalization of the suffix tree to square matrices, with application. SIAM J Comput 24:520–562
4. Giancarlo R, Grossi R (1996) On the construction of classes of suffix trees for square matrices: algorithms and applications. Inf Comput 130:151–182
5. Giancarlo R, Grossi R (1997) Suffix tree data structures for matrices. In: Apostolico A, Galil, Z (eds) Pattern matching algorithms, ch. 11. Oxford University Press, Oxford, pp 293–340
6. Giancarlo R, Guaiana D (1999) On-line construction of two-dimensional suffix trees. J Complex 15:72–127
7. Gonnet GH (1988) Efficient searching of text and pictures. Technical report OED-88-02, University of Waterloo
8. Kim DK, Kim YA, Park K (1998) Constructing suffix arrays for multi-dimensional matrices. In: Proceedings of the 9th symposium on combinatorial pattern matching, Piscataway, pp 249–260
9. Kim DK, Kim YA, Park K (2003) Generalizations of suffix arrays to multi-dimensional matrices. Theor Comput Sci 302:401–416
10. Kim DK, Na JC, Sim JS, Park K (2011) Linear-time construction of two-dimensional suffix trees. Algorithmica 59:269–297
11. Manber U, Myers G (1993) Suffix arrays: a new method for on-line string searches. SIAM J Comput 22:935–948
12. McCreight EM (1976) A space-economical suffix tree construction algorithms. J ACM 23:262–272
13. Na JC, Giancarlo R, Park K (2007) On-line construction of two-dimensional suffix trees in $O(n^2 \log n)$ time. Algorithmica 48:173–186
14. Storer JA (1996) Lossless image compression using generalized LZ1-type methods. In: Proceedings of data compression conference, Snowbird, pp 290–299

Inductive Inference

Sandra Zilles
Department of Computer Science, University of Regina, Regina, SK, Canada

Keywords

Induction; Learning from examples; Recursion theory

Years and Authors of Summarized Original Work

1983; Case, Smith

Problem Definition

The theory of inductive inference is concerned with the capabilities and limitations of machine learning. Here the learning machine, the concepts to be learned, as well as the hypothesis space are modeled in recursion theoretic terms, based on the framework of identification in the limit [1, 9, 15].

Formally, considering recursive functions (mapping natural numbers to natural numbers) as target concepts, a learner (inductive inference machine) is supposed to process, step by step, gradually growing initial segments of the graph of a target function. In each step, the learner outputs a program in some fixed programming system, where successful learning means that the sequence of programs returned in this process eventually stabilizes on some program actually computing the target function.

Case and Smith [3, 4] proposed several variants of this model in order to study the influence that certain constraints or relaxations may have on the capabilities of learners. Their models restrict (i) the number of mind changes (i.e., changes of output programs) a learner is allowed to make during the learning process and (ii) the number of errors the program eventually hypothesized may have when compared to the target function.

One major result of studying the corresponding effects is a hierarchy of inference types culminating in a model general enough to allow for the identification of the whole class of recursive functions by a single inductive inference machine.

Notation

The target concepts for learning in the model discussed below are recursive functions [14] mapping natural numbers to natural numbers. Such functions, as well as partial recursive functions in general, are considered as computable in an arbitrary, but fixed Gödel numbering $\varphi = (\varphi_i)_{i \in \mathbb{N}}$. Here $\mathbb{N} = \{0, 1, 2, \ldots\}$ denotes the set of all natural numbers. $\varphi = (\varphi_i)_{i \in \mathbb{N}}$ is interpreted as a programming system, where the number $i \in \mathbb{N}$ is called a program for the partial recursive function φ_i.

Suppose f and g are partial recursive functions and $n \in \mathbb{N}$. Below $f =^n g$ is written if the set $\{x \in \mathbb{N} | f(x) \neq g(x)\}$ is of cardinality at most n. If the set $\{x \in \mathbb{N} | f(x) \neq g(x)\}$ is finite, this is denoted by $f =^* g$. One considers $*$ as a special symbol for which the $<$-relation is extended by $n < *$ for all $n \in \mathbb{N}$. For any recursive f and any $z \in \mathbb{N}$, let $f[z]$ denote $(z, (f(0), \ldots, f(z)))$ for short.

For further basic recursion theoretic notions, the reader is referred to [14].

Learning Models

Case and Smith [4] build their theory upon the fundamental model of identification in the limit [1, 9]. There a learner can be understood as an algorithmic device, called an inductive inference machine, which, given any "graph segment" $f[z]$ as its input, returns a program $i \in \mathbb{N}$. Such a learner M identifies a recursive function f in the limit, if there is some $j \in \mathbb{N}$ such that

$$\varphi_j = f \text{ and } M(f[z]) = j \text{ for all but finitely many } z \in \mathbb{N}.$$

A class of recursive functions is learnable in the limit, if there is an inductive inference machine identifying each function in the class in the limit. Identification in the limit is called EX-identification, since a program for f is termed an *ex*planation for f.

For instance, the class of all primitive recursive functions is EX-identifiable, whereas the class of all recursive functions is not [9].

The central question discussed by Case and Smith [4] is how the limitations of EX-learners are affected by posing certain requirements on the success criterion, concerning:

- Convergence criteria:
  - e.g., when restricting the number of permitted mind changes
  - e.g., when relaxing the constraints on syntactical convergence of the sequence of programs returned in the learning process
- Accuracy:
  - e.g., when relaxing the number of permitted anomalies in the programs returned eventually

Problem 1 In which way do modifications of EX-identification in terms of accuracy and convergence criteria affect the capabilities of the corresponding learners?

Problem 2 In particular, if inaccuracies are permitted, can EX-learners always refute inaccurate hypotheses?

Problem 3 How much relaxation of the model of EX-identification is needed to achieve learnability of the full class of recursive functions?

Key Results

Accuracy and Convergence Constraints

In order to systematically address these problems, Case and Smith [4] defined inference types reflecting restrictions and relaxations of EX-identification as follows.

Definition 1 Suppose S is a class of recursive functions and $m, n \in \mathbb{N} \cup \{*\}$. S is $\mathrm{EX}_n{}^m$-identifiable if there is an inductive inference machine M, such that for any function $f \in S$, there is some $j \in \mathbb{N}$ satisfying:

- $M(f[z]) = j$ for all but finitely many $z \in \mathbb{N}$.
- $j =^m f$.
- The cardinality of the set $\{z \in \mathbb{N} | M(f[z]) \neq M(f[z+1])\}$ is at most n.

For intuition one may view n as an upper bound on the allowed number of "mind changes" and m as an upper bound on the allowed number of "anomalies."

$\mathrm{EX}_n{}^m$ denotes the set of all classes of recursive functions which are $\mathrm{EX}_n{}^m$-identifiable.

Definition 2 Suppose S is a class of recursive functions and $m \in \mathbb{N} \cup \{*\}$. S is BC^m-identifiable if there is an inductive inference machine M, which, for any function $f \in S$, satisfies:

- $\varphi_{M(f[z])} =^m f$ for all but finitely many $z \in \mathbb{N}$.

BC^m denotes the set of all classes of recursive functions which are BC^m-identifiable. BC is short for behaviorally correct; the difference to EX-learning is that convergence of the sequence of programs returned by the learner is defined only in terms of semantics, no longer in terms of syntax.

The Impact of Accuracy and Convergence Constraints

In general, each permission of mind changes or anomalies increases the capabilities of learners; however, mind changes cannot be traded in for anomalies or vice versa.

Theorem 1 *Let* $a, b, c, d \in \mathbb{N} \cup \{*\}$. *Then* ${EX_b}^a \subseteq {EX_d}^c$ *if and only if* $a \leq c$ *and* $b \leq d$.

Corollary 1 *For any* $m, n \in \mathbb{N}$, *the following inclusions hold.*

1. $EX_n^m \subset EX_n^{m+1} \subset EX_n^*$.
2. $EX_n^m \subset EX_{n+1}^m \subset EX_*^m$.

Theorem 2 *Let* $n \in \mathbb{N}$. *Then* $EX_*^* \subset BC^n \subset BC^{n+1} \subset BC^*$.

These results provide a solution to Problem 1.

Refutability

In particular, refutability demands (in the sense that every incorrect hypothesis should be refutable; see [13]) are not applicable in the theory of inductive inference; see Problem 2.

Formally, Case and Smith [4] consider refutability as a property guaranteed by Popperian machines, the latter being defined as follows:

Definition 3 Suppose M is an inductive inference machine M. M is Popperian if, on any input, M returns a program of a recursive function.

Results thereon include the following:

Theorem 3 *There is an EX-identifiable class S of recursive functions for which there is no Popperian inductive inference machine witnessing its EX-identifiability.*

Corollary 2 *There is an* EX^1*-identifiable class S of recursive functions for which there is no Popperian inductive inference machine witnessing its* EX^1*-identifiability.*

Additionally, in EX^1-identification, Popper's refutability principle cannot be applied even if it concerns only those hypotheses returned in the limit.

Learning All Recursive Functions

Since the results above yield a hierarchy of inference types with strictly growing collections of learnable classes, there is also an implicit answer to Problem 3: the class of recursive functions is neither in ${EX_n}^m$ for any $m, n \in \mathbb{N} \cup \{*\}$ nor in BC^m for any $m \in \mathbb{N}$. In contrast to that, Case and Smith [4] prove:

Theorem 4 *The class of all recursive functions is in* BC^*.

Applications

The work of Case and Smith [4] has been of high impact in learning theory.

A consequence of the discussion of anomalies is that refutability principles in general do not hold for identification in the limit. This result has given rise to later studies on methods and techniques inductive inference machines might apply in order to discover their errors [7] and thus to further insights into the nature of inductive inference.

Concerning the study of mind change hierarchies, among others, their lifting to transfinite ordinal numbers [8] is a notable extension.

Moreover, the theory of learning as proposed by Case and Smith [4] has been applied for the development of the theory of identifying recursive [11] or recursively enumerable [10] languages.

Open Problems

Among the currently open problems in inductive inference, one key challenge is to find a reasonable notion of the complexity of learning problems (i.e., of classes of recursive functions) involving the run-time complexity of learners as well as the number of mind changes required to learn the functions in a class. In particular, special natural classes of functions should be analyzed in terms of such a complexity notion.

Though of course the hierarchies $EX_0{}^m \subset EX_1{}^m \subset EX_2{}^m \subset \ldots$ for any $m \in \mathbb{N}$ reflect some increase of complexity in that sense, a corresponding complexity notion would not address the aspect of run-time complexity of learners. Different complexity notions have been introduced, such as the so-called intrinsic complexity [2, 6] (neglecting run-time complexity) and the "measure under the curve" [5] (respecting the number of examples required, but neglecting the number of mind changes). In particular, for learning deterministic finite automata, different notions of run-time complexity have been discussed [12].

However, the definition of a more capacious complexity notion remains an open issue.

Cross-References

▶ PAC Learning

Recommended Reading

1. Blum L, Blum M (1975) Toward a mathematical theory of inductive inference. Inf Control 28(2):125–155
2. Case J, Kötzing T (2011) Measuring learning complexity with criteria epitomizers. In: Proceedings of the 28th international symposium on theoretical aspects of computer science, Dortmund. Leibniz international proceedings in informatics, pp 320–331
3. Case J, Smith CH (1978) Anomaly hierarchies of mechanized inductive inference. In: Proceedings of the 10th symposium on the theory of computing, San Diego. ACM, New York, pp 314–319
4. Case J, Smith CH (1983) Comparison of identification criteria for machine inductive inference. Theor Comput Sci 25(2):193–220
5. Daley RP, Smith CH (1986) On the complexity of inductive inference. Inf Control 69(1–3):12–40
6. Freivalds R, Kinber E, Smith CH (1995) On the intrinsic complexity of learning. Inf Comput 118(2):208–226
7. Freivalds R, Kinber E, Wiehagen R (1995) How inductive inference strategies discover their errors. Inf Comput 123(1):64–71
8. Freivalds R, Smith CH (1993) On the role of procrastination in machine learning. Inf Comput 107(2):237–271
9. Gold EM (1967) Language identification in the limit. Inf Control 10(5):447–474
10. Kinber EB, Stephan F (1995) Language learning from texts: mindchanges, limited memory, and monotonicity. Inf Comput 123(2):224–241
11. Lange S, Grieser G, Zeugmann T (2005) Inductive inference of approximations for recursive concepts. Theor Comput Sci 348(1):15–40
12. Pitt L (1989) Inductive inference, DFAs, and computational complexity. In: Analogical and inductive inference, 2nd international workshop, Reinhardsbrunn Castle, GDR. Lecture notes in computer science, vol 397. Springer, Berlin, pp 18–44
13. Popper K (1959) The logic of scientific discovery. Harper & Row, New York
14. Rogers H (1967) Theory of recursive functions and effective computability. McGraw-Hill, New York
15. Zeugmann T, Zilles S (2008) Learning recursive functions: a survey. Theor Comput Sci 397:4–56

Influence and Profit

Yuqing Zhu[2,3] and Weili Wu[1,2,3]
[1]College of Computer Science and Technology, Taiyuan University of Technology, Taiyuan, Shanxi Province, China
[2]Department of Computer Science, California State University, Los Angeles, CA, USA
[3]Department of Computer Science, The University of Texas at Dallas, Richardson, TX, USA

Keywords

Data mining; Influence; Profit; Social networks; Viral marketing

Years and Authors of Summarized Original Work

2003; Kempe, Kleinberg, Tardos
2013; Zhu, Lu, Bi, Wu, Jiang, Li

Problem Definition

A social network is a graph of relationships and interactions within a set of individuals. Information can spread within a social network by "word-of-mouth" effects. In other words, information diffuses from individuals to individuals in a social network through the connections between them, and if some information is spread by some initial individuals, many individuals may believe in it due to information diffusion. A social network is denoted as $G=(V,E,w)$, where V is a set of vertexes with size n, $E \subseteq V \times V$ is a set of edges with size m, and $w : E \rightarrow [0,1]$ is the set of all $w(u,v)$ which is the weight of edge (u,v).

Independent Cascade (IC) and Linear Threshold (LT) Models

The IC and LT models [1] are two basic models of influence diffusion in social networks, and there are two vertex stages: *inactive* and *active*. The influence always starts from a set of a set S consists of seeds (initially active nodes). The time is divided into discrete steps $0, 1, 2, \ldots$ Denote S_i the set of active vertexes at step i ($S_0 = S$ and $S_{-1} = \emptyset$). In the IC model, influence propagates as follows: S_i is the union set of S_{i-1} and other vertexes activated by vertexes in $S_{i-1} \setminus S_{i-2}$ in step i. Each node u has only one chance to activate each of its neighbors v with probability $w(u,v)$ when u first becomes active. In the LT model, influence propagates as follows: at the beginning each vertex v picks a threshold θ_v uniformly at random from $[0,1]$ which is the threshold of this vertex becoming active. In each step i, $S_i = S_{i-1} \cup \{v | \sum_{u \in S_{i-1}} w(u,v) \geq \theta_v\}$. Both IC and LT models stop at the step $t+1$ when the process reaches its maximum influence, i.e., $S_{t+1} = S_t$.

Problem 1: Influence Maximization Problem (InfMax) [1]

INPUT: *A social network $G = (V,E,w)$ and k, the number of seeds.*

OUTPUT: *The set S containing k seeds that maximizes the influence $\mathcal{I}(S)$.*

Price-Related Propagation (PR) Frame

Adding monetary factor into the propagation process of the IC and LT models makes this *price related (PR) propagation* frame. In the PR frame, only the individual who adopts a product propagates this product's influence, and the adoption depends on the relationship between the price offered and the individual's valuation about this product. In detail, every vertex u has three stages: *neutral*, *influenced*, and *active*. Vertex u being neutral means it has no idea or positive attitude about this product. When u becomes influenced, u holds a positive attitude to the product but u hasn't adopted this product yet. Only if u further turns into active stage, u adopts the product and propagates the influence by telling its network neighbors. The PR frame separates, holding a positive attitude and propagating influence, in which the two are the same in traditional IC and LT models. This separation comes from the fact that individuals in social networks are independent human beings who not only are influenced by the people around but also have their own judgements. If someone receives some information, surely he or she should first evaluate the information before spreading it.

The PR frame assumes that each individual u has a *valuation* for the product, which is the highest price this individual thinks the product is worth. The rule of judging whether an influenced individual turns into active is the following: only if u is *influenced* and its valuation is higher than the offered price, u will turn *active*, adopt this product, and propagate the influence. The PR frame is an extension to the IC and LT models; it contains the PR-I model based on the IC and the PR-L model based on the LT. The rules of an individual turning from *neutral* to *influenced* in PR-I and PR-L model are the same as the rules of an individual turning from *inactive* to *active* in IC and LT model, respectively. However in the PR frame the *influenced* individuals do not propagate influence, but only the *active* ones do, and an *influence* individual turns to *active* if and only if the offered price is lower than its valuation.

Pricing Strategies in the PR Frame

Since price is vitally significant in the PR frame, we design two strategies to determine the prices offered to the individuals. The first one is *binary pricing* (BYC), in which all chosen seeds are given free samples and all other individuals are charged the same price, and the second one is *panoramic pricing* (PAP), in which prices for individuals including seeds are unconstrained different values that can be any value if needed.

In the PR frame, choosing node u as a seed merely means turning u to *influenced*. However, in BYC, any seed u must further become *active* for each seed is offered a free sample, i.e., the offered price is 0 and no greater than the valuation. In PAP, on the other hand, a seed u may not be *active*.

Price plays a vital role on the influence and profit in the PR frame. High prices may bring high profit but it hinders the influence propagation, and to enlarge the influence, some sacrifice on profit is inevitable. Base on this observation, a parameter $\lambda \in [0, 1)$ is adopted to denote the decision maker's preference toward influence and profit, and the objective is the weighted sum of influence and profit, which we call balanced influence and profit (BIP).

Problem 2: Balanced Influence and Profit Maximization Problem (BIPMax) [2]

INPUT: *A social network $G = (V, E, w)$, the distribution of customer evaluation, and λ the decision maker's preference.*

OUTPUT: *The seed set S and the price $\boldsymbol{p}$ for all individuals that maximize the objective function $\mathcal{B}(S, \boldsymbol{p}) = \lambda \cdot \mathcal{I}(S, \boldsymbol{p}) + (1 - \lambda) \cdot \mathcal{R}(S, \boldsymbol{p})$ where $\mathcal{I}(S, \boldsymbol{p})$ is the influence and $\mathcal{R}(S, \boldsymbol{p})$ is the profit.*

Key Results

Result 1: *InfMax under the IC and LT models is both NP hard.* [1]

Result 2: *BIPMax under the PR-I and PR-L models is both NP-hard.* [2]

The above two results show the difficulties of solving InfMax and BIPMax, respectively. It can be seen that both of them are "hard" to solve. However, approximation algorithms may exist, and the following two properties are used to design and analyze algorithms. Suppose f is a set function on subsets of V.

Submodularity and Monotony

1. **Submodular function.** f is called *submodular* if for every $X \subseteq Y \subseteq V$ and $z \in V \setminus Y$, $f(X \cup \{z\}) - f(X) \geq f(Y \cup \{z\}) - f(Y)$.
2. **Monotone function.** f is called *monotone* if $f(X \cup \{z\}) \geq f(X)$ for any set $X \subset V$ and element $z \in V$.

Result 3: *Influence $\mathcal{I}(S)$ under both IC and LT models is submodular and monotone w.r.t. S.* [1]

Result 4: *BIP $\mathcal{B}(S, \boldsymbol{p})$ under both PR-I and PR-L models is submodular w.r.t. S, if the prices $\boldsymbol{p}$ are fixed and $p_i \geq c$, where p_i is the ith element of $\boldsymbol{p}$ and c is the manufacturing cost of the product.* [2]

Remark 1 $\mathcal{B}(S, \boldsymbol{p})$ under both PR-I and PR-L models is non-monotone w.r.t. S. [2]

Algorithm for InfMax

Nemhauser et al. in [3] showed that greedy hill-climbing algorithm has the approximation ratio with $1 - 1/e$ of maximizing a submodular and monotone set function f. The greedy algorithm of maximizing influence is presented in Algorithm 1: each time the vertex that brings the highest marginal influence will be picked as a new seed, until the desired number of seeds are picked. Hence, according to Result 3, Algorithm 1 has a constant performance ratio $1 - 1/e$ solving InfMax.

Note that computing actual influence as well as marginal influence is #P-hard [1]. To estimate the influence in a reasonable time, Monte Carlo simulation is usually adopted, generating a number of samples and calculating the average value of all samples.

Algorithm 1 Greedy algorithm

$S \leftarrow \emptyset$;
while $|S| < k$ **do**
  $u = \arg\max_{u \in V \setminus S}\{\mathcal{I}(S \cup \{u\}) - \mathcal{I}(S)\}$;
  $S \leftarrow S \cup u$;
end while
output S;

Algorithms for BIPMax

For a non-monotone submodular set function f, Feige et al. [4] devised a deterministic local-search $\frac{1}{3}$-approximation and a randomized $\frac{2}{5}$-approximation algorithm if f is nonnegative; therefore, if the prices $\boldsymbol{p}$ are preknown and fixed, the techniques in [4] may be ideal approximation algorithms. However, usually prices $\boldsymbol{p}$ are to be determined to achieve the maximum BIP, and general algorithms need careful consideration.

To give a better pricing method for BIPMax, both the manufacturing cost and local influence should be considered. The manufacturing cost is denoted by c. Individual v_i's evaluation is a random variable X_i whose cumulative distribution function (CDF) is denoted by F_i. (If v_i is *influenced* and being offered price q, then the probability that v_i turns *active* is $\text{Prob}(x_i \geq q) = 1 - F_i(q)$.) For v_i itself, if it is chosen as the only seed and offered price p, the expected profit solely from v_i is $(1-F_i(p))(p-c)$, what is more, the expected influence to other nodes solely from v_i is $(1-F_i(p))\cdot\mathcal{I}(v_i, \boldsymbol{p})$. However, $\mathcal{I}(v_i, \boldsymbol{p})$ depends on other nodes' prices; to ease the computation, the following simple one-hop estimation is adopted: $\sum_{\forall u|\text{ outneighbor of } v_i} w(v_i, u)/d^{\text{out}}(v_i)$, where $d^{\text{out}}(v_i)$ is the outdegree of v_i. Then the *optimal price* p'_i for v_i is calculated as follows:

$$\tilde{\mathcal{B}}_i(p) = \lambda(1 - F_i(p)) \sum \frac{w(v_i, u)}{d^{\text{out}}(v_i)} + (1 - \lambda)(1 - F_i(p))(p - c), \quad (1)$$

$$p'_i = \arg\min_{p \in [0,1]} \tilde{\mathcal{B}}_i(p). \quad (2)$$

Equation (1) considers both the manufacturing cost and the network structure; however, the price calculated by (2) is still myopic.

Determine the Seeds and Prices Under BYC

In BYC the prices can only be 0 or the full price; for a company this strategy takes the least implementation expense. *ABYC* is the algorithm for BYC. ABYC contains two stages: first offering every individual a same full price and second determining the seeds whom free samples are given to.

Equation (2) is not used in the first stage since the obtained p'_i may vary from v_i. Instead we calculate the *universal optimal price*: $p'_U = \arg\min_{p \in [0,1]} \sum_i^n \tilde{\mathcal{B}}_i(p)$.

Greedy is used in the second stage of determining seeds: every round for each non-seed vertex u we compute the marginal BIP of picking u as a seed, and choose the vertex that provides the highest marginal gain. When no marginal BIP gain can be bought by any vertex, ABYC stops.

Suppose the price vector $\boldsymbol{p} = (p_1, \dots p_n)$; denote $(\boldsymbol{p}_{-i}, q)$ the vector obtained by altering p_i, the ith element of $\boldsymbol{p}$ to q, i.e., $(\boldsymbol{p}_{-i}, q) = (p_1, \dots, p_{i-1}, q, p_{i+1}, \dots, p_n)$.

Algorithm 2 ABYC: the algorithm for BYC

$S \leftarrow \emptyset$, $\boldsymbol{p} \leftarrow \boldsymbol{0}$;
for $\forall v_i \in V$ **do**
  $p_i \leftarrow p'_U$;
end for
while true **do**
  $u \leftarrow \arg\max_{v_i \in V \setminus S}\{\mathcal{B}(S \cup \{v_i\}, (\boldsymbol{p}_{-i}, 0)) - \mathcal{B}(S, \boldsymbol{p})\}$;
  if $\mathcal{B}(S \cup \{u\}, (\boldsymbol{p}_{-i}, 0)) - \mathcal{B}(S, \boldsymbol{p}) > 0$ **then**
    $S \leftarrow S \cup \{u\}$; $\boldsymbol{p} \leftarrow (\boldsymbol{p}_{-i}, 0)$;
  else *break*;
  end if
end while
output $(S, \boldsymbol{p})$;

Determine the Seeds and Prices Under PAP

BYC is easy to implement, however it is too simple and constrained, PAP is much freer where prices are assigned with no constraint. APAP is the algorithm for PAP, and like ABYC it also contains two stages. In the first stage, to obtain p'_i for every v_i (2) is adopted. In the second stage, the vertex with the maximum marginal BIP gain is picked step by step until no positive gain is available.

The computation of the marginal BIP gain under PAP when adding v_i into the seed set S is much more complex comparing to BYC, since when choosing v_i as a new seed a new price may also be offered to it. Suppose the new price for v_i is q, then the marginal BIP gain of adding v_i is: $\mathcal{B}(S \cup \{v_i\}, (\boldsymbol{p}_{-i}, q)) - \mathcal{B}(S, \boldsymbol{p})$. Since $\mathcal{B}(S, \boldsymbol{p})$ is a constant w.r.t. q, $\mathcal{B}(S \cup \{v_i\}, (\boldsymbol{p}_{-i}, q))$ should be maximized. When offering price q to v_i, only two outcomes exist in the sample space, outcome ω_1 where v_i accepts the price and turns *active*, outcome ω_0 where v_i rejects the price, stays *influenced* and never spreads the influence. If ω_1 happens, the influence gain collected from v_i is 1 and the profit gain collected from v_i is $q - c$, suppose the influence from other nodes is I_1 and the profit from other nodes is R_1, then the BIP gain is $g_i(q) = \lambda(I_1 + 1) + (1 - \lambda)(q - c + R_1)$, which is a linear function w.r.t. q. Else if ω_0 happens, the influence gain collected from v_i is 1 and the profit gain collected from v_i is 0, suppose the influence from other nodes is I_0 and the profit from other nodes is R_0, then the BIP gain is $h_i = \lambda(I_0 + 1) + (1 - \lambda)R_0$, a constant independent of q. $\text{Prob}(\omega_1) = 1 - F_i(q)$ and $\text{Prob}(\omega_0) = F_i(q)$. Hence the expected BIP is:

$$\delta_i(q) = g_i(q) \cdot (1 - F_i(q)) + h_i \cdot F_i(q) \quad (3)$$

Algorithm 3 APAP: the algorithm for PAP

$S \leftarrow \emptyset, \boldsymbol{p} \leftarrow \boldsymbol{0}$;
for $\forall v_i \in V$ **do**
  $p_i \leftarrow p_i' = \arg\min_{p \in [0,1]} \tilde{\mathcal{B}}_i(p)$;
end for
while true **do**
  for $\forall v_i \in V \setminus S$ **do**
    $p_i^* \leftarrow \arg\max_{q \in [0,1]} \delta_i(q)$;
  end for
  $u \leftarrow \arg\max_{v_i \in V \setminus S} \{\mathcal{B}(S \cup \{v_i\}, (\boldsymbol{p}_{-i}, p_i^*)) - \mathcal{B}(S, \boldsymbol{p})\}$;
  if $\mathcal{B}(S \cup \{u\}, (\boldsymbol{p}_{-i}, p_i^*)) - \mathcal{B}(S, \boldsymbol{p}) > 0$ **then**
    $S \leftarrow S \cup \{u\}$; $\boldsymbol{p} \leftarrow (\boldsymbol{p}_{-i}, p_i^*)$;
  else *break*;
  end if
end while
output $(S, \boldsymbol{p})$;

To calculate I_1 and R_1, set v_i turns active with probability 1 and run Monte Carlo simulations, and to calculate I_0 and R_0, set v_i turns active with probability 0 and run Monte Carlo simulations. After obtaining I_1, R_1, I_0, and R_0, p_i^* should be computed. If $\delta_i(q)$ is a closed form, then p_i^* is easy to calculate. However $\delta_i(q)$ may not be a close form. For example, if the valuation follows normal distribution, then F_i contains an integral term and $\delta_i(q)$ is not a closed form. In this case, *golden section search* [5] which works fast on finding the extremum of a strictly unimodal function can be used. This technique successively narrows the range inside which the extremum exists to find it. Even if $\delta_i(q)$ is not always unimodal, it is unimodel in subintervals of $[0, 1]$. To reduce error, divide the interval $[0, 1]$ into several small intervals with the same size and pick each small interval's midpoint as a sample q_t. The search starts with the interval that contains the sample $q_0 = \arg\max_t \delta_i(q_t)$ and stops when the interval that contains p^* is narrower than a predefined threshold.

Cross-References

▶ Greedy Approximation Algorithms
▶ Influence and Profit

Recommended Reading

1. Kempe D, Kleinberg J, Tardos É (2003) Maximizing the spread of influence through a social network. In: ACM KDD '03, Washington, DC, pp 137–146
2. Zhu Y, Lu Z, Bi Y, Wu W, Jiang Y, Li D (2013) Influene and profit: two sizes of the coin. In: IEEE ICDM '13, Dallas, 1301–1306
3. Nemhauser GL, Wolsey LA, Fisher ML (1978) An analysis of approximations for maximizing submodular set functions-i. Math Program 14(1):265–294
4. Feige U, Mirrokni VS, Vondrák J (2007) Maximizing non-monotone submodular functions. In: IEEE symposium on foundations of computer science (FOCS'07), Providence, pp 461–471

5. Kiefer L (1953) Sequential minimax search for a maximum. In: Proceedings of the American mathematical society, pp 502–206
6. Domingos P, Richardson M (2001) Mining the network value of customers. In: ACM KDD '01, San Francisco, pp 57–66
7. Richardson M, Domingos P (2002) Mining knowledge-sharing sites for viral marketing. In: ACM KDD '02, Edmonton, pp 61–70
8. Kleinberg R, Leighton T (2003) The value of knowing a demand curve: bounds on regret for online posted-price auctions. In: IEEE FOCS '03, Cambridge, pp 594–628
9. Hartline J, Mirrokni V, Sundararajan M (2008) Optimal marketing strategies over social networks. In: ACM WWW '08, Beijing, pp 189–198
10. Arthur D, Motwani R, Sharma A, Xu Y (2009) Pricing strategies for viral marketing on social networks. In: CoRR. abs/0902.3485
11. Lu W, Lakshmanan LVS (2012) Profit maximization over social networks. In: IEEE ICDM '12, Brussels

Influence Maximization

Zaixin Lu[1] and Weili Wu[2,3,4]
[1]Department of Mathematics and Computer Science, Marywood University, Scranton, PA, USA
[2]College of Computer Science and Technology, Taiyuan University of Technology, Taiyuan, Shanxi Province, China
[3]Department of Computer Science, California State University, Los Angeles, CA, USA
[4]Department of Computer Science, The University of Texas at Dallas, Richardson, TX, USA

Keywords

Approximation algorithm; Influence maximization; NP-hard; Social network

Years and Authors of Summarized Original Work

2011; Lu, Zhang, Wu, Fu, Du
2012; Lu, Zhang, Wu, Kim, Fu

Problem Definition

One of the fundamental problems in social network is influence maximization. Informally, if we can convince a small number of individuals in a social network to adopt a new product or innovation, and the target is to trigger a maximum further adoptions, then which set of individuals should we convince? Consider a social network as a graph $G(V, E)$ consisting of individuals (node set V) and relationships (edge set E); essentially influence maximization comes down to the problem of finding important nodes or structures in graphs.

Influence Diffusion

In order to address the influence maximization problem, first it is needed to understand the influence diffusion process in social networks. In other words, how does the influence propagate over time through a social network? Assume time is partitioned into discrete time slots, and then influence diffusion can be modeled as the process by which activations occur from neighbor to neighbor. In each time slot, all previously activated nodes remain active and others either remain inactive or be activated by their neighbors according to the activation constraints. The whole process runs in a finite number of time slots and stops at a time slot when no more activation occurs. Let S denote the set of initially activated nodes; we denote by $f(S)$ eventually the number of activations, and the target is to maximize $f(S)$ with a limited budget.

Problem (Influence Maximization)

INPUT: A graph $G(V, E)$ where V is the set of individuals and E is the set of relationships, an activation model f, and a limited budget number K.

OUTPUT: A set S of nodes where $S \subseteq V$ such that the final activations $f(S)$ is maximized and $|S| \leq K$.

Activation Models

The influence maximization problem was first proposed by Domingos et al. and Richardson

Influence Maximization, Fig. 1 Pseudo-code: Greedy algorithm

```
Greedy Algorithm
1: let S ← ∅ (S holds the selected nodes);
2: while |S| ≤ K do
3:    find v ∈ (V \ S) such that f(S ∪ {v}) is maximized;
4:    let S ← S ∪ {v};
5: end while
```

et al. in [4] and [8], respectively, in which the social networks are modeled as Markov random field. After that, Kempe et al. ([6] and [7]) further investigated this problem in two models: *Independent Cascade* proposed by Goldenberg et al. ([5] and [11]) and *Linear Threshold* proposed by Granovetter et al. and Schelling et al., respectively, in [9] and [10].

In the *Independent Cascade* model, the activations are independent among different individuals, i.e., each newly activated individual u will have a chance, in the next time slot, to activate his or her neighbors v with certain probability $p(u, v)$ which is independent with other activations. In the *Linear Threshold* model, the activation is based on a threshold manner; the influence from an individual u to another individual v is presented by a weight $w(i, j)$ and the individual v will be activated at the moment when the sum of weights he or she receives from previous activated neighbors exceeds the threshold $t(v)$. It is worthy to note that there are two ways to assign the thresholds to individuals: *random* and *deterministic*. In the random model, the thresholds are randomly selected at uniform during the time, while in the deterministic model, the thresholds are assigned to individuals at the beginning and fixed for all time slots. For the sake of simplicity, they are called *Random Linear Threshold* and *Deterministic Linear Threshold*, respectively.

Key Results

Greedy Algorithm

In [6], it has been found that the activation function f under the *Independent Cascade* model and the *Random Linear Threshold* model is sub-modular. Therefore, the natural greedy algorithm (Fig. 1), which selects the node with the maximum marginal gain repeatedly, achieves a $\left(1 - \frac{1}{e}\right)$-approximation solution. However, the problem of exactly calculating the activation function f in a general graph G under the *Independent Cascade* model or the *Random Linear Threshold* model, respectively, is #P-hard [1, 2], which indicates that the greedy algorithm is not a polynomial time algorithm for the two models. The time complexity directly follows the pseudo-code (Fig. 1). Assume there exists an oracle that can compute the activation function f in τ time, and then the greedy algorithm runs in $O(K|V|\tau)$ time.

In [13], it has been found that the problem of exactly calculating the activation function f given an arbitrary set S under the *Deterministic Linear Threshold* model can be solved in linear time in terms of the number of edges. Therefore, the greedy algorithm runs in $O(K|V||E|)$ time. However, it has no approximation guarantee under this model.

Inapproximation Results

Under the *Independent Cascade* model or the *Random Linear Threshold* model, it can be shown by doing a gap-preserving reduction from the Set Cover problem [3] that $\left(1 - \frac{1}{e}\right)$ is the best possible polynomial time approximation ratio for the influence maximization problem; assume NP $\not\subset$ DTIME $\left(n^{\log\log n}\right)$. Under the *Deterministic Linear Threshold* model, it has been shown that there is no polynomial time $n^{1-\epsilon}$-approximation algorithm for the influence maximization problem unless P = NP where n is the number of nodes and $0 < \epsilon < 1$ [12].

Actually in the case that an individual can be activated after one of his or her neighbors becomes active, the greedy algorithm achieves a polynomial time $(1 - \frac{1}{e})$-approximation solution, and even in the simple case that an individual can be activated when one or two of his or her neighbors become active, the influence maximization problem under the *Deterministic Linear Threshold* model is NP-hard to approximate.

Degree-Bounded Graphs

A graph $G(V, E)$ is a (d_1, d_2)-degree-bounded graph if every node in V has at most d_1 incoming edges and at most d_2 outgoing edges.

For the sake of simplicity, the influence maximization problem over such a degree-bounded graph is called (d_1, d_2)-influence maximization. In [13], it has been found that for any constant $\epsilon \in (0, 1)$, there is no polynomial time $n^{1-\epsilon}$-approximation algorithm for the $(2, 2)$-influence maximization problem under the *Deterministic Linear Threshold* model unless P = NP where n is the number of nodes, which indicates that the influence maximization problem under *Deterministic Linear Threshold* model is NP-hard to approximate to within any nontrivial factor, even if an individual can be activated when at least two of his or her neighbors become active.

Applications

Influence maximization would be of great interest for corporations, such as Facebook, LinkedIn, and Twitter, as well as individuals who desire to spread their products, ideas, etc. The solutions have a wide range of applications in various fields, such as product promotions where corporations want to distribute sample products among customers, political elections where candidates want to spread their popularity or political ideas among voters, and emergency situations where emergency news like sudden earthquake needs to spread to every resident in the community. In addition, the solutions may be also applicable in military defense where malicious information which has already propagated dynamically needs to be blocked.

Cross-Reference

▶ Influence and Profit

Recommended Reading

1. Chen W, Yuan Y, Zhang L (2010) Scalable influence maximization in social networks under the linear threshold model. In: The 2010 international conference on data mining. Sydney, Australia
2. Chen W, Wang C, Wang Y (2009) Scalable influence maximization for prevalent viral marketing in large-scale social networks. In: The 2010 international conference on knowledge discovery and data mining. Washington DC, USA
3. Feige U (1998) A threshold of ln n for approximating set cover. J ACM 45:314–318
4. Domingos P, Richardson M (2001) Mining the network value of customers. In: The 2001 international conference on knowledge discovery and data mining. San Francisco, CA, USA
5. Goldenberg J, Libai B, Muller E (2001) Using complex systems analysis to advance marketing theory development. Acad Mark Sci Rev 9(3):1–18
6. Kempe D, Kleinberg J, Tardos É (2003) Maximizing the spread of influence through a social network. In: The 2003 international conference on knowledge discovery and data mining. Washington DC, USA
7. Kempe D, Kleinberg J, Tardos É (2005) Influential nodes in a diffusion model for social networks. In: The 2005 international colloquium on automata, languages and programming. Lisbon, Portugal
8. Richardson M, Domingos P (2002) Mining knowledge-sharing sites for viral marketing. In: The 2002 international conference on knowledge discovery and data mining. Edmonton, AB, Canada
9. Granovetter M (1978) Threshold models of collective behavior. Am J Sociol 83(6):1420–1443
10. Schelling T (1978) Micromotives and macrobehavior. Norton, New York
11. Goldenberg J, Libai B, Muller E (2001) Talk of the network: a complex systems look at the underlying process of word-of-mouth. Mark Lett 12(3): 211–223
12. Lu Z, Zhang W, Wu W, Kim J, Fu B (2012) The complexity of influence maximization problem in deterministic threshold mode. J Comb Optim 24(3):374–378
13. Lu Z, Zhang W, Wu W, Fu B, Du D (2011) Approximation and inapproximation for the influence maximization problem in social networks under deterministic linear threshold model. In: The 2011 international conference on distributed computing systems workshops. Minneapolis, USA

Intersections of Inverted Lists

Andrew Kane and Alejandro López-Ortiz
David R. Cheriton School of Computer Science, University of Waterloo, Waterloo, ON, Canada

Keywords

Algorithms; Bitvectors; Compression; Efficiency; Information retrieval; Intersection; Optimization; Performance; Query processing; Reordering

Years and Authors of Summarized Original Work

2014; Kane, Tompa

Problem Definition

This problem is concerned with efficiently finding a set of documents that closely match a query within a large corpus of documents (i.e., a search engine). This is accomplished by producing an index offline to the query processing and then using the index to quickly answer the queries. The indexing stage involves splitting the dataset into tokens and then constructing an *inverted index* which maps from each token to the list of document identifiers of the documents that contain that token (a *postings list*). The query can then be executed by converting it to a set of query tokens, using the inverted index to find the corresponding postings lists, and *intersecting* the lists to find the documents contained in all the lists (conjunctive intersection or boolean AND). A subsequent ranking step is used to restrict the conjunctive intersection to a list of top-k results that best answer the query.

Objective

Produce an efficient system to answer queries, where efficiency is a space-time trade-off involving the storage of the inverted index (space) and the intersection of the lists (time). If the inverted index is stored on a slow medium, then efficiency might also include the size of the lists required to answer the query (transfer).

Design Choices

There are many degrees of freedom in designing such a system:

1. Creating a document to internal identifier mapping
2. Encoding of the inverted index mapping
3. Encoding of the postings lists
4. Using auxiliary structures in postings lists
5. Ordering of the internal identifiers in the postings lists
6. Order and method of executing the list intersection

Variants

For queries where the conjunctive intersection is too small, the intersection can be relaxed to find documents containing a weighted portion of the query tokens (see t-threashold or Weak-AND). The query lists can be intersected using any boolean operators, though the most commonly added is the boolean NOT operator which can be used to quickly reduce the number of conjunctive results. The query results can also be reduced by including token offset restrictions, such as ensuring tokens appear as a phrase or within some proximity. While many implementations interleave the conjunctive intersection with the calculation of the ranking, some also use ranking information to prune documents from the intersection or to terminate the query early when the correct results (or good enough results) have been found.

Key Results

Traditionally, inverted indexes were stored on disk, causing the reduction of transfer costs to be the dominant objective. Modern systems often store their inverted indexes in memory meaning that reducing overall index size is important and that implementation details of the intersection algorithms can produce significant performance differences, thus leading to a more subtle space-time trade-off. In either case, the mapping portion of the inverted index (the dictionary or lexicon) can be implemented using a data structure such as a B-tree which is both fast and compact, so we do not examine dictionary implementations in the remainder of this article.

Multi-list Processing

Intersecting multiple lists can be implemented by intersecting the two smallest lists and then intersecting the result with the next smallest list iteratively, thus producing a *set versus set* (svs) or *term-at-a-time* (TAAT) approach. If the lists are in sorted order, then each step of the svs approach uses the *merge* algorithm, which takes each element in the smaller list M and finds it in the larger list N by executing a forward search and then reports any elements that are found. The M list could be encoded differently than the N list, and indeed, after the first svs step, it is the uncompressed result list of the previous step. The sequential processing and memory accesses of the svs approach allows the compiler and CPU to optimize the execution, making this approach extremely fast, even though temporary memory is required for intermediate result accumulators. If the lists are not sorted, then additional temporary memory must be used to intersect the lists using some equality-join algorithm.

Intersecting multiple lists can also be implemented by intersecting all the lists at the same time. If the lists are not in sorted order, using this approach may require a large amount of temporary space for the join structures and the accumulators. If the lists are sorted, then we call this a *document-at-a-time* (DAAT) approach, and it requires very little temporary space: just one pointer per list to keep track of its processing location. The order that the lists are intersected could be static, such as ascending term frequency order (as done with svs), or it could adapt to the processing. All of these non-svs approaches jump among the lists that are stored at different memory locations, so it is more difficult for the compiler and CPU to optimize the execution. In addition, the loop iterating over the lists for each result item and the complications when using different list encodings will slow down non-svs implementations. Despite these limitations, many of the optimizations for svs list intersection can be applied to implementations using non-svs approaches. For systems that return the top-ranked results, a small amount of additional memory is needed for a top-k heap to keep track of the best results. Instead of adding results that match all the terms into a simple array of results, they are added to the heap. At the end of query processing, the content of the heap is output in rank order to form the final top-k query results.

Uncompressed Lists

Storing the lists of document identifiers in an *uncompressed* format simply means using a sequential array of integers. For fast intersection, the integers in these lists are stored in order, thus avoiding join structures and allowing many methods of searching for a particular value in a list. As a result, there are many fast algorithms available for intersecting uncompressed integer lists, but the memory used to store the uncompressed lists is very large and probes into the list can produce wasted or inefficient memory access. All of these uncompressed intersection algorithms rely on random access into the lists, so they are inappropriate for compressed lists. We present only three of the best performing algorithms [2]:

Galloping svs (g-svs): Galloping forward search probes into the list to find a point past the desired value, where the probe distance doubles each time, then the desired location is found using binary search within the last two probe points.

Galloping swapping svs (g-swsvs): In the previous galloping svs algorithm, values from the smaller list are found in the larger list. Galloping swapping svs, however, finds values from the list with the smaller number of remaining integers in the other list, thus potentially swapping the roles of the lists.

Sorted Baeza-Yates using adaptive binary forward search (ab-sBY): The Baeza-Yates algorithm is a divide and conquer approach that finds the median value of the smaller list in the larger list, splits the lists, and recurses. Adding matching values at the end of the recursion produces a sorted result list. The adaptive binary forward search variant uses binary search within the recursed list boundaries, rather than using the original list boundaries.

Compressed Lists

There are a large variety of compression algorithms available for sorted integer lists. The lists are first converted into differences minus one (i.e., deltas or d-gaps) to get smaller values, but this removes the ability to randomly access elements in the list. Next, a variable length encoding is used to reduce the number of bits needed to store the values, often grouping multiple values together to allow word or byte alignment of the groups and faster bulk decoding. The most common list compression algorithms are *Variable byte (vbyte)*, *PForDelta (PFD)*, and *Simple9 (S9)*. Recent work has improved decoding, and delta restore speeds for many list compression algorithms using vectorization [7]. Additional gains are possible by changing delta encoding to act on groups of values, thus improving runtime at the expense of using more space. Another recent approach called quasi-succinct indexing [10] first acts on the values as monotone sequences and then incorporates some delta encoding more deeply in the compression algorithm.

List Indexes

List indexes, also known as skip structures or auxiliary indexes, can be included to jump over values in the postings lists and thus avoid decoding, or even accessing, portions of the lists. The desired jump points can be encoded inline with the lists, but they are better stored in a separate contiguous memory location without compression, allowing fast scanning through the jump points. These list index algorithms can be used with compressed lists by storing the deltas of the jump points, but the block-based structure causes complications if the jump point is not byte or word aligned, as well as block aligned. The actual list values that are found in the skip structures can either be maintained within the original compressed list (overlaid) preserving fast iteration through the list, or the values could be extracted (i.e., removed) from the original compressed lists, giving a space reduction but slower iteration through the list.

A simple list index algorithm ("skipper" [9]) groups by a fixed number of elements storing every X^{th} element in a separate array structure, where X is a constant, so we refer to it as skips(X). When intersecting lists, the skip structure is scanned linearly to find the appropriate jump point into the compressed structure, where the decoding can commence. Using variable length skips is possible, such as tuning the number of skipped values relative to the list size n, perhaps using a multiple of $\sqrt{n}$ or $\log(n)$.

Another type of list index algorithm ("lookup" [9]) groups by a fixed size document identifier range using the top-level bits of the value to index into an array storing the desired location in the encoded list, similar to a segment/offset scheme. Each list can pick the number of bits in order to produce reasonable jump sizes. We use D as the domain size and n as the list size, giving a list's density as $y = \frac{n}{D}$. If we assume randomized data and use the parameter B to tune the system, then by using $\left\lceil \log_2\left(\frac{B}{y}\right)\right\rceil$ bottom level bits will leave between $\frac{B}{2}$ and B entries per segment in expectation. As a result, we call this algorithm segment(B).

Bitvectors

When using a compact domain of integers, as we are, the lists can instead be stored as bitvectors, where the bit number is the integer value and the bit is set if the integer is in the list. For runtime performance benefits, this mapping from the identifier to the bit location can be changed if it remains a one-to-one mapping and is applied to all bitvectors.

If all the lists are stored as bitvectors, conjunctive list intersection can be easily implemented by combining the bitvectors of the query terms using bitwise AND (bvand), with the final step converting the result to a list of integers (bvconvert). Note, except for the last step, the result of each step is a bitvector rather than an uncompressed result list. The bvconvert algorithm can be implemented as a linear scan of the bits of each word, but using a logarithmic check is faster since the bitvectors being converted are typically sparse. Encoding all the lists as bitvectors gives good query runtime, but the space usage is very large since there are many tokens.

To alleviate the space costs of using bitvectors, the lists with density less than a parameter value F can be stored using normal delta compression, resulting in a hybrid bitvector algorithm [4]. This hybrid algorithm intersects the delta-compressed lists using merge and then intersects the remainder with the bitvectors by checking if the elements are contained in the first bitvector (bv-contains), repeating this for each bitvector in the query, with the final remaining values being the query result. Bitvectors are faster than other approaches for dense lists, so this hybrid algorithm is faster than non-bitvector algorithms. It can also be more compact than other compression schemes, because dense lists can be compactly stored as bitvectors. In addition, large overlaid skips can be used in the delta-compressed lists to improve query runtime.

In order to store more postings in the faster bitvector form, a semi-bitvector structure [5] encodes the front portion of a list as a bitvector and the rest using skips over delta compression. By skewing lists to have dense front portions, this approach can improve both space and runtime.

Other Approaches

Quasi-succinct indices [10] store list values in blocks with the lower bits of the values in an array and the higher bits as deltas using unary encoding combined with skips for fast access. This structure produces a good space-time trade-off when the number of higher-level bits is limited. The list intersection implementation can exploit the unary encoding of the higher-level bits by counting the number of ones in machine words to find values quickly. The resultant space-time performance is comparable to various skip-type implementations for conjunctive list intersection, though indexing speed may be slower.

The treap data structure combines the functionality of a binary tree and a binary heap. This treap data structure can implement list intersection [6] by storing each list as a treap where the list values are used as the tree order and the frequencies are used as the heap order. During list intersection, subtrees can be pruned from the processing if the frequency is too low to produce highly ranked results. In order to make this approach viable, low-frequency values are stored in separate lists using delta-compression with skips. This treap and frequency separated index structure can produce some space-time performance improvements compared to existing ranking-based search systems.

Wavelet trees can also be used to implement list intersection [8]. The postings lists are ordered by frequency, and each value is assigned a global location, so that each list can be represented by a range of global locations. A series of bit sequences represents a tree which starts with the frequency-ordered lists of the global locations and translates them into the actual document identifiers. Each level of the tree splits the document identifier domain ranges in half and encodes the edges of the tree using the bit sequences (left as 0 and right as 1). Multiple lists can be intersected by following the translation of their document ranges in this tree of bit sequences, only following the branch if it occurs in all of the list translations. After some careful optimization of the translation code, the wavelet tree data structure results in similar space usage, but faster query runtimes than some existing methods for conjunctive queries.

Ranking

Ranking algorithms are closely guarded trade secrets for large web search companies, so their details are not generally known. Many approaches, however, add term frequencies and/or postings offsets into the postings lists and combine this with corpus statistics to produce good results, as done in the standard BM25 approach. Unfortunately, information such as frequencies cannot be easily added to bitvector structures, thus limiting their use. Data and link analysis can also help by producing a global order, such as PageRank, which can be factored into the ranking function to improve results.

Reordering
Document identifiers in postings lists can be assigned to make the identifier deltas smaller and more compressible. As a first stage, they are assigned to form a compact domain of values, while a second stage renumbers these identifiers to optimize the system performance in a process referred to as document *reordering* [3]. Reordering can improve space usage by placing documents with similar terms close together in the ordering, thus reducing the size of the deltas, which can then be stored more compactly, among other benefits.

Applications

Intersecting inverted lists is at the heart of search engine query processing and top-k operators in databases.

Open Problems

Going forward, main goal is to design list representations which are provably optimal in terms of space usage (entropy) together with algorithms that achieve the optimal trade-off between time and the space used by the given representation.

Experimental Results

Most published works involving intersections of inverted lists prove their results through experimentation. Following this approach, we show the relative performance of the described approaches using an in-memory conjunctive list intersection system that has document ordered postings lists without any ranking-based structures or processing. We run this system on the TREC GOV2 corpus and execute 5,000 of the associated queries. These experiments were executed on an AMD Phenom II X6 1090T 3.6 GHz Processor with 6 GB of memory running Ubuntu Linux 2.6.32-43-server. The code was compiled using the gcc 4.4.3 compiler with the -O3 command line parameter. Our results are presented in Fig. 1 using a space-time (log-log) plot. For the configurations considered, the block size of the encoding always equals the skip size, X, and the bitvector configurations all use $X = 256$.

With compressed lists alone, intersection is slow. On the other hand, uncompressed lists are much larger than compressed ones, but random access allows them to be fast. List indexes when combined with the compressed lists add some space, but their targeted access into the lists allows them to be even faster than the uncompressed algorithms. (For the list indexes, we present the fastest configurations we tested over the parameter ranges of X and B.)

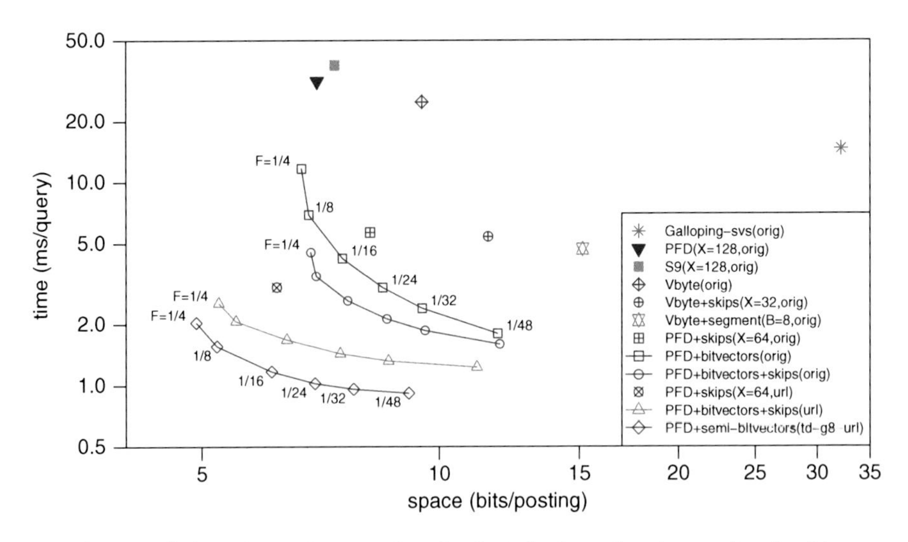

Intersections of Inverted Lists, Fig. 1 Space vs. time (log-log) plot for various intersection algorithms

The performance of list indexes suggests that the benefits of knowing where to probe into the list (i.e., using skips rather than a probing search routine) outweigh the cost of decoding the data at that probe location. Using the hybrid bitvector approach is much faster and somewhat smaller than the other techniques. Adding large overlaid skips to the delta-compressed lists allows the bitvectors + skips algorithm to improve performance.

Reordering the documents to be in URL order gives significant space improvements. This ordering also improves query runtimes significantly, for all combinations of skips and/or bitvectors. Splitting the documents into eight groups by descending number of terms in document, reordering within the groups by URL ordering (td-g8-url), and using semi-bitvectors produce additional improvements in both space and runtime. This demonstration of the superior performance of bitvectors suggests that integrating them into ranking-based systems warrants closer examination.

URLs to Code and Datasets

Several standard datasets and query workloads are available from the Text REtreival Conference (TREC at http://trec.nist.gov). Many implementations of search engines are available in the open source community, including Wumpus (http://www.wumpus-search.org), Zettair (http://www.seg.rmit.edu.au/zettair/), and Lucene (http://lucene.apache.org).

Cross-References

▸ Compressing Integer Sequences

Recommended Reading

1. Anh VN, Moffat A (2005) Inverted index compression using word-aligned binary codes. Inf Retr 8(1):151–166
2. Barbay J, López-Ortiz A, Lu T, Salinger A (2009) An experimental investigation of set intersection algorithms for text searching. J Exp Algorithmics 14:3.7, 1–24
3. Blandford D, Blelloch G (2002) Index compression through document reordering. In: Proceedings of the data compression conference (DCC), Snowbird. IEEE, pp 342–351
4. Culpepper JS, Moffat A (2010) Efficient set intersection for inverted indexing. ACM Trans Inf Syst 29(1):1, 1–25
5. Kane A, Tompa FW (2014) Skewed partial bitvectors for list intersection. In: Proceedings of the 37th ACM international conference on research and development in information retrieval (SIGIR), Gold Coast. ACM, pp 263–272
6. Konow R, Navarro G, Clarke CLA, López-Ortiz A (2013) Faster and smaller inverted indices with treaps. In: Proceedings of the 36th ACM international conference on research and development in information retrieval (SIGIR), Dubin. ACM, pp 193–202
7. Lemire D, Boytsov L (2013) Decoding billions of integers per second through vectorization. Softw Pract Exp. doi: 10.1002/spe.2203. To appear
8. Navarro G, Puglisi SJ (2010) Dual-sorted inverted lists. In: String processing and information retrieval (SPIRE), Los Cabos. Springer, pp 309–321
9. Sanders P, Transier F (2007) Intersection in integer inverted indices. In: Proceedings of the 9th workshop on algorithm engineering and experiments (ALENEX), New Orlean. SIAM, pp 71–83
10. Vigna S (2013) Quasi-succinct indices. In: Proceedings of the 6th international conference on web search and data mining (WSDM), Rome. ACM, pp 83–92
11. Zukowski M, Heman S, Nes N, Boncz P (2006) Super-scalar RAM-CPU cache compression. In: Proceedings of the 22nd international conference on data engineering (ICDE), Atlanta. IEEE, pp 59.1–59.12

Intrinsic Universality in Self-Assembly

Damien Woods
Computer Science, California Institute of Technology, Pasadena, CA, USA

Keywords

Abstract Tile Assembly Model; Intrinsic universality; Self-assembly; Simulation

Years and Authors of Summarized Original Work

2012; Doty, Lutz, Patitz, Schweller, Summers, Woods

2013; Demaine, Patitz, Rogers, Schweller, Summers, Woods

2014; Meunier, Patitz, Summers, Theyssier, Winslow, Woods
2014; Demaine, Demaine, Fekete, Patitz, Schweller, Winslow, Woods

Problem Definition

Algorithmic self-assembly [11] is the idea that small self-assembling molecules can compute as they grow structures. It gives programmers a set of theoretical models in which to specify and design target structures while trying to optimize resources such as number of molecule types or even construction time. The abstract Tile Assembly Model [11] is one such model. An instance of the model is called a tile assembly system and is a triple $\mathcal{T} = (T, \sigma, \tau)$ consisting of a finite set T of square tiles, a seed assembly σ (one or more tiles stuck together), and a temperature $\tau \in \{1, 2, 3, \ldots\}$, as shown in Fig. 1a. Each side of a square tile has a glue (or color) g which in turn has a strength $s \in \{0, 1, 2, \ldots\}$. Growth occurs on the integer plane and begins from a seed assembly (or a seed tile) placed at the origin, as shown in Fig. 1b. A tile sticks to a partially formed assembly if it can be placed next to the assembly in such a way that enough of its glues match the glues of the adjacent tiles on the assembly and the sum of the matching glue strengths is at least the temperature. Growth proceeds one tile at a time, asynchronously and nondeterministically.

Here we discuss recent results and suggest open questions on intrinsic universality and simulation as a method to compare self-assembly models. Figure 2 gives an overview of these and other results. For more details, see [12].

Simulation and Intrinsic Universality

Intuitively, one self-assembly model simulates another if they grow the same structures, via the same dynamical growth processes, possibly with some spatial scaling. Let $\mathcal{S}$ and $\mathcal{T}$ be tile assembly systems of the abstract Tile Assembly Model described above. $\mathcal{S}$ is said to *simulate* $\mathcal{T}$ if the following conditions hold: (1) each tile of $\mathcal{T}$ is represented by one or more $m \times m$ blocks of tiles in $\mathcal{S}$ called *supertiles*, (2) the seed assembly

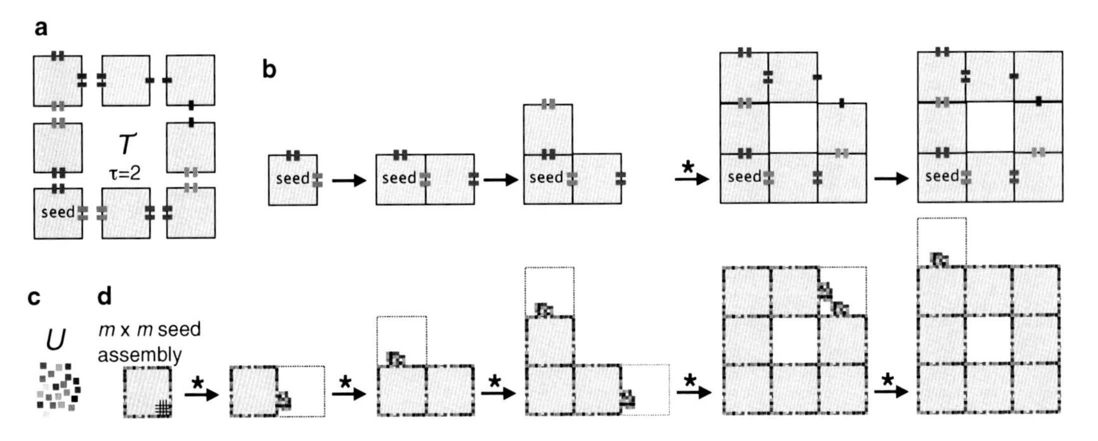

Intrinsic Universality in Self-Assembly, Fig. 1 An instance of the abstract Tile Assembly Model and an example showing simulation and intrinsic universality. (**a**) A tile assembly system $\mathcal{T}$ consists of a tile set, seed tile, and a temperature $\tau \in \mathbb{N}$. Colored glues on the tiles' sides have a natural number strength (shown here as 0, 1, or 2 colored tabs). (**b**) Growth begins from the seed with tiles sticking to the growing assembly if the sum of the strengths of the matching glues is at least τ. (**c**) An intrinsically universal tile set U. (**d**) When initialized with a seed assembly (which encodes $\mathcal{T}$) and at temperature 2, the intrinsically universal tile set simulates the dynamics of $\mathcal{T}$ with each tile placement in $\mathcal{T}$ being simulated by the growth of an $m \times m$ block of tiles. Single tile attachment is denoted by $\rightarrow$, and $\xrightarrow{*}$ denotes multiple tile attachments. Note that both systems have many other growth dynamics that are not shown

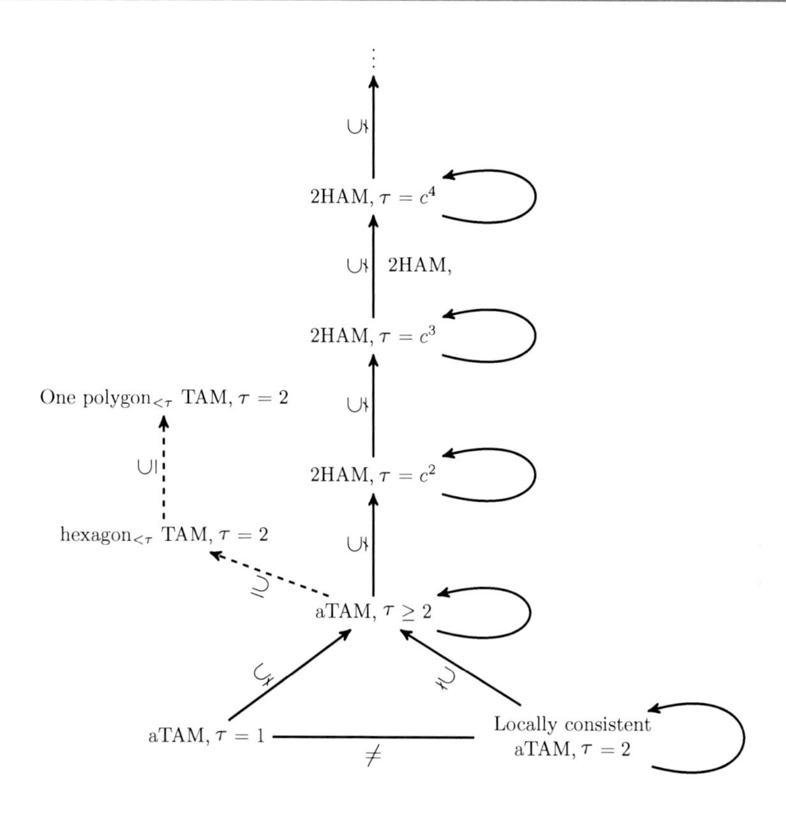

Intrinsic Universality in Self-Assembly, Fig. 2 Classes of tile assembly systems and their relationship with respect to simulation. There is an arrow from B to A if A contains B with respect to simulation: that is, for each tile assembly system $\mathcal{B} \in B$, there is a tile assembly system $\mathcal{A}_{\mathcal{B}} \in A$ that simulates $\mathcal{B}$. *Dashed arrows* denote containment, *solid arrows* denote strict containment, a self-loop denotes the existence of an intrinsically universal tile set for a class and its omission implies that the existence of such a tile set is an open problem. aTAM: abstract Tile Assembly Model (growth from a seed assembly by single tile addition in 2D), τ denotes "temperature." 2HAM: Two-Handed Tile Assembly Model (assemblies of tiles stick together in 2D). A 2HAM temperature hierarchy is shown for some $c \in \{2, 3, 4, \ldots\}$ and, in fact, for each such c the set of temperatures $\{c^i \,|\, i \in \{2, 3, \ldots\}\}$ gives an infinite hierarchy of classes of strictly increasing simulation power in the 2HAM

of $\mathcal{T}$ is represented by the *seed assembly* of $\mathcal{S}$ (one or more connected $m \times m$ supertiles), and (3) via supertile representation every sequence of tile placements in the simulated system $\mathcal{T}$ has a corresponding sequence of supertile placements in the simulator system $\mathcal{S}$, and vice versa. It is worth pointing out that although the intuitive idea of one assembly system simulating another is fairly simple, the formal definition of simulation [10] gets a little technical as the filling out of supertiles in the simulator is an asynchronous and nondeterministic distributed process with many supertiles growing independently and in parallel in the simulator system.

Key Results

The Abstract Tile Assembly Model Is Intrinsically Universal

A class of tile assembly systems C is said to be *intrinsically universal* if there exists a single set of tiles U that simulates any instance of C. For each such simulation, U should be appropriately initialized as an instance (i.e., a tile assembly system) of C itself. Figure 1d illustrates the concept. For example, the abstract Tile Assembly Model has been shown to be intrinsically universal [5]. Specifically, this means that there is a single set of tiles U that when appropriately initialized is capable of simulating an arbitrary tile assembly system $\mathcal{T}$. To program such a simulation, tiles from $\mathcal{T}$ are represented as $m \times m$ supertiles (built from tiles in U), and the seed assembly of $\mathcal{T}$ is represented as a connected assembly $\sigma_{\mathcal{T}}$ of such supertiles. Furthermore, the entire tile assembly system $\mathcal{T}$ (a finite object) is itself encoded in the supertiles of $\sigma_{\mathcal{T}}$ of $\mathcal{U}$. Then if we watch all possible growth dynamics in both $\mathcal{T} = (T, \sigma, \tau)$ and $\mathcal{U} = (U, \sigma_{\mathcal{T}}, 2)$, we get that both systems produce the same set of assemblies via the same dynamics where we use a supertile representation function to map from supertiles over U to tiles from T. It is worth pointing out that in this particular construction [5], the simulating system is always (merely) at temperature $\tau = 2$ no matter how large the temperature ($\tau \geq 1$) of the simulated system.

This intrinsically universal tile set U has the ability to simulate both the geometry and growth order of any tile assembly system. Modulo spatial rescaling U represents the full power and expressivity of the entire abstract Tile Assembly Model.

Noncooperative Assembly Is Weaker than Cooperative Assembly

The temperature 1, or noncooperative, model is a restriction of the abstract Tile Assembly Model. Despite its esoteric name, it models a fundamental and ubiquitous form of growth: asynchronous growing and branching tips in Euclidian space where each new tile is added if it matches on at *least one side*. Separating the power of the noncooperative and cooperative models has presented significant challenge to the community.

Recently it has been shown that the noncooperative model is provably weaker than the full model [10] in that sense that it is not capable of *simulating* arbitrary tile assembly systems. This is the first fully general negative result about temperature 1 that does not assume restrictions on the model nor unproven hypotheses.

An interesting aspect of this result is that it holds for 3D noncooperative systems; they too cannot simulate arbitrary tile assembly systems. This seems quite shocking, given that 3D noncooperative systems are Turing-universal [1]! So in particular, 3D noncooperative systems can simulate 2D (or 3D) cooperative systems by simulating a Turing machine that in turn simulates the cooperative system, but this loose style of simulation ends up destroying the geometry and dynamics of tile assembly by encoding everything as "geometry-less" strings. Hence, Turing-universal algorithmic behavior in self-assembly does not imply the ability to simulate, in a direct geometric fashion, arbitrary algorithmic self-assembly processes.

One Tile to Rule Them All

As an example of a simulation result on a very different model of self-assembly, Demaine, Demaine, Fekete, Patitz, Schweller, Winslow, and Woods [4] describe a sequence of simulations that route from square tiles, to the intrinsically universal tile set, to hexagons (with strength $< \tau$, or weak, glues), to a *single* polygon that is translatable, rotatable, and flipable. Their fixed-sized polygon, when appropriately seeded, simulates any tile assembly system from the abstract Tile Assembly Model. They also show that with translation only (i.e., no rotation), such results are not possible with a small (size ≤ 3) seed (although with larger seeds, a single translation-only polyomino simulates the space-time diagram of a 1D cellular automaton). In the simpler setting of Wang plane tiling, they give an easy method to "compile" any tile set T (on the square or hexagonal lattice) to a single regular polygon that simulates exactly the tilings of T, except with tiny gaps between the polygons.

Two Hands

It has been shown that the two-handed, or hierarchical, model of self-assembly (where large assemblies of tiles may come together in a single step) is not intrinsically universal [3]. Specifically there is no tile set that, in the two-handed model, can simulate all two-handed systems for all temperatures. However, for each $\tau \in \{2, 3, 4, \ldots\}$, there is a tile set U_τ that is intrinsically universal for the class of two-handed systems that work at temperature τ. Also, there is an infinite hierarchy of classes of such systems with each level strictly more powerful than the one below. In fact there are an infinite set of such hierarchies, as described in the caption of Fig. 2. These results give a formalization of the intuition that multiple long-range interactions are more powerful than fewer long-range interactions in the two-handed model.

Open Problems

Gaps in Fig. 2 (i.e., missing solid arrows and missing models) suggest a variety of open questions. Also, it remains as future work to further tease apart the power of restrictions of the abstract Tile Assembly Model, for example, it remains open whether 2D noncooperative systems are intrinsically universal for themselves.

It is an open question whether or not the hexagonal Tile Assembly Model [4], various polygonal Tile Assembly Models [4, 7], the Nubot model [13], and Signal-Passing Tile Assembly Model [6,9] are intrinsically universal. Furthermore, simulation could be used to tease apart the power of subclasses of these models.

Gilbert et al. [7] investigate the computational power of various kinds of polygonal tile assembly systems, showing that regular polygon tiles with >6 sides simulate Turing machines. What is the relationship between tile geometry and simulation power? Do more sides give strictly more simulation power?

A desirable feature of a simulator is not only that it simulates all possible dynamics of some simulated system, but that the probability of a given dynamics is roughly equal in both the simulated system and the simulator. Is there an intrinsically universal tile set with that property? Here, the probability of seeing a given dynamics or assembly in a simulator should be close to that of the simulated system, where "close" means, say, within a factor proportional to the spatial scaling.

Does there exist a tile set U for the abstract Tile Assembly Model, such that for any (adversarially chosen) seed assembly σ, at temperature 2, this tile assembly system simulates some tile assembly system $\mathcal{T}$? Moreover, U should be able to simulate all such members $\mathcal{T}$ of some nontrivial class S. U is a tile set that can do one thing and nothing else: simulate tile assembly systems from the class S. This question about U is inspired by the factor simulation question in CA [2].

Many algorithmic tile assembly systems use cooperative self-assembly to simulate Turing machines in a "zig-zag" fashion, as do a number of experimentally implemented systems. Can the negative result of [10] be extended to show 2D temperature 1 abstract Tile Assembly Model systems do not simulate zig-zag tile assembly systems?

There are a number of future research directions for the two-handed, or hierarchical, self-assembly model. One open question [3, 8] asks whether or not temperature τ two-handed systems can simulate temperature $\tau - 1$ two-handed systems. Another direction involves finding which aspects of the model (e.g., mismatches, excess binding strength, geometric blocking) are required for intrinsic universality at a given temperature, to better understand the intricacies of this very powerful, but natural, model.

Of course, there are many other ways to compare the power of self-assembly models: shape and pattern building, tile complexity, time complexity, determinism versus nondeterminism, and randomized (coin-flipping) algorithms in self-assembly. It remains as important future work to find relationships between these notions on the one hand and intrinsic universality and simulation on the other hand. Can ideas from intrinsic universality be used to answer questions about these notions?

Cross-References

- ▶ Active Self-Assembly and Molecular Robotics With Nubots
- ▶ Combinatorial Optimization and Verification in Self-Assembly
- ▶ Patterned Self-Assembly Tile Set Synthesis
- ▶ Randomized Self-Assembly
- ▶ Robustness in Self-Assembly
- ▶ Self-Assembly at Temperature 1
- ▶ Self-Assembly of Fractals

▶ Self-Assembly of Squares and Scaled Shapes
▶ Self-Assembly with General Shaped Tiles
▶ Temperature Programming in Self-Assembly

Acknowledgments A warm thanks to all of my coauthors on this topic. The author is supported by NSF grants 0832824, 1317694, CCF-1219274, and CCF-1162589.

Recommended Reading

1. Cook M, Fu Y, Schweller RT (2011) Temperature 1 self-assembly: deterministic assembly in 3D and probabilistic assembly in 2D. In: SODA 2011: Proceedings of the 22nd annual ACM-SIAM symposium on discrete algorithms, San Francisco. SIAM, pp 570–589
2. Delorme M, Mazoyer J, Ollinger N, Theyssier G (2011) Bulking II: classifications of cellular automata. Theor Comput Sci 412(30):3881–3905. doi: 10.1016/j.tcs.2011.02.024
3. Demaine ED, Patitz MJ, Rogers TA, Schweller RT, Summers SM, Woods D (2013) The two-handed tile assembly model is not intrinsically universal. In: ICALP: Proceedings of the 40th international colloquium on automata, languages and programming, Part 1, Riga. LNCS, vol 7965. Springer, pp 400–412. arxiv preprint arXiv:1306.6710 [cs.CG]
4. Demaine ED, Demaine ML, Fekete SP, Patitz MJ, Schweller RT, Winslow A, Woods D (2014) One tile to rule them all: simulating any tile assembly system with a single universal tile. In: ICALP: Proceedings of the 41st international colloquium on automata, languages, and programming, Copenhagen. LNCS, vol 8572. Springer, pp 368–379. arxiv preprint arXiv:1212.4756 [cs.DS]
5. Doty D, Lutz JH, Patitz MJ, Schweller RT, Summers SM, Woods D (2012) The tile assembly model is intrinsically universal. In: FOCS: Proceedings of the 53rd annual IEEE symposium on foundations of computer science, New Brunswick, pp 439–446. doi: 10.1109/FOCS.2012.76
6. Fochtman T, Hendricks J, Padilla JE, Patitz MJ, Rogers TA (2014) Signal transmission across tile assemblies: 3D static tiles simulate active self-assembly by 2D signal-passing tiles. Nat Comput 14(2):251–264
7. Gilbert O, Hendricks J, Patitz MJ, Rogers TA (2015) Computing in continuous space with self-assembling polygonal tiles. Tech. rep., arxiv preprint arXiv:1503.00327 [cs.CG]
8. Hendricks J, Patitz MJ, Rogers TA (2015) The simulation powers and limitations of higher temperature hierarchical self-assembly systems. In: MCU: Proceedings of the 7th international conference on machines, computations and universality, North Cyprus, to appear. Tech. rep., arXiv arXiv:1503.04502
9. Jonoska N, Karpenko D (2014) Active tile self-assembly, part 1: universality at temperature 1. Int J Found Comput Sci 25:141–163. doi:10.1142/S0129054114500087
10. Meunier PE, Patitz MJ, Summers SM, Theyssier G, Winslow A, Woods D (2014) Intrinsic universality in tile self-assembly requires cooperation. In: SODA: Proceedings of the ACM-SIAM symposium on discrete algorithms, Portland, pp 752–771. arxiv preprint arXiv:1304.1679 [cs.CC]
11. Winfree E (1998) Algorithmic self-assembly of DNA. PhD thesis, California Institute of Technology
12. Woods D (2015) Intrinsic universality and the computational power of self-assembly. Philos Trans R Soc A Math Phys Eng Sci 373(2046):20140214
13. Woods D, Chen HL, Goodfriend S, Dabby N, Winfree E, Yin P (2013) Active self-assembly of algorithmic shapes and patterns in polylogarithmic time. In: ITCS: Proceedings of the 4th conference on innovations in theoretical computer science, Berkeley. ACM, pp 353–354. arxiv preprint arXiv:1301.2626 [cs.DS]

J

Jamming-Resistant MAC Protocols for Wireless Networks

Andréa W. Richa[1] and Christian Scheideler[2]
[1]School of Computing, Informatics, and Decision Systems Engineering, Ira A. Fulton Schools of Engineering, Arizona State University, Tempe, AZ, USA
[2]Department of Computer Science, University of Paderborn, Paderborn, Germany

Keywords

Adversarial models; Competitive analysis; Jamming; MAC protocol; Wireless communication

Years and Authors of Summarized Original Work

2008; Awerbuch, Richa, Scheideler
2010; Richa, Scheideler, Schmid, Zhang
2011; Richa, Scheideler, Schmid, Zhang
2012; Richa, Scheideler, Schmid, Zhang
2014; Ogierman, Richa, Scheideler, Schmid, Zhang

Motivation

The problem of coordinating the access to a shared medium is a central challenge in wireless networks. In order to solve this problem, a proper medium access control (MAC) protocol is needed. Ideally, such a protocol should not only be able to use the wireless medium as effectively as possible, but it should also be robust against a wide range of interference problems including jamming attacks. Interference problems from outside sources are usually ignored in theory but in practice it is important to take these into account, particularly because the ISM frequency band, which is the standard band used for wireless communication, is one of the most dirty frequency bands as it is affected by many devices like microwaves.

Problem Definition

We model inference from outside sources with the help of an adversary. In the most general model that we have published so far [9], our adversarial model is based on the most widely used model to capture interference problems, which is known as the SINR (signal-to-interference-and-noise ratio) model. In the SINR model, a message sent by node u is correctly received by node v if and only if $P_v(u)/(\mathcal{N} + \sum_{w \in S} P_v(w)) \geq \beta$ where $P_x(y)$ is the received power at node x of the signal transmitted by node y, $\mathcal{N}$ is the background noise, and S is the set of nodes $w \neq u$ that are transmitting at the same time as u. The threshold $\beta > 1$ depends on the desired rate, the modulation scheme, etc. When using the standard model for signal propagation, then this expression results in $(P(u)/d(u,v)^{\alpha})/(\mathcal{N} + \sum_{w \in S}$

M.-Y. Kao (ed.), *Encyclopedia of Algorithms*,
DOI 10.1007/978-1-4939-2864-4

$P(w)/d(w,v)^\alpha) \geq \beta$ where $P(x)$ is the strength of the signal transmitted by x, $d(x,y)$ is the Euclidean distance between x and y, and α is the path-loss exponent. We assume that all nodes transmit with some fixed signal strength P and that $\alpha > 2 + \epsilon$ for some constant $\epsilon > 0$, which is usually the case in an outdoors environment.

In most theory papers on MAC protocols, the background noise $\mathcal{N}$ is either ignored (i.e., $\mathcal{N} = 0$) or assumed to behave like a Gaussian variable. This, however, is an oversimplification of the real world. There are many sources of interference producing a non-Gaussian noise such as electrical devices, temporary obstacles, coexisting networks, or jamming attacks. In order to capture a very broad range of noise phenomena, we model the background noise $\mathcal{N}$ (due to jamming or to environmental noise) with the aid of an adversary $\mathcal{ADV}$ that has a fixed energy budget within a certain time frame for each node v. More precisely, in our case, a message transmitted by a node u will be successfully received by node v if and only if

$$\frac{P/d(u,v)^\alpha}{\mathcal{ADV}(v) + \sum_{w \in S} P/d(w,v)^\alpha} \geq \beta, \quad (1)$$

where $\mathcal{ADV}(v)$ is the current noise level created by the adversary at node v. The goal is to design a MAC protocol that allows the nodes to successfully transmit messages under this model as long as this is in principle possible.

For the formal description and analysis, we assume a synchronized setting where time proceeds in synchronized time steps called *rounds*. In each round, a node u may either transmit a message or sense the channel, but it cannot do both. A node which is sensing the channel may either (i) sense an *idle* channel, (ii) sense a *busy* channel, or (iii) *receive* a packet. In order to distinguish between an idle and a busy channel, the nodes use a fixed noise threshold ϑ: if the measured signal power exceeds ϑ, the channel is considered busy, otherwise idle. Whether a message is successfully received is determined by the SINR rule described above. To leave some chance for the nodes to communicate, we restrict the adversary to be *(B,T)-bounded*: for each node v and time interval I of length T, a (B,T)-*bounded adversary* has an overall noise budget of $B \cdot T$ that it can use to increase the noise level at node v and that it can distribute among the time steps of I as it likes, depending on the current state of the nodes. This adversarial noise model is very general, since in addition to being adaptive, the adversary is allowed to make independent decisions on which nodes to jam at any point in time (provided that the adversary does not exceed its noise budget over a time window of size T).

Our goal is to design a *symmetric local-control* MAC protocol (i.e., there is no central authority controlling the nodes, and all the nodes are executing the same protocol) that has a constant competitive throughput against any (B,T)-bounded adversary as long as certain conditions (that are as general as possible) are met. In order to define what we mean by "competitive," we need some notation. The *transmission range* of a node v is defined as the disk with center v and radius r with $P/r^\alpha \geq \beta\vartheta$. Given a constant $\epsilon > 0$, a time step is called *potentially busy* at some node v if $\mathcal{ADV}(v) \geq (1-\epsilon)\vartheta$ (i.e., only a little bit of additional interference by the other nodes is needed so that v sees a busy channel). For a not potentially busy time step, it is still possible that a message sent by a node u within v's transmission range is successfully received by v. Therefore, as long as the adversary is forced to offer not potentially busy time steps due to its limited budget and every node has a least one other node in its transmission range, it is in principle possible for the nodes to successfully transmit messages. To investigate that formally, we use the following notation. For any time frame F and node v let $f_v(F)$ be the number of time steps in F that are not potentially busy at v and let $s_v(F)$ be the number of time steps in which v successfully receives a message. We call a protocol *c-competitive* for some time frame F if $\sum_{v \in V} s_v(F) \geq c \sum_{v \in V} f_v(F)$. An adversary is *uniform* if at any time step, $\mathcal{ADV}(v) = \mathcal{ADV}(w)$ for all nodes $v,w \in V$, which implies that $f_v(F) = f_w(F)$ for all nodes.

Key Results

We presented a MAC protocol called SADE which can achieve a c-competitive throughput where c only depends on ϵ and the path loss exponent α but not on the size of the network or other network parameters [9]. The intuition behind SADE is simple: each node v maintains a parameter p_v which specifies v's probability of accessing the channel at a given moment of time. That is, in each round, each node u decides to broadcast a message with probability p_v. (This is similar to classical random backoff mechanisms where the next transmission time t is chosen uniformly at random from an interval of size $1/p_v$.) The nodes adapt their p_v values over time in a multiplicative-increase multiplicative-decrease manner, i.e., the value is lowered in times when the channel is utilized (more specifically, we decrease p_v whenever a successful transmission occurs) or increased during times when the channel is idling. However, p_v will never exceed $\hat{p}$, for some sufficiently small constant $\hat{p} > 0$.

In addition to the probability value p_v, each node v maintains a time window estimate T_v and a counter c_v for T_v. The variable T_v is used to estimate the adversary's time window T: a good estimation of T can help the nodes recover from a situation where they experience high interference in the network. In times of high interference, T_v will be increased and the sending probability p_v will be decreased. Now we are ready to describe SADE in full detail.

Initially, every node v sets $T_v := 1$, $c_v := 1$, and $p_v := \hat{p}$. In order to distinguish between idle and busy rounds, each node uses a fixed noise threshold of ϑ.

The SADE protocol works in synchronized rounds. In every round, each node v decides with probability p_v to send a message. If it decides not to send a message, it checks the following two conditions:

- If v successfully receives a message, then $p_v := (1+\gamma)^{-1} p_v$.

(continued)

- If v senses an idle channel (i.e., the total noise created by transmissions of other nodes and the adversary is less than ϑ), then $p_v := \min\{(1+\gamma)p_v, \hat{p}\}, T_v := \max\{1, T_v - 1\}$.

Afterward, v sets $c_v := c_v + 1$. If $c_v > T_v$ then it does the following – v sets $c_v := 1$ – and if there was no idle step among the past T_v rounds, then $p_v := (1+\gamma)^{-1} p_v$ and $T_v := T_v + 2$.

Given that $\gamma \in O(1/(\log T + \log\log n))$, one can show the following theorem, where n is the number of nodes and $N = \max\{n, T\}$.

Theorem 1 *When running* SADE *for at least* $\Omega((T \log N)/\epsilon + (\log N)^4/(\gamma\epsilon)^2)$ *time steps,* SADE *has a* $2^{-O((1/\epsilon)^{2/(\alpha-2)})}$*-competitive throughput for any* $((1-\epsilon)\vartheta, T)$*-bounded adversary as long as (a) the adversary is uniform and the transmission range of every node contains at least one node or (b) there are at least* $2/\epsilon$ *nodes within the transmission range of every node.*

SADE is an adaption of the MAC protocol described in [6] for Unit Disk Graphs that works in more realistic network scenarios considering physical interference. Variants of SADE have also been shown to be successful in other scenarios:

In [7] a variant called ANTIJAM is presented for a simpler wireless model but a more severe adversary called *reactive* adversary, which is an adversary that can base the jamming decision on the actions of the nodes in the current time step and not just the initial state of the system at the current time step. However, the adversary can only distinguish between the cases that at least one node is transmitting or no node is transmitting, i.e., it cannot determine whether a transmitted message is successfully received.

In [8] another variant called COMAC is presented for a simpler wireless model that can handle coexisting networks. Even if these networks cannot exchange any information and the number of these networks is unknown, the protocol is shown to be competitive.

All of these results trace back to a first result in [1] for a very simple wireless model and the case of a single-hop wireless network.

Applications

Practical applications of our results are MAC protocols that are much more robust to outside interference and jamming than the existing protocols like 802.11. In fact, it is known that a much weaker jammer than the ones considered by us already suffices to dramatically reduce the throughput of the standard 802.11 MAC protocol [2].

Open Problems

So far, we have not considered the case of power control and multiple communication channels. Multiple communication channels have been covered in several other works (e.g., [3–5]) but under an adversary that is not as powerful as our adversary. Also, several of our bounds are not tight yet, so it remains to determine tight upper and lower bounds on the competitiveness of MAC protocols within our models.

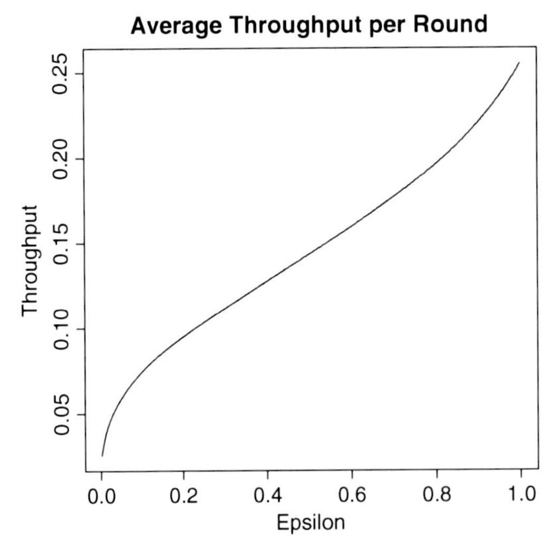

Jamming-Resistant MAC Protocols for Wireless Networks, Fig. 1 The throughput with respect to varying ϵ

Experimental Results

We conducted various simulations to study the robustness of SADE. When varying ϵ, we found out that the worst-case bound of Theorem 1 may be too pessimistic in many scenarios, and the throughput depends to a lesser extent on the constant ϵ. To be more specific, our results suggest that the throughput depends only polynomially on ϵ (cf. the left-most image of Fig. 1), so more work is needed here.

Recommended Reading

1. Awerbuch B, Richa A, Scheideler C (2008) A jamming-resistant MAC protocol for single-hop wireless networks. In: Proc. of the 27th ACM Symp. on Principles of Distributed Computing (PODC), pp. 45–54, Toronto, Canada, 2008.
2. Bayraktaroglu E, King C, Liu X, Noubir G, Rajaraman R, Thapa B (2008) On the performance of IEEE 802.11 under jamming. In. Proc. of the 27th IEEE Conf. on Computer Communications (INFOCOM), pp. 1265–1273, Phoenix, AZ, USA, 2008.
3. Dolev S, Gilbert S, Guerraoui R, Newport C (2007) Gossiping in a multi-channel radio network: an oblivious approach to coping with malicious interference. In: Proc. of the 21st Intl. Symp. on Distributed Computing (DISC), pp. 208–222, 2007.
4. Dolev S, Gilbert S, Guerraoui R, Newport C (2008) Secure communication over radio channels. In: Proc. of the 27th ACM Symp. on Principles of Distributed Computing (PODC), pp. 105–114, Toronto, Canada, 2008.
5. Gilbert S, Guerraoui R, Kowalski D, Newport C (2009) Interference-resilient information exchange. In: Proc. of the 28th IEEE Conf. on Computer Communication (INFOCOM), pp. 2249–2257, Rio de Janeiro, Brazil.
6. Richa A, Scheideler C, Schmid S, Zhang J (2010) A jamming-resistant MAC protocol for multi-hop wireless networks. In: Proc. of the 24th Intl. Symp. on Distributed Computing (DISC), pp. 179–193, Cambridge, MA, USA, 2010.
7. Richa A, Scheideler C, Schmid S, Zhang J (2011) Competitive and fair medium access despite reactive jamming. In: Proc. of the 31st Intl. Conf. on Distributed Computing Systems (ICDCS), pp. 507–516, Minneapolis, MN, USA.
8. Richa A, Scheideler C, Schmid S, Zhang J (2012) Competitive and fair throughput for co-existing networks under adversarial interference. In: Proc. of the 31st ACM Symp. on Principles of Distributed Computing (PODC), pp. 291–300, Madeira, Portugal, 2012.
9. Richa A, Scheideler C, Schmid S, Zhang J (2014) Competitive MAC under adversarial SINR. In: Proc. of the 33rd IEEE Conf. on Computer Communication (INFOCOM), pp. 2751–2759, Toronto, Canada, 2014.

K

k-Best Enumeration

David Eppstein
Donald Bren School of Information and Computer Sciences, Computer Science Department, University of California, Irvine, CA, USA

Keywords

k minimum-weight matchings; k shortest paths; k shortest simple paths; k smallest spanning trees

Years and Authors of Summarized Original Work

1959; Hoffman, Pavley
1963; Clarke, Krikorian, Rausen
1968; Murty
1971; Yen
1972; Lawler
1977; Gabow
1982; Katoh, Ibaraki, Mine
1985; Hamacher, Queyranne
1987; Chegireddy, Hamacher
1993; Frederickson
1997; Eppstein, Galil, Italiano, Nissenzweig
1998; Eppstein
2007; Hershberger, Maxel, Suri
2013; Chen, Kanj, Meng, Xia, Zhang

Problem Definition

K-best enumeration problems are a type of combinatorial enumeration in which, rather than seeking a single best solution, the goal is to find a set of k solutions (for a given parameter value k) that are better than all other possible solutions. Many of these problems involve finding structures in a graph that can be represented by subsets of the graph's edges. In particular, the k shortest paths between two vertices s and t in a weighted network are a set of k distinct paths that are shorter than all other paths, and other problems such as the k smallest spanning trees of a graph or the k minimum weight matchings in a graph are defined in the same way.

Key Results

One of the earliest works in the area of k-best optimization was by Hoffman and Pavley [10] formulating the k-shortest path problem; their paper cites unpublished work by Bock, Kantner, and Hayes on the same problem. Later research by Lawler [12], Gabow [7], and Hamacher and Queyranne [8] described a general approach to k-best optimization, suitable for many of these problems, involving the hierarchical partitioning of the solution space into subproblems. One way

This material is based upon work supported by the National Science Foundation under Grant CCF-1228639 and by the Office of Naval Research under Grant No. N00014-08-1-1015.

M.-Y. Kao (ed.), *Encyclopedia of Algorithms*,
DOI 10.1007/978-1-4939-2864-4

of doing this is to view the optimal solution to a problem as a sequence of edges, and define one subproblem for each edge, consisting of the solutions that first deviate from the optimal solution at that edge. Continuing this subdivision recursively leads to a tree of subproblems, each having a worse solution value than its parent, such that each possible solution is the best solution for exactly one subproblem in the hierarchy. A best-first search of this tree allows the k-best solutions to be found. Alternatively, if both the first and second best solutions can be found, and differ from each other at an edge e, then one can form only two subproblems, one consisting of the solutions that include e and one consisting of the solutions that exclude e. Again, the subdivision continues recursively; each solution (except the global optimum) is the second-best solution for exactly one subproblem, allowing a best-first tree search to find the k-best solutions. An algorithm of Frederickson [6] solves this tree search problem in a number of steps proportional to k times the degree of the tree; each step involves finding the solution (or second-best solution) to a single subproblem.

Probably the most important and heavily studied of the k-best optimization problems is the problem of finding k shortest paths, first formulated by Hoffman and Pavley [10]. In the most basic version of this problem, the paths are allowed to have repeated vertices or edges (unless the input is acyclic, in which case repetitions are impossible). An algorithm of Eppstein [4] solves this version of the problem in the optimal time bound $O(m + n \log n + k)$, where m and n are the numbers of edges and vertices in the given graph; that is, after a preprocessing stage that is dominated by the time to use Dijkstra's algorithm to find a single shortest-path tree, the algorithm takes constant time per path. Eppstein's algorithm follows Hoffman and Pavley in representing a path by its sequence of *deviations*, the edges that do not belong to a tree T of shortest paths to the destination node. The deviation edges that can be reached by a path in T from a given node v are represented as a binary heap (ordered by how much additional length the deviation would cause) and these heaps are used to define a partition of the solution space into subproblems, consisting of the paths that follow a certain sequence of deviations followed by one more deviation from a specified heap. The best path in a subproblem is the one that chooses the deviation at the root of its heap, and the remaining paths can be partitioned into three sub-subproblems, two for the children of the root and one for the paths that use the root deviation but then continue with additional deviations. In this way, Eppstein constructs a tree of subproblems to which Frederickson's tree-searching method can be applied.

In a graph with cycles (or in an undirected graph which, when its edges are converted to directed edges, has many cycles), it is generally preferable to list only the k shortest simple (or loopless) paths, not allowing repetitions within a path. This variation of the k shortest paths problem was formulated by Clarke et al. [3]. Yen's algorithm [14] still remains the one with the best asymptotic time performance, $O(kn(m + n \log n))$; it is based on best-solution partitioning using Dijkstra's algorithm to find the best solution in each subproblem. A more recent algorithm of Hershberger et al. [9] is often faster, but is based on a heuristic that can sometimes fail, causing it to become no faster than Yen's algorithm. In the undirected case, it is possible to find the k shortest simple paths in time $O(k(m + n \log n))$ [11].

Gabow [7] introduced both the problem of finding the k minimum-weight spanning trees of an edge-weighted graph, and the technique of finding a binary hierarchical subdivision of the space of solutions, which he used to solve the problem. In any graph, the best and second-best spanning trees differ only by one edge swap (the removal of one edge from a tree and its replacement by a different edge that reconnects the two subtrees formed by the removal), a property that simplifies the search for a second-best tree as needed for Gabow's partitioning technique. The fastest known algorithms for the k-best spanning trees problem are based on Gabow's partitioning technique, together with dynamic graph data structures that keep track of the best swap in a network as that network undergoes a sequence of edge insertion and deletion operations. To use this technique, one initializes a fully-persistent best-swap data structure (one in which each update creates a new version of the structure without modifying the existing versions, and in which updates may be applied to any version) and associates its initial version with the root of the subproblem tree. Then, whenever an algorithm for selecting the k best nodes of the subproblem tree generates a new node (a subproblem formed by including or excluding an edge from the allowed solutions) the parent node's version of the data structure is updated (by either increasing or decreasing the weight of the edge to force it to be included or excluded in all solutions) and the updated version of the data structure is associated with the child node. In this way, the data structure can be used to quickly find the second-best solution for each of the subproblems explored by the algorithm. Based on this method, the k-best spanning trees of a graph with n vertices and m edges can be found (in an implicit representation based on sequences of swaps rather than explicitly listing all edges in each tree) in time $O(\mathrm{MST}(m,n) + k \min(n,k)^{1/2})$ where $\mathrm{MST}(m,n)$ denotes the time for finding a single minimum spanning tree (linear time, if randomized algorithms are considered) [5].

After paths and spanning trees, probably the next most commonly studied k-best enumeration problem concerns matchings. The problem of finding the k minimum-weight perfect matchings in an edge-weighted graph was introduced by Murty [13]. A later algorithm by Chegireddy and Hamacher [1] solves the problem in time $O(kn^3)$ (where n is the number of vertices in the graph) using the technique of building a binary partition of the solution space. Other problems whose k-best solutions have been studied include the Chinese postman problem, the traveling salesman problem, spanning arborescences in a directed network, the matroid intersection problem, binary search trees and Huffman coding, chess strategies, integer flows, and network cuts.

For many NP-hard optimization problems, where even finding a single best solution is difficult, an approach that has proven very successful is *parameterized complexity*, in which one finds an integer parameter describing the input instance or its solution that is often much smaller than the input size, and designs algorithms whose running time is a fixed polynomial of the input size multiplied by a non-polynomial function of the parameter value. Chen et al. [2] extend this paradigm to k-best problems, showing that, for instance, many NP-hard k-best problems can be solved in polynomial time per solution for graphs of bounded treewidth.

Applications

The k shortest path problem has many applications. The most obvious of these are in the generation of alternative routes, in problems involving communication networks, transportation networks, or building evacuation planning. In bioinformatics, it has been applied to shortest-path formulations of dynamic programming algorithms for biological sequence alignment and also applied in the reconstruction of metabolic pathways, and reconstruction of gene regulation networks. The problem has been used frequently in natural language and speech processing, where a path in a network may represent a hypothesis for the correct decoding of an utterance or piece of writing. Other applications include motion tracking, genealogy, the design of power, communications, and transportation networks, timing analysis of circuits, and task scheduling.

The problem of finding the k-best spanning trees has been applied to point process intensity estimation, the analysis of metabolic pathways, image segmentation and classification, the reconstruction of pedigrees from genetic data, the parsing of natural-language text, and the analysis of electronic circuits.

Cross-References

▸ Minimum Spanning Trees
▸ Single-Source Shortest Paths

Recommended Reading

1. Chegireddy CR, Hamacher HW (1987) Algorithms for finding K-best perfect matchings. Discret Appl Math 18(2):155–165. doi:10.1016/0166-218X(87)90017-5

K

2. Chen J, Kanj IA, Meng J, Xia G, Zhang F (2013) Parameterized top-K algorithms. Theor Comput Sci 470:105–119. doi:10.1016/j.tcs.2012.10.052
3. Clarke S, Krikorian A, Rausen J (1963) Computing the N best loopless paths in a network. J SIAM 11:1096–1102
4. Eppstein D (1998) Finding the k shortest paths. SIAM J Comput 28(2):652–673. doi:10.1137/S0097539795290477
5. Eppstein D, Galil Z, Italiano GF, Nissenzweig A (1997) Sparsification–a technique for speeding up dynamic graph algorithms. J ACM 44(5):669–696. doi:10.1145/265910.265914
6. Frederickson GN (1993) An optimal algorithm for selection in a min-heap. Inf Comput 104(2):197–214. doi:10.1006/inco.1993.1030
7. Gabow HN (1977) Two algorithms for generating weighted spanning trees in order. SIAM J Comput 6(1):139–150. doi:10.1137/0206011
8. Hamacher HW, Queyranne M (1985) K best solutions to combinatorial optimization problems. Ann Oper Res 4(1–4):123–143. doi:10.1007/BF02022039
9. Hershberger J, Maxel M, Suri S (2007) Finding the k shortest simple paths: a new algorithm and its implementation. ACM Trans Algorithms 3(4):A45. doi:10.1145/1290672.1290682
10. Hoffman W, Pavley R (1959) A method for the solution of the Nth best path problem. J ACM 6(4):506–514. doi:10.1145/320998.321004
11. Katoh N, Ibaraki T, Mine H (1982) An efficient algorithm for K shortest simple paths. Networks 12(4):411–427. doi:10.1002/net.3230120406
12. Lawler EL (1972) A procedure for computing the K best solutions to discrete optimization problems and its application to the shortest path problem. Manag Sci 18:401–405. doi:10.1287/mnsc.18.7.401
13. Murty KG (1968) Letter to the editor–an algorithm for ranking all the assignments in order of increasing cost. Oper Res 16(3):682–687. doi:10.1287/opre.16.3.682
14. Yen JY (1971) Finding the K shortest loopless paths in a network. Manag Sci 17:712–716. http://www.jstor.org/stable/2629312

Kernelization, Bidimensionality and Kernels

Daniel Lokshtanov
Department of Informatics, University of Bergen, Bergen, Norway

Keywords

Bidimensionality; Graph algorithms; Kernelization; Parameterized complexity; Polynomial time pre-processing

Years and Authors of Summarized Original Work

2010; Fomin, Lokshtanov, Saurabh, Thilikos

Problem Definition

The theory of bidimensionality simultaneously provides subexponential time parameterized algorithms and efficient approximation schemes for a wide range of optimization problems on planar graphs and, more generally, on classes of graphs excluding a fixed graph H as a minor. It turns out that bidimensionality also provides *linear kernels* for a multitude of problems on these classes of graphs. The results stated here unify and generalize a number of kernelization results for problems on planar graphs and graphs of bounded genus; see [2] for a more thorough discussion.

Kernelization

Kernelization is a mathematical framework for the study of polynomial time preprocessing of instances of computationally hard problems. Let $\mathcal{G}$ be the set of all graphs. A *parameterized graph problem* is a subset Π of $\mathcal{G} \times \mathbb{N}$. An *instance* is a pair $(G,k) \in \mathcal{G} \times \mathbb{N}$. The instance (G,k) is a "yes"-instance of Π if $(G,k) \in \Pi$ and a "no"-instance otherwise. A *strict kernel with ck vertices* for a parameterized graph problem Π and constant $c > 0$ is an algorithm $\mathcal{A}$ with the following properties:

- $\mathcal{A}$ takes as input an instance (G,k), runs in polynomial time, and outputs another instance (G',k').
- (G',k') is a "yes"-instance of Π if and only if (G,k) is.
- $|V(G')| \leq c \cdot k$ and $k' \leq k$.

A *linear kernel* for a parameterized graph problem is a strict kernel with ck vertices for some constant c. We remark that our definition of a linear kernel is somewhat simplified compared to the classic definition [8], but that it is essentially equivalent. For a discussion of the definition of a kernel, we refer to the textbook of Cygan et al. [4].

Graph Classes

Bidimensionality theory primarily concerns itself with graph problems where the input graph is restricted to be in a specific *graph class*. A *graph class* $\mathcal{C}$ is simply a subset of the set $\mathcal{G}$ of all graphs. As an example, the set of all planar graphs is a graph class. Another example of a graph class is the set of all *apex* graphs. Here a graph H is *apex* if H contains a vertex v such that deleting v from H leaves a planar graph. Notice that every planar graph is apex.

A graph H is a *minor* of a graph G if H can be obtained from G by deleting vertices, deleting edges, or contracting edges. Here *contracting* the edge $\{u, v\}$ in G means identifying the vertices u and v and removing all self-loops and double edges. If H can be obtained from G just by contracting edges, then H is a *contraction* of G.

A graph class $\mathcal{C}$ is *minor closed* if every minor of a graph in $\mathcal{C}$ is also in $\mathcal{C}$. A graph class $\mathcal{C}$ is *minor-free* if $\mathcal{C}$ is minor closed and there exists a graph $H \notin \mathcal{C}$. A graph class $\mathcal{C}$ is *apex-minor-free* if $\mathcal{C}$ is minor closed and there exists an apex graph $H \notin \mathcal{C}$. Notice that $H \notin \mathcal{C}$ for a minor closed class $\mathcal{C}$ implies that H cannot be a minor of any graph $G \in \mathcal{C}$.

CMSO Logic

CMSO logic stands for *Counting Monadic Second Order* logic, a formal language to describe properties of graphs. A *CMSO-sentence* is a formula ψ with variables for single vertices, vertex sets, single edges and edge sets, existential and universal quantifiers ($\exists$ and $\forall$), logical connectives $\vee$, $\wedge$ and $\neg$, as well as the following operators:

- $v \in S$, where v is a vertex variable and S is a vertex set variable. The operator returns true if the vertex v is in the vertex set S. Similarly, CMSO has an operator $e \in X$ where e is an edge variable and X is an edge set variable.
- $v_1 = v_2$, where v_1 and v_2 are vertex variables. The operator returns true if v_1 and v_2 are the same vertex of G. There is also an operator $e_1 = e_2$ to check equality of two edge variables e_1 and e_2.
- $\mathbf{adj}(v_1, v_2)$ is defined for vertex variables v_1 and v_2 and returns true if v_1 and v_2 are adjacent in G.
- $\mathbf{inc}(v, e)$ is defined for a vertex variable v and edge variable e. $\mathbf{inc}(v, e)$ returns true if the edge e is incident to the vertex v in G, in other words, if v is one of the two endpoints of e.
- $\mathbf{card}_{p,q}(S)$ is defined for every pair of integers p, q, and vertex or edge set variable S. $\mathbf{card}_{p,q}(S)$ returns true if $|S| \equiv q \mod p$. For an example, $\mathbf{card}_{2,1}(S)$ returns true if $|S|$ is odd.

When we quantify a variable, we need to specify whether it is a vertex variable, edge variable, vertex set variable, or edge set variable. To specify that an existentially quantified variable x is a vertex variable we will write $\exists x \in V(G)$. We will use $\forall e \in E(G)$ to universally quantify edge variables and $\exists X \subseteq V(G)$ to existentially quantify vertex set variables. We will always use lower case letters for vertex and edge variables and upper case letters for vertex set and edge set variables.

A graph G on which the formula ψ is true is said to *model* ψ. The notation $G \models \psi$ means that G models ψ. As an example, consider the formula

$$\psi_1 = \forall v \in V(G)\ \forall x \in V(G)\ \forall y \in V(G)\ \forall z \in V(G) : \\ (x = y) \vee (x = z) \vee (y = z) \vee \neg\mathbf{adj}(v, x) \vee \neg\mathbf{adj}(v, y) \vee \neg\mathbf{adj}(v, z)$$

The formula ψ_1 states that for every four (not necessarily distinct) vertices v, x, y, and z, if x, y, and z are distinct, then v is not adjacent to all of $\{x, y, z\}$. In other words, a graph G models ϕ_1 if and only if the degree of every vertex G is at most 2. CMSO can be used to express many graph properties, such as G having a Hamiltonian cycle, G being 3-colorable, or G being planar.

In CMSO, one can also write formulas where one uses *free variables*. These are variables that are used in the formula but never quantified with an $\exists$ or $\forall$ quantifier. As an example, consider the formula

$$\psi_{DS} = \forall u \in V(G)\, \exists v \in V(G) : (v \in S) \wedge (u = v \vee \mathbf{adj}(u, v))$$

The variable S is a free variable in ψ_{DS} because it is used in the formula, but is never quantified. It does not make sense to ask whether a graph G models ψ_{DS} because when we ask whether the vertex v is in S, the set S is not well defined. However, if the set $S \subseteq V(G)$ is provided together with the graph G, we can evaluate the formula ψ_{DS}. ψ_{DS} will be true for a graph G and set $S \subseteq V(G)$ if, for every vertex $u \in V(G)$, there exists a vertex $v \in V(G)$ such that v is in S and either $u = v$ or u and v are neighbors in G. In other words, the pair (G, S) models ψ_{DS} (written $(G, S) \models \psi_{DS}$) if and only if S is a dominating set in G (i.e., every vertex not in S has a neighbor in S).

CMSO-Optimization Problems

We are now in position to define the parameterized problems for which we will obtain kernelization results. For every CMSO formula ψ with a single free vertex set variable S, we define the following two problems:

ψ-CMSO-Min (Max):

INPUT: *Graph G and integer k.*
QUESTION: *Does there exist a vertex set $S \subseteq V(G)$ such that $(G, S) \models \psi$ and $|S| \leq k$ ($|S| \geq k$ for Max).*

Formally, ψ-CMSO-MIN (MAX) is a parameterized graph problem where the "yes" instances are exactly the pairs (G, k) such that there exists a vertex set S of size at most k (at least k) and $(G, S) \models \psi$. We will use the term *CMSO-optimization problems* to refer to ψ-CMSO-MIN (MAX) for some CMSO formula ψ.

Many well-studied and not so well-studied graph problems are CMSO-optimization problems. Examples include VERTEX COVER, DOMINATING SET, CYCLE PACKING, and the list goes on and on (see [2]). We encourage the interested reader to attempt to formulate the problems mentioned above as CMSO-optimization problems. We will be discussing CMSO-optimization problems *on planar graphs* and on minor-free classes of graphs.

Our results are for problems where the input graph is promised to belong to a certain graph class $\mathcal{C}$. We formalize this by encoding membership in $\mathcal{C}$ in the formula ψ. For an example, ψ_{DS}-CMSO-MIN is the well-studied DOMINATING SET problem. If we want to restrict the problem to planar graphs, we can make a new CMSO logic formula ψ_{planar} such that $G \models \psi_{\text{planar}}$ if and only if G is planar. We can now make a new formula

$$\psi'_{DS} = \psi_{DS} \wedge \psi_{\text{planar}}$$

and consider the problem ψ'_{DS}-CMSO-MIN. Here (G, k) is a "yes" instance if G has a dominating set S of size at most k *and* G is planar. Thus, this problem also forces us to check planarity of G, but this is polynomial time solvable and therefore not an issue with respect to kernelization. In a similar manner, one can restrict any CMSO-optimization problem to a graph class $\mathcal{C}$, as long as there exists a CMSO formula $\psi_{\mathcal{C}}$ such that $G \models \psi_{\mathcal{C}}$ if and only if $G \in \mathcal{C}$. Luckily, such a formula is known to exist for every minor-free class $\mathcal{C}$. We will say that a parameterized problem Π is a problem *on the graph class* $\mathcal{C}$ if, for every "yes" instance (G, k) of Π, the graph G is in $\mathcal{C}$.

For any CMSO-MIN problem Π, we have that $(G, k) \in \Pi$ implies that $(G, k') \in \Pi$ for all $k' \geq k$. Similarly, for a CMSO-MAX problem Π, we have that $(G, k) \in \Pi$ implies that $(G, k') \in \Pi$ for all $k' \leq k$. Thus, the notion of "optimality" is well defined for CMSO-optimization problems. For the problem $\Pi = \psi$-CMSO-MIN, we define

$$OPT_{\Pi}(G) = \min\{k : (G, k) \in \Pi\}.$$

If no k such that $(G,k) \in \Pi$ exists, $OPT_\Pi(G)$ returns $+\infty$. Similarly, for the problem $\Pi = \psi$-CMSO-MAX,

$$OPT_\Pi(G) = \max\{k \ : \ (G,k) \in \Pi\}.$$

If no k such that $(G,k) \in \Pi$ exists, $OPT_\Pi(G)$ returns $-\infty$. We define $SOL_\Pi(G)$ to be a function that given as input a graph G returns a set S of size $OPT_\Pi(G)$ such that $(G,S) \models \psi$ and returns **null** if no such set S exists.

Bidimensionality

For many problems, it holds that contracting an edge cannot increase the size of the optimal solution. We will say that such problems are contraction closed. Formally, a CMSO-optimization problem Π is *contraction closed* if for any G and $uv \in E(G)$, $OPT_\Pi(G/uv) \leq OPT_\Pi(G)$. If contracting edges, deleting edges, and deleting vertices cannot increase the size of the optimal solution, we say that the problem is *minor closed.*

Informally, a problem is *bidimensional* if it is minor closed and the value of the optimum grows with both dimensions of a grid. In other words, on a $(k \times k)$-grid, the optimum should be approximately quadratic in k. To formally define bidimensional problems, we first need to define the $(k \times k)$-grid $\boxplus_k$, as well as the related graph Γ_k.

For a positive integer k, a $k \times k$ *grid*, denoted by $\boxplus_k$, is a graph with vertex set $\{(x,y) \ : \ x,y \in \{1,\ldots,k\}\}$. Thus, $\boxplus_k$ has exactly k^2 vertices. Two different vertices (x,y) and (x',y') are adjacent if and only if $|x-x'|+|y-y'| = 1$. For an integer $k > 0$, the graph Γ_k is obtained from the grid $\boxplus_k$ by adding in every grid cell the diagonal edge going up and to the right and making the bottom right vertex of the grid adjacent to all border vertices. The graph Γ_9 is shown in Fig. 1.

We are now ready to give the definition of bidimensional problems. A CMSO-optimization problem Π is *contraction-bidimensional* if it is contraction closed, and there exists a constant $c > 0$ such that $OPT_\Pi(\Gamma_k) \geq ck^2$. Similarly, Π is *minor-bidimensional* if it is minor closed, and there exists a constant $c > 0$ such that $OPT_\Pi(\boxplus_k) \geq ck^2$.

As an example, the DOMINATING SET problem is contraction-bidimensional. It is easy to verify that contracting an edge may not increase the size of the smallest dominating set of a graph G and that Γ_k does not have a dominating set of size smaller than $\frac{(k-2)^2}{7}$.

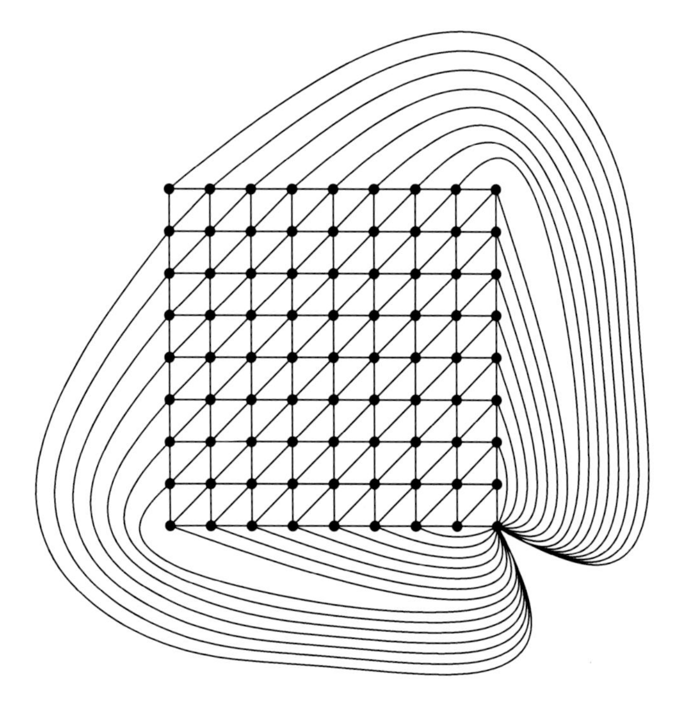

Kernelization, Bidimensionality and Kernels, Fig. 1 The graph Γ_9

Separability

Our kernelization algorithms work by recursively splitting the input instance by small separators. For this to work, the problem has to be somewhat well behaved in the following sense. Whenever a graph is split along a small separator into two independent sub-instances L and R, the size of the optimum solution for the graph $G[L]$ is relatively close to the size of the intersection between L and the optimum solution to the original graph G. We now proceed with a formal definition of what it means for a problem to be well behaved.

For a set $L \subseteq V(G)$, we define $\partial(L)$ to be the set of vertices in L with at least one neighbor outside L. A CMSO-optimization problem Π is *linear separable* if there exists a constant $c \geq 0$ such that for every set $L \subseteq V(G)$, we have

$$|SOL_{\Pi}(G) \cap L| - c \cdot |\partial(L)| \leq OPT_{\Pi}(G[L]) \leq |SOL_{\Pi}(G) \cap L| + c \cdot |\partial(L)|.$$

For a concrete example, we encourage the reader to consider the DOMINATING SET problem and to prove that for DOMINATING SET the inequalities above hold. The crux of the argument is to augment optimal solutions of G and $G[L]$ by adding all vertices in $\partial(L)$ to them.

Key Results

We can now state our main theorem.

Theorem 1 *Let Π be a separable CMSO-optimization problem on the graph class $\mathcal{C}$. Then, if Π is minor-bidimensional and $\mathcal{C}$ is minor-free, or if Π is contraction-bidimensional and $\mathcal{C}$ is apex-minor-free, Π admits a linear kernel.*

The significance of Theorem 1 is that it is, in general, quite easy to formulate graph problems as CMSO-optimization problems and prove that the considered problem is bidimensional and separable. If we are able to do this, Theorem 1 immediately implies that the problem admits a linear kernel on all minor-free graph classes, or on all apex-minor-free graph classes. As an example, the DOMINATING SET problem has been shown to have a linear kernel on planar graphs [1], and the proof of this fact is quite tricky. However, in our examples, we have shown that DOMINATING SET is a CMSO-MIN problem, that it is contraction-bidimensional, and that it is separable. Theorem 1 now implies that DOMINATING SET has a linear kernel not only on planar graphs but on all apex-minor-free classes of graphs! One can go through the motions and use Theorem 1 to give linear kernels for quite a few problems. We refer the reader to [9] for a non-exhaustive list.

We remark that the results stated here are generalizations of results obtained by Bodlaender et al. [2]. Theorem 1 is proved by combining "algebraic reduction rules" (fully developed by Bodlaender et al. [2]) with new graph decomposition theorems (proved in [9]). The definitions here differ slightly from the definitions in the original work [9] and appear here in the way they will appear in the journal version of [9].

Cross-References

▶ Bidimensionality
▶ Data Reduction for Domination in Graphs

Recommended Reading

1. Alber J, Fellows MR, Niedermeier R (2004) Polynomial-time data reduction for dominating set. J ACM 51(3):363–384
2. Bodlaender HL, Fomin FV, Lokshtanov D, Penninkx E, Saurabh S, Thilikos DM (2013) (Meta) Kernelization. CoRR abs/0904.0727. http://arxiv.org/abs/0904.0727
3. Borie RB, Parker RG, Tovey CA (1992) Automatic generation of linear-time algorithms from predicate calculus descriptions of problems on recursively constructed graph families. Algorithmica 7(5&6):555–581
4. Cygan M, Fomin FV, Kowalik Ł, Lokshtanov D, Marx D, Pilipczuk M, Pilipczuk M, Saurabh S (2015, to appear) Parameterized algorithms. Springer, Heidelberg
5. Demaine ED, Hajiaghayi M (2005) Bidimensionality: new connections between FPT algorithms and PTASs. In: Proceedings of the 16th annual ACM-SIAM symposium on discrete algorithms (SODA), Vancouver. SIAM, pp 590–601

6. Demaine ED, Hajiaghayi M (2008) The bidimensionality theory and its algorithmic applications. Comput J 51(3):292–302
7. Demaine ED, Fomin FV, Hajiaghayi M, Thilikos DM (2005) Subexponential parameterized algorithms on graphs of bounded genus and H-minor-free graphs. J ACM 52(6):866–893
8. Downey RG, Fellows MR (2013) Fundamentals of parameterized complexity. Texts in computer science. Springer, London
9. Fomin FV, Lokshtanov D, Saurabh S, Thilikos DM (2010) Bidimensionality and kernels. In: Proceedings of the 20th annual ACM-SIAM symposium on discrete algorithms (SODA), Austin. SIAM, pp 503–510
10. Fomin FV, Lokshtanov D, Raman V, Saurabh S (2011) Bidimensionality and EPTAS. In: Proceedings of the 21st annual ACM-SIAM symposium on discrete algorithms (SODA), San Francisco. SIAM, pp 748–759

Kernelization, Constraint Satisfaction Problems Parameterized above Average

Gregory Gutin
Department of Computer Science, Royal Holloway, University of London, Egham, UK

Keywords

Bikernel; Kernel; MaxCSP; MaxLin; MaxSat

Years and Authors of Summarized Original Work

2011; Alon, Gutin, Kim, Szeider, Yeo

Problem Definition

Let r be an integer, let $V = \{v_1, \ldots, v_n\}$ be a set of variables, each taking values -1 (TRUE) and 1 (FALSE), and let Φ be a set of Boolean functions, each involving at most r variables from V. In the problem MAX-r-CSP, we are given a collection $\mathcal{F}$ of m Boolean functions, each $f \in \mathcal{F}$ being a member of Φ and each with a positive integral weight. Our aim is to find a truth assignment that maximizes the total weight of satisfied functions from $\mathcal{F}$. We will denote the maximum by $\mathrm{sat}(\mathcal{F})$.

Let A be the average weight (over all truth assignments) of satisfied functions. Observe that A is a lower bound for $\mathrm{sat}(\mathcal{F})$. In fact, A is a tight lower bound, whenever the family Φ is closed under replacing each variable by its complement [1]. Thus, it is natural to parameterize MAX-r-CSP as follows (AA stands for *Above Average*).

> MAX-r-CSP-AA
>
> *Instance*: A collection $\mathcal{F}$ of m Boolean functions, each $f \in \mathcal{F}$ being a member of Φ, each with a positive integral weight, and a nonnegative integer k.
> *Parameter*: k.
> *Question*: $\mathrm{sat}(\mathcal{F}) \geq A + k$?

If Φ is the set of clauses with at most r literals, then we get a subproblem of MAX-r-CSP-AA, abbreviated MAX-r-SAT-AA, whose unparameterized version is simply MAX-r-SAT. Assign -1 or 1 to each variable in V randomly and uniformly. Since a clause c of an MAX-r-SAT-AA instance can be satisfied with probability $1 - 2^{r_c}$, where r_c is the number of literals in c, we have $A = \sum_{c \in \mathcal{F}}(1 - 2^{r_c})$. Clearly, A is a tight lower bound.

If Φ is the set S of equations $\prod_{i \in I_j} v_i = b_j$, $j = 1, \ldots, m$, where $v_i, b_j \in \{-1, 1\}$, b_js are constants, $|I_j| \leq r$, then we get a subproblem of MAX-r-CSP-AA, abbreviated MAX-r-LIN2-AA, whose unparameterized version is simply MAX-r-LIN2. Assign -1 or 1 to each variable in V randomly and uniformly. Since each equation of $\mathcal{F}$ can be satisfied with probability $1/2$, we have $A = W/2$, where W is the sum of the weights of equations in $\mathcal{F}$. For an assignment $v = v^0$ of values to the variables, let $\mathrm{sat}(S, v^0)$ denote the total weight of equations of S satisfied by the assignment. The difference $\mathrm{sat}(S, v^0) - W/2$ is called the *excess* of x^0. Let $\mathrm{sat}(S)$ be the maximum of $\mathrm{sat}(S, v^0)$ over all possible assignments v^0.

The following notion was introduced in [1]. Let Π and Π' be parameterized problems. A *bikernel* for Π is a polynomial-time algorithm that maps an instance (I,k) of Π to an instance (I',k') of Π' such that (i) $(I,k) \in \Pi$ if and only if $(I',k') \in \Pi'$ and (ii) $k' \le g(k)$ and $|I'| \le g(k)$ for some function g. The function $g(k)$ is called the *size* of the bikernel. It is known that a decidable problem is fixed-parameter tractable if and only if it admits a bikernel [1]. However, in general a bikernel can have an exponential size, in which case the bikernel may not be useful as a data reduction. A bikernel is called a *polynomial* bikernel if both $f(k)$ and $g(k)$ are polynomials in k.

When $\Pi = \Pi'$ we say that a bikernel for Π is simply a *kernel* of Π. A great deal of research has been devoted to decide whether a problem admits a polynomial kernel.

The following lemma of Alon et al. [1] shows that polynomial bikernels imply polynomial kernels.

Lemma 1 *Let Π, Π' be a pair of decidable parameterized problems such that the nonparameterized version of Π' is in NP and the nonparameterized version of Π is NP-complete. If there is a bikernelization from Π to Π' producing a bikernel of polynomial size, then Π has a polynomial-size kernel.*

Key Results

Following [2], for a Boolean function f of weight $w(f)$ and on $r(f) \le r$ Boolean variables $v_{i_1}, \dots, v_{i_{r(f)}}$, we introduce a polynomial $h_f(v)$, $v = (v_1, \dots, v_n)$ as follows. Let $S_f \subseteq \{-1,1\}^{r(f)}$ denote the set of all satisfying assignments of f. Then

$$h_f(v) = w(f)2^{r-r(f)} \sum_{(a_1,\dots,a_{r(f)})\in S_f} \left[\prod_{j=1}^{r(f)} (1 + v_{i_j} a_j) - 1\right].$$

Let $h(v) = \sum_{f\in\mathcal{F}} h_f(v)$. It is easy to see (cf. [1]) that the value of $h(v)$ at some v^0 is precisely $2^r(U - A)$, where U is the total weight of the functions satisfied by the truth assignment v^0. Thus, the answer to MAX-r-CSP-AA is YES if and only if there is a truth assignment v^0 such that $h(v^0) \ge k2^r$.

Algebraic simplification of $h(v)$ will lead us the following (Fourier expansion of $h(v)$, cf. [7]):

$$h(v) = \sum_{S\in\mathcal{F}} c_S \prod_{i\in S} v_i, \qquad (1)$$

where $\mathcal{F} = \{\emptyset \neq S \subseteq \{1,2,\dots,n\} : c_S \neq 0, |S| \le r\}$. Thus, $|\mathcal{F}| \le n^r$. The sum $\sum_{S\in\mathcal{F}} c_S \prod_{i\in S} v_i$ can be viewed as the excess of an instance of MAX-r-LIN2-AA, and, thus, we can reduce MAX-r-CSP-AA into MAX-r-LIN2-AA in polynomial time (since r is fixed, the algebraic simplification can be done in polynomial time and it does not matter whether the parameter of MAX-r-LIN2-AA is k or $k' = k2^r$). It is proved in [5] that MAX-r-LIN2-AA has a kernel with $O(k^2)$ variables and equations. This kernel is a bikernel from MAX-r-CSP-AA to MAX-r-LIN2-AA. Thus, by Lemma 1, we obtain the following theorem of Alon et al. [1].

Theorem 1 MAX-r-CSP-AA *admits a polynomial-size kernel.*

Applying a reduction from MAX-r-LIN2-AA to MAX-r-SAT-AA in which each monomial in (1) is replaced by 2^{r-1} clauses, Alon et al. [1] obtained the following:

Theorem 2 MAX-r-SAT-AA *admits a kernel with $O(k^2)$ clauses and variables.*

It is possible to improve this theorem with respect to the number of variables in the kernel. The following result was first obtained by Kim and Williams [6] (see also [3]).

Theorem 3 MAX-r-SAT-AA *admits a kernel with $O(k)$ variables.*

Crowston et al. [4] studied the following natural question: How parameterized complexity of MAX-r-SAT-AA changes when r is no longer a constant, but a function $r(n)$ of n. They proved that MAX-$r(n)$-SAT-AA is para-NP-complete for any $r(n) \geq \lceil \log n \rceil$. They also proved that assuming the exponential time hypothesis, MAX-$r(n)$-SAT-AA is not even in XP for any integral $r(n) \geq \log\log n + \phi(n)$, where $\phi(n)$ is any real-valued unbounded strictly increasing computable function. This lower bound on $r(n)$ cannot be decreased much further as they proved that MAX-$r(n)$-SAT-AA is (i) in XP for any $r(n) \leq \log\log n - \log\log\log n$ and (ii) fixed-parameter tractable for any $r(n) \leq \log\log n - \log\log\log n - \phi(n)$, where $\phi(n)$ is any real-valued unbounded strictly increasing computable function. The proofs use some results on MAXLIN2-AA.

Cross-References

▸ Kernelization, MaxLin Above Average
▸ Kernelization, Permutation CSPs Parameterized above Average

Recommended Reading

1. Alon N, Gutin G, Kim EJ, Szeider S, Yeo A (2011) Solving MAX-k-SAT above a tight lower bound. Algorithmica 61:638–655
2. Alon N, Gutin G, Krivelevich M (2004) Algorithms with large domination ratio. J Algorithms 50: 118–131
3. Crowston R, Fellows M, Gutin G, Jones M, Kim EJ, Rosamond F, Ruzsa IZ, Thomassé S, Yeo A (2014) Satisfying more than half of a system of linear equations over GF(2): a multivariate approach. J Comput Syst Sci 80:687–696
4. Crowston R, Gutin G, Jones M, Raman V, Saurabh S (2013) Parameterized complexity of MaxSat above average. Theor Comput Sci 511:77–84
5. Gutin G, Kim EJ, Szeider S, Yeo A (2011) A probabilistic approach to problems parameterized above tight lower bound. J Comput Syst Sci 77: 422–429
6. Kim EJ, Williams R (2012) Improved parameterized algorithms for above average constraint satisfaction. In: IPEC 2011, Saarbrücken. Lecture notes in computer science, vol 7112, pp 118–131
7. O'Donnell R (2008) Some topics in analysis of Boolean functions. Technical report, ECCC report TR08-055. Paper for an invited talk at STOC'08. www.eccc.uni-trier.de/eccc-reports/2008/TR08-055/

Kernelization, Exponential Lower Bounds

Hans L. Bodlaender
Department of Computer Science, Utrecht University, Utrecht, The Netherlands

Keywords

Composition; Compression; Fixed parameter tractability; Kernelization; Lower bounds

Years and Authors of Summarized Original Work

2009; Bodlaender, Downey, Fellows, Hermelin

Problem Definition

Research on kernelization is motivated in two ways. First, when solving a hard (e.g., NP-hard) problem in practice, a common approach is to first *preprocess* the instance at hand before running more time-consuming methods (like integer linear programming, branch and bound, etc.). The following is a natural question. Suppose we use polynomial time for this preprocessing phase: what can be predicted of the size of the instance resulting from preprocessing? The theory of kernelization gives us such predictions. A second motivation comes from the fact that a decidable parameterized problem belongs to the class FPT (i.e., is *fixed parameter tractable*,) if and only if the problem has kernelization algorithm.

A parameterized problem is a subset of $\Sigma^* \times \mathbf{N}$, for some finite set Σ. A *kernelization algorithm* (or, in short *kernel*) for a parameterized problem $Q \subseteq \Sigma^* \times \mathbf{N}$ is an algorithm A that receives as input a pair $(x,k) \in \Sigma^* \times \mathbf{N}$ and outputs a pair $(x',k') = A(x,k)$, such that:

- A uses time, polynomial in $|x| + k$.
- $(x,k) \in Q$, if and only if $(x',k') \in Q$.
- There are functions f, g, such that $|x'| \leq f(k)$ and $k' \leq g(k)$.

In the definition above, f and g give an upper bound on the size, respectively the parameter of the reduced instance. Many well-studied problems have kernels with $k' \leq k$. The running time of an exact algorithm that starts with a kernelization step usually is exponential in the size of the kernel (i.e., $f(k)$), and thus small kernels are desirable. A kernel is said to be *polynomial*, if f and g are bounded by a polynomial. Many well-known parameterized problems have a polynomial kernel, but there are also many for which such a polynomial kernel is not known.

Recent techniques allow us to show, under a complexity theoretic assumption, for some parameterized problems that they do not have a polynomial kernel. The central notion is that of *compositionality*; with the help of transformations and cross compositions, a larger set of problems can be handled.

Key Results

Compositionality

The basic building block of showing that problems do not have a polynomial kernel (assuming $NP \not\subseteq coNP/poly$) is the notion of compositionality. It comes in two types: or-composition and and-composition.

Definition 1 An *or-composition* for a parameterized problem $Q \subseteq \Sigma^* \times \mathbf{N}$ is an algorithm that:

- Receives as input a sequence of instances for Q with the same parameter $(s_1, k), (s_2, k), \ldots, (s_r, k)$
- Uses time, polynomial in $k + \sum_{i=1}^{r} |s_i|$
- Outputs one instance for Q, $(s', k') \in \Sigma^* \times \mathbf{N}$, such that:

  1. $(s', k') \in Q$, if and only if there is an i, $1 \leq i \leq r$, with $(s_i, k) \in Q$.
  2. k' is bounded by a polynomial in k.

The notion of *and-composition* is defined similarly, with the only difference that condition (1) above is replaced by

$(s', k') \in Q$, if and only if for all i, $1 \leq i \leq r$: $(s_i, k) \in Q$.

We define the *classic variant* of a parameterized problem $Q \subseteq \Sigma^* \times \mathbf{N}$ as the decision problem, denoted Q^c where we assume that the parameter is encoded in unary, or, equivalently, an instance (s, k) is assumed to have size $|s| + k$.

Combining results of three papers gives the following results.

Theorem 1 *Let $Q \subseteq \Sigma^* \times \mathbf{N}$ be a parameterized problem. Suppose that the classic variant of Q, Q^c is NP-hard. Assume that $NP \not\subseteq coNP/poly$.*

1. *(Bodlaender et al. [3], Fortnow and Santhanam [12]) If Q has an or-composition, then Q has no polynomial kernel.*
2. *(Bodlaender et al. [3], Drucker [11]) If Q has an and-composition, then Q has no polynomial kernel.*

The condition that $NP \not\subseteq coNP/poly$ is equivalent to $coNP \not\subseteq NP/poly$; if it does not hold, the polynomial time hierarchy collapses to the third level [19].

For many parameterized problems, one can establish (sometimes trivially, and sometimes with quite involved proofs) that they are or-compositional or and-compositional. Taking the disjoint union of instances often gives a trivial composition. A simple example is the LONG PATH problem; it gets as input a pair (G, k) with G an undirected graph and asks whether G has a simple path of length at least k.

Lemma 1 *If $NP \not\subseteq coNP/poly$, then* LONG PATH *has no kernel polynomial in k.*

Proof LONG PATH is well known to be NP-complete. Mapping $(G_1, k), \ldots, (G_r, k)$ to the pair (H, k) with H the disjoint union of $G_1, \ldots, G_r$ is an or-composition. So, the result follows directly as a corollary of Theorem 1. □

The TREEWIDTH problem gets as input a pair (G, k) and asks whether the treewidth of G is most k. As it is NP-hard to decide if the treewidth of a given graph G is at most a given number k [1] and the treewidth of a graph is the maximum of its connected components, taking the disjoint union gives an and-composition for the TREEWIDTH problem and shows that TREEWIDTH has no polynomial kernel unless $NP \subseteq coNP/poly$. Similar proofs work for many more problems. Many problems can be seen to be and- or or-compositional and thus have no polynomial kernels under the assumption that $NP \not\subseteq coNP/poly$. See, e.g., [3,5,9,17].

Transformations

Several researchers observed independently (see [2,5,9]) that transformations can be used to show results for additional problems. The formalization is due to Bodlaender et al. [5].

Definition 2 A *polynomial parameter transformation (ppt)* from parameterized problem $Q \subseteq \Sigma^* \times \mathbf{N}$ to parameterized problem $R \subseteq \Sigma^* \times \mathbf{N}$ is an algorithm A that:

- Has as input an instance of Q, $(s,k) \in \Sigma^* \times \mathbf{N}$.
- Outputs an instance of R, $(s',k') \in \Sigma^* \times \mathbf{N}$.
- $(s,k) \in Q$ if and only if $(s',k') \in R$.
- A uses time polynomial in $|s| + k$.
- k' is bounded by a polynomial in k.

The differences with the well-known polynomial time or Karp reductions from NP-completeness theory are small: note in particular that it is required that the new value of the parameter is polynomially bounded in the old value of the parameter. The following theorem follows quite easily.

Theorem 2 (See [5,6]) *Let R have a polynomial kernel. If there is a ppt from Q to R, and a polynomial time reduction from R to the classic variant of Q, then Q has a polynomial kernel.*

This implies that if we have a ppt from Q to R, Q^c is NP-hard, $R^c \in NP$, then when Q has no polynomial kernel, R has no polynomial kernel.

Cross Composition

Bodlaender et al. [4] introduced the concept of *cross composition*. It gives a more powerful mechanism to show that some problems have no polynomial kernel, assuming $NP \not\subseteq coNP/poly$. We need first the definition of a polynomial equivalence relation.

Definition 3 A *polynomial equivalence relation* is an equivalence relation on Σ^* that can be decided in polynomial time and has for each n, a polynomial number of equivalence classes that contain strings of length at most n.

A typical example may be that strings represent graphs and two graphs are equivalent if and only if they have the same number of vertices and edges.

Definition 4 Let L be a language, R a polynomial equivalence relation, and Q a parameterized problem. An *OR cross composition* of L to Q (w.r.t. R) is an algorithm that:

- Gets as input a sequence of instances $s_1, \ldots, s_r$ of L that belong to the same equivalence class of R.
- Uses time, polynomial in $\sum_{i=1}^{r} |s_i|$.
- Outputs an instance (s',k) of Q.
- k is polynomial in $\max |s_i| + \log k$.
- $(s',k) \in Q$ if and only if there is an i with $s_i \in L$.

The definition for an AND cross composition is similar; the last condition is replaced by

$(s',k) \in Q$ if and only if for all i with $s_i \in L$.

Theorem 3 (Bodlaender et al. [4]) *If we have an OR cross composition, or an AND cross composition from an NP-hard language L into a parameterized problem Q, then Q does not have a polynomial kernel, unless $NP \subseteq coNP/poly$.*

The main differences with or-composition and and-composition are we do not need to start with a collection of instances from Q, but can use a collection of instances of any NP-hard language; the bound on the new value of k usually allows us to restrict to collections of at most 2^k instances, and with the polynomial equivalence relation, we can make assumptions on "similarities" between these instances.

K

For examples of OR cross compositions, and of AND cross compositions, see, e.g., [4, 8, 13, 17].

Other Models and Improvements

Different models of compressibility and stronger versions of the lower bound techniques have been studied, including more general models of compressibility (see [11] and [7]), the use of co-nondeterministic composition [18], weak composition [15], Turing kernelization [16], and a different measure for compressibility based on witness size of problems in NP [14].

Problems Without Kernels

Many parameterized problems are known to be hard for the complexity class $W[1]$. As decidable problems are known to have a kernel, if and only if they are fixed parameter tractable, it follows that $W[1]$-hard problems do not have a kernel, unless $W[1] = FPT$ (which would imply that the exponential time hypothesis does not hold). See, e.g., [10].

Cross-References

▸ Kernelization, Polynomial Lower Bounds
▸ Kernelization, Preprocessing for Treewidth
▸ Kernelization, Turing Kernels

Recommended Reading

1. Arnborg S, Corneil DG, Proskurowski A (1987) Complexity of finding embeddings in a k-tree. SIAM J Algebr Discret Methods 8:277–284
2. Binkele-Raible D, Fernau H, Fomin FV, Lokshtanov D, Saurabh S, Villanger Y (2012) Kernel(s) for problems with no kernel: on out-trees with many leaves. ACM Trans Algorithms 8(5):38
3. Bodlaender HL, Downey RG, Fellows MR, Hermelin D (2009) On problems without polynomial kernels. J Comput Syst Sci 75:423–434
4. Bodlaender HL, Jansen BMP, Kratsch S (2011) Cross-composition: a new technique for kernelization lower bounds. In: Schwentick T, Dürr C (eds) Proceedings 28th international symposium on theoretical aspects of computer science, STACS 2011, Dortmund. Schloss Dagstuhl – Leibniz-Zentrum fuer Informatik, Leibniz International Proceedings in Informatics (LIPIcs), vol 9, pp 165–176
5. Bodlaender HL, Thomassé S, Yeo A (2011) Kernel bounds for disjoint cycles and disjoint paths. Theor Comput Sci 412:4570–4578
6. Bodlaender HL, Jansen BMP, Kratsch S (2012) Kernelization lower bounds by cross-composition. CoRR abs/1206.5941
7. Chen Y, Flum J, Müller M (2011) Lower bounds for kernelizations and other preprocessing procedures. Theory Comput Syst 48(4):803–839
8. Cygan M, Kratsch S, Pilipczuk M, Pilipczuk M, Wahlström M (2012) Clique cover and graph separation: new incompressibility results. In: Czumaj A, Mehlhorn K, Pitts AM, Wattenhofer R (eds) Proceedings of the 39th international colloquium on automata, languages and programming, ICALP 2012, Part I, Warwick. Lecture notes in computer science, vol 7391. Springer, pp 254–265
9. Dom M, Lokshtanov D, Saurabh S (2009) Incompressibility through colors and IDs. In: Albers S, Marchetti-Spaccamela A, Matias Y, Nikoletseas SE, Thomas W (eds) Proceedings of the 36th international colloquium on automata, languages and programming, ICALP 2009, Part I, Rhodes. Lecture notes in computer science, vol 5555. Springer, pp 378–389
10. Downey RG, Fellows MR (2013) Fundamentals of parameterized complexity. Texts in computer science. Springer, London
11. Drucker A (2012) New limits to classical and quantum instance compression. In: Proceedings of the 53rd annual symposium on foundations of computer science, FOCS 2012, New Brunswick, pp 609–618
12. Fortnow L, Santhanam R (2011) Infeasibility of instance compression and succinct PCPs for NP. J Comput Syst Sci 77:91–106
13. Gutin G, Muciaccia G, Yeo: A (2013) (Non-)existence of polynomial kernels for the test cover problem. Inf Process Lett 113:123–126
14. Harnik D, Naor M (2010) On the compressibility of $\mathcal{NP}$ instances and cryptographic applications. SIAM J Comput 39:1667–1713
15. Hermelin D, Wu X (2012) Weak compositions and their applications to polynomial lower bounds for kernelization. In: Rabani Y (ed) Proceedings of the 22nd annual ACM-SIAM symposium on discrete algorithms, SODA 2012, Kyoto. SIAM, pp 104–113
16. Hermelin D, Kratsch S, Soltys K, Wahlström M, Wu X (2013) A completeness theory for polynomial (turing) kernelization. In: Gutin G, Szeider S (eds) Proceedings of the 8th international symposium on parameterized and exact computation, IPEC 2013, Sophia Antipolis. Lecture notes in computer science, vol 8246. Springer, pp 202–215
17. Jansen BMP, Bodlaender HL (2013) Vertex cover kernelization revisited – upper and lower bounds for a refined parameter. Theory Comput Syst 53:263–299
18. Kratsch S (2012) Co-nondeterminism in compositions: a kernelization lower bound for a Ramsey-type

problem. In: Proceedings of the 22nd annual ACM-SIAM symposium on discrete algorithms, SODA 2012, Kyoto, pp 114–122

19. Yap HP (1986) Some topics in graph theory. London mathematical society lecture note series, vol 108. Cambridge University Press, Cambridge

Kernelization, Matroid Methods

Magnus Wahlström
Department of Computer Science, Royal Holloway, University of London, Egham, UK

Keywords

Kernelization; Matroids; Parameterized complexity

Years and Authors of Summarized Original Work

2012; Kratsch, Wahlström

Problem Definition

Kernelization is the study of the power of polynomial-time instance simplification and preprocessing and relates more generally to questions of compact information representation. Given an instance x of a decision problem $\mathcal{P}$, with an associated *parameter* k (e.g., a bound on the solution size in x), a *polynomial kernelization* is an algorithm which in polynomial time produces an instance x' of $\mathcal{P}$, with parameter k', such that $x \in \mathcal{P}$ if and only if $x' \in \mathcal{P}$ and such that both $|x'|$ and k' are bounded by $p(k)$ for some $p(k) = \text{poly}(k)$. A *polynomial compression* is the variant where the output x' is an instance of a new problem $\mathcal{P}'$ (and may not have any associated parameter).

Matroid theory provides the tools for a very powerful framework for kernelization and more general information-preserving sparsification. As an example application, consider the following question. You are given a graph $G = (V, E)$ and two sets $S, T \subseteq V$ of terminal vertices, where potentially $|V| \gg |S|, |T|$. The task is to reduce G to a smaller graph $G' = (V', E')$, with $S, T \subseteq V'$ and $|V'|$ bounded by a function of $|S| + |T|$, such that for any sets $A \subseteq S$, $B \subseteq T$, the minimum (A, B)-cut in G' equals that in G. Here, all cuts are vertex cuts and may overlap A and B (i.e., the terminal vertices are also deletable). It is difficult to see how to do this without using both exponential time in $|S| + |T|$ (due to the large number of choices of A and B) and an exponential dependency of $|V'|$ on $|S|$ and $|T|$ (due to potentially having to include one min cut for every choice of A and B), yet using the appropriate tools from matroid theory, we can in polynomial time produce such a graph G' with $|V'| = O(|S| \cdot |T| \cdot \min(|S|, |T|))$. Call (G, S, T) a *terminal cut system*; we will revisit this example later.

The main power of the framework comes from two sources. The first is a class of matroids known as *gammoids*, which enable the representation of graph-cut properties as linear independence of vectors; the second is a tool known as the *representative sets lemma* (due to Lovász [4] via Marx [5]) applied to such a representation. To describe these closer, we need to review several definitions.

Background on Matroids

We provide only the bare essential definitions; for more, see Oxley [6]. Also see the relevant chapters of Schrijver [8] for a more computational perspective and Marx [5] for a concise, streamlined, and self-contained presentation of the issues most relevant to our concerns. For $s \in \mathbb{N}$, we let $[s]$ denote the set $\{1, \ldots, s\}$.

A *matroid* is a pair $M = (V, \mathcal{I})$ where V is a *ground set* and $\mathcal{I} \subseteq 2^V$ a collection of *independent sets*, subject to three axioms:

1. $\emptyset \in \mathcal{I}$.
2. If $B \in \mathcal{I}$ and $A \subseteq B$, then $A \in \mathcal{I}$.
3. If $A, B \in \mathcal{I}$ and $|A| < |B|$, then there exists some $b \in (B \setminus A)$ such that $A \cup \{b\} \in \mathcal{I}$.

All matroids we deal with will be finite (i.e., have finite ground sets). A set $S \subseteq V$ is *independent* in M if and only if $S \in \mathcal{I}$. A *basis* is a maximal independent set in M; observe that all bases of a matroid have the same cardinality. The *rank* of a set $X \subseteq V$ is the maximum cardinality of an independent set $S \subseteq X$; again, observe that this is well defined.

Linearly Represented Matroids

A prime example of a matroid is a *linear matroid*. Let A be a matrix over some field $\mathbb{F}$, and let V index the column set of A. Let $\mathcal{I}$ contain exactly those sets of columns of A that are linearly independent. Then $M = (V, \mathcal{I})$ defines a matroid, denoted $M(A)$, known as a linear matroid. For an arbitrary matroid M, if M is isomorphic to a linear matroid $M(A)$ (over a field $\mathbb{F}$), then M is *representable* (*over* $\mathbb{F}$), and the matrix A *represents* M. Observe that this is a compact representation, as $|\mathcal{I}|$ would in the general case be exponentially large, while the matrix A would normally have a coding size polynomial in $|V|$. In general, more powerful tools are available for linearly represented matroids than for arbitrary matroids (see, e.g., the MATROID MATCHING problem [8]). In particular, this holds for the representative sets lemma (see below).

Gammoids

The class of matroids central to our concern is the class of *gammoids*, first defined by Perfect [7]. Let $G = (V, E)$ be a (possibly directed) graph, $S \subseteq V$ a set of *source vertices*, and $T \subseteq V$ a set of *sink vertices* (where S and T may overlap). Let $X \subseteq T$ be independent if and only if there exists a collection of $|X|$ pairwise vertex-disjoint directed paths in G, each path starting in S and ending in X; we allow paths to have length zero (e.g., we allow a path from a vertex $x \in S \cap X$ to itself). This notion of independence defines a matroid on the ground set T, referred to as the *gammoid* defined by G, S, and T. By Menger's theorem, the rank of a set $X \subseteq T$ equals the cardinality of an (S, X)-min cut in G.

Gammoids are representable over any sufficiently large field [6], although only randomized procedures for computing a representation are known. An explicit randomized procedure was given in [2], computing a representation of the gammoid (G, S, T) in space (essentially) cubic in $|S|+|T|$. Hence, gammoids imply a polynomial-sized representation of terminal cut systems, as defined in the introduction. This has implications in kernelization [2], though it is not on its own the most useful form, since it is not a representation in terms of graphs.

Representative Sets

Let $M = (V, \mathcal{I})$ be a matroid, and X and Y independent sets in M. We say that Y *extends* X if $X \cup Y$ is independent and $X \cap Y = \emptyset$. The *representative sets lemma* states the following.

Lemma 1 ([4,5]) *Let $M = (V, \mathcal{I})$ be a linearly represented matroid of rank $r + s$, and let $\mathcal{S} = \{S_1, \ldots, S_m\}$ be a collection of independent sets, each of size s. In polynomial time, we can compute a set $\mathcal{S}^* \subseteq \mathcal{S}$ such that $|\mathcal{S}^*| \leq \binom{r+s}{s}$, and for any independent set X, there is a set $S \in \mathcal{S}$ that extends X if and only if there is a set $S' \in \mathcal{S}^*$ that extends X.*

We refer to $\mathcal{S}^*$ as a *representative set* for $\mathcal{S}$ in M. This result is due to Lovász [4], made algorithmic by Marx [5]; recently, Fomin et al. [1] improved the running time and gave algorithmic applications of the result. The power of the lemma is extended by several tools which construct new linearly represented matroids from existing ones; see Marx [5]. For a particularly useful case, for each $i \in [s]$ let $M_i = (V_i, \mathcal{I}_i)$ be a matroid, where each set V_i is a new copy of an original ground set V. Given a representation of these matroids over the same field $\mathbb{F}$, we can form a represented matroid $M = (V_1 \cup \ldots \cup V_s, \mathcal{I}_1 \times \ldots \times \mathcal{I}_s)$ as a *direct sum* of these matroids, where an independent set X in M is the union of an independent set X_i in M_i for each i. For an element $v \in V$, let $v(i)$ denote the copy of v in V_i. Then the set $\{v(1), \ldots, v(s)\}$ extends $X = X_1 \cup \ldots \cup X_s$ if and only if $\{v(i)\}$ extends X_i for each $i \in [s]$. In other words, we have constructed an *AND operation* for the notion of an extending set.

Closest Sets and Gammoids

We need one last piece of terminology. For a (possibly directed) graph $G = (V, E)$ and sets $A, X \subseteq V$, let $R_G(A, X)$ be the set of vertices reachable from A in $G \setminus X$. The set X is *closest to* A if there is no set X' such that $|X'| \leq |X|$ and X' separates X from A, i.e., $X \cap R_G(A, X') = \emptyset$. This is equivalent to X being the unique minimum (A, X)-vertex cut. For every pair of sets $A, B \subseteq V$, there is a unique minimum (A, B)-vertex cut closest to A, which can be computed in polynomial time. Finally, for sets S and X, the set X *pushed towards* S is the unique minimum (S, X)-vertex cut closest to S; this operation is well defined and has no effect if X is already closest to S. The following is central to our applications.

Lemma 2 *Let M be a gammoid defined from a graph $G = (V, E)$ and source set S. Let X be independent in M, and let X' be X pushed towards S. For any $v \in V$, the set $\{v\}$ extends X if and only if $v \in R_G(S, X')$.*

Key Results

The most powerful version of the terminal cut system result is the following.

Theorem 1 *Let $G = (V, E)$ be a (possibly directed) graph, and $X \subseteq V$ a set of vertices. In randomized polynomial time, we can find a set $Z \subseteq V$ of $|Z| = O(|X|^3)$ vertices such that for every partition $X = A \cup B \cup C \cup D$, the set Z contains a minimum (A, B)-vertex cut in the graph $G \setminus D$.*

There is also a variant for cutting into more than two parts, as follows.

Theorem 2 *Let $G = (V, E)$ be an undirected graph, and $X \subseteq V$ a set of vertices. In randomized polynomial time, we can find a set $Z \subseteq V$ of $|Z| = O(|X|^{s+1})$ vertices such that for every partition of X into at most s parts, the set Z contains a minimum solution to the corresponding multiway cut problem.*

We also have the following further kernelization results; see [3] for problem statements.

Theorem 3 *The following problems admit randomized polynomial kernels parameterized by the solution size*: ALMOST 2-SAT, VERTEX MULTICUT *with a constant number of cut requests, and* GROUP FEEDBACK VERTEX SET *with a constant-sized group.*

Applications

We now review the strategy behind kernelization usage of the representative sets lemma.

Representative Sets: Direct Usage

There have been various types of applications of the representative sets lemma in kernelization, from the more direct to the more subtle. We briefly review one more direct and one indirect. The most direct one is for reducing *constraint systems*. We illustrate with the DIGRAPH PAIR CUT problem (which is closely related to a central problem in kernelization [3]). Let $G = (V, E)$ be a digraph, with a source vertex $s \in V$, and let $\mathcal{P} \subseteq V^2$ be a set of pairs. The task is to find a set X of at most k vertices (with $s \notin X$) such that $R_G(s, X)$ does not contain any complete pair from $\mathcal{P}$. We show that it suffices to keep $O(k^2)$ of the pairs $\mathcal{P}$. For this, replace s by a set S of $k + 1$ copies of s, and let M be the gammoid of (G, S, V). By Lemma 2, if X is closest to S and $|X| \leq k$, then $\{u, v\} \subseteq R_G(s, X)$ if and only if both $\{u\}$ and $\{v\}$ extend X in M. Hence, using the direct sum construction, we can construct a representative set $\mathcal{P}^* \subseteq \mathcal{P}$ with $|\mathcal{P}^*| = O(k^2)$ such that for any set X closest to S, the set $R_G(s, X)$ contains a pair $\{u, v\} \in \mathcal{P}$ if and only if it contains a pair $\{u', v'\} \in \mathcal{P}^*$. Furthermore, for an arbitrary set X, pushing X towards S yields a set X' that can only be an improvement on X (i.e., the set of pairs in $R_G(s, X)$ shrinks); hence for any set X with $|X| \leq k$, either pushing X towards S yields a solution to the problem, or there is a pair in $\mathcal{P}^*$ witnessing that X is not a solution. Thus, the set $\mathcal{P}^*$ may be used to replace $\mathcal{P}$, taking the first step towards a kernel for the problem.

Indirect Usage

For more advanced applications, we "force" the lemma to reveal some set Z of special vertices in G, as follows. Let M be a linearly represented matroid, and let $\mathcal{S} = \{S(v) : v \in V\}$ be a collection of subsets of M of bounded size. Assume that we have shown that for every $z \in Z$, there is a carefully chosen set $X(z)$, such that $S(v)$ extends $X(z)$ if and only if $v = z$. Then, necessarily, the representative set $\mathcal{S}^*$ for $\mathcal{S}$ must contain $S(z)$ for every $z \in Z$, by letting $X = X(z)$ in the statement of the lemma. Furthermore, we do not need to provide the set $X(z)$ ahead of time, since the (possibly non-constructive) *existence* of such a set $X(z)$ is sufficient to force $S(z) \in \mathcal{S}^*$. Hence, the set $V^* = \{v \in V : S(v) \in \mathcal{S}^*\}$ must contain Z, among a polynomially bounded number of other vertices. The critical challenge, of course, is to construct the matroid M and sets $S(v)$ and $X(z)$ such that $S(z)$ indeed extends $X(z)$, while $S(v)$ fails to extend $X(z)$ for every $v \neq z$.

We illustrate the application to reducing terminal cut systems. Let $G = (V, E)$ be an undirected graph (the directed construction is similar), with $S, T \subseteq V$, and define a set of vertices Z where $z \in Z$ if and only if there are sets $A \subseteq S$, $B \subseteq T$ such that every minimum (A, B)-vertex cut contains z. We wish to learn Z. Let a *sink-only copy* of a vertex $v \in V$ be a copy v' of v with all edges oriented towards v'. Then the following follows from Lemma 2 and the definition of closest sets.

Lemma 3 *Let $A, B \subseteq V$, and let X be a minimum (A, B)-vertex cut. Then a vertex $v \in V$ is a member of every minimum (A, B)-vertex cut if and only if $\{v'\}$ extends X in both the gammoid (G, A, V) and the gammoid (G, B, V).*

Via a minor modification, we can replace the former gammoid by the gammoid (G, S, V) and the latter by (G, T, V) (for appropriate adjustments to the set X); we can then compute a set V^* of $O(|S| \cdot |T| \cdot k)$ vertices (where k is the size of an (S, T)-min cut) which contains Z. From this, we may compute the sought-after smaller graph G', by iteratively bypassing a single vertex $v \in V \setminus (S \cup T \cup V^*)$ and recomputing V^*, until $V^* \cup S \cup T = V$; observe that bypassing v does not change the size of any (A, B)-min cut. Theorem 1 follows by considering a modification of the graph G, and Theorem 2 follows by a generalization of the above, pushing into s different directions.

Further Applications

A polynomial kernel for MULTIWAY CUT (in the variants with only s terminals or with deletable terminals) essentially follows from the above, but the further kernelization applications in Theorem 3 require a few more steps. However, they follow a common pattern: First, we find an approximate solution X of size poly(k) to "bootstrap" the process; second, we use X to transform the problem into a more manageable form (e.g., for ALMOST 2-SAT, this manageable form is DIGRAPH PAIR CUT); and lastly, we use the above methods to kernelize the resulting problem. This pattern covers the problems listed in Theorem 3.

Finally, the above results have some implications beyond kernelization. In particular, the existence of the smaller graph G' computed for terminal cut systems, and correspondingly an implementation of a gammoid as a graph with poly($|S| + |T|$) vertices, was an open problem, solved in [3].

Cross-References

- ▶ Kernelization, Exponential Lower Bounds
- ▶ Kernelization, Polynomial Lower Bounds
- ▶ Matroids in Parameterized Complexity and Exact Algorithms

Recommended Reading

1. Fomin FV, Lokshtanov D, Saurabh S (2014) Efficient computation of representative sets with applications in parameterized and exact algorithms. In: SODA, Portland, pp 142–151
2. Kratsch S, Wahlström M (2012) Compression via matroids: a randomized polynomial kernel for odd cycle transversal. In: SODA, Kyoto, pp 94–103
3. Kratsch S, Wahlström M (2012) Representative sets and irrelevant vertices: new tools for kernelization. In: FOCS, New Brunswick, pp 450–459

4. Lovász L (1977) Flats in matroids and geometric graphs. In: Proceedings of the sixth British combinatorial conference, combinatorial surveys, Egham, pp 45–86
5. Marx D (2009) A parameterized view on matroid optimization problems. Theor Comput Sci 410(44):4471–4479
6. Oxley J (2006) Matroid theory. Oxford graduate texts in mathematics. Oxford University Press, Oxford
7. Perfect H (1968) Applications of Menger's graph theorem. J Math Anal Appl 22:96–111
8. Schrijver A (2003) Combinatorial optimization: polyhedra and efficiency. Algorithms and combinatorics. Springer, Berlin/New York

Kernelization, Max-Cut Above Tight Bounds

Mark Jones
Department of Computer Science, Royal Holloway, University of London, Egham, UK

Keywords

Kernel; Lambda extendible; Max Cut; Parameterization above tight bound

Years and Authors of Summarized Original Work

2012; Crowston, Jones, Mnich

Problem Definition

In the problem MAX CUT, we are given a graph G with n vertices and m edges, and asked to find a bipartite subgraph of G with the maximum number of edges.

In 1973, Edwards [5] proved that if G is connected, then G contains a bipartite subgraph with at least $\frac{m}{2} + \frac{n-1}{4}$ edges, proving a conjecture of Erdős. This lower bound on the size of a bipartite subgraph is known as the *Edwards-Erdős bound*. The bound is tight – for example, it is an upper bound when G is a clique with odd number of vertices. Thus, it is natural to consider parameterized MAX CUT above this bound, as follows (AEE stands for *Above Edwards-Erdős*).

MAX CUT AEE

Instance: A connected graph G with n vertices and m edges, and a nonnegative integer k.
Parameter: k.
Question: Does G have a bipartite subgraph with at least $\frac{m}{2} + \frac{n-1}{4} + k$ edges?

Mahajan and Raman [6], in their first paper on above-guarantee parameterizations, asked whether this problem is fixed-parameter tractable. As such, the problem was one of the first open problems in above-guarantee parameterizations.

λ-Extendibility and the Poljak-Turzík Bound

In 1982, Poljak and Turzík [8] investigated extending the Edwards-Erdős bound to cases when the desired subgraph is something other than bipartite. To this end, they introduced the notion of λ-extendibility, which generalizes the notion of "bipartiteness." We will define the slightly stronger notion of *strong* λ-extendibility, introduced in [7], as later results use this stronger notion.

Recall that a *block* of a graph G is a maximal 2-connected subgraph of G. The blocks of a graph form a partition of its edges, and a vertex that appears in two or more blocks is a cut-vertex of the graph.

Definition 1 For a family of graphs Π and $0 \leq \lambda \leq 1$, we say Π is strongly λ-extendible if the following conditions are satisfied:

1. If G is connected and $|G| = 1$ or 2, then $G \in \Pi$.
2. G is in Π if and only if each of its blocks is in Π.
3. For any real-valued positive weight function w on the edges of G, if $X \subseteq V(G)$ is such that $G[X]$ is connected and $G[X], G - X \in \Pi$, then G has a subgraph $H \in \Pi$ that uses all the edges of $G[X]$, all the edges of $G - X$, and at least a fraction λ (by weight) of the edges between X and $V(G) \setminus X$.

The definition of λ-extendibility given in [8] is the same as the above, except that the third condition is only required when $|X| = 2$. Clearly strong λ-extendibility implies λ-extendibility; it is an open question whether the converse holds.

The property of being bipartite is strongly λ-extendible for $\lambda = 1/2$. Other strongly λ-extendible properties include being acyclic for directed graphs ($\lambda = 1/2$) and being r-colorable ($\lambda = 1/r$).

Poljak and Turzík [8] extended Edwards' result by showing that for any connected graph G with n vertices and m edges, and any λ-extendible property Π, G contains a subgraph in Π with at least $\lambda m + \frac{1-\lambda}{2}(n-1)$ edges.

Thus, for any λ-extendible property Π, we can consider the following variation of MAX CUT AEE, for any λ-extendible Π (APT stands for *Above Poljak-Turzík*).

Π-SUBGRAPH APT

Instance: A connected graph G with n vertices and m edges, and a nonnegative integer k.
Parameter: k.
Question: Does G have a subgraph in Π with at least $\lambda m + \frac{1-\lambda}{2}(n-1) + k$ edges?

Key Results

We sketch a proof of the polynomial kernel result for MAX CUT AEE, first shown in [2] (although the method described here is slightly different to that in [2]).

For a connected graph G with n vertices and m edges, let $\beta(G)$ denote the maximum number of edges of a bipartite subgraph of G, let $\gamma(G) = \frac{m}{2} + \frac{n-1}{4}$, and let $\varepsilon(G) = \beta(G) - \gamma(G)$. Thus, for an instance (G,k), our aim is to determine whether $\varepsilon(G) \geq k$.

Now consider a connected graph G with a set X of three vertices such that $G' = G - X$ is connected and $G[X] = P_3$, the path with two edges. Note that $G[X]$ is bipartite. Let H' be a subgraph of G' with $\beta(G')$ edges. As bipartiteness is a $1/2$-extendible property, we can create a bipartite subgraph H of G using the edges of H', the edges of $G[X]$, and at least half of the edges between X and $G - X$. It follows that $\beta(G) \geq \beta(G') + \frac{|E(X,V(G)\setminus X)|}{2} + 2$. As $\gamma(G) = \gamma(G') + \frac{|E(X,V(G)\setminus X)|+2}{2} + \frac{3}{4}$, we have that $\varepsilon(G) = \beta(G) - \gamma(G) \geq \beta(G') - \gamma(G') + 2 - \frac{2}{2} - \frac{3}{4} = \varepsilon(G') + \frac{1}{4}$.

Consider a reduction rule in which, if there exists a set X as described above, we delete X from the graph. If we were able to apply such a reduction rule $4k$ times on a graph G, we would end up with a reduced graph G' such that G' is connected and $\varepsilon(G) \geq \varepsilon(G') + \frac{4k}{4} \geq 0 + k$, and therefore we would know that (G,k) is a YES-instance. Of course there may be many graphs for which such a set X cannot be found. However, we can adapt this idea as follows. Given a connected graph G, we recursively calculate a set of vertices $S(G)$ and a rational number $t(G)$ as follows:

- If G is a clique or G is empty, then set $S(G) = \emptyset$ and $t(G) = 0$.
- If G contains a set X such that $|X| = 3$, $G' = G - X$ is connected and $G[X] = P_3$, then set $S(G) = S(G') \cup X$ and set $t(G) = t(G') + \frac{1}{4}$.
- If G contains a cut-vertex v, then there exist non-empty sets of vertices X, Y such that $X \cap Y = \{v\}$, $G[X]$ and $G[Y]$ are connected, and all edges of G are in $G[X]$ or $G[Y]$. Then set $S(G) = S(G[X]) \cup S(G[Y])$ and set $t(G) = t(G[X]) + t(G[Y])$.

It can be shown that for a connected graph G, one of these cases will always hold, and so $S(G)$ and $t(G)$ are well defined. In the first case, we have that $\varepsilon(G) \geq 0$ by the Edwards-Erdős bound. In the second case, we have already shown that $\varepsilon(G) \geq \varepsilon(G') + \frac{1}{4}$. In the third case, we have that $\varepsilon(G) = \varepsilon(G[X]) + \varepsilon(G[Y])$ (note that the union of a bipartite subgraph of $G[X]$ and a bipartite subgraph of $G[Y]$ is a bipartite subgraph of G). It follows that $\varepsilon(G) \geq t(G)$. Note also that $|S(G)| \leq 12t(G)$. If we remove $S(G)$ from G, the resulting graph can be built by joining disjoint graphs at a single vertex, using only cliques as the initial graphs. Thus, $G - S(G)$ has the property that each of its blocks is a clique. We call such a graph a *forest of cliques*.

We therefore get the following lemma.

Lemma 1 ([2]) *Given a connected graph G with n vertices and m edges, and an integer k, we can in polynomial time either decide that (G, k) is a* YES*-instance of* MAX CUT AEE, *or find a set S of at most $12k$ vertices such that $G - S$ is a forest of cliques.*

By guessing a partition of S and then using a dynamic programming algorithm based on the structure of $G - S$, we get a fixed-parameter algorithm.

Theorem 1 ([2]) MAX CUT AEE *can be solved in time $2^{O(k)} \cdot n^4$.*

Using the structure of $G - S$ and the fact that $|S| \leq 12k$, it is possible (using reduction rules) to show first that the number of blocks in $G - S$ must be bounded for any NO-instance, and then that the size of each block must be bounded (see [2]).

Thus, we get a polynomial kernel for MAX CUT AEE.

Theorem 2 ([2]) MAX CUT AEE *admits a kernel with $O(k^5)$ vertices.*

Crowston et al. [3] were later able to improve this to a kernel with $O(k^3)$ vertices.

Extensions to Π-SUBGRAPH APT

A similar approach can be used to show polynomial kernels for Π-SUBGRAPH APT, for other $1/2$-extendible properties. In particular, the property of being an acyclic directed graph is $1/2$-extendible, and therefore every directed graph with n vertices and m arcs has an acyclic subgraph with at least $\frac{m}{2} + \frac{n-1}{4}$ arcs. The problem of deciding whether there exists an acyclic subgraph with at least $\frac{m}{2} + \frac{n-1}{4} + k$ arcs is fixed-parameter tractable, and has a $O(k^2)$-vertex kernel [1].

The notion of a bipartite graph can be generalized in the following way. Consider a graph G with edges labeled either $+$ or $-$. Then we say G is *balanced* if there exists a partition V_1, V_2 of the vertices of G, such that all edges between V_1 and V_2 are labeled $-$ and all other edges are labeled $+$. (Note that if all edges of a graph are labeled $-$, then it is balanced if and only if it is bipartite.) The property of being a balanced graph is $1/2$-extendible, just as the property of being bipartite is. Therefore a graph with n vertices and m vertices, and all edges labeled $+$ or $-$, will have a balanced subgraph with at least $\frac{m}{2} + \frac{n-1}{4}$ edges. The problem of deciding whether there exists a balanced subgraph with at least $\frac{m}{2} + \frac{n-1}{4} + k$ edges is fixed-parameter tractable and has a $O(k^3)$-vertex kernel [3].

Mnich et al. [7] showed that Lemma 1 applies not just for MAX CUT AEE, but for Π-SUBGRAPH APT for any Π which is strongly λ-extendible for some λ (with the bound $12k$ replaced with $\frac{6k}{1-\lambda}$). Thus, Π-SUBGRAPH APT is fixed-parameter tractable as long as it is fixed-parameter tractable on graphs which are close to being a forest of cliques. Using this observation, Mnich et al. showed fixed-parameter tractability for a number of versions of Π-SUBGRAPH APT, including when Π is the family of acyclic directed graphs and when Π is the set of r-colorable graphs.

Crowston et al. [4] proved the existence of polynomial kernels for a wide range of strongly λ-extendible properties:

Theorem 3 ([4]) *Let $0 < \lambda < 1$, and let Π be a strongly λ-extendible property of (possibly oriented and/or labeled) graphs. Then Π-*SUBGRAPH APT *has a kernel on $O(k^2)$ vertices if Condition 1 or 2 holds, and a kernel on $O(k^3)$ vertices if only Condition 3 holds:*

1. *$\lambda \neq \frac{1}{2}$.*
2. *All orientations and labels (if applicable) of the graph K_3 belong to Π.*
3. *Π is a hereditary property of simple or oriented graphs.*

Open Problems

The Poljak-Turzík's bound extends to edge-weighted graphs. The weighted version is as follows: for a graph G with nonnegative real weights on the edges, and a λ-extendible family

K

of graphs Π, there exists a subgraph H of G such that $H \in \Pi$ and H has total weight $\lambda \cdot w(G) + \frac{1-\lambda}{2} \cdot MST(G)$, where $w(G)$ is the total weight of G and $MST(G)$ is the minimum weight of a spanning tree in G.

Thus, we can consider the weighted versions of MAX CUT AEE and Π-SUBGRAPH APT. It is known that a weighted equivalent of Lemma 1 holds (in which all edges in a block of $G - S$ have the same weight), and as a result, the integer-weighted version of MAX CUT AEE can be shown to be fixed-parameter tractable. However, nothing is known about kernelization results for these problems. In particular, it remains an open question whether the integer-weighted version of MAX CUT AEE has a polynomial kernel.

Cross-References

- ▶ Kernelization, Constraint Satisfaction Problems Parameterized above Average
- ▶ Kernelization, MaxLin Above Average

Recommended Reading

1. Crowston R, Gutin G, Jones M (2012) Directed acyclic subgraph problem parameterized above the Poljak-Turzík bound. In: FSTTCS 2012, Hyderabad. LIPICS, vol 18, pp 400–411
2. Crowston R, Jones M, Mnich M (2012) Max-Cut parameterized above the Edwards-Erdős bound. In: ICALP 2012, Warwick. Lecture notes in computer science, vol 7391, pp 242–253
3. Crowston R, Gutin G, Jones M, Muciaccia G (2013) Maximum balanced subgraph problem parameterized above lower bounds. Theor Comput Sci 513:53–64
4. Crowston R, Jones M, Muciaccia G, Philip G, Rai A, Saurabh S (2013) Polynomial kernels for λ-extendible properties parameterized above the Poljak-Turzík bound. In: FSTTCS 2013, Guwahati. LIPICS, vol 24, pp 43–54
5. Edwards CS (1973) Some extremal properties of bipartite subgraphs. Can J Math 25:475–485
6. Mahajan M, Raman V (1999) Parameterizing above guaranteed values: MaxSat and MaxCut. J Algorithms 31(2):335–354
7. Mnich M, Philip G, Saurabh S, Suchý O (2014) Beyond Max-Cut: λ-extendible properties parameterized above the Poljak-Turzík bound. J Comput Syst Sci 80(7):1384–1403
8. Poljak S, Turzík D (1982) A polynomial algorithm for constructing a large bipartite subgraph, with an application to a satisfiability problem. Can J Math 34(4):519–524

Kernelization, MaxLin Above Average

Anders Yeo
Engineering Systems and Design, Singapore University of Technology and Design, Singapore, Singapore
Department of Mathematics, University of Johannesburg, Auckland Park, South Africa

Keywords

Fixed-parameter tractability above lower bounds; Kernelization; Linear equations; M-sum-free sets

Years and Authors of Summarized Original Work

2010; Crowston, Gutin, Jones, Kim, Ruzsa
2011; Gutin, Kim, Szeider, Yeo
2014; Crowston, Fellows, Gutin, Jones, Kim, Rosamond, Ruzsa, Thomassé, Yeo

Problem Definition

The problem MAXLIN2 can be stated as follows. We are given a system of m equations in variables $x_1, \ldots, x_n$ where each equation is $\prod_{i \in I_j} x_i = b_j$, for some $I_j \subseteq \{1, 2, \ldots, n\}$ and $x_i, b_j \in \{-1, 1\}$ and $j = 1, \ldots, m$. Each equation is assigned a positive integral weight w_j. We are required to find an assignment of values to the variables in order to maximize the total weight of the satisfied equations. MAXLIN2 is a well-studied problem, which according to Håstad [8] "is as basic as satisfiability."

Note that one can think of MAXLIN2 as containing equations, $\sum_{i \in I_j} y_i = a_j$ over $\mathbb{F}_2$. This is equivalent to the previous definition by letting $y_i = 0$ if and only if $x_i = 1$ and letting $y_i = 1$ if and only if $x_i = -1$ (and $a_j = 1$ if and only if $b_j = -1$ and $a_j = 0$ if and only if $b_j = 1$). We will however use the original definition as this was the formulation used in [1].

Let W be the sum of the weights of all equations in an instance, S, of MAXLIN2 and let sat(S) be the maximum total weight of equations that can be satisfied simultaneously. To see that $W/2$ is a tight lower bound on sat(S), choose assignments to the variables independently and uniformly at random. Then $W/2$ is the expected weight of satisfied equations (as the probability of each equation being satisfied is $1/2$) and thus $W/2$ is a lower bound. It is not difficult to see that this bound is tight. For example, consider a system consisting of pairs of equations of the form $\prod_{i \in I} x_i = -1$, $\prod_{i \in I} x_i = 1$ of the same weight, for some nonempty sets $I \subseteq \{1, 2, \ldots, n\}$.

As MAXLIN2 is an NP-hard problem, we look for parameterized algorithms. We will give the basic definitions of fixed-parameter tractability (FPT) here and refer the reader to [4, 5] for more information. A *parameterized problem* is a subset $L \subseteq \Sigma^* \times \mathbb{N}$ over a finite alphabet Σ. L is *fixed-parameter tractable* (FPT, for short) if membership of an instance (x, k) in $\Sigma^* \times \mathbb{N}$ can be decided in time $f(k)|x|^{O(1)}$, where f is a function of the parameter k only.

If we set the parameter, k, of an instance, S, of MAXLIN2 to sat(S), then it is easy to see that there exists an $O(f(k)|S|^c)$ algorithm, due to the fact that $k = \text{sat}(S) \geq W/2 \geq |S|/2$. Therefore, this parameter is not of interest (it is never small in practice), and a better parameter would be k, where we want to decide if sat$(S) \geq W/2 + k$. Parameterizing above tight lower bounds in this way was first introduced in 1997 in [11]. This leads us to define the following problem, where AA stands for *Above Average*.

MAXLIN2-AA

Instance: A system S of equations $\prod_{i \in I_j} x_i = b_j$, where $x_i, b_j \in \{-1, 1\}$, $j = 1, \ldots, m$ and where each equation is assigned a positive integral weight w_j and a nonnegative integer k.

Question: sat$(S) \geq W/2 + k$?

The above problem has also been widely studied when the number of variables in each equation is bounded by some constant, say r, which leads to the following problem.

MAX-r-LIN2-AA

Instance: A system S of equations $\prod_{i \in I_j} x_i = b_j$, where $x_i, b_j \in \{-1, 1\}$, $|I_j| \leq r$, $j = 1, \ldots, m$; equation j is assigned a positive integral weight w_j and a nonnegative integer k.

Question: sat$(S) \geq W/2 + k$?

Given a parameterized problem, Π, a *kernel* of Π is a polynomial-time algorithm that maps an instance (I, k) of Π to another instance, (I', k'), of Π such that (i) $(I, k) \in \Pi$ if and only if $(I', k') \in \Pi'$, (ii) $k' \leq f(k)$, and (iii) $|I'| \leq g(k)$ for some functions f and g. The function $g(k)$ is called the *size* of the kernel. It is well known that a problem is FPT if and only if it has a kernel.

A kernel is called a polynomial kernel if both $f(k)$ and $g(k)$ are polynomials in k. A great deal of research has been devoted to finding small-sized kernels and in particular to decide if a problem has a polynomial kernel.

We will show that both the problems stated above are FPT and in fact contain kernels with a polynomial number of variables. The number of equations may be non-polynomial, so these kernels are not real polynomial kernels. The above problems were investigated in a number of papers; see [1–3, 6].

Key Results

We will below outline the key results for both MAXLIN2-AA and MAX-r-LIN2-AA. See [1] for all the details not given here.

MAXLIN2-AA

Recall that MAXLIN2-AA considers a system S of equations $\prod_{i \in I_j} x_i = b_j$, where $x_i, b_j \in \{-1, 1\}$, $j = 1, \ldots, m$ and where each equation is assigned a positive integral weight w_j. Let $\mathcal{F}$ denote the m different sets I_j in the equations of S and let $b_{I_j} = b_j$ and $w_{I_j} = w_j$ for each $j = 1, 2, \ldots, m$.

Let $\varepsilon(x) = \sum_{I \in \mathcal{F}} w_I b_I \prod_{i \in I} x_i$ and note that $\varepsilon(x)$ is the difference between the total weight of satisfied and falsified equations. Crowston et al. [3] call $\varepsilon(x)$ the *excess* and the maximum possible value of $\varepsilon(x)$ the *maximum excess*.

Remark 1 Observe that the answer to MAXLIN2-AA and MAX-r-LIN2-AA is YES if and only if the maximum excess is at least $2k$.

Let A be the matrix over $\mathbb{F}_2$ corresponding to the set of equations in S, such that $a_{ji} = 1$ if $i \in I_j$ and 0, otherwise. Consider the following two reduction rules, where Rule 1 was introduced in [9] and Rule 2 in [6].

Reduction Rule 1 ([9]) *If we have, for a subset I of $\{1, 2, \ldots, n\}$, an equation $\prod_{i \in I} x_i = b'_I$ with weight w'_I, and an equation $\prod_{i \in I} x_i = b''_I$ with weight w''_I, then we replace this pair by one of these equations with weight $w'_I + w''_I$ if $b'_I = b''_I$ and, otherwise, by the equation whose weight is bigger, modifying its new weight to be the difference of the two old ones. If the resulting weight is 0, we delete the equation from the system.*

Reduction Rule 2 ([6]) *Let $t = \mathrm{rank} A$ and suppose columns $a^{i_1}, \ldots, a^{i_t}$ of A are linearly independent. Then delete all variables not in $\{x_{i_1}, \ldots, x_{i_t}\}$ from the equations of S.*

Lemma 1 ([6]) *Let S' be obtained from S by Rule 1 or 2. Then the maximum excess of S' is equal to the maximum excess of S. Moreover, S' can be obtained from S in time polynomial in n and m.*

If we cannot change a weighted system S using Rules 1 and 2, we call it *irreducible*. Let S be an irreducible system of MAXLIN2-AA. Consider the following algorithm introduced in [3]. We assume that, in the beginning, no equation or variable in S is marked.

ALGORITHM $\mathcal{H}$
While the system S is nonempty, do the following:

1. Choose an equation $\prod_{i \in I} x_i = b$ and mark a variable x_l such that $l \in I$.
2. Mark this equation and delete it from the system.
3. Replace every equation $\prod_{i \in I'} x_i = b'$ in the system containing x_l by $\prod_{i \in I \Delta I'} x_i = bb'$, where $I \Delta I'$ is the symmetric difference of I and I' (the weight of the equation is unchanged).
4. Apply Reduction Rule 1 to the system.

The *maximum $\mathcal{H}$-excess* of S is the maximum possible total weight of equations marked by $\mathcal{H}$ for S taken over all possible choices in Step 1 of $\mathcal{H}$. The following lemma indicates the potential power of $\mathcal{H}$.

Lemma 2 ([3]) *Let S be an irreducible system. Then the maximum excess of S equals its maximum $\mathcal{H}$-excess.*

Theorem 1 ([1]) *There exists an $O(n^{2k} (nm)^{O(1)})$-time algorithm for* MAXLIN2-AA$[k]$ *that returns an assignment of excess of at least $2k$ if one exists, and returns* NO *otherwise.*

In order to prove the above, the authors pick n equations $e_1, \ldots, e_n$ such that their rows in A are linearly independent. An assignment of excess at least $2k$ must either satisfy one of these equations or falsify them all. If they are all falsified, then the value of all variables is completely determined. Thus, by Lemma 2, algorithm $\mathcal{H}$ can mark one of these equations, implying a search tree of depth at most $2k$ and width at most k. This implies the desired time bound.

Theorem 2 below is proved using M-sum-free sets, which are defined as follows (see [3]). Let K and M be sets of vectors in $\mathbb{F}_2^n$ such that $K \subseteq M$. We say K is *M-sum-free* if no sum of two or more distinct vectors in K is equal to a vector in M.

Theorem 2 ([1]) *Let S be an irreducible system of* MAXLIN2-AA[k] *and let $k \geq 1$. If $2k \leq m \leq \min\{2^{n/(2k-1)} - 1, 2^n - 2\}$, then the maximum excess of S is at least $2k$. Moreover, we can find an assignment with excess of at least $2k$ in time $O(m^{O(1)})$.*

Using the above, we can solve the problem when $2k \leq m \leq 2^{n/(2k-1)} - 2$ and when $m \geq n^{2k} - 1$ (using Theorem 1). The case when $m < 2k$ immediately gives a kernel and the remaining case when $2^{n/(2k-1)} - 2 \leq m \leq n^{2k} - 2$ can be shown to imply that $n \in O(k^2 \log k)$, thereby giving us the main theorem and corollary of this section.

Theorem 3 ([1]) *The problem* MAXLIN2-AA[k] *has a kernel with at most $O(k^2 \log k)$ variables.*

Corollary 1 ([1]) *The problem* MAXLIN2-AA[k] *can be solved in time $2^{O(k \log k)}(nm)^{O(1)}$.*

MAX-r-LIN2-AA

In [6] it was proved that the problem MAX-r-LIN2-AA admits a kernel with at most $O(k^2)$ variables and equations (where r is treated as a constant). The bound on the number of variables can be improved and it was done by Crowston et al. [3] and Kim and Williams [10]. The best known improvement is by Crowston et al. [1].

Theorem 4 ([1]) *The problem* MAX-r-LIN2-AA *admits a kernel with at most $(2k - 1)r$ variables.*

Both Theorem 4 and a slightly weaker analogous result of the results in [10] imply the following:

Lemma 3 ([1,10]) *There is an algorithm of runtime $2^{O(k)} + m^{O(1)}$ for* MAX-r-LIN2-AA.

Kim and Williams [10] proved that the last result is best possible, in a sense, if the exponential time hypothesis holds.

Theorem 5 ([10]) *If* MAX-3-LIN2-AA *can be solved in $O(2^{\epsilon k}2^{\epsilon m})$ time for every $\epsilon > 0$, then 3-SAT can be solved in $O(2^{\delta n})$ time for every $\delta > 0$, where n is the number of variables.*

Open Problems

The kernel for MAXLIN2-AA contains at most $O(k^2 \log k)$ variables, but may contain an exponential number of equations. It would be of interest to decide if MAXLIN2-AA admits a kernel that has at most a polynomial number of variables and equations.

Cross-References

- ▸ Kernelization, Constraint Satisfaction Problems Parameterized above Average
- ▸ Kernelization, Permutation CSPs Parameterized above Average

Recommended Reading

1. Crowston R, Fellows M, Gutin G, Jones M, Kim EJ, Rosamond F, Ruzsa IZ, Thomassé S, Yeo A (2014) Satisfying more than half of a system of linear equations over GF(2): a multivariate approach. J Comput Syst Sci 80(4):687–696
2. Crowston R, Gutin G, Jones M (2010) Note on Max Lin-2 above average. Inform Proc Lett 110:451–454
3. Crowston R, Gutin G, Jones M, Kim EJ, Ruzsa I (2010) Systems of linear equations over $\mathbb{F}_2$ and problems parameterized above average. In: SWAT 2010, Bergen. Lecture notes in computer science, vol 6139, pp 164–175
4. Downey RG, Fellows MR (2013) Fundamentals of parameterized complexity. Springer, London/Heidelberg/New York
5. Flum J, Grohe M (2006) Parameterized complexity theory. Springer, Berlin
6. Gutin G, Kim EJ, Szeider S, Yeo A (2011) A probabilistic approach to problems parameterized above or below tight bounds. J Comput Syst Sci 77:422–429
7. Gutin G, Yeo A (2012) Constraint satisfaction problems parameterized above or below tight bounds: a survey. Lect Notes Comput Sci 7370:257–286

8. Håstad J (2001) Some optimal inapproximability results. J ACM 48:798–859
9. Håstad J, Venkatesh S (2004) On the advantage over a random assignment. Random Struct Algorithms 25(2):117–149
10. Kim EJ, Williams R (2012) Improved parameterized algorithms for above average constraint satisfaction. In: IPEC 2011, Saarbrücken. Lecture notes in computer science, vol 7112, pp 118–131
11. Mahajan M, Raman V (1999) Parameterizing above guaranteed values: MaxSat and MaxCut. J Algorithms 31(2):335–354. Preliminary version in Electr. Colloq. Comput. Complex. (ECCC), TR-97-033, 1997

Kernelization, Partially Polynomial Kernels

Christian Komusiewicz
Institute of Software Engineering and Theoretical Computer Science, Technical University of Berlin, Berlin, Germany

Keywords

Data reduction; Fixed-parameter algorithms; Kernelization; NP-hard problems

Years and Authors of Summarized Original Work

2011; Betzler, Guo, Komusiewicz, Niedermeier
2013; Basavaraju, Francis, Ramanujan, Saurabh
2014; Betzler, Bredereck, Niedermeier

Problem Definition

In parameterized complexity, each instance (I, k) of a problem comes with an additional parameter k which describes structural properties of the instance, for example, the maximum degree of an input graph. A problem is called *fixed-parameter tractable* if it can be solved in $f(k) \cdot \text{poly}(n)$ time, that is, the super-polynomial part of the running time depends only on k. Consequently, instances of the problem can be solved efficiently if k is small.

One way to show fixed-parameter tractability of a problem is the design of a polynomial-time data reduction algorithm that reduces any input instance (I, k) to one whose size is bounded in k. This idea is captured by the notion of kernelization.

Definition 1 Let (I, k) be an instance of a parameterized problem P, where $I \in \Sigma^*$ denotes the input instance and $k \in \mathbb{N}$ is a parameter. Problem P admits a *problem kernel* if there is a polynomial-time algorithm, called *problem kernelization*, that computes an instance (I', k') of the same problem P such that:

- (I, k) is a yes-instance if and only if (I', k') is a yes-instance, and
- $|I'| + k' \leq g(k)$

for a function g of k only.

Kernelization gives a performance guarantee for the effectiveness of data reduction: instances (I, k) with $|I| > g(k)$ are provably reduced to smaller instances. Thus, one aim in the design of kernelization algorithms is to make the function g as small as possible. In particular, one wants to obtain kernelizations where g is a polynomial function. These algorithms are called *polynomial problem kernelizations*.

For many parameterized problems, however, the existence of such a polynomial problem kernelization is considered to be unlikely (under a standard complexity-theoretic assumption) [4]. Consequently, alternative models of parameterized data reduction, for example Turing kernelization, have been proposed.

The concept of *partial kernelization* offers a further approach to obtain provably useful data reduction algorithms. Partial kernelizations do not aim for a decrease of the instance *size* but for a decrease of some *part* or *dimension* of the instance. For example, if the problem input is a binary matrix with m rows and n columns, the instance size is $\Theta(n \cdot m)$. A partial kernelization can now aim for reducing one dimension of the input, for example the number of rows n. Of course, such a reduction is worthwhile only if we can algorithmically exploit the fact that the number of rows n is small. Hence, the aim is to reduce a dimension of the problem for which there are fixed-parameter algorithms. The dimension can thus be viewed as a secondary parameter.

Altogether, this idea is formalized as follows.

Definition 2 Let (I, k) be an instance of a parameterized problem P, where $I \in \Sigma^*$ denotes the input instance and k is a parameter. Let $d : \Sigma^* \to \mathbb{N}$ be a computable function such that P is fixed-parameter tractable with respect to $d(I)$. Problem P admits a *partial problem kernel* if there is a polynomial-time algorithm, called *partial problem kernelization*, that computes an instance (I', k') of the same problem such that:

- (I, k) is a yes-instance if and only if (I', k') is a yes-instance, and
- $d(I') + k' \leq g(k)$

for a computable function g.

Any parameterized problem P which has a partial kernel for some appropriate dimension d is fixed-parameter tractable with respect to k: First, one may reduce the original input instance (I, k) to the partial kernel (I', k'). In this partial kernel, we have $d(I') \leq g(k)$ and, since P can be solved in $f(d(I')) \cdot \text{poly}(n)$ time, it can thus be solved in $f(g(k)) \cdot \text{poly}(n)$ time.

Using partial problem kernelization instead of classic problem kernelization can be motivated by the following two arguments.

First, the function d in the partial problem kernelization gives us a different goal in the design of efficient data reduction rules. For instance, if the main parameter determining the hardness of a graph problem is the maximum degree, then an algorithm that produces instances whose maximum degree is $O(k)$ but whose size is unbounded might be more useful than an algorithm that produces instances whose size is $O(k^4)$ but the maximum degree is $\Omega(k^2)$.

Second, if the problem does not admit a polynomial-size problem kernel, then it might still admit a *partially polynomial kernel*, that is, a partial kernel in which $d(I') + k' \leq \text{poly}(k)$.

We now give two examples for applications of partial kernelizations.

Key Results

The partial kernelization concept was initially developed to obtain data reduction algorithms for consensus problems, where one is given a collection of combinatorial objects and one is asked to find one object that represents this collection [2].

In the KEMENY SCORE problem, these objects are permutations of a set U and the task is to find a permutation that is close to these permutations with respect to what is called *Kendall's Tau distance*, here denoted by τ. The formal definition of the (unparameterized) problem is as follows.

Input: A multiset $\mathcal{P}$ of permutations of a ground set U and an integer ℓ.
Question: Is there a permutation P such that $\sum_{P' \in \mathcal{P}} \tau(P, P') \leq \ell$?

The parameter k under consideration is the average distance between the input partitions, that is,

$$k := \sum_{\{P, P'\} \subseteq \mathcal{P}} \tau(P, P') / \binom{|\mathcal{P}|}{2}.$$

Observe that, since τ can be computed efficiently, KEMENY SCORE is fixed-parameter tractable with respect to $|U|$: try all possible permutations of U and choose the best one. Hence, if U is small, then the problem is easy. Furthermore, the number of input permutations is not such a crucial feature since KEMENY SCORE is already NP-hard for a constant number of permutations; the partial kernelization thus aims for a reduction of $|U|$ and ignores the—less important—number of input permutations.

This reduction is obtained by removing elements in U that are, compared to the other elements, in roughly the same position in many input permutations. The idea is based on a generalization of the following observation: If some element u is the first element of at least $3|\mathcal{P}|/4$ input permutations, then this element is the first element of an optimal partition. Any instance containing such an element u can thus be reduced to an equivalent with one less element.

By removing such elements, one obtains a sub-instance of the original instance in which every element contributes a value of $16/3$ to the average distance k between the input permutations. This leads to the following result.

Theorem 1 ([3]) KEMENY SCORE *admits a partial kernel with* $|U| \leq 16/3k$.

Further partial kernelizations for consensus problems have been obtained for CONSENSUS CLUSTERING [2, 7] and SWAP MEDIAN PARTITION [2], the partial kernelization for KEMENY SCORE has been experimentally evaluated [3].

Another application of partial kernelization has been proposed for covering problems such as SET COVER [1].

> **Input:** A family $\mathcal{S}$ of subsets of a ground set U.
> **Question:** Is there a subfamily $\mathcal{S}' \subseteq \mathcal{S}$ of size at most ℓ such that every element in U is contained in at least one set of $\mathcal{S}'$?

If $\ell \geq |U|$, then SET COVER has a trivial solution. Thus, a natural parameter is the amount that can be saved compared to this trivial solution, that is, $k := |U| - \ell$. A polynomial problem kernelization for SET COVER parameterized by k is deemed unlikely, again under standard complexity-theoretic assumptions. There is, however, a partially polynomial problem kernel. The dimension d is the universe size $|U|$. SET COVER is fixed-parameter tractable with respect to $|U|$ as it can be solved in $f(|U|) \cdot \text{poly}(n)$ time, for example, by dynamic programming.

The idea behind the partial kernelization is to greedily compute a subfamily $\mathcal{T} \subseteq \mathcal{S}$ of size k. Then, it is observed that either this subfamily has a structure that can be used to efficiently compute a solution of the problem, or $|U| \leq 2k^2 - 2$, or there are elements in U whose removal yields an equivalent instance. Altogether this leads to the following.

Theorem 2 ([1]) SET COVER *admits a partial problem kernel with* $|U| \leq 2k^2 - 2$.

Open Problems

The notion of partial kernelization is quite recent. Hence, the main aim for the near future is to identify further useful applications of the technique. We list some problem areas that contain natural candidates for such applications. Problems that are defined on set families, such as SET COVER, have two obvious dimensions: the number m of sets in the set family and the size n of the universe. Matrix problems also have two obvious dimensions: the number m of rows and the number n of columns. For graph problems, useful dimensions could be identified by examining the so-called parameter hierarchy [6, 8]. Here, the idea is to find dimensions whose value can be much smaller than the number of vertices in the graph. If the size $|I|$ of the instance cannot be reduced to be smaller than $\text{poly}(k)$, then this might be still possible for the smaller dimension $d(I)$. A further interesting research direction could be to study the relationship between partial kernelization and other relaxed notions of kernelization such as Turing kernelization.

For some problems, the existence of partially polynomial kernels has been proven, but it is still unknown whether polynomial kernels exist. One such example is MAXLIN2-AA [5].

Cross-References

- ▶ Kernelization, MaxLin Above Average
- ▶ Kernelization, Polynomial Lower Bounds
- ▶ Kernelization, Turing Kernels

Recommended Reading

1. Basavaraju M, Francis MC, Ramanujan MS, Saurabh S (2013) Partially polynomial kernels for set cover and test cover. In: FSTTCS '13, Guwahati. Schloss Dagstuhl – Leibniz-Zentrum fuer Informatik, LIPIcs, vol 24, pp 67–78
2. Betzler N, Guo J, Komusiewicz C, Niedermeier R (2011) Average parameterization and partial kernelization for computing medians. J Comput Syst Sci 77(4):774–789

3. Betzler N, Bredereck R, Niedermeier R (2014) Theoretical and empirical evaluation of data reduction for exact Kemeny rank aggregation. Auton Agents Multi-Agent Syst 28(5):721–748
4. Bodlaender HL, Downey RG, Fellows MR, Hermelin D (2009) On problems without polynomial kernels. J Comput Syst Sci 75(8):423–434
5. Crowston R, Fellows M, Gutin G, Jones M, Kim EJ, Rosamond F, Ruzsa IZ, Thomassé S, Yeo A (2014) Satisfying more than half of a system of linear equations over GF(2): a multivariate approach. J Comput Syst Sci 80(4):687–696
6. Fellows MR, Jansen BMP, Rosamond FA (2013) Towards fully multivariate algorithmics: parameter ecology and the deconstruction of computational complexity. Eur J Comb 34(3): 541–566
7. Komusiewicz C (2011) Parameterized algorithmics for network analysis: clustering & querying. PhD thesis, Technische Universität Berlin, Berlin
8. Komusiewicz C, Niedermeier R (2012) New races in parameterized algorithmics. In: MFCS '12, Bratislava. Lecture notes in computer science, vol 7464. Springer, pp 19–30

Kernelization, Permutation CSPs Parameterized above Average

Gregory Gutin
Department of Computer Science, Royal Holloway, University of London , Egham, UK

Keywords

Betweenness; Linear ordering; Parameterization above tight bound

Years and Authors of Summarized Original Work

2012; Gutin, van Iersel, Mnich, Yeo

Problem Definition

Let r be an integer and let V be a set of n variables. An *ordering* α is a bijection from V to $\{1, 2, \dots, n\}$; a *constraint* is an ordered r-tuple $(v_1, v_2, \dots, v_r)$ of distinct variables of V; α *satisfies* $(v_1, v_2, \dots, v_r)$ if $\alpha(v_1) < \alpha(v_2) < \cdots < \alpha(v_r)$. An instance of MAX-$r$-LIN-ORDERING consists of a multiset $\mathcal{C}$ of constraints, and the objective is to find an ordering that satisfies the maximum number of constraints. Note that MAX-2-LIN ORDERING is equivalent to the problem of finding a maximum weight acyclic subgraph in an integer-weighted directed graph. Since the FEEDBACK ARC SET problem is NP-hard, MAX-2-LIN ORDERING is NP-hard, and thus MAX-r-LIN-ORDERING is NP-hard for each $r \geq 2$.

Let α be an ordering chosen randomly and uniformly from all orderings and let $c \in \mathcal{C}$ be a constraint. Then the probability that α satisfies c is $1/r!$. Thus the expected number of constraints in $\mathcal{C}$ satisfied by α equals $|\mathcal{C}|/r!$. This is a lower bound on the maximum number of constraints satisfied by an ordering, and, in fact, it is a tight lower bound. This allows us to consider the following parameterized problem (AA stands for *Above Average*).

> MAX-r-LIN-ORDERING-AA
>
> *Instance:* A multiset $\mathcal{C}$ of constraints and a nonnegative integer k.
> *Parameter:* k.
> *Question:* Is there an ordering satisfying at least $|\mathcal{C}|/r! + k$ constraints?

$(1, 2, \dots, r)$ is the identity permutation of the symmetric group $\mathcal{S}_r$. We can extend MAX-r-LIN-ORDERING by considering an arbitrary subset of $\mathcal{S}_r$ rather than just $\{(1, 2, \dots, r)\}$. Instead of describing the extension for each arity $r \geq 2$, we will do it only for $r = 3$, which is our main interest, and leave the general case to the reader.

Let $\Pi \subseteq \mathcal{S}_3 = \{(1,2,3), (1,3,2), (2,1,3), (2,3,1), (3,1,2), (3,2,1)\}$ be arbitrary. For an ordering $\alpha: V \to \{1, 2, \dots, n\}$, a constraint $(v_1, v_2, v_3) \in \mathcal{C}$ is *Π-satisfied by α* if there is a permutation $\pi \in \Pi$ such that $\alpha(v_{\pi(1)}) < \alpha(v_{\pi(2)}) < \alpha(v_{\pi(3)})$. Given Π, the problem Π-CSP is the problem of deciding if there exists an ordering of V that Π-satisfies all the constraints. Every such problem is called a Permutation CSP of arity 3. We will consider the maximization version of these problems, denoted by MAX-Π-

Kernelization, Permutation CSPs Parameterized above Average, Table 1 Permutation CSPs of arity 3 (after symmetry considerations)

$\Pi \subseteq \mathcal{S}_3$	Name	Complexity
$\Pi_0 = \{(123)\}$	3-LIN-ORDERING	Polynomial
$\Pi_1 = \{(123), (132)\}$		Polynomial
$\Pi_2 = \{(123), (213), (231)\}$		Polynomial
$\Pi_3 = \{(132), (231), (312), (321)\}$		Polynomial
$\Pi_4 = \{(123), (231)\}$		NP-comp.
$\Pi_5 = \{(123), (321)\}$	BETWEENNESS	NP-comp.
$\Pi_6 = \{(123), (132), (231)\}$		NP-comp.
$\Pi_7 = \{(123), (231), (312)\}$	CIRCULAR ORDERING	NP-comp.
$\Pi_8 = \mathcal{S}_3 \setminus \{(123), (231)\}$		NP-comp.
$\Pi_9 = \mathcal{S}_3 \setminus \{(123), (321)\}$	NON-BETWEENNESS	NP-comp.
$\Pi_{10} = \mathcal{S}_3 \setminus \{(123)\}$		NP-comp.

CSP, parameterized above the average number of constraints satisfied by a random ordering of V (which can be shown to be a tight bound).

It is easy to see that there is only one distinct Π-CSP of arity 2. Guttmann and Maucher [5] showed that there are in fact only 13 distinct Π-CSPs of arity 3 up to symmetry, of which 11 are nontrivial. They are listed in Table 1 together with their complexity. Some of the problems listed in the table are well known and have special names. For example, the problem for $\Pi = \{(123), (321)\}$ is called the BETWEENNESS problem.

Gutin et al. [4] proved that all 11 nontrivial MAX-Π-CSP problems are NP-hard (even though four of the Π-CSP are polynomial).

Now observe that given a variable set V and a constraint multiset $\mathcal{C}$ over V, for a random ordering α of V, the probability of a constraint in $\mathcal{C}$ being Π-satisfied by α equals $\frac{|\Pi|}{6}$. Hence, the expected number of satisfied constraints from $\mathcal{C}$ is $\frac{|\Pi|}{6}|\mathcal{C}|$, and thus there is an ordering α of V satisfying at least $\frac{|\Pi|}{6}|\mathcal{C}|$ constraints (and this bound is tight). A derandomization argument leads to $\frac{|\Pi|}{6}$-approximation algorithms for the problems MAX-Π-CSP [1]. No better constant factor approximation is possible assuming the Unique Games Conjecture [1].

We will study the parameterization of MAX-Π-CSP above tight lower bound:

> Π-ABOVE AVERAGE (Π-AA)
>
> *Instance*: A finite set V of variables, a multiset $\mathcal{C}$ of ordered triples of distinct variables from V and a nonnegative integer k.
>
> *Parameter*: k.
>
> *Question*: Is there an ordering α of V such that at least $\frac{|\Pi|}{6}|\mathcal{C}| + k$ constraints of $\mathcal{C}$ are Π-satisfied by α?

Key Results

The following is a simple but important observation in [4] allowing one to reduce Π-AA to MAX-3-LIN-ORDERING-AA.

Proposition 1 *Let Π be a subset of $\mathcal{S}_3$ such that $\Pi \notin \{\emptyset, \mathcal{S}_3\}$. There is a polynomial time transformation f from Π-AA to* MAX-3-LIN-ORDERING-AA *such that an instance $(V, \mathcal{C}, k)$ of Π-AA is a Yes-instance if and only if $(V, \mathcal{C}', k) = f(V, \mathcal{C}, k)$ is a Yes-instance of* MAX-3-LIN-ORDERING-AA.

Using a nontrivial reduction from MAX-3-LIN-ORDERING-AA to a combination of MAX-2-LIN-ORDERING-AA and BETWEENNESS-AA and the facts that both problems admit kernels with quadratic numbers of variables and constraints (proved in [3] and [2], respectively),

Gutin et al. [4] showed that MAX-3-LIN-ORDERING-AA also admits a kernel with quadratic numbers of variables and constraints. Kim and Williams [6] partially improved this result by showing that MAX-3-LIN-ORDERING-AA admits a kernel with $O(k)$ variables.

The polynomial-size kernel result for MAX-3-LIN-ORDERING-AA and Proposition 1 imply the following (see [4] for details):

Theorem 1 ([4]) *Let Π be a subset of $\mathcal{S}_3$ such that $\Pi \notin \{\emptyset, \mathcal{S}_3\}$. The problem Π-AA admits a polynomial-size kernel with $O(k^2)$ variables.*

Open Problems

Similar to Proposition 1, it is easy to prove that, for each fixed r every Π-AA can be reduced to LIN-r-ORDERING-AA. Gutin et al. [4] conjectured that for each fixed r the problem MAX-r-LIN-ORDERING-AA is fixed-parameter tractable.

Cross-References

- ▶ Kernelization, Constraint Satisfaction Problems Parameterized above Average
- ▶ Kernelization, MaxLin Above Average

Recommended Reading

1. Charikar M, Guruswami V, Manokaran R (2009) Every permutation CSP of arity 3 is approximation resistant. In: Computational complexity 2009, Paris, pp 62–73
2. Gutin G, Kim EJ, Mnich M, Yeo A (2010) Betweenness parameterized above tight lower bound. J Comput Syst Sci 76:872–878
3. Gutin G, Kim EJ, Szeider S, Yeo A (2011) A probabilistic approach to problems parameterized above tight lower bound. J Comput Syst Sci 77:422–429
4. Gutin G, van Iersel L, Mnich M, Yeo A (2012) All ternary permutation constraint satisfaction problems parameterized above average have Kernels with quadratic number of variables. J Comput Syst Sci 78:151–163
5. Guttmann W, Maucher M (2006) Variations on an ordering theme with constraints. In: 4th IFIP international conference on theoretical computer science-TCS 2006, Santiago. Springer, pp 77–90
6. Kim EJ, Williams R (2012) Improved parameterized algorithms for above average constraint satisfaction. In: IPEC 2011, Saarbrücken. Lecture notes in computer science, vol 7112, pp 118–131

Kernelization, Planar F-Deletion

Neeldhara Misra
Department of Computer Science and Automation, Indian Institute of Science, Bangalore, India

Keywords

Finite-integer index; Meta-theorems; Protrusions; Treewidth

Years and Authors of Summarized Original Work

2012; Fomin, Lokshtanov, Misra, Saurabh
2013; Kim, Langer, Paul, Reidl, Rossmanith, Sau, Sikdar

Problem Definition

Several combinatorial optimization problems on graphs involve identifying a subset of nodes S, of the smallest cardinality, such that the graph obtained after removing S satisfies certain properties. For example, the VERTEX COVER problem asks for a minimum-sized subset of vertices whose removal makes the graph edgeless, while the FEEDBACK VERTEX SET problem involves finding a minimum-sized subset of vertices whose removal makes the graph acyclic. The $\mathcal{F}$-DELETION problem is a generic formulation that encompasses several problems of this flavor.

Let $\mathcal{F}$ be a finite set of graphs. In the $\mathcal{F}$-DELETION problem, the input is an n-vertex graph G and an integer k, and the question is if

G has a subset S of at most k vertices, such that $G - S$ does not contain a graph from $\mathcal{F}$ as a minor. The optimization version of the problem seeks such a subset of the smallest possible size. The PLANAR $\mathcal{F}$-DELETION problem is the version of the problem where $\mathcal{F}$ contains at least one planar graph. The $\mathcal{F}$-DELETION problem was introduced by [3], who gave a non-constructive algorithm running in time $O(f(k) \cdot n^2)$ for some function $f(k)$. This result was improved by [1] to $O(f(k) \cdot n)$, for $f(k) = 2^{2^{O(k \log k)}}$.

For different choices of sets of forbidden minors $\mathcal{F}$, one can obtain various fundamental problems. For example, when $\mathcal{F} = \{K_2\}$, a complete graph on two vertices, this is the VERTEX COVER problem. When $\mathcal{F} = \{C_3\}$, a cycle on three vertices, this is the FEEDBACK VERTEX SET problem. The cases of $\mathcal{F}$ being $\{K_{2,3}, K_4\}$, $\{K_4\}$, $\{\theta_c\}$, and $\{K_3, T_2\}$, correspond to removing vertices to obtain an outerplanar graph, a series-parallel graph, a diamond graph, and a graph of pathwidth one, respectively.

Tools

Most algorithms for the PLANAR $\mathcal{F}$-DELETION problem appeal to the notion of *protrusions* in graphs. An r-protrusion in a graph G is a subgraph H of treewidth at most r such that the number of neighbors of H in $G - H$ is at most r. Intuitively, a protrusion H in a graph G may be thought of as subgraph of small treewidth which is cut off from the rest of the graph by a small separator.

Usually, as a means of preprocessing, protrusions are identified and *replaced* by smaller ones, while maintaining equivalence. The notion of graph replacement in this fashion originates in the work of [4]. The modern notion of protrusion reductions have been employed in various contexts [2, 5, 6, 12]. A widely used method for developing a protrusion replacement algorithm is via the notion of *finite-integer index*. Roughly speaking, this property ensures that graphs can be related under some appropriate notion of equivalence with respect to the problem, and that there are only finitely many equivalence classes. This allows us to identify the class that the protrusion belongs to and replace it with a canonical representative for that class.

Key Results

The algorithms proposed for PLANAR $\mathcal{F}$-DELETION usually have the following ingredients. First, the fact that $\mathcal{F}$ contains a planar graph implies that any YES-instance of the problem must admit a small subset of vertices whose removal leads to a graph of small treewidth. It turns out that such graphs admit a convenient structure from the perspective of the existence of protrusions. In particular, most of the graph can be decomposed into protrusions. From here, there are two distinct themes.

In the first approach, the protrusions are *replaced* by smaller, equivalent graphs. Subsequently, we have a graph that has no large protrusions. For such instances, it can be shown that if there is a solution, there is always one that is incident to a constant fraction of the edges in the graph, and this leads to a randomized algorithm by branching. Notably, the protrusion replacement can be performed by an algorithm that guarantees the removal of a constant fraction of vertices in every application. This helps in ensuring that the overall running time of the algorithm has a linear dependence on the size of the input. This algorithm is limited to the case when all graphs in $\mathcal{F}$ are connected, as is required in demonstrating finite-integer index.

Theorem 1 ([8]) *When every graph in $\mathcal{F}$ is connected, there is a randomized algorithm solving* PLANAR $\mathcal{F}$-DELETION *in time $2^{O(k)} \cdot n$.*

The second approach involves exploring the structure of the instance further. Here, an $O(k)$-sized subset of vertices is identified, with the key property that there is a solution that lives within it. The algorithm then proceeds to exhaustively branch on these vertices. This technique requires a different protrusion decomposition from the previous one. The overall algorithm is implemented using iterative compression. Since the protrusions are not replaced, this algorithm works for all instances of PLANAR $\mathcal{F}$-DELETION, without any further assumptions on the family $\mathcal{F}$. While both approaches lead to algorithms that are single-exponential in k, the latter has a quadratic dependence on the size of the input.

Theorem 2 ([11]) PLANAR-$\mathcal{F}$-DELETION *can be solved in time* $2^{O(k)} \cdot n^2$.

In the context of approximation algorithms, the protrusion replacement is more intricate, because the notion of equivalence is now more demanding. The replacement should preserve not only the exact solutions, but also approximate ones. By appropriately adapting the machinery of replacements with lossless protrusion replacers, the problem admits the following approximation algorithm.

Theorem 3 ([8]) PLANAR $\mathcal{F}$-DELETION *admits a randomized constant ratio approximation algorithm.*

The PLANAR $\mathcal{F}$-DELETION problem also admits efficient preprocessing algorithms. Formally, a kernelization algorithm for the problem takes an instance (G, k) as input and outputs an equivalent instance (H, k') where the size of the output is bounded by a function of k. If the size of the output is bounded by a polynomial function of k, then it is called a polynomial kernel. The reader is referred to the survey [13] for a more detailed introduction to kernelization.

The technique of protrusion replacement was developed and used successfully for kernelization algorithms on sparse graphs [2,5]. These methods were also used for the special case of the PLANAR $\mathcal{F}$-DELETION problem when $\mathcal{F}$ is a graph with two vertices and constant number of parallel edges [6]. In the general setting of PLANAR $\mathcal{F}$-DELETION, kernelization involves anticipating protrusions, that is, identifying subgraphs that become protrusions after the removal of some vertices from an optimal solution. These "near-protrusions" are used to find irrelevant edges, i.e., an edge whose removal does not change the problem, leading to natural reduction rules. The process of finding an irrelevant edge appeals to the well-quasi-ordering of a certain class of graphs as a subroutine.

Theorem 4 ([8]) PLANAR $\mathcal{F}$-DELETION *admits a polynomial kernel.*

Applications

The algorithms for PLANAR $\mathcal{F}$-DELETION apply to any vertex deletion problem that can be described as hitting minor models of some fixed finite family that contains a planar graph.

For a finite set of graphs $\mathcal{F}$, let $\mathcal{G}_{\mathcal{F},k}$ be a class of graphs such that for every $G \in \mathcal{G}_{\mathcal{F},k}$ there is a subset of vertices S of size at most k such that $G \setminus S$ has no minor from $\mathcal{F}$. The following combinatorial result is a consequence of the kernelization algorithm for PLANAR $\mathcal{F}$-DELETION.

Theorem 5 ([8]) *For every set $\mathcal{F}$ that contains a planar graph, every minimal obstruction for $\mathcal{G}_{\mathcal{F},k}$ is of size polynomial in k.*

Kernelization algorithms on apex-free and H-minor-free graphs for all bidimensional problems from [5] can be implemented in linear time by employing faster protrusion reducers. This leads to randomized linear time, linear kernels for several problems.

In the framework for obtaining EPTAS on H-minor-free graphs in [7], the running time of approximation algorithms for many problems is $f(1/\varepsilon) \cdot n^{O(g(H))}$, where g is some function of H only. The only bottleneck for improving polynomial-time dependence is a constant factor approximation algorithm for TREEWIDTH η-DELETION. Using Theorem 3 instead, each EPTAS from [7] runs in time $O(f(1/\varepsilon) \cdot n^2)$. For the same reason, the PTAS algorithms for many problems on unit disk and map graphs from [9] become EPTAS algorithms.

Open Problems

An interesting direction for further research is to investigate PLANAR $\mathcal{F}$-DELETION when none of the graphs in $\mathcal{F}$ is planar. The most interesting case here is when $\mathcal{F} = \{K_5, K_{3,3}\}$, also known as the VERTEX PLANARIZATION problem. The work in [10] demonstrates an algorithm with running time $2^{O(k \log k)}n$, which notably has a linear-time dependence on n. It remains open as to whether VERTEX PLANARIZATION can be solved in $2^{O(k)}n$ time. The question of polynomial kernels in the non-planar setting is also open, in particular, even the specific case of $\mathcal{F} = \{K_5\}$ is unresolved.

Cross-References

▸ Bidimensionality
▸ Kernelization, Preprocessing for Treewidth
▸ Treewidth of Graphs

Recommended Reading

1. Bodlaender HL (1997) Treewidth: algorithmic techniques and results. In: 22nd international symposium on mathematical foundations of computer science (MFCS), Bratislava, vol 1295, pp 19–36
2. Bodlaender HL, Fomin FV, Lokshtanov D, Penninkx E, Saurabh S, Thilikos DM (2009) (Meta) kernelization. In: Proceedings of the 50th annual IEEE symposium on foundations of computer science (FOCS), Atlanta, pp 629–638
3. Fellows MR, Langston MA (1988) Nonconstructive tools for proving polynomial-time decidability. J ACM 35(3):727–739
4. Fellows MR, Langston MA (1989) An analogue of the Myhill-Nerode theorem and its use in computing finite-basis characterizations (extended abstract). In: Proceedings of the 30th annual IEEE symposium on foundations of computer science (FOCS), Research Triangle Park, pp 520–525
5. Fomin FV, Lokshtanov D, Saurabh S, Thilikos DM (2010) Bidimensionality and kernels. In: Proceedings of the twenty-first annual ACM-SIAM symposium on discrete algorithms (SODA), Austin, pp 503–510
6. Fomin FV, Lokshtanov D, Misra N, Philip G, Saurabh S (2011) Hitting forbidden minors: approximation and kernelization. In: Proceedings of the 8th international symposium on theoretical aspects of computer science (STACS), LIPIcs, Dortmund, vol 9, pp 189–200
7. Fomin FV, Lokshtanov D, Raman V, Saurabh S (2011) Bidimensionality and EPTAS. In: Proceedings of the 22nd annual ACM-SIAM symposium on discrete algorithms (SODA), San Francisco. SIAM, pp 748–759
8. Fomin FV, Lokshtanov D, Misra N, Saurabh S (2012) Planar F-deletion: approximation, kernelization and optimal FPT algorithms. In: Proceedings of the 2012 IEEE 53rd annual symposium on foundations of computer science, New Brunswick, pp 470–479
9. Fomin FV, Lokshtanov D, Saurabh S (2012) Bidimensionality and geometric graphs. In: Proceedings of the 23rd annual ACM-SIAM symposium on discrete algorithms (SODA), Kyoto. SIAM, pp 1563–1575
10. Jansen BMP, Lokshtanov D, Saurabh S (2014) A near-optimal planarization algorithm. In: Proceedings of the twenty-fifth annual ACM-SIAM symposium on discrete algorithms, (SODA), Portland, pp 1802–1811
11. Kim EJ, Langer A, Paul C, Reidl F, Rossmanith P, Sau I, Sikdar S (2013) Linear kernels and single-exponential algorithms via protrusion decompositions. In: Proceedings of the 40th international colloquium on automata, languages, and programming (ICALP), Riga, Part I, pp 613–624
12. Langer A, Reidl F, Rossmanith P, Sikdar S (2012) Linear kernels on graphs excluding topological minors. CoRR abs/1201.2780
13. Lokshtanov D, Misra N, Saurabh S (2012) Kernelization – preprocessing with a guarantee. In: Bodlaender HL, Downey R, Fomin FV, Marx D (eds) The multivariate algorithmic revolution and beyond. Bodlaender, HansL. and Downey, Rod and Fomin, FedorV. and Marx, Dániel (eds) Lecture notes in computer science, vol 7370. Springer, Berlin/Heidelberg, pp 129–161

Kernelization, Polynomial Lower Bounds

Stefan Kratsch
Department of Software Engineering and Theoretical Computer Science, Technical University Berlin, Berlin, Germany

Keywords

Kernelization; Parameterized complexity; Satisfiability; Sparsification

Years and Authors of Summarized Original Work

2010; Dell, van Melkebeek

Problem Definition

The work of Dell and van Melkebeek [4] refines the framework for lower bounds for kernelization introduced by Bodlaender et al. [1] and Fortnow and Santhanam [6]. The main contribution is that their results yield a framework for proving polynomial lower bounds for kernelization rather than ruling out all polynomial kernels for a problem; this, for the first time, gives a technique for proving that some polynomial kernelizations are actually best possible, modulo reasonable complexity assumptions. A further important aspect is that, rather than studying kernelization directly, the authors give lower bounds for a far more general oracle communication protocol. In this way, they also obtain strong lower bounds for sparsification, lossy compression (in the sense of Harnik and Naor [7]), and probabilistically checkable proofs (PCPs).

To explain the connection between kernelization and oracle communication protocols, let us first recall the following. A *parameterized problem* is a language $\mathcal{Q} \subseteq \Sigma^* \times \mathbb{N}$; the second component k of instances $(x,k) \in \Sigma^* \times \mathbb{N}$ is called the *parameter*. A *kernelization for $\mathcal{Q}$ with size $h\colon \mathbb{N} \to \mathbb{N}$* is an efficient algorithm that gets as input an instance $(x,k) \in \Sigma^* \times \mathbb{N}$ and returns an equivalent instance (x',k'), i.e., such that $(x,k) \in \mathcal{Q}$ if and only if $(x',k') \in \mathcal{Q}$, with $|x'|, k' \leq h(k)$. If $h(k)$ is polynomially bounded in k, then we also call it a polynomial kernelization.

One way to use a kernelization is to first simplify a given input instance and then solve the reduced instance by any (possibly brute-force) algorithm; together this yields an algorithm for solving the problem in question. If we abstract out the algorithm by saying that the answer for the reduced instance is given by an oracle, then we arrive at a special case of the following communication protocol.

Definition 1 (oracle communication protocol [4]) An *oracle communication protocol* for a language L is a communication protocol for two players. The first player is given the input x and has to run in time polynomial in the length of the input; the second player is computationally unbounded but is not given any part of x. At the end of the protocol, the first player should be able to decide whether $x \in L$. The cost of the protocol is the number of bits of communication from the first player to the second player.

As an example, if $\mathcal{Q}$ has a kernelization with size h, then instances (x,k) can be solved by a protocol of cost $h(k)$. It suffices that the first player can compute a reduced instance (x',k') and send it to the oracle who decides membership of (x',k') in $\mathcal{Q}$; this yields the desired answer for whether $(x,k) \in \mathcal{Q}$. Note that the communication protocol is far more general than kernelization because it makes no assumption about what exactly is sent (or in what encoding). More importantly, it also allows multiple rounds of communication, and the behavior of the oracle could also be active rather than just answering queries for the first player. Thus, the obtained lower bounds for oracle communication protocols are very robust, covering also relaxed forms of kernelization (like bikernels and compressions), and also yield the other mentioned applications.

Key Results

A central result in the work of Dell and van Melkebeek [4] (see also [5]) is the following lemma, called *complementary witness lemma*.

Lemma 1 (complementary witness lemma [4]) *Let L be a language and $t\colon \mathbb{N} \to \mathbb{N} \setminus \{0\}$ be polynomially bounded such that the problem of deciding whether at least one out of $t(s)$ inputs of length at most s belongs to L has an oracle communication protocol of cost $\mathcal{O}(t(s) \log t(s))$, where the first player can be conondeterministic. Then $L \in$* coNP/poly.

A previous work of Fortnow and Santhanam [6] showed that an efficient algorithm for encoding any t instances $x_1, \ldots, x_t$ of size at most s into one instance y of size poly(s) such that $y \in L$ if and only if at least one x_i is in L implies $L \in$ coNP/poly. (We recall that this settled the *OR-distillation conjecture* of Bodlaender et al. [1] and allowed their framework to rule out polynomial kernels under the assumption that $\mathsf{NP} \not\subseteq \mathsf{coNP/poly}$.) Lemma 1 is obtained by a more detailed analysis of this result and requires an encoding of the OR of $t(s)$ instances into one instance of size $\mathcal{O}(t(s)\log t(s))$ rather than allowing only size poly(s) for all values of t. This focus on the number $t(s)$ of instances in relation to the maximum instance size s is the key for getting polynomial lower bounds for kernelization (and other applications). In this overview, we will not discuss the possibility of conondeterministic behavior of the first player, but the interested reader is directed to [9, 10] for applications thereof.

Before outlining further results of Dell and van Melkebeek [4], let us state a lemma that captures one way of employing the complementary witness lemma for polynomial lower bounds for kernelization. The lemma is already implicit in [4] and is given explicitly in follow-up work of Dell and Marx [3] (it can also be found in the current full version [5] of [4]). We recall that $\text{OR}(L)$ refers to the language of all tuples $(x_1, \ldots, x_t)$ such that at least one x_i is contained in L.

Lemma 2 ([3,5]) *Suppose that a parameterized problem Π has the following property for some constant c: For some* NP*-complete language L, there exists a polynomial-time mapping reduction from $\text{OR}(L)$ to Π that maps an instance $(x_1, \ldots, x_t)$ of $\text{OR}(L)$ in which each x_i has size at most s to an instance of Π with parameter $k \leq t^{1/c+o(1)} \cdot \text{poly}(s)$. Then Π does not have a communication protocol of cost $\mathcal{O}(k^{c-\epsilon})$ for any constant $\epsilon > 0$ unless $\mathsf{NP} \subseteq \mathsf{coNP/poly}$, even when the first player is conondeterministic.*

Intuitively, Lemma 2 follows from Lemma 1 because if the reduction and communication protocol in Lemma 2 both exist (for all t), then we can choose $t(s)$ large enough (but polynomially bounded in s) such that for all s we get an oracle communication protocol of cost $\mathcal{O}(t(s))$ as required for Lemma 1. This implies $L \in \mathsf{coNP/poly}$ and, hence, $\mathsf{NP} \subseteq \mathsf{coNP/poly}$ (since L is NP-complete). As discussed earlier, any kernelization yields an oracle communication protocol with cost equal to the kernel size and, thus, this bound carries over directly to kernelization.

Let us now state the further results of Dell and van Melkebeek [4] using the context of Lemma 2. The central result is the following theorem on lower bounds for vertex cover on d-uniform hypergraphs.

Theorem 1 ([4]) *Let $d \geq 2$ be an integer and ϵ a positive real. If $\mathsf{NP} \not\subseteq \mathsf{coNP/poly}$, there is no protocol of cost $\mathcal{O}(n^{d-\epsilon})$ to decide whether a d-uniform hypergraph on n vertices has a vertex cover of at most k vertices, even when the first player is conondeterministic.*

To prove Theorem 1, Dell and van Melkebeek devise a reduction from $\text{OR}(SAT)$ to CLIQUE on d-uniform hypergraphs parameterized by the number of vertices (fulfilling the assumption of Lemma 2 for $c = d$). This reduction relies on an intricate lemma, the *packing lemma*, that constructs a d-uniform hypergraph with t cliques on s vertices each, but having only about $\mathcal{O}(t^{1/d+o(1)} \cdot s)$ vertices and no further cliques of size s. In follow-up work, Dell and Marx [3] give a simpler proof for Theorem 1 without making use of the packing lemma, but use the lemma for another of their results.

Note that the stated bound for VERTEX COVER in d-uniform hypergraphs follows by complementation. Furthermore, since every nontrivial instance has $k \leq n$, this also rules out kernelization to size $\mathcal{O}(k^{d-\epsilon})$. The following lower bound for SATISFIABILITY is obtained by giving a reduction from VERTEX COVER on d-uniform hypergraphs with parameter n. In the reduction, hyperedges of size d are encoded by positive clauses on d variables (one per vertex), and an additional part of the formula (which requires $d \geq 3$) checks that at most k of these variables are set to true.

Theorem 2 ([4]) *Let $d \geq 3$ be an integer and ϵ a positive real. If $\mathsf{NP} \not\subseteq \mathsf{coNP/poly}$, there is no* *protocol of cost $\mathcal{O}(n^{d-\epsilon})$ to decide whether an n-variable d-CNF formula is satisfiable, even when the first player is conondeterministic.*

Finally, the following theorem proves that several known kernelizations for graph modification problems are already optimal. The theorem is proved by a reduction from VERTEX COVER (on graphs) with parameter k that is similar in spirit to the classical result of Lewis and Yannakakis on NP-completeness of the Π-VERTEX DELETION problem for nontrivial hereditary properties Π. Note that Theorem 3 requires that the property Π is not only hereditary, i.e., inherited by *induced* subgraphs, but inherited by *all subgraphs*.

Theorem 3 ([4]) *Let Π be a graph property that is inherited by subgraphs and is satisfied by infinitely many but not all graphs. Let ϵ be a positive real. If* NP $\not\subseteq$ coNP/poly, *there is no protocol of cost $\mathcal{O}(k^{2-\epsilon})$ for deciding whether a graph satisfying Π can be obtained from a given graph by removing at most k vertices.*

As an example, the theorem implies that the FEEDBACK VERTEX SET problem does not admit a kernelization with size $\mathcal{O}(k^{2-\epsilon})$. This is in fact tight since a kernelization by Thomassé [12] achieves $\mathcal{O}(k^2)$ vertices and $\mathcal{O}(k^2)$ edges (cf. [4]); improving to $\mathcal{O}(k^{2-\epsilon})$ edges is ruled out since it would yield an encoding in size $\mathcal{O}(k^{2-\epsilon'})$. Similarly, the well-known kernelization for VERTEX COVER to $2k$ vertices is tight and cannot, in general, be expected to yield instances with less than the trivial $\mathcal{O}(k^2)$ edges.

Applications

Several authors have used the present approach to get polynomial lower bounds for kernelizations of certain parameterized problems; see, e.g., [2, 3, 8, 11]. Similarly, some results make use of conondeterminism [9, 10] and the more general setting of lower bounds for oracle communication protocols [11].

Open Problems

Regarding applications it would be interesting to have more lower bounds that use the full generality of the oracle communication protocols. Furthermore, it is an open problem to relax the assumption of NP $\not\subseteq$ coNP/poly to the minimal P $\neq$ NP.

Recommended Reading

1. Bodlaender HL, Downey RG, Fellows MR, Hermelin D (2009) On problems without polynomial kernels. J Comput Syst Sci 75(8):423–434
2. Cygan M, Grandoni F, Hermelin D (2013) Tight kernel bounds for problems on graphs with small degeneracy – (extended abstract). In: Bodlaender HL, Italiano GF (eds) ESA. Lecture notes in computer science, vol 8125. Springer, pp 361–372
3. Dell H, Marx D (2012) Kernelization of packing problems. In: SODA, Kyoto. SIAM, pp 68–81
4. Dell H, van Melkebeek D (2010) Satisfiability allows no nontrivial sparsification unless the polynomial-time hierarchy collapses. In: Schulman LJ (ed) STOC, Cambridge. ACM, pp 251–260
5. Dell H, van Melkebeek D (2010) Satisfiability allows no nontrivial sparsification unless the polynomial-time hierarchy collapses. Electron Colloq Comput Complex 17:38
6. Fortnow L, Santhanam R (2011) Infeasibility of instance compression and succinct PCPs for NP. J Comput Syst Sci 77(1):91–106
7. Harnik D, Naor M (2010) On the compressibility of $\mathcal{NP}$ instances and cryptographic applications. SIAM J Comput 39(5):1667–1713
8. Hermelin D, Wu X (2012) Weak compositions and their applications to polynomial lower bounds for kernelization. In: SODA, Kyoto. SIAM, pp 104–113
9. Kratsch S (2012) Co-nondeterminism in compositions: a kernelization lower bound for a Ramsey-type problem. In: SODA, Kyoto. SIAM, pp 114–122
10. Kratsch S, Pilipczuk M, Rai A, Raman V (2012) Kernel lower bounds using co-nondeterminism: finding induced hereditary subgraphs. In: Fomin FV, Kaski P (eds) SWAT, Helsinki. Lecture notes in computer science, vol 7357. Springer, pp 364–375
11. Kratsch S, Philip G, Ray S (2014) Point line cover: the easy kernel is essentially tight. In: SODA, Portland. SIAM, pp 1596–1606
12. Thomassé S (2010) A quadratic kernel for feedback vertex set. ACM Trans Algorithms 6(2), 32:1–32:8

Kernelization, Preprocessing for Treewidth

Stefan Kratsch
Department of Software Engineering and Theoretical Computer Science, Technical University Berlin, Berlin, Germany

Keywords

Kernelization; Parameterized complexity; Preprocessing; Structural parameters; Treewidth

Years and Authors of Summarized Original Work

2013; Bodlaender, Jansen, Kratsch

Problem Definition

This work undertakes a theoretical study of preprocessing for the NP-hard TREEWIDTH problem of finding a *tree decomposition* of width at most k for a given graph G. In other words, given G and $k \in \mathbb{N}$, the question is whether G has *treewidth* at most k. Several efficient reduction rules are known that provably preserve the correct answer, and experimental studies show significant size reductions [3, 5]. The present results study these and further newly introduced rules and obtain upper and lower bounds within the framework of *kernelization* from parameterized complexity.

The general interest in computing tree decompositions is motivated by the well-understood approach of using dynamic programming on tree decompositions that is known to allow fast algorithms on graphs of bounded treewidth (but with runtime exponential in the treewidth). A bottleneck for practical applications is the need for finding, as a first step, a sufficiently good tree decomposition; the best known exact algorithm due to Bodlaender [2] runs in time exponential in k^3 and is thus only of theoretical interest. This motivates the use of heuristics and preprocessing to find a reasonably good tree decomposition quickly.

Tree Decompositions and Treewidth

A *tree decomposition* for a graph $G = (V, E)$ consists of a tree $T = (N, F)$ and a family $\mathcal{X} := \{X_i \mid i \in N, X_i \subseteq V\}$. The sets X_i are also called *bags* and the vertices of T are usually referred to as *nodes* to avoid confusion with G; there is exactly one bag X_i associated with each node $i \in N$. The pair $(T, \mathcal{X})$ must fulfill the following three properties: (1) Every vertex of G is contained in at least one bag; (2) For each edge $\{u, v\} \in E$ there must be a bag X_i containing both u and v; (3) For each vertex v of G the set of nodes i of T with $v \in X_i$ induce a (connected) subtree of T. The *width of a tree decomposition* $(T, \mathcal{X})$ is equal to the size of the largest bag $X_i \in \mathcal{X}$ minus one. The *treewidth* of a graph G, denoted $\mathsf{tw}(G)$, is the smallest width taken over all tree decompositions of G.

Parameters

The framework of parameterized complexity allows the study of the TREEWIDTH problem with respect to different *parameters*. A parameter is simply an integer value associated with each problem instance. The *standard parameter* for an optimization problem like TREEWIDTH is the desired solution quality k and we denote this problem by TREEWIDTH(k). Apart from this, *structural parameters* are considered that capture structural aspects of G. For example, the work considers the behavior of TREEWIDTH when the input graph G has a small vertex cover S, i.e., such that deletion of $\ell = |S|$ vertices yields an independent set, with ℓ being used as the parameter. Similarly, several other parameters are discussed, foremost among them the feedback vertex set number and the vertex deletion distance to a single clique; the corresponding vertex sets are called *modulators*, e.g., a feedback vertex set is a modulator to a forest. We denote the arising parameterized problems by TREEWIDTH(vc), TREEWIDTH(fvs), and TREEWIDTH($\mathsf{vc}(\overline{G})$). To decouple the overhead of finding, e.g., a minimum vertex cover for G, all these variants assume that an appropriate modulator is given along with the input and the obtained guarantees are in terms of the size of this modulator. Since all studied parameters can be efficiently approximated to within a constant factor of the optimum, not providing an (optimal) modulator gives only a constant-factor blowup in the obtained results.

Kernelization

A kernelization for a problem with parameter ℓ is an efficient algorithm that given an instance (x, ℓ) returns an equivalent instance (x', ℓ') of size and parameter value ℓ' bounded by some computable function of ℓ. If the bound is polynomial in ℓ then we have a *polynomial kernelization*. Specialized to, for example, TREEWIDTH(vc) a polynomial kernelization would have the following behavior: It gets as input an instance (G, S, k), asking whether the treewidth of G is at most k, where S is a vertex cover for G. In polynomial time it creates an instance (G', S', k') such that: (1) The size of the instance (G', S', k') and the parameter value $|S'|$ are bounded polynomially in k; (2) The set S' is a vertex cover of G'; (3) The graph G has treewidth at most k if and only if G' has treewidth at most k'.

Key Results

The kernelization lower bound framework of Bodlaender et al. [6] together with recent results of Drucker [9] is known to imply that TREEWIDTH(k) admits no polynomial kernelization unless NP $\subseteq$ coNP/poly and the polynomial hierarchy collapses. The present work takes a more detailed look at polynomial kernelization for TREEWIDTH with respect to structural parameters. The results are as follows.

Theorem 1 TREEWIDTH(vc) *i.e., parameterized by vertex cover number, admits a polynomial kernelization to an equivalent instance with* $\mathcal{O}((\mathsf{vc}(G))^3)$ *vertices.*

An interesting feature of this result is that it uses only three simple reduction rules that are well known and often used (cf. [5]). Two rules address so-called simplicial vertices, whose neighborhood is a clique, and a third rule inserts edges between certain pairs of vertices that have a large number of shared vertices. Analyzing these empirically successful rules with respect to the vertex cover number of the input graph yields a kernelization. A fact that nicely complements the observed experimental success.

Theorem 2 TREEWIDTH(fvs) *i.e., parameterized by feedback vertex set number, admits a polynomial kernelization to an equivalent instance with* $\mathcal{O}((\mathsf{fvs}(G))^4)$ *vertices.*

The feedback vertex set number of a graph is upper bounded by its vertex cover number, and forests have feedback vertex set number zero but arbitrarily large vertex cover number. Thus, for large families of input graphs, this second result is stronger. The result again builds on several known reduction rules (including the above ones), among others, for handling vertices that are *almost simplicial*, i.e., all but one neighboring vertex form a clique. On top of these, several new rules are added. One of them addresses a previously uncovered case of almost simplicial vertex removal, namely, when the vertex has degree exactly $k + 1$, where k is the desired treewidth bound. Furthermore, these reduction rules lead to a structure dubbed *clique-seeing paths*, which takes a series of fairly technical rules and analysis to reduce and bound. Altogether, this combination leads to the above result.

Theorem 3 TREEWIDTH($\mathsf{vc}(\overline{G})$) *i.e., parameterized by deletion distance to a single clique, admits no polynomial kernelization unless* NP $\subseteq$ coNP/poly *and the polynomial hierarchy collapses.*

The proof uses the notion of a *cross-composition* introduced by Bodlaender et al. [8], which builds directly on the kernelization lower bound framework of Bodlaender et al. [6] and Fortnow and Santhanam [10]. The cross-composition builds on the proof of NP-

K

completeness of TREEWIDTH by Arnborg et al. [1], which uses a Karp reduction from CUTWIDTH to TREEWIDTH. This construction is extended significantly to yield a cross-composition of CUTWIDTH ON SUBCUBIC GRAPHS (i.e., graphs of maximum degree three) into TREEWIDTH, which, roughly, requires an encoding of many CUTWIDTH instances into a single instance of TREEWIDTH with sufficiently small parameter.

Overall, together with previously known results, the obtained upper and lower bounds for TREEWIDTH cover a wide range of natural parameter choices (see the discussion in [7]). If $\mathcal{C}$ is any graph class that contains all cliques, then the vertex deletion distance of a graph G to $\mathcal{C}$ is upper bounded by $\mathsf{vc}(\overline{G})$. Thus, TREEWIDTH parameterized by distance to $\mathcal{C}$ does not admit a polynomial kernelization unless $\mathsf{NP} \subseteq \mathsf{coNP/poly}$. This includes several well-studied classes like interval graphs, cographs, and perfect graphs. Since TREEWIDTH remains NP-hard on bipartite graphs, the result for parameterization by feedback vertex set number cannot be generalized to vertex deletion to a bipartite graph. It may, however, be possible to generalize this parameter to vertex deletion distance to an outerplanar graph, i.e., planar graphs having an embedding with all vertices appearing on the outer face. Since these graphs generalize forests, this value is upper bounded by the feedback vertex set number.

Theorem 4 WEIGHTED TREEWIDTH(vc) *i.e., parameterized by vertex cover number, admits no polynomial kernelization unless* $\mathsf{NP} \subseteq \mathsf{coNP/poly}$ *and the polynomial hierarchy collapses.*

In the WEIGHTED TREEWIDTH problem, each vertex comes with an integer weight, and the size of a bag in the tree decomposition is defined as the sum of the weights of its vertices. (To note, the present paper uses an extra deduction of one such that treewidth and weighted treewidth coincide for graphs with all vertices having weight one.) The result is proved by a cross-composition from TREEWIDTH (to WEIGHTED TREEWIDTH parameterized by vertex cover number) and complements the polynomial kernelization for the unweighted case. A key idea for the cross-composition is to use a result of Bodlaender and Möhring [4] on the behavior of treewidth under the join operation on graphs. This is combined with replacing all edges (in input graphs and join edges) by using a small number of newly introduced vertices of high weight.

Open Problems

A particular interesting case left open by existing results on polynomial kernelization for structural parameterizations of TREEWIDTH is the vertex deletion distance to outerplanar graphs.

Recommended Reading

1. Arnborg S, Corneil DG, Proskurowski A (1987) Complexity of finding embeddings in a k-tree. SIAM J Algebra Discret 8(2):277–284. doi:10.1137/0608024
2. Bodlaender HL (1996) A linear-time algorithm for finding tree-decom-positions of small treewidth. SIAM J Comput 25(6):1305–1317. doi:10.1145/167088.167161
3. Bodlaender HL, Koster AMCA (2006) Safe separators for treewidth. Discret Math 306(3): 337–350. doi:10.1016/j.disc.2005.12.017
4. Bodlaender HL, Möhring RH (1993) The pathwidth and treewidth of cographs. SIAM J Discret Math 6(2):181–188. doi:10.1137/0406014
5. Bodlaender HL, Koster AMCA, van den Eijkhof F (2005) Preprocessing rules for triangulation of probabilistic networks. Comput Intell 21(3):286–305. doi:10.1111/j.1467-8640.2005.00274.x
6. Bodlaender HL, Downey RG, Fellows MR, Hermelin D (2009) On problems without polynomial kernels. J Comput Syst Sci 75(8):423–434
7. Bodlaender HL, Jansen BMP, Kratsch S (2013) Preprocessing for treewidth: a combinatorial analysis through kernelization. SIAM J Discret Math 27(4):2108–2142
8. Bodlaender HL, Jansen BMP, Kratsch S (2014) Kernelization lower bounds by cross-composition. SIAM J Discret Math 28(1):277–305
9. Drucker A (2012) New limits to classical and quantum instance compression. In: FOCS, New Brunswick. IEEE Computer Society, pp 609–618
10. Fortnow L, Santhanam R (2011) Infeasibility of instance compression and succinct PCPs for NP. J Comput Syst Sci 77(1):91–106

Kernelization, Turing Kernels

Henning Fernau
Fachbereich 4, Abteilung Informatikwissenschaften, Universität Trier, Trier, Germany
Institute for Computer Science, University of Trier, Trier, Germany

Keywords

Karp reduction; Kernelization; Kernel size; Turing kernelization; Turing reduction

Years and Authors of Summarized Original Work

2012; Binkele-Raible, Fernau, Fomin, Lokshtanov, Saurabh
2013; Hermelin, Kratsch, Soltys, Wahlström, Wu
2014; Jansen

Definition and Discussion

The basic definition of the field expresses kernelization as a Karp (many-one) self-reduction. Classical complexity and recursion theory offers quite a lot of alternative and more general notions of reducibilities. The most general notion, that of a Turing reduction, motivates the following definition:

Let (Q, κ) be a parameterized problem over a finite alphabet Σ.

- An *input-bounded oracle* for (Q, κ) is an oracle that, for any given input $x \in \Sigma^*$ of (Q, κ) and any bound t, first checks if $|x|, |\kappa(x)| \leq t$, and if this is certified, it decides in constant time whether the input x is a YES instance of (Q, κ).
- A *Turing kernelization (algorithm)* for (Q, κ) is an algorithm that, provided with access to some input-bounded oracle for (Q, κ), decides on input $x \in \Sigma^*$ in polynomial time whether x is a YES instance of (Q, κ) or not. During its computation, the algorithm can produce (polynomially many) oracle queries x' with bound $t = h(\kappa(x))$, where h is an arbitrary computable function. The function h is referred to as the *size* of the kernel.

If only one oracle access is permitted in a run of the algorithm, we basically get the classical notion of a (many-one or Karp) kernelization.

A more general definition was given in [4], allowing access to a different (auxiliary) problem (Q', κ'). As long as there is a computable reduction from Q' to Q, this does not make much of a difference, as we could translate the queries to Q' into queries of Q. Therefore, we prefer to use the definition given in [1].

Out-Branching: Showing the Difference

In [1], the first example of a natural problem is provided that admits a Turing kernel of polynomial size, but (most likely) no Karp kernel of polynomial size. We provide some details in the following.

Problem Definition

A subdigraph T of a digraph D is an *out-tree* if T is an oriented tree with only one vertex r of indegree zero (called the *root*). The vertices of T of outdegree zero are called *leaves*. If T is a spanning out-tree, i.e., $V(T) = V(D)$, then T is called an *out-branching* of D. The DIRECTED MAXIMUM LEAF OUT-BRANCHING problem is to find an out-branching in a given digraph with the maximum number of leaves. The parameterized version of the DIRECTED MAXIMUM LEAF OUT-BRANCHING problem is k-LEAF OUT-BRANCHING, where for a given digraph D and integer k, it is asked to decide whether D has an out-branching with at least k leaves. If we replace "out-branching" with "out-tree" in the definition of k-LEAF OUT-BRANCHING, we get a problem called k-LEAF

OUT-TREE. The parameterization κ is set to k in both problems. As the two problems are easily translatable into each other, we focus on k-LEAF OUT-BRANCHING as the digraph analogue of the well-known MAXIMUM LEAF SPANNING TREE problem.

Key Results

It is shown that the problem variant where an explicit root is given as additional input, called ROOTED k-LEAF OUT-BRANCHING, admits a polynomial Karp kernel. Alternatively, this variant can be seen as a special case of k-LEAF OUT-BRANCHING by adding one vertex of indegree zero and outdegree one, pointing to the designated root of the original graph. By making a call to this oracle for each of the vertices as potential roots, this provides a Turing kernelization of polynomial size for k-LEAF OUT-BRANCHING. This result is complemented by showing that k-LEAF OUT-TREE has no polynomial Karp kernel unless $coNP \subseteq NP/poly$.

We list the reduction rules leading to the polynomial-size kernel for the rooted version in the following.

Reachability Rule:If there exists a vertex u which is disconnected from the root r, then return NO.

Useless Arc Rule:If vertex u disconnects a vertex v from the root r, then remove the arc vu.

Bridge Rule:If an arc uv disconnects at least two vertices from the root r, contract the arc uv.

Avoidable Arc Rule:If a vertex set S, $|S| \leq 2$, disconnects a vertex v from the root r, $vw \in A(D)$ and $xw \in A(D)$ for all $x \in S$, then delete the arc vw.

Two Directional Path Rule:If there is a path $P = p_1 p_2 \dots p_{l-1} p_l$ with $l = 7$ or $l = 8$ such that

- p_1 and $p_{\text{in}} \in \{p_{l-1}, p_l\}$ are the only vertices with in-arcs from the outside of P
- p_l and $p_{\text{out}} \in \{p_1, p_2\}$ are the only vertices with out-arcs to the outside of P
- The path P is the unique out-branching of $D[V(P)]$ rooted at p_1
- There is a path Q that is the unique out-branching of $D[V(P)]$ rooted at p_{in} and ending in p_{out}
- The vertex after p_{out} on P is not the same as the vertex after p_l on Q

then delete $R = P \setminus \{p_1, p_{\text{in}}, p_{\text{out}}, p_l\}$ and all arcs incident to these vertices from D. Add two vertices u and v and the arc set $\{p_{\text{out}}u, uv, vp_{\text{in}}, p_l v, vu, up_1\}$ to D.

This reduction was simplified and improved in [2] by replacing the rather complicated last reduction rule by a rule that shortens induced bipaths of length four to length two. Here, $P = \{x_1, \dots, x_l\}$, with $l \geq 3$, is an *induced bipath of length* $l - 1$ if the set of arcs neighbored to $\{x_2, \dots, x_{l-1}\}$ in D is exactly $\{(x_i, x_{i+1}), (x_{i+1}, x_i) \mid i \in \{1, \dots, l-1\}\}$. This yielded a Karp kernel with a quadratic number of vertices (measured in terms of the parameter k) for the rooted version. For directed acyclic graphs (DAGs), even a Karp kernel with a linear number of vertices is known for the rooted version [3]. Notice that also for DAGs (in fact, for quite restricted DAGs called *willow graphs*), the unrooted problem versions have no polynomial Karp kernel unless $coNP \subseteq NP/poly$, as suggested by the hardness proof in [1]. Another direction of research is to obtain faster kernelization algorithms, often by restricting the use (and power) of reduction rules. For the k-LEAF OUT-BRANCHING, this was done by Kammer [6].

Hierarchies Based on Turing Kernels

Based on the notion of *polynomial parametric transformation*, in [4] an intertwined WK/MK hierarchy was defined, in analogy to the well-known W/M hierarchy of (hard) parameterized problems. The lowest level (MK[1]) corresponds to (NP) problems with polynomial-size Karp kernels. The second-lowest level is WK[1], and this does not equal MK[1] unless $coNP \subseteq NP/poly$. Typical complete problems for WK[1] are:

- Given a graph G of order n and an integer k, does G contain a clique of size k? Here, the parameterization is $\kappa(G,k) = k \cdot \log(n)$.
- Given a nondeterministic Turing machine M and an integer k, does M stop within k steps? Here, the parameterization is $\kappa(M,k) = k \cdot \log(|M|)$.

As noticed in [4], the CLIQUE problem provides also another (less natural) example of a problem without polynomial-size Karp kernel that has a polynomial-size Turing kernel, taking as parameterization the maximum degree of the input graph.

How Much Oracle Access Is Needed?

The examples we gave so far make use of oracles in a very simple way. More precisely, a very weak notion of truth-table reduction (disjunctive reduction) is applied. The INDEPENDENT SET problem on bull-free graphs [7] seems to provide a first example where the power of Turing reductions is used more extensively, as the oracle input is based on the previous computation of the reduction. Therefore, it could be termed an adaptive kernelization [5]. Yet another way of constructing Turing kernels was described by Jansen [5]. There, in a first step, the instance is decomposed (according to some graph decomposition in that case), and then the fact is used that either a solution is already obtained or it only exists in one of the (small) components of the decomposition. This framework is then applied to deduce polynomial-size Turing kernels, e.g., for the problem of finding a path (or a cycle) of length at least k in a planar graph G, where k is the parameter of the problem.

Open Problems

One of the most simple open questions is whether LONGEST PATH, i.e., the problem of finding a path of length at least k, admits a polynomial-size Turing kernel on general graphs.

Conversely, no tools have been developed so far that allow for ruling out polynomial-size Turing kernels. For the question of practical applications of kernelization, this would be a much stronger statement than ruling out traditional Karp kernels of polynomial size, as a polynomial number of polynomial-size kernels can give a practical solution (see the discussion of k-LEAF OUT-BRANCHING above).

Cross-References

▶ Enumeration of Paths, Cycles, and Spanning Trees
▶ Kernelization, Exponential Lower Bounds

Recommended Reading

1. Binkele-Raible D, Fernau H, Fomin FV, Lokshtanov D, Saurabh S, Villanger Y (2012) Kernel(s) for problems with no kernel: on out-trees with many leaves. ACM Trans Algorithms 8(4):38
2. Daligault J, Thomassé S (2009) On finding directed trees with many leaves. In: Chen J, Fomin FV (eds) Parameterized and exact computation, 4th international workshop, IWPEC, Copenhagen. LNCS, vol 5917. Springer, pp 86–97
3. Daligault J, Gutin G, Kim EJ, Yeo A (2010) FPT algorithms and kernels for the directed k-leaf problem. J Comput Syst Sci 76(2):144–152
4. Hermelin D, Kratsch S, Soltys K, Wahlström M, Wu X (2013) A completeness theory for polynomial (Turing) kernelization. In: Gutin G, Szeider S (eds) Parameterized and exact computation – 8th international symposium, IPEC, Sophia Antipolis. LNCS, vol 8246. Springer, pp 202–215
5. Jansen BMP (2014) Turing kernelization for finding long paths and cycles in restricted graph classes. Tech. Rep. 1402.4718v1, arXiv.CS.DS
6. Kammer F (2013) A linear-time kernelization for the rooted k-leaf outbranching problem. In: Brandstädt A, Jansen K, Reischuk R (eds) Graph-theoretic concepts in computer science – 39th international workshop, WG, Lübeck. LNCS, vol 8165. Springer, pp 310–320
7. Thomassé S, Trotignon N, Vuskovic K (2013) Parameterized algorithm for weighted independent set problem in bull-free graphs. CoRR abs/1310.6205, a conference version appeared at WG (LNCS volume) 2014

K

Kinetic Data Structures

Bettina Speckmann
Department of Mathematics and Computer Science, Technical University of Eindhoven, Eindhoven, The Netherlands

Years and Authors of Summarized Original Work

1999; Basch, Guibas, Hershberger

Problem Definition

Many application areas of algorithms research involve objects in motion. Virtual reality, simulation, air-traffic control, and mobile communication systems are just some examples. Algorithms that deal with objects in motion traditionally discretize the time axis and compute or update their structures based on the position of the objects at every time step. If all objects move continuously then in general their configuration does not change significantly between time steps – the objects exhibit *spatial* and *temporal coherence*. Although *time-discretization* methods can exploit spatial and temporal coherence they have the disadvantage that it is nearly impossible to choose the perfect time step. If the distance between successive steps is too large, then important interactions might be missed, if it is too small, then unnecessary computations will slow down the simulation. Even if the time step is chosen just right, this is not always a satisfactory solution: some objects may have moved only slightly and in such a way that the overall data structure is not influenced.

One would like to use the temporal coherence to detect precisely those points in time when there is an actual change in the structure. The *kinetic data structure* (KDS) framework, introduced by Basch et al. in their seminal paper [2], does exactly that: by maintaining not only the structure itself, but also some additional information, they can determine when the structure will undergo a "real" (combinatorial) change.

Key Results

A kinetic data structure is designed to maintain or monitor a discrete attribute of a set of moving objects, for example, the convex hull or the closest pair. The basic idea is, that although all objects move continuously, there are only certain discrete moments in time when the combinatorial structure of the attribute changes (in the earlier examples, the ordered set of convex-hull vertices or the pair that is closest, respectively). A KDS therefore contains a set of *certificates* that constitutes a proof of the property of interest. Certificates are generally simple inequalities that assert facts like "point c is on the left of the directed line through points a and b." These certificates are inserted in a priority queue (*event queue*) based on their time of expiration. The KDS then performs an event-driven simulation of the motion of the objects, updating the structure whenever an *event* happens, that is, when a certificate fails (see Fig. 1). It is part of the art of designing efficient kinetic data structures to find a small set of simple and easily updatable certificates that serve as a proof of the property one wishes to maintain.

A KDS assumes that each object has a known motion trajectory or *flight plan*, which may be subject to restrictions to make analysis tractable. Two common restrictions would be translation along paths parametrized by polynomials of fixed degree d, or translation and rotation described by algebraic curves. Furthermore, certificates are generally simple algebraic equations, which implies that the failure time of a certificate can be computed as the next largest root of an algebraic expression. An important aspect of kinetic data structures is their on-line character: although the positions and motions (flight plans) of the objects are known at all times, they are not necessarily known far in advance. In particular, any object can change its flight plan at any time. A good KDS should be able to handle such changes in flight plans efficiently.

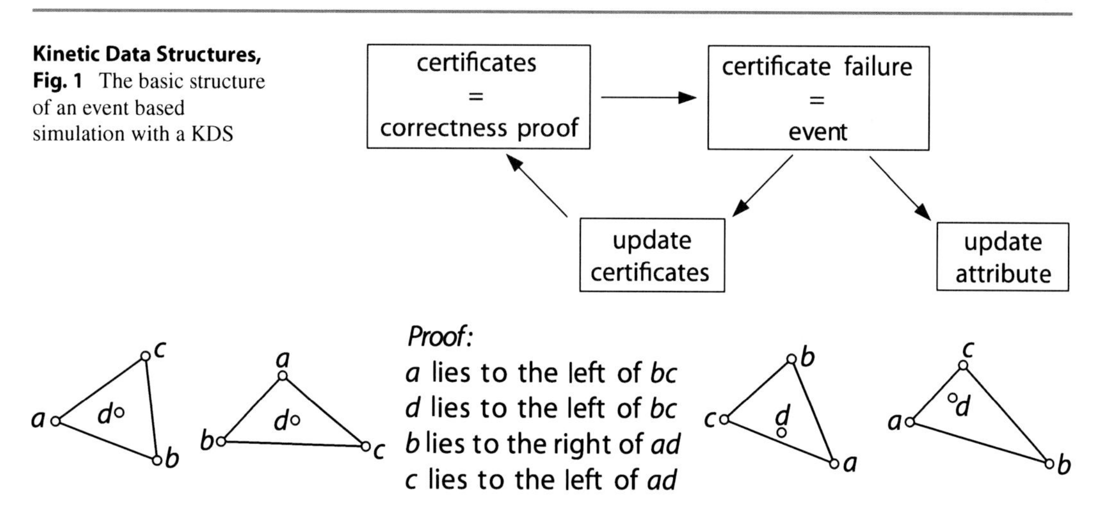

Kinetic Data Structures, Fig. 1 The basic structure of an event based simulation with a KDS

Kinetic Data Structures, Fig. 2 Equivalent convex hull configurations (*left* and *right*), a proof that a, b, and c form the convex hull of S (*center*)

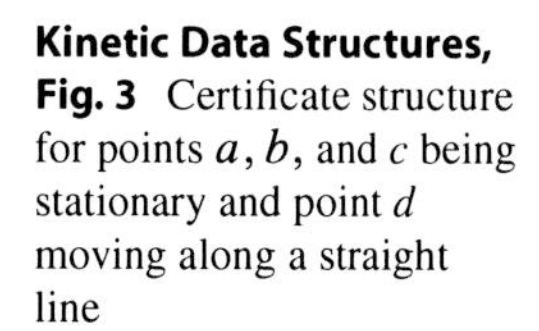

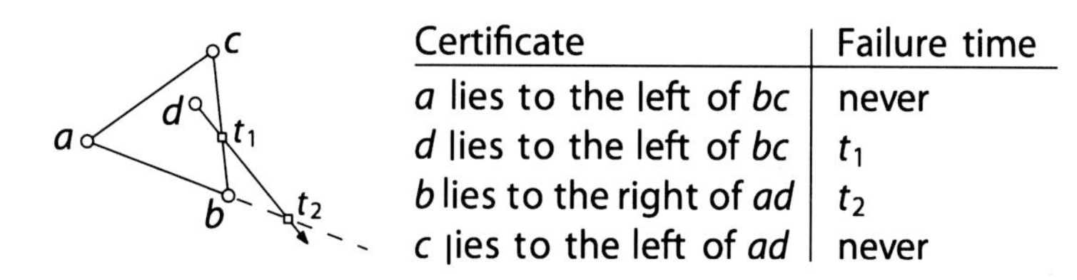

Certificate	Failure time
a lies to the left of bc	never
d lies to the left of bc	t_1
b lies to the right of ad	t_2
c lies to the left of ad	never

Kinetic Data Structures, Fig. 3 Certificate structure for points a, b, and c being stationary and point d moving along a straight line

A detailed introduction to kinetic data structures can be found in Basch's Ph. D. thesis [1] or in the surveys by Guibas [3, 4]. In the following the principles behind kinetic data structures are illustrated by an easy example.

Consider a KDS that maintains the convex hull of a set S of four points a, b, c, and d as depicted in Fig. 2. A set of four simple certificates is sufficient to certify that a, b, and c form indeed the convex hull of S (see Fig. 2 center). This implies, that the convex hull of S will not change under any motion of the points that does not lead to a violation of these certificates. To put it differently, if the points move along trajectories that move them between the configurations depicted in Fig. 2 without the point d ever appearing on the convex hull, then the KDS in principle does not have to process a single event.

Now consider a setting in which the points a, b, and c are stationary and the point d moves along a linear trajectory (Fig. 3 left). Here the KDS has exactly two events to process. At time t_1 the certificate "d is to the left of bc" fails as the point d appears on the convex hull. In this easy setting, only the failed certificate is replaced by "d is to the *right* of bc" with failure time "never", generally processing an event would lead to the scheduling and descheduling of several events from the event queue. Finally at time t_2 the certificates "b is to the right of ad" fails as the point b ceases to be on the convex hull and is replaced by "b is to the *left* of ad" with failure time "never."

Kinetic data structures and their accompanying maintenance algorithms can be evaluated and compared with respect to four desired characteristics.

Responsiveness. One of the most important performance measures for a KDS is the time needed to update the attribute and to repair the certificate set when a certificate fails. A KDS is called *responsive* if this update time is "small", that is, polylogarithmic.

Compactness. A KDS is called *compact* if the number of certificates is near-linear in the total number of objects. Note that this is not necessarily the same as the amount of storage the entire structure needs.

Locality. A KDS is called *local* if every object is involved in only a small number of certificates (again, "small" translates to polylogarithmic). This is important whenever an object changes its flight plane, because one has to recompute the failure times of all certificates this object is involved in, and update the event queue accordingly. Note that a local KDS is always compact, but that the reverse is not necessarily true.

Efficiency. A certificate failure does not automatically imply a change in the attribute that is being maintained, it can also be an *internal event*, that is, a change in some auxiliary structure that the KDS maintains. A KDS is called *efficient* if the worst-case number of events handled by the data structure for a given motion is small compared to the number of combinatorial changes of the attribute (*external events*) that must be handled for that motion.

Applications

The paper by Basch et al. [2] sparked a large amount of research activities and over the last years kinetic data structures have been used to solve various dynamic computational geometry problems. A number of papers deal foremost with the maintenance of discrete attributes for sets of moving points, like the closest pair, width and diameter, clusters, minimum spanning trees, or the constrained Delaunay triangulation. Motivated by ad hoc mobile networks, there have also been a number of papers that show how to maintain the connected components in a set of moving regions in the plane. Major research efforts have also been seen in the study of kinetic binary space partitions (BSPs) and kinetic kd-trees for various objects. Finally, there are several papers that develop KDSs for collision detection in the plane and in three dimensions. A detailed discussion and an extensive list of references can be found in the survey by Guibas [4].

Cross-References

▶ Fully Dynamic Minimum Spanning Trees
▶ Minimum Geometric Spanning Trees

Recommended Reading

1. Basch J (1999) Kinetic data structures. PhD thesis, Stanford University
2. Basch J, Guibas L, Hershberger J (1999) Data structures for mobile data. J Algorithms 31:1–28
3. Guibas L (1998) Kinetic data structures: a state of the art report. In: Proceedings of 3rd workshop on algorithmic foundations of robotics, pp 191–209
4. Guibas L (2004) Modeling motion. In: Goodman J, O'Rourke J (eds) Handbook of discrete and computational geometry, 2nd edn. CRC

Knapsack

Hans Kellerer
Department of Statistics and Operations Research, University of Graz, Graz, Austria

Keywords

Approximation algorithm; Fully polynomial time approximation scheme (FPTAS)

Years and Authors of Summarized Original Work

2000; Ibarra, Kim

Problem Definition

For a given set of items $N = \{1, \ldots, n\}$ with nonnegative integer weights w_j and profits p_j, $j = 1, \ldots, n$, and a knapsack of capacity c, the *knapsack problem* (KP) is to select a subset of the items such that the total profit of the selected items is maximized and the corresponding total weight does not exceed the knapsack capacity c.

Alternatively, a knapsack problem can be formulated as a solution of the following linear integer programming formulation:

$$(KP)\ \text{maximize} \sum_{j=1}^{n} p_i x_j \tag{1}$$

$$\text{subject to} \sum_{j=1}^{n} w_j x_j \leq c, \tag{2}$$

$$x_j \in (0,1),\ j = 1,\ldots,n. \tag{3}$$

The knapsack problem is the simplest nontrivial integer programming model having binary variables, only a single constraint, and only positive coefficients. A large number of theoretical and practical papers have been published on this problem and its extensions. An extensive overview can be found in the books by Kellerer, Pferschy, and Pisinger [4] or Martello and Toth [7].

Adding the integrality condition (3) to the simple linear program (1)–(2) already puts (KP) into the class of $\mathcal{NP}$-hard problems. Thus, (KP) admits no polynomial time algorithms unless $\mathcal{P} = \mathcal{NP}$ holds.

Therefore, this entry will focus on approximation algorithms for (KP). A common method to judge the quality of an approximation algorithm is its worst-case performance. For a given instance I, define by $z^*(I)$ the optimal solution value of (KP) and by $z^H(I)$ the corresponding solution value of a heuristic H. For $\varepsilon \in [0,1[$, a heuristic H is called a $(1-\varepsilon)$*-approximation algorithm* for (KP) if for any instance I

$$z^H(I) \geq (1-\varepsilon)\, z^*(I)$$

holds. Given a parameter ε, a heuristic H is called a *fully polynomial approximation scheme*, or an FTPAS, if H is a $(1-\varepsilon)$-approximation algorithm for (KP) for any $\varepsilon \in [0,1[$, and its running time is polynomial both in the length of the encoded input n and $1/\varepsilon$. The first FTPAS for (KP) was suggested by Ibarra and Kim [1] in 1975. It was among the early FPTASes for discrete optimization problems. It will be described in detail in the following.

Key Results

(KP) can be solved in pseudopolynomial time by a simple dynamic programming algorithm. One possible variant is the so-called *dynamic programming by profits* (DP-Profits). The main idea of DP-Profits is to reach every possible total profit value with a subset of items of minimal total weight. Clearly, the highest total profit value, which can be reached by a subset of weight not greater than the capacity c, will be an optimal solution.

Let $y_j(q)$ denote the minimal weight of a subset of items from $\{1,\ldots,j\}$ with total profit equal to q. To bound the length of every array y_j, an upper bound u on the optimal solution value has to be computed. An obvious possibility would be to use the upper bound $U_{\text{LP}} = \lfloor z^{\text{LP}} \rfloor$ from the solution z^{LP} of the LP-relaxation of (KP) and set $U := U_{\text{LP}}$. It can be shown that U_{LP} is at most twice as large as the optimal solution value z^*. Initializing $y_0(0) := 0$ and $y_0(q) := c+1$ for $q = 1,\ldots,U$, all other values can be computed for $j = 1,\ldots,n$ and $q = 0,\ldots,U$ by using the recursion

$$y_i(q) := \begin{cases} y_{i-1}(q) & \text{if } q < p_j, \\ \min\left(y_{i-1}(q),\ (y_{i-1}(q))\right) & \text{if } q \geq p_j. \end{cases}$$

The optimal solution value is given by $\max\{q \mid y_n(q) \leq c\}$ and the running time of DP-Profits is bounded by $O(nU)$.

Theorem 1 (Ibarra, Kim) *There is an FTPAS for (KP) which runs in $O(n \log n + n/\varepsilon^2)$ time.*

Proof The FTPAS is based on appropriate *scaling* of the profit values p_j and then running DP-Profits with the scaled profit values. Scaling means here that the given profit values p_j are replaced by new profits $\tilde{p}_j$ such that $\tilde{p}_j := \left\lfloor \frac{p_j}{K} \right\rfloor$ for an appropriate chosen constant K.

This scaling can be seen as a partitioning of the profit range into intervals of length K with starting points $0, K, 2K, \ldots$. Naturally, for every profit value p_j, there is some integer value $i \geq 0$ such that p_j falls into the interval $[iK, (i+1)K[$. The scaling procedure generates for every p_j the

K

value $\tilde{p}_j$ as the corresponding index i of the lower interval bound iK.

Running DP-Profits yields a solution set $\tilde{X}$ for the scaled items which will usually be different from the original optimal solution set X^*. Evaluating the original profits of item set $\tilde{X}$ yields the approximate solution value z^H. The difference between z^H and the optimal solution value can be bounded as follows:

$$z^H \geq \sum_{j \in \tilde{X}} K \left\lfloor \frac{p_j}{K} \right\rfloor \geq \sum_{j \in \tilde{X}*} K \left\lfloor \frac{p_j}{K} \right\rfloor$$

$$\geq \sum_{j \in \tilde{X}*} K \left(\frac{p_j}{K} - 1 \right) = z* - |X*|\, K.$$

To get the desired performance guarantee of $1-\varepsilon$, it is sufficient to have

$$\frac{z * - z^H}{z*} \leq \frac{|X*|\, K}{z*} \leq \varepsilon.$$

To ensure this, K has to be chosen such that

$$K \leq \frac{\varepsilon z^*}{|X*|}. \tag{4}$$

Since $n \geq |X^*|$ and $U_{\mathrm{LP}}/2 \leq z^*$, choosing $K := \frac{\varepsilon U_{\mathrm{LP}}}{2n}$ satisfies condition (4) and thus guarantees the performance ratio of $1-\varepsilon$. Substituting U in the $O(nU)$ bound for DP-Profits by U/K yields an overall running time of $O(n^2 \varepsilon)$.

A further improvement in the running time is obtained in the following way. Separate the items into *small* items (having profit $\leq \frac{\varepsilon}{2} U_{\mathrm{LP}}$) and *large* items (having profit $> \frac{\varepsilon}{2} U_{\mathrm{LP}}$). Then, perform DP-Profits for the scaled large items only. To each entry q of the obtained dynamic programming array with corresponding weight $y(q)$, the small items are added to a knapsack with residual capacity $c - y(q)$ in a greedy way. The small items shall be sorted in nonincreasing order of their profit to weight ratio. Out of the resulting combined profit values, the highest one is selected. Since every optimal solution contains at most $2/\varepsilon$ large items, $|X^*|$ can be replaced in (4) by $2/\varepsilon$ which results in an overall running time $O(n \log n + n/\varepsilon^2)$. The memory requirement of the algorithm is $O(n + 1/\varepsilon^3)$.

Two important approximation schemes with advanced treatment of items and algorithmic fine-tuning were presented some years later. The classical paper by Lawler [5] gives a refined scaling resp. partitioning of the items and several other algorithmic improvements which results in a running time $O(n \log(1/\varepsilon) + 1/\varepsilon^4)$. A second paper by Magazine and Oguz [6] contains among other features a partitioning and recombination technique to reduce the space requirements of the dynamic programming procedure. The fastest algorithm is due to Kellerer and Pferschy [2, 3] with running time $O(n \min\{\log n, \log(1/\varepsilon)\} + 1/\varepsilon^2 \log(1/\varepsilon) \cdot \min\{n, 1/\varepsilon \log(1/\varepsilon)\})$ and space requirement $O(n + 1/\varepsilon^2)$.

Applications

(KP) is one of the classical problems in combinatorial optimization. Since (KP) has this simple structure and since there are efficient algorithms for solving it, many solution methods of more complex problems employ the knapsack problem (sometimes iteratively) as a subproblem.

A straightforward interpretation of (KP) is an investment problem. A wealthy individual or institutional investor has a certain amount of money c available which he wants to put into profitable business projects. As a basis for his decisions, he compiles a long list of possible investments including for every investment the required amount w_j and the expected net return p_j over a fixed period. The aspect of risk is not explicitly taken into account here. Obviously, the combination of the binary decisions for every investment such that the overall return on investment is as large as possible can be formulated by (KP).

One may also view the (KP) as a "cutting" problem. Assume that a sawmill has to cut a log into shorter pieces. The pieces must however be cut into some predefined standard-lengths w_j, where each length has an associated selling price p_j. In order to maximize the profit of the log, the sawmill can formulate the problem as a (KP) where the length of the log defines the capacity c.

Among the wide range of "real-world" applications shall be mentioned two-dimensional cutting problems, column generation, separation of cover inequalities, financial decision problems, asset-backed securitization, scheduling problems, knapsack cryptosystems, and most recent combinatorial auctions. For a survey on applications of knapsack problems, the reader is referred to [4].

Recommended Reading

1. Ibarra OH, Kim CE (1975) Fast approximation algorithms for the knapsack and sum of subset problem. J ACM 22:463–468
2. Kellerer H, Pferschy U (1999) A new fully polynomial time approximation scheme for the knapsack problem. J Comb Optim 3:59–71
3. Kellerer H, Pferschy U (2004) Improved dynamic programming in connection with an FPTAS for the knapsack problem. J Comb Optim 8:5–11
4. Kellerer H, Pisinger D, Pferschy U (2004) Knapsack problems. Springer, Berlin
5. Lawler EL (1979) Fast approximation algorithms for knapsack problems. Math Oper Res 4:339–356
6. Magazine MJ, Oguz O (1981) A fully polynomial approximation algorithm for the 0-1 knapsack problem. Eur J Oper Res 8:270–273
7. Martello S, Toth P (1990) Knapsack problems: algorithms and computer implementations. Wiley, Chichester

Knowledge in Distributed Systems

Yoram Moses
Department of Electrical Engineering, Technion – Israel Institute of Technology, Haifa, Israel

Keywords

Common knowledge; Coordinated Attack; Distributed computing; Knowledge; Knowledge gain; Message chain; Potential causality

Years and Authors of Summarized Original Work

1984; Halpern, Moses
1986; Chandy, Misra

Problem Definition

What is the role of knowledge in distributed computing?

Actions taken by a process in a distributed system can only be based on its local information or local *knowledge*. Indeed, in reasoning about distributed protocols, people often talk informally about what processes know about the state of the system and about the progress of the computation. Can the informal reasoning about knowledge in distributed and multi-agent systems be given a rigorous mathematical formulation, and what uses can this have?

Key Results

In [4] Halpern and Moses initiated a theory of knowledge in distributed systems. They suggested that states of knowledge ascribed to groups of processes, especially *common knowledge*, have an important role to play. Knowledge-based analysis of distributed protocols has generalized well-known results and enables the discovery of new ones. These include new efficient solutions to basic problems, tools for relating results in different models, and proving lower bounds and impossibility results. For example, the inability to attain common knowledge when communication is unreliable was established in [4] and shown to imply and generalize the Coordinated Attack problem. Chandy and Misra showed in [1] that in asynchronous systems there is a tight connection between the manner in which knowledge is gained or lost and the message chains that underly Lamport's notion of potential causality.

Modeling Knowledge

In philosophy, knowledge is often modeled by so-called *possible-worlds* semantics. Roughly speaking, at a given "world," an agent will know a given fact to be true precisely if this fact holds at all worlds that the agent "considers possible." In a distributed system, the agents are processes or computing elements. A simple language for reasoning about knowledge is obtained by starting out with a set $\boldsymbol{\Phi} = \{p, q, p', q' \ldots\}$ of propositions, or basic facts. The facts in Φ will depend on the application we wish to study; they may involve statements such as $x = 0$ or $x > y$ concerning values of variables or about other aspects of the computation (e.g., in the analysis of mutual exclusion, a proposition $\mathsf{CS}(i)$ may be used to state that process i is in the critical section). We obtain a logical language $\mathcal{L}_n^K = \mathcal{L}_n^K(\Phi)$ for knowledge, which is a set of formulas, by the following inductive definition. First, $p \in \mathcal{L}_n^K$ for all propositions $p \in \Phi$. Moreover, for all formulas $\varphi, \psi \in \mathcal{L}_n^K$, the language contains the formulas $\neg\varphi$ (standing for "*not* φ"), $\varphi \wedge \psi$ (standing for "φ *and* ψ"), and $K_i\varphi$ ("*process* i *knows* φ"), for every process $i \in \{1, \ldots, n\}$ (Using the operators "$\neg$" and "$\wedge$," we can express all of the Boolean operators. Thus, $\varphi \vee \psi$ ("φ *or* ψ") can be expressed as $\neg(\neg\varphi \wedge \neg\psi)$, while $\varphi \Rightarrow \psi$ ("φ *implies* ψ") is $\neg\varphi \vee \psi$, etc.). The language $\mathcal{L}_n^K$ is the basis of a propositional logic of knowledge. Using it, we can make formulas such as $K_1\mathsf{CS}(1) \wedge K_1K_2\neg\mathsf{CS}(2)$, which states that "*process 1 knows that it is in the critical section, and it knows that process 2 knows that 2 is not in the critical section.*" $\mathcal{L}_n^K$ determines the *syntax* of formulas of the logic. A mathematical definition of what the formulas mean is called its *semantics*.

A process will typically know different things in different computations; even within a computation, its knowledge changes over time as a result of communication and of observing various events. We refer to time t in a run (or computation) r by the pair $(\boldsymbol{r}, \boldsymbol{t})$, which is called a *point*. Formulas are considered to be true or false at a point (r, t), with respect to a set of runs R (we call R a *system*). The set of points of R is denoted by $\mathbf{Pts}(\boldsymbol{R})$.

The definition of knowledge is based on the idea that at any given point (r, t), each process has a well-defined *view*, which depends on i's history up to time t in r. This view may consist of all events that i has observed, or on a much more restricted amount of information, which is considered to be available to i at (r, t). In the language of [3], this view can be thought of as being process i's *local state* at (r, t), which we denote by $\boldsymbol{r_i(t)}$. Intuitively, a process is assumed to be able to distinguish two points iff its local state at one is different from its state in the other. In a given system R, the meaning of the propositions in a set Φ needs to be defined explicitly. This is done by way of an *interpretation* $\pi : \Phi \times \mathsf{Pts}(R) \rightarrow \{\mathsf{True}, \mathsf{False}\}$. The pair $\mathcal{I} = (\boldsymbol{R}, \boldsymbol{\pi})$ is called an *interpreted system*. We denote the fact that φ is satisfied, or true, at a point (r, t) in the system $\mathcal{I}$ by $(\mathcal{I}, r, t) \models \varphi$. Semantics of $\mathcal{L}_n^K(\Phi)$ with respect to an interpreted system $\mathcal{I}$ is given by defining the satisfaction relation "$\models$" defined by induction on the structure of the formulas, as follows:

$$
\begin{array}{lll}
(\mathcal{I}, r, t) \models p & \textit{iff } \pi\big(p, (r, t)\big) = \mathsf{True}, & \text{for a proposition } p \in \Phi \\
(\mathcal{I}, r, t) \models \neg\varphi & \textit{iff } (\mathcal{I}, r, t) \not\models \varphi & \\
(\mathcal{I}, r, t) \models \varphi \wedge \psi & \textit{iff } \text{both } (\mathcal{I}, r, t) \models \varphi \text{ and } (\mathcal{I}, r, t) \models \psi & \\
(\mathcal{I}, r, t) \models K_i\varphi & \textit{iff } (\mathcal{I}, r', t') \models \varphi \text{ whenever } r_i'(t') = r_i(t) \text{ and } (r', t') \in \mathsf{Pts}(R) &
\end{array}
$$

The fourth clause, which defines satisfaction for knowledge formulas, can be applied repeatedly. This gives meaning to formulas involving knowledge about formulas that themselves involve knowledge, such as $K_2\big(\mathsf{CS}(2) \wedge \neg K_1\neg\mathsf{CS}(2)\big)$. Knowledge here is ascribed to processes. The intuition is that the local state captures all of the information available to the process. If there is a scenario leading to another point at which a fact φ is false and the process has the same state as it has not, then the process does not know φ.

This notion of knowledge has fairly strong properties that distinguish it from what one might consider a reasonable notion of, say, human knowledge. For example, it does not depend on computation, thoughts, or a derivation of what the process knows. It is purely "information based." Indeed, any fact that holds at all elements of $\mathsf{Pts}(R)$ (e.g., the protocol that processes are following) is automatically known to all processes. Moreover, it is not assumed that a process can report its knowledge or that its knowledge is explicitly recorded in the local state. This notion of knowledge can be thought of as being ascribed to the processes by an external observer and is especially useful for analysis by a protocol designer.

Common Knowledge and Coordinated Attack

A classic example of a problem for which the knowledge terminology can provide insight is Jim Gray's *Coordinated Attack* problem. We present it in the style of [4]:

The Coordinated Attack Problem

Two divisions of an army are camped on two hilltops, and the enemy awaits in the valley below. Neither general will decide to attack unless he is sure that the other will attack with him, because only a simultaneous attack guarantees victory. The divisions do not initially have plans to attack, and one of the commanding generals wishes to coordinate a simultaneous attack (at some time the next day). The generals can only communicate by means of a messenger. Normally, it takes the messenger 1 h to get from one encampment to the other. However, the messenger can lose his way or be captured by the enemy. Fortunately, on this particular night, everything goes smoothly. How long will it take them to coordinate an attack?

It is possible to show by induction that k trips of the messenger do not suffice, for all $k \geq 0$, and hence the generals will be unable to attack. Gray used this example to illustrate the impact of unreliable communication on the ability to consistently update distinct sites of a distributed database. A much stronger result that generalizes this and applies directly to practical problems can be obtained based on a notion called *common knowledge*. Given a group $G \subseteq \{1, \ldots, n\}$ of processes, we define two new logical operators E_G and C_G, corresponding to *everyone (in G) knows* and *is common knowledge in G*, respectively. We shall denote $E_G^1\varphi = E_G\varphi$ and inductively define $E_G^{k+1}\varphi = E_G(E_G^k\varphi)$. Satisfaction for the new operators is given by

$$(\mathcal{I}, r, t) \models E_G\varphi \ \ \textit{iff} \ \ (\mathcal{I}, r, t) \models K_i\varphi \text{ holds for all } i \in G$$
$$(\mathcal{I}, r, t) \models C_G\varphi \ \ \textit{iff} \ \ (\mathcal{I}, r, t) \models E_G^k\varphi \text{ holds for all } k \geq 1$$

Somewhat surprisingly, common knowledge is not uncommon in practice. People shake hands to signal that they attained common knowledge of an agreement, for example. Similarly, a public announcement to a class or to an audience is considered common knowledge. Indeed, as we now discuss, simultaneous actions can lead to common knowledge.

Returning to the Coordinated Attack problem, consider three propositions attack_A, attack_B, and $\mathsf{delivered}$, corresponding, respectively, to "*general A is attacking*," "*general B is attacking*," and "*at least one message has been delivered*." The fact that the generals do not have a plan to attack can be formalized by saying that at least one of them does not attack unless $\mathsf{delivered}$ is true. Consider a set of runs R consisting of all possible interactions of the generals in the above setting. Suppose that the generals follow the specifications, so they only ever attack simultaneously at points of R. Then, roughly speaking, since the generals' actions depend on their local state, general A knows when attack_A is true. But since they only attack simultaneously and attack_B is true whenever attack_B is true, $K_A\mathsf{attack}_B$ will hold whenever general A attacks. Since B similarly knows when A attacks, $K_BK_A\mathsf{attack}_B$ will hold as well. Indeed, it can be shown that when the generals attack in a system that guarantees that attacks are simultaneous, they must have common knowledge that they are attacking.

Theorem 1 (Halpern and Moses [4]) *Let R be a system with unreliable communication, let $\mathcal{I} = (R, \pi)$, let $(r, t) \in \mathsf{Pts}(R)$, and assume that $|G| > 1$. Then $(\mathcal{I}, r, t) \models \neg C_G \mathsf{delivered}$.*

As in the case of Coordinated Attack, simultaneous actions must be common knowledge when they are performed. Moreover, in cases in which such actions require a minimal amount of communication to be materialize, $C_G \mathsf{delivered}$ must hold when they are performed. Theorem 1 implies that no such actions can be coordinated when communication is unreliable. One immediate consequence is:

Corollary 1 *Under a protocol that satisfies the constraints of the Coordinated Attack problem, the generals never attack.*

The connection between common knowledge and simultaneous actions goes even deeper. It can be shown that when a fact that is not common knowledge to G becomes common knowledge to G, a state transition must occur simultaneously at all sites in G. If simultaneity cannot be coordinated in the system, common knowledge cannot be attained. This raises some philosophical issues: Events and transitions that are viewed as being simultaneous in a system that is modeled at a particular ("coarse") granularity of time will fail to be simultaneous when time is modeled at a finer granularity. As discussed in [4], this is not quite a paradox, since there are many settings in which it is acceptable, and even desirable, to model interactions at a granularity of time in which simultaneous transitions do occur.

A Hierarchy of States of Knowledge and Common Knowledge

Common knowledge is a much stronger state of knowledge than, say, knowledge of an individual process. Indeed, it is best viewed as a state of knowledge of a group. There is an essential difference between E_G^k (everyone knows that everyone knows, for k levels), even for large k, and C_G (common knowledge). Indeed, for every k, there are examples of tasks that can be achieved if $E_G^{k+1}\varphi$ holds but not if $E_G^k\varphi$ does. This suggests the existence of a hierarchy of states of group knowledge, ranging from $E_G\varphi$ to $C_G\varphi$. But it is also possible to define natural states of knowledge for a group that are weaker than these. One is S_G, where $S_G\varphi$ is true if $\bigvee_{i \in G} K_i\varphi$ – *someone in G knows* φ. Even weaker is *distributed knowledge*, denoted by D_G, which is defined by

$$(\mathcal{I}, r, t) \models D_G\varphi \ \textit{iff} \ (\mathcal{I}, r', t') \models \varphi \text{ for all } (r', t') \text{ satisfying } r'_i(t') = r_i(t) \text{ for all } i \in G$$

Roughly speaking, the distributed knowledge of a group corresponds to what follows from the combined information of the group at a given instant. Thus, for example, if all processes start out with initial value 1, they will have distributed knowledge of this fact, even if no single process knows this individually. Halpern and Moses propose a hierarchy of states of group knowledge and suggest that communication can often be seen as the process of moving the state of knowledge up the hierarchy:

$$C_G\varphi \Rightarrow E_G^{k+1}\varphi \Rightarrow E_G^k\varphi \Rightarrow \cdots \Rightarrow E_G\varphi$$

$$\Rightarrow S_G\varphi \Rightarrow D_G\varphi.$$

Knowledge Gain and Loss in Asynchronous Systems

In asynchronous systems there are no guarantees about the pace at which communication is delivered and no guarantees about the relative rates at which processes operate. This motivated Lamport's definition of the happened-before relation among events. It is based on the intuition that in asynchronous systems only information obtained via message chains can affect the activity at a given site. A crisp formalization of this intuition was discovered by Chandy and Misra in [1]:

Theorem 2 (Chandy and Misra) *Let $\mathcal{I}$ be an asynchronous interpreted system, let $\varphi \in \mathcal{L}_n^K$, and let $t' > t$. Then*

Knowledge Gain: *If $(\mathcal{I}, r, t) \not\models K_j \varphi$ and $(\mathcal{I}, r, t') \models K_{i_m} K_{i_{m-1}} \cdots K_{i_1} \varphi$, then there is a message chain through processes $\langle i_1, i_2, \ldots, i_m \rangle$ in r between times t and t'.*

Knowledge Loss: *If $(\mathcal{I}, r, t) \models K_{i_m} K_{i_{m-1}} \cdots K_{i_1} \varphi$ and $(\mathcal{I}, r, t') \not\models \varphi$, then there is a message chain through processes $\langle i_m, i_{m-1}, \ldots, i_1 \rangle$ in r between times t and t'.*

Note that the second clause implies that sending messages can cause a process to lose knowledge about other sites. Roughly speaking, the only way a process can know a nontrivial fact about a remote site is if this fact can only be changed by explicit permission from the process.

Applications and Extensions

The knowledge framework has been used in several ways. We have already seen its use for proving impossibility results in the discussion of the Coordinated Attack example. One interesting use of the formalism is as a tool for expressing *knowledge-based* protocols, in which programs can contain tests such as *if K_i(msg received) then...*, Halpern and Zuck, for example, showed that distinct solutions to the sequence transmission problem under different assumptions regarding communication faults were all implementations of the same knowledge-based protocol [5].

A knowledge-based analysis can lead to the design of efficient, sometimes optimal, distributed protocols. Dwork and Moses analyzed when facts become common knowledge when processes can crash and obtained an efficient and optimal solution to simultaneous consensus in which decisions are taken when initial values become common knowledge [2]. Moses and Tuttle showed that in a slightly harsher failure model, similar optimal solutions exist, but they are not computationally efficient, because computing when values are common knowledge is NP-hard [7]. A thorough exposition of reasoning about knowledge in a variety of fields including distributed systems, game theory, and philosophy appears in [3], while a later discussion of the role of knowledge in coordination, with further references, appears in [6].

Cross-References

- ▸ Byzantine Agreement
- ▸ Causal Order, Logical Clocks, State Machine Replication
- ▸ Distributed Snapshots

K

Recommended Reading

1. Chandy KM, Misra J (1986) How processes learn. Distrib Comput 1(1):40–52
2. Dwork C, Moses Y (1990) Knowledge and common knowledge in a Byzantine environment: crash failures. Inf Comput 88(2):156–186
3. Fagin R, Halpern JY, Moses Y, Vardi MY (2003) Reasoning about knowledge. MIT, Cambridge, MA
4. Halpern JY, Moses Y (1990) Knowledge and common knowledge in a distributed environment. J ACM 37(3):549–587. A preliminary version appeared in Proceeding of the 3rd ACM symposium on principles of distributed computing, 1984
5. Halpern JY, Zuck LD (1992) A little knowledge goes a long way: knowledge-based derivations and correctness proofs for a family of protocols. J ACM 39(3):449–478
6. Moses Y (2015) Relating knowledge and coordinated action: the knowledge of preconditions principle. In: Proceedings of the 15th TARK conference, Pittsburgh, pp 207–215
7. Moses Y, Tuttle MR (1988) Programming simultaneous actions using common knowledge. Algorithmica 3:121–169

L

Large-Treewidth Graph Decompositions

Julia Chuzhoy
Toyota Technological Institute, Chicago, IL, USA

Keywords

Bidimensionality theory; Excluded grid theorem; Erdős-Pósa; Graph decomposition; Treewidth

Years and Authors of Summarized Original Work

2013; Chekuri, Chuzhoy

Problem Definition

Treewidth is an important and a widely used graph parameter. Informally, the treewidth of a graph measures how close the graph is to being a tree. In particular, low-treewidth graphs often exhibit behavior somewhat similar to that of trees, in that many problems can be solved efficiently on such graphs, often by using dynamic programming. The treewidth of a graph $G = (V, E)$ is typically defined via tree decompositions. A tree decomposition for G consists of a tree $T = (V(T), E(T))$ and a collection of sets $\{X_v \subseteq V\}_{v \in V(T)}$ called *bags*, such that the following two properties are satisfied: (i) for each edge $(a, b) \in E$, there is some node $v \in V(T)$ with both $a, b \in X_v$, and (ii) for each vertex $a \in V$, the set of all nodes of T whose bags contain a form a nonempty (connected) subtree of T. The *width* of a given tree decomposition is $\max_{v \in V(T)}\{|X_v| - 1\}$, and the treewidth of a graph G, denoted by $\text{tw}(G)$, is the width of a minimum-width tree decomposition for G.

In large-treewidth graph decompositions, we seek to partition a given graph G into a large number of disjoint subgraphs $G_1, \ldots, G_h$, where each subgraph G_i has a large treewidth. Specifically, if k denotes the treewidth of G, h is the desired number of the subgraphs in the decomposition, and r is the desired lower bound on the treewidth of each subgraph G_i, then we are interested in efficient algorithms that partition any input graph G of treewidth k into h disjoint subgraphs of treewidth at least r each, and in establishing the bounds on h and r in terms of k, for which such a partition exists.

Key Results

The main result of [1] is summarized in the following theorem.

Theorem 1 *There is an efficient algorithm that, given integers $h, r, k \geq 0$, where either $hr^2 \leq k/\text{poly}\log k$ or $h^3 r \leq k/\text{poly}\log k$ holds, and a graph G of treewidth k, computes a partition of G into h disjoint subgraphs of treewidth at least r each.*

M.-Y. Kao (ed.), *Encyclopedia of Algorithms*,
DOI 10.1007/978-1-4939-2864-4

Applications

While low-treewidth graphs can often be handled well by dynamic programming, the major tool for dealing with large-treewidth graphs so far has been the Excluded Grid Theorem of Robertson and Seymour [11]. The theorem states that there is some function $g : \mathbb{Z}^+ \rightarrow \mathbb{Z}^+$, such that for any integer t, every graph of treewidth at least $g(t)$ contains a $(t \times t)$-grid as a minor (we say that a graph H is a minor of G iff we can obtain H from G by a sequence of edge deletions and edge contractions). A long line of work is dedicated to improving the upper and the lower bounds on the function g [2, 6, 7, 9–12]. The best current bounds show that the theorem holds for $g(t) = O(t^{98} \cdot \text{poly} \log(t))$ [2], and the best negative result shows that $g(t) = \Omega(t^2 \log t)$ must hold [12]. Robertson et al. [12] suggest that $g(t) = \Theta(t^2 \log t)$ may be sufficient, and Demaine et al. [5] conjecture that the bound of $g(t) = \Theta(t^3)$ is both necessary and sufficient. Large-treewidth graph decomposition is a tool that allows, in several applications, to bypass the Excluded Grid Theorem while obtaining stronger parameters. Such applications include Erdős-Pósa-type results and fixed-parameter tractable algorithms that rely on the bidimensionality theory. We note that the Excluded Grid Theorem of Robertson and Seymour provides a large-treewidth graph decomposition with weaker bounds. The most recent polynomial bounds for the Excluded Grid Theorem of [2] only ensure that a partition exists for any h, r where $h^{49} r^{98} \leq k/\text{poly} \log k$. Prior to the work of [1], the state-of-the-art bounds for the Grid-Minor Theorem could only guarantee that the partition exists whenever $hr^2 \leq \log k / \log \log k$.

We now provide several examples where the large-treewidth graph decomposition theorem can be used to improve previously known bounds.

Erdős-Pósa-Type Results

A family $\mathcal{F}$ of graphs is said to satisfy the Erdős-Pósa property, iff there is an integer-valued function $f_{\mathcal{F}}$, such that for every graph G, either G contains k disjoint subgraphs isomorphic to members of $\mathcal{F}$, or there is a set S of $f_{\mathcal{F}}(k)$ nodes, such that $G \setminus S$ contains no subgraph isomorphic to a member of $\mathcal{F}$. In other words, S is a cover, or a hitting set, for $\mathcal{F}$ in G. Erdős and Pósa [8] showed such a property when $\mathcal{F}$ is the family of cycles, with $f_{\mathcal{F}}(k) = \Theta(k \log k)$.

The Excluded Grid Theorem has been widely used in proving Erdős-Pósa-type results, where the specific parameters obtained depend on the best known upper bound on the function $g(k)$ in the Excluded Grid Theorem. The parameters in many Erdős-Pósa-type results can be significantly strengthened using Theorem 1, as shown in the following theorem:

Theorem 2 *Let $\mathcal{F}$ be any family of connected graphs and assume that there is an integer r, such that any graph of treewidth at least r is guaranteed to contain a subgraph isomorphic to a member of $\mathcal{F}$. Then $f_{\mathcal{F}}(k) \leq O(kr^2 \text{poly} \log(kr))$.*

Combining Theorem 2 with the best current bound for the Excluded Grid Theorem [2], we obtain the following corollary.

Corollary 1 *Let $\mathcal{F}$ be any family of connected graphs, such that for some integer q, any graph containing a $q \times q$ grid as a minor is guaranteed to contain a subgraph isomorphic to a member of $\mathcal{F}$. Then $f_{\mathcal{F}}(k) \leq O(q^{98} k \text{poly} \log(kq))$.*

For a fixed graph H, let $\mathcal{F}(H)$ be the family of all graphs that contain H as a minor. Robertson and Seymour [11], as one of the applications of their Excluded Grid Theorem, showed that $\mathcal{F}(H)$ has the Erdős-Pósa property iff H is planar. By directly applying Corollary 1, we get the following improved near-linear dependence on k.

Theorem 3 *For any fixed planar graph H, the family $\mathcal{F}(H)$ of graphs has the Erdős-Pósa property with $f_{\mathcal{F}(H)}(k) = O(k \cdot \text{poly} \log(k))$.*

Improved Running Times for Fixed-Parameter Tractability

The theory of bidimensionality [3] is a powerful methodology in the design of fixed-parameter tractable (FPT) algorithms. It led to sub-exponential (in the parameter k) time FPT algorithms for bidimensional parameters in planar graphs and more generally graphs that exclude a fixed graph H as a minor. The theory is based on the Excluded Grid Theorem. However, in general graphs, the weak bounds of the Excluded Grid Theorem meant that one could only derive FPT algorithms with running time of the form $2^{k^c} n^{O(1)}$, for some large constant c, by using the results of Demaine and Hajiaghayi [4], and the recent polynomial bounds for the Excluded Grid Theorem [2]. Using Theorem 1, we can obtain algorithms with running times of the form $2^{k \text{ poly} \log(k)} n^{O(1)}$ for the same class of problems as in [4].

Open Problems

The authors conjecture that there is an efficient algorithm that, given integers k, r, h with $hr \leq k / \text{poly} \log k$, and any graph G of treewidth k, finds a partition of G into h disjoint subgraphs of treewidth at least r each. This remains an open problem.

Experimental Results

None are reported.

URLs to Code and Data Sets

None are reported.

Recommended Reading

1. Chekuri C, Chuzhoy J (2013) Large-treewidth graph decompositions and applications. In: Proceedings of ACM STOC, Palo Alto, pp 291–300
2. Chekuri C, Chuzhoy J (2014) Polynomial bounds for the grid-minor theorem. In: STOC, New York
3. Demaine E, Hajiaghayi M (2007) The bidimensionality theory and its algorithmic applications. Comput J 51(3):292–302
4. Demaine ED, Hajiaghayi M (2007) Quickly deciding minor-closed parameters in general graphs. Eur J Comb 28(1):311–314
5. Demaine E, Hajiaghayi M, Kawarabayashi Ki (2009) Algorithmic graph minor theory: improved grid minor bounds and Wagner's contraction. Algorithmica 54:142–180. http://dx.doi.org/10.1007/s00453-007-9138-y
6. Diestel R (2012) Graph theory. Graduate texts in mathematics, vol 173, 4th edn. Springer, Berlin
7. Diestel R, Jensen TR, Gorbunov KY, Thomassen C (1999) Highly connected sets and the excluded grid theorem. J Comb Theory Ser B 75(1):61–73
8. Erdos P, Pósa L (1965) On independent circuits contained in a graph. Can J Math 17:347–352
9. Kawarabayashi K, Kobayashi Y (2012) Linear min-max relation between the treewidth of H-minor-free graphs and its largest grid minor. In: Proceedings of STACS, Paris
10. Leaf A, Seymour P (2012) Treewidth and planar minors, manuscript. Available at https://web.math.princeton.edu/~pds/papers/treewidth/paper.pdf
11. Robertson N, Seymour PD (1986) Graph minors. V. Excluding a planar graph. J Comb Theory Ser B 41(1):92–114
12. Robertson N, Seymour P, Thomas R (1994) Quickly excluding a planar graph. J Comb Theory Ser B 62(2):323–348

Layout Decomposition for Multiple Patterning

Haitong Tian and Martin D.F. Wong
Department of Electrical and Computer Engineering, University of Illinois at Urbana-Champaign, Urbana, IL, USA

Keywords

Double patterning; Layout decomposition; Lithography; Triple patterning

Years and Authors of Summarized Original Work

2012; Haitong Tian

Problem Definition

As the feature size keeps shrinking, there are increasing difficulties to print circuit patterns using single litho exposure. For 32/22 nm technology nodes, double patterning lithography (DPL) is one of the most promising techniques for the industry. In DPL, a layout is decomposed into two masks, where each feature in the layout is uniquely assigned to one of the masks. By using two masks which go through two separate exposures, better printing resolution can be achieved. For 14/10 nm technology node and beyond, triple patterning lithography (TPL) is one technique to obtain qualified printing results. In TPL, a layout is decomposed into three masks which further enhance the printing resolution. Currently, DPL and TPL are the two most studied multiple patterning techniques for advanced technology nodes [1–7]. Multiple patterning techniques such as quadruple patterning lithography and beyond usually are not investigated because of their increasing mask cost and other technical issues.

For DPL/TPL, there is a minimum coloring distance $d_{\min}$. If the distance of two features is less than $d_{\min}$, they cannot be printed in the same mask. $d_{\min}$ reflects the printing capabilities of current technology and can be redeemed as a constant in the problem. One practical concern of DPL/TPL is feature splitting, in which a feature is split into two or more parts for a legal color assignment. Such a splitting is called a stitch, which increases manufacturing cost and complexity due to additional line ends and more tight overlay control. Therefore, minimizing the number of stitches is a key objective for DPL/TPL decompositions. Other concerns include minimizing design rule violations, maximizing the overlap length, and balancing the usage of different colors. Among these concerns for DPL/TPL, minimizing the number of stitches is the most commonly studied one.

Multiple patterning decomposition is essentially a graph k-coloring problem, where $k = 2$ for DPL and $k = 3$ for TPL. It is well known that 3-coloring problem is NP-Complete, even when the graph is planar. For the general layout, ILP formulations are used in [1–3], and some heuristics are proposed in [5–7]. In reality, many industry designs are based on predesigned standard cells, where the layout is usually in row structures. All the cells are of exactly the same height, with power rails going from the left most of the cell to the right most of it. It is shown that multiple patterning decomposition ($k = 3$) for cell-based row structure layout is polynomial time solvable [4]. The following discussions are based on $k = 3$. The same concept can be easily extended to other multiple patterning techniques such as $k = 2$ or $k > 3$.

Problem: Multiple Patterning Decomposition

Using k colors to represent the k masks, multiple patterning decomposition can be defined as follows:

Input: *Circuit layout and a minimum coloring distance* $d_{\min}$.

Output: *A coloring solution where all features are assigned to one of the k colors.*

Constraint: *Any two features with the distance less than* $d_{\min}$ *cannot be assigned to the same color.*

Key Results

Multiple Patterning Decomposition for Standard Cell Designs

Given a layout, a conflict graph $G = (V, E)$ is constructed, where (I) vertices $V = \{v_1, \ldots, v_n\}$ represent the features in the layout and (II) $E = \{e_1, \ldots, e_m\}$ represent conflicting relationships between the features. A conflict edge exists if the distance of the two features is within $d_{\min}$. Imagine a cutting line that goes vertically across the cell, there are limited number of features that intersect with the cutting line due to the fixed height of the cell. Therefore, the coloring solutions of each cutting can be enumerated in polynomial time. The set of polygons that intersect with the same cutting line is called a cutting line set. An example of conflict graph, cutting line, and cutting line set is shown in Fig. 1.

By using the left boundary of each feature as the cutting line, the solutions of each cut line are computed. Solutions of adjacent cut lines are connected together, which leads to a solution graph. Polygon dummy extension is performed to ensure that the constructed solution graph is legal. For each polygon, its right boundary is

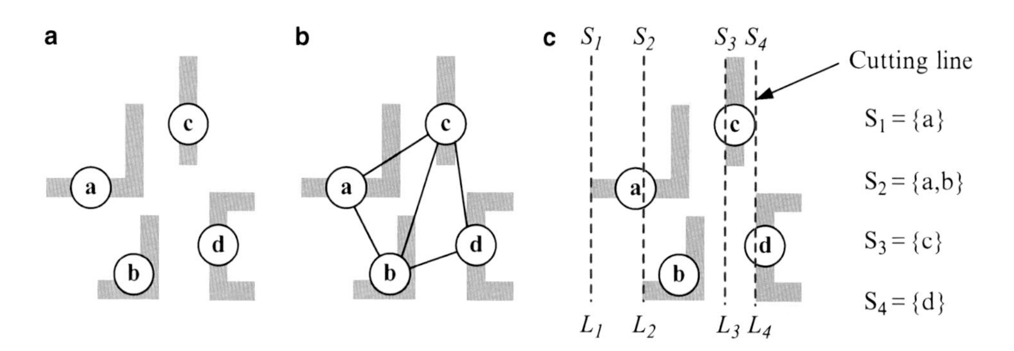

Layout Decomposition for Multiple Patterning, Fig. 1 (**a**) Input layout. (**b**) Conflict graph. (**c**) Cutting lines and the corresponding cutting line sets. There are four cutting lines L_1–L_4 and four cutting line sets S_1–S_4 in this layout

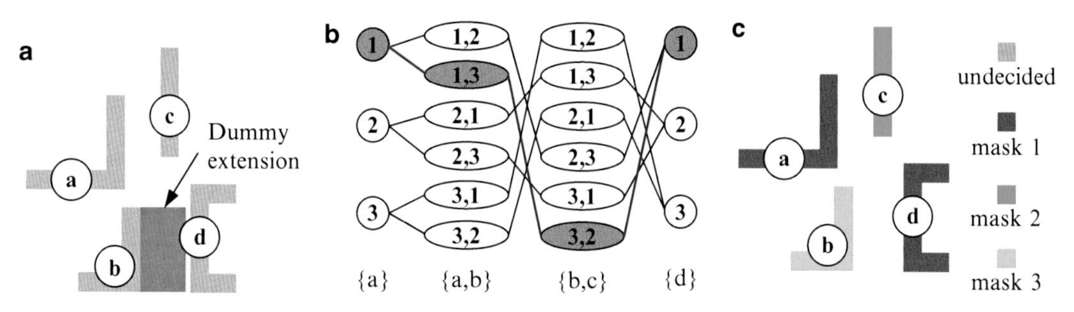

Layout Decomposition for Multiple Patterning, Fig. 2 (**a**) Input layout with polygon dummy extension. (**b**) Solution graph. The highlighted path is a sample decomposition. (**c**) Sample decomposition. Different colors represent different masks

virtually extended to its right most conflicting polygon. After extending the right boundaries of the polygons, it is guaranteed that for any polygon in a cutting line set, all its conflicting polygons (with smaller x coordinates) appear in the previous cutting line set. Therefore, the solution graph can be incrementally constructed and the correctness of the graph is guaranteed.

The solution graph is complete in the sense that it explores all the solution space. It is proven in [4] that every path in the solution graph corresponds to legal TPL decomposition and every legal TPL decomposition corresponds to a path in the solution graph. Figure 2 illustrates the overall flow of their approach.

Minimizing Stitches

The approach can be extended to handle stitches. All legal stitch candidates are computed for the layout, where a polygon feature is decomposed into a set of touching polygons by the stitch candidates. Conflict graph $G = (V, E)$ is constructed to model the rectangular layout, where (I) vertices $V = \{v_1, \ldots, v_n\}$ represent the features in the layout and (II) $E = \{e_1, \ldots, e_m\}$ represent different relationships between the features. There are two types of edges in the graph: conflict edges and stitch edges. A conflict edge exists if the two features do not touch each other and their distance is within $d_{\min}$. A stitch edge exists if the two features touch each other.

A weighted solution graph is constructed, where the weight of an edge denotes the number of stitches needed between the two vertices. A shortest path algorithm is utilized to get the decomposition with optimal number of stitches.

Multiple Patterning Coloring Constraint

In practice, there are additional coloring constraints such as balancing the usage of different masks [4,8] and assigning the same pattern for the same type of cells [9]. For standard cell designs, coloring balancing can be simply achieved by using three global variables when parsing the solution graph [4]. An efficient SAT formulation with limited number of clauses is used to guarantee that the same type of cells has the same coloring decomposition [9].

Applications

Products using DPL in 22 nm technology node are already available in markets. TPL can be used in 14/10 nm technology node.

Open Problems

None is reported.

Experimental Results

The authors in [1] show that as the minimum coloring distance $d_{\min}$ increases, the number of unsolved conflicts increases. They also observe that the placement utilization has a very small impact on the number of unsolved conflicts. The results in [3] show that the speedup techniques can greatly reduce the overall runtime without adversely affecting the quality of the decomposition. Better results on the same benchmarks are reported in [5–7]. The authors in [4] show that their algorithm is able to solve all TPL decomposable benchmarks. For complex layout with stitch candidates, their approach computes a decomposition with the optimal number of stitches.

URLs to Code and Data Sets

The NanGate Open Cell Library can be obtained online for free.

Cross-References

▶ Graph Coloring
▶ Layout Decomposition for Triple Patterning

Recommended Reading

1. Kahng AB, Xu X, Park C-H, Yao H (2008) Layout decomposition for double patterning lithography. In: IEEE/ACM international conference on computer-aided design, San Jose
2. Yuan K, Yang J-S, Pan DZ (2010) Double patterning layout decomposition for simultaneous conflict and stitch minimization. IEEE Trans Comput-Aided Des Integr Circuits Syst 29:185–196
3. Yu B, Yuan K, Zhang B, Ding D, Pan DZ (2011) Layout decomposition for triple patterning lithography. In: IEEE/ACM international conference on computer-aided design, San Jose
4. Tian H, Zhang H, Ma Q, Xiao Z, Wong MDF (2012) A polynomial time triple patterning algorithm for cell based row-structure layout. In: IEEE/ACM international conference on computer-aided design, San Jose
5. Fang S-Y, Chang Y-W, Chen W-Y (2012) A novel layout decomposition algorithm for triple patterning lithography. In: IEEE/ACM proceedings of design automation conference, San Francisco
6. Kuang J, Yang EFY (2013) An efficient layout decomposition approach for triple patterning lithography. In: IEEE/ACM proceedings of design automation conference, Austin
7. Zhang Y, Luk W-S, Zhou H, Yan C, Zeng X (2013) Layout decomposition with pairwise coloring for multiple patterning lithography. In: IEEE/ACM international conference on computer-aided design, San Jose
8. Yu B, Lin Y-H, Luk-Pat G, Ding D, Lucas K, Pan DZ (2013) A high-performance triple patterning layout decomposer with balanced density. In: IEEE/ACM international conference on computer-aided design, San Jose
9. Tian H, Zhang H, Du Y, Xiao Z, Wong MDF (2013) Constrained pattern assignment for standard cell based triple patterning lithography. In: IEEE/ACM international conference on computer-aided design, San Jose

Layout Decomposition for Triple Patterning

Bei Yu and David Z. Pan
Department of Electrical and Computer Engineering, University of Texas, Austin, TX, USA

Keywords

Graph coloring; Layout decomposition; Lithography; Mathematical programming

Years and Authors of Summarized Original Work

2011; Yu, Yuan, Zhang, Ding, Pan

Problem Definition

Layout decomposition is a key stage in triple patterning lithography manufacturing process, where the original designed layout is divided into three masks. There will be three exposure/etching steps, through which the circuit layout can be produced. When the distance between two input features is less than certain minimum distance min_s, they need to be assigned to different masks (colors) to avoid coloring conflict. Sometimes coloring conflict can be resolved by splitting a pattern into two different masks. However, this introduces stitches, which lead to yield loss because of overlay error. Therefore, two of the main objectives in layout decomposition are conflict minimization and stitch minimization. An example of triple patterning layout decomposition is shown in Fig. 1, where all features are divided into three masks without any conflict and one stitch is introduced.

Given an input layout, a *conflict graph* is constructed to transfer initial geometrical relationship into an undirected graph with a set of vertices V and two sets of edges, which are the *conflict edges* (CE) and *stitch edges* (SE), respectively. V has one or more vertices for each polygonal shape and each vertex is associated with a polygonal shape. An edge is in CE iff the two corresponding vertices are within minimum coloring distance min_s. An edge is in SE iff there is a stitch candidate between the two vertices which are associated with the same polygonal shape.

Problem 1 (Layout Decomposition for Triple Patterning)

INPUT: *The decomposition graph where each vertex represents one polygonal shape, and all possible conflicts and stitches are in the conflict edge set CE and the stitch edge set SE, respectively.*

OUTPUT: *A three-color assignment to the conflict graph, such that the weighted cost of conflicts and stitches are minimized. The additional constraints may include color balancing, overlay control, and color preference.*

Key Results

Given an input layout, the conflict graph is constructed. Based on the conflict graph, the layout decomposition for triple patterning can be formulated as an integer linear programming (ILP) formulation [5]. As shown in (1), the objective function in the ILP formulation is to minimize the weighted cost function of conflict and stitch numbers simultaneously:

$$\min \sum_{e_{ij} \in CE} c_{ij} + \alpha \sum_{e_{ij} \in SE} s_{ij} \tag{1}$$

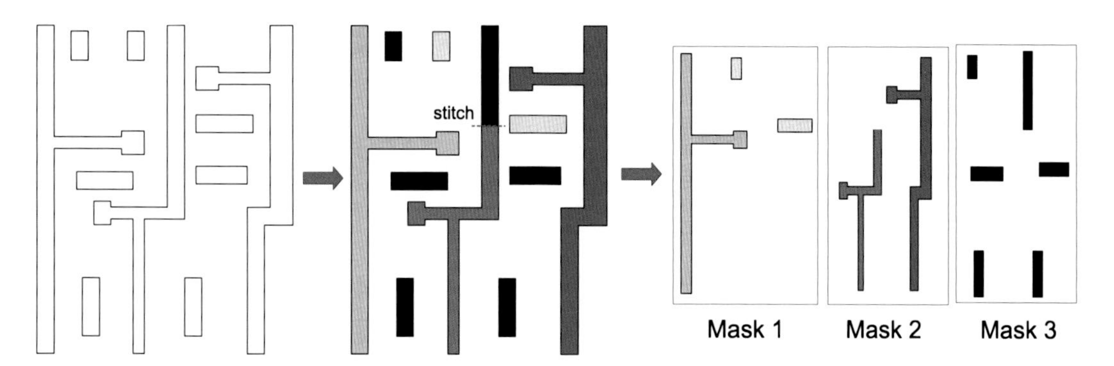

Layout Decomposition for Triple Patterning, Fig. 1 Layout decomposition for triple patterning lithography (TPL)

where α is a parameter for assigning relative cost of stitch versus conflict. Typically, α is much smaller than 1, for example, 0.1, as resolving conflict is the most important objective during layout decomposition. Although the ILP formulation can solve the above layout decomposition problem optimally, it is not scalable to deal with large layouts in modern VLSI designs as the ILP problem is NP-complete.

In [5], a semidefinite programming (SDP)-based algorithm was proposed to achieve good runtime and solution quality. Instead of using a two binary variables to represent three masks, three unit vectors $(1,0), \left(-\frac{1}{2}, \frac{\sqrt{3}}{2}\right)$, and $\left(-\frac{1}{2}, -\frac{\sqrt{3}}{2}\right)$ are proposed to represent them. Note that the angle between any two vectors of the same color is 0, while the angle between any two vectors with different colors is $2\pi/3$. The inner product of two m-dimension vectors $\mathbf{v_i}$ and $\mathbf{v_j}$ is defined as $\mathbf{v_i} \cdot \mathbf{v_j} = \sum_k v_{ik} v_{jk}$. Then for any two vectors $\mathbf{v_i}, \mathbf{v_j} \in \left\{(1,0), \left(-\frac{1}{2}, \frac{\sqrt{3}}{2}\right), \left(-\frac{1}{2}, -\frac{\sqrt{3}}{2}\right)\right\}$, the following property holds:

$$\mathbf{v_i} \cdot \mathbf{v_j} = \begin{cases} 1, & \mathbf{v_i} = \mathbf{v_j} \\ -\frac{1}{2}, & \mathbf{v_i} \neq \mathbf{v_j} \end{cases}$$

Based on the vector representation, the layout decomposition for triple patterning problem can be written as the following vector programming:

$$\min \sum_{e_{ij} \in CE} \frac{2}{3}\left(\mathbf{v_i} \cdot \mathbf{v_j} + \frac{1}{2}\right) + \frac{2\alpha}{3} \sum_{e_{ij} \in SE} (1 - \mathbf{v_i} \cdot \mathbf{v_j}) \tag{2}$$

$$\text{s.t. } \mathbf{v_i} \in \left\{(1,0), \left(-\frac{1}{2}, \frac{\sqrt{3}}{2}\right), \left(-\frac{1}{2}, -\frac{\sqrt{3}}{2}\right)\right\} \tag{2a}$$

It shall be noted that $\mathbf{v_i}$ here is discrete, which is very expensive to solve. Then the discrete vector program is relaxed to the corresponding continuous formulation, which can be solved as a standard semidefinite programming (SDP), as shown below:

$$\min \sum_{e_{ij} \in CE} \frac{2}{3}\left(\mathbf{y_i} \cdot \mathbf{y_j} + \frac{1}{2}\right) + \frac{2\alpha}{3} \sum_{e_{ij} \in SE} (1 - \mathbf{y_i} \cdot \mathbf{y_j}) \tag{3}$$

$$\text{s.t. } \mathbf{y_i} \cdot \mathbf{y_i} = 1, \quad \forall i \in V \tag{3a}$$

$$-\frac{1}{2} \leq \mathbf{y_i} \cdot \mathbf{y_j}, \quad \forall e_{ij} \in CE \tag{3b}$$

$$Y \succeq 0 \tag{3c}$$

The resulting matrix Y, where $y_{ij} = \mathbf{y_i} \cdot \mathbf{y_i}$, essentially provides the relative coloring guidance between two layout features (nodes in the conflict graph). It will be used to guide the final color assignment. If y_{ij} is close to 1, nodes i and j should be in the same mask; if y_{ij} is close to -0.5, nodes i and j tend to be in different masks. The results show that with reasonable threshold such as $0.9 < y_{ij} \leq 1$ for the same mask and $-0.5 \leq y_{ij} < -0.4$ for different masks, more than 80 % of nodes/polygons are decided by the global SDP. For the rest values, heuristic mapping algorithms will be performed to assign all vertices to their final colors.

A set of graph simplification techniques have been proposed to achieve speedup [1,2,5,8]. For example, one technique is called *iterative vertex removal*, where all vertices with degree less than or equal to two are detected and removed temporarily from the conflict graph. After each vertex removal, the degrees of other vertices would be updated. This removing process will continue until all the vertices have degree three or more. All the vertices that are temporarily removed are stored in a stack S. After the color assignment of the remained conflict graph is solved, the removed vertices in S are added back for coloring assignment. For row-based structure layout, specific graph-based algorithms are proposed to provide fast layout decomposition solutions [3,7].

Triple patterning layout decomposition has been actively studied in the last few years, with many interesting results reported. In [5], the performances between ILP- and SDP-based methods were compared. As shown in Fig. 2, SDP-based method can achieve the same optimal solutions

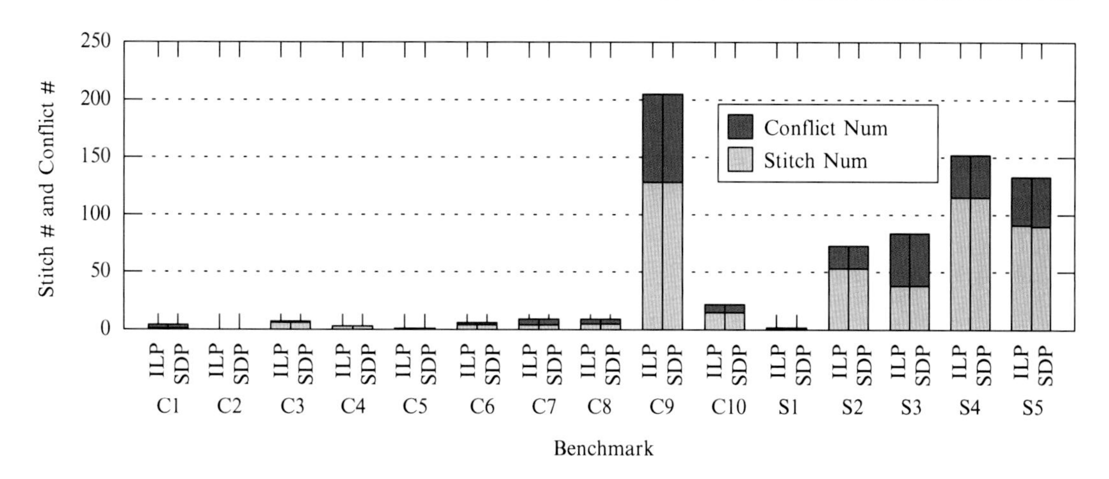

Layout Decomposition for Triple Patterning, Fig. 2 For ISCAS benchmark suite, the results of ILP- and SDP-based methods are very comparable

as obtained by ILP for 14 out of 15 test cases. However, the runtime of ILP-based algorithm is prohibitive when the problem size is big and the layout is dense. Graph simplification techniques are very effective to speed up the layout decomposition process as that can effectively reduce the ILP and SDP problem size. The coloring density balance was integrated into the SDP formulation in [6]. In [4], the SDP framework was further extended to handle quadruple patterning or more general multiple patterning lithography with new vector definition and linear runtime heuristic algorithms.

URLs to Code and Data Sets

Programs and benchmark suites can be found through http://www.cerc.utexas.edu/utda/download/MPLD/.

Cross-References

- ▶ Graph Coloring
- ▶ Greedy Approximation Algorithms

Recommended Reading

1. Fang SY, Chen WY, Chang YW (2012) A novel layout decomposition algorithm for triple patterning lithography. In: IEEE/ACM design automation conference (DAC), San Francisco, pp 1185–1190
2. Kuang J, Young EF (2013) An efficient layout decomposition approach for triple patterning lithography. In: IEEE/ACM design automation conference (DAC), Austin, pp 69:1–69:6
3. Tian H, Zhang H, Ma Q, Xiao Z, Wong M (2012) A polynomial time triple patterning algorithm for cell based row-structure layout. In: IEEE/ACM international conference on computer-aided design (ICCAD), San Jose, pp 57–64
4. Yu B, Pan DZ (2014) Layout decomposition for quadruple patterning lithography and beyond. In: IEEE/ACM design automation conference (DAC), San Francisco
5. Yu B, Yuan K, Zhang B, Ding D, Pan DZ (2011) Layout decomposition for triple patterning lithography. In: IEEE/ACM international conference on computer-aided design (ICCAD), San Jose, pp 1–8
6. Yu B, Lin YH, Luk-Pat G, Ding D, Lucas K, Pan DZ (2013) A high-performance triple patterning layout decomposer with balanced density. In: IEEE/ACM International conference on computer-aided design (ICCAD), San Jose, pp 163–169
7. Yu B, Xu X, Gao JR, Pan DZ (2013) Methodology for standard cell compliance and detailed placement for triple patterning lithography. In: IEEE/ACM international conference on computer-aided design (ICCAD), San Jose, pp 349–356
8. Zhang Y, Luk WS, Zhou H, Yan C, Zeng X (2013) Layout decomposition with pairwise coloring for multiple patterning lithography. In: IEEE/ACM international conference on computer-aided design (ICCAD), San Jose, pp 170–177

Learning Automata

Stefano Varricchio
Department of Computer Science, University of Roma, Rome, Italy

Keywords

Boolean formulae; Computational learning; Formal series; Machine learning; Multiplicity automata; Multivariate polynomials

Years and Authors of Summarized Original Work

2000; Beimel, Bergadano, Bshouty, Kushilevitz, Varricchio

Problem Definition

This problem is concerned with the learnability of *multiplicity automata* in Angluin's *exact learning model* and applications to the learnability of functions represented by small multiplicity automata.

The Learning Model

It is the *exact learning* model [2]: Let f be a *target* function. A learning algorithm may propose to an oracle, in each step, two kinds of queries: *membership queries* (MQ) and *equivalence queries* (EQ). In a MQ it may query for the value of the function f on a particular assignment z. The response to such a query is the value $f(z)$. (If f is Boolean, this is the standard membership query.) In an EQ it may propose to the oracle a hypothesis function h. If h is equivalent to f on all input assignments, then the answer to the query is YES and the learning algorithm succeeds and halts. Otherwise, the answer to the equivalence query is NO and the algorithm receives a *counterexample*, i.e., an assignment z such that $f(z) \neq h(z)$. One says that the learner *learns* a class of functions $\mathcal{C}$, if for every function $f \in \mathcal{C}$ the learner outputs a hypothesis h that is equivalent to f and does so in time polynomial in the "size" of a shortest representation of f and the length of the longest counterexample. The exact learning model is strictly related to the *Probably Approximately Correct* (PAC) model of Valiant [19]. In fact, every equivalence query can be easily simulated by a sample of random examples. Therefore, learnability in the exact learning model also implies learnability in the PAC model with membership queries [2, 19].

Stefano Varricchio: deceased

Multiplicity Automata

Let $\mathcal{K}$ be a field, Σ be an alphabet, and ϵ be the empty string. A *multiplicity automaton* (MA) A of size r consists of $|\Sigma|$ matrices $\{\mu_\sigma : \sigma \in \Sigma\}$, each of which is an $r \times r$ matrix of elements from $\mathcal{K}$ and an r-tuple $\vec{\gamma} = (\gamma_1, \ldots, \gamma_r) \in \mathcal{K}^r$. The automaton A defines a function $f_A : \Sigma^* \rightarrow \mathcal{K}$ as follows. First, define a mapping μ, which associates with every string in Σ^* an $r \times r$ matrix over $\mathcal{K}$, by $\mu(\epsilon) \triangleq \text{ID}$ where ID denotes the *identity matrix*, and for a string $w = \sigma_1\sigma_2 \ldots \sigma_n$, let $\mu(\omega) \triangleq \mu_{\sigma_1} \cdot \mu_{\sigma_2} \cdots \mu_{\sigma_n}$. A simple property of μ is that $\mu(x \circ y) = \mu(x) \cdot \mu(y)$, where $\circ$ denotes concatenation. Now, $f_A(\omega) \triangleq [\mu(\omega)]1 \cdot \vec{\gamma}$ (where $[\mu(w)]_i$ denotes the ith row of the matrix $\mu(w)$). Let $f : \Sigma^* \rightarrow \mathcal{K}$ be a function. Associate with f an infinite matrix F, where each of its rows is indexed by a string $x \in \Sigma^*$ and each of its columns is indexed by a string $y \in \Sigma^*$. The (x, y) entry of F contains the value $f(x \circ y)$. The matrix F is called the *Hankel Matrix* of f. The xth row of F is denoted by F_x. The (x, y) entry of F may be therefore denoted as $F_x(y)$ and as $F_{x,y}$. The following result relates the size of the minimal MA for f to the rank of F (cf. [4] and references therein).

Theorem 1 *Let $f : \Sigma^* \rightarrow \mathcal{K}$ such that $f \not\equiv 0$ and let F be its Hankel matrix. Then, the size r of the smallest multiplicity automaton A such that $f_A \equiv f$ satisfies $r = \text{rank}(F)$ (over the field $\mathcal{K}$).*

Key Results

The learnability of multiplicity automata has been proved in [7] and, independently, in [17]. In what follows, let $\mathcal{K}$ be a field, $f : \Sigma^* \rightarrow \mathcal{K}$

be a function, and F its Hankel matrix such that $r = \text{rank}(F)$ (over $\mathcal{K}$).

Theorem 2 ([4]) *The function f is learnable by an algorithm in time $O(|\Sigma| \cdot r \cdot \mathrm{M}(r) + m \cdot r^3)$ using r equivalence queries and $O((|\Sigma| + \log m) r^2)$ membership queries, where m is the size of the longest counterexample obtained during the execution of the algorithm and $\mathrm{M}(r)$ is the complexity of multiplying two $r \times r$ matrices.*

Some extensions of the above result can be found in [8, 13, 16]. In many cases of interest, the domain of the target function f is not Σ^* but rather Σ^n for some value n, i.e., $f : \Sigma^n \to \mathcal{K}$. The length of counterexamples, in this case, is always n and so $m = n$. Denote by F^d the submatrix of F whose rows are strings in Σ^d and whose columns are strings in Σ^{n-d} and let $r_{\max} = \max_{d=0}^{n} \text{rank}(F^d)$ (where rank is taken over $\mathcal{K}$).

Theorem 3 ([4]) *The function f is learnable by an algorithm in time $O(|\Sigma| rn \cdot M(r_{\max}))$ using $O(r)$ equivalence queries and $O((|\Sigma| + \log n) r \cdot r_{\max})$ membership queries.*

The time complexity of the two above results has been recently further improved [9].

Applications

The results of this section can be found in [3–6]. They show the learnability of various classes of functions as a consequence of Theorems 2 and 3. This can be done by proving that for every function f in the class in question, the corresponding Hankel matrix F has low rank. As is well known, any nondeterministic automaton can be regarded as a multiplicity automaton, whose associated function returns the number of accepting paths of the nondeterministic automaton on w. Therefore, the learnability of multiplicity automata gives a new algorithm for learning deterministic automata and unambiguous automata. (A nondeterministic automata is *unambiguous* if for every $w \in \Sigma^*$, there is at most one accepting path.) The learnability of deterministic automata has been proved in [1]. By [14], the class of deterministic automata contains the class of $O(\log n)$-term DNF, i.e., DNF formulae over n Boolean variables with $O(\log n)$ number of terms. Hence, this class can be learned using multiplicity automata.

Classes of Polynomials

Theorem 4 *Let $p_{i,j} : \Sigma \to \mathcal{K}$ be arbitrary functions of a single variable ($1 \le i \le t$, $1 \le j \le n$). Let $g_i : \Sigma^n \to \mathcal{K}$ be defined by $\prod_{j=1}^{n} p_{i,j}(z_j)$. Finally, let $f : \Sigma^n \to \mathcal{K}$ be defined by $f = \sum_{i=1}^{t} g_i$. Let F be the Hankel matrix corresponding to f and F^d the submatrices defined in the previous section. Then, for every $0 \le d \le n$, $\text{rank}(F^d) \le t$.*

Corollary 1 *The class of functions that can be expressed as functions over $\mathrm{GF}(p)$ with t summands, where each summand T_i is a product of the form $p_{i,1}(x_1) \cdots p_{i,n}(x_n)$ (and $p_{i,j} : \mathrm{GF}(p) \to \mathrm{GF}(p)$ are arbitrary functions), is learnable in time $\mathrm{poly}(n, t, p)$.*

The above corollary implies as a special case the learnability of polynomials over $\mathrm{GF}(p)$. This extends the result of [18] from multi-linear polynomials to arbitrary polynomials. The algorithm of Theorem 3, for polynomials with n variables and t terms, uses $O(nt)$ equivalence queries and $O(t^2 n \log n)$ membership queries. The special case of the above class – the class of polynomials over GF(2) – was known to be learnable before [18]. Their algorithm uses $O(nt)$ equivalence queries and $O(t^3 n)$ membership queries. The following theorem extends the latter result to *infinite* fields, assuming that the functions $p_{i,j}$ are bounded-degree polynomials.

Theorem 5 *The class of functions over a field $\mathcal{K}$ that can be expressed as t summands, where each summand T_i is of the form $p_{i,1}(x_1) \cdots p_{i,n}(x_n)$ and $p_{i,j} : \mathcal{K} \to \mathcal{K}$ are univariate polynomials of degree at most k, is learnable in time $\mathrm{poly}(n, t, k)$. Furthermore, if $|\mathcal{K}| \ge nk + 1$, then this class is learnable from membership queries only in time $\mathrm{poly}(n, t, k)$ (with small probability of error).*

Classes of Boxes

Let $[\ell]$ denote the set $\{0, 1, \ldots, \ell - 1\}$. A box in $[\ell]^n$ is defined by two corners $(a_1, \ldots, a_n)$ and $(b_1, \ldots, b_n)$ (in $[\ell]^n$) as follows:

$$B_{a_1,\ldots,a_n,b_1,\ldots,b_n} = \{(x_1, \ldots, x_n) : \forall i,\, a_i \le x_i \le b_i\}.$$

A box can be represented by its characteristic function in $[\ell]^n$. The following result concerns a more general class of functions.

Theorem 6 *Let $p_{i,j} : \Sigma \to \{0,1\}$ be arbitrary functions of a single variable ($1 \le i \le t$, $1 \le j \le n$). Let $g_i : \Sigma^n \to \{0,1\}$ be defined by $\prod_{j=1}^{n} p_{i,j}(z_j)$. Assume that there is no point $x \in \Sigma^n$ such that $g_i(x) = 1$ for more than s functions g_i. Finally, let $f : \Sigma^n \to \{0,1\}$ be defined by $f = \bigvee_{i=1}^{t} g_i$. Let F be the Hankel matrix corresponding to f. Then, for every field $\mathcal{K}$ and for every $0 \le d \le n$, $\mathrm{rank}(F^d) \le \sum_{i=1}^{s} \binom{t}{i}$.*

Corollary 2 *The class of unions of disjoint boxes can be learned in time $\mathrm{poly}(n, t, \ell)$ (where t is the number of boxes in the target function). The class of unions of $O(\log n)$ boxes can be learned in time $\mathrm{poly}(n, \ell)$.*

Classes of DNF Formulae

The learnability of DNF formulae has been widely investigated. The following special case of Corollary 1 solves an open problem of [18]:

Corollary 3 *The class of functions that can be expressed as exclusive OR of t (not necessarily monotone) monomials is learnable in time $\mathrm{poly}(n, t)$.*

While Corollary 3 does not refer to a subclass of DNF, it already implies the learnability of disjoint (i.e., satisfy-1) DNF. Since DNF is a special case of union of boxes (with $\ell = 2$), one obtains also the learnability of disjoint DNF from Corollary 2. Positive results for satisfy-s DNF (i.e., DNF formulae in which each assignment satisfies at most s terms) can be obtained, with larger values of s. The following two important corollaries follow from Theorem 6. Note that Theorem 6 holds in any field. For convenience (and efficiency), let $\mathcal{K} = \mathrm{GF}(2)$.

Theorem 7 *Let $f = T_1 \vee T_2 \vee \ldots \vee T_t$ be a satisfy-s DNF (i.e., each T_i is a monomial). Let F be the Hankel matrix corresponding to f. Then, $\mathrm{rank}(F^d) \le \sum_{i=1}^{s} \binom{t}{i} \le t^s$.*

Corollary 4 *The class of satisfy-s DNF formulae, for $s = O(1)$, is learnable in time $\mathrm{poly}(n, t)$.*

Corollary 5 *The class of satisfy-s, t-term DNF formulae is learnable in time $\mathrm{poly}(n)$ for the following choices of s and t: (1) $t = O(\log n)$, (2) $t = \mathrm{poly}\log(n)$ and $s = O(\log n / \log \log n)$, (3) $t = 2^{O(\log n / \log \log n)}$ and $s = O(\log \log n)$.*

Classes of Decision Trees

The algorithm of Theorem 3 efficiently learns the class of disjoint DNF formulae. This includes the class of decision trees. Therefore, decision trees of size t on n variables are learnable using $O(tn)$ equivalence queries and $O(t^2 n \log n)$ membership queries. This is better than the best known algorithm for decision trees [11] (which uses $O(t^2)$ equivalence queries and $O(t^2 n^2)$ membership queries). The following results concern more general classes of decision trees.

Corollary 6 *Consider the class of decision trees that compute functions $f : \mathrm{GF}(p)^n \to \mathrm{GF}(p)$ as follows: each node v contains a query of the form "$x_i \in S_v$?" for some $S_v \subseteq \mathrm{GF}(p)$. If $x_i \in S_v$, then the computation proceeds to the left child of v, and if $x_i \notin S_v$ the computation proceeds to the right child. Each leaf ℓ of the tree is marked by a value $\gamma\ell \in \mathrm{GF}(p)$ which is the output on all the assignments which reach this leaf. Then, this class is learnable in time $\mathrm{poly}(n, |L|, p)$, where L is the set of leaves.*

The above result implies the learnability of decision trees with "greater-than" queries in the nodes, solving a problem of [11]. Every decision tree with "greater-than" queries that computes a Boolean function can be expressed as the union of disjoint boxes. Hence, this case can also be derived from Corollary 2. The next theorem will be used to learn more classes of decision trees.

Theorem 8 *Let* $g_i : \Sigma^n \rightarrow \mathcal{K}$ *be arbitrary functions* ($1 \leq i \leq \ell$). *Let* $f : \Sigma^n \rightarrow \mathcal{K}$ *be defined by* $f = \prod_{i=1}^{\ell} g_i$. *Let* F *be the Hankel matrix corresponding to* f, *and* G_i *be the Hankel matrix corresponding to* g_i. *Then,* $\text{rank}(F^d) \leq \prod_{i=1}^{\ell} \text{rank}(G_i^d)$.

This theorem has some interesting applications. The first application states that arithmetic circuits of depth two with multiplication gate of fan-in $O(\log n)$ at the top level and addition gates with unbounded fan-in in the bottom level are learnable.

Corollary 7 *Let* C *be the class of functions that can be expressed in the following way: Let* $p_{i,j} : \Sigma \rightarrow \mathcal{K}$ *be arbitrary functions of a single variable* ($1 \leq i \leq \ell$, $1 \leq j \leq n$). *Let* $\ell = O(\log n)$ *and* $g_i : \Sigma^n \rightarrow \mathcal{K}(1 \leq i \leq \ell)$ *be defined by* $\Sigma_{j=1}^{n} p_{i,j}(z_j)$. *Finally, let* $f : \Sigma^n \rightarrow \mathcal{K}$ *be defined by* $f = \prod_{i=1}^{\ell} g_i$. *Then,* C *is learnable in time* $\text{poly}(n, |\Sigma|)$.

Corollary 8 *Consider the class of decision trees of depth* s, *where the query at each node* v *is a Boolean function* f_v *with* $r_{\max} \leq t$ *(as defined in section "Key Results") such that* $(t+1)^s = \text{poly}(n)$. *Then, this class is learnable in time* $\text{poly}(n, |\Sigma|)$.

The above class contains, for example, all the decision trees of depth $O(\log n)$ that contain in each node a term or a XOR of a subset of variables as defined in [15] (in this case $r_{\max} \leq 2$).

Negative Results

In [4] some limitation of the learnability via the automaton representation has been proved. One can show that the main algorithm does not efficiently learn several important classes of functions. More precisely, these classes contain functions f that have no "small" automaton, i.e., by Theorem 1, the corresponding Hankel matrix F is "large" over every field $\mathcal{K}$.

Theorem 9 *The following classes are not learnable in time polynomial in n and the formula size using multiplicity automata (over any field* $\mathcal{K}$*): DNF, monotone DNF; 2-DNF; read-once DNF;* k*-term DNF, for* $k = \omega(\log n)$*; satisfy-*s *DNF, for* $s = \omega(1)$*; and read-*j *satisfy-*s *DNF, for* $j = \omega(1)$ *and* $s = \Omega(\log n)$.

Some of these classes are known to be learnable by other methods, some are natural generalizations of classes known to be learnable as automata ($O(\log n)$-term DNF [11, 12, 14], and satisfy-s DNF for $s = O(1)$ (Corollary 4)) or by other methods (read-j satisfy-s for $js = O(\log n / \log \log n)$ [10]), and the learnability of some of the others is still an open problem.

Cross-References

▸ Learning Constant-Depth Circuits
▸ Learning DNF Formulas

Recommended Reading

1. Angluin D (1987) Learning regular sets from queries and counterexamples. Inf Comput 75: 87–106
2. Angluin D (1988) Queries and concept learning. Mach Learn 2(4):319–342
3. Beimel A, Bergadano F, Bshouty NH, Kushilevitz E, Varricchio S (1996) On the applications of multiplicity automata in learning. In: Proceedings of the 37th annual IEEE symposium on foundations of computer science, Burlington, Vermont, USA. IEEE Computer Society, Los Alamitos, pp 349–358
4. Beimel A, Bergadano F, Bshouty NH, Kushilevitz E, Varricchio S (2000) Learning functions represented as multiplicity automata. J ACM 47: 506–530
5. Beimel A, Kushilevitz E (1997) Learning boxes in high dimension. In: Ben-David S (ed) 3rd European conference on computational learning theory (EuroCOLT '97), Jerusalem, Israel. Lecture notes in artificial intelligence, vol 1208. Springer, Berlin, pp 3–15. Journal version: Algorithmica 22:76–90 (1998)
6. Bergadano F, Catalano D, Varricchio S (1996) Learning sat-k-DNF formulas from membership queries. In: Proceedings of the 28th annual ACM symposium on the theory of computing, Philadelphia, Pennsylvania, USA. ACM, New York, pp 126–130

7. Bergadano F, Varricchio S (1994) Learning behaviors of automata from multiplicity and equivalence queries. In: Proceedings of 2nd Italian conference on algorithms and complexity, Rome, Italy. Lecture notes in computer science, vol 778. Springer, Berlin, pp 54–62. Journal version: SIAM J Comput 25(6):1268–1280 (1996)
8. Bergadano F, Varricchio S (1996) Learning behaviors of automata from shortest counterexamples. In: EuroCOLT '95, Barcelona, Spain. Lecture notes in artificial intelligence, vol 904. Springer, Berlin, pp 380–391
9. Bisht L, Bshouty NH, Mazzawi H (2006) On optimal learning algorithms for multiplicity automata. In: Proceedings of 19th annual ACM conference on computational learning theory, Pittsburgh, Pennsylvania, USA. Lecture notes in computer science, vol 4005. Springer, Berlin, pp 184–198
10. Blum A, Khardon R, Kushilevitz E, Pitt L, Roth D (1994) On learning read-k-satisfy-j DNF. In: Proceedings of the 7th annual ACM conference on computational learning theory, New Brunswick, New Jersey, USA. ACM, New York, pp 110–117
11. Bshouty NH (1993) Exact learning via the monotone theory. In: Proceedings of the 34th annual IEEE symposium on foundations of computer science, Palo Alto, California, USA. IEEE Computer Society, Los Alamitos, pp 302–311. Journal version: Inf Comput 123(1):146–153 (1995)
12. Bshouty NH (1995) Simple learning algorithms using divide and conquer. In: Proceedings of 8th annual ACM conference on computational learning theory, Santa Cruz, California, USA. ACM, New York, pp 447–453. Journal version: Comput Complex 6:174–194 (1997)
13. Bshouty NH, Tamon C, Wilson DK (1998) Learning matrix functions over rings. Algorithmica 22(1/2):91–111
14. Kushilevitz E (1996) A simple algorithm for learning $O(\log n)$-term DNF. In: Proceedings of 9th annual ACM conference on computational learning theory, Desenzano del Garda, Italy. ACM, New York, pp 266–269. Journal version: Inf Process Lett 61(6):289–292 (1997)
15. Kushilevitz E, Mansour Y (1993) Learning decision trees using the fourier spectrum. SIAM J Comput 22(6):1331–1348
16. Melideo G, Varricchio S (1998) Learning unary output two-tape automata from multiplicity and equivalence queries. In: ALT '98, Otzenhausen, Germany. Lecture notes in computer science, vol 1501. Springer, Berlin, pp 87–102
17. Ohnishi H, Seki H, Kasami T (1994) A polynomial time learning algorithm for recognizable series. IEICE Trans Inf Syst E77-D(10)(5):1077–1085
18. Schapire RE, Sellie LM (1996) Learning sparse multivariate polynomials over a field with queries and counterexamples. J Comput Syst Sci 52(2):201–213
19. Valiant LG (1984) A theory of the learnable. Commun ACM 27(11):1134–1142

Learning Constant-Depth Circuits

Rocco A. Servedio
Computer Science, Columbia University,
New York, NY, USA

Keywords

Boolean Fourier analysis; Learning AC^0 circuits; PAC learning; Uniform-distribution learning

Years and Authors of Summarized Original Work

1993; Linial, Mansour, Nisan

Problem Definition

This problem deals with learning "simple" Boolean functions $f : \{0,1\}^n \to \{-1,1\}$ from uniform random labeled examples. In the basic uniform-distribution PAC framework, the learning algorithm is given access to a *uniform random example oracle* $EX(f,U)$ which, when queried, provides a labeled random example $(x, f(x))$ where x is drawn from the uniform distribution U over the Boolean cube $\{0,1\}^n$. Successive calls to the $EX(f,U)$ oracle yield independent uniform random examples. The goal of the learning algorithm is to output a representation of a hypothesis function $h : \{0,1\}^n \to \{-1,1\}$ which with high probability has high accuracy; formally, for any $\epsilon, \delta > 0$, given ϵ and δ the learning algorithm should output an h which with probability at least $1-\delta$ has $\Pr_{x \in U}[h(x) \neq f(x)] \leq \epsilon$.

Many variants of the basic framework described above have been considered. In the *distribution-independent* PAC learning model, the random example oracle is $EX(f,\mathcal{D})$ where $\mathcal{D}$ is an arbitrary (and unknown to the learner) distribution over $\{0,1\}^n$; the hypothesis h should now have high accuracy with respect to $\mathcal{D}$, i.e., with probability $1-\delta$, it must satisfy

$\Pr_{x \in \mathcal{D}}[h(x) \neq f(x)] \leq \epsilon$. Another variant that has been considered is when the distribution $\mathcal{D}$ is assumed to be an unknown *product distribution*; such a distribution is defined by n parameters $0 \leq p_1, \ldots, p_n \leq 1$, and a draw from $\mathcal{D}$ is obtained by independently setting each bit x_i to 1 with probability p_i. Yet another variant is to consider learning with the help of a *membership oracle*: this is a "black box" oracle $MQ(f)$ for f which, when queried on an input $x \in \{0,1\}^n$, returns the value of $f(x)$. The model of uniform-distribution learning with a membership oracle has been well studied; see e.g. [7, 15].

There are many ways to make precise the notion of a "simple" Boolean function; one common approach is to stipulate that the function be computed by a Boolean circuit of some restricted form. A circuit of *size s and depth d* consists of s AND and OR gates (of unbounded fanin) in which the longest path from any input literal $x_1, \ldots, x_n, \overline{x}_1, \ldots, \overline{x}_n$ to the output node is of length d. Note that a circuit of size s and depth 2 is simply a CNF formula or DNF formula. The complexity class consisting of those Boolean functions computed by poly(n)-size, $O(1)$-depth circuits is known as *nonuniform* AC^0.

Key Results

Positive Results

Linial et al. [16] showed that almost all of the Fourier weight of any constant-depth circuit is on "low-order" Fourier coefficients:

Lemma 1 *Let* $f : \{0,1\}^n \to \{-1,1\}$ *be a Boolean function that is computed by an AND/OR/NOT circuit of size* s *and depth* d*. Then for any integer* $t \geq 0$,

$$\sum_{S \subset \{1,\ldots,n\}, |S| > t} \hat{f}(S)^2 \leq 2s2^{-t^{1/d}/20}.$$

(Hastad [6] has given a refined version of Lemma 1 with slightly sharper bounds; see also [21] for a streamlined proof.) They also showed that any Boolean function can be well approximated by approximating its Fourier spectrum.

Lemma 2 *Let* $f : \{0,1\}^n \to \{-1,1\}$ *be any Boolean Function and let* $g : \{0,1\}^n \to \mathbf{R}$ *be an arbitrary function such that* $\sum_{S \subseteq \{1,\ldots,n\}} (\hat{f}(S) - \hat{g}(S))^2 \leq \epsilon$*. Then* $\Pr_{x \in U}[f(x) \neq \text{sign}(g(x))] \leq \epsilon$.

Using the above two results together with a procedure that estimates all the "low-order" Fourier coefficients, they obtained a quasipolynomial-time algorithm for learning constant-depth circuits:

Theorem 1 *There is an* $n^{O((\log^d(n/\epsilon)))}$*-time algorithm that learns any poly*(n)*-size, depth-d Boolean AND/OR/NOT circuit to accuracy* ϵ *with respect to the uniform distribution, using uniform random examples only.*

Furst et al. [3] extended this result to learning under constant-bounded product distributions. A product distribution $\mathcal{D}$ is said to be *constant bounded* if each of its n parameters $p_1, \ldots, p_n$ is bounded away from 0 and 1, i.e., satisfies $\min\{p_i, 1 - p_i\} = \Theta(1)$.

Theorem 2 *There is an* $n^{O((\log^d(n/\epsilon)))}$*-time algorithm that learns any poly*(n)*-size, depth-d Boolean AND/OR/NOT circuit to accuracy* ϵ *given random examples drawn from any constant-bounded product distribution.*

By combining the Fourier arguments of Linial et al. with hypothesis boosting, Jackson et al. [8] were able to extend Theorem 1 to a broader class of circuits, namely, constant-depth AND/OR/NOT circuits that additionally contain (a limited number of) *majority gates*. A majority gate over r Boolean inputs is a binary gate which outputs "true" if and only if at least half of its r Boolean inputs are set to "true."

Theorem 3 *There is an* $n^{\log^{O(1)}(n/\epsilon)}$*-time algorithm that learns any poly*(n)*-size, constant-depth Boolean AND/OR/NOT circuit that additionally contains polylog*(n) *many majority gates to accuracy* ϵ *with respect to the uniform distribution, using uniform random examples only.*

A *threshold gate* over r Boolean inputs is a binary gate defined by r real weights $w_1, \ldots, w_r$ and a real threshold θ; on input $(x_1, \ldots, x_r)$

L

the value of the threshold gate is 1 if and only if $w_1 x_1 + \cdots + w_r x_r \geq \theta$. Gopalan and Servedio [4] observed that a conjecture of Gotsman and Linial [5] bounding the average sensitivity of low-degree polynomial threshold functions implies Fourier concentration – and hence quasipolynomial time learnability using the original Linial et al. [16] algorithm – for Boolean functions computed by polynomial-size constant-depth AND/OR/NOT circuits augmented with a threshold gate as the topmost (output) gate. Combining this observation with upper bounds on the noise sensitivity of low-degree polynomial threshold functions given in [2], they obtained unconditional sub-exponential time learning for these circuits. Subsequent improvements of these noise sensitivity results, nearly resolving the Gotsman-Linial conjecture and giving stronger unconditional running times for learning these circuits, were given by Kane [10]; see [2, 4, 10] for detailed statements of these results.

Negative Results

Kharitonov [11] showed that under a strong but plausible cryptographic assumption, the algorithmic result of Theorem 1 is essentially optimal. A *Blum integer* is an integer $N = P \cdot Q$ where both P and Q are congruent to 3 modulo 4. Kharitonov proved that if the problem of factoring a randomly chosen n-bit Blum integer is 2^{n^ϵ}-hard for some fixed $\epsilon > 0$, then any algorithm that (even weakly) learns polynomial-size depth-d circuits must run in time $2^{\log^{\Omega(d)} n}$, even if it is only required to learn under the uniform distribution and can use a membership oracle. This implies that there is no polynomial-time algorithm for learning polynomial-size, depth-d circuits (for d larger than some absolute constant).

Using a cryptographic construction of Naor and Reingold [18], Jackson et al. [8] proved a related result for circuits with majority gates. They showed that under Kharitonov's assumption, any algorithm that (even weakly) learns depth-5 circuits consisting of $\log^k n$ many majority gates must run in time $2^{\log^{\Omega(k)} n}$ time, even if it is only required to learn under the uniform distribution and can use a membership oracle.

Applications

The technique of learning by approximating most of the Fourier spectrum (Lemma 2 above) has found many applications in subsequent work on uniform-distribution learning. It is a crucial ingredient in the current state-of-the-art algorithms for learning monotone DNF formulas [20], monotone decision trees [19], and intersections of half-spaces [12] from uniform random examples only. Combined with a membership oracle-based procedure for identifying large Fourier coefficients, this technique is at the heart of an algorithm for learning decision trees [15]; this algorithm in turn plays a crucial role in the celebrated polynomial-time algorithm of Jackson [7] for learning polynomial-size depth-2 circuits under the uniform distribution.

The ideas of Linial et al. have also been applied for the difficult problem of *agnostic learning*. In the agnostic learning framework, there is a joint distribution $\mathcal{D}$ over example-label pairs $\{0,1\}^n \times \{-1,1\}$; the goal of an agnostic learning algorithm for a class $\mathcal{C}$ of functions is to construct a hypothesis h such that $\Pr_{(x,y)\in\mathcal{D}}[h(x) \neq y] \leq \min_{f\in\mathcal{C}} \Pr_{(x,y)\in\mathcal{D}}[f(x) \neq y] + \epsilon$. Kalai et al. [9] gave agnostic learning algorithms for half-spaces and related classes via an algorithm which may be viewed as a generalization of Linial et al.'s algorithm to a broader class of distributions.

Finally, there has been some applied work on learning using Fourier representations as well [17].

Open Problems

Perhaps the most outstanding open question related to this work is whether polynomial-size circuits of depth two – i.e., DNF formulas – can be learned in polynomial time from uniform random examples only. Blum [1] has offered a cash prize for a solution to a restricted version of this problem. A hardness result for learning DNF would also be of great interest; recent work of Klivans and Sherstov [14] gives a hardness result for learning ANDs of majority gates, but hardness for DNF (ANDs of ORs) remains an open question.

Another open question is whether the quasipolynomial-time algorithms for learning constant-depth circuits under uniform distributions and product distributions can be extended to the general distribution-independent model. Known results in complexity theory imply that quasipolynomial-time distribution-independent learning algorithms for constant-depth circuits would follow from the existence of efficient linear threshold learning algorithms with a sufficiently high level of tolerance to "malicious" noise. Currently no nontrivial distribution-independent algorithms are known for learning circuits of depth 3; for depth-2 circuits the best known running time in the distribution-independent setting is the $2^{\tilde{O}(n^{1/3})}$-time algorithm of Klivans and Servedio [13].

A third direction for future work is to extend the results of [8] to a broader class of circuits. Can constant-depth circuits augmented with MOD_p gates be learned in quasipolynomial time? Jackson et al. [8] discusses the limitations of current techniques to address these extensions.

Experimental Results

None to report.

Data Sets

None to report.

URL to Code

None to report.

Cross-References

▶ Cryptographic Hardness of Learning
▶ Learning DNF Formulas
▶ PAC Learning
▶ Statistical Query Learning

Recommended Reading

1. Blum A (2003) Learning a function of r relevant variables (open problem). In: Proceedings of the 16th annual COLT, Washington, DC, pp 731–733
2. Diakonikolas I, Harsha P, Klivans A, Meka R, Raghavendra P, Servedio RA, Tan L-Y (2010) Bounding the average sensitivity and noise sensitivity of polynomial threshold functions. In: Proceedings of the 42nd annual ACM symposium on theory of computing (STOC), Cambridge, pp 533–542
3. Furst M, Jackson J, Smith S (1991) Improved learning of AC^0 functions. In: Proceedings of the 4th annual conference on learning theory (COLT), Santa Cruz, pp 317–325
4. Gopalan P, Servedio R (2010) Learning and lower bounds for AC^0 with threshold gates. In: Proceedings of the 14th international workshop on randomization and computation (RANDOM), Barcelona, pp 588–601
5. Gotsman C, Linial N (1994) Spectral properties of threshold functions. Combinatorica 14(1): 35–50
6. Håstad J (2001) A slight sharpening of LMN. J Comput Syst Sci 63(3):498–508
7. Jackson J (1997) An efficient membership-query algorithm for learning DNF with respect to the uniform distribution. J Comput Syst Sci 55: 414–440
8. Jackson J, Klivans A, Servedio R (2002) Learnability beyond AC^0. In: Proceedings of the 34th annual ACM symposium on theory of computing (STOC), Montréal, pp 776–784
9. Kalai A, Klivans A, Mansour Y, Servedio R (2005) Agnostically learning halfspaces. In: Proceedings of the 46th IEEE symposium on foundations of computer science (FOCS), Pittsburgh, pp 11–20
10. Kane DM (2014) The correct exponent for the Gotsman-Linial conjecture. Comput Complex 23:151–175
11. Kharitonov M (1993) Cryptographic hardness of distribution-specific learning. In: Proceedings of the twenty-fifth annual symposium on theory of computing (STOC), San Diego, pp 372–381
12. Klivans A, O'Donnell R, Servedio R (2004) Learning intersections and thresholds of halfspaces. J Comput Syst Sci 68(4):808–840

L

13. Klivans A, Servedio R (2004) Learning DNF in time $2^{\tilde{O}(n^{1/3})}$. J Comput Syst Sci 68(2):303–318
14. Klivans A, Sherstov A (2006) Cryptographic hardness results for learning intersections of halfspaces. In: Proceedings of the 46th IEEE symposium on foundations of computer science (FOCS), Berkeley
15. Kushilevitz E, Mansour Y (1993) Learning decision trees using the Fourier spectrum. SIAM J Comput 22(6):1331–1348
16. Linial N, Mansour Y, Nisan N (1993) Constant depth circuits, Fourier transform and learnability. J ACM 40(3):607–620
17. Mansour Y, Sahar S (2000) Implementation issues in the Fourier transform algorithm. Mach Learn 40(1):5–33
18. Naor M, Reingold O (1997) Number-theoretic constructions of efficient pseudo-random functions. In: Proceedings of the thirty-eighth annual symposium on foundations of computer science (FOCS), Miami Beach, pp 458–467
19. O'Donnell R, Servedio R (2006) Learning monotone decision trees in polynomial time. In: Proceedings of the 21st conference on computational complexity (CCC), Prague, pp 213–225
20. Servedio R (2004) On learning monotone DNF under product distributions. Inf Comput 193(1):57–74
21. Stefankovic D (2002) Fourier transforms in computer science. PhD thesis, University of Chicago. Masters thesis, TR-2002-03

Learning DNF Formulas

Jeffrey C. Jackson
Department of Mathematics and Computer Science, Duquesne University, Pittsburgh, PA, USA

Keywords

Disjunctive normal form; DNF; Membership oracle; PAC learning; Random walk learning; Smoothed analysis

Years and Authors of Summarized Original Work

1997; Jackson
2005; Bshouty, Mossel, O'Donnell, Servedio
2009; Kalai, Samorodnitsky, Teng

Problem Definition

A disjunctive normal form (DNF) expression is a Boolean expression written as a disjunction of *terms*, where each term is the conjunction of Boolean variables that may or may not be negated. For example, $(v_1 \wedge \overline{v_2}) \vee (v_2 \wedge v_3)$ is a two-term DNF expression over three variables. DNF expressions occur frequently in digital circuit design, where DNF is often referred to as sum of products notation. From a learning perspective, DNF expressions are of interest because they provide a natural representation for certain types of expert knowledge. For example, the conditions under which complex tax rules apply can often be readily represented as DNFs. Another nice property of DNF expressions is their universality: every n-bit Boolean function (the type of function considered in this entry unless otherwise noted) $f : \{0,1\}^n \to \{0,1\}$ can be represented as a DNF expression F over at most n variables.

In the basic *probably-approximately correct (PAC)* learning model [24], n is a fixed positive integer and a *target distribution* $\mathcal{D}$ over $\{0,1\}^n$ is assumed fixed. The learner will have black-box access to an unknown arbitrary Boolean f through an *example oracle* $EX(f, \mathcal{D})$ which, when queried, selects $x \in \{0,1\}^n$ at random according to $\mathcal{D}$ and returns the pair $\langle x, f(x) \rangle$. The DNF learning problem is then to design an algorithm provably meeting the following specifications.

Problem 1 (PAC-DNF)

INPUT: Positive integer n; $\epsilon, \delta > 0$; oracle $EX(f, \mathcal{D})$ for $f : \{0,1\}^n \to \{0,1\}$ expressible as DNF having at most s terms and for $\mathcal{D}$ an arbitrary distribution over $\{0,1\}^n$.

OUTPUT: With probability at least $1 - \delta$ over the random choices made by $EX(f, \mathcal{D})$ and the algorithm (if any), a function $h : \{0,1\}^n \to \{0,1\}$ such that $\Pr_{x \sim \mathcal{D}}[h(x) \neq f(x)] < \epsilon$ and such that $h(x)$ is computable in time polynomial in n and s for each $x \in \{0,1\}^n$.

RUN TIME: Polynomial in n, s, $1/\epsilon$, and $1/\delta$.

The PAC-DNF problem has not been resolved at the time of this writing, and many believe that no algorithm can solve this problem (see, e.g., [2]). However, DNF has been shown to be learnable in several models that relax the PAC assumptions in various ways. In particular, all polynomial-time learning results to date have limited the choice of the target distribution $\mathcal{D}$ to (at most) the class of *constant-bounded product distributions*. For a constant $c \in (0, \frac{1}{2}]$, a c-bounded product distribution $\mathcal{D}_\mu$ is defined by fixing a vector $\mu \in [c, 1-c]^n$ and having $\mathcal{D}_\mu$ correspond to selecting each bit x_i of x independently so that the mean value of x_i is μ_i; mathematically, this distribution function can be written as $\mathcal{D}_\mu(x) \equiv \prod_{i=1}^n (x_i\mu_i + (1-x_i)(1-\mu_i))$. In most learning models, the target distribution is assumed to be selected by an unknown, even adversarial, process from among the allowed distributions; thus, the limitation relative to PAC learning is on the class of allowed distributions, not on how a distribution is chosen. However, in an alternative model of learning from *smoothed product distributions* [17], the mechanism used to choose the target distribution is also constrained as follows. A constant $c \in (0, \frac{1}{2}]$ and an arbitrary vector $\mu \in [2c, 1-2c]^n$ are fixed, a perturbation $\Delta \in [-c, c]^n$ is chosen uniformly at random, and the target distribution is taken to be the c-bounded product distribution $\mathcal{D}_{\mu'}$ such that $\mu' = \mu + \Delta$. The learning algorithm now needs to succeed with only high probability over the choice of Δ (as well as over the usual choices, and in the same run time as for PAC-DNF).

Problem 2 (SMOOTHED-DNF)

INPUT: Same as PAC-DNF except that oracle $EX(f, \mathcal{D}_{\mu'})$ has $\mathcal{D}_{\mu'}$ a smoothed product distribution.

OUTPUT: With probability at least $1-\delta$ over the random choice of μ' and the random choices made by $EX(f, \mathcal{D}_{\mu'})$ and the algorithm (if any), a function $h : \{0,1\}^n \to \{0,1\}$ such that $\Pr_{x \sim \mathcal{D}_{\mu'}}[h(x) \neq f(x)] < \epsilon$.

Various more-informative oracles have also been studied, and we will consider two in particular. A *membership oracle* $MEM(f)$, given x, returns $f(x)$. A *product random walk oracle* $PRW(f, \mathcal{D}_\mu)$ [15] is initialized by selecting an internal state vector x at random according to a fixed arbitrary constant-bounded product distribution $\mathcal{D}_\mu$. Then, on each call to $PRW(f, \mathcal{D}_\mu)$, an $i \in \{1, 2, \ldots, n\}$ is chosen uniformly at random and bit x_i in the internal vector x is replaced with a bit b effectively chosen by flipping a μ_i-biased coin, so that x_i will be 1 with probability μ_i. The oracle then returns the triple $\langle i, x', f(x')\rangle$ consisting of the selected i, the resulting new internal state vector x', and the value that f assigns to this vector. These oracles are used to define two additional DNF learning problems.

Problem 3 (PMEM-DNF)

INPUT: Same as PAC-DNF except that the oracle supplied is $MEM(f)$ and the target distribution is a constant-bounded product distribution $\mathcal{D}_\mu$.

OUTPUT: With probability at least $1-\delta$ over the random choices made by the algorithm, a function h such that $\Pr_{x \sim \mathcal{D}_\mu}[h(x) \neq f(x)] < \epsilon$.

Problem 4 (PRW-DNF)

INPUT: Same as PAC-DNF except that the oracle supplied is $PRW(f, \mathcal{D}_\mu)$ for a constant-bounded product distribution $\mathcal{D}_\mu$.

OUTPUT: With probability at least $1-\delta$ over the random choices made by $PRW(f, \mathcal{D}_\mu)$ and the algorithm (if any), a function h such that $\Pr_{x \sim \mathcal{D}_\mu}[h(x) \neq f(x)] < \epsilon$.

Certain other DNF learning problems and associated results are mentioned briefly in the next section.

Key Results

The first algorithm for efficiently learning arbitrary functions in time polynomial in their DNF size was the Harmonic Sieve [13], which solved the PMEM-DNF problem. This algorithm, like all algorithms for learning arbitrary DNF functions to date, relies in large part on Fourier analysis for its proof of correctness. In particular, a key component of the Sieve involves finding *heavy* (large magnitude) Fourier coefficients of certain functions using a variation on an algorithm discovered by Goldreich and Levin [11] and first employed to obtain a learning result by Kushilevitz and Mansour [22]. The original Sieve also depends on a certain hypothesis boosting algorithm [10]. Subsequent work on the PMEM-DNF problem [5,7,8,18,21] has produced simpler and/or faster algorithms. In the case of [18], the result is somewhat stronger as the approximator h is *reliable*: it rarely produces false positives. The best run time for the PMEM-DNF problem obtained thus far, $\tilde{O}(ns^4/\epsilon)$, is due to Feldman [7].

Using a membership oracle is a form of *active learning* in which the learning algorithm is able to influence the oracle, as opposed to *passive learning*—exemplified by learning from an example oracle—where the learning algorithm merely accepts the information provided by the oracle. Thus, an apparently significant step in the direction of a solution to PAC-DNF occurred when Bshouty et al. [6] showed that PRW-DNF, which is obviously a passive learning problem, can be solved when the product distribution is constrained to be uniform (the distribution is c-bounded for $c = \frac{1}{2}$). Noise sensitivity analysis plays a key role in Bshouty et al.'s result. Jackson and Wimmer [15] subsequently defined the product random walk model and extended the result of [6] to show that PRW-DNF can be solved in general, not merely for uniform random walks. Both results still rely to some extent on the Harmonic Sieve, or more precisely on a slightly modified version called the Bounded Sieve [3].

More recently, the SMOOTHED-DNF problem has been defined and solved by Kalai, Samorodnitsky, and Teng [17]. As an example oracle is used by the algorithm solving this problem, this result can be viewed as a partial solution to PAC-DNF that applies when the target distribution is chosen in a somewhat "friendly," rather than adversarial, way. Their algorithm avoids the Sieve's need for boosting by combining a form of gradient projection optimization [12] with reliable DNF learning [18] in order to produce a good approximator h, given only the heavy Fourier coefficients of the function f to be approximated. Feldman [9] subsequently discovered a simpler algorithm for this problem.

Algorithms for efficiently learning DNF in a few, less studied, models are mentioned only briefly here due to space constraints. These results include uniform-distribution learning of DNF from a quantum example oracle [4] and from two different types of extended statistical queries [3, 14].

Finally, note that although the focus of this entry is on learning arbitrary functions in time polynomial in their DNF size, there is also a substantial literature on polynomial-time learning of restricted classes of functions representable by constrained forms of DNF, such as monotone DNF (functions expressible as a DNF having no negated variables). For the most part, this work predates the algorithms described here. [19] provides a good summary of many early restricted-DNF results. See also Sellie's algorithm [23] that, roughly speaking, efficiently learns with respect to the uniform target distribution most—but not necessarily all—functions representable by polynomial-size DNF given uniform random examples of each function.

Applications

DNF learning algorithms have proven useful as a basis for learning more expressive function-representation classes. In fact, the Harmonic Sieve, without alteration, can be used to learn with respect to the uniform distribution arbitrary functions in time polynomial in their size as a majority of parity functions (see [13] for a definition of the TOP class and a discussion of its superior expressiveness relative to DNF). In another generalization direction, a DNF expression can be viewed as a union of rectangles of the Boolean hypercube. Atici and Servedio [1] have given a generalized version of the Harmonic Sieve that can, among other things, efficiently learn with respect to uniform an interesting subset of the class of unions of rectangles over $\{0, 1, \ldots, b-1\}^n$ for non-constant b. Both of the preceding results use membership queries. A few quasipolynomial-time passive learning results for classes more expressive than DNF in various learning models have also been obtained [15, 16] by building on techniques employed originally in DNF learning algorithms.

Open Problems

As indicated at the outset, a resolution, either positively or negatively, of the PAC-DNF question would be a major step forward. Several other DNF questions are also of particular interest. In the problem of *agnostic learning* [20] of DNF, the goal, roughly, is to efficiently find a function h that approximates arbitrary function $f : \{0, 1\}^n$ nearly as well as the best s-term DNF, for any fixed s. Is DNF agnostically learnable in any reasonable model? Also welcome would be the discovery of efficient algorithms for learning DNF with respect to interesting classes of distributions beyond product distributions. Finally, although monotone DNF is not a universal function class, it should be mentioned that an efficient example-oracle algorithm for monotone DNF, even if restricted to the uniform target distribution, would be considered a breakthrough.

Cross-References

▸ Learning Constant-Depth Circuits
▸ Learning Heavy Fourier Coefficients of Boolean Functions
▸ PAC Learning

Recommended Reading

1. Atici A, Servedio RA (2008) Learning unions of $\omega(1)$-dimensional rectangles. Theoretical Computer Science 405(3):209 – 222
2. Blum A, Furst M, Kearns M, Lipton R (1994) Cryptographic primitives based on hard learning problems. In: Stinson D (ed) Advances in cryptology—CRYPTO '93. Lecture notes in computer science, vol 773, pp 278–291. Springer, Berlin/Heidelberg
3. Bshouty NH, Feldman V (2002) On using extended statistical queries to avoid membership queries. J Mach Learn Res 2:359–395
4. Bshouty NH, Jackson JC (1999) Learning DNF over the uniform distribution using a quantum example oracle. SIAM J Comput 28(3):1136–1153
5. Bshouty NH, Jackson JC, Tamon C (2004) More efficient PAC-learning of DNF with membership queries under the uniform distribution. J Comput Syst Sci 68(1):205–234
6. Bshouty NH, Mossel E, O'Donnell R, Servedio RA (2005) Learning DNF from random walks. J Comput Syst Sci 71(3):250–265
7. Feldman V (2007) Attribute-efficient and non-adaptive learning of parities and DNF expressions. J Mach Learn Res 8:1431–1460
8. Feldman V (2010) Distribution-specific agnostic boosting. In: Proceedings of innovations in computer science, Tsinghua University, Beijing, pp 241–250
9. Feldman V (2012) Learning DNF expressions from Fourier spectrum. In: COLT 2012—the 25th annual conference on learning theory, Edinburgh, pp 17.1–17.19
10. Freund Y (1995) Boosting a weak learning algorithm by majority. Inf Comput 121(2):256–285
11. Goldreich O, Levin LA (1989) A hard-core predicate for all one-way functions. In: Proceedings of the twenty first annual ACM symposium on theory of computing, Seattle, pp 25–32
12. Gopalan P, Kalai AT, Klivans AR (2008) Agnostically learning decision trees. In: Proceedings of the 40th annual ACM symposium on theory of computing, STOC '08, Victoria, pp 527–536. ACM, New York
13. Jackson J (1997) An efficient membership-query algorithm for learning DNF with respect to the uniform distribution. J Comput Syst Sci 55(3):414–440
14. Jackson J, Shamir E, Shwartzman C (1999) Learning with queries corrupted by classification noise. Discret Appl Math 92(2–3):157–175
15. Jackson JC, Wimmer K (2009) New results for random walk learning. In: Dasgupta S, Klivans A (eds) Proceedings of the 22nd annual conference on learning theory, Montreal, pp 267–276. Omnipress, Madison
16. Jackson JC, Klivans AR, Servedio RA (2002) Learnability beyond AC^0. In: STOC '02: Proceedings of the thiry-fourth annual ACM symposium on theory of computing, Montréal, pp 776–784. ACM, New York
17. Kalai AT, Samorodnitsky A, Teng SH (2009) Learning and smoothed analysis. In: Proceedings of the 50th IEEE symposium on foundations of computer science (FOCS), Atlanta, pp 395–404
18. Kalai AT, Kanade V, Mansour Y (2012) Reliable agnostic learning. J Comput Syst Sci 78(5):1481–1495. {JCSS} Special Issue: Cloud Computing 2011
19. Kearns MJ, Vazirani UV (1994) An introduction to computational learning theory. MIT, Cambridge
20. Kearns MJ, Schapire RE, Sellie LM (1994) Toward efficient agnostic learning. Mach Learn 17(2–3):115–141
21. Klivans AR, Servedio RA (2003) Boosting and hard-core set construction. Mach Learn 51(3):217–238
22. Kushilevitz E, Mansour Y (1993) Learning decision trees using the Fourier spectrum. SIAM J Comput 22(6):1331–1348

23. Sellie L (2009) Exact learning of random DNF over the uniform distribution. In: Proceedings of the 41st annual ACM symposium on theory of computing, STOC '09, Bethesda, pp 45–54. ACM, New York
24. Valiant LG (1984) A theory of the learnable. Commun ACM 27(11):1134–1142

Learning Heavy Fourier Coefficients of Boolean Functions

Luca Trevisan
Department of Computer Science, University of California, Berkeley, CA, USA

Keywords

Error-control codes

Years and Authors of Summarized Original Work

1989; Goldreich, Levin

Problem Definition

The Hamming distance $d_H(y, z)$ between two binary strings y and z of the same length is the number of entries in which y and z disagree. A binary error-correcting code of minimum distance d is a mapping $C : \{0,1\}^k \to \{0,1\}^n$ such that for every two distinct inputs $x, x' \in \{0,1\}^k$, the encodings $C(x)$ and $C(x')$ have Hamming distance at least d. Error-correcting codes are employed to transmit information over noisy channels. If a sender transmits an encoding $C(x)$ of a message x via a noisy channel, and the recipient receives a corrupt bit string $y \neq C(x)$, then, provided that y differs from $C(x)$ in at most $(d-1)/2$ locations, the recipient can recover y from $C(x)$. The recipient can do so by searching for the string x that minimizes the Hamming distance between $C(x)$ and y: there can be no other string x' such that $C(x')$ has Hamming distance $(d-1)/2$ or smaller from y, otherwise $C(x)$ and $C(x')$ would be within Hamming distance $d-1$ or smaller, contradicting the above definition. The problem of recovering the message x from the corrupted encoding y is the *unique decoding* problem for the error-correcting code C. For the above-described scheme to be feasible, the decoding problem must be solvable via an efficient algorithm. These notions are due to Hamming [4].

Suppose that C is a code of minimum distance d, and such that there are pairs of encodings $C(x)$, $C(x')$ whose distance is exactly d. Furthermore, suppose that a communication channel is used that could make a number of errors larger than $(d-1)/2$. Then, if the sender transmits an encoded message using C, it is no longer possible for the recipient to uniquely reconstruct the message. If the sender, for example, transmits $C(x)$, and the recipient receives a string y that is at distance $d/2$ from $C(x)$ and at distance $d/2$ from $C(x')$, then, from the perspective of the recipient, it is equally likely that the original message was x or x'. If the recipient knows an upper bound e on the number of entries that the channel has corrupted, then, given the received string y, the recipient can at least compute the list of all strings x such that $C(x)$ and y differ in at most e locations. An error-correcting code $C : \{0,1\}^k \to \{0,1\}^n$ is (e, L)-list decodable if, for every string $y \in \{0,1\}^n$, the set $\{x \in \{0,1\}^k : d_H(C(x), y) \leq e\}$ has cardinality at most L. The problem of reconstructing the list given y and e is the *list-decoding problem* for the code C. Again, one is interested in efficient algorithms for this problem. The notion of list-decoding is due to Elias [1].

A code $C : \{0,1\}^k \to \{0,1\}^n$ is a *Hadamard code* if every two encodings $C(x)$, $C(x')$ differ in precisely $n/2$ locations. In the Computer Science literature, it is common to use the term Hadamard code for a specific construction (the *Reed–Muller code of order 2*) that satisfies the above property. For a string $a \in \{0,1\}^k$, define the function $\ell_a : \{0,1\}^k \to \{0,1\}$ as

$$\ell_a(x) := \sum_i a_i x_i \mod 2 .$$

Observe that, for $a \neq b$, the two functions ℓ_a and ℓ_b differ on precisely $(2^k)/2$ inputs. For $n = 2^k$, the code $H : \{0,1\}^k \to \{0,1\}^n$ maps a message $a \in \{0,1\}^k$ into the n-bit string which is the *truth-table of the function* ℓ_a. That is, if $b_1, \ldots, b_n$ is an enumeration of the $n = 2^k$ elements of $\{0,1\}^k$, and $a \in \{0,1\}^k$ is a message, then the encoding $H(a)$ is the n-bit string that contains the value $\ell_a(b_i)$ in the i-th entry. Note that any two encodings $H(x)$, $H(x')$ differ in precisely $n/2$ entries, and so what was just defined is a Hadamard code. From now on, the term *Hadamard code* will refer exclusively to this construction.

It is known that the Hadamard code $H : \{0,1\}^k \to \{0,1\}^{2^k}$ is $(\frac{1}{2} - \epsilon, \frac{1}{4\epsilon^2})$-list decodable for every $\epsilon > 0$. The Goldreich–Levin results provide efficient list-decoding algorithm.

The following definition of the *Fourier spectrum* of a boolean function will be needed later to state an application of the Goldreich–Levin results to computational learning theory. For a string $a \in \{0,1\}^k$, define the function $\chi_a : \{0,1\}^k \to \{-1,+1\}$ as $\chi_a(x) := (-1)^{\ell_a(x)}$. Equivalently, $\chi_a(x) = (-1)^{\sum_i a_i x_i}$. For two functions $f, g : \{0,1\}^k \to \mathbb{R}$, define their *inner product* as

$$\langle f, g \rangle := \frac{1}{2^k} \sum_x f(x) \cdot g(x) .$$

Then it is easy to see that, for every $a \neq b$, $\langle \chi_a, \chi_b \rangle = 0$, and $\langle \chi_a, \chi_a \rangle = 1$. This means that the functions $\{\chi_a\}_{a \in \{0,1\}^k}$ form an orthonormal basis for the set of all functions $f : \{0,1\}^k \to \mathbb{R}$. In particular, every such function f can be written as a linear combination

$$f(x) = \sum_a \hat{f}(a) \chi_a(x)$$

where the coefficients $\hat{f}(a)$ satisfy $\hat{f}(a) = \langle f, \chi_a \rangle$. The coefficients $\hat{f}(a)$ are called the *Fourier coefficients* of the function f.

Key Results

Theorem 1 *There is a randomized algorithm GL that, given in input an integer k and a parameter $\epsilon > 0$, and given oracle access to a function $f : \{0,1\}^k \to \{0,1\}$, runs in time polynomial in $1/\epsilon$ and in k and outputs, with high probability over its internal coin tosses, a set $S \subseteq \{0,1\}^k$ that contains all the strings $a \in \{0,1\}^k$ such that ℓ_a and f agree on at least a $1/2 + \epsilon$ fraction of inputs.*

Theorem 1 is proved by Goldreich and Levin [3]. The result can be seen as a list-decoding for the Hadamard code $H : \{0,1\}^k \to \{0,1\}^{2^k}$; remarkably, the algorithm runs in time polynomial in k, which is poly-logarithmic in the length of the given corrupted encoding.

Theorem 2 *There is a randomized algorithm KM that given in input an integer k and parameters $\epsilon, \delta > 0$, and given oracle access to a function $f : \{0,1\}^k \to \{0,1\}$, runs in time polynomial in $1/\epsilon$, in $1/\delta$, and in k and outputs a set $S \subseteq \{0,1\}^k$ and a value $g(a)$ for each $a \in S$.*

With high probability over the internal coin tosses of the algorithm,

1. *S contains all the strings $a \in \{0,1\}^k$ such that $|\hat{f}(a)| \geq \epsilon$, and*
2. *For every $a \in S$, $|\hat{f}(a) - g(a)| \leq \delta$.*

Theorem 2 is proved by Kushilevitz and Mansour [5]; it is an easy consequence of the Goldreich–Levin algorithm.

Applications

There are two key applications of the Goldreich–Levin algorithm: one is to cryptography and the other is to computational learning theory.

Application in Cryptography

In cryptography, a *one-way permutation* is a family of functions $\{p_n\}_{n \geq 1}$ such that: (i) for every n, $p_n : \{0,1\}^n \to \{0,1\}^n$ is bijective, (ii) there is a polynomial time algorithm that, given $x \in \{0,1\}^n$, computes $p_n(x)$, and (iii) for every polynomial time algorithm A and polynomial q, and for every sufficiently large n,

$$\mathbb{P}_{x\sim\{0,1\}^n}[A(p_n(x)) = x] \le \frac{1}{q(n)} .$$

That is, even though computing $p_n(x)$ given x is doable in polynomial time, the task of computing x given $p_n(x)$ is intractable. A *hard core predicate* for a one-way permutation $\{p_n\}$ is a family of functions $\{B_n\}_{n\ge1}$ such that: (i) for every n, $B_n : \{0,1\}^n \to \{0,1\}$, (ii) there is a polynomial time algorithm that, given $x \in \{0,1\}^n$, computes $B_n(x)$, and (iii) for every polynomial time algorithm A and polynomial q, and for every sufficiently large n,

$$\mathbb{P}_{x\sim\{0,1\}^n}[A(p_n(x)) = B_n(x)] \le \frac{1}{2} + \frac{1}{q(n)} .$$

That is, even though computing $B_n(x)$ given x is doable in polynomial time, the task of computing $B_n(x)$ given $p_n(x)$ is intractable.

Goldreich and Levin [3] use their algorithm to show that every one-way permutation has a hard-core predicate, as stated in the next theorem.

Theorem 3 *Let $\{p_n\}$ be a one-way permutation; define $\{p'_n\}$ such that $p'_{2n}(x,y) := p_n(x), y$ and let $B_{2n}(x,y) := \sum_i x_i y_i$ mod 2. (For odd indices, let $p'_{2n+1}(z,b) := p'_{2n}(z)$ and $B_{2n+1}(z,b) := B_{2n}(z)$.)*

Then $\{p'_n\}$ is a one-way permutation and $\{B_n\}$ is a hard-core predicate for $\{p'_n\}$.

This result is used in efficient constructions of pseudorandom generators, pseudorandom functions, and private-key encryption schemes based on one-way permutations. The interested reader is referred to Chapter 3 in Goldreich's monograph [2] for more details.

There are also related applications in computational complexity theory, especially in the study of average-case complexity. See [7] for an overview.

Application in Computational Learning Theory

Loosely speaking, in computational learning theory one is given an unknown function $f : \{0,1\}^k \to \{0,1\}$ and one wants to compute, via an efficient randomized algorithm, a representation of a function $g : \{0,1\}^k \to \{0,1\}$ that agrees with f on most inputs. In the *PAC learning* model, one has access to f only via randomly sampled pairs $(x, f(x))$; in the model of *learning with queries*, instead, one can evaluate f at points of one's choice. Kushilevitz and Mansour [5] suggest the following algorithm: using the algorithm of Theorem 2, find a set S of large coefficients and approximations $g(a)$ of the coefficients $\hat{f}(a)$ for $a \in S$. Then define the function $g(x) = \sum_{a\in S} g(a)\chi_a(x)$. If the error caused by the absence of the smaller coefficients and the imprecision in the larger coefficient is not too large, g and f will agree on most inputs. (A technical point is that g as defined above is not necessarily a boolean function, but it can be easily "rounded" to be boolean.) Kushilevitz and Mansour show that such an approach works well for the class of functions f for which $\sum_a |\hat{f}(a)|$ is bounded, and they observe that functions of small *decision tree complexity* fall into this class. In particular, they derive the following result.

Theorem 4 *There is a randomized algorithm that, given in input parameters k, m, ε and δ, and given oracle access to a function $f : \{0,1\}^k \to \{0,1\}$ of decision tree complexity at most m, runs in time polynomial in k, m, $1/\epsilon$ and $\log 1/\delta$ and, with probability at least $1-\delta$ over its internal coin tosses, outputs a circuit computing a function $g : \{0,1\}^k \to \{0,1\}$ that agrees with f on at least a $1-\epsilon$ fraction of inputs.*

Another application of the Kushilevitz–Mansour technique is due to Linial, Mansour, and Nisan [6].

Cross-References

▸ Decoding Reed-Solomon Codes

Recommended Reading

1. Elias P (1957) List decoding for noisy channels. Technical report 335. Research Laboratory of Electronics, MIT, Cambridge

2. Goldreich O (2001) The foundations of cryptography, vol 1. Cambridge University Press, Cambridge
3. Goldreich O, Levin L (1989) A hard-core predicate for all one-way functions. In: Proceedings of the 21st ACM symposium on theory of computing, Seattle, 14–17 May 1989, pp 25–32
4. Hamming R (1950) Error detecting and error correcting codes. Bell Syst Tech J 29:147–160
5. Kushilevitz E, Mansour Y (1993) Learning decision trees using the Fourier spectrum. SIAM J Comput 22(6):1331–1348
6. Linial N, Mansour Y, Nisan N (1993) Constant depth circuits, Fourier transform and learnability. J ACM 40(3):607–620
7. Trevisan L (2004) Some applications of coding theory in computational complexity. Quad Mat 13:347–424, arXiv:cs.CC/0409044

Learning Significant Fourier Coefficients over Finite Abelian Groups

Adi Akavia
Department of Electrical Engineering and Computer Science, MIT, Cambridge, MA, USA

Keywords

Finding heavy fourier coefficients; Learning heavy fourier coefficients

Years and Authors of Summarized Original Work

2003; Akavia, Goldwasser, Safra

Problem Definition

Fourier transform is among the most widely used tools in computer science. Computing the Fourier transform of a signal of length N may be done in time $\Theta(N \log N)$ using the Fast Fourier Transform (FFT) algorithm. This time bound clearly cannot be improved below $\Theta(N)$, because the output itself is of length N. Nonetheless, it turns out that in many applications it suffices to find only the *significant Fourier coefficients*, i.e., Fourier coefficients occupying, say, at least 1 % of the energy of the signal. This motivates the problem discussed in this entry: the problem of efficiently finding and approximating the significant Fourier coefficients of a given signal (SFT, in short). A naive solution for SFT is to first compute the entire Fourier transform of the given signal and then to output only the significant Fourier coefficients; thus yielding no complexity improvement over algorithms computing the entire Fourier transform. In contrast, SFT can be solved far more efficiently in running time $\widetilde{\Theta}(\log N)$ and while reading at most $\widetilde{\Theta}(\log N)$ out of the N signal's entries [2]. This fast algorithm for SFT opens the way to applications taken from diverse areas including computational learning, error correcting codes, cryptography, and algorithms.

It is now possible to formally define the SFT problem, restricting our attention to discrete signals. Use functional notation where a signal is a function $f: G \to \mathbb{C}$ over a finite Abelian group G, its energy is $\|f\|_2^2 \stackrel{\text{def}}{=} 1/|G| \sum_{x \in G} |f(x)|^2$, and its maximal amplitude is $\|f\|_\infty \stackrel{\text{def}}{=} \max\{|f(x)| \mid x \in G\}$. (For readers more accustomed to vector notation, the authors remark that there is a simple correspondence between vector and functional notation. For example, a one-dimensional signal $(v_1, \ldots, v_N) \in \mathbb{C}^{\mathbb{N}}$ corresponds to the function $f: \mathbb{Z}_N \to \mathbb{C}$ defined by $f(i) = v_i$ for all $i = 1, \ldots, N$. Likewise, a two-dimensional signal $M \in \mathbb{C}^{N_1 \times N_2}$ corresponds to the function $f: \mathbb{Z}_{N_1} \times \mathbb{Z}_{N_2} \to \mathbb{C}$ defined by $f(i,j) = M_{ij}$ for all $i = 1, \ldots, N_1$ and $j = 1, \ldots, N_2$.) For ease of presentation assume without loss of generality that $G = \mathbb{Z}_{N_1} \times \mathbb{Z}_{N_2} \times \cdots \times \mathbb{Z}_{N_k}$ for $N_1, \ldots, N_k \in \mathbb{Z}^+$ (i.e., positive integers), and for $\mathbb{Z}_N$ is the additive group of integers modulo N.

The *Fourier transform* of f is the function $\widehat{f}: G \to \mathbb{C}$ defined for each $\alpha = (\alpha_1, \ldots, \alpha_k) \in G$ by

$$\widehat{f}(\alpha) \stackrel{\text{def}}{=} \frac{1}{|G|} \sum_{(x_1,\ldots,x_k) \in G} \left[f(x_1, \ldots, x_k) \prod_{j=1}^{k} \omega_{N_j}^{\alpha_j x_j} \right],$$

where $\omega_{N_j} = \exp(i2\pi/N_j)$ is a primitive root of unity of order N_j. For any $\alpha \in G$, $val_\alpha \in \mathbb{C}$ and $\tau, \varepsilon \in [0,1]$, say that α is a *τ-significant Fourier coefficient* iff $|\widehat{f}(\alpha)|^2 \geq \tau \|f\|_2^2$, and say that val_α is an *ε-approximation for $\widehat{f}(\alpha)$* iff $|val_\alpha - \widehat{f}(\alpha)| < \varepsilon$.

Problem 1 (SFT)
INPUT: Integers $N_1, \ldots, N_k \geq 2$ specifying the group $G = \mathbb{Z}_{N_1} \times \cdots \times \mathbb{Z}_{N_k}$, a threshold $\tau \in (0,1)$, an approximation parameter $\varepsilon \in (0,1)$, and oracle access (Say that an algorithm is given *oracle access* to a function f over G, if it can request and receive the value $f(x)$ for any $x \in G$ in unit time.) to $f: G \to \mathbb{C}$.
OUTPUT: A list of all τ-significant Fourier coefficients of f along with ε-approximations for them.

Key Results

The key result of this entry is an algorithm solving the SFT problem which is much faster than algorithms for computing the entire Fourier transform. For example, for f a Boolean function over $\mathbb{Z}_N$, the running time of this algorithm is $\log N \cdot poly(\log\log N, 1/\tau, 1/\varepsilon)$, in contrast to the $\Theta(N \log N)$ running time of the FFT algorithm. This algorithm is named the SFT algorithm.

Theorem 1 (SFT algorithm [2]) *There is an algorithm solving the SFT problem with running time* $\log|G| \cdot poly(\log\log|G|, \|f\|_\infty/\|f\|_2, 1/\tau, 1/\varepsilon)$ *for* $|G| = \prod_{j=1}^k N_j$ *the cardinality of* G.

Remarks

1. The above result extends to functions f over any finite Abelian group G, as long as the algorithm is given a description of G by its generators and their orders [2].
2. The SFT algorithm reads at most $\log|G| \cdot poly(\log\log|G|, \|f\|_\infty/\|f\|_2, 1/\tau, 1/\varepsilon)$ out of the $|G|$ values of the signal.
3. The SFT algorithm is non adaptive, that is, oracle queries to f are independent of the algorithm's progress.
4. The SFT algorithm is a probabilistic algorithm having a small error probability, where probability is taken over the internal coin tosses of the algorithm. The error probability can be made arbitrarily small by standard amplification techniques.

The SFT algorithm follows a line of works solving the SFT problem for restricted function classes. Goldreich and Levin [9] gave an algorithm for Boolean functions over the group $\mathbb{Z}_2^k = \{0,1\}^k$. The running time of their algorithm is polynomial in $k, 1/\tau$ and $1/\varepsilon$. Mansour [10] gave an algorithm for complex functions over groups $G = \mathbb{Z}_{N_1} \times \cdots \times \mathbb{Z}_{N_k}$ provided that $N_1, \ldots, N_k$ are powers of two. The running time of his algorithm is polynomial in $\log|G|, \log(\max_{\alpha\in G}|\widehat{f}(\alpha)|), 1/\tau$ and $1/\varepsilon$. Gilbert et al. [6] gave an algorithm for complex functions over the group $\mathbb{Z}_N$ for any positive integer N. The running time of their algorithm is polynomial in $\log N$, $\log(\max_{x\in\mathbb{Z}_N} f(x)/\min_{x\in\mathbb{Z}_N} f(x)), 1/\tau$ and $1/\varepsilon$. Akavia et al. [2] gave an algorithm for complex functions over any finite Abelian group. The latter [2] improves on [6] even when restricted to functions over $\mathbb{Z}_N$ in achieving $\log N \cdot poly(\log\log N)$ rather than $poly(\log N)$ dependency on N. Subsequent works [7] improved the dependency of [6] on τ and ε.

Applications

Next, the paper surveys applications of the SFT algorithm [2] in the areas of computational learning theory, coding theory, cryptography, and algorithms.

Applications in Computational Learning Theory

A common task in computational learning is to find a hypothesis h approximating a function f, when given only samples of the function f. Samples may be given in a variety of forms, e.g., via oracle access to f. We consider the following variant of this learning problem: f and h are complex functions over a finite Abelian group $G = \mathbb{Z}_{N_1} \times \cdots \times \mathbb{Z}_{N_k}$, the goal is to find h such that $\|f - h\|_2^2 \leq \gamma \|f\|_2^2$ for $\gamma > 0$ an approximation parameter, and samples of f are given via oracle access.

A straightforward application of the SFT algorithm gives an efficient solution to the above learning problem, provided that there is a small set $\Gamma \subseteq G$ s.t. $\sum_{\alpha \in \Gamma} |\widehat{f}(\alpha)|^2 > (1 - \gamma/3)\|f\|_2^2$. The learning algorithm operates as follows. It first runs the SFT algorithm to find all $\alpha = (\alpha_1, \ldots, \alpha_k) \in G$ that are $\gamma/|\Gamma|$-significant Fourier coefficients of f along with their $\gamma/|\Gamma|\|f\|_\infty$-approximations val_α, and then returns the hypothesis

$$h(x_1, \ldots, x_k) \stackrel{\text{def}}{=} \sum_{\alpha \text{ is } \gamma/|\Gamma|-\text{significant}} val_\alpha \cdot \prod_{j=1}^{k} \omega_{N_j}^{\alpha_j x_j}.$$

This hypothesis h satisfies that $\|f - h\|_2^2 \leq \gamma \|f\|_2^2$. The running time of this learning algorithm and the number of oracle queries it makes is polynomially bounded by $\log |G|$, $\|f\|_\infty / \|f\|_2$, $|\Gamma| \|f\|_\infty / \gamma$.

Theorem 2 *Let* $f : G \to \mathbb{C}$ *be a function over* $G = \mathbb{Z}_{N_1} \times \cdots \times \mathbb{Z}_{N_k}$, *and* $\gamma > 0$ *an approximation parameter. Denote* $t = \min\{|\Gamma| \; |\Gamma \subseteq G$ *s.t.* $\sum_{\alpha \in \Gamma} |\widehat{f}(\alpha)|^2 > (1 - \gamma/3)\|f\|_2^2\}$. *There is an algorithm that given* $N_1, \ldots, N_k$, γ, *and oracle access to f, outputs a (short) description of* $h : G \to \mathbb{C}$ *s.t.* $\|f - h\|_2^2 < \gamma \|f\|_2^2$. *The running time of this algorithm is* $\log |G| \cdot poly(\log \log |G|, \|f\|_\infty / \|f\|_2, t\|f\|_\infty / \gamma)$.

More examples of function classes that can be efficiently learned using our SFT algorithm are given in [3].

Applications in Coding Theory

Error correcting codes encode messages in a way that allows *decoding*, that is, recovery of the original message, even in the presence of noise. When the noise is very high, unique decoding may be infeasible, nevertheless it may still be possible to *list decode*, that is, to find a short list of messages containing the original message. Codes equipped with an efficient list decoding algorithm have found many applications (see [11] for a survey).

Formally, a binary code is a subset $C \subseteq \{0,1\}^*$ of *codewords* each encoding some message. Denote by $C_{N,x} \in \{0,1\}^N$ a codeword of length N encoding a message x. The *normalized Hamming distance* between a codeword $C_{N,x}$ and a received word $w \in \{0,1\}^N$ is $\Delta(C_{N,x}, w) \stackrel{\text{def}}{=} 1/N |\{i \in \mathbb{Z}_N \; | C_{N,x}(i) \neq w(i)\}|$ where $C_{N,x}(i)$ and $w(i)$ are the ith bits of $C_{N,x}$ and w, respectively. Given $w \in \{0,1\}^N$ and a noise parameter $\eta > 0$, the *list decoding* task is to find a list of all messages x such that $\Delta(C_{N,x}, w) < \eta$. The received word w may be given explicitly or implicitly; we focus on the latter where oracle access to w is given. Goldreich and Levin [9] give a list decoding algorithm for Hadamard codes, using in a crucial way their algorithm solving the SFT problem for functions over the Boolean cube.

The SFT algorithm for functions over ZZ_N is a key component in a list decoding algorithm given by Akavia et al. [2]. This list decoding algorithm is applicable to a large class of codes. For example, it is applicable to the code $C^{msb} = \{C_{N,x} : \mathbb{Z}_N \to \{0,1\}\}_{x \in \mathbb{Z}_N^*, N \in \mathbb{Z}^+}$ whose codewords are $C_{N,x}(j) = msb_N(j \cdot x \mod N)$ for $msb_N(y) = 1$ iff $y \geq N/2$ and $msb_N(y) = 0$ otherwise. More generally, this list decoding algorithm is applicable to any *Multiplication code* $C^{\mathcal{P}}$ for $\mathcal{P}$ a family of *balanced* and *well concentrated* functions, as defined below. The running time of this list decoding algorithm is polynomial in $\log N$ and $1/(1 - 2\eta)$, as long as $\eta < \frac{1}{2}$.

Abstractly, the list decoding algorithm of [2] is applicable to any code that is "balanced," "(well) concentrated," and "recoverable," as defined next (and those Fourier coefficients have small greatest common divisor (GCD) with N). A code is *balanced* if $\Pr_{j \in \mathbb{Z}_N}[C_{N,x}(j) = 0] = \Pr_{j \in \mathbb{Z}_N}[C_{N,x}(j) = 1]$ for every codeword $C_{N,x}$. A code is *(well) concentrated* if its codewords can be approximated by a small number of significant coefficients in their Fourier representation (and those Fourier coefficients have small greatest common divisor (GCD) with N). A code is *recoverable* if there is an efficient algorithm mapping each Fourier coefficient α to a short list of codewords for which α is a significant Fourier coefficient. The key property of concentrated codes is that received words w share a significant Fourier coefficient with all close codewords $C_{N,x}$. The high level structure of the list decoding algorithm of [2] is therefore as follows. First it runs the SFT algorithm to find all significant Fourier coefficients α of the received word w. Second for each such α, it runs the recovery algorithm to find all codewords $C_{N,x}$ for which α is significant. Finally, it outputs all those codewords $C_{N,x}$.

Definition 1 (Multiplication codes [2]) Let $\mathcal{P} = \{P_N: \mathbb{Z}_N \to \{0,1\}\}_{N \in \mathbb{Z}^+}$ be a family of functions. Say that $C^{\mathcal{P}} = \{C_{N,x}: \mathbb{Z}_N \to \{0,1\}\}_{x \in \mathbb{Z}_N^*, N \in \mathbb{Z}^+}$ is a *multiplication code for* $\mathcal{P}$ if for every $N \in \mathbb{Z}^+$ and $x \in \mathbb{Z}_N^*$, the encoding $C_{N,x}: \mathbb{Z}_N \to \{0,1\}$ of x is defined by

$$C_{N,x}(j) = P(j \cdot x \bmod N) .$$

Definition 2 (Well concentrated [2]) Let $\mathcal{P} = \{P_N: \mathbb{Z}_N \to \mathbb{C}\}_{N \in \mathbb{Z}^+}$ be a family of functions. Say that $\mathcal{P}$ is *well concentrated* if $\forall N \in \mathbb{Z}^+, \gamma > 0$, $\exists \Gamma \subseteq \mathbb{Z}_N$ s.t. (i) $|\Gamma| \leq poly(\log N/\gamma)$, (ii) $\sum_{\alpha \in \Gamma} |\widehat{P_N}(\alpha)|^2 \geq (1-\gamma)\|P_N\|_2^2$, and (iii) for all $\alpha \in \Gamma$, $gcd(\alpha, N) \leq poly(\log N/\gamma)$ (where $gcd(\alpha, N)$ is the greatest common divisor of α and N).

Theorem 3 (List decoding [2]) *Let* $\mathcal{P} = \{P_N: \mathbb{Z}_N \to \{0,1\}\}_{N \in \mathbb{Z}^+}$ *be a family of efficiently computable* ($\mathcal{P} = \{P_N: \mathbb{Z}_N \to \{0,1\}\}_{N \in \mathbb{Z}^+}$ *is a family of efficiently computable functions if there is an algorithm that given any* $N \in \mathbb{Z}^+$ *and* $x \in \mathbb{Z}_N$ *outputs* $P_N(x)$ *in time* $poly(\log N)$.), *well concentrated, and balanced functions. Let* $C^{\mathcal{P}} = \{C_{N,x}: \mathbb{Z}_N \to \{0,1\}\}_{x \in \mathbb{Z}_N^*, N \in \mathbb{Z}^+}$ *be the multiplication code for* $\mathcal{P}$. *Then there is an algorithm that, given* $N \in \mathbb{Z}_N^+$, $\eta < \frac{1}{2}$ *and oracle access to* $w: \mathbb{Z}_N \to \{0,1\}$, *outputs all* $x \in \mathbb{Z}_N^*$ *for which* $\Delta(C_{N,x}, w) < \eta$. *The running time of this algorithm is polynomial in* $\log N$ *and* $1/(1-2\eta)$.

Remarks

1. The requirement that $\mathcal{P}$ is a family of *efficiently computable* functions can be relaxed. It suffices to require that the list decoding algorithm receives or computes a set $\Gamma \subseteq \mathbb{Z}_N$ with properties as specified in Definition 2.
2. The requirement that $\mathcal{P}$ is a family of *balanced* functions can be relaxed. Denote $bias(\mathcal{P}) = \min_{b \in \{0,1\}} \inf_{N \in \mathbb{Z}^+} \Pr_{j \in \mathbb{Z}_N}[P_N(j) = b]$. Then the list decoding algorithm of [2] is applicable to $C^{\mathcal{P}}$ even when $bias(\mathcal{P}) \neq \frac{1}{2}$, as long as $\eta < bias(\mathcal{P})$.

Applications in Cryptography

Hard-core predicates for one-way functions are a fundamental cryptographic primitive, which is central for many cryptographic applications such as pseudo-random number generators, semantic secure encryption, and cryptographic protocols. Informally speaking, a Boolean predicate P is a *hard-core predicate* for a function f if $P(x)$ is easy to compute when given x, but hard to guess with a non-negligible advantage beyond 50% when given only $f(x)$. The notion of hard-core predicates was introduced by Blum and Micali [2]. Goldreich and Levin [9] showed a randomized hardcore predicate for any one-way function, using in a crucial way their algorithm solving the SFT problem for functions over the Boolean cube.

Akavia et al. [2] introduce a unifying framework for proving that a predicate P is hard-core for a one-way function f. Applying their framework they prove for a wide class of predicates – *segment predicates* – that they are hard-core predicates for various well-known candidate one-way functions. Thus showing new hard-core predicates for well-known one-way function candidates as well as reproving old results in an entirely different way.

Elaborating on the above, a *segment predicate* is any assignment of Boolean values to an arbitrary partition of $\mathbb{Z}_N$ into $poly(\log N)$ segments, or dilations of such an assignment. Akavia et al. [2] prove that any segment predicate is hard-core for any one-way function f defined over $\mathbb{Z}_N$ for which, for a non-negligible fraction of the x's in $\mathbb{Z}_N$, given $f(x)$ and y, one can efficiently compute $f(xy)$ (where xy is multiplication in $\mathbb{Z}_N$). This includes the following functions: the *exponentiation* function $EXP_{p,g}: \mathbb{Z}_p \to \mathbb{Z}_p^*$ defined by $EXP_{p,g}(x) = g^x \bmod p$ for each prime p and a generator g of the group $\mathbb{Z}_p^*$; the *RSA* function $RSA: \mathbb{Z}_N^* \to \mathbb{Z}_N^*$ defined by $RSA(x) = \mathrm{e}^x \bmod N$ for each $N = pq$ a product of two primes p, q, and e co-prime to N; the *Rabin* function $Rabin: \mathbb{Z}_N^* \to \mathbb{Z}_N^*$ defined by $Rabin(x) = x^2 \bmod N$ for each $N = pq$ a product of two primes p, q; and the *elliptic curve log* function defined by $ECL_{a,b,p,Q} = xQ$ for each elliptic curve $E_{a,b,p}(Z_p)$ and Q a point of high order on the curve.

The SFT algorithm is a central tool in the framework of [2]: Akavia et al. take a list decoding methodology, where computing a hard-core predicate corresponds to computing an entry in some error correcting code, predicting a predicate corresponds to access to an entry in a corrupted codeword, and the task of inverting a one-way function corresponds to the task of list decoding a corrupted codeword. The codes emerging in [2] are multiplication codes (see Definition 1 above), which are list decoded using the SFT algorithm.

Definition 3 (Segment predicates [2]) Let $\mathcal{P} = \{P_N: \mathbb{Z}_N \to \{0,1\}\}_{N \in \mathbb{Z}^+}$ be a family of predicates that are non-negligibly far from constant (A family of functions $\mathcal{P} = \{P_N: \mathbb{Z}_N \to \{0,1\}\}_{N \in \mathbb{Z}^+}$ is *non-negligibly far from constant* if $\forall N \in \mathbb{Z}^+$ and $b \in \{0,1\}$, $\Pr_{j \in \mathbb{Z}_N}[P_N(j) = b] \le 1 - poly(1/\log N)$).

- It can be sayed that P_N is a basic t-segment predicate if $P_N(x+1) \neq P_N(x)$ for at most t x's in $\mathbb{Z}_N$.
- It can be sayed that P_N is a t-segment predicate if there exist a basic t-segment predicate P' and $a \in \mathbb{Z}_N$ which is co-prime to N s.t. $\forall x \in \mathbb{Z}_N$, $P_N(x) = P'(x/a)$.
- It can be sayed that $\mathcal{P}$ is a family of segment predicates if $\forall N \in \mathbb{Z}^+$, P_N is a $t(N)$-segment predicate for $t(N) \le poly(\log N)$.

Theorem 4 (Hardcore predicates [2]) *Let $\mathcal{P}$ be a family of segment predicates. Then, $\mathcal{P}$ is hard-core for RSA, Rabin, EXP, ECL, under the assumption that these are one-way functions.*

Application in Algorithms

Our modern times are characterized by information explosion incurring a need for faster and faster algorithms. Even algorithms classically regarded as efficient – such as the FFT algorithm with its $\Theta(N \log N)$ complexity – are often too slow for data-intensive applications, and linear or even sub-linear algorithms are imperative. Despite the vast variety of fields and applications where algorithmic challenges arise, some basic algorithmic building blocks emerge in many of the existing algorithmic solutions. Accelerating such building blocks can therefore accelerate many existing algorithms. One of these recurring building blocks is the Fast Fourier Transform (FFT) algorithm. The SFT algorithm offers a great efficiency improvement over the FFT algorithm for applications where it suffices to deal only with the significant Fourier coefficients. In such applications, replacing the FFT building block with the SFT algorithm accelerates the $\Theta(N \log N)$ complexity in each application of the FFT algorithm to $poly(\log N)$ complexity [1]. Lossy compression is an example of such an application [1, 5, 8]. To elaborate, central component in several transform compression methods (e.g., JPEG) is to first apply Fourier (or Cosine) transform to the signal, and then discard many of its coefficients. To accelerate such algorithms – instead of computing the entire Fourier (or Cosine) transform – the SFT algorithm can be used to directly approximate only the significant Fourier coefficients. Such an accelerated algorithm achieves compression guarantee as good as the original algorithm (and possibly better), but with running time improved to $poly(\log N)$ in place of the former $\Theta(N \log N)$.

Cross-References

- ▶ Abelian Hidden Subgroup Problem
- ▶ Learning Constant-Depth Circuits
- ▶ Learning DNF Formulas
- ▶ Learning Heavy Fourier Coefficients of Boolean Functions
- ▶ Learning with Malicious Noise
- ▶ List Decoding near Capacity: Folded RS Codes
- ▶ PAC Learning
- ▶ Statistical Query Learning

Recommended Reading

1. Akavia A, Goldwasser S (2005) Manuscript submitted as an NSF grant, awarded CCF-0514167
2. Akavia A, Goldwasser S, Safra S (2003) Proving hard-core predicates using list decoding. In: Proceedings of the 44th symposium on foundations of computer science (FOCS'03). IEEE Computer Society, pp 146–157
3. Atici A, Servedio RA (2006) Learning unions of $\omega(1)$-dimensional rectangles. ALT, pp 32–47
4. Blum M, Micali S (1984) How to generate cryptographically strong sequences of pseudo-random bits. SIAM J Comput 4(13):850–864
5. Cormode G, Muthukrishnan S (2006) Combinatorial algorithms for compressed sensing. In: 13th International colloquium on structural information and communication complexity (SIROCCO), Chester, 2–5 July 2006, pp 280–294
6. Gilbert AC, Guha S, Indyk P, Muthukrishnan S, Strauss M (2002) Near-optimal sparse fourier representations via sampling. In: Proceedings of the thiry-fourth annual ACM symposium on theory of computing (STOC'02). ACM, pp 152–161
7. Gilbert AC, Muthukrishnan S, Strauss MJ (2005) Improved time bounds for near-optimal sparse Fourier representation via sampling. In: Proceedings of SPIE Wavelets XI, San Diego
8. Gilbert AC, Strauss MJ, Tropp JA, Vershynin R (2007) One sketch for all: fast algorithms for compressed sensing. In: Proceedings of the 39th ACM symposium on theory of computing (STOC'07)
9. Goldreich O, Levin L (1989) A hard-core predicate for all one-way functions. In: Proceedings of the 27th ACM symposium on theory of computing (STOC'89)
10. Mansour Y (1995) Randomized interpolation and approximation of sparse polynomials. SIAM J Comput 24:357–368
11. Sudan M (2000) List decoding: algorithms and applications. SIGART Newsl 31:16–27

Learning with Malicious Noise

Peter Auer
Chair for Information Technology,
Montanuniversitaet Leoben, Leoben, Austria

Keywords

Agnostic learning; Boosting; Clustering; Computational learning theory; Efficient learning; Malicious noise; PAC (probably approximately correct) learning; Sample complexity; Statistical query learning

Years and Authors of Summarized Original Work

1993; Kearns, Li
1999; Cesa-Bianchi, Dichterman, Fischer, Shamir, Simon

Problem Definition

This problem is concerned with PAC learning of concept classes when training examples are affected by malicious errors. The PAC (probably approximately correct) model of learning (also known as the distribution-free model of learning) was introduced by Valiant [13]. This model makes the idealized assumption that error-free training examples are generated from the same distribution which is then used to evaluate the learned hypothesis. In many environments, however, there is some chance that an erroneous example is given to the learning algorithm. The malicious noise model – again introduced by Valiant [14] – extends the PAC model by allowing example errors of any kind: it makes no assumptions on the nature of the errors that occur. In this sense the malicious noise model is a worst-case model of errors, in which errors may be generated by an adversary whose goal is to foil the learning algorithm. Kearns and Li [8,9] study the maximal malicious error rate such that learning is still possible. They also provide a canonical method to transform any standard learning algorithm into an algorithm which is robust against malicious noise.

Notations Let X be a set of instances. The goal of a learning algorithm is to infer an unknown subset $C \subseteq X$ of instances which exhibit a certain property. Such subsets are called concepts. It is known to the learning algorithm that the correct concept C is from a concept class $\mathcal{C} \subseteq 2^X$, $C \in \mathcal{C}$. Let $C(x)=1$ if $x \in C$ and $C(x)=0$ if $x \notin C$. As input the learning algorithm receives an accuracy parameter $\varepsilon > 0$, a confidence parameter $\delta > 0$, and the malicious noise rate $\beta \geq 0$. The learning algorithm may request a sample of labeled instances $S = \langle (x_1, \ell_1), \ldots, (x_m, \ell_m) \rangle$, $x_i \in X$, and $\ell_i \in \{0,1\}$ and produces a hypothesis $H \subseteq X$. Let $\mathcal{D}$ be the unknown distribution of instances in X. Learning is successful if H misclassifies an example with probability less than ε, $\text{err}_{\mathcal{D}}(C,H) := \mathcal{D}\{x \in X : C(x) \neq H(x)\} < \varepsilon$. A learning algorithm is required to be successful with probability $1-\delta$. The error of a hypothesis H in respect to a sample S of labeled instances is defined as $\text{err}(S,H) := |\{(x,\ell) \in S : H(x) \neq \ell\}|/|S|$.

The VC dimension VC($\mathcal{C}$) of a concept class $\mathcal{C}$ is the maximal number of instances $x_1, \ldots, x_d$ such that $\{(C(x_1), \ldots, C(x_d)) : C \in \mathcal{C}\} = \{0,1\}^d$. The VC dimension is a measure of the difficulty to learn concept class $\mathcal{C}$ [4].

To investigate the computational complexity of learning algorithms, sequences of concept classes with increasing complexity $(X_n, \mathcal{C}_n)_n = \langle (X_1, \mathcal{C}_1), (X_2, \mathcal{C}_2), \ldots \rangle$ are considered. In this case the learning algorithm receives also a complexity parameter n as input.

Generation of Examples In the malicious noise model, the labeled instances (x_i, ℓ_i) are generated independently from each other by the following random process:

(a) Correct examples: with probability $1-\beta$, an instance x_i is drawn from distribution $\mathcal{D}$ and labeled by the correct concept C, $\ell_i = C(x_i)$.

(b) Noisy examples: with probability β, an arbitrary example (x_i, ℓ_i) is generated, possibly by an adversary.

Problem 1 (Malicious Noise Learning of $(X, \mathcal{C})$)

INPUT: Reals $\varepsilon, \delta > 0$, $\beta \geq 0$
OUTPUT: A hypothesis $H \subseteq X$

For any distribution $\mathcal{D}$ on X and any concept $C \in \mathcal{C}$, the algorithm needs to produce with probability $1-\delta$ a hypothesis H such that $\text{err}_{\mathcal{D}}(C,H) < \varepsilon$. The probability $1-\delta$ is taken in respect to the random sample $(x_1, \ell_1), \ldots, (x_m, \ell_m)$ requested by the algorithm. The examples (x_i, ℓ_i) are generated as defined above.

Problem 2 (Polynomial Malicious Noise Learning of $(X_n, \mathcal{C}_n)_n$)

INPUT: Reals $\varepsilon, \delta > 0$, $\beta \geq 0$, integer $n \geq 1$
OUTPUT: A hypothesis $H \subseteq X_n$

For any distribution $\mathcal{D}$ on X_n and any concept $C \in \mathcal{C}_n$, the algorithm needs to produce with probability $1-\delta$ a hypothesis H such that $\text{err}_{\mathcal{D}}(C,H) < \varepsilon$. The computational complexity of the algorithm must be bounded by a polynomial in $1/\varepsilon$, $1/\delta$, and n.

Key Results

Theorem 1 ([9]) *Let $\mathcal{C}$ be a nontrivial concept class with two concepts $C_1, C_2 \in \mathcal{C}$ that are equal on an instance x_1 and differ on another instance x_2, $C_1(x_1) = C_2(x_1)$, and $C_1(x_2) \neq C_2(x_2)$. Then no algorithm can learn $\mathcal{C}$ with malicious noise rate $\beta \geq \frac{\varepsilon}{1+\varepsilon}$.*

Theorem 2 *Let $\Delta > 0$ and $d = \text{VC}(\mathcal{C})$. For a suitable constant κ, any algorithm which requests a sample S of $m \geq \kappa \frac{\varepsilon d \log 1/(\Delta\delta)}{\Delta^2}$ labeled examples and returns a hypothesis $H \in \mathcal{C}$ which minimizes $\text{err}(S,H)$ learns the concept class $\mathcal{C}$ with malicious noise rate $\beta \leq \frac{\varepsilon}{1+\varepsilon} - \Delta$.*

Lower bounds on the number of examples necessary for learning with malicious noise were derived by Cesa-Bianchi et al.

Theorem 3 ([7]) *Let $\Delta > 0$ and $d = \mathrm{VC}(\mathcal{C}) \geq 3$. There is a constant κ such that any algorithm which learns $\mathcal{C}$ with malicious noise rate $\beta = \frac{\varepsilon}{1+\varepsilon} - \Delta$ by requesting a sample and returning a hypothesis $H \in \mathcal{C}$ which minimizes $\mathrm{err}(S, H)$ needs a sample of size at least $m \geq \kappa \frac{\varepsilon d}{\Delta^2}$.*

A general conversion of a learning algorithm for the noise-free model into an algorithm for the malicious noise model was given by Kearns and Li.

Theorem 4 ([9]) *Let $\mathcal{A}$ be a (polynomial-time) learning algorithm which learns concept classes $\mathcal{C}_n$ from $m(\varepsilon, \delta, n)$ noise-free examples, i.e., $\beta = 0$. Then $\mathcal{A}$ can be converted into a (polynomial-time) learning algorithm for $\mathcal{C}_n$ for any malicious noise rate $\beta \leq \frac{\log m(\varepsilon/8, 1/2, n)}{m(\varepsilon/8, 1/2, n)}$.*

The next theorem relates learning with malicious noise to a type of combinatorial optimization problems.

Theorem 5 ([9]) *Let $r \geq 1$ and $\alpha > 0$.*

1. *Let $\mathcal{A}$ be a polynomial-time algorithm which, for any sample S, returns a hypothesis $H \in \mathcal{C}$ with $\mathrm{err}(S, H) \leq r \cdot \min_{C \in \mathcal{C}} \mathrm{err}(S, C)$. Then $\mathcal{A}$ learns concept class $\mathcal{C}$ for any malicious noise rate $\beta \leq \frac{\varepsilon}{(1+\alpha)(1+\varepsilon)r}$ in time polynomial in $1/\epsilon$, $\log 1/\delta$, $\mathrm{VC}(\mathcal{C})$, and $1/\alpha$.*
2. *Let $\mathcal{A}$ be a polynomial-time learning algorithm for concept classes $\mathcal{C}_n$ which tolerates a malicious noise rate $\beta = \frac{\varepsilon}{r}$ and returns a hypothesis $H \in \mathcal{C}_n$. Then $\mathcal{A}$ can be converted into a polynomial-time algorithm which for any sample S, with high probability, returns a hypothesis $H \in \mathcal{C}_n$ such that $\mathrm{err}(S, H) \leq (1 + \alpha)r \cdot \min_{C \in \mathcal{C}} \mathrm{err}(S, C)$.*

The computational hardness of several such related combinatorial optimization problems was shown by Ben-David, Eiron, and Long [3]. Some particular concept classes for which learning with malicious noise has been considered are monomials, CNF and DNF formulas [9, 14], symmetric functions and decision lists [9], multiple intervals on the real line [7], and halfspaces [11].

Applications

Several extensions of the learning model with malicious noise have been proposed, in particular the agnostic learning model [10] and the statistical query model [1]. The following relations between these models and the malicious noise model have been established:

Theorem 6 ([10]) *If concept class $\mathcal{C}$ is polynomial-time learnable in the agnostic model, then $\mathcal{C}$ is polynomial-time learnable with any malicious noise rate $\beta \leq \varepsilon/2$.*

Theorem 7 ([1]) *If $\mathcal{C}$ is learnable from (relative error) statistical queries, then $\mathcal{C}$ is learnable with any malicious noise rate $\beta \leq \varepsilon / \log^p(1/\varepsilon)$ for a suitable large p independent of $\mathcal{C}$.*

Another learning model related to the malicious noise model is learning with nasty noise [6]. In this model examples affected by malicious noise are not chosen at random with probability β, but an adversary might manipulate an arbitrary fraction of βm examples out of a given sample of size m. The malicious noise model was also considered in the context of online learning [2] and boosting [12]. A variant of the malicious noise model for unsupervised learning has been investigated in [5]. In this model, noisy data points are again replaced by arbitrary points possibly generated by an adversary. Still, a correct clustering of the points can learned if the noise rate is moderate.

Cross-References

- ▶ Performance-Driven Clustering
- ▶ Learning DNF Formulas
- ▶ Online Learning and Optimization
- ▶ PAC Learning
- ▶ Statistical Query Learning

Recommended Reading

1. Aslam JA, Decatur SE (1998) Specification and simulation of statistical query algorithms for efficiency and noise tolerance. J Comput Syst Sci 56: 191–208

2. Auer P, Cesa-Bianchi N (1998) On-line learning with malicious noise and the closure algorithm. Ann Math Artif Intell 23:83–99
3. Ben-David S, Eiron N, Long P (2003) On the difficulty of approximately maximizing agreements. J Comput Syst Sci 66:496–514
4. Blumer A, Ehrenfeucht A, Haussler D, Warmuth M (1989) Learnability and the Vapnik-Chervonenkis dimension. J ACM 36:929–965
5. Brubaker SC (2009) Robust PCA and clustering in noisy mixtures. In: Proceedings of 20th annual ACM-SIAM symposium on discrete algorithms, New York, pp 1078–1087
6. Bshouty N, Eiron N, Kushilevitz E (2002) PAC learning with nasty noise. Theor Comput Sci 288: 255–275
7. Cesa-Bianchi N, Dichterman E, Fischer P, Shamir E, Simon HU (1999) Sample-efficient strategies for learning in the presence of noise. J ACM 46: 684–719
8. Kearns M, Li M (1988) Learning in the presence of malicious errors. In: Proceedings of 20th ACM symposium on theory of computing, Chicago, pp 267–280
9. Kearns M, Li M (1993) Learning in the presence of malicious errors. SIAM J Comput 22:807–837
10. Kearns M, Schapire R, Sellie L (1994) Toward efficient agnostic learning. Mach Learn 17: 115–141
11. Klivans AR, Long PM, Servedio RA (2009) Learning halfspaces with malicious noise. Journal of Mach Learn Res 10:2715–2740
12. Servedio RA (2003) Smooth boosting and learning with malicious noise. Journal of Mach Learn Res 4:633–648
13. Valiant L (1984) A theory of the learnable. Commun ACM 27:1134–1142
14. Valiant L Learning (1985) disjunctions of conjunctions. In: Proceedings of 9th international joint conference on artificial intelligence, Los Angeles, pp 560–566

Learning with the Aid of an Oracle

Christino Tamon
Department of Computer Science, Clarkson University, Potsdam, NY, USA

Keywords

Boolean circuits; Disjunctive normal form; Exact learning via queries

Years and Authors of Summarized Original Work

1996; Bshouty, Cleve, Gavaldà, Kannan, Tamon

Problem Definition

In the exact learning model of Angluin [2], a learning algorithm A must discover an unknown function $\mathsf{f} : \{0,1\}^n \to \{0,1\}$ that is a member of a known class C of Boolean functions. The learning algorithm can make at least one of the following types of queries about f:

- Equivalence query $\mathsf{EQ_f(g)}$, for a candidate function g:
  The reply is either "yes," if $\mathsf{g} \Leftrightarrow \mathsf{f}$, or a counterexample a with $\mathsf{g}(a) \neq \mathsf{f}(a)$, otherwise.
- Membership query $\mathsf{MQ_f}(a)$, for some $a \in \{0,1\}^n$:
  The reply is the Boolean value $\mathsf{f}(a)$.
- Subset query $\mathsf{SubQ_f(g)}$, for a candidate function g:
  The reply is "yes," if $\mathsf{g} \Rightarrow \mathsf{f}$, or a counterexample a with $\mathsf{f}(a) < \mathsf{g}(a)$, otherwise.
- Superset query $\mathsf{SupQ_f(g)}$, for a candidate function g:
  The reply is "yes," if $\mathsf{f} \Rightarrow \mathsf{g}$, or a counterexample a with $\mathsf{g}(a) < \mathsf{f}(a)$, otherwise.

A disjunctive normal formula (DNF) is a depth-2 OR-AND circuit whose size is given by the number of its AND gates. Likewise, a conjunctive normal formula (CNF) is a depth-2 AND-OR circuit whose size is given by the number of its OR gates. Any Boolean function can be represented as both a DNF or a CNF formula. A k-DNF is a DNF where each AND gate has a fan-in of at most k; similarly, we may define a k-CNF.

Problem For a given class C of Boolean functions, such as polynomial-size Boolean circuits or disjunctive normal form (DNF) formulas, the goal is to design polynomial-time learning algorithms for any unknown f ∈ C and ask a polynomial number of queries. The output of the learning algorithm should be a function g of polynomial size satisfying $g \Leftrightarrow f$. The polynomial functions bounding the running time, query complexity, and output size are defined in terms of the number of inputs n and the size of the smallest representation (Boolean circuit or DNF) of the unknown function f.

Key Results

One of the main results proved in [5] is that Boolean circuits and disjunctive normal formulas are exactly learnable using equivalence queries and access to an NP oracle.

Theorem 1 *The following tasks can be accomplished with probabilistic polynomial-time algorithms that have access to an* NP *oracle and make polynomially many equivalence queries:*

- *Learning* DNF *formulas of size s using equivalence queries that are depth-*3 AND-OR-AND *formulas of size $O(sn^2/\log^2 n)$.*
- *Learning Boolean circuits of size s using equivalence queries that are circuits of size $O(sn + n \log n)$.*

The idea behind this result is simple. Any class C of Boolean functions is exactly learnable with equivalence queries using the Halving algorithm of Littlestone [11]. This algorithm asks equivalence queries that are the *majority* of candidate functions from C. These are functions in C that are consistent with the counterexamples obtained so far by the learning algorithm. Since each such majority query eliminates at least half of the candidate functions, $\log_2 |C|$ equivalence queries are sufficient to learn any function in C. A problem with using the Halving algorithm here is that the majority query has exponential size. But, it can be shown that a majority of a polynomial number of uniformly random candidate functions is a good enough approximator to the majority of all candidate functions. Moreover, with access to an NP oracle, there is a randomized polynomial time algorithm for generating random uniform candidate functions due to Jerrum, Valiant, and Vazirani [7]. This yields the result.

The next observation is that subset and superset queries are apparently powerful enough to simulate both equivalence queries and the NP oracle. This is easy to see since the tautology test $g \Leftrightarrow 1$ is equivalent to $\mathrm{SubQ_f}(\overline{g}) \wedge \mathrm{SubQ_f}(g)$, for any unknown function f; and, $\mathrm{EQ_f}(g)$ is equivalent to $\mathrm{SubQ_f}(g) \wedge \mathrm{SupQ_f}(g)$. Thus, the following generalization of Theorem 1 is obtained.

Theorem 2 *The following tasks can be accomplished with probabilistic polynomial-time algorithms that make polynomially many subset and superset queries:*

- *Learning* DNF *formulas of size s using equivalence queries that are depth-*3 AND-OR-AND *formulas of size $O(sn^2/\log^2 n)$.*
- *Learning Boolean circuits of size s using equivalence queries that are circuits of size $O(sn + n \log n)$.*

Stronger deterministic results are obtained by allowing more powerful complexity-theoretic oracles. The first of these results employ techniques developed by Sipser and Stockmeyer [12, 13].

Theorem 3 *The following tasks can be accomplished with deterministic polynomial-time algorithms that have access to an Σ_3^p oracle and make polynomially many equivalence queries:*

- *Learning* DNF *formulas of size s using equivalence queries that are depth-*3 AND-OR-AND *formulas of size $O(sn^2/\log^2 n)$.*
- *Learning Boolean circuits of size s using equivalence queries that are circuits of size $O(sn + n \log n)$.*

In the following result, C is an infinite class of functions containing functions of the form $f : \{0,1\}^\star \to \{0,1\}$. The class C is p-evaluatable if the following tasks can be performed in polynomial time:

- Given y, is y a valid representation for any function $f_y \in C$?
- Given a valid representation y and $x \in \{0,1\}^\star$, is $f_y(x) = 1$?

Theorem 4 *Let* C *be any p-evaluatable class. The following statements are equivalent:*

- C *is learnable from polynomially many equivalence queries of polynomial size (and unlimited computational power).*
- C *is learnable in deterministic polynomial time with equivalence queries and access to a* Σ_5^p *oracle.*

For exact learning with membership queries, the following results are proved.

Theorem 5 *The following tasks can be accomplished with deterministic polynomial-time algorithms that have access to an* NP *oracle and make polynomially many membership queries (in n,* DNF *and* CNF *sizes of* f*, where* f *is the unknown function):*

- *Learning monotone Boolean functions.*
- *Learning* $O(\log n)$-CNF $\bigcap$ $O(\log n)$-DNF.

The ideas behind the above result use techniques from [2,4]. For a monotone Boolean function f, the standard closure algorithm uses both equivalence and membership queries to learn f using candidate functions g satisfying g $\Rightarrow$ f. The need for membership can be removed using the following observation. Viewing $\neg$f as a monotone function on the inverted lattice, we can learn f and $\neg$f simultaneously using candidate functions g, h, respectively, that satisfy g $\Rightarrow$ h. The NP oracle is used to obtain an example *a* that either helps in learning f or in learning $\neg$f; when no such example can be found, we have learned f.

Theorem 6 *Any class* C *of Boolean functions that is exactly learnable using a polynomial number of membership queries (and unlimited computational power) is exactly learnable in expected polynomial time using a polynomial number of membership queries and access to an* NP *oracle.*

Moreover, any p-evaluatable class C *that is exactly learnable from a polynomial number of membership queries (and unlimited computational power) is also learnable in deterministic polynomial time using a polynomial number of membership queries and access to a* Σ_5^p *oracle.*

Theorems 4 and 6 showed that information-theoretic learnability using equivalence and membership queries can be transformed into computational learnability at the expense of using the Σ_5^p and NP oracles, respectively.

Applications

The learning algorithm for Boolean circuits using equivalence queries and access to an NP oracle has found an application in complexity theory. Watanabe (see [10]) showed an improvement on a known theorem of Karp and Lipton [8]: if NP has polynomial-size circuits, then the polynomial-time hierarchy PH collapses to $\mathsf{ZPP}^{\mathsf{NP}}$. Subsequently, Aaronson (see [1]) showed that queries to the NP oracle used in the learning algorithm (for Boolean circuits) cannot be parallelized by any relativizing techniques.

Some techniques developed in Theorem 5 for exact learning using membership queries of monotone Boolean functions have found applications in data mining [6].

Open Problems

It is unknown if there are polynomial-time learning algorithms for Boolean circuits and DNF formulas using equivalence queries (without complexity-theoretic oracles). There are strong cryptographic evidence that Boolean circuits are not learnable in polynomial-time (see [3] and the references therein). The best running time for learning DNF formulas is $2^{\tilde{O}(n^{1/3})}$ as given by Klivans and Servedio [9]. It is unclear if membership queries help in this case.

Cross-References

For related learning results, see
▶ Learning DNF Formulas
▶ Learning Automata

Recommended Reading

1. Aaronson S (2006) Oracles are subtle but not malicious. In: Proceedings of the 21st annual IEEE conference on computational complexity (CCC'06), Prague, pp 340–354
2. Angluin D (1988) Queries and concept learning. Mach Learn 2:319–342
3. Angluin D, Kharitonov M (1995) When Won't Membership Queries Help? J Comput Syst Sci 50:336–355
4. Bshouty NH (1995) Exact learning boolean function via the monotone theory. Inf Comput 123:146–153
5. Bshouty NH, Cleve R, Gavaldà R, Kannan S, Tamon C (1996) Oracles and queries that are sufficient for exact learning. J Comput Syst Sci 52(3):421–433
6. Gunopolous D, Khardon R, Mannila H, Saluja S, Toivonen H, Sharma RS (2003) Discovering all most specific sentences. ACM Trans Database Syst 28:140–174
7. Jerrum MR, Valiant LG, Vazirani VV (1986) Random generation of combinatorial structures from a uniform distribution. Theor Comput Sci 43:169–188
8. Karp RM, Lipton RJ (1980) Some connections between nonuniform and uniform complexity classes. In: Proceedings of the 12th annual ACM symposium on theory of computing, Los Angeles, pp 302–309
9. Klivans AR, Servedio RA (2004) Learning DNF in time $2^{\tilde{O}(n^{1/3})}$. J Comput Syst Sci 68:303–318
10. Köbler J, Watanabe O (1998) New collapse consequences of np having small circuits. SIAM J Comput 28:311–324
11. Littlestone N Learning quickly when irrelevant attributes abound: a new linear-threshold algorithm. Mach Learn 2:285–318 (1987)
12. Sipser M (1983) A complexity theoretic approach to randomness. In: Proceedings of the 15th annual ACM symposium on theory of computing, Boston, pp 330–334
13. Stockmeyer LJ (1985) On approximation algorithms for $\#P$. SIAM J Comput 14:849–861

LEDA: a Library of Efficient Algorithms

Christos Zaroliagis
Department of Computer Engineering and Informatics, University of Patras, Patras, Greece

Keywords

LEDA platform for combinatorial and geometric computing

Years and Authors of Summarized Original Work

1995; Mehlhorn, Näher

Problem Definition

In the last forty years, there has been a tremendous progress in the field of computer algorithms, especially within the core area known as *combinatorial algorithms*. Combinatorial algorithms deal with objects such as lists, stacks, queues, sequences, dictionaries, trees, graphs, paths, points, segments, lines, convex hulls, etc, and constitute the basis for several application areas including network optimization, scheduling, transport optimization, CAD, VLSI design, and graphics. For over thirty years, asymptotic analysis has been the main model for designing and assessing the efficiency of combinatorial algorithms, leading to major algorithmic advances.

Despite so many breakthroughs, however, very little had been done (at least until 15 years ago) about the practical utility and assessment of this wealth of theoretical work. The main reason for this lack was the absence of a standard *algorithm library*, that is, of a software library that contains a systematic collection of robust and efficient implementations of algorithms and data structures, upon which other algorithms and data structures can be easily built.

The lack of an algorithm library limits severely the great impact which combinatorial algorithms can have. The continuous re-implementation of basic algorithms and data structures slows down progress and typically discourages people to make the (additional) effort to use an efficient solution, especially if such a solution cannot be re-used. This makes the migration of scientific discoveries into practice a very slow process.

The major difficulty in building a library of combinatorial algorithms stems from the fact that such algorithms are based on complex data types, which are typically not encountered in programming languages (i.e., they are not built-in types). This is in sharp contrast with other computing areas such as statistics, numerical analysis, and linear programming.

Key Results

The currently most successful algorithm library is LEDA (Library for Efficient Data types and Algorithms) [4, 5]. It contains a very large collection of advanced data structures and algorithms for combinatorial and geometric computing. The development of LEDA started in the early 1990s, it reached a very mature state in the late 1990s, and it continues to grow. LEDA has been written in C++ and has benefited considerably from the object-oriented paradigm.

Four major goals have been set in the design of LEDA.

1. *Ease of use*: LEDA provides a sizable collection of data types and algorithms in a form that they can be readily used by non-experts. It gives a precise and readable specification for each data type and algorithm, which is short, general and abstract (to hide the details of implementation). Most data types in LEDA are parameterized (e.g., the dictionary data type works for arbitrary key and information type). To access the objects of a data structure by position, LEDA has invented the *item concept* that casts positions into an abstract form.
2. *Extensibility*: LEDA is easily extensible by means of parametric polymorphism and can be used as a platform for further software development. Advanced data types are built on top of basic ones, which in turn rest on a uniform conceptual framework and solid implementation principles. The main mechanism to extend LEDA is through the so-called LEDA extension packages (LEPs). A LEP extends LEDA into a particular application domain and/or area of algorithms that is not covered by the core system. Currently, there are 15 such LEPs; for details see [1].
3. *Correctness*: In LEDA, programs should give sufficient justification (proof) for their answers to allow the user of a program to easily assess its correctness. Many algorithms in LEDA are accompanied by *program checkers*. A program checker C for a program P is a (typically very simple) program that takes as input the input of P, the output of P, and perhaps additional information provided by P, and verifies that the answer of P in indeed the correct one.
4. *Efficiency*: The implementations in LEDA are usually based on the asymptotically most efficient algorithms and data structures that are known for a problem. Quite often, these implementations have been fine-tuned and enhanced with heuristics that considerably improve running times. This makes LEDA not only the most comprehensive platform for combinatorial and geometric computing, but also a library that contains the currently fastest implementations.

Since 1995, LEDA is maintained by the Algorithmic Solutions Software GmbH [1] which is responsible for its distribution in academia and industry.

Other efforts for algorithm libraries include the Standard Template Library (STL) [7], the Boost Graph Library [2, 6], and the Computational Geometry Algorithms Library (CGAL) [3].

STL [7] (introduced in 1994) is a library of interchangeable components for solving many fundamental problems on sequences of elements, which has been adopted into the C++ standard. It contributed the *iterator concept* which provides an interface between *containers* (an object that stores other objects) and algorithms. Each algorithm in STL is a function template parameterized by the types of iterators upon which it operates. Any iterator that satisfies a minimum set of requirements can be used regardless of the data structure accessed by the iterator. The systematic approach used in STL to build abstractions and interchangeable components is called *generic programming*.

The Boost Graph Library [2, 6] is a C++ graph library that applies the notions of generic programming to the construction of graph algorithms. Each graph algorithm is written not in terms of a specific data structure, but instead in terms of a graph abstraction that can be easily implemented by many different data structures. This offers the programmer the flexibility to use graph algorithms in a wide variety of applications. The first release of the library became available in September 2000.

The Computational Geometry Algorithms Library [3] is another C++ library that focuses on geometric computing only. Its main goal is to provide easy access to efficient and reliable geometric algorithms to users in industry and academia. The CGAL library started in 1996 and the first release was in April 1998.

Among all libraries mentioned above LEDA is by far the best (both in quality and efficiency of implementations) regarding combinatorial computing. It is worth mentioning that the late versions of LEDA have also incorporated the iterator concept of STL.

Finally, a notable effort concerns the Stony Brook Algorithm Repository [8]. This is not an algorithm library, but a comprehensive collection of algorithm implementations for over seventy problems in combinatorial computing, started in 2001. The repository features implementations coded in different programming languages, including C, C++, Java, Fortran, ADA, Lisp, Mathematic, and Pascal.

Applications

An algorithm library for combinatorial and geometric computing has a wealth of applications in a wide variety of areas, including: network optimization, scheduling, transport optimization and control, VLSI design, computer graphics, scientific visualization, computer aided design and modeling, geographic information systems, text and string processing, text compression, cryptography, molecular biology, medical imaging, robotics and motion planning, and mesh partition and generation.

Open Problems

Algorithm libraries usually do not provide an interactive environment for developing and experimenting with algorithms. An important research direction is to add an interactive environment into algorithm libraries that would facilitate the development, debugging, visualization, and testing of algorithms.

Experimental Results

There are numerous experimental studies based on LEDA, STL, Boost, and CGAL, most of which can be found in the world-wide web. Also, the web sites of some of the libraries contain pointers to experimental work.

URL to Code

The afore mentioned algorithm libraries can be downloaded from their corresponding web sites, the details of which are given in the bibliography (Recommended Reading).

Cross-References

► Engineering Algorithms for Large Network Applications
► Experimental Methods for Algorithm Analysis
► Implementation Challenge for Shortest Paths
► Shortest Paths Approaches for Timetable Information
► TSP-Based Curve Reconstruction

Recommended Reading

1. Algorithmic Solutions Software GmbH, http://www.algorithmic-solutions.com/. Accessed Feb 2008
2. Boost C++ Libraries, http://www.boost.org/. Accessed Feb 2008
3. CGAL: Computational Geometry Algorithms Library, http://www.cgal.org/. Accessed Feb 2008
4. Mehlhorn K, Näher S (1995) LEDA: a platform for combinatorial and geometric computing. Commun ACM 38(1):96–102
5. Mehlhorn K, Näher S (1999) LEDA: a platform for combinatorial and geometric computing. Cambridge University Press, Boston
6. Siek J, Lee LQ, Lumsdaine A (2002) The boost graph library. Addison-Wesley, Cambridge
7. Stepanov A, Lee M (1994) The standard template library. In: Technical report X3J16/94–0095, WG21/N0482, ISO Programming Language C++ project. Hewlett-Packard, Palo Alto
8. The Stony Brook Algorithm Repository, http://www.cs.sunysb.edu/~algorith/. Accessed Feb 2008

Lempel-Ziv Compression

Simon J. Puglisi
Department of Computer Science, University of Helsinki, Helsinki, Finland

Keywords

Dictionary-based compression; LZ77; LZ78; LZ compression; LZ parsing

Years and Authors of Summarized Original Work

1976; Lempel, Ziv
1977; Ziv, Lempel
1978; Ziv, Lempel

Problem Definition

Lossless data compression is concerned with compactly representing data in a form that allows the original data to be faithfully recovered. Reduction in space can be achieved by exploiting the presence of repetition in the data.

Many of the main solutions for lossless data compression in the last three decades have been based on techniques first described by Ziv and Lempel [22, 33, 34]. These methods gained popularity in the 1980s via tools like Unix `compress` and the GIF image format, and today they pervade computer software, for example, in the zip, gzip, and lzma compression utilities, in modem compression standards V.42bis and V.44, and as a basis for the compression used in information retrieval systems [8] and Google's BigTable [4] database system. Perhaps the primary reason for the success of Lempel and Ziv's methods is their powerful combination of compression effectiveness and compression/decompression throughput. We refer the reader to [3, 29] for a review of related dictionary-based compression techniques.

Key Results

Let $S[1 \ldots n]$ be a string of n symbols drawn from an alphabet Σ. Lempel-Ziv-based compression algorithms work by parsing S into a sequence of substrings called *phrases* (or factors). To achieve compression, each phrase is replaced by a compact representation, as detailed below.

LZ78

Assume the encoder has already parsed the phrases $S_1, S_2, \ldots, S_{i-1}$, that is, $S = S_1 S_2 \ldots S_{i-1} S'$ for some suffix S' of S. The LZ78 [34] dictionary is the set of strings obtained by adding a single symbol to one of the strings S_j or to the empty string. The next phrase S_i is then the longest prefix of S' that is an element of the dictionary. For example, $S = bbaabbbabbabbaab$ has an LZ78 parsing of b, ba, a, bb, bab, $babb$, aa, b. Clearly, all LZ78 phrases will be distinct, except possibly the final one. Let S_0 denote the empty string. If $S_i = S_j \alpha$, where $0 \le j < i$ and $\alpha \in \Sigma$, the code word emitted by LZ78 for S_i will be the pair (j, α). Thus, if LZ78 parses the string S into y words, its output is bounded by $y \log y + y \log |\Sigma| + \Theta(y)$ bits.

LZ77

The LZ77 parsing algorithm takes a single positive integer parameter w, called the *window size*. Say the encoder has already parsed the phrases $S_1, S_2, \ldots, S_{i-1}$, that is, $S = S_1 S_2 \ldots S_{i-1} S'$ for some suffix S' of S. The next phrase S_i starts at position $q = |S_1 S_2 \ldots S_{i-1}| + 1$ in S and is the shortest prefix of S' that does not have an occurrence starting at any position $q - w \le p_i < q$ in S. Thus defined, LZ77 phrases have the form $t\alpha$, where t (possibly empty) is the longest prefix of S' that also has an occurrence starting at some position $q - w \le p_i < q$ in S and α is the symbol $S[q + |t| + 1]$.

The version of LZ77 described above is often called *sliding window* LZ77: a text window of length w that slides along the string during parsing is used to decide the next phrase. In the so-called infinite window LZ77, we enforce that $q - w$ is always equal to 0 – in other words, throughout parsing the window stretches all the way back to the beginning of the string. Infinite window parsing is more powerful than sliding window parsing, and for the remainder of this entry, the term "LZ77" refers to infinite window LZ77, unless explicitly stated.

For $S = bbaabbbabbabbaab$, the infinite window LZ77 parsing is b, ba, ab, $bbab$, $babbaa$, b. Note that phrases are allowed to overlap their definitions, as is the case with phrase S_5 in our example. Like LZ78, all LZ77 phrases are distinct, with the possible exception of the last phrase. It is easy to see that infinite window LZ77 will always produce a smaller number of phrases than LZ78. If $S_i = t\alpha$ with $\alpha \in \Sigma$, the code word for S_i is the triple (p_i, ℓ_i, α), where p_i is the position of a previous occurrence of t in $S_1 S_2 \ldots S_{i-1}$ and $\ell_i = |t|$.

Finally, it is important to note that for a given phrase $S_i = t\alpha$, there is sometimes more than one previous occurrence of t, leading to a choice of p_i value. If $p_i < q$ is the largest possible for every phrase, then we call the parsing *rightmost*. In their study on the bit complexity of LZ compression, Ferragina et al. [9] showed that the rightmost parsing can lead to encodings asymptotically smaller than what is achievable otherwise.

Compression and Decompression Complexity

The LZ78 parsing for a string of length n can be computed in $O(n)$ time by maintaining the phrases in a try. Doing so allows finding in time proportional to its length the longest prefix of the unparsed portion that is in the dictionary, and so the time overall is linear in n. Compression in sublinear time and space is possible using succinct dynamic tries [16]. Decoding is somewhat symmetric – an explicit try is not needed, just the parent pointers implicitly represented in the encoded pairs.

Recovering the original string S from its LZ77 parsing (i.e., decompression) is very easy: (p_i, ℓ_i, α) is decoded by copying the symbols of the substring $S[p_i \ldots p_i + \ell_i - 1]$ and then appending α. By the definition of the parsing, any symbol we need to copy will have already been decoded (if we copy the strings left to right).

Obtaining the LZ77 parsing in the first place is not as straightforward and has been the subject of intense research since LZ77 was published. Indeed, it seems safe to speculate that LZ78 and sliding window LZ77 were invented primarily because it was initially unclear how infinite window LZ77 could be computed efficiently. Today, parsing is possible in worst-case $O(n)$ time, using little more than $n(\log n + \log|\Sigma|)$ bits of space. Current state-of-the-art methods [14, 15, 18, 19] operate offline, combining the suffix array [24] of the input string with data structures for answering next and previous smaller value queries [10]. For online parsing, the current best algorithm uses $O(n \log n)$ time and $O(n \log|\Sigma|)$ bits of space [32].

Compression Effectiveness

It is well known that LZ converges to the entropy of any ergodic source [6, 30, 31]. However, it is also possible to prove compression bounds on LZ-based schemes without probabilistic assumptions on the input, using the notion of *empirical entropy* [25].

Convergence to Empirical Entropy

For any string S, the kth-order empirical entropy $H_k(S)$ is a lower bound on the compression achievable by any compressor that assigns code words to symbols based on statistics derived from the k letters preceding each symbol in the string. In particular, the output of LZ78 (and so LZ77) is upper-bounded by $|S|H_k(S) + o(|S| \log |\Sigma|)$ bits [20] for any $k = o(\log_{|\Sigma|} n)$.

Relationship to Grammar Compression

The *smallest grammar problem* is the problem of finding the smallest context-free grammar that generates only a given input string S. The size g^* of the smallest grammar is a rather elegant measure of compressibility, and Charikar et al. [5] established that finding it is NP-hard. They also considered several approximation algorithms, including LZ78. The LZ78 parsing of S can be viewed as a context-free grammar in which for each dictionary word $S_i = S_j\alpha$, there is a production rule $X_i = X_j\alpha$. LZ78's approximation ratio is rather bad: $\Omega(n^{2/3}/\log n)$.

Charikar et al. also showed that g^* is at least the number of phrases z of the LZ77 parse of S and used the phrases of the parsing to derive a new grammar compression algorithm with approximation ratio $O(\log(|S|/g^*))$. The same result was discovered contemporaneously by Rytter [28] and later simplified by Jeż [17].

Greedy Versus Non-greedy Parsing

LZ78 and LZ77 are both greedy algorithms: they select, at each step, the longest prefix of the remaining suffix of the input that is in the dictionary. For LZ77, the greedy strategy is optimal in the sense that it yields the minimum number of code words. However, if variable-length codes are used to represent each element of the code word triple, the greedy strategy does not yield an optimal parsing, as Ferragina, Nitto, and Venturini have recently established [9]. For LZ78, greedy parsing does not always produce the minimum number of phrases. Indeed, in the worst case, greedy parsing can produce a factor $O(\sqrt{n})$ more than a simple non-greedy parsing strategy that, instead of choosing the prefix that gives the longest extension in the current iteration, chooses the prefix that gives the longest extension in the next iteration [26]. There are many, many variants on LZ parsing that relax the greedy condition with the aim of reducing the overall encoding size in practice. Several of these non-greedy methods are covered in textbooks (e.g., [3, 29]).

Applications

As outlined at the start of this entry, the major applications of Lempel and Ziv's methods are in the fields of lossless data compression. However, the deep connections of these methods to string data mean the Lempel-Ziv parsing has also found important applications in string processing: the parsing reveals a great deal of information about the repetitive structure of the underlying string, and this can be used to design efficient algorithms and data structures.

Pattern Matching in Compressed Space

A compressed full-text index for a string $S[1 \dots n]$ is a data structure that takes space proportional to the entropy of S while simultaneously supporting efficient queries over S. The supported queries can be relatively simple, such as random access to symbols of S, or more complex, such as reporting all the occurrences of a pattern $P[1 \dots m]$ in S.

Arroyuelo et al. [2] describe a compressed index based on LZ78. For any text S, their index uses $(2+\epsilon)nH_k(S) + o(n \log |\Sigma|)$ bits of space and reports all c occurrences of P in S in $O(m^2 \log m + (m + c) \log n)$ time. Their approach stores two copies of the LZ78 dictionary represented as tries. One try contains the dictionary phrases, and the other contains the reverse phrases. The main trick to pattern matching is to then split the pattern in two (in all m possible ways) and then check for each half in the tries.

Kreft and Navarro [21] describe a compressed index based on the LZ77 parsing. It requires $3z \log n + O(z \log |\Sigma|) + o(n)$ bits of space and supports extraction of ℓ symbols in $O(\ell h)$ time and pattern matching in $O(m^2 h + m \log z + c \log z)$ time, where $h \leq \sqrt{n}$ is the maximum length of a referencing chain in the parsing (a position is copied from another, that one from another, and so on, h times). More recently, Gagie et al. [12] describe a different LZ77-based index with improved query bounds that takes $O(z \log n \log(n/z))$ bits of space and supports extraction of ℓ symbols in $O(\ell + \log n)$ time and pattern matching in $O(m \log m + c \log \log n)$ time. A related technique called *indexing by kernelization*, which does not support access and restricts the maximum pattern length that can be searched for, has recently emerged as a promising practical approach applicable to highly repetitive data [11]. This technique uses an LZ parsing to identify a subsequence (called a *kernel*) of the text that is guaranteed to contain at least one occurrence of every pattern the text contains. This kernel string is then further processed to obtain an index capable of searching the original text.

A related problem is the compressed matching problem, in which the text and the pattern are given together and the text is compressed. The task here is to perform pattern matching in the compressed text without decompressing it. For LZ-based compressors, this problem was first considered by Amir, Benson, and Farach [1]. Considerable progress has been made since then, and we refer the reader to [13, 27] (and to another encyclopedia entry) for an overview of more recent results.

String Alignment

Crochemore, Landau, and Ziv-Ukelson [7] used LZ78 to accelerate sequence alignment: the problem of finding the lowest-cost sequence of edit operations that transforms one string $S[1 \ldots n]$ into another string $T[1 \ldots n]$. Masek and Paterson proposed an $O(n^2/\log n)$ time algorithm that applies when the costs of the edit operations are rational. Crochemore et al.'s method runs in the same time in the worst case, but allows real-valued costs, and obtains an asymptotic speedup when the underlying texts are compressible.

The textbook solution to the string alignment problem runs in $O(n^2)$, using a straightforward dynamic program that computes a matrix $M[1 \ldots n, 1 \ldots n]$. The approach of the faster algorithms is to break the dynamic program matrix into blocks. Masek and Paterson use blocks of uniform size. Crochemore et al. use blocks delineated by the LZ78 parsing, the idea being that whenever they need to solve a block $M[i \ldots i', j \ldots j']$, they can solve it in $O(i'-i+j'-j)$ time by essentially copying their solutions to the previous blocks $M[i \ldots i'-1, j \ldots j']$ and $M[i \ldots i', j \ldots j'-1]$. A similar approach was later used to speed up training of hidden Markov models [23].

Open Problems

Ferragina, Nitto, and Venturini [9] provide an algorithm for computing the rightmost LZ77 parsing that takes $O(n + n \log |\Sigma| / \log \log n)$ time and $O(n)$ words of space to process a string of length n. The existence of an $O(n)$ time algorithm independent of the alphabet size is an open problem.

As mentioned above, the size z of the LZ77 parsing is a lower bound on the size g^* of the smallest grammar for a given string. Proving an asymptotic separation between g^* and z (or, alternatively, finding a way to produce grammars of size z) is a problem of considerable theoretical interest.

URLs to Code and Data Sets

The source code of the gzip tool (based on LZ77) is available at http://www.gzip.org, and the related compression library zlib is available at http://www.zlib.net. Source code for the more efficient compressor LZMA is at: http://www.7-zip.org/sdk.html.

Source code for more recent LZ parsing algorithms (developed in the last 2 years) is available at http://www.cs.helsinki.fi/group/pads/. This code includes the current fastest LZ parsing algorithms for both internal and external memory.

The Pizza&Chili Corpus is a frequently used test data set for LZ parsing algorithms; see http://pizzachili.dcc.uchile.cl/repcorpus.html.

Cross-References

▶ Approximate String Matching
▶ Compressed Suffix Array
▶ Grammar Compression

▶ Pattern Matching on Compressed Text
▶ Suffix Trees and Arrays

Recommended Reading

1. Amir A, Benson G, Farach M (1996) Let sleeping files lie: pattern matching in Z-compressed files. J Comput Syst Sci 52(2):299–307
2. Arroyuelo D, Navarro G, Sadakane K (2012) Stronger Lempel-Ziv based compressed text indexing. Algorithmica 62(1–2):54–101
3. Bell TC, Cleary JG, Witten IH (1990) Text compression. Prentice-Hall, Upper Saddle River
4. Chang F, Dean J, Ghemawat S, Hsieh WC, Wallach DA, Burrows M, Chandra T, Fikes A, Gruber RE (2008) Bigtable: a distributed storage system for structured data. ACM Trans Comp Sys 26(2):1–26
5. Charikar M, Lehman E, Liu D, Panigrahy R, Prabhakaran M, Sahai A, Shelat A (2005) The smallest grammar problem. IEEE Trans Inform Theory 51(7):2554–2576
6. Cover T, Thomas J (1991) Elements of information theory. Wiley, New York
7. Crochemore M, Landau GM, Ziv-Ukelson M (2003) A subquadratic sequence alignment algorithm for unrestricted scoring matrices. SIAM J Comput 32(6):1654–1673
8. Ferragina P, Manzini G (2010) On compressing the textual web. In: Proceedings of the third international conference on web search and web data mining (WSDM) 2010, New York, 4–6 February 2010. ACM, pp 391–400
9. Ferragina P, Nitto I, Venturini R (2013) On the bit-complexity of lempel-Ziv compression. SIAM J Comput 42(4):1521–1541
10. Fischer J, Heun V (2011) Space-efficient preprocessing schemes for range minimum queries on static arrays. SIAM J Comput 40(2):465–492
11. Gagie T, Puglisi SJ (2015) Searching and indexing genomic databases via kernelization. Front Bioeng Biotechnol 3(12). doi:10.3389/fbioe.2015.00012
12. Gagie T, Gawrychowski P, Kärkkäinen J, Nekrich Y, Puglisi SJ (2014) LZ77-based self-indexing with faster pattern matching. In: Proceedings of Latin-American symposium on theoretical informatics (LATIN), Montevideo. Lecture notes in computer science, vol 8392. Springer, pp 731–742
13. Gawrychowski P (2013) Optimal pattern matching in LZW compressed strings. ACM Trans Algorithms 9(3):25
14. Goto K, Bannai H (2013) Simpler and faster Lempel Ziv factorization. In: Proceedings of the 23rd data compression conference (DCC), Snowbird, pp 133–142
15. Goto K, Bannai H (2014) Space efficient linear time Lempel-Ziv factorization for small alphabets. In: Proceedings of the 24th data compression conference (DCC), Snowbird, pp 163–172
16. Jansson J, Sadakane K, Sung W (2007) Compressed dynamic tries with applications to LZ-compression in sublinear time and space. In: Proceedings of 27th FSTTCS, Montevideo. Lecture notes in computer science, vol 4855. Springer, New Delhi, pp 424–435
17. Jez A (2014) A really simple approximation of smallest grammar. In: Kulikov AS, Kuznetsov SO, Pevzner PA (eds) Proceedings of 25th annual symposium combinatorial pattern matching (CPM) 2014, Moscow, 16–18 June 2014. Lecture notes in computer science, vol 8486. Springer, pp 182–191
18. Kärkkäinen J, Kempa D, Puglisi SJ (2013) Linear time Lempel-Ziv factorization: simple, fast, small. In: Proceedings of CPM, Bad Herrenalb. Lecture notes in computer science, vol 7922, pp 189–200
19. Kempa D, Puglisi SJ (2013) Lempel-Ziv factorization: simple, fast, practical. In: Zeh N, Sanders P (eds) Proceedings of ALENEX, New Orleans. SIAM, pp 103–112
20. Kosaraju SR, Manzini G (1999) Compression of low entropy strings with lempel-ziv algorithms. SIAM J Comput 29(3):893–911
21. Kreft S, Navarro G (2013) On compressing and indexing repetitive sequences. Theor Comput Sci 483:115–133
22. Lempel A, Ziv J (1976) On the complexity of finite sequences. IEEE Trans Inform Theory 22(1):75–81
23. Lifshits Y, Mozes S, Weimann O, Ziv-Ukelson M (2009) Speeding up hmm decoding and training by exploiting sequence repetitions. Algorithmica 54(3):379–399
24. Manber U, Myers GW (1993) Suffix arrays: a new method for on-line string searches. SIAM J Comput 22(5):935–948
25. Manzini G (2001) An analysis of the Burrows-Wheeler transform. J ACM 48(3):407–430
26. Matias Y, Sahinalp SC (1999) On the optimality of parsing in dynamic dictionary based data compression. In: Proceedings of the tenth annual ACM-SIAM symposium on discrete algorithms, 17–19 January 1999, Baltimore, pp 943–944
27. Navarro G, Tarhio J (2005) LZgrep: a Boyer-Moore string matching tool for Ziv-Lempel compressed text. Softw Pract Exp 35(12):1107–1130
28. Rytter W (2003) Application of Lempel-Ziv factorization to the approximation of grammar-based compression. Theor Comput Sci 302(1–3):211–222
29. Salomon D (2006) Data compression: the complete reference. Springer, New York/Secaucus
30. Sheinwald D (1994) On the Ziv-Lempel proof and related topics. Proc. IEEE 82:866–871
31. Wyner A, Ziv J (1994) The sliding-window Lempel-Ziv algorithm is asymptotically optimal. Proc IEEE 82:872–877
32. Yamamoto J, I T, Bannai H, Inenaga S, Takeda M (2014) Faster compact on-line Lempel-Ziv factorization. In: Proceedings of 31st international symposium on theoretical aspects of computer science (STACS), Lyon. LIPIcs 25, pp 675–686

33. Ziv J, Lempel A (1977) A universal algorithm for sequential data compression. IEEE Trans Inform Theory 23(3):337–343
34. Ziv J, Lempel A (1978) Compression of individual sequences via variable-rate coding. IEEE Trans Inform Theory 24(5):530–536

Leontief Economy Equilibrium

Yinyu Ye
Department of Management Science and Engineering, Stanford University, Stanford, CA, USA

Keywords

Bimatrix game; Competitive exchange; Computational equilibrium; Leontief utility function

Synonyms

Algorithmic game theory; Arrow-debreu market; Max-min utility; Walras equilibrium

Years and Authors of Summarized Original Work

2005; Codenotti, Saberi, Varadarajan, Ye
2005; Ye

Problem Definition

The Arrow-Debreu exchange market equilibrium problem was first formulated by Léon Walras in 1954 [7]. In this problem, everyone in a population of m traders has an initial endowment of a divisible goods and a utility function for consuming all goods – their own and others'. Every trader sells the entire initial endowment and then uses the revenue to buy a bundle of goods such that his or her utility function is maximized. Walras asked whether prices could be set for everyone's goods such that this is possible. An answer was given by Arrow and Debreu in 1954 [1] who showed that, under mild conditions, such equilibrium would exist if the utility functions were concave. In general, it is unknown whether or not an equilibrium can be computed efficiently; see, e.g., ▶ General Equilibrium.

Consider a special class of Arrow-Debreu's problems, where each of the n traders has exactly one unit of a divisible and distinctive good for trade, and let trader i, $i = 1, \ldots, n$, bring good i, where the class of problems is called the *pairing class*. For given prices p_j on good j, consumer i's maximization problem is

$$\begin{aligned} &\text{maximize } u_i(x_{i1}, \ldots, x_{in}) \\ &\text{subject to } \textstyle\sum_j p_j x_{ij} \leq p_i, \\ &\qquad\qquad x_{ij} \geq 0, \quad \forall j, \end{aligned} \tag{1}$$

where x_{ij} is the quantity of good j purchased by trader i. Let x_i^* denote a maximal solution vector of (1). Then, vector p is called the Arrow-Debreu price equilibrium if there exists an x_i^* for consumer i, $i = 1, \ldots, n$, to clear the market, that is,

$$\sum_i x_i^* = e,$$

where e is the vector of all ones representing available goods on the exchange market.

The Leontief economy equilibrium problem is the Arrow-Debreu equilibrium problem when the utility functions are in the Leontief form:

$$u_i(x_i) = \min_{j:\, h_{ij}>0} \left\{ \frac{x_{ij}}{h_{ij}} \right\},$$

where the Leontief coefficient matrix is given by

$$H = \begin{pmatrix} h_{11} & h_{12} & \ldots & h_{1n} \\ h_{21} & h_{22} & \ldots & h_{2n} \\ \ldots & \ldots & \ldots & \ldots \\ h_{n1} & h_{n2} & \ldots & h_{nn} \end{pmatrix}. \tag{2}$$

Here, one may assume that

Assumption 1 *H has no all-zero row, that is, every trader likes at least one good.*

Key Results

Let u_i be the equilibrium utility value of consumer i and p_i be the equilibrium price for good $i, i = 1, \ldots, n$. Also, let U and P be diagonal matrices whose diagonal entries are u_i's and p_i's, respectively. Then, the Leontief economy equilibrium $p \in R^n$, together with $u \in R^n$, must satisfy

$$\begin{aligned} UHp &= p, \\ P(e - H^T u) &= 0, \\ H^T u &\leq e, \\ u,\ p &\geq 0, \\ p &\neq 0. \end{aligned} \tag{3}$$

One can prove:

Theorem 1 (Ye [8]) *System (3) always has a solution ($u \neq 0, p$) under Assumption 1 (i.e., H has no all-zero row). However, a solution to System (3) may not be a Leontief equilibrium, although every Leontief equilibrium satisfies System (3).*

A solution to System (3) is called a quasi-equilibrium. For example,

$$H^T = \begin{pmatrix} 1\ 2\ 0 \\ 0\ 1\ 2 \\ 0\ 0\ 1 \end{pmatrix}$$

has a quasi-equilibrium $p^T = (1, 0, 0)$ and $u^T = (1, 0, 0)$, but it is not an equilibrium. This is because that trader 3, although with zero budget, can still purchase goods 2 and 3 at zero prices. In fact, check if H has an equilibrium that is an NP-hard problem; see discussion later. However, under certain sufficient conditions, e.g., all entries in H are positive, every quasi-equilibrium is an equilibrium.

Theorem 2 (Ye [8]) *Let $B \subset \{1, 2, \ldots, n\}$, $N = \{1, 2, \ldots, n\} \setminus B$, H_{BB} be irreducible, and u_B satisfy the linear system*

$$H_{BB}^T u_B = e, \quad H_{BN}^T u_B \leq e, \quad \textit{and} \quad u_B > 0.$$

Then the (right) Perron-Frobenius eigenvector p_B of $U_B H_{BB}$ together with $p_N = 0$ will be a solution to System (3). And the converse is also true. Moreover, there is always a rational solution for every such B, that is, the entries of price vector are rational numbers, if the entries of H are rational. Furthermore, the size (bit length) of the solution is bounded by the size (bit length) of H.

The theorem implies that the traders in block B can trade among themselves and keep others goods "free." In particular, if one trader likes his or her own good more than any other good, that is, $h_{ii} \geq h_{ij}$ for all j, then $u_i = 1/h_{ii}$, $p_i = 1$, and $u_j = p_j = 0$ for all $j \neq i$, that is, $B = \{i\}$, makes a Leontief economy equilibrium. The theorem thus establishes, for the first time, a combinatorial algorithm to compute a Leontief economy equilibrium by finding a right block $B \neq \emptyset$, which is actually a nontrivial solution ($u \neq 0$) to an LCP problem

$$H^T u + v = e,\ u^T v = 0,\ 0 \neq u, v \geq 0. \tag{4}$$

If $H > 0$, then any complementary solution $u \neq 0$, together with its support $B = \{j : u_j > 0\}$, of (4) induce a Leontief economy equilibrium that is the (right) Perron-Frobenius eigenvector of $U_B H_{BB}$, and it can be computed in polynomial time by solving a linear equation. Even if $H \not> 0$, any complementary solution $u \neq 0$ and $B = \{j : u_j > 0\}$, as long as H_{BB} is irreducible, induces an equilibrium for System (3). The equivalence between the pairing Leontief economy model and the LCP also implies

Corollary 1 *LCP (4) always has a nontrivial solution $u \neq 0$, where H_{BB} is irreducible with $B = \{j : u_j > 0\}$, under Assumption 1 (i.e., H has no all-zero row).*

If Assumption 1 does not hold, the corollary may not be true; see example below:

$$H^T = \begin{pmatrix} 0\ 2 \\ 0\ 1 \end{pmatrix}.$$

L

Applications

Given an arbitrary bimatrix game, specified by a pair of $n \times m$ matrices A and B, with positive entries, one can construct a Leontief exchange economy with $n + m$ traders and $n + m$ goods as follows. In words, trader i comes to the market with one unit of good i, for $i = 1, \ldots, n + m$. Traders indexed by any $j \in \{1, \ldots, n\}$ receive some utility only from goods $j \in \{n+1, \ldots, n+m\}$, and this utility is specified by parameters corresponding to the entries of the matrix B. More precisely the proportions in which the j-th trader wants the goods are specified by the entries on the jth row of B. Vice versa, traders indexed by any $j \in \{n + 1, \ldots, n + m\}$ receive some utility only from goods $j \in \{1, \ldots, n\}$. In this case, the proportions in which the j-th trader wants the goods are specified by the entries on the j-th column of A.

In the economy above, one can partition the traders in two groups, which bring to the market disjoint sets of goods and are only interested in the goods brought by the group they do not belong to.

Theorem 3 (Codenotti et al. [4]) *Let (A, B) denote an arbitrary bimatrix game, where one assumes, w.l.o.g., that the entries of the matrices A and B are all positive. Let*

$$H^T = \begin{pmatrix} 0 & A \\ B^T & 0 \end{pmatrix}$$

describe the Leontief utility coefficient matrix of the traders in a Leontief economy. There is a one-to-one correspondence between the Nash equilibria of the game (A, B) and the market equilibria H of the Leontief economy. Furthermore, the correspondence has the property that a strategy is played with positive probability at a Nash equilibrium if and only if the good held by the corresponding trader has a positive price at the corresponding market equilibrium.

The theorem implies that finding an equilibrium for Leontief economies is at least as hard as finding a Nash equilibrium for two-player nonzero sum games, a problem recently proven PPAD-complete (Chen and Deng [3]), where no polynomial time approximation algorithm is known today.

Furthermore, Gilboa and Zemel [6] proved a number of hardness results related to the computation of Nash equilibria (NE) for finite games in normal form. Since the NE for games with more than two players can be irrational, these results have been formulated in terms of NP-hardness for multiplayer games, while they can be expressed in terms of NP-completeness for two-player games. Using a reduction to the NE game, Codenotti et al. proved:

Theorem 4 (Codenotti et al. [4]) *It is NP-hard to decide whether a Leontief economy H has an equilibrium.*

On the positive side, Zhu et al. [9] recently proved the following result:

Theorem 5 *Let the Leontief utility matrix H be symmetric and positive. Then there is a fully polynomial time approximation scheme (FPTAS) for approximating a Leontief equilibrium, although the equilibrium set remains non-convex or non-connected.*

Cross-References

- ▶ Approximations of Bimatrix Nash Equilibria
- ▶ Complexity of Bimatrix Nash Equilibria
- ▶ General Equilibrium
- ▶ Non-approximability of Bimatrix Nash Equilibria

Recommended Reading

The reader may want to read Brainard and Scarf [2] on how to compute equilibrium prices in 1891; Chen and Deng [2] on the most recent hardness result of computing the bimatrix game; Cottle et al. [5] for literature on linear complementarity problems; and all references listed in [4] and [8] for the recent literature on computational equilibrium.

1. Arrow KJ, Debreu G (1954) Existence of an equilibrium for competitive economy. Econometrica 22:265–290

2. Brainard WC, Scarf HE (2000) How to compute equilibrium prices in 1891. Cowles Foundation Discussion Paper 1270
3. Chen X, Deng X (2005) Settling the complexity of 2-player Nash-equilibrium. ECCC TR05-140
4. Codenotti B, Saberi A, Varadarajan K, Ye Y (2006) Leontief economies encode nonzero sum two-player games. In: Proceedings SODA, Miami
5. Cottle R, Pang JS, Stone RE (1992) The linear complementarity problem. Academic, Boston
6. Gilboa I, Zemel E (1989) Nash and correlated equilibria: some complexity considerations. Games Econ Behav 1:80–93
7. Walras L (1954) [1877] Elements of pure economics. Irwin. ISBN 0-678-06028-2
8. Ye Y (2005) Exchange market equilibria with Leontief's utility: freedom of pricing leads to rationality. In: Proceedings WINE, Hong Kong
9. Zhu Z, Dang C, Ye Y (2012) A FPTAS for computing a symmetric Leontief competitive economy equilibrium. Math Program 131:113–129

LexBFS, Structure, and Algorithms

Nicolas Trotignon
Laboratoire de l'Informatique du Parallélisme (LIP), CNRS, ENS de Lyon, Lyon, France

Keywords

Classes of graphs; LexBFS; Truemper configurations; Vertex elimination ordering

Years and Authors of Summarized Original Work

2014; Aboulker, Charbit, Trotignon, Vušković

Problem Definition

We provide a general method to prove the existence and compute efficiently elimination orderings in graphs. Our method relies on several tools that were known before but that were not put together so far: the algorithm LexBFS due to Rose, Tarjan, and Lueker, one of its properties discovered by Berry and Bordat, and a local decomposition property of graphs discovered by Maffray, Trotignon, and Vušković.

Terminology

In this paper, all graphs are finite and simple. A graph G *contains* a graph F if F is isomorphic to an induced subgraph of G. A class of graphs is *hereditary* if for every graph G of the class, all induced subgraphs of G belong to the class. A graph G is F*-free* if it does not contain F. When $\mathcal{F}$ is a set of graphs, G is $\mathcal{F}$*-free* if it is F-free for every $F \in \mathcal{F}$. Clearly every hereditary class of graphs is equal to the class of $\mathcal{F}$-free graphs for some $\mathcal{F}$ ($\mathcal{F}$ can be chosen to be the set of all graphs not in the class but all induced subgraphs of which are in the class). The induced subgraph relation is not a well quasi order (contrary, e.g., to the minor relation), so the set $\mathcal{F}$ does not need to be finite.

When $X \subseteq V(G)$, we write $G[X]$ for the subgraph of G induced by X. An ordering $(v_1, \dots, v_n)$ of the vertices of a graph G is an $\mathcal{F}$*-elimination ordering* if for every $i = 1, \dots, n$, $N_{G[\{v_1,\dots,v_i\}]}(v_i)$ is $\mathcal{F}$-free. Note that this is equivalent to the existence, in every induced subgraph of G, of a vertex whose neighborhood is $\mathcal{F}$-free.

Example

Let us illustrate our terminology on a classical example. We denote by S_2 the independent graph on two vertices. A vertex is *simplicial* if its neighborhood is S_2-free, or equivalently is a clique. A graph is *chordal* if it is hole-free, where a *hole* is a chordless cycle of length at least 4.

Theorem 1 (Dirac [6]) *Every chordal graph admits an* $\{S_2\}$*-elimination ordering.*

Theorem 2 (Rose, Tarjan, and Lueker [14]) *There exists a linear-time algorithm that computes an* $\{S_2\}$*-elimination ordering of an input chordal graph.*

LexBFS

To explain the results, we need to define LexBFS. It is a linear time algorithm of Rose, Tarjan, and Lueker [14] whose input is any graph G together with a vertex s and whose output is a linear ordering of the vertices of G starting at s. A linear ordering of the vertices of a graph G is a *LexBFS ordering* if there exists a vertex s of G such that the ordering can be produced by LexBFS when the input is G, s. The order from Theorem 2 is in fact computed by LexBFS. We do not need here to define LexBFS more precisely, because the following result fully characterizes LexBFS orderings.

Theorem 3 (Brandstädt, Dragan, and Nicolai [3]) *An ordering $\prec$ of the vertices of a graph $G = (V, E)$ is a LexBFS ordering if and only if it satisfies the following property: for all $a, b, c \in V$ such that $c \prec b \prec a$, $ca \in E$ and $cb \notin E$ there exists a vertex d in G such that $d \prec c$, $db \in E$ and $da \notin E$.*

Key Results

The following property was introduced by Maffray, Trotignon, and Vušković in [12] (where it was called Property ($\star$)).

Definition 1 Let $\mathcal{F}$ be a set of graphs. A graph G is *locally $\mathcal{F}$-decomposable* if for every vertex v of G, every $F \in \mathcal{F}$ contained in $N(v)$, and every connected component C of $G - N[v]$, there exists $y \in F$ such that y has a non-neighbor in F and no neighbors in C. A class of graphs $\mathcal{C}$ is *locally $\mathcal{F}$-decomposable* if every graph $G \in \mathcal{C}$ is locally $\mathcal{F}$-decomposable.

It is easy to see that if a graph is locally $\mathcal{F}$-decomposable, then so are all its induced subgraphs. Therefore, for all sets of graphs $\mathcal{F}$, the class of graphs that are locally $\mathcal{F}$-decomposable is hereditary. The main result is the following.

Theorem 4 *If $\mathcal{F}$ is a set of non-complete graphs, and G is a locally $\mathcal{F}$-decomposable graph, then every LexBFS ordering of G is an $\mathcal{F}$-elimination ordering.*

First Example of Application

Let us now illustrate how Theorem 4 can be used with the simplest possible set made of non-complete graphs $\mathcal{F} = \{S_2\}$, where S_2 is the independent graph on two vertices. The following is well known and easy to prove.

Lemma 1 *A graph G is locally $\{S_2\}$-decomposable if and only if G is chordal.*

Hence, a proof for Theorems 1 and 2 is easily obtained by using Lemma 1 and Theorem 4.

Sketch of Proof

The proof of Theorem 4 relies mainly on the following.

Theorem 5 (Berry and Bordat [2]) *If G is a non-complete graph and z is the last vertex of a LexBFS ordering of G, then there exists a connected component C of $G - N[z]$ such that for every neighbor x of z, either $N[x] = N[z]$ or $N(x) \cap C \neq \emptyset$.*

Equivalently, if we put z together with its neighbors of the first type, the resultant set of vertices is a clique, a homogeneous set, and its neighborhood is a minimal separator. Such sets are called *moplexes* in [2] and Theorem 5 is stated in term of moplexes in [2]. Note that Theorem 5 can be proved from the following very convenient lemma.

Lemma 2 *Let $\prec$ be a LexBFS ordering of a graph $G = (V, E)$. Let z denote the last vertex in this ordering. Then for all vertices $a, b, c \in V$ such that $c \prec b \prec a$ and $ca \in E$, there exists a path from b to c whose internal vertices are disjoint from $N[z]$.*

Truemper Configurations

To state the next results, we need special types of graphs that are called Truemper configurations. They play an important role in structural graph theory; see [15]. Let us define them. A *3-path configuration* is a graph induced by three internally vertex disjoint paths of length at least 1, $P_1 = x_1 \dots y_1$, $P_2 = x_2 \dots y_2$ and $P_3 = x_3 \dots y_3$, such that either $x_1 = x_2 = x_3$ or x_1, x_2, x_3 are all distinct and pairwise adjacent and either $y_1 = y_2 = y_3$ or y_1, y_2, y_3 are all distinct and pairwise adjacent. Furthermore, the vertices of $P_i \cup P_j$, $i \neq j$ induce a hole. Note that this last condition in the definition implies the following:

- If x_1, x_2, x_3 are distinct (and therefore pairwise adjacent) and y_1, y_2, y_3 are distinct, then the three paths have length at least 1. In this case, the configuration is called a *prism*.
- If $x_1 = x_2 = x_3$ and $y_1 = y_2 = y_3$, then the three paths have length at least 2 (since a path of length 1 would form a chord of the cycle formed by the two other paths). In this case, the configuration is called a *theta*.
- If $x_1 = x_2 = x_3$ and y_1, y_2, y_3 are distinct, or if x_1, x_2, x_3 are distinct and $y_1 = y_2 = y_3$, then at most one of the three paths has length 1, and the others have length at least 2. In this case, the configuration is called a *pyramid*.

A *wheel* (H, v) is a graph formed by a hole H, called the *rim*, and a vertex v, called the *center*, such that the center has at least three neighbors on the rim. A *Truemper configuration* is a graph that is either a prism, a theta, a pyramid, or a wheel.

A *hole* in a graph is a chordless cycle of length at least 4. It is *even* or *odd* according to the parity of the number of its edges. A graph is *universally signable* if it contains no Truemper configuration.

Speeding Up of Known Algorithms

We now state the previously known optimization algorithms for which we get better complexity by applying our method. In each case, we prove the existence of an elimination ordering, compute it with LexBFS, and take advantage of the ordering to solve the problem. Each time, we improve the previously known complexity by at least a factor of n:

- Maximum weighted clique in even-hole-free graphs in time $O(nm)$
- Maximum weighted clique in universally signable graphs in time $O(n + m)$
- Coloring in universally signable graphs in time $O(n + m)$

New Algorithms

We now apply systematically our method to all possible sets made of non-complete graphs of order 3. For each such set $\mathcal{F}$ (there are seven of them), we provide a class with a $\mathcal{F}$-elimination ordering.

To describe the classes of graphs that we obtain, we need to be more specific about wheels. A wheel is a *1-wheel* if for some consecutive vertices x, y, z of the rim, the center is adjacent to y and nonadjacent to x and z. A wheel is a *2-wheel* if for some consecutive vertices x, y, z of the rim, the center is adjacent to x and y and nonadjacent to z. A wheel is a *3-wheel* if for some consecutive vertices x, y, z of the rim, the center is adjacent to x, y and z. Observe that a wheel can be simultaneously a 1-wheel, a 2-wheel, and a 3-wheel. On the other hand, every wheel is a 1-wheel, a 2-wheel, or a 3-wheel. Also, any 3-wheel is either a 2-wheel or a *universal wheel* (i.e., a wheel whose center is adjacent to all vertices of the rim).

Up to isomorphism, there are four graphs on three vertices, and three of them are not complete. These three graphs (namely, the independent graph on three vertices denoted by S_3, the path of length 2 denoted by P_3, and its complement denoted by $\overline{P_3}$) are studied in the next lemma.

Lemma 3 *For a graph G, the following hold:*

(i) G is locally $\{S_3\}$-decomposable if and only if G is {1-wheel, theta, pyramid}-free.
(ii) G is locally $\{P_3\}$-decomposable if and only if G is 3-wheel-free.
(iii) G is locally $\{\overline{P_3}\}$-decomposable if and only if G is {2-wheel, prism, pyramid}-free.

Applying our method then leads to the next result, that is, a description of eight classes of graphs (one of them is the class of chordal graphs, and one of them is the class of universally signable graphs). They are described in Table 1: the second column describes the forbidden induced subgraphs that define the class and the last column describes the neighborhood of the last vertex of a LexBFS ordering.

LexBFS, Structure, and Algorithms, Table 1 Eight classes of graphs

i	Class $\mathcal{C}_i$	$\mathcal{F}_i$	Neighborhood
1	{1-wheel, theta, pyramid}-free		No stable set of size 3
2	3-wheel-free		Disjoint union of cliques
3	{2-wheel, prism, pyramid}-free	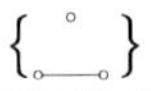	Complete multipartite
4	{1-wheel, 3-wheel, theta, pyramid}-free	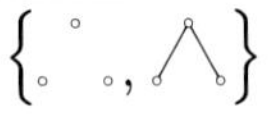	Disjoint union of at most two cliques
5	{1-wheel, 2-wheel, prism, theta, pyramid}-free	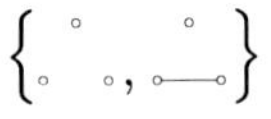	Stable sets of size at most 2 with all possible edges between them
6	{2-wheel, 3-wheel, prism, pyramid}-free		Clique or stable set
7	{wheel, prism, theta, pyramid}-free	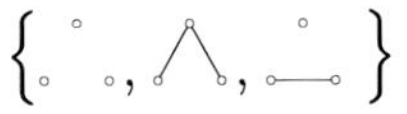	Clique or stable set of size 2
8	hole-free	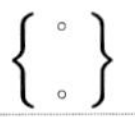	Clique

LexBFS, Structure, and Algorithms, Table 2 Several properties of classes defined in Table 1

i	Max clique	Coloring
1	NP-hard [13]	NP-hard [9]
2	$O(nm)$ [14]	NP-hard [11]
3	$O(nm)$	NP-hard [11]
4	$O(n+m)$	?
5	$O(nm)$	?
6	$O(n+m)$	NP-hard [11]
7	$O(n+m)$	$O(n+m)$
8	$O(n+m)$ [14]	$O(n+m)$ [14]

Theorem 6 *For $i = 1,\ldots,8$, let $\mathcal{C}_i$ and $\mathcal{F}_i$ be the classes defined as in Table 1. For $i = 1,\ldots,8$, the class $\mathcal{C}_i$ is exactly the class of locally $\mathcal{F}_i$-decomposable graphs.*

For each class $\mathcal{C}_i$, we survey in Table 2 the complexity of the maximum clique problem (for which our method provides sometimes a fast algorithm) and of the coloring problem.

Open Problems

Addario-Berry, Chudnovsky, Havet, Reed, and Seymour [1] proved that every even-hole-free graph admits a vertex whose neighborhood is the union of two cliques. We wonder whether this result can be proved by some search algorithm.

Our work suggests that a linear time algorithm for the maximum clique problem might exist in $\mathcal{C}_2$, but we could not find it.

We are not aware of a polynomial time coloring algorithm for graphs in $\mathcal{C}_4$ or $\mathcal{C}_5$.

Since class $\mathcal{C}_1$ generalizes claw-free graphs, it is natural to ask which of the properties of claw-free graphs it has, such as a structural description (see [4]), a polynomial time algorithm for the maximum stable set (see [7]), approximation algorithms for the chromatic number (see [10]), and a polynomial time algorithm for the induced linkage problem (see [8]).

In [5], an $O(nm)$ time algorithm is described for the maximum weighted stable set problem in $\mathcal{C}_7$. Since the class is a simple generalization of chordal graphs, we wonder whether a linear time algorithm exists.

Recommended Reading

1. Addario-Berry L, Chudnovsky M, Havet F, Reed B, Seymour P (2008) Bisimplicial vertices in even-hole-free graphs. J Comb Theory Ser B 98(6):1119–1164
2. Berry A, Bordat JP (1998) Separability generalizes dirac's theorem. Discret Appl Math 84(1–3):43–53
3. Brandstädt A, Dragan F, Nicolai F (1997) LexBFS-orderings and powers of chordal graphs. Discret Math 171(1–3):27–42
4. Chudnovsky M, Seymour P (2008) Clawfree graphs. IV. Decomposition theorem. J Comb Theory Ser B 98(5):839–938

5. Conforti M, Cornuéjols G, Kapoor A, Vušković K (1997) Universally signable graphs. Combinatorica 17(1):67–77
6. Dirac G (1961) On rigid circuit graphs. Abhandlungen aus dem Mathematischen Seminar der Universität Hamburg 25:71–76
7. Faenza Y, Oriolo G, Stauffer G (2011) An algorithmic decomposition of claw-free graphs leading to an $O(n^3)$-algorithm for the weighted stable set problem. In: SODA, San Francisco, pp 630–646
8. Fiala J, Kamiński M, Lidický B, Paulusma D (2012) The k-in-a-path problem for claw-free graphs. Algorithmica 62(1–2):499–519
9. Holyer I (1981) The NP-completeness of some edge-partition problems. SIAM J Comput 10(4): 713–717
10. King A (2009) Claw-free graphs and two conjectures on ω, δ, and χ. PhD thesis, McGill University
11. Maffray F, Preissmann M (1996) On the NP-completeness of the k-colorability problem for triangle-free graphs. Discret Math 162:313–317
12. Maffray F, Trotignon N, Vušković K (2008) Algorithms for square-$3PC(\cdot,\cdot)$-free Berge graphs. SIAM J Discret Math 22(1):51–71
13. Poljak S (1974) A note on the stable sets and coloring of graphs. Commentationes Mathematicae Universitatis Carolinae 15:307–309
14. Rose D, Tarjan R, Lueker G (1976) Algorithmic aspects of vertex elimination on graphs. SIAM J Comput 5:266–283
15. Vušković K (2013) The world of hereditary graph classes viewed through Truemper configurations. In: Blackburn SR, Gerke S, Wildon M (eds) Surveys in combinatorics. London mathematical society lecture note series, vol 409. Cambridge University Press, Cambridge, pp 265–325

Linearity Testing/Testing Hadamard Codes

Sofya Raskhodnikova[1] and Ronitt Rubinfeld[2,3]
[1]Computer Science and Engineering Department, Pennsylvania State University, University Park, State College, PA, USA
[2]Massachusetts Institute of Technology (MIT), Cambridge, MA, USA
[3]Tel Aviv University, Tel Aviv-Yafo, Israel

Keywords

Error-correcting codes; Group homomorphism; Linearity of functions; Property testing; Sublinear-time algorithms

Years and Authors of Summarized Original Work

1993; Blum, Luby, Rubinfeld

Problem Definition

In this article, we discuss the problem of testing linearity of functions and, more generally, testing whether a given function is a group homomorphism. An algorithm for this problem, given by [9], is one of the most celebrated property testing algorithms. It is part of or is a special case of many important property testers for algebraic properties. Originally designed for program checkers and self-correctors, it has found uses in probabilistically checkable proofs (PCPs), which are an essential tool in proving hardness of approximation.

We start by formulating an important special case of the problem, testing the linearity of Boolean functions. A function $f : \{0,1\}^n \to \{0,1\}$ is *linear* if for some $a_1, a_2, \ldots, a_n \in \{0,1\}$,

$$f(x_1, x_2, \ldots, x_n) = a_1 x_1 + a_2 x_2 + \cdots a_n x_n.$$

The operations in this definition are over $\mathbb{F}_2$. That is, given vectors $\mathbf{x} = (x_1, \ldots, x_n)$ and $\mathbf{y} = (y_1, \ldots, y_n)$, where $x_1, \ldots, x_n, y_1, \ldots, y_n \in \{0,1\}$, the vector $\mathbf{x} + \mathbf{y} = (x_1 + y_1 \bmod 2, \ldots, x_n + y_n \bmod 2)$. There is another, equivalent definition of linearity of Boolean functions over $\{0,1\}^n$: a function f is *linear* if for all $x, y \in \{0,1\}^n$,

$$f(x) + f(y) = f(x + y).$$

A generalization of a linear function, defined above, is a group homomorphism. Given two finite groups, $(G, \circ)$ and $(H, \star)$, a *group homomorphism* from G to H is a function $f : G \to H$ such that for all elements $x, y \in G$,

$$f(x) \star f(y) = f(x \circ y).$$

We would like to test (approximately) whether a given function is linear or, more generally, is a group homomorphism. Next, we define the property testing framework [12, 23]. Linearity testing was the first problem studied in this framework. The linearity tester of [9] actually preceded the definition of this framework and served as an inspiration for it. Given a proximity parameter $\epsilon \in (0,1)$, a function is ϵ-far from satisfying a specific property $\mathcal{P}$ (such as being linear or being a group homomorphism) if it has to be modified on at least an ϵ fraction of its domain in order to satisfy $\mathcal{P}$. A function is ϵ-close to $\mathcal{P}$ if it is not ϵ-far from it. A tester for property $\mathcal{P}$ gets a parameter $\epsilon \in (0,1)$ and an oracle access to a function f. It must accept with probability (The choice of error probability in the definition of the tester is arbitrary. Using standard techniques, a tester with error probability 1/3 can be turned into a tester with error probability $\delta \in (0,1/3)$ by repeating the original tester $O(\log \frac{1}{\delta})$ times and taking the majority answer.) at least 2/3 if the function f satisfies property $\mathcal{P}$ and reject with probability at least 2/3 if f is ϵ-far from satisfying $\mathcal{P}$. Our goal is to design an efficient tester for group homomorphism.

Alternative Formulation

Another way of viewing the same problem is in terms of error-correcting codes. Given a function $f : G \to H$, we can form a codeword corresponding to f by listing the values of f on all points in the domain. The *homomorphism* code is the set of all codewords that correspond to homomorphisms from G to H. This is an error-correcting code with large distance because, for two different homomorphisms $f, g : G \to H$, the fraction of points $x \in G$ on which $f(x) = g(x)$ is at most 1/2. In the special case when G is $\{0,1\}^n$ and H is $\{0,1\}$, we get the Hadamard code. Our goal can be formulated as follows: design an efficient algorithm that tests whether a given string is a codeword of a homomorphism code (or ϵ-far from it).

Key Results

The linearity (homomorphism) tester designed by [9] repeats the following test several times, until the desired success probability is reached, and accepts iff all iterations accept.

Algorithm 1: BLR Linearity (Homomorphism) Test

input : Oracle access to an unknown function $f : G \to H$.

1 Pick $x, y \in G$ uniformly and independently at random.
2 Query f on x, y, and $x + y$ to find out $f(x)$, $f(y)$, and $f(x + y)$.
3 **Accept** if $f(x) + f(y) = f(x + y)$; otherwise, **reject**.

Blum et al. [9] and Ben-Or et al. [7] showed that $O(1/\epsilon)$ iterations of the BLR test suffice to get a property tester for group homomorphism. (The analysis in [9] worked for a special case of the problem, and [7] extended it to all groups). It is not hard to prove that $\Omega(1/\epsilon)$ queries are required to test for linearity and, in fact, any nontrivial property, so the resulting tester is optimal in terms of the query complexity and the running time.

Lots of effort went into understanding the rejection probability of the BLR test for functions that are ϵ-far from homomorphisms over various groups and, especially, for the case $F = \{0,1\}^n$ (see [17] and references therein). A nice exposition of the analysis for the latter special case, which follows the Fourier-analytic approach of [5], can be found in the book by [21].

Several works [8, 14, 24–26] showed how to reduce the number of random bits required by homomorphism tests. In the natural implementation of the BLR test, $2\log|G|$ random bits per iteration are used to pick x and y. Shpilka and Wigderson [25] gave a homomorphism test for general groups that needs only $(1+o(1))\log_2|G|$ random bits.

The case when G is a subset of an infinite group, f is a real-valued function, and the oracle query to f returns a finite-precision approximation to $f(x)$ has been considered in [2, 10, 11, 19, 20]. These works gave testers with query complexity independent of the domain size (see [18] for a survey).

Applications

Self-Testing/Correcting Programs

The linearity testing problem was motivated in [9] by applications to self-testing and self-correcting of programs. Suppose you are given a program that is known to be correct on most inputs but has not been checked (or, perhaps, is even known to be incorrect) on remaining inputs. A *self-tester* for f is an algorithm that can quickly verify whether a given program that supposedly computes f is correct on most inputs, without the aid of another program for f that has already been verified. A *self-corrector* for f is an algorithm that takes a program that correctly computes f on most inputs and uses it to correctly compute f on all inputs.

Blum et al. [9] used their linearity test to construct self-testers for programs intended to compute various homomorphisms. Such functions include integer, polynomial, matrix, and modular multiplication and division. Once it is verified that a program agrees on most inputs with some homomorphism, the task of determining whether it agrees with the *correct* homomorphism on most inputs becomes much easier.

For programs intended to compute homomorphisms, it is easy to construct self-correctors: Suppose a program outputs $f(x)$ on input x, where f agrees on most inputs with a homomorphism g. Fix a constant c. Consider the algorithm that, on input x, picks $c \log 1/\delta$ values y from the domain G uniformly at random, computes $f(x+y) - f(y)$, and outputs the value that is seen most often, breaking ties arbitrarily. If f is $\frac{1}{8}$-close to g, then, since both y and $x+y$ are uniformly distributed in G, it is the case that for at least 3/4 of the choices of y, both $g(x+y) = f(x+y)$ and $g(y) = f(y)$, in which case $f(x+y) - f(y) = g(x)$. Thus, it is easy to show that there is a constant c such that if f is $\frac{1}{8}$-close to a homomorphism g, then for all x, the above algorithm outputs $g(x)$ with probability at least $1 - \delta$.

Probabilistically Checkable Proofs

We discussed an equivalent formulation of the linearity testing problem in terms of testing whether a given string is a codeword of a Hadamard code. This formulation has been used in proofs of hardness of approximation of some NP-hard problems and to construct PCP systems that can be verified with a few queries (see, e.g., [3, 13]).

The BLR Test as a Building Block

The BLR test has been generalized and extended in many ways, as well as used as a building block in other testers. One generalization, particularly useful in PCP constructions, is to testing if a given function is a polynomial of low degree (see, e.g., [1, 15, 16]). Other generalizations include tests for long codes [6, 13] and tests of linear consistency among multiple functions [4]. An example of an algorithm that uses the BLR test as a building block is a tester by [22] for the *singleton* property of functions $f : \{0,1\}^n \to \{0,1\}$, namely, the property that the function $f(x) = x_i$ for some $i \in [1, n]$.

Open Problems

We discussed that the BLR test can be used to check whether a given string is a Hadamard codeword or far from it. For which other codes can such a check be performed efficiently? In other words, which codes are locally testable? We refer the reader to the entry ▶ Locally Testable Codes.

Which other properties of functions can be efficiently tested in the property testing model? Some examples are given in the entries ▶ Testing Juntas and Related Properties of Boolean Functions and ▶ Monotonicity Testing. Testing properties of graphs is discussed in the entries ▶ Testing Bipartiteness in the Dense-Graph Model and ▶ Testing Bipartiteness of Graphs in Sublinear Time.

Cross-References

▶ Learning Heavy Fourier Coefficients of Boolean Functions
▶ Locally Testable Codes
▶ Monotonicity Testing

► Quantum Error Correction
► Testing Bipartiteness in the Dense-Graph Model
► Testing Bipartiteness of Graphs in Sublinear Time
► Testing if an Array Is Sorted
► Testing Juntas and Related Properties of Boolean Functions

Acknowledgments The first author was supported in part by NSF award CCF-1422975 and by NSF CAREER award CCF-0845701.

Recommended Reading

1. Alon N, Kaufman T, Krivilevich M, Litsyn S, Ron D (2003) Testing low-degree polynomials over GF(2). In: Proceedings of RANDOM'03, Princeton, pp 188–199
2. Ar S, Blum M, Codenotti B, Gemmell P (1993) Checking approximate computations over the reals. In: Proceedings of the Twenty-Fifth Annual ACM Symposium on the Theory of Computing, San Diego, pp 786–795
3. Arora S, Lund C, Motwani R, Sudan M, Szegedy M (1998) Proof verification and the hardness of approximation problems. J ACM 45(3): 501–555
4. Aumann Y, Håstad J, Rabin MO, Sudan M (2001) Linear-consistency testing. J Comput Syst Sci 62(4):589–607
5. Bellare M, Coppersmith D, Håstad J, Kiwi M, Sudan M (1996) Linearity testing over characteristic two. IEEE Trans Inf Theory 42(6):1781–1795
6. Bellare M, Goldreich O, Sudan M (1998) Free bits, PCPs, and nonapproximability—towards tight results. SIAM J Comput 27(3):804–915
7. Ben-Or M, Coppersmith D, Luby M, Rubinfeld R (2008) Non-Abelian homomorphism testing, and distributions close to their self-convolutions. Random Struct Algorithms 32(1):49–70
8. Ben-Sasson E, Sudan M, Vadhan S, Wigderson A (2003) Randomness-efficient low degree tests and short PCPs via epsilon-biased sets. In: Proceedings of the Thirty-Fifth Annual ACM Symposium on the Theory of Computing, San Diego, pp 612–621
9. Blum M, Luby M, Rubinfeld R (1993) Self-testing/correcting with applications to numerical problems. JCSS 47:549–595
10. Ergun F, Kumar R, Rubinfeld R (2001) Checking approximate computations of polynomials and functional equations. SIAM J Comput 31(2):s 550–576
11. Gemmell P, Lipton R, Rubinfeld R, Sudan M, Wigderson A (1991) Self-testing/correcting for polynomials and for approximate functions. In: Proceedings of the Twenty-Third Annual ACM Symposium on Theory of Computing, New Orleans, pp 32–42
12. Goldreich O, Goldwasser S, Ron D (1998) Property testing and its connection to learning and approximation. J ACM 45(4):653–750
13. Håstad J (2001) Some optimal in approximability results. J ACM 48(4):798–859
14. Hastad J, Wigderson A (2003) Simple analysis of graph tests for linearity and PCP. Random Struct Algorithms 22(2):139–160
15. Jutla CS, Patthak AC, Rudra A, Zuckerman D (2009) Testing low-degree polynomials over prime fields. Random Struct Algorithms 35(2): 163–193
16. Kaufman T, Ron D (2006) Testing polynomials over general fields. SIAM J Comput 36(3):779–802
17. Kaufman T, Litsyn S, Xie N (2010) Breaking the epsilon-soundness bound of the linearity test over GF(2). SIAM J Comput 39(5):1988–2003
18. Kiwi M, Magniez F, Santha M (2001) Exact and approximate testing/correcting of algebraic functions: a survey. Electron. Colloq. Comput. Complex. 8(14). http://dblp.uni-trier.de/db/journals/eccc/eccc8.html#ECCC-TR01-014
19. Kiwi M, Magniez F, Santha M (2003) Approximate testing with error relative to input size. JCSS 66(2):371–392
20. Magniez F (2005) Multi-linearity self-testing with relative error. Theory Comput Syst 38(5):573–591
21. O'Donnell R (2014) Analysis of Boolean Functions. Cambridge University Press, New York
22. Parnas M, Ron D, Samorodnitsky A (2002) Testing basic Boolean formulae. SIAM J Discret Math 16(1):20–46
23. Rubinfeld R, Sudan M (1996) Robust characterizations of polynomials with applications to program testing. SIAM J Comput 25(2):252–271
24. Samorodnitsky A, Trevisan L (2000) A PCP characterization of NP with optimal amortized query complexity. In: Proceedings of the Thirty-Second Annual ACM Symposium on Theory of Computing, Portland, pp 191–199
25. Shpilka A, Wigderson A (2006) Derandomizing homomorphism testing in general groups. SIAM J Comput 36(4):1215–1230
26. Trevisan L (1998) Recycling queries in PCPs and in linearity tests. In: Proceedings of the Thirtieth Annual ACM Symposium on the Theory of Computing, Dallas, pp 299–308

Linearizability

Maurice Herlihy
Department of Computer Science, Brown University, Providence, RI, USA

Keywords

Atomicity

Years and Authors of Summarized Original Work

1990; Herlihy, Wing

Problem Definition

An *object* in languages such as Java and C++ is a container for data. Each object provides a set of *methods* that are the only way to to manipulate that object's internal state. Each object has a *class* which defines the methods it provides and what they do.

In the absence of concurrency, methods can be described by a pair consisting of a *precondition* (describing the object's state before invoking the method) and a *postcondition*, describing, once the method returns, the object's state and the method's return value. If, however, an object is shared by concurrent threads in a multiprocessor system, then method calls may overlap in time, and it no longer makes sense to characterize methods in terms of pre- and post-conditions.

Linearizability is a correctness condition for concurrent objects that characterizes an object's concurrent behavior in terms of an "equivalent" sequential behavior. Informally, the object behaves "as if" each method call takes effect instantaneously at some point between its invocation and its response. This notion of correctness has some useful formal properties. First, it is *non-blocking*, which means that linearizability as such never requires one thread to wait for another to complete an ongoing method call. Second, it is *local*, which means that an object composed of linearizable objects is itself linearizable. Other proposed correctness conditions in the literature lack at least one of these properties.

Notation

An execution of a concurrent system is modeled by a *history*, a finite sequence of method *invocation* and *response events*. A *subhistory* of a history H is a subsequence of the events of H. A method invocation is written as $\langle x.m(a^*)A\rangle$, where x is an object, m a method name, a^* a sequence of arguments, and A a thread. A method response is written as $\langle x{:}t(r^*)A\rangle$ where t is a termination condition and r^* is a sequence of result values.

A response *matches* an invocation if their objects and thread names agree. A *method call* is a pair consisting of an invocation and the next matching response. An invocation is *pending* in a history if no matching response follows the invocation. If H is a history, *complete*(H) is the subsequence of H consisting of all matching invocations and responses. A history H is *sequential* if the first event of H is an invocation, and each invocation, except possibly the last, is immediately followed by a matching response.

Let H be a a history. The *thread subhistory* $H|P$ is the subsequence of events in H with thread name P. The *object subhistory* $H|x$ is similarly defined for an object x. Two histories H and H' are *equivalent* if for every thread A, $H|A = H'|A$. A history H is *well-formed* if each thread subhistory $H|A$ of H is sequential. Notice that thread subhistories of a well-formed history are always sequential, but object subhistories need not be.

A *sequential specification* for an object is a prefix-closed set of sequential object histories that defines that object's *legal* histories. A sequential history H is *legal* if each object subhistory is legal. A method is *total* if it is defined for every object state, otherwise it is *partial*. (For example, a *deq()* method that blocks on an empty queue is partial, while one that throws an exception is total.)

A history H defines an (irreflexive) partial order $\rightarrow_H$ on its method calls: $m_0 \rightarrow_H m_1$ if the result event of m_0 occurs before the invocation event of m_1. If H is a sequential history, then $\rightarrow_H$ is a total order.

Let H be a history and x an object such that $H|x$ contains method calls m_0 and m_1. A call $m_0 \rightarrow_x m_1$ if m_0 precedes m_1 in $H|x$. Note that $\rightarrow_x$ is a total order.

Informally, linearizability requires that each method call appear to "take effect" instantaneously at some moment between its invocation and response. An important implication of this definition is that method calls that do not overlap cannot be reordered:

linearizability preserves the "real-time" order of method calls. Formally,

Definition 1 A history H is *linearizable* if it can be extended (by appending zero or more response events) to a history H' such that:

- L1 *complete*(H') is equivalent to a legal sequential history S, and
- L2 If method call m_0 precedes method call m_1 in H, then the same is true in S.

S is called a *linearization* of H. (H may have multiple linearizations.) Informally, extending H to H' captures the idea that some pending invocations may have taken effect even though their responses have not yet been returned to the caller.

Key Results

The Locality Property

A property is *local* if all objects collectively satisfy that property provided that each individual object satisfies it.

Linearizability is local:

Theorem 1 *H is linearizable if and only if $H|x$ is linearizable for ever object x.*

Proof The "only if" part is obvious.

For each object x, pick a linearization of $H|x$. Let R_x be the set of responses appended to $H|x$ to construct that linearization, and let $\rightarrow_x$ be the corresponding linearization order. Let H' be the history constructed by appending to H each response in R_x.

The $\rightarrow_H$ and $\rightarrow_x$ orders can be "rolled up" into a single partial order. Define the relation $\rightarrow$ on method calls of $complete(H')$: For method calls m and $\bar{m}$, $m \rightarrow \bar{m}$ if there exist method calls $m_0, \ldots, m_n$, such that $m = m_0, \bar{m} = m_n$, and for each i between 0 and $n-1$, either $m_i \rightarrow_x m_{i+1}$ for some object x, or $m_i \rightarrow_H m_{i+1}$.

It turns out that $\rightarrow$ is a partial order. Clearly, $\rightarrow$ is transitive. It remains to be shown that $\rightarrow$ is anti-reflexive: for all x, it is false that $x \rightarrow x$.

The proof proceeds by contradiction. If not, then there exist method calls $m_0, \ldots, m_n$, such that $m_0 \rightarrow m_1 \rightarrow \cdots \rightarrow m_n, m_n \rightarrow m_0$, and each pair is directly related by some $\rightarrow_x$ or by $\rightarrow_H$.

Choose a cycle whose length is minimal. Suppose all method calls are associated with the same object x. Since $\rightarrow_x$ is a total order, there must exist two method calls m_{i-1} and m_i such that $m_{i-1} \rightarrow_H m_i$ and $m_i \rightarrow_x m_{i-1}$, contradicting the linearizability of x.

The cycle must therefore include method calls of at least two objects. By reindexing if necessary, let m_1 and m_2 be method calls of distinct objects. Let x be the object associated with m_1. None of $m_2, \ldots, m_n$ can be a method call of x. The claim holds for m_2 by construction. Let m_i be the first method call in $m_3, \ldots, m_n$ associated with x. Since m_{i-1} and m_i are unrelated by $\rightarrow_x$, they must be related by $\rightarrow_H$, so the response of m_{i-1} precedes the invocation of m_i. The invocation of m_2 precedes the response of m_{i-1}, since otherwise $m_{i-1} \rightarrow_H m_2$, yielding the shorter cycle $m_2, \ldots, m_{i-1}$. Finally, the response of m_1 precedes the invocation of m_2, since $m_1 \rightarrow_H m_2$ by construction. It follows that the response to m_1 precedes the invocation of m_i, hence $m_1 \rightarrow_H m_i$, yielding the shorter cycle $m_1, m_i, \ldots, m_n$.

Since m_n is not a method call of x, but $m_n \rightarrow m_1$, it follows that $m_n \rightarrow_H m_1$. But $m_1 \rightarrow_H m_2$ by construction, and because $\rightarrow_H$ is transitive, $m_n \rightarrow_H m_2$, yielding the shorter cycle $m_2, \ldots, m_n$, the final contradiction. □

Locality is important because it allows concurrent systems to be designed and constructed in a modular fashion; linearizable objects can be implemented, verified, and executed independently. A concurrent system based on a non-local correctness property must either rely on a centralized scheduler for all objects, or else satisfy additional constraints placed on objects to ensure that they follow compatible scheduling protocols. Locality should not be taken for granted; as discussed below, the literature includes proposals for alternative correctness properties that are not local.

The Non-blocking Property

Linearizability is a *non-blocking* property: a pending invocation of a total method is never required to wait for another pending invocation to complete.

Theorem 2 *Let inv(m) be an invocation of a total method. If* $\langle x\ invP\rangle$ *is a pending invocation in a linearizable history H, then there exists a response* $\langle x\ resP\rangle$ *such that* $H \cdot \langle x\ resP\rangle$ *is linearizable.*

Proof Let S be any linearization of H. If S includes a response $\langle x\ resP\rangle$ to $\langle x\ invP\rangle$, the proof is complete, since S is also a linearization of $H \cdot \langle x\ resP\rangle$. Otherwise, $\langle x\ invP\rangle$ does not appear in S either, since linearizations, by definition, include no pending invocations. Because the method is total, there exists a response $\langle x\ resP\rangle$ such that

$$S' = S \cdot \langle x\ invP\rangle \cdot \langle x\ res\ P\rangle$$

is legal. S', however, is a linearization of $H \cdot \langle x\ resP\rangle$, and hence is also a linearization of H. □

This theorem implies that linearizability by itself never forces a thread with a pending invocation of a total method to block. Of course, blocking (or even deadlock) may occur as artifacts of particular implementations of linearizability, but it is not inherent to the correctness property itself. This theorem suggests that linearizability is an appropriate correctness condition for systems where concurrency and real-time response are important. Alternative correctness conditions, such as serializability [1] do not share this non-blocking property.

The non-blocking property does not rule out blocking in situations where it is explicitly intended. For example, it may be sensible for a thread attempting to dequeue from an empty queue to block, waiting until another thread enqueues an item. The queue specification captures this intention by making the `deq()` method's specification partial, leaving it's effect undefined when applied to an empty queue. The most natural concurrent interpretation of a partial sequential specification is simply to wait until the object reaches a state in which the method is defined.

Other Correctness Properties

Sequential Consistency [4] is a weaker correctness condition that requires Property *L1* but not *L2*: method calls must appear to happen in some one-at-a-time, sequential order, but calls that do not overlap can be reordered. Every linearizable history is sequentially consistent, but not vice versa. Sequential consistency permits more concurrency, but it is not a local property: a system composed of multiple sequentially-consistent objects is not itself necessarily sequentially consistent.

Much work on databases and distributed systems uses *serializability* as the basic correctness condition for concurrent computations. In this model, a *transaction* is a "thread of control" that applies a finite sequence of methods to a set of objects shared with other transactions. A history is *serializable* if it is equivalent to one in which transactions appear to execute sequentially, that is, without interleaving. A history is *strictly serializable* if the transactions' order in the sequential history is compatible with their precedence order: if every method call of one transaction precedes every method call of another, the former is serialized first. (Linearizability can be viewed as a special case of strict serializability where transactions are restricted to consist of a single method applied to a single object.)

Neither serializability nor strict serializability is a local property. If different objects serialize transactions in different orders, then there may be no serialization order common to all objects. Serializability and strict serializability are *blocking* properties: Under certain circumstances, a transaction may be unable to complete a pending method without violating serializability. A *deadlock* results if multiple transactions block one another. Such transactions must be rolled back and restarted, implying that additional mechanisms must be provided for that purpose.

Applications

Linearizability is widely used as the basic correctness condition for many concurrent data structure algorithms [5], particularly for lock-free and wait-free data structures [2]. Sequential consistency is widely used for describing low-level systems such as hardware memory interfaces. Serializability and strict serializability are widely used for database systems in which it must be easy for application programmers to preserve complex application-specific invariants spanning multiple objects.

Open Problems

Modern multiprocessors often support very weak models of memory consistency. There are many open problems concerning how to model such behavior, and how to ensure linearizable object implementations on top of such architectures.

Cross-References

- ▸ Concurrent Programming, Mutual Exclusion
- ▸ Registers

Recommended Reading

The notion of Linearizability is due to Herlihy and Wing [3], while Sequential Consistency is due to Lamport [4], and serializability to Eswaran et al. [1].

1. Eswaran KP, Gray JN, Lorie RA, Traiger IL (1976) The notions of consistency and predicate locks in a database system. Commun ACM 19(11):624–633. doi:10.1145/360363.360369
2. Herlihy M (1991) Wait-free synchronization. ACM Trans Program Lang Syst (TOPLAS) 13(1): 124–149
3. Herlihy MP, Wing JM (1990) Linearizability: a correctness condition for concurrent objects. ACM Trans Program Lang Syst (TOPLAS) 12(3): 463–492
4. Lamport L (1979) How to make a multiprocessor computer that correctly executes multiprocess programs. IEEE Trans Comput C-28(9):690
5. Vafeiadis V, Herlihy M, Hoare T, Shapiro M (2006) Proving correctness of highly-concurrent linearisable objects. In: PPoPP'06: proceedings of the eleventh ACM SIGPLAN symposium on principles and practice of parallel programming, pp 129–136. doi:10.1145/1122971.1122992

List Decoding near Capacity: Folded RS Codes

Atri Rudra
Department of Computer Science and Engineering, State University of New York, Buffalo, NY, USA

Keywords

Decoding; Error correction

Years and Authors of Summarized Original Work

2006; Guruswami, Rudra

Problem Definition

One of the central trade-offs in the theory of error-correcting codes is the one between the amount of redundancy needed and the fraction of errors that can be corrected. (This entry deals with the adversarial or worst-case model of errors–no assumption is made on how the errors and error locations are distributed beyond an upper bound on the total number of errors that may be caused.) The redundancy is measured by the *rate* of the code, which is the ratio of the the number of information symbols in the message to that in the codeword – thus, for a code with encoding function $E : \Sigma^k \rightarrow \Sigma^n$, the rate equals k/n. The *block length* of the code equals n, and Σ is its *alphabet*.

The goal in *decoding* is to find, given a noisy received word, the actual codeword that it could have possibly resulted from. If the target is to correct a fraction ρ of errors (ρ will be called the error-correction radius), then this amounts to finding codewords within (normalized Hamming) distance ρ from the received word. We are guaranteed that there will be a unique such codeword provided the distance between *every* two distinct codewords is at least 2ρ, or in other words the relative distance of the code is at least 2ρ. However, since the relative distance δ of a code must satisfy $\delta \leq 1 - R$ where R is the rate of the code (by the Singleton bound), if one insists on an unique answer, the best trade-off between ρ and R is $\rho = \rho_U(R) = (1 - R)/2$. But this is an overly pessimistic estimate of the error-correction radius, since the way Hamming spheres pack in space, for *most* choices of the received word there will be at most one codeword within distance ρ from it even for ρ much greater than $\delta/2$. Therefore, *always* insisting on a unique answer will preclude decoding most such received words owing to a few pathological received words that have more than one codeword within distance roughly $\delta/2$ from them.

A notion called list decoding, that dates back to the late 1950s [1, 9], provides a clean way to get around this predicament, and yet deal with worst-case error patterns. Under list decoding, the decoder is required to output a list of all codewords within distance ρ from the received word. Let us call a code C (ρ, L) *-list decodable* if the number of codewords within distance ρ of any received word is at most L. To obtain better trade-offs via list decoding, (ρ, L)-list decodable codes are needed where L is bounded by a polynomial function of the block length, since this an *a priori* requirement for polynomial time list decoding. How large can ρ be as a function of R for which such (ρ, L)-list decodable codes exist? A standard random coding argument shows that $\rho \geq 1 - R - o(1)$ can be achieved over large enough alphabets, cf. [2, 10], and a simple counting argument shows that ρ must be at most $1 - R$. Therefore the *list decoding capacity*, i.e., the information-theoretic limit of list decodability, is given by the trade-off $\rho_{\text{cap}}(R) = 1 - R = 2\rho_U(R)$. Thus list decoding holds the promise of correcting *twice* as many errors as unique decoding, for *every* rate. The above-mentioned list decodable codes are non-constructive. In order to realize the potential of list decoding, one needs explicit constructions of such codes, and on top of that, polynomial time algorithms to perform list decoding.

Building on works of Sudan [8], Guruswami and Sudan [6] and Parvaresh and Vardy [7], Guruswami and Rudra [5] present codes that get arbitrarily close to the list decoding capacity $\rho_{\text{cap}}(R)$ for every rate. In particular, for every $1 > R > 0$ and every $\epsilon > 0$, they give *explicit* codes of rate R together with polynomial time list decoding algorithm that can correct up to a fraction $1 - R - \epsilon$ of errors. These are the first explicit codes (with efficient list decoding algorithms) that get arbitrarily close to the list decoding capacity for *any* rate.

Description of the Code

Consider a Reed–Solomon (RS) code $C = \mathsf{RS}_{\mathbb{F},\mathbb{F}^*}[n, k]$ consisting of evaluations of degree k polynomials over some finite field $\mathbb{F}$ at the set $\mathbb{F}^*$ of nonzero elements of $\mathbb{F}$. Let $q = |\mathbb{F}| = n + 1$. Let γ be a generator of the multiplicative group $\mathbb{F}^*$, and let the evaluation points be ordered as $1, \gamma, \gamma^2, \ldots, \gamma^{n-1}$. Using all nonzero field elements as evaluation points is one of the most commonly used instantiations of Reed–Solomon codes.

Let $m \geq 1$ be an integer parameter called the *folding parameter*. For ease of presentation, it will assumed that m divides $n = q - 1$.

Definition 1 (Folded Reed–Solomon Code) The m-folded version of the RS code C, denoted $\mathsf{FRS}_{\mathbb{F},\gamma,m,k}$, is a code of block length $N = n/m$ over $\mathbb{F}^m$. The encoding of a message $f(X)$, a polynomial over $\mathbb{F}$ of degree at most k, has as its j'th symbol, for $0 \leq j < n/m$, the m-tuple $(f(\gamma^{jm}), f(\gamma^{jm+1}), \cdots, f(\gamma^{jm+m-1}))$. In other words, the codewords of $C' = \mathsf{FRS}_{\mathbb{F},\gamma,m,k}$ are in one-one correspondence with those of the RS code C and are obtained by bundling together consecutive m-tuple of symbols in codewords of C.

Key Results

The following is the main result of Guruswami and Rudra.

Theorem 1 ([5]) *For every $\epsilon > 0$ and $0 < R < 1$, there is a family of folded Reed–Solomon codes that have rate at least R and which can be list decoded up to a fraction $1 - R - \epsilon$ of errors in time (and outputs a list of size at most) $(N/\epsilon^2)^{O(\epsilon^{-1}\log(1/R))}$ where N is the block length of the code. The alphabet size of the code as a function of the block length N is $(N/\epsilon^2)^{O(1/\epsilon^2)}$.*

The result of Guruswami and Rudra also works in a more general setting called *list recovering*, which is defined next.

Definition 2 (List Recovering) A code $C \subseteq \Sigma^n$ is said to be (ζ, l, L)-list recoverable if for every sequence of sets $S_1,\cdots,S_n$ where each $S_i \subseteq \Sigma$ has at most l elements, the number of codewords $c \in C$ for which $c_i \in S_i$ for at least ζn positions $i \in \{1, 2, \ldots, n\}$ is at most L.

A code $C \subseteq \Sigma^n$ is said to (ζ, l)-list recoverable in polynomial time if it is $(\zeta, l, L(n))$-list recoverable for some polynomially bounded function $L(\cdot)$, and moreover there is a polynomial time algorithm to find the at most $L(n)$ codewords that are solutions to any $(\zeta, l, L(n))$-list recovering instance.

Note that when $l = 1$, $(\zeta, 1, \cdot)$-list recovering is the same as list decoding up to a $(1 - \zeta)$ fraction of errors. Guruswami and Rudra have the following result for list recovering.

Theorem 2 ([5]) *For every integer $l \geq 1$, for all R, $0 < R < 1$ and $\epsilon > 0$, and for every prime p, there is an explicit family of folded Reed–Solomon codes over fields of characteristic p that have rate at least R and which can be $(R + \epsilon, l)$-list recovered in polynomial time. The alphabet size of a code of block length N in the family is $(N/\epsilon^2)^{O(\epsilon^{-2}\log l/(1-R))}$.*

Applications

To get within ϵ of capacity, the codes in Theorem 1 have alphabet size $N^{\Omega(1/\epsilon^2)}$ where N is the block length. By concatenating folded RS codes of rate close to 1 (that are list recoverable) with suitable inner codes followed by redistribution of symbols using an expander graph (similar to a construction for linear-time unique decodable codes in [3]), one can get within ϵ of capacity with codes over an alphabet of size $2^{O(\epsilon^{-4}\log(1/\epsilon))}$. A counting argument shows that codes that can be list decoded efficiently to within ϵ of the capacity need to have an alphabet size of $2^{\Omega(1/\epsilon)}$.

For binary codes, the list decoding capacity is known to be $\rho_{\text{bin}}(R) = H^{-1}(1 - R)$ where $H(\cdot)$ denotes the binary entropy function. No explicit constructions of binary codes that approach this capacity are known. However, using the Folded RS codes of Guruswami Rudra in a natural concatenation scheme, one can obtain polynomial time constructable binary codes of rate R that can be list decoded up to a fraction $\rho_{\text{Zyab}}(R)$ of errors, where $\rho_{\text{Zyab}}(R)$ is the "Zyablov bound".

See [5] for more details.

Open Problems

The work of Guruswami and Rudra could be improved with respect to some parameters. The size of the list needed to perform list decoding to a radius that is within ϵ of capacity grows as $N^{O(\epsilon^{-1}\log(1/R))}$ where N and R are the block length and the rate of the code respectively. It remains an open question to bring this list size down to a constant independent of n (the existential random coding arguments work with a list size of $O(1/\epsilon)$). The alphabet size needed to approach capacity was shown to be a constant independent of N. However, this involved a brute-force search for a rather large (inner) code, which translates to a construction time of about $N^{O(\epsilon^{-2}\log(1/\epsilon))}$ (instead of the ideal construction time where the exponent of N does not depend on ϵ). Obtaining a "direct" algebraic construction over a constant-sized alphabet, such as the generalization of the Parvaresh-Vardy framework to algebraic-geometric codes in [4], might help in addressing these two issues.

Finally, constructing binary codes that approach list decoding capacity remains open.

Cross-References

- ▶ Decoding Reed–Solomon Codes
- ▶ Learning Heavy Fourier Coefficients of Boolean Functions
- ▶ LP Decoding

Recommended Reading

1. Elias P (1957) List decoding for noisy channels. Technical report 335, Research Laboratory of Electronics MIT
2. Elias P (1991) Error-correcting codes for list decoding. IEEE Trans Inf Theory 37:5–12
3. Guruswami V, Indyk P (2005) Linear-time encodable/decodable codes with near-optimal rate. IEEE Trans Inf Theory 51(10):3393–3400
4. Guruswami V, Patthak A (2006) Correlated Algebraic-Geometric codes: improved list decoding over bounded alphabets. In: Proceedings of the 47th annual IEEE symposium on foundations of computer science (FOCS), Oct 2006, Berkley, pp 227–236
5. Guruswami V, Rudra A (2006) Explicit capacity-achieving list-decodable codes. In: Proceedings of the 38th annual ACM symposium on theory of computing, May 2006, Seattle, pp 1–10
6. Guruswami V, Sudan M (1999) Improved decoding of Reed–Solomon and algebraic-geometric codes. IEEE Trans Inf Theory 45:1757–1767
7. Parvaresh F, Vardy A (2005) Correcting errors beyond the Guruswami–Sudan radius in polynomial time. In: Proceedings of the 46th annual IEEE symposium on foundations of computer science, Pittsburgh, pp 285–294
8. Sudan M (1997) Decoding of Reed-Solomon codes beyond the error-correction bound. J Complex 13(1):180–193
9. Wozencraft JM (1958) List decoding. Quarterly progress report, research laboratory of electronics. MIT 48:90–95
10. Zyablov VV, Pinsker MS (1981) List cascade decoding. Probl Inf Transm 17(4):29–34 (in Russian); pp 236–240 (in English) (1982)

List Ranking

Riko Jacob[1,2], Ulrich Meyer[3], and Laura Toma[4]
[1]Institute of Computer Science, Technical University of Munich, Munich, Germany
[2]IT University of Copenhagen, Copenhagen, Denmark
[3]Department of Computer Science, Goethe University Fankfurt am Main, Frankfurt, Germany
[4]Department of Computer Science, Bowdoin College, Brunswick, ME, USA

Keywords

3-coloring; Euler Tour; External memory algorithms; Graph algorithms; Independent set; PRAM algorithms; Time-forward processing

Years and Authors of Summarized Original Work

1995; Chiang, Goodrich, Grove, Tamassia, Vengroff, Vitter

Problem Definition

Let L be a linked list of n vertices $x_1, x_2, \ldots, x_n$ such that every vertex x_i stores a pointer $succ(x_i)$ to its successor in L. As with any linked list, we assume that no two vertices have the same successor, any vertex can reach the tail of the list by following successor pointers, and we denote the *head* of the list the vertex that no other vertex in L points to and the *tail* the vertex whose successor is $null$. Given the head x_h of L, the *list-ranking* problem is to find the *rank*, or distance, of each vertex x_i in L from the head of L: that is, $rank(x_h) = 0$ and $rank(succ(x_i)) = rank(x_i) + 1$; refer to Fig. 1. A generalization of this problem is to consider that each vertex x_i stores, in addition to $succ(x_i)$, a weight $w(x_i)$; in this case the list is given as a set of tuples $\{(x_i, w(x_i), succ(x_i))\}$ and we want to compute

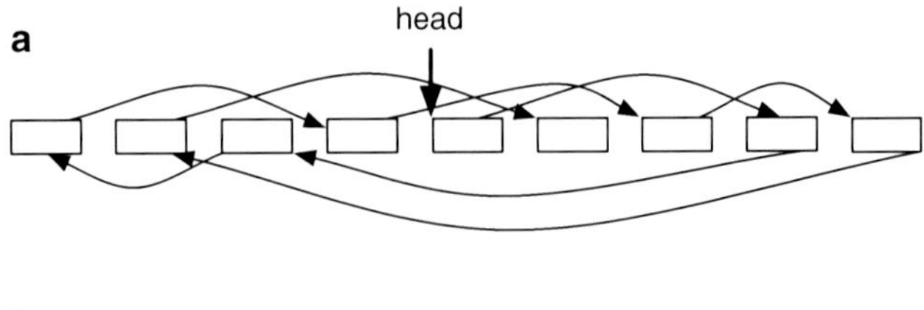

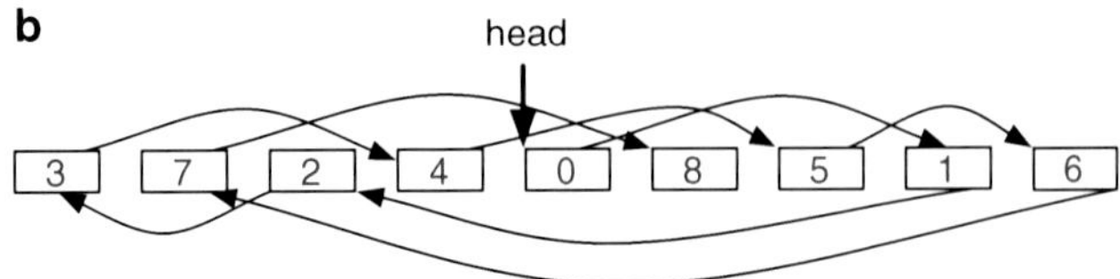

List Ranking, Fig. 1 (**a**) An instance of the LR problem, with the ranks shown in (**b**)

$rank(x_h) = w(x_h)$, and $rank(succ(x_i)) = rank(x_i) + w(succ(x_i))$.

Key Results

List ranking is one of the fundamental problems in the external memory model (EM or I/O-model) which requires nontrivial techniques and illustrates the differences (and connection) between the models of computation, namely, the random access machine (RAM), its parallel version PRAM, and the parallel external memory model (PEM). It also illustrates how ideas from parallel algorithms are used in serial external memory algorithms and the idea of using geometrically decreasing sizes to get an algorithm that in total is as fast as sorting. Furthermore list ranking is the main ingredient in the Euler Tour technique, which is one of the main techniques for obtaining I/O-efficient solutions for fundamental problems on graphs and trees.

In internal memory, list ranking can be solved in $O(n)$ time with a straightforward algorithm that starts from the head of the list and follows successor pointers. In external memory, the same algorithm may use $\Omega(n)$ I/Os – the intuition is that in the worst case, the vertices are arranged in such an order that following a successor pointer will always require loading a new block from disk (one I/O). If the vertices were arranged in order of their ranks, then traversing the list would be trivial, but arranging them in this order would require knowing their ranks, which is exactly the problem we are trying to solve.

The external memory model (see also chapter ▸ I/O-Model) has been extended to a parallel version, the PEM model, which models a private cache shared memory architecture, where the shared memory (external memory) is organized in blocks of size B, each cache (internal memory) has size M, and there are P processors. The cost of executing an algorithm is the number of parallel I/Os. When we use asymptotic notation, we think of M, B, and P as arbitrary nondecreasing functions depending on n, the number of elements constituting the input. Similar to the PRAM, there are different versions of the model, depending on the possibility of concurrent read and write. In the following we assume a Concurrent Read Exclusive Write (CREW) policy [2]. For $M = 2, B = 1$ the PEM model is a PRAM model, and for $P = 1$ the EM model.

Similar to the EM model (see chapter ▸ External Sorting and Permuting), sorting is an important building block for the PEM having complexity $\text{sort}(n) = O(\frac{n}{PB} \log_{\text{d}} \frac{n}{B})$ for $\text{d} = \max\{2, \min\{\frac{n}{PB}, \frac{M}{B}\}\}$ if $P \leq \frac{n}{B}$ [7].

The complexity of permuting in the PEM model is that of sorting unless the direct algorithm with $O(\frac{n}{P})$ I/Os is faster, i.e., for B smaller than the logarithmic term.

In the RAM model, permuting and list ranking have scanning complexity, while (comparison-based) sorting is more expensive. In the PRAM model, this changes slightly: permuting still has scanning complexity, while list ranking has sorting complexity (like any function where a single output depends on all inputs it needs $n/P + \log n$ time). In the external memory model, list ranking has permuting complexity which is higher than scanning unless $B = O(1)$, and usually it is sorting complexity. In contrast, for many processors and large B (more precisely $P = M = 2B = \sqrt{n}$) and a restricted model where the input is only revealed to the algorithm if certain progress has been made, list ranking has complexity $\Omega(\log^2 n)$ [8].

List-Ranking Algorithms in Parallel External Memory

A solution for list ranking that runs in $O(\text{sort}(n))$ I/Os was described by Chiang et al. [4]. The general idea is based on a PRAM algorithm by Cole and Vishkin and consists of the following steps:

1. Find an independent set I of L (a set of vertices such that no two vertices in I are adjacent in L) consisting of $\Theta(n)$ vertices.
2. Compute a new list $L - I$ by removing the vertices in I from L; that is, all vertices x in I are bridged out: let y be the vertex with $succ(y) = x$, and then we set $succ(y) := succ(x)$; additionally the weight of x is added to the weight of $succ(x)$. This ensures that the rank of any vertex in $L - I$ is the same as its rank in L.
3. Compute the ranks recursively on $L - I$.
4. Compute the ranks of the vertices in I from the ranks of the neighbors in $L - I$.

The key idea of the algorithm is finding an independent set of size $\Omega(c \cdot n)$ for some constant $c \in (0, 1)$ and thus recursing on a list of size $O((1 - c) \cdot n)$. The first step, finding a large independent set, can be performed in $O(\text{sort}(n))$ I/Os and is described in more detail below. The second and fourth step can be performed in a couple of scanning and sorting passes in overall $O(\text{sort}(n))$ I/Os. For example, to update the weights of vertices in $L - I$, it suffices to sort the vertices in I by their successor, sort the vertices in $L - I$ by their vertex ID, and then scan the two sorted lists, and for each pair (x', x) in I with $x = succ(x')$, and $(x, succ(x))$ in $L - I$, we update the weight of x to include the deleted vertex x': $w(x) = w(x) + w(x')$. Similarly, to update the successors of vertices in $L - I$, it suffices to sort I by vertex ID, sort the vertices in $L - I$ by their successor, and then scan the two sorted lists. Once the ranks of the vertices in $L - I$ are computed, it suffices to sort I by vertex ID, sort $L - I$ by successor, and then scan the two lists to update the rank of each vertex $x \in I$. Overall, the I/O-complexity of this list-ranking algorithm is given by the recurrence $T(n) \leq O(\text{sort}(n)) + T(c \cdot n)$, for some constant $c \in (0, 1)$, with solution $O(\text{sort}(n))$ I/Os. This is due to the convexity of the I/O behavior of sorting, i.e., $\text{sort}(c \cdot n) \leq c \cdot \text{sort}(n)$. The algorithm can be used in the parallel setting as well because the scanning here works on pairs of elements that are stored in neighboring cells. For moderately large number of processors, sorting the original instance still dominates the overall running times, and more processors lead to an additional log factor in the number of parallel I/Os [8].

Computing a Large Independent Set of L

There are several algorithms for finding an independent set of a list L that run in $O(\text{sort}(n))$ I/Os. The simplest one is a randomized algorithm by Chiang et al. [4], based on a PRAM algorithm by Anderson and Miller. The idea is to flip a coin for each vertex and then select the vertices whose coin came up heads and their successor's coin came up tail. This produces an independent set of expected size $(n - 1)/4 = \Theta(n)$.

In the serial setting, a different way to compute an independent set of L is to 3-color the vertices in L (assign one of the three colors to each vertex such that no two adjacent vertices have same color) and then pick the most popular of the three colors, an independent set of at least $n/3$ vertices. This can be implemented in $O(\text{sort}(n))$ I/Os using time-forward processing [4]. A more direct algorithm, also based on time-forward processing, is described in [11].

In the parallel setting, time-forward processing is not available. Instead deterministic coin tossing can be used. With permuting complexity, any k-coloring of a list can be transformed to a $\log k$ coloring by what is known as deterministic coin tossing [5]. This immediately leads to a parallel deterministic algorithm with an additional

$\log^* n$ factor in the number of parallel I/Os. Alternatively, as long as sorting is still convex in n, a technique called delayed pointer processing [3] can be used. This can be understood as a parallel version of time-forward processing for a DAG of depth $\log \log n$.

Alternatively, all the CREW-PRAM list-ranking algorithm can be executed on the PEM leading to one parallel I/O per step.

Lower Bounds

There is a permuting complexity lower bound for list ranking, showing that the above-explained algorithm is asymptotically optimal for a large range of the parameters. This was sketched in [4] for the serial case and made precise in [8] by the indivisibility assumption that edges have to be treated as atoms and extended to the parallel case. Observe that in the PEM model, superlinear speedups in the number of processors are possible (the overall available fast memory increases), and hence a lower bound for the serial case does not imply a good lower bound the parallel case.

If the number of processors is high (or for parallel computational models where permuting is easy like BSP or map-reduce), list ranking seems to become more difficult than permuting. All known algorithms are a factor $O(\log n)$ more expensive than permuting. One attempt at an explanation is a lower bound in the mentioned setting where the instance is only gradually revealed (depending on the algorithm already having solved certain other parts of the instance) [8]. For this particular parameter setting, the lower bound shows that the described sorting-based algorithm is optimal.

Applications

List ranking in external memory is particularly useful in connection with *Euler tours* [4]. An Euler tour of an undirected tree $T = (V, E)$ is a traversal of T that visits every edge twice, once in each direction. Such a traversal is represented as a linear list L of edges and can be obtained I/O-efficiently as follows: after fixing an order of the edges $\{v, w_1\}, \ldots, \{v, w_k\}$ incident to each node v of T, we set the successor of $\{w_i, v\}$ in L to be $\{v, w_{i+1}\}$ and the successor of $\{w_k, v\}$ in L to be $\{v, w_1\}$. We break the resulting circular list at some root node r by choosing an edge $\{v, r\}$ with successor $\{r, w\}$, setting the successor of $\{v, r\}$ to be null and marking $\{r, w\}$ to be the first edge of the traversal list L. List ranking on L is then applied in order to lay out the Euler tour on disk in a way that about B consecutive list elements are kept in each block. As an Euler tour reflects the structure of its underlying tree T, many properties of T can be derived from a few scanning and sorting steps on the edge sequence of the Euler tour once it has been stored in a way suitable for I/O-efficient traversal. In fact, this technique is not restricted to external memory but has already been used earlier [9] for parallel (PRAM) algorithms, where the scanning steps are replaced by parallel prefix computations.

Classic tree problems solved with the Euler tour technique include *tree rooting* (finding the parent-child direction of tree edges after a vertex of an unrooted and undirected tree has been chosen to become the root), assigning *pre-/post-/inorder numbers*, and computing *node levels* or *number of descendants*. Euler tours and hence list ranking are also useful for non-tree graphs. For example, they are a basic ingredient of a *clustering* preprocessing step [1] for I/O-efficient breadth first search (BFS) on sparse undirected graphs G: after obtaining a spanning tree T for G, an Euler tour around T is used in order to deterministically obtain low diameter clusters of G.

Experimental Results

Despite their theoretical sorting complexity I/O bound, external-memory list-ranking implementations based on independent set removal suffer from non-negligible constant factors. For small to medium input sizes featuring $n < M^2/(4 \cdot B)$, Sibeyn modifies his connected components algorithm [10] in order to solve practical list-ranking problems in scanning complexity with a small constant factor ($22 \cdot n/B$ I/Os). The algorithm splits the input list into at most $M/(2 \cdot B)$ subproblems of $M/2$ consecutive node indices each and processes these subproblems in two passes (the first one running from high to low index ranges and the second one vice versa).

For all nodes of the current range, Sibeyn's algorithm follows the links leading to the nodes of the same sublists and updates the information on their final node and the number of links to it. For all nodes with links running outside the current sublist, the required information is requested in a batched fashion from the subproblems containing the nodes to which they are linked. Phase-one-requests from and phase-two-answers to the sublists are processed only when the wave through the data hits the corresponding subproblem. Due to the overall size restriction, a buffer of size $\Theta(B)$ can be kept in main memory for each subproblem in order to facilitate I/O-efficient information transfer between the subproblems.

The implementation of Sibeyn's algorithm in the STXXL [6] framework has been used as a building block in the engineering of many graph traversal algorithms [1]. For example, the improved clustering preprocessing based on list ranking and Euler tours helped in reducing the I/O wait time for BFS by up to two orders of magnitude compared to a previous clustering method.

Cross-References

- ▶ Cache-Oblivious Model
- ▶ External Sorting and Permuting
- ▶ I/O-Model

Recommended Reading

1. Ajwani D, Meyer U (2009) Design and engineering of external memory traversal algorithms for general graphs. In: Lerner J, Wagner D, Zweig KA (eds) Algorithmics of large and complex networks. Springer, Berlin/Heidelberg, pp 1–33
2. Arge L, Goodrich M, Nelson M, Sitchinava N (2008) Fundamental parallel algorithms for private-cache chip multiprocessors. In: SPAA 2008, pp 197–206
3. Arge L, Goodrich M, Sitchinava N (2010) Parallel external memory graph algorithms. In: IPDPS. IEEE, pp 1–11. http://dx.doi.org/10.1109/IPDPS.2010.5470440
4. Chiang Y, Goodrich M, Grove E, Tamassia R, Vengroff D, Vitter J (1995) External memory graph algorithms. In: Proceedings of the 6th annual symposium on discrete algorithms (SODA), San Francisco, pp 139–149
5. Cole R, Vishkin U (1986) Deterministic coin tossing with applications to optimal parallel list ranking. Inf Control 70(1):32–53
6. Dementiev R, Kettner L, Sanders P (2008) STXXL: standard template library for XXL data sets. Software: Pract Exp 38(6): 589–637
7. Greiner G (2012) Sparse matrix computations and their I/O complexity. Dissertation, Technische Universität München, München. http://nbn-resolving.de/urn/resolver.pl?urn:nbn:de:bvb:91-diss-20121123-1113167-0-6
8. Jacob R, Lieber T, Sitchinava N (2014) On the complexity of list ranking in the parallel external memory model. In: Proceedings 39th international symposium on mathematical foundations of computer science (MFCS'14), Budapest. LNCS, vol 8635. Springer, pp 384–395
9. JáJá J (1992) An introduction to parallel algorithms. Addison-Wesley, Reading
10. Sibeyn J (2004) External connected components. In: Proceedings of the 9th Scandinavian workshop on algorithm theory (SWAT), Lecture Notes in Computer Science, vol 3111. Humlebaek, pp 468–479. http://link.springer.com/chapter/10.1007/978-3-540-27810-8_40
11. Zeh N (2002) I/O-efficient algorithms for shortest path related problems. Phd thesis, School of Computer Science, Carleton University

List Scheduling

Leah Epstein
Department of Mathematics, University of Haifa, Haifa, Israel

Keywords

Online scheduling on identical machines

Years and Authors of Summarized Original Work

1966; Graham

Problem Definition

The paper of Graham [8] was published in the 1960s. Over the years, it served as a common example of online algorithms (though the original algorithm was designed as a simple approximation heuristic). The following basic setting is considered.

A sequence of n jobs is to be assigned to m identical machines. Each job should be assigned to one of the machines. Each job has a size associated with it, which can be seen as its processing time or its load. The load of a machine is the sum of sizes of jobs assigned to it. The goal is to minimize the maximum load of any machine, also called the makespan. We refer to this problem as JOB SCHEDULING.

If jobs are presented one by one and each job needs to be assigned to a machine in tur, without any knowledge of future jobs, the problem is called online. Online algorithms are typically evaluated using the (absolute) *competitive ratio*, which is similar to the *approximation ratio* of approximation algorithms. For an algorithm $\mathcal{A}$, we denote its cost by $\mathcal{A}$ as well. The cost of an optimal offline algorithm that knows the complete sequence of jobs is denoted by OPT. The competitive ratio of an algorithm $\mathcal{A}$ is the infimum $\mathcal{R} \geq 1$ such that for any input, $\mathcal{A} \leq \mathcal{R} \cdot \text{OPT}$.

Key Results

In paper [8], Graham defines an algorithm called LIST SCHEDULING (LS). The algorithm receives jobs one by one. Each job is assigned in turn to a machine which has a minimal current load. Ties are broken arbitrarily.

The main result is the following:

Theorem 1 LS *has a competitive ratio of* $2 - \frac{1}{m}$.

Proof Consider a schedule created for a given sequence. Let ℓ denote a job that determines the makespan (that is, the last job assigned to a machine i that has a maximum load), let L denote its size, and let X denote the total size of all other jobs assigned to i. At the time when L was assigned to i, this was a machine of minimum load. Therefore, the load of each machine is at least X. The makespan of an optimal schedule (i.e., a schedule that minimizes the makespan) is the cost of an optimal offline algorithm and thus is denoted by OPT. Let P be the sum of all job sizes in the sequence.

The two following simple lower bounds on OPT can be obtained:

$$\text{OPT} \geq L. \tag{1}$$

$$\text{OPT} \geq \frac{P}{m} \geq \frac{m \cdot X + L}{m} = X + \frac{L}{m}. \tag{2}$$

Inequality (1) follows from the fact that $\{\text{OPT}\}$ needs to run job ℓ and thus at least one machine has a load of at least L. The first inequality in (2) is due to the fact that at least one machine receives at least a fraction $\frac{1}{m}$ of the total size of jobs. The second inequality in (2) follows from the comments above on the load of each machine.

This proves that the makespan of the algorithm, $X + L$ can be bounded as follows:

$$X+L \leq \text{OPT} + \frac{m-1}{m} L \leq \text{OPT} + \frac{m-1}{m} \text{OPT} = (2 - 1/m)\,\text{OPT}. \tag{3}$$

The first inequality in (3) follows from (2) and the second one from (1).

To show that the analysis is tight, consider $m(m-1)$ jobs of size 1 followed by a single job of size m. After the smaller jobs arrive, LS obtains a balanced schedule in which every machine has a load of $m-1$. The additional job increases the makespan to $2m-1$. However, an optimal offline solution would be to assign the smaller jobs to $m-1$ machines and the remaining job to the remaining machine, getting a load of m.

A natural question was whether this bound is best possible. In a later paper, Graham [9] showed that applying LS with a sorted sequence of jobs (by nonincreasing order of sizes) actually gives a better upper bound of $\frac{4}{3} - \frac{1}{3m}$ on the approximation ratio. A polynomial time approximation scheme was given by Hochbaum and Shmoys in [10]. This is the best offline result one could hope for as the problem is known to be NP hard in the strong sense.

As for the online problem, it was shown in [5] that no (deterministic) algorithm has a smaller competitive ratio than $2 - \frac{1}{m}$ for the cases $m = 2$ and $m = 3$. On the other hand, it was shown in a sequence of papers that an algorithm with a smaller competitive ratio can be found for any $m \geq 4$, and even algorithms with a competitive ratio that does not approach 2 for large m were designed.

The best such result is by Fleischer and Wahl [6], who designed a 1.9201-competitive algorithm. Lower bounds of 1.852 and 1.85358 on the competitive ratio of any online algorithm were shown in [1,7]. Rudin [13] claimed a better lower bound of 1.88.

Applications

As the study of approximation algorithms and specifically online algorithms continued, the analysis of many scheduling algorithms used similar methods to the proof above. Below, several variants of the problem where almost the same proof as above gives the exact same bound are mentioned.

Load Balancing of Temporary Tasks

In this problem, the sizes of jobs are seen as loads. Time is a separate axis. The input is a sequence of events, where every event is an arrival or a departure of a job. The set of active jobs at time t is the set of jobs that have already arrived at this time and have not departed yet. The cost of an algorithm at a time t is its makespan at this time. The cost of an algorithm is its maximum cost over time. It turns out that the analysis above can be easily adapted for this model as well. It is interesting to note that in this case, the bound $2 - \frac{1}{m}$ is actually best possible, as shown in [2].

Scheduling with Release Times and Precedence Constraints

In this problem, the sizes represent processing times of jobs. Various versions have been studied. Jobs may have designated release times, which are the times when these jobs become available for execution. In the online scenario, each job arrives and becomes known to the algorithm only at its release time. Some precedence constraints may also be specified, defined by a partial order on the set of jobs. Thus, a job can be run only after its predecessors complete their execution. In the online variant, a job becomes known to the algorithm only after its predecessors have been completed. In these cases, LS acts as follows. Once a machine becomes available, a waiting job that arrived earliest is assigned to it. (If there is no waiting job, the machine is idle until a new job arrives).

The upper bound of $2 - \frac{1}{m}$ on the competitive ratio can be proved using a relation between the cost of an optimal schedule and the amount of time when at least one machine is idle (See [14] for details).

This bound is tight for several cases. For the case where there are release times, no precedence constraints, and processing times (sizes) are not known upon arrival, Shmoys, Wein, and Williamson [15] proved a lower bound of $2 - \frac{1}{m}$. For the case where there are only precedence constraints (no release times, and sizes of jobs are known upon arrival), a lower bound of the same value appeared in [4]. Note that the case with clairvoyant scheduling (i.e., sizes of jobs are known upon arrival), release times, and no precedence constraints is not settled. For $m = 2$, it was shown by Noga and Seiden [11] that the tight bound is $(5 - \sqrt{5})/2 \approx 1.38198$, and the upper bound is achieved using an algorithm that applies waiting with idle machines rather than scheduling a job as soon as possible, as done by LS.

Open Problems

The most challenging open problem is to find the best possible competitive ratio for this basic online problem of job scheduling. The gap between the upper bound and the lower bound is not large, yet it seems very difficult to find the exact bound. A possibly easier question would be to find the best possible competitive ratio for $m = 4$. A lower bound of $\sqrt{3} \approx 1.732$ has been shown by [12], and the currently known upper bound is 1.733 by [3]. Thus, it may be the case that this bound would turn out to be $\sqrt{3}$.

Cross-References

▶ Approximation Schemes for Makespan Minimization
▶ Robust Scheduling Algorithms

Recommended Reading

1. Albers S (1999) Better bounds for online scheduling. SIAM J Comput 29(2):459–473
2. Azar Y, Epstein L (2004) On-line load balancing of temporary tasks on identical machines. SIAM J Discret Math 18(2):347–352
3. Chen B, van Vliet A, Woeginger GJ (1994) New lower and upper bounds for on-line scheduling. Oper Res Lett 16:221–230
4. Epstein L (2000) A note on on-line scheduling with precedence constraints on identical machines. Inf Process Lett 76:149–153
5. Faigle U, Kern W, Turán G (1989) On the performane of online algorithms for partition problems. Acta Cybern 9:107–119
6. Fleischer R, Wahl M (2000) On-line scheduling revisited. J Sched 3:343–353
7. Gormley T, Reingold N, Torng E, Westbrook J (2000) Generating adversaries for request-answer games. In: Proceedings of the 11th symposium on discrete algorithms (SODA2000), San Francisco, pp 564–565
8. Graham RL (1966) Bounds for certain multiprocessing anomalies. Bell Syst Technol J 45:1563–1581
9. Graham RL (1969) Bounds on multiprocessing timing anomalies. SIAM J Appl Math **17**, 263–269
10. Hochbaum DS, Shmoys DB (1987) Using dual approximation algorithms for scheduling problems: theoretical and practical results. J ACM 34(1):144–162
11. Noga J, Seiden SS (2001) An optimal online algorithm for scheduling two machines with release times. Theor Comput Sci 268(1):133–143
12. Rudin JF III, Chandrasekaran R (2003) Improved bounds for the online scheduling problem. SIAM J Comput 32:717–735
13. Rudin JF III (2001) Improved bounds for the online scheduling problem. Ph.D. thesis, The University of Texas at Dallas
14. Sgall J (1998) On-line scheduling. In: Fiat A, Woeginger GJ (eds) Online algorithms: the state of the art. Springer, Berlin/New York, pp 196–231
15. Shmoys DB, Wein J, Williamson DP (1995) Scheduling parallel machines on-line. SIAM J Comput 24:1313–1331

Local Alignment (with Affine Gap Weights)

Henry Leung
Department of Computer Science, The University of Hong Kong, Hong Kong, China

Keywords

Affine gap penalty; Local alignment; Mapping; Pairwise alignment

Years and Authors of Summarized Original Work

1986; Altschul, Erickson

Problem Definition

The pairwise local alignment problem is concerned with identification of a pair of similar substrings from two molecular sequences. This problem has been studied in computer science for four decades. However, most problem models were generally not biologically satisfying or interpretable before 1974. In 1974, Sellers developed a metric measure of the similarity between molecular sequences. [9] generalized this metric to include deletions and insertions of arbitrary length which represent the minimum number of mutational events required to convert one sequence into another.

Given two sequences S and T, a pairwise alignment is a way of inserting space characters '_' in S and T to form sequences S' and T' respectively with the same length. There can be different alignments of two sequences. The score of an alignment is measured by a scoring metric $\delta(x, y)$. At each position i where both x and

y are not spaces, the similarity between $S'[i]$ and $T'[i]$ is measured by $\delta(S'[i], T'[j])$. Usually, $\delta(x, y)$ is positive when x and y are the same and negative when x and y are different. For positions with consecutive space characters, the alignment scores of the space characters are not considered independently; this is because inserting or deleting a long region in molecular sequences is more likely to occur than inserting or deleting several short regions. Smith and Waterman use an affine gap penalty to model the similarity at positions with space characters. They define a consecutive substring with spaces in S' or T' as a gap. For each length l gap, they give a linear penalty $W_k = W_s + l \times W_p$ for some predefined positive constants W_s and W_p. The score of an alignment is the sum of the score at each position i minus the penalties of each gap. For example, the alignment score of the following alignment is $\delta(G, G) + \delta(C, C) + \delta(C, C) + \delta(U, C) + \delta(G, G) - (W_s + 2 \times W_p)$.

$$S : GCCAUUG$$

$$T : GCC\_\_CG$$

The optimal global alignment of sequences S and T is the alignment of S and T with the maximum alignment score.

Sometimes we want to know whether sequences S and T contain similar substrings instead of whether S and T are similar. In this case, they solve the pairwise local alignment problem, which wants to find a substring U in S and another substring V in T such that the global alignment score of U and V is maximized.

Pairwise Local Alignment Problem

Input: Two sequences $S[1 \ldots n]$ and $T[1 \ldots m]$.

Output: A substring U in S and a substring V in T such that the optimal global alignment of U and V is maximized.

O(mn) time and O(mn) space algorithm is based on dynamic programming.

The pairwise local alignment problem can be solved in O(mn) time and O(mn) space by dynamic programming. The algorithm needs to fill in the 4 $m \times n$ tables H, H_N, H_S, and H_T, where each entry takes constant time. The individual meanings of these 4 tables are as follows.

$H(i, j)$: maximum score of the global alignment of U and V over all suffixes U in $S[1 \ldots i]$ and all suffixes V in $T[1 \ldots j]$.

$H_N(i, j)$: maximum score of the global alignment of U and V over all suffixes U in $S[1 \ldots i]$ and all suffixes V in $T[1 \ldots j]$, with the restriction that $S[i]$ and $T[j]$ must be aligned.

$H_S(i, j)$: maximum score of the global alignment of U and V over all suffixes U in $S[1 \ldots i]$ and all suffixes V in $T[1 \ldots j]$, with $S[j]$ aligned with a space character.

$H_T(i, j)$: maximum score of the global alignment of U and V over all suffixes U in $S[1 \ldots i]$ and all suffixes V in $T[1 \ldots j]$, with $T[j]$ aligned with a space character.

The optimal local alignment score of S and T will be $\max\{H(i, j)\}$, and the local alignment of S and T can be found by tracking back table H.

In the tables, each entry can be filled in by the following recursion in constant time.

Basic Step

$$H(i,0) = H(0, j) = 0, \qquad 0 \leqslant i \leqslant n,\ 0 \leqslant i \leqslant m$$

$$H_N(i,0) = H_N(0, j) = -\infty, \qquad 0 \leqslant i \leqslant n,\ 0 \leqslant i \leqslant m$$

$$H_s(i,0) = H_T(0, j) = W_s + W_p, \ 0 \leqslant i \leqslant n,\ 0 \leqslant i \leqslant m$$

$$H_s(0, j) = H_T(i,0) = -\infty, \qquad 0 \leqslant i \leqslant n,\ 0 \leqslant i \leqslant m$$

Recursion Step

$$H(i,j) = \max\{H_N(i,j), H_s(i,j), H_T(i,j), 0\},$$
$$1 \leqslant i \leqslant n, \ 1 \leqslant i \leqslant m$$
$$H_N(i,j) = H(i-1, j-1) + \delta(S[i], T[j]),$$
$$1 \leqslant i \leqslant n, \ 1 \leqslant i \leqslant m$$
$$H_s(i,j) = \max\{H(i-1,j) - (W_s + W_p),$$
$$H_S(i-1,j) - W_p\},$$
$$1 \leqslant i \leqslant n, \ 1 \leqslant i \leqslant m$$
$$H_T(i,j) = \max\{H(i, j-1) - (W_s + W_p),$$
$$H_T(i, j-1) - W_p\},$$
$$1 \leqslant i \leqslant n, \ 1 \leqslant i \leqslant m$$

Applications

Local alignment with affine gap penalty can be used for protein classification, phylogenetic footprinting, and identification of functional sequence elements.

URL to Code

http://bioweb.pasteur.fr/seqanal/interfaces/water.html

Recommended Reading

1. Allgower EL, Schmidt PH (1985) An algorithm for piecewise-linear approximation of an implicitly defined manifold. SIAM J Num Anal 22:322–346
2. Altschul SF, Gish W, Miller W, Myers EW, Lipman DJ (1990) Basic local alignment search tool. J Mol Biol 215:403–410
3. Chao KM, Miller W (1995) Linear-space algorithms that build local alignments from fragments. Algorithmica 13:106–134
4. Gusfield D (1999) Algorithms on strings, trees and sequences. Cambridge University Press, Cambridge. ISBN:052158519
5. Ma B, Tromp J, Li M (2002) PatternHunter: faster and more sensitive homology search. Bioinformatics 18:440–445
6. Myers EW, Miller W (1988) Optimal alignments in linear space. Bioinformatics 4:11–17
7. Needleman SB, Wunsch CD (1970) A general method applicable to the search for similarities in the amino acid sequence of two proteins. J Mol Biol 48:443–453
8. Pearson WR, Lipman DJ (1988) Improved tools for biological sequence comparison. Proc Natl Acad Sci USA 85:2444–2448
9. Sellers PH (1974) On the theory and computation of evolutionary distances. SIAM J Appl Math 26:787–793
10. Smith TF, Waterman MS (1981) Identification of common molecular subsequences. J Mol Biol 147:195–197

Local Alignment (with Concave Gap Weights)

S.M. Yiu
Department of Computer Science, University of Hong Kong, Hong Kong, China

Keywords

Pairwise local alignment; Sequence alignment

Years and Authors of Summarized Original Work

1988; Miller, Myers

Problem Definition

This work of Miller and Myers [11] deals with the problem of pairwise sequence alignment in which the distance measure is based on the gap penalty model. They proposed an efficient algorithm to solve the problem when the gap penalty is a concave function of the gap length.

Let X and Y be two strings (sequences) of alphabet Σ. The pairwise alignment $\mathcal{A}$ of X and Y maps X, Y into strings X', Y' that may contain spaces (not in Σ) such that (1) $|X'| = |Y'| = \ell$; (2) removing spaces from X' and Y' returns X and Y, respectively; and (3) for any $1 \leq i \leq \ell$, $X'[i]$ and $Y'[i]$ cannot be both spaces where $X'[i]$ denotes the ith character in X'.

To evaluate the quality of an alignment, there are many different measures proposed (e.g., edit distance, scoring matrix [12]). In this work, they consider the *gap penalty* model.

A *gap* in an alignment $\mathcal{A}$ of X and Y is a maximal substring of contiguous spaces in either X' or Y'. There are gaps and aligned characters (both $X'[i]$ and $Y'[i]$ are not spaces) in an alignment. The score for a pair of aligned characters is based on a distance function $\delta(a, b)$ where $a, b \in \Sigma$. Usually δ is a metric, but this assumption is not required in this work. The penalty of a gap of length k is based on a nonnegative function $W(k)$. The score of an alignment is the sum of the scores of all aligned characters and gaps. An alignment is *optimal* if its score is the minimum possible.

The penalty function $W(k)$ is *concave* if $\Delta W(k) \geq \Delta W(k+1)$ for all $k \geq 1$, where $\Delta W(k) = W(k+1) - W(k)$.

The penalty function $W(k)$ is *affine* if $W(k) = a + bk$ where $a and b$ are constants. Affine function is a special case of concave function. The problem for affine gap penalty has been considered in [1, 7].

The penalty function $W(k)$ is a *P-piece affine curve* if the domain of W can be partitioned into P intervals, $(\tau_1 = 1, \chi_1), (\tau_2, \chi_2), \ldots, (\tau_p, \chi_p = \infty)$, where $\tau_i = \chi_{i-1} + 1$ for all $1 < i \leq p$, such that for each interval, the values of W follow an affine function. More precisely, for any $k \in (\tau_i, \chi_i)$, $W(k) = a_i + b_i k$ for some constants a_i, b_i.

Problem

Input: Two strings X and Y, the scoring function δ, and the gap penalty function $W(k)$.

Output: An optimal alignment of X and Y.

Key Results

Theorem 1 *If $W(k)$ is concave, they provide an algorithm for computing an optimal alignment that runs in $O(n^2 \log n)$ time where n is the length of each string and uses $O(n)$ expected space.*

Corollary 1 *If $W(k)$ is an affine function, the same algorithm runs in $O(n^2)$ time.*

Theorem 2 *For some special types of gap penalty functions, the algorithm can be modified to run faster.*

- *If $W(k)$ is a P-piece affine curve, the algorithm can be modified to run in $O(n^2 \log P)$ time.*
- *For logarithmic gap penalty function, $W(k) = a + b \log k$, the algorithm can be modified to run in $O(n^2)$ time.*
- *If $W(k)$ is a concave function when $k > K$, the algorithm can be modified to run in $O(K + n^2 \log n)$ time.*

Applications

Pairwise sequence alignment is a fundamental problem in computational biology. Sequence similarity usually implies functional and structural similarity. So, pairwise alignment can be used to check whether two given sequences have similar functions or structures and to predict functions of newly identified DNA sequence. One can refer to Gusfield's book for some examples on the importance of sequence alignment (pp. 212–214 of [8]).

The alignment problem can be further divided into the *global* alignment problem and the *local* alignment problem. The problem defined here is the global alignment problem in which the whole input strings are required to align with each other. On the other hand, for local alignment, the main interest lies in identifying a substring from each of the input strings such that the alignment score of the two substrings is the minimum among all possible substrings. Local alignment is useful in aligning sequences that are not similar, but contain a region that are highly conserved (similar). Usually this region is a functional part (domain) of the sequences. Local alignment is particularly useful in comparing proteins. Proteins in the same family from different species usually have some functional domains that are highly conserved while the other parts are not similar at all. Examples are the homeobox genes [4] for which the protein sequences are quite different in each species except the functional domain *homeodomain*.

Conceptually, the alignment score is used to capture the evolutionary distance between the two given sequences. Since a gap of more than one space can be created by a single mutational event, considering a gap of length k as a unit instead of k different point mutation may be more appropriate in some cases. However, which gap penalty function should be used is a difficult question to answer and sometimes depends on the actual applications. Most applications, such as BLAST, uses the affine gap penalty which is still the dominate model in practice. On the other hand, Benner et al. [2] and Gu and Li [9] suggested to use the logarithmic gap penalty in some cases. Whether using a concave gap penalty function in general is meaningful is still an open issue.

Open Problem

Note that the results of this paper have been independently obtained by Galil and Giancarlo [6], and for affine gap penalty, Gotoh [7] also gave an $O(n^2)$ algorithm for solving the alignment problem. In [5], Eppstein gave a faster algorithm that runs in $O(n^2)$ time for solving the same sequence alignment problem with concave gap penalty function. Whether a subquadratic algorithm exists for solving this problem remains open. As a remark, subquadratic algorithms do exist for solving the sequence alignment problem if the measure is not based on the gap penalty model, but is computed as $\sum_{i=1}^{\ell} \delta(X1'[i], Y'[i])$ based only on a scoring function $\delta(a, b)$ where $a, b \in \Sigma \cup \{\_\}$ where '_' represents the space [3, 10].

Experimental Results

They have performed some experiments to compare their algorithm with Waterman's $O(n^3)$ algorithm [13] on a number of different concave gap penalty functions. Artificial sequences are generated for the experiments. Results from their experiments lead to their conjectures that Waterman's method runs in $O(n^3)$ time when the two given strings are very similar or the score for mismatch characters is small and their algorithm runs in $O(n^2)$ time if the range of the function $W(k)$ is not functionally dependent on n.

Cross-References

▸ Local Alignment (with Affine Gap Weights)

Recommended Reading

1. Altschul SF, Erickson BW (1986) Optimal sequence alignment using affine gap costs. Bull Math Biol 48:603–616
2. Benner SA, Cohen MA, Gonnet GH (1993) Empirical and structural models for insertions and deletions in the divergent evolution of proteins. J Mol Biol 229:1065–1082
3. Crochemore M, Landau GM, Ziv-Ukelson M (2003) A subquadratic sequence alignment algorithm for unrestricted scoring matrices. SIAM J Comput 32(6):1654–1673
4. De Roberts E, Oliver G, Wright C (1990) Homeobox genes and the vertibrate body plan. Sci Am 263(1):46-52
5. Eppstein D (1990) Sequence comparison with mixed convex and concave costs. J Algorithms 11(1):85–101
6. Galil Z, Giancarlo R (1989) Speeding up dynamic programming with applications to molecular biology. Theor Comput Sci 64:107–118
7. Gotoh O (1982) An improved algorithm for matching biological sequences. J Mol Biol 162:705–708
8. Gusfield D (1997) Algorithms on strings, trees, and sequences: computer science and computational biology. Cambridge University Press, Cambridge
9. Li W-H, Gu X (1995) The size distribution of insertions and deletions in human and rodent pseudogenes suggests the logarithmic gap penalty for sequence alignment. J Mol Evol 40:464–473
10. Masek WJ, Paterson MS (1980) A fater algorithm for computing string edit distances. J Comput Syst Sci 20:18–31
11. Miller W, Myers EW (1988) Sequence comparison with concave weighting functions. Bull Math Biol 50(2):97–120
12. Sankoff D, Kruskal JB (1983) Time warps, strings edits, and macromolecules: the theory and practice of sequence comparison. Addison-Wesley, Reading
13. Waterman MS (1984) Efficient sequence alignment algorithms. J Theor Biol 108:333–337

Local Approximation of Covering and Packing Problems

Fabian Kuhn
Department of Computer Science, ETH Zurich, Zurich, Switzerland

Synomyms

Distributed approximation of covering and packing problems

Years and Authors of Summarized Original Work

2003–2006; Kuhn, Moscibroda, Nieberg, Wattenhofer

Problem Definition

A *local algorithm* is a distributed algorithm on a network with a running time which is independent or almost independent of the network's size or diameter. Usually, a distributed algorithm is called local if its time complexity is at most polylogarithmic in the size n of the network. Because the time needed to send information from one node of a network to another is at least proportional to the distance between the two nodes, in such an algorithm, each node's computation is based on information from nodes in a close vicinity only. Although all computations are based on local information, the network as a whole typically still has to achieve a global goal. Having local algorithms is inevitable to obtain time-efficient distributed protocols for large-scale and dynamic networks such as peer-to-peer networks or wireless ad hoc and sensor networks.

In [2, 6, 7], Kuhn, Moscibroda, and Wattenhofer describe upper and lower bounds on the possible trade-off between locality (time complexity) of distributed algorithms and the quality (approximation ratio) of the achievable solution for an important class of problems called covering and packing problems. Interesting covering and packing problems in the context of networks include minimum dominating set, minimum vertex cover, maximum matching, as well as certain flow maximization problems. All the results given in [2, 6, 7] hold for general network topologies. Interestingly, it is shown by Kuhn, Moscibroda, Nieberg, and Wattenhofer in [3, 4, 5] that covering and packing problems can be solved much more efficiently when assuming that the network topology has special properties which seem realistic for wireless networks.

Distributed Computation Model

In [2, 3, 4, 5, 6, 7], the network is modeled as an undirected and except for [5] unweighted graph $G = (V, E)$. Two nodes $u, v \in V$ of the network are connected by an edge $(u, v) \in E$ whenever there is a direct bidirectional communication channel connecting u and v. In the following, the number of nodes and the maximal degree of G are denoted by $n = |V|$ and by Δ.

For simplicity, communication is assumed to be synchronous. That is, all nodes start an algorithm simultaneously and time is divided into rounds. In each round, every node can send an arbitrary message to each of its neighbors and perform some local computation based on the information collected in previous rounds. The *time complexity* of a synchronous distributed algorithm is the number of rounds until all nodes terminate.

Local distributed algorithms in the described synchronous model have first been considered in [8] and [9]. As an introduction to the above and similar distributed computation models, it is also recommended to read [11].

Distributed Covering and Packing Problems

A fractional covering problem (P) and its dual fractional packing problem (D), are linear programs (LPs) of the canonical forms

$$\begin{array}{llll} \min & \boldsymbol{c}^{\mathrm{T}}\boldsymbol{x} & & \\ \text{s.t.} & A\cdot\boldsymbol{x} \geq \boldsymbol{b} & \text{(P)} \\ & \boldsymbol{x} \geq \boldsymbol{0} & \end{array} \qquad \begin{array}{llll} \max & \boldsymbol{b}^{\mathrm{T}}\boldsymbol{y} & \\ \text{s.t.} & A^{\mathrm{T}}\cdot\boldsymbol{y} \leq \boldsymbol{c} & \text{(D)} \\ & \boldsymbol{y} \geq \boldsymbol{0} & \end{array}$$

where all a_{ij}, b_i, and c_i are non-negative. In a distributed context, finding a small (weighted) dominating set or a small (weighted) vertex cover of the network graph are the most important covering problems. A dominating set of a graph G is a subset S of its nodes such that all nodes of G either are in S or have a neighbor in S. The dominating set problem can be formulated as covering integer LP by setting A to be the adjacency matrix with 1s in the diagonal, by setting $\boldsymbol{b}$ to be a vector with all 1s and if $\boldsymbol{c}$ is the weight vector. A vertex cover is a subset of the nodes such that all edges are covered. Packing problems occur in a broad range of resource allocation problems. As an example, in [1] and [10], the problem of assigning flows to a given fixed set of paths is described. Another common packing problem is (weighted) maximum matching, the problem of finding a largest possible set of pairwise non-adjacent edges.

While computing a dominating set, vertex cover, or matching of the network graph are inherently distributed tasks, general covering and packing LPs have no immediate distributed meaning. To obtain a distributed version of these LPs, two dual LPs (P) and (D) are mapped to a bipartite network as follows. For each primal variable x_i and for each dual variable y_j, there are nodes v_i^p and v_j^d, respectively. There is an edge between two nodes v_i^p and v_j^d whenever $a_{ji} \neq 0$, i.e., there is an edge if the ith variable of an LP occurs in its jth inequality.

In most real-world examples of distributed covering and packing problems, the network graph is of course not equal to the described bipartite graph. However, it is usually straightforward to simulate an algorithm which is designed for the above bipartite network on the actual network graph without affecting time and message complexities.

Bounded Independence Graphs

In [3, 4, 5], local approximation algorithms for covering and packing problems for graphs occuring in the context of wireless ad hoc and sensor networks are studied. Because of scale, dynamism and the scarcity of resources, these networks are a particular interesting area to apply local distributed algorithms.

Wireless networks are often modeled as *unit disk graphs* (UDGs): Nodes are assumed to be in a two-dimensional Euclidean plane and two nodes are connected by an edge iff their distance is at most 1. This certainly captures the inherent geometric nature of wireless networks. However, unit disk graphs seem much too restrictive to accurately model real wireless networks. In [3, 4, 5], Kuhn et. al. therefore consider two generalizations of the unit disk graph model, *bounded independent graphs* (BIGs) and *unit ball graphs* (UBGs). A BIG is a graph where all local independent sets are of bounded size. In particular, it is assumed that there is a function $I(r)$ which upper bounds the size of the largest independent set of every r-neighborhood in the graph. Note that the value of $I(r)$ is independent of n, the size of the network. If $I(r)$ is a polynomial in r, a BIG is said to be polynomially bounded. UDGs are BIGs with $I(r) \in O(r^2)$. UBGs are a natural generalization of UDGs. Given some underlying metric space (V, d) two nodes $u, v \in V$ are connected by an edge iff $d(u, v) \leq 1$. If the metric space (V, d) has constant doubling dimension, (The doubling dimension of a metric space is the logarithm of the maximal number of balls needed to cover a ball $B_r(x)$ in the metric space with balls $B_{r/2}(y)$ of half the radius), a UBG is a polynomially bounded BIG.

Key Results

The first algorithms to solve general distributed covering and packing LPs appear in [1, 10]. In [1], it is shown that it is possible to find a solution which is within a factor of $1+\varepsilon$ of the optimum in $O(\log^3(\rho n)/\varepsilon^3)$ rounds where ρ is the ratio between the largest and the smallest non-zero coefficient of the LPs. The result of [1] is improved and generalized in [6, 7] where the following result is proven:

Theorem 1 *In k rounds, (P) and (D) can be approximated by a factor of* $(\rho\Delta)^{O(1/\sqrt{k})}$

using messages of size at most $O(\log(\rho\Delta))$. *An* $(1+\varepsilon)$*-approximation can be found in time* $O(\log^2(\rho\Delta)/\varepsilon^4)$.

The algorithm underlying Theorem 1 needs only small messages of size $O(\log(\rho\Delta))$ and extremely simple and efficient local computations. If larger messages and more complicated (but still polynomial) local computations are allowed, it is possible to improve the result of Theorem 1:

Theorem 2 *In k rounds, LPs of the form (P) or (D) can be approximated by a factor of* $O(n^{O(1/k)})$. *This implies that a constant approximation can be found in time* $O(\log n)$.

Theorems 1 and 2 only give bounds on the quality of distributed solutions of covering and packing LPs. However, many of the practically relevant problems are integer versions of covering and packing LPs. Combined with simple randomized rounding schemes, the following upper bounds for dominating set, vertex cover, and matching are proven in [6, 7]:

Theorem 3 *Let* Δ *be the maximal degree of the given network graph. In k rounds, minimum dominating set can be approximated by a factor of* $O(\Delta^{O(1/\sqrt{k})} \cdot \log \Delta)$ *in expectation by using messages of size* $O(\Delta)$. *Without bound on the message size, an expected approximation ratio of* $O(n^{O(1/k)} \cdot \log \Delta)$ *can be achieved. Minimum vertex cover and maximum matching can both be approximated by a factor of* $O(\Delta^{1/k})$ *in k rounds.*

In [2, 7], it is shown that the upper bounds on the trade-offs between time complexity and approximation ratio given by Theorems 1–3 are almost optimal:

Theorem 4 *In k rounds, it is not possible to approximate minimum vertex cover better than by factors of* $\Omega(\Delta^{1/k}/k)$ *and* $\Omega(n^{\Omega(1/k^2)}/k)$. *This implies time lower bounds of* $\Omega(\log \Delta/\log\log \Delta)$ *and* $\Omega(\sqrt{\log n/\log\log n})$ *for constant or even poly-logarithmic approximation ratios. The same bounds hold for minimum dominating set, for maximum matching, as well as for the underlying LPs.*

While Theorem 4 shows that the results given by Theorems 1–3 are close to optimal for worst-case network topologies, the problems might be much simpler if restricted to networks which actually occur in reality. In fact, it is shown in [3, 4, 5] that the above results can indeed be improved if the network graph is assumed to be a BIG or a UBG with constant doubling dimension. In [5], the following result for UBGs is proven:

Theorem 5 *Assume that the network graph* $G = (V, E)$ *is a UBG with underlying metric* (V, d). *If* (V, d) *has constant doubling dimension and if all nodes know the distances to their neighbors in G up to a constant factor, it is possible to find constant approximations for minimum dominating set, minimum vertex cover, maximum matching, as well as for general LPs of the forms (P) and (D) in* $O(\log^* n)$ *rounds. (The log-star function* $\log^* n$ *is an extremely slowly increasing function which gives the number of times the logarithm has to be taken to obtain a number smaller than 1.)*

While the algorithms underlying the results of Theorems 1 and 2 for solving covering and packing LPs are deterministic or straightforward to be derandomized, all known efficient algorithms to solve minimum dominating set and more complicated integer covering and packing problems are randomized. Whether there are good deterministic local algorithms for dominating set and related problems is a long-standing open question. In [3], it is shown that if the network is a BIG, efficient deterministic distributed algorithms exist:

Theorem 6 *On a BIG it is possible to find constant approximations for minimum dominating set, minimum vertex cover, maximum matching, as well as for LPs of the forms (P) and (D) deterministically in* $O(\log \Delta \cdot \log^* n)$ *rounds.*

In [4], it is shown that on polynomially bounded BIGs, one can even go one step further and efficiently find an arbitrarily good approximation by a distributed algorithm:

Theorem 7 *On a polynomially bounded BIG, there is a local approximation scheme which*

L

computes a $(1+\varepsilon)$*-approximation for minimum dominating set in time* $O(\log\Delta\log^*(n)/\varepsilon + 1/\varepsilon^{O(1)})$. *If the network graph is a UBG with constant doubling dimension and nodes know the distances to their neighbors, a* $(1+\varepsilon)$*-approximation can be computed in* $O(\log^*(n)/\varepsilon + 1/\varepsilon^{O(1)})$ *rounds.*

Applications

The most important application environments for local algorithms are large-scale decentralized systems such as wireless ad hoc and sensor networks or peer-to-peer networks. On such networks, only local algorithms lead to scalable systems. Local algorithms are particularly well-suited if the network is dynamic and computations have to be repeated frequently.

A particular application of the minimum dominating set problem is the task of clustering the nodes of wireless ad hoc or sensor networks. Assigning each node to an adjacent node in a dominating set induces a simple clustering of the nodes. If the nodes of the dominating set (i.e., the cluster centers) are connected with each other by using additional nodes, the resulting structure can be used as a backbone for routing.

Open Problems

There are a number of open problems related to the distributed approximation of covering and packing problems in particular and to distributed approximation algorithms in general. The most obvious open problem certainly is to close the gaps between the upper bounds of Theorems 1, 2, and 3 and the lower bounds of Theorem 4. It would also be interesting to see how well other optimization problems can be approximated in a distributed manner. In particular, the distributed complexity of more general classes of linear programs remains completely open. A very intriguing unsolved problem is to determine to what extent randomization is needed to obtain time-efficient distributed algorithms. Currently, the best determinic algorithms for finding a dominating set of reasonable size and for many other problems take time $2^{O(\sqrt{\log n})}$ whereas the time complexity of the best randomized algorithms usually is at most polylogarithmic in the number of nodes.

Cross-References

▶ Fractional Packing and Covering Problems
▶ Maximum Matching
▶ Randomized Rounding

Recommended Reading

1. Bartal Y, Byers JW, Raz D (1997) Global optimization using local information with applications to flow control. In: Proceedings of the 38th IEEE symposium on the foundations of computer science (FOCS), pp 303–312
2. Kuhn F, Moscibroda T, Wattenhofer R (2004) What cannot be computed locally! In: Proceedings of the 23rd ACM symposium on principles of distributed computing (PODC), pp 300–309
3. Kuhn F, Moscibroda T, Nieberg T, Wattenhofer R (2005) Fast deterministic distributed maximal independent set computation on growth-bounded graphs. In: Proceedings of the 19th international conference on distributed computing (DISC), pp 273–287
4. Kuhn F, Moscibroda T, Nieberg T, Wattenhofer R (2005) Local approximation schemes for ad hoc and sensor networks. In: Proceedings of the 3rd joint workshop on foundations of mobile computing (DIALM-POMC), pp 97–103
5. Kuhn F, Moscibroda T, Wattenhofer R (2005) On the locality of bounded growth. In: Proceedings of the 24th ACM symposium on principles of distributed computing (PODC), pp 60–68
6. Kuhn F, Wattenhofer R (2005) Constant-time distributed dominating set approximation. Distrib Comput 17(4):303–310
7. Kuhn F, Moscibroda T, Wattenhofer R (2006) The price of being near-sighted. In: Proceedings of the 17th ACM-SIAM symposium on discrete algorithms (SODA), pp 980–989
8. Linial N (1992) Locality in distributed graph algorithms. SIAM J Comput 21(1):193–201
9. Naor M, Stockmeyer L (1993) What can be computed locally? In: Proceedings of the 25th annual ACM symposium on theory of computing (STOC), pp 184–193
10. Papadimitriou C, Yannakakis M (1993) Linear programming without the matrix. In: Proceedings of the 25th ACM symposium on theory of computing (STOC), pp 121–129
11. Peleg D (2000) Distributed computing: a locality-sensitive approach. SIAM

Local Computation in Unstructured Radio Networks

Thomas Moscibroda
Systems and Networking Research Group, Microsoft Research, Redmond, WA, USA

Keywords

Coloring unstructured radio networks; Maximal independent sets in radio networks

Years and Authors of Summarized Original Work

2005; Moscibroda, Wattenhofer

Problem Definition

In many ways, familiar distributed computing communication models such as the *message passing model* do not describe the harsh conditions faced in wireless ad hoc and sensor networks closely enough. Ad hoc and sensor networks are multi-hop radio networks and hence, messages being transmitted may interfere with concurrent transmissions leading to collisions and packet losses. Furthermore, the fact that all nodes share the same wireless communication medium leads to an inherent broadcast nature of communication. A message sent by a node can be received by all nodes in its transmission range. These aspects of communication are modeled by the *radio network model*, e.g., [2].

Definition 1 (Radio Network Model) In the radio network model, the wireless network is modeled as a graph $G = (V, E)$. In every time-slot, a node $u \in V$ can either send or not send a message. A node v, $(u, v) \in E$, receives the message if and only *exactly one* of its neighbors has sent a message in this time-slot.

While communication primitives such as broadcast, wake-up, or gossiping, have been widely studied in the literature on radio networks (e.g., [1, 2, 8]), less is known about the computation of *local network coordination structures* such as clusterings or colorings. The most basic notion of a clustering in wireless networks boils down to the graph-theoretic notion of a dominating set.

Definition 2 (Minimum Dominating Set (MDS)) Given a graph $G = (V, E)$. A dominating set is a subset $S \subseteq V$ such that every node is either in S or has at least one neighbor in S. The minimum dominating set problem asks for a dominating set S of minimum cardinality.

A dominating set $S \subseteq V$ in which no two neighboring nodes are in S is a *maximal independent set (MIS)*. The distributed complexity of computing a MIS in the message passing model has been of fundamental interest to the distributed computing community for over two decades (e.g., [11–13]), but much less is known about the problem's complexity in radio network models.

Definition 3 (Maximal Independent Set (MIS)) Given a graph $G = (V, E)$. An independent set is a subset of pair-wise non-adjacent nodes in G. A maximal independent set in G is an independent set $S \subseteq V$ such that for every node $u \notin S$, there is a node $v \in \Gamma(u)$ in S.

Another important primitive in wireless networks is the *vertex coloring* problem, because associating different colors with different time slots in a time-division multiple access (TDMA) scheme; a correct coloring corresponds to a medium access control (MAC) layer without *direct interference*, that is, no two neighboring nodes send at the same time.

Definition 4 (Minimum Vertex Coloring) Given a graph $G = (V, E)$. A correct vertex coloring for G is an assignment of a color $c(v)$ to each node $v \in V$, such that $c(u) \neq c(v)$ any two adjacent nodes $(u, v) \in E$. A minimum vertex coloring is a correct coloring that minimizes the number of used colors.

In order to capture the especially harsh characteristics of wireless multi-hop networks immediately after their deployment, the unstructured radio network model makes additional assumptions. In particular, a new notion of asynchronous wake-up is considered, because, in a wireless, multi-hop environment, it is realistic to assume that some nodes join the network (e.g., become deployed, or switched on) later than others. Notice that this is different from the notion of asynchronous wake-up defined and studied in [8] and subsequent work, in which nodes are assumed to be "woken up" by incoming messages.

Definition 5 (Unstructured Radio Network Model) In the *unstructured radio network model*, the wireless network is modeled as a unit disk graph (UDG) $G = (V, E)$. In every time-slot, a node $u \in V$ can either send or not send a message. A node v, $(u, v) \in E$, receives the message if and only *exactly one* of its neighbors has sent a message in this time-slot. Additionally, the following assumptions are made:

- *Asynchronous wake-up:* New nodes can wake up/join in *asynchronously* at any time. Before waking-up, nodes do neither receive nor send any messages.
- *No global clock:* Nodes only have access to a local clock that starts increasing after wake-up.
- *No collision detection:* Nodes cannot distinguish between the event of a collision and no message being sent. Moreover, a sending node does not know how many (if any at all!) neighbors have received its transmission correctly.
- *Minimal global knowledge:* At the time of their wake-up, nodes have no information about their neighbors in the network and they do not whether some neighbors are already awake, executing the algorithm. However, nodes know an upper bound for the maximum number of nodes $n = |V|$.

The measure that captures the efficiency of an algorithm defined in the unstructured radio network model is its *time-complexity*. Since every node can wake up at a different time, the time-complexity of an algorithm is defined as the maximum number of time-slots between a node's wake-up and its final, irrevocable decision.

Definition 6 (Time Complexity) The *running time* T_v of a node $v \in V$ is defined as the number of time slots between v's *waking up* and the time v makes an irrevocable *final decision* on the outcome of its protocol (e.g., whether or not it joins the dominating set in a clustering algorithm, or which color to take in a coloring algorithm, etc.). The *time complexity* $T(\mathcal{Q})$ of algorithm $\mathcal{Q}$ is defined as the maximum running time over all nodes in the network, i.e., $T(\mathcal{Q}) := \max_{v \in V} T_v$.

Key Results

Naturally, algorithms for such uninitialized, chaotic networks have a different flavor compared to "traditional" algorithms that operate on a given network graph that is static and well-known to all nodes. Hence, the algorithmic difficulty of the following algorithms partly stems from the fact that since nodes wake up asynchronously and do not have access to a global clock, the different phases of the algorithm may be arbitrarily intertwined or shifted in time. Hence, while some nodes may already be in an advanced stage of the algorithm, there may be nodes that have either just woken up, or that are still in early stage. It was proven in [9] that even in single-hop networks (G is the complete graph), no efficient algorithms exist if nodes have no knowledge on n.

Theorem 1 *If nodes have no knowledge of n, every (possibly randomized) algorithm requires* *up to $\Omega(n/\log n)$ time slots before at least one node can send a message in single-hop networks.*

In single-hop networks, and if n is globally known, [8] presented a randomized algorithm that selects a unique leader in time $O(n \log n)$, with high probability. This result has subsequently been improved to $O(\log^2 n)$ by Jurdziński and Stachowiak [9]. The generalized wake-up problem in multi-hop radio network was first studied in [4].

The complexity of local network structures such as clusterings or colorings in unstructured multi-hop radio networks was first studied in [10]: A good approximation to the minimum dominating set problem can be computed in polylogarithmic time.

Theorem 2 *In the unstructured radio network model, an expected O(1)-approximation to the dominating set problem can be computed in expected time $O(\log^2 n)$. That is, every node decides whether to join the dominating set within $O(\log^2 n)$ time slots after its wake-up.*

In a subsequent paper [18], it has been shown that the running time of $O(\log^2 n)$ is sufficient even for computing the more sophisticated MIS structure. This result is asymptotically optimal because – improving on a previously known bound of $\Omega(\log^2 n/\log\log n)$ [9] –, a corresponding lower bound of $\Omega(\log^2 n)$ has been proven in [6].

Theorem 3 *With high probability, a maximal independent set (MIS) can be computed in expected time $O(\log^2 n)$ in the unstructured radio network model. This is asymptotically optimal.*

It is interesting to compare this achievable upper bound on the harsh unstructured radio network model with the best known time lower bounds in message passing models: $\Omega(\log^* n)$ in unit disk graphs [12] and $\Omega(\sqrt{\log n/\log\log n})$ in general graphs [11]. Also, a time bound of $O(\log^2 n)$ was also proven in [7] in a radio network model without asynchronous wake-up and in which nodes have a-priori knowledge about their neighborhood.

Finally, it is also possible to efficiently color the nodes of a network as shown in [17], and subsequently improved and generalized in Chap. 12 of [15].

Theorem 4 *In the unstructured radio network model, a correct coloring with at most $O(\Delta)$ colors can be computed in time $O(\Delta \log n)$ with high probability.*

Similar bounds for a model with collision detection mechanisms are proven in [3].

Applications

In wireless ad hoc and sensor networks, local network coordination structures find important applications. In particular, clusterings and colorings can help in facilitating the communication between adjacent nodes (MAC layer protocols) and between distant nodes (routing protocols), or to improve the energy efficiency of the network.

The following mentions two specific examples of applications: Based on the MIS algorithms of Theorem 3, a protocol is presented in [5], which efficiently constructs a *spanner*, i.e., a more sophisticated initial infrastructure that helps in structuring wireless multi-hop network. In [16], the same MIS algorithm is used as an ingredient for a protocol that minimizes the energy consumption of wireless sensor nodes during the *deployment phase*, a problem that has been first studied in [14].

L

Recommended Reading

1. Alon N, Bar-Noy A, Linial N, Peleg D (1991) A Lower Bound for Radio Broadcast. J Comput Syst Sci 43:290–298
2. Bar-Yehuda R, Goldreich O, Itai A (1987) On the time-complexity of broadcast in radio networks: an exponential gap between determinism randomization. In: Proceedings of the 6th symposium on principles of distributed computing (PODC), pp 98–108

3. Busch R, Magdon-Ismail M, Sivrikaya F, Yener B (2004) Contention-free MAC protocols for wireless sensor networks. In: Proceedings of the 18th annual conference on distributed computing (DISC)
4. Chrobak M, Gąsieniec L, Kowalski, D (2004) The wake-up problem in multi-hop radio networks. In: Proceedings of the 15th ACM-SIAM symposium on discrete algorithms (SODA), pp 992–1000
5. Farach-Colton M, Fernandes RJ, Mosteiro MA (2005) Bootstrapping a hop-optimal network in the weak sensor model. In: Proceedings of the 13th European symposium on algorithms (ESA), pp 827–838
6. Farach-Colton M, Fernandes RJ, Mosteiro MA (2006) Lower bounds for clear transmissions in radio networks. In: Proceedings of the 7th Latin American symposium on theoretical informatics (LATIN), pp 447–454
7. Gandhi R, Parthasarathy S (2004) Distributed algorithms for coloring and connected domination in wireless ad hoc networks. In: Foundations of software technology and theoretical computer science (FSTTCS), pp 447–459
8. Gąsieniec L, Pelc A, Peleg D (2000) The wakeup problem in synchronous broadcast systems (extended abstract). In: Proceedings of the 19th ACM symposium on principles of distributed computing (PODC), pp 113–121
9. Jurdziński T, Stachowiak G (2002) Probabilistic algorithms for the wakeup problem in single-hop radio networks. In: Proceedings of the 13th Annual International Symposium on Algorithms and Computation (ISAAC), pp 535–549
10. Kuhn F, Moscibroda T, Wattenhofer R (2004) Initializing newly deployed ad hoc and sensor networks. In: Proceedings of the 10th Annual International Conference on Mobile Computing and Networking (MOBICOM), pp 260–274
11. Kuhn F, Moscibroda T, Wattenhofer R (2004) What cannot be computed locally! In: Proceedings of 23rd annual symposium on principles of distributed computing (PODC), pp 300–309
12. Linial N (1992) Locality in distributed graph algorithms. SIAM J Comput 21(1):193–201
13. Luby M (1986) A simple parallel algorithm for the maximal independent set problem. SIAM J Comput 15:1036–1053
14. McGlynn MJ, Borbash SA (2001) Birthday protocols for low energy deployment and flexible neighborhood discovery in ad hoc wireless networks. In: Proceedings of the 2nd ACM international symposium on mobile ad hoc networking & computing (MOBIHOC)
15. Moscibroda T (2006) Locality, scheduling, and selfishness: algorithmic foundations of highly decentralized networks. Doctoral thesis Nr. 16740, ETH Zurich
16. Moscibroda T, von Rickenbach P, Wattenhofer R (2006) Analyzing the energy-latency trade-off during the deployment of sensor networks. In: Proceedings of the 25th joint conference of the IEEE computer and communications societies (INFOCOM)
17. Moscibroda T, Wattenhofer R (2005) Coloring unstructured radio networks. In: Proceedings of the 17th ACM symposium on parallel algorithms and architectures (SPAA), pp 39–48
18. Moscibroda T, Wattenhofer R (2005) Maximal independent sets in radio networks. In: Proceedings of the 23rd ACM symposium on principles of distributed computing (PODC), pp 148–157

Local Reconstruction

Comandur Seshadhri
Sandia National Laboratories, Livermore, CA, USA
Department of Computer Science, University of California, Santa Cruz, CA, USA

Keywords

Data reconstruction; Property testing; Sublinear algorithms

Years and Authors of Summarized Original Work

2010; Saks, Seshadhri

Problem Definition

Consider some massive dataset represented as a function $f : D \mapsto R$, where D is discrete and R is an arbitrary range. This dataset could be as varied as an array of numbers, a graph, a matrix, or a high-dimensional function. Datasets are often useful because they possess some property of interest. An array might be sorted, a graph might be connected, a matrix might be orthogonal, or a function might be convex. These properties are critical to the use of the dataset. Yet, due to unavoidable errors (say, in storing the dataset), these properties might not hold any longer. For example, a sorted array could become unsorted because of roundoff errors.

Can we find a function $g : D \mapsto R$ that satisfies the property and is "sufficiently close" to f? Let us formalize this question. Let $\mathcal{P}$ denote a property, which we define as a subset of functions. We define a *distance* between functions, $\mathsf{dist}(f, g) = |\{x | f(x) \neq g(x)\}|/|D|$. In words, this is the fraction of domain points where f and g differ (the relative Hamming distance). This definition naturally extends to properties: $\mathsf{dist}(f, \mathcal{P}) = \min_{h \in \mathcal{P}} \mathsf{dist}(f, h)$. This is the minimum amount of change f must undergo to have property $\mathcal{P}$. Our aim is to construct $g \in \mathcal{P}$ such that $\mathsf{dist}(f, g)$ is "small." The latter can be quantified by comparing with the baseline, $\mathsf{dist}(f, \mathcal{P})$.

The *offline reconstruction* problem involves explicitly constructing g from f. But this is prohibitively expensive if f is a large dataset. Instead, we wish to *locally* construct g, meaning we want to quickly compute $g(x)$ (for $x \in D$) without constructing g completely.

Local filters [13]: A *local filter* for property $\mathcal{P}$ is an algorithm A satisfying the following conditions. The filter has *oracle access* to function $f : D \mapsto R$, meaning that it can access $f(x)$ for any $x \in D$. Each such access is called a *lookup*. The filter takes as input an auxiliary random seed ρ and $x \in D$. For fixed f, ρ, A runs deterministically on input x to produce an output $A_{f,\rho}(x)$. Note that $A_{f,\rho}$ specifies a function on domain D, which will be the desired function g.

1. For each f and ρ, $A_{f,\rho}$ always satisfies $\mathcal{P}$.
2. For each f, with high probability over ρ, the function $A_{f,\rho}$ is suitably close to f.
3. For each x, $A_{f,\rho}(x)$ can be computed with very few lookups.
4. The size of the random seed ρ should be much smaller than D.

Let g be $A_{f,\rho}$. Condition 2 has been formalized in at least two different ways. The original definition demanded that $\mathsf{dist}(f, g) \leq c \cdot \mathsf{dist}(f, \mathcal{P})$, where c is a fixed constant [13]. Other results only enforce Condition 2 when $f \in \mathcal{P}$ [3, 9]. One could imagine desiring $|\mathsf{dist}(f, g) - \mathsf{dist}(f, \mathcal{P})| < \delta$, for input parameter δ. Conditions 3 and 4 typically demand that the lookup complexity and random seed lengths are $o(|D|)$ (sometimes we desire them to be $\mathrm{poly}(\log |D|)$ or even constant).

Connections with Other Models

The notion of reconstruction through filters was first proposed by Ailon et al. in [1], though the requirements were less stringent. There is a sequence $x_1, x_2, \ldots$ of domain points generated online. Given x_i, the filter outputs value $g(x_i)$. The filter is allowed to store previous outputs to ensure consistency. Saks and Seshadhri [13] defined local filters to address two concerns with this model. First, the storage of all previous queries and answers is a massive space overhead. Second, different runs of the filter construct different g's (because the filter is randomized). So we cannot instantiate multiple copies of the filter to handle queries independently and consistently.

Independent of this line of work, Brakerski defined *local restorers* [6], which are basically equivalent to filters with an appropriate setting of Conditions 1 and 2. A major generalization of local filters, called *local computation algorithms*, was subsequently proposed by Rubinfeld et al. [12]. This model considers computation on a large input where the output itself is large (e.g., one may want a maximal independent set of a massive graph). The aim is to locally compute the output, by an algorithm akin to a filter.

Depending on how Property 2 is instantiated, filters can easily be used for tolerant testing and distance approximation [11]. If the filter ensures that $\mathsf{dist}(f, g)$ is comparable to $\mathsf{dist}(f, \mathcal{P})$, then it suffices to estimate $\mathsf{dist}(f, g)$ for distance approximation.

A special case of local reconstruction that has received extensive attention is decoding of error correcting codes. Here, f is some string, and $\mathcal{P}$ is the set of all valid code words. In local decoding, there is either *one* correct output or a sparse list of possible correct outputs. For general properties, there may be many (possibly infinitely many) ways to construct a valid reconstruction g. This creates challenges for designing filters. Once the random seed is fixed, all query answers provided by the filter must be consistent with a *single* function having the property.

Key Results

Over the past decade, there have been numerous results on local reconstruction, over a variety of domains.

Monotonicity

The most studied property for local reconstruction is monotonicity [1, 3, 5, 13]. Consider $f : [n]^d \mapsto \mathbb{R}$, where d, n are positive integers and $[n] = \{1, 2, \ldots, n\}$. The domain is equipped with the natural coordinate-wise partial order. Namely, $x \prec y$ if $x \neq y$ and all coordinates of x are less than or equal those of y. A function f is monotone if: $\forall x \prec y$, $f(x) \leq f(y)$. When $d = 1$, monotone functions are exactly sorted arrays.

Most initial work on local filters focused exclusively on monotonicity. There exists a filter of running time $(\log n)^{O(d)}$ with $\mathsf{dist}(f, g) \leq 2^{O(d)}\mathsf{dist}(f, \mathcal{P})$ [13]. There are nearly matching lower bounds that are extremely strong; even for relaxed versions of Condition 2 [3].

Lipschitz Continuity

Let n, d be positive integers and c be a positive real number. A function $f : [n]^d \mapsto \mathbb{R}$ is c-Lipschitz if $\forall x, y, |f(x) - f(y)| \leq c\|x - y\|_1$. This is a fundamental property of functions and appears in functional analysis, statistics, and optimization. Recently, Lipschitz continuity was studied from a property testing perspective by Jha and Raskhodnikova [9]. Most relevant to this essay, they gave an application of local Lipschitz filters for differential privacy. The guarantees on their filter are analogous to monotonicity (with a weaker form of Property 2). Awasthi, Jha, Molinaro, and Raskhodnikova [3] gave matching lower bounds for these reconstruction problems.

Dense Graph Properties

Dense graphs are commonly studied in property testing, where the input is given as a binary adjacency matrix. Brakerski's work on local restorers provides filters for bipartiteness and existence of large cliques. Large classes of dense graphs are known to be tolerant testable. These results have been extended to local filters for hypergraph properties by Austin and Tao [2]. This work was developed independently of all the previous work on filters, and their algorithms are called "repair" algorithms.

Connectivity Properties of Sparse Graphs

In the sparse graph setting, the input G is an adjacency list of a bounded-degree graph. Filters have been given for several properties regarding connectivity. Campagna, Guo, and Rubinfeld [7] provide reconstructors for k-connectivity and the property of having low diameter. Local reconstructors for the property of expansion were given by Kale and Seshadhri [10].

Convexity in 2, 3-Dimensions

Chazelle and Seshadhri [8] studied reconstruction in the geometric setting. They focus on convex polygon and 3D convex polytope reconstruction. These results were in the online filter setting of [1], though their 3D result is a local filter.

Open Problems

For any property tester, one can ask if the associated property has a local filter. Given the breadth of this area, we cannot hope to give a good summary of open problems. Nonetheless, we make a few suggestions.

The Curse of Dimensionality for Monotonicity and Lipschitz

Much work has gone into understanding local filters for monotonicity, but it is not clear how to remove the exponential dependence on d. Can we find a reasonable setting for filters where a $\text{poly}(d, \log n)$ lookup complexity is possible? One possibility is requiring only "additive error" in Condition 2. For some parameter $\delta > 0$, we only want $|\mathsf{dist}(f, g) - \mathsf{dist}(f, \mathcal{P})| \leq \delta$. Is there a filter with lookup complexity $\text{poly}(d, \log n, 1/\delta)$? This definition would avoid previous lower bounds [3].

Filters for Convexity

Filters for convexity could have a great impact on optimization. A local filter would implicitly represent a close enough convex function to an input non-convex function, which would be extremely useful for (say) minimization. For this application, it would be essential to handle high-dimensional functions. Unfortunately, there are no known property testers for convexity in this setting, so designing local filters is a distant goal.

Filters for Properties of Bounded-Degree Graphs

The large class of minor-closed properties (such as planarity) is known to be testable for bounded-degree graphs [4]. Can we get local filters for these properties? This appears to be a challenging question, since even approximating the distance to planarity is a difficult problem. Nonetheless, the right relaxations of filter conditions could lead to positive results for filters.

Cross-References

▸ Monotonicity Testing
▸ Testing if an Array Is Sorted

Recommended Reading

1. Ailon N, Chazelle B, Comandur S, Liu D (2008) Property-preserving data reconstruction. Algorithmica 51(2):160–182
2. Austin T, Tao T (2010) Testability and repair of hereditary hypergraph properties. Random Struct Algorithms 36(4):373–463
3. Awasthi P, Jha M, Molinaro M, Raskhodnikova S (2012) Limitations of local filters of lipschitz and monotone functions. In: Workshop on approximation, randomization, and combinatorial optimization. Algorithms and techniques (RANDOM-APPROX), pp 374–386
4. Benjamini I, Schramm O, Shapira A (2010) Every minor-closed property of sparse graphs is testable. Adv Math 223(6):2200–2218
5. Bhattacharyya A, Grigorescu E, Jha M, Jung K, Raskhodnikova S, Woodruff D (2012) Lower bounds for local monotonicity reconstruction from transitive-closure spanners. SIAM J Discret Math 26(2):618–646
6. Brakerski Z (2008) Local property restoring. Manuscript
7. Campagna A, Guo A, Rubinfeld R (2013) Local reconstructors and tolerant testers for connectivity and diameter. In: Workshop on approximation, randomization, and combinatorial optimization. Algorithms and techniques (RANDOM-APPROX), pp 411–424
8. Chazelle B, Seshadhri C (2011) Online geometric reconstruction. J ACM 58(4):14
9. Jha M, Raskhodnikova S (2013) Testing and reconstruction of lipschitz functions with applications to data privacy. SIAM J Comput 42(2):700–731
10. Kale S, Seshadhri C (2011) Testing expansion in bounded degree graphs. SIAM J Comput 40(3):709–720
11. Parnas M, Ron D, Rubinfeld R (2006) Tolerant property testing and distance approximation. J Comput Syst Sci 6(72):1012–1042
12. Rubinfeld R, Tamir G, Vardi S, Xie N (2011) Fast local computation algorithms. In: Proceedings of innovations in computer science, pp 223–238
13. Saks M, Seshadhri C (2006) Local monotonicity reconstruction. SIAM J Comput 39(7):2897–2926

Local Search for K-medians and Facility Location

Kamesh Munagala
Levine Science Research Center, Duke University, Durham, NC, USA

Keywords

Clustering; Facility location; k-Means; k-Medians; k-Medioids; Point location; Warehouse location

Years and Authors of Summarized Original Work

2001; Arya, Garg, Khandekar, Meyerson, Munagala, Pandit

Problem Definition

Clustering is a form of *unsupervised learning*, where the goal is to "learn" useful patterns in a data set $\mathcal{D}$ of size n. It can also be thought of as a data compression scheme where a large data set is represented using a smaller collection of "representatives". Such a scheme is characterized by specifying the following:

1. A *distance* metric $\mathbf{d}$ between items in the data set. This metric should satisfy the triangle inequality: $\mathbf{d}(i, j) \leq \mathbf{d}(j, k) + \mathbf{d}(k, i)$ for any three items $i, j, k \in \mathcal{D}$. In addition, $\mathbf{d}(i, j) = \mathbf{d}(j, i)$ for all $i, j \in S$ and $\mathbf{d}(i, i) = 0$. Intuitively, if the distance between two items is smaller, they are more similar. The items are usually points in some high dimensional Euclidean space $\mathcal{R}^d$. The commonly used distance metrics include the Euclidean and Hamming metrics, and the cosine metric measuring the angle between the vectors representing the items.
2. The output of the clustering process is a partitioning of the data. This chapter deals with *center-based* clustering. Here, the output is a smaller set $C \subset \mathcal{R}^d$ of *centers* which best represents the input data set $S \subset \mathcal{R}^d$. It is typically the case that $|C| \ll |\mathcal{D}|$. Each item $j \in \mathcal{D}$ is *mapped to or approximated by* the the closest center $i \in C$, implying $\mathbf{d}(i, j) \leq \mathbf{d}(i', j)$ for all $i' \in C$. Let $\sigma : \mathcal{D} \to C$ denote this mapping. This is intuitive since closer-by (similar) items will be mapped to the same center.
3. A measure of the *quality* of the clustering, which depends on the desired output. There are several commonly used measures for the quality of clustering. In each of the clustering measures described below, the goal is to choose C such that $|C| = k$ and the objective function $f(C)$ is minimized.

k-center: $f(C) = \max_{j \in \mathcal{D}} \mathbf{d}(j, \sigma(j))$.
k-median: $f(C) = \sum_{j \in \mathcal{D}} \mathbf{d}(j, \sigma(j))$.
k-means: $f(C) = \sum_{j \in \mathcal{D}} \mathbf{d}(j, \sigma(j))^2$.

All the objectives described above are NP-HARD to optimize in general metric spaces $\mathbf{d}$, leading to the study of heuristic and approximation algorithms. In the rest of this chapter, the focus is on the k-median objective. The approximation algorithms for k-median clustering are designed for $\mathbf{d}$ being a general possibly non-Euclidean metric space. In addition, a collection $\mathcal{F}$ of possible center locations is given as input, and the set of centers C is restricted to $C \subseteq \mathcal{F}$. From the perspective of approximation, the restriction of the centers to a finite set $\mathcal{F}$ is not too restrictive – for instance, the optimal solution which is restricted to $\mathcal{F} = \mathcal{D}$ has objective value at most a factor 2 of the optimal solution which is allowed arbitrary $\mathcal{F}$. Denote $|\mathcal{D}| = n$, and $|\mathcal{F}| = m$. The running times of the heuristics designed will be polynomial in m n, and a parameter $\varepsilon > 0$. The metric space $\mathbf{d}$ is now defined over $\mathcal{D} \cup \mathcal{F}$.

A related problem to k-medians is its Lagrangean relaxation, called FACILITY LOCATION. In this problem, there is a again collection $\mathcal{F}$ of possible center locations. Each location $i \in \mathcal{F}$ has a location cost r_i. The goal is to choose a collection $C \subseteq \mathcal{F}$ of centers and construct the mapping $\sigma : S \to C$ from the items to the centers such that the following function is minimized:

$$f(C) = \sum_{j \in \mathcal{D}} \mathbf{d}(j, \sigma(j)) + \sum_{i \in C} r_i .$$

The facility location problem effectively gets rid of the hard bound k on the number of centers in k-medians, and replaces it with the center cost term $\sum_{i \in C} r_i$ in the objective function, thereby making it a Lagrangean relaxation of the k-median problem. Note that the costs of centers can now be non-uniform.

The approximation results for both the k-median and facility location problems carry over as is to the weighted case: Each item $j \in \mathcal{D}$ is allowed to have a non-negative weight w_j. In the objective function $f(C)$, the term $\sum_{j \in \mathcal{D}} \mathbf{d}(j, \sigma(j))$ is replaced with $\sum_{j \in \mathcal{D}} w_j \cdot \mathbf{d}(j, \sigma(j))$. The weighted case is especially relevant to the FACILITY LOCATION problem where the item weights signify user demands for a resource, and the centers denote locations of the resource. In the remaining discussion, "items" and "users" are used interchangably to denote members of the set $\mathcal{D}$.

Key Results

The method of choice for solving both the k-median and FACILITY LOCATION problems are the class of local search heuristics, which run in "local improvement" steps. At each step t, the heuristic maintains a set C_t of centers. For the k-median problem, this collection satisfies $|C_t| = k$. A local improvement step first generates a collection of new solutions $\mathcal{E}_{t+1}$ from C_t. This is done such that $|\mathcal{E}_{t+1}|$ is polynomial in the input size. For the k-median problem, in addition, each $C \in \mathcal{E}_{t+1}$ satisfies $|C| = k$. The improvement step sets $C_{t+1} = \text{argmin}_{C \in \mathcal{E}_{t+1}} f(C)$. For a pre-specified parameter $\varepsilon > 0$, the improvement iterations stop at the first step T where $f(C_T) \geq (1-\varepsilon) f(C_{T-1})$.

The key design issue is the specification of the start set C_0, and the construction of $\mathcal{E}_{t+1}$ from C_t. The key analysis issues are bounding the number of steps T till termination, and the quality of the final solution $f(C_T)$ against the optimal solution $f(C^*)$. The ratio $(f(C_T))/(f(C^*))$ is termed the "locality gap" of the heuristic.

Since each improvement step reduces the value of the solution by at least a factor of $(1-\varepsilon)$, the running time in terms of number of improvement steps is given by the following expression (here D is the ratio of the largest to smallest distance in the metric space over $\mathcal{D} \cup \mathcal{F}$).

$$T \leq \log_{1/(1-\varepsilon)}\left(\frac{f(C_0)}{f(C_T)}\right) \leq \frac{\log\left(\frac{f(C_0)}{f(C_T)}\right)}{\varepsilon} \leq \frac{\log(nD)}{\varepsilon}$$

which is polynomial in the input size. Each improvement step needs computation of $f(C)$ for $C \in \mathcal{E}_t$. This is polynomial in the input size since $|\mathcal{E}_t|$ is assumed to be polynomial.

k-Medians

The first local search heuristic with provable performance guarantees is presented in the work of Arya et al. [1]. The is the natural p-swap heuristic: Given the current center set C_t of size k, the set $\mathcal{E}_{t+1}$ is defined by:

$$\mathcal{E}_{t+1} = \{(C_t \setminus \mathcal{A}) \cup \mathcal{B}, \quad \text{where } \mathcal{A} \subseteq C_t, \mathcal{B} \subseteq \mathcal{F} \setminus C_t, |\mathcal{A}| = |\mathcal{B}| \leq p\}.$$

The above simply means swap at most p centers from C_t with the same number of centers from $\mathcal{F} \setminus C_t$. Recall that $|\mathcal{D}| = n$ and $|\mathcal{F}| = m$. Clearly, $|\mathcal{E}_{t+1}| \leq (k(m-k))^p \leq (km)^p$. The start set C_0 is chosen arbitrarily. The value p is a parameter which affects the running time and the approximation ratio. It is chosen to be a constant, so that $|\mathcal{E}_t|$ is polynomial in m.

Theorem 1 ([1]) *The p-swap heuristic achieves locality gap* $(3 + 2/p) + \varepsilon$ *in running time* $O(nk(\log(nD))/\varepsilon(mk)^p)$. *Furthermore, for every p there is a k-median instance where the p-swap heuristic has locality gap exactly* $(3 + 2/p)$.

Setting $p = 1/\varepsilon$, the above heuristic achieves a $3 + \varepsilon$ approximation in running time $\tilde{O}(n(mk)^{O(1/\varepsilon)})$.

Facility Location

For this problem, since there is no longer a constraint on the number of centers, the local improvement step needs to be suitably modified. There are two local search heuristics both of which yield a locality gap of $3 + \varepsilon$ in polynomial time.

The "add/delete/swap" heuristic proposed by Kuehn and Hamburger [10] either adds a center to C_t, drops a center from C_t, or swaps a center in C_t with one in $\mathcal{F} \setminus C_t$. The start set C_0 is again arbitrary.

$$\mathcal{E}_{t+1} = \{(C_t \setminus \mathcal{A}) \cup \mathcal{B}, \text{ where } \mathcal{A} \subseteq C_t, \mathcal{B} \subseteq \mathcal{F} \setminus C_t, |\mathcal{A}| = 0, |\mathcal{B}| = 1 \text{ or } |\mathcal{A}| = 1, |\mathcal{B}| = 0, \text{ or } |\mathcal{A}| = 1, |\mathcal{B}| = 1\}$$

Clearly, $|\mathcal{E}_{t+1}| = O(m^2)$, making the running time polynomial in the input size and $1/\varepsilon$. Korupolu, Plaxton, and Rajaraman [9] show that this heuristic achieves a locality gap of at most $5 + \varepsilon$. Arya et al. [1] strengthen this analysis to show that this heuristic achieves a locality gap of $3 + \varepsilon$, and that bound this is tight in the sense that there are instances where the locality gap is exactly 3.

The "add one/delete many" heuristic proposed by Charikar and Guha [2] is slightly more involved. This heuristic adds one facility and drops all facilities which become irrelevant in the new solution.

$$\mathcal{I}_{t+1} = \{(C_t \cup \{i\}) \setminus I(i), \text{ where } i \in \mathcal{F} \setminus C_t, I(i) \subseteq C_t\}$$

The set $I(i)$ is computed as follows: Let W denote the set of items closer to i than to their assigned centers in C_t. These items are ignored from the computation of $I(i)$. For every center $s \in C_t$, let U_s denote all items which are assigned to s. If $f_s + \sum_{j \in U_s \setminus W} d_j \mathbf{d}(j,s) > \sum_{j \in U_s \setminus W} d_j \mathbf{d}(j,i)$, then it is cheaper to remove location s and reassign the items in $U_s \setminus W$ to i. In this case, s is placed in $I(i)$. Let N denote $m + n$. Computing $I(i)$ is therefore a $O(N)$ time greedy procedure, making the overall running time polynomial. Charikar and Guha [2] show the following theorem:

Theorem 2 ([2]) *The local search heuristic which attempts to add a random center $i \notin C_t$ and remove set $I(i)$, computes a $3+\varepsilon$ approximation with high probability within $T = O(N \log N(\log N + 1/\varepsilon))$ improvement steps, each with running time $O(N)$.*

Capacitated Variants

Local search heuristics are also known for capacitated variants of the k-median and facility location problems. In this variant, each possible location $i \in \mathcal{F}$ can serve at most u_i number of users. In the soft capacitated variant of facility location, some $r_i \geq 0$ copies can be opened at $i \in \mathcal{F}$ so that the facility cost is $f_i r_i$ and the number of users served is at most $r_i u_i$. The optimization goal is now to decide the value of r_i for each $i \in \mathcal{F}$ so that the assignment of users to the centers satisfies the capacity constraints at each center, and the cost of opening the centers and assigning the users is minimized. For this variant, Arya et al. [1] show a local search heuristic with a locality gap of $4+\varepsilon$.

In the version of facility location with hard capacities, location $i \in \mathcal{F}$ has a hard bound u_i on the number of users that can be assigned here. If all the capacities u_i are equal (uniform case), Korupolu, Plaxton, and Rajaraman [9] present an elegant local search heuristic based on solving a transshipment problem which achieves a $8+\varepsilon$ locality gap. The analysis is improved by Chudak and Williamson [4] to show a locality gap $6+\varepsilon$. The case of non-uniform capacities requires significantly new ideas – Pál, Tardos, and Wexler [14] present a network flow based local search heuristic that achieves a locality gap of $9+\varepsilon$. This bound is improved to $8+\varepsilon$ by Mahdian and Pál [12], who generalize several of the local search techniques described above in order to obtain a constant factor approximation for the variant of facility location where the facility costs are arbitrary non-decreasing functions of the demands they serve.

Related Algorithmic Techniques

Both the k-median and facility location problems have a rich history of approximation results. Since the study of uncapacitated facility location was initiated by Cornuejols, Nemhauser, and Wolsey [5], who presented a natural linear programming (LP) relaxation for this problem, several constant-factor approximations have been designed via several techniques, ranging from rounding of the LP solution [11, 15], local search [2, 9], the primal-dual schema [7], and dual fitting [6]. For the k-median problem, the first constant factor approximation [3] of $6\frac{2}{3}$ was obtained by rounding the natural LP relaxation via a generalization of the filtering technique in [11]. This result was subsequently improved to a 4 approximation by Lagrangean relaxation and the primal-dual schema [2, 7], and finally to a $(3+\varepsilon)$ approximation via local search [1].

Applications

The facility location problem has been widely studied in operations research [5, 10], and forms a fundamental primitive for several resource location problems. The k-medians and k-means metrics are widely used in clustering, or unsupervised learning. For clustering applications, several heuristic improvements to the basic local search framework have been proposed: k-Medioids [8] selects a random input point and replaces it with one of the existing centers if there is an improvement; the CLARA [8] implementation of k-Medioids chooses the centers from a random sample of the input points to speed up the computation; the CLARANS [13] heuristic draws a fresh random sample of feasible centers before each improvement step to further improve the efficiency.

Cross-References

▶ Facility Location

Recommended Reading

1. Arya V, Garg N, Khandekar R, Meyerson A, Munagala K, Pandit V (2004) Local search heuristics for k-median and facility location problems. SIAM J Comput 33(3):544–562
2. Charikar M, Guha S (2005) Improved combinatorial algorithms for facility location problems. SIAM J Comput 34(4):803–824
3. Charikar M, Guha S, Tardos É, Shmoys DB (1999) A constant factor approximation algorithm for the k-median problem(extended abstract). In: STOC '99: proceedings of the thirty-first annual ACM symposium on theory of computing, Atlanta, 1–4 May 1999, pp 1–10
4. Chudak FA, Williamson DP (2005) Improved approximation algorithms for capacitated facility location problems. Math Program 102(2):207–222
5. Cornuejols G, Nemhauser GL, Wolsey LA (1990) The uncapacitated facility location problem. In: Discrete location theory. Wiley, New York, pp 119–171
6. Jain K, Mahdian M, Markakis E, Saberi A, Vazirani VV (2003) Greedy facility location algorithms analyzed using dual fitting with factor-revealing LP. J ACM 50(6):795–824
7. Jain K, Vazirani VV (2001) Approximation algorithms for metric facility location and k-median problems using the primal-dual schema and Lagrangian relaxation. J ACM 48(2):274–296
8. Kaufman L, Rousseeuw PJ (1990) Finding groups in data: an introduction to cluster analysis. Wiley, New York
9. Korupolu MR, Plaxton CG, Rajaraman R (1998) Analysis of a local search heuristic for facility location problems. In: SODA '98: proceedings of the ninth annual ACM-SIAM symposium on discrete algorithms, San Francisco, 25–26 Jan 1998, pp 1–10
10. Kuehn AA, Hamburger MJ (1963) A heuristic program for locating warehouses. Manag Sci 9(4):643–666
11. Lin JH, Vitter JS (1992) ε-approximations with minimum packing constraint violation (extended abstract). In: STOC '92: proceedings of the twenty-fourth annual ACM symposium on theory of computing, Victoria, pp 771–782
12. Mahdian M, Pál M (2003) Universal facility location. In: European symposium on algorithms, Budapest, 16–19 Sept 2003, pp 409–421
13. Ng RT, Han J (1994) Efficient and effective clustering methods for spatial data mining. In: Proceedings of the symposium on very large data bases (VLDB), Santiago de Chile, 12–15 Sept 1994, pp 144–155
14. Pál M, Tardos É, Wexler T (2001) Facility location with nonuniform hard capacities. In: Proceedings of the 42nd annual symposium on foundations of computer science, Las Vegas, 14–17 Oct 2001, pp 329–338
15. Shmoys DB, Tardos É, Aardal K (1997) Approximation algorithms for facility location problems. In: Proceedings of the twenty-ninth annual ACM symposium on theory of computing, El Paso, 4–6 May 1997, pp 265–274

L

Locality in Distributed Graph Algorithms

Pierre Fraigniaud
Laboratoire d'Informatique Algorithmique: Fondements et Applications, CNRS and University Paris Diderot, Paris, France

Keywords

Coloring; Distributed computing; Maximal independent set; Network computing

Years and Authors of Summarized Original Work

1992; Linial
1995; Naor, Stockmeyer
2013; Fraigniaud, Korman, Peleg

Problem Definition

In the context of *distributed network computing*, an important concern is the ability to design *local* algorithms, that is, distributed algorithms in which every node (Each node is a computing entity, which has the ability to exchange messages with its neighbors in the network along its communication links.) of the network can deliver its result after having consulted only nodes in its vicinity. The word "vicinity" has a rather vague interpretation in general. Nevertheless, the objective is commonly to design algorithms in which every node outputs after having exchanged information with nodes at constant distance from it (i.e., at distance independent of the number of nodes n in the networks) or at distance at most polylogarithmic in n, but certainly significantly smaller than n or than the diameter of the network.

The *tasks* to be solved by distributed algorithms acting in networks can be formalized as follows. The network itself is modeled by an undirected connected graph G with node set $V(G)$ and edge set $E(G)$, without loops and double edges. In the sequel, by *graph* we are only referring to this specific type of graphs. Nodes are *labeled* by a function $\ell : V \rightarrow \{0,1\}^*$ that assigns to every node v its label $\ell(v)$. A pair (G,ℓ), where G is a graph and ℓ is a labeling of G, is called *configuration*, and a collection $\mathcal{L}$ of configurations is called a *distributed language*. A typical example of a distributed language is: $\mathcal{L}_{\text{properly colored}} = \{(G,\ell) : \ell(v) \neq \ell(v') \text{ for all } \{v,v'\} \in E(G)\}$.

Unless specified otherwise, we are always assuming that the considered languages are decidable in the sense of classical (sequential) computability theory. To every distributed language $\mathcal{L}$ can be associated a *construction* task which consists in computing the appropriate labels for a given network (Here, we are restricting ourselves to *input-free* construction tasks, but the content of this chapter can be generalized to tasks with inputs, in which case the labels are input-output pairs, and, given the inputs, the nodes must produce the appropriate outputs to fit in the considered language.):

Problem 1 (Construction Task for $\mathcal{L}$)

INPUT: A graph G (in which nodes haves no labels);

OUTPUT: A label $\ell(v)$ at each node v, such that $(G,\ell) \in \mathcal{L}$.

For instance, the construction task for $\mathcal{L}_{\text{properly colored}}$ consists, for each node of a graph G, to output a color so that any two adjacent nodes do not output the same color. To every distributed language $\mathcal{L}$ can also be associated a *decision* task, which consists in having nodes deciding whether any given configuration (G,ℓ) is in $\mathcal{L}$ (in this case, every node v is given its label $\ell(v)$ as inputs). This type of tasks finds applications whenever it is desired to check the correctness of a solution produced by another algorithm or, say, by some black box that may act incorrectly. The decision *rule*, motivated by various considerations including *termination detection*, is as follows: if $(G,\ell) \in \mathcal{L}$, then all nodes must *accept* the configuration, while if $(G,\ell) \notin \mathcal{L}$, then at least one node must *reject* that configuration. In other words:

Problem 2 (Decision Task for $\mathcal{L}$)

INPUT: A configuration (G,ℓ) (i.e., each node $v \in V(G)$ has a label $\ell(v)$);

OUTPUT: A boolean $b(v)$ at each node v such that:

$$(G,\ell) \in \mathcal{L} \iff \bigwedge_{v \in V(G)} b(v) = \textit{true}.$$

For instance, a decision algorithm for $\mathcal{L}_{\text{properly colored}}$ consists, for each node v, of a graph G with input some color $\ell(v)$, to accept if all its neighbors have colors distinct from $\ell(v)$, and to reject otherwise. Finally, the third type of tasks can be associated to distributed languages, called *verification* tasks, which can also be seen as a *nondeterministic* variant of the decision tasks. In the context of verification, in addition to its label $\ell(v)$, every node $v \in V(G)$ is also given a *certificate* $c(v)$. This provides G with a global distributed certificate $c : V(G) \rightarrow \{0,1\}^*$

that is supposed to attest to the fact that the labels are correct. If this is indeed the case, i.e., if $(G, \ell) \in \mathcal{L}$, then all nodes must accept the instance (provided with the due certificate). Note that a verification algorithm is allowed to reject a configuration $(G, \ell) \in \mathcal{L}$ in case the certificate is not appropriate for that configuration since for every configuration $(G, \ell) \in \mathcal{L}$, one just asks for the existence of *at least one* appropriate certificate. In addition, to prevent the nodes to be fooled by some certificate on an illegal instance, it is also required that if $(G, \ell) \notin \mathcal{L}$, then for *every* certificate, at least one node must reject that configuration. In other words:

Problem 3 (Verification Task for $\mathcal{L}$)

INPUT: A configuration (G, ℓ), and a distributed certificate c;
OUTPUT: A boolean $b(v, c)$ at each node v, which may indeed depend on c, such that:

$$(G, \ell) \in \mathcal{L} \iff \bigvee_{c' \in \{0,1\}^*} \bigwedge_{v \in V(G)} b(v, c') = \text{true}.$$

For instance, cycle-freeness cannot be locally decided, as even cycles and paths cannot be locally distinguished. However, cycle-freeness can be locally verified, using certificates on $O(\log n)$ bits, as follows. The certificate of node v in a cycle-free graph G is its distance in G to some fixed node $v_0 \in V(G)$. The verification algorithm essentially checks that every node v with $c(v) > 0$ has a unique neighbor v' with $c(v') = c(v) - 1$ and all its other neighbors w with $c(w) = c(v) + 1$, while a node v with $c(v) = 0$ checks that all its neighbors w satisfy $c(w) = 1$. If G has a cycle, then no certificates can pass these tests. As in sequential computability theory, the terminology "verification" comes from the fact that a distributed certificate can be viewed as a (distributed) *proof* that the current configuration is in the language, and the role of the algorithm is to verify this proof. The ability to simultaneously construct a labeling ℓ for G as well as a proof c certifying the correctness of ℓ is a central notion in the design of distributed *self-stabilizing* algorithms – in which variables can be transiently corrupted.

Locality in distributed graph algorithms is dealing with the design and analysis of distributed network algorithms solving any of the above three kinds of tasks.

Computational Model

The study of local algorithms is usually tackled in the framework of the so-called LOCAL model, formalized and thoroughly studied in [13]. In this model, every node v is a Turing machine which is given an *identity*, i.e., a nonnegative integer $\text{id}(v)$. All identities given to the nodes of any given network are pairwise distinct. All nodes execute the same algorithm. They wake up simultaneously, and the computation performs in synchronous *rounds*, where each round consists in three phases executed by each node: (1) send a message to all neighboring nodes in the network, (2) receive the messages sent by the neighboring nodes in the network, and (3) perform some individual computation. The complexity of an algorithm in the LOCAL model is measured in term of *number of rounds* until all nodes terminate. This number of rounds is actually simply the maximum, taken over all nodes in the network, of the distance at which information is propagated from a node in the network. In fact, an algorithm performing in t rounds can be rewritten into an algorithm in which every node, first, collects all data from the nodes at distance at most t from it in G and, second, performs some individual computation on these data.

Observe that the LOCAL model is exclusively focusing on the locality issue and ignores several aspects of the computation. In particular, it is synchronous and fault-free. Also, the model is oblivious to the amount of individual computation performed at each node, and it is oblivious to the amount of data that are transmitted between neighbors at each round. An important consequence of these facts is that lower bounds derived in this model are very robust, in the sense that they are not resulting from clock drifts, crashes, nor from any kind of limitation on the individual computation or on the volume of transmitted data. Instead, lower bounds in the LOCAL model result solely from the fact that every node is unaware of what is lying beyond a certain horizon in the network and must cope with this uncertainty (Most upper bounds are however based on algorithms that perform polynomial-time individual computations at each node and exchange only a polylogarithmic amount of bits between nodes.).

Note also that the identities given to the nodes may impact the result of the computation. In particular the label ℓ produced by a construction algorithm may not only depend on G but also on the identity assignment id : $V(G) \rightarrow \mathbb{N}$. The same holds for decision and verification algorithms, in which the accept/reject decision at a node may be impacted by its identity (thus, for an illegal configuration, the nodes that reject may differ depending on the identity assignment to the nodes). However, in the case of verification, it is desirable that the certificates given to the nodes do not depend on their identities, but solely on the current configuration. Indeed, the certificates should rather be solely depending on the given configuration with respect to the considered language and should better not depend on implementation factors such as, say, the IP address given to a computer (The theory of *proof-labeling scheme* [7] however refers to scenarios in which it is fully legitimate that certificates may also depend on the node identities).

Classical Tasks

Many tasks investigated in the framework of network computing are related to classical graph problems, including computing proper colorings, independent sets, matchings, dominating sets, etc. Optimization problems are however often weakened. For instance, the coloring problem considered in the distributed setting is typically $(\Delta + 1)$-coloring, where Δ denotes the maximum node degree of the current network. Similarly, instead of looking for a minimum dominating set, or for a maximum independent set, one typically looks for dominating sets (resp., independent sets) that are minimal (resp., maximal) for inclusion. There are at least two reasons for such relaxations, besides the fact that the relaxed versions are sequentially solvable in polynomial time by simple greedy algorithms while the original versions are NP-hard. First, one can trivially locally decide whether a solution of the aforementioned relaxed problems satisfies the constraints of the relaxed variants, which yield the question of whether one can also construct their solutions locally (Instead, problems like minimum-weight spanning tree construction cannot be checked locally as the presence of an edge in the solution may depend of another edge, arbitrarily far in the network.). Second, these relaxed problems already involve one of the most severe difficulties distributed computing has to cope with, that is, *symmetry breaking*.

Key Results

In this section, we say that a distributed algorithm is *local* if and only if it performs in a constant number of rounds in the LOCAL model, and we are interested in identifying distributed languages that are locally constructible, locally decidable, and/or locally verifiable.

Local Algorithms

Naor and Stockmeyer [11] have thoroughly studied the distributed languages that can be locally constructed. They established that it is TM-undecidable whether a distributed language can be locally constructed, and this holds even if one restricts the problem to distributed languages that can be locally decided (On the other hand, it appears to be not easy to come up with a nontrivial example of a distributed language that can be constructed locally. One such nontrivial example is given in [11]: *weak coloring*, in which every non-isolated node must have at least one neighbor colored differently, is locally constructible for a large class of graphs. This problem is related to some *resource allocation* problem.). The crucial notion of *order-invariant* algorithms, defined as algorithms such that the output at every node is identical for every two identity assignments that preserve the relative ordering of the identities, was also introduced in [11]. Using Ramsey theory, it is proved that in networks with constant maximum degree, for every locally decidable distributed language $\mathcal{L}$ with constant-size labels, if $\mathcal{L}$ can be constructed by a local algorithm, then $\mathcal{L}$ can also be constructed by a local order-invariant algorithm. This result has many important consequences. One is for instance the impossibility to solve $(\Delta + 1)$-coloring and maximal independent set (MIS) in a constant number of rounds. This follows from the fact that a t-round order-invariant algorithm cannot solve these problems in rings where nodes are consecutively labeled from 1 to n, because adjacent nodes with identities in $[t + 1, n - t - 1]$ must produce the same output. Another important consequence of the restriction to order-invariant algorithms is the derandomization theorem in [11] stating that, in constant degree graphs, for every locally decidable distributed language $\mathcal{L}$ with constant-size label, if $\mathcal{L}$ can be constructed by a randomized Monte Carlo local algorithm, then $\mathcal{L}$ can also be constructed by a deterministic local algorithm.

The distributed languages that can be locally decided, or verified, have been studied by Fraigniaud, Korman, and Peleg in [6]. Several complexity classes are defined and separated, and complete languages are identified for the local reduction. It is also shown in [6] that the class of all distributed languages that can be locally verified by a randomized Monte Carlo algorithm with success probability $\frac{\sqrt{5}-1}{2}$ includes *all* distributed languages. The impact of randomization is however somehow limited, at least for the class of distributed languages closed under node deletion. Indeed, [6] establishes that for any such language $\mathcal{L}$, if $\mathcal{L}$ can be locally decided by a randomized Monte Carlo algorithm with success probability greater than $\frac{\sqrt{5}-1}{2}$, then $\mathcal{L}$ can be locally decided by a deterministic algorithm. Finally, [6] additionally discusses the power of *oracles* providing nodes with information about the current network, like, typically, its number of nodes.

Almost Local Algorithms

Linial [8] proved that constructing a $(\Delta + 1)$-coloring, or, equivalently, a MIS, requires $\Omega(\log^* n)$ rounds, where $\log^* x$ is the number of times one should iterate the log function, starting from x, for reaching a value less than 1. The $\log^*$ function grows quite slowly (e.g., $\log^* 10^{100} = 5$), but is not constant. This lower bound holds even for n-node rings in which identities are in $[1, n]$, nodes know n, and nodes share a consistent notion of clockwise direction. Linial's lower bound is tight, as a 3-coloring algorithm performing in $O(\log^* n)$ rounds can be obtained by adapting the algorithm by Cole and Vishkin [5] originally designed for the PRAM model to the setting of the lower bound. Also, Linial [8] describes a $O(\log^* n)$-round algorithm for Δ^2-coloring. Note that the $\Omega(\log^* n)$-round lower bound for $(\Delta + 1)$-coloring extends to randomized Monte Carlo algorithms [10]. On the other hand, the best known upper bounds on the number of rounds to solve $(\Delta + 1)$-coloring in arbitrary graphs are $2^{O(\sqrt{\log n})}$ for deterministic algorithms [12] and expected $O(\log n)$ for randomized Las Vegas algorithms [1, 9].

By expressing the complexity of local algorithms in terms of both the size n of the network and its max-degree Δ, one can distinguish the impact of these two parameters. For instance, Linial's $O(\log^* n)$-round algorithm for Δ^2-coloring [8] can be adapted to produce an $O(\Delta^2 + \log^* n)$-round algorithm for $(\Delta + 1)$-coloring. This bound has been improved by a series of contributions, culminating to the currently best known algorithm for $(\Delta + 1)$-coloring performing in $O(\Delta + \log^* n)$ rounds [3]. Also, there is a randomized $(\Delta + 1)$-coloring algorithms performing in expected $O\left(\log \Delta + \sqrt{\log n}\right)$ rounds [14]. This algorithm was recently improved to another algorithm performing in $O\left(\log \Delta + e^{O(\sqrt{\log \log n})}\right)$ rounds [4].

Additional Results

The reader is invited to consult the monograph [2] for more inputs on local distributed graph coloring algorithms, the survey [15] for a detailed survey on local algorithms, as well as the textbook [13] for the design of distributed graph algorithms in various contexts, including the LOCAL model.

Open Problems

As far as local construction tasks are concerned, in a way similar to what happens in sequential computing, the theory of local distributed computing lacks lower bounds. (Celebrated Linial's lower bound [8] is actually one of the very few examples of a nontrivial lower bound for local computation). As a consequence, one observes large gaps between the numbers of rounds of the best known lower bounds and of the best known algorithms. This is typically the case for $(\Delta + 1)$-coloring. One of the most important open problems in this field is in fact to close these gaps for coloring as well as for many other graph problems. Similarly, although studied in depth by, e.g., Naor and Stockmeyer [11], the power of randomization is still not fully understood in the context of local computation. In general, the best known randomized algorithms are significantly faster than the best known deterministic algorithms, as witnessed by the case of $(\Delta + 1)$-coloring. Nevertheless, it is not known whether this is just an artifact of a lack of knowledge or an intrinsic separation between the two classes of algorithms.

In the context of local decision and verification tasks, the interplay between the ability to decide or verify locally and the ability to search (i.e., construct) locally is not fully understood. The completeness notions for local decision in [6] do not seem to play the same role as the completeness notions in classical complexity theory. In particular, in the context of local computing, one has not yet observed phenomena similar to self-reduction for NP-complete problems. Yet, the theory of local decision and verification is in its infancy, and it may be too early for drawing conclusions about its impact on distributed local search. An intriguing question is related to generalizing decision and verification tasks in a way similar to the polynomial hierarchy in sequential computing, by adding more alternating quantifiers in the specification of Problem 3. For instance, it would then be interesting to figure out whether each level of the hierarchy has a "natural" language as representative.

Recommended Reading

1. Alon N, Babai L, Itai A (1986) A fast and simple randomized parallel algorithm for the maximal independent set problem. J. Algorithms 7(4): 567–583
2. Barenboim L, Elkin M (2013) Distributed graph coloring: fundamentals and recent developments. Synthesis lectures on distributed computing theory. Morgan & Claypool Publishers
3. Barenboim L, Elkin M, Kuhn F (2014) Distributed (Delta+1)-coloring in linear (in delta) time. SIAM J Comput 43(1):72–95
4. Barenboim L, Elkin M, Pettie S, Schneider J (2012) The locality of distributed symmetry breaking. In: Proceedings of the 53rd IEEE symposium on foundations of computer science (FOCS), New Brunswick, pp 321–330
5. Cole R, Vishkin U (1986) Deterministic coin tossing and accelerating cascades: micro and macro techniques for designing parallel algorithms. In: Proceedings of the 18th ACM symposium on theory of computing (STOC), Berkeley, pp 206–219
6. Fraigniaud P, Korman A, Peleg D (2013) Towards a complexity theory for local distributed computing. J ACM 60(5):35
7. Korman A, Kutten S, Peleg D (2010) Proof labeling schemes. Distrib Comput 22(4):215–233
8. Linial N (1992) Locality in distributed graph algorithms. SIAM J Comput 21(1):193–201
9. Luby M (1986) A simple parallel algorithm for the maximal independent set problem. SIAM J Comput 15:1036–1053
10. Naor M (1991) A lower bound on probabilistic algorithms for distributive ring coloring. SIAM J Discret Math 4(3):409–412
11. Naor M, Stockmeyer L (1995) What can be computed locally? SIAM J Comput 24(6): 1259–1277
12. Panconesi A, Srinivasan A (1996) On the complexity of distributed network decomposition. J Algorithms 20(2):356–374
13. Peleg D (2000) Distributed computing: a locality-sensitive approach. SIAM, Philadelphia
14. Schneider J, Wattenhofer R (2010) A new technique for distributed symmetry breaking. In: Proceedings of the 29th ACM symposium on principles of distributed computing (PODC), Zurich, pp 257–266
15. Suomela J (2013) Survey of local algorithms. ACM Comput Surv 45(2):24

Locally Decodable Codes

Shubhangi Saraf
Department of Mathematics and Department of Computer Science, Rutgers University, Piscataway, NJ, USA

Keywords

Error correcting codes; Locally decodable codes; Sublinear time algorithms

Years and Authors of Summarized Original Work

2000; Katz, Trevisan
2002; Goldreich, Karloff, Schulman, Trevisan
2004; Kerenedis, de Wolf
2007; Woodruff
2007; Raghavendra
2008; Yekhanin
2009; Efremenko
2010; Dvir, Gopalan, Yekhanin
2010; Woodruff
2011; Kopparty, Saraf, Yekhanin
2013; Hemenway, Ostrovsky, Wootters
2013; Guo, Kopparty, Sudan
2015; Kopparty, Meir, Ron-Zewi, Saraf

Problem Definition

Classical error-correcting codes allow one to encode a k-bit message $\mathbf{x}$ into an n-bit codeword $C(\mathbf{x})$, in such a way that $\mathbf{x}$ can still be accurately recovered even if $C(\mathbf{x})$ gets corrupted in a small number of coordinates. The traditional way to recover even a small amount of information contained in $\mathbf{x}$ from a corrupted version of $C(\mathbf{x})$ is to run a traditional decoder for C, which would read and process the entire corrupted codeword, to recover the entire original message $\mathbf{x}$. The required information or required piece of $\mathbf{x}$ can then be read off. In the current digital age where huge amounts of data need to be encoded and decoded, even running in linear time to read the entire encoded data might be too wasteful, and the need for *sublinear* algorithms for error correction is greater than ever. Specially if one is only interested in recovering a single bit or a few bits of $\mathbf{x}$, it is possible to have codes with much more efficient decoding algorithms, which allow for the decoding to take place by only reading a sublinear number of code positions. Such codes are known as locally decodable codes (LDCs). Locally decodable codes allow reconstruction of an arbitrary bit $\mathbf{x}_i$ with high probability by only reading $t \ll k$ randomly chosen coordinates of (a possibly corrupted) $C(\mathbf{x})$.

The two main interesting parameters of a locally decodable code are (1) the codeword length n (as a function of the message length k) which measures the amount of redundancy that is introduced into the message by the encoder and (2) the query complexity of local decoding which counts the number of bits that need to be read from a (corrupted) codeword in order to recover a single bit of the message. Ideally, one would like to have both of these parameters as small as possible. One cannot, however, simultaneously minimize both of them; there is a trade-off. On one end of the spectrum, we have LDCs with the codeword length close to the message length, decodable with somewhat large query complexity. Such codes are useful for data storage and transmission. On the other end we have LDCs where the query complexity is a small constant, but the codeword length is large compared to the message length. Such codes find applications in complexity theory, derandomization, and cryptography (and this was the reason they were originally studied [15]). The true shape of the trade-off between the codeword length and the query complexity of LDCs is not known. Determining it is a major open problem (see [23] for an excellent recent survey of the LDC literature).

Natural variants of locally decodable codes are *locally correctable codes* (LCCs) where every symbol of the true codeword can be recovered with high probability by reading only a small number of locations of the corrupted codeword. When the underlying code is linear, it is known that every LCC is also an LDC.

Notation and Formal Definition

For a set Σ and two vectors $c, c' \in \Sigma^n$, we define the relative Hamming distance between c and c', which we denote by $\Delta(c, c')$, to be the fraction of coordinates where they differ: $\Delta(c, c' = \Pr_{i \in [n]}[c_i \neq c'_i]$.

An error-correcting code of length n over the alphabet Σ is a subset $\mathcal{C} \subseteq \Sigma^n$. The *rate* of $\mathcal{C}$ is defined to be $\frac{\log |\mathcal{C}|}{n \log |\Sigma|}$. The (minimum) distance of $\mathcal{C}$ is defined to be the smallest $\delta > 0$ such that for every distinct $c_1, c_2 \in \mathcal{C}$, we have $\Delta(c_1, c_2) \geq \delta$.

For an algorithm A and a string r, we will use A^r to represent that A is given query access to r.

We now define locally decodable codes.

Definition 1 (Locally Decodable Code) Let $\mathcal{C} \subseteq \Sigma^n$ be a code with $|\mathcal{C}| = |\Sigma|^k$. Let $E : \Sigma^k \to \mathcal{C}$ be a bijection (we refer to E as the encoding map for $\mathcal{C}$; note that k/n equals the rate of the code $\mathcal{C}$). We say that $(\mathcal{C}, E)$ is locally decodable from a δ'-fraction of errors with t queries if there is a randomized algorithm A, such that:

- **Decoding:** Whenever a message $x \in \Sigma^k$ and a received word $r \in \Sigma^n$ are such that $\Delta(r, E(x)) < \delta'$, then, for each $i \in [k]$,

$$\Pr[\mathsf{A}^r(i) = x_i] \geq 2/3.$$

- **Query complexity** t**:** Algorithm $\mathsf{A}^r(i)$ always makes at most t queries to r.

Key Results

Locally decodable codes have been implicitly studied in coding theory for a long time, starting with Reed's "majority-logic decoder" for binary Reed-Muller codes [18]. They were first formally defined by Katz and Trevisan [15] (see also [19]). Since then, the quest for understanding locally decodable codes has generated many developments.

As mentioned before, there are two main regimes in which LDCs have been studied. The first regime, on which most prior work has focused, is the regime where the query complexity is small – even a constant. In this setting, the most interesting question is to construct codes with rate as large as possible (though it is known that some significant loss in rate is inevitable). The second regime, which has been the focus of more recent work, is the high-rate regime. In this regime, one insists on the rate of the code being very close to 1 and then tries to obtain as low-query complexity as possible. We now discuss the known results in both regimes.

Low-Query Regime

A significant body of work on LDCs has focused on local decoding with a constant number of queries. Local decoding with 2 queries is almost fully understood. The Hadamard code is a 2-query LDC with codeword length $n = 2^k$. Moreover, it was shown in [9, 14] that any 2-query locally decodable code (that is decodable from some small fixed constant fraction of errors) must have $n = 2^{\Omega(k)}$.

For a long time, it was generally believed that for decoding with any constant number of queries, this exponential blowup is needed: a k-bit message must be encoded into at least $\exp(k^{\epsilon})$ bits, for some constant $\epsilon > 0$. This is precisely the trade-off exhibited by the Reed-Muller codes, which were believed to be optimal. Recently, in a surprising and beautiful sequence of works [6, 8, 17, 22], a new family of codes called matching vector codes was constructed, and they were shown to have local decoding parameters surprisingly much better than that of Reed-Muller codes! This family of codes gives constant query locally decodable codes which encode k bits into as few as $n = \exp(\exp(\log^{\epsilon}(k)))$ bits for some small (Parameter ϵ can be chosen arbitrarily close to 0 by increasing the number of queries as a function of ϵ.) constant $\epsilon > 0$ and are locally decodable from some fixed small constant fraction of errors.

There has also been considerable work [9, 14, 15, 20, 21] on lower bounds on the length of low-query locally decodable codes. However, there is still a huge gap in our understanding of the best rate achievable for any query complexity that is at least 3. For instance, for 3-query

LDCs that decode from a fixed small constant fraction of errors, the best lower bounds we know are of the form $n = \Omega(k^2)$ [21]. The best upper bounds on the other hand only give constructions with $n = \exp(\exp(\log^{O(1)}(k)))$. For codes locally decodable for general query complexity t, it is known [15] that $n = k^{1+\Omega(1/t)}$ (again, a really long way off from the upper bounds). Thus, in particular, for codes of constant rate, local decoding requires at least $\Omega(\log k)$ queries.

High-Rate Regime

In the high-rate regime, one fixes the rate to be constant, i.e., $n = O(k)$ or even $(1+\alpha)k$, for some small constant α and then tries to construct locally decodable codes with the smallest query complexity possible. Reed-Muller codes in this regime demonstrate the following setting of parameters: for any constant $\epsilon > 0$, there is a family of Reed-Muller codes of rate $= \exp(1/\epsilon)$ that is decodable with n^ϵ queries. Till recently this was the best trade-off for parameters known in this regime, and in fact we did not know any family of codes of rate $> 1/2$ with any sublinear query complexity.

In the last few years, three very different families of codes have been constructed [10–12] that go well beyond the parameters of Reed-Muller codes. These codes show that for arbitrary $\alpha, \epsilon > 0$, and for every finite field $\mathbb{F}$, for infinitely many n, there is a linear code over $\mathbb{F}$ of length n with rate $1-\alpha$, which is locally decodable (and even locally correctable) from a constant fraction (This constant is positive and is a function of only α and ϵ.) of errors with $O(n^\epsilon)$ queries. Codes with such parameters were in fact conjectured not to exist, and this conjecture, if it were true, would have yielded progress on some well-known open questions in arithmetic circuit complexity [7].

Even more recently, it was shown that one can achieve even subpolynomial query complexity while keeping rate close to 1. In [13], it was shown that, for any $\alpha > 0$ and for every finite field $\mathbb{F}$, for infinitely many n, there is a code over $\mathbb{F}$ of length n with rate $1-\alpha$, which is locally decodable (and even locally correctable) from a constant fraction (This constant is positive and is a function of only α.) of errors with $2^{O(\sqrt{\log n \log\log n})}$ queries.

On the lower bound front, all we know are the $n = k^{1+\Omega(1/t)}$ lower bounds by [15]. Thus, in particular, it is conceivable that for any small constant α, one could have a family of codes of rate $1-\alpha$ that are decodable with $O(\log n)$ queries.

Applications

In theoretical computer science, locally decodable codes have played an important part in the proof-checking revolution of the early 1990s [1,2,4,16], as well as in other fundamental results in hardness amplification and pseudorandomness [3, 5, 19]. Variations of locally decodable codes are also beginning to find practical applications in data storage and data retrieval.

Open Problems

In the constant query regime, the most important question is to get the rate to be as large as possible, and the major open question in this direction is the following:

Question 1 Do there exist LDCs with polynomial rate, i.e., $n = k^{O(1)}$, that are decodable with $O(1)$ queries?

In the high-rate regime, the best query lower bounds we know are just logarithmic, and the main challenge is to construct codes with improved query complexity, hopefully coming close to the best lower bounds we can prove.

Question 2 Do there exist LDCs of rate $1-\alpha$ or even $\Omega(1)$ that are decodable with $poly(\log n)$ queries?

Given the recent constructions of new families of codes in both the high-rate and low-query regimes with the strengthened parameters, it doesn't seem all that farfetched to imagine that we might soon be able to give much better answers to the above questions.

Recommended Reading

1. Arora S, Lund C, Motwani R, Sudan M, Szegedy M (1998) Proof verification and the hardness of approximation problems. J ACM 45(3):501–555
2. Arora S, Safra S (1998) Probabilistic checking of proofs: a new characterization of NP. J ACM 45(1):70–122
3. Arora S, Sudan M (2003) Improved low-degree testing and its applications. Combinatorica 23:365–426
4. Babai L, Fortnow L, Levin L, Szegedy M (1991) Checking computations in polylogarithmic time. In: 23rd ACM symposium on theory of computing (STOC), New Orleans, pp 21–31
5. Babai L, Fortnow L, Nisan N, Wigderson A (1993) BPP has subexponential time simulations unless EXPTIME has publishable proofs. Comput Complex 3:307–318
6. Dvir Z, Gopalan P, Yekhanin S (2010) Matching vector codes. In: 51st IEEE symposium on foundations of computer science (FOCS), Las Vegas, pp 705–714
7. Dvir Z (2010) On matrix rigidity and locally self-correctable codes. In: 26th IEEE computational complexity conference (CCC), Cambridge, pp 291–298
8. Efremenko K (2009) 3-query locally decodable codes of subexponential length. In: 41st ACM symposium on theory of computing (STOC), Bethesda, pp 39–44
9. Goldreich O, Karloff H, Schulman L, Trevisan L (2002) Lower bounds for locally decodable codes and private information retrieval. In: 17th IEEE computational complexity conference (CCC), Montréal, pp 175–183
10. Guo A, Kopparty S, Sudan M (2013) New affine-invariant codes from lifting. In: ACM innovations in theoretical computer science (ITCS), Berkeley, pp 529–540
11. Hemenway B, Ostrovsky R, Wootters M (2013) Local correctability of expander codes. In: 40th international colloquium on automata, languages, and programming (ICALP), Riga, pp 540–551
12. Kopparty S, Saraf S, Yekhanin S (2011) High-rate codes with sublinear-time decoding. In: 43rd ACM symposium on theory of computing (STOC), San Jose, pp 167–176
13. Kopparty S, Meir O, Ron-Zewi N, Saraf S (2015) High rate locally-correctable and locally-testable codes with sub-polynomial query complexity. In: Electronic colloquium on computational complexity (ECCC). TR15-068
14. Kerenidis I, de Wolf R (2004) Exponential lower bound for 2-query locally decodable codes via a quantum argument. J Comput Syst Sci 69:395–420
15. Katz J, Trevisan L (2000) On the efficiency of local decoding procedures for error-correcting codes. In: 32nd ACM symposium on theory of computing (STOC), Portland, pp 80–86
16. Lipton R (1990) Efficient checking of computations. In: 7th international symposium on theoretical aspects of computer science (STACS), Rouen. Lecture notes in computer science, vol 415, pp 207–215. Springer, Berlin/Heidelberg
17. Raghavendra P (2007) A note on Yekhanin's locally decodable codes. In: Electronic colloquium on computational complexity (ECCC). TR07-016
18. Reed IS (1954) A class of multiple-error-correcting codes and the decoding scheme. IEEE Trans Inf Theory 4:38–49
19. Sudan M, Trevisan L, Vadhan S (1999) Pseudorandom generators without the XOR lemma. In: 31st ACM symposium on theory of computing (STOC), Atlanta, pp 537–546
20. Woodruff D (2007) New lower bounds for general locally decodable codes. In: Electronic colloquium on computational complexity (ECCC). TR07-006
21. Woodruff D (2010) A quadratic lower bound for three-query linear locally decodable codes over Any Field. In: Approximation, randomization, and combinatorial optimization. Algorithms and techniques, 13th international workshop, APPROX 2010, and 14th international workshop, RANDOM, Barcelona, pp 766–779
22. Yekhanin S (2008) Towards 3-query locally decodable codes of subexponential length. J ACM 55:1–16
23. Yekhanin S (2012) Locally decodable codes. Found Trends Theor Comput Sci 6(3):139–255

Locally Testable Codes

Prahladh Harsha
Tata Institute of Fundamental Research, Mumbai, Maharashtra, India

Keywords

Error-correcting codes; Locally checkable; PCPs; Property testing

Years and Authors of Summarized Original Work

1990; Blum, Luby, Rubinfeld
2002; Goldreich, Sudan

Problem Definition

Locally testable codes (LTC) are error-correcting codes that support algorithms which can distinguish valid codewords from words that are "far" from all codewords by probing a given word only at a sublinear (typically constant) number of locations. LTCs are useful in the following scenario. Suppose data is transmitted by encoding it using a LTC. Then, one could check if the received data is nearly uncorrupted or has been considerably corrupted by making very few probes into the received data.

An error-correcting code $\mathcal{C} : \{0,1\}^k \to \{0,1\}^n$ is a function mapping k-bit messages to n-bit codewords. The ratio k/n is referred to as the rate of the code $\mathcal{C}$. The Hamming distance between two n-bit strings x and y, denoted by $\Delta(x, y)$, is the number of locations where x and y disagree, i.e., $\Delta(x, y) = \{i \in [n] \mid x_i \neq y_i\}$. The relative distance between x and y, denoted by $\delta(x, y)$, is the normalized distance, i.e., $\delta(x, y) = \Delta(x, y)/n$. The distance of the code $\mathcal{C}$, denote by $d(\mathcal{C})$, is the minimum Hamming distance between two distinct codewords, i.e., $d(\mathcal{C}) = \min_{x \neq y} \Delta(\mathcal{C}(x), \mathcal{C}(y))$. The distance of a string w from the code $\mathcal{C}$, denoted by $\Delta(w, \mathcal{C})$, is the distance of the nearest codeword to w, i.e., $\Delta(w, \mathcal{C}) = \min_x \Delta(w, \mathcal{C}(x))$. The relative distance of a code and the relative distance of a string w to the code are the normalized versions of the corresponding distances.

Definition 1 (locally testable code (LTC)) A code $\mathcal{C} : \{0,1\}^k \to \{0,1\}^n$ is said to be (q, δ, ε)-locally testable if there exists a probabilistic oracle algorithm T, also called a tester that, on oracle access to an input string $w \in \{0,1\}^n$, makes at most q queries (A *query* models a probe into the input string w in which one symbol (here a bit) of w is read.) to the string w and has the following properties:

Completeness: For every message $x \in \{0,1\}^k$, with probability 1 (over the tester's internal random coins), the tester T accepts the word $\mathcal{C}(x)$. Equivalently,

$$\forall x \in \{0,1\}^k, \qquad \Pr[T^{\mathcal{C}(x)} \text{ accepts}] = 1.$$

Soundness: For every string $w \in \{0,1\}^n$ such that $\Delta(w, \mathcal{C}) \geq \delta n$, the tester T rejects the word w with probability at least ε (despite reading only q bits of the word w). Equivalently,

$$\forall w \in \{0,1\}^n,$$
$$\Delta(w, \mathcal{C}) \geq \delta n \implies \Pr[T^w \text{ accepts}] \leq 1 - \varepsilon.$$

Local testability was first studied in the context of program checking by Blum, Luby, and Rubinfeld [8] who showed that the Hadamard code is locally testable (Strictly, speaking Blum et al. only showed that "Reed-Muller codes of order 1," a strict subclass of Hadamard codes are locally testable, while later Kaufman and Litsyn [15] demonstrated the local testability of the entire class of Hadamard codes.) and Gemmell et al. [11] who showed that the Reed-Muller codes are locally testable. The notion of LTCs is implicit in the work on locally checkable proofs by Babai et al. [2] and subsequent works on PCP. The explicit definition appeared independently in the works of Rubinfeld and Sudan [17], Friedl and Sudan [10], Arora's PhD thesis [1] (under the name of "probabilistically checkable proofs"), and Spielman's PhD thesis [18] (under the name of "checkable codes"). A formal study of LTCs was initiated by Goldreich and Sudan [14].

The following variants of the above definition of locally testability have also been studied.

- 2-sided vs. 1-sided error: The above definition of LTCs has perfect completeness, in the sense that every valid codeword is accepted with probability exactly 1. The tester, in this case, is said to have 1-sided error. A 2-sider error tester, on the other hand, accepts valid codewords with probability at least c for some $c \in (1 - \varepsilon, 1]$. However, most constructions of LTCs have perfect completeness.
- Strong/robust LTCs: The soundness requirement in Definition 1 can be strengthened in the following sense. We can require that there exists a constant $\rho \in (0, 1)$ such that for every string $w \in \{0,1\}^n$ which satisfies $\Delta(w, \mathcal{C}) \geq d$, we have

$$\Pr[T^w \text{ accepts}] \leq 1 - \frac{\rho d}{n}.$$

In other words, non-codewords which are at least a certain minimum distance from the code are not only rejected with probability at least ε but are in fact rejected with probability proportional to the distance of the non-codeword from the code. Codes that have such testers are called (q, ρ)-strong locally testable codes. They are sometimes also referred to as (q, ρ)-robust locally testable codes. Most constructions of LTCs satisfy the stronger soundness requirement.

- Adaptive vs. nonadaptive: The q queries of the tester T could either be adaptive or nonadaptive. Almost all constructions of LTCs are nonadaptive.
- Tolerant testers: Tolerant LTCs are codes with testers that accept not only valid codewords but also words which are close to the code, within a particular tolerance parameter δ' for $\delta' \leq \delta$.

LTCs are closely related to probabilistically checkable proofs (PCPs). Most known constructions of PCPs yield LTCs with similar parameters. In fact, there is a generic transformation to convert a PCP of proximity (which is a PCP with more requirements) into an LTC with comparable parameters [7,19]. See a survey by Goldreich [13] for the interplay between PCP and LTC constructions.

Locally decodable codes (LDCs), in contrast to LTCs, are codes with sublinear time decoders. Informally, such decoders can recover each message entry with high probability by probing the word at a sublinear (even constant) number of locations provided that the codeword has not been corrupted at too many locations. Observe that LTCs distinguish codewords from words that are far from the code while LDCs allow decoding from words that are close to the code.

Key Results

Local Testability of Hadamard Codes

As a first example, we present the seminal result of Blum, Luby, and Rubinfeld [8] that showed that Hadamard codes are locally testable. In this setting, we are given a string $f \in \{0,1\}^{2^k}$ and we would like to test if f is a Hadamard codeword. It will be convenient to view the 2^k-bit long string f as a function $f : \{0,1\}^k \to \{0,1\}$. In this alternate view, Hadamard codewords (strictly speaking, "Reed-Muller codewords of order 1") correspond to linear functions, i.e., they satisfy $\forall x, y \in \{0,1\}^k, f(x) + f(y) = f(x+y)$ (Here addition "+" refers to bitwise xor. In other words, $b_1 + b_2 := b_1 \oplus b_2$ for $b_1, b_2 \in \{0,1\}$ and $x + y = (x_1, x_2 \ldots, x_k) + (y_1, y_2 \ldots, y_k) := (x_1 \oplus y_1, x_2 \oplus y_2 \ldots, x_k \oplus y_k)$ for $x, y \in \{0,1\}^k$). The following test is due to Blum, Luby, and Rubinfeld. The accompanying theorem shows that this is in fact a robust characterization of linear functions.

BLR-Test

Input: Parameter k and oracle access to $f : \{0,1\}^k \to \{0,1\}$:

1. Choose $x, y \in_R \{0,1\}^k$ uniformly at random.
2. Query f at locations x, y and $x + y$.
3. Accept iff $f(x) + f(y) = f(x + y)$.

Clearly, the BLR-test always accepts all linear functions (i.e., Hadamard codewords). Blum, Luby, and Rubinfeld (with subsequent improvements due to Coppersmith) showed that if f has relative distance at least δ from all linear functions, then BLR-test rejects with probability at least $\min\{\delta/2, 2/9\}$. Their result was more general in the sense that it applied to all additive groups and not just $\{0,1\}$. For the special case of $\{0,1\}$, Bellare et al. [3] obtained the following stronger result:

Theorem 1 ([3,8]) *If f is at relative distance at least δ from all linear functions, then the BLR-test rejects f with probability at least δ.*

Local Testability of Reed-Muller Codes

Rubinfeld and Sudan [17] considered the problem of local testability of the Reed-Muller codes. Here, we consider codes over non-Boolean alphabets and the natural extension of LTCs to this non-Boolean setting. Given a field $\mathbb{F}$ and parameters d and m (where $d+1<|\mathbb{F}|$), the Reed-Muller code consists of codewords which are evaluations of m-variate polynomials of total degree at most d. Let $\alpha_0, \alpha_1, \alpha_2, \ldots, \alpha_{d+1}$ be $(d+2)$ distinct elements in the field $\mathbb{F}$ (recall that $d+1<|\mathbb{F}|$). The following test checks if a given $f: \mathbb{F}^m \rightarrow \mathbb{F}$ is close to the Reed-Muller code.

RS-test

Input: Field $\mathbb{F}$, parameter d, and oracle access to $f: \mathbb{F}^m \rightarrow \mathbb{F}$:

1. Choose $x, y \in_R \mathbb{F}^m$ uniformly at random.
2. Query f at locations $(x+\alpha_i \cdot y), i = 1, \ldots, d+1$.
3. Interpolate to construct a univariate polynomial $q: \mathbb{F} \rightarrow \mathbb{F}$ of degree at most d such that $q(\alpha_i) = f(x+\alpha_i \cdot y), i = 1, \ldots, d+1$.
4. If $q(\alpha_0) = f(x+\alpha_0 \cdot y)$ accept, else reject.

The above test checks that the restriction of the function f to the line $l(t) = x + ty$ is a univariate polynomial of degree at most d. Clearly, multivariate polynomials of degree at most d are always accepted by the RS-test.

Theorem 2 ([17]) *There exists a constant c such that if $|\mathbb{F}| \geq 2d+2$ and $\delta < c/d^2$, then the following holds for every (positive) integer m. If f has relative distance at least δ from all m-variate polynomials of total degree at most d, then the RS-test rejects f with probability at least $\delta/2$.*

Open Problems

The Hadamard code is testable with three queries but has inverse exponential rate, whereas the Reed-Muller code (for certain setting of parameters d, m, and $\mathbb{F}$) has polylogarithmic query complexity and inverse polynomial rate. We can ask if there exist codes good with respect to both parameters. In other words, do there exist codes with inverse polynomial rate and linear distance which are testable with a constant number of queries? Such a construction, with nearly linear rate, was obtained by Ben-Sasson and Sudan [5] and Dinur [9].

Theorem 3 ([5,9]) *There exists a constant q and an* explicit *family of codes $\{C_k\}_k$ where $C_k: \{0,1\}^k \rightarrow \{0,1\}^{k \cdot \text{poly} \log n}$ that have linear distance and are q-locally testable.*

This construction is obtained by combining the algebraic PCP of proximity-based constructions due to Ben-Sasson and Sudan [5] with the gap amplification technique of Dinur [9]. Meir [16] obtained a LTC with similar parameters using a purely combinatorial construction, albeit a non-explicit one.

It is open if this construction can be further improved. In particular, it is open if there exist codes with constant rate and linear relative distance (such codes are usually referred to as *good codes*) that are constant query locally testable. We do not know of even a non-explicit code with such properties. To the contrary, it is known that random low-density parity check matrix (LDPC) codes are in fact not locally testable [6]. For a more detailed survey on LTCs, their constructions, and limitations, the interested reader is directed to excellent surveys by Goldreich [13], Trevisan [19], and Ben-Sasson [4].

Cross-References

▶ Linearity Testing/Testing Hadamard Codes
▶ Locally Decodable Codes

Recommended Reading

1. Arora S (1994) Probabilistic checking of proofs and the hardness of approximation problems. PhD thesis, University of California, Berkeley
2. Babai L, Fortnow L, Levin LA, Szegedy M (1991) Checking computations in polylogarithmic time. In: Proceedings of the 23rd ACM symposium on theory of computing (STOC), pp 21–31. doi:10.1145/103418.103428
3. Bellare M, Coppersmith D, Håstad J, Kiwi MA, Sudan M (1996) Linearity testing in characteristic two. IEEE Trans Inf Theory 42(6):1781–1795. doi:10.1109/18.556674, (Preliminary version in *36th FOCS*, 1995)

4. Ben-Sasson E (2010) Limitation on the rate of families of locally testable codes. In: [12], pp 13–31. doi:10.1007/978-3-642-16367-8_3
5. Ben-Sasson E, Sudan M (2008) Short PCPs with polylog query complexity. SIAM J Comput 38(2):551–607. doi:10.1137/050646445, (Preliminary version in *37th STOC*, 2005)
6. Ben-Sasson E, Harsha P, Raskhodnikova S (2005) Some 3CNF properties are hard to test. SIAM J Comput 35(1):1–21. doi:10.1137/S0097539704445445, (Preliminary version in *35th STOC*, 2003)
7. Ben-Sasson E, Goldreich O, Harsha P, Sudan M, Vadhan S (2006) Robust PCPs of proximity, shorter PCPs and applications to coding. SIAM J Comput 36(4):889–974. doi:10.1137/S0097539705446810, (Preliminary version in *36th STOC*, 2004)
8. Blum M, Luby M, Rubinfeld R (1993) Self-testing/correcting with applications to numerical problems. J Comput Syst Sci 47(3):549–595. doi:10.1016/0022-0000(93)90044-W, (Preliminary version in *22nd STOC*, 1990)
9. Dinur I (2007) The PCP theorem by gap amplification. J ACM 54(3):12. doi:10.1145/1236457.1236459, (Preliminary version in *38th STOC*, 2006)
10. Friedl K, Sudan M (1995) Some improvements to total degree tests. In: Proceedings of the 3rd Israel symposium on theoretical and computing systems, pp 190–198. Corrected version available online at http://theory.lcs.mit.edu/~madhu/papers/friedl.ps
11. Gemmell P, Lipton RJ, Rubinfeld R, Sudan M, Wigderson A (1991) Self-testing/correcting for polynomials and for approximate functions. In: Proceedings of the 23rd ACM symposium on theory of computing (STOC), pp 32–42. doi:10.1145/103418.103429
12. Goldreich O (ed) (2010) Property testing (Current research and surverys). Lecture notes in computer science, vol 6390. Springer, Berlin
13. Goldreich O (2010) Short locally testable codes and proofs: a survey in two parts. In: [12], pp 65–104. doi:10.1007/978-3-642-16367-8_6
14. Goldreich O, Sudan M (2006) Locally testable codes and PCPs of almost linear length. J ACM 53(4):558–655. doi:10.1145/1162349.1162351, (Preliminary Verison in *43rd FOCS*, 2002)
15. Kaufman T, Litsyn S (2005) Almost orthogonal linear codes are locally testable. In: Proceedings of the 46th IEEE symposium on foundations of computer science (FOCS), pp 317–326. doi:10.1109/SFCS.2005.16
16. Meir O (2009) Combinatorial construction of locally testable codes. SIAM J Comput 39(2):491–544. doi:10.1137/080729967, (Preliminary version in *40th STOC*, 2008)
17. Rubinfeld R, Sudan M (1996) Robust characterizations of polynomials with applications to program testing. SIAM J Comput 25(2):252–271. doi:10.1137/S0097539793255151, (Preliminary version in *23rd STOC*, 1991 and *3rd SODA*, 1992)
18. Spielman DA (1995) Computationally efficient error-correcting codes and holographic proofs. PhD thesis, Massachusetts Institute of Technology. http://www.cs.yale.edu/homes/spielman/Research/thesis.html
19. Trevisan L (2004) Some applications of coding theory in computational complexity. Quaderni di Matematica 13:347–424. cs/0409044

Low Stretch Spanning Trees

Michael Elkin
Department of Computer Science, Ben-Gurion University, Beer-Sheva, Israel

Keywords

Spanning trees with low average stretch

Years and Authors of Summarized Original Work

2005; Elkin, Emek, Spielman, Teng

Problem Definition

Consider a weighted connected multigraph $G = (V, E, \omega)$, where ω is a function from the edge set E of G into the set of positive reals. For a path P in G, the *weight* of P is the sum of weights of edges that belong to the path P. For a pair of vertices $u, v \in V$, the *distance* between them in G is the minimum weight of a path connecting u and v in G. For a spanning tree T of G, the stretch of an edge $(u,v) \in E$ is defined by

$$\mathrm{stretch}_T(u, v) = \frac{\mathrm{dist}_T(u, v)}{\mathrm{dist}_G(u, v)},$$

and the average stretch over all edges of E is

$$\text{avestr}(G,T) = \frac{1}{|E|} \sum_{(u,v)\in E} \text{stretch}_T(u,v).$$

The average stretch of a multigraph $G = (V, E, \omega)$ is defined as the smallest average stretch of a spanning tree T of G, avestr(G, T). The average stretch of a positive integer n, avestr(n), is the maximum average stretch of an n-vertex multigraph G. The problem is to analyze the asymptotic behavior of the function avestr(n).

A closely related (dual) problem is to construct a probability distribution $\mathcal{D}$ of spanning trees for G, so that

$$\text{expstr}(G,\mathcal{D}) = \max_{e=(u,v)\in E} \mathbb{E}_{T\in\mathcal{D}}(\text{stretch}_T(u,v))$$

is small as possible. Analogously, expstr$(G) = \min_{\mathcal{D}}\{\text{expstr}(G,\mathcal{D})\}$, where the minimum is over all distributions $\mathcal{D}$ of spanning trees of G, and expstr$(n) = \max_G\{\text{expstr}(G)\}$, where the maximum is over all n-vertex multigraphs.

By viewing the problem as a 2-player zero-sum game between a tree player that aims to minimize the payoff and an edge player that aims to maximize it, it is easy to see that for every positive integer n, avestr(n) = expstr(n) [3]. The probabilistic version of the problem is, however, particularly convenient for many applications.

Key Results

The problem was studied since 1960s [9, 14, 16, 17]. A major progress in its study was achieved by Alon et al. [3], who showed that

$$\Omega(\log n) = \text{avestr}(n) = \text{expstr}(n) = \exp\left(O\left(\sqrt{\log n \cdot \log\log n}\right)\right).$$

Elkin et al. [10] improved the upper bound and showed that

$$\text{avestr}(n) = \text{expstr}(n) = O\left(\log^2 n \cdot \log\log n\right).$$

Applications

One application of low-stretch spanning trees is for solving symmetric diagonally dominant linear systems of equations. Boman and Hendrickson [6] were the first to discover the surprising relationship between these two seemingly unrelated problems. They applied the spanning trees of [3] to design solvers that run in time $m^{3/2}2^{O(\sqrt{\log n \log\log n})\log(1/\epsilon)}$. Spielman and Teng [15] improved their results by showing how to use the spanning trees of [3] to solve diagonally dominant linear systems in time

$$m2^{O(\sqrt{\log n \log\log n})}\log(1/\epsilon).$$

By applying the low-stretch spanning trees developed in [10], the time for solving these linear systems reduces to

$$m\log^{O(1)} n \log(1/\epsilon),$$

and to $O(n(\log n \log\log n)^2 \log(1/\epsilon))$ when the systems are planar. Applying a recent reduction of Boman, Hendrickson, and Vavasis [7], one obtains a $O(n(\log n \log\log n)^2 \log(1/\epsilon))$ time algorithm for solving the linear systems that arise when applying the finite element method to solve two-dimensional elliptic partial differential equations.

Chekuri et al. [8] used low-stretch spanning trees to devise an approximation algorithm for nonuniform buy-at-bulk network design problem. Their algorithm provides a first polylogarithmic approximation guarantee for this problem.

Abraham et al. [2] use a technique of Star decomposition introduced by Elkin et al. [10] to construct embeddings with a constant average stretch, where the average is over all *pairs of vertices*, rather than over all edges. The result of Abraham et al. [2] was, in turn, already used in a yet more recent work of Elkin et al. [11] on fundamental circuits.

Open Problems

Abraham and Neiman [1] subsequently devised an algorithm for constructing a spanning tree with average stretch $O(\log n \log\log n)$. The most evident open problem is to close the gap between this algorithm and the $\Omega(\log n)$ lower bound. Another intriguing subject is the study of low-stretch spanning trees for various restricted families of graphs. Progress in this direction was recently achieved by Emek and Peleg [12] that constructed low-stretch spanning trees with average stretch $O(\log n)$ for unweighted series-parallel graphs. Discovering other applications of low-stretch spanning trees is another promising venue of study.

Finally, there is a closely related relaxed notion of low-stretch *Steiner* or *Bartal* trees. Unlike a spanning tree, a Steiner tree does not have to be a subgraph of the original graph, but rather is allowed to use edges and vertices that were not present in the original graph. It is, however, required that the distances in the Steiner tree will be no smaller than the distances in the original graph. Low-stretch Steiner trees were extensively studied [4, 5, 13]. Fakcharoenphol et al. [13] devised a construction of low-stretch Steiner trees with an average stretch of $O(\log n)$. It is currently unknown whether the techniques used in the study of low-stretch Steiner trees can help in improving the bounds for the low-stretch spanning trees.

Cross-References

▸ Approximating Metric Spaces by Tree Metrics

Recommended Reading

1. Abraham I, Neiman O (2012) Using petal-decompositions to build low-stretch spanning trees. In: Proceedings of the 44th annual ACM symposium on theory of computing, New York, June 2012, pp 395–406
2. Abraham I, Bartal Y, Neiman O (2007) Embedding metrics into ultrametrics and graphs into spanning trees with constant average distortion. In: Proceedings of the 18th ACM-SIAM symposium on discrete algorithms, New Orleans, Jan 2007
3. Alon N, Karp RM, Peleg D, West D (1995) A graph-theoretic game and its application to the k-server problem. SIAM J Comput 24(1):78–100. Also available technical report TR-91-066, ICSI, Berkeley (1991)
4. Bartal Y (1996) Probabilistic approximation of metric spaces and its algorithmic applications. In: Proceedings of the 37th annual symposium on foundations of computer science, Berlington, Oct 1996, pp 184–193
5. Bartal Y (1998) On approximating arbitrary metrices by tree metrics. In: Proceedings of the 30th annual ACM symposium on theory of computing, Dallas, 23–26 May 1998, pp 161–168
6. Boman E, Hendrickson B (2001) On spanning tree preconditioners. Manuscript, Sandia National Lab
7. Boman E, Hendrickson B, Vavasis S (2004) Solving elliptic finite element systems in near-linear time with suppost preconditioners. Manuscript, Sandia National Lab and Cornell. http://arXiv.org/abs/cs/0407022. Accessed 9 July 2004
8. Chekuri C, Hagiahayi MT, Kortsarz G, Salavatipour M (2006) Approximation algorithms for non-uniform buy-at-bulk network design. In: Proceedings of the 47th annual symposium on foundations of computer science, Berkeley, Oct 2006, pp 677–686
9. Deo N, Prabhu GM, Krishnamoorthy MS (1982) Algorithms for generating fundamental cycles in a graph. ACM Trans Math Softw 8:26–42
10. Elkin M, Emek Y, Spielman D, Teng S-H (2005) Lower-stretch spanning trees. In: Proceedings of the 37th annual ACM symposium on theory of computing (STOC'05), Baltimore, May 2005, pp 494–503
11. Elkin M, Liebchen C, Rizzi R (2007) New length bounds for cycle bases. Inf Process Lett 104(5): 186–193
12. Emek Y, Peleg D (2006) A tight upper bound on the probabilistic embedding of series-parallel graphs. In: Proceedings of symposium on discrete algorithms (SODA'06), Miami, Jan 2006, pp 1045–1053
13. Fakcharoenphol J, Rao S, Talwar K (2003) A tight bound on approximating arbitrary metrics by tree metrics. In: Proceedings of the 35th annual ACM symposium on theory of computing, San Diego, June 2003, pp 448–455
14. Horton JD (1987) A polynomial-time algorithm to find the shortest cycle basis of a graph. SIAM J Comput 16(2):358–366
15. Spielman D, Teng S-H (2004) Nearly-linear time algorithm for graph partitioning, graph sparsification, and solving linear systems. In: Proceedings of the 36th annual ACM symposium on theory of computing (STOC'04), Chicago, June 2004, pp 81–90

16. Stepanec GF (1964) Basis systems of vector cycles with extremal properties in graphs. Uspekhi Mat Nauk 19:171–175. (In Russian)
17. Zykov AA (1969) Theory of finite graphs. Nauka, Novosibirsk. (In Russian)

Lower Bounds Based on the Exponential Time Hypothesis: Edge Clique Cover

Michał Pilipczuk
Institute of Informatics, University of Warsaw, Warsaw, Poland
Institute of Informatics, University of Bergen, Bergen, Norway

Keywords

Cocktail party graph; Edge clique cover; Exponential Time Hypothesis; Parameterized complexity

Years and Authors of Summarized Original Work

2008; Gramm, Guo, Hüffner, Niedermeier
2013; Cygan, Pilipczuk, Pilipczuk

The Exponential Time Hypothesis and Its Consequences

In 2001, Impagliazzo, Paturi, and Zane [5, 6] introduced the *Exponential Time Hypothesis* (ETH): a complexity assumption saying that there exists a constant $c > 0$ such that no algorithm for 3-SAT can achieve the running time of $\mathcal{O}(2^{cn})$, where n is the number of variables of the input formula. In particular, this implies that there is no *subexponential-time* algorithm for 3-SAT, that is, one with running time $2^{o(n)}$. The key result of Impagliazzo, Paturi, and Zane is the *Sparsification Lemma*, proved in [6]. Without going into technical details, the Sparsification Lemma provides a reduction that allows us to assume that the input instance of 3-SAT is sparse in the following sense: the number of clauses is linear in the number of variables. Thus, a direct consequence is that, assuming ETH, there is a constant $c > 0$ such that there is no algorithm for 3-SAT with running time $\mathcal{O}(2^{c(n+m)})$. Hence, an algorithm with running time $2^{o(n+m)}$ is excluded in particular.

After the introduction of the Exponential Time Hypothesis and the Sparsification Lemma, it turned out that ETH can be used as a robust assumption for proving sharp lower bounds on the time complexity of various computational problems. For many classic NP-hard graph problems, like VERTEX COVER, 3-COLORING, or HAMILTONIAN CYCLE, the known NP-hardness reductions from 3-SAT are *linear*, i.e., they transform an instance of 3-SAT with n variables and m clauses into an instance of the target problem whose total size is $\mathcal{O}(n + m)$. Consequently, if any of these problems admitted an algorithm with running time $2^{o(N+M)}$, where N and M are the numbers of vertices and edges of the graph, respectively, then the composition of the reduction and such an algorithm would yield an algorithm for 3-SAT with running time $2^{o(n+m)}$, thus contradicting ETH. As all these problems indeed can be solved in time $2^{\mathcal{O}(N)}$, this shows that the single-exponential running time is essentially optimal. The same problems restricted to planar graphs have NP-hardness reductions from 3-SAT with a quadratic size blowup, which excludes the existence of $2^{o(\sqrt{N+M})}$ algorithms under ETH. Again, this is matched by $2^{\mathcal{O}(\sqrt{N})}$ algorithms obtained using the Lipton-Tarjan planar separator theorem.

Of particular interest to us are applications in parameterized complexity. Recall that a parameterized problem is *fixed-parameter tractable* (FPT) if there is an algorithm solving it in time $f(k) \cdot n^c$, where n is the total input size, k is the parameter, f is some computable function, and c is a universal constant. Observe that if we provide a reduction from 3-SAT to a parameterized problem where the output parameter depends linearly on $n + m$, then assuming ETH we exclude the existence of a *subexponential parameterized algorithm*, i.e., one with running time $2^{o(k)} \cdot n^{\mathcal{O}(1)}$. If the dependence of the output parameter on $n + m$ is different, then we obtain a lower bound for a different function f. This idea has been successfully applied for many various parameterizations and different running times. For example, Lokshtanov et al. [8] introduced a framework for proving lower bounds excluding the running time of the form $2^{o(k \log k)} \cdot n^{\mathcal{O}(1)}$. The framework can be used to show the optimality of known FPT algorithm for several important problems, with a notable example of CLOSEST STRING.

More information on lower bounds based on ETH can be found in the survey of Lokshtanov et al. [7] or in the PhD thesis of the current author [9].

Problem Definition

In the EDGE CLIQUE COVER problem, we are given a graph G and an integer k, and the question is whether one can find k complete subgraphs $C_1, C_2, \ldots, C_k$ of G such that $E(G) = \bigcup_{i=1}^{k} E(C_i)$. In other words, we have to cover the whole edge set of G using k complete subgraphs of G. Such a selection of k complete subgraphs is called an *edge clique cover*.

The study of the parameterized complexity of EDGE CLIQUE COVER was initiated by Gramm et al. [3]. The main observation of Gramm et al. is the applicability of the following data reduction rule: as long as in G there exists a pair of *perfect twins* (i.e., adjacent vertices having exactly the same neighborhood), then it is safe to remove one of them (and decrement the parameter k in the case when these twins form an isolated edge in G). Once there are no perfect twins and no isolated vertices in the graph (the latter ones can be also safely removed), then one can easily show the following: there is no edge clique cover of size less than $\log |V(G)|$. Consequently, instances with $k < \log |V(G)|$ can be discarded as no-instances, and we are left with instances satisfying $|V(G)| \leq 2^k$. In the language of parameterized complexity, this is called a *kernel* with 2^k vertices. By applying a standard covering dynamic programming algorithm on this kernel, we obtain an FPT algorithm for EDGE CLIQUE COVER with running time $2^{2^{\mathcal{O}(k)}} + |V(G)|^{\mathcal{O}(1)}$; the second summand is the time needed to apply the data reduction rule exhaustively.

Given the striking simplicity of the approach of Gramm et al. [3], the natural open question was whether the obtained double-exponential running time of the algorithm for EDGE CLIQUE COVER could be improved.

Key Results

This question has been resolved by Cygan et al. [1], who showed that, under ETH, the running time obtained by Gramm et al. [3] is essentially optimal. More precisely, they proved the following result:

Lemma 1 *There exists a polynomial-time algorithm that, given an instance φ of 3-SAT with n variables and m clauses, constructs an equivalent* EDGE CLIQUE COVER *instance (G, k) with $k = \mathcal{O}(\log n)$ and $|V(G)| = \mathcal{O}(n + m)$.*

Thus, by considering a composition of the reduction of Lemma 1 with a hypothetical algorithm for EDGE CLIQUE COVER with running time $2^{2^{o(k)}} \cdot |V(G)|^{\mathcal{O}(1)}$, we obtain the following lower bound:

Theorem 1 *Unless ETH fails, there is no algorithm for* EDGE CLIQUE COVER *with running time $2^{2^{o(k)}} \cdot |V(G)|^{\mathcal{O}(1)}$.*

Curiously, Lemma 1 can be also used to show that the kernelization algorithm of Gramm et al. [3] is also essentially optimal. More precisely, we have the following theorem.

Theorem 2 *There exists a universal constant $\varepsilon > 0$ such that, unless* P = NP, *there is no constant λ and a polynomial-time algorithm $\mathcal{A}$ that takes an instance (G, k) of* EDGE CLIQUE COVER *and outputs an equivalent instance (G', k') of* EDGE CLIQUE COVER *with binary encoding of length at most $\lambda \cdot 2^{\varepsilon k}$.*

The idea of the proof of Theorem 2 is to consider the composition of three algorithms: (i) the reduction of Lemma 1, (ii) a hypothetical algorithm as in the statement of Theorem 2 for a very small $\varepsilon > 0$, and (iii) a polynomial-time reduction from EDGE CLIQUE COVER to 3-SAT, whose existence follows from the fact that the former problem is in NP and the latter one is NP-hard. Since the constants hidden in the bounds for algorithms (i) and (iii) are universal, for some very small $\varepsilon > 0$ this composition would result in an algorithm that takes any instance of 3-SAT on n variables and shrinks it to an instance that has total bitsize $o(n)$, i.e., *sublinear* in n. Hence, by applying this algorithm multiple times, we would eventually shrink the instance at hand to constant size, and then we could solve it by brute force. As all the algorithms involved run in polynomial time, this would imply that P $=$ NP.

We remark that Theorem 2 was not observed in the original work of Cygan et al. [1], but its proof can be found in the PhD thesis of the current author [9], and it will appear in the upcoming journal version of [1]. Also, in an earlier work, Cygan et al. [2] proved a weaker statement that EDGE CLIQUE COVER does not admit a polynomial kernel unless NP $\subseteq$ coNP/poly. Theorem 2 shows that even a subexponential kernel is unlikely under the weaker assumption of P $\neq$ NP.

Let us now shed some light on the proof of Lemma 1, which is the crucial technical ingredient of the results. The main idea is to base the reduction on the analysis of the *cocktail party graph*: for an integer $n > 1$, the cocktail party graph H_{2n} is defined as a complete graph on $2n$ vertices with a perfect matching removed. Observe that H_{2n} does not contain any perfect twins, so it is immune to the data reduction rule of Gramm et al. [3] and the minimum size of an edge clique cover in H_{2n} is at least $1 + \log n$. On the other hand, it is relatively easy to construct a large family of edge clique covers of H_{2n} that have size $2\lceil \log n \rceil$. Actually, the question of determining the minimum size of an edge clique cover in H_{2n} was studied from a purely combinatorial point of view: Gregory and Pullman [4] proved that it is equal to $\inf\{k : n \leq \binom{k-1}{\lceil k/2 \rceil}\}$, answering an earlier question of Orlin. Note that this value is larger than $1 + \log n$ only by an additive factor of $\mathcal{O}(\log \log n)$.

Thus, the cocktail party graph provides a natural example of a hard instance where the parameter is logarithmic. The crux of the construction is to start with the cocktail party graph H_{2n} and, by additional gadgeteering, force the solution inside it to belong to the aforementioned family of edge clique covers of size $2\lceil \log n \rceil$. The behavior of these edge clique covers (called *twin covers*) can be very well understood, and we can encode the evaluation of the variables of the input 3-SAT formula as a selection of a twin cover to cover H_{2n}. In order to verify that the clauses of the input formula are satisfied, we construct additional clause gadgets. This involves only a logarithmic number of additional cliques and is based on careful constructions using binary encodings.

Discussion

After announcing the results of Cygan et al. [1], there was some discussion about their actual meaning. For instance, some authors suggested that the surprisingly high lower bound for EDGE CLIQUE COVER may be a argument against the plausibility of the Exponential Time Hypothesis. Our view on this is quite different: the double-exponential lower bound suggests that EDGE CLIQUE COVER is an inherently hard problem, even though it may not seem as such at first glance. The relevant parameter in this problem is not really the number of cliques k, but rather 2^k, the number of possible different neighborhoods that can arise in a graph that is a union of k complete graphs. The lower bound of Cygan et al. [1] intuitively shows that one cannot expect to significantly reduce the number of neighborhoods that needs to be considered.

Recommended Reading

1. Cygan M, Pilipczuk M, Pilipczuk M (2013) Known algorithms for edge clique cover are probably optimal. In: Proceedings of the twenty-fourth annual ACM-SIAM symposium on discrete algorithms, SODA 2013, New Orleans, 6–8 Jan 2013, pp 1044–1053
2. Cygan M, Kratsch S, Pilipczuk M, Pilipczuk M, Wahlström M (2014) Clique cover and graph separation: new incompressibility results. TOCT 6(2):6
3. Gramm J, Guo J, Hüffner F, Niedermeier R (2008) Data reduction and exact algorithms for clique cover. ACM J Exp Algorithmics 13:article 2.2
4. Gregory DA, Pullman NJ (1982) On a clique covering problem of Orlin. Discret Math 41(1): 97–99
5. Impagliazzo R, Paturi R (2001) On the complexity of k-SAT. J Comput Syst Sci 62(2):367–375
6. Impagliazzo R, Paturi R, Zane F (2001) Which problems have strongly exponential complexity? J Comput Syst Sci 63(4):512–530
7. Lokshtanov D, Marx D, Saurabh S (2011) Lower bounds based on the exponential time hypothesis. Bull EATCS 105:41–72
8. Lokshtanov D, Marx D, Saurabh S (2011) Slightly superexponential parameterized problems. In: Proceedings of the twenty-second annual ACM-SIAM symposium on discrete algorithms, SODA 2011, San Francisco, 23–25 Jan 2011. SIAM, pp 760–776
9. Pilipczuk M (2013) Tournaments and optimality: new results in parameterized complexity. PhD thesis, University of Bergen, Norway. Available at the webpage of the author

Lower Bounds for Dynamic Connectivity

Mihai Pătraşcu
Computer Science and Artificial Intelligence Laboratory (CSAIL), Massachusetts Institute of Technology (MIT), Cambridge, MA, USA

Keywords

Dynamic trees

Years and Authors of Summarized Original Work

2004; Pătraşcu, Demaine

Problem Definition

The dynamic connectivity problem requests maintenance of a graph G subject to the following operations:

insert(u, v): insert an undirected edge (u, v) into the graph.
delete(u, v): delete the edge (u, v) from the graph.
connected(u, v): test whether u and v lie in the same connected component.

Let m be an upper bound on the number of edges in the graph. This entry discusses cell-probe lower bounds for this problem. Let t_u be the complexity of `insert` and `delete` and t_q the complexity of `query`.

The Partial-Sums Problem

Lower bounds for dynamic connectivity are intimately related to lower bounds for another classic problem: maintaining partial sums. Formally, the problem asks one to maintain an array $A[1..n]$ subject to the following operations:

update(k, Δ): let $A[k] \leftarrow \Delta$.
sum(k): returns the partial sum $\sum_{i=1}^{k} A[i]$.
testsum(k, σ): returns a boolean value indicating whether $\mathrm{sum}(k) = \sigma$.

To specify the problem completely, let elements $A[i]$ come from an arbitrary group G containing at least 2^δ elements. In the cell-probe model with b-bit cells, let t_u^Σ be the complexity of `update` and t_q^Σ the complexity of `testsum` (which is also a lower bound on `sum`).

The tradeoffs between t_u^Σ and t_q^Σ are well understood for all values of b and δ. However, this entry only considers lower bounds under the standard assumptions that $b = \Omega(\lg n)$ and $t_u \geq t_q$. It is standard to assume $b = \Omega(\lg n)$ for upper bounds in the RAM model; this assumption also means that the lower bound applies to the pointer machine. Then, Pătraşcu and Demaine [6] prove:

Theorem 1 *The complexity of the partial-sums problems satisfies:* $t_q^\Sigma \cdot \lg(t_u^\Sigma / t_q^\Sigma) = \Omega(\delta/b \cdot \lg n)$.

Observe that this matches the textbook upper bound using augmented trees. One can build a balanced binary tree over $A[1], \ldots, A[n]$ and store in every internal node the sum of its subtree. Then, updates and queries touch $O(\lg n)$ nodes (and spend $O(\lceil \delta/b \rceil)$ time in each one due to the size of the group). To decrease the query time, one can use a B-tree.

Relation to Dynamic Connectivity

We now clarify how lower bounds for maintaining partial sums imply lower bounds for dynamic connectivity. Consider the partial-sums problem over the group $G = S_n$, i.e., the permutation group on n elements. Note that $\delta = \lg(n!) = \Omega(n \lg n)$. It is standard to set $b = \Theta(\lg n)$, as this is the natural word size used by dynamic connectivity upper bounds. This implies $t_q^\Sigma \lg(t_u^\Sigma / t_q^\Sigma) = \Omega(n \lg n)$.

The lower bound follows from implementing the partial-sums operations using dynamic connectivity operations. Refer to Fig. 1. The vertices of the graph form an integer grid of size $n \times n$. Each vertex is incident to at most two edges, one edge connecting to a vertex in the previous column and one edge connecting to a vertex in the next column. Point (x, y_1) in the grid is connected to point $(x+1, A[x](y_1))$, i.e.,the edges between two adjacent columns describe the corresponding permutation from the partial-sums vector.

To implement `update` (x, π), all the edges between column x and $x+1$ are first deleted and then new edges are inserted according to π. This gives $t_u^\Sigma = O(2n \cdot t_u)$. To implement `testsum` (x, π), one can use n `connected` queries between the pairs of points $(1, y) \rightsquigarrow (x+1, \pi(y))$. Then, $t_q^\Sigma = O(n \cdot t_q)$. Observe that the `sum` query cannot be implemented as easily. Dynamic connectivity is the main motivation to study the `testsum` query.

The lower bound of Theorem 1 translates into $nt_q \cdot \lg(2nt_u/nt_q) = \Omega(n \lg n)$; hence $t_q \lg(t_u/t_q) = \Omega(\lg n)$. Note that this lower bound implies $\max\{t_u, t_q\} = \Omega(\lg n)$. The best known upper bound (using amortization and randomization) is $O(\lg n (\lg \lg n)^3)$ [9]. For any $t_u = \Omega(\lg n (\lg \lg n)^3)$, the lower bound tradeoff is known to be tight. Note that the graph in the lower bound is always a disjoint union of paths. This implies optimal lower bounds for two important special cases: dynamic trees [8] and dynamic connectivity in plane graphs [2].

Key Results

Understanding Hierarchies

Epochs

To describe the techniques involved in the lower bounds, first consider the `sum` query and assume $\delta = b$. In 1989, Fredman and Saks [3] initiated the study of dynamic cell-probe lower bounds, essentially showing a lower bound of $t_q^\Sigma \lg t_u^\Sigma = \Omega(\lg n)$. Note that this implies $\max\{t_q^\Sigma, t_u^\Sigma\} = \Omega(\lg n / \lg \lg n)$.

At an intuitive level, their argument proceeded as follows. The hard instance will have n random updates, followed by one random query. Leave $r \geq 2$ to be determined. Looking *back* in time

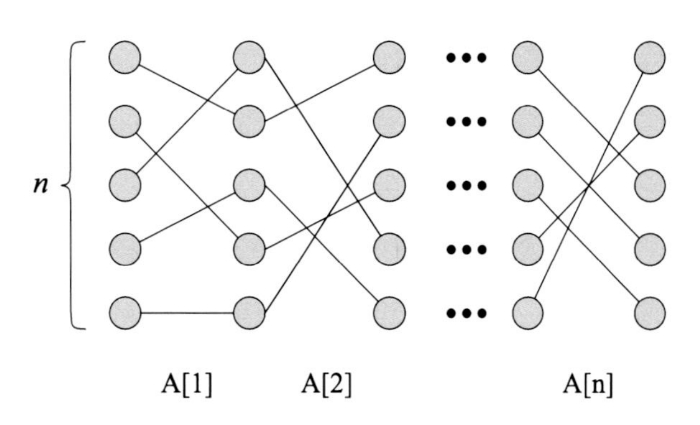

Lower Bounds for Dynamic Connectivity, Fig. 1 Constructing an instance of dynamic connectivity that mimics the partial-sums problem

from the query, one groups the updates into exponentially growing epochs: the latest r updates are epoch 1, the earlier r^2 updates are epoch 2, etc. Note that epoch numbers increase going back in time, and there are $O(\log_r n)$ epochs in total.

For some epoch i, consider revealing to the query all updates performed in all epochs different from i. Then, the query reduces to a partial-sums query among the updates in epoch i. Unless the query is to an index below the minimum index updated in epoch i, the answer to the query is still uniformly random, i.e., has δ bits of entropy. Furthermore, even if one is given, say, $r^i\delta/100$ bits of information about epoch i, the answer still has $\Omega(\delta)$ bits of entropy on average. This is because the query and updates in epoch i are uniformly random, so the query can ask for any partial sum of these updates, uniformly at random. Each of the r^i partial sums is an independent random variable of entropy δ.

Now one can ask how much information is available to the query. At the time of the query, let each cell be associated with the epoch during which it was last written. Choosing an epoch i uniformly at random, one can make the following intuitive argment:

1. No cells written by epochs $i+1, i+2, \ldots$ can contain information about epoch i, as they were written in the past.
2. In epochs $1, \ldots, i-1$, a number of $bt_u^\Sigma \cdot \sum_{j=1}^{i-1} r^j \leq bt_u^\Sigma \cdot 2r^{i-1}$ bits were written. This is less than $r^i\delta/100$ bits of information for $r > 200t_u^\Sigma$ (recall the assumption $\delta = b$). By the above, this implies the query answer still has $\Omega(\delta)$ bits of entropy.
3. Since i is uniformly random among $\Theta(\log_r n)$ epochs, the query makes an expected $O(t_q^\Sigma / \log_r n)$ probes to cells from epoch i. All queries that make no cell probes to epoch i have a fixed answer (entropy 0), and all other queries have answers of entropy $\leq \delta$. Since an average query has entropy $\Omega(\delta)$, a query must probe a cell from epoch i with constant probability. That means $t_q^\Sigma / \log_r n = \Omega(1)$, and $\sum = \Omega(\log_r n) = \Omega(\lg n / \lg t_u^\Sigma)$.

One should appreciate the duality between the proof technique and the natural upper bounds based on a hierarchy. Consider an upper bound based on a tree of degree r. The last r random updates (epoch 1) are likely to be uniformly spread in the array. This means the updates touch different children of the root. Similarly, the r^2 updates in epoch 2 are likely to touch every node on level 2 of the tree, and so on. Now, the lower bound argues that the query needs to traverse a root-to-leaf path, probing a node on every level of the tree (this is equivalent to one cell from every epoch).

Time Hierarchies

Despite considerable refinement to the lower bound techniques, the lower bound of $\Omega(\lg n / \lg\lg n)$ was not improved until 2004. Then, Pătraşcu and Demaine [6] showed an optimal bound of $t_q^\Sigma \lg(t_u^\Sigma / t_q^\Sigma) = \Omega(\lg n)$, implying $\max\{t_u^\Sigma, t_q^\Sigma\} = \Omega(\lg n)$. For simplicity, the discussion below disregards the tradeoff and just sketches the $\Omega(\lg n)$ lower bound.

Pătraşcu and Demaine's [6] counting technique is rather different from the epoch technique; refer to Fig. 2. The hard instance is a sequence of n operations alternating between updates and queries. They consider a balanced binary tree over the time axis, with every leaf being an operation. Now for every node of the tree, they propose to count the number of cell probes made in the right subtree to a cell written in the left subtree. Every probe is counted exactly once, for the lowest common ancestor of the read and write times.

Now focus on two sibling subtrees, each containing k operations. The $k/2$ updates in the left subtree, and the $k/2$ queries in the right subtree, are expected to interleave in index space. Thus, the queries in the right subtree ask for $\Omega(k)$ different partial sums of the updates in the left subtree. Thus, the right subtree "needs" $\Omega(k\delta)$ bits of information about the left subtree, and this information can only come from cells written in the left subtree and read in the right one. This implies

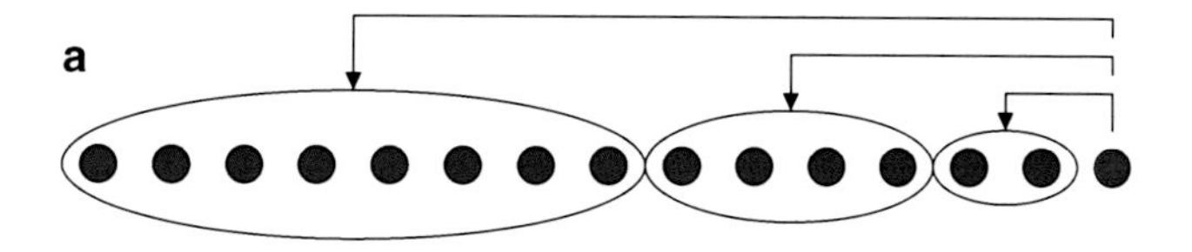

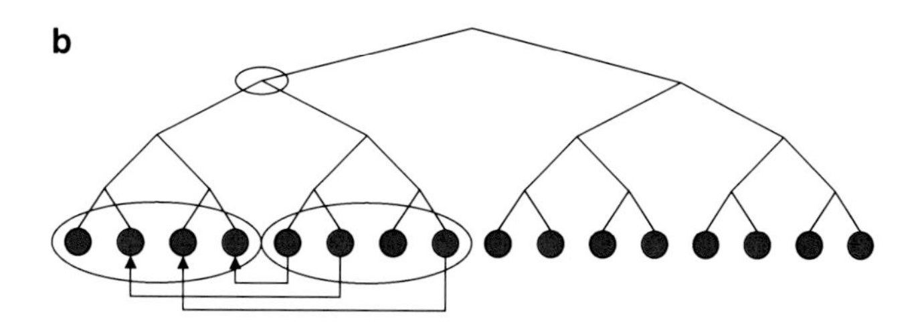

Lower Bounds for Dynamic Connectivity, Fig. 2 Analysis of cell probes in the **a** epoch-based and **b** time-hierarchy techniques

a lower bound of $\Omega(k)$ probes, associated with the parent of the sibling subtrees. This bound is linear in the number of leaves, so summing up over the tree, one obtains a total $\Omega(n \lg n)$ lower bound, or $\Omega(\lg n)$ cost per operation.

An Optimal Epoch Construction

Rather surprisingly, Pătraşcu and Tarniţă [7] managed to reprove the optimal tradeoff of Theorem 1 with minimal modifications to the epoch argument. In the old epoch argument, the information revealed by epochs $1, \ldots, i-1$ about epoch i was bounded by the number of cells written in these epochs. The key idea is that an equally good bound is the number of cells read during epochs $1, \ldots, i-1$ and written during epoch i.

In principle, all cell reads from epoch $i-1$ could read data from epoch i, making these two bounds identical. However, one can randomize the epoch construction by inserting the query after an unpredictable number of updates. This randomization "smooths" out the distribution of epochs from which cells are read, i.e., a query reads $O(t_q^\Sigma / \log_r n)$ cells from every epoch, in expectation over the randomness in the epoch construction. Then, the $O(r^{i-1})$ updates in epochs $1, \ldots, i-1$ only read $O(r^{i-1} \cdot t_u^\Sigma / \log_r n)$ cells from epoch i. This is not enough information if $r \gg t_u^\Sigma / \log_r n = \Theta(t_u^\Sigma / t_q^\Sigma)$, which implies $t_q^\Sigma = \Omega(\log_r n) = \Omega(\lg n / \lg(t_u^\Sigma / t_q^\Sigma))$.

Technical Difficulties

Nondeterminism

The lower bounds sketched above are based on the fact that the `sum` query needs to output $\Omega(\delta)$ bits of information about every query. If dealing with the *decision* `testsum` query, an argument based on output entropy can no longer work.

The most successful idea for decision queries has been to convert them to queries with nonboolean output, in an extended cell-probe model that allows nondeterminism. In this model, the query algorithm is allowed to spawn an arbitrary number of computation threads. Each thread can make t_q cell probes, after with it must either terminate with a 'reject' answer, or return an answer to the query. All nonrejecting threads must return the same output. In this model, a query with arbitrary output is equivalent to a decision query, because one can just nondeterministically guess the answer, and then verify it.

By the above, the challenge is to prove good lower bounds for `sum` even in the nondeterminstic model. Nondeterminism shakes our view that when analyzing epoch i, only cell probes to epoch i matter. The trouble is that the query may not know *which* of its probes are actually to epoch i. A probe that reads a cell from a previous epoch provides at least some information about epoch i: no update in the epoch decided to overwrite the cell. Earlier this was not a problem because the goal was only to rule out the case that there are *zero* probes to epoch i. Now, however, different threads can probe any cell in memory, and one cannot determine which threads actually avoid probing anything in epoch i. In other words, there is a covert communication channel between epoch i and the query in which the epoch can use the choice of which cells to write in order to communicate information to the query.

There are two main strategies for handling nondeterministic query algorithms. Husfeldt and Rauhe [4] give a proof based on some interesting observations about the combinatorics of nondeterministic queries. Pătraşcu and Demaine [6] use the power of nondeterminism itself to output a small certificate that rules out useless cell probes. The latter result implies the optimal lower bound of Theorem 1 for `testsum` and, thus, the logarithmic lower bound for dynamic connectivity.

Alternative Histories

The framework described above relies on fixing all updates in epochs different from i to an average value and arguing that the query answer still has a lot of variability, depending on updates in epoch i. This is true for aggregation problems but not for search problems. If a searched item is found with equal probability in any epoch, then fixing all other epochs renders epoch i irrelevant with probability $1 - 1/(\log_r n)$.

Alstrup et al. [1] propose a very interesting refinement to the technique, proving $\Omega(\lg n/\lg\lg n)$ lower bounds for an impressive collection of search problems. Intuitively, their idea is to consider $O(\log_r n)$ alternative histories of updates, chosen independently at random. Epoch i is relevant in at least one of the histories with constant probability. On the other hand, even if one knows what epochs 1 through $i-1$ learned about epoch i in *all histories*, answering a random query is still hard.

Bit-Probe Complexity

Intuitively, if the word size is $b = 1$, the lower bound for connectivity should be roughly $\Omega(\lg^2 n)$, because a query needs $\Omega(\lg n)$ bits from every epoch. However, ruling out anything except zero probes to an epoch turns out to be difficult, for the same reason that the nondeterministic case is difficult. Without giving a very satisfactory understanding of this issue, Pătraşcu and Tarniţă [7] use a large bag of tricks to show an $\Omega((\lg n/\lg\lg n)^2)$ lower bound for dynamic connectivity. Furthermore, they consider the partial-sums problem in $\mathbb{Z}_2$ and show an $\Omega(\lg n/\lg\lg\lg n)$ lower bound, which is a triply-logarithmic factor away from the upper bound!

Applications

The lower bound discussed here extends by easy reductions to virtually all natural fully dynamic graph problems [6].

Open Problems

By far, the most important challenge for future research is to obtain a lower bound of $\omega(\lg n)$ per operation for some dynamic data structure in the cell-probe model with word size $\Theta(\lg n)$. Miltersen [5] specifies a set of technical conditions for what qualifies as a solution to such a challenge. In particular, the problem should be a *dynamic language membership* problem.

For the partial-sums problem, though `sum` is perfectly understood, `testsum` still lacks tight bounds for certain ranges of parameters [6]. In addition, obtaining tight bounds in the bit-probe model for partial sums in $\mathbb{Z}_2$ appears to be rather challenging.

Recommended Reading

1. Alstrup S, Husfeldt T, Rauhe T (1998) Marked ancestor problems. In: Proceedings of the 39th IEEE symposium on foundations of computer science (FOCS), pp 534–543
2. Eppstein D, Italiano GF, Tamassia R, Tarjan RE, Westbrook JR, Yung M (1992) Maintenance of a minimum spanning forest in a dynamic planar graph. J Algorithm 13:33–54. See also SODA'90

3. Fredman ML, Saks ME (1989) The cell probe complexity of dynamic data structures. In: Proceedings of the 21st ACM symposium on theory of computing (STOC), pp 345–354
4. Husfeldt T, Rauhe T (2003) New lower bound techniques for dynamic partial sums and related problems. SIAM J Comput 32:736–753. See also ICALP'98
5. Miltersen PB (1999) Cell probe complexity – a survey. In: 19th conference on the foundations of software technology and theoretical computer science (FSTTCS) (Advances in Data Structures Workshop)
6. Pătraşcu M, Demaine ED (2006) Logarithmic lower bounds in the cell-probe model. SIAM J Comput 35:932–963. See also SODA'04 and STOC'04
7. Pătraşcu M, Tarniţă C (2007) On dynamic bit-probe complexity. Theor Comput Sci 380:127–142. See also ICALP'05
8. Sleator DD, Tarjan RE (1983) A data structure for dynamic trees. J Comput Syst Sci 26:362–391, See also STOC'81
9. Thorup M (2000) Near-optimal fully-dynamic graph connectivity. In: Proceedings of the 32nd ACM symposium on theory of computing (STOC), pp 343–350

Lower Bounds for Online Bin Packing

Rob van Stee
University of Leicester, Leicester, UK

Keywords

Bin packing; Competitive analysis; Lower bounds; Online algorithms

Years and Authors of Summarized Original Work

1992; van Vliet
2012; Balogh, Békési, Galambos

Problem Definition

In the online bin packing problem, a sequence of *items* with sizes in the interval $(0, 1]$ arrive one by one and need to be packed into *bins*, so that each bin contains items of total size at most 1. Each item must be irrevocably assigned to a bin before the next item becomes available. The algorithm has no knowledge about future items. There is an unlimited supply of bins available, and the goal is to minimize the total number of used bins (bins that receive at least one item).

The most common performance measure for online bin packing algorithms is the asymptotic performance ratio, or asymptotic competitive ratio, which is defined as

$$R_{\mathrm{ASY}}(A) := \limsup_{n\to\infty}\left\{\max_{L}\left\{\frac{A(L)}{n}\,\middle|\,\mathrm{OPT}(L)=n\right\}\right\}. \quad (1)$$

Hence, for any input L, the number of bins used by an online algorithm A is compared to the optimal number of bins needed to pack the same input. Note that calculating the optimal number of bins might take exponential time; moreover, it requires that the entire input is known in advance.

Key Results

Yao showed that no online algorithm has performance ratio less than $\frac{3}{2}$ [7]. The following construction is very important in the context of proving lower bounds for online algorithms. Start with an item of type 1 and size $1/2 + \varepsilon$, for some very small $\varepsilon > 0$. Now, in each step, add an item of the largest possible size of the form $1/s + \varepsilon$ that can fit with all previous items into a single bin. That is, the second item has size $1/3 + \varepsilon$, the third item has size $1/7 + \varepsilon$, etc. To be more precise, it can be shown that the sizes in this input sequence are given by $1/t_i + \varepsilon$ $(i \geq 1)$, where t_i is defined by

$$t_1 = 2, \quad t_{i+1} = t_i(t_i - 1) + 1 \quad i \geq 1.$$

The first few numbers of this sequence are $2, 3, 7, 43, 1{,}807$. This sequence was first examined by Sylvester [5].

Since we allow an additive constant to the competitive ratio, in order to turn the above set of items into an input that can be used to prove a lower bound for any online algorithm, we need to repeat each item in the input N times for some arbitrarily large N. To summarize the preceding discussion, the input has the following form:

$$N \times \left(\frac{1}{2}+\varepsilon\right), N \times \left(\frac{1}{3}+\varepsilon\right), N \times \left(\frac{1}{7}+\varepsilon\right), N \times \left(\frac{1}{43}+\varepsilon\right), N \times \left(\frac{1}{1{,}807}+\varepsilon\right), \ldots, \quad (2)$$

where the items are given to the algorithm in order of *nondecreasing* size.

Brown and Liang independently gave a lower bound to 1.53635 [2, 3], using the sequence (2). Van Vliet showed how to use the input defined by (2) to prove a lower bound of 1.54014. Van Vliet set up a linear programming formulation to define all possible online algorithms for this input to prove the lower bound.

This works by characterizing online algorithms by which *patterns* they use and by how frequently they use them. A pattern is a multiset of items which fit together in one bin (have total size at most 1). As N tends to infinity, an online algorithm can be fully characterized by the fraction of bins that it packs according to each pattern.

As an example, consider the two largest item sizes in (2). The only valid patterns are $(1,0)$, $(1,1)$, $(0,2)$, where (x,y) means that there are x items of size $\frac{1}{2}+\varepsilon$ and y items of size $\frac{1}{3}+\varepsilon$ in the bin. The N smallest items arrive first, and the online algorithm will pack them into bins with patterns $(1,1)$ and $(0,2)$ (where the first choice means that one item is now packed into it, and in the future only one item of size $\frac{1}{2}+\varepsilon$ is possibly packed with it). Say it uses x_1 bins with pattern $(1,1)$ and x_2 with pattern $(0,2)$; then we must have $x_1+2x_2 \geq N$ or $x_1/N+2x_2/N \geq 1$. In the linear program, we use variables $x_i' = x_i/N$, thus eliminating the appearance of the number N altogether.

The input (2) appeared to be "optimal" to make life hard for online algorithms: the smallest items arrive first, and the input is constructed in such a way that each item is as large as possible given the larger items (that are defined first). Intuitively, larger items are more difficult to handle by online algorithms than smaller items. Surprisingly, however, in 2012 Balogh et al. [1] managed to prove a lower bound of $248/161 \approx 1.54037$ using a slight modification of the input. Instead of using $1/43$ as the fourth item size, they use $1/49$ and then continue the construction in the same manner as before. We get the following input:

$$N \times \left(\frac{1}{2}+\varepsilon\right), N \times \left(\frac{1}{3}+\varepsilon\right), N \times \left(\frac{1}{7}+\varepsilon\right), N \times \left(\frac{1}{49}+\varepsilon\right), N \times \left(\frac{1}{343}+\varepsilon\right), \ldots, \quad (3)$$

For the input that consists only of the first four phases of this input, the resulting lower bound is now slightly lower than before, but this is more than compensated for by the next items.

Open Problems

Other variations of the input sequence (2) do not seem to give better lower bounds. Yet there is still a clear gap to the best known upper bound of 1.58889 by Seiden [4]. Can we give a stronger lower bound using some other construction?

Cross-References

▶ Bin Packing with Cardinality Constraints
▶ Current Champion for Online Bin Packing
▶ Harmonic Algorithm for Online Bin Packing

Recommended Reading

1. Balogh J, Békési J, Galambos G (2012) New lower bounds for certain bin packing algorithms. Theor Comput Sci 440–441:1–13
2. Brown DJ (1979) A lower bound for on-line one-dimensional bin packing algorithms. Technical report R-864, Coordinated Science Laboratory, Urbana
3. Liang FM (1980) A lower bound for online bin packing. Inf Process Lett 10:76–79
4. Seiden SS (2002) On the online bin packing problem. J ACM 49(5):640–671
5. Sylvester JJ (1880) On a point in the theory of vulgar fractions. Am J Math 3:332–335
6. van Vliet A (1992) An improved lower bound for on-line bin packing algorithms. Inf Process Lett 43:277–284
7. Yao AC-C (1980) New algorithms for bin packing. J ACM 27:207–227

Lowest Common Ancestors in Trees

Martín Farach-Colton
Department of Computer Science,
Rutgers University, Piscataway, NJ, USA

Keywords

Least common ancestor; Nearest common ancestor; Range minimum query; Succinct structures

Years and Authors of Summarized Original Work

1984; Gabow, Bentley, Tarjan
1984; Harel, Tarjan
1989; Berkman, Breslauer, Galil, Schieber, Vishkin
2000; Bender, Farach-Colton

Problem Definition

One of the most fundamental algorithmic problems on trees is how to find the *lowest common ancestor (LCA)* of a pair of nodes. The LCA of nodes u and v in a tree is the shared ancestor of u and v that is located farthest from the root. More formally, the *lowest common ancestor (LCA)* problem is:

Preprocess: A rooted tree T having n nodes.
Query: For nodes u and v of tree T, query $\text{LCA}_T(u, v)$ returns the least common ancestor of u and v in T, that is, it returns the node farthest from the root that is an ancestor of both u and v. (When the context is clear, we drop the subscript T on the LCA.)

The goal is to optimize both the preprocessing time and the query time. We will therefore refer to the running time of an algorithm with preprocessing time $T_P(N)$ and query time of $T_Q(N)$ as having run time $\langle T_P(N), T_Q(N)\rangle$.

The LCA problem has been studied intensively both because it is inherently beautiful algorithmically and because fast algorithms for the LCA problem can be used to solve other algorithmic problems.

Key Results

In [7], Harel and Tarjan showed the surprising result that LCA queries can be answered in constant time after only linear preprocessing of the tree T. This result was simplified over the course of several papers, and current solutions are based on combinations of four themes:

1. The LCA problem is equivalent to the range minimum query (RMQ) problem, defined below, in that they can be reduced to each other in linear preprocessing time and constant query time. Thus, an optimal solution for one yields an optimal solution for the other.
2. The LCA of certain trees, notably complete binary trees and trees that are linear paths, can be computed quickly. General trees can be decomposed into special trees. Similarly, the RMQ of certain classes of arrays can be computed quickly.
3. Nodes can be labeled to capture information about their position in the tree. These labels can be used to compute the label of the LCA of two nodes.

Harel and Tarjan [7] showed that LCA computation has a lower bound of $\Omega(\log\log n)$ on a pointer machine. Therefore, the fast algorithms presented here will all require operations on $O(\log n)$-bit words. We will see how to use this assumption not only to make queries $O(1)$ time but to improve preprocessing from $O(n \log n)$ to $O(n)$ via Four Russians encoding of small problem instances.

Below, we explore each of these themes for LCA computation.

RMQ

The *range minimum query (RMQ) problem*, which seems quite different from the LCA problem, is, in fact, intimately linked. It is defined as:

Preprocess: A length n array A of numbers.

Query: For indices i and j between 1 and n, query $\text{RMQ}_A(x, y)$ returns the index of the smallest element in the subarray $A[i \ldots j]$. (When the context is clear, we drop the subscript A on the RMQ.)

The following two lemmas give linear reductions between LCA and RMQ.

Reducing LCA to RMQ

Lemma 1 ([3]) *If there is a* $\langle f(n), g(n)\rangle$*-time solution for RMQ, then there is a* $\langle f(2n-1)+O(n), g(2n-1)+O(1)\rangle$*-time solution for LCA.*

Proof Let T be the input tree. The reduction relies on one key observation:

Observation 1 *The LCA of nodes u and v is the shallowest node encountered between the visits to u and to v during a depth-first search traversal of T.*

Therefore, the reduction proceeds as follows.

1. Let array $E[1,\ldots,2n-1]$ store the nodes visited in an Euler tour of the tree T. (The Euler tour of T is the sequence of nodes we obtain if we write down the label of each node each time it is visited during a DFS. The array of the Euler tour has length $2n-1$ because we start at the root and subsequently output a node each time we traverse an edge. We traverse each of the $n-1$ edges twice, once in each direction.) That is, $E[i]$ is the label of the ith node visited in the Euler tour of T.
2. Let the *level* of a node be its distance from the root. Compute the Level Array $L[1,\ldots,2n-1]$, where $L[i]$ is the level of node $E[i]$ of the Euler tour.
3. Let the *representative* of a node in an Euler tour be the index of the first occurrence of the node in the tour (In fact, any occurrence of i will suffice to make the algorithm work, but we consider the first occurrence for the sake of concreteness.); formally, the representative of i is $\text{argmin}_j\{E[j] = i\}$. Compute the Representative Array $R[1,\ldots,n]$, where $R[i]$ is the index of the representative of node i.

Each of these three steps takes $O(n)$ time, yielding $O(n)$ total time. To compute $\text{LCA}_T(x, y)$, we note the following:

- The nodes in the Euler tour between the first visits to u and to v are $E[R[u],\ldots,R[v]]$ (or $E[R[v],\ldots,R[u]]$).
- The shallowest node in this subtour is at index $\text{RMQ}_L(R[u], R[v])$, since $L[i]$ stores the level of the node at $E[i]$ and the RMQ will thus report the position of the node with minimum level.
- The node at this position is $E[\text{RMQ}_L(R[u], R[v])]$, which is thus the output of $\text{LCA}_T(u, v)$, by Observation 1.

Thus, we can complete our reduction by preprocessing Level Array L for RMQ. As promised, L is an array of size $2n-1$, and building it takes time $O(n)$. The total preprocessing is $f(2n-1)+O(n)$. To calculate the query time, observe that an LCA query in this reduction uses one RMQ query in L and three array references at $O(1)$ time each, for a total of $g(2n-1)+O(1)$ time, and we have completed the proof of the reduction. ■

Reducing RMQ to LCA

Lemma 2 ([6]) *If there is a* $\langle f(n), f(1)\rangle$ *solution for LCA, then there is a* $\langle f(n)+O(n), g(n)+O(1)\rangle$ *solution for RMQ.*

Proof Let $A[1, \ldots, n]$ be the input array.

The Cartesian tree of an array is defined as follows. The root of a Cartesian tree is the minimum element of the array, and the root is labeled with the position of this minimum. Removing the root element splits the array into two pieces. The left and right children of the root are the recursively constructed Cartesian trees of the left and right subarrays, respectively.

A Cartesian tree can be built in linear time as follows. Suppose C_i is the Cartesian tree of $A[1, \ldots, i]$. To build C_{i+1}, we notice that node $i+1$ will belong to the rightmost path of C_{i+1}, so we climb up the rightmost path of C_i until finding the position where $i + 1$ belongs. Each comparison either adds an element to the rightmost path or removes one, and each node can only join the rightmost path and leave it once. Thus, the total time to build C_n is $O(n)$.

The reduction is as follows.

- Let C be the Cartesian tree of A. Recall that we associate with each node in C the index i corresponding to $A[i]$.

Claim $\text{RMQ}_A(i, j) = \text{LCA}_C(i, j)$.

Proof Consider the least common ancestor, k, of i and j in the Cartesian tree C. In the recursive description of a Cartesian tree, k is the first node that separates i and j. Thus, in the array A, element $A[k]$ is between elements $A[i]$ and $A[j]$. Furthermore, $A[k]$ must be the smallest such element in the subarray $A[i, \ldots, j]$ since, otherwise, there would be a smaller element k' in $A[i, \ldots, j]$ that would be an ancestor of k in C, and i and j would already have been separated by k'.

More concisely, since k is the first element to split i and j, it is between them because it splits them, and it is minimal because it is the first element to do so. Thus, it is the RMQ. ■

We can complete our reduction by preprocessing the Cartesian tree C for LCA. Tree C takes time $O(n)$ to build, and because C is an n node tree, LCA preprocessing takes $f(n)$ time, for a total of $f(n) + O(n)$ time. The query then takes $f(n) + O(1)$, and we have completed the proof of the reduction. ■

An Algorithm for RMQ

Observe that RMQ has a solution with complexity $\langle O(n^2), O(1)\rangle$: build a table storing answers to all of the $\binom{n}{2}$ possible queries. To achieve $O(n^2)$ preprocessing rather than the $O(n^3)$ naive preprocessing, we apply a trivial dynamic program. Notice that answering an RMQ query now requires just one array lookup.

To improve the $\langle O(n^2), O(1)\rangle$-time brute-force table algorithm for RMQ to $\langle O(n \log n), O(1)\rangle$, precompute the result of all queries with a range size that is a power of two. That is, for every i between 1 and n and every j between 1 and $\log n$, find the minimum element in the block starting at i and having length 2^j, that is, compute $M[i, j] = \text{argmin}_{k=i \ldots i+2^j-1}\{A[k]\}$. Table M therefore has size $O(n \log n)$, and it can be filled in time $O(n \log n)$ by using dynamic programming. Specifically, find the minimum in a block of size 2^j by comparing the two minima of its two constituent blocks of size 2^{j-1}. More formally, $M[i, j] = M[i, j - 1]$ if $A[M[i, j - 1]] \leq M[i + 2^{j-1} - 1, j - 1]$, and $M[i, j] = M[i + 2^{j-1} - 1, j - 1]$ otherwise.

How do we use these blocks to compute an arbitrary $\text{RMQ}(i, j)$? We select two overlapping blocks that entirely cover the subrange: let 2^k be the size of the largest block that fits into the range from i to j, that is, let $k = \lfloor \log(j - i) \rfloor$. Then $\text{RMQ}(i, j)$ can be computed by comparing the minima of the following two blocks: i to $i + 2^k - 1$ $(M(i, k))$ and $j - 2^k + 1$ to j $(M(j - 2^k + 1, k))$. These values have already been computed, so we can find the RMQ in constant time.

This gives the *Sparse Table (ST)* algorithm for RMQ, with complexity $\langle O(n \log n), O(1)\rangle$.

LCA on Special Trees and RMQ on Special Arrays

In this section, we consider special cases of LCA and RMQ that have fast solutions. These can be used to build optimal algorithms.

Paths and Balanced Binary Trees

If a tree is a path, that is, every node has outdegree 1, then computing the LCA is quite trivial. In that case, the depth of each node can be computed in

$O(n)$ time, and the LCA of two nodes is the node with smaller depth.

For a complete binary tree, the optimal algorithm is somewhat more involved. In this case, each node can be assigned a label of length $O(\log n)$ from which the LCA can be computed in constant time. This labeling idea can be extended to general trees, as we will see in section "Labeling Schemes."

Consider the following node labeling: for any node v of depth $d(v)$, the label $\mathcal{L}(v)$ is obtained by assigning to the first $d(v)$ bits of the code the left-right path from the root, where a left edge is coded with a 0 and a right edge is coded with a 1. The $d(v) + 1^{st}$ bit is 1, and all subsequent bits, up to $1 + \log n$ are 0.

Now let $x = \mathcal{L}(u)$ XOR $\mathcal{L}(v)$ and let $w = \text{LCA}(u, v)$. The first $d(w)$ bits of x are 0, since u and v share the same path until then. The next bit differs so the $d(w) + 1$st bit of x is the first bit that is 1. Thus, by computing $\lfloor \log x \rfloor$, we find the depth of w. Then we can construct the label of w by taking the first $d(w)$ bits of $\mathcal{L}(u)$, then a 1 at position $d(w) + 1$, and then 0s. All these operations take constant time, and all labels can be computed in linear time.

The first optimal LCA algorithm, by Harel and Tarjan [7], decomposed arbitrary trees into paths and balanced binary trees, using optimal labeling algorithms for each part. Such labeling schemes have been used as components in many of the subsequent algorithms. It turns out that there is an $O(\log n)$-bit labeling scheme for arbitrary trees where the LCA can be computed in constant time just from labels. This algorithm will be discussed in section "Labeling Schemes."

An $\langle O(n), O(1)\rangle$-Time Algorithm for ±1RMQ

We have already seen an $\langle O(n \log n), O(1)\rangle$-time algorithm for RMQ, which thus yields an LCA algorithm of the same complexity. However, it is possible to do better, via a simple observation, plus the Four Russians technique.

Consider the RMQ problem generated by the reduction given in Lemma 1. The level tour of a tree is not an arbitrary instance of RMQ. Rather, we note that all entries are integers and adjacent entries differ by one. We call this special case ±1RMQ.

If we can show an $\langle O(n), O(1)\rangle$-time algorithm for ±1RMQ, we directly get an algorithm of the same complexity for LCA, by Lemma 1, but we also get an algorithm of the same complexity for general RMQ by Lemma 2. Thus, to solve an arbitrary RMQ problem optimally, first compute the Cartesian tree and then the Euler and Level tours, thus reducing an arbitrary RMQ to a ±1RMQ in linear time.

In order to improve the preprocessing of ±1RMQ, we will use a table lookup technique to precompute answers on small subarrays, for a log-factor speedup. To this end, partition A into blocks of size $\frac{\log n}{2}$. Define an array $A'[1, \ldots, 2n/\log n]$, where $A'[i]$ is the minimum element in the ith block of A. Define an equal size array B, where $B[i]$ is a position in the ith block in which value $A'[i]$ occurs. Recall that RMQ queries return the position of the minimum and that the LCA to RMQ reduction uses the position of the minimum, rather than the minimum itself. Thus, we will use array B to keep track of where the minima in A' came from.

The ST algorithm runs on array A' in time $\langle O(n), O(1)\rangle$. Having preprocessed A' for RMQ, consider how we answer any query $\text{RMQ}(i, j)$ in A. The indices i and j might be in the same block, so we have to preprocess each block to answer RMQ queries. If $i < j$ are in different blocks, then we can answer the query $\text{RMQ}(i, j)$ as follows. First compute the values:

1. The minimum from i forward to the end of its block
2. The minimum of all the blocks between i's block and j's block
3. The minimum from the beginning of j's block to j

The query will return the position of the minimum of the three values computed. The second minimum is found in constant time by an RMQ on A', which has been preprocessed using the ST algorithm. But we need to know how to answer range minimum queries inside blocks to compute the first and third minima and thus to finish off the algorithm. Thus, the in-block queries are needed whether i and j are in the same block or not.

Therefore, we focus now only on in-block RMQs. If we simply performed RMQ preprocessing on each block, we would spend too much time in preprocessing. If two blocks were identical, then we could share their preprocessing. However, it is too much to hope for that blocks would be so repeated. The following observation establishes a much stronger shared-preprocessing property.

Observation 2 *If two arrays* $X[1,\ldots,k]$ *and* $Y[1,\ldots,k]$ *differ by some fixed value at each position, that is, there is a* c *such that* $X[i] = Y[i] + c$ *for every* i*, then all RMQ answers will be the same for* X *and* Y*. In this case, we can use the same preprocessing for both arrays.*

Thus, we can *normalize* a block by subtracting its initial offset from every element. We now use the ± 1 property to show that there are very few kinds of normalized blocks.

Lemma 3 *There are* $O(\sqrt{n})$ *kinds of normalized blocks.*

Proof Adjacent elements in normalized blocks differ by $+1$ or -1. Thus, normalized blocks are specified by a ± 1 vector of length $(1/2 \cdot \log n) - 1$. There are $2^{(1/2 \cdot \log n) - 1} = O(\sqrt{n})$ such vectors. ■

We are now basically done. We create $O(\sqrt{n})$ tables, one for each possible normalized block. In each table, we put all $(\frac{\log n}{2})^2 = O(\log^2 n)$ answers to all in-block queries. This gives a total of $O(\sqrt{n}\log^2 n)$ total preprocessing of normalized block tables and $O(1)$ query time. Finally, compute, for each block in A, which normalized block table it should use for its RMQ queries. Thus, each in-block RMQ query takes a single table lookup.

Overall, the total space and preprocessing used for normalized block tables and A' tables is $O(n)$ and the total query time is $O(1)$. This gives an optimal algorithm for LCA and RMQ. This algorithm was first presented as a PRAM algorithm by Berkman et al. [3]. Although this algorithm is quite simple, and easily implementable, for many years after its publication, LCA computation was still considered to be too complicated to implement. The algorithm presented here is somewhat streamlined compared to Berkman et al.'s algorithm, because they had the added goal of making a parallelizable algorithm. Thus, for example, they used two levels of encoding to remove the log factor from the preprocessing, with the first level breaking the RMQ into blocks of size $\log n$ and in the second level breaking the blocks up into mini-blocks of size $\log\log n$. Similarly, the sparse table algorithm was somewhat different and required binary-tree LCA as a subroutine.

It is possible, even probably, that the slight complexities of the PRAM version of this algorithm obscured its elegance. This theory was tested by Bender and Farach-Colton [2], who presented the sequential version of the same algorithm with the simplified Sparse Table and RMQ blocking scheme presented here, with the goal of establishing the practicality of LCA computation. This seems to have done the trick, and many variants and implementations of this algorithm now exist.

Labeling Schemes

We have already seen a labeling scheme that allows for the fast computation of LCA on complete binary trees. But labels can be used to solve the LCA problem on arbitrary trees, as shown by Alstrup et al. [1].

To be specific, the goal is to assign to every node an $O(\log n)$-bit label so that the label of the LCA of two nodes can be computed in constant time from their labels. We have seen how to do this for a complete binary tree, but the problem there was simplified by the fact that the depth of such a tree is $O(\log n)$ and the branching factor is two. Here, we consider the general case of arbitrary depth and degree.

Begin by decomposing the tree into *heavy paths*. To do so, let the weight $w(v)$ of any node v be the number of nodes in the subtree rooted at that node. All edges between a node and its heaviest child are called *heavy edges* and all other edges are called *light edges*.

Ties are broken arbitrarily. Note that there are $O(\log n)$ light edges on any root-leaf path.

A path to a node can be specified by alternately specifying how far one traverses a heavy path before exiting to a light edge and the rank of the light edge at the point of exit. Each such code takes $O(\log n)$ bits yielding $O(\log^2 n)$ bits in total.

To reduce the number of bits to $O(\log n)$, Alstrup et al. applied alphabetic codes, which preserve lexicographic ordering but are of variable length. In particular, they used a code with the following properties:

- Let $Y = y_1, y_2, \ldots, y_k$ be a sequence of integers with $\sum_{i=1}^{k} y_i = s$.
- There exists an alphabetic sequence $B = b_1, b_2, \ldots, b_k$ for Y such that, for all i, $|b_i| \leq \lceil \log s \rceil - \lfloor \log y_i \rfloor$.

The idea now is that large trees get short codes and small trees get large codes. By cleverly building alphabetic codes that depend on the size of the trees, the code lengths telescope, giving a final code of length $O(\log n)$. We refer the reader to the paper or to the excellent presentation by Bille [4] for details.

Succinct Representations

There have been several extensions of the LCA/RMQ problem. The one that has received the most attention is that of succinct structures to compute the LCA or RMQ. These structures are succinct because they use $O(n)$ bits of extra space, as opposed to the structures above, which use $\Omega(n)$ words of memory, or $\Omega(n \log n)$ bits.

The first succinct solution for LCA was due to Sadakane [10], who gave an optimal LCA algorithm using $2n + O(n(\log \log n)^2/\log n)$ bits. The main approach of this algorithm is to replace the Sparse Table algorithm with a more bit-efficient variant.

The first succinct solution for RMQ was also due to Sadakane [9]. His algorithm takes $4n + o(n)$ bits. The main idea, once again, follows the Sparse Table algorithm. By using a ternary Cartesian tree which stores values not in internal nodes but in leaves, the preorders of nodes coincide the orders of the values.

The current best solution is by Fischer [5], who uses $2n + o(n)$ bits, which is shown to be optimal up to lower-order terms. The structure using the least known $o(n)$-bit term [8] uses $2n + O(n/\log^c n)$ bits, for any constant c. All the succinct solutions are $O(1)$ time.

Cross-References

▸ Compressed Range Minimum Queries

Recommended Reading

1. Alstrup S, Gavoille C, Kaplan H, Rauhe T (2004) Nearest common ancestors: a survey and a new algorithm for a distributed environment. Theory Comput Syst 37(3):441–456
2. Bender MA, Farach-Colton M (2000) The LCA problem revisited. In: Proceedings of Latin American theoretical informatics (LATIN), Montevideo, pp 88–94
3. Berkman O, Breslauer D, Galil Z, Schieber B, Vishkin U (1989) Highly parallelizable problems. In: Proceedings of the 21st annual ACM symposium on theory of computing, New Orleans, pp 309–319
4. Bille P (2014) Nearest common ancestors. http://massivedatasets.files.wordpress.com/2014/02/nearestcommonancestors_2014.pdf
5. Fischer J (2010) Optimal succinctness for range minimum queries. In: Proceedings of LATIN, Oaxaca, pp 158–169
6. Gabow HN, Bentley JL, Tarjan RE (1984) Scaling and related techniques for geometry problems. In: Proceedings of the 16th annual ACM symposium on theory of computing, New York, vol 67, pp 135–143
7. Harel D, Tarjan RE (1984) Fast algorithms for finding nearest common ancestors. SIAM J Comput 13(2):338–355
8. Navarro G, Sadakane K (2014) Fully functional static and dynamic succinct trees. ACM Trans Algorithms 10(3)
9. Sadakane K (2002) Space-efficient data structures for flexible text retrieval systems. In: International symposium on algorithms and computation (ISAAC), Vancouver, pp 14–24
10. Sadakane K (2002) Succinct representations of lcp information and improvements in the compressed suffix arrays. In: Proceedings of the 13th annual ACM-SIAM symposium on discrete algorithms, San Francisco, pp 225–232

LP Based Parameterized Algorithms

M.S. Ramanujan
Department of Informatics, University of Bergen, Bergen, Norway

Keywords

Above guarantee parameterizations; FPT algorithms; Linear programming

Years and Authors of Summarized Original Work

2013; Cygan, Pilipczuk, Pilipczuk, Wojtaszczyk
2014; Lokshtanov, Narayanaswamy, Raman, Ramanujan, Saurabh
2014; Wahlstrom

Problem Definition

Linear and integer programs have played a crucial role in the theory of approximation algorithms for combinatorial optimization problems. While they have also been central in identifying polynomial time solvable problems, it is only recently that these tools have been put to use in designing exact algorithms for NP-complete problems. Following the paradigm of above-guarantee parameterization in fixed-parameter tractability, these efforts have focused on designing algorithms where the exponential component of the running time depends only on the excess of the solution above the optimum value of a linear program for the problem.

Method Description

The linear program obtained from a given integer linear program (ILP) by relaxing the integrality conditions on the variables is called the *standard relaxation* of the ILP or the standard LP. Similarly, the linear program obtained from the ILP by restricting the domain of the variables to the set of all half integers is called the *half-integral relaxation* of the ILP or the half-integral LP. The standard LP is said to have a half-integral optimum if the optimum values of the standard relaxation and half-integral relaxation coincide.

The VERTEX COVER problem provides one of the simplest illustrations of this method. In this problem, the objective is to find a minimum-sized subset of vertices whose removal makes a given graph edgeless. In the well-known integer linear programming formulation (ILP) for VERTEX COVER, given a graph G, a *feasible solution* is defined as a function $x : V \rightarrow \{0, 1\}$ satisfying the edge constraints $x(u) + x(v) \geq 1$ for every edge (u, v). The objective of the linear program is to minimize $\Sigma_{u \in V} x(u)$ over all feasible solutions x. The value of the optimum solution to this ILP is denoted by $vc(G)$. In the standard relaxation of the above ILP, the constraint $x(v) \in \{0, 1\}$ is replaced with $x(v) \geq 0$, for all $v \in V$. For a graph G, this relaxation is denoted by LPVC(G), and the minimum value of LPVC(G) is denoted by $vc^*(G)$.

It is known that LPVC(G) has a half-integral optimum [10] and that LPVC(G) is *persistent* [11], that is, if a variable is assigned 0 (respectively 1) in an optimum solution to the standard LP, then it can be discarded from (respectively included into) an optimum vertex cover of G. Based on the persistence of LPVC(G), a polynomial time preprocessing procedure for VERTEX COVER immediately follows. More precisely, as long as there is an optimum solution to the standard LP which assigns 0 or 1 to a vertex of G, one may discard or include this vertex in the optimum solution. When this procedure cannot be executed any longer, an arbitrary vertex of G is selected, and the algorithm branches into 2 exhaustive cases based on this vertex being included or excluded in an optimum vertex cover of G. Standard analysis shows that this is an algorithm running in time $O(4^{(vc(G)-vc^*(G))}|G|^{O(1)})$.

Key Results

This method was first used in the context of fixed-parameter tractability by Guillemot [5]

who used it to give FPT algorithms for path-transversal and cycle-transversal problems. Subsequently, Cygan et al. [4] improved upon this result to give an FPT algorithm for the MULTIWAY CUT problem parameterized above the LP value. In this problem, the objective is to find a smallest set of vertices which pair-wise separates a given subset of vertices of a graph. As a consequence of this algorithm, they were able to obtain the current fastest FPT algorithm for MULTIWAY CUT parameterized by the solution size.

Theorem 1 ([4]) *There is an algorithm for* MULTIWAY CUT *running in time* $O(2^k|G|^{O(1)})$, *where k is the size of the solution.*

Following this work, Narayanaswamy et al. [9] and Lokshtanov et al. [8] considered the VERTEX COVER problem and built upon these methods with several additional problem-specific steps to obtain improved FPT algorithms for several problems, with the most notable among them being the ODD CYCLE TRANSVERSAL problem – the problem of finding a smallest set of vertices to delete in order to obtain a bipartite graph. These results were the first improvements over the very first FPT algorithm for this problem given by Reed, Smith, and Vetta [12].

Theorem 2 ([8]) *There is an algorithm for* ODD CYCLE TRANSVERSAL *running in time* $O(2.32^k|G|^{O(1)})$, *where k is the size of the solution.*

Iwata et al. [7] applied this method to several problems to improve the polynomial dependence of the running times on the input size. Using network-flow-based *linear time* algorithms for solving the half-integral Vertex Cover LP (LPVC), they obtained an FPT algorithm for ODD CYCLE TRANSVERSAL with a linear dependence on the input size. Most recently, using tools from the theory of constraint satisfaction, Wahlstrom [13] extended this approach to a much broader class of problems with half-integral LPs and obtained improved FPT algorithms for a number of problems including node-deletion UNIQUE LABEL COVER and GROUP FEEDBACK VERTEX SET. The UNIQUE LABEL COVER problem plays a central role in the theory of approximation and was studied from the point of view of parameterized complexity by Chitnis et al. [2]. The GROUP FEEDBACK VERTEX SET problem is a generalization of the classical FEEDBACK VERTEX SET problem. The fixed-parameter tractability of this problem was proved in [5] and [3].

Theorem 3 ([2]) *There is an algorithm for node-deletion* UNIQUE LABEL COVER *running in time* $O(|\Sigma|^{2k}|G|^{O(1)})$ *and an algorithm for* GROUP FEEDBACK VERTEX *running in time* $O(4^k|G|^{O(1)})$. *In the first case,* Σ *denotes the size of the alphabet, and in either case, k denotes the size of the solution.*

Applications

This method relies crucially on the half integrality of a certain LP for the problem at hand. The most well-known problems with this property are VERTEX COVER, MULTIWAY CUT, and certain problems for which Hochbaum [6] defined a particular kind of ILPs, referred to as IP2. However, the work of Wahlstrom [13] lifts this approach to a more general class of problems by interpreting a half-integral relaxation as a polynomial-time solvable problem on the discrete search space of $\{0, \frac{1}{2}, 1\}$.

Open Problems

A primary challenge here is to build upon these LP-based tools to design an FPT algorithm for ODD CYCLE TRANSVERSAL with a provably optimal dependence on the parameter under appropriate complexity theoretic assumptions.

Experimental Results

Experimental results comparing algorithms for VERTEX COVER based on this method with other state-of-the art empirical methods are given in [1].

Cross-References

▸ Kernelization, Max-Cut Above Tight Bounds
▸ Kernelization, MaxLin Above Average

Recommended Reading

1. Akiba T, Iwata Y (2015) Branch-and-reduce exponential/FPT algorithms in practice: a case study of vertex cover. In: Proceedings of the seventeenth workshop on algorithm engineering and experiments, ALENEX 2015, San Diego, 5 Jan 2015, pp 70–81
2. Chitnis RH, Cygan M, Hajiaghayi M, Pilipczuk M, Pilipczuk M (2012) Designing FPT algorithms for cut problems using randomized contractions. In: 53rd annual IEEE symposium on foundations of computer science, FOCS 2012, New Brunswick, 20–23 Oct 2012, pp 460–469
3. Cygan M, Pilipczuk M, Pilipczuk M (2012) On group feedback vertex set parameterized by the size of the cutset. In: Graph-theoretic concepts in computer science – 38th international workshop, WG 2012, Jerusalem, 26–28 June 2012, revised selected papers, pp 194–205
4. Cygan M, Pilipczuk M, Pilipczuk M, Wojtaszczyk JO (2013) On multiway cut parameterized above lower bounds. Trans Comput Theory 5(1):3
5. Guillemot S (2011) FPT algorithms for path-transversal and cycle-transversal problems. Discret Optim 8(1):61–71
6. Hochbaum DS (2002) Solving integer programs over monotone inequalities in three variables: a framework for half integrality and good approximations. Eur J Oper Res 140(2):291–321
7. Iwata Y, Oka K, Yoshida Y (2014) Linear-time FPT algorithms via network flow. In: Proceedings of the twenty-fifth annual ACM-SIAM symposium on discrete algorithms, SODA 2014, Portland, 5–7 Jan 2014, pp 1749–1761
8. Lokshtanov D, Narayanaswamy NS, Raman V, Ramanujan MS, Saurabh S (2014) Faster parameterized algorithms using linear programming. ACM Trans Algorithms 11(2):15
9. Narayanaswamy NS, Raman V, Ramanujan MS, Saurabh S (2012) LP can be a cure for parameterized problems. In: 29th international symposium on theoretical aspects of computer science, STACS 2012, Paris, 29th Feb – 3rd Mar 2012, pp 338–349
10. Nemhauser GL, Trotter LE (1974) Properties of vertex packing and independence system polyhedra. Math Program 6:48–61
11. Nemhauser GL, Trotter LE (1975) Vertex packings: structural properties and algorithms. Math Program 8:232–248
12. Reed BA, Smith K, Vetta A (2004) Finding odd cycle transversals. Oper Res Lett 32(4):299–301
13. Wahlström M (2014) Half-integrality, LP-branching and FPT algorithms. In: Proceedings of the twenty-fifth annual ACM-SIAM symposium on discrete algorithms, SODA 2014, Portland, 5–7 Jan 2014, pp 1762–1781

LP Decoding

Jonathan Feldman
Google, Inc., New York, NY, USA

Keywords

Belief propagation; Error-correcting codes; LDPC codes; Low-density parity-check codes; LP decoding; Pseudocodewords

Years and Authors of Summarized Original Work

2002 and later; Feldman, Karger, Wainwright

Problem Definition

Error-correcting codes are fundamental tools used to transmit digital information over unreliable channels. Their study goes back to the work of Hamming and Shannon, who used them as the basis for the field of information theory. The problem of decoding the original information up to the full error-correcting potential of the system is often very complex, especially for modern codes that approach the theoretical limits of the communication channel.

LP decoding [4, 5, 8] refers to the application of *linear programming* (LP) *relaxation* to the problem of decoding an error-correcting code. Linear programming relaxation is a standard technique in approximation algorithms and operations research, and is central to the study of efficient algorithms to find good (albeit suboptimal) solutions to very difficult optimization problems [13]. LP decoders have tight combinatorial characterizations of decoding success that can be used to analyze error-correcting performance.

The codes for which LP decoding has received the most attention are *low-density parity-check* (LDPC) codes [9], due to their excellent error-correcting performance. The LP decoder is particularly attractive for analysis of these codes because the standard message-passing algorithms such as *belief propagation* (see [15]) used for decoding are often difficult to analyze, and indeed the performance of LP decoding is closely tied to these methods.

Error-Correcting Codes and Maximum-Likelihood Decoding

This section begins with a very brief overview of error-correcting codes, sufficient for formulating the LP decoder. Some terms are not defined for space reasons; for a full treatment of error-correcting codes in context, the reader is referred to textbooks on the subject (e.g., [11]).

A *binary error-correcting code* is a subset $C \subseteq \{0,1\}^n$. The *rate* of the code C is $r = \log(|C|)/n$. A *linear* binary code is a linear subspace of $\{0,1\}^n$. A *codeword* is a vector $y \in C$. Note that 0^n is always a codeword of a linear code, a fact that will be useful later. When the code is used for communication, a codeword $\dot{y} \in C$ is transmitted over a *noisy channel*, resulting in some *received word* $\hat{y} \in \Sigma^n$, where Σ is some alphabet that depends on the channel model. Generally in LP decoding a *memoryless, symmetric* channel is assumed. One common such channel is the *binary symmetric channel (BSC) with parameter* p, which will be referred to as BSC_p, where $0 < p < 1/2$. In the BSC_p, the alphabet is $\Sigma = \{0,1\}$, and for each i, the received symbol $\hat{y}_i$ is equal to $\dot{y}_i$ with probability p, and $\hat{y}_i = 1 - \dot{y}_i$ otherwise. Although LP decoding works with more general channels, this chapter will focus on the BSC_p.

The *maximum-likelihood (ML) decoding problem* is the following: given a received word $\hat{y} \in \{0,1\}^n$, find the codeword $y^* \in C$ that is most likely to have been sent over the channel. Defining the vector $\gamma \in \{-1,+1\}^n$ where $\gamma_i = 1 - 2\hat{y}_i$, it is easy to show:

$$y^* = \arg\min_{y \in C} \sum_i \gamma_i y_i. \tag{1}$$

The complexity of the ML decoding problem depends heavily on the code being used. For simple codes such as a *repetition code* $C = \{0^n, 1^n\}$, the task is easy. For more complex (and higher-rate) codes such as LDPC codes, ML decoding is NP-hard [1].

LP Decoding

Since ML decoding can be very hard in general, one turns to sub-optimal solutions that can be found efficiently. LP decoding, instead of trying to solve (1), relaxes the constraint $y \in C$, and instead requires that $y \in \mathcal{P}$ for some succinctly describable linear polytope $\mathcal{P} \subseteq [0,1]^n$, resulting in the following linear program:

$$y_{\text{LP}} = \arg\min_{y \in \mathcal{P}} \sum_{i=1}^{n} \gamma_i y_i. \tag{2}$$

It should be the case that the polytope includes all the codewords, and does not include any integral non-codewords. As such, a polytope $\mathcal{P}$ is called *proper* for code C if $\mathcal{P} \cap \{0,1\}^n = C$.

The LP decoder works as follows. Solve the LP in (2) to obtain $y_{\text{LP}} \in [0,1]^n$. If y_{LP} is integral (i.e., all elements are 0 or 1), then output y_{LP}. Otherwise, output "error". By the definition of a proper polytope, if the LP decoder outputs a codeword, it is guaranteed to be equal to the ML codeword y^*. This fact is known as the *ML certificate* property.

Comparing with ML Decoding

A successful decoder is one that outputs the original codeword transmitted over the channel, and so the quality of an algorithm is measured by the likelihood that this happens. (Another common non-probabilistic measure is the *worst-case* performance guarantee, which measures how many bit-flips an algorithm can tolerate and still be guaranteed to succeed.) Note that y^* is the one *most likely* to be the transmitted codeword $\dot{y}$, but it is not always the case that $y^* = \dot{y}$. However, no

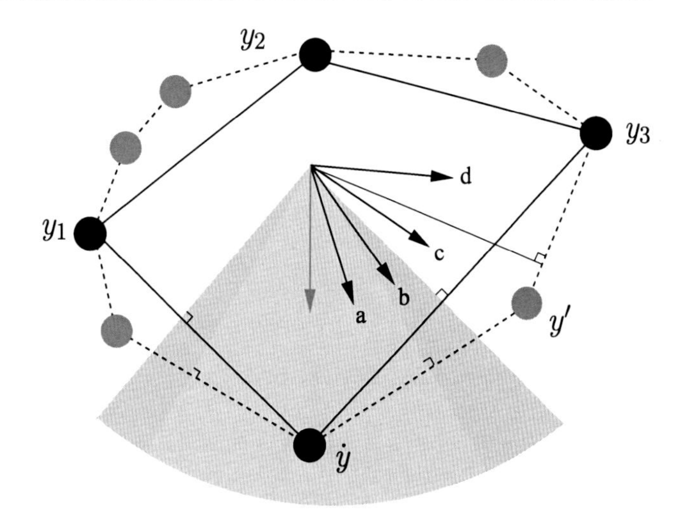

LP Decoding, Fig. 1 A decoding polytope $\mathcal{P}$ (*dotted line*) and the convex hull C (*solid line*) of the codewords $\dot{y}$, y_1, y_2, and y_3. Also shown are the four possible cases (a–d) for the objective function, and the normal cones to both $\mathcal{P}$ and C

decoder can perform better than an ML decoder, and so it is useful to use ML decoding as a basis for comparison.

Figure 1 provides a geometric perspective of LP decoding, and its relation to exact ML decoding. Both decoders use the same LP objective function, but over different constraint sets. In exact ML decoding, the constraint set is the convex hull C of codewords (i.e., the set of points that are convex combinations of codewords from C), whereas relaxed LP decoding uses the larger polytope $\mathcal{P}$. In Fig. 1, the four arrows labeled (a)–(d) correspond to different "noisy" versions of the LP objective function. (a) If there is very little noise, then the objective function points to the transmitted codeword $\dot{y}$, and thus both ML decoding and LP decoding succeed, since both have the transmitted codeword $\dot{y}$ as the optimal point. (b) If more noise is introduced, then ML decoding succeeds, but LP decoding fails, since the fractional vertex y' is optimal for the relaxation. (c) With still more noise, ML decoding fails, since y_3 is now optimal; LP decoding still has a fractional optimum y', so this error is in some sense "detected". (d) Finally, with a lot of noise, both ML decoding and LP decoding have y_3 as the optimum, and so both methods fail and the error is "undetected". Note that in the last two cases (c, d), when ML decoding fails, the failure of the LP decoder is in some sense the fault of the code itself, as opposed to the decoder.

Normal Cones and C-Symmetry

The (negative) *normal cones* at $\dot{y}$ (also called the *fundamental cone* [10]) is defined as follows:

$$N_{\dot{y}}(\mathcal{P})=\{\gamma\in\mathbb{R}^n:\sum_i \gamma_i(y_i-\dot{y}_i)\geq 0 \text{ for all } y\in\mathcal{P}\},$$

$$N_{\dot{y}}(C)=\{\gamma\in\mathbb{R}^n:\sum_i \gamma_i(y_i-\dot{y}_i)\geq 0 \text{ for all } y\in C\}.$$

Note that $N_{\dot{y}}(\mathcal{P})$ corresponds to the set of cost vectors γ such that $\dot{y}$ is an optimal solution to (2). The set $N_{\dot{y}}(C)$ has a similar interpretation as the set of cost vectors γ for which $\dot{y}$ is the ML codeword. Since $\mathcal{P}\subset C$, it is immediate from the definition that $N_y(C)\supset N_y(\mathcal{P})$ for all $y\in C$. Fig. 1 shows these two cones and their relationship.

The success probability of an LP decoder is equal to the total probability mass of $N_{\dot{y}}(\mathcal{P})$, under the distribution on cost vectors defined by the channel. The success probability of ML decoding is similarly related to the probability mass in the normal cone $N_y(C)$. Thus, the discrepancy between the normal cones of $\mathcal{P}$ and C is a measure of the gap between exact ML and relaxed LP decoding.

This analysis is specific to a particular transmitted codeword $\dot{y}$, but one would like to apply it in general. When dealing with linear codes, for most decoders one can usually assume that an arbitrary codeword is transmitted, since the decision region for decoding success is symmetric. The same holds true for LP decoding (see [4] for proof), as long as the polytope $\mathcal{P}$ is *C-symmetric*, defined as follows:

Definition 1 A proper polytope $\mathcal{P}$ for the binary code C is C-**symmetric** if, for all $y \in \mathcal{P}$ and $\dot{y} \in C$, it holds that $y' \in \mathcal{P}$, where $y'_i = |y_i - \dot{y}_i|$.

Using a Dual Witness to Prove Error Bounds

In order to prove that LP decoding succeeds, one must show that $\dot{y}$ is the optimal solution to the LP in (2). If the code C is linear, and the relaxation is proper and C-symmetric, one can assume that $\dot{y} = 0^n$, and then show that 0^n is optimal. Consider the *dual* of the decoding LP in (2). If there is a feasible point of the dual LP that has the same cost (i.e., zero) as the point 0^n has in the primal, then 0^n must be an optimal point of the decoding LP. Therefore, to prove that the LP decoder succeeds, it suffices to exhibit a zero-cost point in the dual. (Actually, since the existence of the zero-cost dual point only proves that 0^n is one of possibly many primal optima, one needs to be a bit more careful, a minor issue deferred to more complete treatments of this material.)

Key Results

LP decoders have mainly been studied in the context of Low-Density Parity-Check codes [9], and their generalization to expander codes [12]. LP decoders for Turbo codes [2] have also been defined, but the results are not as strong. This summary of key results gives bounds on the *word error rate (WER)*, which is the probability, over the noise in the channel, that the decoder does not output the transmitted word. These bounds are relative to specific *families* of codes, which are defined as infinite set of codes of increasing length whose rate is bounded from below by some constant. Here the bounds are given in asymptotic form (without constants instantiated), and only for the binary symmetric channel.

Many other important results that are not listed here are known for LP decoding and related notions. Some of these general areas are surveyed in the next section, but there is insufficient space to reference most of them individually; the reader is referred to [3] for a thorough bibliography.

Low-Density Parity-Check Codes

The polytope $\mathcal{P}$ for LDPC codes, first defined in [4, 8, 10], is based on the underlying *Tanner graph* of the code, and has a linear number of variables and constraints. If the Tanner graph expands sufficiently, it is known that LP decoding can correct a constant fraction of errors in the channel, and thus has an inverse exponential error rate. This was proved using a dual witness:

Theorem 1 ([6]) *For any rate $r > 0$, there is a constant $\epsilon > 0$ such that there exists a rate r family of low-density parity-check codes with length n where the LP decoder succeeds as long as at most ϵn bits are flipped by the channel. This implies that there exists a constant $\epsilon' > 0$ such that the word error rate under the BSC_p with $p < \epsilon'$ is at most $2^{-\Omega(n)}$.*

Expander Codes

The *capacity* of a communication channel bounds from above the rate one can obtain from a family of codes and still get a word error rate that goes to zero as the code length increases. The notation C_p is used to denote the capacity of the BSC_p. Using a family of codes based on expanders [12], LP decoding can achieve rates that approach capacity. Compared to LDPC codes, however, this comes at the cost of increased decoding complexity, as the size of the LP is exponential in the gap between the rate and capacity.

Theorem 2 ([7]) *For any $p > 0$, and any rate $r < C_p$, there exists a rate r family of expander codes with length n such that the word error rate of LP decoding under the BSC_p is at most $2^{-\Omega(n)}$.*

Turbo Codes

Turbo codes [2] have the advantage that they can be encoded in linear time, even in a streaming fashion. *Repeat-accumulate* codes are a simple form of Turbo code. The LP decoder for Turbo codes and their variants was first defined in [4, 5], and is based on the *trellis* structure of the component *convolutional* codes. Due to certain properties of turbo codes it is impossible to prove bounds for turbo codes as strong as the ones for LDPC codes, but the following is known:

Theorem 3 ([5]) *There exists a rate* $1/2 - o(1)$ *family of repeat-accumulate codes with length n, and a constant* $\epsilon > 0$*, such that under the* BSC_p *with* $p < \epsilon$*, the LP decoder has a word error rate of at most* $n^{-\Omega(1)}$.

Applications

The application of LP decoding that has received the most attention so far is for LDPC codes. The LP for this family of codes not only serves as an interesting alternative to more conventional iterative methods [15], but also gives a useful tool for analyzing those methods, an idea first explored in [8, 10, 14]. Iterative methods such as *belief propagation* use local computations on the Tanner graph to update approximations of the marginal probabilities of each code bit. In this type of analysis, the vertices of the polytope $\mathcal{P}$ are referred to as *pseudocodewords*, and tend to coincide with the fixed points of this iterative process. Other notions of pseudocodeword-like structures such as *stopping sets* are also known to coincide with these polytope vertices. Understanding these structures has also inspired the design of new codes for use with iterative and LP decoding. (See [3] for a more complete bibliography of this work).

The decoding method itself can be extended in many ways. By adding redundant information to the description of the code, one can derive tighter constraint sets to improve the error-correcting performance of the decoder, albeit at an increase in complexity. Adaptive algorithms that try to add constraints "on the fly" have also been explored, using branch-and-bound or other techniques. Also, LP decoding has inspired the use of other methods from optimization theory in decoding error-correcting codes. (Again, see [3] for references.)

Open Problems

The LP decoding method gives a simple, efficient and analytically tractable approach to decoding error-correcting codes. The results known to this point serve as a proof of concept that strong bounds are possible, but there are still important questions to answer. Although LP decoders can achieve capacity with decoding time polynomial in the length of the code, the complexity of the decoder still depends exponentially on the gap between the rate and capacity (as is the case for all other known provably efficient capacity-achieving decoders). Decreasing this dependence would be a major accomplishment, and perhaps LP decoding could help. Improving the fraction of errors correctable by LP decoding is also an important direction for further research.

Another interesting question is whether there exist constant-rate linear-distance code families for which one can formulate a polynomial-sized exact decoding LP. Put another way, is there a constant-rate linear-distance family of codes whose convex hulls have a polynomial number of facets? If so, then LP decoding would be equivalent to ML decoding for this family. If not, this is strong evidence that suboptimal decoding is inevitable when using good codes, which is a common belief.

An advantage to LP decoding is the *ML certificate* property mentioned earlier, which is not enjoyed by most other standard suboptimal decoders. This property opens up the possibility for a wide range of heuristics for improving decoding performance, some of which have been analyzed, but largely remain wide open.

LP decoding has (for the most part) only been explored for LDPC codes under memoryless symmetric channels. The LP for turbo codes has been defined, but the error bounds proved so far are not a satisfying explanation of the excellent performance observed in practice. Other codes and channels have gotten little, if any, attention.

Cross-References

▶ Decoding Reed–Solomon Codes
▶ Learning Heavy Fourier Coefficients of Boolean Functions
▶ Linearity Testing/Testing Hadamard Codes
▶ List Decoding near Capacity: Folded RS Codes

Recommended Reading

1. Berlekamp E, McEliece R, van Tilborg H (1978) On the inherent intractability of certain coding problems. IEEE Trans Inf Theory 24:384–386
2. Berrou C, Glavieux A, Thitimajshima P (1993) Near Shannon limit error-correcting coding and decoding: turbo-codes. In: Proceedings of the IEEE international conference on communications (ICC), Geneva, 23–26 May 1993, pp 1064–1070
3. Boston N, Ganesan A, Koetter R, Pazos S, Vontobel P Papers on pseudocodewords. HP Labs, Palo Alto. http://www.pseudocodewords.info
4. Feldman J (2003) Decoding error-correcting codes via linear programming. Ph.D. thesis, Massachusetts Institute of Technology
5. Feldman J, Karger DR (2002) Decoding turbo-like codes via linear programming. In: Proceedings of the 43rd annual IEEE symposium on foundations of computer science (FOCS), Vancouver, 16–19 Nov 2002
6. Feldman J, Malkin T, Servedio RA, Stein C, Wainwright MJ (2004) LP decoding corrects a constant fraction of errors. In: Proceedings of the IEEE international symposium on information theory, Chicago, 27 June – 2 July 2004
7. Feldman J, Stein C (2005) LP decoding achieves capacity. In: Symposium on discrete algorithms (SODA '05), Vancouver, Jan 2005
8. Feldman J, Wainwright MJ, Karger DR (2003) Using linear programming to decode linear codes. In: 37th annual conference on information sciences and systems (CISS '03), Baltimore, 12–14 Mar 2003
9. Gallager R (1962) Low-density parity-check codes. IRE Trans Inf Theory (IT) 8:21–28
10. Koetter R, Vontobel P (2003) Graph covers and iterative decoding of finite-length codes. In: Proceedings of the 3rd international symposium on turbo codes and related topics, Brest, Sept 2003, pp 75–82
11. MacWilliams FJ, Sloane NJA (1981) The theory of error correcting codes. North-Holland, Amsterdam
12. Sipser M, Spielman D (1996) Expander codes. IEEE Trans Inf Theory 42:1710–1722
13. Vazirani VV (2003) Approximation algorithms. Springer, Berlin
14. Wainwright M, Jordan M (2003) Variational inference in graphical models: the view from the marginal polytope. In: Proceedings of the 41st Allerton conference on communications, control, and computing, Monticello Oct 2003
15. Wiberg N (1996) Codes and decoding on general graphs. Ph.D. thesis, Linkoping University

M

Majority Equilibrium

Qizhi Fang
School of Mathematical Sciences, Ocean University of China, Qingdao, Shandong Province, China

Keywords

Condorcet winner

Years and Authors of Summarized Original Work

2003; Chen, Deng, Fang, Tian

Problem Definition

Majority rule is arguably the best decision mechanism for public decision-making, which is employed not only in public management but also in business management. The concept of majority equilibrium captures such a democratic spirit in requiring that no other solutions would please more than half of the voters in comparison to it. The work of Chen, Deng, Fang, and Tian [1] considers a public facility location problem decided via a voting process under the majority rule on a discrete network. This work distinguishes itself from previous work by applying the computational complexity approach to the study of majority equilibrium. For the model with a single public facility located in trees, cycles, and cactus graphs, it is shown that the majority equilibrium can be found in linear time. On the other hand, when the number of public facilities is taken as the input size (not a constant), finding a majority equilibrium is shown to be $\mathcal{NP}$-hard.

Consider a network $G = ((V, \omega), (E, l))$ with vertex and edge- weight functions $\omega : V \to R^+$ and $l : E \to R^+$, respectively. Each vertex $i \in V$ represents a community, and $\omega(i)$ represents the number of voters that reside there. For each $e \in E$, $l(e) > 0$ represents the length of the road $e = (i, j)$ connecting two communities i and j. For two vertices $i, j \in V$, the distance between i and j, denoted by $d_G(i, j)$, is the length of a shortest path joining them. The location of a public facility such as a library, community center, etc., is to be determined by the public via a voting process under the majority rule. Here, each member of the community desires to have the public facility close to himself, and the decision has to be agreed upon by a majority of the voters. Denote the vertex set of G by $V = \{v_1, v_2, \ldots, v_n\}$. Then each $v_i \in V$ has a preference order $\geq_i$ on V induced by the distance on G. That is, $x \geq_i y$ if and only if $d_G(v_i, x) \leq d_G(v_i, y)$ for two vertices $x, y \in V$; similarly, $x >_i y$ if and only if $d_G(v_i, x) < d_G(v_i, y)$. Based on such a preference profile, four types of majority equilibrium, called Condorcet winners, are defined as follows.

M.-Y. Kao (ed.), *Encyclopedia of Algorithms*,
DOI 10.1007/978-1-4939-2864-4

Definition 1 Let $v_0 \in V$, then v_0 is called:

1. *A weak quasi-Condorcet winner*, if for every $u \in V$ distinct of v_0,

$$\omega(\{v_i \in V : v_0 \geq_i u\}) \geq \sum_{v_i \in V} \omega(v_i)/2;$$

2. *A strong quasi-Condorcet winner*, if for every $u \in V$ distinct of v_0,

$$\omega(\{v_i \in V : v_0 \geq_i u\}) > \sum_{v_i \in V} \omega(v_i)/2;$$

3. *A weak Condorcet winner*, if for every $u \in V$ distinct of v_0,

$$\omega(\{v_i \in V : v_0 \underset{i}{>} u\}) \geq \omega(\{v_i \in V : u \underset{i}{>} v_0\});$$

4. *A strong Condorcet winner*, if for every $u \in V$ distinct of v_0,

$$\omega(\{v_i \in V : v_0 \underset{i}{>} u\}) > \omega(\{v_i \in V : u \underset{i}{>} v_0\}).$$

Under the majority voting mechanism described above, the problem is to develop efficient ways for determining the existence of Condorcet winners and finding such a winner when one exists.

Problem 1 (Finding Condorcet Winners) INPUT: A network $G = ((V, w), (E, l))$. OUTPUT: A Condorcet winner $v \in V$ or nonexistence of Condorcet winners.

Key Results

The mathematical results on the Condorcet winners depend deeply on the understanding of combinatorial structures of underlying networks. Theorems 1–3 below are given for weak quasi-Condorcet winners in the model with a single facility to be located. Other kinds of Condorcet winners can be discussed similarly.

Theorem 1 *Every tree has one weak quasi-Condorcet winner or two adjacent weak quasi-Condorcet winners, which can be found in linear time.*

Theorem 2 *Let C_n be a cycle of order n with vertex-weight function $\omega : V(C_n) \to R^+$. Then $v \in V(C_n)$ is a weak quasi-Condorcet winner of C_n if and only if the weight of each $\lfloor \frac{n+1}{2} \rfloor$-interval containing v is at least $\frac{1}{2} \sum_{v \in C_n} \omega(v)$. Furthermore, the problem of finding a weak quasi-Condorcet winner of C_n is solvable in linear time.*

Given a graph $G = (V, E)$, a vertex v of G is a *cut vertex* if $E(G)$ can be partitioned into two nonempty subsets E_1 and E_2 such that the induced graphs $G[E_1]$ and $G[E_2]$ have just the vertex v in common. A *block* of G is a connected subgraph of G that has no cut vertices and is maximal with respect to this property. Every graph is the union of its blocks. A graph G is called a *cactus graph*, if G is a connected graph in which each block is an edge or a cycle.

Theorem 3 *The problem of finding a weak quasi-Condorcet winner of a cactus graph with vertex-weight function is solvable in linear time.*

In general, the problem can be extended to the cases where a number of public facilities are required to be located during one voting process, and the definitions of Condorcet winners can also be extended accordingly. In such cases, the public facilities may be of the same type or different types; and the utility functions of the voters may be of different forms.

Theorem 4 *If there are a bounded constant number of public facilities to be located at one voting process under the majority rule, then the problem of finding a Condorcet winner (any of the four types) can be solved in polynomial time.*

Theorem 5 *If the number of public facilities to be located is not a constant but considered as the input size, the problem of finding a Condorcet winner is $\mathcal{NP}$-hard; and the corresponding decision problem: deciding whether a candidate set of public facilities is a Condorcet winner is co-$\mathcal{NP}$-complete.*

Applications

Damange [2] first reviewed continuous and discrete spatial models of collective choice, aiming at characterizing the public facility location problem as a result of the pubic voting process. Although the network models in Chen et al. [1] have been studied for some problems in economics [3,4], the main point of Chen et al.'s work is the computational complexity and algorithmic approach. This approach can be applied to more general public decision-making processes.

For example, consider a public road repair problem, pioneered by Tullock [5] to study redistribution of tax revenue under a majority rule system. An edge-weighted graph $G = (V, E, w)$ represents a network of local roads, where the weight of each edge represents the cost of repairing the road. There is also a distinguished vertex $s \in V$ representing the entry point to the highway system. The majority decision problem involves a set of agents $A \subseteq V$ situated at vertices of the network who would choose a subset F of edges. The cost of repairing F, which is the sum of the weights of edges in F, will be shared by all n agents, each an n-th of the total. In this model, a majority stable solution under the majority rule is a subset $F \subseteq E$ that connects s to a subset $A_1 \subset A$ of agents with $|A_1| > |A|/2$ such that no other solution H connecting s to a subset of agents $A_2 \subset A$ with $|A_2|A|/2$ satisfies the conditions that $\sum_{e \in H} w(e) \leq \sum_{e \in F} w(e)$, and for each agent in A_2, its shortest path to s in solution H is not longer than that in solution F, and at least one of the inequalities is strict. It is shown in Chen et al. [1] that for this model, finding a majority equilibrium is $\mathcal{NP}$-hard for general networks and is polynomially solvable for tree networks.

Cross-References

- ▶ General Equilibrium
- ▶ Leontief Economy Equilibrium
- ▶ Local Search for K-medians and Facility Location

Recommended Reading

1. Chen L, Deng X, Fang Q, Tian F (2002) Majority equilibrium for public facility allocation. Lect Notes Comput Sci 2697:435–444
2. Demange G (1983) Spatial models of collective choice. In: Thisse JF, Zoller HG (eds) Locational analysis of public facilities. North-Holland Publishing Company, Amsterdam
3. Hansen P, Thisse JF (1981) Outcomes of voting and planning: condorcet, weber and rawls locations. J Publ Econ 16:1–15
4. Schummer J, Vohra RV (2002) Strategy-proof location on a network. J Econ Theory 104:405–428
5. Tullock G (1959) Some problems of majority voting. J Polit Econ 67:571–579

Manifold Reconstruction

Siu-Wing Cheng
Department of Computer Science and Engineering, Hong Kong University of Science and Technology, Hong Kong, China

Keywords

Čzech complex; Delaunay complex; Homology; Homeomorphism; Implicit function; Voronoi diagram

Years and Authors of Summarized Original Work

2005; Cheng, Dey, Ramos
2008; Niyogi, Smale, Weinberger
2014; Boissonnat, Ghosh
2014; Cheng, Chiu

Problem Definition

With the widespread of sensing and Internet technologies, a large number of numeric attributes for a physical or cyber phenomenon can now be collected. If each attribute is viewed as a coordinate, an instance in the collection can be viewed as a point in $\mathbb{R}^d$ for some large d. When the physical or cyber phenomenon is governed by only a few latent parameters, it is often postulated that the data points lie on some unknown smooth compact manifold $\mathcal{M}$ of dimension k, where $k \ll d$. The goal is to reconstruct a faithful representation of $\mathcal{M}$ from the data points. Reconstruction problem are ill-posed in general. Therefore, the data points are assumed to be dense enough so that it becomes theoretically possible to obtain a faithful reconstruction. The quality of the reconstruction is measured in several ways: the Hausdorff distance between the reconstruction and $\mathcal{M}$, the deviation between the normal spaces of the reconstruction and $\mathcal{M}$ at nearby points, and whether the reconstruction and $\mathcal{M}$ are topologically equivalent.

Key Results

It is clear that more data points are needed in some parts of $\mathcal{M}$ than others, and this is captured well by the concepts of *medial axis* and *local feature size*. The medial axis of $\mathcal{M}$ is the closure of the set of points in $\mathbb{R}^d$ that are at the same distances from two or more closest points in $\mathcal{M}$. For every point $x \in \mathcal{M}$, its local feature $f(x)$ is the distance from x to the medial axis of $\mathcal{M}$. The input set S of data points in $\mathcal{M}$ is an *ϵ-sample* if for every point $x \in \mathcal{M}$, $d(x, S) \leq \epsilon f(x)$. The input set S is a *uniform* ϵ-sample if for every point $x \in \mathcal{M}$, $d(x, S) \leq \varepsilon$, assuming that the ambient space has been scaled such that $\min_{x \in \mathcal{M}} f(x) = 1$. Furthermore, for every pair of points $p, q \in S$, if $d(p,q) \geq \delta f(p)$ or $d(p,q) \geq \delta$ for some constant δ, then we call S an (ϵ, δ)-sample or a uniform (ϵ, δ)-sample, respectively. Most reconstruction results in the literature are about what theoretical guarantees can be offered when ϵ is sufficiently small.

Cocone Complex

Cheng, Dey, and Ramos [7] gave the first proof that a faithful homeomorphic simplicial complex can be constructed from the data points. First, the dimension k of $\mathcal{M}$ and the tangent spaces at the data points are estimated using known algorithms in the literature (e.g., [4, 8, 10]). Let θ be some appropriately small constant angle. The *θ-cocone* at a point $p \in S$ is the subset of points $z \in \mathbb{R}^d$ such that pz makes an angle at most θ with the estimated tangent space at p. Consider the Delaunay triangulation of S. Let σ be a Delaunay simplex and let V_σ be its dual Voronoi cell. The *cocone complex* $\mathcal{K}$ of S is the collection of Delaunay simplices σ such that V_σ intersects the θ-cocones of the vertices of σ. It turns out that if the simplices in the Delaunay triangulation of S have bounded aspect ratio, then $\mathcal{K}$ is a k-dimensional simplicial complex homeomorphic to $\mathcal{M}$. Moreover, the Hausdorff distance between $\mathcal{K}$ and $\mathcal{M}$ is at most ε times the local feature size, and the angle deviation between the normal spaces of $\mathcal{K}$ and $\mathcal{M}$ at nearby points is $O(\epsilon)$. A difficult step of the algorithm is the generalization of sliver removal [6] in $\mathbb{R}^3$ to $\mathbb{R}^d$ in order to ensure that the simplices have bounded aspect ratio. The intuition is simple, but the analysis is quite involved. Figure 1 shows a sliver $pqrs$. View the Delaunay triangulation as a weighted Delaunay triangulation with all weights

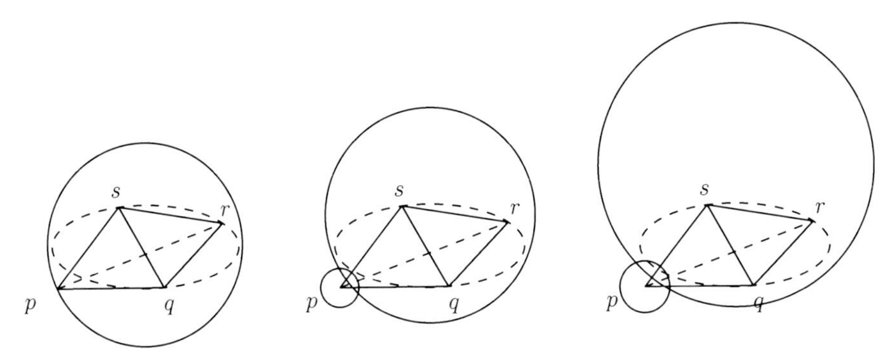

Manifold Reconstruction, Fig. 1 As the weight of p increases, the orthoball becomes larger and moves away from p

equal to zero. If the weight of p is increased, the orthoball of $pqrs$ – the ball at zero power distances from all its vertices – moves away from p and becomes larger. Therefore, the ball will become so large that it contains some other points in S, thereby eliminating the sliver $pqrs$ from the weighted Delaunay triangulation.

Theorem 1 ([7]) *There exist constants ϵ and δ such that if S is an (ϵ, δ)-sample of a smooth compact manifold of dimension k in $\mathbb{R}^d$, a simplicial complex $\mathcal{K}$ can be constructed such that:*

- *$\mathcal{K}$ is homeomorphic to $\mathcal{M}$.*
- *Let τ be a j-dimensional simplex of $\mathcal{K}$. Let p be a vertex of τ. For every point $q \in \mathcal{M}$, if $d(p,q) = O(\epsilon f(p))$, then for every normal vector $\mathbf{n}$ at q, τ has a normal vector $\mathbf{v}$ that makes an $O(\epsilon)$ angle with $\mathbf{n}$.*
- *For every point x in $\mathcal{K}$, its distance from the nearest point $y \in \mathcal{M}$ is $O(\epsilon f(y))$.*

The running time of the algorithm is exponential in d.

Boissonnat, Guibas, and Oudot [3] show that the Voronoi computation can be avoided if one switches to the *weighted witness complex*. Moreover, given an ϵ-sample without the lower bound on the interpoint distances, they can construct a family of plausible reconstructions of dimensions $1, 2, \ldots, k$ and let the user choose an appropriate one. The sliver issue is also encountered [3] and resolved in an analogous manner.

Čech Complex

Betti numbers are informative topological invariants of the shape. There are $d + 1$ betti numbers, β_i for $i \in [0, d]$. The zeroth betti number β_0 is the number of connected components in $\mathcal{M}$. In three dimensions (i.e., $d = 3$), β_2 is the number of voids in $\mathcal{M}$ (i.e., bounded components in $\mathbb{R}^3 \setminus \mathcal{M}$), and β_1 is the number of independent one-dimensional cycles in $\mathcal{M}$ that cannot be contracted within $\mathcal{M}$ to a single point. Two cycles are *homologous* if one can be deformed into the other continuously. Two overlapping cycles can be combined by eliminating the overlap. A set of cycles are independent if no cycle can be obtained by continuously deforming and combining some other cycles in the set. A tunnel is physically accommodated in the complement of $\mathcal{M}$, and its existence is witnessed by a one-dimensional cycle in $\mathcal{M}$ that "circles around" the tunnel (i.e., cannot be contracted in $\mathcal{M}$ to a single point). Therefore, the number of independent one-dimensional cycles that cannot be contracted to a single point measures the number of "independent tunnels." In fact, voids live in the complement of $\mathcal{M}$ too and their number is measured by the number of independent 2-dimensional cycles that cannot be contracted to a single point. In general, β_i is the number of independent i-dimensional cycles that cannot be contracted in $\mathcal{M}$ to a single point. Alternatively, β_i is the rank of the ith homology group. Note that $\beta_i = 0$ for $i > k$ as $\mathcal{M}$ is k-dimensional.

Numerical procedures are known to compute the betti numbers of a simplicial complex (e.g., [11]) and only the incidence relations among the elements in the complex are needed. Therefore, given a homeomorphic simplicial complex $\mathcal{K}$ of $\mathcal{M}$, the betti numbers of $\mathcal{K}$ and hence of $\mathcal{M}$ can be computed. In fact, $\mathcal{K}$ and $\mathcal{M}$ have the same homology (groups). But requiring a homeomorphic complex is an overkill. Let $\mathcal{B}$ be a set of balls of the same radius r. For every subset of balls in $\mathcal{B}$, connect the ball centers in the subset to form a simplex if these balls have a nonempty common intersection. The resulting collection of simplices is known as a *Čech complex* $\mathcal{C}$. Niyogi, Smale, and Weinberger [12] proved that if S is a dense sample from a uniform probability distribution on $\mathcal{M}$ and r is set appropriately, then $\mathcal{C}$ has the same homology as $\mathcal{M}$ with high probability.

Theorem 2 ([12]) *Let S be a set of n points drawn in i.i.d. fashion according to the uniform probability measure on $\mathcal{M}$. Assume that $r < 1/2$. There exists constants α_1 and α_2 depending on $\mathcal{M}$ such that if $n > \alpha_1(\log \alpha_2 + \log(1/\delta))$, then $\mathcal{C}$ has the same homology as $\mathcal{M}$ with probability greater than $1 - \delta$.*

In general the Čech complex contains many simplices. Niyogi, Smale, and Weinberger

M

showed that the same result also holds when the probability distribution has support near $\mathcal{M}$ instead of exactly on $\mathcal{M}$. Subsequently, similar results have also been obtained for the *Vietoris-Rips complex* [1].

Tangential Delaunay Complex

Although the cocone complex gives a homeomorphic reconstruction, the running time is exponential in d. It is natural to ask whether the running time can be made to depend exponentially on k instead. Boissonnat and Ghosh [2] answered this question affirmatively by introducing a new local Delaunay reconstruction.

Let S be a dense (ϵ, δ)-sample of $\mathcal{M}$. Suppose that the tangent spaces at the data points in S have been estimated with an $O(\epsilon)$ angular error. Take a point $p \in S$. Let H_p be the estimated tangent space at p. Let V_p be the Voronoi cell owned by p. Identify the set star(p) of Delaunay simplices that are incident to p and whose dual Voronoi faces intersect H_p. The collection of all such stars form the *tangential Delaunay complex*. The bisectors between p and the other points in S intersect H_p in $(k-1)$-dimensional affine subspaces. These affine subspaces define a Voronoi diagram in H_p, and star(p) is determined by the cell owned by p in this Voronoi diagram in H_p. Therefore, no data structure of dimension higher than k is needed in the computation.

The tangential Delaunay complex is not a triangulated manifold in general though due to some inconsistencies. For example, if σ is a simplex in star(p), it is not necessarily true that σ is in star(q) for another vertex q of σ. Such inconsistencies can be removed by assigning weights to the points in S appropriately as in the case of eliminating slivers from the cocone complex.

Theorem 3 ([2]) *The guarantees in Theorem 1 can be obtained by an algorithm that runs in $O(dn^2 + d2^{O(k^2)}n)$ time, where n is the number of data points.*

More theoretical guarantees are provided in [2].

Implicit Function

The complexes in the previous methods are either very large or not so easy to compute in practice. An alternative approach is to approximate $\mathcal{M}$ by the zero-set of an implicit function $\varphi : \mathbb{R}^d \to \mathbb{R}^{d-k}$ that is defined using the data points in S, assuming that S is a uniform ϵ-sample:

$$\varphi(x) = \sum_{p \in S} \omega(x,p) \cdot B^t_{\varphi,x} \cdot (x - p),$$

- Let $c \geq 3$. Define $\omega(x,p) = \gamma(d(x,p)) / \sum_{p \in S} \gamma(d(x,q))$, where

$$\gamma(s) = \begin{cases} \left(1 - \frac{s}{kc\epsilon}\right)^{2k} \left(\frac{2s}{c\epsilon} + 1\right), & \text{if } s \in [0, kc\epsilon], \\ 0, & \text{if } s > kc\epsilon. \end{cases}$$

  Notice that $\varphi(x)$ depends only on the points in S within a distance $kc\epsilon$ from x.
- For every point $p \in S$, let T_p be a $d \times k$ matrix with orthonormal columns such that its column space is an approximate tangent space at p with angular error $O(\epsilon)$. For every point $x \in \mathbb{R}^d$, let $C_x = \sum_{p \in S} \omega(x,p) \cdot T_p T_p^t$. Therefore, the space L_x spanned by the eigenvectors of C_x corresponding to the smallest $d-k$ eigenvalues is a "weighted average" of the approximate normal spaces at the data points. Define $B_{\varphi,x}$ to be a $d \times (d-k)$ matrix with linearly independent columns such that its column space is L_x. It turns out the zero-set of φ is independent of the choices of $B_{\varphi,x}$.

The weight function ω makes local reconstruction possible without a complete sampling of $\mathcal{M}$. Moreover, the construction of φ is computational less intensive than the construction of a complex. The following guarantees are offered.

Theorem 4 ([5]) *Let $\hat{\mathcal{M}}$ be the set of points at distance ϵ^τ or less from $\mathcal{M}$ for any fixed $\tau \in (1,2)$. Let S_φ denote the zero-set of φ. Let ν denote the map that sends a point in $\mathbb{R}^d$ to the nearest point in $\mathcal{M}$.*

- *For a small enough ϵ, the restriction of ν to $S_\varphi \cap \hat{\mathcal{M}}$ is a homeomorphism between $S_\varphi \cap \hat{\mathcal{M}}$ and $\mathcal{M}$.*

- *For every $x \in S_\varphi \cap \hat{\mathcal{M}}$, the angle between the normal space of $\mathcal{M}$ at $\nu(x)$ and the normal space of S_φ at x is $O(\epsilon^{(\tau-1)/2})$.*

A provably good iterative projection operator is also known for φ [9].

Cross-References

- ▶ Curve Reconstruction
- ▶ Delaunay Triangulation and Randomized Constructions
- ▶ Surface Reconstruction
- ▶ Voronoi Diagrams and Delaunay Triangulations

Recommended Reading

1. Attali D, Lieutier A, Salinas D (2013) Vietoris-Rips complexes also provide topologically correct reconstructions of sampled shapes. Comput Geom Theory Appl 46(4):448–465
2. Boissonnat J-D, Ghost A (2014) Manifold reconstruction using tangential Delaunay complexes. Discret Comput Geom 51(1):221–267
3. Boissonnat J-D, Guibas LJ, Oudot S (2009) Manifold reconstruction in arbitrary dimensions using witness complexes. Discret Comput Geom 42(1):37–70
4. Cheng S-W, Chiu, M-K (2009) Dimension detection via slivers. In: Proceedings of the ACM-SIAM symposium on discrete algorithms, New York, pp 1001–1010
5. Cheng S-W, Chiu M-K (2014) Implicit manifold reconstruction. In: Proceedings of the ACM-SIAM symposium on discrete algorithms, Portland, pp 161–173
6. Cheng S-W, Dey TK, Edelsbrunner H, Facello MA, Teng S-H (2000) Sliver exudation. J ACM 17(1–2):51–68
7. Cheng S-W, Dey TK, Ramos EA (2005) Manifold reconstruction from point samples. In: Proceedings of the ACM-SIAM symposium on discrete algorithms, Vancouver, pp 1018–1027
8. Cheng S-W, Wang Y, Wu Z (2008) Provable dimension detection using principal component analysis. Int J Comput Geom Appl 18(5):415–440
9. Chiu M-K (2013) Manifold reconstruction from discrete point sets. PhD dissertation, Hong Kong University of Science and Technology
10. Dey TK, Giesen J, Goswami S, Zhao W (2003) Shape dimension and approximation from samples. Discret Comput Geom 29(3):419–343
11. Friedman J (1998) Computing Betti numbers via combinatorial laplacians. Algorithmica 21(4): 331–346
12. Niyogi P, Smale S, Weinberger S (2008) Finding the homology of submanifolds with high confidence from random samples. Discret Comput Geom 39(1):419–441

Market Games and Content Distribution

Vahab S. Mirrokni
Theory Group, Microsoft Research, Redmond, WA, USA

Keywords

Congestion games; Market sharing games; Stable matching; Valid-Utility games

Years and Authors of Summarized Original Work

2005; Mirrokni

Problem Definition

This chapter studies market games for their performance and convergence of the equilibrium points. The main application is the content distribution in cellular networks in which a service provider needs to provide data to users. The service provider can use several cache locations to store and provide the data. The assumption is that cache locations are selfish agents (resident subscribers) who want to maximize their own profit. Most of the results apply to a general framework of monotone two-sided markets.

Uncoordinated Two-Sided Markets

Various economic interactions can be modeled as two-sided markets. A two-sided market consists of two disjoint groups of agents: active agents and passive agents. Each agent has a preference list over the agents of the other side, and can be matched to one (or many) of the agents in the other side. A central solution concept to these markets are *stable matchings*, introduced by Gale and Shapley [5]. It is well known that stable matchings can be achieved using a centralized polynomial-time algorithm. Many markets, however, do not have any centralized matching mechanism to match agents. In those markets, matchings are formed by actions of self-interested agents. Knuth [9] introduced uncoordinated two-sided markets. In these markets, cycles of better or best responses exist, but random better response and best response dynamics converge to a stable matching with probability one [2, 10, 14]. Our model for content distribution corresponds to a special class of uncoordinated two-sided markets that is called *the distributed caching games*.

Before introducing the distributed caching game as an uncoordinated two-sided market, the distributed caching problem and some game theoretic notations are defined.

Distributed Caching Problem

Let U be a set of n cache locations with given available capacities A_i and given available band widths B_i for each cache location i. There are k request types; (Request type can be thought of as different files that should be delivered to clients.) each request type t has a size a_t $(1 \leq t \leq k)$. Let H be a set of m requests with a reward R_j, a required bandwidth b_j, a request type t_j for each request j, and a cost c_{ij} for connecting each cache location i to each request j. The profit of providing request j by cache location i is $f_{ij} = R_j - c_{ij}$. A cache location i can service a set of requests S_i, if it satisfies the *bandwidth constraint*: $\sum_{j \in S_i} b_j \leq B_i$, and the *capacity constraint*: $\sum_{t \in \{t_j | j \in S_i\}} a_t \leq A_i$ (this means that the sum of the sizes of the request types of the requests in cache location i should be less than or equal to the available capacity of cache location i). A set S_i of requests is feasible for cache location i if it satisfies both of these constraints. The goal of the DCP problem is to find a feasible assignment of requests to cache locations to maximize the total profit; i.e., the total reward of requests that are provided minus the connection costs of these requests.

Strategic Games

A *strategic game* $\mathcal{G}$ is defined as a tuple $\mathcal{G}(U, \{F_i | i \in U\}, \{\alpha_i() | i \in U\})$ where (i) U is the set of n players or agents, (ii) F_i is a family of feasible *(pure) strategies* or *actions* for player i and (iii) $\alpha_i : \Pi_{i \in U} F_i \rightarrow \mathbb{R}^+ \cup \{0\}$ is the (private) *payoff* or *utility* function for agent i, given the set of strategies of all players. Player i's strategy is denoted by $s_i \in F_i$. A *strategy profile* or a (strategy) *state*, denoted by $S = (s_1, s_2, \ldots, s_n)$, is a vector of strategies of players. Also let $S \oplus s'_i := (s_1, \ldots, s_{i-1}, s'_i, s_{i+1}, \ldots, s_k)$.

Best-Response Moves

In a *non-cooperative* game, each agent wishes to maximize its own payoff. For a strategy profile $S = (s_1, s_2, \ldots, s_n)$, a *better response move* of player i is a strategy s'_i such that $\alpha_i(S \oplus s'_i) \geq \alpha_i(S)$. In a *strict better response move*, the above inequality is strict. Also, for a strategy profile $S = (s_1, s_2, \ldots, s_n)$ a *best response* of player i in S is a better response move $s^*_i \in F_i$ such that for any strategy $s_i \in F_i$, $\alpha_i(S \oplus s^*_i) \geq \alpha_i(S \oplus s_i)$.

Nash Equilibria

A pure strategy Nash equilibrium (PSNE) of a strategic game is a strategy profile in which each player plays his best response.

State Graph

The state graph, $\mathcal{D} = (\mathcal{F}, \mathcal{E})$, of a strategic game $\mathcal{G}$, is an arc-labeled directed graph, where the vertex set $\mathcal{F}$ corresponds to the set of strategy profiles or states in $\mathcal{G}$, and there is an arc from state S to state S' with label i if the only difference between S and S' is in the strategy of player i; and player i plays one of his best responses in strategy profile S'. A *best-response walk* is a directed walk in the state graph.

Price of Anarchy

Given a strategic game, $\mathcal{G}(U, \{F_i | i \in U\}, \{\alpha() | i \in U\})$, and a maximization social function $\gamma : \Pi_{i \in U} F_i \rightarrow \mathbb{R}$, the price of anarchy, denoted by $\mathsf{poa}(\mathcal{G}, \gamma)$, is the worst ratio between the social value of a pure Nash equilibrium and the optimum.

Distributed Caching Games

The distributed caching game can be formalized as a two-sided market game: active agents correspond to n resident subscribers or cache locations, and passive agents correspond to m requests from transit subscribers. Formally, given an instance of the DCP problem, a strategic game $\mathcal{G}(U, \{F_i | i \in U\}, \{\alpha_i | i \in U\})$ is defined as follows. The set of players (or active agents) U is the set of cache locations. The family of feasible strategies F_i of a cache location i is the family of subsets s_i of requests such that $\sum_{j \in s_i} b_j \leq B_i$ and $\sum_{t \in \{t_j | j \in s_i\}} a_t \leq A_i$. Given a vector $S = (s_1, s_2, \ldots, s_n)$ of strategies of cache locations, the *favorite* cache locations for request j, denoted by $\mathsf{FAV}(j)$, is the set of cache locations i such that $j \in s_i$ and f_{ij} has the maximum profit among the cache locations that have request j in their strategy set, i.e., $f_{ij} \geq f_{i'j}$ for any i' such that $j \in s_{i'}$. For a strategy profile $S = (s_1, \ldots, s_n)$ $\alpha_i(S) = \sum_{j : i \in \mathsf{FAV}(j)} f_{ij} / |\mathsf{FAV}(j)|$. Intuitively, the above definition implies that the profit of each request goes to the cache locations with the minimum connection cost (or equivalently with the maximum profit) among the set of cache locations that provide this request. If more than one cache location have the maximum profit (or minimum connection cost) for a request j, the profit of this request is divided equally between these cache locations. The payoff of a cache location is the sum of profits from the requests it actually serves. A player i *serves* a request j if $i \in \mathsf{FAV}(j)$. The social value of strategy profile S, denoted by $\gamma(S)$, is the sum of profits of all players. This value $\gamma(S)$ is a measure of the efficiency of the assignment of requests and request types to cache locations.

Special Cases

In this paper, the following variants and special cases of the DCP problem are also studied: The CapDCP problem is a special case of DCP problem without bandwidth constraints. The BanDCP problem is a special case of DCP problem without capacity constraints. In the uniform BanDCP problem, the bandwidth consumption of all requests is the same. In the uniform CapDC problem, the size of all request types is the same.

Many-to-One Two-Sided Markets with Ties

In the distributed caching game, active and passive agents correspond to cache locations and requests respectively. The set of feasible strategies for each active agent correspond to a set of solutions to a packing problem. Moreover, the preferences of both active and passive agents is determined from the profit of requests to cache locations. In many-to-one two-sided markets, the preference of passive and active agents as well as the feasible family of strategies are arbitrary. The preference list of agents may have ties as well.

Monotone and Matroid Markets

In *monotone many-to-one two-sided markets*, the preferences of both active and passive agents are determined based on payoffs $p_{ij} = p_{ji}$ for each active agent i and passive agent j (similar to the DCP game). An agent i prefers j to j' if $p_{ij} > p_{ij'}$. In *matroid two-sided markets*, the feasible set of strategies of each active agent is the set of independent sets of a matroid. Therefore, uniform BanDCP game is a matroid two-sided market game.

Key Results

In this section, the known results for these problems are summarized.

Centralized Approximation Algorithm

The distributed caching problem generalizes the multiple knapsack problem and the generalized assignment problem [3] and as a result is an APX-hard problem.

Theorem 1 ([4]) *There exists a linear programming based* $1 - \frac{1}{e}$*-approximation algorithm and a local search* $\frac{1}{2}$*-approximation algorithm for the* DCP *problem.*

The $1 - \frac{1}{e}$-approximation for this problem is based on rounding an exponentially large configuration linear program [4]. On the basis of some reasonable complexity theoretic assumptions, this approximation factor of $1 - \frac{1}{e}$ is tight for this problem. More formally,

Theorem 2 ([4]) *For any $\epsilon > 0$, there exists no $1 - \frac{1}{e} - \epsilon$-approximation algorithm for the* DCP *problem unless $NP \subseteq DTIME(n^{O(\log \log n)})$.*

Price of Anarchy

Since the DCP game is a strategic game, it possesses mixed Nash equilibria [12]. The DCP game is a valid-utility game with a submodular social function as defined by Vetta [16]. This implies that the performance of any mixed Nash equilibrium of this game is at least $\frac{1}{2}$ of the optimal solution.

Theorem 3 ([4, 11]) *The* DCP *game is a valid-utility game and the price of anarchy for mixed Nash equilibria is $\frac{1}{2}$. Moreover, this result holds for all monotone many-to-one two-sided markets with ties.*

A direct proof of the above price of anarchy bound for the DCP game can be found in [11].

Pure Nash Equilibria: Existence and Convergence

This part surveys known results for existence and convergence of pure Nash equilibria.

Theorem 4 ([11]) *There are instances of the* IBDC *game that have no pure Nash equilibrium.*

Since, IBDC is a special case of CapDCP, the above theorem implies that there are instances of the CapDCP game that have no pure Nash equilibrium. In the above theorem, the bandwidth consumption of requests are not uniform, and this was essential in finding the example. The following gives theorems for the uniform variant of these games.

Theorem 5 ([1, 11]) *Any instance of the uniform* BanDCP *game does not contain any cycle of strict best-response moves, and thus possess a pure Nash equilibrium. On the other hand, there are instances of the uniform* CapDCP *game with no pure Nash equilibria.*

The above result for the uniform BanDCP game can be generalized to matroid two-sided markets with ties as follows.

Theorem 6 ([1]) *Any instance of the monotone matroid two-sided market game with ties is a potential game, and possess pure Nash equilibria. Moreover, any instance of the matroid two-sided market game with ties possess pure Nash equilibria.*

Convergence Time to Equilibria

This section proves that there are instances of the uniform CapDCP game in which finding a pure Nash equilibrium is PLS-hard [8]. The definition of PLS-hard problems can be found in papers by Yannakakis et al. [8, 15].

Theorem 7 ([11]) *There are instances of the uniform* CapDCP *game with pure Nash equilibria (It is also possible to say that finding a* sink equilibrium *is PLS-hard. A sink equilibrium is a set of strategy profiles that is closed under best-response moves. A pure equilibrium is a sink equilibrium with exactly one profile. This equilibrium concept is formally defined in [7].) for which finding a pure Nash equilibrium is PLS-hard.*

Using the above proof and a result of Schaffer and Yannakakis [13, 15], it is possible to show that in some instances of the uniform CapDCP game, there are states from which all paths of best responses have exponential length.

Corollary 1 ([11]) *There are instances of the uniform* CapDCP *game that have pure Nash equilibria with states from which any sequence of best-response moves to any pure Nash equilibrium (or sink equilibrium) has an exponential length.*

The above theorems show exponential convergence to pure Nash equilibria in general DCP games. For the special case of the uniform BanDCP game, the following is a positive result for the convergence time to equilibria.

Theorem 8 ([2]) *The expected convergence time of a random best-response walk to pure Nash equilibria in matroid monotone two-sided markets (without ties) is polynomial.*

Since the uniform BanDCP game is a special case of matroid monotone two-sided markets with ties, the above theorem indicates that for the BanDCP game with no tie in the profit of requests, the convergence time of a random best-response walk is polynomial. Finally, we state a theorem about the convergence time of the general (non-monotone) matroid two-sided market games.

Theorem 9 ([2]) *In the matroid two-sided markets (without ties), a random best response dynamic of players may cycle, but it converges to a Nash equilibrium with probability one. However, it may take exponential time to converge to a pure Nash equilibrium.*

Pure Nash equilibria of two-sided market games correspond to stable matchings in two-sided markets and vice-versa [2]. The fact that better response dynamics of players in two-sided market games may cycle, but will converge to a stable matching has been proved in [9, 14]. Ackermann et al. [2] extend these results for best-response dynamics, and show an exponential lower bound for expected convergence time to pure Nash equilibria.

Applications

The growth of the Internet, the World Wide Web, and wide-area wireless networks allow an increasing number of users to access vast amounts of information in different geographic areas. As one of the most important functions of the service provider, content delivery can be performed by caching popular items in cache locations close to the users. Performing such a task in a decentralized manner in the presence of self-interested entities in the system can be modeled as an uncoordinated two-sided market game.

The 3G subscriber market can be categorized into groups with shared interest in location-based services, e.g., the preview of movies in a theater or scenes of the mountain nearby. Since the 3G radio resources are limited, it is expensive to repeatedly transmit large quantities of data over the air interface from the base station (BS). It is more economical for the service provider to offload such repeated requests on to the ad-hoc network comprised of its subscribers where some of them recently acquired a copy of the data. In this scenario, the goal for the service provider is to give incentives for peer subscribers in the system to cache and forward the data to the requesting subscribers. Since each data item is large in size and transit subscribers are mobile, we assume that the data transfer occurs in a close range of a few hops.

In this setting, envision a system consisting of two groups of subscribers: resident and transit subscribers. Resident subscribers are less mobile and mostly confined to a certain geographical area. Resident subscribers have incentives to cache data items that are specific to this geographical region since the service provider gives monetary rewards for satisfying the queries of transit subscribers. Transit subscribers request their favorite data items when they visit a particular region. Since the service provider does not have knowledge of the spatial and temporal distribution of requests, it is difficult if not impossible for the provider to stipulate which subscriber should cache which set of data items. Therefore, the decision of what to cache is left to each individual subscriber. The realization of this content distribution system depends on two main issues. First, since subscribers are selfish agents, they may act to increase their individual payoff and decrease the performance of the system. Here, we provide a framework for which we can prove that in an equilibrium situation of this framework, we use the performance of the system efficiently. The second issue is that the payoff of each request for each agent must be a function of the set of agents that have this request in their strategy, since these agents compete on this request and the profit of this request should be divided among these agents in an appropriate way. Therefore, each selfish agent may change the set of items it cached in response to the set of items cached by others. This model leads to a non-cooperative caching scenario that can be modeled on a two-sided market game, studied and motivated in the context of market sharing games and distributed caching games [4, 6, 11].

Open Problems

It is known that there exist instances of the distributed caching game with no pure Nash equilibria. It is also known that best response dynamics of players may take exponential time to converge to pure Nash equilibria. An interesting question is to study the performance of sink equilibria [7, 11] or the price of sinking [7, 11] for these games. The distributed caching game is a valid-utility game. Goemans, Mirrokni, and Vetta [7] show that despite the price of anarchy of $\frac{1}{2}$ for valid-utility games, the performance of sink equilibria (or price of sinking) for these games is $\frac{1}{n}$. We conjecture that the price of sinking for DCP games is a constant. Moreover, it is interesting to show that after polynomial rounds of best responses of players the approximation factor of the solution is a constant. We know that one round of best responses of players is not sufficient to get constant-factor solutions. It might be easier to show that after a polynomial number of random best responses of players, the expected total profit of players is at least a constant factor of the optimal solution. Similar positive results for sink equilibria and random best responses of players are known for congestion games [7, 11].

The complexity of verifying if a given state of the distributed caching game is in a sink equilibrium or not is an interesting question to explore. Also, given a distributed caching game (or a many-to-one two-sided market game), an interesting problem is to check if the set of all sink equilibria is pure Nash equilibria or not. Finally, an interesting direction of research is to classify classes of two-sided market games for which pure Nash equilibria exists or best-response dynamics of players converge to a pure Nash equilibrium.

Cross-References

▸ Stable Marriage
▸ Stable Marriage with Ties and Incomplete Lists
▸ Best Response Algorithms for Selfish Routing

Recommended Reading

1. Ackermann H, Goldberg P, Mirrokni V, Röglin H, Vöcking B (2007) A unified approach to congestion games and twosidedmarkets. In: 3rd workshop of internet economics (WINE), pp 30–41
2. Ackermann H, Goldberg P, Mirrokni V, Röglin H, Vöcking B (2008) Uncoordinated two-sided markets. ACM Electronic Commerce (ACMEC)
3. Chekuri C, Khanna S (2000) A PTAS for the multiple knapsack problem. In: Proceedings of the 11th annual ACM-SIAM symposium on discrete algorithms (SODA), pp 213–222
4. Fleischer L, Goemans M, Mirrokni VS, Sviridenko M (2006) Tight approximation algorithms for maximum general assignment problems. In: Proceedings of the 16th annual ACM–SIAM symposium on discrete algorithms (SODA), pp 611–620
5. Gale D, Shapley L (1962) College admissions and the stability of marriage. Am Math Mon 69:9–15
6. Goemans M, Li L, Mirrokni VS, Thottan M (2004) Market sharing games applied to content distribution in ad-hoc networks. In: Proceedings of the 5th ACM international symposium on mobile ad hoc networking and computing (MobiHoc), pp 1020–1033
7. Goemans M, Mirrokni VS, Vetta A (2005) Sink equilibria and convergence. In: Proceedings of the 46th conference on foundations of computer science (FOCS), pp 123–131
8. Johnson D, Papadimitriou CH, Yannakakis M (1988) How easy is local search? J Comput Syst Sci 37:79–100
9. Knuth D (1976) Marriage Stables et leurs relations avec d'autres problèmes Combinatories. Les Presses de l'Université de Montréal
10. Kojima F, Unver Ü (2006) Random paths to pairwise stability in many-to-many matching problems: a study on market equilibration. Int J Game Theory
11. Mirrokni VS (2005) Approximation algorithms for distributed and selfish agents. Ph.D. thesis, Massachusetts Institute of Technology
12. Nash JF (1951) Non-cooperative games. Ann Math 54:268–295

13. Papadimitriou CH, Schaffer A, Yannakakis M (1990) On the complexity of local search. In: Proceedings of the 22nd symposium on theory of computing (STOC), pp 438–445
14. Roth AE, Vande Vate JH (1990) Random paths to stability in two sided matching. Econometrica 58(6):1475–1480
15. Schaffer A, Yannakakis M (1991) Simple local search problems that are hard to solve. SIAM J Comput 20(1):56–87
16. Vetta A (2002) Nash equilibria in competitive societies, with applications to facility location, traffic routing and auctions. In: Proceedings of the 43rd symposium on foundations of computer science (FOCS), pp 416–425

Matching in Dynamic Graphs

Surender Baswana[1], Manoj Gupta[2], and Sandeep Sen[2]
[1]Department of Computer Science and Engineering, Indian Institute of Technology (IIT), Kanpur, India
[2]Indian Institute of Technology (IIT) Delhi, Hauz Khas, New Delhi, India

Keywords

Dynamic graph; Matching; Maximum matching; Maximal matching; Randomized algorithms

Years and Authors of Summarized Original Work

2011; Baswana, Gupta, Sen

Problem Definition

Let $G = (V, E)$ be an undirected graph on $n = |V|$ vertices and $m = |E|$ edges. A matching in G is a set of edges $\mathcal{M} \subseteq E$ such that no two edges in $\mathcal{M}$ share any vertex. Matching has been one of the most well-studied problems in algorithmic graph theory for decades [4]. A matching $\mathcal{M}$ is called *maximum matching* if the number of edges in $\mathcal{M}$ is maximum. The fastest known algorithm for maximum matching, due to Micali and Vazirani [5], runs in $O(m\sqrt{n})$. A matching is said to be *maximal* if it is not strictly contained in any other matching. It is well known that a maximal matching achieves a factor 2 approximation of the maximum matching.

Key Result

We address the problem of maintaining maximal matching in a fully dynamic environment – allowing updates in the form of both insertion and deletion of edges. Ivković and Llyod [3] designed the first fully dynamic algorithm for maximal matching with $O((n+m)^{0.7072})$ update time. In this entry, we present a fully dynamic algorithm for maximal matching that achieves $O(\log n)$ expected amortized time per update.

Ideas Underlying the Algorithm

We begin with some terminologies and notations that will facilitate our description and also provide some intuition behind our approach. Let $\mathcal{M}$ denote a matching in the given graph at any instant – an edge $(u, v) \in \mathcal{M}$ is called a *matched* edge where u is referred to as a *mate* of v and vice versa. An edge in $E \backslash \mathcal{M}$ is an *unmatched* edge. A vertex x is *matched* if there exists an edge $(x, y) \in \mathcal{M}$; otherwise it is *free* or *unmatched*.

In order to maintain a maximal matching, it suffices to ensure that there is no edge (u, v) in the graph such that both u and v are free with respect to the matching $\mathcal{M}$. Therefore, a natural technique for maintaining a maximal matching is to keep track of each vertex if it is matched or free. When an edge (u, v) is inserted, we add (u, v) to the matching if u and v are free. For the case when an unmatched edge (u, v) is deleted, no action is required. Otherwise, for both u and v, we search their neighborhoods for any free vertex and update the matching accordingly. It follows that each update takes $O(1)$ computation time except when it involves deletion of a matched edge; in this case the computation time is of the order of the sum of the degrees of the two endpoints of the deleted edge. So this trivial algorithm is quite efficient for *small* degree vertices, but could be expensive for *large* degree vertices. An alternate approach could be to match a free vertex u with a randomly chosen neighbor, say v. Following the standard adversarial model, it can be observed that an expected $\deg(u)/2$ edges incident to u will be deleted before deleting the matched edge (u, v). So the expected amortized cost per edge deletion for u is roughly $O\left(\frac{\deg(u)+\deg(v)}{\deg(u)/2}\right)$. If $\deg(v) < \deg(u)$, this cost is $O(1)$. But if $\deg(v) \gg \deg(u)$, then it can be as bad as the trivial algorithm. To circumvent this problem, we introduce an important notion, called *ownership* of edges. Intuitively, we assign an edge to that endpoint which has *higher* degree.

The idea of choosing a random mate and the trivial algorithm described above can be combined together to design a simple algorithm for maximal matching. This algorithm maintains a partition of the vertices into two levels. Level 0 consists of vertices which own *fewer* edges, and we handle the updates there using the trivial algorithm. Level 1 consists of vertices (and their mates) which own *larger* number of edges, and we use the idea of random mate to handle their updates. This 2 – LEVEL algorithm achieves $O(\sqrt{n})$ expected amortized time per update. A careful analysis of the 2 – LEVEL algorithm suggests that a *finer* partition of vertices could help in achieving a faster update time. This leads to our $\log_2 n$ – LEVEL algorithm that achieves expected amortized $O(\log n)$ time per update.

Our algorithm uses randomization very crucially in order to handle the updates efficiently. The matching maintained (based on the random bits) by the algorithm at any stage is not known to the adversary for it to choose the updates adaptively. This oblivious adversarial model is no different from randomized data structures like universal hashing.

The 2-LEVEL Algorithm

The algorithm maintains a partition of the set of vertices into two levels. Each edge present in the graph will be owned by one or both of its endpoints as follows. If both the endpoints of an edge are at level 0, then it is owned by both of them. Otherwise it will be owned by exactly that endpoint which lies at a higher level. If both the endpoints are at level 1, the tie will be broken suitably by the algorithm. Let $\mathcal{O}_u$ denote the set of edges owned by a vertex u at any instant of the algorithm. With a slight abuse of the notation, we will also use $\mathcal{O}_u$ to denote $\{v | (u, v) \in \mathcal{O}_u\}$. As the algorithm proceeds, the vertices will make transition from one level to another and the ownership of edges will also change accordingly.

The algorithm maintains the following three invariants after each update:

1. Every vertex at level 1 is matched. Every free vertex at level 0 has all its neighbors matched.
2. Every vertex at level 0 owns less than $\sqrt{n}$ edges at any stage.
3. Both endpoints of every matched edge are at the same level.

It follows from the first invariant that the matching $\mathcal{M}$ is maximal at each stage. The second and third invariants help in incorporating the two ideas of our algorithm efficiently.

Handling Insertion of an Edge

Let (u, v) be the edge being inserted. If either u or v are at level 1, there is no violation of any invariant. However, if both u and v are at level 0, then we proceed as follows. Both u and v become the owner of the edge (u, v). If u and v are free, then we add (u, v) to $\mathcal{M}$. Notice that the insertion of (u, v) also leads to increase of $|\mathcal{O}_u|$ and $|\mathcal{O}_v|$ by one and so may lead to violation of Invariant 2. We process the vertex that owns more edges; let u be that vertex. If $|\mathcal{O}_u| = \sqrt{n}$, then Invariant 2 has got violated. In order to restore it, u moves to level 1 and gets matched to some vertex, say y, selected uniformly at random from $\mathcal{O}_u$. Vertex y also moves to level 1 to satisfy Invariant 3. If w and x were, respectively, the earlier mates of u and y at level 0, then the matching of u with y has rendered w and x free. Both w and x search for free neighbors at level 0 and update the matching accordingly. It is easy to observe that in all these cases, it takes $O(\sqrt{n})$ time to handle an edge insertion.

Handling Deletion of an Edge

Let (u, v) be an edge that is deleted. If $(u, v) \notin \mathcal{M}$, all the invariants are still valid. Let us consider the more important case of $(u, v) \in \mathcal{M}$ – the deletion of (u, v) has caused u and v to become free. Therefore, the first invariant might have got violated for u and v. If edge (u, v) was at level 0, then both u and v search for a free neighbor and update the matching accordingly. This takes $O(\sqrt{n})$ time. If edge (u, v) was at level 1, then u (similarly v) is processed as follows.

First, u disowns all its edges whose other endpoint is at level 1. If $|\mathcal{O}_u|$ is still greater than or equal to $\sqrt{n}$, then u stays at level 1 and selects a random mate from $\mathcal{O}_u$. However, if $|\mathcal{O}_u|$ has fallen below $\sqrt{n}$, then u moves to level 0 and gets matched to a free neighbor (if any). For each neighbor of u at level 0, the transition of u from level 1 to 0 is, effectively, like insertion of a new edge. This transition leads to an increase in the number of owned edges by each neighbor of u at level 0. As a result, the second invariant for each such neighbor at level 0 may get violated if the number of edges it owns now becomes $\sqrt{n}$. To take care of these scenarios, we proceed as follows. We scan each neighbor of u at level 0, and for each neighbor w, with $|\mathcal{O}_w| = \sqrt{n}$, a mate is selected randomly from $\mathcal{O}_w$ and w is moved to level 1 along with its mate. This concludes the deletion procedure of edge (u, v).

Analysis of the Algorithm

It may be noted that, unlike insertion, the deletion of an edge could potentially lead to moving of many vertices from level 0 to 1 and this may involve significant computation. However, we will show that the expected amortized computation per update is $O(\sqrt{n})$.

We analyze the algorithm using the concept of *epochs*.

Definition 1 At any time t, let (u, v) be any edge in $\mathcal{M}$. Then the **epoch** of (u, v) at time t is the maximal time interval containing t during which $(u, v) \in \mathcal{M}$.

The entire life span of an edge (u, v) can be viewed as a sequence of epochs when it is matched, separated by periods when it is unmatched. Any edge update that does not change the matching is processed in O(1) time. An edge update that changes the matching results in the start of new epoch(s) or the termination of some existing epoch(s). And it is only during the creation or termination of an epoch that significant computation is involved. For the purpose of analyzing the update time (when matching is affected), we assign the computation performed to the corresponding epochs created or terminated. It is easy to see that the computation associated with an epoch at level 0 is $O(\sqrt{n})$. The computation associated with an epoch at level 1 is of the order of sum of the degrees of the endpoints of the corresponding matched edge which may be $\Omega(n)$. When a vertex moves from level 0 to 1, although it owns $\sqrt{n}$ edges, this may grow later to $O(n)$. So the computation associated with an epoch at level 1 can be quite high. We will show that the expected number of such epochs that get terminated during any arbitrary sequence of edge updates will be relatively small. The following lemma plays a key role.

Lemma 1 *The deletion of an edge (u, v) at level 1 terminates an epoch with probability $\leq 1/\sqrt{n}$.*

Proof The deletion of edge (u, v) will lead to termination of an epoch only if $(u, v) \in \mathcal{M}$. If edge (u, v) was owned by u at the time of its deletion, note that u owned at least $\sqrt{n}$ edges at the moment of start of its epoch. Since u selected its matched edge uniformly at random from these edges, the (conditional) probability is $\frac{1}{\sqrt{n}}$. The same argument applies if v was the owner, so (u, v) is a matched edge at the time of deletion of (u, v) with probability at most $1/\sqrt{n}$. □

Consider any sequence of m edge updates. We analyze the computation associated with all the epochs that get terminated during these m updates. It follows from Lemma 1 and the linearity of expectation that the expected number of epochs terminated at level 1 will be $m/\sqrt{n}$. As discussed above, computation associated with each epoch at level 1 is $O(n)$. So the expected computation associated with the termination of all epochs at level 1 is $O(m\sqrt{n})$. The number of epochs destroyed at level 0 is trivially bounded by

$O(m)$. Each epoch at level 0 has $O(\sqrt{n})$ computation associated with it, so the total computation associated with these epochs is also $O(m\sqrt{n})$. We conclude the following.

Theorem 1 *Starting with a graph on n vertices and no edges, we can maintain maximal matching for any sequence of m updates in expected* $O(m\sqrt{n})$ *time.*

The $\log_2 n$ – LEVEL Algorithm

The key idea for improving the update time lies in the second invariant of our 2 – LEVEL algorithm. Let $\alpha(n)$ be the threshold for the maximum number of edges that a vertex at level 0 can own. Consider an epoch at level 1 associated with some edge, say (u, v). The computation associated with this epoch is of the order of the number of edges u and v own which can be $\Theta(n)$ in the worst case. However, the expected duration of the epoch is of the order of the minimum number of edges u can own at the time of its creation, i.e., $\Theta(\alpha(n))$. Therefore, the expected amortized computation per edge deletion at level 1 is $O(n/\alpha(n))$. Balancing this with the $\alpha(n)$ update time at level 0 yields $\alpha(n) = \sqrt{n}$.

In order to improve the running time of our algorithm, we need to decrease the ratio between the maximum and the minimum number of edges a vertex can own during an epoch at any level. It is this ratio that determines the expected amortized time per edge deletion. This observation leads us to a finer partitioning of the ownership classes. When a vertex creates an epoch at level i, it owns at least 2^i edges, and during the epoch, it is allowed to own at most $2^{i+1} - 1$ edges. As soon as it owns 2^{i+1} edges, it migrates to a higher level. Notice that the ratio of maximum to minimum edges owned by a vertex during an epoch gets reduced from $\sqrt{n}$ to a constant leading to about $\log_2 n$ levels. Though the $\log_2 n$–LEVEL algorithm can be seen as a natural generalization of our 2 – LEVEL algorithm, there are many intricacies that make the algorithm and its analysis quite involved. For example, a single edge update may lead to a sequence of falls and rise of many vertices across the levels of the data structure. Moreover, there may be several vertices trying to fall or rise at any time while processing an update. Taking a top-down approach in processing these vertices simplifies the description of the algorithm. The analysis of the algorithm becomes easier when we analyze each level separately. This analysis at any level is quite similar to the analysis of LEVEL–1 in our 2–LEVEL algorithm. We recommend the interested reader to refer to the journal version of this paper in order to fully comprehend the algorithm and its analysis. The final result achieved by our $\log_2 n$ – LEVEL algorithm is stated below.

Theorem 2 *Starting with a graph on n vertices and no edges, we can maintain maximal matching for any sequence of m updates in expected* $O(m \log n)$ *time.*

Using standard probability tools, it can be shown that the bound on the update time as stated in Theorem 2 holds with high probability, as well as with limited independence.

Open Problems

There have been new results on maintaining approximate weighted matching [2] and $(1 + \epsilon)$-approximate matching [1, 6] for $\epsilon < 1$. The interested reader should study these results. For any $\epsilon < 1$, whether it is possible to maintain $(1 + \epsilon)$-approximate matching in poly-logarithmic update time is still an open problem.

Recommended Reading

1. Anand A, Baswana S, Gupta M, Sen S (2012) Maintaining approximate maximum weighted matching in fully dynamic graphs. In: FSTTCS, Hyderabad, pp 257–266
2. Gupta M, Peng R (2013) Fully dynamic (1+e)-approximate matchings. In: FOCS, Berkeley, pp 548–557
3. Ivkovic Z, Lloyd EL (1994) Fully dynamic maintenance of vertex cover. In: WG '93: proceedings of the 19th international workshop on graph-theoretic concepts in computer science, Utrecht. Springer, London, pp 99–111
4. Lovasz L, Plummer M (1986) Matching theory. AMS Chelsea Publishing/North-Holland, Amsterdam/New York
5. Micali S, Vazirani VV (1980) An $O(\sqrt{(|V|)}|E|)$ algorithm for finding maximum matching in general graphs. In: FOCS, Syracuse, pp 17–27

6. Neiman O, Solomon S (2013) Simple deterministic algorithms for fully dynamic maximal matching. In: STOC, Palo Alto, pp 745–754

Matching Market Equilibrium Algorithms

Ning Chen[1] and Mengling Li[2]
[1]Division of Mathematical Sciences, School of Physical and Mathematical Sciences, Nanyang Technological University, Singapore, Singapore
[2]Division of Mathematical Sciences, Nanyang Technological University, Singapore, Singapore

Keywords

Competitive equilibrium; Matching market; Maximum competitive equilibrium; Minimum competitive equilibrium

Years and Authors of Summarized Original Work

1971; Shapley, Shubik
1982; Kelso, Crawford
1986; Demange, Gale and Sotomayor

Problem Definition

The study of matching market equilibrium was initiated by Shapley and Shubik [13] in an assignment model. A classical instance of the matching market involves a set B of n unit-demand buyers and a set Q of m indivisible items, where each buyer wants to buy at most one item and each item can be sold to at most one buyer. Each buyer i has a valuation $v_{ij} \geq 0$ for each item j, representing the maximum amount that i is willing to pay for item j. Each item j has a reserve price $r_j \geq 0$, below which it won't be sold. Without loss of generality, one can assume there is a null item whose value is zero to all buyers and whose price is always zero.

An output of the matching market is a tuple $(\mathbf{p}, \mathbf{x})$, where $\mathbf{p} = (p_1, \ldots, p_m) \geq 0$ is a *price* vector with p_j denoting the price charged for item j and $\mathbf{x} = (x_1, \ldots, x_n) \geq 0$ is an *allocation* vector with x_i denoting the item that i wins. If i does not win any item, $x_i = \emptyset$. An output is essentially a bipartite matching with monetary transfers between the matched parties (buyers and items). A *feasible* price vector is any vector $\mathbf{p} \geq \mathbf{r} = (r_1 \ldots r_m)$. Given an outcome $(\mathbf{p}, \mathbf{x})$, the *utility* (*payoff*) to a buyer i who gets item j and pays p_j is $u_i(\mathbf{px}) = v_{ij} - p_j$ (assume linear surplus) and $u_i(\mathbf{px}) = 0$ if i gets nothing. At price $\mathbf{p}$, the *demand set* for buyer i is $D_i(\mathbf{p}) = \{j \in \arg\max_{j'}(v_{ij'} - p_{j'}) \,|\, v_{ij} - p_j \geq 0\}$

A tuple $(\mathbf{p}, \mathbf{x})$ is called a *competitive equilibrium* if:

- For any item j, $p_j = r_j$ if no one wins j in allocation $\mathbf{x}$.
- If buyer i wins item j ($x_i = j$), then $j \in D_i(\mathbf{p})$.
- If buyer i does not win any item ($x_i = \emptyset$), then for every item j, $v_{ij} - p_j \leq 0$.

The first condition above is a market clearance (efficiency) condition, which says that all unallocated items are priced at the given reserve prices. The second and third conditions ensure envy-freeness (fairness), implying that each buyer is allocated with an item that maximizes his utility at these prices. In a market competitive equilibrium, all items with prices higher than the reserve prices are sold out and everyone gets his maximum utility at the corresponding allocation.

An outcome $(\mathbf{p}, \mathbf{x})$ is called a *minimum competitive equilibrium* if it is a competitive equilibrium and for any other competitive equilibrium $(\mathbf{p}', \mathbf{x}')$, $p_j \leq p'_j$ for every item j. It represents the interests of all buyers in terms of their total payment. Similarly, an outcome $(\mathbf{p}, \mathbf{x})$ is called a *maximum competitive equilibrium* if it is a competitive equilibrium and for any other competitive equilibrium $(\mathbf{p}', \mathbf{x}')$, $p_j \geq p'_j$ for every item j. It represents the interests of all sellers in terms of total payment received. Maximum and minimum equilibria represent the contradictory interests of the two parties in a two-sided matching market at the two extremes.

M

The solution concept of competitive equilibrium is closely related to the well-established stability solution concept initiated by Gale and Shapley [9] in pure two-sided matching markets without money. To define stability in a multi-item matching market with money, we need to have a definition of preferences. Buyers prefer items with larger utility ($v_{ij} - p_j$), and sellers (items) prefer buyers with larger payment. Given an outcome $(\mathbf{p}, \mathbf{x})$, where p_j is the price of item j and x_i is the allocation of buyer i, we say (i, j) is a *blocking pair* if there is p'_j such that $p'_j > p_j$ (item j can receive more payment) and the utility that i obtains in $(\mathbf{p}, \mathbf{x})$ is less than $v_{ij} - p'_j$ (by payment p'_j to item j buyer i can get more utility). An outcome $(\mathbf{p}, \mathbf{x})$ is *stable* if it has no blocking pairs, that is, no unmatched buyer-seller pair can mutually benefit by trading with each other instead of their current partners.

Key Results

For any given matching market, Shapley and Shubik [13] formulate an efficient matching market outcome as a linear program of maximizing the social welfare of the allocation. The duality theorem then shows the existence of competitive equilibrium. They also prove that there is a unique minimum (maximum) equilibrium price vector.

Theorem 1 (Shapley and Shubik [13]) *A matching market competitive equilibrium always exits. The set of competitive equilibrium price vectors* $\mathbf{p}$ *form a lattice in* $\mathbb{R}^m_{\geq 0}$.

Shapley and Shubik [13] also establish the connections between stability and competitive equilibrium.

Theorem 2 (Shapley and Shubik [13]) *In a multi-item market, an outcome* $(\mathbf{p}, \mathbf{x})$ *is a competitive equilibrium if and only if it is stable.*

However, they did not define an adjustment process like the deferred-acceptance algorithm by Gale and Shapley [9]. Crawford and Knoer [5] study a more generalized setup with firms and workers, where firms can hire multiple workers while each worker can be employed by at most one firm. It is essentially a many-to-one matching framework, where firms can be considered as buyers with multi-unit demand while buyers can be considered as items. They describe a salary adjustment process, which is essentially a version of the deferred-acceptance algorithm. They also show that for arbitrary capacities of the firms (buyers), when ties are ruled out, it always converges to a minimum competitive equilibrium. They provide an alternative proof of Shapley and Shubik's [13] result and allow significant generalization of it. However, the analysis in Crawford and Knoer [5] is flawed by two unnecessarily restrictive assumptions. Later, Kelso and Crawford [10] relax these assumptions and propose a modification to the salary adjustment process, which works as follows (firms are essentially buyers with multi-unit demand and workers are the items):

```
Salary Adjustment Process
(Kelso-Crawford [10])

1. Firms begin by facing a set of
   initially very low salaries
   (reserve prices).
2. Firms make offers to their most
   preferred set of workers. Any offer
   previously made by a firm to a
   worker that was not rejected must
   be honored.
3. Each worker who receives one or
   more offers rejects all but his
   favorite, which he tentatively
   accepts.
4. Offers not rejected in previous
   periods remain in force. For
   each rejected offer a firm made,
   increase the feasible salary
   for the rejecting worker. Firms
   continue to make offers to their
   favorite sets of workers.
5. The process stops when no
   rejections are made. Workers then
   accept the offers that remain in
   force from the firms they have not
   rejected.
```

An important assumption underlying the algorithm is that workers are gross substitutes from the standpoint of the firm. That is, when the price of one worker goes up, demand for another worker should not go down. As the salary adjustment process goes on, for firms, the set of feasible offers is reduced as some offers are rejected. For workers, the set of offers grows as more offers come up. Therefore, there is a monotonicity of offer sets in opposite directions. Alternatively, we can have a worker-proposing algorithm, where workers all begin by offering their services at the highest possible wage at their most preferred firm.

Theorem 3 (Kelso and Crawford [10]) *The salary adjustment process converges to a minimum competitive equilibrium, provided that all workers are gross substitutes from each firm's standpoint. In other words the final competitive equilibrium allocation is weakly preferred by every firm to any other equilibrium allocation.*

Taking a different perspective, Demange, Gale, and Sotomayor [7] propose an ascending auction-based algorithm (called "exact auction mechanism" in their original paper) that converges to a minimum competitive equilibrium for the original one-to-one matching problem. It is a variant of the so-called Hungarian method by Kuhn [11] for solving the optimal assignment problem. The algorithm works as the following:

Exact Auction Mechanism (Demange-Gale-Sotomayor [7])

---

1. Assume (for simplicity) that valuations v_{ij} are all integers.
2. Set the initial price vector, $\mathbf{p}^0$, to the reserve prices, $\mathbf{p}^0 = \mathbf{r}$
3. At round t when current prices are $\mathbf{p}^t$, each buyer i declares his demand set, $D_i(\mathbf{p}^t)$.
4. If there is no over-demanded set of items, terminate the process. A market equilibrium at prices $\mathbf{p}^t$ exists, and a corresponding allocation can be found by maximum matching.
5. If there exits over-demanded set(s), find a minimal over-demanded set S, and for all $j \in S$, $p_j^{t+1} = p_j^t + 1$. Set $t = t + 1$, and go to step 3.

Theorem 4 (Demange, Gale, and Sotomayor [7]) *The exact auction mechanism always finds a competitive equilibrium. Moreover, the equilibrium it finds is the minimum competitive equilibrium.*

This auction outcome can be computed efficiently since we can compute (minimal) over-demanded sets in P-time. It is not necessary to increment prices by one unit in each iteration. Instead, one can raise prices in the over-demanded set until the demand set of one of the respective bidders enlarges. It turns out that the payments (minimum price equilibrium) are precisely the VCG payments. Therefore, the mechanism is incentive compatible, and for every buyer, it is a dominant strategy to specify his true valuations. Moreover, this mechanism is even group strategy-proof, meaning that no strict subset of buyers who can collude have an incentive to misrepresent their true valuations.

Demange, Gale, and Sotomayor [7] also propose an approximation algorithm, called "approximate auction mechanism" for computing a minimum competitive equilibrium. It is a version of the deferred-acceptance algorithm proposed by Crawford and Knoer [5], which in turn is a special case of the algorithm of Kelso and Crawford [10]. The algorithm works as follows.

Approximate Auction Mechanism (Demange-Gale-Sotomayor [7])

---

1. Set the initial price vector, $\mathbf{p}^0$, to the reserve prices, $\mathbf{p}^0 = \mathbf{r}$
2. At round t when current prices are $\mathbf{p}^t$, each buyer i may bid for any item. When he does so, he is committed to that item, which means he commits himself to possibly buying the item at the announced price. The item is (tentatively) assigned to that bidder.
3. At this point, any uncommitted bidder may:
   - Bid for some unassigned item, in which case he becomes committed to it at its initial price.
   - Bid for an assigned item, in which case he becomes committed to that item, its price increase by some fixed amount δ, and the bidder to whom it was assigned becomes uncommitted
   - Drop out of the bidding.
4. When there are no more uncommitted bidders, the auction terminates. Each committed bidder buys the item assigned to him at its current price.

This approximate auction mechanism would be appealing to the buyers since it does not require them to decide in advance exactly what their bidding behavior will be. Instead, at each stage, a buyer can make use of present and past stages of the auction to decide his next bid. If buyers behave in accordance with linear valuations, the final price will differ from the minimum equilibrium price by at most $k\delta$ units, where k is the minimum of the number of items and bidders. Thus, by making δ (the unit by which bids are increased) sufficiently small, one can get arbitrarily close to the minimum equilibrium price.

Theorem 5 (Demange, Gale, and Sotomayor [7]) *Under the approximate auction mechanism, the final price of an item will differ from the minimum equilibrium price by at most* $k\delta$*, where* $k = \min(m, n)$.

The mechanisms discussed so far (approximately) compute a minimum competitive equilibrium. These approaches can be easily transformed to compute a maximum competitive equilibrium. Chen and Deng [3] discuss a combinatorial algorithm which iteratively increases prices to converge to a maximum competitive equilibrium starting from an arbitrary equilibrium.

Applications

The assignment model is used by Becker [2] to study marriage and household economics. Based on the fact that stable outcomes all correspond to optimal assignments, he studies which men are matched to which women under different assumptions of the assignment matrix.

Extensions

The existence of competitive equilibrium has later been established by Crawford and Knoer [5], Gale [8], and Quinnzi [12] for more general utility functions rather than the linear surplus, provided $u_{ij}(\cdot)$ is strictly decreasing and continuous everywhere.

Under a minimum competitive equilibrium mechanism, it is a dominant strategy for every buyer to report his true valuation. On the other hand, under a maximum competitive equilibrium mechanism, while the sellers will be truthful, it is possible that some buyer bids a false value to obtain more utility. In a recent study, Chen and Deng [3] show the convergence from the maximum competitive equilibrium toward the minimum competitive equilibrium in a deterministic and dynamic setting.

Another strand of recent studies focus on the assignment model with budget constraints, which is applicable to many marketplaces such as online and TV advertising markets. An extra budget constraint introduces discontinuity in the utility function, which fundamentally changes the properties of competitive equilibria. In such setups, a competitive equilibrium does not always exist. Aggarwal et al. [1] study the problem of computing a weakly stable matching in the assignment model with quasi-linear utilities subject to a budget constraint. However, a weakly stable matching does not possess the envy-freeness property of a competitive equilibrium. Chen et al. [4] establish a connection between competitive equilibrium in the assignment model with budgets and strong stability. Then they give a strong polynomial time algorithm for deciding existence of and computing a minimum competitive equilibrium for a general class of utility functions in the assignment model with budgets.

Cross-References

▸ Hospitals/Residents Problem
▸ Optimal Stable Marriage
▸ Stable Marriage
▸ Stable Marriage and Discrete Convex Analysis
▸ Stable Marriage with Ties and Incomplete Lists

Recommended Reading

1. Aggarwal G, Muthukrishnan S, Pal D, Pal M (2009) General auction mechanism for search advertising. In: WWW, Madrid, pp 241–250

2. Becker GS (1981) A treatise on the family. Harvard University Press, Cambridge
3. Chen N, Deng X (2011) On Nash dynamics of matching market equilibria. CoRR abs/1103.4196
4. Chen N, Deng X, Ghosh A (2010) Competitive equilibria in matching markets with budgets. ACM SIGecom Exch 9(1):1–5
5. Crawford VP, Knoer EM (1981) Job matching with heterogeneous firms and workers. Econometrica 49(2):437–450
6. Demange G, Gale D (1985) The strategy structure of two-sided matching markets. Econometrica 53(4):873–883
7. Demange G, Gale D, Sotomayor M (1986) Multi-item auctions. J Pol Econ 94(4):863–872
8. Gale D (1984) Equilibrium in a discrete exchange economy with money. Int J Game Theory 13:61–64
9. Gale D, Shapley LS (1962) College admissions and the stability of marriage. Am Math Mon 69(1):9–15
10. Kelso AS, Crawford VP (1982) Job matching, coalition formation, and gross substitutes. Econometrica 50(6):1483–1504
11. Kuhn HW (1955) The Hungarian method for the assignment problem. Nav Reserv Logist Q 2(1):83–97
12. Quinnzi M (1984) Core and competitive equilibrium with indivisibilities. Int J Game Theory 13:41–60
13. Shapley LS, Shubik M (1971) The assignment game I: the core. Int J Game Theory 1(1):110–130

Matroids in Parameterized Complexity and Exact Algorithms

Fahad Panolan[1] and Saket Saurabh[1,2]
[1]Institute of Mathematical Sciences, Chennai, India
[2]University of Bergen, Bergen, Norway

Keywords

Exact algorithms; Matroids; Parameterized algorithms; Representative families

Years and Authors of Summarized Original Work

2009; Marx
2014; Fomin, Lokshtanov, Saurabh
2014; Fomin, Lokshtanov, Panolan, Saurabh
2014; Shachnai, Zehavi

Problem Definition

In recent years matroids have been used in the fields of parameterized complexity and exact algorithms. Many of these works mainly use a computation of *representative families*. Let $M = (E, \mathcal{I})$ be a matroid and $\mathcal{S} = \{S_1, \ldots, S_t\} \subseteq \mathcal{I}$ be a family of independent sets of size p. A subfamily $\hat{\mathcal{S}} \subseteq \mathcal{S}$ is called a *q-representative family* for $\mathcal{S}$ (denoted by $\hat{\mathcal{S}} \subseteq^q_{rep} \mathcal{S}$), if for every $Y \subseteq E$ of size at most q, if there exists a set $S \in \mathcal{S}$ disjoint from Y with $S \cup Y \in \mathcal{I}$, then there exists a set $\hat{S} \in \hat{\mathcal{S}}$ disjoint from Y with $\hat{S} \cup Y \in \mathcal{I}$. The basic algorithmic question regarding representative families is, given a matroid $M = (E, \mathcal{I})$, a family $\mathcal{S} \subseteq \mathcal{I}$ of independent sets of size p and a positive integer q, compute $\hat{\mathcal{S}} \subseteq^q_{rep} \mathcal{S}$ of size as small as possible in time as fast as possible.

The Two-Families Theorem of Bollobás [1] for extremal set systems implies that every family of independent sets of size p in a uniform matroid has a q-representative family with at most $\binom{p+q}{p}$ sets. The generalization of Two-Families Theorem to subspaces of a vector space by Lovász [5] implies that every family of independent sets of size p in a linear matroid has a q-representative family with at most $\binom{p+q}{p}$ sets. In fact one can show that the cardinality $\binom{p+q}{p}$ of a q-representative family of a family of independent sets of size p is optimal. It is important to note that the size of the q-representative family of a family of sets of size p in a uniform or linear matroid only depends on p and q and not on the cardinality of ground set of the matroid, and this fact is used to design parameterized and exact algorithms.

Key Results

For uniform matroids, Monien [8] gave an algorithm for computing a q-representative family of size at most $\sum_{i=0}^{q} p^i$ in time $\mathcal{O}(pq \cdot \sum_{i=0}^{q} p^i \cdot t)$ and Marx [6] gave another algorithm, for computing a q-representative families of size at most $\binom{p+q}{p}$ in time $\mathcal{O}(p^q \cdot t^2)$. For uniform matroids, Fomin et al. [2] proved the following theorem.

Theorem 1 ([2, 9]) *Let $\mathcal{S} = \{S_1, \ldots, S_t\}$ be a family of sets of size p over a universe of size n and let $0 < x < 1$. For a given q, a q-representative family $\widehat{\mathcal{S}} \subseteq^q_{rep} \mathcal{S}$ with at most $x^{-p}(1-x)^{-q} \cdot 2^{o(p+q)} \log n$ sets can be computed in time $\mathcal{O}((1-x)^{-q} \cdot 2^{o(p+q)} \cdot t \cdot \log n)$.*

In [3], Fomin, Lokshtanov, and Saurabh proved Theorem 1 for $x = \frac{p}{p+q}$. That is, a q-representative family $\hat{\mathcal{S}} \subseteq^q_{rep} \mathcal{S}$ with at most $\binom{p+q}{p} \cdot 2^{o(p+q)} \cdot \log n$ sets can be computed in time $\mathcal{O}((\frac{p+q}{q})^q \cdot 2^{o(p+q)} \cdot t \cdot \log n)$. Later Fomin et al. [2] observed that the proof in [3] can be modified to work for every $0 < x < 1$ and allows an interesting trade-off between the size of the computed representative families and the time taken to compute them, and this trade-off can be exploited algorithmically to speed up "representative families based" algorithms. Independently, at the same time, Shachnai and Zehavi [9] also observed that the proof in [3] can be generalized to get Theorem 1. We would like to mention that in fact a variant of Theorem 1 is proved in [2, 9] which computes a weighted q-representative family. The proof of Theorem 1 uses algorithmic variant of "random permutation" proof of Bollobás Lemma and an efficient construction of a variant of universal sets called n-p-q-separating collections.

For linear matroids, Marx [7] showed that Lovász's proof can be transformed into an algorithm computing a q-representative family:

Theorem 2 ([7]) *Given a linear representation A_M of a matroid $M = (E, \mathcal{I})$, a family $\{S_1, \ldots, S_t\}$ of independent sets of size p and a positive integer q, there is an algorithm which computes $\hat{\mathcal{S}} \subseteq^q_{rep} \mathcal{S}$ of size $\binom{p+q}{p}$ in time $2^{\mathcal{O}(p \log(p+q))} \cdot \binom{p+q}{p}^{\mathcal{O}(1)} (||A_M||t)^{\mathcal{O}(1)}$ where $||A_M||$ is the size of A_M.*

Fomin, Lokshtanov, and Saurabh [3] gave an efficient computation of representative families in linear matroids:

Theorem 3 ([3]) *Let $M = (E, \mathcal{I})$ be a linear matroid of rank $p + q = k$ given together with its representation matrix A_M over a field $\mathbb{F}$. Let $\mathcal{S} = \{S_1, \ldots, S_t\}$ be a family of independent sets of size p. Then $\hat{\mathcal{S}} \subseteq^q_{rep} \mathcal{S}$ with at most $\binom{p+q}{p}$ sets can be computed in $\mathcal{O}\left(\binom{p+q}{p} t p^\omega + t \binom{p+q}{q}^{\omega-1}\right)$ operations over $\mathbb{F}$, where $\omega < 2.373$ is the matrix multiplication exponent.*

We would like to draw attention of the reader that in Theorems 3 and 2, the cardinality of the computed q-representative family is optimal and polynomial in p and q if one of p or q is a constant, unlike in Theorem 1. As in the case of Theorem 1, Theorem 3 is also proved for a weighted q-representative family.

Most of the algorithms using representative families are dynamic programming algorithms. A class of families which often arise in dynamic programming are product families. A family $\mathcal{F}$ is called *product* of two families $\mathcal{A}$ and $\mathcal{B}$, where $\mathcal{A}$ and $\mathcal{B}$ are families of independent sets in a matroid $M = (E, \mathcal{I})$, if $\mathcal{F} = \{A \cup B \mid A \in \mathcal{A}, B \in \mathcal{B}, A \cap B = \emptyset, A \cup B \in \mathcal{I}\}$. Fomin et al. [2] gave two algorithms to compute q-representative family of a product family $\mathcal{F}$, one in case of uniform matroid and other in case of linear matroid. These algorithms significantly outperform the naive way of computing the product family $\mathcal{F}$ first and then a representative family of it.

Applications

Representative families are used to design efficient algorithms in parameterized complexity and exact algorithms.

Parameterized and Exact Algorithms

In this subsection we list some of the parameterized and exact algorithms obtained using representative families.

1. ℓ-MATROID INTERSECTION. In this problem we are given ℓ matroids $M_1 = (E, \mathcal{I}_1), \ldots, M_\ell = (E, \mathcal{I}_\ell)$ along with their linear representations $A_{M_1}, \ldots, A_{M_\ell}$ over the same field $\mathbb{F}$ and a positive integer k. The objective is to find a k element subset of E, which is independent in all the matroids $M_1, \ldots, M_\ell$. Marx [7] gave a randomized

algorithm for the problem running in time $f(k,l)\left(\sum_{i=1}^{\ell} ||A_{M_i}||\right)^{\mathcal{O}(1)}$, where f is a computable function. By giving an algorithm for deterministic truncation of linear matroids, Lokshtanov et al. [4] gave a deterministic algorithm for the problem running in time $2^{\omega k \ell}\left(\sum_{i=1}^{\ell} ||A_{M_i}||\right)^{\mathcal{O}(1)}$, where ω is the matrix multiplication exponent.

2. LONG DIRECTED CYCLE. In the LONG DIRECTED CYCLE problem, we are interested in finding a cycle of length at least k in a directed graph. Fomin et al. [2] and Shachnai and Zehavi [9] gave an algorithm of running time $\mathcal{O}(6.75^{k+o(k)} mn^2 \log^2 n)$ for this problem.
3. SHORT CHEAP TOUR. In this problem we are given an undirected n-vertex graph G, $w: E(G) \rightarrow \mathbb{N}$ and an integer k. The objective is to find a path of length k with minimum weight. Fomin et al. [2] and Shachnai and Zehavi [9] gave a $\mathcal{O}(2.619^k n^{\mathcal{O}(1)} \log W)$ time algorithm for SHORT CHEAP TOUR, where W is the largest edge weight in the given input graph.
4. MULTILINEAR MONOMIAL DETECTION. Here the input is an arithmetic circuit C over $\mathbb{Z}^+$ representing a polynomial $P(X)$ over $\mathbb{Z}^+$. The objective is to test whether $P(X)$ construed as a sum of monomials contain a multilinear monomial of degree k. For this problem Fomin et al. [2] gave an algorithm of running time $\mathcal{O}(3.8408^k 2^{o(k)} s(C) n \log^2 n)$, where $s(C)$ is the size of the circuit.
5. MINIMUM EQUIVALENT GRAPH(MEG). In this problem we are seeking a spanning subdigraph D' of a given n-vertex digraph D with as few arcs as possible in which the reachability relation is the same as in the original digraph D. Fomin, Lokshtanov, and Saurabh [3] gave the first single-exponential exact algorithm, i.e., of running time $2^{\mathcal{O}(n)}$, for the problem.
6. **Dynamic Programming Over Graphs of Bounded Treewidth.** Fomin et al. [2] gave algorithms with running time $\mathcal{O}\left((1+2^{\omega-1}\cdot 3)^{\mathbf{tw}}\mathbf{tw}^{\mathcal{O}(1)} n\right)$ for FEEDBACK VERTEX SET and STEINER TREE, where $\mathbf{tw}$ is the treewidth of the input graph, n is the number of vertices in the input graph, and ω is the matrix multiplication exponent.

Open Problems

1. Can we improve the running time for the computation of representative families in linear matroids or in specific matroids like graphic matroids?

Cross-References

▸ Shadowless Solutions for Fixed-Parameter Tractability of Directed Graphs
▸ Treewidth of Graphs

Recommended Reading

1. Bollobás B (1965) On generalized graphs. Acta Math Acad Sci Hungar 16:447–452
2. Fomin FV, Lokshtanov D, Panolan F, Saurabh S (2014) Representative sets of product families. In: Proceedings of 22nd Annual European Symposium on Algorithms (ESA 2014), Wroclaw, 8–10 Sept 2014, vol 8737, pp 443–454. doi:10.1007/978-3-662-44777-2_37
3. Fomin FV, Lokshtanov D, Saurabh S (2014) Efficient computation of representative sets with applications in parameterized and exact algorithms. In: Proceedings of the Twenty-Fifth Annual ACM-SIAM Symposium on Discrete Algorithms (SODA 2014), Portland, 5–7 Jan 2014, pp 142–151. doi:10.1137/1.9781611973402.10
4. Lokshtanov D, Misra P, Panolan F, Saurabh S (2015) Deterministic truncation of linear matroids. In: Proceedings of 42nd International Colloquium on Automata, Languages, and Programming (ICALP 2015), Kyoto, 6–10 July 2015, Part I, pp 922–934. doi:10.1007/978-3-662-47672-7_75
5. Lovász L (1977) Flats in matroids and geometric graphs. In: Combinatorial surveys (Proceedings of the Sixth British Combinatorial Conference, Royal Holloway College, Egham). Academic, London, pp 45–86
6. Marx D (2006) Parameterized coloring problems on chordal graphs. Theor Comput Sci 351(3):407–424
7. Marx D (2009) A parameterized view on matroid optimization problems. Theor Comput Sci 410(44):4471–4479
8. Monien B (1985) How to find long paths efficiently. In: Analysis and design of algorithms for combinatorial problems (Udine, 1982). North-Holland mathe-

matics studies, vol 109. North-Holland, Amsterdam, pp 239–254. doi:10.1016/S0304-0208(08)73110-4
9. Shachnai H, Zehavi M (2014) Representative families: a unified tradeoff-based approach. In: Proceedings of 22nd Annual European Symposium on Algorithms (ESA 2014), Wroclaw, 8–10 Sept 2014, vol 8737, pp 786–797. doi:10.1007/978-3-662-44777-2_65

Max Cut

Alantha Newman
CNRS-Université Grenoble Alpes and G-SCOP, Grenoble, France

Keywords

Approximation algorithms; Graph partitioning

Synonyms

Maximum bipartite subgraph

Year and Authors of Summarized Original Work

1994; 1995; Goemans, Williamson

Problem Definition

Given an undirected edge-weighted graph, $G = (V, E)$, the maximum cut problem (MAX CUT) is to find a bipartition of the vertices that maximizes the weight of the edges crossing the partition. If the edge weights are non-negative, then this problem is equivalent to finding a maximum weight subset of the edges that forms a bipartite subgraph, i.e., the maximum bipartite subgraph problem. All results discussed in this article assume non-negative edge weights. MAX CUT is one of Karp's original NP-complete problems [20]. In fact, it is NP-hard to approximate to within a factor better than $\frac{16}{17}$ [17, 35].

For nearly 20 years, the best-known approximation factor for MAX CUT was half, which can be achieved by a very simple algorithm: form a set S by placing each vertex in S with probability half. Since each edge crosses the cut $(S, V \setminus S)$ with probability half, the expected value of this cut is half the total edge weight. This implies that for any graph, there exists a cut with value at least half of the total edge weight. In 1976, Sahni and Gonzalez presented a deterministic half-approximation algorithm for MAX CUT, which is essentially a de-randomization of the aforementioned randomized algorithm [31]: iterate through the vertices and form sets S and $\bar{S}$ by placing each vertex in the set that maximizes the weight of cut $(S, \bar{S})$ thus far. After each iteration of this process, the weight of this cut will be at least half of the weight of the edges with both endpoints in $S \cup \bar{S}$.

This simple half-approximation algorithm uses the fact that for any graph with non-negative edge weights, the total edge weight of a given graph is an upper bound on the value of its maximum cut. There exist classes of graphs for which a maximum cut is arbitrarily close to half the total edge weight, i.e., graphs for which this "trivial" upper bound can be close to twice the true value of an optimal solution. An example of such a class of graphs is complete graphs on n vertices, K_n. In order to obtain an approximation factor better than half, one must be able to compute an upper bound on the value of a maximum cut that is better, i.e., smaller, than the trivial upper bound for such classes of graphs.

Linear Programming Relaxations

For many optimization (maximization) problems, linear programming has been shown to yield better (upper) bounds on the value of an optimal solution than can be obtained via combinatorial methods. There are several well-studied linear programming relaxations for MAX CUT. For example, a classical integer program has a variable x_e for each edge and a constraint for each odd cycle, requiring that an odd cycle C contribute at most $|C| - 1$ edges to an optimal solution.

$$\max \sum_{e \in E} w_e x_e$$

$$\sum_{e \in C} x_e \leq |C| - 1 \quad \forall \text{ odd cycles } C$$

$$x_e \in \{0, 1\}.$$

The last constraint can be relaxed so that each x_e is required to lie between 0 and 1, but need not be integral, i.e., $0 \leq x_e \leq 1$. Although this relaxation may have exponentially many constraints, there is a polynomial-time separation oracle (equivalent to finding a minimum weight odd cycle), and thus, the relaxation can be solved in polynomial time [14]. Another classical integer program contains a variable x_{ij} for each pair of vertices. In any partition of the vertices, either zero or two edges from a three-cycle cross the cut. This requirement is enforced in the following integer program. If edge $(i, j) \notin E$, then w_{ij} is set to 0.

$$\max \sum_{i,j \in V} w_{ij} x_{ij}$$

$$x_{ij} + x_{jk} + x_{ki} \leq 2 \quad \forall i, j, k \in V$$

$$x_{ij} + x_{jk} - x_{ki} \geq 0 \quad \forall i, j, k \in V$$

$$x_{ij} \in \{0, 1\}.$$

Again, the last constraint can be relaxed so that each x_{ij} is required to lie between 0 and 1. In contrast to the aforementioned cycle-constraint-based linear program, this linear programming relaxation has a polynomial number of constraints.

Both of these relaxations actually have the same optimal value for any graph with non-negative edge weights [3,26,30]. (For a simplified proof of this, see [25].) Poljak showed that the integrality gap for each of these relaxations is arbitrarily close to 2 [26]. In other words, there are classes of graphs that have a maximum cut containing close to half of the edges, but for which each of the above relaxations yields an upper bound close to all the edges, i.e., no better than the trivial "all-edges" bound. In particular, graphs with a maximum cut close to half the edges and with high girth can be used to demonstrate this gap. A comprehensive look at these linear programming relaxations is contained in the survey of Poljak and Tuza [30].

Another natural integer program uses variables for vertices rather than edges:

$$\max \sum_{(i,j) \in E} w_{ij} \left(x_i(1-x_j)+x_j(1-x_i)\right) \quad (1)$$

$$x_i \in \{0, 1\} \qquad \forall i \in V. \quad (2)$$

Replacing (2) with $x_i \in [0, 1]$ results in a nonlinear relaxation that is actually just as hard to solve as the integer program. This follows from the fact that any fractional solution can be rounded to obtain an integer solution with at least the same value. Indeed, for any vertex $h \in V$ with fractional value x_h, we can rewrite the objective function (1) as follows. Edges adjacent to vertex h are denoted by $\delta(h)$. For ease of notation, let us momentarily assume the graph is unweighted, although the argument works for non-negative edge weights.

$$\sum_{(i,j) \in E \setminus \delta(h)} x_i(1 - x_j) + x_j(1 - x_i) +$$

$$x_h \overbrace{\sum_{j \in \delta(h)} (1 - x_j)}^{A} + (1 - x_h) \overbrace{\sum_{j \in \delta(h)} x_j}^{B}. \quad (3)$$

If $A \geq B$, we round x_ℓ to 1, otherwise we round it to 0. Repeating this process for all vertices results in an integral solution whose objective value is no less than the objective value of the initial fractional solution.

Eigenvalue Upper Bounds

Delorme and Poljak [8] presented an eigenvalue upper bound on the value of a maximum cut, which was a strengthened version of a previous eigenvalue bound considered by Mohar and Poljak [24]. Computing Delorme and Poljak's upper bound is equivalent to solving an eigenvalue minimization problem. They showed that their bound is computable in polynomial time with arbitrary precision. In a series of work, Delorme, Poljak and Rendl showed that this upper bound behaves "differently" from the linear programming-based upper bounds. For example, they studied classes of sparse random graphs (e.g., $G(n, p)$ with $p = 50/n$) and showed that their upper bound is close to optimal on these graphs [9]. Since graphs of this type can also be used to demonstrate an integrality gap arbitrarily close to 2 for the aforementioned linear programming relaxations, their work highlighted contrasting behavior between these two upper bounds. Further computational experiments on other classes of graphs gave more evidence that the bound was indeed stronger than previously studied bounds [27, 29]. Delorme and Poljak conjectured that the five cycle demonstrated the worst-case behavior for their bound: a ratio of $\frac{32}{25+5\sqrt{5}} \approx 0.88445$ between their bound and the optimal integral solution. However, they could not prove that their bound was strictly less than twice the value of a maximum cut in the worst case.

Key Result

In 1994, Goemans and Williamson presented a randomized 0.87856-approximation algorithm for MAX CUT [12]. Their breakthrough work was based on rounding a semidefinite programming relaxation and was the first use of semidefinite programming in approximation algorithms. Poljak and Rendl showed that the upper bound provided by this semidefinite relaxation is equivalent to the eigenvalue bound of Delorme and Poljak [28]. Thus, Goemans and Williamson proved that the eigenvalue bound of Delorme and Poljak is no more than 1.138 times the value of a maximum cut.

A Semidefinite Relaxation

MAX CUT can be formulated as the following quadratic integer program, which is NP-hard to solve. Each vertex $i \in V$ is represented by a variable y_i, which is assigned either 1 or -1 depending on which side of the cut it appears.

$$\max \frac{1}{2} \sum_{(i,j)\in E} w_{ij}(1 - y_i y_j)$$
$$y_i \in \{-1, 1\} \qquad \forall i \in V.$$

Goemans and Williamson considered the following relaxation of this integer program, in which each vertex is represented by a unit vector.

$$\max \frac{1}{2} \sum_{(i,j)\in E} w_{ij}(1 - v_i \cdot v_j)$$
$$v_i \cdot v_i = 1 \qquad \forall i \in V$$
$$v_i \in \mathcal{R}^n \qquad \forall i \in V.$$

They showed that this relaxation is equivalent to a semidefinite program. Specifically, consider the following semidefinite relaxation:

$$\max \frac{1}{2} \sum_{(i,j)\in E} w_{ij}(1 - y_{ij})$$
$$y_{ii} = 1 \qquad \forall i \in V$$
$$Y \qquad \text{positive semidefinite.}$$

The equivalence of these two relaxations is due to the fact that a matrix Y is positive semidefinite if and only if there is a matrix B such that $B^T B = Y$. The latter relaxation can be solved to within arbitrary precision in polynomial time via the ellipsoid algorithm, since it has a polynomial-time separation oracle [15]. Thus, a solution to the first relaxation can be obtained by finding a solution to the second relaxation and finding a matrix B such that $B^T B = Y$. If the columns of B correspond to the vectors $\{v_i\}$, then $y_{ij} = v_i \cdot v_j$, yielding a solution to the first relaxation.

Random-Hyperplane Rounding

Goemans and Williamson showed how to round the semidefinite programming relaxation of MAX CUT using a new technique that has since become known as "random-hyperplane rounding" [12]. First obtain a solution to the first relaxation, which consists of a set of unit vectors $\{v_i\}$, one vector for each vertex. Then choose a random vector $r \in \mathcal{R}^n$ in which each coordinate of r is chosen from the standard normal distribution. Finally, set $S = \{i \mid v_i \cdot r \geq 0\}$ and output the cut $(S, V \setminus S)$.

The probability that a particular edge $(i, j) \in E$ crosses the cut is equal to the probability that the dot products $v_i \cdot r$ and $v_j \cdot r$ differ in sign. This probability is exactly equal to θ_{ij}/π, where θ_{ij} is the angle between vectors v_i and v_j. Thus, the expected weight of edges crossing the cut is equal to $\sum_{(i,j)\in E} \theta_{ij}/\pi$. How large is this compared to the objective value given by the semidefinite programming relaxation, i.e., what is the approximation ratio?

Define α_{gw} as the worst-case ratio of the expected contribution of an edge to the cut, to its contribution to the objective function of the semidefinite programming relaxation. In other words: $\alpha_{gw} = \min_{0\leq\theta\leq\pi} \frac{2}{\pi}\frac{\theta}{1-\cos\theta}$. It can be shown that $\alpha_{gw} > 0.87856$. Thus, the expected value of a cut is at least $\alpha_{gw} \cdot SDP_{OPT}$, resulting in an approximation ratio of at least 0.87856 for MAX CUT. The same analysis applies to weighted graphs with non-negative edge weights.

This algorithm was de-randomized by Mahajan and Hariharan [23]. Goemans and Williamson also applied their random-hyperplane rounding techniques to give improved approximation guarantees for other problems such as MAX-DICUT and MAX-2SAT.

Integrality Gap and Hardness

Karloff showed that there exist graphs for which the best hyperplane is only a factor α_{gw} of the maximum cut [19], showing that there are graphs for which the analysis in [12] is tight. Since the optimal SDP value for such graphs equals the optimal value of a maximum cut, these graphs cannot be used to demonstrate an integrality gap. However, Feige and Schechtman showed that there exist graphs for which the maximum cut is a α_{gw} fraction of the SDP bound [10], thereby establishing that the approximation guarantee of Goemans and Williamson's algorithm matches the integrality gap of their semidefinite programming relaxation. Recently, Khot, Kindler, Mossel, and O'Donnell [22] showed that if the Unique Games Conjecture of Khot [21] is true, then it is NP-hard to approximate MAX CUT to within any factor larger than α_{gw}.

Better-than-Half Approximations Without SDPs

Since Goemans and Williamson presented an α_{gw}-approximation algorithm for MAX CUT, it has been an open question if one can obtain a matching approximation factor or even an approximation ratio of $\frac{1}{2} + \epsilon$ for some constant $\epsilon > 0$ without using SDPs. Trevisan presented an algorithm based on spectral partitioning with running time $\tilde{O}(n^2)$ and an approximation guarantee of 0.531 [34], which was subsequently improved to 0.614 by Soto [32].

Applications

The work of Goemans and Williamson paved the way for the further use of semidefinite programming in approximation algorithms, particularly for graph partitioning problems. Methods based on the random-hyperplane technique have been successfully applied to many optimization problems that can be categorized as partition problems. A few examples are 3-COLORING [18], MAX-3-CUT [7,11,13], MAX-BISECTION [16], CORRELATION CLUSTERING [5,33], and SPARSEST CUT [2]. Additionally, some progress has been made in extending semidefinite programming techniques outside the domain of graph partitioning to problems such as BETWEENNESS [6], BANDWIDTH [4], and LINEAR EQUATIONS mod p [1].

Cross-References

- ▸ Exact Algorithms for Maximum Two-Satisfiability
- ▸ Graph Bandwidth
- ▸ Graph Coloring
- ▸ Sparsest Cut

Recommended Reading

1. Andersson G, Engebretsen L, Håstad J (2001) A new way to use semidefinite programming with applications to linear equations mod p. J Algorithms 39:162–204

M

2. Arora S, Rao S, Vazirani U (2004) Expander flows, geometric embeddings and graph partitioning. In: Proceedings of the 36th annual symposium on the theory of computing (STOC), Chicago, pp 222–231
3. Barahona F (1993) On cuts and matchings in planar graphs. Math Program 60:53–68
4. Blum A, Konjevod G, Ravi R, Vempala S (2000) Semi-definite relaxations for minimum bandwidth and other vertex-ordering problems. Theor Comput Sci 235:25–42
5. Charikar M, Guruswami V, Wirth A (2003) Clustering with qualitative information. In: Proceedings of the 44th annual IEEE symposium on foundations of computer science (FOCS), Boston, pp 524–533
6. Chor B, Sudan M (1998) A geometric approach to betweeness. SIAM J Discret Math 11:511–523
7. de Klerk E, Pasechnik DV, Warners JP (2004) On approximate graph colouring and MAX-k-CUT algorithms based on the θ function. J Comb Optim 8(3):267–294
8. Delorme C, Poljak S (1993) Laplacian eigenvalues and the maximum cut problem. Math Program 62:557–574
9. Delorme C, Poljak S (1993) The performance of an eigenvalue bound in some classes of graphs. Discret Math 111:145–156. Also appeared in Proceedings of the conference on combinatorics, Marseille, 1990
10. Feige U, Schechtman G (2002) On the optimality of the random hyperplane rounding technique for MAX-CUT. Random Struct Algorithms 20(3):403–440
11. Frieze A, Jerrum MR (1997) Improved approximation algorithms for MAX-k-CUT and MAX BISECTION. Algorithmica 18:61–77
12. Goemans MX, Williamson DP (1995) Improved approximation algorithms for maximum cut and satisfiability problems using semidefinite programming. J ACM 42:1115–1145
13. Goemans MX, Williamson DP (2004) Approximation algorithms for MAX-3-CUT and other problems via complex semidefinite programming. STOC 2001 Spec Issue J Comput Syst Sci 68:442–470
14. Grötschel M, Lovász L, Schrijver A (1981) The ellipsoid method and its consequences in combinatorial optimization. Combinatorica 1:169–197
15. Grötschel M, Lovász L, Schrijver A (1988) Geometric algorithms and combinatorial optimization. Springer, Berlin
16. Halperin E, Zwick U (2002) A unified framework for obtaining improved approximation algorithms for maximum graph bisection problems. Random Struct Algorithms 20(3):382–402
17. Håstad J (2001) Some optimal inapproximability results. J ACM 48:798–869
18. Karger DR, Motwani R, Sudan M (1998) Improved graph coloring via semidefinite programming. J ACM 45(2):246–265
19. Karloff HJ (1999) How good is the Goemans-Williamson MAX CUT algorithm? SIAM J Comput 29(1):336–350
20. Karp RM (1972) Reducibility among combinatorial problems. In: Complexity of computer computations. Plenum, New York, pp 85–104
21. Khot S (2002) On the power of unique 2-prover 1-round games. In: Proceedings of the 34th annual symposium on the theory of computing (STOC), Montreal, pp 767–775
22. Khot S, Kindler G, Mossel E, O'Donnell R (2004) Optimal inapproximability results for MAX CUT and other 2-variable CSPs? In: Proceedings of the 45th annual IEEE symposium on foundations of computer science (FOCS), Rome, pp 146–154
23. Mahajan R, Hariharan R (1995) Derandomizing semidefinite programming based approximation algorithms. In: Proceedings of the 36th annual IEEE symposium on foundations of computer science (FOCS), Milwaukee, pp 162–169
24. Mohar B, Poljak S (1990) Eigenvalues and the max-cut problem. Czech Math J 40(115): 343–352
25. Newman A (2004) A note on polyhedral relaxations for the maximum cut problem. Unpublished manuscript
26. Poljak S (1992) Polyhedral and eigenvalue approximations of the max-cut problem. Sets Graphs Numbers Colloqiua Mathematica Societatis Janos Bolyai 60:569–581
27. Poljak S, Rendl F (1994) Node and edge relaxations of the max-cut problem. Computing 52:123–137
28. Poljak S, Rendl F (1995) Nonpolyhedral relaxations of graph-bisection problems. SIAM J Optim 5:467–487
29. Poljak S, Rendl F (1995) Solving the max-cut using eigenvalues. Discret Appl Math 62(1–3):249–278
30. Poljak S, Tuza Z (1995) Maximum cuts and large bipartite subgraphs. DIMACS Ser Discret Math Theor Comput Sci 20:181–244
31. Sahni S, Gonzalez T (1976) P-complete approximation problems. J ACM 23(3):555–565
32. Soto JA (2015) Improved analysis of a max-cut algorithm based on spectral partitioning. SIAM J Discret Math 29(1):259–268
33. Swamy C (2004) Correlation clustering: maximizing agreements via semidefinite programming. In: Proceedings of 15th annual ACM-SIAM symposium on discrete algorithms (SODA), New Orleans, pp 526–527
34. Trevisan L (2012) Max cut and the smallest eigenvalue. SIAM J Comput 41(6):1769–1786
35. Trevisan L, Sorkin GB, Sudan M, Williamson DP (2000) Gadgets, approximation, and linear programming. SIAM J Comput 29(6): 2074–2097

Max Leaf Spanning Tree

Frances Rosamond
Parameterized Complexity Research Unit,
University of Newcastle, Callaghan, NSW,
Australia

Keywords

Connected dominating set; Extremal structure; Maximum leaf spanning tree

Years and Authors of Summarized Original Work

2005; Estivill-Castro, Fellows, Langston, Rosamond

Problem Definition

The MAX LEAF SPANNING TREE problem asks us to find a spanning tree with at least k leaves in an undirected graph. The decision version of parameterized MAX LEAF SPANNING TREE is the following:

> MAX LEAF SPANNING TREE
> INPUT: A connected graph G, and an integer k.
> PARAMETER: An integer k.
> QUESTION: Does G have a spanning tree with at least k leaves?

The parameterized complexity of the nondeterministic polynomial-time complete MAX LEAF SPANNING TREE problem has been extensively studied [2, 3, 9, 11] using a variety of kernelization, branching and other fixed-parameter tractable (FPT) techniques. The authors are the first to propose an extremal structure method for hard computational problems. The method, following in the sense of Grothendieck and in the spirit of the graph minors project of Robertson and Seymour, is that a mathematical project should unfold as a series of small steps in an overall trajectory that is described by the appropriate "mathematical machine." The authors are interested in statements of the type: Every connected graph on n vertices that satisfies a certain set of properties has a spanning tree with at least k leaves, and this spanning tree can be found in time $O(f(k) + n^c)$, where c is a constant (independent of k) and f is an arbitrary function.

In parameterized complexity, the value k is called the *parameter* and is used to capture some structure of the input or other aspect of the computational objective. For example, k might be the number of edges to be deleted in order to obtain a graph with no cycles, or k might be the number of DNA sequences to be aligned in an alignment, or k may be the maximum type-declaration nesting depth of a compiler, or $k = 1/\epsilon$ may be the parameterization in the analysis of approximation, or k might be a composite of several variables.

There are two important ways of comparing FPT algorithms, giving rise to two FPT *races*. In the "$f(k)$" race, the competition is to find ever more slowing growing parameter functions $f(k)$ governing the complexity of FPT algorithms. The "kernelization race" refers to the following lemma stating that a problem is in FPT if and only if the input can be preprocessed (*kernelized*) in "ordinary" polynomial time into an instance whose size is bounded by a function of k only.

Lemma 1 *A parameterized problem* Π *is in FPT if and only if there is a polynomial-time transformation (in both n and k) that takes* (x, k) *to* (x', k') *such that:*

(1) (x, k) *is a yes-instance of* Π *if and only if* (x', k') *is a yes-instance of* Π,
(2) $k' \leq k$, *and*
(3) $|x'| \leq g(k)$ *for some fixed function* g.

In the situation described by the lemma, say that we can *kernelize* to instances of size at most $g(k)$. Although the two races are often closely related, the result is not always the same. The current best FPT algorithm for MAX LEAF is due to Bonsma [1] (following the extremal structure approach outlined by the authors) with a running time of $O^*(8.12^k)$ to determine whether a graph G on n vertices has a spanning tree with at least k leaves; however the authors present the FPT algorithm with the smallest kernel size.

The authors list five independent deliverables associated to the extremal structure theory, and illustrate all of the objectives for the MAX LEAF problem. The five objectives are:

(A) Better FPT algorithms as a result of deeper structure theory, more powerful reduction rules associated with that structure theory, and stronger inductive proofs of improved kernelization bounds.
(B) Powerful preprocessing *(data reduction/kernelization)* rules and *combinations* of rules that can be used regardless of whether the parameter is small and that can be combined with other approaches, such as approximation and heuristics. These are usually easy to program.
(C) Gradients and transformation rules for local search heuristics.
(D) Polynomial-time approximation algorithms and performance bounds proved in a systematic way.
(E) Structure to exploit for solving other problems.

Key Results

The key results are programmatic, providing a *method of extremal structure* as a systematic method for designing FPT algorithms. The five interrelated objectives listed above are surveyed, and each is illustrated using the MAX LEAF SPANNING TREE problem.

Objective A: FPT Algorithms

The objective here is to find polynomial-time preprocessing (*kernelization*) rules where $g(k)$ is as small as possible. This has a direct payoff in terms of program objective B.

Rephrased as a structure theory question, the crucial issue is: *What is the structure of graphs that do not have a subgraph with k leaves?*

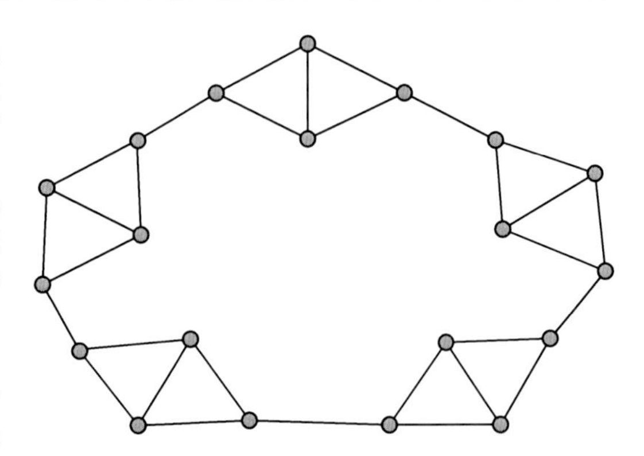

Max Leaf Spanning Tree, Fig. 1 Reduction rules were developed in order to reduce this Kleitman–West graph structure

A graph theory result due to Kleitman and West shows that a graph of minimum degree at least 3, that excludes a k-leaf subgraph, has at most $4(k-3)$ vertices. Figure 1 shows that this is the best possible result for this hypothesis. However, investigating the structure using extremal methods reveals the need for the reduction rule of Fig. 2. About 20 different polynomial-time reduction rules (some much more complex and "global" in structure than the simple local reduction rule depicted) are sufficient to kernelize to a graph of minimum degree 2 having at most $3.5k$ vertices.

In general, an instance of a parameterized problem consists of a pair (x, k) and a "boundary" which is located by holding x fixed and varying k and regarding whether the outcome of the decision problem is *yes* or *no*. Of interest is the boundary when x is reduced. A typical boundary lemma looks like the following.

Lemma 2 *Suppose* (G, k) *is a reduced instance of* MAX LEAF, WITH (G, k) *a yes-instance and* $(G, k+1)$ *a no-instance. Then* $|G| \leq ck$. *(Here* c *is a small constant that becomes clarified during the investigation.)*

A proof of a boundary lemma is by minimum counterexample. A counterexample would be a graph such that (1) (G, k) is reduced, (2) (G, k) is a *yes*-instance of MAX LEAF, (3) $(G, k+1)$ is a *no*-instance, and (4) $|G| > ck$.

The proof of a boundary lemma unfolds gradually. Initially, it is not known what bound will k' = k-1

Max Leaf Spanning Tree, Fig. 2 A reduction rule for the Kleitman–West graph

eventually succeed and it is not known exactly what is meant by *reduced*. In the course of an attempted proof, these details are worked out. As the arguments unfold, structural situations will suggest new reduction rules. Strategic choices involved in a boundary lemma include:

(1) Determining the polarity of the boundary, and setting up the boundary lemma.
(2) Choosing a witness structure.
(3) Setting inductive priorities.
(4) Developing a series of structural claims that describe the situation at the boundary.
(5) Discovering reduction rules that can act in polynomial-time on relevant structural situations at the boundary.
(6) As the structure at the boundary becomes clear, filling in the blank regarding the kernelization bound.

The overall structure of the argument is "by minimum counterexample" according to the priorities established by choice 3, which generally make reference to choice 2. The proof proceeds by a series of small steps consisting of structural claims that lead to a detailed structural picture at the "boundary"– and thereby to the bound on the size of G that is the conclusion of the lemma. The complete proof assembles a series of claims made against the witness tree, various sets of vertices, and inductive priorities and sets up a master inequality leading to a proof by induction, and a $3.5k$ problem kernel.

Objective B: Polynomial-Time Preprocessingand Data-Reduction Routines

The authors have designed a table for tracing each possible *boundary state* for a possible solution. Examples are given that show the surprising power of cascading data-reduction rules on real input distributions and that describe a variety of mathematical phenomena relating to reduction rules. For example, some reduction rules, such as the *Kleitman–West dissolver rule* for MAX LEAF (Fig. 2), have a fixed "boundary size" (in this case 2), whereas crown-type reduction rules do not have a fixed boundary size.

Objective C: Gradients and Solution Transformations for Local Search

A generalization of the usual setup for local search is given, based on the mathematical power of the more complicated gradient in obtaining superior kernelization bounds. Idea 1 is that local search be conducted based on maintaining a "current witness structure" rather than a full solution (spanning tree). Idea 2 is to use the list of inductive priorities to define a "better solution" gradient for the local search.

Objective D: Polynomial-Time Approximation Algorithms

The polynomial-time extremal structure theory leads directly to a constant-factor p-time approximation algorithm for MAX LEAF. First, reduce G using the kernelization rules. The rules are approximation-preserving. Take *any* tree T (not necessarily spanning) in G. If all of the structural claims hold, then (by the boundary lemma arguments) the tree T must have at least n/c leaves for $c = 3.75$. Therefore, lifting T back along the reduction path, we obtain a c-approximation.

If at least one of the structural claims does not hold, then the tree T can be improved against one of the inductive priorities. Notice that each claim is proved by an argument that can be interpreted as a polynomial-time routine that improves T, when the claim is contradicted.

These consequences can be applied to the original T (and its successors) only a polynomial number of times (determined by the list of

Max Leaf Spanning Tree, Table 1 The complexity ecology of parameters

	TW	BW	VC	DS	G	ML
TW	*FPT*	*W*[1]-hard	*FPT*	*FPT*	?	*FPT*
BW	*FPT*	*W*[1]-hard	*FPT*	*FPT*	?	*FPT*
VC	*FPT*	?	*FPT*	*FPT*	?	*FPT*
DS	?	?	*W*[1]-hard	*W*[1]-hard	?	?
G	*W*[1]-hard	*W*[1]-hard	*W*[1]-hard	*W*[1]-hard	*FPT*	?
ML	*FPT*	?	*FPT*	*FPT*	*FPT*	?

inductive priorities) until one arrives at a tree T' for which all of the various structural claims hold. At that point, we must have a c-approximate solution.

Objective E: Structure To Exploitin The Ecology of Complexity

The objective here is to understand how every input-governing problem parameter affects the complexity of every other problem. As a small example, consider Table 1 using the shorthand TW is TREEWIDTH, BW is BANDWIDTH, VC is VERTEX COVER, DS is DOMINATING SET, G is GENUS and ML is MAX LEAF. The entry in the second row and fourth column indicates that there is an *FPT* algorithm to optimally solve the DOMINATING SET problem for a graph G of bandwidth at most k. The entry in the fourth row and second column indicates that it is unknown whether BANDWIDTH can be solved optimally by an *FPT* algorithm when the parameter is a bound on the domination number of the input.

MAX LEAF applies to the last row of the table. For graphs of *max leaf number* bounded by k, the maximum size of an independent set can be computed in time $O^*(2.972^k)$ based on a reduction to a kernel of size at most $7k$. There is a practical payoff for using the output of one problem as the input to another.

Applications

The MAX LEAF SPANNING TREE problem has motivations in computer graphics for creating triangle strip representations for fast interactive rendering [5]. Other applications are found in the area of traffic grooming and network design, such as the design of optical networks and the utilization of wavelengths in order to minimize network cost, either in terms of the line-terminating equipment deployed or in terms of electronic switching [6]. The minimum-energy problem in wireless networks consists of finding a transmission radius vector for all stations in such a way that the total transmission power of the whole network is the least possible. A restricted version of this problem is equivalent to the MAX LEAF SPANNING TREE problem [7]. Finding spanning trees with many leaves is equivalent to finding small connected dominating sets and is also called the MINIMUM CONNECTED DOMINATING problem [13].

Open Problems

Branching Strategies

While extremal structure is in some sense *the right way* to design an FPT algorithm, this is not the only way. In particular, the recipe is silent on what to do with the kernel. An open problem is to find general strategies for employing "parameter-appropriate structure theory" in branching strategies for sophisticated problem kernel analysis.

Turing Kernelizability

The polynomial-time transformation of (x, k) to the simpler reduced instance (x', k') is a many:1 transformation. One can generalize the notion of many:1 reduction to Turing reduction. How should the quest for p-time extremal theory unfold under this "more generous" FPT?

Algorithmic Forms of The Boundary Lemma Approach

The hypothesis of the boundary lemma that (G, k) is a *yes*-instance implies that there exists a witness structure to this fact. There is no assumption that one has algorithmic access to this structure, and when reduction rules are discovered, these have to be transformations that can be applied to (G, k) and a structure that can be discovered in (G, k) in polynomial time. In other words, reduction rules cannot be *defined* with respect to the witness structure. Is it possible to describe more general approaches to kernelization where the witness structure used in the proof of the boundary lemma is polynomial-time computable, and this structure provides a conditional context for some reduction rules? How would this change the extremal method recipe?

Problem Annotation

One might consider a generalized MAX LEAF problem where vertices and edges have various annotations as to whether they *must* be leaves (or internal vertices) in a solution, etc. Such a generalized form of the problem would generally be expected to be "more difficult" than the vanilla form of the problem. However, several of the "best known" FPT algorithms for various problems, are based on these generalized, annotated forms of the problems. Examples include PLANAR DOMINATING SET and FEEDBACK VERTEX SET [4]. Should annotation be part of the recipe for the best possible polynomial-time kernelization?

Cross-References

▸ Connected Dominating Set
▸ Data Reduction for Domination in Graphs

Recommended Reading

1. Bonsma P (2006) Spanning trees with many leaves: new extremal results and an improved FPT algorithm, vol 1793. Memorandum Department of Applied Mathematics, University of Twente, Enschede
2. Bonsma P, Brueggemann T, Woeginger G (2003) A faster FPT algorithm for finding spanning trees with many leaves. In: Proceedings of MFCS 2003. Lecture notes in computer science, vol 2747. Springer, Berlin, pp 259–268
3. Downey RG, Fellows MR (1999) Parameterized complexity, Monographs in computer science. Springer, New York
4. Dehne F, Fellows M, Langston M, Rosamond F, Stevens K (2005) An $O(2^{O(k)}n^3)$ FPT algorithm for the undirected feedback vertex set problem. In: Proceedings COCOON 2005. Lecture notes in computer science, vol 3595. Springer, Berlin, pp 859–869
5. Diaz-Gutierrez P, Bhushan A, Gopi M, Pajarola R (2006) Single-strips for fast interactive rendering. J Vis Comput 22(6):372–386
6. Dutta R, Savage C (2005) A Note on the complexity of converter placement supporting broadcast in WDM optical networks. In: Proceedings of the international conference on telecommunication systems-modeling and analysis, Dallas, Nov 2005, pp 23–31. ISBN: 0-9716253-3-6. American Telecommunication Systems Management Association, Nashville
7. Egecioglu O, Gonzalez T (2001) Minimum-energy broadcast in simple graphs with limited node power. In: Proceedings of the IASTED international conference on parallel and distributed computing and systems (PDCS), Anaheim, Aug 2001, pp 334–338
8. Estivill-Castro V, Fellows MR, Langston MA, Rosamond FA (2005) FPT is P-time extremal structure I. In: Algorithms and complexity in Durham 2005. Texts in algorithmics, vol 4. Kings College Publications, London, pp 1–41
9. Fellows M, Langston M (1992) On well-partial-order theory and its applications to combinatorial problems of VLSI design. SIAM J Discret Math 5:117–126
10. Fellows M (2003) Blow-ups, win/win's and crown rules: some new directions in FPT. In: Proceedings of the 29th workshop on graph theoretic concepts in computer science (WG 2003). Lecture notes in computer science, vol 2880. Springer, Berlin, pp 1–12
11. Fellows M, McCartin C, Rosamond F, Stege U (2000) Coordinatized kernels and catalytic reductions: an improved FPT algorithm for max leaf spanning tree and other problems. In: Proceedings of the 20th conference on foundations of software technology and theoretical computer science (FST-TCS 2000). Lecture notes in theoretical computer science, vol 1974. Springer, Berlin, pp 240–251
12. Kleitman DJ, West DB (1991) Spanning trees with many leaves. SIAM J Discret Math 4:99–106
13. Kouider M, Vestergaard PD (2006) Generalized connected domination in graphs. Discret Math Theory Comput Sci (DMTCS) 8:57–64
14. Lu H-I, Ravi R (1998) Approximating maximum leaf spanning trees in almost linear time. J Algorithm 29:132–141

15. Niedermeier R (2006) Invitation to fixed parameter algorithms. Lecture series in mathematics and its applications. Oxford University Press, Oxford
16. Prieto-Rodriguez E (2005) Systematic kernelization in FPT algorithm design. Dissertation, School of Electrical Engineering and Computer Science, University of Newcastle
17. Solis-Oba R (1998) 2-approximation algorithm for finding a spanning tree with the maximum number of leaves. In: Proceedings of the 6th annual European symposium on algorithms (ESA'98). Lecture notes in computer science, vol 1461. Springer, Berlin, pp 441–452

Maximizing the Minimum Machine Load

Csanad Imreh
Institute of Informatics, University of Szeged, Szeged, Hungary

Keywords

Approximation algorithms; Multiprocessor systems; Scheduling

Years and Authors of Summarized Original Work

1982; Deuermeyer, Friesen, Langston
1997; Woeginger
1998; Azar, Epstein

Problem Definition

In a scheduling problem we have to find an optimal schedule of jobs. Here we consider the parallel machines case, where m machines are given, and we can use them to schedule the jobs. In the most fundamental model, each job has a known processing time, and to schedule the job we have to assign it to a machine, and we have to give its starting time and a completion time, where the difference between the completion time and the starting time is the processing time. No machine may simultaneously run two jobs. If no further assumptions are given then the machines can schedule the jobs assigned to them without an idle time and the total time required to schedule the jobs on a machine is the sum of the processing times of the jobs assigned to it. We call this value the load of the machine.

Concerning the machine environment three different models are considered. If the processing time of a job is the same for each machine, then we call the machines identical machines. If each machine has a speed s_i, the jobs have a processing weight p_j and the processing time of job j on the i-th machine is p_j/s_i, then we call the machines related machines. If the processing time of job j is given by an arbitrary positive vector $P_j = (p_j(1), \ldots, p_j(m))$, where the processing time of the job on the i-th machine is $p_j(i)$, then we call the machines unrelated machines. Here we consider the identical machine case unless it is stated otherwise.

Many objective functions are considered for scheduling problems. Here we consider only such models where the goal is the maximization of the minimal load which problem was proposed in [4]. We note that the most usual objective function is minimizing the maximal load which is called makespan. This objective is the dual of the makespan in some sense but both objective functions require to balance the loads of the machines.

A straightforward reduction from the well-known NP-hard partition problem shows that the investigated problem is NP-hard. Therefore one main research question is to develop polynomial time approximation algorithms which cannot ensure the optimal solution but always give a solution which is not much worse than the optimal one. These approximation algorithms are usually evaluated by the approximation ratio. In case of maximization problems an algorithm is called c-approximation if the objective value given by the algorithm is at least c-times as large than the optimal objective value. If we have a polynomial time $1-\varepsilon$-approximation algorithm for every $\varepsilon > 0$, then this class of algorithms is called polynomial approximation scheme (PTAS in short). If these algorithms are also polynomial in $1/\varepsilon$, then we call them fully polynomial approximation scheme (FPTAS in short).

Key Results

Approximation Algorithms

Since we plan to balance the load on the machines, a straightforward idea to find a solution is to use some greedy method to schedule the jobs. If we schedule the jobs one by one, then a greedy algorithm assigns the job to the machine with the smallest load. Unfortunately if we schedule the jobs in arbitrary order, then this algorithm does not have constant approximation ratio. In the worst inputs the problem is caused by the large jobs which are scheduled at the end of the input. Therefore, the next idea is to order the jobs by decreasing size and schedule them one by one assigning the actual job to the machine with the smallest load. This algorithm is called LPT (longest processing time) and analyzed in [4] and [3]. The first analysis was presented in [4], where the authors proved that the algorithm is 3/4-approximation and also proved that no greater approximation ratio can be given for the algorithm as the number of machines tends to ∞. In [3] a more sophisticated analysis is given, the authors proved that the exact competitive ratio is $(3m - 1)/(4m - 2)$. Later in [7] a PTAS was presented for the problem. The time complexity of the algorithm is $O(c_\varepsilon \cdot n \cdot m)$ where c_ε is a constant which depends exponentially on ε. Thus the presented class of algorithms is not an FPTAS. But it worths noting that we cannot expect an FPTAS for the problem. It belongs to the class of strongly NP-complete problems; thus an FPTAS would yield P = NP. The case of unrelated machines is much more difficult. In [2] it is proved that no better than 1/2 approximation algorithm exists unless P = NP; therefore we cannot expect a PTAS in this case.

Online and Semi-online Problem

In many applications we do not have a priori knowledge about the input and the algorithms must make their decision online based only on the past information. These algorithms are called online algorithms. In scheduling problems this means that the jobs arrive one by one and the algorithm has to schedule the arriving job without any knowledge about the further ones. This area is very similar to the area of approximation algorithms, again we cannot expect algorithms which surely find optimal solutions. In approximation algorithms the problem is that we do not have exponential time for computation; in online algorithms it is the lack of information. The algorithms are also analyzed by a similar method, but in the area of online algorithms it is called competitive ratio. For maximization problems an online algorithm is called c-competitive if the objective value given by the algorithm is at least c-times as large than the optimal objective value.

The online version of scheduling maximizing the minimal load is studied in [1]. The most straightforward online algorithm is the above-mentioned List algorithm which assigns the actual job to the machine with the smallest load. It is $1/m$-competitive and it is easy to see (considering m jobs of size 1 and if they are assigned to different machines $m-1$ further jobs of size m) that no better deterministic online algorithm can be given. In [1] randomized algorithms are studied, the authors presented an $1/O(\sqrt{m}\log m)$-competitive randomized algorithm and proved that no randomized algorithm can have better competitive ratio than $1/\Omega(\sqrt{m})$. The case of related machines is also studied and it is proved that no algorithm exists which has a competitive ratio depending on the number of machines.

In semi-online problems usually some extra information is given to the algorithm. The first such model is also studied in [1]. The authors studied the version where the optimal value is known in advance and they presented an $m/(2m - 1)$-competitive algorithm and they proved that if $m = 2$ or $m = 3$ then no semi-online algorithm in this model with better competitive ratio exists. In case of related machines and known optimal value, an $1/m$-competitive algorithm is given. Several further semi-online version is studied in the literature. In [5] the semi-online version where the maximal job size is known in advance, in [6] the version where total processing time of all jobs and the largest processing time is known in advance is studied.

Applications

The first paper [4] mentions an application as the motivation of the model. It is stated that the problem was motivated by modeling the sequencing of maintenance actions for modular gas turbine aircraft engines. If a fleet of M identical machines (engines) are given and they must be kept operational for as long as possible, then we obtain the objective to maximize the minimal load.

Cross-References

▸ List Scheduling

Recommended Reading

1. Azar Y, Epstein L (1998) On-line machine covering. J Sched 1(2):67–77
2. Bezáková I, Dani V (2005) Nobody left behind: fair allocation of indivisible goods. ACM SIGecom Exch 5.3
3. Csirik J, Kellerer H, Woeginger GJ (1992) The exact LPT-bound for maximizing the minimum completion time. Oper Res Lett 11:281–287
4. Deuermeyer BL, Friesen DK, Langston MA (1982) Scheduling to maximize the minimum processor finish time in a multiprocessor system. SIAM J Discret Methods 3:190–196
5. He Y, Zhang GC (1999) Semi on-line scheduling on two identical machines. Computing 62:179–187
6. Tan Z, Wu Y (2007) Optimal semi-online algorithms for machine covering. Theor Comput Sci 372: 69–80
7. Woeginger GJ (1997) A polynomial time approximation scheme for maximizing the minimum machine completion time. Oper Res Lett 20(4):149–154

Maximum Agreement Subtree (of 2 Binary Trees)

Ramesh Hariharan
Strand Life Sciences, Bangalore, India

Keywords

Isomorphism; Tree agreement

Years and Authors of Summarized Original Work

1996; Cole, Hariharan
2002; Farach, Hariharan, Przytycka, Thorup

Problem Definition

Consider two rooted trees T_1 and T_2 with n leaves each. The internal nodes of each tree have at least two children each. The leaves in each tree are labeled with the same set of labels, and further, no label occurs more than once in a particular tree. An *agreement subtree* of T_1 and T_2 is defined as follows. Let L_1 be a subset of the leaves of T_1 and let L_2 be the subset of those leaves of T_2 which have the same labels as leaves in L_1. The subtree of T_1 *induced* by L_1 is an agreement subtree of T_1 and T_2 if and only if it is *isomorphic* to the subtree of T_2 induced by L_2. The maximum agreement subtree problem (henceforth called *MAST*) asks for the largest agreement subtree of T_1 and T_2.

The terms *induced subtree* and *isomorphism* used above need to be defined. Intuitively, the subtree of T induced by a subset L of the leaves of T is the topological subtree of T restricted to the leaves in L, with branching information relevant to L preserved. More formally, for any two leaves a, b of a tree T, let $\text{lca}_T(a,b)$ denote their lowest common ancestor in T. If $a = b$, $\text{lca}_T(a,b) = a$. The *subtree* U of T *induced* by a subset L of the leaves is the tree with leaf set L and interior node set $\{\text{lca}_T(a,b) \mid a,b \in L\}$ inheriting the ancestor relation from T, that is, for all $a, b \in L$, $\text{lca}_U(a,b) = \text{lca}_T(a,b)$.

Intuitively, two trees are isomorphic if the children of each node in one of the trees can be reordered so that the leaf labels in each tree occur in the same order and the shapes of the two trees become identical. Formally, two trees U_1 and U_2 with the same leaf labels are said to be *isomorphic* if there is a 1–1 mapping μ between their nodes, mapping leaves to leaves with the same labels, and such that for any two different leaves a, b of U_1, $\mu(\text{lca}_{U1}(a,b)) = \text{lca}_{U2}(\mu(a), \mu(b))$.

Key Results

Previous Work

Finden and Gordon [8] gave a heuristic algorithm for the *MAST* problem on binary trees which had an $O(n^5)$ running time and did not guarantee an optimal solution. Kubicka, Kubicki, and McMorris [12] gave an $O(n^{(5+\epsilon)\log n})$ algorithm for the same problem. The first polynomial time algorithm for this problem was given by Steel and Warnow [14]; it had a running time of $O(n^2)$. Steel and Warnow also considered the case of nonbinary and unrooted trees. Their algorithm takes $O(n^2)$ time for fixed-degree rooted and unrooted trees and $O(n^{4.5}\log n)$ for arbitrary-degree rooted and unrooted trees. They also give a linear reduction from the rooted to the unrooted case. Farach and Thorup gave an $O\left(nc^{\sqrt{\log n}}\right)$ time algorithm for the *MAST* problem on binary trees; here c is a constant greater than 1. For arbitrary-degree trees, their algorithm takes $O\left(n^2 c^{\sqrt{\log n}}\right)$ time for the unrooted case [6] and $O(n^{1.5}\log n)$ time for the rooted case [7]. Farach, Przytycka, and Thorup [4] obtained an $O(n\log^3 n)$ algorithm for the *MAST* problem on binary trees. Kao [11] obtained an algorithm for the same problem which takes $O(n\log^2 n)$ time. This algorithm takes $O(\min\{nd^2\log d\log^2 n, nd^{\frac{3}{2}}\log^3 n\})$ for degree d trees.

The *MAST* problem for more than two trees has also been studied. Amir and Keselman [1] showed that the problem is *NP*-hard for even 3 unbounded degree trees. However, polynomial bounds are known [1, 5] for three or more bounded degree trees.

Our Contribution

An $O(n\log n)$ algorithm for the *MAST* problem for two binary trees is presented here. This algorithm is obtained by improving upon the $O(n\log^3 n)$ algorithm from [4] (in fact, the final journal version [3] combines both papers). The $O(n\log^3 n)$ algorithm of [4] can be viewed as taking the following approach (although the authors do not describe it this way). It identifies two special cases and then solves the general case by interpolating between these cases.

Special Case 1

The internal nodes in both trees form a path. The *MAST* problem reduces to essentially a size n Longest Increasing Subsequence problem in this case. As is well known, this can be solved in $O(n\log n)$ time.

Special Case 2

Both trees T_1 and T_2 are complete binary trees. For each node v in T_2, only certain nodes u in T_1 can be usefully mapped to v, in the sense that the subtree of T_1 rooted at u and the subtree of T_2 rooted at v have a nonempty agreement subtree. There are $O(n\log^2 n)$ such pairs (u, v). This can be seen as follows. Note that for (u, v) to be such a pair, the subtree of T_1 rooted at u and the subtree of T_2 rooted at v must have a leaf label in common. For each label, there are only $O(\log^2 n)$ such pairs obtained by pairing each ancestor of the leaf with this label in T_1 with each ancestor of the leaf with this label in T_2. The total number of interesting pairs is thus $O(n\log^2 n)$. For each pair, computing the *MAST* takes $O(1)$ time, as it is simply a question of deciding the best way of pairing their children.

The interpolation process takes a centroid decomposition of the two trees and compares pairs of centroid paths, rather than individual nodes as in the complete tree case. The comparison of a pair of centroid paths requires finding matchings with special properties in appropriately defined bipartite graphs, a nontrivial generalization of the Longest Increasing Subsequence problem. This process creates $O(n\log^2 n)$ interesting (u, v) pairs, each of which takes $O(\log n)$ time to process.

This work provides two improvements, each of which gains a $\log n$ factor.

Improvement 1

The complete tree special case is improved to $O(n\log n)$ time as follows. A pair of nodes $(u, \nu), u \in T_1, \nu \in T_2$, is said to be *interesting* if there is an agreement subtree mapping u to ν. As is shown below, for complete trees, the total number of interesting pairs (u, v) is just $O(n \log n)$. Consider a node v in T_2. Let L_2 be the set of leaves which are descendants of v. Let L_1 be the set of leaves in T_1 which have the same labels as the leaves in L_2. The only nodes that may be mapped to v are the nodes u in the subtree of T_1 induced by L_1. The number of such nodes u is O(size of the subtree of T_2 rooted at v). The total number of interesting pairs is thus the sum of the sizes of all subtrees of T_2, which is $O(n \log n)$.

This reduces the number of interesting pairs (u, v) to $O(n \log n)$. Again, processing a pair takes $O(1)$ time (this is less obvious, for identifying the descendants of u which root the subtrees with which the two subtrees of v can be matched is nontrivial). Constructing the above induced subtree itself can be done in $O(|L_1|)$ time, as will be detailed later. The basic tool here is to preprocess trees T_1 and T_2 in $O(n)$ time so that the least common ancestor queries can be answered in $O(1)$ time.

Improvement 2

As in [4], when the trees are not complete binary trees, the algorithm takes centroid paths and matches pairs of centroid paths. The $O(\log n)$ cost that the algorithm in [4] incurs in processing each such interesting pair of paths arises when there are large (polynomial in n size) instances of the generalized Longest Increasing Subsequence problem. At first sight, it is not clear that large instances of these problems can be created for sufficiently many of the interesting pairs; unfortunately, this turns out to be the case. However, these problem instances still have some useful structure. By using (static) weighted trees, pairs of interesting vertices are processed in $O(1)$ time per pair, on the average, as is shown by an appropriately parametrized analysis.

The Multiple Degree Case

The techniques can be generalized to higher degree bounds $d > 2$ by combining it with techniques from [6, Sect. 2] for unbounded degrees. This appears to yield an algorithm with running time $O(\min\{n\sqrt{d}\log^2 n, nd \log n \log d\})$. The conjecture, however, is that there is an algorithm with running time $O(n\sqrt{d}\log n)$.

Applications

Motivation

The *MAST* problem arises naturally in biology and linguistics as a measure of consistency between two evolutionary trees over species and languages, respectively. An evolutionary tree for a set of *taxa*, either species or languages, is a rooted tree whose leaves represent the taxa and whose internal nodes represent ancestor information. It is often difficult to determine the true phylogeny for a set of taxa, and one way to gain confidence in a particular tree is to have different lines of evidence supporting that tree. In the biological taxa case, one may construct trees from different parts of the DNA of the species. These are known as *gene trees*. For many reasons, these trees need not entirely agree, and so one is left with the task of finding a consensus of the various gene trees. The maximum agreement subtree is one method of arriving at such a consensus. Notice that a gene is usually a binary tree, since DNA replicates by a binary branching process. Therefore, the case of binary trees is of great interest.

Another application arises in automated translation between two languages (Grishman and Yangarber, NYU, Private Communication). The two trees are the parse trees for the same meaning sentences in the two languages. A complication that arises in this application (due in part to imperfect dictionaries) is that words need not be uniquely matched, i.e., a word at the leaf of one tree could match a number (usually small) of words at the leaves of the other tree. The aim is to find a maximum agreement subtree; this is done with the goal of improving context-using dictionaries for automated translation. So long as each word in one tree has only a constant number of matches in the other tree (possibly with differing weights), the algorithm given here can be used, and its performance remains $O(n \log n)$. More generally, if there are m word matches in all, the performance becomes $O((m + n) \log n)$. Note, however, that if there are two collections of equal-meaning words in the two trees of sizes k_1 and k_2 respectively, they induce $k_1 k_2$ matches.

Cross-References

▶ Maximum Agreement Subtree (of 3 or More Trees)
▶ Maximum Agreement Supertree

Recommended Reading

1. Amir A, Keselman D (1997) Maximum agreement subtree in a set of evolutionary trees. SIAM J Comput 26(6):1656–1669
2. Cole R, Hariharan R (1996) An $O(n \log n)$ algorithm for the maximum agreement subtree problem for binary trees. In: Proceedings of 7th ACM-SIAM SODA, Atlanta, pp 323–332
3. Cole R, Farach-Colton M, Hariharan R, Przytycka T, Thorup M (2000) An $O(n \log n)$ algorithm for the maximum agreement subtree problem for binary trees. SIAM J Comput 30(5):1385–1404
4. Farach M, Przytycka T, Thorup M (1995) The maximum agreement subtree problem for binary trees. In: Proceedings of 2nd ESA
5. Farach M, Przytycka T, Thorup M (1995) Agreement of many bounded degree evolutionary trees. Inf Process Lett 55(6):297–301
6. Farach M, Thorup M (1995) Fast comparison of evolutionary trees. Inf Comput 123(1):29–37
7. Farach M, Thorup M (1997) Sparse dynamic programming for evolutionary-tree comparison. SIAM J Comput 26(1):210–230
8. Finden CR, Gordon AD (1985) Obtaining common pruned trees. J Classific 2:255–276
9. Fredman ML (1975) Two applications of a probabilistic search technique: sorting $X + Y$ and building balanced search trees. In: Proceedings of the 7th ACM STOC, Albuquerque, pp 240–244
10. Harel D, Tarjan RE (1984) Fast algorithms for finding nearest common ancestors. SIAM J Comput 13(2):338–355
11. Kao M-Y (1998) Tree contractions and evolutionary trees. SIAM J Comput 27(6):1592–1616
12. Kubicka E, Kubicki G, McMorris FR (1995) An algorithm to find agreement subtrees. J Classific 12:91–100
13. Mehlhorn K (1977) A best possible bound for the weighted path length of binary search trees. SIAM J Comput 6(2):235–239
14. Steel M, Warnow T (1993) Kaikoura tree theorems: computing the maximum agreement subtree. Inf Process Lett 48:77–82

Maximum Agreement Subtree (of 3 or More Trees)

Teresa M. Przytycka
Computational Biology Branch, NCBI, NIH, Bethesda, MD, USA

Keywords

Tree alignment

Years and Authors of Summarized Original Work

1995; Farach, Przytycka, Thorup

Problem Definition

The maximum agreement subtree problem for k trees (k-MAST) is a generalization of a similar problem for two trees (MAST). Consider a tuple of k rooted leaf-labeled trees $(T_1, T_2 \ldots T_k)$. Let $A = \{a_1, a_2, \ldots a_n\}$ be the set of leaf labels. Any subset $B \subseteq A$ uniquely determines the so-called topological restriction $T|B$ of the three T to B. Namely, $T|B$ is the topological subtree of T spanned by all leaves labeled with elements from B and the lowest common ancestors of all pairs of these leaves. In particular, the ancestor relation in $T|B$ is defined so that it agrees with the ancestor relation in T. A subset B of A such $T^1 \mid B, \ldots, T^k \mid B$ are isomorphic is called an *agreement set.*

Problem 1 (k-MAST) INPUT: A tuple $\vec{T} = (T^1, \ldots, T^k)$ of leaf-labeled trees, with a common set of labels $A = \{a_1, \ldots, a_n\}$, such that for each tree T^i there exists one-to-one mapping between the set of leaves of that tree and the set of labels A.

OUTPUT: k-MAST($\vec{T}$) is equal to the maximum cardinality agreement set of $\vec{T}$.

Key Results

In the general setting, k-MAST problem is NP-complete for $k \geq 3$ [1]. Under the assumption that the degree of at least one of the trees is bounded, Farach et al. proposed an algorithm leading to the following theorem:

Theorem 1 *If the degree of the trees in the tuple* $\vec{T} = (T^1, \ldots, T^k)$ *is bounded by* d*, then the* k-MAST$(\vec{T})$ *can be computed in* $O(kn^3 + n^d)$ *time.*

In what follows, the problem is restricted to finding the cardinality of the maximum agreement set rather than the set itself. The extension of this algorithm to an algorithm that finds the agreement set (and subsequently the agreement subtree) within the same time bounds is relatively straightforward.

Recall that the classical $O(n^2)$ dynamic programming algorithm for MAST of two binary trees [11] processes all pairs of internal nodes of the two trees in a bottom-up fashion. For each pair of such nodes, it computes the MAST value for the subtrees rooted at this pair. There are $O(n^2)$ pairs of nodes, and the MAST value for the subtrees rooted at a given pair of nodes can be computed in constant time from MAST values of previously processed pairs of nodes.

To set the stage for the more general case, let k-MAST$(\vec{v})$ be the solution to the k-MAST problem for the subtrees of $T^1(v_1), \ldots, T^k(v_k)$ where $T^i(v_i)$ is the subtree if T^i rooted at v_i. If, for all i, u_i is a strict ancestor of v_i in T^i, then, $\vec{v}$ is *dominated* by $\vec{u}$ (denoted $\vec{v} \prec \vec{u}$).

A naive extension of the algorithm for two trees to an algorithm for k trees would require computing k-MAST$(\vec{v})$ for all possible tuples $\vec{v}$ by processing these tuples in the order consistent with the domination relation. The basic idea that allows to avoid $\Omega(n^k)$ complexity is to replace the computation of k-MAST$(\vec{v})$ with the computation of a related value, mast$(\vec{v})$, defined to be the size of the maximum agreement set for the subtrees of $(T^1, \ldots, T^k)$ rooted at $(v_1, \ldots v_k)$ subject to the additional restriction that the agreement subtrees themselves are rooted at $v_1, \ldots v_k$, respectively. Clearly

$$k\text{-MAST}(T^1, \ldots, T^k) = \max_{\vec{v}} \text{mast}(\vec{v}).$$

The benefit of computing mast rather than k-MAST follows from the fact that most of mast values are zero and it is possible to identify (very efficiently) $\vec{v}$ with nonzero mast values.

Remark 1 If mast$(\vec{v}) > 0$ then $\vec{v} = (\text{lca}^{T^1}(a,b), \ldots \text{lca}^{T^k}(a,b))$ for some leaf labels a, b where $\text{lca}^{T_i}(a,b)$ is the lowest common ancestor of leaves labeled by a and b in the tree T^i.

A tuple $\vec{v}$ such that $\vec{v} = (\text{lca}^{T^1}(a,b), \ldots \text{lca}^{T^k}(a,b))$ for some $a, b \in A$ is called an *lca-tuple*. By Remark 1 it suffices to compute mast values for the lca-tuples only. Just like in the naive approach, mast$(\vec{v})$ is computed from mast values of other lca-tuples dominated by $\vec{v}$. Another important observation is that only some lca-tuples dominated by $\vec{v}$ are needed to compute mast$(\vec{v})$. To capture this, Farach et al. define the so-called proper domination relation (introduced formally below) and show that the mast value for any lca-tuple $\vec{v}$ can be computed from mast values of lca-tuples properly dominated by $\vec{v}$ and that the proper domination relation has size $O(n^3)$.

Proper Domination Relation

Index the children of each internal node of any tree in an arbitrary way. Given a pair $\vec{v}, \vec{w}$ of lca-tuples such that $\vec{w} \prec \vec{v}$ the corresponding domination relation has associated with it *direction* $\vec{\delta}_{\vec{w} \prec \vec{v}} = (\delta_1, \ldots, \delta_k)$ where w_i descends from the child of v_i indexed with δ_i. Let $v_i(j)$ be the child of v_i with index j. The direction domination is termed *active* is if the subtrees rooted at the $v_1(\delta_1), \ldots, v_k(\delta_k)$ have at least one leaf label in common. Note that each leaf label can witness only one active direction, and consequently each lca-tuple can have at most n active domination directions. Two directions $\vec{\delta}_{\vec{w} \prec \vec{v}}$ and $\vec{\delta}_{\vec{u} \prec \vec{v}}$ are called *compatible* if and only if the direction vectors differ in all coordinates.

Definition 1 $\vec{v}$ properly denominates $\vec{u}$ (denoted $\vec{u} < \vec{v}$) if $\vec{v}$ dominates $\vec{u}$ along an active direction $\vec{\delta}$ and there exists another tuple $\vec{w}$ which is also dominated by $\vec{v}$ along an active direction $\vec{\delta}_{\perp}$ compatible with δ.

From the definition of proper domination and from the fact that each leaf label can witness only one active domination direction, the following observations can be made:

Remark 2 The strong domination relation $<$ on lca-tuples has size $O(n^3)$. Furthermore, the relation can be computed in $O(kn^3)$ time.

Remark 3 For any lca-tuple $\vec{v}$, if $\mathrm{mast}(\vec{v}) > 0$ then either $\vec{v}$ is an lca-tuple composed of leaves with the same label or $\vec{v}$ properly dominates some lca-tuple.

It remains to show how the values $\mathrm{mast}(\vec{v})$ are computed. For each lca-tuple $\vec{v}$, the so-called compatibility graph $G(\vec{v})$ is constructed. The nodes of $G(\vec{v})$ are active directions from $\vec{v}$ and there is an edge between two such nodes if and only if corresponding directions are compatible. The vertices of $G(\vec{v})$ are weighted and the weight of a vertex corresponding to an active direction $\vec{\delta}$ equals the maximum mast value of a lca-tuple dominated by $\vec{v}$ along the this direction. Let $\mathrm{MWC}(G(\vec{v}))$ be the maximum weight clique in $G(\vec{v})$.

The bottom-up algorithm for computing nonzero mast values based on the following recursive dependency whose correctness follows immediately from the corresponding definitions and Remark 3:

Lemma 1 *For any lca-tuple* $\vec{v}$

$$\mathrm{mast}(\vec{v}) = \max \begin{cases} 1 & \text{if all elements of } \vec{v} \text{ are leaves} \\ \mathrm{MWC}(G(\vec{v})) & \text{otherwise} \end{cases} . \quad (1)$$

The final step is to demonstrate that once the lca-tuples and the strong domination relation is precomputed, the computation all nonzero mast values can be preformed in $O(n^d)$ time. This is done by generating all possible cliques for all $G(\vec{v})$. Using the fact that the degree of at least one tree is bounded by d, one can show that all the cliques can be generated in $O\left(\sum_{l \le d} \binom{n}{l}\right) = O(d^3(ne/d)^d)$ time and that there is $O(d(ne/d)^d)$ of them [6].

Applications

The k-MAST problem is motivated by the need to compare evolutionary trees. Recent advances in experimental techniques in molecular biology provide diverse data that can be used to construct evolutionary trees. This diversity of data combined with the diversity of methods used to construct evolutionary trees often leads to the situation when the evolution of the same set of species is explained by different evolutionary trees. The maximum agreement subtree problem has emerged as a measure of the agreement between such trees and as a method to identify subtree which is common for these trees. The problem was first defined by Finden and Gordon in the context of two binary trees [7]. These authors also gave a heuristic algorithm to solve the problem. The $O(n^2)$ dynamic programming algorithm for computing MAST values for two binary trees has been given in [11]. Subsequently, a number of improvements leading to fast algorithms for computing MAST value of two trees under various assumptions about rooting and tree degrees [5, 8, 10] and references therein.

The MAST problem for three or more unbounded degree trees is NP-complete [1]. Amir and Keselman report an $O(kn^{d+1} + n^{2d})$ time algorithm for the agreement of k bounded degree trees. The work described here provides a $O(kn^3 + n^d)$ for the case where the number of trees is k and the degree of at least one tree is bounded by d. For $d = 2$ the complexity of the algorithm is dominated by the first term. An $O(kn^3)$ algorithm for this case was also given by Bryant [4] and $O(n^2 \log^{k-1} n)$ implementation of this algorithm was proposed in [9].

k-MAST problem is a fixed parameter tractable in p, the smallest number of leaf labels such that the removal of the corresponding leaves

M

produces agreement (see [2] and references therein). The approximability of the MAST and related problem has been studied in [3] and references therein.

Cross-References

▶ Maximum Agreement Subtree (of 2 Binary Trees)
▶ Maximum Agreement Supertree
▶ Maximum Compatible Tree

Acknowledgments This work was supported by the Intramural Research Program of the National Institutes of Health, National Library of Medicine.

Recommended Reading

1. Amir A, Keselman D (1997) Maximum agreement subtree in a set of evolutionary trees: metrics and efficient algorithms. SIAM J Comput 26(6):1656–1669
2. Berry V, Nicolas F (2006) Improved parameterized complexity of the maximum agreement subtree and maximum compatible tree problems. IEEE/ACM Trans Comput Biol Bioinform 3(3):289–302
3. Berry V, Guillemot S, Nicolas F, Paul C (2005) On the approximation of computing evolutionary trees. In: COCOON, Kunming, pp 115–125
4. Bryand D (1997) Building trees, hunting for trees, and comparing trees: theory and methods in phylogenetic analysis. Ph.D. thesis, Department of Mathematics, University of Canterbury
5. Cole R, Farach-Colton M, Hariharan R, Przytycka T, Thorup M (2001) An $o(n \log n)$ algorithm for the maximum agreement subtree problem for binary trees. SIAM J Comput 1385–1404
6. Farach M, Przytycka TM, Thorup M (1995) On the agreement of many trees. Inf Process Lett 55(6):297–301
7. Finden CR, Gordon AD (1985) Obtaining common pruned trees. J Classif 2:255–276
8. Kao M-Y, Lam T-W, Sung W-K, Ting H-F (2001) An even faster and more unifying algorithm for comparing trees via unbalanced bipartite matchings. J Algorithms 40(2):212–233
9. Lee C-M, Hung L-J, Chang M-S, Tang C-Y (2004) An improved algorithm for the maximum agreement subtree problem. In: BIBE, Taichung, p 533
10. Przytycka TM (1998) Transforming rooted agreement into unrooted agreement. J Comput Biol 5(2):335–349
11. Steel MA, Warnow T (1993) Kaikoura tree theorems: computing the maximum agreement subtree. Inf Process Lett 48(2):77–82

Maximum Agreement Supertree

Jesper Jansson[1] and Wing-Kin Sung[2]
[1]Laboratory of Mathematical Bioinformatics, Institute for Chemical Research, Kyoto University, Gokasho, Uji, Kyoto, Japan
[2]Department of Computer Science, National University of Singapore, Singapore, Singapore

Keywords

Fixed-parameter tractability; Maximum agreement supertree; NP-hardness; Phylogenetic tree; Rooted triplet

Years and Authors of Summarized Original Work

2005; Jansson, Ng, Sadakane, Sung
2007; Berry, Nicolas
2010; Guillemot, Berry
2011; Hoang, Sung

Problem Definition

A *phylogenetic tree* is a rooted, unordered tree whose leaves are distinctly labeled and whose internal nodes have degree at least two. By distinctly labeled, we mean that no two leaves in the tree have the same label. Let T be a phylogenetic tree with a leaf label set S. For any subset S' of S, *the topological restriction of T to S'* (denoted by $T \mid S'$) is the tree obtained from T by deleting all nodes which are not on any path from the root to a leaf in S' along with their incident edges and then contracting every edge between a node having just one child and its child. See Fig. 1 for an illustration. For any phylogenetic tree T, denote its set of leaf labels by $\Lambda(T)$.

The *maximum agreement supertree problem (MASP)* [12] is defined as follows.

Problem 1 Let $\mathcal{T} = \{T_1, T_2, \ldots, T_k\}$ be an input set of phylogenetic trees, where the sets $\Lambda(T_i)$ may overlap. The maximum agreement supertree problem (MASP) asks for a phylogenetic

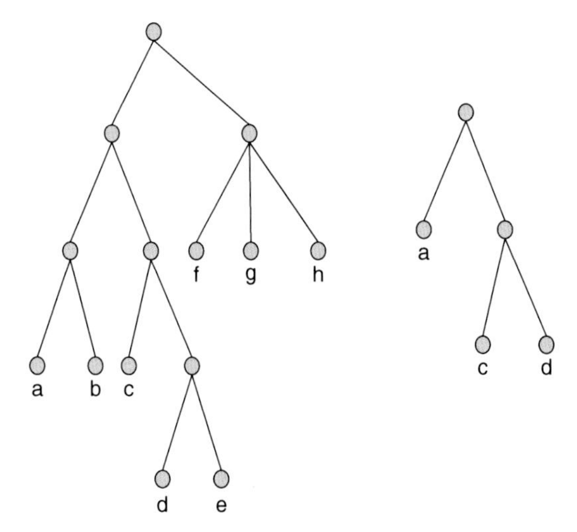

Maximum Agreement Supertree, Fig. 1 Let T be the phylogenetic tree on the *left*. Then $T \mid \{a, c, d\}$ is the phylogenetic tree shown on the *right*

tree Q with leaf label set $\Lambda(Q) \subseteq \bigcup_{T_i \in \mathcal{T}} \Lambda(T_i)$ such that $|\Lambda(Q)|$ is maximized and for each $T_i \in \mathcal{T}$, it holds that $T_i \mid \Lambda(Q)$ is isomorphic to $Q \mid \Lambda(T_i)$.

The following notation is used below: $n = \left|\bigcup_{T_i \in \mathcal{T}} \Lambda(T_i)\right|$, $k = |\mathcal{T}|$, and $D = \max_{T_i \in \mathcal{T}}\{\deg(T_i)\}$, where $\deg(T_i)$ is the degree of T_i (i.e., the maximum number of children of any node belonging to T_i).

A problem related to MASP is the *maximum compatible supertree problem (MCSP)* [2]:

Problem 2 Let $\mathcal{T} = \{T_1, T_2, \ldots, T_k\}$ be an input set of phylogenetic trees, where the sets $\Lambda(T_i)$ may overlap. The maximum compatible supertree problem (MCSP) asks for a phylogenetic tree W with leaf label set $\Lambda(W) \subseteq \bigcup_{T_i \in \mathcal{T}} \Lambda(T_i)$ such that $|\Lambda(W)|$ is maximized and for each $T_i \in \mathcal{T}$, it holds that $T_i \mid \Lambda(W)$ can be obtained from $W \mid \Lambda(T_i)$ by applying a series of edge contractions.

For information about MCSP, refer to [2, 11].

Key Results

The special case of the maximum agreement supertree problem in which $\Lambda(T_1) = \Lambda(T_2) \ldots = \Lambda(T_k)$ has been well studied in the literature and is also known as the *maximum agreement subtree problem (MAST)*. By utilizing known results for MAST, several results can be obtained for various special cases of MASP. Firstly, it is known that MAST can be solved in $O(\sqrt{D}n \log(2n/D))$ time when $k = 2$ (see [13]) or in $O(kn^3 + n^D)$ time when $k \geq 3$ (see [4, 6]), which leads to the following theorems.

Theorem 1 ([12]) *When $k = 2$, MASP can be solved in $O(T_{MAST} + n)$ time, where T_{MAST} is the time required to solve MAST for two $O(n)$-leaf trees. Note that $T_{MAST} = O(\sqrt{D}\, n \log(2n/D))$.*

Theorem 2 ([2]) *For any fixed $k \geq 3$, if every leaf appears in either 1 or k trees, MASP can be solved in $O(T'_{MAST} + kn)$ time, where T'_{MAST} is the time required to solve MAST for $\{T_1|L, T_2|L, \ldots, T_k|L\}$, where $L = \bigcap_{T_i \in \mathcal{T}} \Lambda(T_i)$. Note that $T'_{MAST} = O(k|L|^3 + |L|^D)$.*

On the negative side, the maximum agreement supertree problem is NP-hard in general, as shown by the next theorem. (A *rooted triplet* is a binary phylogenetic tree with exactly three leaves.)

Theorem 3 ([2, 12]) *For any fixed $k \geq 3$, MASP with unbounded D is NP-hard. Furthermore, MASP with unbounded k remains NP-hard even if restricted to rooted triplets, i.e., $D = 2$.*

The inapproximability results for MAST by Hein et al. [9] and Gąsieniec et al. [7] immediately carry over to MASP with unbounded D as follows.

Theorem 4 ([2, 12]) *cannot be approximated within a factor of $2^{\log^{\delta} n}$ in polynomial time for any constant $\delta < 1$, unless $NP \subseteq DTIME[2^{\text{polylog}\, n}]$, even when restricted to $k = 3$. Also, MASP cannot be approximated within a factor of n^{ε} for any constant ε where $0 \leq \varepsilon < \frac{1}{9}$ in polynomial time unless $P = NP$, even for instances containing only trees of height 2.*

Although MASP is difficult to approximate in polynomial time, a simple approximation algorithm based on a technique from [1] achieves an approximation factor that is close to the bounds given in Theorem 4.

Theorem 5 ([12]) *MASP can be approximated within a factor of* $\frac{n}{\log n}$ *in* $O(n^2) \cdot \min\{O(k \cdot (\log\log n)^2), O(k + \log n \cdot \log\log n)\}$ *time. MASP restricted to rooted triplets can be approximated within a factor of* $\frac{n}{\log n}$ *in* $O(k + n^2 \log^2 n)$ *time.*

Fixed-parameter tractable algorithms for solving MASP also exist. In particular, for *binary* phylogenetic trees, Jansson et al. [12] first gave an $O(k(2n^2)^{3k^2})$-time algorithm. Later, Guillemot and Berry [8] improved the time complexity to $O((8n)^k)$. Hoang and Sung [11] further improved the time complexity to $O((6n)^k)$, as summarized in Theorem 6.

Theorem 6 ([11]) *MASP restricted to* $D = 2$ *can be solved in* $O((6n)^k)$ *time.*

For the case where each tree in $\mathcal{T}$ has degree at most D, Hoang and Sung [11] gave the following fixed-parameter polynomial-time solution.

Theorem 7 ([11]) *MASP restricted to phylogenetic trees of degree at most D can be solved in* $O((kD)^{kD+3}(2n)^k)$ *time.*

For unbounded n, k, and D, Guillemot and Berry [8] proposed a solution that is efficient when the input trees are similar.

Theorem 8 ([8]) *MASP can be solved in* $O((2k)^p kn^2)$ *time, where p is an upper bound on the number of leaves that are missing from* $\bigcup_{T_i \in \mathcal{T}} \Lambda(T_i)$ *in a MASP solution.*

Applications

One challenge in phylogenetics is to develop good methods for merging a collection of phylogenetic trees on overlapping sets of taxa into a single supertree so that no (or as little as possible) branching information is lost. Ideally, the resulting supertree can then be used to deduce evolutionary relationships between taxa which do not occur together in any one of the input trees. Supertree methods are useful because most individual studies investigate relatively few taxa [15] and because sample bias leads to certain taxa being studied much more frequently than others [3]. Also, supertree methods can combine trees constructed for different types of data or under different models of evolution. Furthermore, although computationally expensive methods for constructing reliable phylogenetic trees are infeasible for large sets of taxa, they can be applied to obtain highly accurate trees for smaller, overlapping subsets of the taxa which may then be merged using computationally less intense, supertree-based techniques (see, e.g., [5, 10, 14]).

Since the set of trees which is to be combined may in practice contain contradictory branching structure (e.g., if the trees have been constructed from data originating from different genes or if the experimental data contains errors), a supertree method needs to specify how to resolve conflicts. One intuitive idea is to identify and remove a smallest possible subset of the taxa so that the remaining taxa can be combined without conflicts. In this way, one would get an indication of which ancestral relationships can be regarded as resolved and which taxa need to be subjected to further experiments. The above biological problem can be formalized as MASP.

Open Problems

An open problem is to improve the time complexity of the currently fastest algorithms for solving MASP. Moreover, the existing fixed-parameter polynomial-time algorithms for MASP are not practical, so it could be useful to provide heuristics that work well on real data.

Cross-References

- ▸ Maximum Agreement Subtree (of 2 Binary Trees)
- ▸ Maximum Agreement Subtree (of 3 or More Trees)
- ▸ Maximum Compatible Tree

Acknowledgments JJ was funded by the Hakubi Project at Kyoto University and KAKENHI grant number 26330014.

Recommended Reading

1. Akutsu T, Halldórsson MM (2000) On the approximation of largest common subtrees and largest common point sets. Theor Comput Sci 233(1–2): 33–50
2. Berry V, Nicolas F (2007) Maximum agreement and compatible supertrees. J Discret Algorithms 5(3):564–591
3. Bininda-Emonds ORP, Gittleman JL, Steel MA (2002) The (super)tree of life: procedures, problems, and prospects. Annu Rev Ecol Syst 33: 265–289
4. Bryant D (1997) Building trees, hunting for trees, and comparing trees: theory and methods in phylogenetic analysis. PhD thesis, University of Canterbury, Christchurch
5. Chor B, Hendy M, Penny D (2007) Analytic solutions for three taxon ML trees with variable rates across sites. Discret Appl Math 155(6–7): 750–758
6. Farach M, Przytycka T, Thorup M (1995) On the agreement of many trees. Inf Process Lett 55(6):297–301
7. Gąsieniec L, Jansson J, Lingas A, Östlin A (1999) On the complexity of constructing evolutionary trees. J Comb Optim 3(2–3):183–197
8. Guillemot S, Berry V (2010) Fixed-parameter tractability of the maximum agreement supertree problem. IEEE/ACM Trans Comput Biol Bioinform 7(2):342–353
9. Hein J, Jiang T, Wang L, Zhang K (1996) On the complexity of comparing evolutionary trees. Discret Appl Math 71(1–3):153–169
10. Henzinger MR, King V, Warnow T (1999) Constructing a tree from homeomorphic subtrees, with applications to computational evolutionary biology. Algorithmica 24(1):1–13
11. Hoang VT, Sung W-K (2011) Improved algorithms for maximum agreement and compatible supertrees. Algorithmica 59(2):195–214
12. Jansson J, Ng JHK, Sadakane K, Sung W-K (2005) Rooted maximum agreement supertrees. Algorithmica 43(4):293–307
13. Kao M-Y, Lam T-W, Sung W-K, Ting H-F (2001) An even faster and more unifying algorithm for comparing trees via unbalanced bipartite matchings. J Algorithms 40(2):212–233
14. Kearney P (2002) Phylogenetics and the quartet method. In: Current topics in computational molecular biology. MIT, Cambridge pp 111–133
15. Sanderson MJ, Purvis A, Henze C (1998) Phylogenetic supertrees: assembling the trees of life. TRENDS Ecol Evol 13(3):105–109

Maximum Cardinality Stable Matchings

Eric McDermid
Cedar Park, TX, USA

Keywords

Approximation algorithm; Lower bounds; Matching; NP-hard; Preferences; Stability; Ties; UGC-hard; Upper bounds

Years and Authors of Summarized Original Work

2002; Manlove, Irving, Iwama, Miyazaki, Morita
2003; Halld'orsson, Iwama, Miyazaki, Yanagisawa
2004; Halld'orsson, Iwama, Miyazaki, Yanagisawa
2004; Iwama, Miyazaki, Okamoto
2007; Halld'orsson, Iwama, Miyazaki, Yanagisawa
2007; Iwama, Miyazaki, Yamauchi
2007; Yanagisawa
2008; Irving, Manlove
2008; Iwama, Miyazaki, Yamauchi
2009; McDermid
2011; Kir'aly
2013; Kir'aly
2014; Huang and Kavitha
2014; Iwama, Miyazaki, and Yanagisawa
2014; Paluch
2014; Radnai
2015; Dean, Jalasutram

Problem Definition

The input to an instance of the classical *stable marriage problem* consists of a set of n men and n women. Additionally, each person provides a strictly ordered preference list of the opposite set. The goal is to find a complete matching of men to women that is also *stable*, i.e., a matching having the property that there does not exist a man and a woman who prefer each other over their matched assignment. In their seminal work, Gale and Shapley [2] showed that every instance of the stable marriage problem admits at least one stable matching and showed that one can be found in polynomial time (see entry ▶ Stable Marriage). In fact, stable marriage instances can have exponentially many stable matchings [18].

More general settings arise when relaxations of Gale and Shapley's original version are permitted. In the *stable marriage problem with incomplete lists* (SMI), men and women may deem arbitrary members of the opposite set *unacceptable*, prohibiting the pair from being matched together. In the *stable marriage problem with ties* (SMT), preference lists need not be strictly ordered but may instead contain subsets of agents all having the same rank. Instances of SMT and SMI always admit a stable matching, and, crucially, all stable matchings for a fixed instance have the same cardinality. Interestingly, when both ties and incomplete lists are allowed (denoted SMTI, see entry ▶ Stable Marriage with Ties and Incomplete Lists), stable matchings again exist but can differ in cardinality. How can we find one of maximum cardinality?

Key Results

Benchmark Results

Manlove et al. [20] established two key benchmarks for the problem of computing a maximum cardinality stable matching (MAX-SMTI). First, they showed that the problem is NP-hard under the following two simultaneous restrictions. One set of agents, say, the men, all have strictly ordered preference lists, while each woman's preference list is either strictly ordered or is a tie of length two. Second, they showed that MAX-SMTI is approximable within a factor of 2 by arbitrarily breaking the ties and finding any stable matching in the resulting SMI instance.

Since then, researchers have focused on improving the approximability bounds for MAX-SMTI. The severity of the restrictions in Manlove et al.'s hardness results has led researchers to study not only the general version but a number of special cases of MAX-SMTI as well.

Upper Bounds: The General Case

For the general case of MAX-SMTI, Iwama et al. [12] gave a $2-c\left(\frac{\log n}{n}\right)$ approximation algorithm, where c is a positive constant. This algorithm was subsequently improved to yield a performance guarantee of $2 - \frac{c'}{\sqrt{n}}$, where c' is a positive constant which is at most $1/4\sqrt{6}$ [14]. The first approximation algorithm to achieve a constant performance guarantee better than two was given by Iwama et al. [13], establishing a performance ratio of 15/8. Next, Király [16] devised a new approximation algorithm with a bound of 5/3. Finally, McDermid [21] obtained 3/2, which is currently the best known approximation ratio. Later, Paluch [22] and Király [17] also obtained approximation algorithms with the same performance guarantee of 3/2; however, their algorithms have the advantage of running in linear time. Király's has the extra benefit of requiring only "local" preference list information. The following theorem summarizes the best known upper bound for the general case.

Theorem 1 *There is a 3/2-approximation algorithm for MAX-SMTI.*

Upper Bounds: Special Cases

The special case of MAX-SMTI that has received the most attention is that in which ties may only appear in one set only. We let 1S-MAX-SMTI denote this problem. Halldórsson et al. [3] gave a $(2/(1 + T^{-2}))$-approximation algorithm for 1S-MAX-SMTI, where T is the length of the longest tie. This bound was improved to 13/7 for MAX-SMTI instances in which ties are restricted to be of size at most 2 [3]. They later showed that 10/7 is achievable for 1S-MAX-SMTI [4] via a randomized approximation algorithm. Irving and Manlove [11] described a 5/3-approximation algorithm for 1S-MAX-SMTI instances in which lists may have at most one tie that may only appear at the end of the preference list. One of the most important results in this area was that of Király [16], who provided a particularly simple and elegant 3/2-approximation algorithm with an equally transparent analysis for 1S-MAX-SMTI (with no further restrictions on the problem). Since then, further improvements have been obtained for 1S-MAX-SMTI by Iwama, Miyazaki, and Yanagisawa [15], who exploited a linear programming relaxation to obtain a 25/17-approximation. Huang and Kavitha [10] used different techniques to improve upon this, giving a linear-time algorithm with a ratio of 22/15. Radnai [23] tightened their analysis to show that 41/28 is in fact achieved. Finally, Dean and Jalasutram [1] showed that the algorithm given in [15] actually achieves 19/13 through an analysis using a factor-revealing LP. The following theorem summarizes the best known upper bounds for the special cases of MAX-SMTI.

Theorem 2 *There is a 19/13-approximation algorithm for 1S-MAX-SMTI. When all ties have length at most two, there is a (randomized) 10/7-approximation algorithm for MAX-SMTI.*

Lower Bounds

The best lower bounds on approximability are due to Yanagisawa [24] and Iwama et al. [5]. Yanagisawa [24] showed that MAX-SMTI is NP-hard to approximate within 33/29 and UGC-hard to approximate within 4/3. 1S-MAX-SMTI was shown by Iwama et al. [5] to be NP-hard to approximate within 21/19 and UGC-hard to approximate within 5/4. The next theorem summarizes these results.

Theorem 3 *It is NP-hard to approximate MAX-SMTI (1S-MAX-SMTI) within* 33/29 (21/19). *It is UGC-hard to approximate MAX-SMTI (1S-MAX-SMTI) within* 4/3 (5/4).

Applications

Stable marriage research is a fascinating subset of theoretical computer science not only for its intrinsic interest but also for its widespread application to real-world problems. Throughout the world, centralized matching schemes are used in various contexts such as the assignment of students to schools and graduating medical students to hospitals. We direct the reader to [19, *Section 1.3.7*] for a comprehensive overview (see also entry ▸ Hospitals/Residents Problem). Perhaps the most famous of these is the National Resident Matching Program (NRMP) [7] in the United States, which allocates over 35,000 graduating medical students to their first job at a hospital. Similar schemes exist in Canada [8], Scotland [9], and Japan [6]. In one way or another, all of these matching schemes require one or both of the sets involved to produce preference lists ranking the other set. Methods similar to the Gale-Shapley algorithm are then used to create the assignments.

Both economists and computer scientists alike have influenced the design and implementation of such matching schemes. In fact, the 2012 Nobel Prize for Economic Sciences was awarded to Alvin Roth and Lloyd Shapley, in part for their contribution to the widespread deployment of matching algorithms in practical settings. Researchers Irving and Manlove at the School of Computing Science at the University of Glasgow led the design and implementation of algorithms for the *Scottish Foundation Allocation Scheme* that have been used by NHS Education for Scotland to assign graduating medical students to hospital programs [19, *Section 1.3.7*]. This setting has actually yielded true instances of MAX-SMTI, as hospital programs have been allowed to have ties in their preference lists.

Recommended Reading

1. Dean B, Jalasutram R (2015, to appear) Factor revealing LPs and stable matching with ties and incomplete lists. In: Proceedings of MATCH-UP 2015: the 3rd international workshop on matching under preferences, Glasgow
2. Gale D, Shapley LS (1962) College admissions and the stability of marriage. Am Math Mon 69:9–15
3. Halldórsson MM, Iwama K, Miyazaki S, Yanagisawa H (2003) Improved approximation of the stable marriage problem. In: Proceedings of ESA 2003: the 11th annual European symposium on algorithms, Budapest. Lecture notes in computer science, vol 2832. Springer, pp 266–277
4. Halldórsson MM, Iwama K, Miyazaki S, Yanagisawa H (2004) Randomized approximation of the stable marriage problem. Theor Comput Sci 325(3):439–465
5. Halldórsson M, Iwama K, Miyazaki S, Yanagisawa H (2007) Improved approximation of the stable marriage problem. ACM Trans Algorithms 3(3):30-es

M

6. http://www.jrmp.jp (Japan Resident Matching Program website)
7. http://www.nrmp.org (National Resident Matching Program website)
8. http://www.carms.ca (Canadian Resident Matching Service website)
9. http://www.nes.scot.nhs.uk/sfas (Scottish Foundation Allocation Scheme website)
10. Huang C-C, Kavitha T (2014) An improved approximation algorithm for the stable marriage problem with one-sided ties. In: Proceedings of IPCO 2014, the 17th conference on integer programming and combinatorial optimization, Bonn. Lecture notes in computer science, vol 8494. Springer, pp 297–308
11. Irving RW, Manlove D (2008) Approximation algorithms for hard variants of the stable marriage and hospitals/residents problem. J Comb Optim 16(3):279–292
12. Iwama K, Miyazaki S, Okamoto K (2004) A $\left(2 - c\frac{\log n}{n}\right)$-approximation algorithm for the stable marriage problem. In: Proceedings of SWAT 2004: the 9th Scandinavian workshop on algorithm theory, Humlebaek. Lecture notes in computer science, vol 3111. Springer, pp 349–361
13. Iwama K, Miyazaki S, Yamauchi N (2007) A 1.875-approximation algorithm for the stable marriage problem. In: Proceedings of SODA 2007: the eighteenth ACM/SIAM symposium on discrete algorithms, New Orleans, pp 288–297
14. Iwama K, Miyazaki S, Yamauchi N (2008) A $\left(2 - c\frac{1}{\sqrt{n}}\right)$-approximation algorithm for the stable marriage problem. Algorithmica 51(3):342–356
15. Iwama K, Miyazaki S, Yanagisawa H (2014) A 25/17-approximation algorithm for the stable marriage problem with one-sided ties. Algorithmica 68:758–775
16. Király Z (2011) Better and simpler approximation algorithms for the stable marriage problem. Algorithmica 60(1):3–20
17. Király Z (2013) Linear time local approximation algorithm for maximum stable marriage. MDPI Algorithms 6(3):471–484
18. Knuth DE (1976) Mariages stables. Les Presses de L'Université de Montréal, Montréal
19. Manlove D (2013) Algorithmics of matching under preferences. World Scientific, Hackensack
20. Manlove DF, Irving RW, Iwama K, Miyazaki S, Morita Y (2002) Hard variants of stable marriage. Theor Comput Sci 276(1–2):261–279
21. McDermid EJ (2009) A 3/2-approximation algorithm for general stable marriage. In: Proceedings of ICALP 2009: the 36th international colloquium on automata, languages and programming, Rhodes. Lecture notes in computer science, vol 5555. Springer, pp 689–700
22. Paluch KE (2014) Faster and simpler approximation of stable matchings. MDPI Algorithms 7(2): 189–202
23. Radnai A (2014) Approximation algorithms for the stable matching problem. Master's thesis, Eötvös Loránd University
24. Yanagisawa H (2007) Approximation algorithms for stable marriage problems. PhD thesis, School of Informatics, Kyoto University

Maximum Compatible Tree

Vincent Berry
Institut de Biologie Computationnelle,
Montpellier, France

Keywords

Consensus of trees; Maximum compatible tree; Pattern matching on trees; Phylogenetics; Trees

Years and Authors of Summarized Original Work

2001; Ganapathy, Warnow

Problem Definition

This problem is a pattern matching problem on leaf-labeled trees. Each input tree is considered as a branching pattern inducing specific groups of leaves. Given a set of input trees with identical leaf sets, the goal is to find the largest subset of leaves on the branching pattern of which the input trees do not disagree. A *maximum compatible tree* is a tree on such a leaf set and with a branching pattern respecting that of each input tree (see below for a formal definition). The maximum compatible tree problem (MCT) is to find such a tree or, equivalently, its leaf set. The main motivation for this problem is in phylogenetics, to measure the similarity between evolutionary trees or to represent a consensus of a set of trees. The problem was introduced in [10] and [11, under the MRST acronym]. Previous related works concern the well-known maximum agreement subtree problem (MAST).

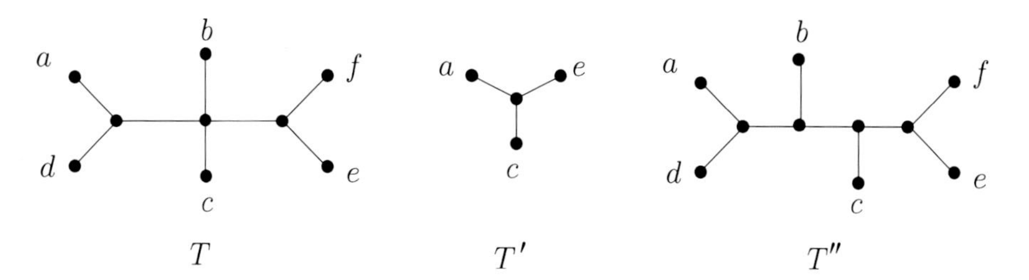

Maximum Compatible Tree, Fig. 1 Three unrooted trees. A tree T, a tree T' such that $T' = T|\{a, c, e\}$, and a tree T'' such that $T'' \trianglerighteq T$

Solving MAST is finding the largest subset of leaves on which all input trees *exactly* agree. The difference between MAST and MCT is that MAST seeks a tree whose branching information is isomorphic to that of a subtree in each of the input trees, while MCT seeks a tree that contains the branching information (i.e., groups) of the corresponding subtree of each input tree. This difference allows the tree obtained for MCT to be more informative, as it can include branching information present in one input tree but not in the others, as long as this information is *compatible* (in the sense of [14]) with the latters. Both problems are equivalent when all input trees are binary. Ganapathy and Warnow [6] were the first to give an algorithm to solve MCT in its general form. Their algorithm relies on a simple dynamic programming approach similar to a work on MAST [13] and has a running time exponential in the number of input trees and in the maximum degree of a node in the input trees. Later, [1] proposed a fixed-parameter algorithm using one parameter only. Approximation results have also been obtained, [3,7] proposing low-cost polynomial-time algorithms that approximate the complement of MCT within a constant factor.

Notations Trees considered here are evolutionary trees (*phylogenies*). Such a tree T has its leaf set $L(T)$ in bijection with a label set and is either rooted, in which case all internal nodes have at least two children each, or unrooted, in which case internal nodes have a degree of at least three. Given a set L of labels and a tree T, the *restriction* of T to L, denoted $T|L$, is the tree obtained in taking the smallest induced subgraph of T that connects leaves with labels in $L \cap L(T)$ and then removing any degree-two (non-root) node to make the tree homeomorphically irreducible. Two trees T, T' are *isomorphic*, denoted $T = T'$, if and only if there is a graph isomorphism $T \mapsto T'$ preserving leaf labels (and the roots if both trees are rooted). A tree T *refines* a tree T', denoted $T \trianglerighteq T'$, whenever T can be transformed into T' by collapsing some of its internal edges (*collapsing* an edge means removing it and merging its extremities). See Fig. 1 for examples of these relations between trees. Note that a tree T properly refining another tree T' agrees with the entire evolutionary history of T' while containing additional information absent from T': at least one high-degree node of T' is replaced in T by several nodes of lesser degree; hence, T contains more information than T' on which species belong together.

Given a collection $\mathcal{T} = \{T_1, T_2, \ldots, T_k\}$ of input trees with identical leaf sets L, a tree T with leaves in L is said to be *compatible with* $\mathcal{T}$ if and only if $\forall T_i \in \mathcal{T}, T \trianglerighteq T_i|L(T)$. If there is a tree T compatible with $\mathcal{T}$ such that $L(T) = L$, then the collection $\mathcal{T}$ is said to be *compatible*. Knowing whether a collection is compatible is a problem for which linear-time algorithms have been known for a long time (e.g., [9]). The MAXIMUM COMPATIBLE TREE problem is a natural optimization version of this problem to deal with incompatible collections of trees.

Problem 1 (MAXIMUM COMPATIBLE TREE – MCT)

INPUT: A collection $\mathcal{T}$ of trees with the same leaf sets.

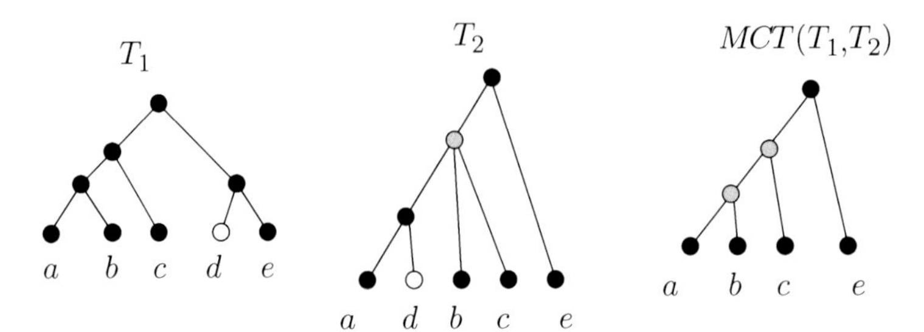

Maximum Compatible Tree, Fig. 2 An incompatible collection of two input trees $\{T_1, T_2\}$ and their maximum compatible tree, $T = MCT(T_1, T_2)$. Removing the leaf d renders the input trees compatible, hence $L(T) = \{a, b, c, e\}$. Here, T strictly refines T_2 restricted to $L(T)$, which is expressed by the fact that a node in T_2 (the *blue* one) has its child subtrees distributed between several connected nodes of T (*blue nodes*). Note also that here $|MCT(T_1, T_2)| > |MAST(T_1, T_2)|$

OUTPUT: A tree compatible with $\mathcal{T}$ having the largest number of leaves. Such a tree is denoted $MCT(\mathcal{T})$.

See Fig. 2 for an example. Note that $\forall \mathcal{T}$, $|MCT(\mathcal{T})| \geq |MAST(\mathcal{T})|$ and that MCT is equivalent to MAST when the input trees are binary. Note also that instances of MCT and MAST can have several optimum solutions.

Key Results

Exact Algorithms

The MCT problem was shown to be NP-hard on 6 trees by [10] and then on 2 trees by [11]. The NP-hardness holds as long as one of the input trees is not of bounded degree. For two bounded-degree trees, Hein et al. [11] mention a polynomial-time algorithm based on *aligning* trees. Ganapathy and Warnow propose an exponential algorithm for solving MCT in the general case [6]. Given two trees T_1, T_2, they show how to compute a binary MCT of any pair of subtrees $(S_1 \in T_1, S_2 \in T_2)$ by dynamic programming. Subtrees whose root is of high degree are handled by considering every possible partition of the roots's children in two sets. This leads the complexity bound to have a term exponential in d, the maximum degree of a node in the input trees. When dealing with k input trees, k-tuples of subtrees are considered, and the simultaneous bipartitions of the roots's children for k subtrees are considered. Hence, the complexity bound is also exponential in k.

Theorem 1 ([6]) *Let L be a set of n leaves. The* MCT *problem for a collection of k rooted trees on L in which each tree has degree at most $d + 1$ can be solved in $O(2^{2kd} n^k)$ time.*

The result easily extends to unrooted trees by considering each of the n leaves in turn as a possible root for all trees of the collection.

Theorem 2 ([6]) *Given a collection of k unrooted trees with degree at most $d + 1$ on an n-leaf set, the* MCT *problem can be solved in $O(2^{2kd} n^{k+1})$.*

Let $\mathcal{T}$ be a collection on a set L of n leaves, [1] considered the following decision problem denoted MCT_p: given $\mathcal{T}$ and $p \in [0, n]$, does $|MCT(\mathcal{T})| \geq n - p$?

Theorem 3 ([1])

1. *MCT_p on rooted trees can be solved in $O(\min\{3^p kn, 2.27^p + kn^3\})$ time.*
2. *MCT_p on unrooted trees can be solved in $O\big((p + 1) \times \min\{3^p kn, 2.27^p + kn^3\}\big)$ time.*

The $3^p kn$ term comes from an algorithm that first identifies in $O(kn)$ time a 3-leaf set S

on which the input trees conflict and then recursively obtains a maximum compatible tree T_1, resp. T_2, T_3 for each of the three collections $\mathcal{T}_1$, resp. $\mathcal{T}_2, \mathcal{T}_3$ obtained by removing from the input trees a leaf in S and lastly returning the $\mathcal{T}_i$ such that $|\text{MCT}(\mathcal{T}_i)|$ is maximum (with $i \in [1,3]$). The $2.27^p + kn^3$ term comes from an algorithm using a reduction of MCT to 3-HITTING SET. Negative results have been obtained by Guillemot and Nicolas concerning the fixed-parameter tractability of MCT with regard to the maximum degree D of the input trees:

Theorem 4 ([8])

1. MCT *is* $W[1]$*-hard with respect to* D.
2. MCT *cannot be solved in* $O(N^{o(2^{D/2})})$ *time unless SNP* $\subseteq$ *SE, where N denotes the input length, i.e.,* $N = O(kn)$.

The MCT problem also admits a variant that deals with *supertrees*, i.e., trees having different (but overlapping) sets of leaves. The resulting problem is $W[2]$-hard with respect to p [2].

Approximation Algorithms

The idea of locating and then eliminating successively all the conflicts between the input trees has also led to approximation algorithms for the *complement* version of the MCT problem, denoted CMCT. Let L be the leaf set of each tree in an input collection $\mathcal{T}$; CMCT aims at selecting the smallest number of leaves $S \subseteq L$ such that the collection $\{T_i|(L - S) : T_i \in \mathcal{T}\}$ is compatible.

Theorem 5 ([7]) *Given a collection* $\mathcal{T}$ *of* k *rooted trees on an* n*-leaf set* L*, there is a 3-approximation algorithm for* CMCT *that runs in* $O(k^2n^2)$ *time.*

The running time of this algorithm was later improved:

Theorem 6 ([3, 5]) *There is an* $O(kn + n^2)$ *time 3-approximation algorithm for* CMCT *on a collection of* k *rooted trees with* n *leaves.*

Note also that working on rooted or unrooted trees does not change the achievable approximation ratio for CMCT [3].

Applications

In bioinformatics, the MCT problem (and similarly MAST) is used to reach different practical goals. The first motivation is to measure the similarity of a set of trees. These trees can represent RNA secondary structures [11, 12] or estimates of a phylogeny inferred from different datasets composed of molecular sequences (e.g., genes) [14]. The gap between the size of a maximum compatible tree and the number of input leaves indicates the degree of disimilarity of the input trees. Concerning the phylogenetic applications, quite often some edges of the trees inferred from the datasets have been collapsed due to insufficient statistical support, resulting in some higher-degree nodes in the trees considered by MCT. Each such node does not indicate a multi-speciation event but rather the uncertainty with respect to the branching pattern to be chosen for its child subtrees. In such a situation, the MCT problem is to be preferred to MAST, as it correctly handles high-degree nodes, enabling them to be resolved according to branching information present in other input trees. As a result, more leaves are conserved in the output tree; hence, a larger degree of similarity is detected between the input trees. Note also that a low similarity value between the input trees can be due to horizontal gene transfers. When these events are not too numerous, identifying species subject to such effects is done by first suspecting leaves discarded from a maximum compatible tree.

The shape of a maximum compatible tree, i.e., not just its size, also has an application in systematic biology to obtain a consensus of a set of phylogenies that are optimal for some tree-building criterion. For instance, the maximum parsimony and maximum likelihood criteria can provide several dozens (sometimes hundreds) of optimal or near-optimal trees. In practice, these trees are first grouped into islands of neighboring trees, and a consensus tree is obtained for each island by resorting to a classical consensus tree method, e.g., the majority-rule or strict consensus. The trees representing the islands form a collection of which a consensus is then sought. However, consensus methods keeping all input leaves tend to create poorly resolved trees. An alternative approach lies in proposing a representative tree that contains a largest possible subset of leaves on the position of which the trees of the collection agree. Again, MCT is more suited than MAST as the input trees can contain some high-degree nodes, with the same meaning as discussed above.

Open Problems

A direction for future work would be to examine the variant of MCT where some leaves are imposed in the output tree. This question arises when a biologist wants to ensure that the species central to his study are contained in the output tree. For MAST on two trees, this constrained variant of the problem was shown in a natural way to be of the same complexity as the regular version [4]. For MCT however, such a constraint can lead to several optimization problems that need to be sorted out. Another important work to be done is a set of experiments to measure the range of parameters for which the algorithms proposed to solve or approximate MCT are useful.

URLs to Code and Datasets

A Perl program can be asked to the author of this entry.

Cross-References

▶ Maximum Agreement Subtree (of 2 Binary Trees)
▶ Maximum Agreement Subtree (of 3 or More Trees)

Recommended Reading

1. Berry V, Nicolas F (2006) Improved parametrized complexity of the maximum agreement subtree and maximum compatible tree problems. IEEE/ACM Trans Comput Biol Bioinformatics 3(3):289–302
2. Berry V, Nicolas F (2007) Maximum agreement and compatible supertrees. J Discret Algorithms 5(3):564–591
3. Berry V, Guillemot S, Nicolas F, Paul C (2005) On the approximation of computing evolutionary trees. In: Wang L (ed) Proceedings of the 11th annual international conference on computing and combinatorics (COCOON'05), Shanghai. LNCS, vol 3595. Springer, pp 115–125
4. Berry V, Peng ZS, Ting HF (2008) From constrained to unconstrained maximum agreement subtree in linear time. Algorithmica 50(3):369–385
5. Berry V, Guillemot S, Nicolas F, Paul C (2009) Linear time 3-approximation for the mast problem. ACM Trans. Algorithms 5(2):23:1–23:18
6. Ganapathy G, Warnow TJ (2001) Finding a maximum compatible tree for a bounded number of trees with bounded degree is solvable in polynomial time. In: Gascuel O, Moret BME (eds) Proceedings of the 1st international workshop on algorithms in bioinformatics (WABI'01), Aarhus, pp 156–163
7. Ganapathy G, Warnow TJ (2002) Approximating the complement of the maximum compatible subset of leaves of k trees. In: Proceedings of the 5th international workshop on approximation algorithms for combinatorial optimization (APPROX'02), Rome, pp 122–134
8. Guillemot S, Nicolas F (2006) Solving the maximum agreement subtree and the maximum compatible tree problems on many bounded degree trees. In: Lewenshtein M, Valiente G (eds) Proceedings of the 17th combinatorial pattern matching symposium (CPM'06), Barcelona. LNCS, vol 4009. Springer, pp 165–176
9. Gusfield D (1991) Efficient algorithms for inferring evolutionary trees. Networks 21:19–28
10. Hamel AM, Steel MA (1996) Finding a maximum compatible tree is NP-hard for sequences and trees. Appl Math Lett 9(2):55–59
11. Hein J, Jiang T, Wang L, Zhang K (1996) On the complexity of comparing evolutionary trees. Discr Appl Math 71(1–3):153–169
12. Jiang T, Wang L, Zhang K (1995) Alignment of trees – an alternative to tree edit. Theor Comput Sci 143(1):137–148
13. Steel MA, Warnow TJ (1993) Kaikoura tree theorems: computing the maximum agreement subtree. Inf Process Lett 48(2):77–82
14. Swofford D, Olsen G, Wadell P, Hillis D (1996) Phylogenetic inference. In: Hillis D, Moritz D, Mable B (eds) Molecular systematics, 2nd edn. Sinauer Associates, Sunderland, pp 407–514

Maximum Lifetime Coverage

Weili Wu[1,2,3] and Ling Ding[4]
[1]College of Computer Science and Technology, Taiyuan University of Technology, Taiyuan, Shanxi Province, China
[2]Department of Computer Science, California State University, Los Angeles, CA, USA
[3]Department of Computer Science, The University of Texas at Dallas, Richardson, TX, USA
[4]Institute of Technology, University of Washington Tacoma, Tacoma, WA, USA

Keywords

Algorithm; Approximation; Constant; Coverage; Lifetime; Wireless sensor network

Years and Authors of Summarized Original Work

2005; Cardei, Mihaela; Thai, My T.; Li, Yingshu; Wu, Weili
2012; Ding, Ling; Wu, Weili; Willson, James; Wu, Lidong; Lu, Zaixin; Lee, Wonjun

Problem Definition

Energy resources are very limited in wireless sensor networks since the wireless devices are small and battery powered. There are two ways to deploy wireless sensor networks. One is deterministic [1] deployment, and the other one is stochastic or random deployment [2]. In deterministic deployment, the goal is to minimize the number of sensors. In the latter one, the goal is to improve coverage ratio. Normally, in random or stochastic deployment, one target is covered by several sensors. It is unnecessary and a waste of energy to activate all sensors around the target to monitor it. We can prolong the coverage duration through making sleep/activate schedules in wireless sensor networks when we don't have abundant energy resources. In Fig. 1, t_1, t_2, and t_3 are three targets. s_1, s_2, and s_3 are three sensors.

Maximum Lifetime Coverage, Fig. 1 Network model and comparison of disjoint and non-disjoint coverage [3]

s_1 can cover t_1 and t_2. s_2 can cover t_1 and t_3. s_3 can cover t_2 and t_3. Assume each sensor can be active for 1 h. s_1 and s_2 can collaborate to cover all targets, and the coverage duration will be 1 h. After 1 h, there are no enough sensors to cover all targets. The coverage lifetime in this case is 1 h, and s_3 is sleep within this 1 h. There can be another coverage choice. s_1 and s_2 collaborate for 0.5 h to cover all targets while s_3 sleeps. s_2 and s_3 collaborate for 0.5 h while s_1 sleeps. s_1 and s_3 collaborate for 0.5 h while s_2 sleeps. The total coverage lifetime will become 1.5 h. The problem is how to divide sensors into groups and how long each group should work to prolong the coverage lifetime. One sensor can appear in several groups. But the total active time of one sensor should satisfy its battery capacity.

We model a wireless sensor network as a graph $G(S, T, E, W, L)$. The sensor set is denoted as S. T represents the set of targets in the network. If one target $t \in T$ can be covered by $s \in S$, then there is an edge (s, t) in G. In Fig. 1b, there is an edge between t_1 and s_1 since t_1 is in s_1's sensing range. All of these edges are stored in E. Heterogeneous sensors are considered. Different sensors may have different energy consumption to do the same tasks. In general, we have different weights of sensors. W denotes the weights of all sensors, and L denotes the energy capacity of all sensors in G. Based on the definition of G, the formal definition of MLCP is defined as follow.

Formal Definition of MLCP

Definition 1 (MLCP[3]) MLCP is that given $G = (\mathcal{S}, \mathcal{T}, \mathcal{E}, \mathcal{W}, \mathcal{L})$, find a set of sensor subsetsets and duration of each subset $(S_1, L_1), (S_2, L_2), \ldots, (S_k, L_k)$ in G to maximize $\sum_{i=1}^{k} L_i$, where S_i represents the sensor subset in G and L_i represents the time duration of S_i, satisfying:

1. $\forall i \in \{1, 2, \ldots, k\}$, S_i satisfies full coverage. $\forall\ t \in \mathcal{T}$ and $\forall S_i$, $\exists s \in S_i$ satisfying $(t, s) \in \mathcal{E}$.
2. For each sensor, the total active time should be smaller or equal to its power constraint.

It is a long-standing open problem whether Maximum Lifetime Coverage Problem (MLCP) has a polynomial-time constant-approximation algorithm [3].

Based on primal-dual method (PD method), *M*inimum *W*eight *S*ensor *C*overage *P*roblem (MWSCP) is used to help to solve MLCP. The formal definition of MWSCP is given in Definition 2.

Definition 2 (MWSCP[3]) MWSCP is to find a sensor subset SC in $G = (\mathcal{S}, \mathcal{T}, \mathcal{E}, \mathcal{W}, \mathcal{L})$ to minimize $\sum_{s \in SC} w(s)$, where $w(s)$ represents the weight of sensor s, such that $\forall\ t \in \mathcal{T}$, $\exists s \in SC$ satisfying $(t, s) \in \mathcal{E}$.

Key Results

1. Heuristic linear programming without performance guarantee [4]
2. Pure heuristic algorithm better than heuristic linear programming
3. One 4-approximation algorithms for MLCP with improvement from $1 + \ln n$, where n is the number of sensors
4. One 4-approximation algorithms for MWSCP

Integer Linear Programming and Heuristic Algorithms Proposed by Cardei et al. [4]

Cardei et al. prove that MLCP is NP-hard. Two algorithms are proposed [4]. The first algorithm is to model MLCP as an integer linear programming. To solve it, the authors first relax the integer variables to real values and get the optimal solution to the relaxed linear programming. Find the maximum time duration of each group based on the optimal solution to the relaxed linear programming. Update all sensors' remaining battery capacity. A new MLCP is formed with different remaining sensor power abilities after previous round. Finally, a maximum lifetime will be achieved in the network.

The second algorithm is a heuristic algorithm. Find a sensor group which can cover all targets. The lifetime of this group is determined by the minimum power ability of sensors in the group. Update all sensors' energy level and choose sensor group again till no such group can be found. The final lifetime is the sum of time duration of all sensor groups.

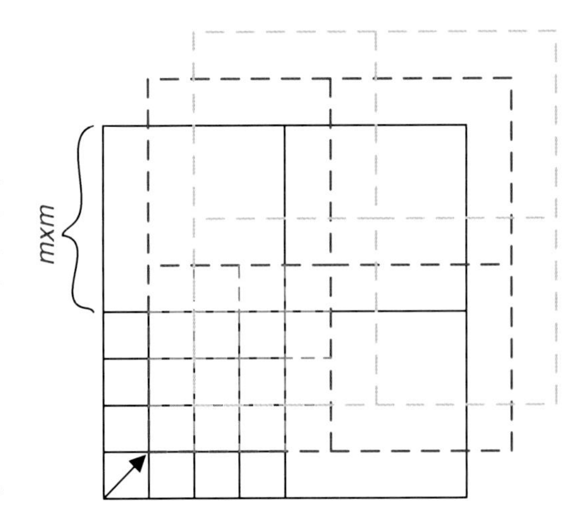

Maximum Lifetime Coverage, Fig. 2 Double partition and shifting [3]

Performance-Guaranteed Approximation Algorithm Proposed by Ling et al. [3]

Ling et al. use primal-dual method to solve MCLP. The primal problem is MWSCP [5]. To get a constant-approximation algorithm for MWSCP, double partition and shifting are used. As shown in Fig. 2, the area is divided into cells with size $\frac{m \times r}{\sqrt{2}} \times \frac{m \times r}{\sqrt{2}}$, where r represents sensing range of sensors and m is a predetermined value. $r = 1$ in Fig. 2. Each cell is further divided into

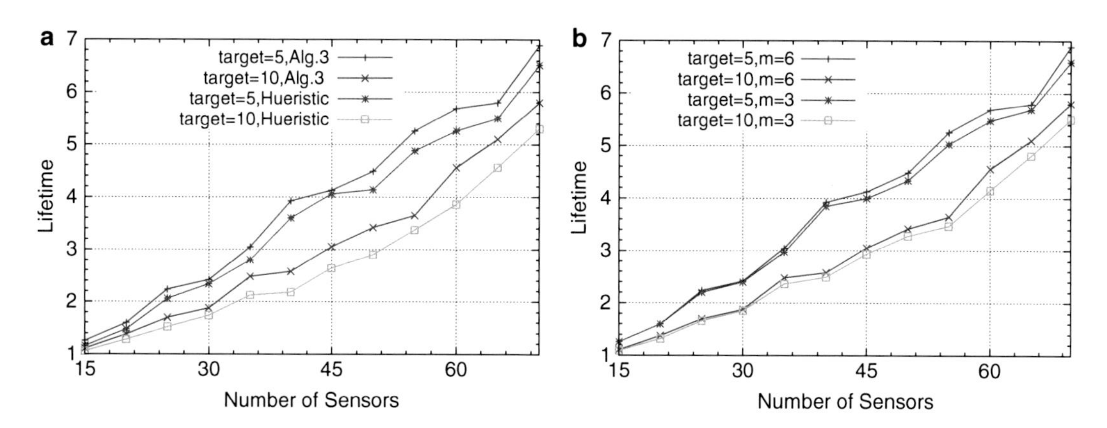

Maximum Lifetime Coverage, Fig. 3 Comparison of lifetime. (**a**) Lifetimes among different algorithms. (**b**) Lifetimes among different partitions [3]

small squares of size $\frac{r}{\sqrt{2}} \times \frac{r}{\sqrt{2}}$ which means there will be $m \times m$ small squares in each cell. After twice partition, there will be m horizontal strips and m vertical strips in each cell. Using dynamic programming, the optimal solution to MWSCP can be found in each cell. Combined all the optimal solutions to each cell, we can get an approximation algorithm to MWSCP for the whole area. To solve the sensors on the border of each cell, the small squares are shifted to purple position in Fig. 2 and then to yellow position. The shift will stop after m times. The final result will be achieved by taking average of the solutions to black partition, purple partition, yellow partition, and other shifts. It is proved in [3] that the final result of MWSCP has a constant performance ratio 4. The running time of the algorithm is determined by the predetermined value m. If m is big, the result will be more precise, but the running time is high. If m is small, the running time is small, but the performance will drop.

Based on the idea of primal-dual method, the solution to MWSCP will help to derive the solution to MCLP with the same performance ratio. The solution to MWSCP is derived iteration by iteration. In each iteration, find the MWSCP firstly and then determine the time duration of the sensor set. Update the lifetime and weight of each sensors. The algorithm will stop if there is no such a sensor set exists.

Experimental Results

Cardei et al. demonstrate that their pure heuristic algorithm outperforms their heuristic linear programming algorithm in running time and lifetime.

Ding et al. [3] conducts their experimental comparisons in an area of $6\sqrt{2} * 6\sqrt{2}$ and $m = 6$. They deploy sensors and targets randomly in that area. All sensors have the same sensing range of 2. The initial power capacity is 1 of each sensor. To show the density's effect on the performance, they increase the sensors from 15 to 70 by 5 and increase the number of targets from 5 to 10. Ling et al. compare their algorithm to Cardei's pure heuristic algorithm. If there are more sensors and the number of targets is fixed, the lifetime will be increased because there are more sensor groups (Fig. 3).

Recommended Reading

1. Bai X, Xuan D, Yun Z, Lai TH, Jia W (2008) Complete optimal deployment patterns for full-coverage and k-connectivity (k<=6) wireless sensor networks. In: Proceedings of ACM MobiHoc, Hong Kong
2. Balister P, Bollobas B, Sarkar A, Kumar S (2007) Reliable density estimates for coverage and connectivity in thin strips of finite length. In: Proceedings of ACM MobiCom, Montréal
3. Ding L, Wu W, Willson J, Wu L, Lu Z, Lee W (2012) Constant-approximation for target coverage problem in wireless sensor networks. In: INFOCOM, Orlando, pp 1584–1592

4. Cardei M, Thai MT, Li Y, Wu W (2005) Energy-efficient target coverage in wireless sensor networks. In: Proceedings of IEEE INFOCOM, Miami
5. Huang Y, Gao X, Zhang Z, Wu W (2009) A better constant-factor approximation for weighted dominating set in unit disk graph. J Comb Optim 18(2):179–194

Maximum Matching

Marcin Mucha
Faculty of Mathematics, Informatics and Mechanics, Institute of Informatics, Warsaw, Poland

Keywords

Algebraic graph algorithms; Fast matrix multiplication; Matching in graphs

Years and Authors of Summarized Original Work

2004; Mucha, Sankowski

Problem Definition

Let $G = (V, E)$ be an undirected graph, and let $n = |V|$, $m = |E|$. A *matching* in G is a subset $M \subseteq E$, such that no two edges of M have a common endpoint. A *perfect matching* is a matching of cardinality $n/2$. The most basic matching related problems are finding a *maximum matching* (i.e., a matching of maximum size) and, as a special case, finding a *perfect matching* if one exists. One can also consider the case where a weight function $w : E \to R$ is given and the problem is to find a *maximum weight matching*.

The maximum matching and maximum weight matching are two of the most fundamental algorithmic graph problems. They have also played a major role in the development of combinatorial optimization and algorithmics. An excellent account of this can be found in a classic monograph [11] by Lovász and Plummer devoted entirely to matching problems. A more up-to-date but also more technical discussion of the subject can be found in [19].

Classical Approach

Solving the maximum matching problem in time polynomial in n is a highly nontrivial task. The first such solution was given by Edmonds [3] in 1965 and has time complexity $O(n^3)$. Edmond's ingenious algorithm uses a combinatorial approach based on augmenting paths and blossoms. Several improvements followed, culminating in the algorithm with complexity $O(m\sqrt{n})$ given by Micali and Vazirani [12] in 1980 (a complete proof of the correctness of this algorithm was given much later by Vazirani [21], a nice exposition of the algorithm and its generalization to the weighted case can be found in a work of Gabow and Tarjan [4]). Beating this bound proved very difficult, several authors managed to achieve only a logarithmic speed-up for certain values of m and n. All these algorithms essentially follow the combinatorial approach introduced by Edmonds.

The maximum matching problem is much simpler for bipartite graphs. The complexity of $O(m\sqrt{n})$ was achieved for this case already in 1971 by Hopcroft and Karp [7], while the key ideas of the first polynomial algorithms date back to the 1920s and the works of König and Egerváry (see [11] and [19]).

Algebraic Approach

Around the time Micali and Vazirani introduced their matching algorithm, Lovász gave a randomized (Monte Carlo) reduction of the problem of testing whether a given n-vertex graph has a perfect matching to the problem of computing a certain determinant of a $n \times n$ matrix. Using the Hopcroft-Bunch fast Gaussian elimination algorithm [1], this determinant can be computed in time $MM(n) = O(n^{\omega})$ – time required to multiply two $n \times n$ matrices. Since $\omega < 2.38$ (see [2, 20]), for dense graphs, this algorithm is asymptotically faster than the matching algorithm of Micali and Vazirani.

However, Lovász's algorithm only tests for perfect matching, it does not find it. Using it to find perfect/maximum matchings in a straightforward fashion yields algorithm with complexity $O(mn^{\omega}) = O(n^{4.38})$. A major open problem in the field was thus: can maximum matchings be actually found in $O(n^{\omega})$ time?

The first step in this direction was taken in 1989 by Rabin and Vazirani [16]. They showed that maximum matchings can be found in time $O(n^{\omega+1}) = O(n^{3.38})$.

Key Results

The following theorems state the key results of [13].

Theorem 1 *Maximum matching in a n-vertex graph G can be found in $O(n^3)$ time (Las Vegas) by performing Gaussian elimination on a certain matrix related to G.*

Theorem 2 *Maximum matching in an n-vertex bipartite graph can be found in $\tilde{O}(n^{\omega})$ time (Las Vegas) by performing a Hopcroft-Bunch fast Gaussian elimination on a certain matrix related to G.*

Theorem 3 *Maximum matching in an n-vertex graph can be found in $\tilde{O}(n^{\omega})$ time (Las Vegas).*

Note: $\tilde{O}$ notation suppresses polylogarithmic factors, so $\tilde{O}(f(n))$ means $O(f(n)\log^k(n))$ for some k.

Let us briefly discuss these results. Theorem 1 shows that effective matching algorithms can be simple. This is in large contrast to augmenting paths-/blossoms-based algorithms which are generally regarded as quite complicated.

The other two theorems show that, for dense graphs, the algebraic approach is asymptotically faster than the combinatorial one.

The algorithm for the bipartite case is very simple. It's only nonelementary part is the fast matrix multiplication algorithm used as black box by the Hopcroft-Bunch algorithm. The general algorithm, however, is complicated and uses strong structural results from matching theory. A natural question is whether or not it is possible to give a simpler and/or purely algebraic algorithm. This has been positively answered by Harvey [5].

Several other related results followed. Mucha and Sankowski [14] showed that maximum matchings in planar graphs can be found in time $\tilde{O}(n^{\omega/2}) = O(n^{1.19})$ which is currently fastest known. Yuster and Zwick [22] extended this to any excluded minor class of graphs. Harvey [6] described a significantly simpler and purely algebraic version of the algorithm for general graphs. Sankowski [17] gave an RNC work-efficient matching algorithm (see also Mulmuley et al. [15] and Karp et al. [9] for earlier, less efficient RNC matching algorithms, and Karloff [8] for a description of a general technique for making such algorithm Las Vegas). He also generalized Theorem 2 to the case of weighted bipartite graphs with integer weights from $[0, \ldots, W]$, showing that in this case maximum weight matchings can be found in time $\tilde{O}(Wn^{\omega})$ (see [18]).

Applications

The maximum matching problem has numerous applications, both in practice and as a subroutine in other algorithms. A nice discussion of practical applications can be found in the monograph [11] by Lovász and Plummer. It should be noted, however, that algorithms based on fast matrix multiplication are completely impractical, so the results discussed here are not really useful in these applications.

On the theoretical side, faster maximum (weight) matching algorithms yield faster algorithms for related problems: disjoint $s - t$ paths problem, the minimum (weight) edge cover problem, the (maximum weight) b-matching problem, the (maximum weight) b-factor problem, the maximum (weight) T-join, or the Chinese postman problem. For detailed discussion of all these applications, see [11] and [19].

The algebraic algorithm of Theorem 1 also has a significant educational value. The combinatorial algorithms for the general maximum matching problem are generally regarded too complicated for an undergraduate course. That is definitely not the case with the algebraic $O(n^3)$ algorithm.

Open Problems

One of the most important open problems in the area is generalizing the results discussed above to weighted graphs. Sankowski [18] gives a $\tilde{O}(Wn^{\omega})$ algorithm for bipartite graphs with integer weights from the interval $[0 \ldots W]$. The complexity of this algorithm is really bad in terms of W. No effective algebraic algorithm is known for general weighted graphs.

Another interesting but most likely very hard problem is the derandomization of the algorithms discussed.

Cross-References

▶ All Pairs Shortest Paths via Matrix Multiplication
▶ Assignment Problem

Recommended Reading

1. Bunch J, Hopcroft J (1974) Triangular factorization and inversion by fast matrix multiplication. Math Comput 125:231–236
2. Coppersmith D, Winograd S (1987) Matrix multiplication via arithmetic progressions. In: Proceedings of the 19th annual ACM conference on theory of computing (STOC), New York, pp 1–6
3. Edmonds J (1965) Paths, trees, and flowers. Can J Math 17:449–467
4. Gabow HN, Tarjan RE (1991) Faster scaling algorithms for general graph matching problems. J ACM 38(4):815–853
5. Harvey N (2006) Algebraic structures and algorithms for matching and matroid problems. In: Proceedings of the 47th annual IEEE symposium on foundations of computer science (FOCS), Berkeley
6. Harvey JAN (2009) Algebraic algorithms for matching and matroid problems. SIAM J Comput 39(2):679–702
7. Hopcroft JE, Karp RM (1973) An $O(n^{5/2})$ Algorithm for maximum matchings in bipartite graphs. SIAM J Comput 2:225–231
8. Karloff H (1986) A Las Vegas RNC algorithm for maximum matching. Combinatorica 6:387–391
9. Karp R, Upfal E, Widgerson A (1986) Constructing a perfect matching is in random NC. Combinatorica 6:35–48
10. Lovász L (1979) On determinants, matchings and random algorithms. In: Budach L (ed) Fundamentals of computation theory (FCT'79), Wendisch-Rietz, pp 565–574. Akademie-Verlag, Berlin
11. Lovász L, Plummer MD (1986) Matching theory. Akadémiai Kiadó – North Holland, Budapest
12. Micali S, Vazirani VV (1980) An $O(\sqrt{V}E)$ algorithm for finding maximum matching in general graphs. In: Proceedings of the 21st annual IEEE symposium on foundations of computer science (FOCS), Syracuse, pp 17–27
13. Mucha M, Sankowski P (2004) Maximum matchings via Gaussian elimination. In: Proceedings of the 45th annual IEEE symposium on foundations of computer science (FOCS), Rome, pp 248–255
14. Mucha M, Sankowski P (2006) Maximum matchings in planar graphs via Gaussian elimination. Algorithmica 45:3–20
15. Mulmuley K, Vazirani UV, Vazirani VV (1987) Matching is as easy as matrix inversion. In: Proceedings of the 19th annual ACM conference on theory of computing, New York, pp 345–354. ACM
16. Rabin MO, Vazirani VV (1989) Maximum matchings in general graphs through randomization. J Algorithms 10:557–567
17. Sankowski P (2005) Processor efficient parallel matching. In: Proceeding of the 17th ACM symposium on parallelism in algorithms and architectures (SPAA), Las Vegas, pp 165–170
18. Sankowski P (2006) Weighted bipartite matching in matrix multiplication time. In: Proceedings of the 33rd international colloquium on automata, languages and programming, Venice, pp 274–285
19. Schrijver A (2003) Combinatorial optimization: polyhedra and efficiency. Springer, Berlin/Heidelberg
20. Vassilevska Williams V (2012) Multiplying matrices faster than Coppersmith-Winograd. In: Proceedings of the 44th symposium on theory of computing conference (STOC), New York, pp 887–898
21. Vazirani VV (1994) A theory of alternating paths and blossoms for proving correctness of the $O(\sqrt{V}E)$ maximum matching algorithm. Combinatorica 14(1):71–109
22. Yuster R, Zwick U (2007) Maximum matching in graphs with an excluded minor. In: Proceedings of the ACM-SIAM symposium on discrete algorithms (SODA), New Orleans

Maximum-Average Segments

Kun-Mao Chao
Department of Computer Science and Information Engineering, National Taiwan University, Taipei, Taiwan

Keywords

Maximum-average segment

Years and Authors of Summarized Original Work

1994; Huang

Problem Definition

Given a sequence of numbers, $A = \langle a_1, a_2, \ldots, a_n \rangle$, and two positive integers L, U, where $1 \leq L \leq U \leq n$, the maximum-density segment problem is to find a consecutive subsequence, i.e., a segment or substring, of A with length at least L and at most U such that the average value of the numbers in the subsequence is maximized.

Key Results

If there is no length constraint, then obviously the maximum-density segment is the maximum number in the sequence. Let's first consider the problem where only the length lower bound L is imposed. By observing that the length of the shortest maximum-density segment with length at least L is at most $2L - 1$, Huang [9] gave an $O(nL)$-time algorithm. Lin et al. [13] proposed a new technique, called the *right-skew decomposition*, to partition each suffix of A into *right-skew* segments of strictly decreasing averages. The right-skew decomposition can be done in $O(n)$ time, and it can answer, for each position i, a consecutive subsequence of A starting at that position such that the average value of the numbers in the subsequence is maximized. On the basis of the right-skew decomposition, Lin et al. [13] devised an $O(n \log L)$-time algorithm for the maximum-density segment problem with a lower bound L, which was improved to $O(n)$ time by Goldwasser et al. [8]. Kim [11] gave another $O(n)$-time algorithm by reducing the problem to the maximum-slope problem in computation geometry. As for the problem which takes both L and U into consideration, Chung and Lu [6] bypassed the construction of the right-skew decomposition and gave an $O(n)$-time algorithm.

It should be noted that a closely related problem in data mining, which basically deals with a binary sequence, was independently formulated and studied by Fukuda et al. [7].

An Extension to Multiple Segments

Given a sequence of numbers, $A = \langle a_1, a_2, \ldots, a_n \rangle$, and two positive integers L and k, where $k \leq \frac{n}{L}$, let $d(A[i, j])$ denote the *density* of segment $A[i, j]$, defined as $(a_i + a_{i+1} + \cdots + a_j)/(j - i + 1)$. The problem is to find k disjoint segments $\{s_1, s_2, \ldots, s_k\}$ of A, each has a length of at least L, such that $\sum_{1 \leq i \leq k} d(s_i)$ is maximized. Chen et al. [5] proposed an $O(nkL)$-time algorithm and an improved $O(nL + k^2 L^2)$-time algorithm was given by Bergkvist and Damaschke [2]. Liu and Chao [14] gave an $O(n + k^2 L \log L)$-time algorithm.

Applications

In all organisms, the GC base composition of DNA varies between 25–75 %, with the greatest variation in bacteria. Mammalian genomes typically have a GC content of 45–50 %. Nekrutenko and Li [15] showed that the extent of the compositional heterogeneity in a genomic sequence strongly correlates with its GC content. Genes are found predominantly in the GC-richest isochore classes. Hence, finding GC-rich regions is an important problem in gene recognition and comparative genomics.

M

Given a DNA sequence, one would attempt to find segments of length at least L with the highest C+G ratio. Specifically, each of nucleotides C and G is assigned a score of 1, and each of nucleotides A and T is assigned a score of 0.

DNA sequence: ATGACTCGAGCTCGTCA
Binary sequence: 00101011011011010 The maximum-average segments of the binary sequence correspond to those segments with the highest GC ratio in the DNA sequence. Readers can refer to [1, 3, 4, 11–13, 16–18] for more variants and applications.

Open Problems

The best asymptotic time bound of the algorithms for the multiple maximum-density segments problem is $O(n + k^2 L \log L)$. Can this problem be solved in $O(n)$ time?

Cross-References

▶ Maximum-Sum Segments

Recommended Reading

1. Arslan A, Eğecioğlu Ö, Pevzner P (2001) A new approach to sequence comparison: normalized sequence alignment. Bioinformatics 17:327–337
2. Bergkvist A, Damaschke P (2005) Fast algorithms for finding disjoint subsequences with extremal densities. In: Proceedings of the 16th annual international symposium on algorithms and computation, Sanya. Lecture notes in computer science, vol 3827, pp 714–723
3. Burton BA (2011) Searching a bitstream in linear time for the longest substring of any given density. Algorithmica 61:555–579
4. Burton BA, Hiron M (2013) Locating regions in a sequence under density constraints. SIAM J Comput 42:1201–1215
5. Chen YH, Lu HI, Tang CY (2005) Disjoint segments with maximum density. In: Proceedings of the 5th annual international conference on computational science, Atlanta, pp 845–850
6. Chung K-M, Lu H-I (2004) An optimal algorithm for the maximum-density segment problem. SIAM J Comput 34:373–387
7. Fukuda T, Morimoto Y,Morishita S, Tokuyama T (1999) Mining optimized association rules for numeric attributes. J Comput Syst Sci 58:1–12
8. Goldwasser MH, Kao M-Y, Lu H-I (2005) Linear-time algorithms for computing maximum-density sequence segments with bioinformatics applications. J Comput Syst Sci 70:128–144
9. Hsieh Y-H, Yu C-C, Wang B-F (2008) Optimal algorithms for the interval location problem with range constraints on length and average. IEEE/ACM Trans Comput Biol Bioinform 5:281–290
10. Huang X (1994) An algorithm for identifying regions of a DNA sequence that satisfy a content requirement. Comput Appl Biosci 10:219–225
11. Kim SK (2003) Linear-time algorithm for finding a maximum-density segment of a sequence. Inf Process Lett 86:339–342
12. Lin Y-L, Huang X, Jiang T, Chao K-M (2003) MAVG: locating non-overlapping maximum average segments in a given sequence. Bioinformatics 19:151–152
13. Lin Y-L, Jiang T, Chao K-M (2002) Efficient algorithms for locating the length-constrained heaviest segments with applications to biomolecular sequence analysis. J Comput Syst Sci 65:570–586
14. Liu H-F, Chao K-M (2006) On locating disjoint segments with maximum sum of densities. In: Proceedings of the 17th annual international symposium on algorithms and computation, Kolkata. Lecture notes in computer science, vol 4288, pp 300–307
15. Nekrutenko A, Li WH (2000) Assessment of compositional heterogeneity within and between eukaryotic genomes. Genome Res 10:1986–1995
16. Stojanovic N, Dewar K (2005) Identifying multiple alignment regions satisfying simple formulas and patterns. Bioinformatics 20:2140–2142
17. Stojanovic N, Florea L, Riemer C, Gumucio D, Slightom J, Goodman M, Miller W, Hardison R (1999) Comparison of five methods for finding conserved sequences in multiple alignments of gene regulatory regions. Nucl Acid Res 19:3899–3910
18. Zhang Z, Berman P, Wiehe T, Miller W (1999) Post-processing long pairwise alignments. Bioinformatics 15:1012–1019

Maximum-Sum Segments

Kun-Mao Chao
Department of Computer Science and Information Engineering, National Taiwan University, Taipei, Taiwan

Keywords

Shortest path; Longest path

Years and Authors of Summarized Original Work

2002; Lin, Jiang, Chao

Problem Definition

Given a sequence of numbers, $A = \langle a_1, a_2, \ldots, a_n \rangle$, and two positive integers L, U, where $1 \le L \le U \le n$, the maximum-sum segment problem is to find a consecutive subsequence, i.e., a segment or substring, of A with length at least L and at most U such that the sum of the numbers in the subsequence is maximized.

Key Results

The maximum-sum segment problem without length constraints is linear-time solvable by using Kadane's algorithm [2]. Huang extended the recurrence relation used in [2] for solving the maximum-sum segment problem and derived a linear-time algorithm for computing the maximum-sum segment with length at least L. Lin et al. [13] proposed an $O(n)$-time algorithm for the maximum-sum segment problem with both L and U constraints, and an online version was given by Fan et al. [10].

An Extension to Multiple Segments

Computing the k largest sums over all possible segments is a natural extension of the maximum-sum segment problem. This extension has been considered from two perspectives, one of which allows the segments to overlap, while the other disallows.

Linear-time algorithms for finding all the nonoverlapping maximal segments were given in [5, 15]. On the other hand, one may focus on finding the k maximum-sum segments whose overlapping is allowed. A naïve approach is to choose the k largest from the sums of all possible contiguous subsequences which requires $O(n^2)$ time. Bae and Takaoka [1] presented an $O(kn)$-time algorithm for the k maximum segment problem. Liu and Chao [14] noted that the k maximum-sum segment problem can be solved in $O(n + k)$ time [9] and gave an $O(n + k)$-time algorithm for the length-constrained k maximum-sum segment problem.

Applications

The algorithms for the maximum-sum segment problem have applications in finding GC-rich regions in a genomic DNA sequence, postprocessing sequence alignments, and annotating multiple sequence alignments. Readers can refer to [3–8, 11, 13, 15–18] for more variants and applications.

Open Problems

It would be interesting to consider the higher dimensional cases.

Cross-References

▸ Maximum-Average Segments

Recommended Reading

1. Bae SE, Takaoka T (2004) Algorithms for the problem of k maximum sums and a VLSI algorithm for the k maximum subarrays problem. In: Proceedings of the 7th international symposium on parallel architectures, algorithms and networks, Hong Kong, pp 247–253
2. Bentley J (1986) Programming pearls. Addison-Wesley, Reading
3. Burton BA (2011) Searching a bitstream in linear time for the longest substring of any given density. Algorithmica 61:555–579
4. Burton BA, Hiron M (2013) Locating regions in a sequence under density constraints. SIAM J Comput 42:1201–1215
5. Chen K-Y, Chao K-M (2004) On the range maximum-sum segment query problem. In: Proceedings of the 15th international symposium on algorithms and computation, Hong Kong. LNCS, vol 3341, pp 294–305
6. Chen K-Y, Chao K-M (2005) Optimal algorithms for locating the longest and shortest segments satisfying a sum or an average constraint. Inf Process Lett 96:197–201

7. Cheng C-H, Chen K-Y, Tien W-C, Chao K-M (2006) Improved algorithms for the k maximum-sum problems. Theor Comput Sci 362:162–170
8. Csűrös M (2004) Maximum-scoring segment sets. IEEE/ACM Trans Comput Biol Bioinform 1:139–150
9. Eppstein D (1998) Finding the k shortest paths. SIAM J Comput 28:652–673
10. Fan T-H, Lee S, Lu H-I, Tsou T-S, Wang T-C, Yao A (2003) An optimal algorithm for maximum-sum segment and its application in bioinformatics. In: Proceedings of the eighth international conference on implementation and application of automata, Santa Barbara. LNCS, vol 2759, pp 251–257
11. Hsieh Y-H, Yu C-C, Wang B-F (2008) Optimal algorithms for the interval location problem with range constraints on length and average. IEEE/ACM Trans Comput Biol Bioinform 5:281–290
12. Huang X (1994) An algorithm for identifying regions of a DNA sequence that satisfy a content requirement. Comput Appl Biosci 10:219–225
13. Lin Y-L, Jiang T, Chao K-M (2002) Efficient algorithms for locating the length-constrained heaviest segments with applications to biomolecular sequence analysis. J Comput Syst Sci 65:570–586
14. Liu H-F, Chao K-M (2008) Algorithms for finding the weight-constrained k longest paths in a tree and the length-constrained k maximum-sum segments of a sequence. Theor Comput Sci 407:349–358
15. Ruzzo WL, Tompa M (1999) A linear time algorithm for finding all maximal scoring subsequences. In: Proceedings of the 7th international conference on intelligent systems for molecular biology, Heidelberg, pp 234–241
16. Stojanovic N, Dewar K (2005) Identifying multiple alignment regions satisfying simple formulas and patterns. Bioinformatics 20:2140–2142
17. Stojanovic N, Florea L, Riemer C, Gumucio D, Slightom J, Goodman M, Miller W, Hardison R (1999) Comparison of five methods for finding conserved sequences in multiple alignments of gene regulatory regions. Nucleic Acids Res 19:3899–3910
18. Zhang Z, Berman P, Wiehe T, Miller W (1999) Post-processing long pairwise alignments. Bioinformatics 15:1012–1019

Max-Min Allocation

Deeparnab Chakrabarty
Microsoft Research, Bangalore, Karnataka, India

Keywords

Approximation algorithms; Linear programs; Scheduling and resource allocation

Years and Authors of Summarized Original Work

2006; Bansal, Sviridenko
2007; Asadpour, Saberi
2008; Feige
2009; Chakrabarty, Chuzhoy, Khanna

Problem Definition

The max-min allocation problem has the following setting. There is a set A of m agents and a set I of n items. Each agent $i \in A$ has utility $u_{ij} \in \mathbb{R}_{\geq 0}$ for item $j \in I$. Given a subset of items $S \subseteq I$, the utility of this set to agent i is denoted as $u_i(S) := \sum_{j \in S} u_{ij}$. The max-min allocation problem is to find an allocation of items to agents such that the minimum utility among the agents is maximized. That is, $\min_{i \in A} u_i(S_i)$ is maximized, where $S_i \subseteq I$ is the set of items allocated to agent i and $S_i \cap S_{i'} = \emptyset$.

The problem naturally arises as an approach to maximize fairness. Fairness is an important concept arising in numerous settings ranging from border disputes in political science to frequency allocations in spectrum auctions. Max-min fairness is one of the standard notions of fairness and has been an object of study for decades [6]. Most of the older works, however, have focussed on divisible settings, that is, situations where the resource can be infinitely divided and allocated. Furthermore, the computational perspective, that is, how *efficiently* can one find a fair allocation, has not been a primary viewpoint. The max-min allocation problem is a combinatorial allocation problem where the items cannot be divided, and the interest is in designing polynomial time algorithms to obtain fair, or near-fair, allocations.

Key Results

The max-min allocation problem is NP-hard and the focus is on designing approximation algorithms. Let OPT be the optimum value of a certain instance. A ρ-approximate solution, for $\rho > 1$, is an allocation where each agent gets utility at least OPT/ρ. A ρ-approximation algorithm returns a ρ-approximate solution given any instance. An algorithm is a polynomial time approximation scheme (PTAS) if for any constant $\varepsilon > 0$, it returns an $(1+\varepsilon)$-approximate solution in polynomial time.

Woeginger [11] obtained a PTAS for the max-min allocation problem when the utility of an item is the *same* for all agents. Bezakova and Dani [5] gave the first nontrivial $(n - m + 1)$-approximation algorithm for the general max-min allocation problem and also showed that it is NP-hard to obtain a better than 2-approximation algorithm for the problem. The latter result remains the best hardness known till date.

Bansal and Sviridenko [3] introduced a restricted version of the max-min allocation problem which they called the *Santa Claus* problem. In this version, each item has an inherent utility u_j, however, it can only be allocated to an agent in a certain subset $A_j \subseteq A$. Equivalently, for each item j, $u_{ij} \in \{u_j, 0\}$. Bansal and Sviridenko [3] described an $O(\log\log m/\log\log\log m)$-approximation algorithm for the Santa Claus problem. Soon after, Feige [8] described an algorithm which *estimates* the value of the optimum of the Santa Claus problem up to $O(1)$-factor in polynomial time, although at the time no efficient algorithm was known to construct the allocation. Following constructive versions of the Lovasz local lemma due to Moser and Tardos [10] and Haeupler et al. [9], there now exists a polynomial time $O(1)$-approximation algorithm for the Santa Claus problem. The constant, however, is necessarily quite large and to our knowledge has not been explicitly specified in any published work. In contrast, Asadpour et al. [2] described a local search algorithm which returns a 4-approximate solution to the Santa Claus problem; however, it is not known whether the procedure terminates in polynomial time or not.

Asadpour and Saberi [1] described a polynomial time $O(\sqrt{m}\log^3 m)$ approximation algorithm for the general max-min allocation problem. Bateni et al. [4] obtained an $O(m^\varepsilon)$-approximation algorithm running in $m^{O(1/\varepsilon)}$ time for certain special cases of the max-min allocation problem; in their special cases, utilities u_{ij} lay in the set $\{0, 1, \infty\}$, and furthermore, for each item j there exists at most one agent i with $u_{ij} = \infty$. Chakrabarty et al. [7] designed an $O(n^\varepsilon)$-approximation algorithm for the general max-min allocation problem which runs in $n^{O(1/\varepsilon)}$-time, for any $\varepsilon > \frac{9\log\log n}{\log n}$. This implies *quasi-polynomial* time $O(\log^{10} n)$-approximation algorithm and $O(m^\varepsilon)$-approximation algorithm, for any constant $\varepsilon > 0$, for the max-min allocation problem. An algorithm runs in quasi-polynomial time, if the logarithm of its running time is upper bounded by a polynomial in the bit length of the data.

Sketch of the Techniques

Almost all algorithms for the max-min allocation problem follow by rounding linear programming (LP) relaxations of the problem. One starts with a guess T of the optimum OPT. Using this, one writes an LP which has a feasible solution if $\mathsf{OPT} \geq T$. The nontrivial part is to round this LP solution to obtain an allocation with every agent getting utility $\geq T/\rho$. Since, by doing a binary search over the guesses, one can get T very close to OPT, the rounding step implies a ρ-approximation algorithm. Henceforth, we assume that T has been guessed to be OPT, and furthermore, by scaling all utility values appropriately, we assume $\mathsf{OPT} = 1$.

The first LP relaxation one may think of is the following. First one clips each utility value at 1; $u_{ij} = \min(1, u_{ij})$. If $\mathsf{OPT} = 1$, then the following LP is feasible.

$$\sum_{j \in I} u_{ij} x_{ij} \geq 1, \qquad \forall i \in A \qquad (1)$$

$$\sum_{i \in A} x_{ij} = 1, \qquad \forall j \in I \qquad (2)$$

The first inequality states that every agent gets utility at least $\mathsf{OPT} = 1$, and the second states that each item is allocated. It is not hard to find instances, and in fact Santa Claus instances, where the LP is feasible for $\mathsf{OPT} = 1$, but in any allocation, some agent will obtain utility at most $1/m$. In other words, the *integrality gap* of this LP relaxation is (at least) m.

There is a considerably stronger LP relaxation which is called the *configuration* LP relaxation. The variables in this LP are of the form $y_{i,C}$ where $i \in A$ and $C \subseteq I$ where $\sum_{j \in C} u_{ij} \geq 1$. (i, C) is called a feasible configuration in this case. $\mathcal{C}$ denotes the collection of all feasible configurations.

$$\sum_C y_{i,C} = 1, \quad \forall i \in A \tag{3}$$

$$\sum_{i \in A} \sum_{C:(i,C) \in \mathcal{C}, j \in C} y_{i,C} = 1, \quad \forall j \in I \tag{4}$$

The first inequality states that each agent precisely gets one subset of items, and the second states that each item is in precisely one feasible configuration. Although the LP has possibly exponentially variables, the dual has only polynomially many variables and can be solved to arbitrary accuracy via the ellipsoid method. We refer the reader to [3] for details. Bansal and Sviridenko [3] show that in the Santa Claus problem, a solution above LP can be rounded to give an allocation where every agent obtains utility $\rho = \Omega\left(\frac{\log \log \log n}{\log \log n}\right)$. To do this, the authors partition the items into *big*, if $u_j \geq \rho$ and *small* otherwise. A solution is ρ-approximate if any agent gets either one big item or $\geq 1/\rho$ small items. The big items are taken care of via a "matching like" procedure, while the small items are allocated by randomized rounding. Bansal and Sviridenko [3] use the Lovasz local lemma (LLL) to analyze the randomized algorithm. Feige [8] uses a more sophisticated randomized rounding in phases along with an LLL analysis to obtain a constant factor approximation to the value of the optimum. At the time, no algorithmic proofs of LLL were known; however, following works of [9, 10], a polynomial time $O(1)$-approximation algorithm is now known for the Santa Claus problem.

In the general max-min allocation, the main problem in generalizing the above technique is that the same item could be big for one agent and small for another agent. Asadpour and Saberi [1] give a two-phase rounding algorithm. In the first phase, a random matching is obtained between agents and *their* big items with roughly the property that each item is allocated with the same probability that the configuration LP prescribes. In the second phase, each agent i randomly selects a set C of items with probability $y_{i,C}$. Since there are m agents, at most m items are allocated in the first phase. This allows [1] to argue that, with high probability, there is enough (roughly $1/\sqrt{m}$) utility remaining among the unmatched items in C. Finally, they also show that the same item is not "claimed" by not more than $O(\log m)$ agents.

The integrality gap of the configuration LP is $\Omega(\sqrt{m})$. Therefore a new LP relaxation is required to go beyond the Asadpour-Saberi result. We now briefly sketch the technique of Chakrabarty et al. [7]. First, they show that any instance can be "reduced" to a canonical instance where *agents* are either heavy or light. Heavy agents have utility 1 for a subset of big items and 0 for the rest. Light agents have a unique *private* item which give them utility 1, and the rest of the items either are small and give utility $1/K$ or give utility 0. Here $K \approx n^{\varepsilon}$ is a large integer. The LP of [7] is parametrized by a maximum matching M between heavy agents and their big items. If all heavy agents are matched, then there is nothing to be done since light agents can allocate their private item. Otherwise, there is a *reassignment* strategy: where a light agent is allocated K small items upon which he "frees" his private item, which is then again allocated to another agent and so on, till an unmatched heavy agent gets a big item. This reassignment can be seen as a directed in-arborescence whose depth can be argued is at most $1/\varepsilon$ since at each level we encounter roughly K new light agents. The LP encodes this reassignment as a flow with a variable for each flow path of length at most $1/\varepsilon$; this implies the number of variables is at most $n^{O(1/\varepsilon)}$ and a similar number of constraints. Therefore, the LP can be solved in $n^{O(1/\epsilon)}$ time which dominates the running time. If the instance has $\mathsf{OPT} = 1$, then the LP has a feasible solution. One would then expect that given such a feasible solution, one can obtain an allocation with every heavy agent getting a big item and each light agent getting either his private item or n^{ε} small items. Unfortunately, this may not be true. What Chakrabarty et al. [7] show is that if the LP has a feasible solution, then a "partial" allocation can be found where some light agents obtain sufficiently many small items, and their private items can be used to obtain a larger matching M' among heavy agents and big items. The process is then repeated iteratively, with a new LP at each step guiding the partial allocation, till one matches every heavy agent.

Summary

In summary, the best polynomial time algorithms for the general max-min allocation problem, as of the date this article is written, have approximation factors which are a polynomial in the input data. On the other hand, even a 2-approximation for the problem has not been ruled out. Closing this gap is a confounding problem in the area of approximation algorithms. A constant factor approximation algorithm is known for the special case called the Santa Claus problem; even here, getting a polynomial time algorithm achieving a "small" constant factor is an interesting problem.

Recommended Reading

1. Asadpour A, Saberi A (2010) An approximation algorithm for max-min fair allocation of indivisible goods. SIAM J Comput 39(7):2970–2989
2. Asaspour A, Feige U, Saberi A (2012) Santa Claus meets hypergraph matching. ACM Trans Algorithms 8(3):24
3. Bansal N, Sviridenko M (2006) The Santa Claus problem. In: ACM symposium on theory of computing (STOC), Seattle
4. Bateni M, Charikar M, Guruswami V (2009) Maxmin allocation via degree lower-bounded arborescences. In: ACM symposium on theory of computing (STOC), Bethesda
5. Bezakova I, Dani V (2005) Allocating indivisible goods. SIGecom Exch 5(3):11–18
6. Brams S, Taylor A (1996) Fair division: from cake-cutting to dispute resolution. Cambridge University Press, Cambridge/New York
7. Chakrabarty D, Chuzhoy J, Khanna S (2009) On allocations that maximize fairness. In: Proceedings, IEEE symposium on foundations of computer science (FOCS), Atlanta
8. Feige U (2008) On allocations that maximize fairness. In: Proceedings, ACM-SIAM symposium on discrete algorithms (SODA), San Francisco
9. Haeupler B, Saha B, Srinivasan A (2011) New constructive aspects of the lovász local lemma. J ACM 58(6)
10. Moser R, Tardos G (2010) A constructive proof of the general lovász local lemma. J ACM 57(2)
11. Woeginger G (1997) A polynomial-time approximation scheme for maximizing the minimum machine completion time. Oper Res Lett 20(4):149–154

Mechanism Design and Differential Privacy

Kobbi Nisim[1] and David Xiao[2]
[1]Department of Computer Science, Ben-Gurion University, Beer Sheva, Israel
[2]CNRS, Université Paris 7, Paris, France

Keywords

Differential privacy; Mechanism design; Privacy-aware mechanism design; Purchasing privacy

Years and Authors of Summarized Original Work

2007; McSherry, Talwar
2011; Ghosh, Roth
2012; Nissim, Orlandi, Smorodinsky
2012; Fleischer, Lyu
2012; Ligett, Roth
2013; Chen, Chong, Kash, Moran, Vadhan
2014; Nissim, Vadhan, Xiao

Problem Definition

Mechanism design and private data analysis both study the question of performing computations over data collected from individual agents while satisfying additional restrictions. The focus in mechanism design is on performing computations that are compatible with the incentives of the individual agents, and the additional restrictions are toward motivating agents to participate in the computation (individual rationality) and toward having them report their true data (incentive compatibility). The focus in private data analysis is on performing computations that limit the information leaked by the output on each individual agent's sensitive data, and the additional restriction is on the influence each agent may have on the outcome distribution (differential privacy). We refer the reader to the sections on *algorithmic game theory* and on *differential privacy* for further details and motivation.

Incentives and privacy. In real-world settings, incentives influence how willing individuals are to part with their private data. For example, an agent may be willing to share her medical data with her doctor, because the utility from sharing is greater than the loss of utility from privacy concerns, while she would probably not be willing to share the same information with her accountant.

Furthermore, privacy concerns can also cause individuals to misbehave in otherwise incentive-compatible, individually rational mechanisms. Consider for example a second-price auction: the optimal strategy in terms of payoff is to truthfully report valuations, but an agent may consider misreporting (or abstaining) because the outcome reveals the valuation of the second-price agent, and the agent does not want to risk their valuation being revealed. In studies based on sensitive information, e.g., a medical study asking individuals to reveal whether they have syphilis, a typical individual with syphilis may be less likely to participate than a typical individual without the disease, thereby skewing the overall sample. The bias may be reduced by offering appropriate compensation to participating agents.

The framework. Consider a setting with n individual agents, and let $x_i \in X$ be the private data of agent i for some type set X. Let $f : X^n \to Y$ be a function of the joint inputs of the agents $\mathbf{x} = (x_1, \ldots, x_n)$. Our goal is to build a mechanism M that computes $f(\mathbf{x})$ accurately and is compatible with incentives and privacy as we will now describe.

We first fix a function v that models the gain in utility that an agent derives from the *outcome* of the mechanism. We restrict our attention to a setting where this value can only depend on the agent's data and the outcome y of the mechanism:

$$v_i = v(x_i, y).$$

We also fix a function λ that models the loss in utility that an agent incurs because information about her private data is leaked by the outcome of the mechanism. Importantly, λ depends on the mechanism M, as the computation M performs determines the leakage. The loss can also depend on how much the agent values privacy, described by a parameter p_i (a real number in our modeling), on the actual data of all the individuals, on the outcome, as well as other parameters such as the strategy of the agent:

$$\lambda_i = \lambda(M, p_i, \mathbf{x}_{-i}, x_i, y, \ldots).$$

The overall utility that agent i derives from participating in the computation of M is

$$u_i = v_i - \lambda_i. \quad (1)$$

With this utility function in mind, our goal will be to construct truthful mechanisms M that compute f accurately. We note that in Eq. 1 we typically think about both v_i and λ_i as positive quantities, but we do not exclude either of them being negative, so either quantity may result in a gain or a loss in utility.

We can now define the mechanism $M : X^n \times \mathbb{R}^n \to Y$ to be a randomized function taking as inputs the private inputs of the agents $\mathbf{x}$ and their privacy valuations $\mathbf{p}$ and returns a value in the set Y.

Modeling the privacy loss. In order to analyze specific mechanisms, we will need to be able to control the privacy loss λ. Toward this end, we will need to assume that λ has some structure, and so we now discuss the assumptions we make and their justifications.

One view of privacy loss is to consider a framework of *sequential games*: an individual is not only participating in mechanism M, but she will also participate in other mechanisms $M', M'', \ldots$ in the future, and each participation will cause her to gain or lose in utility. Because her inputs to these functions may be correlated, revealing her private inputs in M may cause her to obtain less utility in the future. For example, an individual may hesitate to participate in a medical study because doing so might reveal she has a genetic predisposition to a certain disease, which may increase her insurance premiums in the future. This view is general and can formalize many of the concerns we typically associate with privacy: discrimination because of medical conditions, social ostracism, demographic profiling, etc.

The main drawback of this view is that it is difficult to know what the future mechanisms $M', M'', \ldots$ may be. However, if M is differentially private, then participating in M entails a guarantee that remains meaningful *even without knowing the future mechanisms*. To see this, we will use the following definition that is equivalent to the definition of ϵ-differential privacy [3]:

Definition 1 (Differential privacy) A (randomized) mechanism $M : X^n \to Y$ is ϵ-differentially private if for all $\mathbf{x}, \mathbf{x}' \in X^n$ that differ on one entry, and for all $g : Y \to [0, \infty)$, it holds that

$$\mathrm{Exp}[g(M(\mathbf{x}))] \leq e^\epsilon \cdot \mathrm{Exp}[g(M(\mathbf{x}'))],$$

where the expectation is over the randomness introduced by the mechanism M.

Note that $e^\epsilon \approx 1 + \epsilon$ for small ϵ; thus, if $g(y)$ models the expected utility of an individual tomorrow given that the result of $M(x) = y$ today, then by participating in a differentially private mechanism, the individual's utility will change by at most ϵ.

Fact 1. Let $g : Y \to [-1, 1]$. If M is ϵ-differential private, then $\mathrm{Exp}[g(M(\mathbf{x}'))] - \mathrm{Exp}[g(M(\mathbf{x}))] \leq 2(e^\epsilon - 1) \approx 2\epsilon$ for all $\mathbf{x}, \mathbf{x}' \in X^n$ that differ on one entry.

To see why this is true, let $g_-(y) = \mathbf{max}(0, -g(y))$ and $g_+(y) = \mathbf{max}(0, g(y))$. From Definition 1 and the bound on the outcome of g, we get that $\mathrm{Exp}[g_+(M(\mathbf{x}'))] - \mathrm{Exp}[g_+(M(\mathbf{x}))] \leq (e^\epsilon - 1) \cdot \mathrm{Exp}[g_+(M(\mathbf{x}))] \leq e^\epsilon - 1$ and, similarly, $\mathrm{Exp}[g_-(M(\mathbf{x}))] - \mathrm{Exp}[g_-(M(\mathbf{x}'))] \leq e^\epsilon - 1$. As $g(y) = g_+(y) - g_-(y)$, we conclude that $\mathrm{Exp}[g(M(\mathbf{x}'))] - \mathrm{Exp}[g(M(\mathbf{x}))] \leq 2(e^\epsilon - 1)$.

With this in mind, we typically view λ as being "bounded by differential privacy" in the sense that if M is ϵ-differentially private, then $|\lambda_i| \leq p_i \cdot \epsilon$, where p_i (a positive real number) is an upper bound on the maximum value of $2|g(y)|$. In certain settings we make even more specific assumptions about λ_i, and these are discussed in the sequel.

Generic Problems

We will discuss two generic problems for which key results will be given in the next section:

Privacy-aware mechanism design. Given an optimization problem $q : X^n \times Y \to \mathbb{R}$, construct a privacy-aware mechanism whose output $\hat{y}$ approximately maximizes $q(\mathbf{x}, \cdot)$. Using the terminology above, this corresponds to setting $f(\mathbf{x}) = \mathbf{argmax}_y q(\mathbf{x}, y)$, and the mechanism is said to compute $f()$ with accuracy α if (with high probability) $q(\mathbf{x}, f(\mathbf{x})) - q(\mathbf{x}, \hat{y}) \leq \alpha$. We mention two interesting instantiations of $q()$. When $q(\mathbf{x}, y) = \sum_i v(x_i, y)$, the problem is of maximizing social welfare. When x_i corresponds to how agent i values a digital good and $Y = \mathbb{R}^+$ is interpreted as a price for the good, setting $q(\mathbf{x}, y) = y \cdot |i : x_i \geq y|$ corresponds to maximizing the revenue from the good.

Purchasing privacy. Given a function $f : X^n \to Y$, construct a mechanism computing payments to agents for eliciting permission to use (some of) the entries of $\mathbf{x}$ in an approximation for $f(\mathbf{x})$. Here it is assumed that the agents cannot lie about their private values (but can misreport their privacy valuations). We will consider two variants of the problem. In the *insensitive value model*, agents only care about the privacy of their private values $\mathbf{x}$. In the *sensitive value model*, agents also care about the privacy of their

M

privacy valuations $\mathbf{p}$, e.g., because there may be a correlation between x_i and p_i.

Basic Differentially Private Mechanisms

We conclude this section with two differentially private mechanisms that are used in the constructions presented in the next section.

The Laplace mechanism [3]. The Laplace distribution with parameter $1/\epsilon$, denoted Lap$(1/\epsilon)$, is a continuous probability distribution with zero mean and variance $2/\epsilon$. The probability density function of Lap$(1/\epsilon)$ is $h(z) = \frac{\epsilon}{2}e^{-\epsilon|z|}$. For $\Delta \geq 0$ we get $\Pr_{Z\sim \text{Lap}(1/\epsilon)}[|Z| > \Delta] = e^{-\epsilon\Delta}$.

Fact 2. The mechanism M_{Lap} that on input $\mathbf{x} \in \{0,1\}^n$ outputs $y = \#\{i : x_i = 1\} + Z$ where $Z \sim \text{Lap}(1/\epsilon)$ is ϵ-differentially private. From the properties of the Laplace distribution, we get that

$$\Pr_{y\sim M_{\text{Lap}}(\mathbf{x})}[|y - \#\{i : x_i = 1\}| > \Delta] \leq e^{-\epsilon\Delta}.$$

The exponential mechanism [8]. Consider the optimization problem defined by $q : X^n \times Y \to \mathbb{R}$, where q satisfies $|q(\mathbf{x}, y) - q(\mathbf{x}', y)| \leq 1$ for all $y \in Y$ and all $\mathbf{x}, \mathbf{x}'$ that differ on one entry.

Fact 3. The mechanism M_{Exp} that on input $\mathbf{x} \in X^n$ outputs $y \in Y$ chosen according to

$$\Pr[y = t] = \frac{\exp\left(\frac{\epsilon}{2}q(\mathbf{x}, t)\right)}{\sum_{\ell\in Y}\exp\left(\frac{\epsilon}{2}q(\mathbf{x}, \ell)\right)} \tag{2}$$

is ϵ-differentially private. Moreover,

$$\Pr_{y\sim M_{\text{Exp}}(\mathbf{x})}[q(\mathbf{x}, y) \geq \text{opt}(\mathbf{x}) - \Delta] \geq 1 - |Y| \cdot \exp(-\epsilon \cdot \Delta/2), \tag{3}$$

where $\text{opt}(\mathbf{x}) = \mathbf{max} y \in Y(q(\mathbf{x}, y))$.

Notation. For two n-entry vectors $\mathbf{x}, \mathbf{x}'$, we write $\mathbf{x} \sim_i \mathbf{x}'$ to denote that they agree on all but the i-th entry. We write $\mathbf{x} \sim \mathbf{x}'$ if $\mathbf{x} \sim_i \mathbf{x}'$ for some i.

Key Results

The work of McSherry and Talwar [8] was first to realize a connection between differential privacy and mechanism design. They observed that (with bounded utility from the outcome) a mechanism that preserves ϵ-differential privacy is also ϵ-truthful, yielding ϵ-truthful mechanisms for approximately maximizing social welfare or revenue. Other works in this vein – using differential privacy but without incorporating the effect of privacy loss directly into the agent's utility function – include [6, 10, 12].

Privacy-Aware Mechanism Design

The mechanisms of this section share the following setup assumptions:

Optimization problem. $q : X^n \times Y \to [0, n]$ and a utility function $U : X \times Y \to [0, 1]$.

Input. n players each having an input $x_i \in X$ and a privacy valuation p_i. The players may misreport x_i.

Output. The mechanism outputs an element $y \in Y$ approximately maximizing $q(\mathbf{x}, y)$.

Utility. Each player obtains utility $U(x_i, y) - \lambda_i$ where the assumptions on how the privacy loss λ_i behaves vary for the different mechanisms below and are detailed in their respective sections.

Accuracy. Let $\text{opt}(\mathbf{x}) = \mathbf{max}_{y\in Y}(q(\mathbf{x}, y))$. A mechanism is (Δ, δ)-accurate for all $\mathbf{x}$ if it chooses $y \in Y$ such that $\Pr[\text{opt}(\mathbf{x}) - q(\mathbf{x}, y) \leq \Delta] \geq 1 - \delta$ where the probability is taken over the random coins of the mechanism. (One can also define accuracy in terms of $\text{opt}(\mathbf{x}) - \text{Exp}[q(\mathbf{x}, y)]$.)

Worst-Case Privacy Model

In the worst-case privacy model, the privacy loss of mechanism M is only assumed to be upper bounded by the mechanisms' privacy parameter, as in the discussion following Fact 1 [9]:

$$0 \leq \lambda_i \leq p_i \cdot \epsilon \quad \text{where} \quad \epsilon = \mathbf{max}_{x'\sim x, y\in Y} \ln \frac{\Pr[M(x) = y]}{\Pr[M(x') = y]}. \tag{4}$$

Nissim, Orlandi, and Smorodinsky [9] give a generic construction of privacy-aware mechanisms assuming an upper bound on the privacy loss as in Eq. 4. The fact λ_i is only upper bounded and excludes the possibility of punishing misreporting via privacy loss (compare with Algorithms 3 and 4 below), and hence, the generic construction resorts to a somewhat nonstandard modeling from [10]. To illustrate the main components of the construction, we present a specific instantiation in the context of pricing a digital good, where such a nonstandard modeling is not needed.

Algorithm 1 (ApxOptRev)

Auxiliary input: privacy parameter ϵ, probability $0 < \eta < 1$.
Input: $\mathbf{x}' = (x'_1, \ldots, x'_n) \in X^n$.

ApxOptRev executes M_1 with probability $1 - \eta$ and M_2 otherwise, where M_1, M_2 are:

M_1: Choose $y \in Y$ using the exponential mechanism, M_{Exp} (Fact 3), i.e.,

$$\Pr[y = t] = \frac{\exp\left(\frac{\epsilon}{2} \cdot t \cdot |\{i : x'_i \geq t\}|\right)}{\sum_{\ell \in Y} \exp\left(\frac{\epsilon}{2} \cdot \ell \cdot |\{i : x'_i \geq \ell\}|\right)}.$$

M_2: Choose $y \in Y$ uniformly at random.

Pricing a digital good. An auctioneer selling digital good wishes to design a single price mechanism that would (approximately) optimize her revenue. Every agent i has a valuation $x_i \in X = \{0, 0.01, 0.02, \ldots, 1\}$ for the good and privacy preference p_i. Agents are asked to declare x_i to the mechanism, which chooses a price $y \in Y = \{0.01, 0.02, \ldots, 1\}$. Let x'_i be the report of agent i. If $x'_i < y$, then agent i does not pay nor receives the good and hence gains zero utility, i.e., $v_i = 0$. If $x'_i \geq y$, then agent i gets the good and pays y and hence gains in utility. We let this gain be $v_i = x'_i - y + 0.005$, where the additional 0.005 can be viewed as modeling a preference to receive the good (technically, this breaks the tie between the cases $x'_i = y$ and $x'_i = y - 1$). To summarize,

$$v(x_i, x'_i, y) = \begin{cases} x_i - y + 0.005 & \text{if } y < x'_i \\ 0 & \text{otherwise} \end{cases}$$

The privacy loss for agent i is from the information that may be potentially leaked on x'_i via the chosen price y. The auctioneer's optimal revenue is $\text{opt}(\mathbf{x}) = \mathbf{max}_{t \in Y}(t \cdot |\{i : x_i \geq t\}|)$, and the revenue she obtains when the mechanism chooses price y is $y \cdot |\{i : x'_i \geq y\}|$. The mechanism is presented in Algorithm 1.

Agent utility. To analyze agent behavior, compare the utility of a misreporting agent to a truthful agent. (i) As Algorithm 1 is ϵ-differentially private, by our assumption on λ_i, by misreporting agent i may reduce her disutility due to information leakage by *at most* $p_i \cdot \epsilon$. (ii) Note that $v(x_i, x'_i, y) \leq v(x_i, x_i, y)$. Using this and Fact 1, we can bound the expected gain due to misreporting in M_1 as follows:

$$\text{Exp}_{y \sim M_1(\mathbf{x}'_{-i}, x'_i)}[v(x_i, x'_i, y)] - \text{Exp}_{y \sim M_1(\mathbf{x}'_{-i}, x_i)}[v(x_i, x_i, y)] \leq$$
$$\text{Exp}_{y \sim M_1(\mathbf{x}'_{-i}, x'_i)}[v(x_i, x'_i, y)] - \text{Exp}_{y \sim M_1(\mathbf{x}'_{-i}, x_i)}[v(x_i, x'_i, y)] \leq 2 \cdot \epsilon.$$

(iii) On the other hand, in M_2, agent i loses at least $g = 0.01 \cdot 0.005$ in utility whenever $x'_i \neq x'_i$; this is because y falls in the set $\{x_i + 0.01, \ldots, x'_i\}$ with probability $x'_i - x_i \geq 0.01$ when $x_i < x'_i$, in which case she loses at least 0.005 in utility and, similarly, y falls in the set $\{x'_i, \ldots, x_i - 0.01\}$ with probability $x_i - x'_i \geq 0.01$ when $x_i < x'_i$, in which case she loses at least 0.005 in utility.

We hence get that agent i strictly prefers to report truthfully when

$$2 \cdot \epsilon - \eta \cdot g + p_i \cdot \epsilon < 0. \tag{5}$$

Designer utility. Let m be the number of agents for which Eq. 5 does not hold. We have $\text{opt}(\mathbf{x}') \geq \text{opt}(\mathbf{x}) - m$, and hence, using Fact 3, we get that

M

$$\Pr_{y \sim \text{ApxOptRev}(\mathbf{x}')}\left[y \cdot \left|\{i : x_i' \geq y\}\right| < \text{opt}(\mathbf{x}) - m - \Delta\right] \leq |Y| \cdot \exp(-\epsilon\Delta/2) + \eta$$
$$= 100 \cdot \exp(-\epsilon\Delta/2) + \eta.$$

We omit from this short summary the discussion of how to choose the parameters ϵ and η (this choice directly affects m). One possibility is to assume the p_i has nice properties [9].

Per-Outcome Privacy Model

In the output specific model, the privacy loss of mechanism M is evaluated on a per-output basis [2]. Specifically, on output $y \in Y$ is assumed that

$$|\lambda_i(\mathbf{x}, y)| \leq p_i \cdot F_i(\mathbf{x}, y) \text{ where } F_i(\mathbf{x}, y) = \mathbf{max}_{\mathbf{x}', \mathbf{x}'' \sim_i \mathbf{x}} \ln \frac{\Pr[M(\mathbf{x}') = y]}{\Pr[M(\mathbf{x}'') = y]}. \quad (6)$$

To interpret Eq. 6, consider an Bayesian adversary that has a prior belief μ on x_i and fix $\mathbf{x}_{-i}$. After seeing $y = M(\mathbf{x}_{-i}, x_i)$, the Bayesian adversary updates her belief to μ'. For every event E defined over x_i, we get that

$$\mu'(E) = \mu(E|M(\mathbf{x}_{-i}, x_i) = y)$$
$$= \mu(E) \cdot \frac{\Pr[M(\mathbf{x}_{-i}, x_i) = y|E]}{\Pr[M(\mathbf{x}_{-i}, x_i) = y]}$$
$$\in \mu'(E) \cdot e^{\pm F_i(\mathbf{x}, y)}.$$

This suggests that λ_i models harm that is "continuous" in the change in adversarial belief about i, in the sense that a small adversarial change in belief entails small harm. (Note, however, that this argument is restricted to adversarial beliefs on x_i *given* $\mathbf{x}_{-i}$.)

Comparison with Worst-Case Privacy

Note that if M is ϵ-differentially private, then $F_i(\mathbf{x}, y) \leq \epsilon$ for all $\mathbf{x}, y$. Equation 6 can hence be seen as a variant of Eq. 4 where the fixed value ϵ is replaced with the output specific $F_i(\mathbf{x}, y)$. One advantage of such a per-outcome model is that the typical gain from misreporting is significantly smaller than ϵ. In fact, for all $\mathbf{x} \in X^n$ and $x_i' \in X$,

$$\left|\text{Exp}_{y \sim \text{M}(\mathbf{x})}\left[F_i(\mathbf{x}, y)\right] - \text{Exp}_{y \sim \text{M}(\mathbf{x}_{-i}, x_i')}\left[F_i(\mathbf{x}, y)\right]\right| = O(\epsilon^2).$$

On the other hand, the modeled harm is somewhat weaker, as (by Fact 1) Eq. 4 also captures harm that is not continuous in beliefs (such as decisions based on the belief crossing a certain threshold).

Assuming privacy loss is bounded as in Eq. 6, Chen, Chong, Kash, Moran, and Vadhan [2] construct truthful mechanisms for an election between two candidates, facility location, and a VCG mechanism for public projects (the latter uses payments). Central to the constructions is the observation that F_i is large exactly when agent i has influence on the outcome of $M()$. To illustrate the main ideas in the construction, we present here the two-candidate election mechanism.

Two-candidate election. Consider the setting of an election between two candidates. Every agent i has a preference $x_i \in X = \{A, B\}$ and privacy preference p_i. Agents are asked to declare x_i to the mechanism, which chooses an outcome $y \in Y = \{A, B\}$. The utility of agent i is then

$$v(x_i, y) = \begin{cases} 1 \text{ if } x = y \\ 0 \text{ otherwise} \end{cases}$$

The privacy loss for agent i is from the information that may be potentially leaked on her reported x_i' via the outcome y. The designer's goal is to (approximately) maximize the agents' social

Algorithm 2 ApxMaj

Auxiliary input: privacy parameter ϵ.
Input: $\mathbf{x}' = (x'_1, \ldots, x'_n) \in X^n$.

ApxMaj performs the following:

1. Sample a value Z from Lap$(1/\epsilon)$.
2. Choose $y = A$ if $|\{j : x'_j = A\}| > |\{j : x'_j = B\}| + Z$ and $y = B$ otherwise.

welfare (i.e., total utility from the outcome). The mechanism is presented in Algorithm 2.

Agent utility. To analyze agent behavior, we compare the utility of a misreporting agent to a truthful agent. Notice that once the noise Z is fixed if agent i affects the outcome, then her disutility from information leakage is at most $p_i \cdot \epsilon$ and her utility from the outcome decreases by 1. If agent i cannot affect the outcome, then misreporting does not change either. We hence get that agent i strictly prefers to report truthfully when

$$p_i \cdot \epsilon < 1. \tag{7}$$

Note that by our analysis, Eq. 7 implies *universal truthfulness* – agent i prefers to report truthfully for every choice of the noise Z. In contrast, Eq. 5 only implies truthfulness in expectation.

Social welfare. Letting m be the number of agents for which Eq. 7 does not hold, and using Fact 2, we get that Algorithm ApxMaj maximizes social welfare up to error $m + \frac{\log 1/\delta}{\epsilon}$ with probability $1 - \delta$. As in the previous section, we omit from this short summary the discussion of how to choose ϵ (this choice affects m and hence the accuracy of the mechanism).

Purchasing Privacy

The mechanisms of this section share the following setup assumptions, unless noted otherwise:

Input. n players each having a data bit $x_i \in \{0, 1\}$ and a privacy valuation $p_i > 0$. The players may misreport p_i but cannot misreport x_i. We will assume for convenience of notation that $p_1 \leq p_2 \leq \ldots \leq p_n$.

Intermediate outputs. The mechanism selects a subset of participating players $S \subset [n]$ and a scaling factor t and a privacy parameter ϵ.

Output. The mechanism uses the Laplace mechanism to output an estimate $s = \frac{1}{t}\left(\sum_{i \in [S]} x_i + Z\right)$ where $Z \sim$ Lap$(1/\epsilon)$ and payments v_i for $i \in [n]$.

Utility. Each player obtains utility $v_i - \lambda_i$ where the assumptions on how the privacy loss λ_i behaves vary for the different mechanisms below and are detailed in their respective sections.

Accuracy. A mechanism is α-accurate if $\Pr[|s - f(x)| \leq \alpha n] \geq 2/3$ where the probability is taken over the random coins of the mechanism.

We focus on designing mechanisms that approximate the sum function $f(x) = \sum_{i=1}^{n} x_i$ where each $x_i \in \{0, 1\}$, which has been the most widely studied function in this area. As one can see from the above setup assumptions, the crux of the mechanism design problem is in selecting the set S, choosing a privacy parameter ϵ, and computing payments for the players. We note that several of the works we describe below generalize beyond the setting we describe here (i.e., computing different f, fewer assumptions, etc.). The following presentation was designed to give a unified overview (sacrificing some generality), but to preserve the essence both of the challenges posed by the problem of purchasing private data and each mechanism's idea in addressing the challenges.

Insensitive Valuation Model

In the insensitive valuation model, the privacy loss λ_i of a mechanism M is assumed to be [5]

$$\lambda_i = p_i \cdot \epsilon_i \text{ where } \epsilon_i = \mathbf{max}_{x, x' \sim_i x, p, s} \ln \frac{\Pr[M(x, p) = s]}{\Pr[M(x', p) = s]}. \tag{8}$$

It is named the insensitive valuation model because ϵ_i only measures the effect on privacy of changing player i's data bit, but not the effect of changing that player's privacy valuation.

M

Algorithm 3 (FairQuery)

Auxiliary input: budget constraint $B > 0$.

1. Let $k \in [n]$ be the largest integer such that $p_k(n-k) \leq B/k$.
2. Select $S = \{1, \ldots, k\}$ and set $\epsilon = \frac{1}{n-k}$. Set the scaling factor $t = 1$.
3. Set payments $v_i = 0$ for all $i > k$ and $v_i = \min\{\frac{B}{k}, p_{k+1}\epsilon\}$ for all $i \leq k$.

Algorithm 4 MinCostAuction

Auxiliary input: accuracy parameter $\alpha \in (0, 1)$.

1. Set $\alpha' = \frac{\alpha}{1/2+\ln 3}$ and $k = (1-\alpha')n$.
2. Select $S - \{1, \ldots, k\}$ and $\epsilon = \frac{1}{n-k}$. Set the scaling factor $t = 1$.
3. Set payments $v_i = 0$ for $i > k$ and $v_i = p_{k+1}\epsilon$ for all $i \leq k$.

Mechanisms. Two mechanisms are presented in the insensitive value model in [5], listed in Algorithms 3 and 4. Algorithm 3 (FairQuery) is given a hard budget constraint and seeks to optimize accuracy under this constraint; Algorithm 4 (MinCostAuction) is given a target accuracy requirement and seeks to minimize payouts under these constraints.

Guarantees. Algorithms 3 and 4 are individually rational and truthful. Furthermore, Algorithm 3 achieves the best possible accuracy (up to constant factors) for the class of *envy-free* and individually rational mechanisms, where the sum of payments to players does not exceed B. Algorithm 4 achieves the minimal payout (up to constant factors) for the class of *envy-free* and individually rational mechanisms that achieve α-accuracy.

Sensitive Value Model

Ghosh and Roth [5] also defined the sensitive value model where λ_1 is as in Eq. 8, except that ϵ_i is defined to equal

$$\mathbf{max}_{x,p,(x',p')\sim_i(x,p),s} \ln \frac{\Pr[M(x,p)=s]}{\Pr[M(x',p')=s]}. \tag{9}$$

Namely, we also measure the effect on the outcome distribution of the change in a single player's privacy valuation. It was shown in [5] and subsequent generalizations [11] that in this model and various generalizations where the privacy valuation itself is sensitive, it is impossible to build truthful, individually rational, and accurate mechanisms with worst-case guarantees and making finite payments. To bypass these impossibility results, several relaxations were introduced.

Bayesian relaxation [4]. Fleischer and Lyu use the sensitive notion of privacy loss given in Eq. 9. In order to bypass the impossibility results about sensitive values, they assume that the mechanism designer has knowledge of prior distributions P^0, P^1 for the privacy valuations. They assume that all players with data bit b have privacy valuation sampled independently according to P^b, namely, that $p_i \stackrel{R}{\leftarrow} P^{x_i}$, independently for all i. Their mechanism is given in Algorithm 5.

Algorithm 5 Bayesian mechanism from [4]

Auxiliary input: privacy parameter ϵ.

1. Compute $c = 1 - \frac{2}{\epsilon^2 n}$. Compute α_b for $b \in \{0, 1\}$ such that $\Pr_{p \stackrel{R}{\leftarrow} P^b}[p \leq \alpha_b] = c$.
2. Set S be the set of players i such that $p_i \leq \alpha_{x_i}$. Set the scaling factor $t = c$.
3. For each player $i \in S$, pay $\epsilon\alpha_{x_i}$. Pay the other players 0.

Algorithm 5 is truthful and individually rational. Assuming that the prior beliefs are correct, the mechanism is $O(\frac{1}{\epsilon n})$-accurate. The key use of knowledge of the priors is in accuracy: the probability of a player participating is c independent of its data bit.

Take-it-or-leave-it mechanisms [7]. Ligett and Roth put forward a setting where the privacy loss is decomposed into two parts

$$\lambda_i = \lambda_i^p + \lambda_i^x,$$

where λ_i^p is the privacy loss incurred by leaking information of whether or not an individual is selected to participate (i.e., whether individual i is in the set S), and where λ_i^x is the privacy loss incurred by leaking information about the actual data bit.

The interpretation is that a surveyor approaches an individual and offers them v_i to participate. The individual cannot avoid responding to this question and so unavoidably incurs a privacy loss λ_i^p without compensation. If he chooses to participate, then he loses an additional λ_i^x, but in this case he receives v_i in payment. While this is the framework we have been working in all along, up until now we have not distinguished between these two sources of privacy loss, rather considering only the overall loss. By explicitly separating them, [7] can make more precise statements about how incentives relate to each source of privacy loss.

In this model the participation decision of an individual is a function (only) of its privacy valuation, and so we define

$$\lambda_i^p = p_i \epsilon_i^p \text{ where } \epsilon_i^p = \mathbf{max}_{x,p,p' \sim_i p, s} \ln \frac{\Pr[M(x,p)=s]}{\Pr[M(x,p')=s]}. \quad (10)$$

We define $\lambda_i^x = p_i \epsilon_i^x$ where ϵ_i^x is as in the insensitive model, Eq. 8. The mechanism is given in Algorithm 6.

Algorithm 6 is α-accurate. It is not individually rational since players cannot avoid the take-it-or-leave-it offer, which leaks information about their privacy valuation that is not compensated. However, it is "one-sided truthful" in the sense that rational players will accept any offer v_i satisfying $v_i \geq \lambda_i^p - \lambda_i^x$. [7] also proves that for appropriately chosen η, the total payments made by Algorithm 6 are not much more than that of the optimal envy-free mechanism making the same take-it-or-leave-it offers to every player.

Algorithm 6 (Take-it-or-leave-it mechanism [7])

Auxiliary input: accuracy parameter $\alpha \in (0,1)$, payment increment $\eta > 0$.

1. Set $j = 1$ and $\epsilon = \alpha$.
2. Repeat the following:
   (a) Set $E_j = 100(\log j + 1)/\alpha^2$ and $S_j = \emptyset$.
   (b) For $i = 1$ until E_j:
      i. Sample without replacement $i \xleftarrow{R} [n]$.
      ii. Offer player i a payment of $(1+\eta)^j$.
      iii. If player i accepts, set $S_j = S_j \cup \{i\}$.
   (c) Sample $\nu \xleftarrow{R} \Lambda(1/\epsilon)$. If $|S_j| + \nu \geq (1-\alpha/8)E_j$, then break and output selected set $S = S_j$, privacy parameter ϵ, and normalizing factor $t = E_j$. For every $j' \leq j$, pay $(1+\eta)^{j'}$ to each player that accepted in round j' and pay 0 to all other players.
   (d) Otherwise, increment j and continue.

Monotonic valuations [11]. Nissim, Vadhan, and Xiao [11] study a relaxation of sensitive values that they call *monotonic valuations*, where it is assumed that

$$\lambda_i(x,p) \leq p_i \cdot \epsilon_i^{\text{mon}}(x,p) \text{ where } \epsilon_i^{\text{mon}}(x,p) = \mathbf{max}_{(x',p') \sim_i^{\text{mon}} (x,p), s} \ln \frac{\Pr[M(x,p)=s]}{\Pr[M(x',p')=s]}. \quad (11)$$

Here, $(x',p') \sim_i^{\text{mon}} (x,p)$ denotes that $(x',p'),(x,p)$ are identical in all entries except the i'th entry, and in the i'th entry, it holds that either $x_i > x_i'$ and $p_i \geq p_i'$ both hold or $x_i < x_i'$ and $p_i \leq p_i'$ both hold.

The intuition behind the definition is that for many natural settings, $x_i = 1$ is more sensitive than $x_i = 0$ (e.g., if x_i represents whether an individual tested positive for syphilis), and it is therefore reasonable to restrict attention to the case where the privacy valuation when $x_i = 1$ is at least the privacy valuation when $x_i = 0$.

There are two other aspects in which this notion is unlike those used in the earlier works on purchasing privacy: (i) the definition may depend on the input, so the privacy loss may be smaller on some inputs than others, and (ii) we assume only an upper bound on the privacy loss, since ϵ_i^{mon} does not say *which* information is leaked about player i, and so it may be that the harm done to player i is not as severe as ϵ_i^{mon} would suggest. The mechanism is given in Algorithm 7.

Algorithm 7 (Mechanism for monotonic valuations [11])

Auxiliary inputs: budget constraint $B > 0$, privacy parameter $\epsilon > 0$.

1. Set $\tau = \frac{B}{2\epsilon n}$.
2. Output selected set $S = \{i \mid p_i \leq \tau\}$, output privacy parameter ϵ, and scaling factor $t = 1$.
3. Pay B/n to players in S, pay 0 to others.

Algorithm 7 is individually rational for all players and truthful for all players satisfying $p_i \leq \tau$. Assuming all players are rational, on inputs where there are h players having $p_i > \tau$, the mechanism is $(O(\frac{1}{\epsilon n}) + h)$-accurate. The accuracy guarantee holds regardless of how the players with $p_i > \tau$ report their privacy valuations.

Recommended Reading

1. Alpert CJ, Chan T, Kahng AB, Markov IL, Mulet P (1998) Faster minimization of linear wirelength for global placement. IEEE Trans CAD 17(1):3–13
2. Chen Y, Chong S, Kash IA, Moran T, Vadhan SP (2013) Truthful mechanisms for agents that value privacy. In: EC 2013, Philadelphia, pp 215–232
3. Dwork C, McSherry F, Nissim K, Smith A (2006) Calibrating noise to sensitivity in private data analysis. In: TCC 2006, New York, pp 265–284
4. Fleischer L, Lyu Y-H (2012) Approximately optimal auctions for selling privacy when costs are correlated with data. In: EC 2012, Valencia, pp 568–585
5. Ghosh A, Roth A (2011) Selling privacy at auction. In: EC 2011, San Jose, pp 199–208
6. Huang Z, Kannan S (2012) The exponential mechanism for social welfare: private, truthful, and nearly optimal. In: FOCS 2012, New Brunswick, pp 140–149
7. Ligett K, Roth A (2012) Take it or leave it: running a survey when privacy comes at a cost. In: WINE 2012, Liverpool, pp 378–391
8. McSherry F, Talwar K (2007) Mechanism design via differential privacy. In: FOCS 2007, Providence, pp 94–103
9. Nissim K, Orlandi C, Smorodinsky R (2012) Privacy-aware mechanism design. In: EC 2012, Valencia, pp 774–789
10. Nissim K, Smorodinsky R, Tennenholtz M (2012) Approximately optimal mechanism design via differential privacy. In: ITCS 2012, Boston, pp 203–213
11. Nissim K, Vadhan SP, Xiao D (2014) Redrawing the boundaries on purchasing data from privacy-sensitive individuals. In: ITCS 2014, Princeton, pp 411–422
12. Xiao D (2013) Is privacy compatible with truthfulness? In: ITCS 2013, Berkeley, pp 67–86

Mellor-Crummey and Scott Mutual Exclusion Algorithm

Danny Hendler
Department of Computer Science, Ben-Gurion University of the Negev, Beer-Sheva, Israel

Keywords

Critical section; Local spinning; MCS lock; Queue lock; Remote memory references (RMRs)

Years and Authors of Summarized Original Work

1991; Mellor-Crummey, Scott

Problem Definition

Mutual exclusion is a fundamental concurrent programming problem (see ▶ Concurrent Programming, Mutual Exclusion entry), in which a set of processes must coordinate their access to a *critical section* so that, at any point in time, at most a single process is in the critical section.

To a large extent, shared-memory mutual exclusion research focused on *busy-waiting* mutual exclusion, in which, while waiting for the critical section to be freed, processes repeatedly test the values of shared-memory variables. A significant portion of this research over the last two decades was devoted to *local-spin algorithms* [2], in which all busy-waiting is done by means of read-only loops that repeatedly test locally accessible variables.

Local-Spin Algorithms and the RMRs Metric

A natural way to measure the time complexity of algorithms in shared-memory multiprocessors is to count the number of memory accesses they require. This measure is problematic for busy-waiting algorithms because, in this case, a process may perform an unbounded number of memory accesses while busy-waiting for another process holding the lock. Moreover, Alur and Taubenfeld [1] have shown that even the first process to enter the critical section can be made to perform an unbounded number of accesses.

As observed by Anderson, Kim, and Herman [6], "most early shared-memory algorithms employ... busy-waiting loops in which many shared variables are read and written... Under contention, such busy-waiting loops generate excessive traffic on the processors-to-memory interconnection network, resulting in poor performance."

Contemporary shared-memory mutual exclusion research focuses on *local-spin* algorithms, which avoid this problem as they busy-wait by means of performing read-only loops that repeatedly test locally accessible variables (see, e.g., [4, 5, 7, 9, 11, 13, 14]). The performance of these algorithms is measured using the remote memory references (RMRs) metric.

The classification of memory accesses into local and remote depends on the type of multiprocessor. In the *distributed shared-memory* (DSM) model, each shared variable is local to exactly one processor and remote to all others. In the *cache-coherent* (CC) model, each processor maintains local copies of shared variables inside a cache; the consistency of copies in different caches is ensured by a coherence protocol. At any given time, a variable is local to a processor if the coherence protocol guarantees that the corresponding cache contains an up-to-date copy of the variable and is remote otherwise.

Anderson was the first to present a local-spin mutual exclusion algorithm using only reads and writes with bounded RMR complexity [3]. In his algorithm, a process incurs $O(n)$ RMRs to enter and exit its critical section, where n is the maximum number of processes participating in the algorithm. Yang and Anderson improved on that and presented an $O(\log n)$ RMRs mutual exclusion algorithm based on reads and writes [20]. This is asymptotically optimal under both the CC and DSM models [7].

Read-Modify-Write Operations

The system's hardware or operating system provides *primitive operations* (or simply *operations*) that can be applied to shared variables. The simplest operations, which are always assumed, are the familiar *read* and *write* operations. Modern architectures provide stronger *read-modify-write* operations (a.k.a. *fetch-and-*Φ operations). The most notable of these is *compare and swap* (abbreviated CAS), which takes three arguments: an address of a shared variable, an expected value, and a new value. The CAS operation atomically does the following: if the variable stores the expected value, it is replaced with the new value; otherwise, it is unchanged. The success or failure of the CAS operation is then reported back to the program. It is crucial that this operation is executed atomically; thus, an algorithm can read a datum from memory, modify it, and write it back only if no other process modified it in the meantime.

Another widely implemented RMW operation is the *swap* operation, which takes two arguments: an address of a shared variable and a new value. When applied, it atomically stores the new value to the shared variable and returns the previous value. The CAS operation may be viewed as a conditional version of swap, since it performs a swap operation only if the value of the variable to which it is applied is the expected value.

Architectures supporting strong RMW operations admit implementations of mutual exclusion that are more efficient in terms of their RMR complexity as compared with architectures that support only read and write operations. In work that preceded the introduction of the MCS lock, Anderson [2] and Graunke and Thakkar [12] presented lock algorithms, using strong RMW operations such as CAS and swap, that incur only a constant number of RMRs on CC multiprocessors. However, these algorithms are not local spin on DSM multiprocessors and the amount of pre-allocated memory per lock is linear in the number of processes that may use it.

Key Results

Mellor-Crummey and Scott's algorithm [18] is the first local-spin mutual exclusion algorithm in which processes incur only a constant number of RMRs to enter and exit the critical section, in both CC and DSM multiprocessors. The amount of memory that needs to be pre-allocated by locks using this algorithm (often called *MCS locks*) is constant rather than a function of the maximum number of processes that may use the lock. Moreover, MCS locks guarantee a strong notion of fairness called *first-in, first-out* (FIFO, a.k.a. *first-come-first-served*). Informally, FIFO ensures that processes succeed in capturing the lock in the order in which they start waiting for it (see [15] for a more formal definition of the FIFO property).

Algorithm 1 Mellor-Crummey and Scott algorithm

```
1  Qnode: structure {bit locked, Qnode* next};
2  shared Qnode nodes[0 ... n − 1], Qnode* tail
   ch605:initially null;
3  local Qnode* myNode initially &nodes[i],
   successor, pred;
   Entry code for process i
5  myNode.next ← null;
6  pred ← swap(tail, myNode);
7  if pred ≠ null then
8  |   myNode.locked ← true;
9  |   pred.next ← myNode;
10 |   repeat while myNode.locked = true;
   end
   Critical Section;
   Exit code for process i;
14 if myNode.next = null then
15 |   if compare-and-swap(tail, myNode, null) =
   |   false then
16 |   |   repeat while myNode.next = null ;
17 |   |   successor ← myNode.next;
18 |   |   succcessor.locked ← false ;
   |   end
   else
21 |   successor ← myNode.next;
22 |   succcessor.locked ← false;
   end
```

The Algorithm

Pseudocode of the algorithm is presented in Algorithm 1. The key data structure used by the algorithm is the *nodes* array (statement **2**), where entry i is owned by process i, for $i \in \{0, \ldots, n-1\}$. This array represents a queue of processes waiting to enter the critical section. Each array entry is a Qnode structure (statement **1**), comprising a *next* pointer to the structure of the next process in the queue and a *locked* flag on which a process waits until it is signaled by its predecessor. Shared variable *tail* points to the end of the queue and either stores a pointer to the structure of the last process in the queue or is null if the queue is empty.

Before entering the critical section, a process p first initializes the *next* pointer of its Qnode structure to null (statement **5**), indicating that it is about to become the last process in the queue. It then becomes the last queue process by atomically swapping the values of *tail* and its local Qnode structure (statement **6**); the previous value of *tail* is stored to local variable *pred*. If the queue was previously empty (statement **7**), p enters the critical section immediately. Otherwise, p initializes its Qnode structure (statement **8**), writes a pointer to its Qnode structure to the *next* field of its predecessor's Qnode structure (statement **9**), and then busy-waits until it is signaled by its predecessor (statement **10**).

To exit the critical section, process p first checks whether its queue successor (if any) has set the *next* pointer of its Qnode structure (statement **14**), in which case p signals its successor to enter the critical section (statements **21–22**). Even if no process has set p's *next* pointer yet, it is still possible that p does have a successor q that executed the swap of statement **6** but did not yet update p's *next* pointer in statement **9**. Also in this case, p must signal q before it is allowed to exit the critical section. To determine whether or not this is the case, p attempts to perform a

CAS operation that will swap the value of *tail* back to null if p is the single queue process. If the CAS fails, then p does have a successor and must therefore wait until its *next* pointer is updated by the successor. Once this happens, p signals its successor and exits (statements **16–18**).

Mellor-Crummey and Scott's paper won the 2006 Edsger W. Dijkstra Prize in Distributed Computing. Quoting from the prize announcement, the MCS lock is "...probably the most influential practical mutual exclusion algorithm of all time."

Cross-References

- ▶ Concurrent Programming, Mutual Exclusion
- ▶ Transactional Memory
- ▶ Wait-Free Synchronization

Further Reading

For a comprehensive discussion of local-spin mutual exclusion algorithms, the reader is referred to the excellent survey by Anderson, Kim, and Herman [6]. Craig, Landin, and Hagersten [8, 17] presented another queue lock – the CLH lock. The algorithm underlying CLH locks is simpler than the MCS algorithm and, unlike MCS, only requires the swap strong synchronization operation. On the downside, CLH locks are not local spin on DSM multiprocessors.

In many contemporary multiprocessor architectures, processors are organized in clusters and intercluster communication is much slower than intra-cluster communication. *Hierarchical locks* take into account architecture-dependent considerations such as inter- and intra-cluster latencies and may thus improve lock performance on nonuniform memory architectures (NUMA). The idea underlying hierarchical locks is that intercluster lock transfers should be favored. A key challenge faced by hierarchical lock implementations is that of ensuring fairness.

Radovic and Hagersten presented the first hierarchical lock algorithms [19], based on the idea that a process busy-waiting for a lock should back off for a short duration if the lock is held by a process from its own cluster and for a much longer duration otherwise; this is a simple way of ensuring that intra-cluster lock transfers become more likely. Several works pursued this line of research by presenting alternative NUMA-aware lock algorithms (e.g., [10, 16]). For alternatives to lock-based concurrent programming, the reader is referred to [Wait-free Synchronization, Transactional Memory].

Recommended Reading

1. Alur R, Taubenfeld G (1992) Results about fast mutual exclusion. In: Proceedings of the 13th IEEE real-time systems symposium, Phoenix, pp 12–21
2. Anderson TE (1990) The performance of spin lock alternatives for shared-memory multiprocessors. IEEE Trans Parallel Distrib Syst 1(1):6–16
3. Anderson JH (1993) A fine-grained solution to the mutual exclusion problem. Acta Inf 30(3):249–265
4. Anderson J, Kim YJ (2002) An improved lower bound for the time complexity of mutual exclusion. Distrib Comput 15(4):221–253
5. Kim Y-J, Anderson JH (2006) Nonatomic mutual exclusion with local spinning. Distrib Comput 19(1):19–61
6. Anderson J, Kim YJ, Herman T (2003) Shared-memory mutual exclusion: major research trends since 1986. Distrib Comput 16:75–110
7. Attiya H, Hendler D, Woelfel P (2008) Tight RMR lower bounds for mutual exclusion and other problems. In: STOC, Victoria, pp 217–226
8. Craig T (1993) Building FIFO and priority-queuing spin locks from atomic swap. Technical report
9. Danek R, Golab WM (2010) Closing the complexity gap between FCFS mutual exclusion and mutual exclusion. Distrib Comput 23(2):87–111
10. Dice D, Marathe VJ, Shavit N (2012) Lock cohorting: a general technique for designing NUMA locks. In: PPOPP, New Orleans, pp 247–256
11. Golab WM, Hendler D, Woelfel P (2010) An $O(1)$ RMRS leader election algorithm. SIAM J Comput 39(7):2726–2760
12. Graunke G, Thakkar SS (1990) Synchronization algorithms for shared-memory multiprocessors. IEEE Comput 23(6):60–69
13. Kim YJ, Anderson JH (2007) Adaptive mutual exclusion with local spinning. Distrib Comput 19(3):197–236
14. Kim YJ, Anderson JH (2012) A time complexity lower bound for adaptive mutual exclusion. Distrib Comput 24(6):271–297
15. Lamport L (1974) A new solution of Dijkstra's concurrent programming problem. Commun ACM 17(8):453–455

16. Luchangco V, Nussbaum D, Shavit N (2006) A hierarchical CLH queue lock. In: Euro-Par, Dresden, pp 801–810
17. Magnusson P, Landin A, Hagersten E (1994) Queue locks on cache coherent multiprocessors. In: IPPS, Cancún. IEEE Computer Society, pp 165–171
18. Mellor-Crummey JM, Scott ML (1991) Algorithms for scalable synchronization on shared-memory multiprocessors. ACM Trans Comput Syst 9(1):21–65. doi:10.1145/103727.103729, http://doi.acm.org/10.1145/103727.103729
19. Radovic Z, Hagersten E (2003) Hierarchical backoff locks for nonuniform communication architectures. In: HPCA, Anaheim, pp 241–252
20. Yang JH, Anderson J (1995) A fast, scalable mutual exclusion algorithm. Distrib Comput 9(1):51–60

Memory-Constrained Algorithms

Matias Korman
Graduate School of Information Sciences,
Tohoku University, Miyagi, Japan

Keywords

Connectivity; Constant workspace; Logspace; Multi-pass algorithms; One-pass algorithms; Selection; Sorting; Stack algorithms; Streaming; Time-space trade-off; Undirected graphs

Years and Authors of Summarized Original Work

1980; Munro, Patterson
2000; Reingold
2013; Barba, Korman, Langerman, Sadakane, Silveira

Problem Definition

This field of research evolves around the design of algorithms in the presence of memory constraints. Research on this topic has been going on for over 40 years [16]. Initially, this was motivated by the high cost of memory space. Afterward, the topic received a renewed interest with appearance of smartphones and other types of handheld devices for which large amounts of memory are either expensive or not desirable.

Although many variations of this principle exist, the general idea is the same: the input is in some kind of read-only data structure, the output must be given in a write-only structure, and in addition to these two structures, we can only use a fixed amount of memory to compute the solution. This memory should be enough to cover all space requirements of the algorithm (including the variables directly used by the algorithm, space needed to make recursion, invoking procedures, *etc.*). In the following we list the most commonly considered limitations for both the input and the workspace.

Considerations on the Input

One of the most restrictive models that has been considered is the *one-pass* (or *streaming*) model. In this setting the elements of the input can only be scanned once in a sequential fashion. Given the limitations, the usual aim is to approximate the solution and ideally obtain some kind of worst-case approximation ratio with respect to the optimal solution.

The natural extension of the above constraint is the *multi-pass* model, in which the input can be scanned sequentially a constant number of times. In here we look for trade-off between the number of passes and either the size of the workspace or the quality of the approximation.

The next natural step is to allow input to be scanned any number of times and even allowing *random access* to the input values. Research for this model focuses on either computability (i.e., determining whether or not a particular problem is solvable with a workspace of fixed size) or the design of efficient algorithms whose running time is not much worse (when compared to the case in which no space constraints exist).

A more generous model considered is the *in-place* one. In this model, the values of the input can be rearranged (or sometimes even overwritten). Note that the input need not be recoverable after an execution of the algorithm. By making an appropriate permutation of the input, we can usually encode different data structures. Thus, algorithms under this model often achieve the running times comparable to those in unconstrained settings.

Considerations on the Workspace

The most natural way to measure the space required by the algorithms is simply the number of bits used. On many cases it is simpler to count the number of *words* (i.e., the minimum amount of space needed to store a variable, a pointer to some position in the input, or simply a counter) used by the algorithm. It is widely accepted that a word needs $\Theta(\log n)$ bits; thus it is easy to alternate between both approaches.

Most of the literature focuses in the case in which the workspace can only fit $O(\log n)$ bits. This workspace (combined with random access to the input) defines the heavily studied *log-space* complexity class within computational complexity. Within this field the main focus of research is to determine whether or not a problem can be solved (without considering other properties such as the running time of the algorithm). Due to the logarithmic bit-word equivalence, an algorithm that uses $O(\log n)$ bits is also referred to as a *constant workspace* algorithm.

There has also been an interest in the design of algorithms whose workspace depends on some parameter determined by the user. In this case the aim is to obtain an algorithm whose running time decreases as the space increases (this is often referred to as a *time-space trade-off*).

Key Results

Selection and Sorting in Multi-pass Models

One of the most studied problems under the multi-pass model is *sorting*. That is, given a list of n distinct numbers, how fast can we sort them? How many passes of the input are needed when our total amount of memory is limited? Whenever workspace is not large enough to hold the sorted list, the aim is to simply report the values of the input in ascending order.

The first time-space trade-off algorithm for sorting under the multi-pass model was given by Munro and Paterson [13], where several upper and lower space bounds were given (as a function on the number of times we can scan the input). The bounds were afterward improved and extended for the random access model: it is known that the running time of an algorithm that uses $O(b)$ bits must be at least $\Omega(n^2/b)$ [8]. Matching upper bounds for the case in which $b \in \Omega(\log n) \cap O(n/\log n)$ were shown by Pagter and Rauhe [15] (where b denotes the size of the workspace, in bits).

Selection

Another closely related topic is *selection*. In addition to the list of n numbers, we are given an integer $k \leq n$. The aim is to compute the number whose rank is k (i.e., the kth smallest value). It is well-known that this problem can be solved in linear time when no space constraints exist [7]. Munro and Patterson [13] presented a time-space trade-off algorithm for the multi-pass model. For any $w \in \Omega(\log^2 n) \cap O(n/\log n)$, the algorithm runs in $O(n \log_w n + n \log w)$ time and uses $O(w)$ words of space.

The algorithm stores two values – called *filters* – for which we know that the element to select lies in between (initially, the filters are simply set to $\pm\infty$, respectively). Thus, the aim is to iteratively scan the input shrinking the distance between the filters. At each iteration we look for an element whose rank is as close as possible to k (ignoring elements that do not lie within the filters). Once we have a candidate, we can compute its exact rank in linear time by comparing its value with the other elements of the input and update either the upper or lower filter accordingly. The process is repeated until $O(w)$ elements remain between the filters.

The key of this algorithm lies in a good choice of an approximation so that a large amount of values are filtered. The method of Munro and Patterson [13] first constructs a sample of the input as follows: for a block of up to $\frac{w}{\log n}$ elements, its sample simply consists of these elements sorted in increasing value. For larger blocks B (say, $2^i \frac{w}{\log n}$ elements for some $i \in \{1, \ldots, \lceil \log(n/w) \rceil\}$), partition the block into two equally sized sub-blocks and construct their samples inductively. Then, the sample of B is created by selecting one every other element of each of the samples of the two sub-blocks and sorting the obtained list. The sample of the whole input can be constructed in a bottom-up fashion that uses at most $O(\frac{w}{\log n})$ words in each level (thus, $O(w)$ in total). Once we have computed the sample of the input, we can extract its approximation by selecting the corresponding value within the sample.

Randomized Algorithms

The previous approach can be drastically simplified under randomized environments. Simply select an element of the input uniformly at random, compute its rank, and update one of the filters accordingly. With high probability after a constant number of iterations, a constant fraction of the input will be discarded. Thus, overall $O(\log n)$ iterations (and a constant number of words) will be needed to find the solution. Chan [9] improved this idea, obtaining an algorithm that runs in $O(n \log \log_w n)$ time and uses $O(w)$ words (for any $w \leq n$).

Improvements

For most values of w, the algorithm of Munro and Patterson is asymptotically tight, even if we allow random access to the input. Thus, further research focused in extending the range space for which optimality is known. Frederickson [11] increased the optimality range for the selection problem to any $w \in \Omega(\log n^2) \cap O(2^{\log n/\log^* n})$. Recently, Elmasry et al. [10] gave a linear time algorithm that only uses $O(n)$ bits (i.e., they preserve the linear running time of [7] and reduce the size of the workspace by a logarithmic factor). Raman and Ramnath [17] used a similar approximate median approach for the case in which $o(\log n)$ words fit in the workspace.

Undirected Graph Connectivity in the Random Access Model

Given an undirected graph $G = (V, E)$ and two vertices $s, t \in V$, the *undirected s–t connectivity problem* is to decide whether or not there exists a path from s to t in G. This problem can be easily solved in linear time using linear space (with either breadth-first search or width-first search schemes) in unrestricted models. However, determining the existence of a deterministic algorithm that only uses $O(\log n)$ bits space was a long-standing open problem in the field of complexity theory.

Problem Background

Aleliunas et al. [1] showed that the problem can be easily solved with a randomized algorithm. Essentially they show that a sufficiently long random walk will traverse all vertices of a connected graph. Thus, if we start at s and do not reach t after a polynomially bounded number of steps, we can conclude that with high probability, s and t are not connected.

The connectivity problem can also be solved with a nondeterministic logspace algorithm (where the certificate is simply the path connecting s and t). Thus, Savitch's theorem [19] can transform it to a deterministic algorithm that uses $O(\log^2 n)$ bits (and superpolynomial time). The space requirements were afterward reduced to $O(\log^{3/2} n)$ [14] and $O(\log^{4/3} n)$ [2]. Recently, Reingold [18] positively answered the question by giving a deterministic logspace algorithm. Although no discussion on the running time is explicitly mentioned, it is well known that Reingold's algorithm runs in polynomial time. This is due to the fact that a Turing machine with a logarithmic space constraint can have at most $2^{O(\log n)} = O(n^{O(1)})$ different configurations.

Reingold's Algorithm

Conceptually, the algorithm aims to transform G into a graph in which all vertices have degree three. This is done by virtually replacing each vertex of degree $k > 3$ by a cycle of k vertices each of which is adjacent to one of the neighbors of the previous graph (and adding self-loops to vertices of low degree).

The algorithm then combines the *squaring* and the *zig-zag product* operations. The squaring operation connects vertices whose distance in the original graph is at most two, while the zig-zag product between two graphs G and H essentially replaces every vertex of G with a copy of H (and connects vertices of two copies of H if and only if the original vertices were adjacent in G).

Intuitively speaking, the squaring operation will reduce the diameter while the zig-zag product keeps the degree of the vertices bounded (for this algorithm, H consists of a sparse graph of constant degree and small diameter). After repeating this process a logarithmic number of times, the resulting graph has bounded degree, logarithmic diameter and preserves the connectivity between the corresponding vertices. In particular, we can determine the connectivity between u and v by exhaustively looking through all paths of logarithmic length starting from u. Since each vertex has bounded degree, the paths can be encoded without exceeding the space bounds. The algorithm will stop as soon as v is found in any of these paths or after all paths have been explored.

Even though we cannot store the transformation of G explicitly, the only operation that is needed during the exhaustive search is to determine the ith clockwise neighbor of a vertex after j transformation steps have been done on G (for some i and $j \in O(\log n)$). Reingold provided a method to answer such queries using constant number of bits on the graph resulting after doing $j-1$ transformation steps on G. Thus, by repeating the process inductively, we can find the solution to our query without exceeding the space bounds.

Other Models of Note

The study of memory constrained algorithms has received a lot of interest by the computational geometry community. Most of them use random access to the input, use a constant number of words, and aim to reporting fundamental geometric data structures. For example, Jarvis march [12] (also known as the *gift-wrapping algorithm*) computes the convex hull of a set of n points in $O(nh)$ time (where h is the number of vertices on the convex hull). Asano and Rote [3] showed how to compute the Voronoi diagram and minimum spanning tree of a given set of points in $O(n^2)$ and $O(n^3)$ time, respectively. Similarly, several time-space trade-off algorithms have been designed for classic problems within a simple polygon, such as triangulation [4], shortest path computation [4], or visibility [5]. These algorithms use properties of the problem considered so as to somehow compute the solution using local information whenever possible. In most cases, this ends up in an algorithm that is completely different from those used when no memory constraints exist.

Compressed Stack

A different approach was taken by Barba et al. [6], where the class of *stack algorithms* is considered. This class consists of deterministic algorithms that have a one-pass access to the input and can use a constant number of words. In addition, they allow the usage of a stack so as to store elements of the input. At any instant of time, only the top element of the stack is available to the algorithm. Note that with this additional stack, we can store up to $\Theta(n)$ values of the input. Hence, the model is not strictly speaking memory constrained.

Although this model may seem a bit artificial, Barba et al. give several examples of well-known programs that fit into this class of algorithms (such as computing the convex hull of a simple polygon or computing the visibility of a point inside a simple polygon). More interestingly, they show how to remove the auxiliary stack, effectively transforming any stack algorithm into a memory constrained algorithm that uses $O(w)$ words (for any $w \in \{1, \ldots, n\}$).

Block Reconstruction

The key to this approach lies in the fact that portions of the stack can be *reconstructed* efficiently: let $a_i, a_{i+1}, \ldots, a_j$ be a set of consecutive elements of the input (for some $i < j$) such that we know that a_i and a_j are in the stack after a_j has been processed. Then, we can determine which values in between a_i and a_j are also in the stack by re-executing the algorithm restricting the input to the $a_i, \ldots, a_j$ interval (thus, taking time proportional to $O(j-i)$).

Using this idea, we can avoid explicitly storing the whole stack by using a *compressed stack* data structure that never exceeds the size of the workspace. For the particular case in which $w = \Theta(\sqrt{n})$, the algorithm virtually partitions the input into blocks of size $\sqrt{n}$. The invariant of the data structure is that the portions of the stack corresponding to the last two blocks that have pushed elements into the stack are stored, whereas for any other block, we only store the first and last elements that are in the stack, if any. By using a charging scheme, they show that each block triggers at most one reconstruction, and each reconstruction takes time proportional to the size of the destroyed block. In all, the compressed stack data structure reduces the size of the workspace to $\Theta(\sqrt{n})$ without asymptotically increasing the running time.

In the general case, the input is split into p equally sized blocks (where $p = \max\{2, w/\log n\}$), and each block is further subpartitioned into blocks until the blocks of the lowermost level can be explicitly stored (or the recursion exceeds size of the workspace). The smaller the size of the workspace, the higher the number of levels it will have and thus more time will be spent in reconstructing blocks. This creates a time-space trade-off for any stack algorithm whose running time is $O(\frac{n^2 \log n}{2^w})$ time for any workspace of $w \in o(\log n)$ words. For larger workspaces, the algorithm runs in $O(n^{1+1/\log p})$ time and uses $O(p \log_p n)$ words (for $2 \leq p \leq n$).

Cross-References

- ▶ Exact Algorithms and Time/Space Tradeoffs
- ▶ Memoryless Gathering of Mobile Robotic Sensors
- ▶ Rank and Select Operations on Bit Strings

Recommended Reading

1. Aleliunas R, Karp RM, Lipton R, Lovasz L, Rackoff C (1979) Random walks, universal traversal sequences, and the complexity of maze problems. In: FOCS, San Juan, pp 218–223
2. Armoni R, Ta-Shma A, Widgerson A, Zhou S (2000) An $o(\log(n)^{4/3})$ space algorithm for (s,t) connectivity in undirected graphs. J ACM 47(2):294–311
3. Asano T, Rote G (2009) Constant-working-space algorithms for geometric problems. In: CCCG, Vancouver, pp 87–90
4. Asano T, Buchin K, Buchin M, Korman M, Mulzer W, Rote G, Schulz A (2012) Memory-constrained algorithms for simple polygons. Comput Geom Theory Appl 46(8):959–969
5. Barba L, Korman M, Langerman S, Silveira R (2014) Computing the visibility polygon using few variables. Comput Geom Theory Appl 47(9):918–926
6. Barba L, Korman M, Langerman S, Sadakane K, Silveira R (2015) Space-time trade-offs for stack-based algorithms. Algorithmica 72(4):1097–1129
7. Blum M, Floyd RW, Pratt VR, Rivest RL, Tarjan RE (1973) Time bounds for selection. J Comput Syst Sci 7(4):448–461
8. Borodin A, Cook S (1982) A time-space tradeoff for sorting on a general sequential model of computation. SIAM J Comput 11:287–297
9. Chan TM (2010) Comparison-based time-space lower bounds for selection. ACM Trans Algorithms 6:1–16
10. Elmasry A, Juhl DD, Katajainen J, Satti SR (2013) Selection from read-only memory with limited workspace. In: COCOON, Hangzhou, pp 147–157
11. Frederickson GN (1987) Upper bounds for time-space trade-offs in sorting and selection. J Comput Syst Sci 34(1):19–26
12. Jarvis R (1973) On the identification of the convex hull of a finite set of points in the plane. Inf Process Lett 2(1):18–21
13. Munro JI, Paterson M (1980) Selection and sorting with limited storage. Theor Comput Sci 12:315–323
14. Nisan N, Szemeredi E, Wigderson A (1992) Undirected connectivity in $o(log^{1.5}n)$ space. In: FOCS, Pittsburg, pp 24–29
15. Pagter J, Rauhe T (1998) Optimal time-space tradeoffs for sorting. In: FOCS, Palo Alto, pp 264–268
16. Pohl I (1969) A minimum storage algorithm for computing the median. IBM technical report RC2701
17. Raman V, Ramnath S (1998) Improved upper bounds for time-space tradeoffs for selection with limited storage. In: SWAT '98, Stockholm, pp 131–142
18. Reingold O (2008) Undirected connectivity in log-space. J ACM 55(4):1–24
19. Savitch WJ (1970) Relationships between nondeterministic and deterministic tape complexities. J Comput Syst Sci 4(2):177–192

Memoryless Gathering of Mobile Robotic Sensors

Paola Flocchini
School of Electrical Engineering and Computer Science, University of Ottawa, Ottawa, ON, Canada

Keywords

Gathering; Rendezvous; Sensors aggregation

Years and Authors of Summarized Original Work

1999; Ando, Oasa, Suzuki, Yamashita
2005; Flocchini, Prencipe, Santoro, Widmayer

Problem Definition

The Model: A *mobile robotic sensor* (or simply *sensor*) is modeled as a computational unit with sensorial capabilities: it can perceive the spatial environment within a fixed distance $V > 0$, called *visibility range*, it has its own local working memory, and it is capable of performing local computations [3,4].

Each sensor is a point in its own local Cartesian coordinate system (not necessarily consistent with the others), of which it perceives itself as the center. A sensor can move in any direction, but it may be stopped before reaching its destination, e.g. because of limits to its motion energy; however, it is assumed that the distance traveled in a move by a sensor is not infinitesimally small (unless it brings the sensor to its destination).

The sensors have no means of direct communication: any communication occurs in an implicit manner, by observing the other sensors' positions. Moreover, they are *autonomous* (i.e., without a central control) *identical* (i.e., they execute the same protocol), and *anonymous* (i.e., without identifiers that can be used during the computation).

The sensors can be *active* or *inactive*. When *active*, a sensor performs a *Look-Compute-Move* cycle of operations: it first observes the portion of the space within its visibility range obtaining a snapshot of the positions of the sensors in its range at that time (*Look*); using the snapshot as an input, the sensor then executes the algorithm to determine a destination point (*Compute*); finally, it moves toward the computed destination, if different from the current location (*Move*). After that, it becomes *inactive* and stays idle until the next activation. Sensors are *oblivious*: when a sensor becomes active, it does not remember any information from previous cycles. Note that several sensors could actually occupy the same point; we call *multiplicity detection* the ability of a sensor to see whether a point is occupied by a single sensor or by more than one.

Depending on the degree of synchronization among the cycles of different sensors, three submodels are traditionally identified: *synchronous*, *semi-synchronous*, and *asynchronous*. In the *synchronous* (FSYNC) and in the *semi-synchronous* (SSYNC) models, there is a global clock tick reaching all sensors simultaneously, and a sensor's cycle is an instantaneous event that starts at a clock tick and ends by the next. In FSYNC, at each clock tick all sensors become active, while in SSYNC only a subset of the sensors might be active in each cycle. In the *asynchronous* model (ASYNC), there is no global clock and the sensors do not have a common notion of time. Furthermore, the duration of each activity (or inactivity) is finite but unpredictable. As a result, sensors can be seen while moving, and computations can be made based on obsolete observations.

Let $S(t) = \{s_1(t), \ldots, s_n(t)\}$ denote the set of the n sensors' at time t. When no ambiguity arises, we shall omit the temporal index t. Moreover, with an abuse of notation we indicate by s_i both a sensor and its position. Let $S_i(t)$ denote the set of sensors that are within distance V from s_i at time t, that is, the set of sensors that are visible from s_i. At any point in time t, the sensors induce a *visibility graph* $G(t) = (N, E(t))$ defined as follows: $N = S$ and $\forall r, s \in N$, $(r, s) \in E(t)$ iff r and s are at distance no more than the visibility range V.

The Problem: In this setting, one of the most basic coordination and synchronization task is *Gathering*: the sensors, initially placed in arbitrary distinct positions in a 2-dimensional space, must congregate at a single location (the choice of the location is not predetermined) within finite time. In the following, we assume $n > 2$. A problem closely related to *Gathering* is *Convergence*, where the sensors need to be arbitrarily close to a common location, without the requirement of ever reaching it. A special type of convergence (also called *Near-Gathering* or *collisionless con-*

vergence) requires the sensors to converge without ever colliding with each other.

Key Results

Basic Impossibility Results

First of all, neither *Convergence* nor *Gathering* can be achieved from arbitrary initial placements if the initial visibility graph $G(0)$ is not connected. So, in all the literature, $G(0)$ is always assumed to be connected. Furthermore, if the sensors have *neither* agreement on the coordinate system *nor* multiplicity detection, then *Gathering* is not solvable in SSYNC (and thus in ASYNC), regardless of the range of visibility and the amount of memory that they may have.

Theorem 1 ([8]) *In absence of multiplicity detection and of any agreement on the coordinate systems, Gathering is deterministically unsolvable in* SSYNC.

Given this impossibility result, the natural question is whether the problem can be solved with common coordinate systems.

Gathering with Common Coordinate Systems

Assuming that the sensors agree on a common coordinate system, *Gathering* has been shown to be solvable under the weakest of the three schedulers (ASYNC) [2].

Let $\mathcal{R}$ be the rightmost vertical axis where some sensor initially lie. The idea of the algorithm is to make the sensors move toward $\mathcal{R}$, in such a way that, after a finite number of steps, they will reach it and gather at the bottommost position occupied by a sensor at that time. Let the *Look* operation of sensor s_i at time t return $S_i(t)$. The computed destination point of s_i depends on the positions of the visible sensors. Once the computation is completed, s_i moves toward its destination (but it may stop before the destination is reached). Informally,l

- If s_i sees sensors to its left or above on the vertical axis passing through its position (this axis will be referred to as *its vertical axis*), it does not move.
- If s_i sees sensors only below on its vertical axis, it moves down toward the nearest sensor.
- If s_i sees sensors only to its right, it moves horizontally toward the vertical axis of the nearest sensor.
- If s_i sees sensors both below on its vertical axis and on its right, it computes a destination point and performs a diagonal move to the right and down, as explained below.

To describe the diagonal movement we introduce some notation (refer to Fig. 1). Let $\overline{AA'}$ be the vertical diameter of $S_i(t)$ with A' the top and A the bottom end point; let $\mathcal{L}_i$ denote the topologically open region (with respect to $\overline{AA'}$) inside $S_i(t)$ and to the right of s_i and let $S = \overline{s_i A}$ and $S' = \overline{s_i A'}$, where neither S' and S include s_i. Let Ψ be the vertical axis of the horizontally closest sensor (if any) in $\mathcal{L}_i$.

```
Diagonal_Movement(Ψ)
  B := upper intersection between S_i(t) and Ψ;
  C := lower intersection between S_i(t) and Ψ;
  2β = A ŝ_i B;
  If β < 60° then (B,Ψ) := Rotate(s_i, B);
  H := Diagonal_Destination(Ψ,A,B);
  Move towards H.
```

where `Rotate()` and `Diagonal_Destination()` are as follows:

- `Rotate(`s_i, B`)` rotates the segment $\overline{s_i B}$ in such a way that $\beta = 60°$ and returns the new position of B and Ψ. This angle choice ensures that the destination point is not outside the circle.
- `Diagonal_Destination(`Ψ, A, B`)` computes the destination of s_i as follows: the direction of s_i's movement is given by the perpendicular to the segment $\overline{AB}$; the destination of s_i is the point H on the intersection of the direction of its movement and of the axis Ψ.

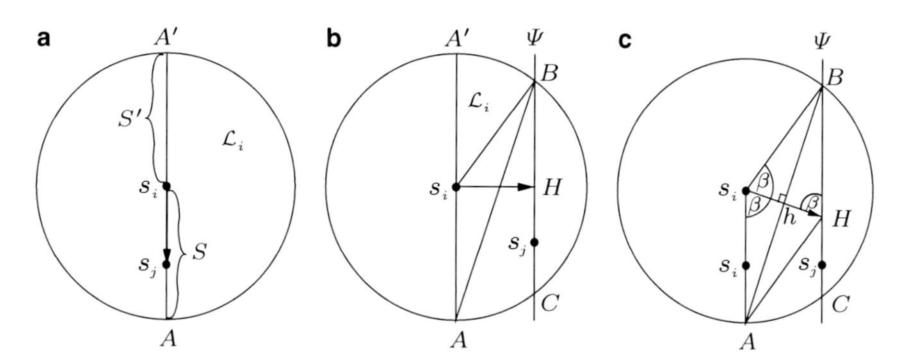

Memoryless Gathering of Mobile Robotic Sensors, Fig. 1 From [4]: (**a**) Notation. (**b**) Horizontal move. (**c**) Diagonal move

Memoryless Gathering of Mobile Robotic Sensors, Fig. 2 From [4]: Notation for algorithm CONVERGENCE [1]

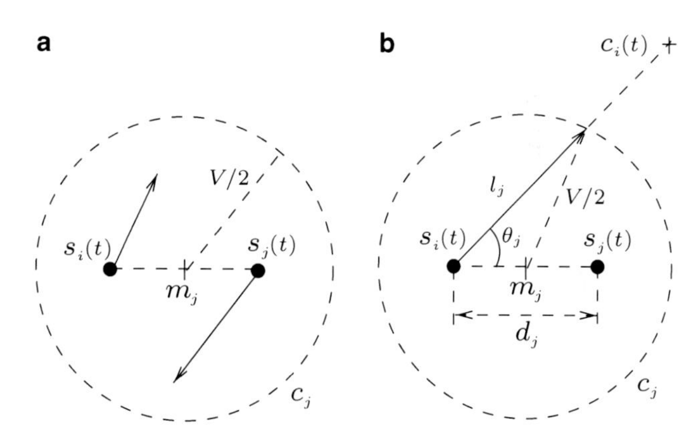

Theorem 2 ([2]) *With common coordinate systems, Gathering is possible in* ASYNC.

Gathering has been shown to be possible in SSYNC also when compasses are unstable for some arbitrary long periods, provided they have a common clockwise notion, and that they eventually stabilize, and assuming the total number of sensors is known [9].

Convergence and Near-Gathering

Convergence in SSYNC

The impossibility result does not apply to the case of *Convergence*. In fact, it is possible to solve it in SSYNC in the basic model (i.e., without common coordinate systems) [1].

Let $SC_i(t)$ denote the smallest enclosing circle of the set $\{s_j(t)|s_j \in S(t)\}$ of positions of sensors in $S(t)$; let $c_i(t)$ be the center of $SC_i(t)$.

CONVERGENCE
Assumptions: SSYNC.

If $S_i(t) = \{s_i\}$, **then** gathering is completed.
$\forall s_j \in S_i(t) \setminus \{s_i\}$:

$d_j := dist(s_i(t), s_j(t))$,
$\theta_j := c_i(t)\widehat{s_i(t)}s_j(t)$,
$l_j := (d_j/2)\cos\theta_j + \sqrt{(V/2)^2 - ((d_j/2)\sin\theta_j)^2}$,

$limit := \min_{s_j \in S_i(t)\setminus\{s_i\}}\{l_j\}$,
$goal := dist(s_i(t), c_i(t))$,
$D := \min\{goal, limit\}$,
$p :=$ point on $\overline{s_i(t)c_i(t)}$ at distance D from $s_i(t)$.
Move towards p.

Everytime s_i becomes active, it moves toward $c_i(t)$, but only up to a certain distance. Specifically, if s_i does not see any sensor other than itself, then s_i does not move. Otherwise, its destination is the point p on the segment

$\overline{s_i(t)c_i(t)}$ that is closest to $c_i(t)$ and that satisfies the following condition: For every sensor $s_j \in S(t)$, p lies in the disk $\mathcal{C}_j$ whose center is the midpoint m_j of $s_i(t)$ and $s_j(t)$ and whose range is $V/2$ (see Fig. 2). This condition ensures that s_i and s_j will still be visible after the movement of s_i, and possibly of s_j.

Theorem 3 ([1]) *Convergence is possible in* SSYNC.

Convergence in ASYNC

Convergence has been shown to be possible also in ASYNC, but under special schedulers: *partial* ASYNC [6] and *1-fair* ASYNC [5]. In *partial* ASYNC the time spent by a sensor performing the *Look*, *Compute*, and *Sleep* operations is bounded by a globally predefined amount and the time spent in the *Move* operation by a locally predefined amount; in 1-fair ASYNC between two successive activations of each sensor, all the other sensors are activated at most once. Finally, *Convergence* has been studied also in presence of perception inaccuracies (radial errors in locating a sensor) and it has been shown how to reach convergence in FSYNC for small inaccuracies.

Near-Gathering

Slight modifications can make the algorithm of [1] described above collisionless, thus solving also the *collisionless Convergence* problem (i.e., *Near-Gathering*) in SSYNC. *Near-Gathering* can be achieved also in ASYNC, with two additional assumptions [7]: 1) the sensors must partially agree on a common coordinate system (one axis is sufficient) and 2) the initial visibility graph must be *well connected*, that is, the subgraph of the visibility graph that contains only the edges corresponding to sensors at distance *strictly smaller* than V must be connected.

Open Problems

The existing results for *Gathering* and *Convergence* leave several problems open. For example, *Gathering* in SSYNC (and thus ASYNC) has been proven impossible when neither multiplicity detection nor an orientation are available. While common orientation suffices, it is not known whether the presence of multiplicity detection alone is sufficient to solve the problem. Also, the impossibility result does not apply to FSYNC; however no algorithm is known in such a setting that does not make use of orientation. Finally, it is not known whether *Convergence* (collisionless or not) is solvable in ASYNC without additional assumptions: so far no algorithm has been provided nor an impossibility proof.

Recommended Reading

1. Ando H, Oasa Y, Suzuki I, Yamashita M (1999) A distributed memoryless point convergence algorithm for mobile robots with limited visibility. IEEE Trans Robot Autom 15(5):818–828
2. Flocchini P, Prencipe G, Santoro N, Widmayer P (2005) Gathering of asynchronous mobile robots with limited visibility. Theor Comput Sci 337: 147–168
3. Flocchini P, Prencipe G, Santoro N (2011) Computing by mobile robotic sensors. In: Nikoletseas S, Rolim J (eds) Theoretical aspects of distributed computing in sensor networks, chap 21. Springer, Heidelberg. ISBN:978-3-642-14849-1
4. Flocchini P, Prencipe G, Santoro N (2012) Distributed computing by oblivious mobile robots. Morgan & Claypool, San Rafael
5. Katreniak B (2011) Convergence with limited visibility by asynchronous mobile robots. In: 18th international colloquium on structural information and communication complexity (SIROCCO), Gdańsk, Poland, pp 125–137
6. Lin J, Morse A, Anderson B (2007) The multi-agent rendezvous problem. Part 2: the asynchronous case. SIAM J Control Optim 46(6): 2120–2147
7. Pagli L, Prencipe G, Viglietta G (2012) Getting close without touching. In: 19th international colloquium on structural information and communication complexity (SIROCCO), Reykjavík, Iceland, pp 315–326
8. Prencipe G (2007) Impossibility of gathering by a set of autonomous mobile robots. Theor Comput Sci 384(2–3):222–231
9. Souissi S, Défago X, Yamashita M (2009) Using eventually consistent compasses to gather memoryless mobile robots with limited visibility. ACM Trans Auton Adapt Syst 4(1):1–27

Meshing Piecewise Smooth Complexes

Tamal Krishna Dey
Department of Computer Science and Engineering, The Ohio State University, Columbus, OH, USA

Keywords

Delaunay mesh; Delaunay refinement; Piecewise smooth complex; Topology; Volume mesh

Years and Authors of Summarized Original Work

2006; Boissonnat, Oudot
2007; Cheng, Dey, Ramos
2007: Cheng, Dey, Levine
2012; Cheng, Dey, Shewchuk

Problem Definition

The class of piecewise smooth complex (PSC) includes geometries that go beyond smooth surfaces. They contain polyhedra, smooth and non-smooth surfaces with or without boundaries, and more importantly non-manifolds. Thus, provable mesh generation algorithms for this domain extend the scope of mesh generation to a wide variety of domains. Just as in surface mesh generation, we are required to compute a set of points on the input complex and then connect them with a simplicial complex which is *geometrically* close and is *topologically* equivalent to the input. One challenge that makes this task harder is that the PSCs allow arbitrary small input angles, a notorious well-known hurdle for mesh generation.

A PSC is a set of *cells*, each being a smooth, connected manifold, possibly with boundary. The 0-cells, 1-cells, and 2-cells are called corners, ridges, and patches, respectively. A PSC could also contain 3-cells that designate regions to be meshed with tetrahedra, if we are interested in a volume mesh.

Definition 1 (ridge; patch) A *ridge* is a closed, connected subset of a smooth 1-manifold without boundary in $\mathbb{R}^3$. A *patch* is a 2-manifold that is a closed, connected subset of a smooth 2-manifold without boundary in $\mathbb{R}^3$.

Definition 2 A *piecewise smooth complex* $\mathcal{S}$ is a finite set of vertices, ridges, and patches that satisfy the following conditions.

1. The boundary of each cell in $\mathcal{S}$ is a union of cells in $\mathcal{S}$.
2. If two distinct cells c_1 and c_2 in $\mathcal{S}$ intersect, their intersection is a union of cells in $\mathcal{S}$ included in the boundary of c_1 or c_2.

Our goal is to generate a triangulation of a PSC. Element quality is not a primary issue here though a good radius-edge ratio can be obtained by additional refinement except near the small input angles. The definition below makes our notion of triangulation of a PSC precise. Recall that $|T|$ denotes the underlying space of a complex T (Fig. 1).

Definition 3 (triangulation of a piecewise smooth complex) A simplicial complex T is a *triangulation* of a PSC $\mathcal{S}$ if there is a homeomorphism h from $|\mathcal{S}|$ to $|T|$ such that $h(v) = v$ for each vertex $v \in \mathcal{S}$ and for each cell $\xi \in \mathcal{S}$, there is a subcomplex $T_\xi \subseteq T$ such that h is a homeomorphism from ξ to $|T_\xi|$.

Key Results

To generate a mesh for a PSC with theoretical guarantees, we use Delaunay refinement as in the smooth surface meshing. For a point set $P \subset \mathbb{R}^3$, let Vor P and Del P denote the Voronoi diagram and Delaunay triangulation of P, respectively. The restricted Delaunay complex as in the smooth surface meshing plays an important role in sampling the PSCs.

M

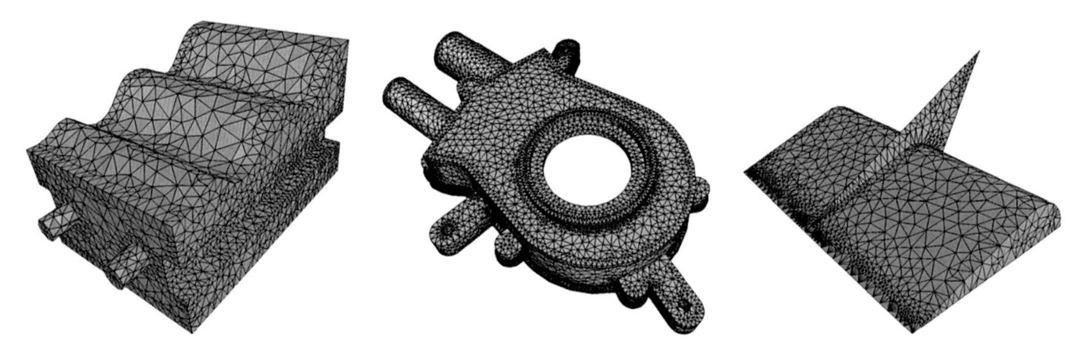

Meshing Piecewise Smooth Complexes, Fig. 1 Example meshes of PSC: (left) a piecewise smooth surface, a non-manifold, a surface with non-trivial topology

Let V_ξ denote the dual Voronoi face of a Delaunay simplex ξ in Del P. The restricted Voronoi face of V_ξ with respect to $\mathbb{X} \subset \mathbb{R}^3$ is the intersection $V_\xi|_\mathbb{X} = V_\xi \cap \mathbb{X}$. The *restricted Voronoi diagram* and *restricted Delaunay triangulation* of P with respect to $\mathbb{X}$ are

$$\text{Vor } P|_\mathbb{X} = \{V_\xi|_\mathbb{X} \mid V_\xi|_\mathbb{X} \neq \emptyset\} \text{ and Del } P|_\mathbb{X} = \{\xi \mid V_\xi|_\mathbb{X} \neq \emptyset\} \text{ respectively.}$$

In words, Del $P|_\mathbb{X}$ consists of those Delaunay simplices in Del P whose dual Voronoi face intersects $\mathbb{X}$. We call these simplices *restricted.* For a restricted triangle, its dual Voronoi edge intersects the domain in a single or multiple points. These are the centers of *surface Delaunay balls* that circumscribe the vertices of the triangle.

In smooth surface meshing, the restricted triangles violating certain desirable properties are refined by the addition of the surface Delaunay ball centers. It turns out that this process cannot continue forever because the inserted points maintain a fixed lower bound on their distances from the existing points. This argument breaks down if non-smoothness is allowed. In particular, ridges and corners where several patches meet cause the local feature size to be zero in which case inserted points with a lower bound on local feature size may come arbitrarily close to each other. Nevertheless, Boissonnat and Oudot [1] showed that the Delaunay refinement that they proposed for smooth surfaces can be extended to a special class of piecewise smooth surfaces called *Lipschitz surfaces*. Their algorithm may break down for domains with small angles. The analysis requires that the input angles subtended by the tangent planes of the patches meeting at the ridges or a corner are close to 180°.

The first guaranteed algorithm for PSCs with small angles is due to Cheng, Dey, and Ramos [3]. They introduced the idea of using weighted vertices as protecting balls in a weighted Delaunay triangulation. In this triangulation each point p is equipped with a weight w_p which can also be viewed as a ball $\hat{p} = B(p, w_p)$ centered at p with radius w_p. The squared weighted distance between two points (p, w_p) and (q, w_q) is measured as $\|p-q\|^2 - w_p^2 - w_q^2$. Notice that the weight can be zero in which case the weighted point is a regular point. One can define a Voronoi diagram and its dual Delaunay triangulation for a weighted point set just like their Euclidean counterparts by replacing Euclidean distances with weighted distances. To emphasize the weighted case, let us denote a weighted point set P with $P[w]$ and its weighted Delaunay triangulation with Del $P[w]$.

The algorithm of Cheng, Dey, and Ramos [3] has two phases, the *protection* phase and the *refinement* phase. In the protection phase, it computes a set of protecting balls centered at the ridges and corners of the input PSC. The union of the protecting balls cover the ridges and corners completely. Let $P[w]$ be the weighted points representing the protecting balls. The algorithm computes Del $P[w]$ and the restricted Delaunay triangles in Del $P[w]|_S$. In the refinement phase, it refines the restricted triangles if either they have a large surface Delaunay ball (albeit in the weighted sense), or they fail to form a topological disk around each vertex on each patch adjoining the vertex. The algorithm guarantees that the final mesh is a triangulation of the input PSC and the two are related by a homeomorphism. The proof of the homeomorphism uses an extension of the topological ball property of Edelsbrunner and Shah [6] to accommodate PSCs.

The algorithm of Cheng et al. [3] requires difficult geometric computations such as feature size estimates. Cheng, Dey, and Levine [2] simplified some of these computations at the expense of weakening the topological guarantees. Like the smooth surface meshing algorithm in [4], they guarantee that each input patch and ridge is approximated with output simplices that form a manifold of appropriate dimension. The algorithm is supplied with a user-specified size parameter. If this size parameter is small enough, the output is a triangulation of the PSC in the sense of Definition 3. In a subsequent paper, Dey and Levine [5] proposed a strategy to combine the protection phase with refinement phase which allowed to adjust the ball sizes on the fly rather than computing them beforehand by estimating feature sizes. Cheng, Dey, and Shewchuk [4] refined this strategy even further to have an improved algorithm with detailed analysis.

Theorem 1 ([4]) *There is a Delaunay refinement algorithm that runs with a parameter $\lambda > 0$ on an input PSC $\mathcal{S}$ and outputs a mesh $T = \text{Del}\, P[w]|_{\mathcal{S}}$ with the following guarantees:*

1. *For each patch $\sigma \in \mathcal{S}$, $|\text{Del}\, P[w]|_\sigma|$ is a 2-manifold with boundary, and every vertex in $\text{Del}\, P[w]|_\sigma$ lies on σ. The boundary of $|\text{Del}\, P[w]|_\sigma|$ is homeomorphic to $\text{Bd}\,\sigma$, the boundary of σ, and every vertex in $\text{Del}\, P[w]|_{\text{Bd}\,\sigma}$ lies on $\text{Bd}\,\sigma$.*
2. *If λ is sufficiently small, then T is a triangulation of $\mathcal{S}$ (recall Definition 3). Furthermore, there is a homeomorphism h from $|\mathcal{S}|$ to $|T|$ such that for every i-dimensional cell $\xi \in \mathcal{S}_i$ with $i \in [0, 2]$, h is a homeomorphism from ξ to $|\text{Del}\, P[w]|_\xi|$, every vertex in $\text{Del}\, P[w]|_\xi$ lies on ξ, and the boundary of $|\text{Del}\, P[w]|_\xi|$ is $|\text{Del}\, P[w]|_{\text{Bd}\,\xi}|$*

Notice that the above guarantee specifies that the homeomorphism between the input and the output respects the stratification of corners, ridges, and patches and thus preserves these features. Once a mesh for the surface patches is completed, Delaunay refinement algorithms can further refine the mesh to improve the quality of the surface triangles or the tetrahedra they enclose. The algorithm can only attack triangles and tetrahedra with large orthoradius-edge ratios; some simplices with large circumradius-edge ratios may survive. The tetrahedral refinement algorithm should be careful in that if inserting a vertex at the circumcenter of a poor-quality tetrahedron destroys some surface triangle, the algorithm simply should opt not to insert the new vertex. This approach has the flaw that tetrahedra with large radius-edge ratios sometimes survive near the boundary.

URLs to Code and Data Sets

CGAL(http://cgal.org), a library of geometric algorithms, contains software for mesh generation of piecewise smooth surfaces. The DelPSC software that implements the PSC meshing algorithm as described in [4] is available from http://web.cse.ohio-state.edu/~tamaldey/delpsc.html.

Cross-References

▶ 3D Conforming Delaunay Triangulation
▶ Smooth Surface and Volume Meshing

Recommended Reading

1. Boissonnat J-D, Oudot S (2006) Provably good sampling and meshing of Lipschitz surfaces. In: Proceedings of the 22nd annual symposium on computational geometry, Sedona, pp 337–346
2. Cheng S-W, Dey TK, Levine J (2007) A practical Delaunay meshing algorithm for a large class of domains.

In: Proceedings of the 16th international meshing roundtable, Seattle, pp 477–494
3. Cheng S-W, Dey TK, Ramos EA (2007) Delaunay refinement for piecewise smooth complexes. In: Proceedings of the 18th annual ACM-SIAM symposium on discrete algorithms, New Orleans, pp 1096–1105
4. Cheng S-W, Dey TK, Shewchuk JR (2012) Delaunay mesh generation. CRC, Boca Raton
5. Dey TK, Levine JA (2009) Delaunay meshing of piecewise smooth complexes without expensive predicates. Algorithms 2(4):1327–1349
6. Edelsbrunner H, Shah N (1997) Triangulating topological spaces. Int J Comput Geom Appl 7:365–378

Message Adversaries

Michel Raynal
Institut Universitaire de France and IRISA, Université de Rennes, Rennes, France

Keywords

Dynamic transmission failure; Mobile failure; Synchronous dynamic network; Ubiquitous failures

Years and Authors of Summarized Original Work

1989; Santoro, Widmayer

Problem Definition: The Notion of a Message Adversary

Message adversaries have been introduced by N. Santoro and P. Widmayer in a paper titled *Time is not a healer* [15] to model and understand what they called *dynamic* transmission failures in the context of synchronous systems. Then, they extended their approach in [16] where they used the term *ubiquitous* failures. The terms *heard-of communication* [5], *transient link failure* [17], and *mobile failure* [12] have later been used by other authors to capture similar network behaviors in synchronous or asynchronous systems.

The aim of this approach is to consider message losses as a normal link behavior (as long as messages are correctly transmitted). The notion of a message adversary is of a different nature than the notion of the fair link assumption. A fair link assumption is an assumption on each link taken separately, while the message adversary notion considers the network as a whole; its aim is not to build a reliable network but to allow the statement of connectivity requirements that must be met for problems to be solved. Message adversaries allow us to consider topology changes not as anomalous network behaviors, but as an essential part of the deep nature of the system.

Reliable Synchronous Systems

A fully connected synchronous system is made up of n computing entities (processes) denoted $p_1, \ldots, p_n$, where each pair of processes is connected by a bidirectional link. The progress of the processes is ruled by a global clock which generates a sequence of rounds. During a round, each process (a) first sends a message to all the other processes (broadcast), (b) then receives a message from each other process, and (c) finally executes a local computation. The fundamental property of a synchronous system is that the messages sent at a round r are received during the very same round r. As we can see, this type of synchrony is an abstraction that encapsulates (and hides) specific timing assumptions (there are bounds on message transfer delays and processing times, and those are known and used by the underlying system level to generate the sequence of synchronized rounds [13]).

In the case of a reliable synchronous system, both processes and links are reliable, i.e., no process deviates from its specification, and all the messages that are sent, and only them, are received (exactly once) by each process.

Message Adversary

A *message adversary* is a daemon which, at each round, can suppress messages (hence, these messages are never received). The adversary is not prevented from having a read access to the local states of the processes at the beginning of each round.

It is possible to associate a directed graph G^r with each round r. Its vertices are the processes, and there is an edge from p_i to p_j if the message sent at round r by p_i to p_j is not suppressed by the adversary. There is a priori no relation on the consecutive graphs G^r, G^{r+1}, etc. As an example, the daemon can define G^{r+1} from the local states of the processes at the end of round r.

Let $\mathcal{SMP}_n[\text{adv} : \text{AD}]$ denote the synchronous system whose communications are under the control of an adversary-denoted AD. $\mathcal{SMP}_n[\text{adv} : \emptyset]$ denotes the synchronous system in which the adversary has no power (it can suppress no message), while $\mathcal{SMP}_n[\text{adv} : \infty]$ denotes the synchronous system in which the adversary can suppress all the messages at every round. It is easy to see that, from a message adversary and computability point of view, $\mathcal{SMP}_n[\text{adv} : \emptyset]$ is the most powerful synchronous system model, while $\mathcal{SMP}_n[\text{adv} : \infty]$ is the weakest. More generally, the more constrained the message adversary AD, the more powerful the synchronous system.

Key Results

Key Results in Synchronous Systems

The Spanning Tree Adversary

Let TREE be the message adversary defined by the following constraint: at every round r, the graph G^r is an undirected spanning tree, i.e., the adversary cannot suppress the two messages – one in each direction – sent on the edges of G^r. Let $\mathcal{SMP}_n[\text{adv} : \text{TREE}]$ denote the corresponding synchronous system. As already indicated, for any r and $r' \neq r$, G^r and $G^{r'}$ are not required to be related; they can be composed of totally different sets of links.

Let us assume that each process p_i has an initial input v_i. It is shown in [11] that $\mathcal{SMP}_n[\text{adv} : \text{TREE}]$ allows the processes to compute any computable function on their inputs, i.e., functions on the vector $[v_1, \ldots, v_n]$.

Solving this problem amounts to ensure that each input v_i attains each process p_j despite the fact that the spanning tree can change arbitrarily from a round to the next one. This follows from the following observation. At any round r, the set of processes can be partitioned into two subsets: the set yes_i which contains the processes that have received v_i, and the set no_i which contains the processes that have not yet received v_i. As G^r is an undirected spanning tree (the tree is undirected because no message is suppressed on each of its edges), it follows that there is an edge of G^r that connects a process of the set yes_i to a process that belongs to the set no_i. So during round r, there at least one process of the set no_i which receives a copy of v_i and will consequently belong to the set yes_i of the next round. It follows that at most $(n-1)$ rounds are necessary for v_i to attain all the processes.

Consensus in the Presence of Message Adversaries

In the consensus problem, each process proposes a value and has to decide a value v such that v was proposed by a process, and no two processes decide different values. This problem is addressed in [2, 6, 8, 12, 17] in the message adversary context.

Impossibility Agreement-Related Results

As presented in [15], the k-process agreement problem (which must not be confused with the k-set agreement problem) is defined as follows. Each process p_i proposes an input value $v_i \in \{0, 1\}$, and at least k processes have to decide the same proposed value v. Let us observe that this problem can be trivially solved without any communication when $k \leq \lceil \frac{n}{2} \rceil$ (namely, each process decides its input value). Let us also notice that if $k > \lceil \frac{n}{2} \rceil$, there is at most one decision value.

It is shown in [15] that the k-process agreement problem cannot be solved for $k > \lceil \frac{n}{2} \rceil$, if the message adversary is allowed to suppress up to $(n-1)$ messages at every round. Other impossibility results are presented in [16]. More results can be found in [3].

M

d-Solo Executions

A process runs *solo* when it computes its local output without receiving any information from other processes, either because they crashed or they are too slow. This corresponds to a message adversary that suppresses all the messages sent to some subset of processes (which consequently run solo). The computability power of models in which several processes may run solo is addressed in [9]. This paper introduces the concept of *d-solo* model, $1 \leq d \leq n$ (synchronous round-based wait-free models where up to d processes may run solo in each round), and characterizes the colorless tasks that can be solved in a *d-solo* model. Among other results, this paper shows that the (d, ϵ)*-solo approximate agreement* task (which generalizes ϵ-approximate agreement) can be solved in the d-solo model, but cannot be solved in the $(d + 1)$-solo model. Hence, the *d-solo* models define a strict hierarchy.

Key Results in Asynchronous Systems

In a very interesting way, message adversaries allows the establishment of equivalences between synchronous systems and asynchronous systems.

These equivalences, which are depicted in Fig. 1 (from [14]), concern tasks. A task is the distributed analogous of a mathematical function [10], in the sense that each process has an input and must compute an output, and the processes need to cooperate to compute their individual outputs (this cooperation is inescapable, which makes the task *distributed*).

$A \simeq_T B$ means that any task that can be computed in the model A can be computed in the model B and vice versa. An arrow from A to B means that, from a task solvability point of view, the model A is strictly stronger than the model B (there are tasks solvable in A and not in B, and all the tasks solvable in B are solvable in A).

Let $\mathcal{ARW}_{n,n-1}[fd : \emptyset]$ (resp. $\mathcal{AMP}_{n,n-1}[fd : \emptyset]$) denote the basic asynchronous read/write (resp. message-passing) system model in which up to $(n - 1)$ processes may crash (premature halting). These models are also called wait-free models. The notation $[fd : \emptyset]$ means that these systems are not enriched with additional computational power. Differently, $\mathcal{ARW}_{n,n-1}[fd : \text{FD}]$ (resp. $\mathcal{AMP}_{n,n-1}[fd : \text{FD}]$) denotes $\mathcal{ARW}_{n,n-1}[fd : \emptyset]$ (resp. $\mathcal{AMP}_{n,n-1}[fd : \emptyset]$) enriched with a failure detector FD (see below).

- The message adversary-denoted TOUR (four tournament) has been introduced By Afek and Gafni in [1]. At any round, this adversary can suppress one message on each link but not both: for any pair of processes (p_i, p_j), either the message from p_i to p_j or the message from p_j to p_i or none of them can be suppressed.

  The authors have shown the following model equivalence: $\mathcal{SMP}_n[\text{adv} : \text{TOUR}] \simeq_T \mathcal{ARW}_{n,n-1}[fd : \emptyset]$. This is an important result as it established for the first time a

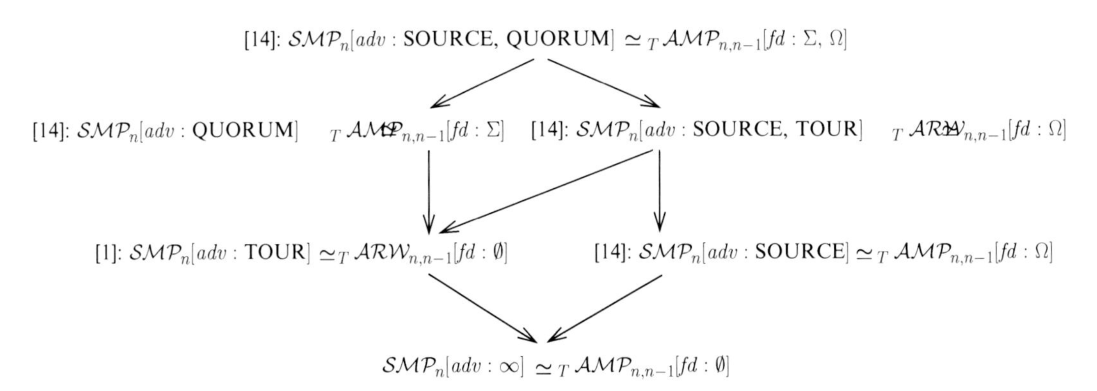

Message Adversaries, Fig. 1 A message adversary-hierarchy based on task equivalence and failure detectors

very strong relation linking message-passing synchronous systems where no process crashes, but messages can be lost according to the adversary TOUR, with the basic asynchronous wait-free read/write model.

- The other model equivalences, from a task solvability point of view, are due to Raynal and Stainer [14], who considered two failure detectors and introduced two associated message adversaries.

  A failure detector is an oracle that provides processes with information on failures. The failure detector Ω, called "eventual leader," was introduced in [4]. It is the failure detector that provides the minimal information on failures that allow consensus to be solved in $\mathcal{ARW}_{n,n-1}[fd : \emptyset]$ and in $\mathcal{AMP}_{n,n-1}[fd : \emptyset]$ where a majority of processes do not crash. The failure detector Σ, which is called "quorum", was introduced in [7]. It is the failure detector that provides the minimal information on failures that allow a read/write register to be built on top of an asynchronous message-passing system prone to up to $(n-1)$ process crashes.

  The message adversary SOURCE is constrained by the following property: there is a process p_x and a round r such that, at any round $r' > r$, the adversary does not suppress the message sent by p_x to the other processes. The message adversary QUORUM captures the message patterns that allow to obtain the quorums defined by Σ.

  As indicated, the corresponding model equivalences are depicted on Fig. 1. As an example, when considering distributed tasks, the synchronous message-passing model where no process crashes and where the message adversary is constrained by SOURCE and TOUR ($\mathcal{SMP}_n$[adv : SOURCE, TOUR]) and the basic wait-free asynchronous read/write model enriched with Ω ($\mathcal{ARW}_{n,n-1}[fd : \Omega]$) have the same computability power.

  When looking at the figure, it is easy to see that the suppression of the constraint TOUR from the model $\mathcal{SMP}_n$[adv : SOURCE, TOUR] gives the model $\mathcal{SMP}_n$ [adv : SOURCE], which is equivalent to $\mathcal{AMP}_{n,n-1}[fd : \Omega]$. Hence, suppressing TOUR weakens the model from $\mathcal{ARW}_{n,n-1}[fd : \Omega]$ to $\mathcal{AMP}_{n,n-1}[fd : \Omega]$ (i.e., from asynchronous read/write with Ω to asynchronous message-passing with Ω).

Applications

Message adversaries are important because they allow network changes in synchronous systems to be easily captured. Message losses are no longer considered as link or process failures, but as a normal behavior generated by process mobility and wireless links. Message adversaries provide us with a simple way to state assumptions (and sometimes minimal assumptions) on link connectivity which allow distributed computing problems to be solved in synchronous systems. They also allow the statement of equivalences relating (a) synchronous systems weakened by dynamically changing topology and (b) asynchronous read/write (or message-passing) systems enriched with distinct types of failure detectors.

M

Cross-References

▸ Distributed Computing for Enumeration
▸ Failure Detectors
▸ Locality in Distributed Graph Algorithms

Recommended Reading

1. Afek Y, Gafni E (2013) Asynchrony from synchrony. In: Proceedings of the international conference on distributed computing and networking (ICDCN'13), Mumbai. LNCS, vol 7730. Springer, pp 225–239
2. Biely M, Robinson P, Schmid U (2012) Agreement in directed dynamic networks. In: Proceedings of the 19th international colloquium on structural information and communication complexity (SIROCCO'12), Reykjavik. LNCS, vol 7355. Springer, pp 73–84
3. Casteigts P, Flocchini P, Godard E, Santoro N, Yamashita M (2013) Expressivity of time-varying

graphs. In: Proceedings of the 19th international symposium on fundamentals of computation theory (FST'13), Liverpool. LNCS, vol 8070. Springer, pp 95–106
4. Chandra T, Hadzilacos V, Toueg S (1996) The weakest failure detector for solving consensus. J ACM 43(4):685–722
5. Charron-Bost B, Schiper A (2009) The *heard-of* model: computing in distributed systems with benign faults. Distrib Comput 22(1):49–71
6. Coulouma E, Godard E (2013) A characterization of dynamic networks where consensus is solvable. In: Proceedings of the 19th international colloquium on structural information and communication complexity (SIROCCO'13), Ischia. LNCS, vol 8179. Springer, pp 24–35
7. Delporte-Gallet C, Fauconnier H, Guerraoui R (2010) Tight failure detection bounds on atomic object implementations. J ACM 57(4):Article 22
8. Godard E, Peters GP (2011) Consensus vs broadcast in communication networks with arbitrary mobile omission faults. In: Proceedings of the 17th international colloquium on structural information and communication complexity (SIROCCO'11), Gdansk. LNCS, vol 6796. Springer, pp 29–41
9. Herlihy M, Rajsbaum S, Raynal M, Stainer J (2014) Computing in the presence of concurrent solo executions. In: Proceedings of the 11th Latin-American theoretical informatics symposium (LATIN'2014), Montevideo. LNCS, vol 8392. Springer, pp 214–225
10. Herlihy MP, Shavit N (1999) The topological structure of asynchronous computability. J ACM 46(6):858–923
11. Kuhn F, Lynch NA, Oshman R (2010) Distributed computation in dynamic networks. In: Proceedings of the 42nd ACM symposium on theory of computing (STOC'10), Cambridge. ACM, pp 513–522
12. Moses Y, Rajsbaum S (2002) A layered analysis of consensus. SIAM J Comput 31:989–1021
13. Raynal M (2010) Fault-tolerant agreement in synchronous message-passing systems. Morgan & Claypool Publishers, 165p. ISBN:978-1-60845-525-6
14. Raynal M, Stainer J (2013) Synchrony weakened by message adversaries vs asynchrony restricted by failure detectors. In: Proceedings of the 32nd ACM symposium on principles of distributed computing (PODC '13), Montréal. ACM, pp 166–175
15. Santoro N, Widmayer P (1989) Time is not a healer. In: Proceedings of the 6th annual symposium on theoretical aspects of computer science (STACS'89), Paderborn. LNCS, vol 349. Springer, pp 304–316
16. Santoro N, Widmayer P (2007) Agreement in synchronous networks with ubiquitous faults. Theor Comput Sci 384(2–3):232–249
17. Schmid U, Weiss B, Keidar I (2009) Impossibility results and lower bounds for consensus under link failures. SIAM J Comput 38(5):1912–1951

Metric TSP

Markus Bläser
Department of Computer Science, Saarland University, Saarbrücken, Germany

Keywords

Metric traveling salesman problem; Metric traveling salesperson problem

Years and Authors of Summarized Original Work

1976; Christofides

Problem Definition

The *Traveling Salesman Problem (TSP)* is the following optimization problem:

Input: A complete loopless undirected graph $G=(V,E,w)$ with a weight function w: $E \to \mathbb{Q}_{\geq 0}$ that assigns to each edge a non-negative weight.

Feasible solutions: All Hamiltonian tours, i.e., the subgraphs H of G that are connected, and each node in them that has degree two.

Objective function: The weight function $w(H) = \sum_{e \in H} w(e)$ of the tour.

Goal: Minimization.

The TSP is an NP-hard optimization problem. This means that a polynomial time algorithm for the TSP does not exist unless P = NP. One way out of this dilemma is provided by *approximation algorithms*. A polynomial time algorithm for the TSP is called an α-approximation algorithm if the tour H produced by the algorithm fulfills $w(H) \leq \alpha \cdot \mathrm{OPT}(G)$. Here OPT($G$) is the weight of a minimum weight tour of G. If G is clear from the context, one just writes OPT. An α-approximation algorithm always produces a feasible solution whose objective value is at most a factor of α away from the optimum value. α is also called the approximation factor or performance guarantee. α does not need to be a constant; it can be a function that depends on the size of the instance or the number of nodes n.

If there exists a polynomial time approximation algorithm for the TSP that achieves an exponential approximation factor in n, then $\mathsf{P} = \mathsf{NP}$ [6]. Therefore, one has to look at restricted instances. The most natural restriction is the *triangle inequality*, that means,

$$w(u, v) \leq w(u, x) + w(x, v) \quad \text{for all } u, v, x \in V.$$

The corresponding problem is called the *Metric TSP*. For the Metric TSP, approximation algorithms that achieve a constant approximation factor exist. Note that for the Metric TSP, it is sufficient to find a tour that visits each vertex *at least* once: Given such a tour, we can find a Hamiltonian tour of no larger weight by skipping every vertex that we already visited. By the triangle inequality, the new tour cannot get heavier.

Key Results

A simple 2-approximation algorithm for the Metric TSP is the *tree doubling algorithm*. It uses minimum spanning trees to compute Hamiltonian tours. A *spanning tree* T of a graph $G = (V, E, w)$ is a connected acyclic subgraph of G that contains each node of V. The weight $w(T)$ of such a spanning tree is the sum of the weights of the edges in it, i.e., $w(T) = \sum_{e \in T} w(e)$. A spanning tree is called a minimum spanning tree if its weight is minimum among all spanning trees of G. One can efficiently compute a minimum spanning tree, for instance via Prim's or Kruskal's algorithm, see e.g., [5].

The tree doubling algorithm seems to be folklore. The next lemma is the key for proving the upper bound on the approximation performance of the tree doubling algorithm.

Lemma 1 *Let T be a minimum spanning tree of $G = (V, E, w)$. Then $w(T) \leq$ OPT.*

Proof If one deletes any edge of a Hamiltonian tour of G, one gets a spanning tree of G. □

Algorithm 1 Tree doubling algorithm

Input: a complete loopless edge weighted undirected graph $G = (V, E, w)$ with weight function $w : E \to \mathbb{Q}_{\geq 0}$ that fulfills the triangle inequality

Output: a Hamiltonian tour of G that is a 2" approximation

1: Compute a minimum spanning tree T of G.
2: Duplicate each edge of T and obtain a Eulerian multigraph T'.
3: Compute a Eulerian tour of T' (for instance via a depth first search in T). Whenever a node is visited in the Eulerian tour that was already visited, this node is skipped and one proceeds with the next unvisited node along the Eulerian cycle. (This process is called *shortcutting*.) Return the resulting Hamiltonian tour H.

Theorem 2 *Algorithm 1 alwaysreturns a Hamiltonian tour whose weight is at most twice the weight of an optimum tour. Its running time is polynomial.*

Proof By Lemma 1, $w(T) \leq$ OPT. Since one duplicates each edge of T, the weight of T' equals $w(T') = 2w(T) \leq 2$OPT. When taking shortcuts in step 3, a path in T' is replaced by a single edge. By the triangle inequality, the sum of the weights of the edges in such a path is at least the weight of the edge it is replaced by. (Here, the algorithm breaks down for arbitrary weight functions.) Thus $w(H) \leq w(T')$. This proves the claim about the approximation performance.

The running time is dominated by the time needed to compute a minimum spanning tree. This is clearly polynomial. □

Christofides' algorithm (Algorithm 2) is a clever refinement of the tree doubling algorithm. It first computes a minimum spanning tree. On the nodes that have an odd degree in T, it then computes a minimum weight perfect matching. A matching M of G is called a matching on $U \subseteq V$ if all edges of M consist of two nodes from U. Such a matching is called *perfect* if every node of U is incident with an edge of M.

Lemma 3 *Let$U \subseteq V, \#U$ even. Let M be a minimum weight perfect matching on U. Then $w(M) \leq$ OPT$/2$.*

M

Algorithm 2 Christofides' algorithm

Input: a complete loopless edge weighted undirected graph $G = (V, E, w)$ with weight function $w : E \to \mathbb{Q}_{\geq 0}$ that fulfills the triangle inequality
Output: a Hamiltonian tour of G that is a 3/2" approximation
1: Compute a minimum spanning tree T of G.
2: Let $U \subseteq V$ be the set of all nodes that have odd degree in T. In G, compute a minimum weight perfect matching M on U.
3: Compute a Eulerian tour of $T \cup M$ (considered as a multigraph).
4: Take shortcuts in this Eulerian tour to a Hamiltonian tour H.

Proof Let H be an optimum Hamiltonian tour of G. One takes shortcuts in H to get a tour H' on $G|_U$ as follows: H induces a permutation of the nodes in U, namely the order in which the nodes are visited by H. One connects the nodes of U in the order given by the permutation. To every edge of H' corresponds a path in H connecting the two nodes of this edge. By the triangle inequality, $w(H') \leq w(H)$. Since $\#U$ is even, H' is the union of two matchings. The lighter one of these two has a weight of at most $w(H')/2 \leq \mathrm{OPT}/2$. □

One can compute a minimum weight perfect matching in time $O(n^3)$, see for instance [5].

Theorem 4 *Algorithm 2 is a 3/2-approximation algorithm with polynomial running time.*

Proof First observe that the number of odd degree nodes of the spanning tree is even, since the sum of the degrees of all nodes equals $2(n-1)$, which is even. Thus a perfect matching on U exists. The weight of the Eulerian tour is obviously $w(T) + w(M)$. By Lemma 1, $w(T) \leq \mathrm{OPT}$. By Lemma 3, $w(M) \leq \mathrm{OPT}/2$. The weight $w(H)$ of the computed tour H is at most the weight of the Eulerian tour by the triangle inequality, i.e., $w(H) \leq \frac{3}{2}\mathrm{OPT}$. Thus the algorithm is a 3/2-approximation algorithm. Its running time is $O(n^3)$. □

Applications

Experimental analysis shows that Christofides' algorithm itself deviates by 10 % to 15 % from the optimum tour [3]. However, it can serve as a good starting tour for other heuristics like the Lin–Kernigham heuristic.

Open Problems

The analysis of Algorithm 2 is tight; an example is the metric completion of the graph depicted in Fig. 1. The unique minimum spanning tree consists of all solid edges. It has only two nodes of odd degree. The edge between these two nodes has weight $(1+\epsilon)(n+1)$. No shortcuts are needed, and the weight of the tour produced by the algorithm is $\approx 3n$. An optimum tour consists of all dashed edges plus the leftmost and rightmost solid edge. The weight of this tour is $(2n-1)(1+\epsilon) + 2 \approx 2n$.

The question whether there is an approximation algorithm with a better performance guarantee is a major open problem in the theory of approximation algorithms.

Held and Karp [2] design an LP based algorithm that computes a lower bound for the weight of an optimum TSP tour. It is conjectured that the weight of an optimum TSP tour is at most a factor of 4/3 larger than this lower bound, but this conjecture is unproven for more than three decades. An algorithmic proof of this conjecture

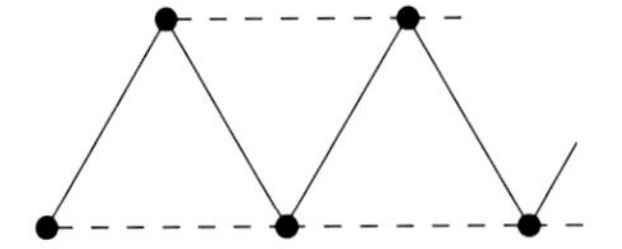
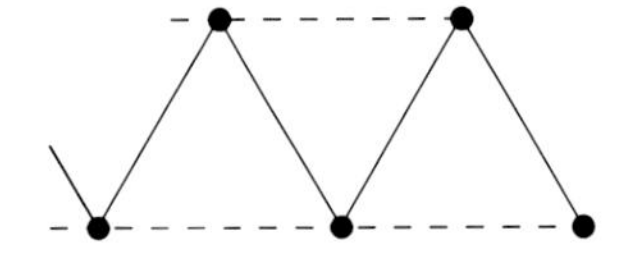

Metric TSP, Fig. 1 A tight example for Christofides' algorithm. There are $2n + 1$ nodes. *Solid edges* have a weight of one, *dashed ones* have a weight of $1 + \epsilon$

would yield an 4/3-approximation algorithm for the Metric TSP.

Experimental Results

See e.g., [3], where a deviation of 10 % to 15 % of the optimum (more precisely of the Held–Karp bound) is reported for various sorts of instances.

Data Sets

The webpage of the 8th DIMACS implementation challenge, www.research.att.com/~dsj/chtsp/, contains a lot of instances.

Cross-References

▶ Minimum Spanning Trees

Recommended Reading

Christofides never published his algorithm. It is usually cited as one of two technical reports from Carnegie Mellon University, TR 388 of the Graduate School of Industrial Administration (now Tepper School of Business) and CS-93-13. None of them seem to be available at Carnegie Mellon University anymore [Frank Balbach, personal communication, 2006]. A one-page abstract was published in a conference record. But his algorithm quickly found his way into standard textbooks on algorithm theory, see [7] for a recent one.

1. Christofides N (1976) Worst case analysis of a new heuristic for the traveling salesman problem, Technical Report 388, Graduate School of Industrial Administration, Carnegie-Mellon University, Pittsburgh. Also: Carnegie-Mellon University Technical Report CS-93-13, 1976. Abstract in Traub JF (ed) Symposium on new directions and recent results in algorithms and complexity, pp 441. Academic, New York (1976)
2. Held M, Karp RM (1970) The traveling salesman problem and minimum spanning trees. Oper Res 18:1138–1162
3. Johnson DS, McGeoch LA (2002) Experimental analysis of heuristics for the STSP. In: Gutin G, Punnen AP (eds) The traveling salesman problem and its variations. Kluwer, Dordrecht
4. Lawler EL, Lenstra JK, Rinnooy Kan AHG, Shmoys DB (eds) (1985) The traveling salesman problem. A guided tour of combinatorial optimization. Wiley, Chichester
5. Papadimitriou C, Steiglitz K (1982) Combinatorial optimization: algorithms and complexity. Prentice-Hall, Englewood Cliffs
6. Sahni S, Gonzalez T (1976) P-complete approximation problems. J ACM 23:555–565
7. Vazirani VV (2001) Approximation algorithms. Springer, Berlin
8. Traveling Salesman Problem (2006). www.tsp.gatech.edu. Accessed 28 Mar 2008

Metrical Task Systems

Manor Mendel
Department of Mathematics and Computer Science, The Open University of Israel, Raanana, Israel

Keywords

MTS

Years and Authors of Summarized Original Work

1992; Borodin, Linial, Saks

Problem Definition

Metrical task systems (MTS), introduced by Borodin, Linial, and Saks [5], is a cost minimization problem defined on a metric space (X, d_X) and informally described as follows: A given *system* has a set of internal states X. The aim of the system is to serve a given sequence of tasks. The servicing of each task has a certain cost that depends on the task and the state of the system. The system may switch states before serving the task, and the total cost for servicing the task is the sum of the service cost of the task in the new state and the distance between the states in a metric space defined on the set of states. Following Manasse, McGeoch, and Sleator [11], an extended model is considered here, in which the set of allowable tasks may be restricted.

Notation

Let T^* denote the set of finite sequences of elements from a set T. For $x, y \in T^*$, $x \circ y$ is the concatenation of the sequences x and y, and $|x|$ is the length of the sequence x.

Definition 1 (Metrical Task System) Fix a metric space (X, d_X). Let $\Gamma = \{(r_x)_{x \in X} : \forall x \in X, r(x) \in [0, \infty]\}$ be the set of all possible tasks. Let $T \subseteq \Gamma$ be a subset of tasks, called *allowable tasks*.
MTS((X, d_X), T, $a_0 \in X$):
INPUT: A finite sequence of tasks $\tau = (\tau_1, \ldots, \tau_m) \in T^*$.
OUTPUT: A sequence of points $a = (a_1, \ldots, a_m) \in X^*$, $|a| = |\tau|$.
OBJECTIVE: minimize

$$\mathsf{cost}(\tau, a) = \sum_{i=1}^{m} (d_X(a_{i-1}, a_i) + \tau_i(a_i)).$$

When $T = \Gamma$, the MTS problem is called *general*.

When X is finite and the task sequence $\tau \in T^*$ is given in advance, a dynamic programming algorithm can compute an optimal solution in space $O(|X|)$ and time $O(|\tau| \cdot |X|)$. MTS, however, is most interesting in an online setting, where the system must respond to a task τ_i with a state $a_i \in X$ without knowing the future tasks in τ. Formally,

Definition 2 (Online algorithms for MTS) A deterministic algorithm for a MTS((X, d_X), T, a_0) is a mapping $S : T^* \to X^*$ such that for every $\tau \in T$, $|S(\tau)| = |\tau|$. A deterministic algorithm $S : T^* \to X^*$ is called *online* if for every $\tau, \sigma \in T^*$, there exists $a \in X^*$, $|a| = |\sigma|$ such that $S(\tau \circ \sigma) = S(\tau) \circ a$. A randomized online algorithm is a probability distribution over deterministic online algorithms.

Online algorithms for MTS are evaluated using *(asymptotic) competitive analysis*, which is, roughly speaking, the worst ratio of the algorithm's cost to the optimal cost taken over all possible task sequences.

Definition 3 A randomized online algorithm R for MTS((X, d_X), a_0) is called c-competitive (against oblivious adversaries) if there exists $b = b(X) \in \mathbb{R}$ such that for any task sequence $\tau \in T^*$, and any point sequence $a \in X^*$, $|a| = |\tau|$,

$$\mathbb{E}[\mathsf{cost}(\tau, R(\tau))] \le c \cdot \mathsf{cost}(\tau, a) + b,$$

where the expectation is taken over the distribution R.

The competitive ratio of an online algorithm R is the infimum over $c \ge 1$ for which R is c-competitive. The deterministic [respectively, randomized] competitive ratio of MTS((X, d_X), T, a_0) is the infimum over the competitive ratios of all deterministic [respectively, randomized] online algorithms for this problem. Note that because of the existential quantifier over b, the asymptotic competitive ratio (both randomized and deterministic) of a MTS((X, d_X), T, a_0) is independent of a_0, and it can therefore be dropped from the notation.

Key Results

Theorem 1 ([5]) *The deterministic competitive ratio of the general MTS problem on any n-point metric space is* $2n - 1$.

In contrast to the deterministic case, the understanding of randomized algorithms for general MTS is not complete, and generally no sharp bounds such as Theorem 1 are known.

Theorem 2 ([5, 10]) *The randomized competitive ratio of the general MTS problem on n-point uniform space (where all distances are equal) is at least* $H_n = \sum_{i=1}^{n-1} i^{-1}$, *and at most* $(1 + o(1))H_n$.

The best bounds currently known for general n-point metrics are proved in two steps: First the given metric is approximated by an *ultrametric*, and then a bound on the competitive ratio of general MTS on ultrametrics is proved.

Theorem 3 ([8, 9]) *For any n-point metric space (X, d_X), there exists an $O(\log^2 n \log\log n)$ competitive randomized algorithm for the general MTS on (X, d_X).*

The metric approximation component in the proof of Theorem 3 is called *probabilistic embedding*. An optimal $O(\log n)$ probabilistic embedding is shown by Fakcheroenphol, Rao and Talwar before [8] improving on results by Alon, Karp, Peleg, and West and by Bartal, where this notion was invented. A different type of metric approximation with better bounds for metrics of low *aspect ratio* is given in [3].

Fiat and Mendel [9] show a $O(\log n \log\log n)$ competitive algorithm for n-point ultrametrics, improving (and using) a result of Bartal, Blum, Burch, and Tomkins [1], where the first polylogarithmic (or even sublinear) competitive randomized algorithm for general MTS on general metric spaces is presented.

Theorem 4 ([2, 12]) *For any n-point metric space (X, d_X), the randomized competitive ratio of the general MTS on (X, d_X) is at least $\Omega(\log n / \log\log n)$.*

The metric approximation component in the proof of Theorem 4 is called *Ramsey subsets*. It was first used in this context by Karloff, Rabani, and Ravid, later improved by Blum, Karloff, Rabani and Saks, and Bartal, Bollobás, and Mendel [2]. A tight result on Ramsey subsets is proved by Bartal, Linial, Mendel, and Naor. For a simpler (and stronger) proof, see [12].

A lower bound of $\Omega(\log n / \log\log n)$ on the competitive ratio of any randomized algorithm for general MTS on n-point ultrametrics is proved in [2], improving previous results of Karloff, Rabani, and Ravid, and Blum, Karloff, Rabani and Saks.

The last theorem is the only one not concerning general MTSs.

Theorem 5 ([6]) *It is PSPACE hard to determine the competitive ratio of a given MTS instance $((X, d_X), a_0 \in X, T)$, even when d_X is the uniform metric. On the other hand, when d_X is uniform, there is a polynomial time deterministic online algorithm for MTS $((X, d_X), a_0 \in X, T)$ whose competitive ratio is $O(\log |X|)$ times the deterministic competitive ratio of the MTS((X, d_X), a_0, T). Here it is assumed that the instance ((X, d_X), a_0, T) is given explicitly.*

Applications

Metrical task systems were introduced as an abstraction for online computation, they generalize many concrete online problems such as paging, weighted caching, k-server, and list update. Historically, it served as an indicator for a general theory of competitive online computation.

The main technical contribution of the MTS model is the development of the work function algorithm used to prove the upper bound in Theorem 1. This algorithm was later analyzed by Koutsoupias and Papadimitriou in the context of the k-server problem, and was shown to be $2k - 1$ competitive. Furthermore, although the MTS model generalizes the k-server problem, the general MTS problem on the n-point metric is essentially equivalent to the $(n - 1)$-server problem on the same metric [2]. Hence, lower bounds on the competitive ratio of general MTS imply lower bounds for the k-server problem, and algorithms for general MTS may constitute a first step in devising an algorithm for the k-server problem, as is the case with the work function algorithm.

The metric approximations used in Theorem 3, and Theorem 4 have found other algorithmic applications.

Open Problems

There is still an obvious gap between the upper bound and lower bound known on the randomized competitive ratio of general MTS on general finite metrics. It is known that, contrary to the deterministic case, the randomized competitive ratio is *not* constant across all metric spaces of the same size. However, in those cases where exact bounds are known, the competitive ratio is $\Theta(\log n)$. An obvious conjecture is that the randomized competitive is $\Theta(\log n)$ for any n-point metric. Arguably, the simplest classes of metric spaces for which no upper bound on the randomized competitive ratio better than $O(\log^2 n)$ is known, are paths and cycles.

Also lacking is a "middle theory" for MTS. On the one hand, general MTS are understood fairly well. On the other hand, specialized MTS such as list update, deterministic k-server algorithms, and deterministic weighted-caching, are also understood fairly well, and have a much better competitive ratio than the corresponding general MTS. What may be missing are "in between" models of MTS that can explain the low competitive ratios for some of the concrete online problems mentioned above.

It would be also nice to strengthen Theorem 5, and obtain a polynomial time deterministic online algorithm whose competitive ratio on any MTS instance on *any* n-point metric space is at most poly-log(n) times the deterministic competitive ratio of that MTS instance.

Cross-References

- ▶ Algorithm DC-TREE for k-Servers on Trees
- ▶ Approximating Metric Spaces by Tree Metrics
- ▶ Online List Update
- ▶ Online Paging and Caching
- ▶ Ski Rental Problem
- ▶ Work-Function Algorithm for k-Servers

Recommended Reading

1. Bartal Y, Blum A, Burch C, Tomkins A (1997) A polylog(n)-competitive algorithm for metrical task systems. In: Proceedings of the 29th annual ACM symposium on the theory of computing. ACM, New York, pp 711–719
2. Bartal Y, Bollobás B, Mendel M (2006) Ramsey-type theorems for metric spaces with applications to online problems. J Comput Syst Sci 72:890–921
3. Bartal Y, Mendel M (2004) Multiembedding of metric spaces. SIAM J Comput 34:248–259
4. Borodin A, El-Yaniv R (1998) Online computation and competitive analysis. Cambridge University Press, Cambridge, UK
5. Borodin A, Linial N, Saks ME (1992) An optimal on-line algorithm for metrical task system. J ACM 39:745–763
6. Burley WR, Irani S (1997) On algorithm design for metrical task systems. Algorithmica 18:461–485
7. Chrobak M, Larmore LL (1998) Chapter 4, Metrical task systems, the server problem and the work function algorithm. In: Fiat A, Woeginger GJ (eds) Online algorithms. The state of the art. LNCS, vol 1442. Springer, London, pp 74–96
8. Fakcharoenphol J, Rao S, Talwar K (2004) A tight bound on approximating arbitrary metrics by tree metrics. J Comput Syst Sci 69:485–497
9. Fiat A, Mendel M (2003) Better algorithms for unfair metrical task systems and applications. SIAM J Comput 32:1403–1422
10. Irani S, Seiden SS (1998) Randomized algorithms for metrical task systems. Theor Comput Sci 194:163–182
11. Manasse MS, McGeoch LA, Sleator DD (1990) Competitive algorithms for server problems. J Algorithms 11:208–230
12. Mendel M, Naor A (2007) Ramsey partitions and proximity data structures. J Eur Math Soc 9(2):253–275

Min-Hash Sketches

Edith Cohen
Tel Aviv University, Tel Aviv, Israel
Stanford University, Stanford, CA, USA

Keywords

Approximate distinct counting; Bottom-k sketches; k-mins sketches; k-partition sketches; Similarity estimation; Summary structures

Years and Authors of Summarized Original Work

1985; Flajolet, Martin
1997; Broder
1997; Cohen

Problem Definition

MINHASH sketches (also known as min-wise sketches) are randomized summary structures of subsets which support set union operations and approximate processing of cardinality and similarity queries.

Set-union support, also called *mergeability*, means that a sketch of the union of two sets can be computed from the sketches of the two sets. In particular, this applies when the second set is a single element. The queries supported by MINHASH sketches include cardinality (of a subset from its sketch) and similarity (of two subsets from their sketches).

Sketches are useful for massive data analysis. Working with sketches often means that instead of explicitly maintaining and manipulating very large subsets (or equivalently 0/1 vectors), we can instead maintain the much smaller sketches and can still query properties of these subsets.

We denote the universe of elements by U and its size by $n = |U|$. We denote by $S(X)$ the sketch of the subset $X \subset U$.

Set Operations

- **Inserting an element:** Given a set X and element $y \in U$, a sketch $S(X \cup \{y\})$ can be computed from $S(X)$ and y.
- **Merging two sets:** For two (possibly overlapping) sets X and Y, we can obtain a sketch of their union $S(X \cup Y)$ from $S(X)$ and $S(Y)$.

Support for insertion makes the sketches suitable for streaming, where elements (potentially with repeated occurrences) are introduced sequentially. Support for merges is important for parallel or distributed processing: We can sketch a data set that has multiple parts by sketching each part and combining the sketches. We can also compute the sketches by partitioning the data into parts, sketching each of the parts concurrently, and finally merging the sketches of the parts to obtain a sketch of the full data set.

Queries

From the sketches of subsets, we would like to (approximately) answer queries on the original data. More precisely, for a set pf subsets $\{X_i\}$, we are interested in estimating a function $f(\{X_i\})$. To do this, we apply an *estimator* $\hat{f}$ to the respective set of sketches $\{S(X_i)\}$.

We would like our estimators to have certain properties: When estimating nonnegative quantities (such as cardinalities or similarities), we would want the estimator to be nonnegative as well. We are often interested in unbiased estimators and always in *admissible* estimators, which are Pareto optimal in terms of variance (variance on one instance cannot be improved without increasing variance on another). We also seek good *concentration*, meaning that the probability of error decreases exponentially with the relative error.

We list some very useful queries that are supported by MINHASH sketches:

- **Cardinality:** The number of elements in the set $f(X) = |X|$.
- **Similarity:** The Jaccard coefficient $f(X, Y) = |X \cap Y|/|X \cup Y|$, cosine similarity $f(X, Y) = |X \cap Y|/\sqrt{|X||Y|}$, or cardinality of the union $f(X, Y) = |X \cup Y|$.
- **Complex relations:** Cardinality of the union of multiple sets $|\bigcup_i X_i|$, number of elements occurring in at least 2 sets $|\{j \mid \exists i_1 \neq i_2, j \in X_{i_1} \cap X_{i_2}\}|$, set differences, etc.
- **Domain queries:** When elements have associated metadata (age, topic, activity level), we can include this information in the sketch, which becomes a random sample of the set. Including this information allows us to process domain queries, which depend on the matadata. For example, "the number of Californians in the union of two (or more) sets."

Key Results

MINHASH sketches had been proposed as summary structures which satisfy the above requirements. There are multiple variants of the MINHASH sketch, which are optimized for different applications. The common thread is that the elements $x \in U$ of the universe U are assigned random *rank* values $r(x)$ (which are typically produced by a random hash function). The MINHASH sketch $S(X)$ of a set X includes order statistics (maximum, minimum, or top-/bottom-k values) of the set of ranks $\{r(x) \mid x \in X\}$. Note that when we sketch multiple sets, the same random rank assignment is common to all sketches (we refer to this as *coordination*).

Before stating precise definitions for the different MINHASH sketch structures, we provide some intuition for the power of order statistics. We first consider cardinality estimation. The minimum rank value $\min_{x \in X} r(x)$ is the minimum of $|X|$ independent random variables, and therefore, its expectation should be smaller when the cardinality $|X|$ is larger. Thus, the minimum rank carries information on $|X|$. We next consider the sketches of two sets X and Y. Recall that they are computed with respect to the same assignment r. Therefore, the minimum rank values carry information on the similarity of the sets: in particular, when the sets are more similar, their minimum ranks are more likely to be equal. Finally, the minimum rank element of a set is a random sample from the set and, therefore, as such, can support estimation of statistics of the set.

The variations of MINHASH sketches differ in the particular structure: how the rank assignment is used and the domain and distribution of the ranks $r(x) \sim D$.

Structure

MINHASH sketches are parameterized by an integer $k \geq 1$, which controls a trade-off between the size of the sketch representation and the accuracy of approximate query results.

MINHASH sketches come in the following three common flavors:

- A *k-mins sketch* [6, 13] includes the smallest rank in each of k independent rank assignments. There are k different rank functions r_i and the sketch $S(X) = (\tau_1, \ldots, \tau_k)$ has $\tau_i = \min_{y \in X} r_i(y)$. When viewed as a sample, it corresponds to sampling k times with replacement.
- A *k-partition sketch* [13, 14, 18], which in the context of cardinality estimation is called *stochastic averaging*, uses a single rank assignment together with a uniform at random mapping of items to k buckets. We use $b : U \to [k]$ for the bucket mapping and r for the rank assignment. The sketch $(\tau_1, \ldots, \tau_k)$ then includes the item with minimum rank in each bucket. That is, $\tau_i = \min_{y \in X | b(y)=i} r_i(y)$. If the set is empty, the entry is typically defined as the supremum of the domain of r.
- A *bottom-k sketch* [4, 6] $\tau_1 < \cdots < \tau_k$ includes the k items with smallest rank in $\{r(y) \mid y \in X\}$. Interpreted as a sample, it corresponds to sampling k elements without replacement. Related uses of the same method include KMV sketch [2], coordinated order samples [3, 19, 21], or conditional random sampling [17].

Note that all three flavors are the same when $k = 1$.

With all three flavors, the sketch represents k random elements of D. When viewed as random samples, MINHASH sketches of different subsets X are *coordinated*, since they are generated using the same random rank assignments to the domain U. The notion of coordination is very powerful. It means that similar subsets have similar sketches (a locality sensitive hashing property). It also allows us to support merges and similarity estimation much more effectively. Coordination in the context of survey sampling was introduced in [3] and was applied for sketching data in [4,6].

Rank Distribution

Since we typically use a random hash function $H(x) \sim D$ to generate the ranks, it always suffices to store element identifiers instead of ranks, which means the representation of each rank value is $\lceil \log_2 n \rceil$ bits and the bit size of the sketch is at most $k \log n$. This representation size is necessary when we want to support domain queries – the sketch of each set should identify the element associated with each included rank, so we can retrieve the metadata needed to evaluate a selection predicate.

For the applications of estimating cardinalities or pairwise similarities, however, we can work with ranks that are not unique to elements and, in particular, come from a smaller discrete domain. Working with smaller ranks allows us to use sketches of a much smaller size and also replace the dependence of the sketch on the domain size ($O(\log n)$ per entry) by dependence on the subset sizes. In particular, we can support cardinality estimation and similarity estimation of subsets of size at most m with ranks of size $O(\log\log m)$. Since the k rank values used in the sketch are typically highly correlated, the sketch $S(X)$ can be stored using $O(\log\log m + k)$ bits in expectation ($O(\log\log m + k \log k)$ bits for similarity). This is useful when we maintain sketches of many sets and memory is at a premium, as when collecting traffic statistics in IP routers.

For analysis, it is often convenient to work with continuous ranks, which without loss of generality are $r \sim U[0,1]$ [6], since there is a monotone (order preserving) transformation from any other continuous distribution. Using ranks of size $O(\log n)$ is equivalent to working with continuous ranks.

In practice, we work with discrete ranks, for example, values restricted to $1/2^i$ for integral $i > 0$ [13] or more generally using a *base* $b > 1$ and using $1/b^i$. This is equivalent to drawing a continuous rank and rounding it down to the largest discrete point of the form $1/b^i$.

Streaming: Number of Updates

Consider now maintaining a MINHASH sketch in the streaming model. We maintain a sketch S of the elements X that we had seen until now. When we process a new element y, then if $y \in X$, the sketch is not modified. We can show that the number of times the sketch is modified is in expectation at most $k \ln n$, where $n = |X|$ is the number of distinct elements in the prefix. We provide the argument for bottom-k sketches. It is similar with other flavors. The probability that a new element has a rank value that is smaller than the kth smallest rank in X is the probability that it is in one of the first k positions in a permutation of size $n + 1$. That is, the probability is 1 if $n < k$ and is $k/(n+1$ otherwise. Summing over new distinct elements $n = 1, \ldots, |X|$, we obtain $\sum_{i=1}^{n} k/i \leq k \ln n$.

Inserting an Element

We now consider inserting an element y, that is, obtaining a sketch $S(X \cup \{y\})$ from $S(X)$ and y. The three sketch flavors have different properties and trade-offs in terms of insertion costs. We distinguish between insertions that result in an actual update of the sketch and insertions where the sketch is not modified.

- k-mins sketch: We need to generate the rank of y, $r_i(y)$, in each of k different assignments (k hash computations). We can then compare, coordinate-wise, each rank with the respective one in the sketch, taking the minimum of the two values. This means that each insertion, whether the sketch is updated or not, results in $O(k)$ operations.
- Bottom-k sketch: We apply our hash function to generate $r(y)$. We then compare $r(y)$ with τ_k. If the sketch contains fewer than k ranks ($|S| < k$ or τ_k is the rank domain supremum), then $r(y)$ is inserted to S.

  Otherwise, the sketch is updated only if $r(y) < \tau_k$. In this case, the largest sketch entry τ_k is discarded and $r(y)$ is inserted to the sketch S. When the sketch is not modified, the operation is $O(1)$. Otherwise, it can be $O(\log k)$.
- k-partition sketch: We apply the hash functions to y to determine the bucket $b(y) \in [k]$ and the rank $r(y)$. To determine if an update is needed, we compare $r(y)$ and $\tau_{b(y)}$. If the latter is empty ($\tau_{b(y)}$ is the domain supremum) or if $r(y) < \tau_{b(y)}$, we assign $\tau_{b(y)} \leftarrow r(y)$.

Merging

We now consider computing the sketch $S(X \cup Y)$ from the sketches $S(X)$ and $S(Y)$ of two sets X, Y.

For k-mins and k-partition sketches, the sketch of $S(X \cup Y)$ is simply the coordinate-wise minimum $(\min\{\tau_1, \tau_1'\}, \ldots, \min\{\tau_k, \tau_k'\})$

of the sketches $S(X) = (\tau_1, \ldots, \tau_k)$ and $S(Y) = (\tau'_1, \ldots, \tau'_k)$. For bottom-$k$ sketches, the sketch of $S(X \cup Y)$ includes the k smallest rank values in $S(X) \cup S(Y)$.

Estimators

Estimators are typically specifically derived for a given sketch flavor and rank distribution.

Cardinality estimators were pioneered by Flajolet and Martin [13], continuous ranks were considered in [6], and lower bounds were presented in [1]. State-of-the-art practical solutions include [7, 14]. Cardinality estimation can be viewed in the context of the theory of point estimation: estimating the parameter of a distribution (the cardinality) from the sketch (the "outcome"). Estimation theory implies that current estimators are optimal (minimize variance) for the sketch [7]. Recently, *historic inverse probability (HIP)* estimators were proposed, which apply with all sketch types and improve variance by maintaining an approximate count alongside the MINHASH sketch [7], which is updated when the sketch is modified.

Estimators for set relations were first considered in [6] (cardinality of union, by computing a sketch of the union and applying a cardinality estimator) and [4] (the Jaccard coefficient, which is the ratio of intersection to union size). The Jaccard coefficient can be estimated on all sketch flavors (when ranks are not likely to have collisions) by simply looking at the respective ratio in the sketches themselves. In general, many set relations can be estimated from the sketches, and state-of-the-art derivation is given in [8, 10].

Applications

Approximate distinct counters are widely used in practice. Applications include statistics collection at IP routers and counting distinct search queries [15].

An important application of sketches, and their first application to estimate set relations, was introduced in [6]. Given a directed graph, and a node v, we can consider the set $R(v)$ of all reachable nodes. It turns out that sketches $S(R(v))$ for all nodes v in a graph can be computed very efficiently, in nearly linear time. The approach naturally extends to sketching neighborhoods in a graph. The sketches of nodes support efficient estimation of important graph properties, such as the distance distribution, node similarities (compare their relations to other nodes), and influence of a set of nodes [11, 12].

Similarity estimation using sketches was applied to identify near-duplicate Web pages by sketching "shingles" that are consecutive lists or words [4]. Since then, MINHASH sketches are extensively applied for similarity estimation of text documents and other entities.

Extensions

Weighted Elements

MINHASH sketches are summaries of sets or of 0/1 vectors. In many applications, each element $x \in U$ has a different intrinsic nonnegative weight $w(x) > 0$, and queries are formulated with respect to these weights: Instead of cardinality estimates we can consider the respective weighted sum $\sum_{x \in X} w(x)$. Instead of the Jaccard for the similarity of two sets X and Y, we may be interested in the weighted version $\sum_{x \in X \cap Y} w(x) / \sum_{x \in X \cup Y} w(x)$. When this is the case, to obtain more accurate query results, we use sketches so that the inclusion probability of an element increases with its weight. The sketch in this case would correspond to a weighted sample. This is implemented by using ranks which are drawn from a distribution that depends on the weights $w(y)$ [6, 9, 19–21].

Hash Functions

We assumed here the availability of truly random hash function. In practice, observed performance is consistent with this assumption. We mention however that the amount of independence needed was formally studied using *min-wise independent* hash functions [5, 16].

Cross-References

- ▶ All-Distances Sketches of graphs.
- ▶ Coordinated Sampling. They are also a building block of
- ▶ MIN-HASH Sketches can be viewed as

Recommended Reading

1. Alon N, Matias Y, Szegedy M (1999) The space complexity of approximating the frequency moments. J Comput Syst Sci 58:137–147
2. Bar-Yossef Z, Jayram TS, Kumar R, Sivakumar D, Trevisan L (2002) Counting distinct elements in a data stream. In: RANDOM, Cambridge. ACM
3. Brewer KRW, Early LJ, Joyce SF (1972) Selecting several samples from a single population. Aust J Stat 14(3):231–239
4. Broder AZ (1997) On the resemblance and containment of documents. In: Proceedings of the compression and complexity of sequences, Salerno. IEEE, pp 21–29
5. Broder AZ, Charikar M, Frieze AM, Mitzenmacher M (2000) Min-wise independent permutations. J Comput Syst Sci 60(3):630–659
6. Cohen E (1997) Size-estimation framework with applications to transitive closure and reachability. J Comput Syst Sci 55:441–453
7. Cohen E (2014) All-distances sketches, revisited: HIP estimators for massive graphs analysis. In: PODS, Snowbird. ACM. http://arxiv.org/abs/1306.3284
8. Cohen E (2014) Estimation for monotone sampling: competitiveness and customization. In: PODC, Paris. ACM. http://arxiv.org/abs/1212.0243, full version http://arxiv.org/abs/1212.0243
9. Cohen E, Kaplan H (2007) Summarizing data using bottom-k sketches. In: PODC, Portland. ACM
10. Cohen E, Kaplan H (2009) Leveraging discarded samples for tighter estimation of multiple-set aggregates. In: SIGMETRICS, Seattle. ACM
11. Cohen E, Delling D, Fuchs F, Goldberg A, Goldszmidt M, Werneck R (2013) Scalable similarity estimation in social networks: closeness, node labels, and random edge lengths. In: COSN, Boston. ACM
12. Cohen E, Delling D, Pajor T, Werneck RF (2014) Sketch-based influence maximization and computation: scaling up with guarantees. In: CIKM. ACM. http://research.microsoft.com/apps/pubs/?id=226623, full version http://research.microsoft.com/apps/pubs/?id=226623
13. Flajolet P, Martin GN (1985) Probabilistic counting algorithms for data base applications. J Comput Syst Sci 31:182–209
14. Flajolet P, Fusy E, Gandouet O, Meunier F (2007) Hyperloglog: the analysis of a near-optimal cardinality estimation algorithm. In: Analysis of algorithms (AOFA), Juan des Pins
15. Heule S, Nunkesser M, Hall A (2013) HyperLogLog in practice: algorithmic engineering of a state of the art cardinality estimation algorithm. In: EDBT, Genoa
16. Indyk P (1999) A small approximately min-wise independent family of hash functions. In: Proceedings of the 10th ACM-SIAM symposium on discrete algorithms, Baltimore. ACM-SIAM
17. Li P, Church KW, Hastie T (2008) One sketch for all: theory and application of conditional random sampling. In: NIPS, Vancouver
18. Li P, Owen AB, Zhang CH (2012) One permutation hashing. In: NIPS, Lake Tahoe
19. Ohlsson E (1998) Sequential poisson sampling. J Off Stat 14(2):149–162
20. Rosén B (1972) Asymptotic theory for successive sampling with varying probabilities without replacement, I. Ann Math Stat 43(2):373–397. http://www.jstor.org/stable/2239977
21. Rosén B (1997) Asymptotic theory for order sampling. J Stat Plan Inference 62(2):135–158

Minimal Dominating Set Enumeration

Mamadou Moustapha Kanté and Lhouari Nourine
Clermont-Université, Université Blaise Pascal, LIMOS, CNRS, Aubière, France

Keywords

Dominating set; Enumeration; Transversal hypergraph

Years and Authors of Summarized Original Work

2011–2014; Kanté, Limouzy, Mary, Nourine, Uno

Problem Definition

Let G be a graph on n vertices and m edges. An edge is written xy (equivalently yx). A *dominating set* in G is a set of vertices D

such that every vertex of G is either in D or is adjacent to some vertex of D. It is said to be *minimal* if it does not contain any other dominating set as a proper subset. For every vertex x, let $N[x]$ be $\{x\} \cup \{y | xy \in E\}$ and for every $S \subseteq V$ let $N[S] := \bigcup_{x \in S} N[x]$. For $S \subseteq V$ and $x \in S$ we call any $y \in N[x] \setminus N[S \setminus x]$, a *private neighbor of x with respect to S*. The set of minimal dominating sets of G is denoted by $\mathcal{D}(G)$. We are interested in an output-polynomial algorithm for enumerating $\mathcal{D}(G)$, i.e., listing, without repetitions, all the elements of $\mathcal{D}(G)$ in time bounded by $p\left(n+m, \sum_{D \in \mathcal{D}(G)} |D|\right)$ (DOM-ENUM for short).

It is easy to see that DOM-ENUM is a special case of HYPERGRAPH DUALIZATION. Let $\mathcal{N}(G)$, called the *closed neighborhood hypergraph*, be the hypergraph with hyperedges $\{N[x] | x \in V\}$. It is easy to see that D is a dominating set of G if and only if D is a transversal of $\mathcal{N}(G)$. Hence, DOM-ENUM is a special case of HY-PERGRAPH DUALIZATION. For several graph classes their closed neighborhood hypergraphs are subclasses of hypergraph classes where an output-polynomial algorithm is known for HYPERGRAPH DUALIZATION, e.g., minor-closed classes of graphs, graphs of bounded degree, graphs of bounded conformality, graphs of bounded degeneracy, graphs of logarithmic degeneracy [11, 12, 19]. So, DOM-ENUM seems more tractable than HYPERGRAPH DUALIZATION since there exist families of hypergraphs that are not closed neighborhoods of graphs [1].

Key Results

Contrary to several special cases of HYPERGRAPH DUALIZATION in graphs, (e.g., enumeration of maximal independent sets, enumeration of spanning forests, etc.) DOM-ENUM is equivalent to HYPERGRAPH DUALIZATION. Indeed, it is proved in [14] that with every hypergraph $\mathcal{H}$, one can associate a co-bipartite graph $\mathcal{B}(\mathcal{H})$ such that every minimal dominating set of $\mathcal{B}(\mathcal{H})$ is either a transversal of $\mathcal{H}$ or has size at most 2. A consequence is that there exists a polynomial delay polynomial space algorithm for HYPERGRAPH DUALIZATION if and only if there exists one for DOM-ENUM, even in co-bipartite graphs. The reduction is moreover asymptotically tight (with respect to polynomial delay reductions as defined in [19]) in the sense that there exist hypergraphs $\mathcal{H}$ such that for every graph G we cannot have $tr(\mathcal{H}) = \mathcal{D}(G)$ [14]. This intriguing result has the advantage of bringing tools from graph structural theory to tackle the difficult and widely open problem HYPERGRAPH DUALIZATION. Furthermore, until recently the most graph classes where DOM-ENUM is known to be tractable were those for which closed neighborhood hypergraphs were subclasses of some of the tractable hypergraph classes for HYPERGRAPH DUALIZATION. We will give examples of graph classes where graph theory helps a lot to solve DOM-ENUM, and sometimes allows to introduce new techniques for the enumeration.

It is widely known now that every monadic second-order formula can be checked in polynomial time in graph classes of bounded *clique-width* [3,20]. Courcelle proved in [2] that one can also enumerate, with linear delay linear space, the solutions of every monadic second-order formula. Since one can express in monadic second-order logic that a subset D of vertices is a minimal dominating set, DOM-ENUM has a linear delay linear space in graph classes of bounded clique-width. The algorithm by Courcelle is quite ingenious: it firsts constructs a DAG, some subtrees of which correspond to the positive runs of the tree-automata associated with the formula on the given graph and then enumerate these subtrees.

Many graph classes do not have bounded clique-width (interval graphs, permutation graphs, unit-disk graphs, etc.) and many such graph classes have nice structures that helped in the past for solving combinatorial problems, e.g., the clique-tree of chordal graphs, permutation models, etc. For some of these graph classes structural results can help to solve DOM-ENUM.

A common tool in enumeration area is the *parsimonious reduction*. One wants to enumerate a set of objects $\mathcal{O}$ and instead constructs a bijective function $b : \mathcal{O} \rightarrow \mathcal{T}$ such that there is an efficient algorithm to enumerate $\mathcal{T}$. For instance it is proved in [11, 14] that every minimal dominating set D of a split graph G can be characterized by $D \cap C(G)$ where $C(G)$ is the clique of G. A consequence is that in a split graph G there is a bijection between $\mathcal{D}(G)$ and the set $\{S \subseteq C(G) | \forall x \in S, x \text{ has a private neighbor}\}$, and since this latter set is an independent system, DOM-ENUM in split graphs admits a linear delay polynomial space algorithm.

One can obtain other parsimonious reductions using graph structures. For instance, it is easy to check that every minimal dominating set in an interval graph is a collection of paths. Moreover, using the interval model (and ordering intervals from their left endpoints) every minimal dominating set can be constructed greedily by keeping track of the last two chosen vertices. Indeed it is proved in [13] that with every interval graph G one can associate a DAG, the maximal paths of which are in bijection with the minimal dominating sets of G. The nodes of the DAG are pairs (x, y) such that $x < y$ and such that x and y can be both in a minimal dominating set, and the arcs are $((x, y), (y, z))$ such that (1) $\{x, y, z\}$ can be in a minimal dominating set, (2) there is no vertex between y and z that is not dominated by y or z, sources are pairs (x, y) where every interval before x is dominated by x, and sinks are pairs (x, y) where every interval after y is dominated by y. This reduction to maximal paths of a DAG can be adapted to several other graph classes having a linear structure similar to the interval model, e.g. permutation graphs, circular-arc graphs [13]. In general, if for every graph G in a graph class $\mathcal{C}$ one can associate an ordering of the vertices such that for every subset $S \subseteq V$ the possible ways to extend S into a minimal dominating set depends only on the last k vertices of S, for some fixed constant k depending only on $\mathcal{C}$, then for every $G \in \mathcal{C}$ the enumeration of $\mathcal{D}(G)$ can be reduced to the enumeration of paths in a DAG as for interval graphs and thus DOM-ENUM is tractable in $\mathcal{C}$ [19]. This seems for instance to be the case for d-trapezoid graphs.

Parsimonious reductions between graph classes can be also defined. For instance, the *completion* of a graph G, i.e., the set of edges that can be added to G without changing $\mathcal{D}(G)$ are characterized in [11, 14], this characterization lead the authors to prove that the completion of every P_6-free chordal graph is a split graph, which results in a linear delay polynomial space algorithm for DOM-ENUM in P_6-free chordal graphs.

The techniques developed by the HYPERGRAPH DUALIZATION community combined with graph structural theory can give rise to new tractable cases of DOM-ENUM. For instance, the main drawback of Berge's algorithm is that at some level computed transversals are not necessarily subsets of solutions and this prevents from obtaining an output-polynomial algorithm since the computed set may be arbitrary large compared to the solution set [21]. One way to overcome this difficulty consists in choosing some levels $l_1 \ldots, l_k$ of Berge's algorithm such that every computed set at level l_j is a subset of a solution at level l_{j+1}. A difficulty with that scheme is to compute all the descendants in level l_{j+1} of a transversal in level l_j. This idea combined with the structure of minimal dominating sets in line graphs is used to derive a polynomial delay polynomial space algorithm for DOM-ENUM in line graphs [15]. A consequence is that there is a polynomial delay polynomial space algorithm to list the set of *minimal edge dominating sets* in graphs.

Another famous technique in enumeration area is the *back tracking*. Start from the empty set, and in each iteration choose a vertex x and partition the problem into two sub-problems: the enumeration of minimal dominating sets containing x and the enumeration of those not containing x, at each step we have a set X to include in the solution and a set Y not to include. If at each step one can solve the EXTENSION PROBLEM, i.e., whether there is a minimal dominating set containing X and not intersecting Y, then DOM-ENUM admits a polynomial delay polynomial space algorithm. However, the EXTENSION PROBLEM is NP-complete in general [19] and even in split graphs [16]. But, sometimes structure helps. For instance, in split graphs whenever $X \cup Y \subseteq C(G)$, the EXTENSION PROBLEM is polynomial [11, 14] and was the key for the linear delay algorithm. Another special case of the EXTENSION PROBLEM is proved to be polynomial in chordal graphs using the *clique tree* of chordal graphs and is also the key to prove that DOM-ENUM in chordal graphs admits a polynomial delay polynomial space algorithm [16]. The algorithm uses deeply the clique tree and is a nested combination of several enumeration algorithms.

Open Problems

1. The first major challenge is to find an output-polynomial algorithm for DOM-ENUM, even in co-bipartite graphs. One way to address this problem is to understand the structure of minimal dominating sets in a graph. Failing to solve this problem, can graphs help to improve the quasi-polynomial time algorithm by Fredman and Khachiyan [7]?
2. Until now if the techniques used to solve DOM-ENUM in many graph classes are well-known, deep structural theory of graphs is not used and the used graph structures are more or less ad hoc. Can we unify all these results and obtain at the same time new positive results? Indeed, there are several well-studied graph classes where the status of DOM-ENUM is still open: bipartite graphs, unit-disk graphs, graphs of bounded expansion to cite a few. Are developed tools sufficient to address these graph classes?
3. There are several well-studied variants of the dominating set problem, in particular *total dominating set* and *connected dominating set* (see the monographs [9, 10]). It is proved in [14] that the enumeration of minimal total dominating sets and minimal connected dominating sets in split graphs is equivalent to HYPERGRAPH DUALIZATION. This is somehow surprising and we do not yet understand why such small variations make the problem difficult even in split graphs. Can we explain this situation?
4. From [14] we know that the enumeration of minimal connected dominating sets is harder than HYPERGRAPH DUALIZATION. Are both problems equivalent? Can we find a graph class $\mathcal{C}$ where each graph in $\mathcal{C}$ has a non-exponential number of minimal connected dominating sets, but minimum connected dominating set is NP-complete? Notice that if a class of graphs $\mathcal{C}$ has a polynomially bounded number of minimal separators, then the enumeration of minimal connected dominating sets can be reduced to DOM-ENUM [14].
5. A related question to DOM-ENUM is a tight bound for the number of minimal dominating sets in graphs. The best upper bound is $O(1.7159^n)$ and the best lower bound is $15^{n/6}$ [6]. For several graph classes, tight bounds were obtained [4, 8]. Prove that $15^{n/6}$ is the upper bound or find the tight bound.
6. Another related subject to DOM-ENUM is the counting of (minimal) dominating sets in time polynomial in the input graph. If the counting of dominating sets is a #P-hard problem and have been investigated in the past [5, 17, 18], not so much is known for the counting of minimal dominating sets, one can cite few examples: graphs of bounded clique-width [2], and interval, permutation and circular-arc graphs [13]. If we define for G the *minimal domination polynomial* $MD(G, x)$ that is the generating function of its minimal dominating sets, for which graph classes this polynomial can be computed? Does it have a (linear) recursive definition? For which values x can we evaluate it?

Cross-References

▶ Beyond Hypergraph Dualization
▶ Efficient Polynomial Time Approximation Scheme for Scheduling Jobs on Uniform Processors
▶ Reverse Search; Enumeration Algorithms

Recommended Reading

1. Brandstädt A, Van Le B, Spinrad JP (1999) Graph classes a survey. SIAM monographs on discrete mathematics and applications. SIAM, Philadelphia
2. Courcelle B (2009) Lineai delay enumeration and monadic second-order logic. Discret Appl Math 157:2675–2700
3. Courcelle B, Makowsky JA, Rotics U (2000) Linear time solvable optimization problems on graphs of bounded clique-width. Theory Comput Syst 33(2):125–150
4. Couturier J-F, Heggernes P, van 't Hof P, Kratsch D (2013) Minimal dominating sets in graph classes: combinatorial bounds and enumeration. Theor Comput Sci 487:82–94
5. Dohmen K, Tittmann P (2012) Domination reliability. Electron J Comb 19(1):P15
6. Fomin FV, Grandoni F, Pyatkin AV, Stepanov AA (2008) Combinatorial bounds via measure and conquer: bounding minimal dominating sets and applications. ACM Trans Algorithms 5(1)
7. Fredman ML, Khachiyan L (1996) On the complexity of dualization of monotone disjunctive normal forms. J Algorithms 21(3):618–628
8. Golovach PA, Heggernes P, Kanté MM, Kratsch D, Villanger Y (2014) Minimal dominating sets in interval graphs and trees. Submitted
9. Haynes TW, Hedetniemi ST, Slater PJ (1998) Fundamentals of domination in graphs. Volume 208 of pure and applied mathematics. Marcel Dekker, New York
10. Haynes TW, Hedetniemi ST, Slaterv PJ (1998) Domination in graphs: advanced topics. Volume 209 of pure and applied mathematics. Marcel Dekker, New York
11. Kanté MM, Limouzy V, Mary A, Nourine L (2011) Enumeration of minimal dominating sets and variants. In: FCT 2011, Oslo, pp 298–309
12. Kanté MM, Limouzy V, Mary A, Nourine L (2012) On the neighbourhood helly of some graph classes and applications to the enumeration of minimal dominating sets. In: ISAAC 2012, Taipei, pp 289–298
13. Kanté MM, Limouzy V, Mary A, Nourine L, Uno T (2013) On the enumeration and counting of minimal dominating sets in interval and permutation graphs. In: ISAAC 2013, Hong Kong, pp 339–349
14. Kanté MM, Limouzy V, Mary A, Nourine L (2014) On the enumeration of minimal dominating sets and related notions. Accepted for publication at SIAM Journal on Discrete Mathematics
15. Kanté MM, Limouzy V, Mary A, Nourine L, Uno T (2014) Polynomial delay algorithm for listing minimal edge dominating sets in graphs. In: CoRR. abs/1404.3501
16. Kanté MM, Limouzy V, Mary A, Nourine L, Uno T (2014) A polynomial delay algorithm for enumerating minimal dominating sets in chordal graphs. CoRRabs/1404.3501
17. Kijima S, Okamoto Y, Uno T (2011) Dominating set counting in graph classes. In: COCOON 2011, Dallas, pp 13–24
18. Kotek T, Preen J, Simon F, Tittmann P, Trinks M (2012) Recurrence relations and splitting formulas for the domination polynomial. Electron J Comb 19(3):P47
19. Mary A (2013) Énumération des Dominants Minimaux d'un graphe. PhD thesis, Université Blaise Pascal
20. Oum S, Seymour PD (2006) Approximating clique-width and branch-width. J Comb Theory Ser B 96(4):514–528
21. Takata K (2007) A worst-case analysis of the sequential method to list the minimal hitting sets of a hypergraph. SIAM J Discret Math 21(4):936–946

Minimal Perfect Hash Functions

Paolo Boldi and Sebastiano Vigna
Dipartimento di Informatica, Università degli Studi di Milano, Milano, Italy

Keywords

Minimal perfect hash functions

Years and Authors of Summarized Original Work

1984; Fredman, Komlós
1984; Mehlhorn
1996; Majewski, Wormald, Havas, Czech
2001; Hagerup, Tholey
2004; Chazelle, Kilian, Rubinfeld, Tal
2007; Botelho, Pagh, Ziviani
2009; Belazzougui, Botelho, Dietzfelbinger

Problem Definition

A minimal perfect hash function (MPHF) is a (data structure providing a) bijective map from a set S of n keys to the set of the first n natural numbers. In the static case (i.e., when the set S is known in advance), there is a wide spectrum of solutions available, offering different trade-offs in

M

terms of construction time, access time, and size of the data structure.

Problem Formulation

Let $[x]$ denote the set of the first x natural numbers. Given a positive integer $u = 2^w$, and a set $S \subseteq [u]$ with $|S| = n$, a function $h : S \to [m]$ is *perfect* if and only if it is injective and *minimal* if and only if $m = n$. An (M)PHF is a data structure that allows one to evaluate a (minimal) perfect function of this kind.

When comparing different techniques for building MPHFs, one should be aware of the trade-offs between construction time, evaluation time, and space needed to store the function. A general tenet is that evaluation should happen in constant time (with respect to n), whereas construction is only required to be feasible in a practical sense.

Space is often the most important aspect when a construction is taken into consideration; usually space is computed in an exact (i.e., non-asymptotic) way. Some exact space lower bounds for this problem are known (they are pure space bounds and do not consider evaluation time): Fredman and Komlós proved [4] that no MPHF can occupy less than $n \log e + \log \log u + O(\log n)$ bits, as soon as $u \geq n^{2+\epsilon}$; this bound is essentially tight [9, Sect. III.2.3, Thm. 8], disregarding evaluation time.

Key Results

One fundamental question is how close to the space lower bound $n \log e + \log \log u$ one can stay if the evaluation must be performed in constant time. The best theoretical results in this direction are given in [6], where an $n \log e + \log \log u + O(n(\log \log n)^2 / \log n + \log \log \log u)$ technique is provided (optimal up to an additive factor) whose construction takes linear time in expectation. The technique is only of theoretical relevance, though, as it yields a low number of bits per key only for unrealistically large values of n.

We will describe two practical solutions: the first one provides a structure that is simple, constant time, and asymptotically space optimal (i.e., $O(n)$); its actual space requirement is about twice the lower bound. The second one can potentially approach the lower bound, even if in practice this would require an unfeasibly long construction time; nonetheless, it provides the smallest known practical data structure – it occupies about 1.44 times the lower bound.

We present the two constructions in some detail below; they both use the idea of building an MPHF out of a PHF that we explain first.

From a PHF to a MPHF

Given a set $T \subseteq [m]$ of size $|T| = n$, define $\text{rank}_T : [m] \to [n]$ by letting

$$\text{rank}_T(p) = |\{i \in T \mid i < p\}|.$$

Clearly, every PHF $g : S \to [m]$ can be combined with $\text{rank}_{g(S)} : [m] \to [n]$ to obtain an MPHF. Jacobson [7] offers a constant-time implementation for the rank data structure that uses $o(m)$ additional bits besides the set T represented as an array of m bits; furthermore, constant-time solutions exist that require as little as $O(n/(\log n)^c)$ (for any desired c) over the information-theoretical lower bound $\log \binom{m}{n}$ [10].

For practical solutions, see [5, 11]. For very sparse sets T, the Elias-Fano scheme can be rewarding in terms of space, but query time becomes $O(\log(m/n))$.

The Hypergraph-Based Construction

We start by recalling the hypergraph-based construction presented in [8]. Their method, albeit originally devised only for order-preserving MPHF, can be used to store compactly an *arbitrary* r-bit function $f : S \to [2^r]$. The construction draws three hash functions $h_0, h_1, h_2 : S \to [\gamma n]$ (with $\gamma \approx 1.23$) and builds a 3-hypergraph with one hyperedge $(h_0(x), h_1(x), h_2(x))$ for every $x \in S$. With positive probability, this hypergraph does not have a nonempty 2-core, that is, its hyperedges can be sorted in such a way that every hyperedge contains (at least) a vertex that never appeared before, called the *hinge*. Equivalently, the set of equations (in the variables a_i)

$$f(x) = \left(a_{h_0(x)} + a_{h_1(x)} + a_{h_2(x)}\right) \bmod 2^r$$

has a solution that can be found by a hypergraph-peeling process in time $O(n)$. Storing the function consists in storing γn integers of r bits each (the array a_i), so $\gamma r n$ bits are needed (excluding the bits needed for the hash functions); function evaluation takes constant time.

In [3] the authors (which were not aware of [8]) present a "mutable Bloomier filter," which is formed by a PHF and a data storage indexed by the output of the PHF. The idea is to let $r = 2$ and to decide f *after* the hinges have been successfully determined, letting $f(x)$ be the index of the hinge of the hyperedge associated with x, that is, the index $i \in \{0, 1, 2\}$ such that $h_i(x)$ is the hinge of $(h_0(x), h_1(x), h_2(x))$. This way, the function $g : S \to [m]$ defined by $g(x) = h_{f(x)}(x)$ is a PHF, and it is stored in $2\gamma n$ bits.

The fact that combining such a construction with a ranking data structure might actually provide an MPHF was noted in [2] (whose authors did not know [3]). An important implementation trick that makes it possible to get $\approx$2.65 bits per key is the fact that $r = 2$, but actually we need to store three values. Thus, when assigning values to hinges, we can use 3 (which modulo 3 is equivalent to zero) in place of 0: in this way, hinges are exactly associated to those a_i that are nonzero, which makes it possible to build a custom ranking structure that does not use an additional bit vector, but rather ranks directly nonzero pairs of bits.

The "Hash, Displace, and Compress" Construction

A completely different approach is suggested in [1]: once more, they first build a PHF $h : S \to [m]$ where $m = (1 + \epsilon)n$ for some $\epsilon > 0$. The set S is first divided into r buckets by means of a first-level hash function $g : S \to [r]$; the r buckets $g^{-1}(0), \ldots, g^{-1}(r-1)$ are sorted by their cardinalities, with the largest buckets first.

Let $B_0, \ldots, B_{r-1}$ be the buckets and let $\phi_0, \phi_1, \phi_2, \ldots$ be a sequence of independent fully random hash functions $S \to [m]$. For every $i = 0, \ldots, r - 1$, the construction algorithm determines the smallest index p_i such that ϕ_{p_i} is injective when applied to B_i and moreover $\phi_{p_i}(B_i)$ is disjoint from $\cup_{j<i}\phi_{p_j}(B_j)$. A careful analysis shows that this construction takes linear time in expectation (the choice of r impacts on construction time) and that the expected p_i is bounded by a constant, so the indices can be stored in $O(\log(1/\epsilon)n)$ space.

In practice, if r is chosen so that the average bucket size is $\approx$5, it is possible to obtain an MPHF using $\approx$2.05 bits per key with a construction time that is still feasible, albeit an order of magnitude larger than the hypergraph-based construction.

The authors of [1] also discuss a variant that can directly build MPHFs, but the construction time is no longer linear in expectation; moreover, from a practical viewpoint it is useful to enlarge slightly the buckets so that they have a prime size (this makes it easier to generate a good sequence of hash functions [1]).

Open Problems

Improving construction and query time in practice and getting closer to the space lower bound keeping the construction feasible are the main open problems about MPHFs, as there are already known constructions that close the gap asymptotically.

URLs to Code and Data Sets

The Sux4J library (http://sux4j.di.unimi.it/) provides Java implementations of the methods we discussed. The CMPH library (http://cmph.sourceforge.net/) provides C implementations.

Cross-References

▸ Monotone Minimal Perfect Hash Functions
▸ Rank and Select Operations on Bit Strings

Recommended Reading

1. Belazzougui D, Botelho FC, Dietzfelbinger M (2009) Hash, displace, and compress. In: Fiat A, Sanders P (eds) Algorithms – ESA 2009, 17th annual European

symposium, Copenhagen, 7–9 Sept 2009, proceedings, pp 682–693
2. Botelho FC, Pagh R, Ziviani N (2007) Simple and space-efficient minimal perfect hash functions. In: Dehne FKHA, Sack JR, Zeh N (eds) Proceedings of the WADS 2007, 10th international workshop on algorithms and data structures, Halifax. Lecture notes in computer science, vol 4619. Springer, pp 139–150
3. Chazelle B, Kilian J, Rubinfeld R, Tal A (2004) The Bloomier filter: an efficient data structure for static support lookup tables. In: Munro JI (ed) Proceedings of the fifteenth annual ACM-SIAM symposium on discrete algorithms, SODA 2004, New Orleans. SIAM, pp 30–39
4. Fredman ML, Komlós J (1984) On the size of separating systems and families of perfect hash functions. SIAM J Algebr Discret Methods 5(1):61–68
5. Gog S, Petri M (2014) Optimized succinct data structures for massive data. Softw Pract Exp 44(11):1287–1314
6. Hagerup T, Tholey T (2001) Efficient minimal perfect hashing in nearly minimal space. In: Ferreira A, Reichel H (eds) STACS 2001, 18th annual symposium on theoretical aspects of computer science, Dresden, 15–17 Feb 2001, proceedings, pp 317–326
7. Jacobson G (1989) Space-efficient static trees and graphs. In: 30th annual symposium on foundations of computer science (FOCS '89), Research Triangle Park. IEEE Computer Society, pp 549–554
8. Majewski BS, Wormald NC, Havas G, Czech ZJ (1996) A family of perfect hashing methods. Comput J 39(6):547–554
9. Mehlhorn K (1984) Data structures and algorithms 1: sorting and searching. EATCS monographs on theoretical computer science, vol 1. Springer, Berlin/New York
10. Patrascu M (2008) Succincter. In: 49th annual IEEE symposium on foundations of computer science, Philadelphia. IEEE Computer Society, pp 305–313
11. Vigna S (2008) Broadword implementation of rank/select queries. In: McGeoch CC (ed) 7th international workshop on experimental algorithms, WEA 2008, Provincetown. Lecture notes in computer science, vol 5038. Springer, pp 154–168

Minimum Bisection

Robert Krauthgamer
Weizmann Institute of Science, Rehovot, Israel
IBM Almaden Research Center, San Jose, CA, USA

Keywords

Graph bisection

Years and Authors of Summarized Original Work

1999; Feige, Krauthgamer

Problem Definition

Overview

Minimum bisection is a basic representative of a family of discrete optimization problems dealing with partitioning the vertices of an input graph. Typically, one wishes to minimize the number of edges going across between the different pieces, while keeping some control on the partition, say by restricting the number of pieces and/or their size. (This description corresponds to an edge-cut of the graph; other variants correspond to a vertex-cut with similar restrictions.) In the minimum bisection problem, the goal is to partition the vertices of an input graph into two equal-size sets, such that the number of edges connecting the two sets is as small as possible.

In a seminal paper in 1988, Leighton and Rao [14] devised for MINIMUM-BISECTION a logarithmic-factor bicriteria approximation algorithm. (A bicriteria approximation algorithm partitions the vertices into two sets each containing at most 2/3 of the vertices, and its value, i.e., the number of edges connecting the two sets, is compared against that of the best partition into equal-size sets.) Their algorithm has found numerous applications, but the question of finding a true approximation with a similar factor remained open for over a decade later. In 1999, Feige and Krauthgamer [6] devised the first polynomial-time algorithm that approximates this problem within a factor that is polylogarithmic (in the graph size).

Cuts and Bisections

Let $G = (V, E)$ be an undirected graph with $n = |V|$ vertices, and assume for simplicity that n is even. For a subset S of the vertices, let $\bar{S} = V \setminus S$. The *cut* (also known as *cutset*) $(S, \bar{S})$ is defined as the set of all edges with one endpoint in S and one endpoint in $\bar{S}$. These edges are said to *cross* the cut, and the two sets S and $\bar{S}$ are called the two *sides* of the cut.

Assume henceforth that G has nonnegative edge-weights. (In the unweighted version, every edge has a unit weight.) The *cost* of a cut $(S, \bar{S})$ is then defined to be the total edge-weight of all the edges crossing the cut.

A cut $(S, \bar{S})$ is called a *bisection* of G if its two sides have equal cardinality, namely $|S| = |\bar{S}| = n/2$. Let $b(G)$ denote the minimum cost of a bisection of G.

Problem 1 (MINIMUM-BISECTION)
INPUT: An undirected graph G with nonnegative edge-weights.
OUTPUT: A bisection $(S, \bar{S})$ of G that has minimum cost.

This definition has a crucial difference from the classical MINIMUM-CUT problem (see e.g., [10] and references therein), namely, there is a restriction on the sizes of the two sides of the cut. As it turns out, MINIMUM-BISECTION is NP-hard (see [9]), while MINIMUM-CUT can be solved in polynomial time.

Balanced Cuts and Edge Separators

The above rather basic definition of minimum bisection can be extended in several ways. Specifically, one may require only an upper bound on the size of each side. For $0 < \beta < 1$, a cut $(S, \bar{S})$ is called β-balanced if $\max\{|S|, |\bar{S}|\} \leq \beta n$. Note the latter requirement implies $\min\{|S|, |\bar{S}|\} \geq (1 - \beta)n$. In this terminology, a bisection is a 1/2-balanced cut.

Problem 2 (β-BALANCED-CUT)
INPUT: An undirected graph G with nonnegative edge-weights.
OUTPUT: A β-balanced cut $(S, \bar{S})$ of G with $\max\{|S|, |\bar{S}|\} \leq \beta n$, that has cost as small as possible.

The special case of $\beta = 2/3$ is commonly refered to as the EDGE-SEPARATOR problem.

In general, the sizes of the two sides may be specified in advance arbitrarily (rather than being equal); in this case the input contains a number k, and the goal is to find a cut $(S, \bar{S})$ such that $|S| = k$. One may also wish to divide the graph into more than two pieces of equal size and then the input contains a number $r \geq 2$, or alternatively, to divide the graph into r pieces of whose sizes are $k_1, \ldots, k_r$, where the numbers k_i are prescribed in the input; in either case, the goal is to minimize the number of edges crossing between different pieces.

Problem 3 (PRESCRIBED-PARTITION)
INPUT: An undirected graph $G = (V, E)$ with nonnegative edge-weights, and integers $k_1, \ldots, k_r$ such that $\sum_i k_i = |V|$.
OUTPUT: A partition $V = V_1 \cup \cdots \cup V_r$ of G with $|V_i| = k_i$ for all i, such that the total edge-weight of edges whose endpoints lie in different sets V_i is as small as possible.

Key Results

The main result of Feige and Krauthgamer [6] is an approximation algorithm for MINIMUM-BISECTION. The approximation factor they originally claimed is $O(\log^2 n)$, because it used the algorithm of Leighton and Rao [14]; however, by using instead the algorithm of [2], the factor immediately improves to $O(\log^{1.5} n)$.

Theorem 1 Minimum-Bisection *can be approximated in polynomial time within $O(\log^{1.5} n)$ factor. Specifically, the algorithm produces for an input graph G a bisection $(S, \bar{S})$ whose cost is at most $O(\log^{1.5} n) \cdot b(G)$.*

The algorithm immediately extends to similar results for related and/or more general problems that are defined above.

Theorem 2 β-Balanced-Cut *(and in particular Edge-Separator) can be approximated in polynomial time within $O(\log^{1.5} n)$ factor.*

Theorem 3 Prescribed-Partition *can be approximated in time $n^{O(r)}$ to within $O(\log^{1.5} n)$ factor.*

For all three problems above, the approximation ratio improves to $O(\log n)$ for the family of graphs excluding a fixed minor (which includes in particular planar graphs). For simplicity, this result is stated for Minimum-Bisection.

Theorem 4 *In graphs excluding a fixed graph as a minor (e.g., planar graphs), the problems (i)* Minimum-Bisection, *(ii)* β-Balanced-Cut, *and (iii)* Prescribed-Partition *with fixed r can all be approximated in polynomial time within factor* $O(\log n)$.

It should be noted that all these results can be generalized further, including vertex-weights and terminals-vertices ($s - t$ pairs), see [Sect. 5 in 6].

Related Work

A bicriteria approximation algorithm for β-balanced cut returns a cut that is β'-balanced for a predetermined $\beta' > \beta$. For bisection, for example, $\beta = 1/2$ and typically $\beta' = 2/3$.

The algorithms in the above theorems use (in a black-box manner) an approximation algorithm for a problem called minimum quotient-cuts (or equivalently, sparsest-cut with uniform-demands). For this problem, the best approximation currently known is $O(\sqrt{\log n})$ for general graphs due to Arora, Rao, and Vazirani [2], and $O(1)$ for graphs excluding a fixed minor due to Klein, Plotkin, and Rao [13]. These approximation algorithms for minimum quotient-cuts immediately give a polynomial time bicriteria approximation (sometimes called pseudo-approximation) for MINIMUM-BISECTION. For example, in general graphs the algorithm is guaranteed to produce a 2/3-balanced cut whose cost is at most $O(\sqrt{\log n}) \cdot b(G)$. Note however that a 2/3-balanced cut does not provide a good approximation for the value of $b(G)$. For instance, if G consists of three disjoint cliques of equal size, an optimal 2/3-balanced cut has no edges, whereas $b(G) = \Omega(n^2)$. For additional related work, including approximation algorithms for dense graphs, for directed graphs, and for other graph partitioning problems, see [Sect. 1 in 6] and the references therein.

Applications

One major motivation for MINIMUM-BISECTION, and graph partitioning in general, is a divide-and-conquer approach to solving a variety of optimization problems, especially in graphs, see e.g., [15, 16]. In fact, these problems arise naturally in a wide range of practical settings such as VLSI design and image processing; sometimes, the motivation is described differently, e.g., as a clustering task.

Another application of MINIMUM-BISECTION is in assignment problems, of a form that is common in parallel systems and in scientific computing: jobs need to be assigned to machines in a balanced way, while assigning certain pairs of jobs the same machine, as much as possible. For example, consider assigning n jobs to 2 machines, when the amount of communication between every two jobs is known, and the goal is to have equal load (number of jobs) on each machine, and bring to minimum the total communication that goes between the machines. Clearly, this last problem can be restated as MINIMUM-BISECTION in a suitable graph.

It should be noted that in many of these settings, a true approximation is not absolutely necessary, and a bicriteria approximation may suffice. Nevertheless, the algorithms stated in section "Key Results" have been used to design algorithms for other problems, such as (1) an approximation algorithm for minimum bisection in k-uniform hypergraphs [3]; (2) an approximation algorithm for a variant of the minimum multicut problem [17]; and (3) an algorithm that efficiently certifies the unsatisfiability of random $2k$-SAT with sufficiently many clauses [5].

From a practical perspective, numerous heuristics (algorithms without worst-case guarantees) for graph partitioning have been proposed and studied, see [1] for an extensive survey. For example, one of the most famous heuristics is Kerninghan and Lin's local search heuristic for minimum bisection [11].

Open Problems

Currently, there is a large gap between the $O(\log^{1.5} n)$ approximation ratio for MINIMUM-BISECTION achieved by Theorem 1 and the hardness of approximation results known for it. As mentioned above, MINIMUM-BISECTION is known to be NP-hard (see [9]).

The problem is not known to be APX-hard but several results provide evidence towards this possibility. Bui and Jones [4] show that for every fixed $\epsilon > 0$, it is NP-hard to approximate the minimum bisection within an *additive* term of $n^{2-\epsilon}$. Feige [7] showed that if refuting 3SAT is hard on average on a natural distribution of inputs, then for every fixed $\varepsilon > 0$ there is no $4/3 - \varepsilon$ approximation algorithm for minimum bisection. Khot [12] proved that minimum bisection does not admit a polynomial-time approximation scheme (PTAS) unless NP has randomized sub-exponential time algorithms.

Taking a broader perspective, currently there is a (multiplicative) gap of $O(\log n)$ between the approximation ratio for MINIMUM-BISECTION and that of minimum quotient-cuts (and thus also to the factor achieved by bicriteria approximation). It is interesting whether this gap can be reduced, e.g., by using the algorithm of [2] in a non-black box manner.

The vertex-cut version of MINIMUM-BISECTION is defined as follows: the goal is to partition the vertices of the input graph into $V = A \cup B \cup S$ with $|S|$ as small as possible, under the constraints that $\max\{|A|, |B|\} \leq n/2$ and no edge connects A with B. It is not known whether a polylogarithmic factor approximation can be attained for this problem. It should be noted that the same question regarding the directed version of MINIMUM-BISECTION was answered negatively by Feige and Yahalom [8].

Cross-References

See entry on the paper by Arora, Rao, and Vazirani [2].

▶ Separators in Graphs
▶ Sparsest Cut

Recommended Reading

1. Alpert CJ, Kahng AB (1995) Recent directions in netlist partitioning: a survey. Integr VLSI J 19(1–2):1–81
2. Arora S, Rao S, Vazirani U (2004) Expander flows, geometric embeddings, and graph partitionings. In: 36th annual symposium on the theory of computing, Chicago, June 2004, pp 222–231
3. Berman P, Karpinski M (2003) Approximability of hypergraph minimum bisection. ECCC report TR03-056, Electronic Colloquium on Computational Complexity, vol 10
4. Bui TN, Jones C (1992) Finding good approximate vertex and edge partitions is NP-hard. Inform Process Lett 42(3):153–159
5. Coja-Oghlan A, Goerdt A, Lanka A, Schädlich F (2004) Techniques from combinatorial approximation algorithms yield efficient algorithms for random 2k-SAT. Theory Comput Sci 329(1–3):1–45
6. Feige U (2002) Relations between average case complexity and approximation complexity. In: 34th annual ACM symposium on the theory of computing, Montréal, 19–21 May 2002, pp 534–543
7. Feige U, Krauthgamer R (2006) A polylogarithmic approximation of the minimum bisection. SIAM Rev 48(1):99–130, Previous versions appeared in Proceedings of 41st FOCS, 1999; and in SIAM J Comput 2002
8. Feige U, Yahalom O (2003) On the complexity of finding balanced oneway cuts. Inf Process Lett 87(1):1–5
9. Garey MR, Johnson DS (1979) Computers and intractability: a guide to the theory of NP-completeness. W.H. Freeman
10. Karger DR (2000) Minimum cuts in near-linear time. J ACM 47(1):46–76
11. Kernighan BW, Lin S (1970) An efficient heuristic procedure for partitioning graphs. Bell Syst Tech J 49(2):291–307
12. Khot S (2004) Ruling out PTAS for graph min-bisection, densest subgraph and bipartite clique. In: 45th annual IEEE symposium on foundations of computer science, Georgia Institute of Technology, Atlanta, 17–19 Oct 2004, pp 136–145
13. Klein P, Plotkin SA, Rao S (1993) Excluded minors, network decomposition, and multicommodity flow. In: 25th annual ACM symposium on theory of computing, San Diego, 16–18 May 1993, pp 682–690
14. Leighton T, Rao S (1999) Multicommodity max-flow min-cut theorems and their use in designing approximation algorithms. J ACM 46(6):787–832, 29th FOCS, 1988
15. Lipton RJ, Tarjan RE (1980) Applications of a planar separator theorem. SIAM J Comput 9(3): 615–627
16. Rosenberg AL, Heath LS (2001) Graph separators, with applications. Frontiers of computer science. Kluwer/Plenum, New York
17. Svitkina Z, Tardos É (2004) Min-Max multiway cut. In: 7th international workshop on approximation algorithms for combinatorial optimization (APPROX), Cambridge, 22–24 Aug 2004, pp 207–218

Minimum Congestion Redundant Assignments

Dimitris Fotakis[1] and Paul (Pavlos) Spirakis[2,3,4]
[1]Department of Information and Communication Systems Engineering, University of the Aegean, Samos, Greece
[2]Computer Engineering and Informatics, Research and Academic Computer Technology Institute, Patras University, Patras, Greece
[3]Computer Science, University of Liverpool, Liverpool, UK
[4]Computer Technology Institute (CTI), Patras, Greece

Keywords

Maximum fault-tolerant partition; Minimum fault-tolerant congestion

Years and Authors of Summarized Original Work

2002; Fotakis, Spirakis

Problem Definition

This problem is concerned with the most efficient use of redundancy in load balancing on faulty parallel links. More specifically, this problem considers a setting where some messages need to be transmitted from a source to a destination through some faulty parallel links. Each link fails independently with a given probability, and in case of failure, none of the messages assigned to it reaches the destination. (This assumption is realistic if the messages are split into many small packets transmitted in a round-robin fashion. Then the successful delivery of a message requires that all its packets should reach the destination.) An assignment of the messages to the links may use redundancy, i.e., assign multiple copies of some messages to different links. The reliability of a redundant assignment is the probability that every message has a copy on some active link, thus managing to reach the destination. Redundancy increases reliability, but also increases the message load assigned to the links. A good assignment should achieve high reliability and keep the maximum load of the links as small as possible.

The reliability of a redundant assignment depends on its structure. In particular, the reliability of different assignments putting the same load on every link and using the same number of copies for each message may vary substantially (e.g., compare the reliability of the assignments in Fig. 1). The crux of the problem is to find an efficient way of exploiting redundancy in order to achieve high reliability and low maximum load. (If one does not insist on minimizing the maximum load, a reliable assignment is constructed by assigning every message to the most reliable links.)

The work of Fotakis and Spirakis [1] formulates the scenario above as an optimization problem called *Minimum Fault-Tolerant Congestion* and suggests a simple and provably efficient approach of exploiting redundancy. This approach naturally leads to the formulation of another interesting optimization problem, namely that of computing an efficient fault-tolerant partition of a set of faulty parallel links. [1] presents polynomial-time approximation algorithms for computing a fault-tolerant partition of the links and proves that combining fault-tolerant partitions with standard load balancing algorithms results in a good approximation to Minimum Fault-Tolerant Congestion. To the best knowledge of the entry authors, this work is the first to consider the approximability of computing a redundant assignment that minimizes the maximum load of the links subject to the constraint that random faults should be tolerated with a given probability.

Notations and Definitions

Let M denote a set of m faulty parallel links connecting a source s to a destination t, and let J denote a set of n messages to be transmitted from s to t. Each link i has a rational capacity $c_i \geq 1$ and a rational failure probability $f_i \in (0, 1)$. Each message j has a rational

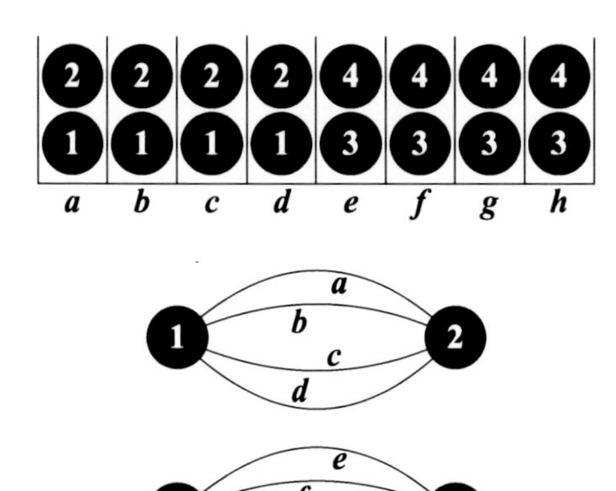

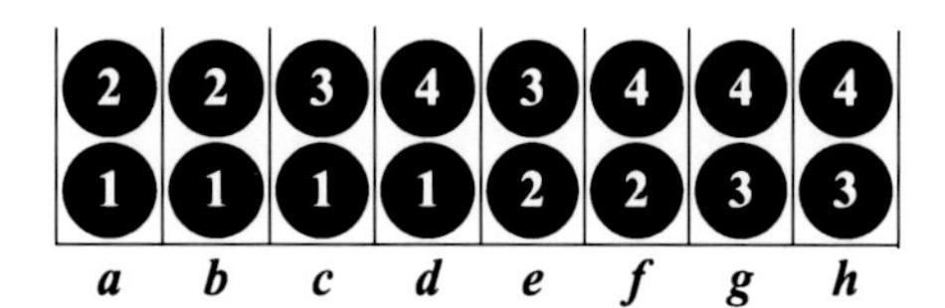

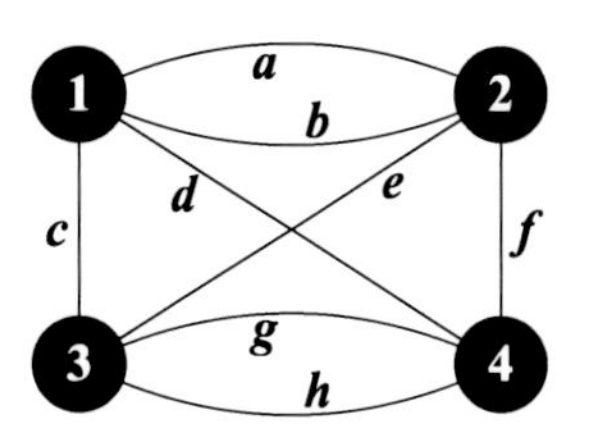

Minimum Congestion Redundant Assignments, Fig. 1 Two redundant assignments of 4 unit size messages to 8 identical links. Both assign every message to 4 links and 2 messages to every link. The corresponding graph is depicted below each assignment. The assignment on the *left* is the most reliable 2-partitioning assignment ϕ_2. Lemma 3 implies that for every failure probability f, ϕ_2 is at least as reliable as any other assignment ϕ with $\mathrm{Cong}(\phi) \leq 2$. For instance, ϕ_2 is at least as reliable as the assignment on the *right*. Indeed the reliability of the assignment on the right is $1-4f^4+2f^6+4f^7-3f^8$, which is bounded from above by $\mathrm{Rel}(\phi_2) = 1-2f^4+f^8$ for all $f \in [0,1]$

size $s_j \geq 1$. Let $f_{\max} \equiv \max_{i \in M}\{f_i\}$ denote the failure probability of the most unreliable link. Particular attention is paid to the special case of identical capacity links, where all capacities are assumed to be equal to 1.

The reliability of a set of links M', denoted $\mathrm{Rel}(M')$, is the probability that there is an active link in M'. Formally, $\mathrm{Rel}(M') \equiv 1 - \prod_{i \in M'} f_i$. The reliability of a collection of disjoint link subsets $\mathcal{M} = \{M_1, \ldots, M_\nu\}$, denoted $\mathrm{Rel}(\mathcal{M})$, is the probability that there is an active link in every subset of $\mathcal{M}$. Formally,

$$\mathrm{Rel}(\mathcal{M}) \equiv \prod_{\ell=1}^{\nu} \mathrm{Rel}(M_\ell) = \prod_{\ell=1}^{\nu} \left(1 - \prod_{i \in M_\ell} f_i \right).$$

A redundant assignment $\phi : J \mapsto 2^M \setminus \emptyset$ is a function that assigns every message j to a nonempty set of links $\phi(j) \subseteq M$. An assignment ϕ is feasible for a set of links M' if for every message j, $\phi(j) \cap M' \neq \emptyset$. The *reliability* of an assignment ϕ, denoted $\mathrm{Rel}(\phi)$, is the probability that ϕ is feasible for the actual set of active links. Formally,

$$\mathrm{Rel}(\phi) \equiv \sum_{\substack{M' \subseteq M \\ \forall j \in J,\, \phi(j) \cap M' \neq \emptyset}} \left(\prod_{i \in M'} (1-f_i) \prod_{i \in M \setminus M'} f_i \right)$$

The *congestion* of an assignment ϕ, denoted $\mathrm{Cong}(\phi)$, is the maximum load assigned by ϕ to a link in M. Formally,

$$\mathrm{Cong}(\phi) \equiv \max_{i \in M} \left\{ \sum_{j : i \in \phi(j)} \frac{s_j}{c_i} \right\}.$$

Problem 1 (Minimum Fault-Tolerant Congestion)

INPUT: *A set of faulty parallel links* $M = \{(c_1, f_1), \ldots, (c_m, f_m)\}$, *a set of messages* $J = \{s_1, \ldots, s_n\}$, *and a rational number* $\epsilon \in (0, 1)$.

OUTPUT: *A redundant assignment* $\phi : J \mapsto 2^M \setminus \emptyset$ *with* $\mathrm{Rel}(\phi) \geq 1 - \epsilon$ *that minimizes* $\mathrm{Cong}(\phi)$.

Minimum Fault-Tolerant Congestion is **NP**-hard because it is a generalization of minimizing makespan on (reliable) parallel machines. The decision version of Minimum Fault-Tolerant Congestion belongs to **PSPACE**, but it is not clear whether it belongs to **NP**. The reason is that computing the reliability of a redundant assignment and deciding whether it is a feasible solution is **#P**-complete.

The work of Fotakis and Spirakis [1] presents polynomial-time approximation algorithms for Minimum Fault-Tolerant Congestion based on a simple and natural class of redundant assignments whose reliability can be computed easily. The high level idea is to separate the reliability aspect from load balancing. Technically, the set of links is partitioned in a collection of disjoint subsets $\mathcal{M} = \{M_1, \ldots, M_\nu\}$ with $\text{Rel}(\mathcal{M}) \geq 1 - \epsilon$. Every subset $M_\ell \in \mathcal{M}$ is regarded as a *reliable link* of *effective capacity* $c(M_\ell) \equiv \min_{i \in M_\ell}\{c_i\}$. Then any algorithm for load balancing on reliable parallel machines can be used for assigning the messages to the subsets of $\mathcal{M}$, thus computing a redundant assignment ϕ with $\text{Rel}(\phi) \geq 1 - \epsilon$.

The assignments produced by this approach are called *partitioning assignments*. More precisely, an assignment $\phi : J \mapsto 2^M \setminus \emptyset$ is a *ν-partitioning assignment* if for every pair of messages j, j', either $\phi(j) = \phi(j')$ or $\phi(j) \cap \phi(j') = \emptyset$, and ϕ assigns the messages to ν different link subsets.

Computing an appropriate fault-tolerant collection of disjoint link subsets is an interesting optimization problem by itself. A feasible solution $\mathcal{M}$ satisfies the constraint that $\text{Rel}(\mathcal{M}) \geq 1 - \epsilon$. For identical capacity links, the most natural objective is to maximize the number of subsets in $\mathcal{M}$ (equivalently, the number of reliable links used by the load balancing algorithm). For arbitrary capacities, this objective generalizes to maximizing the total effective capacity of $\mathcal{M}$.

Problem 2 (Maximum Fault-Tolerant Partition)

INPUT: A set of faulty parallel links $M = \{(c_1, f_1), \ldots, (c_m, f_m)\}$, and a rational number $\epsilon \in (0, 1)$.

OUTPUT: A collection $\mathcal{M} = \{M_1, \ldots, M_\nu\}$ of disjoint subsets of M with $\text{Rel}(\mathcal{M}) \geq 1 - \epsilon$ that maximizes $\sum_{\ell=1}^{\nu} c(M_\ell)$.

The problem of Maximum Fault-Tolerant Partition is **NP**-hard. More precisely, given m identical capacity links with rational failure probabilities and a rational number $\epsilon \in (0, 1)$, it is **NP**-complete to decide whether the links can be partitioned into sets M_1 and M_2 with $\text{Rel}(M_1) \cdot \text{Rel}(M_2) \geq 1 - \epsilon$.

Key Results

Theorem 1 *There is a 2-approximation algorithm for Maximum Fault-Tolerant Partition of identical capacity links. The time complexity of the algorithm is* $O((m - \sum_{i \in M} \ln f_i) \ln m)$.

Theorem 2 *For every constant* $\delta > 0$, *there is a* $(8+\delta)$*-approximation algorithm for Maximum Fault-Tolerant Partition of capacitated links. The time complexity of the algorithm is polynomial in the input size and* $1/\delta$.

To demonstrate the efficiency of the partitioning approach for Maximum Fault-Tolerant Congestion, Fotakis and Spirakis prove that for certain instances, the reliability of the most reliable partitioning assignment bounds from above the reliability of any other assignment with the same congestion (see Fig. 1 for an example).

Lemma 3 *For any positive integers* Λ, ν, μ *and any rational* $f \in (0, 1)$, *let* ϕ *be a redundant assignment of* $\Lambda\nu$ *unit size messages to* $\nu\mu$ *identical capacity links with failure probability f. Let* ϕ_ν *be the ν-partitioning assignment that assigns* Λ *messages to each of* ν *disjoint subsets consisting of* μ *links each. If* $\text{Cong}(\phi) \leq \Lambda = \text{Cong}(\phi_\nu)$, *then* $\text{Rel}(\phi) \leq (1 - f^\mu)^\nu = \text{Rel}(\phi_\nu)$.

Based on the previous upper bound on the reliability of any redundant assignment, [1] presents polynomial-time approximation algorithms for Maximum Fault-Tolerant Congestion.

Theorem 4 *There is a quasi-linear-time 4-approximation algorithm for Maximum Fault-Tolerant Congestion on identical capacity links.*

Theorem 5 *There is a polynomial-time* $2\lceil \ln(m/\epsilon)/\ln(1/f_{\max}) \rceil$*-approximation algorithm for Maximum Fault-Tolerant Congestion on instances with unit size messages and capacitated links.*

Applications

In many applications dealing with faulty components (e.g., fault-tolerant network design, fault-tolerant routing), a combinatorial structure (e.g., a graph, a hypergraph) should optimally tolerate random faults with respect to a given property (e.g., connectivity, nonexistence of isolated points). For instance, Lomonosov [5] derived tight upper and lower bounds on the probability that a graph remains connected under random edge faults. Using the bounds of Lomonosov, Karger [3] obtained improved theoretical and practical results for the problem of estimating the reliability of a graph. In this work, Lemma 3 provides a tight upper bound on the probability that isolated nodes do not appear in a not necessarily connected hypergraph with $\Lambda\nu$ nodes and $\nu\mu$ "faulty" hyperedges of cardinality Λ.

More precisely, let ϕ be any assignment of $\Lambda\nu$ unit size messages to $\nu\mu$ identical links that assigns every message to μ links and Λ messages to every link. Then ϕ corresponds to a hypergraph H_ϕ, where the set of nodes consists of $\Lambda\nu$ elements corresponding to the unit size messages and the set of hyperedges consists of $\nu\mu$ elements corresponding to the identical links. Every hyperedge contains the messages assigned to the corresponding link and has cardinality Λ (see Fig. 1 for a simple example with $\Lambda = 2$, $\nu = 2$, and $\mu = 4$). Clearly, an assignment ϕ is feasible for a set of links $M' \subseteq M$ iff the removal of the hyperedges corresponding to the links in $M \setminus M'$ does not leave any isolated nodes (For a node v, let $\deg_H(v) \equiv |\{e \in E(H) : v \in e\}|$. A node v is isolated in H if $\deg_H(v) = 0$) in H_ϕ. Lemma 3 implies that the hypergraph corresponding to the most reliable ν-partitioning assignment maximizes the probability that isolated nodes do not appear when hyperedges are removed equiprobably and independently.

The previous work on fault-tolerant network design and routing mostly focuses on the worst-case fault model, where a feasible solution must tolerate any configuration of a given number of faults. The work of Gasieniec et al. [2] studies the fault-tolerant version of minimizing congestion of virtual path layouts in a complete ATM network. In addition to several results for the worst-case fault model, [2] constructs a virtual path layout of logarithmic congestion that tolerates random faults with high probability. On the other hand, the work of Fotakis and Spirakis shows how to construct redundant assignments that tolerate random faults with a probability given as part of the input and achieve a congestion close to optimal.

Open Problems

An interesting research direction is to determine the computational complexity of Minimum Fault-Tolerant Congestion and related problems. The decision version of Minimum Fault-Tolerant Congestion is included in the class of languages decided by a polynomial-time non-deterministic Turing machine that reduces the language to a single call of a **#P** oracle. After calling the oracle once, the Turing machine rejects if the oracle's outcome is less than a given threshold and accepts otherwise. This class is denoted $\mathbf{NP}^{\#\mathbf{P}[1,\mathrm{comp}]}$ in [1]. In addition to Minimum Fault-Tolerant Congestion, $\mathbf{NP}^{\#\mathbf{P}[1,\mathrm{comp}]}$ includes the decision version of Stochastic Knapsack considered in [4]. A result of Toda and Watanabe [6] implies that $\mathbf{NP}^{\#\mathbf{P}[1,\mathrm{comp}]}$ contains the entire Polynomial Hierarchy. A challenging open problem is to determine whether the decision version of Minimum Fault-Tolerant Congestion is complete for $\mathbf{NP}^{\#\mathbf{P}[1,\mathrm{comp}]}$.

A second direction for further research is to consider the generalizations of other fundamental optimization problems (e.g., shortest paths, minimum connected subgraph) under random faults. In the fault-tolerant version of minimum connected subgraph for example, the input consists of a graph whose edges fail independently with given probabilities, and a rational number $\epsilon \in (0, 1)$. The goal is to compute a spanning subgraph with a minimum number of edges whose reliability is at least $1 - \epsilon$.

M

Cross-References

- ▶ Approximation Schemes for Bin Packing
- ▶ Bin Packing
- ▶ Connectivity and Fault Tolerance in Random Regular Graphs
- ▶ List Scheduling

Recommended Reading

1. Fotakis D, Spirakis P (2002) Minimum congestion redundant assignments to tolerate random faults. Algorithmica 32:396–422
2. Gasieniec L, Kranakis E, Krizanc D, Pelc A (1996) Minimizing congestion of layouts for ATM networks with faulty links. In: Penczek W, Szalas A (eds) Proceedings of the 21st international symposium on mathematical foundations of computer science. Lecture notes in computer science, vol 1113. Springer, Berlin, pp 372–381
3. Karger D (1999) A randomized fully polynomial time approximation scheme for the all-terminal network reliability problem. SIAM J Comput 29:492–514
4. Kleinberg J, Rabani Y, Tardos E (2000) Allocating bandwidth for bursty connections. SIAM J Comput 30:191–217
5. Lomonosov M (1974) Bernoulli scheme with closure. Probl Inf Transm 10:73–81
6. Toda S, Watanabe O (1992) Polynomial-time 1-turing reductions from #PH to #P. Theor Comput Sci 100:205–221

Minimum Connected Sensor Cover

Lidong Wu[1] and Weili Wu[2,3,4]
[1]Department of Computer Science, The University of Texas, Tyler, TX, USA
[2]College of Computer Science and Technology, Taiyuan University of Technology, Taiyuan, Shanxi Province, China
[3]Department of Computer Science, California State University, Los Angeles, CA, USA
[4]Department of Computer Science, The University of Texas at Dallas, Richardson, TX, USA

Keywords

Approximation algorithm; Connected sensor cover; Performance ratio

Years and Authors of Summarized Original Work

2013; Wu, Du, Wu, Li, Lv, Lee

Problem Definition

Nowadays, the sensor exists everywhere. The wireless sensor network has been studied extensively. In view of this type of networks, there are two most important properties, coverage and connectivity. In fact, the sensor is often used for collecting information, and hence, its sensing area has to cover the target (points or area). Usually, for a wireless sensor, its sensing area is a disk with the center at the sensor. The radius of this disk is called the sensing radius. After information is collected, the sensor has to send to central station for analysis. This requires all active sensors to form a connected communication network. Actually, every sensor has also a communication function, and it can send information to other sensors located in its communication area, which is also a disk with center at the sensor. The radius of the communication disk is called the communication radius.

The sensor is often very small and energy is often supplied with batteries. Therefore, energy efficiency is a big issue in the study of wireless sensor networks. A sensor network is said to be *homogeneous* if all sensors in the network have the same size of sensing radius and the same size of communication radius. For a homogeneous wireless sensor network, the energy consumption can be measured by the number of active sensors. The minimum connected sensor cover problem is a classic optimization problem based on the above consideration in the study of wireless sensor networks, which is described as follows.

Consider a homogeneous wireless sensor network in the Euclidean plane. Given a connected target area Ω, find the minimum number of sensors satisfying the following two conditions:

[Coverage] The target area Ω is covered by selected sensors.

[Connectivity] Selected sensors induce a connected network.

A subset of sensors is called a *sensor cover* if it satisfies the coverage condition and called a *connected sensor cover* if it satisfies both the coverage condition and the connectivity condition.

The minimum connected sensor cover problem is NP-hard. The study on approximation solutions of this problem has attracted many researchers.

Key Results

The minimum connected sensor cover problem was first proposed by Gupta, Das, and Gu [8]. They presented a greedy algorithm with performance ratio $O(r \ln n)$ where n is the number of sensors and r is the link radius of the sensor network, i.e., for any two sensors with overlapping sensing disks, their hop distance in the communication network is at most r.

Zhang and Hou [12] studied a special case that the communication radius is at least twice of the sensing radius, and they showed that in this case, the coverage of a connected region implied the connectivity. In this case, they presented a polynomial-time constant-approximation.

Das and Gupta [13] and Xing et al. [11] explored more about the relationship between coverage and connectivity. Bai et al. [2] studied a sensor deployment problem regarding the coverage and connectivity. Alam and Haas [1] studied the minimum connected sensor cover problem in three-dimensional sensor networks.

Funke et al. [5] allow sensors to vary their sensing radius. With variable sensing radius and communication radius, Zhou, Das, and Gupta [14] designed a polynomial-time approximation with performance ratio $O(\log n)$. Chosh and Das [6] designed a greedy approximation using the maximal independent set and Voronoi diagram. They determined the size of connected sensor cover produced by their algorithm. However, no comparison with optimal solution, that is, no analysis on approximation performance ratio, is given. In fact, none of the above efforts give an improvement on the approximation performance ratio $O(r \log n)$ given by Gupta, Das, and Gu [8].

Wu et al. [10] made the first improvement. They present two polynomial-time approximations. The first one has performance ratio $O(r)$. This approximation is designed based on a polynomial-time constant-approximation [4] or a polynomial-time approximation scheme [9] for the minimum target coverage problem as follows: given a homogeneous set of sensors and a set of target points in the Euclidean plane, find the minimum number of sensors covering all given target points.

The $O(r)$-approximation consists of three steps. In the first step, it replaces the target area by $O(n^2)$ target points such that the target area is covered by a subset of sensors if and only if those target points are all covered by this subset of sensors. In the second step, it computes a constant-approximation solution, say c-approximation solution S for the minimum target coverage problem with those target points as input. Since the optimal solution for the minimum connected sensor cover problem must be a feasible solution for the minimum target coverage problem. $|S|$ is within a factor c of the size of a minimum connected sensor cover, i.e., $|S| \leq c \cdot \text{opt}_{\text{mcsc}}$.

In the third step, the algorithm employs a polynomial-time 1.39-approximation algorithm for the network Steiner tree problem [3], and apply this algorithm on the input consisting of a graph with unit weight for each edge, which is the communication network of sensors, and a terminal set S. Note that in a graph with unit weight for each edge, the total edge weight of a tree is the number of vertices in the tree minus one. Therefore, the result obtained in the third step is a connected sensor cover with cardinality upper bounded by $|S|$ plus $(1 + 1.39 \times$ (size of minimum network Steiner tree on S)).

To estimate the size of minimum network Steiner tree on S, consider a minimum connected sensor cover S^*. Let T be a spanning tree in the subgraph induced by S^*. For each sensor $s \in S$, there must exist a sensor $s' \in S^*$ with sensing disk overlapping with the sensing disk of s. Therefore, there is a path P_s, with distance at most r, connecting s and s' in the communication network. Clearly, $T \cup (\cup_{s \in S} P_s)$ is a Steiner tree on S. Hence, the size of minimum network

Steiner tree on S is at most $|S| - 1 + |S| \cdot r$. Therefore, the connected sensor cover obtained in the third step has size at most

$$\begin{aligned} |S| + 1 + 1.39 \cdot (|S|(r+1) - 1) &\leq |S|(1.39r + 2.39) \\ &\leq c(1.39r + 2.39) \cdot \text{opt}_{\text{mcsc}} \\ &= O(r) \cdot \text{opt}_{\text{mcsc}}. \end{aligned}$$

The second polynomial-time approximation designed by Wu et al. [10] is a random algorithm with performance ratio $O(\log^3 n)$. This approximation is obtained by the following two observations: (1) The minimum connected sensor cover problem is a special case of the minimum connected set cover problem. (2) The minimum connected set-cover problem has a close relationship to the group Steiner tree. Therefore, some results on group Steiner trees can be transformed into connected sensor covers.

In conclusion, Wu et al. [10] obtained the following:

Theorem 1 *There exists a polynomial-time $O(r)$-approximation for the minimum connected sensor cover problem. There exists also a polynomial-time random algorithm with performance $O(\log^3 n)$ for the minimum connected sensor cover problem.*

Open Problems

For two approximations in Theorem 1, one has performance ratio $O(r)$ independent from n and the other one has performance ratio $O(\log^3 n)$ independent from r. This fact suggests that either n or r is closely related or there exists a polynomial-time constant-approximation. Therefore, we have the following conjecture:

Conjecture 1 There exists a polynomial-time $O(1)$-approximation for the minimum connected sensor cover problem.

Cross-References

- ▸ Connected Set-Cover and Group Steiner Tree
- ▸ Performance-Driven Clustering

Recommended Reading

1. Alam SMN, Haas ZJ (2006) Coverage and connectivity in three-dimensional networks. In: MobiCom'06, Los Angeles
2. Bai X, Kumar S, Xuan D, Yun Z, Lai TH (2006) DEploying wireless sensors to achieve both coverage and connectivity. In: MobiHoc'06, Florence, pp 131–142
3. Byrka J, Grandoni F, Rothvoss T, Sanita L (2010) An improved LP-based approximation for Steiner tree. In: STOC'10, Cambridge, 5–8 June, pp 583–592
4. Ding L, Wu W, Willson JK, Wu L, Lu Z, Lee W (2012) Constant-approximation for target coverage problem in wireless sensor networks. In: Proceedings of the 31st annual joint conference of IEEE communication and computer society (INFOCOM), Miami
5. Funke S, Kesselman A, Kuhn F, Lotker Z, Segal M (2007) Improved approximation algorithms for connected sensor cover. Wirel Netw 13:153–164
6. Ghosh A, Das SK (2005) A distributed greedy algorithm for connected sensor cover in dense sensor networks. LNCS 3560:340–353
7. Ghosh A, Das SK (2006) Coverage and connectivity issues in wireless sensor networks. In: Shorey R, Ananda AL, Chan MC, Ooi WT (eds) Mobile, wireless, and sensor networks: technology, applications, and future directions. Wiley, Hoboken, pp 221–256
8. Gupta H, Das SR, Gu Q (2003) Connected sensor cover: self-organization of sensor networks for efficient query execution. In: MobiHoc'03, Annapolis, pp 189–200
9. Mustafa N, Ray S (2009) PTAS for geometric hitting set problems via local search. In: Proceedings of the SoCG 2009, Aarhus, pp 17–22. ACM, New York
10. Wu L, Du H, Wu W, Li D, Lv J, Lee W (2013) Approximations for minimum connected sensor cover. In: INFOCOM, Turin
11. Xing G, Wang X, Zhang Y, Liu C, Pless R, Gill C (2005) Integrated coverage and connectivity configuration for energy conservation in sensor networks. ACM Trans Sens Netw 1(1):36–72
12. Zhang H, Hou JC (2005) Maintaining sensing coverage and connectivity in large sensor networks. Ad Hoc Sens Wirel Netw 1:89–124
13. Zhou Z, Das S, Gupta H (2004) Connected k-coverage problem in sensor networks. In: Proceedings of the 13th international conference on computer communications and networks, Chicago, pp 373–378
14. Zhou Z, Das SR, Gupta H (2009) Variable radii connected sensor cover in sensor networks. ACM Trans Sens Netw 5(1):1–36

Minimum Energy Broadcasting in Wireless Geometric Networks

Christoph Ambühl
Department of Computer Science, University of Liverpool, Liverpool, UK

Keywords

Energy-efficient broadcasting in wireless networks

Years and Authors of Summarized Original Work

2005; Ambühl

Problem Definition

In the most common model for wireless networks, stations are represented by points in $\mathbb{R}^d$. They are equipped with a omnidirectional transmitter and receiver which enables them to communicate with other stations. In order to send a message from a station s to a station t, station s needs to emit the message with enough power such that t can receive it. It is usually assumed that the power required by a station s to transmit data directly to station t is $\|st\|^\alpha$, for some constant $\alpha \geq 1$, where $\|st\|$ denotes the distance between s and t.

Because of the omnidirectional nature of the transmitters and receivers, a message sent by a station s with power r^α can be received by all stations within a disc of radius r around s. Hence the energy required to send a message from a station s directly to a set of stations S' is determined by $\max_{v \in S'} \|sv\|^\alpha$.

An instance of the *minimum energy broadcast routing problem in wireless networks* (MEBR) consists of a set of stations S and a constant $\alpha \geq 1$. One of the stations in S is designated as the source station s_0. The goal is to send a message at minimum energy cost from s_0 to all other stations in S. This operation is called *broadcast*.

In the case $\alpha = 1$, the optimal solution is to send the message directly from s_0 to all other stations. For $\alpha > 1$, sending the message via intermediate stations which forward it to other stations is often more energy efficient.

A solution of the MEBR instance can be described in terms of a so-called *broadcast tree*. That is, a directed spanning tree of S which contains directed paths from s_0 to all other vertices. The solution described by a broadcast tree T is the one in which every station forwards the message to all its out-neighbors in T.

Problem 1 (MEBR)
INSTANCE: A set S of points in R^d, $s_0 \in S$ designated as the source, and a constant α.
SOLUTION: A broadcast tree T of S.
MEASURE: The objective is to minimize the total energy needed to broadcast a message from s_0 to all other nodes,which can be expressed by

$$\sum_{u \in S} \max_{v \in \delta(u)} \|uv\|^\alpha, \tag{1}$$

where $\delta(u)$ denotes the set of out-neighbors of station u in T.

The MEBR problem is known to be NP-hard for $d \geq 2$ and $\alpha > 1$ [2]. APX-hardness is known for $d \geq 3$ [5].

Key Results

Numerous heuristics have been proposed for this problem. Only a few of them have been analyzed theoretically. The one which attains the best approximation guarantee is the so-called MST-heuristic [12].

MST-HEURISTIC: Compute a minimum spanning tree of S (*mst(S)*) and turn it into an broadcast tree by directing the edges.

Theorem 1 ([1]) *In the Euclidean plane, the MST-heuristic is a 6 approximation algorithm for MEBR for all $\alpha \geq 2$.*

M

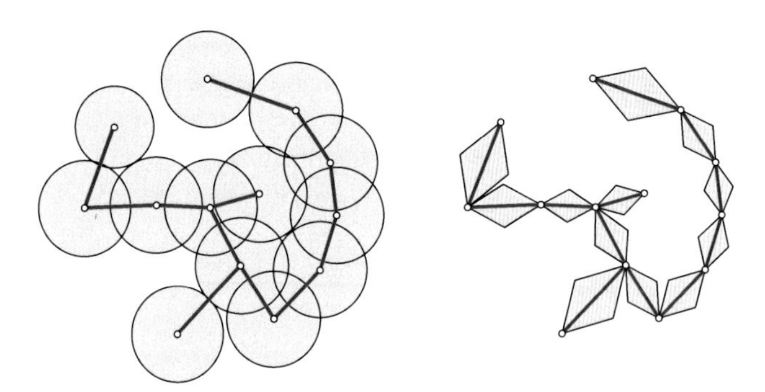

Minimum Energy Broadcasting in Wireless Geometric Networks, Fig. 1 Illustration of the first and second approach for bounding *w(S)*. In both approaches, *w(S)* is bounded in terms of the total area covered by the shapes

Theorem 2 ([9]) *In the Euclidean three-dimensional space, the MST-heuristic is a 18.8 approximation algorithm for MEBR for all* $\alpha \geq 3$.

For $\alpha < d$, the MST-heuristic does not provide a constant approximation ratio. The d-dimensional kissing numbers represent lower bounds for the performance of the MST-heuristic. Hence the analysis for $d = 2$ is tight, whereas for $d = 3$ the lower bound is 12.

Analysis

The analysis of the MST-heuristic is based on good upper bounds for

$$w(S) := \sum_{e \in mst(S)} \|e\|^{\alpha}, \quad (2)$$

which obviously is an upper bound on (1). The radius of an instance of MEBR is the distance between s_0 to the station furthest from s_0. It turns out that the MST-heuristic performs worst on instances of radius 1 whose optimal solution is to broadcast the message directly from s_0 to all other stations. Since the optimal value for such instances is 1, the approximation ratio follows from good upper bounds on *w(S)* for instances with radius 1.

The rest of this section focuses on the case $d = \alpha = 2$. There are two main approaches for upper bounding *w(S)*. In both approaches, *w(S)* is upper bounded in terms of the area of particular kinds of shapes associated with either the stations or with the edges of the MST.

In the first approach, the shapes are disks of radius $m/2$ placed around every station of S, where m is the length of the longest edge of *mst(S)*. Let A denote the area covered by the disks. One can prove $w(S) \leq \frac{4}{\pi}\left(A - \pi(m/2)^2\right)$. Assuming that S has radius 1, one can prove $w(S) \leq 8$ quite easily [4]. This approach can even be extended to obtain $w(S) \leq 6.33$ [8], and it can be generalized for $d \geq 2$.

In the second approach [7, 11], *w(S)* is expressed in terms of shapes associated with the edges of *mst(S)*, e.g., diamond shapes as shown on the right of Fig. 1. The area of a diamond for an edge e is equal to $\|e\|^2/(2\sqrt{3})$. Since one can prove that the diamonds never intersect, one obtains $w(S) = A/(2\sqrt{3})$. For instances with radius 1, one can get $w(S) \leq 12.15$.

For the 2-dimensional case, one can even obtain a matching upper bound [1]. The shapes used in this proof are equilateral triangles, arranged in pairs along every edge of the MST. As can be seen on the left of Fig. 2, these shapes do intersect. Still one can obtain a good upper bound on their total area by means of the convex hull of S:

Let the *extended convex hull of S* be the convex hull of S extended by equilateral triangles along the border of the convex hull. One can prove that the total area generated by the equilateral triangle shapes along the edges of *mst(S)* is upper bounded by the area of the extended convex hull

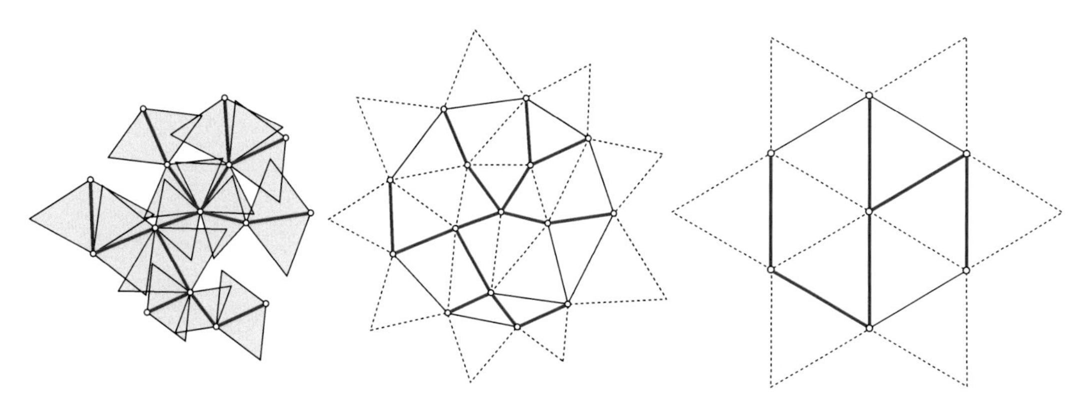

Minimum Energy Broadcasting in Wireless Geometric Networks, Fig. 2 Illustration of the tight bound for $d = 2$. The total area of the equilateral triangles on the left is bounded by its extended convex hull shown in the middle. The point set that maximizes area of the extended convex hull is the star shown on the right

of S. By showing that for instances of radius 1 the area of the extended convex hull is maximized by the point configuration shown on the right of Fig. 2, the matching upper bound of 6 can be established.

Applications

The MEBR problem is a special case of a large class of problems called *range assignment problems*. In all these problems, the goal is to assign a range to each station such that a certain type of communication operation such as broadcast, all-to-1 (gathering), all-to-all (gossiping), can be accomplished. See [3] for a survey on range assignment problems.

It is worth noting that the problem of upper bounding $w(S)$ has already been considered in different contexts. The idea of using diamond shapes to upper bound the length of an MST has already been used by Gilbert and Pollak in [6]. Steele [10] makes use of space filling curves to bound $w(S)$.

Open Problems

An obvious open problem is to close the gap in the analysis of the MST-heuristic for the three dimensional case. This might be very difficult, as the lower bound from the kissing number might not be tight.

Even in the plane, the approximation ratio of the MST-heuristic is quite large. It would be interesting to see a different approach for the problem, maybe based on LP-rounding. It is still not known whether MEBR is APX-hard for instances in the Euclidean plane. Hence there might exist a PTAS for this problem.

Cross-References

▶ Broadcasting in Geometric Radio Networks
▶ Deterministic Broadcasting in Radio Networks
▶ Geometric Spanners
▶ Minimum Geometric Spanning Trees
▶ Minimum Spanning Trees
▶ Randomized Broadcasting in Radio Networks
▶ Randomized Gossiping in Radio Networks

Recommended Reading

1. Ambühl C (2005) An optimal bound for the MST algorithm to compute energy efficient broadcast trees in wireless networks. In: Proceedings of 32th international colloquium on automata, languages and programming (ICALP). Lecture notes in computer science, vol 3580. Springer, Berlin, pp 1139–1150
2. Clementi A, Crescenzi P, Penna P, Rossi G, Vocca P (2001) On the complexity of computing minimum energy consumption broadcast subgraphs. In: Proceedings of the 18th annual symposium on theoretical aspects of computer science (STACS), pp 121–131

3. Clementi A, Huiban G, Penna P, Rossi G, Verhoeven Y (2002) Some recent theoretical advances and open questions on energy consumption in ad-hoc wireless networks. In: Proceedings of the 3rd workshop on approximation and randomization algorithms in communication networks (ARACNE), pp 23–38
4. Flammini M, Klasing R, Navarra A, Perennes S (2004) Improved approximation results for the minimum energy broadcasting problem. In: Proceedings of the 2004 joint workshop on foundations of mobile computing
5. Fuchs B (2006) On the hardness of range assignment problems. In: Proceedings of the 6th Italian conference on algorithms and complexity (CIAC), pp 127–138
6. Gilbert EN, Pollak HO (1968) Steiner minimal trees. SIAM J Appl Math 16:1–29
7. Klasing R, Navarra A, Papadopoulos A, Perennes S (2004) Adaptive broadcast consumption (ABC), a new heuristic and new bounds for the minimum energy broadcast routing problem. In: Proceedings of the 3rd IFIP-TC6 international networking conference (NETWORKING), pp 866–877
8. Navarra A (2005) Tighter bounds for the minimum energy broadcasting problem. In: Proceedings of the 3rd international symposium on modeling and optimization in mobile, ad-hoc and wireless networks (WiOpt), pp 313–322
9. Navarra A (2006) 3-D minimum energy broadcasting. In: Proceedings of the 13th colloquium on structural information and communication complexity (SIROCCO), pp 240–252
10. Steele JM (1989) Cost of sequential connection for points in space. Oper Res Lett 8:137–142
11. Wan P-J, Calinescu G, Li X-Y, Frieder O (2002) Minimum-energy broadcasting in static ad hoc wireless networks. Wirel Netw 8:607–617
12. Wieselthier JE, Nguyen GD, Ephremides A (2002) Energy-efficient broadcast and multicast trees in wireless networks. Mob Netw Appl 7:481–492

Minimum Energy Cost Broadcasting in Wireless Networks

Peng-Jun Wan, Xiang-Yang Li, and Ophir Frieder
Department of Computer Science, Illinois Institute of Technology, Chicago, IL, USA

Keywords

MEB; Minimum energy broadcast; MST

Years and Authors of Summarized Original Work

2001; Wan, Calinescu, Li, Frieder

Problem Definition

Ad hoc wireless networks have received significant attention in recent years due to their potential applications in battlefield, emergency disaster relief and other applications [11, 15]. Unlike wired networks or cellular networks, no wired backbone infrastructure is installed in ad hoc wireless networks. A communication session is achieved either through a single-hop transmission if the communication parties are close enough, or through relaying by intermediate nodes otherwise. Omni-directional antennas are used by all nodes to transmit and receive signals. They are attractive in their broadcast nature. A single transmission by a node can be received by many nodes within its vicinity. This feature is extremely useful for multicasting/broadcasting communications. For the purpose of energy conservation, each node can dynamically adjust its transmitting power based on the distance to the receiving node and the background noise. In the most common power-attenuation model [10], the signal power falls as $\frac{1}{r^\kappa}$, where r is the distance from the transmitter antenna and κ is a real *constant* between 2 and 4 dependent on the wireless environment. Assume that all receivers have the same power threshold for signal detection, which is typically normalized to one. With these assumptions, the power required to support a link between two nodes separated by a distance r is r^κ. A key observation here is that relaying a signal between two nodes may result in lower total transmission power than communicating over a large distance due to the nonlinear power attenuation. They assume the network nodes are given as a finite point (The terms node, point and vertex are interchangeable here: node is a network term, point is a geometric term, and vertex is a graph term.) set P in a two-dimensional plane. For any real number κ, they use $G^{(\kappa)}$ to denote the weighted complete graph over P in which the weight of an edge e is $\|\mathbf{e}\|^{\kappa}$.

The minimum-energy unicast routing is essentially a shortest-path problem in $G^{(\kappa)}$. Consider any unicast path from a node $\mathbf{p} = \mathbf{p}_0 \in P$ to another node $\mathbf{q} = \mathbf{p}_m \in P$: $\mathbf{p}_0\mathbf{p}_1 \cdots \mathbf{p}_{m-1}\mathbf{p}_m$. In this path, the transmission power of each node $\mathbf{p}_i$, $0 \leq i \leq m-1$, is $\|\mathbf{p}_i\mathbf{p}_{i+1}\|^{\kappa}$ and the transmission power of $\mathbf{p}_m$ is zero. Thus the total transmission energy required by this path is $\sum_{i=0}^{m-1} \|\mathbf{p}_i\mathbf{p}_{i+1}\|^{\kappa}$, which is the total weight of this path in G^{κ}. So by applying any shortest-path algorithm such as the Dijkstra's algorithm [5], one can solve the minimum-energy unicast routing problem.

However, for broadcast applications (in general multicast applications), Minimum-Energy Routing is far more challenging. Any broadcast routing is viewed as an arborescence (a directed tree) T, rooted at the source node of the broadcasting, that spans all nodes. Use $f_T(\mathbf{p})$ to denote the transmission power of the node $\mathbf{p}$ required by T. For any leaf node $\mathbf{p}$ of T, $f_T(\mathbf{p}) = 0$. For any internal node $\mathbf{p}$ of T,

$$f_T(\mathbf{p}) = \max_{\mathbf{pq} \in T} \|\mathbf{pq}\|^{\kappa},$$

in other words, the κ-th power of the longest distance between $\mathbf{p}$ and its children in T. The total energy required by T is $\sum_{\mathbf{p} \in P} f_T(\mathbf{p})$. Thus the minimum-energy broadcast routing problem is different from the conventional link-based minimum spanning tree (MST) problem. Indeed, while the MST can be solved in polynomial time by algorithms such as Prim's algorithm and Kruskal's algorithm [5], it is NP-hard [4] to find the minimum-energy broadcast routing tree for nodes placed in two-dimensional plane. In its general graph version, the minimum-energy broadcast routing can also be shown to be NP-hard [7], and even worse, it can not be approximated within a factor of $(1-\epsilon)\log\Delta$, unless $NP \subseteq DTIME\left[n^{O(\log\log n)}\right]$, by an approximation-preserving reduction from the Connected Dominating Set problem [8], where Δ is the maximal degree and ϵ is any arbitrary small positive constant.

Three greedy heuristics have been proposed for the minimum-energy broadcast routing problem by [15]. The MST heuristic first applies the Prim's algorithm to obtain a MST, and then orient it as an arborescence rooted at the source node. The SPT heuristic applies the Dijkstra's algorithm to obtain a SPT rooted at the source node. The BIP heuristic is the node version of Dijkstra's algorithm for SPT. It maintains, throughout its execution, a single arborescence rooted at the source node. The arborescence starts from the source node, and new nodes are added to the arborescence one at a time on the minimum incremental cost basis until all nodes are included in the arborescence. The incremental cost of adding a new node to the arborescence is the minimum additional power increased by some node in the current arborescence to reach this new node. The implementation of BIP is based on the standard Dijkstra's algorithm, with one fundamental difference on the operation whenever a new node q is added. Whereas the Dijkstra's algorithm updates the node weights (representing the current knowing distances to the source node), BIP updates the cost of each link (representing the incremental power to reach the head node of the directed link). This update is performed by subtracting the cost of the added link pq from the cost of every link qr that starts from q to a node r not in the new arborescence.

Key Results

The performance of these three greedy heuristics have been evaluated in [15] by simulation studies. However, their analytic performances in terms of the approximation ratio remained open until [13]. The work of Wan et al. [13] derived the bounds on their approximation ratios.

Let us begin with the SPT algorithm. Let ϵ be a sufficiently small positive number. Consider m nodes $\mathbf{p}_1, \mathbf{p}_2, \cdots, \mathbf{p}_m$ evenly distributed on a cycle of radius 1 centered at a node $\mathbf{o}$. For $1 \leq i \leq m$, let $\mathbf{q}_i$ be the point in the line segment $\mathbf{op}_i$ with $\|\mathbf{oq}_i\| = \epsilon$. They consider a broadcasting from the node $\mathbf{o}$ to these $n = 2m$ nodes $\mathbf{p}_1, \mathbf{p}_2, \cdots, \mathbf{p}_m, \mathbf{q}_1, \mathbf{q}_2, \cdots, \mathbf{q}_m$. The SPT is the superposition of paths $\mathbf{oq}_i\mathbf{p}_i$, $1 \leq i \leq m$. Its total energy consumption is $\epsilon^2 + m(1-\epsilon)^2$. On the other hand, if the transmission power of node $\mathbf{o}$ is set to 1, then the signal can reach all other points. Thus the minimum energy consumed by all broadcasting methods is at most 1. So the approximation ratio of SPT is at least $\epsilon^2 + m(1-\epsilon)^2$. As $\epsilon \longrightarrow 0$, this ratio converges to $\frac{n}{2} = m$.

They [13] also proved that

Theorem 1 *The approximation ratio of MST is at least 6 for any $\kappa \geq 2$.*

Theorem 2 *The approximation ratio of BIP is at least $\frac{13}{3}$ for any $\kappa = 2$.*

They then derived the upper bounds by extensively using the geometric structures of Euclidean MSTs (EMST). They first observed that as long as the cost of a link is an increasing function of the Euclidean length of the link, the set of MSTs of any point set *coincides* with the set of Euclidean MSTs of the same point set. They proved a key result about an upper bound on the parameter $\sum_{e \in MST(P)} \|e\|^2$ for any finite point set P inside a disk with radius one.

Theorem 3 *Let c be the supreme of $\sum_{e \in MST(P)} \|e\|^2$ over all such point sets P. Then $6 \leq c \leq 12$.*

The following lemma proved in [13] is used to bound the energy cost for broadcast when each node can dynamically adjust its power.

Lemma 4 For any point set P in the plane, the total energy required by any broadcasting among P is at least $\frac{1}{c} \sum_{e \in MST(P)} \|e\|^{\kappa}$.

Lemma 5 For any broadcasting among a point set P in a two-dimensional plane, the total energy required by the arborescence generated by the BIP algorithm is at most $\sum_{e \in MST(P)} \|e\|^{\kappa}$.

Thus, they conclude the following two theorems.

Theorem 6 *The approximation ratio of EMST is at most c, and therefore is at most 12.*

Theorem 7 *The approximation ratio of BIP is at most c, and therefore is at most 12.*

Later, Wan et al. [14] studied the energy efficient multicast for wireless networks when each node can dynamically adjust its power. Given a set of receivers Q, the problem Min-Power Asymmetric Multicast seeks, for any given communication session, an arborescence T of minimum total power which is rooted at the source node s and reaches all nodes in Q. As a generalization of Min-Power Asymmetric Broadcast Routing, Min-Power Asymmetric Multicast Routing is also NP-hard. Wieselthier et al. [15] adapted their three broadcasting heuristics to three multicasting heuristics by a technique of pruning, which was called as pruned minimum spanning tree (P-MST), pruned shortest-path tree (P-SPT), and pruned broadcasting incremental power (P-BIP), respectively in [14]. The idea is as follows. They first obtain a spanning tree rooted at the source of a given multicast session by applying any of the three broadcasting heuristics. They then eliminate from the spanning arborescence all nodes which do not have any descendant in Q. They [14] show by constructing examples that all structures P-SPT, P-MST and P-BIP could have approximation ratio as large as $\Theta(n)$ in the worst case for multicast. They then further proposed a multicast scheme with a constant approximation ratio on the total energy consumption. Their protocol for Min-Power Asymmetric Multicast Routing is based on Takahashi-Matsuyama Steiner tree heuristic [12]. Initially, the multicast tree T contains only the source node. At each iterative step, the multicast tree T is grown by one path from some node in T to some destination node from Q that is not yet in the tree T. The path must have the least total power among all such paths from a node in T to a node in $Q - T$. This procedure is repeated until all required nodes are included in T. This heuristic is referred to as Shortest Path First (SPF).

Theorem 8 *For asymmetric multicast communication, the approximation ratio of SPF is between 6 and 2c, which is at most 24.*

Applications

Broadcasting and multicasting in wireless ad hoc networks are critical mechanisms in various applications such as information diffusion, wireless networks, and also for maintaining consistent global network information. Broadcasting is often necessary in MANET routing protocols. For example, many unicast routing protocols such as Dynamic Source Routing (DSR), Ad Hoc On Demand Distance Vector (AODV), Zone Routing Protocol (ZRP), and Location Aided Routing (LAR) use broadcasting or a derivation of it to establish routes. Currently, these protocols all rely on a simplistic form of broadcasting called *flooding*, in which each node (or all nodes in a localized area) retransmits each received unique packet exactly one time. The main problems with flooding are that it typically causes unproductive and often harmful bandwidth congestion, as well as inefficient use of node resources. Broadcasting is also more efficient than sending multiple copies the same packet through unicast. It is highly important to use power-efficient broadcast algorithms for such networks since wireless devises are often powered by batteries only.

Open Problems

There are some interesting questions left for further study. For example, the exact value of the constant c remains unsolved. A tighter upper bound on c can lead to tighter upper bounds on the approximation ratios of both the link-based MST heuristic and the BIP heuristic. They conjecture that the exact value for c is 6, which seems to be true based on their extensive simulations. The second question is what is the approximation lower bound for minimum energy broadcast? Is there a PTAS for this problem?

So far, all the known theoretically good algorithms either assume that the power needed to support a link uv is proportional to $\|uv\|^{\kappa}$ or is a fixed cost that is independent of the neighboring nodes that it will communicate with. In practice, the energy consumption of a node is neither solely dependent on the distance to its farthest neighbor, nor totally independent of its communication neighbor. For example, a more general power consumption model for a node u would be $c_1 + c_2 \cdot \|uv\|^{\kappa}$ for some constants $c_1 \geq 0$ and $c_2 \geq 0$ where v is its farthest communication neighbor in a broadcast structure. No theoretical result is known about the approximation of the optimum broadcast or multicast structure under this model. When $c_2 = 0$, this is the case where all nodes have a fixed power for communication. Minimizing the total power used by a reliable broadcast tree is equivalent to the minimum connected dominating set problem (MCDS), i.e., minimize the number of nodes that relay the message, since all relaying nodes of a reliable broadcast form a connected dominating set (CDS). Notice that recently a PTAS [2] has been proposed for MCDS in UDG graph.

Another important question is how to find efficient broadcast/multicast structures such that the delay from the source node to the last node receiving message is bounded by a predetermined value while the total energy consumption is minimized. Notice that here the delay of a broadcast/multicast based on a tree is not simply the height of the tree: many nodes cannot transmit simultaneously due to the interference.

Cross-References

▶ Broadcasting in Geometric Radio Networks
▶ Deterministic Broadcasting in Radio Networks

Recommended Reading

1. Ambühl C (2005) An optimal bound for the MST algorithm to compute energy efficient broadcast trees in wireless networks. In: Proceedings of 32th international colloquium on automata, languages and programming (ICALP). LNCS, vol 3580, pp 1139–1150
2. Cheng X, Huang X, Li D, Du D-Z (2003) Polynomial-time approximation scheme for minimum connected dominating set in ad hoc wireless networks. Networks 42:202–208
3. Chvátal V (1979) A greedy heuristic for the set-covering problem. Math Oper Res 4(3):233–235

M

4. Clementi A, Crescenzi P, Penna P, Rossi G, Vocca P (2001) On the complexity of computing minimum energy consumption broadcast subgraphs. In: 18th annual symposium on theoretical aspects of computer science. LNCS, vol 2010, pp 121–131
5. Cormen TJ, Leiserson CE, Rivest RL (1990) Introduction to algorithms. MIT/McGraw-Hill, Columbus
6. Flammini M, Navarra A, Klasing R, Pérennes A (2004) Improved approximation results for the minimum energy broadcasting problem. DIALM-POMC. ACM, New York, pp 85–91
7. Garey MR, Johnson DS (1979) Computers and intractability: a guide to the theory of NP-completeness. W.H. Freeman and Company, New York
8. Guha S, Khuller S (1998) Approximation algorithms for connected dominating sets. Algorithmica 20:347–387
9. Preparata FP, Shamos MI (1985) Computational geometry: an introduction. Springer, New York
10. Rappaport TS (1996) Wireless communications: principles and practices. Prentice Hall/IEEE, Piscataway
11. Singh S, Raghavendra CS, Stepanek J (1999) Power-aware broadcasting in mobile ad hoc networks. In: Proceedings of IEEE PIMRC'99, Osaka
12. Takahashi H, Matsuyama A (1980) An approximate solution for Steiner problem in graphs. Math Jpn 24(6):573–577
13. Wan P-J, Calinescu G, Li X-Y, Frieder O (2002) Minimum-energy broadcast routing in static ad hoc wireless networks. ACM Wirel Netw 8(6):607–617, Preliminary version appeared in IEEE INFOCOM (2000)
14. Wan P-J, Calinescu G, Yi C-W (2004) Minimum-power multicast routing in static ad hoc wireless networks. IEEE/ACM Trans Netw 12:507–514
15. Wieselthier JE, Nguyen GD, Ephremides A (2000) On the construction of energy-efficient broadcast and multicast trees in wireless networks. IEEE Infocom 2:585–594

Minimum Flow Time

Nikhil Bansal
Eindhoven University of Technology, Eindhoven, The Netherlands

Keywords

Response time; Sojourn time

Years and Authors of Summarized Original Work

1997; Leonardi, Raz

Problem Definition

The problem is concerned with efficiently scheduling jobs on a system with multiple resources to provide a good quality of service. In scheduling literature, several models have been considered to model the problem setting, and several different measures of quality have been studied. This note considers the following model: There are several identical machines, and jobs are released over time. Each job is characterized by its size, which is the amount of processing it must receive to be completed, and its release time, before which it cannot be scheduled. In this model, Leonardi and Raz studied the objective of minimizing the average flow time of the jobs, where the flow time of a job is the duration of time since it is released until its processing requirement is met. Flow time is also referred to as response time or sojourn time and is a very natural and commonly used measure of the quality of a schedule.

Notations

Let $\mathcal{J} = \{1, 2, \ldots, n\}$ denote the set of jobs in the input instance. Each job j is characterized by its release time r_j and its processing requirement p_j. There is a collection of m identical machines, each having the same processing capability. A schedule specifies which job executes at what time on each machine. Given a schedule, the completion time c_j of a job is the earliest time at which job j receives p_j amount of service. The flow time f_j of j is defined as $c_j - r_j$. A schedule is said to be preemptive if a job can be interrupted arbitrarily and its execution can be resumed later from the point of interruption without any penalty. A schedule is non-preemptive if a job cannot be interrupted once it is started. In the context of multiple machines, a schedule is said to be migratory if a job can be moved from one machine to another during its execution without any penalty. In the off-line model, all the jobs J are given in advance. In scheduling algorithms, the online model is usually more realistic than the off-line model.

Key Results

For a single machine, it is a folklore result that the Shortest Remaining Processing Time (SRPT) policy that at any time works on the job with the least remaining processing time is optimal for minimizing the average flow time. Note that SRPT is an online algorithm and is a preemptive scheduling policy.

If no preemption is allowed, Kellerer, Tautenhahn, and Woeginger [6] gave an $O(n^{1/2})$ approximation algorithm for minimizing the flow time on a single machine and also showed that no polynomial time algorithm can have an approximation ratio of $n^{1/2-\varepsilon}$ for any $\varepsilon > 0$ unless P = NP.

Leonardi and Raz [8] gave the first nontrivial results for minimizing the average flow time on multiple machines. Later, a simpler presentation of this result was given by Leonardi [7]. The main result of [8] is the following.

Theorem 1 ([8]) *On multiple machines, the SRPT algorithm is* $O(\min(\log(n/m), \log P))$ *competitive for minimizing average flow time, where* P *is the maximum to minimum job size ratio.*

They also gave a matching lower bound (up to constant factors) on the competitive ratio.

Theorem 2 ([8]) *For the problem of minimizing flow time on multiple machines, any online algorithm has a competitive ratio of* $\Omega(\min(\log(n/m), \log P))$*, even when randomization is allowed.*

Note that minimizing the average flow time is equivalent to minimizing the total flow time. Suppose each job pays \$1 at each time unit it is alive (i.e., unfinished), then the total payment received is equal to the total flow time. Summing up the payment over each time step, the total flow time can be expressed as the summation over the number of unfinished jobs at each time unit. As SRPT works on jobs that can be finished as soon as possible, it seems intuitively that it should have the least number of unfinished jobs at any time. While this is true for a single machine, it is not true for multiple machines (as shown in an example below). The main idea of [8] was to show that at any time, the number of unfinished jobs under SRPT is "essentially" no more than $O(\min(\log P))$ times that under any other algorithm. To do this, they developed a technique of grouping jobs into a logarithmic number of classes according to their remaining sizes and arguing about the total unfinished work in these classes. This technique has found a lot of uses since then to obtain other results. To obtain a guarantee in terms of n, some additional ideas are required.

The instance below shows how SRPT could deviate from optimum in the case of multiple machines. This instance is also the key component in the lower bound construction in Theorem 2 above. Suppose there are two machines, and three jobs of size 1, 1, and 2 arrive at time $t = 0$. SRPT would schedule the two jobs of size 1 at $t = 0$ and then work on size 2 job at time $t = 1$. Thus, it has one unit of unfinished work at $t = 2$. However, the optimum could schedule the size 2 job at time 0 and finish all these jobs by time 2. Now, at time $t = 2$, three more jobs with sizes 1/2, 1/2, and 1 arrive. Again, SRPT will work on size 1/2 jobs first, and it can be seen that it will have two unfinished jobs with remaining work 1/2 each at $t = 3$, whereas the optimum can finish all these jobs by time 3. This pattern is continued by giving three jobs of size 1/4, 1/4, and 1/2 at $t = 3$ and so on. After k steps, SRPT will have k jobs with sizes $1/2, 1/4, 1/8, \ldots, 1/2^{k-2}, 1/2^{k-1}, 1/2^{k-1}$, while the optimum has no jobs remaining. Now the adversary can give 2 jobs of size $1/2^k$ each every $1/2^k$ time units for a long time, which implies that SRPT could be $\Omega(\log P)$ worse than optimum.

Leonardi and Raz also considered off-line algorithms for the non-preemptive setting in their paper.

Theorem 3 (**[8]**) *There is a polynomial time off-line algorithm that achieves an approximation ratio of $O(n^{1/2} \log n/m)$ for minimizing average flow time on m machines without preemption.*

To prove this result, they give a general technique to convert a preemptive schedule to a non-preemptive one at the loss of an $O(n^{1/2})$ factor in the approximation ratio. They also showed an almost matching lower bound. In particular,

Theorem 4 (**[8]**) *No polynomial time algorithm for minimizing the total flow time on multiple machines without preemption can have an approximation ratio of $O(n^{1/3-\varepsilon})$ for any $\varepsilon > 0$, unless $P = NP$.*

Extensions

Since the publication of these results, they have been extended in several directions. Recall that SRPT is both preemptive and migratory. Awerbuch, Azar, Leonardi, and Regev [2] gave an online scheduling algorithm that is nonmigratory and still achieves a competitive ratio of $O(\min(\log(n/m), \log P))$. Avrahami and Azar [1] gave an even more restricted $O(\min(\log P, \log(n/m)))$ competitive online algorithm. Their algorithm, in addition to being nonmigratory, dispatches a job immediately to a machine upon its arrival. Recently, Garg and Kumar [4, 5] have extended these results to a setting where machines have nonuniform speeds. Other related problems and settings such as stretch minimization (defined as the flow time divided by the size of a job), weighted flow time minimization, general cost functions such as weighted norms, and the non-clairvoyant setting where the size of a job is not unknown upon its arrival have also been investigated. The reader is referred to the relatively recent survey [9] for more details.

Applications

The flow time measure considered here is one of the most widely used measures of quality of service, as it corresponds to the amount of time one has to wait to get the job done. The scheduling model considered here arises very naturally when there are multiple resources and several agents that compete for service from these resources. For example, consider a computing system with multiple homogeneous processors where jobs are submitted by users arbitrarily over time. Keeping the average response time low also keeps the frustration levels of the users low. The model is not necessarily limited to computer systems. At a grocery store, each cashier can be viewed as a machine, and the users lining up to check out can be viewed as jobs. The flow time of a user is time spent waiting until she finishes her transaction with the cashier. Of course, in many applications, there are additional constraints such as it may be infeasible to preempt jobs or, if customers expect a certain fairness, such people might prefer to be serviced in a first-come-first-served manner at a grocery store.

Open Problems

The online algorithm of Leonardi and Raz is also the best-known off-line approximation algorithm for the problem. In particular, it is not known whether an $O(1)$ approximation exists even for the case of two machines. Settling this would be very interesting. In related work, Bansal [3] considered the problem of finding nonmigratory schedules for a constant number of machines. He gave an algorithm that produces a $(1 + \varepsilon)$-approximate solution for any $\varepsilon > 0$ in time $n^{O(\log n/\varepsilon 2)}$. This suggests the possibility of a polynomial time approximation scheme for the problem, at least for the case of a constant number of machines.

Cross-References

- ▶ Flow Time Minimization
- ▶ Multilevel Feedback Queues
- ▶ Shortest Elapsed Time First Scheduling

Recommended Reading

1. Avrahami N, Azar Y (2003) Minimizing total flow time and completion time with immediate dispacthing. In: Proceedings of 15th SPAA, pp 11–18

2. Awerbuch B, Azar Y, Leonardi S, Regev O (2002) Minimizing the flow time without migration. SIAM J Comput 31:1370–1382
3. Bansal N (2005) Minimizing flow time on a constant number of machines with preemption. Oper Res Lett 33:267–273
4. Garg N, Kumar A (2006) Better algorithms for minimizing average flow-time on related machines. In: Proceesings of ICALP, pp 181–190
5. Garg N, Kumar A (2006) Minimizing average flow time on related machines. In: ACM symposium on theory of compuring (STOC), pp 730–738
6. Kellerer H, Tautenhahn T, Woeginger GJ (1999) Approximability and nonapproximability results for minimizing total flow time on a single machine. SIAM J Comput 28:1155–1166
7. Leonardi S (2003) A simpler proof of preemptive flow-time approximation. In: Bampis E (ed) Approximation and on-line algorithms. Lecture notes in computer science. Springer, Berlin
8. Leonardi S, Raz D (1997) Approximating total flow time on parallel machines. In: ACM symposium on theory of computing (STOC), pp 110–119
9. Pruhs K, Sgall J, Torng E (2004) Online scheduling. In: Handbook on scheduling: algorithms, models and performance analysis. CRC, Boca Raton. Symposium on theory of computing (STOC), 1997, pp 110–119

Minimum Geometric Spanning Trees

Christos Levcopoulos
Department of Computer Science, Lund University, Lund, Sweden

Keywords

EMST; Euclidean minimum spanning trees; Minimum length spanning trees; Minimum weight spanning trees; MST

Years and Authors of Summarized Original Work

1999; Krznaric, Levcopoulos, Nilsson

Problem Definition

Let S be a set of n points in d-dimensional real space where $d \geq 1$ is an integer constant. A *minimum spanning tree* (MST) of S is a connected acyclic graph with vertex set S of minimum total edge length. The length of an edge equals the distance between its endpoints under some metric. Under the so-called L_p metric, the distance between two points x and y with coordinates $(x_1, x_2, \ldots, x_d)$ and $(y_1, y_2, \ldots, y_d)$, respectively, is defined as the pth root of the sum $\sum_{i=1}^{d} |x_i - y_i|^p$.

Key Results

Since there is a very large number of papers concerned with geometric MSTs, only a few of them will be mentioned here.

In the common Euclidean L_2 metric, which simply measures straight-line distances, the MST problem in two dimensions can be solved optimally in time $O(n \log n)$, by using the fact that the MST is a subgraph of the Delaunay triangulation of the input point set. The latter is in turn the dual of the Voronoi diagram of S, for which there exist several $O(n \log n)$-time algorithms. The term "optimally" here refers to the algebraic computation tree model. After computation of the Delaunay triangulation, the MST can be computed in only $O(n)$ additional time, by using a technique by Cheriton and Tarjan [6].

Even for higher dimensions, i.e., when $d > 2$, it holds that the MST is a subgraph of the dual of the Voronoi diagram; however, this fact cannot be exploited in the same way as in the two-dimensional case, because this dual may contain $\Omega(n^2)$ edges. Therefore, in higher dimensions, other geometric properties are used to reduce the number of edges which have to be considered. The first subquadratic-time algorithm for higher dimensions was due to Yao [15]. A more efficient algorithm was later proposed by Agarwal et al. [1]. For $d = 3$, their algorithm runs in randomized expected time $O((n \log n)^{4/3})$ and for $d \geq 4$, in expected time $O\left(n^{2-2/(\lceil d/2 \rceil)+1+\epsilon}\right)$, where ε stands for an arbitrarily small positive constant.

The algorithm by Agarwal et al. builds on exploring the relationship between computing an MST and finding a closest pair between n red points and m blue points, which is called the *bichromatic closest pair* problem. They

M

showed that if $T_d(n,m)$ denotes the time to solve the latter problem, then an MST can be computed in $O(T_d(n,n)\log^d n)$ time. Later, Callahan and Kosaraju [4] improved this bound to $O(T_d(n,n)\log n)$. Both methods achieve running time $O(T_d(n,n))$, if $T_d(n,n) = \Omega(n^{1+\alpha})$, for some $\alpha > 0$. Finally, Krznaric et al. [11] showed that the two problems, i.e., computing an MST and computing the bichromatic closest pair, have the same worst-case time complexity (up to constant factors) in the commonly used algebraic computation tree model and for any fixed L_p metric. The hardest part to prove is that an MST can be computed in time $O(T_d(n,n))$. The other part, which is that the bichromatic closest pair problem is not harder than computing the MST, is easy to show: if one first computes an MST for the union of the $n + m$ red and blue points, one can then find a closest bichromatic pair in linear time, because at least one such pair has to be connected by some edge of the MST.

The algorithm proposed by Krznaric et al. [11] is based on the standard approach of joining trees in a forest with the shortest edge connecting two different trees, similar to the classical Kruskal's and Prim's MST algorithms for graphs. To reduce the number of candidates to be considered as edges of the MST, the algorithm works in a sequence of phases, where in each phase only edges of equal or similar length are considered, within a factor of 2.

The initial forest is the set S of points, that is, each point of the input constitutes an individual edgeless tree. Then, as long as there is more than one tree in the forest, two trees are merged by producing an edge connecting two nodes, one from each tree. After this procedure, the edges produced comprise a single tree that remains in the forest, and this tree constitutes the output of the algorithm.

Assume that the next edge that the algorithm is going to produce has length l. Each tree T in the forest is partitioned into groups of nodes, each group having a specific node representing the group. The representative node in such a group is called a *leader*. Furthermore, every node in a group including the leader has the property that it lies within distance $\epsilon \cdot l$ from its leader, where ε is a real constant close to zero.

Instead of considering all pairs of nodes which can be candidates for the next edge to produce, first, only pairs of leaders are considered. Only if a pair of leaders belong to different trees and the distance between them is approximately l, then the closest pair of points between their two respective groups is computed, using the algorithm for the bichromatic closest pair problem.

Also, the following invariant is maintained: for any phase producing edges of length $\Theta(l)$ and for any leader, there is only a constant number of other leaders at distance $\Theta(l)$. Thus, the total number of pairs of leaders to consider is only linear in the number of leaders.

Nearby leaders for any given leader can be found efficiently by using bucketing techniques and data structures for dynamic closest pair queries [3], together with extra artificial points which can be inserted and removed for probing purposes at various small boxes at distance $\Theta(l)$ from the leader. In order to maintain the invariant, when moving to subsequent phases, one reduces the number of leaders accordingly, as pairs of nearby groups merge into single groups. Another tool which is also needed to consider the right types of pairs is to organize the groups according to the various directions in which there can be new candidate MST edges adjacent to nodes in the group. For details, please see the original paper by Krznaric et al. [11].

There is a special version of the bichromatic closest point problem which was shown by Krznaric et al. [11] to have the same worst-case time complexity as computing an MST, namely, the problem for the special case when both the set of red points and the set of blue points have a very small diameter compared with the distance between the closest bichromatic pair. This ratio can be made arbitrarily small by choosing a suitable ε as the parameter for creating the groups and leaders mentioned above. This fact was exploited in order to derive more efficient algorithms for the three-dimensional case.

For example, in the L_1 metric, it is possible to build in time $O(n\log n)$ a special kind of a planar Voronoi diagram for the blue points on a plane separating the blue from the red points having the following property: for each query point q in the half-space including the red points, one can use this Voronoi diagram to find in time $O(\log n)$ the blue point which is closest to q under the L_1 metric. (This planar Voronoi diagram can be seen as defined by the vertical projections of the blue points onto the plane containing the diagram, and the size of a Voronoi cell depends on the distance between the corresponding blue point and the plane.) So, by using subsequently every red point as a query point for this data structure, one can solve the bichromatic closest pair problem for such well-separated red-blue sets in total $O(n \log n)$ time.

By exploiting and building upon this idea, Krznaric et al. [11] showed how to find an MST of S in optimal $O(n \log n)$ time under the L_1 and L_∞ metrics when $d = 3$. This is an improvement over previous bounds due to Gabow et al. [10] and Bespamyatnikh [2], who proved that, for $d = 3$, an MST can be computed in $O(n \log n \log \log n)$ time under the L_1 and L_∞ metrics.

The main results of Krznaric et al. [11] are summarized in the following theorem.

Theorem *In the algebraic computation tree model, for any fixed L_p metric and for any fixed number of dimensions, computing the MST has the same worst-case complexity, within constant factors, as solving the bichromatic closest pair problem. Moreover, for three-dimensional space under the L_1 and L_∞ metrics, the MST (as well as the bichromatic closest pair) can be computed in optimal $O(n \log n)$ time.*

Approximate and Dynamic Solutions

Callahan and Kosaraju [4] showed that a spanning tree of length within a factor $1 + \epsilon$ from that of an MST can be computed in time $O(n(\log n + \epsilon^{-d/2} \log \epsilon^{-1}))$. Approximation algorithms with worse trade-off between time and quality had earlier been developed by Clarkson [7], Vaidya [14] and Salowe [13]. In addition, if the input point set is supported by certain basic data structures, then the approximate length of an MST can be computed in randomized sublinear time [8]. Eppstein [9] gave fully dynamic algorithms that maintain an MST when points are inserted or deleted.

Applications

MSTs belong to the most basic structures in computational geometry and in graph theory, with a vast number of applications.

Open Problems

Although the complexity of computing MSTs is settled in relation to computing bichromatic closest pairs, this means also that it remains open for all cases where the complexity of computing bichromatic closest pairs remains open, e.g., when the number of dimensions is greater than 3.

Experimental Results

Narasimhan and Zachariasen [12] have reported experiments with computing geometric MSTs via well-separated pair decompositions. More recent experimental results are reported by Chatterjee, Connor, and Kumar [5].

Cross-References

▸ Minimum Spanning Trees
▸ Parallel Connectivity and Minimum Spanning Trees
▸ Randomized Minimum Spanning Tree
▸ Steiner Trees

Recommended Reading

1. Agarwal PK, Edelsbrunner H, Schwarzkopf O, Welzl E (1991) Euclidean minimum spanning trees and bichromatic closest pairs. Discret Comput Geom 6:407–422
2. Bespamyatnikh S (1997) On constructing minimum spanning trees in R_1^k. Algorithmica 18(4): 524–529
3. Bespamyatnikh S (1998) An optimal algorithm for closest-pair maintenance. Discret Comput Geom 19(2):175–195
4. Callahan PB, Kosaraju SR (1993) Faster algorithms for some geometric graph problems in higher dimensions. In: SODA, Austin, pp 291–300

5. Chatterjee S, Connor M, Kumar P (2010) Geometric minimum spanning trees with GeoFilterKruskal. In: Experimental Algorithms: proceedings of SEA, Ischia Island. LNCS, vol 6049, pp 486–500
6. Cheriton D, Tarjan RE (1976) Finding minimum spanning trees. SIAM J Comput 5(4):724–742
7. Clarkson KL (1984) Fast expected-time and approximation algorithms for geometric minimum spanning trees. In: Proceedings of STOC, Washington, DC, pp 342–348
8. Czumaj A, Ergün F, Fortnow L, Magen A, Newman I, Rubinfeld R, Sohler C (2005) Approximating the weight of the Euclidean minimum spanning tree in sublinear time. SIAM J Comput 35(1):91–109
9. Eppstein D (1995) Dynamic Euclidean minimum spanning trees and extrema of binary functions. Discret Comput Geom 13:111–122
10. Gabow HN, Bentley JL, Tarjan RE (1984) Scaling and related techniques for geometry problems. In: STOC, Washington, DC, pp 135–143
11. Krznaric D, Levcopoulos C, Nilsson BJ (1999) Minimum spanning trees in d dimensions. Nord J Comput 6(4):446–461
12. Narasimhan G, Zachariasen M (2001) Geometric minimum spanning trees via well-separated pair decompositions. ACM J Exp Algorithms 6:6
13. Salowe JS (1991) Constructing multidimensional spanner graphs. Int J Comput Geom Appl 1(2):99–107
14. Vaidya PM (1988) Minimum spanning trees in k-Dimensional space. SIAM J Comput 17(3):572–582
15. Yao AC (1982) On constructing minimum spanning trees in k-Dimensional spaces and related problems. SIAM J Comput 11(4):721–736

Minimum k-Connected Geometric Networks

Artur Czumaj[1] and Andrzej Lingas[2]
[1]Department of Computer Science, Centre for Discrete Mathematics and Its Applications, University of Warwick, Coventry, UK
[2]Department of Computer Science, Lund University, Lund, Sweden

Keywords

Geometric networks; Geometric optimization; k-connectivity; Network design; Survivable network design

Synonyms

Euclidean graphs; Geometric graphs

Years and Authors of Summarized Original Work

2000; Czumaj, Lingas

Problem Definition

The following classical optimization problem is considered: for a given undirected weighted geometric network, find its minimum-cost subnetwork that satisfies a priori given multi-connectivity requirements. This problem restricted to *geometric networks* is considered in this entry.

Notations

Let $G = (V, E)$ be a geometric network, whose vertex set V corresponds to a set of n points in $\mathbb{R}^d$ for certain integer d, $d \geq 2$ and whose edge set E corresponds to a set of straight-line segments connecting pairs of points in V. G is called complete if E connects all pairs of points in V.

The cost $\delta(x, y)$ of an edge connecting a pair of points $x, y \in \mathbb{R}^d$ is equal to the Euclidean distance between points x and y, that is, $\delta(x, y) = \sqrt{\sum_{i=1}^{d}(x_i - y_i)^2}$, where $x = (x_1, \dots, x_d)$ and $y = (y_1, \dots, y_d)$. More generally, the cost $\delta(x, y)$ could be defined using other norms, such as ℓ_p norms for any $p > 1$, i.e., $\delta(x, y) = \left(\sum_{i=1}^{p}(x_i - y_i)^p\right)^{1/p}$. The cost of the network is equal to the sum of the costs of the edges of the network, $\text{cost}(G) = \sum_{(x,y)\in E} \delta(x, y)$.

A network $G = (V, E)$ *spans* a set S of points if $V = S$. $G = (V, E)$ is k*-vertex connected* if for any set $U \subseteq V$ of fewer than k vertices, the network $(V \setminus U,\ E \cap ((V \setminus U) \times (V \setminus U)))$ is connected. Similarly, G is k*-edge connected* if for any set $\mathcal{E} \subseteq E$ of fewer than k edges, the network $(V, E \setminus \mathcal{E})$ is connected.

The *(Euclidean) minimum-cost k-vertex-connected spanning network problem*: for a given set S of n points in the Euclidean space $\mathbb{R}^d$, find a minimum-cost k-vertex-connected Euclidean network spanning points in S.

The *(Euclidean) minimum-cost k-edge-connected spanning network problem*: for a given set S of n points in the Euclidean space $\mathbb{R}^d$, find a minimum-cost k-edge-connected Euclidean network spanning points in S.

A variant that allows *parallel edges* is also considered:

The *(Euclidean) minimum-cost k-edge-connected spanning multi-network problem*: for a given set S of n points in the Euclidean space $\mathbb{R}^d$, find a minimum-cost k-edge-connected Euclidean multi-network spanning points in S (where the multi-network can have parallel edges).

The concept of minimum-cost k-connectivity naturally extends to include that of *Euclidean Steiner k-connectivity* by allowing the use of additional vertices, called *Steiner points*. For a given set S of points in $\mathbb{R}^d$, a geometric network G is a *Steiner k-vertex connected (or Steiner k-edge connected) for S* if the vertex set of G is a *superset* of S and for every pair of points from S there are k internally vertex-disjoint (edge-disjoint, respectively) paths connecting them in G.

The *(Euclidean) minimum-cost Steiner k-vertex/edge connectivity* problem: find a minimum-cost network on a superset of S that is Steiner k-vertex/edge connected for S.

Note that for $k = 1$, it is simply the *Steiner minimal tree* problem, which has been very extensively studied in the literature (see, e.g., [15]).

In a more general formulation of multi-connectivity graph problems, nonuniform connectivity constraints have to be satisfied.

The *survivable network design* problem: for a given set S of points in $\mathbb{R}^d$ and a connectivity requirement function $r : S \times S \to \mathbb{N}$, find a minimum-cost geometric network spanning points in S such that for any pair of vertices $p, q \in S$ the subnetwork has $r_{p,q}$ internally vertex-disjoint (or edge-disjoint, respectively) paths between p and q.

In many applications of this problem, often regarded as the most interesting ones [10, 14], the connectivity requirement function is specified with the help of a one-argument function which assigns to each vertex p its connectivity type $r_v \in \mathbb{N}$. Then, for any pair of vertices $p, q \in S$, the connectivity requirement $r_{p,q}$ is simply given as $\min\{r_p, r_q\}$ [13, 14, 18, 19]. This includes the *Steiner tree problem* (see, e.g., [2]), in which $r_p \in \{0, 1\}$ for any vertex $p \in S$.

A *polynomial-time approximation scheme (PTAS)* is a family of algorithms $\{\mathcal{A}_\varepsilon\}$ such that, for each fixed $\varepsilon > 0$, $\mathcal{A}_\varepsilon$ runs in time polynomial in the size of the input and produces a $(1 + \varepsilon)$-approximation.

Related Work

For a very extensive presentation of results concerning problems of finding minimum-cost k-vertex- and k-edge-connected spanning subgraphs, nonuniform connectivity, connectivity augmentation problems, and geometric problems, see [1, 3, 4, 12, 16].

Despite the practical relevance of the multi-connectivity problems for geometrical networks and the vast amount of practical heuristic results reported (see, e.g., [13, 14, 18, 19]), very little theoretical research had been done towards developing efficient approximation algorithms for these problems until a few years ago. This contrasts with the very rich and successful theoretical investigations of the corresponding problems in general metric spaces and for general weighted graphs. And so, until 1998, even for the simplest and most fundamental multi-connectivity problem that of finding a minimum-cost 2-vertex-connected network spanning a given set of points in the Euclidean plane, obtaining approximations achieving better than a $\frac{3}{2}$ ratio had been elusive (the ratio $\frac{3}{2}$ is the best polynomial-time approximation ratio known for general networks whose weights satisfy the triangle inequality [9]; for other results, see, e.g., [5, 16]).

Key Results

The first result is an extension of the well-known $\mathcal{NP}$-hardness result of minimum-cost 2-connectivity in general graphs (see, e.g., [11]) to geometric networks.

Theorem 1 *The problem of finding a minimum-cost 2-vertex-/2-edge-connected geometric network spanning a set of n points in the plane is $\mathcal{NP}$-hard.*

Next result shows that if one considers the minimum-cost multi-connectivity problems in an enough high dimension, the problems become APX-hard.

Theorem 2 ([6]) *There exists a constant $\xi > 0$ such that it is $\mathcal{NP}$-hard to approximate within $1 + \xi$ the minimum-cost 2-connected geometric network spanning a set of n points in $\mathbb{R}^{\lceil \log_2 n \rceil}$*

This result extends also to any ℓ_p norm.

Theorem 3 ([6]) *For integer $d \geq \log n$ and for any fixed $p \geq 1$, there exists a constant $\xi > 0$ such that it is $\mathcal{NP}$-hard to approximate within $1 + \xi$ the minimum-cost 2-connected network spanning a set of n points in the ℓ_p metric in $\mathbb{R}^d$.*

Since the minimum-cost multi-connectivity problems are hard, the research turned into the study of approximation algorithms. By combining some of the ideas developed for the polynomial-time approximation algorithms for TSP due to Arora [2] (see also [17]) together with several new ideas developed specifically for the multi-connectivity problems in geometric networks, Czumaj and Lingas obtained the following results.

Theorem 4 ([6, 7]) *Let k and d be any integers, $k, d \geq 2$, and let ε be any positive real. Let S be a set of n points in $\mathbb{R}^d$. There is a randomized algorithm that in time $n \cdot (\log n)^{(kd/\varepsilon)^{\mathcal{O}(d)}} \cdot 2^{2^{(kd/\varepsilon)^{\mathcal{O}(d)}}}$ with probability at least 0.99 finds a k-vertex-connected (or k-edge-connected) spanning network for S whose cost is at most $(1 + \varepsilon)$-time optimal.*

Furthermore, this algorithm can be derandomized in polynomial time to return a k-vertex-connected (or k-edge-connected) spanning network for S whose cost is at most $(1 + \varepsilon)$ times the optimum.

Observe that when all d, k, and ε are constant, then the running times are $n \cdot \log^{\mathcal{O}(1)} n$.

The results in Theorem 4 give a PTAS for small values of k and d.

Theorem 5 (PTAS for vertex/edge connectivity [6, 7]) *Let $d \geq 2$ be any constant integer. There is a certain positive constant $c < 1$ such that for all k such that $k \leq (\log \log n)^c$, the problems of finding a minimum-cost k-vertex-connected spanning network and a k-edge-connected spanning network for a set of points in $\mathbb{R}^d$ admit PTAS.*

The next theorem deals with multi-networks where feasible solutions are allowed to use parallel edges.

Theorem 6 ([7]) *Let k and d be any integers, $k, d \geq 2$, and let ε be any positive real. Let S be a set of n points in $\mathbb{R}^d$. There is a randomized algorithm that in time $n \cdot \log n \cdot (d/\varepsilon)^{\mathcal{O}(d)} + n \cdot 2^{2^{(k^{\mathcal{O}(1)} \cdot (d/\varepsilon)^{\mathcal{O}(d^2)})}}$, with probability at least 0.99 finds a k-edge-connected spanning multi-network for S whose cost is at most $(1 + \varepsilon)$ times the optimum. The algorithm can be derandomized in polynomial time.*

Combining this theorem with the fact that parallel edges can be eliminated in case $k = 2$, one obtains the following result for 2-connectivity in networks.

Theorem 7 (Approximation schemes for 2-connected graphs, [7]) *Let d be any integer, $d \geq 2$, and let ε be any positive real. Let S be a set of n points in $\mathbb{R}^d$. There is a randomized algorithm that in time $n \cdot \log n \cdot (d/\varepsilon)^{\mathcal{O}(d)} + n \cdot 2^{(d/\varepsilon)^{\mathcal{O}(d^2)}}$ with probability at least 0.99 finds a 2-vertex-connected (or 2-edge-connected) spanning network for S whose cost is at most $(1 + \varepsilon)$ times the optimum. This algorithm can be derandomized in polynomial time.*

For constant d, the running time of the randomized algorithms is $n \log n \cdot (1/\varepsilon)^{\mathcal{O}(1)} + 2^{(1/\varepsilon)^{\mathcal{O}(1)}}$.

Theorem 8 ([8]) *Let d be any integer, $d \geq 2$, and let ε be any positive real. Let S be a set* *of n points in* $\mathbb{R}^d$. *There is a randomized algorithm that in time* $n \cdot \log n \cdot (d/\varepsilon)^{\mathcal{O}(d)} + n \cdot 2^{(d/\varepsilon)^{\mathcal{O}(d^2)}} + n \cdot 2^{2^{d^{d^{\mathcal{O}(1)}}}}$ *with probability at least* 0.99 *finds a Steiner 2-vertex-connected (or 2-edge-connected) spanning network for S whose cost is at most* $(1+\varepsilon)$ *times the optimum. This algorithm can be derandomized in polynomial time.*

Theorem 9 ([8]) *Let d be any integer,* $d \geq 2$, *and let* ε *be any positive real. Let S be a set of n points in* $\mathbb{R}^d$. *There is a randomized algorithm that in time* $n \cdot \log n \cdot (d/\varepsilon)^{\mathcal{O}(d)} + n \cdot 2^{(d/\varepsilon)^{\mathcal{O}(d^2)}} + n \cdot 2^{2^{d^{d^{\mathcal{O}(1)}}}}$ *with probability at least* 0.99 *gives a* $(1+\varepsilon)$*-approximation for the geometric network survivability problem with* $r_v \in \{0,1,2\}$ *for any* $v \in V$. *This algorithm can be derandomized in polynomial time.*

Applications

Multi-connectivity problems are central in algorithmic graph theory and have numerous applications in computer science and operation research, see, e.g., [1, 12, 14, 19]. They also play very important role in the design of networks that arise in practical situations, see, e.g., [1, 14]. Typical application areas include telecommunication, computer, and road networks. Low degree connectivity problems for geometrical networks in the plane can often closely *approximate* such practical connectivity problems (see, e.g., the discussion in [14, 18, 19]). The survivable network design problem in geometric networks also arises in many applications, e.g., in telecommunication, communication network design, VLSI design, etc. [13, 14, 18, 19].

Open Problems

The results discussed above lead to efficient algorithms only for small connectivity requirements k; the running time is polynomial only for the value of k up to $(\log\log n)^c$ for certain positive constant $c < 1$. It is an interesting open problem if one can obtain polynomial-time approximation scheme algorithms also for large values of k.

It is also an interesting open problem if the multi-connectivity problems in geometric networks can have practically fast approximation schemes.

Cross-References

▸ Euclidean Traveling Salesman Problem
▸ Minimum Geometric Spanning Trees

Recommended Reading

1. Ahuja RK, Magnanti TL, Orlin JB, Reddy MR (1995) Applications of network optimization. In: Ball MO, Magnanti TL, Monma CL and Nemhauser GL (eds) Handbooks in operations research and management science, vol 7: network models chapter 1. North-Holland, pp 1–83
2. Arora S (1998) Polynomial time approximation schemes for Euclidean traveling salesman and other geometric problems. J ACM 45(5):753–782
3. Berger A, Czumaj A, Grigni M, Zhao H (2005) Approximation schemes for minimum 2-connected spanning subgraphs in weighted planar graphs. In: Proceedings of the 13th Annual European Symposium on Algorithms (ESA), Palma de Mallorca, pp 472–483
4. Borradaile G, Klein PN (2008) The two-edge connectivity survivable network problem in planar graphs. In: Proceedings of the 35th Annual International Colloquium on Automata Languages and Programming (ICALP), Reykjavik, pp 485–501
5. Cheriyan J, Vetta A (2005) Approximation algorithms for network design with metric costs. In: Proceedings of the 37th Annual ACM Symposium on Theory of Computing (STOC), Baltimore, pp 167–175
6. Czumaj A, Lingas A (1999) On approximability of the minimum-cost k-connected spanning subgraph problem. In: Proceedings of the 10th Annual ACM-SIAM Symposium on Discrete Algorithms (SODA), Baltimore, pp 281–290
7. Czumaj A, Lingas A (2000) Fast approximation schemes for Euclidean multi-connectivity problems. In: Proceedings of the 27th Annual International Colloquium on Automata, Languages and Programming (ICALP), Geneva, pp 856–868
8. Czumaj A, Lingas A, Zhao H (2002) Polynomial-time approximation schemes for the Euclidean survivable network design problem. In: Proceedings of the 29th Annual International Colloquium on Automata, Languages and Programming (ICALP), Málaga, pp 973–984
9. Frederickson GN, JáJá J (1982) On the relationship between the biconnectivity augmentation and

M

traveling salesman problem. Theor Comput Sci 19(2):189–201
10. Gabow HN, Goemans MX, Williamson DP (1998) An efficient approximation algorithm for the survivable network design problem. Math Progr Ser B 82(1–2):13–40
11. Garey MR, Johnson DS (1979) Computers and intractability: a guide to the theory of NP-completeness. Freeman, New York
12. Goemans MX, Williamson DP (1996) The primal-dual method for approximation algorithms and its application to network design problems. In: Hochbaum D (ed) Approximation algorithms for $\mathcal{NP}$-hard problems, chapter 4. PWS, Boston, pp 144–191
13. Grötschel M, Monma CL, Stoer M (1992) Computational results with a cutting plane algorithm for designing communication networks with low-connectivity constraints. Oper Res 40(2):309–330
14. Grötschel M, Monma CL, Stoer M (1995) Design of survivable networks. In: Handbooks in operations research and management science, vol 7: network models, chapter 10. North-Holland, pp 617–672
15. Hwang FK, Richards DS, Winter P (1992) The Steiner tree problem. North-Holland, Amsterdam
16. Khuller S (1996) Approximation algorithms for finding highly connected subgraphs. In: Hochbaum D (ed) Approximation algorithms for $\mathcal{NP}$-hard problems, chapter 6. PWS, Boston, pp 236–265
17. Mitchell JSB (1999) Guillotine subdivisions approximate polygonal subdivisions: a simple polynomial-time approximation scheme for geometric TSP, k-MST, and related problems. SIAM J Comput 28(4):1298–1309
18. Monma CL, Shallcross DF (1989) Methods for designing communications networks with certain two-connected survivability constraints. Oper Res 37(4):531–541
19. Stoer M (1992) Design of survivable networks. Springer, Berlin

Minimum Spanning Trees

Seth Pettie
Electrical Engineering and Computer Science (EECS) Department, University of Michigan, Ann Arbor, MI, USA

Keywords

Minimal spanning tree; Minimum weight spanning tree; Shortest spanning tree

Years and Authors of Summarized Original Work

2002; Pettie, Ramachandran

Problem Definition

The *minimum spanning tree* (MST) problem is, given a connected, weighted, and undirected graph $G = (V, E, w)$, to find the tree with minimum total *weight* spanning all the vertices V. Here, $w : E \rightarrow \mathbb{R}$ is the weight function. The problem is frequently defined in geometric terms, where V is a set of points in d-dimensional space and w corresponds to Euclidean distance. The main distinction between these two settings is the form of the input. In the graph setting, the input has size $O(m+n)$ and consists of an enumeration of the $n = |V|$ vertices and $m = |E|$ edges and edge weights. In the geometric setting, the input consists of an enumeration of the coordinates of each point ($O(dn)$ space): all $\binom{V}{2}$ edges are implicitly present and their weights implicit in the point coordinates. See [16] for a discussion of the Euclidean minimum spanning tree problem.

History

The MST problem is generally recognized [7, 12] as one of the first combinatorial problems studied specifically from an algorithmic perspective. It was formally defined by Borůvka in 1926 [1] (predating the fields of computability theory and combinatorial optimization and even much of graph theory), and since his initial algorithm, there has been a sustained interest in the problem. The MST problem has motivated research in matroid optimization [3] and the development of efficient data structures, particularly priority queues (aka heaps) and disjoint set structures [2, 18].

Related Problems

The MST problem is frequently contrasted with the *traveling salesman* and *minimum Steiner tree* problems [6]. A Steiner tree is a tree that may span any *superset* of the given points; that is, additional points may be introduced that reduce the weight of the minimum spanning tree. The traveling salesman problem asks for a tour (cycle) of the vertices with minimum total length. The generalization of the MST problem to directed graphs is sometimes called the *minimum branching* [5]. Whereas the undirected and directed versions of the MST problem are solvable in polynomial time, traveling salesman and minimum Steiner tree are NP-complete [6].

Optimality Conditions

A *cut* is a partition (V', V'') of the vertices V. An edge (u, v) *crosses* the cut (V', V'') if $u \in V'$ and $v \in V''$. A sequence $(v_0, v_1, \ldots, v_{k-1}, v_0)$ is a *cycle* if $(v_i, v_{i+1(\bmod k)}) \in E$ for $0 \leq i < k$.

The correctness of all MST algorithms is established by appealing to the dual *cut* and *cycle* properties, also known as the blue rule and red rule [18].

Cut Property An edge is in some minimum spanning tree if and only if it is the lightest edge crossing some cut.

Cycle Property An edge is not in any minimum spanning tree if and only if it is the sole heaviest edge on some cycle.

It follows from the cut and cycle properties that if the edge weights are unique, then there is a unique minimum spanning tree, denoted $\mathbf{MST}(G)$. Uniqueness can always be enforced by breaking ties in any consistent manner. MST algorithms frequently appeal to a useful corollary of the cut and cycle properties called the *contractibility* property. Let $G \backslash C$ denote the graph derived from G by contracting the subgraph C, that is, C is replaced by a single vertex c and all edges incident to exactly one vertex in C become incident to c; in general, $G \backslash C$ may have more than one edge between two vertices.

Contractibility Property If C is a subgraph such that for all pairs of edges e and f with exactly one endpoint in C, there exists a path $P \subseteq C$ connecting e f with each edge in P lighter than either e or f, then C is *contractible*. For any contractible C, it holds that $\mathbf{MST}(G) = \mathbf{MST}(C) \cup \mathbf{MST}(G \backslash C)$.

The Generic Greedy Algorithm

Until recently, all MST algorithms could be viewed as mere variations on the following generic greedy MST algorithm. Let $\mathcal{T}$ consist initially of n trivial trees, each containing a single vertex of G. Repeat the following step $n-1$ times. Choose any $T \in \mathcal{T}$ and find the minimum weight edge (u, v) with $u \in T$ and v in a different tree, say $T' \in \mathcal{T}$. Replace T and T' in $\mathcal{T}$ with the single tree $T \cup \{(u, v)\} \cup T'$. After $n-1$ iterations, $\mathcal{T} = \{\mathbf{MST}(G)\}$. By the cut property, every edge selected by this algorithm is in the MST.

Modeling MST Algorithms

Another corollary of the cut and cycle properties is that the set of minimum spanning trees of a graph is determined solely by the relative order of the edge weights – their specific numerical values are not relevant. Thus, it is natural to model MST algorithms as *binary decision trees*, where nodes of the decision tree are identified with edge weight comparisons and the children of a node correspond to the possible outcomes of the comparison. In this decision tree model, a trivial lower bound on the *time* of the optimal MST algorithm is the *depth* of the optimal decision tree.

Key Results

The primary result of [14] is an explicit MST algorithm that is *provably* optimal even though its asymptotic running time is currently unknown.

Theorem 1 *There is an explicit, deterministic minimum spanning tree algorithm whose running time is on the order of $D_{MST}(m, n)$, where m is the number of edges, n the number of vertices, and $D_{MST}(m, n)$ the maximum depth of an optimal decision tree for any m-edge n-node graph.*

It follows that the Pettie-Ramachandran algorithm [14] is asymptotically no worse than *any* MST algorithm that deduces the solution through edge weight comparisons. The best known upper bound on $D_{\text{MST}}(m, n)$ is $O(m\alpha(m, n))$, due to Chazelle [2]. It is trivially $\Omega(m)$.

M

Let us briefly describe how the Pettie-Ramachandran algorithm works. An (m,n) instance is a graph with m edges and n vertices. Theorem 1 is proved by giving a linear time *decomposition* procedure that reduces any (m,n) instance of the MST problem to instances of size $(m^*,n^*),(m_1,n_1),\ldots,(m_s,n_s)$, where $m = m^* + \sum_i m_i$, $n = \sum_i n_i$, $n^* \le n/\log\log\log n$, and each $n_i \le \log\log\log n$. The (m^*,n^*)instance can be solved in $O(m+n)$ time with existing MST algorithms [2]. To solve the other instances, the Pettie-Ramachandran algorithm performs a brute-force search to find the minimum depth decision tree for *every* graph with at most log log log n vertices. Once these decision trees are found, the remaining instances are solved in $O(\sum_i D_{\mathrm{MST}}(m_i,n_i)) = O(D_{\mathrm{MST}}(m,n))$ time. Due to the restricted size of these instances ($n_i \le \log\log\log n$), the time for a brute-force search is a negligible $o(n)$. The decomposition procedure makes use of Chazelle's *soft heap* [2] (an approximate priority queue) and an extension of the contractibility property.

Approximate Contractibility Let G' be derived from G by *increasing* the weight of some edges. If C is contractible w.r.t. G', then $\mathbf{MST}(G) = \mathbf{MST}(\mathbf{MST}(C) \cup \mathbf{MST}(G \backslash C) \cup E^*)$, where $E^*$is the set of edges with increased weights.

A secondary result of [14] is that the running time of the optimal algorithm is actually linear on nearly every graph topology, under any permutation of the edge weights.

Theorem 2 *Let G be selected uniformly at random from the set of all n-vertex, m-edge graphs. Then regardless of the edge weights,* **MST** (G) *can be found in* $O(m+n)$ *time with probability* $1 - 2^{-\Omega(m/\alpha 2)}$, *where* $\alpha = \alpha(m,n)$ *is the slowly growing inverse Ackermann function.*

Theorem 1 should be contrasted with the results of Karger, Klein, and Tarjan [9] and Chazelle [2] on the randomized and deterministic complexity of the MST problem.

Theorem 3 ([9]) *The minimum spanning forest of a graph with m edges can be computed by a randomized algorithm in* $O(m)$ *time with probability* $1 - 2^{-\Omega(m)}$.

Theorem 4 ([2]) *The minimum spanning tree of a graph can be computed in* $O(m\alpha(m,n))$ *time by a deterministic algorithm, where* α *is the inverse Ackermann function.*

Applications

Borůvka [1] invented the MST problem while considering the practical problem of electrifying rural Moravia (present-day Czech Republic) with the shortest electrical network. MSTs are used as a starting point for heuristic approximations to the optimal traveling salesman tour and optimal Steiner tree, as well as other network design problems. MSTs are a component in other graph optimization algorithms, notably the single-source shortest path algorithms of Thorup [19] and Pettie-Ramachandran [15]. MSTs are used as a tool for visualizing data that is presumed to have a tree structure; for example, if a matrix contains dissimilarity data for a set of species, the minimum spanning tree of the associated graph will presumably group closely related species; see [7]. Other modern uses of MSTs include modeling physical systems [17] and image segmentation [8]; see [4] for more applications.

Open Problems

The chief open problem is to determine the *deterministic* complexity of the minimum spanning tree problem. By Theorem 1, this is tantamount to determining the decision tree complexity of the MST problem.

Experimental Results

Moret and Shapiro [11] evaluated the performance of greedy MST algorithms using a variety of priority queues. They concluded that the best MST algorithm is Jarník's [7] (also attributed to Prim and Dijkstra; see [3, 7, 12]) as implemented with a pairing heap [13]. Katriel, Sanders, and Träff [10] designed and implemented a non-greedy randomized MST algorithm based on that of Karger et al. [9]. They concluded that on moderately dense graphs, it runs substantially faster than the greedy algorithms tested by Moret and Shapiro.

Cross-References

▶ Randomized Minimum Spanning Tree

Recommended Reading

1. Borůvka O (1926) O jistém problému minimálním. Práce Moravské Přírodovědecké Společnosti 3:37–58. In Czech
2. Chazelle B (2000) A minimum spanning tree algorithm with inverse-Ackermann type complexity. J ACM 47(6):1028–1047
3. Cormen TH, Leiserson CE, Rivest RL, Stein C (2001) Introduction to algorithms. MIT, Cambridge
4. Eppstein D, Geometry in action: minimum spanning trees. http://www.ics.uci.edu/~eppstein/gina/mst.html. Last downloaded Nov 2014
5. Gabow HN, Galil Z, Spencer TH, Tarjan RE (1986) Efficient algorithms for finding minimum spanning trees in undirected and directed graphs. Combinatorica 6:109–122
6. Garey MR, Johnson DS (1979) Computers and intractability: a guide to NP-completeness. Freeman, San Francisco
7. Graham RL, Hell P (1985) On the history of the minimum spanning tree problem. Ann Hist Comput 7(1):43–57
8. Ion A, Kropatsch WG, Haxhimusa Y (2006) Considerations regarding the minimum spanning tree pyramid segmentation method. In: Proceedings of the 11th workshop structural, syntactic, and statistical pattern recognition (SSPR), Hong Kong. LNCS, vol 4109. Springer, Berlin, pp 182–190
9. Karger DR, Klein PN, Tarjan RE (1995) A randomized linear-time algorithm for finding minimum spanning trees. J ACM 42:321–329
10. Katriel I, Sanders P, Träff JL (2003) A practical minimum spanning tree algorithm using the cycle property. In: Proceedings of the 11th annual European symposium on algorithms, Budapest. LNCS, vol 2832. Springer, Berlin, pp 679–690
11. Moret BME, Shapiro HD (1994) An empirical assessment of algorithms for constructing a minimum spanning tree. In: Computational support for discrete mathematics. DIMACS series in discrete mathematics and theoretical computer science, vol 15. American Mathematical Society, Providence
12. Pettie S (2003) On the shortest path and minimum spanning tree problems. Ph.D. thesis, The University of Texas, Austin
13. Pettie S (2005) Towards a final analysis of pairing heaps. In: Proceedings of the 46th annual symposium on foundations of computer science (FOCS), Pittsburgh, pp 174–183
14. Pettie S, Ramachandran V (2002) An optimal minimum spanning tree algorithm. J ACM 49(1):16–34
15. Pettie S, Ramachandran V (2005) A shortest path algorithm for real-weighted undirected graphs. SIAM J Comput 34(6):1398–1431
16. Preparata FP, Shamos MI (1985) Computational geometry. Springer, New York
17. Subramaniam S, Pope SB (1998) A mixing model for turbulent reactive flows based on euclidean minimum spanning trees. Combust Flame 115(4):487–514
18. Tarjan RE (1983) Data structures and network algorithms. CBMS-NSF regional conference series in applied mathematics, vol 44. SIAM, Philadelphia
19. Thorup M (1999) Undirected single-source shortest paths with positive integer weights in linear time. J ACM 46(3):362–394

Minimum Weight Triangulation

Christos Levcopoulos
Department of Computer Science, Lund University, Lund, Sweden

Keywords

Minimum length triangulation

Years and Authors of Summarized Original Work

1998; Levcopoulos, Krznaric

Problem Definition

Given a set S of n points in the Euclidean plane, a *triangulation* T of S is a maximal set of nonintersecting straight-line segments whose endpoints are in S. The *weight* of T is defined as the total Euclidean length of all edges in T. A triangulation that achieves minimum weight is called a *minimum weight triangulation*, often abbreviated MWT, of S.

Key Results

Since there is a very large number of papers and results dealing with minimum weight triangulation, only relatively very few of them can be mentioned here.

Mulzer and Rote have shown that MWT is NP-hard [12]. Their proof of NP-completeness is not given explicitly; it relies on extensive calculations which they performed with a computer. Remy and Steger have shown a quasi-polynomial time approximation scheme for MWT [13]. These results are stated in the following theorem:

Theorem 1 *The problem of computing the MWT (minimum weight triangulation) of an input set S of n points in the plane is NP-hard. However, for any constant $\epsilon > 0$, a triangulation of S achieving the approximation ratio of $1 + \epsilon$, for an arbitrarily small positive constant ϵ, can be computed in time $n^{\mathcal{O}(\log^8 n)}$.*

The complexity status of the symmetric problem of finding the *maximum* weight triangulation is still open, but there exists a quasi-polynomial time approximation scheme for it [10].

The Quasi-Greedy Triangulation Approximates the MWT

Levcopoulos and Krznaric showed that a triangulation of total length within a constant factor of MWT can be computed in polynomial time for arbitrary point sets [7]. The triangulation achieving this result is a modification of the so-called *greedy* triangulation. The greedy triangulation starts with the empty set of diagonals and keeps adding a shortest diagonal not intersecting the diagonals which have already been added, until a full triangulation is produced. The greedy triangulation has been shown to approximate the minimum weight triangulation within a constant factor, unless a special case arises where the greedy diagonals inserted are "climbing" in a special, very unbalanced way along a relatively long concave chain containing many vertices and with a large empty space in front of it, at the same time blocking visibility from another, opposite concave chain of many vertices. In such "bad" cases, the worst-case ratio between the length of the greedy and the length of the minimum weight triangulation is shown to be $\Theta\left(\sqrt{n}\right)$. To obtain a triangulation which always approximates the MWT within a constant factor, it suffices to take care of this special bad case in order to avoid the unbalanced "climbing," and replace it by a more balanced climbing along these two opposite chains. Each edge inserted in this modified method is still almost as short as the shortest diagonal, within a factor smaller than 1.2. Therefore, the modified triangulation which always approximates the MWT is named the *quasi-greedy* triangulation. In a similar way as the original greedy triangulation, the quasi-greedy triangulation can be computed in time $O(n \log n)$ [8]. Gudmundsson and Levcopoulos [5] showed later that a variant of this method can also be parallelized, thus achieving a constant factor approximation of MWT in $O(\log n)$ time, using $O(n)$ processors in the CRCW PRAM model. Another by-product of the quasi-greedy triangulation is that one can easily select in linear time a subset of its edges to obtain a convex partition which is within a constant factor of the minimum length convex partition of the input point set. This last property was crucial in the proof that the quasi-greedy triangulation approximates the MWT. The proof also uses an older result that the (original, unmodified) greedy triangulation of any convex polygon approximates the minimum weight triangulation [9]. Some of the results from [7] and from [8] can be summarized in the following theorem:

Theorem 2 *Let S be an input set of n points in the plane. The quasi-greedy triangulation of S, which is a slightly modified version of the greedy triangulation of S, has total length within a constant factor of the length of the MWT (minimum weight triangulation) of S and can be computed in time $O(n \log n)$. Moreover, the* *(unmodified) greedy triangulation of S has length within $O\left(\sqrt{n}\right)$ of the length of MWT of S, and this bound is asymptotically tight in the worst case.*

Computing the Exact Minimum Weight Triangulation

Below, three approaches to compute the exact MWT are shortly discussed. These approaches assume that it is numerically possible to efficiently compare the total length of sets of line segments in order to select the set of smallest weight. This is a simplifying assumption, since this is an open problem per se. However, the problem of computing the exact MWT remains NP-hard even under this assumption [12]. The three approaches differ with respect to the creation and selection of subproblems, which are then solved by dynamic programming.

The first approach, sketched by Lingas [11], employs a general method for computing optimal subgraphs of the complete Euclidean graph. By developing this approach, it is possible to achieve subexponential time $2^{O(\sqrt{n}\log n)}$. The idea is to create the subproblems which are solved by dynamic programming. This is done by trying all (suitable) planar separators of length $O\left(\sqrt{n}\right)$, separating the input point set in a balanced way, and then to proceed recursively within the resulting subproblems.

The second approach uses fixed-parameter algorithms. So, for example, if there are only $O(\log n)$ points in the interior of the convex hull of S, then the MWT of S can be computed in polynomial time [4]. This approach extends also to compute the minimum weight triangulation under the constraint that the outer boundary is not necessarily the convex hull of the input vertices; it can be an arbitrary polygon. Some of these algorithms have been implemented; see Grantson et al. [2] for a comparison of some implementations. These dynamic programming approaches take typically cubic time with respect to the points of the boundary but exponential time with respect to the number of remaining points. So, for example, if k is the number of hole points inside the boundary polygon, then an algorithm, which has also been implemented, can compute the exact MWT in time $O(n^3 \cdot 2^k \cdot k)$ [2].

In an attempt to solve larger problems, a different approach uses properties of MWT which usually help to identify, for random point sets, many edges that *must* be, respectively *cannot* be, in MWT. One can then use dynamic programming to fill in the remaining MWT edges. For random sets consisting of tens of thousands of points from the uniform distribution, one can thus compute the exact MWT in minutes [1].

Applications

The problem of computing a triangulation arises, for example, in finite element analysis, terrain modeling, stock cutting, and numerical approximation [3, 6]. The *minimum weight* triangulation has attracted the attention of many researchers, mainly due to its natural definition of optimality, and because it has proved to be a challenging problem over the past 30 years, with unknown complexity status until the end of 2005.

Open Problems

All results mentioned leave open problems. For example, can one find a simpler proof of NP-completeness, which can be checked without running computer programs? It would be desirable to improve the approximation constant which can be achieved in polynomial time (to simplify the proof, the constant shown in [7] is not explicitly calculated and it would be relatively large, if the proof is not refined). The time bound for the approximation scheme could hopefully be improved. It could also be possible to refine the software which computes efficiently the exact MWT for large random point sets, so that it can handle efficiently a wider range of input, i.e., not only completely random point sets. This could perhaps be done by combining this software with implementations of fixed-parameter algorithms, as the ones reported in [2, 4], or with other approaches. It is also open whether or not the subexponential exact method can be further improved.

Experimental Results

Please see the last paragraph under the section about key results.

URL to Code

Link to code used to compare some dynamic programming approaches in [2]: http://fuzzy.cs.unimagdeburg.de/~borgelt/pointgon.html

Cross-References

- ▶ Greedy Set-Cover Algorithms
- ▶ Minimum Geometric Spanning Trees
- ▶ Minimum k-Connected Geometric Networks

Recommended Reading

1. Beirouti R, Snoeyink J (1998) Implementations of the LMT heuristic for minimum weight triangulation. In: Symposium on computational geometry, Minneapolis, 7–10 June 1998, pp 96-105
2. Borgelt C, Grantson M, Levcopoulos C (2008) Fixed parameter algorithms for the minimum weight triangulation problem. Int J Comput Geom Appl 18(3):185–220
3. de Berg M, van Kreveld M, Overmars M, Schwarzkopf O (2000) Computational geometry – algorithms and applications, 2nd edn. Springer, Heidelberg
4. Grantson M, Borgelt C, Levcopoulos C (2005) Minimum weight triangulation by cutting out triangles. In: Proceedings of the 16th annual international symposium on algorithms and computation (ISAAC 2005), Sanya. Lecture notes in computer science, vol 3827. Springer, Heidelberg, pp 984–994
5. Gudmundsson J, Levcopoulos C (2000) A parallel approximation algorithm for minimum weight triangulation. Nord J Comput 7(1):32–57
6. Hjelle Ø, Dæhlen M (2006) Triangulations and applications. In: Mathematics and visualization, vol IX. Springer, Heidelberg. ISBN:978-3-540-33260-2
7. Levcopoulos C, Krznaric D (1998) Quasi-greedy triangulations approximating the minimum weight triangulation. J Algorithms 27(2):303–338
8. Levcopoulos C, Krznaric D (1999) The greedy triangulation can be computed from the Delaunay triangulation in linear time. Comput Geom 14(4):197–220
9. Levcopoulos C, Lingas A (1987) On approximation behavior of the greedy triangulation for convex polygons. Algorithmica 2:15–193
10. Levcopoulos C, Lingas A (2014) A note on a QPTAS for maximum weight triangulation of planar point sets. Inf Process Lett 114:414–416
11. Lingas A (1998) Subexponential-time algorithms for minimum weight triangulations and related problems. In: Proceedings 10th Canadian conference on computational geometry (CCCG), McGill University, Montreal, 10–12 Aug 1998
12. Mulzer W, Rote G (2006) Minimum-weight triangulation is NP-hard. In: Proceedings of the 22nd annual ACM symposium on computational geometry (SoCG'06), Sedona. ACM, New York. The journal version in J ACM 55(2):Article No. 11 (2008)
13. Remy J, Steger A (2006) A quasi-polynomial time approximation scheme for minimum weight triangulation. In: Proceedings of the 38th ACM symposium on theory of computing (STOC'06), Seattle. ACM, New York. The journal version in J ACM 56(3):Article No. 15 (2009)

Minimum Weighted Completion Time

V.S. Anil Kumar[1], Madha V. Marathe[2], Srinivasan Parthasarathy[2], and Aravind Srinivasan[3]
[1]Virginia Bioinformatics Institute, Virginia Tech, Blacksburg, VA, USA
[2]IBM T.J. Watson Research Center, Hawthorne, NY, USA
[3]Department of Computer Science, University of Maryland, College Park, MD, USA

Keywords

Average weighted completion time

Years and Authors of Summarized Original Work

1999; Afrati et al.

Problem Definition

The minimum weighted completion time problem involves (i) a set J of n jobs, a positive weight w_j for each job $j \in J$, and a release date r_j before which it cannot be scheduled; (ii) a set of m machines, each of which can process at most one job at any time; and (iii) an arbitrary set of positive values $\{p_{i,j}\}$, where $p_{i,j}$ denotes the time to process job j on machine i. A schedule involves assigning jobs to machines and choosing an order in which they are processed. Let C_j denote the completion time of job j for a given schedule. The *weighted completion time* of a schedule is defined as $\sum_{j \in J} w_j C_j$, and the goal is to compute a schedule that has the minimum weighted completion time.

In the scheduling notation introduced by Graham et al. [8], a scheduling problem is denoted by a 3-tuple $\alpha|\beta|\gamma$, where α denotes the machine environment, β denotes the additional constraints on jobs, and γ denotes the objective function. In this article, we will be concerned with the α-values 1, P, R, and Rm, which respectively denote one machine, identical parallel machines (i.e., for a fixed job j and for each machine i, $p_{i,j}$ equals a value p_j that is independent of i), unrelated machines (the $p_{i,j}$'s are dependent on both job i and machine j), and a fixed number m (not part of the input) of unrelated machines. The field β takes on the values r_j, which indicates that the jobs have release dates, and the value *pmtn*, which indicates that preemption of jobs is permitted. Further, the value *prec* in the field β indicates that the problem may involve precedence constraints between jobs, which poses further restrictions on the schedule. The field γ is either $\sum w_j C_j$ or $\sum C_j$, which denote total weighted and total (unweighted) completion times, respectively.

Some of the simpler classes of the weighted completion time scheduling problems admit optimal polynomial-time solutions. They include the problem $P||\sum C_j$, for which the *shortest-job-first* strategy is optimal, the problem $1\ ||\sum w_j C_j$, for which Smith's rule [14] (scheduling jobs in their nondecreasing order of p_j/w_j values) is optimal, and the problem $R||\sum C_j$, which can be solved via matching techniques [3, 10]. With the introduction of release dates, even the simplest classes of the weighted completion time minimization problem becomes strongly nondeterministic polynomial-time (NP)-hard. In this article, we focus on the work of Afrati et al. [1], whose main contribution is the design of polynomial-time approximation schemes (PTASs) for several classes of scheduling problems to minimize weighted completion time *with release dates*. Prior to this work, the best solutions for minimizing weighted completion time with release dates were all $O(1)$-approximation algorithms (e.g., [5, 6, 12]); the only known PTAS for a strongly NP-hard problem involving weighted completion time was due to Skutella and Woeginger [13], who developed a PTAS for the problem $P||\sum w_j C_j$. For an excellent survey on the minimum weighted completion time problem, we refer the reader to Chekuri and Khanna [4]. Another important objective is the flow time, which is a generalization of completion time; a recent breakthrough of Bansal and Kulkarni shows how to approximate the total flow time and maximum flow time to within polylgarithmic factors [2].

Key Results

Afrati et al. [1] were the first to develop PTASs for weighted completion time problems involving release dates. We summarize the running times of these PTASs in Table 1.

The results presented in Table 1 were obtained through a careful sequence of input transformations followed by dynamic programming. The input transformations ensure that the input becomes *well structured* at a slight loss in optimality, while dynamic programming allows efficient enumeration of all the near-optimal solutions to the well-structured instance.

The first step in the input transformation is *geometric rounding*, in which the processing times and release dates are converted to powers of $1 + \epsilon$ with at most $1 + \epsilon$ loss in the overall performance. More significantly, the step (i)

Minimum Weighted Completion Time, Table 1 Summary of results of Afrati et al. [1]

Problem	Running time of polynomial-time approximation schemes
$1\,\lvert r_j \rvert \sum w_j C_j$	$O\left(2^{poly(\frac{1}{\epsilon})} n + n \log n\right)$
$P\lvert r_j \rvert \sum w_j C_j$	$O\left((m+1)^{poly(\frac{1}{\epsilon})}\, n + n \log n\right)$
$P\lvert r_j, pmtn \rvert \sum w_j C_j$	$O\left(2^{poly(\frac{1}{\epsilon})} n + n \log n\right)$
$Rm\,\lvert r_j \rvert \sum w_j C_j$	$O\left(f\left(m, \frac{1}{\epsilon}\right) poly\,(n)\right)$
$Rm\,\lvert r_j, pmtn \rvert \sum w_j C_j$	$O\left(f\left(m, \frac{1}{\epsilon}\right) n + n \log n\right)$
$Rm\,\vert\vert \sum w_j C_j$	$O\left(f\left(m, \frac{1}{\epsilon}\right) n + \log n\right)$

ensures that there are only a small number of distinct processing times and release dates to deal with, (ii) allows time to be broken into geometrically increasing intervals, and (iii) aligns release dates with start and end times of intervals. These are useful properties that can be exploited by dynamic programming.

The second step in the input transformation is *time stretching*, in which small amounts of idle time are added throughout the schedule. This step also changes completion times by a factor of at most $1 + O(\epsilon)$ but is useful for *cleaning up* the scheduling. Specifically, if a job is *large* (i.e., occupies a large portion of the interval where it executes), it can be pushed into the idle time of a later interval where it is small. This ensures that most jobs have small sizes compared with the length of the intervals where they execute, which greatly simplifies schedule computation. The next step is *job shifting*. Consider a partition of the time interval $[0,\infty)$ into intervals of the form $I_x = [(1+\epsilon)^x, (1+\epsilon)^{x+1})$ for integral values of x. The job-shifting step ensures that there is a slightly suboptimal schedule in which every job j gets completed within $O(\log_{1+\epsilon}(1+\frac{1}{\epsilon}))$ intervals after r_j. This has the following nice property: If we consider blocks of intervals $\mathcal{B}_0, \mathcal{B}_1, \ldots$, with each block $\mathcal{B}_i$ containing $O(\log_{1+\epsilon}(1+\frac{1}{\epsilon}))$ consecutive intervals, then a job j starting in block $\mathcal{B}_i$ completes within the next block. Further, the other steps in the job-shifting phase ensure that there are not too many *large* jobs which spill over to the next block; this allows the dynamic programming to be done efficiently.

The precise steps in the algorithms and their analysis are subtle, and the above description is clearly an oversimplification. We refer the reader to [1] or [4] for further details.

Applications

A number of optimization problems in parallel computing and operations research can be formulated as machine scheduling problems. When precedence constraints are introduced between jobs, the weighted completion time objective can generalize the more commonly studied makespan objective and hence is important.

Open Problems

Some of the major open problems in this area are to improve the approximation ratios for scheduling on unrelated or related machines for jobs with precedence constraints. The following problems in particular merit special mention. The best known solution for the $1 \mid prec \mid \sum w_j C_j$ problem is the 2-approximation algorithm due to Hall et al. [9]; improving upon this factor is a major open problem in scheduling theory. The problem $R\,\lvert prec \rvert \sum_j w_j C_j$ in which the precedence constraints form an arbitrary acyclic graph is especially open – the only known results in this direction are when the precedence constraints form chains [7] or trees [11].

The other open direction is inapproximability – there are significant gaps between the known approximation guarantees and hardness factors for various problem classes. For instance, the

$R||\sum w_j C_j$ and $R|r_j|\sum w_j C_j$ are both known to be approximable-hard, but the best known algorithms for these problems (due to Skutella [12]) have approximation ratios of 3/2 and 2, respectively. Closing these gaps remains a significant challenge.

Cross-References

- ▶ Approximation Schemes for Makespan Minimization
- ▶ Flow Time Minimization
- ▶ List Scheduling
- ▶ Minimum Flow Time

Acknowledgments The Research of V.S. Anil Kumar and M.V. Marathe was supported in part by NSF Award CNS-0626964. A. Srinivasan's research was supported in part by NSF Award CNS-0626636.

Recommended Reading

1. Afrati FN, Bampis E, Chekuri C, Karger DR, Kenyon C, Khanna S, Milis I, Queyranne M, Skutella M, Stein C, Sviridenko M (1999) Approximation schemes for minimizing average weighted completion time with release dates. In: Proceedings of the foundations of computer science, pp 32–44
2. Bansal N, Kulkarni J (2014) Minimizing flow-time on unrelated machines. arXiv:1401.7284
3. Bruno JL, Coffman EG, Sethi R (1974) Scheduling independent tasks to reduce mean finishing time. Commun ACM 17:382–387
4. Chekuri C, Khanna S (2004) Approximation algorithms for minimizing weighted completion time. In: Leung JY-T (eds) Handbook of scheduling: algorithms, models, and performance analysis. CRC, Boca Raton
5. Chekuri C, Motwani R, Natarajan B, Stein C (2001) Approximation techniques for average completion time scheduling. SIAM J Comput 31(1):146–166
6. Goemans M, Queyranne M, Schulz A, Skutella M, Wang Y (2002) Single machine scheduling with release dates. SIAM J Discret Math 15:165–192
7. Goldberg LA, Paterson M, Srinivasan A, Sweedyk E (2001) Better approximation guarantees for job-shop scheduling. SIAM J Discret Math 14:67–92
8. Graham RL, Lawler EL, Lenstra JK, Rinnooy Kan AHG (1979) Optimization and approximation in deterministic sequencing and scheduling: a survey. Ann Discret Math 5:287–326
9. Hall LA, Schulz AS, Shmoys DB, Wein J (1997) Scheduling to minimize average completion time: off-line and on-line approximation algorithms. Math Oper Res 22(3):513–544
10. Horn W (1973) Minimizing average flow time with parallel machines. Oper Res 21:846–847
11. Kumar VSA, Marathe MV, Parthasarathy S, Srinivasan A (2005) Scheduling on unrelated machines under tree-like precedence constraints. In: APPROX-RANDOM, pp 146–157
12. Skutella M (2001) Convex quadratic and semidefinite relaxations in scheduling. J ACM 46(2):206–242
13. Skutella M, Woeginger GJ (1999) A PTAS for minimizing the weighted sum of job completion times on parallel machines. In: Proceedings of the 31st annual ACM symposium on theory of computing (STOC'99), pp 400–407
14. Smith WE (1956) Various optimizers for single-stage production. Nav Res Log Q 3:59–66

Min-Sum Set Cover and Its Generalizations

Sungjin Im
Electrical Engineering and Computer Sciences (EECS), University of California, Merced, CA, USA

Keywords

Approximation; Covering problems; Greedy algorithm; Latency; Randomized rounding; Set cover; Submodular

Years and Authors of Summarized Original Work

2004; Feige, Lovász, Tetali
2010; Bansal, Gupta, Krishnaswamy
2011; Azar, Gamzu

Problem Definition

The min sum set cover (MSSC) problem is a latency version of the set cover problem. The input to MSSC consists of a collection of sets $\{S_i\}_{i\in[m]}$ over a universe of elements $[n] := \{1, 2, 3, \ldots, n\}$. The goal is to schedule elements, one at a time, to hit all sets as early on average as possible. Formally, we would like to find a permutation $\pi : [n] \to [n]$ of the elements $[n]$ ($\pi(i)$ is the ith element in the ordering) such that the average (or equivalently total) cover time of the sets $\{S_i\}_{i \in [m]}$ is minimized. The cover time of a set S_i is defined as the earliest time t such that $\pi(t) \in S_i$. For convenience, we will say that we schedule/process element $\pi(i)$ at time i.

Since MSSC was introduced in [4], several generalizations have been studied. Here we discuss two of them. In the generalized min sum set cover (GMSSC) problem [2], each set S_i has a requirement κ_i. In this generalization, a set S_i is covered at the first time t when κ_i elements are scheduled from S_i, i.e., $|\{\pi(1), \pi(2), \ldots, \pi(t)\} \cap S_i| \geq \kappa_i$. Note that MSSC is a special case of GMSSC when $\kappa_i = 1$ for all $i \in [n]$.

Another interesting generalization is submodular ranking (SR) [1]. In SR, each set S_i is replaced with a nonnegative and monotone submodular function $f_i : 2^{[n]} \to [0, 1]$ with $f_i([n]) = 1$; function f is said to be submodular if $f(A \cup B) + f(A \cap B) \leq f(A) + f(B)$ for all $A, B \subseteq [n]$ and monotone if $f(A) \leq f(B)$ for all $A \subseteq B$. The cover time of each function f_i is now defined as the earliest time t such that $f_i(\{\pi(1), \pi(2), \ldots, \pi(t)\}) = 1$. Note that GMSSC is a special case of SR when $f_i(A) = \min\{|S_i \cap A|/\kappa_i, 1\}$. Also it is worth noting that SR generalizes set cover.

Key Results

We summarize main results known for MSSC, GMSSC, and SR.

Theorem 1 ([4]) *There is a 4-approximation for MSSC, and there is a matching lower bound $4 - \epsilon$ unless $P = NP$.*

Interestingly, the tight 4-approximation was achieved by a very simple greedy algorithm that schedules an element at each time that covers the largest number of uncovered sets. The analysis in [4] introduced the notion of "histograms"; see below for more detail.

Theorem 2 ([3, 6, 7]) *There is an $O(1)$-approximation for GMSSC.*

Azar and Gamzu gave the first nontrivial approximation for GMSSC whose guarantee was $O(\log \max_i \kappa_i)$ [2]. The analysis was also based on histograms and was inspired by the work in [4]. Bansal et al. [3] showed that the analysis of the greedy algorithm in [2] is essentially tight and used a linear programming relaxation and randomized rounding to give the first $O(1)$-approximation for GMSSC; the precise approximation factor obtained was 485. The LP used in [3] was a time-indexed LP strengthened with knapsack covering inequalities. The rounding procedure combined threshold rounding and randomized "boosted-up" independent rounding. The approximation was later improved to 28 by [7], subsequently to 12.4 by [6]. The key idea for these improvements was to use α-point rounding to resolve conflicts between elements, which is popular in the scheduling literature.

Theorem 3 ([1, 5]) *There is an $O(\log(1/\epsilon))$-approximation for SR where ϵ is the minimum marginal positive increase of any function f_i.*

Note that this result immediately implies an $O(\log \max_i \kappa_i)$-approximation for GMSSC. The algorithm in [1] is an elegant greedy algorithm which schedules an element e at time t with the maximum $\sum_i (f_i(A \cup \{e\}) - f_i(A))/(1 - f_i(A))$ – here A denotes all elements scheduled by time $t - 1$, and if $f_i(A) = 1$, then f_i is excluded from the summation. Note that this algorithm becomes the greedy algorithm in [4] for the special case of MSSC. The analysis was also based on histograms. Later, Im et al. [5] gave an alternative analysis of this greedy algorithm which was inspired by the analysis of other latency problems. We note that the algorithm that schedules element e that gives the maximum total marginal increase of $\{f_i\}$ has a very poor approximation guarantee, as observed in [1].

As we discussed above, there are largely three analysis techniques used in this line of work: histogram-based analysis, latency argument-based analysis, and LP rounding. We will sketch these techniques following [3–5] closely – we chose these papers since they present the techniques in a relatively simpler way, though they do not necessarily give the best approximation guarantees or most general results. We begin with the analysis tools developed for greedy algorithms. To present key ideas more transparently, we will focus on MSSC.

Histogram-Based Analysis

We sketch the analysis of the 4-approximation in [4]. Let R_t denote the uncovered sets at time t and N_t the sets that are first covered at time t. Observe that $\sum_{t\in[n]} |R_t|$ is the algorithm's total cover time. In the analysis, we represent the optimal and the algorithm's solutions using histograms. First, in the optimal solution's histogram sets are ordered in increasing order of their cover times, and set S_i has width 1 and height equal to its cover time. In the algorithm's solution, as before, sets are ordered in increasing order of their cover times, but set S_i has height equal to $|R_t|/|N_t|$ where t is S_i's cover time. Here, the increase in the algorithm's objective at time t is uniformly distributed to sets N_t that are newly covered at time t. Note that the areas of both histograms are equal to the optimal cost and the algorithm's cost, respectively. Then we can show that after shrinking the algorithm's histogram by a factor of 2, both horizontally and vertically, one can place it completely inside the optimal solution's histogram. This analysis is very simple and is based on a clever observation on the greedy solution's structure. This type of analysis was also used in [1,2].

Latency Argument-Based Analysis

This analysis does not seem to yield tight approximation guarantees, but could be more flexible since it does not compare two histograms directly. The key idea is to show that if we can't charge the number of uncovered sets in our algorithm's schedule at time t to the analogous number in the optimal schedule, then our algorithm must have covered a lot of sets during the time interval $[t/2, t]$. In other words, if our algorithm didn't make enough progress recently, then our algorithm's current status can be shown to be comparable to the optimal solution's status. Intuitively, if our algorithm is not comparable to the optimal solution, then the algorithm can nearly catch up with the optimal solution by following the choices the optimal solution has made. For technical reasons, we may have to compare our algorithm's status to the optimal solution's earlier status. This analysis is easily generalized to GMSSC, SR, and more general metric settings [5].

We now discuss the linear programming-based approach. Bansal et al. discussed why greedy algorithms are unlikely to yield an $O(1)$-approximation for GMSSC [3].

LP and Randomized Rounding

Consider the following time-indexed integer program (IP) used in [3]: variable x_{et} is an indicator variable that is 1 if element e is scheduled at time t, otherwise 0. Variable y_{it} is 1 if S_i is covered by time t, otherwise 0. The IP is relaxed into an LP by allowing x, y to be fractional.

$$
\begin{aligned}
\min \quad & \sum_{t\in[n]} \sum_{i\in[m]} (1 - y_{it}) & \\
\text{s.t.} \quad & \sum_{t\in[n]} x_{et} = 1 & \forall e \in [n] \\
& \sum_{e\in[n]} x_{et} = 1 & \forall t \in [n] \\
& \sum_{e\in S_i \setminus A} \sum_{1\le t'<t} x_{et'} \ge (\kappa_i - |A|) \cdot y_{it} & \forall i \in [m], A \subseteq S_i, t \in [n] \\
& 0 \le y \le 1 & \\
& x \ge 0. &
\end{aligned}
$$

M

Note that for *integral* solutions, the objective is exactly the total cover time since each set S_i uncovered at time t adds 1 to the objective. The first two constraints say that every element must be scheduled and exactly one element must be scheduled at a time. If we use the most natural constraint $\sum_{e \in S_i} \sum_{1 \leq t' < t} x_{et'} \geq \kappa_i y_{it}$ (it says that if set S_i is covered by time t, then there must be at least κ_i elements scheduled from S_i by time t), the LP has a large integrality gap [3]. Hence, [3] strengthened the LP with the above knapsack covering inequalities. There is an easy separation oracle for the last constraint; hence we can solve the LP in polynomial time. The analysis is done by showing that the expected cover time of each set S_i is at most $O(1)$ factor larger than the earliest time τ when the set S_i is covered by the LP by at least a half, i.e., $y_{i\tau} \geq 1/2$. This is sufficient to give an $O(1)$-approximation since the LP pays at least $\tau/2$ for set S_i.

Applications

The MSSC problem and its closely related problems have various applications in adaptive query processing and distributed resource allocation problems. Also GMSSC has applications in Web page ranking and broadcast scheduling. For details, see [1, 4]. Perhaps it would be no stretch to say that min sum set cover problems are at least loosely connected to all problems whose goal is to satisfy multiple demands with the overall minimum latency.

Open Problems

An outstanding open problem is to settle the approximability of GMSSC. As mentioned before, GMSSC captures MSSC (all $\kappa_i = 1$), for which there is a tight 4-approximation known [4]. The other extreme case is when $\kappa_i = |S_i|$ for all i. This problem is essentially equivalent to a classic precedence constrained scheduling problem $1|\text{prec}|\sum_j w_j C_j$ for which there are several 2-approximations known; see [3] for pointers. However, the current best approximation factor known for GMSSC is 12.4. Im et al. conjectured that GMSSC admit a 4-approximation [6].

Recommended Reading

1. Azar Y, Gamzu I (2011) Ranking with submodular valuations. In: SODA, San Francisco, pp 1070–1079
2. Azar Y, Gamzu I, Yin X (2009) Multiple intents reranking. In: STOC, Bethesda, pp 669–678
3. Bansal N, Gupta A, Krishnaswamy R (2010) A constant factor approximation algorithm for generalized min-sum set cover. In: SODA, Austin, pp 1539–1545
4. Feige U, Lovász L, Tetali P (2004) Approximating min sum set cover. Algorithmica 40(4):219–234
5. Im S, Nagarajan V, van der Zwaan R (2012) Minimum latency submodular cover. In: ICALP (1), Warwick, pp 485–497
6. Im S, Sviridenko M, Zwaan R (2014) Preemptive and non-preemptive generalized min sum set cover. Math Program 145(1–2):377–401
7. Skutella M, Williamson DP (2011) A note on the generalized min-sum set cover problem. Oper Res Lett 39(6):433–436

Misra-Gries Summaries

Graham Cormode
Department of Computer Science, University of Warwick, Coventry, UK

Keywords

Approximate counting; Frequent items; Streaming algorithms

Years and Authors of Summarized Original Work

1982; Misra, Gries

Problem Definition

The frequent items problem is to process a stream of items and find all items occurring more than a given fraction of the time. It is one of the most heavily studied problems in data stream algorithms, dating back to the 1980s. Many applications rely directly or indirectly on finding the frequent items, and implementations are in use in large-scale industrial systems. Informally, given a sequence of items, the problem is simply to find those items which occur most frequently.

Typically, this is formalized as finding all items whose frequency exceeds a specified fraction of the total number of items. Variations arise when the items have weights and further when these weights can also be negative.

Definition 1 Given a stream $\mathcal{S}$ of n items $t_1 \ldots t_n$, the frequency of an item i is $f_i = |\{j | t_j = i\}|$. The exact ϕ-frequent items comprise the set $\{i | f_i > \phi n\}$.

Example The stream $\mathcal{S} = (a, b, a, c, c, a, b, d)$ has $f_a = 3, f_b = 2, f_c = 2, f_d = 1$. For $\phi = 0.2$, the frequent items are a, b, and c.

A streaming algorithm which solves this problem must use a linear amount of space, even for large values of ϕ: given an algorithm that claims to solve this problem, we could insert a set S of N items, where every item has frequency 1. Then, we could also insert N copies of item i. If i is then reported as a frequent item (occurring more than 50 % of the time), then $i \in S$, else $i \notin S$. Consequently, since set membership requires $\Omega(N)$ space, $\Omega(N)$ space is also required to solve the frequent items problem. Instead, an approximate version is defined based on a tolerance for error ϵ.

Definition 2 Given a stream $\mathcal{S}$ of n items, the ϵ-approximate frequent items problem is to return a set of items F so that for all items $i \in F$, $f_i > (\phi - \epsilon)n$, and there is no $i \notin F$ such that $f_i > \phi n$.

Since the exact ($\epsilon = 0$) frequent items problem is hard in general, we will use "frequent items" or "the frequent items problem" to refer to the ϵ-approximate frequent items problem. A related problem is to estimate the frequency of items on demand.

Definition 3 Given a stream $\mathcal{S}$ of n items defining frequencies f_i as above, the frequency estimation problem is to process a stream so that, given any i, an $\hat{f}_i$ is returned satisfying $\hat{f}_i \leq f_i \leq \hat{f}_i + \epsilon n$.

Key Results

The problem of frequent items dates back at least to a problem first studied by Moore in 1980 [5]. It was published as a "problem" in the Journal of Algorithms in the June 1981 issue [17], to determine if there was a majority choice in a list of n votes.

Preliminaries: The Majority Algorithm

In addition to posing the majority question as a problem, Moore also invented the MAJORITY algorithm along with Boyer in 1980, described in a technical report from early 1981 [4]. A similar solution with proof of the optimal number of comparisons was provided by Fischer and Salzburg [9]. MAJORITY can be stated as follows: store the first item and a counter, initialized to 1. For each subsequent item, if it is the same as the currently stored item, increment the counter. If it differs and the counter is zero, then store the new item and set the counter to 1; else, decrement the counter. After processing all items, the algorithm guarantees that if there is a majority vote, then it must be the item stored by the algorithm. The correctness of this algorithm is based on a pairing argument: if every non-majority item is paired with a majority item, then there should still remain an excess of majority items. Although not posed as a streaming problem, the algorithm has a streaming flavor: it takes only one pass through the input (which can be ordered arbitrarily) to find a majority item. To verify that the stored item really is a majority, a second pass is needed to simply count the true number of occurrences of the stored item.

Algorithm 1: MISRA-GRIES(k)

```
n ← 0; T ← ∅;
for each i :
do { n ← n + 1;
     if i ∈ T
       then c_i ← c_i + 1;
     else if |T| < k − 1
       then { T ← T ∪ {i};
              c_i ← 1;
     else for all j ∈ T
       do { c_j ← c_j − 1;
            if c_j = 0
              then T ← T\{j};
```

M

Misra-Gries Summary

The Misra-Gries summary is a simple algorithm that solves the frequent items problem. It can be viewed as a generalization of MAJORITY to track multiple frequent items.

Instead of keeping a single counter and item from the input, the MISRA-GRIES summary stores $k - 1$ (item, counter) pairs. The natural generalization of MAJORITY is to compare each new item against the stored items T and increment the corresponding counter if it is among them. Else, if there is some counter with count zero, it is allocated to the new item and the counter set to 1. If all $k - 1$ counters are allocated to distinct items, then all are decremented by 1. A grouping argument is used to argue that any item which occurs more than n/k times must be stored by the algorithm when it terminates. Example pseudocode to illustrate this algorithm is given in Algorithm 1, making use of set notation to represent the operations on the set of stored items T: items are added and removed from this set using set union and set subtraction, respectively, and we allow ranging over the members of this set (thus, implementations will have to choose appropriate data structures which allow the efficient realization of these operations). We also assume that each item j stored in T has an associated counter c_j. For items not stored in T, then c_j is defined as 0 and does not need to be explicitly stored.

This n/k generalization was first proposed by Misra and Gries [16]. The time cost of the algorithm is dominated by the $O(1)$ dictionary operations per update and the cost of decrementing counts. Misra and Gries use a balanced search tree and argue that the decrement cost is amortized $O(1)$; Karp et al. propose a hash table to implement the dictionary [11]; and Demaine et al. show how the cost of decrementing can be made worst case $O(1)$ by representing the counts using offsets and maintaining multiple linked lists [8].

Bose et al. [3] observed that executing this algorithm with $k = 1/\epsilon$ ensures that the count associated with each item on termination is at most ϵn below the true value. The bounds on the accuracy of the structure were tightened by Berinde et al. to show that the error depends only on the "tail": the total weight of items outside the top-k most frequent, rather than the total weight of all items [2]. This gives a stronger accuracy guarantee when the input distribution is skewed, for example, if the frequencies follow a Zipfian distribution. They also show that the algorithm can be altered to tolerate updates with weights, rather than assuming that each item has equal unit weight.

A similar data structure called SPACESAVING was introduced by Metwally et al. [15]. This structure also maintains a set of items and counters, but follows a different set of update rules. Recently, it was shown that the SPACESAVING structure is isomorphic to MISRA-GRIES: the state of both structures can be placed in correspondence as each update arrives [1]. The different representations reflect that SPACESAVING maintains an upper bound on the count of stored items, while MISRA-GRIES keeps a lower bound. In studies, the upper bound tends to be closer to the true count, but it is straightforward to switch between the two representations.

Moreover, Agarwal et al. [1] showed that the MISRA-GRIES summary is *mergeable*. That is, two summaries of different inputs of size k can be combined together to obtain a new summary of size k that summarizes the union of the two inputs. This merging can be done repeatedly, to summarize arbitrarily many inputs in arbitrary configurations. This allows the summary to be used in distributed and parallel environments.

Lastly, the concept behind the algorithm of tracking information on k representative elements has inspired work in other settings. Liberty [12] showed how this can be used to track an approximation to the best k-rank summary of a matrix, using k rows. This was extended by Ghashami and Phillips [10] to offer better accuracy by keeping more rows.

Applications

The question of tracking approximate counts for a large number of possible objects arises in a number of settings. Many applications have arisen in the context of the Internet, such as tracking the most popular source, destinations, or source-destination pairs (those with the highest amount of traffic) or tracking the most popular objects, such as the most popular queries to a search engine, or the most popular pieces of content in a large content host. It forms the basis of other problems, such as finding the frequent itemsets within a stream of transactions: those subsets of items which occur as a subset of many transactions. Solutions to this problem have used ideas similar to the count and prune strategy of the Misra-Gries summary to find approximate frequent itemsets [14]. Finding approximate counts of items is also needed within other stream algorithms, such as approximating the entropy of a stream [6].

Experimental Results

There have been a number of experimental studies of Misra-Gries and related algorithms, for a variety of computing models. These have shown that the algorithm is accurate and fast to execute [7, 13].

URLs to Code and Data Sets

Code for this algorithm is widely available:

http://www.cs.rutgers.edu/~muthu/massdal-code-index.html
http://hadjieleftheriou.com/sketches/index.html
https://github.com/cpnielsen/twittertrends

Cross-References

▸ Count-Min Sketch

Recommended Reading

1. Agarwal P, Cormode G, Huang Z, Phillips J, Wei Z, Yi K (2012) Mergeable summaries. In: ACM principles of database systems, Scottsdale
2. Berinde R, Cormode G, Indyk P, Strauss M (2009) Space-optimal heavy hitters with strong error bounds. In: ACM principles of database systems, Providence
3. Bose P, Kranakis E, Morin P, Tang Y (2003) Bounds for frequency estimation of packet streams. In: SIROCCO, Umeå
4. Boyer B, Moore J (1981) A fast majority vote algorithm. Technical report ICSCA-CMP-32, Institute for Computer Science, University of Texas
5. Boyer RS, Moore JS (1991) MJRTY – a fast majority vote algorithm. In: Bledsoe WW, Boyer RS (eds) Automated reasoning: essays in honor of Woody Bledsoe. Automated reasoning series. Kluwer Academic, Dordrecht/Boston, pp 105–117
6. Chakrabarti A, Cormode G, McGregor A (2007) A near-optimal algorithm for computing the entropy of a stream. In: ACM-SIAM symposium on discrete algorithms, New Orleans
7. Cormode G, Hadjieleftheriou M (2009) Finding the frequent items in streams of data. Commun ACM 52(10):97–105
8. Demaine E, López-Ortiz A, Munro JI (2002) Frequency estimation of internet packet streams with limited space. In: European symposium on algorithms (ESA), Rome
9. Fischer M, Salzburg S (1982) Finding a majority among n votes: solution to problem 81-5. J Algorithms 3(4):376–379
10. Ghashami M, Phillips JM (2014) Relative errors for deterministic low-rank matrix approximations. In: ACM-SIAM symposium on discrete algorithms, Portland, pp 707–717
11. Karp R, Papadimitriou C, Shenker S (2003) A simple algorithm for finding frequent elements in sets and bags. ACM Trans Database Syst 28: 51–55
12. Liberty E (2013) Simple and deterministic matrix sketching. In: ACM SIGKDD, Chicago, pp 581–588
13. Manerikar N, Palpanas T (2009) Frequent items in streaming data: an experimental evaluation of the state-of-the-art. Data Knowl Eng 68(4): 415–430
14. Manku G, Motwani R (2002) Approximate frequency counts over data streams. In: International conference on very large data bases, Hong Kong, pp 346–357
15. Metwally A, Agrawal D, Abbadi AE (2005) Efficient computation of frequent and top-k elements in data streams. In: International conference on database theory, Edinburgh
16. Misra J, Gries D (1982) Finding repeated elements. Sci Comput Program 2:143–152
17. Moore J (1981) Problem 81-5. J Algorithms 2:208–209

Mobile Agents and Exploration

Evangelos Kranakis[1] and Danny Krizanc[2]
[1]Department of Computer Science, Carleton, Ottawa, ON, Canada
[2]Department of Computer Science, Wesleyan University, Middletown, CT, USA

Keywords

Distributed algorithms; Graph exploration; Mobile agent; Navigation; Rendezvous; Routing; Time/Memory tradeoffs

Years and Authors of Summarized Original Work

1952; Shannon

Problem Definition

How can a network be explored efficiently with the help of mobile agents? This is a very broad question and to answer it adequately it will be necessary to understand more precisely what mobile agents are, what kind of networked environment they need to probe, and what complexity measures are interesting to analyze.

Mobile Agents

Mobile agents are autonomous, intelligent computer software that can move within a network. They are modeled as automata with limited memory and computation capability and are usually employed by another entity (to which they must report their findings) for the purpose of collecting information. The actions executed by the mobile agents can be discrete or continuous and transitions from one state to the next can be either deterministic or non-deterministic, thus giving rise to various natural complexity measures depending on the assumptions being considered.

Network Model

The network model is inherited directly from the theory of distributed computing. It is a connected graph whose vertices comprise the computing nodes and edges correspond to communication links. It may be static or dynamic and its resources may have various levels of accessibility. Depending on the model being considered, nodes and links of the network may have distinct labels. A particularly useful abstraction is an anonymous network whereby the nodes have no identities, which means that an agent cannot distinguish two nodes except perhaps by their degree. The outgoing edges of a node are usually thought of as distinguishable but an important distinction can be made between a globally consistent edge-labeling versus a locally independent edge-labeling.

Efficiency Measures for Exploration

Efficiency measures being adopted involve the time required for completing the exploration task, usually measured either by the number of edge traversals or nodes visited by the mobile agent. The interplay between time required for exploration and memory used by the mobile agent (*time/memory tradeoffs*) are key parameters considered for evaluating algorithms. Several researchers impose no restrictions on the memory but rather seek algorithms minimizing exploration time. Others, investigate the minimum size of memory which allows for exploration of a given type of network (e.g., tree) of given (known or unknown) size, regardless of the exploration time. Finally, several researchers consider time/memory tradeoffs.

Main Problems

Given a model for both the agents and the network, the graph exploration problem is that of designing an algorithm for the agent that allows it to visit all of the nodes and/or edges of the network. A closely related problem is where the domain to be explored is presented as a region of the plane with obstacles and exploration becomes visiting all unobstructed portions of the region in the sense of visibility. Another related problem is that of rendezvous where two or more agents are required to gather at a single node of a network.

Key Results

Claude Shannon [17] is credited with the first finite automaton algorithm capable of exploring an arbitrary maze (which has a range of 5×5 squares) by trial and error means. Exploration problems for mobile agents have been extensively studied in the scientific literature and the reader will find a useful historical introduction in Fraigniaud et al. [11].

Exploration in General Graphs

The network is modeled as a graph and the agent can move from node to node only along the edges. The graph setting can be further specified in two different ways. In Deng and Papadimitriou [8] the agent explores strongly connected directed graphs and it can move only in the direction from head to tail of an edge, but not vice-versa. At each point, the agent has a map of all nodes and edges visited and can recognize if it sees them again. They minimize the ratio of the total number of edges traversed divided by the optimum number of traversals, had the agent known the graph. In Panaite and Pelc [15] the explored graph is undirected and the agent can traverse edges in both directions. In the graph setting it is often required that apart from completing exploration the agent has to draw a map of the graph, i.e., output an isomorphic copy of it. Exploration of directed graphs assuming the existence of labels is investigated in Albers and Henzinger [1] and Deng and Papadimitriou [8]. Also in Panaite and Pelc [15], an exploration algorithm is proposed working in time $e + O(n)$, where is n the number of nodes and e the number of links. Fraigniaud et al. [11] investigate memory requirements for exploring unknown graphs (of unknown size) with unlabeled nodes and locally labeled edges at each node. In order to explore all graphs of diameter D and max degree d a mobile agent needs $\Omega(D \log d)$ memory bits even when exploration is restricted to planar graphs. Several researchers also investigate exploration of anonymous graphs in which agents are allowed to drop and remove pebbles. For example in Bender et al. [4] it is shown that one pebble is enough for exploration, if the agent knows an upper bound on the size of the graph, and $\Theta(\log \log n)$ pebbles are necessary and sufficient otherwise.

Exploration in Trees

In this setting it is assumed the agent can distinguish ports at a node (locally), but there is no global orientation of the edges and no markers available. *Exploration with stop* is when the mobile agent has to traverse all edges and stop at some node. For *exploration with return* the mobile agent has to traverse all edges and stop at the starting node. In *perpetual exploration* the mobile agent has to traverse all edges of the tree but is not required to stop. The upper and lower bounds on memory for the exploration algorithms analyzed in Diks et al. [9] are summarized in the table, depending on the knowledge that the mobile agent has. Here, n is the number of nodes of the tree, $N \geq n$ is an upper bound known to the mobile agent, and d is the maximum degree of a node of the tree.

Exploration	Knowledge	Lower bounds	Upper bounds
Perpetual	$\emptyset$	None	$O(\log d)$
w/Stop	$n \leq N$	$\Omega(\log \log \log n)$	$O(\log N)$
w/Return	$\emptyset$	$\Omega(\log n)$	$O(\log^2 n)$

Exploration in a Geometric Setting

Exploration in a geometric setting with unknown terrain and convex obstacles is considered by Blum et al. [5]. They compare the distance walked by the agent (or robot) to the length of the shortest (obstacle-free) path in the scene and describe and analyze robot strategies that minimize this ratio for different kinds of scenes. There is also related literature for exploration in more general settings with polygonal and rectangular obstacles by Deng et al. [7] and Bar-Eli et al. [3], respectively. A setting that is important in wireless networking is when nodes are aware of their location. In this case, Kranakis et al. [12] give efficient algorithms for navigation, namely compass routing and face routing that guarantee delivery in Delaunay and arbitrary planar geometric graphs, respectively, using only local information.

Rendezvous

The rendezvous search problem differs from the exploration problem in that it concerns two searchers placed at different nodes of a graph that want to minimize the time required to rendezvous (usually) at the same node. At any given time the mobile agents may occupy a vertex of the graph and can either stay still or move from vertex to vertex. It is of interest to minimize the time required to rendezvous. A natural extension of this problem is to study multi-agent mobile systems. More generally, given a particular agent model and network model, a set of agents distributed arbitrarily over the nodes of the network are said to rendezvous if executing their programs after some finite time they all occupy the same node of the network at the same time. Of special interest is the highly symmetric case of anonymous agents on an anonymous network and the simplest interesting case is that of two agents attempting to rendezvous on a ring network. In particular, in the model studied by Sawchuk [16] the agents cannot distinguish between the nodes, the computation proceeds in synchronous steps, and the edges of each node are oriented consistently. The table summarizes time/memory tradeoffs known for six algorithms (see Kranakis et al. [13] and Flocchini et al. [10]) when the k mobile agents use indistinguishable pebbles (one per mobile agent) to mark their position in an n node ring.

Memory	Time	Memory	Time
$O(k \log n)$	$O(n)$	$O(\log n)$	$O(n)$
$O(\log n)$	$O(kn)$	$O(\log k)$	$O(n)$
$O(k \log \log n)$	$O\left(\frac{n \log n}{\log \log n}\right)$	$O(\log k)$	$O(n \log k)$

Kranakis et al. [14] show a striking computational difference for rendezvous in an oriented, synchronous, $n \times n$ torus when the mobile agents may have more indistinguishable tokens. It is shown that two agents with a constant number of unmovable tokens, or with one movable token each cannot rendezvous if they have $o(\log n)$ memory, while they can perform rendezvous with detection as long as they have one unmovable token and $O(\log n)$ memory. In contrast, when two agents have two movable tokens each then rendezvous (respectively, rendezvous with detection) is possible with constant memory in a torus. Finally, two agents with three movable tokens each and constant memory can perform rendezvous with detection in a torus. If the condition on synchrony is dropped the rendezvous problem becomes very challenging. For a given initial location of agents in a graph, De Marco et al. [6] measure the performance of a rendezvous algorithm as the number of edge traversals of both agents until rendezvous is achieved. If the agents are initially situated at a distance D in an infinite line, they give a rendezvous algorithm with cost $O(D|L_{\min}|^2)$ when D is known and $O((D + |L_{\max}|)^3)$ if D is unknown, where $|L_{\min}|$ and $|L_{\max}|$ are the lengths of the shorter and longer label of the agents, respectively. These results still hold for the case of the ring of unknown size but then they also give an optimal algorithm of cost $O(n|L_{\min}|)$, if the size n of the ring is known, and of cost $O(n|L_{\max}|)$, if it is unknown. For arbitrary graphs, they show that rendezvous is feasible if an upper bound on the size of the graph is known and they give an optimal algorithm of cost $O(D|L_{\min}|)$ if the topology of the graph and the initial positions are known to the agents.

Applications

Interest in mobile agents has been fueled by two overriding concerns. First, to simplify the complexities of distributed computing, and second to overcome the limitations of user interface approaches. Today they find numerous applications in diverse fields such as distributed problem solving and planning (e.g., task sharing and coordination), network maintenance (e.g., daemons in networking systems for carrying out tasks like monitoring and surveillance), electronic commerce and intelligence search (e.g., data mining and surfing crawlers to find products and services from multiple sources), robotic exploration (e.g., rovers, and other mobile platforms that can explore potentially dangerous environments or even enhance planetary extravehicular activity), and distributed rational decision making (e.g., auction protocols, bargaining, decision making). The interested reader can find useful information in several articles in the volume edited by Weiss [18].

Open Problems

Specific directions for further research would include the study of time/memory tradeoffs in search game models (see Alpern and Gal [2]). Multi-agent systems are particularly useful for content-based searches and exploration, and further investigations in this area would be fruitful. Memory restricted mobile agents provide a rich model with applications in sensor systems. In the geometric setting, navigation and routing in a three dimensional environment using only local information is an area with many open problems.

Cross-References

▸ Deterministic Searching on the Line
▸ Robotics
▸ Routing

Recommended Reading

1. Albers S, Henzinger MR (2000) Exploring unknown environments. SIAM J Comput 29:1164–1188
2. Alpern S, Gal S (2003) The theory of search games and rendezvous. Kluwer, Norwell
3. Bar-Eli E, Berman P, Fiat A, Yan R (1994) On-line navigation in a room. J Algorithms 17:319–341
4. Bender MA, Fernandez A, Ron D, Sahai A, Vadhan S (1998) The power of a pebble: exploring and mapping directed graphs. In: Proceedings of the 30th annual symposium on theory of computing, Dallas, 23–26 May 1998, pp 269–278
5. Blum A, Raghavan P, Schieber B (1997) Navigating in unfamiliar geometric terrain. SIAM J Comput 26:110–137
6. De Marco G, Gargano L, Kranakis E, Krizanc D, Pelc A, Vaccaro U (2006) Asynchronous deterministic rendezvous in graphs. Theor Comput Sci 355: 315–326
7. Deng X, Kameda T, Papadimitriou CH (1998) How to learn an unknown environment I: the rectilinear case. J ACM 45:215–245
8. Deng X, Papadimitriou CH (1999) Exploring an unknown graph. J Graph Theory 32:265–297
9. Diks K, Fraigniaud P, Kranakis E, Pelc A (2004) Tree exploration with little memory. J Algorithms 51: 38–63
10. Flocchini P, Kranakis E, Krizanc D, Santoro N, Sawchuk C (2004) Multiple mobile agent rendezvous in the ring. In: Proceedings of the LATIN 2004, Bueons Aires, 5–8 Apr 2004. LNCS, vol 2976, pp 599–608
11. Fraigniaud P, Ilcinkas D, Peer G, Pelc A, Peleg D (2005) Graph exploration by a finite automaton. Theor Comput Sci 345:331–344
12. Kranakis E, Singh H, Urrutia J (1999) Compass routing in geometric graphs. In: Proceedings of the 11th Canadian conference on computational geometry (CCCG-99), Vancouver, 15–18 Aug 1999, pp 51–54
13. Kranakis E, Krizanc D, Santoro N, Sawchuk C (2003) Mobile agent rendezvous search problem in the ring. In: Proceedings of the international conference on distributed computing systems (ICDCS), Providence, 19–22 May 2003, pp 592–599
14. Kranakis E, Krizanc D, Markou E (2006) Mobile agent rendezvous in a synchronous torus. In: Correa J, Hevia A, Kiwi M (eds) Proceedings of LATIN 2006, 7th Latin American symposium, Valdivia, 20–24 March 2006. SVLNCS, vol 3887, pp 653–664
15. Panaite P, Pelc A (1999) Exploring unknown undirected graphs. J Algorithms 33:281–295
16. Sawchuk C (2004) Mobile agent rendezvous in the ring. PhD thesis, Carleton University, Ottawa
17. Shannon C (1951) Presentation of a maze solving machine, in cybernetics, circular, causal and feedback machines in biological and social systems. In: von Feerster H, Mead M, Teuber HL (eds) Trans. 8th Conf, New York, March 15–16, 1951, pp 169–181. Josiah Mary Jr. Foundation, New York (1952)
18. Weiss G (ed) (1999) Multiagent systems: a modern approach to distributed artificial intelligence. MIT, Cambridge, MA

Model Checking with Fly-Automata

Bruno Courcelle and Iréne Durand
Laboratoire Bordelais de Recherche en Informatique (LaBRI), CNRS, Bordeaux University, Talence, France

Keywords

Automaton on terms; Clique-width; Fly-automaton; Fixed-parameter tractable algorithm; Model checking; Monadic second-order logic; Tree-width

Years and Authors of Summarized Original Work

2012, 2013; Courcelle, Durand

Problem Definition

The verification of monadic second-order (MSO) graph properties, equivalently, the model-checking problem for MSO logic over finite binary relational structures, is fixed-parameter tractable (FPT) where the parameter consists of the formula that expresses the property and the tree-width or the clique-width of the input graph or structure. How to build usable algorithms for this problem? The proof of the general theorem (an *algorithmic meta-theorem*, cf. [12]) is based on the description of the input by algebraic terms and the construction of finite automata that accept the terms describing the satisfying inputs. But these automata are in practice much too large to be constructed [11, 14]. A typical number of states is $2^{2^{10}}$, and lower bounds match this number. Can one use automata and overcome this difficulty?

Key Results

We propose to use *fly-automata* (*FA*) [3]. They are automata whose states are *described* and not *listed* and whose transitions are *computed on the fly* and not *tabulated*. When running on a term of size 1,000, a fly-automaton with $2^{2^{10}}$ states computes only 1,000 transitions if it is deterministic. FA can have infinitely many states. For example, a state can record, among other things, the (unbounded) number of occurrences of a particular symbol in the input term. FA can thus check certain graph properties that are *not monadic second-order expressible*. An example is *regularity*, the fact that all vertices have the same degree. Furthermore, an FA equipped with an *output function* that maps the set of accepting states to an effectively given domain $\mathcal{D}$ can compute a value, for example, the number of k-colorings of the given graph G or the minimum cardinality of one of the k color classes if G is k-colorable (this number measures how close this graph is to be $(k-1)$-colorable). We have implemented and tested an FA that computes the number of 3-colorings of a graph.

Tree-width and *clique-width* are graph complexity measures that serve as parameters in many FPT algorithms [7, 8, 10]. Both are based on hierarchical decompositions of graphs that can be expressed by terms written with the operation symbols of appropriate *graph algebras* [6]. The model-checking automata take such terms as inputs. We will present results concerning graphs of bounded clique-width. The similar results for graphs of bounded tree-width reduce to them as we will explain at the end of this section.

Graph Algebras and Monadic Second-Order Logic

Graphs are finite, undirected, and without loops and multiple edges. The extension to directed graphs, possibly with loops and/or labels, is straightforward. A graph G is identified with the relational structure $\langle V_G, edg_G \rangle$ where edg_G is a binary symmetric relation representing adjacency.

Rather than giving a formal definition of *monadic second-order* (MSO) logic, we present the closed formula expressing 3-colorability (an NP-complete property). It is $\exists X, Y.Col(X, Y)$ where $Col(X, Y)$ is the formula

$$X \cap Y = \emptyset \wedge \forall u, v.\{edg(u, v) \Longrightarrow [\neg(u \in X \wedge v \in X) \wedge \neg(u \in Y \wedge v \in Y) \wedge \neg(u \notin X \cup Y \wedge v \notin X \cup Y)]\}.$$

This formula expresses that X, Y and $V_G - (X \cup Y)$ are the three color classes of a 3-coloring. The corresponding colors are respectively 1, 2, and 3.

Definition 1 (The graph algebra $\mathcal{G}$)

(a) We will use $\mathbb{N}_+$ as a set of labels called *port labels*. A *p-graph* is a triple $G = \langle V_G, edg_G, \pi_G \rangle$ where π_G is a mapping: $V_G \rightarrow \mathbb{N}_+$. If $\pi_G(x) = a$, we say that x is an *a-port*. The set $\pi(G)$ of port labels of G is its *type*. By using a default label, say 1, we make every nonempty graph into a p-graph of type $\{1\}$.

(b) We let F_k be the following finite set of *operations* on p-graphs of type included in $C := \{1, \ldots, k\} \subseteq \mathbb{N}_+$:
- The binary symbol $\oplus$ denotes the union of two *disjoint* p-graphs,
- The unary symbol $relab_{a \rightarrow b}$ denotes the *relabelling* that changes every port label a into b (where $a, b \in C$),
- The unary symbol $add_{a,b}$, for $a < b$, $a, b \in C$, denotes the *edge addition* that adds an edge between every a-port x and every b-port y (unless there is already an edge between them, our graphs have no multiple edges),
- For each $a \in C$, the nullary symbol $\mathbf{a}$ denotes an isolated a-port.

(c) Every term t in $T(F_k)$ (the set of finite terms written with F_k) is called a *k-expression*. Its *value* is a p-graph, $val(t)$, that we now define. For each position u of t (equivalently, each node u of the syntax tree of t), we define a p-graph $val(t)/u$, whose vertex set is the set of leaves of t below u. The definition of $val(t)/u$ is, for fixed t, by bottom-up induction on u:
- If u is an occurrence of $\mathbf{a}$, then $val(t)/u$ has vertex u as an a-port and no edge,
- If u is an occurrence of $\oplus$ with sons u_1 and u_2, then $val(t)/u := val(t)/u_1 \oplus val(t)/u_2$ (note that $val(t)/u_1$ and $val(t)/u_2$ are disjoint),
- If u is an occurrence of $relab_{a \rightarrow b}$ with son u_1, then $val(t)/u := relab_{a \rightarrow b}(val(t)/u_1)$,
- If u is an occurrence of $add_{a,b}$ with son u_1, then $val(t)/u := add_{a,b}(val(t)/u_1)$.

Finally, $val(t) := val(t)/root_t$. Its vertex set is the set of all leaves (occurrences of nullary symbols). For an example, let

$$t := add^1_{b,c}(add^2_{a,b}(\mathbf{a}^3 \oplus^4 \mathbf{b}^5) \oplus^6 relab^7_{b \rightarrow c}$$
$$(add^8_{a,b}(\mathbf{a}^9 \oplus^{10} \mathbf{b}^{11})))$$

where the superscripts 1–11 number the positions of t. The p-graph $val(t)$ is $3_a - 5_b - 11_c - 9_a$ where the subscripts a, b, c indicate the port labels. (For clarity, port labels are letters in examples.) If $u := 2$ and $w := 8$, then $t/u = t/w = add_{a,b}(\mathbf{a} \oplus \mathbf{b})$; however, $val(t)/u$ is the p-graph $3_a - 5_b$ and $val(t)/w$ is $9_a - 11_b$, isomorphic to $val(t)/u$.

(d) The *clique-width* of a graph G, denoted by $cwd(G)$, is the least integer k such that G is isomorphic to $val(t)$ for some t in $T(F_k)$. We denote by $\mathcal{G}_k$ the set $val(T(F_k))$ of p-graphs that are the value of a term over F_k. We let F be the union of the sets F_k and $\mathcal{G}$ be the union of the sets $\mathcal{G}_k$. Every p-graph is isomorphic to a graph in $\mathcal{G}$, hence, has a clique-width.

(e) An *F-congruence* is an equivalence relation $\approx$ on p-graphs such that:
- Two isomorphic p-graphs are equivalent, and
- If $G \approx G'$ and $H \approx H'$, then $\pi(G) = \pi(G')$, $G \oplus H \approx G' \oplus H'$, $add_{a,b}(G) \approx add_{a,b}(G')$ and $relab_{a \rightarrow b}(G) \approx relab_{a \rightarrow b}(G')$.

(f) A set of graphs L is *recognizable* if it is a union of classes of an F-congruence such that, for each finite type $C \subseteq \mathbb{N}_+$, the number of equivalence classes of p-graphs of type C is finite.

Definition 2 (Fly-automata)

(a) Let H be a finite or countable, effectively given, signature. A *fly-automaton over H* (in short, an *FA over H*) is a 4-tuple $\mathcal{A} = \langle H, Q_{\mathcal{A}}, \delta_{\mathcal{A}}, Acc_{\mathcal{A}} \rangle$ such that $Q_{\mathcal{A}}$ is the finite or countable, effectively given, set of *states*; $Acc_{\mathcal{A}}$ is the set of *accepting states,* a decidable subset of $Q_{\mathcal{A}}$; and $\delta_{\mathcal{A}}$ is a computable function that defines the *transition rules*: for each tuple $(f, q_1, \ldots, q_m)$ with $q_1, \ldots, q_m \in Q_{\mathcal{A}}$, $f \in H$, $\rho(f) = m \geq 0$, $\delta_{\mathcal{A}}(f, q_1, \ldots, q_m)$ is a finite set of states. We write $f[q_1, \ldots, q_m] \rightarrow q$ (and $f \rightarrow q$ if f is nullary) to mean that $q \in \delta_{\mathcal{A}}(f, q_1, \ldots, q_m)$. We say that $\mathcal{A}$ is *finite* if H and $Q_{\mathcal{A}}$ are finite.

(b) *Runs* and *recognized languages* are defined as usual; see [1]. A *deterministic* FA $\mathcal{A}$ (by "deterministic" we mean "deterministic and complete") has a unique run on each term t, and $q_{\mathcal{A}}(t)$ is the state reached at the root of t. The mapping $q_{\mathcal{A}}$ is computable, and the membership in $L(\mathcal{A})$ of a term $t \in T(H)$ is decidable.

(c) Every FA $\mathcal{A}$ that is not deterministic can be *determinized* by an easy extension of the usual construction, see [3]; it is important that the sets $\delta_{\mathcal{A}}(f, q_1, \ldots, q_m)$ be finite.

(d) A deterministic FA over H *with output function* is a 4-tuple $\mathcal{A} = \langle H, Q_{\mathcal{A}}, \delta_{\mathcal{A}}, Out_{\mathcal{A}} \rangle$ that is a deterministic FA where $Acc_{\mathcal{A}}$ is replaced by a total and computable *output function* $Out_{\mathcal{A}}$: $Q_{\mathcal{A}} \to \mathcal{D}$ such that $\mathcal{D}$ is an effectively given domain. The *function computed by* $\mathcal{A}$ is $Comp(\mathcal{A}) : T(H) \to \mathcal{D}$ such that $Comp(\mathcal{A})(t) := Out_{\mathcal{A}}(q_{\mathcal{A}}(t))$.

Example 1 The number of accepting runs of an automaton.

Let $\mathcal{A} = \langle H, Q_{\mathcal{A}}, \delta_{\mathcal{A}}, Acc_{\mathcal{A}} \rangle$ be a nondeterministic FA. We construct a deterministic FA $\mathcal{B}$ that computes the number of accepting runs of $\mathcal{A}$ on any term in $T(H)$. As set of states $Q_{\mathcal{B}}$, we take the set of finite subsets of $Q_{\mathcal{A}} \times \mathbb{N}_+$. The transitions are defined so that $\mathcal{B}$ reaches state α at the root of $t \in T(H)$ if and only if α is the finite set of pairs $(q, n) \in Q_{\mathcal{A}} \times \mathbb{N}_+$ such that n is the number of runs of $\mathcal{A}$ that reach state q at its root. This number is finite and α can be seen as a partial function: $Q_{\mathcal{A}} \to \mathbb{N}_+$ having a finite domain. For a symbol f of arity 2, $\mathcal{B}$ has the transition: $f[\alpha, \beta] \to \gamma$ where γ is the set of pairs (q, n) such that n is the sum of the integers $n_p.n_r$ over all pairs $(p, r) \in Q_{\mathcal{A}} \times Q_{\mathcal{A}}$ such that $(p, n_p) \in \alpha$, $(r, n_r) \in \beta$ and $q \in \delta_{\mathcal{A}}(f, p, r)$. The transitions for other symbols are defined similarly. The function $Out_{\mathcal{A}}$ maps a state α to the sum of the integers n such that $(q, n) \in \alpha \cap (Acc_{\mathcal{A}} \times \mathbb{N}_+)$.□

Example 2 An FA for checking 3-colorability.

In order to construct an FA that accepts the terms $t \in T(F)$ such that $val(t)$ is 3-colorable, we first construct an FA $\mathcal{A}$ for the property $Col(X, Y)$. For this purpose, we transform F into $F^{(2)}$ by replacing each nullary symbol **a** by the four nullary symbols $(\mathbf{a}, ij)$, $i, j \in \{0, 1\}$. A term $t \in T(F^{(2)})$ defines, first, the graph $val(t')$ where t' is obtained from t by removing the Booleans i, j from the nullary symbols and, second, the pair (V_X, V_Y) such that V_X is the set of vertices u (leaves of t) that are occurrences of $(\mathbf{a}, 1j)$ for some **a** and j and V_Y is the set of those that are occurrences of $(\mathbf{a}, i1)$ for some **a** and i. The set of terms $t \in T(F^{(2)})$ such that $Col(V_X, V_Y)$ holds in $val(t')$ is defined by a deterministic FA $\mathcal{A}$ that we now specify. Its states are *Error* and the finite subsets of $\mathbb{N}_+ \times \{1, 2, 3\}$. Their meanings are as follows:

- At position u of t, the automaton reaches state *Error* if and only if $val(t')/u$ has a vertex in $V_X \cap V_Y$ or an edge between two vertices, either both in V_X or both in V_Y or both in $V_G - (V_X \cup V_Y)$, hence of the same color, respectively 1, 2, or 3;
- It reaches state $\alpha \subseteq \mathbb{N}_+ \times \{1, 2, 3\}$ if and only if these conditions do not hold and α is the set of pairs (a, i) such that $val(t')/u$ has an a-port of color i.

All states except *Error* are accepting. Here are the transitions of $\mathcal{A}$:

$$(\mathbf{a}, 00) \to \{(a, 3)\}, (\mathbf{a}, 10) \to \{(a, 1)\},$$
$$(\mathbf{a}, 01) \to \{(a, 2)\}, (\mathbf{a}, 11) \to Error.$$

For $\alpha, \beta \subseteq \mathbb{N}_+ \times \{1, 2, 3\}$, $\mathcal{A}$ has transitions:

$$\oplus [\alpha, \beta] \to \alpha \cup \beta,$$
$$add_{a,b}[\alpha] \to Error, \text{ if } (a, i) \text{ and } (b, i) \text{ belong to } \alpha \text{ for some } i = 1, 2, 3,$$
$$add_{a,b}[\alpha] \to \alpha, \text{ otherwise },$$
$$relab_{a \to b}[\alpha] \to \beta, \text{ obtained by replacing } a \text{ by } b \text{ in each pair of } \alpha.$$

Its other transitions are $\oplus[\alpha,\beta] \to \mathit{Error}$ if α or β is *Error*, $add_{a,b}[\mathit{Error}] \to \mathit{Error}$, and $relab_{a\to b}[\mathit{Error}] \to \mathit{Error}$.

This FA checks $Col(X,Y)$. To check, $\exists X,Y.Col(X,Y)$, we build a nondeterministic FA $\mathcal{B}$ by deleting the state *Error*, by replacing the first three rules of $\mathcal{A}$ by $\mathbf{a} \to \{(a,3)\}, \mathbf{a} \to \{(a,1)\}, \mathbf{a} \to \{(a,2)\}$, and by deleting those that yield *Error*. All states are accepting, but on some terms, no run can reach the root, and these terms are rejected. Furthermore, the construction of Example 1 shows how to make $\mathcal{B}$ into a deterministic FA that computes the number of 3-colorings, because the 3-colorings of $val(t)$ are in bijection with the accepting runs of $\mathcal{B}$ on t. □

Recognizability Theorem: The set of graphs that satisfy a closed MSO formula φ is F-recognizable.

Weak Recognizability Theorem: For every closed MSO formula φ, for every k, the set of graphs in $\mathcal{G}_k$ that satisfy φ is F_k-recognizable.

About proofs: The Recognizability Theorem is Theorem 5.68 of [6]. Its proof shows that the equivalence defined by the fact that the two considered p-graphs have the same type and satisfy the same closed MSO formulas of quantifier height at most that of φ satisfies the conditions of Definition 1(f). (These formulas have unary predicates for expressing port labels.) The Weak Recognizability Theorem follows from the former one. It can be proved directly by constructing an FA over F [3]. (We construct a single FA, not a particular FA for each subsignature F_k as in Theorem 6.35 of [6].) This construction can be implemented, at least in a number of nontrivial cases. The proof of the strong theorem does not provide any usable automaton.

Counting and Optimizing Automata

Let $P(X_1,\ldots,X_s)$ be an MSO property of vertex sets $X_1,\ldots,X_s$. We denote $(X_1,\ldots,X_s)$ by $\overline{X}$ and $t \models P(\overline{X})$ means that $\overline{X}$ satisfies P in the graph $val(t)$ defined by a term t. We are interested not only to check the validity of $\exists\overline{X}.P(\overline{X})$ but also to compute from a term t the following values:

> $\#\overline{X}.P(\overline{X})$, defined as the number of assignments $\overline{X}$ such that $t \models P(\overline{X})$,
> $\mathrm{Sp}\overline{X}.P(\overline{X})$, the *spectrum* of $P(\overline{X})$, defined as the set of tuples of the form $(|X_1|,\ldots,|X_s|)$ such that $t \models P(\overline{X})$,
> $\mathrm{MSp}\overline{X}.P(\overline{X})$, the *multispectrum* of $P(\overline{X})$, defined as the multiset of tuples $(|X_1|,\ldots,|X_s|)$ such that $t \models P(\overline{X})$.

These computations can be done by FA. The construction for $\#\overline{X}.P(\overline{X})$ is similar to that of Example 1. We obtain in this way FPT or XP algorithms [8, 10].

Edge Set Quantifications and Tree-Width

The two recognizability theorems and the corresponding constructions of FA yielding FPT and XP algorithms hold and can be done for graphs of bounded tree-width and MSO formulas with edge set quantifications: it suffices to replace a graph G by its incidence graph $Inc(G)$, a bipartite graph whose vertices are those of G and its edges, to observe that the clique-width of $Inc(G)$ is bounded in terms of the tree-width of G and that an MSO formula with edge set quantifications over G can be translated into an MSO formula over $Inc(G)$. Another approach is in [2].

Beyond MS Logic

The property that the considered graph is the union of two disjoint regular graphs with possibly some edges between these two subgraphs is not MSO expressible but can be checked by an FA. An FA can also compute the minimal number of edges between X and $V_G - X$ such that $G[X]$ and $G[V_G - X]$ are connected, when such a set X exists.

Open Problems

The parsing problem for graphs of clique-width at most k is NP-complete (with k in the input) [9]. Good heuristics remain to be developed.

Experimental Results

These constructions have been implemented and tested [3–5]. We have computed the number of optimal colorings of graphs of clique-width at most 8 for which the chromatic polynomial is known, which allows to verify the correctness of the automaton. We can verify in, respectively, 35 and 105 min that the 20×20 and the 6×60 grids are 3-colorable. In 29 min, we can verify that the McGee graph (24 vertices) given by a term over F_{10} is acyclically 3-colorable.

A different approach using games is presented in [13].

Recommended Reading

1. Comon H. et al (2007) Tree automata techniques and applications. http://tata.gforge.inria.fr/
2. Courcelle B. (2012) On the model-checking of monadic second-order formulas with edge set quantifications. Discrete Appl. Math. 160:866–887
3. Courcelle B., Durand I (2012) Automata for the verification of monadic second-order graph properties. J. Appl. Logic 10:368–409
4. Courcelle B., Durand I (submitted for publication, 2013) Computations by fly-automata beyond monadic second-order logic. http://arxiv.org/abs/1305.7120
5. Courcelle B., Durand I. (2013) Model-checking by infinite fly-automata. In: Proceedings of the 5th conference on algebraic informatics, Porquerolles, France. Lecture Notes in Computer Science, Vol. 8080, pp 211–222
6. Courcelle B., Engelfriet J. (2012) Graph structure and monadic second-order logic, a language theoretic approach. Vol. 138 of Encyclopedia of Mathematics and its Application. Cambridge University Press, Cambridge, UK
7. Courcelle B, Makowsky J, Rotics U Linear-time solvable optimization problems on graphs of bounded clique-width. Theory Comput. Syst. 33:125–150 (2000)
8. Downey R., Fellows M. (1999) Parameterized complexity. Springer, New York
9. Fellows M., Rosamond F., Rotics U., Szeider S. (2009) Clique-width is NP-complete. SIAM J. Discret. Math. 23:909–939
10. Flum J., Grohe M. (2006) Parametrized complexity theory. Springer, Berlin, Heidelberg
11. Frick M., Grohe M. (2004) The complexity of first-order and monadic second-order logic revisited. Ann Pure Appl. Logic 130:3–31
12. Grohe M., Kreutzer S. (2011) Model theoretic methods in finite combinatorics. In: Grohe M, Makowsky J (eds) Contemporary mathematics, Vol. 558. American Mathematical Society, Providence, Rhode Island, pp 181–206
13. Kneis J., Langer A., Rossmanith P. (2011) Courcelle's theorem – a game-theoretic approach. Discret Optim 8:568–594
14. Reinhardt K. (2002) The complexity of translating logic to finite automata. In: Graedel E. *et al.* (eds) Automata, logics, and infinite games: a guide to current research. Lecture Notes in Computer Science, Vol. 2500. Springer, Berlin, Heidelberg, pp 231–238

Modularity Maximization in Complex Networks

My T. Thai
Department of Computer and Information Science and Engineering, University of Florida, Gainesville, FL, USA

Keywords

Approximation algorithms; Community structure; Complexity; Modularity clustering

Years and Authors of Summarized Original Work

2013; Dinh, Thai
2013; Dinh, Nguyen, Alim, Thai
2013; DasGupta, Desai
2014; Dinh, Thai

Problem Definition

Many complex networks of interests such as the Internet, social, and biological networks exhibit the community structure where nodes are naturally clustered into tightly connected communities, with only sparser connections between them. The modularity maximization is concerned with finding such community structures in a given complex network.

Consider a network represented as an undirected graph $G = (V, E)$ consisting of $n = |V|$ vertices and $m = |E|$ edges. The *adjacency matrix* of G is denoted by $A = (A_{ij})$, where A_{ij} is the weight of edge (i, j) and $A_{ij} = 0$ if $(i, j) \notin E$. We also denote the (weighted) degree of vertex i, the total weights of edges incident at i, by $\mathsf{deg}(i)$ or, in short, d_i.

Community structure (CS) is a division of the vertices in V into a collection of disjoint subsets of vertices $\mathcal{C} = \{C_1, C_2, \ldots, C_l\}$ (with **unspecified** l) where $\bigcup_{i=1}^{l} C_i = V$. Each subset $C_i \subseteq V$ is called a *community*, and we wish to have more edges connecting vertices in the same communities than edges that connect vertices in different communities. The *modularity* [7] of $\mathcal{C}$ is the fraction of the edges that fall within the given communities minus the expected number of such fraction if edges were distributed at random. The randomization of the edges is done so as to preserve the degree of each vertex. If vertices i and j have degrees d_i and d_j, then the expected number of edges between i and j is $\frac{d_i d_j}{2M}$. Thus, the modularity, denoted by Q, is then

$$Q(\mathcal{C}) = \frac{1}{2M} \sum_{ij} \left(A_{ij} - \frac{d_i d_j}{2M} \right) \delta_{ij} \qquad (1)$$

where M is the total edge weights and the element δ_{ij} of the membership matrix $\boldsymbol{\delta}$ is defined as

$$\delta_{ij} = \begin{cases} 1, & \text{if } i \text{ and } j \text{ are in the same community} \\ 0, & \text{otherwise} \end{cases}$$

The modularity values can be either positive or negative, and the higher (positive) modularity values indicate stronger community structure. Therefore, the *maximizing modularity problem* asks us to find a division $\mathcal{C}$ which maximizes the modularity value $Q(\mathcal{C})$.

Key Results

Computational Complexity

This problem is different from the partition problem as we do not know the total number of partitions beforehand. That being said, l is unspecified. Somewhat surprisingly, modularity maximization is still NP-complete on trees, one of the simplest graph classes.

Theorem 1 *Modularity maximization on trees is NP-complete.*

The proof has been presented in [3], reducing from the subset-sum problem. Furthermore, for dense graphs, namely, for the complements of 3-regular graphs, DasGupta and Desai have provided a $(1+\epsilon)$-inapproximability of the modularity maximization problem [1], stated as follows.

Theorem 2 *It is NP-hard to approximate the modularity maximization problem on $(n - 4)$–regular graphs within a factor of $i + \epsilon$ for some constant $\epsilon > 0$.*

The proof has been presented in [1], reducing from the maximum-cardinality-independent set problem for 3-regular graphs (3-MIS). The basic intuition behind this proof is that large-size cliques must be properly contained within the communities.

Exact Solutions

Although the problem is in NP class, efficient algorithms to obtain optimal solutions for small size networks are still of interest. Dinh and Thai have presented an exact algorithm with a running time of $O(n^5)$ to the problem on uniform-weighted trees [3]. The algorithm is based on the dynamic programming, which exploits the relationship between maximizing modularity and minimizing the sum of squares of component volumes, where volume of a component S is defined as $\mathsf{vol}(S) = \sum_{v \in S} d_v$.

When the input graph is not a tree, an exact solution based on integer linear programming (ILP) is provided by Dinh and Thai [3]. Note that in the ILP for modularity maximization, there is a triangle inequality $x_{ij} + xjk - x_{ik} \geq 0$ to guarantee the values of x_{ij} be consistent to each other. Here $x_{ij} = 0$ if i and j are in the same community; otherwise $x_{ij} = 0$. Therefore, the ILP has $3\binom{n}{3} = \theta(n^3)$ constraints, which is about half a million constraints for a network of 100 vertices. As a consequence, the sizes of solved instances were limited to few hundred nodes. Along this direction, Dinh and Thai have pre-

M

sented a sparse metric, which reduces the number of constraints to $O(n^2)$ in sparse networks where $m = O(n)$ [3].

Approximation Algorithms

When G is a tree, the problem can be solved by a polynomial time approximation scheme (PTAS) with a running time of $O(n^{\epsilon+1})$ for $\epsilon > 0$ [3]. The PTAS is solely based on the following observation. Removing $k-1$ edges in G will yield k connected communities and $Q_k \geq (1 - \frac{1}{k})Q_{\text{opt}}$ where Q_k is the maximum modularity of a community structure with k communities and Q_{opt} is the optimal solution.

When G having the degree distribution follows the power law, i.e., the fraction of nodes in the network having k degrees is proportional to $k^{-\gamma}$, where $1 < \gamma \leq 4$, the problem can be approximated to a constant factor for $\gamma > 2$ and up to an $O(1/\log n)$ when $1 < \gamma \leq 2$ [2]. The details of this algorithm, namely, low-degree following (LDF), is presented in Algorithm 1.

Algorithm 1 Low-degree following algorithm (parameter $d_0 \in \mathbb{N}^+$)

1. $L := \emptyset, M := \emptyset, O := \emptyset, p_i = 0\ \forall i = 1..n$
2. **for each** vertex $i \in V$ **do**
3. **if** $(k_i \leq d_0) \& (i \notin L \cup M)$ **then**
4. **if** $N(i) \setminus M \neq \emptyset$ **then**
5. Select a vertex $j \in N(i) \setminus M$
6. Let $M = M \cup \{i\}, L = L \cup \{j\}, p_i = j$
7. **else**
8. Select a vertex $t \in N(i)$
9. $O = O \cup \{i\}, p_i = t$
10. $\mathcal{L} = \emptyset$
11. **for each** vertex $i \in V \setminus (M \cup O)$ **do**
12. $C_i = \{i\} \cup \{j \in M \mid p_j = i\} \cup \{t \in O \mid p_{p_t} = i\}$
13. $\mathcal{L} = \mathcal{L} \cup \{C_i\}$
14. Return $\mathcal{L}$

The selection of d_0 is important to derive the approximation factor as d_0 needs to be a sufficient large constant that is still relative small to n when n tends to infinity. In an actual implementation of the algorithm, Dinh and Thai have designed an automatic selection of d_0 to maximize Q. LDF can be extended to solve the problem in directed graphs [2].

Theorem 3 ([1]) *There exists an $O(\log d)$-approximation for d-regular graphs with $d < \frac{n}{2\ln n}$ and $O(\log d_{\max})$-approximation for weighted graphs $d_{\max} < \frac{\sqrt{n}}{16\ln n}$.*

Modularity in Dynamic Networks

Networks in real life are very dynamic, which requires us to design an adaptive approximation algorithm to solve the problem. In this approach, the community structure (CS) at time t is detected based on the community structure at time $t-1$ and the changes in the network, instead of recomputing it from scratch. Indeed, the above LDF algorithm can be enhanced to cope with this situation [4]. At first LDF is run to find the base CS at time 0. Then at each time step, we adaptively follow and unfollow the nodes that violate the condition 3 in Algorithm 1.

Applications

Finding community structures has applications in vast domains. For example, a community in biology networks often consists of proteins, genes, or subunits with functional similarity. Thus, finding communities can help to predict unknown proteins. Likewise, in online social networks, a community can be a group of users having common interests; therefore, obtaining CS can help to predict user interests. Furthermore, detecting CS finds itself extremely useful in deriving social-based solutions for many network problems, such as forwarding and routing strategies in communication networks, Sybil defense, worm containment on cellular networks, and sensor reprogramming. In the network visualization perspective, finding CS helps display core network components and their mutual interactions, hence presents a more compact and understandable description of the network as a whole.

Recommended Reading

1. DasGupta B Desai D (2013) On the complexity of Newman's community finding approach for biological and social networks. J Comput Syst Sci 79:50–67
2. Dinh TN, Thai MT (2013) Community detection in scale-free networks: approximation algorithms for

maximizing modularity. IEEE J Sel Areas Commun Spec Issue Netw Sci (JSAC) 31(6):997–1006
3. Dinh TN, Thai MT (2014) Towards optimal community detection: from trees to general weighted networks. Internet Math. doi:10.1080/15427951.2014.950875
4. Dinh TN, Nguyen NP, Alim MA, Thai MT (2013) A near-optimal adaptive algorithm for maximizing modularity in dynamic scale-free networks. J Comb Optim (JOCO). doi:10.1007/s10878-013-9665-1
5. Fortunato S (2010) Community detection in graphs. Phys Rep 486(3-5):75-174
6. Fortunato S Barthelemy M (2007) Resolution limit in community detection. Proc Natl Acad Sci 104(1):36-41
7. Newman MEJ (2006) Modularity and community structure in networks. Proc Natl Acad Sci 103:8577–8582
8. Nguyen NP, Dinh TN, Tokala S, Thai MT (2011) Overlapping communities in dynamic networks: their detection and mobile applications. In: Proceedings of ACM international conference on mobile computing and networking (MobiCom), Las Vegas

Monotone Minimal Perfect Hash Functions

Paolo Boldi and Sebastiano Vigna
Dipartimento di Informatica, Università degli Studi di Milano, Milano, Italy

Keywords

Minimal perfect hash functions; Succinct data structures

Years and Authors of Summarized Original Work

2009; Belazzougui, Boldi, Pagh, Vigna
2011; Belazzougui, Boldi, Pagh, Vigna

Problem Definition

A minimal perfect hash function is a (data structure providing a) bijective map from a set S of n keys to the set of the first n natural numbers. In the static case (i.e., when the set S is known in advance), there is a wide spectrum of solutions available, offering different trade-offs in terms of construction time, access time, and size of the data structure.

An important distinction is whether any bijection will be suitable or whether one wants it to respect some specific property. A *monotone minimal perfect hash function (MMPHF)* is one where the keys are bit vectors and the function is required to preserve their lexicographic order.

Sometimes in the literature, this situation is identified with the one in which the set S has some predefined linear order and the bijection is required to respect the order: in this case, one should more precisely speak of *order-preserving* minimal perfect hash functions; for this scenario, a ready-made $\Omega(n \log n)$ space lower bound is trivially available (since all the $n!$ possible key orderings must be representable). This is not true in the monotone case, so the distinction between monotone and order preserving is not moot.

Problem Formulation

Let $[x]$ denote the set of the first x natural numbers. Given a positive integer $u = 2^w$, and a set $S \subseteq [u]$ with $|S| = n$, a function $h : S \to [m]$ is *perfect* iff it is injective, *minimal* iff $m = n$, and *monotone* iff $x \leq y$ implies $h(x) \leq h(y)$. In the following, we assume to work in the RAM model with words of length w. For simplicity, we will describe only the case in which the keys have fixed length w; the results can be extended to the general case [2].

Key Results

One key building block that is needed to describe the possible approaches to the problem is that of storing an *arbitrary* r-bit function $h : S \to [2^r]$ in a succinct way (i.e., using $rn + o(n)$ bits and guaranteeing constant access time). A practical (although slightly less efficient) method for storing a succinct function $f : S \to [2^r]$ was presented in [8]. The construction draws three hash functions $h_0, h_1, h_2 : S \to [\gamma n]$ (with $\gamma \approx 1.23$) and builds a 3-hypergraph with one hyperedge $(h_0(x), h_1(x), h_2(x))$ for every $x \in$

	B_0		B_1		B_2		B_3
s_0	0001001000000	s_3	0010011000000	s_6	0010011010100	s_9	0010011110110
s_1	0010010101100	s_4	0010011001000	s_7	0010011010101	s_{10}	0100100010000
$\mathbf{s_2}$	**0010010101110**	$\mathbf{s_5}$	**0010011010010**	$\mathbf{s_8}$	**0010011010110**		

Monotone Minimal Perfect Hash Functions, Fig. 1 A toy example: $S = \{s_0, \ldots, s_{10}\}$ is divided into three buckets of size three (except for the last one that contains just two elements), whose delimiters $D = \{s_2, s_5, s_8\}$ appear in boldface

d_0 :

s_0	2
s_1	2
s_2	2
s_3	8
s_4	8
s_5	8
s_6	11
s_7	11
s_8	11
s_9	1
s_{10}	1

d_1 :

00	0
00100110	1
00100110101	2
0	3

Monotone Minimal Perfect Hash Functions, Fig. 2 Bucketing with longest common prefix for the set S of Fig. 1: d_0 maps each element x of S to the length of the longest common prefix of the bucket to which x belongs; d_1 maps each longest common prefix to the bucket index

S. With positive probability, this hypergraph does not have a nonempty 2-core; or, equivalently, the set of equations (in the variables a_i)

$$f(x) = \left(a_{h_0(x)} + a_{h_1(x)} + a_{h_2(x)}\right) \bmod 2^r$$

has a solution that can be found by a hypergraph-peeling process in time $O(n)$. Storing the function amounts to storing γn integers of r bits each (the array a_i), so $\gamma r n$ bits are needed (excluding the bits required to store the hash functions), and it is possible to improve this amount to $\gamma n + rn$ using standard techniques [2]; function evaluation takes constant time (one essentially just needs to evaluate three hash functions). We shall refer to this data structure as an *MWHC function* (from the name of the authors). Alternative, asymptotically more efficient data structures for the same purpose are described in [3,4].

Trivial Solution

MWHC functions can themselves be used to store a MMPHF (setting $r = \lceil \log n \rceil$ and using the appropriate function f). This idea (requiring $\gamma n \lceil \log n \rceil$ bits) is also implied in the original paper [8], where the authors actually present their construction as a solution to the order-preserving minimal perfect hash function problem.

A Constant-Time, $O(n \log w)$-Space Solution

More sophisticated approaches are based on the general technique of *bucketing*: a *distributor* function $f : S \to [m]$ is stored, which divides the set of keys into m buckets respecting the lexicographic order; then, for every bucket $i \in [m]$, a MMPHF g_i on $f^{-1}(i)$ is stored as a succinct function. Different choices for the distributor and the bucket sizes provide different space/time trade-offs.

A constant-time solution can be obtained using buckets of equal size $b = O(\log n)$. Let $B_0, \ldots, B_{m-1}$ be the unique order-preserving partition of S into buckets of size b and p_i be the longest common prefix of the keys in B_i; Fig. 1 shows an example with $b = 3$. It is easy to see that the p_i's are all distinct, which allows us to build a succinct function d_1 mapping $p_i \mapsto i$. Moreover, we can store a succinct function $d_0 : S \to [w]$ mapping x to the length of the longest common prefix p_i of the bucket B_i containing x. The two functions together form the distributor (given x one applies d_1 to the prefix of x of length $d_0(x)$); see Fig. 2. The space required by d_0 and d_1 is $O(n \log w)$ and $O((n/b)\log(n/b)) = O(n)$, respectively. The functions g_i's require $\sum_i O(|B_i| \log b) = O(n \log \log n) = O(n \log w)$ bits. The overall space requirement is thus $O(n \log w)$; optimal values for b when using MWHC functions are computed in [2].

A $O(\log w)$-Time, $O(n \log \log w)$-Space Solution

In search for a better space bound, we note that an obvious alternative approach to the bucketing

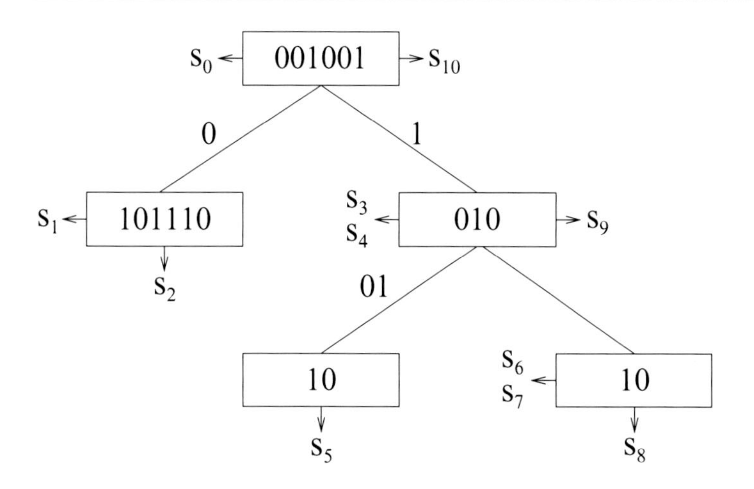

Monotone Minimal Perfect Hash Functions, Fig. 3 The standard compacted trie built from the set D of Fig. 1. This data structure can be used to rank arbitrary elements of the universe with respect to D: when the trie is visited with an element not in D, the visit will terminate at an *exit node*, determining that the given element is to the left (i.e., smaller than) or to the right (i.e., larger than) all the leaves that descend from that node. The picture shows, for each element of S, the node where the visit would end

problem is by *ranking*. Given a set of strings X, a ranking data structure provides, for each string $s \in [u]$, the number of strings in X that are smaller than s, that is, $|\{x \in X \mid x < s\}|$. If you consider the set D of *delimiters* (i.e., the set containing the last string of each bucket), a distributor can be built using any data structure that provides the rank of an arbitrary string with respect to D.

For instance, a trivial way to obtain such rank information is to build a compacted trie [7] containing the strings in D (see Fig. 3). Much more sophisticated data structures are obviously available today (e.g., [6]), but they all fail to meet their purpose in our case. They occupy too much space, and we do not really need to rank *all* possible strings: we just need to rank strings in S. We call this problem the *relative ranking problem*: given sets $D \subseteq S$, we want to rank a string s w.r.t. D under the condition that s belongs to S.

The relative ranking problem can be approached in different ways: in particular, a static *probabilistic z-fast trie* [1] provides a $O(n)$-space, $O(\log w)$-time solution for buckets of size $O(\log w)$ (the trie can actually provide a wrong output for a small set of keys; their correct output needs to be stored separately – see [1] for details). The functions g_i's require $O(n \log \log w)$ bits, which dominates the space bound.

Different Approaches

Other theoretical and practical solutions, corroborated with experimental data, were described in [2].

Open Problems

Currently, the main open problem about monotone minimal perfect hashing is that all known lower bounds are trivial; in particular, the lower bound $n \log e + \log w - O(\log n)$ by Fredman and Komlós (provided that $u = 2^w \geq n^{2+\epsilon}$ for some fixed $\epsilon > 0$) for minimal perfect hash functions [5] is of little help – it is essentially independent from the size of the universe. We already know that the dependence on the size of the universe can be really small (as small as $O(n \log \log w)$ bits), but it is currently conjectured that there is no monotone minimal perfect hashing scheme whose number of bits per key is constant in the size of the universe.

URLs to Code and Datasets

The most comprehensive implementation of MMPHFs is contained in the Sux4J Java free library (http://sux4j.di.unimi.it/). A good collection of datasets is available by the LAW (http://law.di.unimi.it/) under the form of list of URLs from web crawls and list of ids from social networks (e.g., Wikipedia): these are typical use cases of a MMPHF.

Cross-References

▶ Minimal Perfect Hash Functions

Recommended Reading

1. Belazzougui D, Boldi P, Pagh R, Vigna S (2009) Monotone minimal perfect hashing: searching a sorted table with $O(1)$ accesses. In: Proceedings of the 20th annual ACM-SIAM symposium on discrete mathematics (SODA). ACM, New York, pp 785–794
2. Belazzougui D, Boldi P, Pagh R, Vigna S (2011) Theory and practice of monotone minimal perfect hashing. ACM J Exp Algorithmics 16(3):3.2:1–3.2:26
3. Charles DX, Chellapilla K (2008) Bloomier filters: a second look. In: Halperin D, Mehlhorn K (eds) Algorithms – ESA 2008, Proceedings of the 16th annual European symposium, Karlsruhe, 15–17 Sept 2008. Lecture notes in computer science, vol 5193. Springer, pp 259–270
4. Dietzfelbinger M, Pagh R (2008) Succinct data structures for retrieval and approximate membership (extended abstract). In: Aceto L, Damgård I, Goldberg LA, Halldórsson MM, Ingólfsdóttir A, Walukiewicz I (eds) Proceedings of the 35th international colloquium on automata, languages and programming, ICALP 2008, Part I: Track A: algorithms, automata, complexity, and games. Lecture notes in computer science, vol 5125. Springer, pp 385–396
5. Fredman ML, Komlós J (1984) On the size of separating systems and families of perfect hash functions. SIAM J Algebraic Discret Methods 5(1):61–68
6. Gupta A, Hon WK, Shah R, Vitter JS (2007) Compressed data structures: dictionaries and data-aware measures. Theor Comput Sci 387(3):313–331
7. Knuth DE (1997) The art of computer programming. Sorting and searching, vol 3, 2nd edn. Addison-Wesley, Reading
8. Majewski BS, Wormald NC, Havas G, Czech ZJ (1996) A family of perfect hashing methods. Comput J 39(6):547–554

Monotonicity Testing

Deeparnab Chakrabarty
Microsoft Research, Bangalore, Karnataka, India

Keywords

Boolean functions; Monotonicity; Property testing

Years and Authors of Summarized Original Work

1999; Dodis, Goldreich, Lehman, Raskhodnikova, Ron, Samorodnitsky
2000; Goldreich, Goldwasser, Lehman, Ron, Samorodnitsky
2013; Chakrabarty, Seshadhri

Problem Definition

A real-valued function $f : D \to \mathbb{R}$ defined over a partially ordered set (poset) D is *monotone* if $f(x) \leq f(y)$ for any two points $x \prec y$. In this article, we focus on the poset induced by the coordinates of a d-dimensional, n-hypergrid, $[n]^d$, where $x \prec y$ iff $x_i \leq y_i$ for all integers $1 \leq i \leq d$. Here, we have used $[n]$ as a shorthand for $\{1, \ldots, n\}$. The hypercube, $\{0, 1\}^d$, and the n-line, $[n]$, are two special cases of this.

Monotonicity testing is the algorithmic problem of deciding whether a given function is monotone. The algorithm has query access to the function, which means that it can query f at any domain point x and obtain the value of $f(x)$. The performance of the algorithm is measured by the number of queries it makes. Although the running time of the algorithm is also important, we ignore this parameter in this article because in most relevant cases it is of the same order of magnitude as the query complexity. We desire algorithms whose query complexity is bounded polynomially in d and $\log n$.

Monotonicity is a fundamental property. In one dimension, monotone functions correspond to sorted arrays. In many applications it may be useful to know if an outcome of a procedure monotonically changes with some or all of its attributes. In learning theory, for instance, monotone concepts are known to require fewer samples to learn; it is useful to test beforehand if a concept is monotone or not.

Property Testing Framework

It is not too hard to see that, without any extra assumptions, monotonicity testing is infeasible unless almost the entire input is accessed. Therefore, the problem is studied under the property testing framework where the goal is to distinguish monotone functions from those which are "far" from monotone. A function f is said to be ϵ-far from being monotone if it needs to be modified on at least ϵ-fraction of the domain points to make it monotone.

Formally, the monotonicity testing problem is defined as follows. Given an input distance parameter ϵ, the goal is to design a $q(d, n, \epsilon)$-query tester which is a randomized algorithm that queries the function on at most $q(d, n, \epsilon)$ points and satisfies the following requirements:

(a) If the function is monotone, the algorithm **accepts**.
(b) If the function is ϵ-far from being monotone, the algorithm **rejects**.

The tester can err in each of the above cases but the error probability should be at most $1/3$. If a tester never rejects a monotone function, then it is called a *one-sided error* tester. If the queries made by the tester do not depend on the function values returned by the previous queries, then the tester is called a *nonadaptive* tester. The function q is called the *query complexity* of the monotonicity tester, and one is interested in (ideally matching) upper and lower bounds for it.

Key Results

Monotonicity testing was first studied by Ergun et al. [7] for functions defined on the n-line, that is, the case when $d = 1$. The authors designed an $O(\epsilon^{-1} \log n)$-query tester.

Goldreich et al. [9] studied monotonicity testing over the hypercube $\{0, 1\}^n$ and designed an $O(\epsilon^{-1} d)$-query tester for *Boolean* functions. These are functions whose range is $\{0, 1\}$. This tester repeats the following *edge test* $O(\epsilon^{-1} d)$ times: sample an edge of the hypercube uniformly at random, query the function values at its endpoints, and check if these two points violate monotonicity. That is, if $x \prec y$ and $f(x) > f(y)$, where x, y are the queried points. If at any time a violation is found, the tester rejects; otherwise it accepts. Clearly, this tester is nonadaptive and with one-sided error. Goldreich et al. [9] also demonstrated an $O(\epsilon^{-1} d \log n)$-query tester for Boolean functions defined over the d-dimensional n-hypergrid. This tester also defines a distribution over comparable pairs (not necessarily adjacent) of points, that is, points x, y with $x \prec y$. In each iteration, it samples a pair from the distribution, queries the function value on these points, and checks for a violation of monotonicity. Such testers are called *pair testers*.

Goldreich et al. [9] showed a *range reduction theorem* which states that if there exists a pair tester for Boolean functions over the hypergrid whose query complexity has linear dependence on ϵ, that is $q(d, n, \epsilon) = \epsilon^{-1} q(d, n)$, then such a tester could be extended to give an $O(\epsilon^{-1} q(d, n)|\mathbf{R}|)$-query tester for real-valued functions, where $\mathbf{R}$ is the range of the real-valued function. Note that $|\mathbf{R}|$ could be as large as 2^n. Dodis et al. [6] obtained a stronger range reduction theorem and showed an $O(\epsilon^{-1} q(d, n) \log |\mathbf{R}|)$-query tester under the same premise. This implied an $O(\epsilon^{-1} d \log n \log |\mathbf{R}|)$-query monotonicity tester for the hypergrid.

Recently, Chakrabarty and Seshadhri [3] removed the dependence on the range size exhibiting an $O(\epsilon^{-1} d \log n)$-query monotonicity tester for any real-valued function. In fact, the tester in their paper is the same as the tester defined in [9] alluded to above. In a separate paper, the same authors [4] complemented the above result by showing that any tester (adaptive and with two-sided error) for monotonicity with distance parameter ϵ must make $\Omega(\epsilon^{-1} d \log n)$-queries.

Sketch of the Techniques

In this section, we sketch some techniques used in the results mentioned above. For simplicity, we restrict ourselves to functions defined on the hypercube.

To analyze their tester, Goldreich et al. [9] used the following *shifting* technique from combinatorics to convert a Boolean function f to a monotone function. Pick any dimension i and look at all the violated edges whose endpoints differ precisely in this dimension. If $(x \prec y)$ is a violation, then since f is Boolean, it must be that $f(x) = 1$ and $f(y) = 0$. For every such violation, redefine $f(x) = 0$ and $f(y) = 1$. Note that once dimension i is treated so, there is no violation across dimension i. The crux of the argument is that for any other dimension j, the number of violated edges across that dimension can only decrease. Therefore, once all dimensions are treated, the function becomes monotone, and the number of points at which it has been modified is at most twice the number of violated edges in the beginning. In other words, if f is ϵ-far, then the number of violated edges is at least $\epsilon 2^{d-1}$. Therefore, the edge test succeeds with probability at least ϵ/d since the total number of edges is $d2^{d-1}$.

The fact that treating one dimension does not increase the number of violated edges across any other dimension crucially uses the fact that the function is Boolean and is, in general, not true for all real-valued functions. The range reduction techniques of [6,9] convert a real-valued function to a collection of Boolean ones with certain properties, and the size of the collection, which is $|\mathbf{R}|$ in [9] and $O(\log |\mathbf{R}|)$ in [6], appears as the extra multiplicative term in the query complexity.

The technique of Chakrabarty and Seshadhri [3] to handle general real-valued functions departs from the above in that it does not directly fix a function to make it monotone. Rather, they use the connection between the distance to monotonicity and matchings in the *violation graph* G_f which was defined by [6]. The vertices of G_f are the domain points, and (x, y) is an edge in G_f if and only if (x, y) is a violation to monotonicity of f. A folklore result, which appears in print in [8], is that the distance to monotonicity of f is precisely the cardinality of the *minimum vertex cover* of G_f divided by the domain size. This, in turn, implies that if f is ϵ-far, then *any* maximal matching in G_f must have cardinality at least $\epsilon 2^{d-1}$.

Chakrabarty and Seshadhri [3] introduce the notion of the *weighted* violation graph G_f where the weight of $(x \prec y)$ is defined as $(f(x) - f(y))$. They prove that the number of violated edges of the hypercube is at least the size of the *maximum weight* matching M^* in G_f. Since M^* is also maximal, this shows that the number of violated edges is at least $\epsilon 2^{d-1}$. The proof of [3] follows by charging each matched pair of points in M^* differing in the ith dimension to a *distinct* violated edge across the ith dimension.

Boolean Monotonicity Testing

Let us go back to Boolean functions defined over the hypercube. Recall that Goldreich et al. [9] designed an $O(\epsilon^{-1} d)$-query tester for such functions. Furthermore, their analysis is tight: there are functions for which the edge tester's success probability is $\Theta(\epsilon/d)$. The best known lower bound [8], however, is that any nonadaptive tester with one-sided error with respect to a specific constant distance parameter requires $\Omega(\sqrt{d})$ queries, and any adaptive, one-sided error tester required $\Omega(\log d)$-queries.

Chakrabarty and Seshadhri [2] obtained the first $o(d)$-query tester: they describe an $O(d^{7/8}\epsilon^{-3/2} \ln(1/\epsilon))$ query tester for Boolean functions defined over the hypercube. The tester is a combination of the edge test and what the authors call the *path test* – in each step one of the two is performed with probability 1/2. The path test is the following. Orient all edges in the hypercube to go from the point with fewer 1s to the one with more 1s. Sample a random *directed* path from the all-0s point to the all-1s point. Sample two points x, y on this path which (a) have "close" to $d/2$ ones, and (b) are "sufficiently" far away from each other. Query $f(x), f(y)$ and check for a violation to monotonicity.

Chakrabarty and Seshadhri [2] obtain the following result for the path tester. If in the hypercube (and not the violation graph), there exists a matching of violated edges whose cardinality equals $\sigma 2^d$, then the path test catches a violation with probability roughly $\Omega(\sigma^3/\sqrt{d})$. Although this is good if σ is large, say a constant, the analysis above does not give better testers for functions with small σ. The authors circumvent this via the following *dichotomy* theorem. Given a function f with distance parameter ϵ, let the *number* of violated edges be $\delta 2^n$; from the result of Goldreich et al. [9], we know that $\delta \geq \epsilon$. Chakrabarty and Seshadhri [2] prove that if for any function f the quantity σ is small, then δ must be large. In particular, they prove that $\delta\sigma \geq \epsilon^2/32$. Therefore, for *any* function, either the edge test or the path test succeeds with probability $\omega(1/d)$.

Very recent work settles the question of nonadaptive, Boolean monotonicity testing. Chen, De, Servedio, and Tan [5] prove that any nonadaptive, monotonicity tester for Boolean functions needs to make $\Omega(d^{\frac{1}{2}-c})$-queries for any constant $c > 0$, even if it is allowed to have *two-sided error*. Khot, Minzer, and Safra [10] generalize the dichotomy theorem of Chakrabarty and Seshadhri [2] to obtain a $O(d^{1/2}\epsilon^{-2})$-query, nonadaptive monotonicity tester with one-sided error.

Open Problems

The query complexity of *adaptive* monotonicity testers for Boolean valued functions is not well understood. The best upper bounds are that of nonadaptive testers, while the best lower bound is only $\Omega(\log d)$. Understanding whether adaptivity helps in Boolean monotonicity testing or not is an interesting open problem. Recent work of Berman, Raskhodnikova, and Yaroslavtsev [1] sheds some light: they show that any nonadaptive, one-sided error monotonicity tester for functions $f : [n] \times [n] \rightarrow \{0,1\}$, that is, functions defined over the two-dimensional grid, requires $\Omega(\frac{1}{\epsilon}\log(\frac{1}{\epsilon}))$ queries where ϵ is the distance parameter; on the other hand, they also demonstrate an *adaptive*, one-sided error monotonicity tester for such functions which makes only $O(\frac{1}{\epsilon})$ queries.

In this article, we only discussed the poset defined by the n-hypergrid. The best known tester for functions defined over a general N-element poset is an $O(\sqrt{N/\epsilon})$ tester, while the best lower bound is $\Omega(N^{\frac{1}{\log\log N}})$ for nonadaptive testers. Both results are due to Fischer et al. [8], and closing this gap is a challenging problem.

Cross-References

▸ Linearity Testing/Testing Hadamard Codes
▸ Testing if an Array Is Sorted
▸ Testing Juntas and Related Properties of Boolean Functions

Recommended Reading

1. Berman P, Raskhodnikova S, Yaroslavtsev G (2014) L_p Testing. In: Proceedings, ACM symposium on theory of computing (STOC), New York
2. Chakrabarty D, Seshadhri C (2013) A $o(n)$ monotonicity tester for Boolean functions over the hypercube. In: Proceedings, ACM symposium on theory of computing (STOC), Palo Alto
3. Chakrabarty D, Seshadhri C (2013) Optimal bounds for monotonicity and Lipschitz testing over hypercubes and hypergrids. In: Proceedings, ACM symposium on theory of computing (STOC), Palo Alto
4. Chakrabarty D, Seshadhri C (2013) An optimal lower bound for monotonicity testing over hypergrids. In: Proceedings, international workshop on randomization and computation (RANDOM), Berkeley
5. Chen X, De A, Servedio R, Tan LY (2015) Boolean function monotonicity testing requires (almost) $n^{1/2}$ non-adaptive queries. In: Proceedings, ACM symposium on theory of computing (STOC), Portland
6. Dodis Y, Goldreich O, Lehman E, Raskhodnikova S, Ron D, Samorodnitsky A (1999) Improved testing algorithms for monotonicity. In: Proceedings, international workshop on randomization and computation (RANDOM), Berkeley
7. Ergun F, Kannan S, Kumar R, Rubinfeld R, Viswanathan M (2000) Spot-checkers. J Comput Syst Sci 60(3):717–751
8. Fischer E, Lehman E, Newman I, Raskhodnikova S, Rubinfeld R, Samorodnitsky A (2002) Monotonicity testing over general poset domains. In: Proceedings, ACM symposium on theory of computing (STOC), Montreal

9. Goldreich O, Goldwasser S, Lehman E, Ron D, Samorodnitsky A (2000) Testing monotonicity. Combinatorica 20:301–337
10. Khot S, Minzer D, Safra S (2015) On monotonicity testing and boolean isoperimetric type theorems. In: Electornic colloquium on computational complexity (ECCC), TR15-011

Multi-armed Bandit Problem

Nicolò Cesa-Bianchi
Dipartimento di Informatica, Università degli Studi di Milano, Milano, Italy

Keywords

Adaptive allocation; Regret minimization; Repeated games; Sequential experiment design

Years and Authors of Summarized Original Work

2002; Auer, Cesa-Bianchi, Freund, Schapire
2002; Auer, Cesa-Bianchi, Fischer

Problem Definition

A multi-armed bandit is a sequential decision problem defined on a set of actions. At each time step, the decision maker selects an action from the set and obtains an observable payoff. The goal is to maximize the total payoff obtained in a sequence of decisions. The name *bandit* refers to the colloquial term for a slot machine ("one-armed bandit" in American slang) and to the decision problem, faced by a casino gambler, of choosing which slot machine to play next. Bandit problems naturally address the fundamental trade-off between exploration and exploitation in sequential experiments. Indeed, the decision maker must use a strategy (called allocation policy) able to balance the exploitation of actions that did well in the past with the exploration of actions that might give higher payoffs in the future. Although the original motivation came from clinical trials [14] (when different treatments are available for a certain disease and one must decide which treatment to use on the next patient), bandits have often been used in industrial applications, for example, to model the sequential allocation of a unit resource to a set of competing tasks.

Definitions and Notation

A bandit problem with $K \geq 2$ actions is specified by the processes $\langle X_{i,t} \rangle$ that, at each time step $t = 1, 2, \ldots$, assign a payoff $X_{i,t}$ to each action $i = 1, \ldots, K$. An allocation policy selects at time t an action $I_t \in \{1, \ldots, K\}$, possibly using randomization, and receives the associated payoff $X_{I_t,t}$. Note that the index I_t of the action selected by the allocation policy at time t can only depend on the set $X_{I_1,1}, \ldots, X_{I_{t-1},t-1}$ of previously observed payoffs (and on the policy's internal randomization). It is this information constraint that creates the exploration vs. exploitation dilemma at the core of bandit problems.

The performance of an allocation policy over a horizon of T steps is typically measured against that of the policy that consistently plays the optimal action for this horizon. This notion of performance, called regret, is formally defined by

$$R_T = \max_{i=1,\ldots,K} \mathbb{E}\left[\sum_{t=1}^{T} X_{i,t} - \sum_{t=1}^{T} X_{I_t,t}\right] . \quad (1)$$

The expectation in (1) is taken with respect to both the policy's internal randomization and the potentially stochastic nature of the payoff processes. Whereas we focus here on payoff processes that are either deterministic or stochastic i.i.d., other choices have been also considered. Notable examples are the Markovian payoff processes [8] or the more general Markov decision processes studied in reinforcement learning.

If the processes $\langle X_{i,t} \rangle$ are stochastic i.i.d. with unknown expectations $\mu_1, \ldots, \mu_K$, as in Robbins' original formalization of the bandit problem [13], then

$$\max_{i=1,\dots,K} \mathbb{E}\left[\sum_{t=1}^{T} X_{i,t}\right] = T\left(\max_{i=1,\dots,K} \mu_i\right) = T\mu^*$$

where $\mu^* = \max_i \mu_i$ is the highest expected payoff. In this case, the regret (1) becomes the stochastic regret

$$R_T^{\mathrm{IID}} = T\mu^* - \sum_{t=1}^{T} \mathbb{E}[\mu_{I_t}] . \quad (2)$$

On the other hand, if the payoff processes $\langle X_{i,t}\rangle$ are fixed, deterministic sequences of unknown numbers $x_{i,t}$, then (1) becomes the nonstochastic regret

$$R_T^{\mathrm{DET}} = \max_{i=1,\dots,K} \sum_{t=1}^{T} x_{i,t} - \sum_{t=1}^{T} \mathbb{E}[x_{I_t,t}] \quad (3)$$

where the expectation is only with respect to the internal randomization used by the allocation policy. This nonstochastic version of the bandit problem is directly inspired by the problem of playing repeatedly an unknown game – see the pioneering work of Hannan [9] and Blackwell [5] on repeated games and also the recent literature on online learning.

The analyses in [2, 3] assume bounded payoffs, $X_{i,t} \in [0,1]$ for all i and t. Under this assumption, $R_T = \mathcal{O}(T)$ irrespective of the allocation policy being used. The main problem is to determine the optimal allocation policies (the ones achieving the slowest regret growth) for the stochastic and the deterministic case. The parameters that are typically used to express the regret bounds are the horizon T and the number K of actions.

Key Results

Consider first the stochastic i.i.d. bandits with $K \geq 2$ actions and expected payoffs $\mathbb{E}[X_{i,t}] = \mu_i$ for $i = 1,\dots,K$ and $t \geq 1$. Also, let $\Delta_i = \mu^* - \mu_i$ (where, as before, $\mu^* = \max_{i=1,\dots,K} \mu_i$). A simple instance of stochastic bandits are the Bernoulli bandits, where payoffs $X_{i,1}, X_{i,2}, \dots$ for each action are i.i.d. Bernoulli random variables. Lai and Robbins [11] prove the following asymptotic lower bound on R_T for Bernoulli bandits. Let $N_{i,T} = \left|\{t = 1,\dots,T\}\, I_t = i\right|$ be the number of times the allocation policy selected action i within horizon T, and let $\mathrm{KL}(\mu,\mu')$ be the Kullback-Leibler divergence between two Bernoulli random variables of parameter μ and μ'.

Theorem 1 ([11]) *Consider an allocation policy that, for any Bernoulli bandit with $K \geq 2$ actions, for any action i with $\mu_i < \mu^*$, and for any $a > 0$, satisfies $\mathbb{E}[N_{i,T}] = o(T^a)$. Then, for any choice of $\mu_1,\dots,\mu_K$,*

$$\liminf_{T\to\infty} \frac{R_T^{\mathrm{IID}}}{\ln T} \geq \sum_{i:\Delta_i>0} \frac{\Delta_i}{\mathrm{KL}(\mu_i,\mu^*)} .$$

This shows that when $\mu_1,\dots,\mu_K$ are fixed and $T \to \infty$, no policy can have a stochastic regret growing slower than $\Theta(K \ln T)$ in a Bernoulli bandit. For any fixed horizon T, the following stronger lower bound holds.

Theorem 2 ([3]) *For all $K \geq 2$ and any horizon T, there exist $\mu_1,\dots,\mu_K$ such that any allocation policy for the Bernoulli bandit with K actions suffers a stochastic regret of at least*

$$R_T^{\mathrm{IID}} \geq \frac{1}{20}\sqrt{KT} . \quad (4)$$

Note that Yao's minimax principle immediately implies the lower bound $\Omega\left(\sqrt{KT}\right)$ on the nonstochatic regret R_T^{DET}.

Starting with [11], several allocation policies have been proposed that achieve a stochastic regret of optimal order $K \ln T$. The UCB algorithm of [2] is a simple policy achieving this goal nonasymptotically. At each time step t, UCB selects the action i, maximizing $\overline{X}_{i,t} + C_{i,t}$. Here $\overline{X}_{i,t}$ is the sample average of payoffs in previous selections of i, and $C_{i,t}$ is an upper bound on the length of the confidence interval for the estimate of μ_i at confidence level $1 - \frac{1}{t}$.

Theorem 3 ([2]) *There exists an allocation policy that for any Bernoulli bandit with $K \geq 2$ actions satisfies*

$$R_T^{\mathrm{IID}} \leq \sum_{i:\Delta_i>0} \left(\frac{8}{\Delta_i} \ln T + 2 \right) \qquad \textit{for all } T.$$

The same result also applies to any stochastic i.i.d. bandit with payoffs bounded in $[0,1]$. A comparison with Theorem 2 can be made by removing the dependence on Δ_i in the upper bound of Theorem 3. Once rewritten in this way, the bound becomes $R_T^{\mathrm{IID}} \leq 2\sqrt{KT \ln T}$, showing optimality (up to log factors) of UCB even when the values $\mu_1, \dots, \mu_K$ can be chosen as a function of a target horizon T.

A bound on the nonstochastic regret matching the lower bound of Theorem 2 up to log factors is obtained via the randomized Exp3 policy introduced in [3]. At each time step t, Exp3 selects each action i with probability proportional to $\exp(\eta_t \widehat{X}_{i,t})$, where $\widehat{X}_{i,t}$ is an importance sampling estimate of the cumulative payoff $x_{i,1} + \cdots + x_{i,t-1}$ (recall that $x_{i,t}$ is only observed if $I_t = i$) and the parameter η_t is set to $\sqrt{(\ln K)/(tK)}$.

Theorem 4 ([3]) *For any bandit problem with $K \geq 2$ actions and deterministic payoffs $\langle x_{i,t} \rangle$, the regret of the Exp3 algorithm satisfies $R_T^{\mathrm{DET}} \leq 2\sqrt{KT \ln K}$ for all T.*

Variants

The bandit problem has been extended in several directions. For example, in the pure exploration variant of stochastic bandits [6], a different notion of regret (called simple regret) is used. In this setting, at the end of each step t, the policy has to output a recommendation J_t for the index of the optimal action. The simple regret of the policy for horizon T is then defined by $r_T = \mu^* - \mathbb{E}\,\mu_{J_T}$.

The term adaptive adversary is used to denote a generalized nonstochastic bandit problem where the payoff processes of all actions are represented by a deterministic sequence $f_1, f_2, \dots$ of functions. The payoff at time t of action i is then defined by $f_t(I_1, \dots, I_{t-1}, i)$, where $I_1, \dots, I_{t-1}$ is the sequence of actions selected by the policy up to time $t-1$. In the presence of an adaptive adversary, the appropriate performance measure is policy regret [1].

In the setting of contextual bandits [15], at each time step the allocation policy has access to side information (e.g., in the form of a feature vector). Here regret is not measured with respect to the best action, but rather with respect to the best mapping from side information to actions in a given class of such mappings.

If the set of action in a regular bandit is very large, possibly infinite, then the regret can be made small by imposing dependencies on the payoffs. For instance, the payoff at time t of each action a is defined by $f_t(a)$, where f_t is an unknown function. Control on the regret is then achieved by making specific assumptions on the space of actions a and on the payoff functions f_t (e.g., linear, Lipschitz, smooth, convex, etc.) – see, e.g., [4, 7] for early works in this direction.

Applications

Bandit problems have an increasing number of industrial applications particularly in the area of online services, where one can benefit from adapting the service to the individual sequence of requests. A prominent example of bandit application is online advertising. This is the problem of deciding which advertisement to display on the web page delivered to the next visitor of a website. A related problem is website optimization, which deals with the task of sequentially choosing design elements (font, images, layout) of the web page to be displayed to the next visitor. Another important application area is that of personalized recommendation systems, where the goal is to choose what to show from multiple content feeds in order to match the user's interest. In these applications, the payoff is associated with the users's actions, e.g., click-throughs or other desired behaviors – see, e.g., [12]. Bandits have been also applied to the problem of source routing, where a sequence of packets must be routed from a source host to a destination host in a given network, and the network protocol allows to choose a specific source-destination path for each packet to be sent. The (negative) payoff is the time it takes to deliver a packet and depends additively on the congestion of the edges in the chosen path – see, e.g., [4]. A further application area is computer game playing, where each move is chosen by simulating and evaluating many possible game continuations after the move. Algorithms for bandits (more specifically, for a tree-based version of the bandit problem) can be used to explore more efficiently the huge tree of game continuations by focusing on the most promising subtrees. This idea has been successfully implemented in the MoGo player, which plays Go at the world-class level. MoGo is based on the UCT strategy for hierarchical bandits [10], which in turn builds on the UCB allocation policy.

Cross-References

▸ Online Learning and Optimization
▸ Reinforcement Learning

Recommended Reading

1. Arora R, Dekel O, Tewari A (2009) Online bandit learning against an adaptive adversary: from regret to policy regret. In: Proceedings of the 29th international conference on machine learning, Montreal
2. Auer P, Cesa-Bianchi N, Fischer P (2002) Finite-time analysis of the multiarmed bandit problem. Mach Learn J 47(2–3):235–256
3. Auer P, Cesa-Bianchi N, Freund Y, Schapire R (2002) The nonstochastic multiarmed bandit problem. SIAM J Comput 32(1):48–77
4. Awerbuch B, Kleinberg R (2004) Adaptive routing with end-to-end feedback: distributed learning and geometric approaches. In: Proceedings of the 36th annual ACM symposium on theory of computing, Chicago. ACM, pp 45–53
5. Blackwell D (1956) An analog of the minimax theorem for vector payoffs. Pac J Math 6:1–8
6. Bubeck S, Munos R, Stoltz G (2009) Pure exploration in multi-armed bandits problems. In: Proceedings of the 20th international conference on algorithmic learning theory, Porto
7. Flaxman AD, Kalai AT, McMahan HB (2005) Online convex optimization in the bandit setting: gradient descent without a gradient. In: Proceedings of the 16th annual ACM-SIAM symposium on discrete algorithms, Philadelphia. Society for Industrial and Applied Mathematics, pp 385–394
8. Gittins J, Glazebrook K, Weber R (2011) Multi-armed bandit allocation indices, 2nd edn. Wiley, Hoboken
9. Hannan J (1957) Approximation to Bayes risk in repeated play. Contrib. Theory Games 3:97–139
10. Kocsis L, Szepesvari C (2006) Bandit based Monte-Carlo planning. In: Proceedings of the 15th European conference on machine learning, Vienna, pp 282–293
11. Lai TL, Robbins H (1985) Asymptotically efficient adaptive allocation rules. Adv Appl Math 6: 4–22
12. Li L, Chu W, Langford J, Schapire R (2010) A contextual-bandit approach to personalized news article recommendation. In: Proceedings of the 19th international conference on world wide web, Raleigh
13. Robbins H (1952) Some aspects of the sequential design of experiments. Bull Am Math Soc 58:527–535
14. Thompson W (1933) On the likelihood that one unknown probability exceeds another in view of the evidence of two samples. Bull Am Math Soc 25: 285–294
15. Wang CC, Kulkarni S, Poor H (2005) Bandit problems with side observations. IEEE Trans Autom Control 50(3):338–355

Multicommodity Flow, Well-linked Terminals and Routing Problems

Chandra Chekuri
Department of Computer Science, University of Illinois, Urbana-Champaign, Urbana, IL, USA
Department of Mathematics and Computer Science, The Open University of Israel, Raanana, Israel

Keywords

All-or-nothing multicommodity flow problem; Edge disjoint paths problem; Maximum edge disjoint paths problem; Node disjoint paths problem

Years and Authors of Summarized Original Work

2005; Chekuri, Khanna, Shepherd

Problem Definition

Three related optimization problems derived from the classical edge disjoint paths problem (EDP) are described. An instance of EDP consists of an undirected graph $G = (V, E)$ and a multiset $\mathcal{T} = \{s_1t_1, s_2t_2, \ldots, s_kt_k\}$ of k node pairs. EDP is a decision problem: can the pairs in $\mathcal{T}$ be connected (alternatively routed) via edge-disjoint paths? In other words, are there paths $P_1, P_2, \ldots, P_k$ such that for $1 \leq i \leq k$, P_i is path from s_i to t_i, and no edge $e \in E$ is in more than one of these paths? EDP is known to be NP-Complete. This article considers there maximization problems related to EDP.

- **Maximum Edge-Disjoint Paths Problem (MEDP).** Input to MEDP is the same as for EDP. The objective is to *maximize* the number of pairs in $\mathcal{T}$ that can be routed via edge-disjoint paths. The output consists of a subset $S \subseteq \{1, 2, \ldots, k\}$ and for each $i \in S$ a path P_i connecting s_i to t_i such that the paths are edge-disjoint. The goal is to maximize $|S|$.
- **Maximum Edge-Disjoint Paths Problem with Congestion (MEDPwC).** MEDPwC is a relaxation of MEDP. The input, in addition to G and the node pairs, contains an integer congestion parameter c. The output is the same for MEDP; a subset $S \subseteq \{1, 2, \ldots, k\}$ and for each $i \in S$ a path P_i connecting s_i to t_i. However, the paths $P_i, 1 \leq i \leq k$ are not required to be edge-disjoint. The relaxed requirement is that for each edge $e \in E$, the number of paths for the routed pairs that contain e is at most c. Note that MEDPwC with $c = 1$ is the same as MEDP.
- **All-or-Nothing Multicommodity Flow Problem (ANF).** ANF is a different relaxation of MEDP obtained by relaxing the notion of routing. A pair s_it_i is now said to be routed if a unit flow is sent from s_i to t_i (potentially on multiple paths). The input is the same as for MEDP. The output consists of a subset $S \subseteq \{1, 2, \ldots, k\}$ such that there is a feasible multicommodity flow in G that routes one unit of flow for each pair in S. The goal is to maximize $|S|$.

In the rest of the article, graphs are assumed to be undirected multigraphs. Given a graph $G = (V, E)$ and $S \subset V$, let $\delta_G(S)$ denote the set of edges with exactly one end point in S. Let n denote the number of vertices in the input graph.

Key Results

A few results in the broader literature are reviewed in addition to the results from [5]. EDP is NP-Complete when k is part of the input. A highly non-trivial result of Robertson and Seymour yields a polynomial time algorithm when k is a fixed constant.

Theorem 1 ([16]) *There is a polynomial time algorithm for EDP when k is a fixed constant independent of the input size.*

Using Theorem 1 it is easy to see that MEDP and MEDPwC have polynomial time algorithms for fixed k. The same holds for ANF by simple enumeration since the decision version is polynomial-time solvable via linear programming.

The focus of this article is on the case when k is part of the input, and in this setting, all three problems considered are NP-hard. The starting point for most approximation algorithms is the natural multicommodity flow relaxation given below. This relaxation is valid for both MEDP and ANF. The end points of the input pairs are referred to as *terminals* and let X denote the set of terminals. To describe the relaxation as well as simplify further discussion, the following simple assumption is made without loss of generality; each node in the graph participates in at most one of the input pairs. This assumption implies that the input pairs induce a matching M on the terminal set X. Thus the input for the problem can alternatively given as a triple (G, X, M).

For the given instance (G, X, M), let $\mathcal{P}_i$ denote the set of paths joining s_i and t_i in G and let $\mathcal{P} = \cup_i \mathcal{P}_i$. The LP relaxation has the following variables. For each path $P \in \mathcal{P}$ there is a variable $f(P)$ which is the amount of flow sent on P. For each pair $s_i t_i$ there is a variable x_i to indicate the total flow that is routed for the pair.

$$(\text{MCF} - \text{LP}) \max \sum_{i=1}^{k} x_i \quad \text{s.t}$$

$$x_i - \sum_{P \in \mathcal{P}_i} f(P) = 0 \quad 1 \le i \le k$$

$$\sum_{P: e \in P} f(P) \le 1 \quad \forall e \in E$$

$$x_i, f(P) \in [0,1] 1 \le i \le k, P \in \mathcal{P}$$

The above path formulation has an exponential (in n) number of variables, however it can still be solved in polynomial time. There is also an equivalent compact formulation with a polynomial number of variables and constraints. Let OPT denote the value of an optimum solution to a given instance. Similarly, let OPT-LP denote the value of an optimum solution the LP relaxation for the given instance. It can be seen that OPT-LP $\ge$ OPT. It is known that the integrality gap of (MCF-LP) is $\Omega(\sqrt{n})$ [10]; that is, there is an infinite family of instances such that $OPT - LP/OPT = \Omega(\sqrt{n}$. The current best approximation algorithm for MEDP is given by the following theorem.

Theorem 2 ([7]) *The integrality gap of (MCF-LP) for MEDP is $\Theta(\sqrt{n})$ and there is an $O(\sqrt{n})$ approximation for MEDP.*

For MEDPwC the approximation ratio improves with the congestion parameter c.

Theorem 3 ([18]) *There is an $O(n^{1/c})$ approximation for MEDPwC with congestion parameter c. In particular there is a polynomial time algorithm that routes $\Omega(\text{OPT} - \text{LP}/n^{1/c})$ pairs with congestion at most c.*

The above theorem is established via randomized rounding of a solution to (MCF-LP). Similar results, but via simpler combinatorial algorithms, are obtained in [2, 15].

In [5] a new framework was introduced to obtain approximation algorithm for routing problems in undirected graphs via (MCF-LP). A key part of the framework is the so-called well-linked decomposition that allows a reduction of an arbitrary instance to an instance in which the terminals satisfy a strong property.

Definition 1 Let $G = (V, E)$ be a graph. A subset $X \subseteq V$ is *cut-well-linked* in G if for every $S \subset V, |\delta_G(S)| \ge \min\{|S \cap X|, |(V \setminus S) \cap X|\}$. X is *flow-well-linked* if there exists a feasible fractional multicommodity flow in G for the instance in which there is a demand of $1/|X|$ for each unordered pair $uv, u, v \in X$.

The main result in [5] is the following.

Theorem 4 ([5]) *Let (G, X, M) be an instance of MEDP or ANF and let* OPT-LP *be the value of an optimum solution to (MCF-LP) on (G, X, M). There there is a polynomial time algorithm that obtains a collection of instances $(G_1, X_1, M_1), (G_2, X_2, M_2), \ldots, (G_h, X_h, M_h)$ with the following properties:*

- *The graphs $G_1, G_2, \ldots, G_h$ are node-disjoint induced subgraphs of G. For $1 \le i \le h, X_i \subseteq X$ and $M_i \subseteq M$.*
- *For $1 \le i \le h$, X_i is flow-well-linked in G_i.*
- $\sum_{i=1}^{h} |X_i| = \Omega(\text{OPT} - \text{LP}/\log^2 n)$.

For planar graphs and graphs that exclude a fixed minor, the above theorem gives a stronger guarantee: $\sum_{i=1}^{h} |X_i| = \Omega(\text{OPT} - \text{LP}/\log n)$. A well-linked instance satisfies a strong symmetry property based on the following observation. If A is flow-well-linked in G then for *any* matching J on X, OPT-LP on the instance (G, A, J) is $\Omega(|A|)$. Thus the particular matching M of a given well-linked instance (G, X, M) is essentially irrelevant. The second part of the framework in [5] consists of exploiting the well-linked property of the instances produced by the decomposition procedure. At a high level this is done by showing that if G has a well-linked set X, then it contains a "crossbar" (a routing structure) of size $\Omega(|X|/\text{poly}(\log n))$. See [5] for more precise definitions. Techniques for the second part vary based on the problem as well as the family of graphs in question. The following results are obtained using Theorem 4 and other non-trivial ideas for the second part [3–5, 8].

Theorem 5 ([5]) *There is an $O(\log^2 n)$ approximation for ANF. This improves to an $O(\log n)$ approximation in planar graphs.*

Theorem 6 ([5]) *There is an $O(\log n)$ approximation for MEDPwC in planar graphs for $c \geq 2$. There is an $O(\log n)$ approximation for ANF in planar graphs.*

Theorem 7 ([8]) *There is an $O(r \log n \log r)$ approximation for MEDP in graphs of treewidth at most r.*

Generalizations and Variants

Some natural variants and generalizations of the problems mentioned in this article are obtained by considering three orthogonal aspects: (i) node disjointness instead of edge-disjointness, (ii) capacities on the edges and/or nodes, and (iii) demand values on the pairs (each pair $s_i t_i$ has an integer demand d_i and the objective is to route d_i units of flow between s_i and t_i). Results similar to those mentioned in the article are shown to hold for these generalizations and variants [5]. Capacities and demand values on pairs are somewhat easier to handle while node-disjoint problems often require additional non-trivial ideas. The reader is referred to [5] for more details.

For some special classes of graphs (trees, expanders and grids to name a few), constant factor or poly-logarithmic approximation ratios are known for MEDP.

Applications

Flow problems are at the core of combinatorial optimization and have numerous applications in optimization, computer science and operations research. Very special cases of EDP and MEDP include classical problems such as single-commodity flows, and matchings in general graphs, both of which have many applications. EDP and variants arise most directly in telecommunication networks and VLSI design. Since EDP captures difficult problems as special cases, there are only a few algorithmic tools that can address the numerous applications in a unified fashion. Consequently, empirical research tends to focus on application specific approaches to obtain satisfactory solutions. The flip side of the difficulty of EDP is that it offers a rich source of problems, the study of which has led to important algorithmic advances of broad applicability, as well as fundamental insights in graph theory, combinatorial optimization, and related fields.

Open Problems

A number of very interesting open problems remain regarding the approximability of the problems discussed in this article. Table 1 gives the best known upper and lower bounds on the approximation ratio as well as integrality gap of (MCF-LP). All the inapproximability results in Table 1, and the integrality gap lower bounds for MEDPwC and ANF, are from [1]. The inapproximability results are based on the assumption that NP $\not\subseteq$ ZTIME($n^{\text{poly}(\log n)}$). Closing the gaps between the lower and upper bounds are the major open problems.

Multicommodity Flow, Well-linked Terminals and Routing Problems, Table 1 Known bounds for MEDP, ANF and MEDPwC in general undirected graphs. The best upper bound on the approximation ratio is the same as the upper bound on the integrality gap of (MCF-LP)

	Integrality gap of (MCF-LP)		Approximation ratio
	Upper bound	Lower bound	Lower bound
MEDP	$O(\sqrt{n})$	$\Omega(\sqrt{n})$	$\Omega(\log^{1/2-\epsilon} n)$
MEDPwC	$O(n^{1/c})$	$\Omega(\log^{(1-\epsilon)/(c+1)} n)$	$\Omega(\log^{(1-\epsilon)/(c+1)} n)$
ANF	$O(\log^2 n)$	$\Omega(\log^{1/2-\epsilon} n)$	$\Omega(\log^{1/2-\epsilon} n)$

Cross-References

▶ Randomized Rounding
▶ Separators in Graphs
▶ Treewidth of Graphs

Recommended Reading

The limited scope of this article does not do justice to the large literature on EDP and related problems. In addition to the articles cited in the main body of the article, the reader is referred to [6, 9, 11–14, 17] for further reading and pointers to existing literature.

1. Andrews M, Chuzhoy J, Khanna S, Zhang L (2005) Hardness of the undirected edge-disjoint paths problem with congestion. In: Proceedings of the IEEE FOCS, pp 226–244
2. Azar Y, Regev O (2006) Combinatorial algorithms for the unsplittablae flow problem. Algorithmica 44(1):49–66, Preliminary version in Proceedings of the IPCO 2001
3. Chekuri C, Khanna S, Shepherd FB (2004) Edge-disjoint paths in Planar graphs. In: Proceedings of the IEEE FOCS, pp 71–80
4. Chekuri C, Khanna S, Shepherd FB (2004) The all-or-nothing multicommodity flow problem. In: Proceedings of the ACM STOC, pp 156–165
5. Chekuri C, Khanna S, Shepherd FB (2005) Multicommodity flow, well-linked terminals, and routing problems. In: Proceedings of the ACM STOC, pp 183–192
6. Chekuri C, Khanna S, Shepherd FB (2006) Edge-disjoint paths in planar graphs with constant congestion. In: Proceedings of the ACM STOC, pp 757–766
7. Chekuri C, Khanna S, Shepherd FB (2006) An $O(\sqrt{n})$ approximation and integrality gap for disjoint paths and UFP. Theory Comput 2:137–146
8. Chekuri C, Khanna S, Shepherd FB (2007) A note on multiflows and treewidth. Algorithmica. Published online
9. Frank A (1990) Packing paths, cuts, and circuits – a survey. In: Korte B, Lovász L, Prömel HJ, Schrijver A (eds) Paths, flows and VLSI-layout. Springer, Berlin, pp 49–100
10. Garg N, Vazirani V, Yannakakis M (1997) Primal-dual approximation algorithms for integral flow and multicut in trees. Algorithmica 18(1):3–20, Preliminary version appeared in Proceedings of the ICALP 1993
11. Guruswami V, Khanna S, Rajaraman R, Shepherd FB, Yannakakis M (2003) Near-optimal hardness results and approximation algorithms for edge-disjoint paths and related problems. J Comput Syst Sci 67:473–496, Preliminary version in Proceedings of the ACM STOC 1999
12. Kleinberg JM (1996) Approximation algorithms for disjoint paths problems. PhD thesis, MIT, Cambridge, MA
13. Kleinberg JM (2005) An approximation algorithm for the disjoint paths problem in even-degree planar graphs. In: Proceedings of the IEEE FOCS, pp 627–636
14. Kolliopoulos SG (2007) Edge disjoint paths and unsplittable flow. In: Handbook on approximation algorithms and metaheuristics. Chapman & Hall/CRC Press computer & science series, vol 13. Chapman Hall/CRC Press
15. Kolliopoulos SG, Stein C (2004) Approximating disjoint-path problems using greedy algorithms and packing integer programs. Math Program A 99:63–87, Preliminary version in Proceedings of the IPCO 1998
16. Robertson N, Seymour PD (1995) Graph minors XIII. The disjoint paths problem. J Comb Theory B 63(1):65–110
17. Schrijver A (2003) Combinatorial optimization: polyhedra and efficiency. Springer, Berlin
18. Srinivasan A (1997) Improved approximations for edge-disjoint paths, unsplittable flow, and related routing problems. In: Proceedings of the IEEE FOCS, pp 416–425

M

Multicut

Shuchi Chawla
Department of Computer Science, University of Wisconsin–Madison, Madison, WI, USA

Years and Authors of Summarized Original Work

1993; Garg, Vazirani, Yannakakis
1996; Garg, Vazirani, Yannakakis

Problem Definition

The Multicut problem is a natural generalization of the s-t mincut problem – given an undirected capacitated graph $G = (V, E)$ with k pairs of vertices $\{s_i, t_i\}$; the goal is to find a subset of edges of the smallest total capacity whose removal from G disconnects s_i from t_i for every

$i \in \{1, \cdots, k\}$. However, unlike the Mincut problem which is polynomial-time solvable, the Multicut problem is known to be NP-hard and APX-hard for $k \geq 3$ [6].

This problem is closely related to the Multi-Commodity Flow problem. The input to the latter is a capacitated network with k commodities (source-sink pairs); the goal is to route as much total flow between these source-sink pairs as possible while satisfying capacity constraints. The maximum multi-commodity flow in a graph can be found in polynomial time via linear programming, and there are also several combinatorial FPTASes known for this problem [7, 9, 11].

It is immediate from the definition of Multicut that the multicommodity flow in a graph is bounded above by the capacity of a minimum multicut in the graph. When there is a single commodity to be routed, the **max-flow min-cut** theorem of Ford and Fulkerson [8] states that the converse also holds: the maximum s-t flow in a graph is exactly equal to the minimum s-t cut in the graph. This duality between flows and cuts in a graph has many applications and, in particular, leads to a simple algorithm for finding the minimum cut in a graph.

Given its simplicity and elegance, several attempts have been made to extend this duality to other classes of flow and partitioning problems. Hu showed, for example, that the min-multicut equals the maximum multi-commodity flow when there are only two commodities in the graph [12]. Unfortunately, this property does not extend to graphs with more than two commodities. The focus has therefore been on obtaining approximate max-multicommodity flow min-multicut theorems. Such theorems would also imply a polynomial-time algorithm for approximately computing the capacity of the minimum multicut in a graph.

Key Results

Garg, Vazirani and Yannakakis [10] were the first to obtain an approximate max-multicommodity flow min-multicut theorem. They showed that the maximum multicommodity flow in a graph is always at least an $O(\log k)$ fraction of the minimum multicut in the graph. Moreover, their proof of this result is constructive. That is, they also provide an algorithm for computing a multicut for a given graph with capacity at most $O(\log k)$ times the maximum multicommodity flow in the graph. This is the best approximation algorithm known to date for the Multicut problem.

Theorem 1 *Let M denote the minimum multicut in a graph with k commodities and f denote the maximum multicommodity flow in the graph. Then*

$$\frac{M}{O(\log k)} \leq f \leq M .$$

Moreover, there is a polynomial time algorithm for finding an $O(\log k)$-approximate multicut in a graph.

Furthermore, they show that this theorem is tight to within constant factors. That is, there are families of graphs in which the gap between the maximum multicommodity flow and minimum multicut is $\Theta(\log k)$.

Theorem 2 *There exists a infinite family of multicut instances $\{(G_k, P_k)\}$ such that for all k, the graph $G_k = (V_k, E_k)$ contains k vertices and $P_k \subseteq V_k \times V_k$ is a set of $\Omega(k^2)$ source-sink pairs. Furthermore, the maximum multicommodity flow in the instance (G_k, P_k) is $O(k/\log k)$ and the minimum multicut is $\Omega(k)$.*

Garg et al. also consider the Sparsest Cut problem which is another partitioning problem closely related to Multicut, and provided an approximation algorithm for this problem. Their results for Sparsest Cut have subsequently been improved upon [3, 15]. The reader is referred to the entry on ▶ Sparsest Cut for more details.

Applications

A key application of the Multicut problem is to the 2CNF ≡ Deletion problem. The latter is a constraint satisfaction problem in which given a weighted set of clauses of the form $P \equiv Q$, where P and Q are literals, the goal is to delete a minimum weight set of clauses so that the remaining set is satisfiable. The 2CNF $\equiv$ Deletion problem models a number of partitioning problems, for example the Minimum Edge-Deletion Graph Bipartization problem – finding the minimum weight set of edges whose deletion makes a graph bipartite. Klein et al. [14] showed that the 2CNF $\equiv$ Deletion problem reduces in an approximation preserving way to Multicut. Therefore, a ρ-approximation to Multicut implies a ρ-approximation to 2CNF $\equiv$ Deletion. (See the survey by Shmoys [16] for more applications.)

Open Problems

There is a big gap between the best-known algorithm for Multicut and the best hardness result (APX-hardness) known for the problem. Improvements in either direction may be possible, although there are indications that the $O(\log k)$ approximation is the best possible. In particular, Theorem 2 implies that the integrality gap of the natural linear programming relaxation for Multicut is $\Theta(\log k)$. Although improved approximations have been obtained for other partitioning problems using semi-definite programming instead of linear programming, Agarwal et al. [1] showed that similar improvements cannot be achieved for Multicut – the integrality gap of the natural SDP-relaxation for Multicut is also $\Theta(\log k)$. On the other hand, there are indications that the APX-hardness is not tight. In particular, assuming the so-called Unique Games conjecture, it has been shown that Multicut cannot be approximated to within any constant factor [4, 13]. In light of these negative results, the main open problem related to this work is to obtain a super-constant hardness for the Multicut problem under a standard assumption such as $P \neq NP$.

The Multicut problem has also been studied in directed graphs. The best known approximation algorithm for this problem is an $O(n^{11/23} \log^{O(1)} n)$-approximation due to Aggarwal, Alon and Charikar [2], while on the hardness side, Chuzhoy and Khanna [5] show that there is no $2^{\Omega(\log^{1-\epsilon} n)}$ approximation, for any $\epsilon > 0$, unless NP$\subseteq$ZPP. Chuzhoy and Khanna also exhibit a family of instances for which the integrality gap of the natural LP relaxation of this problem (which is also the gap between the maximum directed multicommodity flow and the minimum directed multicut) is $\Omega(n^{1/7})$.

Cross-References

▶ Sparsest Cut

Recommended Reading

1. Agarwal A, Charikar M, Makarychev K, Makarychev Y (2005) O($\sqrt{\log n}$) approximation algorithms for Min UnCut, Min 2CNF deletion, and directed cut problems. In: Proceedings of the 37th ACM symposium on theory of computing (STOC), Baltimore, pp 573–581
2. Aggarwal A, Alon N, Charikar M (2007) Improved approximations for directed cut problems. In: Proceedings of the 39th ACM symposium on theory of computing (STOC), San Diego, pp 671–680
3. Arora S, Satish R, Vazirani U (2004) Expander flows, geometric embeddings, and graph partitionings. In: Proceedings of the 36th ACM symposium on theory of computing (STOC), Chicago, pp 222–231
4. Chawla S, Krauthgamer R, Kumar R, Rabani Y, Sivakumar D (2005) On the hardness of approximating sparsest cut and multicut. In: Proceedings of the 20th IEEE conference on computational complexity (CCC), San Jose, pp 144–153
5. Chuzhoy J, Khanna S (2007) Polynomial flow-cut gaps and hardness of directed cut problems. In: Proceedings of the 39th ACM symposium on theory of computing (STOC), San Diego, pp 179–188
6. Dahlhaus E, Johnson DS, Papadimitriou CH, Seymour PD, Yannakakis M (1994) The complexity of multiterminal cuts. SIAM Comput J 23(4):864–894
7. Fleischer L (1999) Approximating fractional multicommodity flow independent of the number of commodities. In: Proceedings of the 40th IEEE symposium on foundations of computer science (FOCS), New York, pp 24–31
8. Ford LR, Fulkerson DR (1956) Maximal flow through a network. Can J Math 8:399–404
9. Garg N, Könemann J (1998) Faster and simpler algorithms for multicommodity flow and other fractional packing problems. In: Proceedings of the 39th IEEE symposium on foundations of computer science (FOCS), pp 300–309

M

10. Garg N, Vazirani VV, Yannakakis M (1996) Approximate max-flow min-(multi)cut theorems and their applications. SIAM Comput J 25(2):235–251
11. Grigoriadis MD, Khachiyan LG (1996) Coordination complexity of parallel price-directive decomposition. Math Oper Res 21:321–340
12. Hu TC (1963) Multi-commodity network flows. Oper Res 11(3):344–360
13. Khot S, Vishnoi N (2005) The unique games conjecture, integrality gap for cut problems and the embeddability of negative-type metrics into l_1 In: Proceedings of the 46th IEEE symposium on foundations of computer science (FOCS), pp 53–62
14. Klein P, Agrawal A, Ravi R, Rao S (1990) Approximation through multicommodity flow. In: Proceedings of the 31st IEEE symposium on foundations of computer science (FOCS), pp 726–737
15. Linial N, London E, Rabinovich Y (1995) The geometry of graphs and some of its algorithmic applications. Combinatorica 15(2):215–245, Also in Proceedings of 35th FOCS, pp 577–591 (1994)
16. Shmoys DB (1997) Cut problems and their application to divide-and-conquer. In: Hochbaum DS (ed) Approximation algorithms for NP-hard problems. PWS, Boston, pp 192–235

Multidimensional Compressed Pattern Matching

Amihood Amir
Department of Computer Science, Bar-Ilan University, Ramat-Gan, Israel
Department of Computer Science, Johns Hopkins University, Baltimore, MD, USA

Keywords

Multidimensional compressed search; Pattern matching in compressed images; Two-dimensional compressed matching

Years and Authors of Summarized Original Work

2003; Amir, Landau, Sokol

Problem Definition

Let c be a given compression algorithm, and let $c(D)$ be the result of c compressing data D. The *compressed search problem with compression algorithm c* is defined as follows.

INPUT: Compressed text $c(T)$ and pattern P.
OUTPUT: All locations in T where pattern P occurs.

A compressed matching algorithm is *optimal* if its time complexity is $O(|c(T)|)$.

Although optimality in terms of time is always important, when dealing with compression, the criterion of **extra space** is perhaps more important (Ziv, Personal communication, 1995). Applications employ compression techniques specifically because there is a limited amount of available space. Thus, it is not sufficient for a compressed matching algorithm to be optimal in terms of time, it must also satisfy the given space constraints. Space constraints may be due to limited amount of disk space (e.g., on a server), or they may be related to the size of the memory or cache. Note that if an algorithm uses as little extra space as the size of the cache, the runtime of the algorithm is also greatly reduced as no cache misses will occur [13]. It is also important to remember that in many applications, e.g., LZ compression on strings, the *compression ratio* – $|S|/|c(S)|$ – is a small constant. In a case where the compression ratio of the given text is a constant, an optimal compressed matching performs no better than the naive algorithm of decompressing the text. However, if the constants hidden in the *"big O"* are smaller than the compression ratio, then the compressed matching does offer a practical benefit. If those constants are larger than the optimal the compressed search algorithm may, in fact, be using more space than the uncompressed text.

Definition 1 (inplace) A compressed matching is said to be *inplace* if the extra space used is proportional to the input size of the pattern.

Note that this definition encompasses the compressed matching model (e.g., [2]) where the pattern is input in uncompressed form, as well as the *fully compressed* model [10], where the pattern is input in compressed form. The *inplace* requirement allows the extra space to be the input size of the pattern, whatever that size may be. However, in many applications the compression ratio is a constant; therefore, a stronger space constraint is defined.

Definition 2 Let $\mathcal{AP}$ be the set of all patterns of size m, and let $c(\mathcal{AP})$ be the set of all compressed images of $\mathcal{AP}$. Let m' be the length of the smallest pattern in $c(\mathcal{AP})$. A compressed matching algorithm with input pattern P of length m is called *strongly inplace* if the amount of extra space used is proportional to m'.

The problem as defined above is equally applicable to textual (one-dimensional), image (two-dimensional), or any type of data, such as bitmaps, concordances, tables, XML data, or any possible data structure.

The compressed matching problem is considered crucial in image databases, since they are highly compressible. The initial definition of the compressed matching paradigm was motivated by the *two dimensional run-length compression*. This is the compression used for fax transmissions. The run-length compression is defined as follows.

Let $S = s_1 s_2 \cdots s_n$ be a string over some alphabet Σ. The *run-length compression* of string S is the string $S' = \sigma_1^{r_1} \sigma_2^{r_2} \cdots \sigma_k^{r_k}$ such that (1) $\sigma_i \neq \sigma_{i+1}$ for $1 \leq i < k$ and (2) S can be described as the concatenation of k *segments*, the symbol σ_1 repeated r_1 times, the symbol σ_2 repeated r_2 times, ..., and the symbol σ_k repeated r_k times. The *two-dimensional run-length compression* is the concatenation of the run-length compression of all the matrix rows (or columns).

The *two-dimensional run-length compressed matching problem* is defined as follows:

INPUT: Text array T of size $n \times n$, and pattern array P of size $m \times m$ *both* in two-dimensional run-length compressed form.

OUTPUT: All locations in T of occurrences of P. Formally, the output is the set of locations (i, j) such that $T[i+k, j+l] = P[k+1, l+1] k, l = 0 \ldots m-1$.

Another ubiquitous lossless two-dimensional compression is CompuServe's GIF standard, widely used on the World Wide Web. It uses LZW [19] (a variation of LZ78) on the image linearized row by row.

The *two-dimensional LZ compression* is formally defined as follows. Given an image $T[1 \ldots n, 1 \ldots n]$, create a string $T_{lin}\ [1 \ldots n^2]$ by concatenating all rows of T. Compressing T_{lin} with one-dimensional LZ78 yields the two-dimensional LZ compression of the image T.

The *two-dimensional LZ compressed matching problem* is defined as follows:

INPUT: Text array T of size $n \times n$, and pattern array P of size $m \times m$ *both* in two-dimensional LZ compressed form.

OUTPUT: All locations in T of occurrences of P. Formally, the output is the set of locations (i, j) such that $T[i+k, j+l] = P[k+1, l+1]$ $k, l = 0 \ldots m-1$.

Key Results

The definition of compressed search first appeared in the context of searching for two dimensional run-length compression [1, 2]. The following result was achieved there.

Theorem 1 (Amir and Benson [3]) *There exists an $O(|c(T)| \log |c(T)|)$ worst-case time solution to the compressed search problem with the two dimensional run-length compression algorithm.*

The above mentioned paper did not succeed in achieving either an optimal or an inplace algorithm. Nevertheless, it introduced the notion of *two-dimensional periodicity*. As in strings, periodicity plays a crucial rôle in two-dimensional string matching, and its advent has provided solutions to many longstanding open problems of two-dimensional string matching. In [5], it was used to achieve the first linear-time, alphabet-independent, two-dimensional text scanning. Later, in [4, 16] it was used in two different ways for a linear-time witness table construction. In [7] it was used to achieve the first parallel, time and work optimal, CREW algorithm for text scanning. A simpler variant of periodicity was used by [11] to obtain a constant-time CRCW algorithm for text scanning. A recent further attempt has been made [17] to generalize periodicity analysis to higher dimensions.

The first optimal two-dimensional compressed search algorithm was the following.

Theorem 2 (Amir et al. [6]) *There exists an* $O(|c(T)|)$ *worst-case time solution to the compressed search problem with the two-dimensional run-length compression algorithm.*

Optimality was achieved by a concept the authors called *witness-free dueling*. The paper proved new properties of two-dimensional periodicity. This enables duels to be performed in which no witness is required. At the heart of the dueling idea lies the concept that two overlapping occurrences of a pattern in a text can use the content of a predetermined text position or witness in the overlap to eliminate one of them. Finding witnesses is a costly operation in a compressed text; thus, the importance of witness-free dueling.

The original algorithm of Amir et al. [6] takes time $O(|c(T)| + |P| \log \sigma)$, where σ is $\min(|P|, |\Sigma|)$, and Σ is the alphabet. However with the witness table construction of Galil and Park [12] the time is reduced to $O(|c(T)| + |P|)$. Using known techniques, one can modify their algorithm so that its extra space is $O(|P|)$. This creates an optimal algorithm that is also inplace, provided the pattern is input in uncompressed form. With use of the run-length compression, the difference between $|P|$ and $|c(P)|$ can be quadratic. Therefore it is important to seek an inplace algorithm.

Theorem 3 (Amir et al. [9]) *There exists an* $O(|c(T)| + |P| \log \sigma)$ *worst-case time solution to the compressed search problem with the two-dimensional run-length compression algorithm, where* σ *is* $\min(|P|, |\Sigma|)$*, and* Σ *is the alphabet, for all patterns that have no trivial rows (rows consisting of a single repeating symbol). The amount of space used is* $O(|c(P)|)$.

This algorithm uses the framework of the noncompressed two dimensional pattern matching algorithm of [6]. The idea is to use the **dueling** mechanism defined by Vishkin [18]. Applying the dueling paradigm directly to run-length compressed matching has previously been considered impossible since the location of a witness in the compressed text cannot be accessed in constant time. In [9], a way was shown in which a witness *can* be accessed in (amortized) constant time, enabling a relatively straightforward application of the dueling paradigm to compressed matching.

A strongly inplace compressed matching algorithm exists for the two-dimensional LZ compression, but its preprocessing is not optimal.

Theorem 4 (Amir et al. [8]) *There exists an* $O(|c(T)| + |P|^3 \log \sigma)$ *worst-case time solution to the compressed search problem with the two-dimensional LZ compression algorithm, where* σ *is* $\min(|P|, |\Sigma|)$*, and* Σ *is the alphabet. The amount of space used is* $O(m)$*, for an* $m \times m$ *size pattern.* $O(m)$ *is the best compression achievable for any* $m \times m$ *sized pattern under the two-dimensional LZ compression.*

The algorithm of [8] can be applied to any two-dimensional compressed text, in which the compression technique allows sequential decompression in small space.

Applications

The problem has many applications since two-dimensional data appears in many different types of compression. The two compressions discussed here are the run-length compression, used by fax transmissions, and the LZ compression, used by GIF.

Open Problems

Any lossless two-dimensional compression used, especially one with a large compression ratio, presents the problem of enabling the search without uncompressing the data for saving of both time and space.

Searching in two-dimensional lossy compressions will be a major challenge. Initial steps in this direction can be found in [14, 15], where JPEG compression is considered.

Cross-References

► Multidimensional Compressed Pattern Matching
► Multidimensional String Matching

Recommended Reading

1. Amir A, Benson G (1992) Efficient two dimensional compressed matching. In: Proceeding of data compression conference, Snow Bird, pp 279–288
2. Amir A, Benson G (1992) Two-dimensional periodicity and its application. In: Proceeding of 3rd symposium on discrete algorithms, Orlando, pp 440–452
3. Amir A, Benson G (1998) Two-dimensional periodicity and its application. SIAM J Comput 27(1):90–106
4. Amir A, Benson G, Farach M (1992) The truth, the whole truth, and nothing but the truth: alphabet independent two dimensional witness table construction. Technical Report GITCC-92/52, Georgia Institute of Technology
5. Amir A, Benson G, Farach M (1994) An alphabet independent approach to two dimensional pattern-matching. SIAM J Comput 23(2):313–323
6. Amir A, Benson G, Farach M (1997) Optimal two-dimensional compressed matching. J Algorithms 24(2):354–379
7. Amir A, Benson G, Farach M (1998) Optimal parallel two dimensional text searching on a crew pram. Inf Comput 144(1):1–17
8. Amir A, Landau G, Sokol D (2003) Inplace 2D matching in compressed images. J Algorithms 49(2):240–261
9. Amir A, Landau G, Sokol D (2003) Inplace run-length 2D compressed search. Theory Comput Sci 290(3):1361–1383
10. Berman P, Karpinski M, Larmore L, Plandowski W, Rytter W (1997) On the complexity of pattern matching for highly compressed two dimensional texts. In: Proceeding of 8th annual symposium on combinatorial pattern matching (CPM 97). LNCS, vol 1264. Springer, Berlin, pp 40–51
11. Crochemore M, Galil Z, Gasieniec L, Hariharan R, Muthukrishnan S, Park K, Ramesh H, Rytter W (1993) Parallel two-dimensional pattern matching. In: Proceeding of 34th annual IEEE FOCS, pp 248–258
12. Galil Z, Park K (1996) Alphabet-independent two-dimensional witness computation. SIAM J Comput 25(5):907–935
13. Hennessy JL, Patterson DA (1996) Computer architecture: a quantitative approach, 2nd edn. Morgan Kaufmann, San Francisco
14. Klein ST, Shapira D (2005) Compressed pattern matching in JPEG images. In: Proceeding Prague stringology conference, pp 125–134
15. Klein ST, Wiseman Y (2003) Parallel huffman decoding with applications to JPEG files. Comput J 46(5):487–497
16. Park K, Galil Z (1992) Truly alphabet-independent two-dimensional pattern matching. In: Proceeding 33rd IEEE FOCS, pp 247–256
17. Régnier M, Rostami L (1993) A unifying look at d-dimensional periodicities and space coverings. In: 4th symposium on combinatorial pattern matching, p 15
18. Vishkin U (1985) Optimal parallel pattern matching in strings. In: Proceeding 12th ICALP, pp 91–113
19. Welch TA (1984) A technique for high-performance data compression. IEEE Comput 17:8–19

Multidimensional String Matching

Juha Kärkkäinen
Department of Computer Science, University of Helsinki, Helsinki, Finland

Keywords

Approximate string matching; Combinatorial pattern matching; Image matching; Multidimensional array matching; Multidimensional string algorithms; Rotation invariance; Scaling invariance

Years and Authors of Summarized Original Work

1977; Bird
1978; Baker
1991; Amir, Landau
1994; Amir, Benson, Farach
1999; Kärkkäinen, Ukkonen
2000; Baeza-Yates, Navarro
2002; Fredriksson, Navarro, Ukkonen
2006; Amir, Kapah, Tsur
2009; Hundt, Liśkiewicz, Nevries
2010; Amir, Chencinski

Problem Definition

Given two two-dimensional arrays, the *text* $T[1 \ldots n, 1 \ldots n]$ and the *pattern* $P[1 \ldots m, 1 \ldots m]$, $m \leq n$, both with element values from *alphabet* Σ of size σ, the basic *two-dimensional string matching* (2DSM) problem is to find all *occurrences* of P in T, i.e., all $m \times m$ subarrays of T that are identical to P. In addition to the basic problem, several types of generalizations are considered: *approximate matching* (allow local errors), *invariant matching* (allow global transformations), and *multidimensional matching*.

In approximate matching, an occurrence is a subarray S of the text, whose *distance* $d(S, P)$ from the pattern does not exceed a threshold k. Different distance measures lead to different variants of the problem. When no distance is explicitly mentioned, the *Hamming distance*, the number of mismatching elements, is assumed.

For one-dimensional strings, the most common distance is the Levenshtein distance, the minimum number of insertions, deletions, and substitutions for transforming one string into the other. A simple generalization to two dimensions is the *Krithivasan–Sitalakshmi (KS) distance*, which is the sum of row-wise Levenshtein distances. Baeza-Yates and Navarro [6] introduced several other generalizations, one of which, the *RC distance*, is defined as follows. A two-dimensional array can be decomposed into a sequence of rows and columns by removing either the last row or the last column from the array until nothing is left. Different decompositions are possible depending on whether a row or a column is removed at each step. The RC distance is the minimum cost of transforming a decomposition of one array into a decomposition of the other, where the minimum is taken over all possible decompositions as well as all possible transformations. A transformation consists of insertions, deletions, and modifications of rows and columns. The cost of inserting or deleting a row/column is the length of the row/column, and the cost of modification is the Levenshtein distance between the original and the modified row/column.

The invariant matching problems search for occurrences that match the pattern after some global transformation of the pattern. In the *scaling invariant matching* problem, an occurrence is a subarray that matches the pattern scaled by some factor. If only integral scaling factors are allowed, the definition of the problem is obvious. For real-valued scaling, a refined model is needed, where the text and pattern elements, called *pixels* in this case, are unit squares on a plane. Scaling the pattern means stretching the pixels. An occurrence is a matching M between text pixels and pattern pixels. The scaled pattern is placed on top of the text with one corner aligned, and each text pixel $T[r, s]$, whose center is covered by the pattern, is matched with the covering pattern pixel $P[r', s']$, i.e., $([r, s], [r', s']) \in M$.

In the *rotation invariant matching* problem, too, an occurrence is a matching between text pixels and pattern pixels. This time the center of the pattern is placed at the center of a text pixel, and the pattern is rotated around the center. The matching is again defined by which pattern pixels cover which text pixel centers.

All the problems can be generalized to more than two dimensions. In the d-dimensional problem, the text is an n^d array and the pattern an m^d array. The focus is on two dimensions, but multidimensional generalizations of the results are mentioned when they exist.

Many other variants of the problems are omitted here due to a lack of space. Some of them as well as some of the results in this entry are surveyed by Amir [1]. A wider range of problems as well as traditional image processing techniques for solving them can be found in [9].

Key Results

The classical solution to the 2DSM problem by Bird [8] and independently by Baker [7] reduces the problem to one-dimensional string matching. It has two phases:

1. Find all occurrences of pattern rows on the text rows and mark them. This takes $\mathcal{O}(n^2 \log \min(m, \sigma))$ time using the Aho–Corasick algorithm. On an integer alphabet $\Sigma = \{0, 1, \ldots, \sigma - 1\}$, the time can be improved to $\mathcal{O}(n^2 + m^2 \min(m, \sigma) + \sigma)$ using $\mathcal{O}(m^2 \min(m, \sigma) + \sigma)$ space.
2. The pattern is considered a sequence of m rows and each $n \times m$ subarray of the text a sequence of n rows. The Knuth–Morris–Pratt string matching algorithm is used for finding

the occurrences of the pattern in each subarray. The algorithm makes $\mathcal{O}(n)$ row comparisons for each of the $n - m + 1$ subarrays. With the markings from Step 1, a row comparison can be done in constant time, giving $\mathcal{O}(n^2)$ time complexity for Step 2.

The time complexity of the Bird–Baker algorithm is linear if the alphabet size σ is constant. The algorithm of Amir, Benson, and Farach [4] (with improvements by Galil and Park [13]) achieves linear time independent of the alphabet size using a quite different kind of algorithm based on string matching by duels and two-dimensional periodicity.

Theorem 1 (Bird [8]; Baker [7]; Amir, Benson, and Farach [4]) *The 2DSM problem can be solved in the optimal $\mathcal{O}(n^2)$ worst-case time.*

The Bird–Baker algorithm generalizes straightforwardly into higher dimensions by repeated application of Step 1 to reduce a problem in d dimensions into $n - m + 1$ problems in $d - 1$ dimensions. The time complexity is $\mathcal{O}(dn^d \log m^d)$. The Amir–Benson–Farach algorithm has been generalized to three dimensions with the time complexity $\mathcal{O}(n^3)$ [14].

The average-case complexity of the 2DSM problem was studied by Kärkkäinen and Ukkonen [16], who proved a lower bound and gave an algorithm matching the bound.

Theorem 2 (Kärkkäinen and Ukkonen [16]) *The 2DSM problem can be solved in the optimal $\mathcal{O}(n^2(\log_\sigma m)/m^2)$ average-case time.*

The result (both lower and upper bound) generalizes to the d-dimensional case with the $\Theta(n^d \log_\sigma m/m^d)$ average-case time complexity.

Amir and Landau [3] give algorithms for approximate 2DSM problems for both the Hamming distance and the KS distance. The RC model was developed and studied by Baeza–Yates and Navarro [6].

Theorem 3 (Amir and Landau [3]; Baeza–Yates and Navarro [6]) *The approximate 2DSM problem can be solved in $\mathcal{O}(kn^2)$ worst-case time for the Hamming distance, in $\mathcal{O}(k^2n^2)$ worst-case time for the KS distance, and in $\mathcal{O}(k^2mn^2)$ worst-case time for the RC distance.*

The results for the KS and RC distances generalize to d dimensions with the time complexities $\mathcal{O}(k(k + d)n^d)$ and $\mathcal{O}(d!m^{2d}n^d)$, respectively.

Approximate matching algorithms with good average-case complexity are described by Kärkkäinen and Ukkonen [16] for the Hamming distance and by Baeza–Yates and Navarro [6] for the KS and RC distances.

Theorem 4 (Kärkkäinen and Ukkonen [16]; Baeza–Yates and Navarro [6]) *The approximate 2DSM problem can be solved in $\mathcal{O}(kn^2(\log_\sigma m)/m^2)$ average-case time for the Hamming and KS distances and in $\mathcal{O}(n^2/m)$ average-case time for the RC distance.*

The results for the Hamming and the RC distance have d-dimensional generalizations with the time complexities $\mathcal{O}(kn^d(\log_\sigma m^d)/m^d)$ and $\mathcal{O}(kn^d/m^{d-1})$, respectively.

The scaling and rotation invariant 2DSM problems involve a continuous valued parameter (scaling factor or rotation angle). However, the corresponding matching between text and pattern pixels changes only at certain points, and there are only $\mathcal{O}(nm)$ effectively distinct scales and $\mathcal{O}(m^3)$ effectively distinct rotation angles. A separate search for each distinct scale or rotation would give algorithms with time complexities $\mathcal{O}(n^3m)$ and $\mathcal{O}(n^2m^3)$, but faster algorithms exist.

Theorem 5 (Amir and Chencinski [2]; Amir, Kapah, and Tsur [5]) *The scaling invariant 2DSM problem can be solved in $\mathcal{O}(n^2m)$ worst-case time, and the rotation invariant 2DSM problem in $\mathcal{O}(n^2m^2)$ worst-case time.*

Fast average-case algorithms for the rotation invariant problem are described by Fredriksson, Navarro, and Ukkonen [12]. They also consider approximate matching versions.

Theorem 6 (Fredriksson, Navarro, and Ukkonen [12]) *The rotation invariant 2DSM problem can be solved in the optimal $\mathcal{O}(n^2(\log_\sigma m)/m^2)$ average-case time. The rotation invariant approximate 2DSM problem can be solved in the optimal $\mathcal{O}(n^2(k + \log_\sigma m)/m^2)$ average-case time.*

M

Fredriksson, Navarro, and Ukkonen [12] also consider rotation invariant matching in d dimensions.

Hundt, Liśkiewicz, and Nevries [15] show that there are $\mathcal{O}(n^4m^2)$ effectively distinct combinations of scales and rotations and give an $\mathcal{O}(n^6m^2)$ time algorithm for finding the best match under a distance that generalizes the Hamming distance implying the following result.

Theorem 7 (Hundt, Liśkiewicz, and Nevries [15]) *The scaling and rotation invariant 2DSM and approximate 2DSM problems can be solved in* $\mathcal{O}(n^6m^2)$ *time.*

Applications

The main application area is pattern matching in images, particularly applications where the point of view in the image is well defined, such as aerial and astronomical photography, optical character recognition, and biomedical imaging. Even three-dimensional problems arise in biomedical applications [10].

Open Problems

There may be some room for improving the results under the combined scaling and rotation invariance using techniques similar to those in Theorems 5 and 6. Many combinations of the different variants of the problem have not been studied. With rotation invariant approximate matching under the RC distance even the problem needs further specification.

Experimental Results

No conclusive results exist though some experiments are reported in [10, 11, 15, 16].

Cross-References

- ▶ Approximate String Matching
- ▶ Indexed Two-Dimensional String Matching
- ▶ String Matching

Recommended Reading

1. Amir A (2005) Theoretical issues of searching aerial photographs: a bird's eye view. Int J Found Comput Sci 16(6):1075–1097
2. Amir A, Chencinski E (2010) Faster two dimensional scaled matching. Algorithmica 56(2): 214–234
3. Amir A, Landau GM (1991) Fast parallel and serial multidimensional approximate array matching. Theor Comput Sci 81(1):97–115
4. Amir A, Benson G, Farach M (1994) An alphabet independent approach to two-dimensional pattern matching. SIAM J Comput 23(2):313–323
5. Amir A, Kapah O, Tsur D (2006) Faster two-dimensional pattern matching with rotations. Theor Comput Sci 368(3):196–204
6. Baeza-Yates R, Navarro G (2000) New models and algorithms for multidimensional approximate pattern matching. J Discret Algorithms 1(1):21–49
7. Baker TP (1978) A technique for extending rapid exact-match string matching to arrays of more than one dimension. SIAM J Comput 7(4): 533–541
8. Bird RS (1977) Two dimensional pattern matching. Inf Process Lett 6(5):168–170
9. Brown LG (1992) A survey of image registration techniques. ACM Comput Surv 24(4):325–376
10. Fredriksson K, Ukkonen E (2000) Combinatorial methods for approximate pattern matching under rotations and translations in 3D arrays. In: Proceedings of the 7th international symposium on string processing and information retrieval, A Coruña. IEEE Computer Society, pp 96–104
11. Fredriksson K, Navarro G, Ukkonen E (2002) Faster than FFT: rotation invariant combinatorial template matching. In: Pandalai S (ed) Recent research developments in pattern recognition, vol II. Transworld Research Network, Trivandrum, pp 75–112
12. Fredriksson K, Navarro G, Ukkonen E (2005) Sequential and indexed two-dimensional combinatorial template matching allowing rotations. Theor Comput Sci 347(1–2):239–275
13. Galil Z, Park K (1996) Alphabet-independent two-dimensional witness computation. SIAM J Comput 25(5):907–935
14. Galil Z, Park JG, Park K (2004) Three-dimensional periodicity and its application to pattern matching. SIAM J Discret Math 18(2): 362–381
15. Hundt C, Liśkiewicz M, Nevries R (2009) A combinatorial geometrical approach to two-dimensional robust pattern matching with scaling and rotation. Theor Comput Sci 410(51): 5317–5333
16. Kärkkäinen J, Ukkonen E (1999) Two- and higher-dimensional pattern matching in optimal expected time. SIAM J Comput 29(2): 571–589

Multilevel Feedback Queues

Nikhil Bansal
Eindhoven University of Technology,
Eindhoven, The Netherlands

Keywords

Fairness; Low sojourn times; Scheduling with unknown job sizes

Years and Authors of Summarized Original Work

1968; Coffman, Kleinrock

Problem Definition

The problem is concerned with scheduling dynamically arriving jobs in the scenario when the processing requirements of jobs are unknown to the scheduler. This is a classic problem that arises, for example, in CPU scheduling, where users submit jobs (various commands to the operating system) over time. The scheduler is only aware of the existence of the job and does not know how long it will take to execute, and the goal is to schedule jobs to provide good quality of service to the users. Formally, this note considers the average flow time measure, defined as the average duration of time since a job is released until its processing requirement is met.

Notations

Let $\mathcal{J} = \{1, 2, \ldots, n\}$ denote the set of jobs in the input instance. Each job j is characterized by its release time r_j and its processing requirement p_j. In the online setting, job j is revealed to the scheduler only at time r_j. A further restriction is the *non-clairvoyant* setting, where only the existence of job j is revealed at r_j, in particular the scheduler does not know p_j until the job meets its processing requirement and leaves the system. Given a schedule, the completion time c_j of a job is the earliest time at which job j receives p_j amount of service. The flow time f_j of j is defined as $c_j - r_j$. A schedule is said to be preemptive, if a job can be interrupted arbitrarily, and its execution can be resumed later from the point of interruption without any penalty. It is well known that preemption is necessary to obtain reasonable guarantees even in the offline setting [4].

There are several natural non-clairvoyant algorithms such as first come first served, processor sharing (work on all current unfinished jobs at equal rate), and shortest elapsed time first (work on job that has received least amount of service thus far). Coffman and Kleinrock [2] proposed another natural algorithm known as the multilevel feedback queueing (MLF). MLF works as follows: there are queues $Q_0, Q_1, Q_2, \ldots$ and thresholds $0 < t_0 < t_1 < t_2 \ldots$. Initially upon arrival, a job is placed in Q_0. When a job in Q_i receives t_i amount of cumulative service, it is moved to Q_{i+1}. The algorithm at any time works on the lowest numbered nonempty queue. Coffman and Kleinrock analyzed MLF in a queuing theoretic setting, where the jobs arrive according to a Poisson process and the processing requirements are chosen identically and independently from a known probability distribution.

Recall that the online shortest remaining processing time (SRPT) algorithm that at any time works on the job with the least remaining processing time produces an optimum schedule. However, SRPT requires the knowledge of job sizes and hence is not non-clairvoyant. Since a non-clairvoyant algorithm only knows a lower bound on a jobs size (determined by the amount of service it has received thus far), MLF tries to mimic SRPT by favoring jobs that have received the least service thus far.

Key Results

While non-clairvoyant algorithms have been studied extensively in the queuing theoretic setting for many decades, this notion was considered relatively recently in the context of competitive analysis by Motwani, Phillips, and Torng [5]. As in traditional competitive analysis, a non-clairvoyant algorithm is called c-competitive if for every input instance, its performance is no worse than c times that of the optimum offline solution for that instance. Motwani, Phillips, and Torng showed the following.

Theorem 1 ([5]) *For the problem of minimizing average flow time on a single machine, any deterministic non-clairvoyant algorithm must have a competitive ratio of at least* $\Omega(n^{1/3})$*, and any randomized algorithm must have a competitive ratio of at least* $\Omega(\log n)$*, where* n *is number of jobs in the instance.*

It is not too surprising that any deterministic algorithm must have a poor competitive ratio. For example, consider MLF where the thresholds are powers of 2, i.e., $1, 2, 4, \ldots$. Say $n = 2^k$ jobs of size $2^k + 1$ each arrive at times $0, 2^k, 2 \cdot 2^k, \ldots, (2^k - 1)2^k$, respectively. Then, it is easily verified that the average flow time under MLF is $\Omega(n^2)$, where as the average flow time is under the optimum algorithm is $\Omega(n)$.

Note that MLF performs poorly on the above instance since all jobs are stuck till the end with just one unit of work remaining. Interestingly, Kalyanasundaram and Pruhs [3] designed a randomized variant of MLF (known as RMLF) and proved that its competitive ratio is almost optimum. For each job j, and for each queue Q_i, the RMLF algorithm sets a threshold $t_{i,j}$ randomly and independently according to a truncated exponential distribution. Roughly speaking, setting a random threshold ensures that if a job is stuck in a queue, then its remaining processing is a reasonable fraction of its original processing time.

Theorem 2 ([3]) *The RMLF algorithm is* $O(\log n \log \log n)$ *competitive against an oblivious adversary. Moreover, the RMLF algorithm is* $O(\log n \log \log n)$ *competitive even against an adaptive adversary provided the adversary chooses all the job sizes in advance.*

Later, Becchetti and Leonardi [1] showed that in fact the RMLF is optimally competitive up to constant factors. They also analyzed RMLF on identical parallel machines.

Theorem 3 ([1]) *The RMLF algorithm is* $O(\log n)$ *competitive for a single machine. For multiple identical machines, RMLF achieves a competitive ratio of* $O(\log n \log (\frac{n}{m}))$*, where* m *is the number of machines.*

Applications

MLF and its variants are widely used in operating systems [6, 7]. These algorithms are not only close to optimum with respect to flow time but also have other attractive properties such as the amortized number of preemptions is logarithmic (preemptions occur only if a job arrives or departs or moves to another queue).

Open Problems

It is not known whether there exists a $o(n)$-competitive deterministic algorithm. It would be interesting to close the gap between the upper and lower bounds for this case. Often in real systems, even though the scheduler may not know the exact job size, it might have some information about its distribution based on historical data. An interesting direction of research could be to design and analyze algorithms that use this information.

Cross-References

▶ Flow Time Minimization
▶ Minimum Flow Time
▶ Shortest Elapsed Time First Scheduling

Recommended Reading

1. Becchetti L, Leonardi S (2004) Nonclairvoyant scheduling to minimize the total flow time on single and parallel machines. J ACM (JACM) 51(4):517–539
2. Coffman EG, Kleinrock L (1968) Feedback queueing models for time-shared systems. J ACM (JACM) 15(4):549–576
3. Kalyanasundaram B, Pruhs K (2003) Minimizing flow time nonclairvoyantly. J ACM (JACM) 50(4):551–567

4. Kellerer H, Tautenhahn T, Woeginger GJ (1999) Approximability and nonapproximability results for minimizing total flow time on a single machine. SIAM J Comput 28(4):1155–1166
5. Motwani R, Phillips S, Torng E (1994) Nonclairvoyant scheduling. Theor Comput Sci 130(1): 17–47
6. Nutt G (1999) Operating system projects using windows NT. Addison Wesley, Reading
7. Tanenbaum AS (1992) Modern operating systems. Prentice-Hall, Upper Saddle River

Multilinear Monomial Detection

Ioannis Koutis
Computer Science Department, University of Puerto Rico-Rio Piedras, San Juan, PR, USA

Keywords

Algebraic methods; Color coding; Parameterized algorithms; Subgraph containment

Years and Authors of Summarized Original Work

2008; Koutis
2009; Williams

Problem Definition

The topic of this article is the parameterized multilinear monomial detection problem:

k-MLD: Given an arithmetic circuit C representing a polynomial $P(X)$ over $\mathbb{Z}_+$, decide whether $P(X)$ construed as a sum of monomials contains a multilinear monomial of degree k.

An *arithmetic circuit* is a directed acyclic graph with nodes corresponding to addition and multiplication gates, sources labeled with variables from a set X or positive integers, and one special terminal corresponding to the output gate. A *monomial* of degree k is a product of exactly k variables from X, and it is called *multilinear* if these k variables are distinct.

The k-MLD problem is arguably a fundamental problem, encompassing as special cases many natural and well-studied parameterized problems. Along with the algorithm for its solution, k-MLD provides a general framework for designing parameterized algorithms [11, 15]. The framework has yielded the fastest known algorithms for many parameterized problems, including all parameterized decision problems that were previously known to be solvable via dynamic programming combined with the color-coding method [2].

Key Results

Theorem 1 *The k-MLD problem can be solved by a randomized algorithm with one-sided error in $O^*(2^k)$ time and polynomial space. (The $O^*(\cdot)$ notation hides factors polynomial in the size of the input.)*

The algorithm claimed in Theorem 1 always reports the correct answer when the input polynomial does not contain a degree-k multilinear monomial. In the opposite case, it reports a correct answer with probability at least $1/4$.

Overview of the Algorithm

The algorithm utilizes a set of commutative matrices $\mathcal{M} \subseteq \mathbb{Z}_2^{2^k \times 2^k}$, with the following properties:

(i) For each $M \in \mathcal{M}$, we have $M^2 = \mathbf{0} \; mod \; 2$.
(ii) If $M_1, \ldots, M_k$ are randomly selected matrices from $\mathcal{M}$, then their product is equal to the "all-ones" matrix $1^{2^k \times 2^k}$, with probability at least $1/4$, and $0 \; mod \; 2$ otherwise.

The construction of $\mathcal{M}$ is based on matrix representations of the group $\mathbb{Z}_2^k$, i.e., the abelian group of k-dimensional 0–1 vectors with addition mod 2 [11].

To simplify our discussion, let us assume that all monomials in $P(X)$ have total degree k. The role of $\mathcal{M}$ is then almost self-evident: evaluating

$P(X)$ on a random assignment $\bar{X} : X \to \mathcal{M}$ will annihilate (mod 2) all non-multilinear monomials in $P(X)$ due to property (i). On the other hand, each degree-k multilinear monomial will "survive" the assignment with constant probability, due to property (ii).

However, property (ii) clearly does not suffice for $P(X)$ to evaluate to nonzero (with some probability) in the presence of multilinear monomials. The main reason is that the coefficients of all multilinear monomials in $P(X)$ may be equal to 0 mod 2. The solution is the introduction of a new set A of "fingerprint" variables in order to construct an extended polynomial $\tilde{P}(X, A)$ over $\mathbb{Z}_2$. The key property of $\tilde{P}(X, A)$ is that its multilinear monomials are in a one-to-one correspondence with *copies* of the multilinear monomials in $P(X)$. Specifically, each copy of a multilinear monomial $\mu(X)$ of $P(X)$ gets its own distinct multilinear monomial of the form $q(A)\mu(X)$ in $\tilde{P}(X, A)$. The extended polynomial is constructed by applying simple operations on C. In most cases it is enough to attach a distinct "multiplier" variable from A to each edge of C; in the general case, some more work is needed that may increase the size of C by a quadratic factor in the worst case. We can then consider what is the effect of evaluating $\tilde{P}(X, A)$ on $\bar{X}$: (i) If $P(X)$ does not contain any degree-k multilinear monomial, then each monomial of $\tilde{P}(X, A)$ contains a squared variable from X. Hence $\tilde{P}(\bar{X}, A)$ is equal to $\mathbf{0}\ mod\ 2$. On the other hand, if $P(X)$ does contain a degree-k multilinear monomial, then, by construction, the diagonal entries of $\tilde{P}(\bar{X}, A)$ are all equal to a *nonzero* polynomial $Q(A)$, with probability at least $1/4$. Due to its size, we cannot afford to write down $Q(A)$, but we do have "black-box" access to it via evaluating it. We can thus test it for identity with zero via the Schwartz-Zippel Lemma [14]. This requires only a single evaluation of $Q(A)$ on a random assignment $\bar{A} : A \to GF[2^{\log_2 k+10}]$. Overall, the algorithm returns a "yes" if and only if $P(\bar{X}, \bar{A}) \neq \mathbf{0}\ mod\ 2$.

By the properties of $\mathcal{M}$, it suffices to compute the trace of $P(\bar{X}, \bar{A})$ or equivalently the sum of its eigenvalues. As observed in [11, 13], this can be done with 2^k evaluations of $P(X, A)$ over the ring of polynomials $\mathbb{Z}[A]$. This yields the $O^*(2^k)$ time and polynomial space claims.

The construction of an extended polynomial $\tilde{P}(X, A)$ was used in [11, 15] for two special cases, but it can be generalized to arbitrary polynomials as claimed in [13]. The idea to use the Schwartz-Zippel Lemma in order to test $Q(A)$ for identity with zero appeared in [15].

A Negative Result

The matrices in $\mathcal{M}$ together with multiplication and addition modulo 2 form a commutative matrix algebra $\mathcal{A}$ which has 2^{2^k} elements. Computations with this algebra require time at least 2^k, since merely describing one element of $\mathcal{A}$ requires 2^k bits. One may ask whether there is another significantly smaller algebra that can replace $\mathcal{A}$ in this evaluation-based framework and yield a faster algorithm. The question was answered in the negative in [13], for the special but important case when $k = |X|$. Concretely, it was shown that there is no evaluation-based algorithm that can detect a multilinear term of degree n in an n-variate polynomial in $o(2^n/\sqrt{n})$ time.

A Generalization

The CONSTRAINED k-MLD problem is a generalization of k-MLD that was introduced in [12]. The set X is a union of t *mutually disjoint* sets of variables $X_1, \ldots, X_t$, each associated with a positive number μ_i. The sets X_i and the numbers μ_i, for $i = 1, \ldots, t$ are part of the input. A multilinear monomial is *permissible* if it contains at most μ_i variables from X_i, for all i. The problem is then defined as follows:

> CONSTRAINED k-MLD: Given an arithmetic circuit C representing a polynomial $P(X)$ over $\mathbb{Z}_+$, decide whether $P(X)$ construed as a sum of monomials contains a permissible multilinear monomial of degree k.

Theorem 2 *The* CONSTRAINED k-MLD *problem can be solved by a randomized algorithm with one-sided error in $O^*(2^k)$ time and polynomial space.*

Theorem 2 was shown in [5]. It is worth noting that the algorithm in the proof of Theorem 2 does not rely on matrix algebras. Thus, it provides an alternative proof for Theorem 1.

Applications

Many parameterized problems are reducible to the k-MLD problem. For several of these problems, the fastest – in terms of the exponential dependence on k – known algorithms are composed by a relatively simple reduction to a k-MLD instance and a subsequent invocation of the algorithm for its solution. Such problems include the k-TREE problem on directed graphs and certain packing problems [11, 13, 15].

The k-MLD framework provides an exponentially faster alternative to the color-coding method [2] for parameterized decision problems. As noted in [8], color-coding-based algorithms consist canonically of a random assignment of k colors to entities of the input (e.g., to the vertices of graph) followed by a dynamic programming procedure. From the steps of the dynamic programming procedure, one can delineate the construction of a k-MLD instance. This task was, for example, undertaken in [9], giving improved algorithms for all subgraph containment problems previously solved via color coding.

The algebraic language provided by the k-MLD framework has simplified the design of parameterized algorithms, yielding faster algorithms even for problems for which the applicability of color coding was not apparent due to the more complicated underlying dynamic programming procedures. This includes partial graph domination problems [13] and parameterized string problems in computational biology [6].

All *algebraizations* used in [11, 13] construct polynomials whose monomials are in one-to-one correspondence with potential solutions of the underlying combinatorial problem (e.g., length-k walks). The actual solutions (e.g., k-paths) are mapped to multilinear monomials, while nonsolutions are mapped to non-multilinear monomials. Andreas Björklund introduced a significant departure from this approach [4]. He viewed modulo 2 computations as a resource rather as a nuisance and worked on sharper algebraizations using appropriate determinant polynomials. These polynomials do contain multilinear monomials corresponding to nonsolutions, unlike previous algebraizations; however, these come *in pairs* and they cancel out modulo 2. On the other hand, with the appropriate use of "fingerprint" variables, the multilinear monomials corresponding to valid solutions appear with a coefficient of 1. Björklund's novel ideas led initially to a faster algorithm for d-dimensional matching problems [4] and subsequently to the breakthrough $O(1.67^n)$ time algorithm for the HAMILTONIAN PATH problem on n-vertex graphs [3], breaking below the $O^*(2^n)$ barrier and marking the first progress in the problem after nearly 50 years of stagnation.

Modulo 2 cancelations were also explicitly exploited in the design of the first single-exponential time algorithms for various graph connectivity problems parameterized by *treewidth* [7]. For example, the HAMILTONIAN PATH problem can now be solved in $O^*(4^t)$ time for n-vertex graphs, assuming a tree decomposition of width at most t is given as input. The previously fastest algorithm runs in $O^*(t^{O(t)})$ time.

Open Problems

The following problems are open:

- Color coding stood until recently as the fastest *deterministic* algorithm for the k-MLD problem, running in $O^*((2e)^k)$ time, the currently fastest deterministic algorithm in $O^*(2.85^k)$ time [10]. Is there a deterministic algorithm for k-MLD that runs in $O^*(2^k)$ time?
- The known deterministic algorithms can also solve the *weighted* version of k-PATH which asks for a k-path of minimum weight, in $O^*(2.85^k \log W)$ time, where W is the largest edge weight in the graph. The algorithm for k-MLD can be adapted to solve weighted k-PATH in $O^*(2^k W)$ time. Is there an $O^*(2^k \log W)$ time algorithm for weighted

k-PATH and, more generally, for weighted versions of k-MLD?

- Color coding with balanced hashing families has been used to approximately count k-paths in a graph, in $O^*((2e)^k)$ time [1]. Is there an $O^*(2^k)$ approximate counting algorithm?
- Finally, is there an algorithm for k-MLD that runs in $O^*(2^{\epsilon k})$ time, for any $\epsilon < 1$? Such an algorithm would constitute major progress in exact and parameterized algorithms for NP-hard problems.

Cross-References

▶ Color Coding
▶ Treewidth of Graphs

Recommended Reading

1. Alon N, Gutner S (2009) Balanced hashing, color coding and approximate counting. In: Chen J, Fomin FV (eds) Parameterized and exact computation. Springer, Berlin/Heidelberg, pp 1–16
2. Alon N, Yuster R, Zwick U (1995) Color coding. J ACM 42(4):844–856
3. Björklund A (2010) Determinant sums for undirected hamiltonicity. In: 51th annual IEEE symposium on foundations of computer science, FOCS 2010, Las Vegas, 23–26 Oct 2010, pp 173–182
4. Björklund A (2010) Exact covers via determinants. In: 27th international symposium on theoretical aspects of computer science, STACS, Nancy, pp 95–106
5. Björklund A, Kaski P, Kowalik L (2015) Constrained multilinear detection and generalized graph motifs. Algorithmica 1–21. doi:10.1007/s00453-015-9981-1
6. Blin G, Bonizzoni P, Dondi R, Sikora F (2012) On the parameterized complexity of the repetition free longest common subsequence problem. Inf Process Lett 112(7):272–276
7. Cygan M, Nederlof J, Pilipczuk M, Pilipczuk M, van Rooij JMM, Wojtaszczyk JO (2011) Solving connectivity problems parameterized by treewidth in single exponential time. In: IEEE 52nd annual symposium on foundations of computer science, FOCS, Palm Springs, pp 150–159
8. Downey RG, Fellows MR (2013) Fundamentals of parameterized complexity. Texts in computer science. Springer. doi:10.1007/978-1-4471-5559-1
9. Fomin FV, Lokshtanov D, Raman V, Saurabh S, Rao BVR (2012) Faster algorithms for finding and counting subgraphs. J Comput Syst Sci 78(3):698–706. doi:10.1016/j.jcss.2011.10.001
10. Fomin FV, Lokshtanov D, Saurabh S (2014) Efficient computation of representative sets with applications in parameterized and exact algorithms. In: Proceedings of the twenty-fifth annual ACM-SIAM symposium on discrete algorithms. SIAM, Philadelphia, pp 142–151
11. Koutis I (2008) Faster algebraic algorithms for path and packing problems. In: Proceedings of the 35th international colloquium on automata, languages and programming, Part I. Springer, Berlin/Heidelberg, pp 575–586
12. Koutis I (2012) Constrained multilinear detection for faster functional motif discovery. Inf Process Lett 112(22):889–892
13. Koutis I, Williams R (2009) Limits and applications of group algebras for parameterized problems. In: Proceedings of the 36th international colloquium on automata, languages and programming: Part I (ICALP '09). Springer, Berlin/Heidelberg, pp 653–664
14. Motwani R, Raghavan P (1997) Randomized algorithms. In: Tucker AB (ed) The computer science and engineering handbook. CRC Press, Boca Raton, pp 141–161
15. Williams R (2009) Finding paths of length k in $O^*(2^k)$ time. Inf Process Lett 109:315–318

Multiple String Matching

Maxime Crochemore[1,2,3] and Thierry Lecroq[4]
[1]Department of Computer Science, King's College London, London, UK
[2]Laboratory of Computer Science, University of Paris-East, Paris, France
[3]Université de Marne-la-Vallée, Champs-sur-Marne, France
[4]Computer Science Department and LITIS Faculty of Science, Université de Rouen, Rouen, France

Keywords

Failure function; Pattern matching; Shift function; String matching; Trie; DAWG

Years and Authors of Summarized Original Work

1975; Aho, Corasick
1979; Commentz-Walter

1999; Crochemore, Czumaj, Gąsieniec, Lecroq, Plandowski, Rytter

Problem Definition

Given a finite set of k *pattern strings* $\mathcal{P} = \{P^1, P^2, \ldots, P^k\}$ and a *text string* $T = t_1 t_2 \ldots t_n$, T and the P^is being sequences over an alphabet Σ of size σ, the *multiple string matching (MSM)* problem is to find one or, more generally, all the text positions where a P^i occurs in T. More precisely the problem is to compute the set $\{j \mid \exists i, P^i = t_j t_{j+1} \ldots t_{j+|P^i|-1}\}$, or equivalently the set $\{j \mid \exists i, P^i = t_{j-|P^i|+1} t_{j-|P^i|+2} \ldots t_j\}$. Note that reporting all the occurrences of the patterns may lead to a quadratic output (e.g., when P^is and T are drawn from a one-letter alphabet). The length of the shortest pattern in $\mathcal{P}$ is denoted by ℓmin. This problem is an extension of the exact string matching problem.

Both worst- and average-case complexities are considered. For the latter one assumes that pattern and text are randomly generated by choosing each character uniformly and independently from Σ. For simplicity and practicality the assumption $|P^i| = o(n)$ is made, for $1 \leq i \leq k$, in this entry.

Key Results

A first solution to the multiple string matching problem consists in applying an exact string matching algorithm for locating each pattern in $\mathcal{P}$. This solution has an $O(kn)$ worst-case time complexity. There are more efficient solutions along two main approaches. The first one, due to Aho and Corasick [1], is an extension of the automaton-based solution for matching a single string. The second approach, initiated by Commentz-Walter [5], extends the Boyer-Moore algorithm to several patterns.

The Aho-Corasick algorithm first builds a trie $T(\mathcal{P})$, a digital tree recognizing the patterns of $\mathcal{P}$. The trie $T(\mathcal{P})$ is a tree whose edges are labeled by letters and whose branches spell the patterns of $\mathcal{P}$. A node p in the trie $T(\mathcal{P})$ is associated with the unique word w spelled by the path of $T(\mathcal{P})$ from its root to p. The root itself is identified with the empty word ε. Notice that if w is a node in $T(\mathcal{P})$, then w is a prefix of some $P^i \in \mathcal{P}$. If in addition $a \in \Sigma$, then $child(w, a)$ is equal to wa if wa is a node in $T(\mathcal{P})$; it is equal to NIL otherwise.

During a second phase, when patterns are added to the trie, the algorithm initializes an output function *out*. It associates the singleton $\{P^i\}$ with the nodes P^i $(1 \leq i \leq k)$ and associates the empty set with all other nodes of $T(\mathcal{P})$.

Finally, the last phase of the preprocessing consists in building a failure link for each node of the trie and simultaneously completing the output function. The failure function *fail* is defined on nodes as follows (w is a node): $fail(w) = u$ where u is the longest proper suffix of w that belongs to $T(\mathcal{P})$. Computation of failure links is done during a breadth-first traversal of $T(\mathcal{P})$. Completion of the output function is done while computing the failure function *fail* using the following rule: if $fail(w) = u$, then $out(w) = out(w) \cup out(u)$.

To stop going back with failure links during the computation of the failure links, and also to overpass text characters for which no transition is defined from the root during the searching phase, a loop is added on the root of the trie for these symbols. This finally produces what is called a pattern matching machine or an Aho-Corasick automaton (see Fig. 1).

After the preprocessing phase is completed, the searching phase consists in parsing the text T with $T(\mathcal{P})$. This starts at the root of $T(\mathcal{P})$ and uses failure links whenever a character in T does not match any label of outgoing edges of the current node. Each time a node with a nonempty output is encountered, this means that the patterns of the output have been discovered in the text, ending at the current position. Then, the position is output.

Theorem 1 (Aho and Corasick [1]) *After preprocessing $\mathcal{P}$, searching for the occurrences of the strings of $\mathcal{P}$ in a text T can be done in time $O(n \times \log \sigma)$. The running time of the associated preprocessing phase is $O(|\mathcal{P}| \times \log \sigma)$. The extra memory space required for both operations is $O(|\mathcal{P}|)$.*

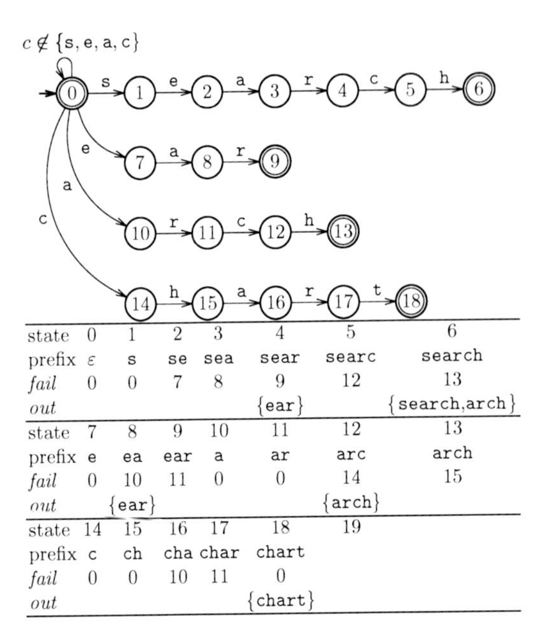

state	0	1	2	3	4	5	6
prefix	ε	s	se	sea	sear	searc	search
fail	0	0	7	8	9	12	13
out					{ear}		{search,arch}
state	7	8	9	10	11	12	13
prefix	e	ea	ear	a	ar	arc	arch
fail	0	10	11	0	0	14	15
out		{ear}				{arch}	
state	14	15	16	17	18	19	
prefix	c	ch	cha	char	chart		
fail	0	0	10	11	0		
out					{chart}		

Multiple String Matching, Fig. 1 The pattern matching machine or Aho-Corasick automaton for the set of strings {search, ear, arch, chart}

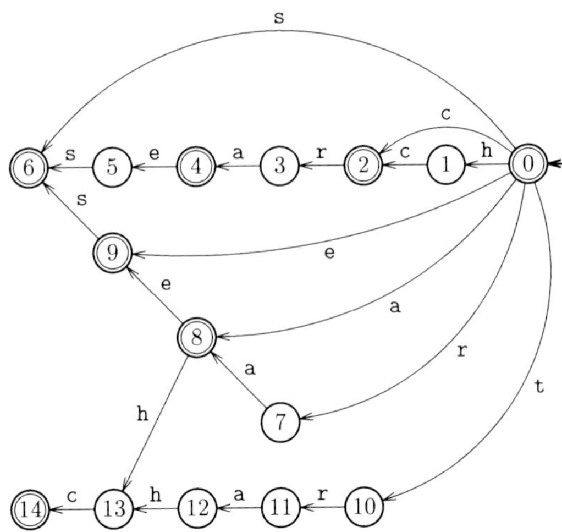

Multiple String Matching, Fig. 2 An example of DAWG, index structure used for matching the set of strings {search, ear, arch, chart}. The automaton accepts the reverse prefixes of the strings

The Aho-Corasick algorithm is actually a generalization to a finite set of strings of the Morris-Pratt exact string matching algorithm.

Commentz-Walter [5] generalized the Boyer-Moore exact string matching algorithm to multiple string matching. Her algorithm builds a trie for the reverse patterns in $\mathcal{P}$ together with two shift tables and applies a right to left scan strategy. However, it is intricate to implement and has a quadratic worst-case time complexity.

The DAWG-match algorithm [6] is a generalization of the BDM exact string matching algorithms. It consists in building an exact indexing structure for the reverse strings of $\mathcal{P}$ such as a factor automaton or a generalized suffix tree, instead of just a trie as in the Aho-Corasick and Commentz-Walter algorithms (see Fig. 2). The overall algorithm can be made optimal by using both an indexing structure for the reverse patterns and an Aho-Corasick automaton for the patterns. Then, searching involves scanning some portions of the text from left to right and some other portions from right to left. This enables to skip large portions of the text T.

Theorem 2 (Crochemore et al. [6]) *The DAWG-match algorithm performs at most $2n$ symbol comparisons. Assuming that the sum of the length of the patterns in $\mathcal{P}$ is less than ℓmin^k, the DAWG-match algorithm makes on average $O((n \log_\sigma \ell min)/\ell min)$ inspections of text characters.*

The bottleneck of the DAWG-match algorithm is the construction time and space consumption of the exact indexing structure. This can be avoided by replacing the exact indexing structure by a factor oracle for a set of strings. A factor oracle is a simple automaton that may recognize a few additional strings comparing to exact indexing structure. When the factor oracle is used alone, it gives the Set Backward Oracle Matching (SBOM) algorithm [2]. It is an exact algorithm that behaves almost as well as the DAWG-match algorithm.

The bit-parallelism technique can be used to simulate the DAWG-match algorithm. It gives the MultiBNDM algorithm of Navarro and Raffinot [9]. This strategy is efficient when $k \times \ell min$ bits fit in a few computer words. The prefixes of strings of $\mathcal{P}$ of length ℓmin are packed together in a bit vector. Then, the search is similar to the

BNDM exact string matching and is performed for all the prefixes at the same time.

The use of the generalization of the bad-character shift alone as done in the Horspool exact string matching algorithm gives poor performances for the MSM problem due to the high probability of finding each character of the alphabet in one of the strings of $\mathcal{P}$.

The algorithm of Wu and Manber [13] considers blocks of length ℓ. Blocks of such a length are hashed using a function h into values less than *maxvalue*. Then *shift*$[h(B)]$ is defined as the minimum between $|P^i| - j$ and $\ell min - \ell + 1$ with $B = p^i_{j-\ell+1} \cdots p^i_j$ for $1 \leq i \leq k$ and $1 \leq j \leq |P^i|$. The value of ℓ varies with the minimum length of the strings in $\mathcal{P}$ and the size of the alphabet. The value of *maxvalue* varies with the memory space available.

The searching phase of the algorithm consists in reading blocks B of length ℓ. If *shift*$[h(B)] > 0$, then a shift of length *shift*$[h(B)]$ is applied. Otherwise, when *shift*$[h(B)] = 0$, the patterns ending with block B are examined one by one in the text. The first block to be scanned is $t_{\ell min-\ell+1} \ldots t_{\ell min}$. This method is incorporated in the *agrep* command [12].

Recent works have been devoted to multiple string matching on packed strings where each symbol is encoded using $\log \sigma$ bits. In this context, Belazzougui [3] gave an efficient algorithm that works in $O(n + ((\log k + \log \ell min + \log\log|\mathcal{P}|)/\ell min + (\log\sigma)/\omega) + occ)$ where ω is the size of the machine word and occ is the number of occurrences of patterns of $\mathcal{P}$ in T. On average it is possible to solve the problem in $O(n/\ell min)$ time using $O(|\mathcal{P}|\log|\mathcal{P}|)$ bits of space [4].

Applications

MSM algorithms serve as basis for multidimensional pattern matching and approximate pattern matching with wildcards. The problem has many applications in computational biology, database search, bibliographic search, virus detection in data flows, and several others.

Experimental Results

The book of G. Navarro and M. Raffinot [10] is a good introduction to the domain. It presents experimental graphics that report experimental evaluation of multiple string matching algorithms for different alphabet sizes, pattern lengths, and sizes of pattern set.

URLs to Code and Data Sets

Well-known packages offering efficient MSM are *agrep* (https://github.com/Wikinaut/agrep) and *grep* with the -F option (http://www.gnu.org/software/grep/grep.html).

Cross-References

- ▸ Multidimensional String Matching is the case where the text dimension is greater than one.
- ▸ Regular Expression Matching is the more complex case where the pattern can be a regular expression;
- ▸ String Matching is the version where a single pattern is searched for in a text;
- ▸ Suffix Trees and Arrays refers to the case where the text can be preprocessed;

Further information can be found in the three following books: [7, 8] and [11].

Recommended Reading

1. Aho AV, Corasick MJ (1975) Efficient string matching: an aid to bibliographic search. C ACM 18(6):333–340
2. Allauzen C, Crochemore M, Raffinot M (1999) Factor oracle: a new structure for pattern matching. In: SOFSEM'99. LNCS, vol 1725, Milovy, Czech Republic, pp 291–306
3. Belazzougui D (2012) Worst-case efficient single and multiple string matching on packed texts in the word-ram model. J Discret Algorithms 14:91–106
4. Belazzougui D, Raffinot M (2013) Average optimal string matching in packed strings. In: Spirakis PG, Serna MJ (eds) Proceedings of the 8th international conference on algorithms and complexity (CIAC

M

2013), Barcelona. Lecture notes in computer science, vol 7878. Springer, Barcelona, Spain, pp 37–48
5. Commentz-Walter B (1979) A string matching algorithm fast on the average. In: Proceedings of ICALP'79. Lecture notes in computer science vol 71. Springer, Graz, Austria, pp 118–132
6. Crochemore M, Czumaj A, Gąsieniec L, Lecroq T, Plandowski W, Rytter W (1999) Fast practical multi-pattern matching. Inf Process Lett 71(3–4):107–113
7. Crochemore M, Hancart C, Lecroq T (2007) Algorithms on strings. Cambridge University Press, Cambridge, New York
8. Gusfield D (1997) Algorithms on strings, trees and sequences. Cambridge University Press, Cambridge, New York
9. Navarro G, Raffinot M (2000) Fast and flexible string matching by combining bit-parallelism and suffix automata. ACM J Exp Algorithms 5:4
10. Navarro G, Raffinot M (2002) Flexible pattern matching in strings – practical on-line search algorithms for texts and biological sequences. Cambridge University Press, Cambridge
11. Smyth WF (2002) Computing patterns in strings. Addison Wesley Longman, Harlow
12. Wu S, Manber U (1992) Agrep – a fast approximate pattern-matching tool. In: Proceedings of USENIX winter 1992 technical conference. USENIX Association, San Francisco, CA, pp 153–162
13. Wu S, Manber U (1994) A fast algorithm for multi-pattern searching. Report TR-94-17, Department of Computer Science, University of Arizona, Tucson

Multiple Unit Auctions with Budget Constraint

Tian-Ming Bu
Software Engineering Institute, East China Normal University, Shanghai, China

Keywords

Auction design; Optimal mechanism design

Years and Authors of Summarized Original Work

2005; Borgs, Chayes, Immorlica, Mahdian, Saberi
2006; Abrams

Problem Definition

In this problem, an auctioneer would like to sell an idiosyncratic commodity with m copies to n bidders, denoted by $i = 1, 2, \ldots, n$. Each bidder i has two kinds of privately known information: $t_i^u \in \mathbb{R}^+$, $t_i^b \in \mathbb{R}^+$. t_i^u represents the price buyer i is willing to pay for per copy of the commodity and t_i^b represents i's budget.

Then a *one-round sealed-bid* auction proceeds as follows. *Simultaneously*, all the bidders submit their bids to the auctioneer. When receiving the reported unit value vector $\mathbf{u} = (u_1, \ldots, u_n)$ and the reported budget vector $\mathbf{b} = (b_1 \ldots, b_n)$ of bids, the auctioneer computes and outputs the allocation vector $\mathbf{x} = (x_1, \ldots, x_n)$ and the price vector $\mathbf{p} = (p_1, \ldots, p_n)$. Each element of the allocation vector indicates the number of copies allocated to the corresponding bidder. If bidder i receives x_i copies of the commodity, he pays the auctioneer $p_i x_i$. Then bidder i's total payoff is $(t_i^u - p_i)x_i$ if $x_i p_i \leq t_i^b$ and $-\infty$ otherwise. Correspondingly, the revenue of the auctioneer is $\mathcal{A}(\mathbf{u}, \mathbf{b}, m) = \sum_i p_i x_i$.

If each bidder submits his privately *true* unit value t_i^u and budget t_i^b to the auctioneer, the auctioneer can determine the single price $p_{\mathcal{F}}$ (i.e., $\forall i,\ p_i = p_{\mathcal{F}}$) and the allocation vector which maximize the auctioneer's revenue. This optimal single price revenue is denoted by $\mathcal{F}(\mathbf{u}, \mathbf{b}, m)$.

Interestingly, in this problem, we assume bidders have free will and have complete knowledge of the auction mechanism. Bidders would just report the bid (maybe different from his corresponding privately true values) which could maximize his payoff according to the auction mechanism.

So the objective of the problem is to design a *truthful* auction satisfying *voluntary participation* to raise the auctioneer's revenue as much as possible. An auction is *truthful* if for every bidder i, bidding his true valuation would maximize his payoff, regardless of the bids submitted by the other bidders [7,8]. An auction satisfies *voluntary participation* if each bidder's payoff is guaranteed to be nonnegative if he reports his bid truthfully. The success of the auction $\mathcal{A}$ is determined by competitive ratio β which is defined as the upper bound of $\frac{\mathcal{F}(\mathbf{u},\mathbf{b},m)}{\mathcal{A}(\mathbf{u},\mathbf{b},m)}$ [6]. Clearly, the smaller competitive ratio β is, the better the auction $\mathcal{A}$ is.

Definition 1 (Multiple-Unit Auctions with Budget Constraint)

INPUT: the number of copies m, the submitted unit value vector $\mathbf{u}$, and the submitted budget vector $\mathbf{b}$.

OUTPUT: the allocation vector $\mathbf{x}$ and the price vector $\mathbf{p}$.

CONSTRAINTS:

(a) Truthful;
(b) Voluntary participation;
(c) $\sum_i x_i \leq m$.

Key Results

Let $b_{\max}$ denote the largest budget among the bidders receiving copies in the optimal solution and define $\alpha = \frac{\mathcal{F}}{b_{\max}}$.

Theorem 1 ([3]) *A truthful auction satisfying voluntary participation with competitive ratio* $1/\max_{0<\delta<1}\left\{(1-\delta)(1-2e^{-\frac{\alpha\delta^2}{36}})\right\}$ *can be designed.*

Theorem 2 ([1]) *A truthful auction satisfying voluntary participation with competitive ratio* $\frac{4\alpha}{\alpha-1}$ *can be designed.*

Theorem 3 ([1]) *If* α *is known in advance, then a truthful auction satisfying voluntary participation with competitive ratio* $\frac{(x\alpha+1)\alpha}{(x\alpha-1)^2}$ *can be designed, where* $x = \frac{\alpha-1+((\alpha-1)^2-4\alpha)^{1/2}}{2\alpha}$.

Theorem 4 ([1]) *For any truthful randomized auction* $\mathcal{A}$ *satisfying voluntary participation, the competitive ratio is at least* $2-\epsilon$ *when* $\alpha \geq 2$.

Applications

This problem is motivated by the development of IT industry and the popularization of auctions, especially, auctions on the Internet. Multiple copy auctions of relatively low-value goods, such as the auction of online ads for search terms to bidders with budget constraints, is assuming a very important role. Companies such as Google and Yahoo!'s revenue depends almost on certain types of auctions. There are many papers including [2,4,5] which focus on different facets of the same model.

Cross-References

▸ Competitive Auction

Recommended Reading

1. Abrams Z (2006) Revenue maximization when bidders have budgets. In: Proceedings of the seventeenth annual ACM-SIAM symposium on discrete algorithms (SODA-06), Miami, pp 1074–1082
2. Bhattacharya S, Conitzer V, Munagala K, Xia L (2010) Incentive compatible budget elicitation in multi-unit auctions. In: Proceedings of the twenty-first annual ACM-SIAM symposium on discrete algorithms (SODA-10), Austin, pp 554–572
3. Borgs C, Chayes JT, Immorlica N, Mahdian M, Saberi A (2005) Multi-unit auctions with budget-constrained bidders. In: ACM conference on electronic commerce (EC-05), Vancouver, pp 44–51
4. Bu TM, Qi Q, Sun AW (2008) Unconditional competitive auctions with copy and budget constraints. Theor Comput Sci 393(1–3):1–13
5. Dobzinski S, Lavi R, Nisan N (2012) Multi-unit auctions with budget limits. Games Econ Behav 74(2):486–503
6. Fiat A, Goldberg AV, Hartline JD, Karlin AR (2002) Competitive generalized auctions. In: Proceedings of the 34th annual ACM symposium on theory of computing (STOC-02), New York, pp 72–81
7. Nisan N, Ronen A (1999) Algorithmic mechanism design. In: Proceedings of the thirty-first annual ACM symposium on theory of computing (STOC-99), New York, pp 129–140
8. Parkes DC (2004) Chapter 2: iterative combinatorial auctions. PhD thesis, University of Pennsylvania

Multiplex PCR for Gap Closing (Whole-Genome Assembly)

Vera Asodi
Center for the Mathematics of Information, California Institute of Technology, Pasadena, CA, USA

Keywords

Multiplex PCR; Whole genome assemble

M

Years and Authors of Summarized Original Work

2002; Alon, Beigel, Kasif, Rudich, Sudakov

Problem Definition

This problem is motivated by an important and timely application in computational biology that arises in whole-genome shotgun sequencing. Shotgun sequencing is a high throughput technique that has resulted in the sequencing of a large number of bacterial genomes as well as Drosophila (fruit fly) and Mouse and the celebrated Human genome (at Celera) (see, e.g., [8]). In all such projects, one is left with a collection of DNA fragments. These fragments are subsequently assembled, in-silico, by a computational algorithm. The typical assembly algorithm repeatedly merges overlapping fragments into longer fragments called contigs. For various biological and computational reasons some regions of the DNA cannot covered by the contigs. Thus, the contigs must be ordered and oriented and the gaps between them must be sequenced using slower, more tedious methods. For further details see, e.g., [3]. When the number of gaps is small (e.g., less than ten) biologists often use combinatorial PCR. This technique initiates a set of "bi-directional molecular walks" along the gaps in the sequence; these walks are facilitated by PCR. In order to initiate the molecular walks biologists use primers. Primers are designed so that they bind to unique (with respect to the entire DNA sequence) templates occurring at the end of each contig. A primer (at the right temperature and concentration) anneals to the designated unique DNA substring and promotes copying of the template starting from the primer binding site, initiating a one-directional walk along the gap in the DNA sequence. A PCR reaction occurs, and can be observed as a DNA ladder, when two primers that bind to positions on two ends of the same gap are placed in the same test tube.

If there are N contigs, the combinatorial (exhaustive) PCR technique tests all possible pairs (quadratically many) of $2N$ primers by placing two primers per tube with the original uncut DNA strand. PCR products can be detected using gels or they can be read using sequencing technology or DNA mass-spectometry. When the number of gaps is large, the quadratic number of PCR experiments is prohibitive, so primers are pooled using $K > 2$ primers per tube; this technique is called multiplex PCR [4]. This problem deals with finding optimal strategies for pooling the primers to minimize the number of biological experiments needed in the gap-closing process.

This problem can be modeled as the problem of identifying or learning a hidden matching given a vertex set V and an allowed query operation: for a subset $F \subseteq V$, the query Q_F is "does F contain at least one edge of the matching"? In this formulation each vertex represents a primer, an edge of the matching represents a reaction, and the query represents checking for a reaction when a set of primers are combined in a test tube. The objective is to identify the matching asking as few queries as possible, that is performing as few tests as possible. For further discussion of this model see [3, 7].

This problem is of interest even in the deterministic, fully non-adaptive case. A family $\mathcal{F}$ of subsets of a vertex set V solves the matching problem on V if for any two distinct matchings M_1 and M_2 on V there is at least one $F \in \mathcal{F}$ that contains an edge of one of the matchings and does not contain any edge of the other. Obviously, any such family enables learning an unknown matching deterministically and non-adaptively, by asking the questions Q_F for each $F \in \mathcal{F}$. The objective here is to determine the minimum possible cardinality of a family that solves the matching problem on a set of n vertices.

Other interesting variants of this problem are when the algorithm may be randomized, or when it is adaptive, that is when the queries are asked in k rounds, and the queries of each round may depend on the answers from the previous rounds.

Key Results

In [2], the authors study the number of queries needed to learn a hidden matching in several models. Following is a summary of the main results presented in this paper.

The trivial upper bound on the size of a family that solves the matching problem on n vertices is $\binom{n}{2}$, achieved by the family of all pairs of vertices. Theorem 1 shows that in the deterministic non-adaptive setting one cannot do much better than this, namely, that the trivial upper bound is tight up to a constant factor. Theorem 2 improves this upper bound by showing a family of approximately half that size that solves the matching problem.

Theorem 1 *For every $n > 2$, every family $\mathcal{F}$ that solves the matching problem on n vertices satisfies*

$$|\mathcal{F}| \geq \frac{49}{153}\binom{n}{2}.$$

Theorem 2 *For every n there exists a family of size*

$$\left(\frac{1}{2}+o(1)\right)\binom{n}{2}$$

that solves the matching problem on n vertices.

Theorem 3 shows that one can do much better using randomized algorithms. That is, one can learn a hidden matching asking only $O(n \log n)$ queries, rather than order of n^2. These randomized algorithms make no errors, however, they might ask more queries with some small probability.

Theorem 3 *The matching problem on n vertices can be solved by probabilistic algorithms with the following parameters:*

- *2 rounds and $(1/(2\ln 2))n \log n(1+o(1)) \approx 0.72n \log n$ queries*
- *1 round and $(1/\ln 2)n \log n(1+o(1)) \approx 1.44n \log n$ queries.*

Finally, Theorem 4 considers adaptive algorithms. In this case there is a tradeoff between the number of queries and the number of rounds. The more rounds one allows, the fewer tests are needed, however, as each round can start only after the previous one is completed, this increases the running time of the entire procedure.

Theorem 4 *For all $3 \leq k \leq \log n$, there is a deterministic k-round algorithm for the matching problem on n vertices that asks*

$$O\left(n^{1+\frac{1}{2(k-1)}}(\log n)^{1+\frac{1}{k-1}}\right)$$

queries per round.

Applications

As described in section "Problem Definition", this problem was motivated by the application of gap closing in whole-genome sequencing, where the vertices correspond to primers, the edges to PCR reactions between pairs of primers that bind to the two ends of a gap, and the queries to tests in which a set of primers are combined in a test tube.

This gap-closing problem can be stated more generally as follows. Given a set of chemicals, a guarantee that each chemical reacts with at most one of the others, and an experimental mechanism to determine whether a reaction occurs when several chemicals are combined in a test tube, the objective is to determine which pairs of chemicals react with each other with a minimum number of experiments.

Another generalization which may have more applications in molecular biology is when the hidden subgraph is not a matching but some other fixed graph, or a family of graphs. The paper [2], as well as some other related works (e.g., [1, 5, 6]), consider this generalization for other graphs. Some of these generalizations have other specific applications in molecular biology.

Open Problems

- Determine the smallest possible constant c such that there is a deterministic non-adaptive algorithm for the matching problem on n vertices that performs $c\binom{n}{2}(1+o(1))$ queries.

- Find more efficient deterministic k-round algorithms or prove lower bounds for the number of queries in such algorithms.
- Find efficient algorithms and prove lower bounds for the generalization of the problem to graphs other than matchings.

Recommended Reading

1. Alon N, Asodi V (2004) Learning a hidden subgraph. In: ICALP. LNCS, vol 3142, pp 110–121; Also: SIAM J Discr Math 18:697–712 (2005)
2. Alon N, Beigel R, Kasif S, Rudich S, Sudakov B (2002) Learning a hidden matching. In: Proceedings of the 43rd IEEE FOCS, pp 197–206; Also: SIAM J Comput 33:487–501 (2004)
3. Beigel R, Alon N, Apaydin MS, Fortnow L, Kasif S (2001) An optimal procedure for gap closing in whole genome shotgun sequencing. In: Proceedings of RECOMB. ACM, pp 22–30
4. Burgart LJ, Robinson RA, Heller MJ, Wilke WW, Iakoubova OK, Cheville JC (1992) Multiplex polymerase chain reaction. Mod Pathol 5:320–323
5. Grebinski V, Kucherov G (1997) Optimal query bounds for reconstructing a Hamiltonian cycle in complete graphs. In: Proceedings of 5th Israeli symposium on theoretical computer science, pp 166–173
6. Grebinski V, Kucherov G (1998) Reconstructing a Hamiltonian cycle by querying the graph: application to DNA physical mapping. Discret Appl Math 88:147–165
7. Tettelin H, Radune D, Kasif S, Khouri H, Salzberg S (1999) Pipette optimal multiplexed PCR: efficiently closing whole genome shotgun sequencing project. Genomics 62:500–507
8. Venter JC, Adams MD, Sutton GG, Kerlavage AR, Smith HO, Hunkapiller M (1998) Shotgun sequencing of the human genome. Science 280:1540–1542

Multitolerance Graphs

George B. Mertzios
School of Engineering and Computing Sciences, Durham University, Durham, UK

Keywords

Intersection model; Maximum clique; Minimum coloring; Maximum-weight independent set; Multitolerance graphs; Tolerance graphs

Years and Authors of Summarized Original Work

2011; Mertzios

Problem Definition

Tolerance graphs model interval relations in such a way that intervals can tolerate a certain degree of overlap without being in conflict. A graph $G = (V, E)$ on n vertices is a *tolerance* graph if there exists a collection $I = \{I_v \mid v \in V\}$ of closed intervals on the real line and a set $t = \{t_v \mid v \in V\}$ of positive numbers, such that for any two vertices $u, v \in V$, $uv \in E$ if and only if $|I_u \cap I_v| \geq \min\{t_u, t_v\}$, where $|I|$ denotes the length of the interval I.

Tolerance graphs have been introduced in [3], in order to generalize some of the well-known applications of interval graphs. If in the definition of tolerance graphs we replace the operation "min" between tolerances by "max," we obtain the class of *max-tolerance* graphs [7]. Both tolerance and max-tolerance graphs have attracted many research efforts (e.g., [4, 5, 7–10]) as they find numerous applications, especially in bioinformatics, constraint-based temporal reasoning, and resource allocation problems, among others [4, 5, 7, 8]. In particular, one of their applications is in the comparison of DNA sequences from different organisms or individuals by making use of a software tool like BLAST [1].

In some circumstances, we may want to treat different parts of the genomic sequences in BLAST nonuniformly, since for instance some of them may be biologically less significant or we have less confidence in the exact sequence due to sequencing errors in more error-prone genomic regions. That is, we may want to be more tolerant at some parts of the sequences than at others. This concept leads naturally to the notion of *multitolerance* (known also as *bitolerance*) graphs [5, 11]. The main idea is to allow two different tolerances to each interval, one to the left and one to the right side, respectively. Then, every interval tolerates in its interior part the intersection with other intervals by an amount that is a convex combination of these two border tolerances.

Formally, let $I = [l, r]$ be a closed interval on the real line and $l_t, r_t \in I$ be two numbers between l and r, called *tolerant points*; note that it is not necessary that $l_t \leq r_t$. For every $\lambda \in [0, 1]$, we define the interval $I_{l_t,r_t}(\lambda) = [l + (r_t - l)\lambda, l_t + (r - l_t)\lambda]$, which is the convex combination of $[l, l_t]$ and $[r_t, r]$. Furthermore, we define the set $\mathcal{I}(I, l_t, r_t) = \{I_{l_t,r_t}(\lambda) \mid \lambda \in [0, 1]\}$ of intervals. That is, $\mathcal{I}(I, l_t, r_t)$ is the set of all intervals that we obtain when we linearly transform $[l, l_t]$ into $[r_t, r]$. For an interval I, the *set of tolerance intervals* τ of I is defined either as $\tau = \mathcal{I}(I, l_t, r_t)$ for some values $l_t, r_t \in I$ of tolerant points or as $\tau = \{\mathbb{R}\}$. A graph $G = (V, E)$ is a *multitolerance* graph if there exists a collection $I = \{I_v \mid v \in V\}$ of closed intervals and a family $t = \{\tau_v \mid v \in V\}$ of sets of tolerance intervals, such that for any two vertices $u, v \in V$, $uv \in E$ if and only if there exists an element $Q_u \in \tau_u$ with $Q_u \subseteq I_v$ or there exists an element $Q_v \in \tau_v$ with $Q_v \subseteq I_u$. Then, the pair $\langle I, t \rangle$ is called a *multitolerance representation* of G. Tolerance graphs are a special case of multitolerance graphs.

Note that, in general, the adjacency of two vertices u and v in a multitolerance graph G depends on both sets of tolerance intervals τ_u and τ_v. However, since the real line $\mathbb{R}$ is not included in any finite interval, if $\tau_u = \{\mathbb{R}\}$ for some vertex u of G, then the adjacency of u with another vertex v of G depends *only* on the set τ_v of v. If G has a multitolerance representation $\langle I, t \rangle$, in which $\tau_v \neq \{\mathbb{R}\}$ for every $v \in V$, then G is called a *bounded multitolerance* graph. Bounded multitolerance graphs coincide with trapezoid graphs, i.e., the intersection graphs of trapezoids between two parallel lines L_1 and L_2 on the plane, and have received considerable attention in the literature [5, 11]. However, the trapezoid intersection model cannot cope with general multitolerance graphs, in which it can be $\tau_v = \{\mathbb{R}\}$ for some vertices v. Therefore, the only way until now to deal with general multitolerance graphs was to use the inconvenient multitolerance representation, which uses an infinite number of tolerance intervals.

Key Results

In this entry we introduce the first nontrivial intersection model for general multitolerance graphs, given by objects in the 3-dimensional space, called *trapezoepipeds*. This *trapezoepiped representation* unifies in a simple and intuitive way the widely known trapezoid representation for bounded multitolerance graphs and the parallelepiped representation for tolerance graphs [9]. The main idea is to exploit the third dimension to capture the information of the vertices with $\tau_v = \{\mathbb{R}\}$ as the set of tolerance intervals. This intersection model can be constructed efficiently (in linear time), given a multitolerance representation.

Apart from being important on its own, the trapezoepiped representation can be also used to design efficient algorithms and structural results. Given a multitolerance graph with n vertices and m edges, we present algorithms that compute a minimum coloring and a maximum clique in $O(n \log n)$ time (which turns out to be *optimal*), and a maximum-weight independent set in $O(m + n \log n)$ time (where $\Omega(n \log n)$ is a lower bound for the complexity of this problem [2]). Moreover, a variation of this algorithm can compute a maximum-weight independent set in optimal $O(n \log n)$ time, when the input is a tolerance graph, thus closing the complexity gap of [9].

Given a multitolerance representation of a graph $G = (V, E)$, vertex $v \in V$ is called *bounded* if $\tau_v = \mathcal{I}(I_v, l_{t_v}, r_{t_v})$ for some values $l_{t_v}, r_{t_v} \in I_v$. Otherwise, v is *unbounded*. V_B and V_U are the sets of bounded and unbounded vertices in V, respectively. Clearly $V = V_B \cup V_U$.

Definition 1 For a vertex $v \in V_B$ (resp. $v \in V_U$) in a multitolerance representation of G, the values $t_{v,1} = l_{t_v} - l_v$ and $t_{v,2} = r_v - r_{t_v}$ (resp. $t_{v,1} = t_{v,2} = \infty$) are the *left tolerance* and the *right tolerance* of v, respectively. Moreover, if $v \in V_U$, then $t_v = \infty$ is the *tolerance* of v.

It can be easily seen by Definition 1 that if we set $t_{v,1} = t_{v,2}$ for every vertex $v \in V$, then we obtain a tolerance representation, in which

$t_{v,1} = t_{v,2}$ is the (unique) tolerance of v. Let now L_1 and L_2 be two parallel lines at unit distance in the plane.

Definition 2 Given an interval $I_v = [l_v, r_v]$ and tolerances $t_{v,1}, t_{v,2}$, $\overline{T}_v$ is the trapezoid in $\mathbb{R}^2$ defined by the points c_v, b_v on L_1 and a_v, d_v on L_2, where $a_v = l_v$, $b_v = r_v$, $c_v = \min\{r_v, l_v + t_{v,1}\}$, and $d_v = \max\{l_v, r_v - t_{v,2}\}$. The values $\phi_{v,1} = \operatorname{arc\,cot}(c_v - a_v)$ and $\phi_{v,2} = \operatorname{arc\,cot}(b_v - d_v)$ are the *left slope* and the *right slope* of $\overline{T}_v$, respectively. Moreover, for every unbounded vertex $v \in V_U$, $\phi_v = \phi_{v,1} = \phi_{v,2}$ is the *slope* of $\overline{T}_v$.

Note that, in Definition 2, the endpoints a_v, b_v, c_v, d_v of any trapezoid $\overline{T}_v$ (on the lines L_1 and L_2) lie on the plane $z = 0$ in $\mathbb{R}^3$. Therefore, since we assumed that the distance between the lines L_1 and L_2 is one, these endpoints of $\overline{T}_v$ correspond to the points $(a_v, 0, 0)$, $(b_v, 1, 0)$, $(c_v, 1, 0)$, and $(d_v, 0, 0)$ in $\mathbb{R}^3$, respectively. For the sake of presentation, we may not distinguish in the following between these points in $\mathbb{R}^3$ and the corresponding real values a_v, b_v, c_v, d_v, whenever this slight abuse of notation does not cause any confusion.

We are ready to give the main definition of this entry, namely, the *trapezoepiped representation*. For a set X of points in $\mathbb{R}^3$, denote by $H_{\text{convex}}(X)$ the *convex hull* defined by the points of X. That is, $\overline{T}_v = H_{\text{convex}}(a_v, b_v, c_v, d_v)$ for every vertex $v \in V$ by Definition 2, where a_v, b_v, c_v, d_v are points of the plane $z = 0$ in $\mathbb{R}^3$.

Definition 3 (trapezoepiped representation) Let $G = (V, E)$ be a multitolerance graph with a multitolerance representation $\{I_v = [a_v, b_v], \tau_v \mid v \in V\}$ and $\Delta = \max\{b_v \mid v \in V\} - \min\{a_v \mid v \in V\}$ be the greatest distance between two interval endpoints. For every vertex $v \in V$, the *trapezoepiped* T_v of v is the convex set of points in $\mathbb{R}^3$ defined as follows:

(a) If $t_{v,1}, t_{v,2} \leq |I_v|$ (i.e., v is bounded), then $T_v = H_{\text{convex}}(\overline{T}_v, a'_v, b'_v, c'_v, d'_v)$.
(b) If $t_v = t_{v,1} = t_{v,2} = \infty$ (i.e., v is unbounded), then $T_v = H_{\text{convex}}(a'_v, c'_v)$.

Where $a'_v = (a_v, 0, \Delta - \cot\phi_{v,1})$, $b'_v = (b_v, 1, \Delta - \cot\phi_{v,2})$, $c'_v = (c_v, 1, \Delta - \cot\phi_{v,1})$, and $d'_v = (d_v, 0, \Delta - \cot\phi_{v,2})$. The set of trapezoepipeds $\{T_v \mid v \in V\}$ is a *trapezoepiped representation* of G (Fig. 1).

Theorem 1 *Let $G = (V, E)$ be a multitolerance graph with a multitolerance representation $\{I_v = [a_v, b_v], \tau_v \mid v \in V\}$. Then for every $u, v \in V$, $uv \in E$ if and only if $T_u \cap T_v \neq \emptyset$.*

Efficient Algorithms

As one of our main tools towards providing efficient algorithms on multitolerance graphs, we refine Definition 3 by introducing the notion of a *canonical* trapezoepiped representation. A trapezoepiped representation R of a multitolerance graph $G = (V, E)$ is called *canonical* if the following is true: for every unbounded vertex $v \in V_U$ in R, if we replace T_v by $H_{\text{convex}}(\overline{T}_v, a'_v, c'_v)$ in R, we would create a new edge in G. Note that replacing T_v by $H_{\text{convex}}(\overline{T}_v, a'_v, c'_v)$ in R is equivalent to replacing in the corresponding multitolerance representation of G the infinite tolerance $t_v = \infty$ by the finite tolerances $t_{v,1} = t_{v,2} = |I_v|$, i.e., to making v a bounded vertex. Clearly, every trapezoepiped representation R can be transformed to a canonical one by iteratively replacing unbounded vertices with bounded ones (without introducing new edges), as long as this is possible. Using techniques from computational geometry, we can prove the next theorem.

Theorem 2 *Every trapezoepiped representation of a multitolerance graph G with n vertices can be transformed to a canonical representation of G in $O(n \log n)$ time.*

The main idea for the proof of Theorem 2 is the following. We associate with every unbounded vertex $v \in V_U$ an (appropriately defined) point p_v and with every bounded vertex $u \in V_B$ three points $p_{u,1}, p_{u,2}, p_{u,3}$ in the plane. Furthermore we associate with every bounded vertex $u \in V_B$ the two line segments $\ell_{u,1}$ and $\ell_{u,2}$ in the plane, which have the points $\{p_{u,1}, p_{u,2}\}$ and $\{p_{u,2}, p_{u,3}\}$ as endpoints, respectively. We can prove that an unbounded vertex $v \in V_U$ can

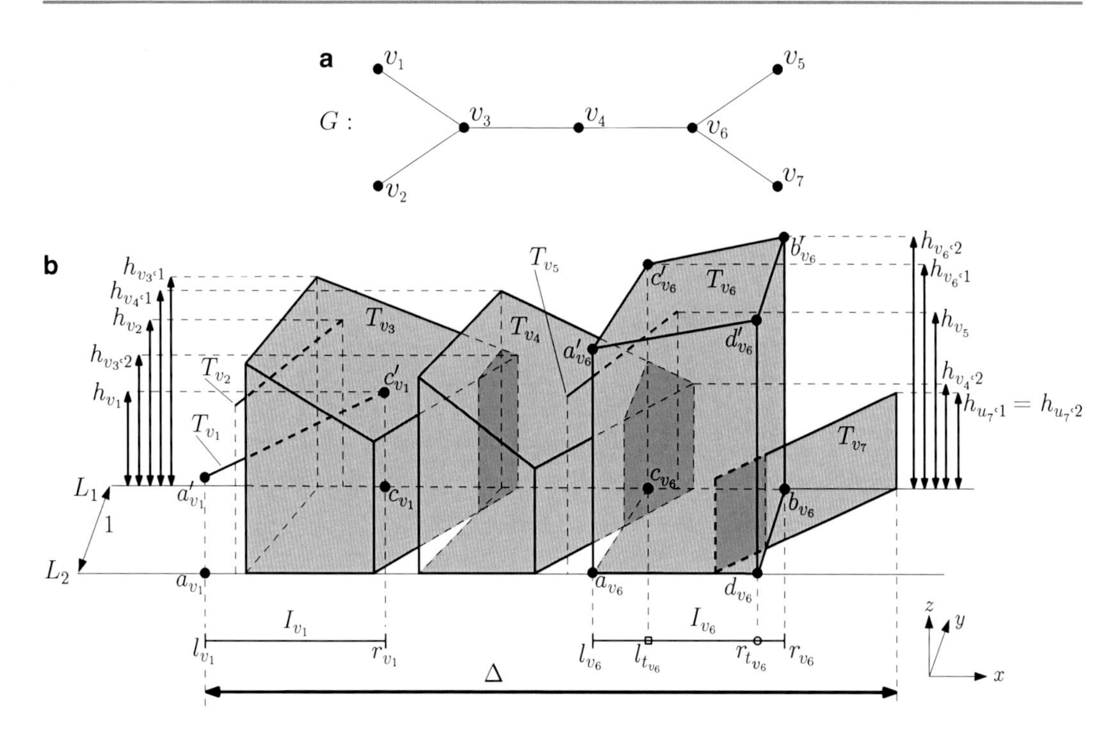

Multitolerance Graphs, Fig. 1 (a) A multitolerance graph G and (b) a trapezoepiped representation R of G. Here, $h_{v_i,j} = \Delta - \cot \phi_{v_i,j}$ for every bounded vertex $v_i \in V_B$ and $j \in \{1,2\}$, while $h_{v_i} = \Delta - \cot \phi_{v_i}$ for every unbounded vertex $v_i \in V_U$

be replaced by a bounded vertex without introducing a new edge if and only if, in the above construction, the point p_v lies above the *lower envelope* $\text{Env}(L)$ of the line segments $L = \{\ell_{u,1}, \ell_{u,2} : u \in V_B\}$. Since $|L| = O(n)$, we can compute $\text{Env}(L)$ in $O(n \log n)$ time using the algorithm of [6].

In the resulting canonical representation R' of G, for every unbounded vertex $v \in V_U$, there exists at least one bounded vertex $u \in V_B$ such that $uv \notin E$ and T_v lies "above" T_u in R'. Moreover, we can prove that in this case $N(v) \subseteq N(u)$, and thus there exists a minimum coloring of G where u and v have the same color. The main idea for our (optimal) $O(n \log n)$-time minimum coloring algorithm is the following. We first compute in $O(n \log n)$ time a minimum coloring of the induced subgraph $G[V_B]$ using the coloring algorithm of [2] for trapezoid graphs. Then, given this coloring, we assign in linear time a color to all unbounded vertices. Furthermore, using Theorem 2, the maximum clique algorithm of [2] for trapezoid graphs, and the fact that multitolerance graphs are perfect, we provide an (optimal) $O(n \log n)$-time maximum clique algorithm for multitolerance graphs.

Our $O(m + n \log n)$-time maximum-weight independent set algorithm for multitolerance graphs is based on dynamic programming. During its execution, the algorithm uses binary search trees to maintain two finite sets M and H of $O(n)$ weighted markers each, which are appropriately sorted on the real line. For the case where the input graph G is a tolerance graph, this algorithm can be slightly modified to compute a maximum-weight independent set in (optimal) $O(n \log n)$ time, thus closing the complexity gap of [9].

Classification of Multitolerance Graphs

Apart from its use in devising efficient algorithms, the trapezoepiped representation proved useful also in classifying multitolerance graphs inside the hierarchy of perfect graphs that is given in [5, Figure 2.8]. The resulting hierarchy of classes of perfect graphs is *complete*, i.e., all inclusions are strict.

Open Problems

The trapezoepiped representation provides geometric insight for multitolerance graphs, and it can be expected to prove useful in deriving new algorithmic as well as structural results. It remains open to close the gap between the lower bound of $\Omega(n \log n)$ and the upper bound of $O(m + n \log n)$ for the weighted independent set on general multitolerance graphs. Furthermore, interesting open problems for further research include the weighted clique problem, the Hamiltonian cycle problem, the dominating set problem, as well as the recognition problem of general multitolerance graphs.

Recommended Reading

1. Altschul SF, Gish W, Miller W, Myers EW, Lipman DJ (1990) Basic local alignment search tool. J Mol Biol 215(3):403–410
2. Felsner S, Müller R, Wernisch L (1997) Trapezoid graphs and generalizations, geometry and algorithms. Discret Appl Math 74:13–32
3. Golumbic MC, Monma CL (1982) A generalization of interval graphs with tolerances. In: Proceedings of the 13th Southeastern conference on combinatorics, graph theory and computing, Boca Raton. Congressus Numerantium, vol 35, pp 321–331
4. Golumbic MC, Siani A (2002) Coloring algorithms for tolerance graphs: reasoning and scheduling with interval constraints. In: Proceedings of the joint international conferences on artificial intelligence, automated reasoning, and symbolic computation (AISC/Calculemus), Marseille, pp 196–207
5. Golumbic MC, Trenk AN (2004) Tolerance graphs. Cambridge studies in advanced mathematics. Cambridge University Press, Cambridge
6. Hershberger J (1989) Finding the upper envelope of n line segments in $O(n \log n)$ time. Inf Process Lett 33(4):169–174
7. Kaufmann M, Kratochvil J, Lehmann KA, Subramanian AR (2006) Max-tolerance graphs as intersection graphs: cliques, cycles, and recognition. In: Proceedings of the 17th annual ACM-SIAM symposium on discrete algorithms (SODA), Miami, pp 832–841
8. Lehmann KA, Kaufmann M, Steigele S, Nieselt K (2006) On the maximal cliques in c-max-tolerance graphs and their application in clustering molecular sequences. Algorithms Mol Biol 1:9
9. Mertzios GB, Sau I, Zaks S (2009) A new intersection model and improved algorithms for tolerance graphs. SIAM J Discret Math 23(4):1800–1813
10. Mertzios GB, Sau I, Zaks S (2011) The recognition of tolerance and bounded tolerance graphs. SIAM J Comput 40(5):1234–1257
11. Parra A (1998) Triangulating multitolerance graphs. Discret Appl Math 84(1–3):183–197

Multiway Cut

Gruia Calinescu
Department of Computer Science, Illinois Institute of Technology, Chicago, IL, USA

Keywords

Multiterminal cut

Years and Authors of Summarized Original Work

1998; Calinescu, Karloff, Rabani

Problem Definition

Given an undirected graph with edge costs and a subset of k nodes called *terminals*, a *multiway cut* is a subset of edges whose removal disconnects each terminal from the rest. MULTIWAY CUT is the problem of finding a multiway cut of minimum cost.

Previous Work

Dahlhaus, Johnson, Papadimitriou, Seymour, and Yannakakis [6] initiated the study of MULTIWAY CUT and proved that MULTIWAY CUT is MAX SNP-hard even when restricted to instances with three terminals and unit edge costs. Therefore, unless $P = NP$, there is no polynomial-time approximation scheme for MULTIWAY CUT. For $k = 2$, the problem is identical to the undirected version of the extensively studied s–t min-cut problem of Ford and Fulkerson, and thus has polynomial-time algorithms (see, e.g., [1]). Prior to this paper, the best (and essentially the only) approximation algorithm for $k \geq 3$ was due to the above-mentioned paper of Dahlhaus et al. They give a very simple combinatorial *isolation heuristic* that achieves an approximation ratio of $2(1 - 1/k)$. Specifically, for each terminal i, find a minimum-cost cut separating i from the remaining terminals, and then output the union of the $k - 1$ cheapest of the k cuts. For $k = 4$ and for $k = 8$, Alon (see [6]) observed that the isolation heuristic can be modified to give improved ratios of 4/3 and 12/7, respectively.

In special cases, far better results are known. For fixed k in planar graphs, the problem is solvable in polynomial time [6]. For trees and 2-trees, there are linear-time algorithms [5]. For dense unweighted graphs, there is a polynomial-time approximation scheme [2, 8].

Key Results

Theorem 1 ([3]) *There is a deterministic polynomial time algorithm that finds a multiway cut of cost at most* $(1.5 - 1/k)$ *times the optimum multiway cut.*

The approximation algorithm from Theorem 1 is based on a novel linear programming relaxation described later. On the basis of the same linear program, the approximation ratio was subsequently improved to 1.3438 by Karger, Klein, Stein, Thorup, and Young [10]. For three terminals, [10] and Cheung, Cunningham, and Tang [4] give very different 12/11-approximation algorithms.

Two variations of the problem have been considered in the literature: Garg, Vazirani, and Yannakakis [9] obtain a $(2 - 2/k)$-approximation ratio for the node-weighted version, and Naor and Zosin [11] obtain 2-approximation for the case of directed graphs. It is known that any approximation ratio for these variations translates immediately into the same approximation ratio for VERTEX COVER, and thus it is hard to get any significant improvement over the approximation ratio of 2.

The algorithm from Theorem 1 appears next, giving a flavor of how this result is obtained. The complete proof of the approximation ratio is not long and appears in [3] or the book [12].

Notation

Let $G = (V, E)$ be an undirected graph on $V = \{1, 2, \ldots, n\}$ in which each edge $uv \in E$ has a non-negative cost $c(u, v) = c(v, u)$, and let $T = \{1, 2, \ldots, k\} \subseteq V$ be a set of *terminals*. MULTIWAY CUT is the problem of finding a minimum cost set $C \subseteq E$ such that in $(V, E \setminus C)$, each of the terminals $1, 2, \ldots, k$ is in a different component. Let MWC = MWC(G) be the value of the optimal solution to MULTIWAY CUT.

Δ_k denotes the $(k-1)$-simplex, i.e., the $(k-1)$-dimensional convex polytope in $\mathbb{R}^k$ given by $\{x \in \mathbb{R}^k | (x \geq 0) \wedge (\sum_i x_i = 1)\}$.

For $x \in \mathbb{R}^k$, $\|x\|$ is its L_1 norm: $\|x\| = \sum_i |x_i|$. For $j = 1, 2, \ldots, k$, $e^j \in \mathbb{R}^k$ denotes the unit vector given by $(e^j)_j = 1$ and $(e^j)_i = 0$ for all $i \neq j$.

LP-Relaxation

The *simplex* relaxation for MULTIWAY CUT with edge costs has as variables k-dimensional real vectors x^u, defined for each vertex $u \in V$:

$$\text{Minimize } \frac{1}{2} \sum_{uv \in E} c(u, v) \cdot \|x^u - x^v\|$$

Subject to:

$$x^u \in \Delta_k \quad \forall u \in V$$

$$x^t = e^t \quad \forall t \in T.$$

In other words, the terminals stay at the vertices of the $(k-1)$-simplex, and the other nodes anywhere in the simplex, and measure an edge's length by the total variation distance between its endpoints. Clearly, placing all nodes at simplex vertices gives an integral solution: the lengths of edges are either 0 (if both endpoints are at the same vertex) or 1 (if the endpoints are at different vertices), and the removal of all unit length edges disconnects the graph into at least k components, each containing at most one terminal.

To solve this relaxation as a linear program, new variables are introduced: y^{uv}, defined for all $uv \in E$, and x_i^u, defined for all $u \in V$ and $i \in T$.

Also new variables are y_i^{uv}, defined for all $i \in T$ and $uv \in E$. Then one writes the linear program:

$$\text{Minimize } \frac{1}{2}\sum_{uv\in E} c(u,v)y^{uv}$$

Subject to :

$$x^u \in \Delta_k \qquad \forall u \in V$$

$$x^t = e^t \qquad \forall t \in T$$

$$y^{uv} = \sum_{i\in T} y_i^{uv} \qquad \forall uv \in E$$

$$y_i^{uv} \geq x_i^u - x_i^v \quad \forall uv \in E\ ,\ i \in T$$

$$y_i^{uv} \geq x_i^v - x_i^u \quad \forall uv \in E\ ,\ i \in T\ .$$

It is easy to see that this linear program optimally solves the simplex relaxation above, by noticing that an optimal solution to the linear program can be assumed to put $y_i^{uv} = |x_i^u - x_i^v|$ and $y^{uv} = \|x^u - x^v\|$. Thus, solving the simplex relaxation can be done in polynomial time. This is the first step of the approximation algorithm. Clearly, the value Z^* of this solution is a lower bound on the cost of the minimum multiway cut MWC.

The second step of the algorithm is a rounding procedure which transforms a feasible solution of the simplex relaxation into an integral feasible solution. The rounding procedure below differs slightly from the one given in [3], but can be proven to give exactly the same solution. This variant is easier to present, although if one wants to prove the approximation ratio then the only way we know of is by showing that indeed this variant gives the same solution as the more complicated algorithm given in [3].

Rounding

Set $B(i,\rho) = \{u \in V \mid x_i^u > 1-\rho\}$, the set of nodes suitably "close" to terminal i in the simplex. Choose a permutation $\sigma = \langle \sigma_1, \sigma_2, \ldots, \sigma_k\rangle$ to be either $\langle 1,2,3,\ldots,k-1,k\rangle$ or $\langle k-1,k-2,k-3,\ldots,1,k\rangle$ with probability $1/2$ each. Independently, choose $\rho \in (0,1)$ uniformly at random. Then, process the terminals in the order $\sigma(1), \sigma(2), \sigma(3), \ldots, \sigma(k)$.

Algorithm 1 The rounding procedure

1: Let $\sigma = \langle 1,\ldots,k-3,k-2,k-1,k\rangle$ or $\langle k-1,k-2,k-3,\ldots,1,k\rangle$, each with prob. 1/2
2: Let ρ be a random real in $(0,1)$ /* See the paragraph below. */
3: **for** $j = 1$ **to** $k-1$ **do**
4:     **for** all u such that $x^u \in B(\sigma_j,\rho) \setminus \cup_{i:i<j} B(\sigma_i,\rho)$ **do**
5:         $\bar{x}^u := e^{\sigma_j}$ /* assign node u to terminal σ_j */
6:     **end for**
7: **end for**
8: **for** all u such that $x^u \notin \cup_{i:i<k} B(\sigma_i,\rho)$ **do**
9:     $\bar{x}^u := e^k$
10: **end for**

For each j from 1 to $k-1$, place the nodes that remain in $B(\sigma_j,\rho)$ at e^{σ_j}. Place whatever nodes remain at the end at e^k. The following code specifies the rounding procedure more formally. $\bar{x}$ denotes the rounded (integral) solution.

To derandomize and implement this algorithm in polynomial time, one tries both permutations σ and at most $k(n+1)$ values of ρ. Indeed, for any permutation σ, two different values of ρ, $\rho_1 < \rho_2$, produce combinatorially distinct solutions only if there is a terminal i and a node u such that $x_i^u \in (1-\rho_2, 1-\rho_1]$. Thus, there are at most $k(n+1)$ "interesting" values of ρ, which can be determined easily by sorting the nodes according to each coordinate separately. The resulting discrete sample space for (σ,ρ) has size at most $2k(n+1)$, so one can search it exhaustively.

The analysis of the algorithm, however, is based on the randomized algorithm above, as the proof shows that the expected total cost of edges whose endpoints are at different vertices of Δ_k in the rounded solution $\bar{x}$ is at most $1.5\,Z^*$. To get an $(1.5 - 1/k)Z^*$ upper bound, one must rename the terminals such that terminal k maximizes a certain quantity given by the simplex relaxation, or alternatively randomly pick a terminal as the last element of the permutation (the order of the first $k-1$ terminals does not matter as long as both the increasing and the decreasing permutations are tried by the rounding procedure). Exhaustive search of the sample space produces one integral solution whose cost does not exceed the average.

Applications

MULTIWAY CUT is used in Computer Vision, but unless one can solve the instance exactly, algorithms for the generalization METRIC LABELING are needed. MULTIWAY CUT has applications in parallel and distributed computing, as well as in chip design.

Open Problems

The improvements of [10, 4] are based on better rounding procedures and both compare the integral solution obtained to Z^*. This leads to the natural question: what is the supremum, over multiway cut instances G, of $Z^*(G)/\mathrm{MWC}(G)$. This supremum is called *integrality gap* or *integrality ratio*. For three terminals, [10] and [4] show that the integrality gap is exactly 12/11, while for general k, Freund and Karloff [7] give a lower bound of 8/7. The best-known upper bound is 1.3438, achieved by an approximation algorithm of [10].

Cross-References

▸ Multicut
▸ Sparsest Cut

Recommended Reading

1. Ahuja RK, Magnanti TL, Orlin JB (1993) Network flows. Prentice Hall, Englewood Cliffs
2. Arora S, Karger D, Karpinski M (1999) Polynomial time approximation schemes for dense instances of NP-hard problems. J Comput Syst Sci 58(1):193–210, Preliminary version in STOC 1995
3. Calinescu G, Karloff HJ, Rabani Y (1998) An improved approximation algorithm for multiway cut. In: ACM symposium on theory of computing, pp 48–52. Journal version in J Comput Syst Sci 60:564–574 (2000)
4. Cheung K, Cunningham WH, Tang L (2006) Optimal 3-terminal cuts and linear programming. Math Prog 105:389–421, Preliminary version in IPCO 1999
5. Chopra S, Rao MR (1991) On the multiway cut polyhedron. Networks 21:51–89
6. Dahlhaus E, Johnson DS, Papadimitriou CH, Seymour PD, Yannakakis M (1994) The complexity of multiterminal cuts. SIAM J Comput 23:864–894, Preliminary version in STOC 1992, An extended abstract was first announced in 1983
7. Freund A, Karloff H (2000) A lower bound of 8/(7+1/k-1) on the integrality ratio of the Calinescu–Karloff–Rabani relaxation for multiway cut. Inf Process Lett 75:43–50
8. Frieze A, Kannan R (1996) The regularity lemma and approximation schemes for dense problems. In: Proceedings of 37th IEEE FOCS. IEEE Computer Society, Los Alamitos, pp 12–20
9. Garg N, Vazirani VV, Yannakakis M (2004) Multiway cuts in node weighted graphs. J Algorithms 50(1):49–61, Preliminary version in ICALP 1994
10. Karger DR, Klein P, Stein C, Thorup M, Young NE (2004) Rounding algorithms for a geometric embedding of minimum multiway cut. Math Oper Res 29(3):436–461, Preliminary version in STOC 1999
11. Naor JS, Zosin L (2001) A 2-approximation algorithm for the directed multiway cut problem. SIAM J Comput 31(2):477–492, Preliminary version in FOCS 1997
12. Vazirani VV (2001) Approximation algorithms. Springer, Berlin/Heidelberg/New York

Musite: Tool for Predicting Protein Phosphorylation Sites

Qiuming Yao[1], Jianjiong Gao[2], and Dong Xu[3]
[1]University of Missouri, Columbia, MO, USA
[2]Computational Biology Center, Memorial Sloan-Kettering Cancer Center, New York, NY, USA
[3]Bond Life Sciences Center, University of Missouri, Columbia, MO, USA

Keywords

Amino acid frequency; Disorder score; K nearest neighbors; Phosphorylation site prediction; Phosphoproteomics; Support vector machine; Substitution matrix

Years and Authors of Summarized Original Work

2010; Gao, Thelen, Dunker, Xu
2012; Yao, Gao, Bollinger, Thelen, Xu

Problem Definition

Protein phosphorylation plays an important role in various biological functions and cellular processes. Identifying potential phosphorylation sites in a protein often helps to reveal functional details at the molecular level and was always performed by in vivo or in vitro experiments. Since the last decade, bioinformatics has been contributing significantly in characterizing protein structures and functionalities solely from its primary information, which also sheds light on phosphorylation site prediction. As per our expectation, in silico prediction should not only provide an alternative way to identify protein phosphorylation sites at lower cost but also with much higher throughput (e.g., proteome-wide screening), so that biologists can quickly pinpoint the potential sites for further experiments from a long list of targets. Therefore, it is soon becoming valuable and imperative to build such a bioinformatics tool or framework that can predict general and kinase-specific phosphorylation sites in proteins.

In definition, protein phosphorylation prediction is a computational approach to determine whether a certain amino acid in a protein sequence can be potentially phosphorylated or not. More specifically, given a protein (or peptide) sequence $P = [a_i]\ i = 1, \ldots n$ (where n is the sequence length, with amino acid a_i at ith position), the prediction algorithm is to tell if each of a_i (especially when a_i is serine, threonine, or tyrosine) can be phosphorylated in P or not. In a kinase-specific format, this question is asked with a proposed kinase name or kinase family. Moreover, this question can also be asked in a species-specific or condition-specific format.

Key Results

Machine Learning Approach

The algorithm we designed to fulfill this prediction task is able to resolve the association between phosphorylation and sequence information from the experimentally identified phosphorylation sites. Thus, it is formulated as a machine learning approach rather than an ab initio method. The collected experimental data need to be split into training and testing set to generate, tune, and validate our machine learning models. These machine learning models are then capable of predicting general or kinase-specific phosphorylation site for proteins with unknown sites. The general or specific prediction models are very dependent on the correspondent data sets in training. For example, kinase-specific prediction requires kinase-specific training data sets. After different preprocessing steps applied on data sets for general or specific purposes, the prediction models are generated from the same machine learning method (i.e., support vector machine [1–3] in our framework).

Technically speaking, per site prediction can be modeled as a binary classification problem, where the class label Y is either $+1$ for identified phosphorylation site or -1 for unidentified site, with X as its feature vector. A machine learning model can be considered as a map function from feature space X to the class label, i.e., $M : g(X) \rightarrow Y$, obtained from the training data set$\{(X_1, Y_1), (X_2, Y_2) \ldots, (X_m, Y_m)\}$. The prediction for the unknown X^* is simply calculated through $Y^* = g(X^*)$.

In our case, X_i is a feature vector from the protein sequence, extracted from a flanking peptide surrounding i-th amino acid a_i. The flanking sequence is often centralized by a_i and is a substring of the original protein sequence, denoted as $p(a_i) = [a_{i-w}, \ldots a_i, \ldots, a_{i+w}]$, where w is called the window size.

Support vector machine (SVM) then generates our prediction model M by maximizing the margin of the classification boundary.

Features

K nearest neighbor (KNN) scores, disorder scores, and amino acid information are used as the features in our SVM-based machine learning approach.

For amino acid a_i, **k** nearest neighbors are defined as the top k most similar peptides (within smallest distances) to the target peptide $p(a_i)$ in the training data set. Besides size k, the neighborhood can be also defined as a certain percentage of the whole training data set (e.g., 1 % of the total population). KNN score is then the ratio between the numbers of positive and negative sites within this predefined neighborhood.

Notice that the peptide similarity needs to be defined and normalized. In a more clear illustration, the two flanking sequences centralized by amino acid a_i and a_j are represented as $p(a_i) = [a_{i-w}, a_{i-w+1}, \ldots, a_i, \ldots, a_{i+w-1}, a_{i+w}]$ and $p(a_j) = [a_{j-w}, a_{j-w+1}, \ldots, a_j, \ldots, a_{j+w-1}, a_{j+w}]$, respectively.

The distance $D(p(a_i), p(a_j))$ between peptides $p(a_i)$ and $p(a_j)$ is calculated by

$$D(p(a_i), p(a_j)) = 1 - \frac{\sum_{k=-w}^{w} S(a_{i+k}, a_{j+k})}{2w+1}$$

where w is the window size, and the function S(.) is to calculate the amino acid similarity between a_i and a_j based on the normalized amino acid substitution matrix Q. More specifically,

$$S(a_i, a_j) = \frac{Q(a_i, a_j) - \min(Q)}{\max(Q) - \min(Q)}$$

where a_i and a_j are two amino acids, Q is the substitution matrix, and max(Q) and min(Q) represent the maximal and minimal values in the matrix Q. By default, BLOSUM62 is used as the most general substitution matrix. In fact, Q can also be directly calculated from the training data set and then KNN score is very specific to the training samples.

Disorder score per amino acid site reflects the stability of the local structure and is calculated by VSL2B [4]. By considering the disorder property as a more neighborhood-dependent and continuous feature, we correct (smooth) the disorder score at the position a_i using the mean value across the flanking peptide $p(a_i)$, i.e.,

$$\begin{aligned} \text{Disorder}(a_i) &= \text{average}(p(a_i)) \\ &= \frac{1}{2w+1} \sum_{k=-w}^{w} \text{disorder}(a_{i+k}) \end{aligned}$$

Amino acid information for flanking sequence $p(a_i)$ can refer to both composition and position information. At one extreme, amino acid frequency reflects composition information but no position information. The amino acid preference in phosphorylated peptides [5] can be revealed by this frequency feature. The size of this frequency vector is 20, which stores the normalized counts for each amino acid type within the range of the flanking peptide $p(a_i)$. On the other extreme, amino acid binary vector can provide position-specific information by bookkeeping a 0-1 vector for each amino acid at each position. The length of amino acid binary vector is 20*w, much longer than the frequency, which may potentially cause over-fitting in machine learning when the sample size is small. Therefore, selecting the right way to encode and represent the amino acid information is a trade-off between the losslessness of the positional information and the length of the feature vector.

Bootstrap and Aggregation

Since the under-identified phosphorylation sites (negative data) are always overwhelming the identified ones (positive data) in the training data set, we resolved this problem with bootstrap procedures to avoid the potential bias in the final classifier due to this unbalancing situation. The bootstrap step is a randomized resampling to get a balanced training data each time, which is repeated many times in order to explore the whole sample space thoroughly. So we will get many models based on the actual number of bootstrap steps, such as $M_1, \ldots, M_k$, and k could be up to thousands. Then, we do the final classification based on the voting or mean value from these many models, i.e., $G : G[g_1(X), \ldots g_k(X)]$, where G is called the aggregation step. The aggregated model is thus unbiased despite the imbalance of the labels in the training data.

Cross Validation

With the trained model, the testing result is often displayed as a trade-off between specificity and sensitivity, e.g., by a receiver-operating characteristic (ROC) curve. Specificity and sensitivity are defined as follows:

$$\text{specificity} = \frac{TN}{TN + FP}$$

$$\text{sensitivity} = \frac{TP}{TP + FN}$$

where TN represents true negative, FP false positive, TP true positive, and FN false negative.

Cross validation is a way to measure if the power of the phosphorylation prediction model trained from the known data can be well extended to the unknown. Usually, the cross validation can be performed with leave-one-out strategy (for small data set, i.e., kinase-specific data set) or from non-overlapped testing and training sets with x folds settings (for general phosphorylation site).

Musite as a Toolkit

Musite is an open-source software toolkit designed for large-scale phosphorylation prediction for both general and kinase-specific cases [6, 7]. The framework is quite flexible, so that user can take advantage of different preprocessing steps for specifying training or testing data, as well as picking different features and tuning parameters. By default, Musite provides general phosphorylation prediction models, several popular kinase-specific models, and multiple species-specific predictions (e.g., a plant-specific tool was build using our in-house plant protein phosphorylation database P^{3}DB [8–10]). Moreover, trained with users' specific data sets, Musite is also capable of generating customized models to do precise prediction particularly on their own research focus.

Applications

This tool or framework can be used as a quick filter on a long list of candidate proteins for experimental biologists to narrow down the phosphorylation sites to perform biochemical assay. It can also help to evaluate or compare the experimental observations in discovery studies. On the other hand, the computational experts can use this tool to do comparative studies, by fast and cheap computer screening across multiple proteomes. This tool can be easily and freely incorporated into any translational bioinformatics pipeline to characterize or annotate protein functionality within large scale proteomics studies.

Open Problems

1. The current version is an alignment free method. Is it possible or necessary to consider alignment, i.e., allowing indels for peptides similarity calculation?
2. The current features are more or less local. Are there any feasible features representing long distance association with phosphorylation site?
3. Can we extract any interesting biological rules from the machine learning models for general or specific phosphorylation events?

URLs to Code

The source code can be downloaded from SourceForge.
http://musite.sourceforge.net/
The online prediction Web services are available at:
http://musite.net/
http://p3db.org/prediction.php

Cross-References

▶ Support Vector Machines

Recommended Reading

1. Schölkopf B, Burges CJC, Smola AJ (1999) Advances in kernel methods support vector learning. MIT, Cambridge
2. Wang L (2005) Support vector machines: theory and applications. In: Studies in fuzziness and soft computing. Springer, Berlin
3. Vapnik VN (2000) The nature of statistical learning theory, 2nd edn. Springer, New York
4. Obradovic Z, Peng K, Vucetic S, Radivojac P, Dunker AK (2005) Exploiting heterogeneous sequence properties improves prediction of protein disorder. Proteins 61(Suppl 7):176–182

5. Iakoucheva LM, Radivojac P, Brown CJ, O'Connor TR, Sikes JG, Obradovic Z, Dunker AK (2004) The importance of intrinsic disorder for protein phosphorylation. Nucleic Acids Res 32(3):1037–1049
6. Yao Q, Gao J, Bollinger C, Thelen JJ, Xu D (2012) Predicting and analyzing protein phosphorylation sites in plants using musite. Front Plant Science 3:186
7. Gao J, Thelen JJ, Dunker AK, Xu D (2010) Musite, a tool for global prediction of general and kinase-specific phosphorylation sites. Mol Cell Proteomics (MCP) 9(12):2586–2600
8. Gao J, Agrawal GK, Thelen JJ, Xu D (2009) P3DB: a plant protein phosphorylation database. Nucleic Acids Research 37(Database issue): D960–D962
9. Yao Q, Bollinger C, Gao J, Xu D, Thelen JJ (2012) P3DB: an integrated database for plant protein phosphorylation. Front Plant Sci 3:206
10. Yao Q, Ge H, Wu S, Zhang N, Chen W, Xu C, Gao J, Thelen JJ, Xu D (2014) P3DB 3.0: from plant phosphorylation sites to protein networks. Nucleic Acids Res 42(Database issue): D1206–D1213

M

N

Nash Equilibria and Dominant Strategies in Routing

Weizhao Wang[1], Xiang-Yang Li[2], and Xiaowen Chu[3]
[1]Google Inc., Irvine, CA, USA
[2]Department of Computer Science, Illinois Institue of Technology, Chicago, IL, USA
[3]Department of Computer Science, Hong Kong Baptist University, Hong Kong, China

Keywords

BB; Nash; Strategyproof; Truthful

Years and Authors of Summarized Original Work

2005; Wang, Li, Chu

Problem Definition

This problem is concerned with the multicast routing and cost sharing in a selfish network composed of relay terminals and receivers. This problem is motivated by the recent observation that the selfish behavior of the network could largely degraded existing system performance, even dysfunction. The work of Wang, Li and Chu [7] first presented some negative results of the strategyproof mechanism in multicast routing and sharing, and then proposed a new solution based on Nash Equilibrium that could greatly improve the performance.

Wang, Li and Chu modeled a network by a link weighted graph $G = (V, E, \mathbf{c})$, where V is the set of all nodes and $\mathbf{c}$ is the cost vector of the set E of links. For a multicast session, let Q denote the set of all receivers. In game theoretical networking literatures, usually there are two models for the multicast cost/payment sharing.

Axiom Model (AM) All receivers must receive the service, or equivalently, each receiver has an infinity valuation [3]. In this model, a sharing method ξ computes how much each receiver should pay when the receiver set is R and cost vector is $\mathbf{c}$.

Valuation Model (VM) There is a set $Q = \{q_1, q_2, \cdots, q_r\}$ of r possible receivers. Each receiver $q_i \in Q$ has a valuation $\boldsymbol{\eta}_i$ for receiving the service. Let $\boldsymbol{\eta} = (\eta_1, \eta_2, \ldots, \eta_r)$ be the valuation vector and $\boldsymbol{\eta}_R$ be the valuation vector of a set $R \subseteq Q$ of receivers. In this model, they are interested in a sharing mechanism $\mathcal{S}$ consisting of a *selection scheme* $\sigma(\boldsymbol{\eta}, \mathbf{c})$ and a *sharing method* $\xi(\boldsymbol{\eta}, \mathbf{c})$. $\sigma_i(\boldsymbol{\eta}, \mathbf{c})$ denotes whether receiver i receives the service or not, and $\xi_i(\boldsymbol{\eta}, \mathbf{c})$ computes how much the receiver q_i should pay for the multicast service. Let $\mathbb{P}(\boldsymbol{\eta}, \mathbf{c})$ be the total payment for providing the service to the receiver set.

In the valuation model, a receiver who is willing to receive the service is not guaranteed to receive the service. For notational simplicity, $\sigma(\boldsymbol{\eta}, \mathbf{c})$ is used to denote the set of actual receivers. Under the Valuation Model, a *fair*

M.-Y. Kao (ed.), *Encyclopedia of Algorithms*,
DOI 10.1007/978-1-4939-2864-4

Algorithm 1 The multicast system $\Psi^{\mathsf{DM}} = (\mathcal{M}^{\mathsf{DM}}, \mathcal{S}^{\mathsf{DM}})$ based on multicast tree LCPT

1: Compute path $\mathsf{LCP}(s, q_j, \mathbf{d})$ and set $\phi_j = \frac{\omega(\mathrm{B}_{mm}(s,q_j,\mathbf{d}),\mathbf{d})}{r}$ for every $q_j \in Q$.
2: Set $\mathcal{O}_i^{\mathsf{DM}}(\eta, \mathbf{d}) = 0$ and $\mathcal{P}_i^{\mathsf{DM}}(\eta, \mathbf{d}) = 0$ for each link $e_i \notin \mathsf{LCP}(s, q_j, \mathbf{d})$.
3: **for** each receiver q_j **do**
4: **if** $\eta_j \geq \phi_j$ **then**
5: Receiver q_j is granted the service and charged $\xi_j^{\mathsf{DM}}(\eta, \mathbf{d})$, set $R = R \cup q_j$.
6: **else**
7: Receiver q_j is not granted the service and is charged 0.
8: **end if**
9: **end for**
10: Set $\mathcal{O}_i^{\mathsf{DM}}(\eta, \mathbf{d}) = 1$ and $\mathcal{P}_i^{\mathsf{DM}}(\eta, \mathbf{d}) = \mathcal{P}_i^{\mathsf{LCPT}}(\eta_R^{=\infty}, \mathbf{d})$ for each link $e_i \in \mathsf{LCPT}(R, \mathbf{d})$.

Algorithm 2 FPA Mechanism $\mathcal{M}^{\mathsf{AUC}}$

1: Each terminal bids a price b_i.
2: Every link sends a unit size dummy packet with property $\rho = \tau \cdot (n \cdot b_u - \sum_{e_i \in G} b_i)$ and receives payment $f_i(s, q_1, \mathbf{b}) = \tau \cdot \left[b_u \cdot (n \cdot b_u - \sum_{e_j \in G - e_i} b_j) - \frac{h_i^2}{2} \right]$. Here, b_u is the maximum cost any link can declare.
3: Compute the unique path $\mathsf{LCP}(s, q_1, \mathbf{b}')$ by applying certain fixed tie-breaking rule consistently.
4: Each terminal bids again for a price b_i'.
5: **for** each link e_i **do**
6: It is select to relay the packet and receives payment b_i' if and only if e_i is on path $\mathsf{LCP}(s, q_1, \mathbf{b}')$.
7: **end for**

sharing according to the following criteria is studied.

- **Budget Balance**: For the receiver set $R = \sigma(\eta, \mathbf{c})$, $\mathbb{P}(\eta, \mathbf{c}) = \sum_{q_i \in Q} \xi_i(\eta, \mathbf{c})$. If $\alpha \cdot \mathbb{P}(\eta, \mathbf{c}) \leq \sum_{i \in R} \xi_i(\eta, \mathbf{c}) \leq \mathbb{P}(\eta, \mathbf{c})$, for some given parameter $0 < \alpha \leq 1$, then $\mathcal{S} = (\sigma, \xi)$ is called α-budget-balance. If budget balance is not achievable, then a sharing scheme $\mathcal{S}$ may need to be α-budget-balance instead of budget balance.
- **No Positive Transfer** (NPT): Any receiver q_i's sharing should not be negative.
- **Free Leaving**: (FR) The potential receivers who do not receive the service should not pay anything.
- **Consumer Sovereignty** (CS): For any receiver q_i, if η_i is sufficiently large, then q_i is guaranteed to be an actual receiver.
- **Group-Strategyproof** (GS): Assume that η is the valuation vector and $\eta' \neq \eta$. If $\xi_i(\eta', \mathbf{c}) \geq \xi_i(\eta, \mathbf{c})$ for each $q_i \in \eta$, then $\xi_i(\eta', \mathbf{c}) = \xi_i(\eta, \mathbf{c})$.

Notations

The path with the lowest cost between two odes s and t is denoted as $\mathsf{LCP}(s, t, \mathbf{c})$, and its cost is dented as $|\mathsf{LCP}(s, t, \mathbf{c})|$. Given a simple path P in the graph G with cost vector $\mathbf{c}$, the sum of the cost of links on path P is denoted as $|\mathsf{P}(\mathbf{c})|$. For a simple path $P = v_i \rightsquigarrow v_j$, if $\mathsf{LCP}(s, t, \mathbf{c}) \bigcap P = \{v_i, v_j\}$, then P is called a *bridge* over $\mathsf{LCP}(s, t, \mathbf{c})$. This bridge P *covers* link e_k if $e_k \in \mathsf{LCP}(v_i, v_j, \mathbf{c})$. Given a link $e_i \in \mathsf{LCP}(s, t, \mathbf{c})$, the path with the minimum cost that covers e_i is denoted as $\mathrm{B}_{\min}(e_i, \mathbf{c})$. The bridge $\mathrm{B}_{mm}(s, t, \mathbf{c}) = \max_{e_i \in \mathsf{LCP}(s,t,\mathbf{c})} \mathrm{B}_{\min}(e_i, \mathbf{c})$ is the *max-min cover* of the path $\mathsf{LCP}(s, t, \mathbf{c})$.

A bridge set $\mathcal{B}$ is a *bridge cover* for $\mathsf{LCP}(s, t, \mathbf{c})$, if for every link $e_i \in \mathsf{LCP}(s, t, \mathbf{c})$, there exists a bridge $\mathrm{B} \in \mathcal{B}$ such that $e_i \in \mathsf{LCP}(v_{s(\mathrm{B})}, v_{t(\mathrm{B})}, \mathbf{c})$. The *weight* of a bridge cover $\mathcal{B}(s, t, \mathbf{c})$ is defined as $|\mathcal{B}(s, t, \mathbf{c})| = \sum_{B \in \mathcal{B}(s,t,\mathbf{c})} \sum_{e_i \in B} c_i$. A bridge cover is a *least bridge cover* (LB), denoted by $\mathbb{LB}(s, t, \mathbf{c})$, if it has the smallest weight among all bridge covers that cover $\mathsf{LCP}(s, t, \mathbf{c})$.

Key Results

Theorem 1 *If $\Psi = (\mathcal{M}, \mathcal{S})$ is an α-stable multicast system, then $\alpha \leq 1/n$.*

Theorem 2 *Multicast system Ψ^{DM} is $1/(r \cdot n)$-stable, where r is the number of receivers.*

Theorem 1 gives an upper bound for α for any α-stable unicast system Ψ. It is not difficult to observe that even the receivers are cooperative, Theorem 1 still holds. Theorem 2 showed that

Algorithm 3 FPA based unicast system

1: Execute Line $1 - 3$ in Algorithm 2.
2: Compute $\mathbb{LB}(s, q_1, \mathbf{b})$, and set $\phi = \frac{|\mathbb{LB}(s,q_1,\mathbf{b})|}{2}$.
3: If $\phi \leq \eta_1$ then set $\sigma_1^{\mathsf{AU}}(\eta_1, \widetilde{\mathbf{b}}) = 1$ and $\xi_1^{\mathsf{AU}}(\eta_1, \widetilde{\mathbf{b}}) = \phi$. Every relay link on LCP is selected and receives an extra payment b_i'.
4: For each link $e_i \notin \mathrm{LCP}(s, q_1, \mathbf{b}')$, it receives a payment $\mathcal{P}_i^{\mathsf{AU}}(\eta_1, \widetilde{\mathbf{b}}) - \gamma \cdot (b_i' - b_i)^2$.

there exists a multicast system is $1/(r \cdot n)$-stable. When $r = 1$, the problem become traditional unicast system and the bound is tight. When relaxing the dominant strategy to the Nash Equilibria requirement, a First Price Auction (FPA) mechanism is proposed by Wang et al. under the Axiom Model that has many nice properties.

Theorem 3 *There exists NE for FPA mechanism $\mathcal{M}^{\mathsf{AUC}}$ and for any NE, (a) each link bids his true cost as the first bid b_i, (b) the actual shortest path is always selected, (c) the total cost for different NE differs at most 2 times.*

Based on the FPA Mechanism Ψ^{AUC}, Wang, Li and Chu design a unicast system as follows.

Theorem 4 *The FPA based unicast system not only has Nash Equilibria, but also is $\frac{1}{2}$-NE-stable with ϵ additive, for any given ϵ.*

By treating each receiver as a separate receiver and applying the similar process as in the unicast system, Wang, Li and Chu extended the unicast system to a multicast system.

Theorem 5 *The FPA based multicast system not only has Nash Equilibria, but also is $1/(2 \cdot r)$-NE-stable with ϵ additive, for any given ϵ.*

Applications

More and more research effort has been done to study the non-cooperative games recently. Among these various forms of games, the unicast/multicast routing game [2, 5, 6] and multicast cost sharing game [1, 3, 4] have received a considerable amount of attentions over the past few year due to its application in the Internet. However, both unicast/multicast routing game and multicast cost sharing game are one folded: the unicast/multicast routing game does not take the receivers into account while the multicast cost sharing game does not treat the links as non-cooperative. In this paper, they study the scenario, which was called *multicast system*, in which both the links and the receivers could be non-cooperative. Solving this problem paving a way for the real world commercial multicast and unicast application. A few examples are, but not limited to, the multicast of the video content in wireless mesh network and commercial WiFi system; the multicast routing in the core Internet.

Open Problems

A number of problems related to the work of Wang, Li and Chu [7] remain open. The first and foremost, the upper bound and lower bound on α still have a gap of r if the multicast system is α-stable; and a gap of $2r$ if the multicast system is α-Nash stable.

The second, Wang, Li and Chu only showed the existence of the Nash Equilibrium under their systems. They have not characterized the convergence of the Nash Equilibrium and the strategies of the user, which are not only interesting but also important problems.

Cross-References

▸ Non-approximability of Bimatrix Nash Equilibria

Recommended Reading

1. Feigenbaum J, Papadimitriou CH, Shenker S (2001) Sharing the cost of multicast transmissions. J Comput Syst Sci 63:21–41
2. Kao M-Y, Li X-Y, Wang W (2005) Towards truthful mechanisms for binary demand games: a general framework. In: ACM EC, Vancouver, pp 213–222
3. Herzog S, Shenker S, Estrin D (1997) Sharing the "cost" of multicast trees: an axiomatic analysis. IEEE/ACM Trans Netw 5:847–860

4. Moulin H, Shenker S (2001) Strategyproof sharing of submodular costs: budget balance versus efficiency. Econ Theory 18:511–533
5. Wang W, Li X-Y, Sun Z, Wang Y (2005) Design multicast protocols for non-cooperative networks. In: Proceedings of the 24th IEEE INFOCOM, Miami, vol 3, pp 1596–1607
6. Wang W, Li X-Y, Wang Y (2004) Truthful multicast in selfish wireless networks. In: Proceedings of the 10th ACM MOBICOM, Philadelphia, pp 245–259
7. Wang W, Li X-Y, Chu X (2005) Nash equilibria, dominant strategies in routing. In: Workshop for Internet and network economics (WINE). Lecture notes in computer science, vol 3828. Springer, Hong Kong, pp 979–988

Nearest Neighbor Interchange and Related Distances

Bhaskar DasGupta[1], Xin He[2], Ming Li[3], John Tromp[4], and Louxin Zhang[5]
[1]Department of Computer Science, University of Illinois, Chicago, IL, USA
[2]Department of Computer Science and Engineering, The State University of New York, Buffalo, NY, USA
[3]David R. Cheriton School of Computer Science, University of Waterloo, Waterloo, ON, Canada
[4]CWI, Amsterdam, The Netherlands
[5]Department of Mathematics, National University of Singapore, Singapore, Singapore

Keywords

Comparison of phylogenies; Network models of evolution

Years and Authors of Summarized Original Work

1997; DasGupta, He, Jiang, Li, Tromp, Zhang

Problem Definition

In this entry, the authors state results on some transformation-based distances for *evolutionary trees*. Several distance models for evolutionary trees have been proposed in the literature. Among them, the best known is perhaps the *nearest neighbor interchange* (nni) distance introduced independently in [10] and [9]. The authors will focus on the nni distance and a closely related distance called the *subtree-transfer* distance originally introduced in [5, 6]. Several papers that involved DasGupta, He, Jiang, Li, Tromp, and Zhang essentially showed the following results:

- A correspondence between the nni distance and the linear-cost subtree-transfer distance on unweighted trees.
- Computing the nni distance is NP-hard, but admits a fixed-parameter tractability and a logarithmic ratio approximation algorithms.
- A 2-approximation algorithm for the linear-cost subtree-transfer distance on weighted evolutionary trees.

The authors first define the nni and linear-cost subtree-transfer distances for unweighted trees. Then the authors extend the nni and linear-cost subtree-transfer distances to weighted trees. For the purpose of this entry, an evolutionary tree (also called *phylogeny*) is an *unordered* tree, has uniquely labeled leaves and unlabeled interior nodes, can be *unrooted* or *rooted*, can be *unweighted* or *weighted*, and has all internal nodes of degree 3.

Unweighted Trees

An nni operation swaps two subtrees that are separated by an internal edge (u, v), as shown in Fig. 1.

The nni operation is said to *operate* on this internal edge. The nni distance, $D_{\text{nni}}(T_1, T_2)$, between two trees T_1 and T_2 is defined as the *minimum* number of nni operations required to transform one tree into the other.

An nni operation can also be viewed as moving a subtree past a neighboring internal node. A more general operation is to transfer a subtree from one place to another arbitrary place. Figure 2 shows such a *subtree-transfer* operation.

The subtree-transfer distance between two trees T_1 and T_2 is the minimum number of

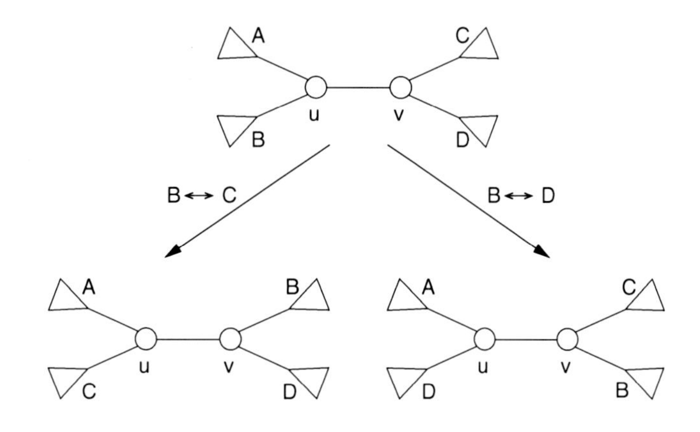

Nearest Neighbor Interchange and Related Distances, Fig. 1 The two possible nni operations on an internal edge (u, v): exchange $B \leftrightarrow C$ or $B \leftrightarrow D$

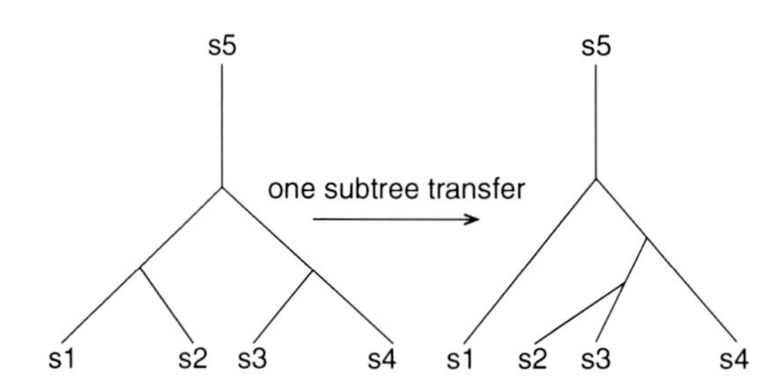

Nearest Neighbor Interchange and Related Distances, Fig. 2 An example of subtree-transfer

subtrees one needs to move to transform T_1 into T_2 [5–7]. It is sometimes appropriate in practice to discriminate among subtree-transfer operations as they occur with different frequencies. In this case, one can charge each subtree-transfer operation a cost equal to the distance (the number of nodes passed) that the subtree has moved in the current tree. The *linear-cost* subtree-transfer distance, $D_{\mathrm{lcst}}(T_1, T_2)$, between two trees T_1 and T_2 is then the minimum total cost required to transform T_1 into T_2 by subtree-transfer operations [1, 3].

Weighted Trees

Both the linear-cost subtree-transfer and nni models can be naturally extended to weighted trees. The extension for nni is straightforward: an nni operation is simply charged a cost equal to the weight of the edge it operates on. For feasibility of weighted nni transformation between two given weighted trees T_1 and T_2, one also requires that the following conditions are satisfied: (1) for each leaf label a, the weight of the edge in T_1 incident on a is the same as the weight of the edge in T_2 incident on a and (2) the multisets of weights of internal edges of T_1 and T_2 are the same (Fig. 3).

In the case of linear-cost subtree-transfer, although the idea is immediate, i.e., a moving subtree should be charged for the weighted distance it travels, the formal definition needs some care and is given below. Consider (unrooted) trees in which each edge e has a weight $w(e) \geq 0$. To ensure feasibility of transforming a tree into another, one requires the total weight of all edges to equal one. A subtree-transfer is now defined as follows. Select a subtree S of T at a given node u and select an edge $e \notin S$. Split the edge e into two edges e_1 and e_2 with weights $w(e_1)$ and $w(e_2)$ ($w(e_1), w(e_2) \geq 0, w(e_1)+w(e_2) = w(e)$), and move S to the common end point of e_1 and e_2. Finally, merge the two remaining edges $e\prime$ and $e\prime\prime$ adjacent to u into one edge with weight $w(e') + w(e'')$. The cost of this subtree-transfer is the total weight of all the edges over which S is moved. Figure 3 gives an example. The edge-weights of the given tree are normalized so that their total sum is 1. The subtree S is transferred to split the edge e_4 to e_6 and e_7 such that $w(e_6)$, $w(e_7) \geq 0$ and $w(e_6) + w(e_7) = w(e_4)$; finally, the two edges e_1 and e_2 are merged to e_5 such that $w(e_5) = w(e_1) + w(e_2)$. The cost of transferring S is $w(e_2) + w(e_3) + w(e_6)$.

Note that for weighted trees, the linear-cost subtree-transfer model is more general than the nni model in the sense that one can slide a subtree along an edge with subtree-transfers. Such an operation is not realizable with nni moves.

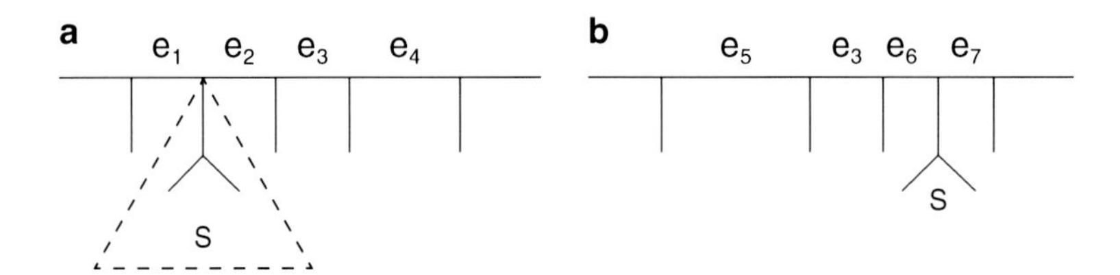

Nearest Neighbor Interchange and Related Distances, Fig. 3 Subtree-transfer on weighted phylogenies. Tree (**b**) is obtained from tree (**a**) with one subtree-transfer

Key Results

Let T_1 and T_2 be the two trees, each with n nodes, that are being used in the distance computation.

Theorem 1 ([1,2,4]) *Assume that T_1 and T_2 are unweighted. Then, the following results hold:*

- $D_{\text{nni}}(T_1, T_2) = D_{\text{lcst}}(T_1, T_2)$.
- *Computing $D_{\text{nni}}(T_1, T_2)$ is NP-complete.*
- *Suppose that $D_{\text{nni}}(T_1, T_2) \leq d$. Then, an optimal sequence of* nni *operations transforming T_1 into T_2 can be computed in $O(n^2 \log n + n \cdot 2^{23d/2})$ time.*
- *$D_{\text{nni}}(T_1, T_2)$ can be approximated to within a factor of* $\log n + O(1)$ *in polynomial time.*

Theorem 2 ([1–4]) *Assume that T_1 and T_2 are weighted. Then, the following results hold:*

- *$D_{\text{nni}}(T_1, T_2)$ can be approximated to within a factor of* $6 + 6 \log n$ *in $O(n^2 \log n)$ time.*
- *Assume that T_1 and T_2 are allowed to have leaves that are not necessarily uniquely labeled. Then, computing $D_{\text{lcst}}(T_1, T_2)$ is NP-hard.*
- *$D_{\text{lcst}}(T_1, T_2)$ can be approximated to within a factor of 2 in $O(n^2 \log n)$ time.*

Applications

The results reported here are on transformation-based distances for evolutionary trees. Such a tree can be *rooted* if the evolutionary origin is known and can be *weighted* if the evolutionary length on each edge is known. Reconstructing the correct evolutionary tree for a set of species is one of the fundamental yet difficult problems in evolutionary genetics. Over the past few decades, many approaches for reconstructing evolutionary trees have been developed, including (not exhaustively) parsimony, compatibility, distance, and maximum likelihood approaches. The outcomes of these methods usually depend on the data and the amount of computational resources applied. As a result, in practice they often lead to different trees on the same set of species [8]. It is thus of interest to compare evolutionary trees produced by different methods or by the same method on different data.

Another motivation for investigating the linear-cost subtree-transfer distance comes from the following motivation. When *recombination* of DNA sequences occurs in an evolution, two sequences meet and generate a new sequence, consisting of genetic material taken left of the recombination point from the first sequence and right of the point from the second sequence [5, 6]. From a phylogenetic viewpoint, before the recombination, the ancestral material on the present sequence was located on two sequences, one having all the material to the left of the recombination point and another having all the material to the right of the breaking point. As a result, the evolutionary history can no longer be described by a single tree. The recombination event partitions the sequences into two neighboring regions. The history for the left and the right regions could be described by separate evolutionary trees. The recombination makes the two evolutionary trees describing neighboring regions differ. However, two neighbor trees cannot be arbitrarily different, one must be obtainable from the other by a *subtree-transfer operation*. When more than one recombination occurs, one can describe an evolutionary history using a list of evolutionary trees, each corresponds to some region of the sequences and each can be obtained by several subtree-transfer operations from its predecessor [6]. The computation of a linear-cost subtree-transfer distance is useful in reconstructing such a list of trees based on parsimony [5, 6].

Open Problems

1. Is there a constant ratio approximation algorithm for the nni distance on unweighted evolutionary trees or is the $O(\log n)$-approximation the best possible?
2. Is the linear-cost subtree-transfer distance NP-hard to compute on weighted evolutionary trees if leaf labels are not allowed to be nonunique?
3. Can one improve the approximation ratio for linear-cost subtree-transfer distance on weighted evolutionary trees?

Cross-References

- ▸ Algorithms for Combining Rooted Triplets into a Galled Phylogenetic Network
- ▸ Maximum Agreement Subtree (of 2 Binary Trees)
- ▸ Maximum Agreement Subtree (of 3 or More Trees)
- ▸ Phylogenetic Tree Construction from a Distance Matrix

Recommended Reading

1. DasGupta B, He X, Jiang T, Li M, Tromp J, Zhang L (1997) On distances between phylogenetic trees. In: 8th annual ACM-SIAM symposium on discrete algorithms, New Orleans, pp 427–436
2. DasGupta B, He X, Jiang T, Li M, Tromp J, Wang L, Zhang L (1998) Computing distances between evolutionary trees. In: Du DZ, Pardalos PM (eds) Handbook of combinatorial optimization, vol 2. Kluwer Academic, Norwell, pp 35–76
3. DasGupta B, He X, Jiang T, Li M, Tromp J (1999) On the linear-cost subtree-transfer distance. Algorithmica 25(2):176–195
4. DasGupta B, He X, Jiang T, Li M, Tromp J, Zhang L (2000) On computing the nearest neighbor interchange distance. In: Du DZ, Pardalos PM, Wang J (eds) Proceedings of the DIMACS workshop on discrete problems with medical applications. DIMACS series in discrete mathematics and theoretical computer science, vol 55. American Mathematical Society, Providence, Rhode Island, USA, pp 125–143
5. Hein J (1990) Reconstructing evolution of sequences subject to recombination using parsimony. Math Biosci 98:185–200
6. Hein J (1993) A heuristic method to reconstruct the history of sequences subject to recombination. J Mol Evol 36:396–405
7. Hein J, Jiang T, Wang L, Zhang K (1996) On the complexity of comparing evolutionary trees. Discret Appl Math 71:153–169
8. Kuhner M, Felsenstein J (1994) A simulation comparison of phylogeny algorithms under equal and unequal evolutionary rates. Mol Biol Evol 11(3):459–468
9. Moore GW, Goodman M, Barnabas J (1973) An iterative approach from the standpoint of the additive hypothesis to the dendrogram problem posed by molecular data sets. J Theor Biol 38:423–457
10. Robinson DF (1971) Comparison of labeled trees with valency three. J Comb Theory Ser B 11:105–119

N

Negative Cycles in Weighted Digraphs

Christos Zaroliagis
Department of Computer Engineering and Informatics, University of Patras, Patras, Greece

Years and Authors of Summarized Original Work

1994; Kavvadias, Pantziou, Spirakis, Zaroliagis

Problem Definition

Let $G = (V, E)$ be an n-vertex, m-edge directed graph (digraph), whose edges are associated with a real-valued cost function $wt : E \to \mathbb{R}$. The cost, $wt(P)$, of a path P in G is the sum of the costs of the edges of P. A simple path C whose starting and ending vertices coincide is called a cycle. If $wt(C) < 0$, then C is called a *negative cycle*. The goal of the negative cycle problem is to detect whether there is such a cycle in a given digraph G with real-valued edge costs, and if indeed exists to output the cycle.

The negative cycle problem is closely related to the shortest path problem. In the latter, a minimum cost path between two vertices s and t is sought. It is easy to see that an s-t shortest path exists if and only if no s-t path in G contains a negative cycle [1, 13]. It is also well-known that shortest paths from a given vertex s to all other vertices form a tree called *shortest path tree* [1, 13].

Key Results

For the case of general digraphs, the best algorithm to solve the negative cycle problem (or to compute the shortest path tree, if such a cycle does not exist) is the classical Bellman–Ford algorithm that takes $O(nm)$ time (see e.g., [1]). Alternative methods with the same time complexity are given in [4, 7, 12, 13]. Moreover, in [11, Chap. 7] an extension of the Bellman–Ford algorithm is described which, in addition to detecting and reporting the existing negative cycles (if any), builds a shortest path tree rooted a some vertex s reaching those vertices u whose shortest s-u path does not contain a negative cycle. If edge costs are integers larger than $-L$ ($L \geq 2$), then a better algorithm was given in [6] that runs in $O(m\sqrt{n}\log L)$ time, and it is based on bit scaling.

A simple deterministic algorithm that runs in $O(n^2 \log n)$ expected time with high probability is given in [10] for a large class of input distributions, where the edge costs are chosen randomly according to the endpoint-independent model (this model includes the common case where all edge costs are chosen independently from the same distribution).

Better results are known for several important classes of sparse digraphs (i.e., digraphs with $m = O(n)$ edges) such as planar digraphs, outerplanar digraphs, digraphs of small genus, and digraphs of small treewidth.

For general sparse digraphs, an algorithm is given in [8] that solves the negative cycle problem in $O(n + \tilde{\gamma}^{1.5} \log \tilde{\gamma})$ time, where $\tilde{\gamma}$ is a topological measure of the input sparse digraph G, and whose value varies from 1 up to $\Theta(n)$. Informally, $\tilde{\gamma}$ represents the minimum number of outerplanar subgraphs, satisfying certain separation properties, into which G can be decomposed. In particular, $\tilde{\gamma}$ is proportional to $\gamma(G) + q$, where G is supposed to be embedded into an orientable surface of genus $\gamma(G)$ so as to minimize the number q of faces that collectively cover all vertices. For instance, if G is outerplanar, then $\tilde{\gamma} = 1$, which implies an optimal $O(n)$ time algorithm for this case. The algorithm in [8] does not require such an embedding to be provided by the input. In the same paper, it is shown that random $G_{n,p}$ graphs with threshold function $1/n$ are planar with probability one and have an expected value for $\tilde{\gamma}$ equal to $O(1)$. Furthermore, an efficient parallelization of the algorithm on the CREW PRAM model of computation is provided in [8].

Better bounds for planar digraphs are as follows. If edge costs are integers, then an algorithm running in $O(n^{4/3} \log(nL))$ time is given in [9]. For real edge costs, an $O(n \log^3 n)$-time algorithm was given in [5].

An optimal $O(n)$-time algorithm is given in [3] for the case of digraphs with small treewidth (and real edge costs). Informally, the treewidth t of a graph G is a parameter which measures how close is the structure of G to a tree. For instance, the class of graphs of small treewidth includes series-parallel graphs ($t = 2$) and outerplanar graphs ($t = 2$). An optimal parallel algorithm for the same problem, on the EREW PRAM model of computation, is provided in [2].

Applications

Finding negative cycles in a digraph is a fundamental combinatorial and network optimization problem that spans a wide range of applications including: shortest path computation, two dimensional package element, minimum cost flows, minimal cost-to-time ratio, model verification, compiler construction, software engineering,

VLSI design, scheduling, circuit production, constraint programming and image processing. For instance, the isolation of negative feedback loops is imperative in the design of VLSI circuits. It turns out that such loops correspond to negative cost cycles in the so-called amplifier-gain graph of the circuit. In constraint programming, it is required to check the feasibility of sets of constraints. Systems of difference constraints can be represented by constraint graphs, and one can show that such a system is feasible if and only if there are no negative cost cycles in its corresponding constraint graph. In zero-clairvoyant scheduling, the problem of checking whether there is a valid schedule in such a scheduling system can be reduced to detecting negative cycles in an appropriately defined graph. For further discussion on these and other applications see [1, 12, 14].

Open Problems

The negative cycle problem is closely related to the shortest path problem. The existence of negative edge costs makes the solution of the negative cycle problem or the computation of a shortest path tree more difficult and thus more time consuming compared to the time required to solve the shortest path tree problem in digraphs with non-negative edge costs. For instance, for digraphs with real edge costs, compare the $O(nm)$-time algorithm in the former case with the $O(m + n \log n)$-time algorithm for the latter case (Dijkstra's algorithm implemented with an efficient priority queue; see e.g., [1]).

It would therefore be interesting to try to reduce the gap between the above two time complexities, even for special classes of graphs or the case of integer costs.

The only case where these two complexities coincide concerns the digraphs of small treewidth [3], making it the currently most general such class of graphs. For planar digraphs, the result in [5] is only a polylogarithmic factor away from the $O(n)$-time algorithm in [9] that computes a shortest path tree when the edge costs are non-negative.

Experimental Results

An experimental study for the negative cycle problem is conducted in [4]. In that paper, several methods that combine a shortest path algorithm (based on the Bellman–Ford approach) with a cycle detection strategy are investigated, along with some new variations of them. It turned out that the performance of algorithms for the negative cycle problem depends on the number and the size of the negative cycles. This gives rise to a collection of problem families for testing negative cycle algorithms.

A follow-up of the above study is presented in [14], where two new heuristics are introduced and are incorporated on three of the algorithms considered in [4] (the original Bellman–Ford and the variations in [13] and [7]), achieving dramatic improvements. The data sets considered in [14] are those in [4].

Data Sets

Data set generators and problem families are described in [4], and are available from http://www.avglab.com/andrew/soft.html.

URL to Code

The code used in [4] is available from http://www.avglab.com/andrew/soft.html.

Cross-References

▸ All Pairs Shortest Paths in Sparse Graphs
▸ All Pairs Shortest Paths via Matrix Multiplication
▸ Single-Source Shortest Paths

Recommended Reading

1. Ahuja R, Magnanti T, Orlin J (1993) Network flows. Prentice-Hall, Englewood Cliffs
2. Chaudhuri S, Zaroliagis C (1998) Shortest paths in digraphs of small treewidth. Part II: optimal parallel algorithms. Theor Comput Sci 203(2):205–223

3. Chaudhuri S, Zaroliagis C (2000) Shortest paths in digraphs of small treewidth. Part I: sequential algorithms. Algorithmica 27(3):212–226
4. Cherkassky BV, Goldberg AV (1999) Negative-cycle detection algorithms. Math Program 85:277–311
5. Fakcharoenphol J, Rao S (2001) Planar graphs, negative weight edges, shortest paths, and near linear time. In: Proceedings of 42nd IEEE symposium on foundations of computer science (FOCS 2001). IEEE Computer Society, Los Alamitos, pp 232–241
6. Goldberg AV (1995) Scaling algorithms for the shortest paths problem. SIAM J Comput 24:494–504
7. Goldberg AV, Radzik T (1993) A heuristic improvement of the Bellman-Ford algorithm. Appl Math Lett 6(3):3–6
8. Kavvadias D, Pantziou G, Spirakis P, Zaroliagis C (1994) Efficient sequential and parallel algorithms for the negative cycle problem. In: Algorithms and computation (ISAAC'94). Lecture notes computer science, vol 834. Springer, Heidelberg, pp 270–278
9. Klein P, Rao S, Rauch M, Subramanian S (1997) Faster shortest path algorithms for planar graphs. J Comput Syst Sci 5(1):3–23
10. Kolliopoulos SG, Stein C (1998) Finding real-valued single-source shortest paths in $o(n^3)$ expected time. J Algorithm 28:125–141
11. Mehlhorn K, Näher S (1999) LEDA: a platform for combinatorial and geometric computing. Cambridge University Press, Cambridge
12. Spirakis P, Tsakalidis A (1986) A very fast, practical algorithm for finding a negative cycle in a digraph. In: Proceedings of 13th ICALP, pp 397–406
13. Tarjan RE (1983) Data structures and network algorithms. SIAM, Philadelphia
14. Wong CH, Tam YC (2005) Negative cycle detection problem. In: Algorithms – ESA 2005. Lecture notes in computer science, vol 3669. Springer, Heidelberg, pp 652–663

Network Creation Games

Erik D. Demaine[1], Mohammad Taghi Hajiaghayi[2], Hamid Mahini[2], and Morteza Zadimoghaddam[3]
[1]MIT Computer Science and Artificial Intelligence Laboratory, Cambridge, MA, USA
[2]Department of Computer Science, University of Maryland, College Park, MD, USA
[3]Google Research, New York, NY, USA

Keywords

Game theory; Network formation; Price of anarchy; Small-world phenomenon

Years and Authors of Summarized Original Work

2015; Demaine, Hajiaghayi, Mahini, Zadimoghaddam

Problem Definition

Over the last few decades, a wide variety of networks have emerged. The general structure of these networks including their global connectivity properties has been studied extensively. On the other hand, strategic aspects of them are also very interesting to explore by considering the nodes as independent agents. The exciting area of network creation games attempts to understand how real-world networks (such as the Internet) develop when multiple independent agents (e.g., ISPs) build pieces of the network to selfishly improve their own objective functions which heavily depend on their connectivity properties.

We start by elaborating on these connectivity objectives and its relation to the global design and structure of the network. Network design is a fundamental family of problems at the intersection between computer science and operations research, amplified in importance by the sustained growth of computer networks such as the Internet. Traditionally, the goal is to find a minimum-cost (sub) network that satisfies some specified property such as k-connectivity or connectivity on terminals (as in the classic Steiner tree problem). Such a formulation captures the (possibly incremental) creation cost of the network but does not incorporate the cost of actually using the network. By contrast, network routing has the goal of optimizing the usage cost of the network but assumes that the network has already been created. The network creation game attempts to unify the network design and network routing problems by modeling both creation and usage costs. Specifically, each node in the system is an independent selfish agent that can create a link (edge) to any other node, at a cost of α. In addition to these creation costs, each node incurs a usage cost related to the distances to the other nodes. In the model introduced by

Fabrikant, Luthra, Maneva, Papadimitriou, and Shenker [11], the usage cost incurred by a node is the sum of distances to all other nodes. Equivalently, if we divide the cost (and thus α) by the number n of nodes, the usage cost is the average distance to other nodes. In another natural model, the usage cost incurred by a node is the maximum distance to all other nodes: this model captures the worst-case instead of average-case behavior of routing. To model the dominant behavior of large-scale networking scenarios such as the Internet, we consider each node to be an agent (player) [12] that selfishly tries to minimize its own creation and usage costs [1, 6, 11]. In this context, the price of anarchy [14, 15, 17] is the worst possible ratio of the total cost found by some independent selfish behavior and the optimal total cost possible by a centralized, social welfare-maximizing solution. The price of anarchy is a well-studied concept in algorithmic game theory for problems such as load balancing, routing, and network design; see, e.g., [1, 3–7, 11, 15, 16].

Equilibria To model the dominant behavior of large-scale networking scenarios such as the Internet, we consider the case where every node (player) selfishly tries to minimize its own creation and usage cost. This game-theoretic setting naturally leads to the various kinds of equilibria and the study of their structure. Two frequently considered notions are Nash equilibrium, where no player can change its strategy (which edges to buy) to locally improve its cost, and strong Nash equilibrium, where no coalition of players can change their collective strategy to locally improve the cost of each player in the coalition. Nash equilibria capture the combined effect of both selfishness and lack of coordination, while strong Nash equilibria separate these issues, enabling coordination and capturing the specific effect of selfishness. However, the notion of strong Nash equilibrium is extremely restrictive in our context, because all players can simultaneously change their entire strategies, abusing the local optimality intended by original Nash equilibria and effectively forcing globally near-optimal solutions. Thus it makes sense to focus on weaker notions of equilibria.

Structure of equilibria What structural properties can be predicted about equilibria in network creation games? For example, Fabrikant et al. [11] conjectured that equilibrium graphs in the unilateral model were all trees, but this is not always the case as shown by Albers et al. [1] One particularly interesting structural feature is whether all equilibrium graphs have small diameter (say, polylogarithmic), analogous to the small-world phenomenon. A closely related issue is the price of anarchy, that is, the worst possible ratio of the total cost of an equilibrium (found by independent selfish behavior) and the optimal total cost possible by a centralized solution (maximizing social welfare). The price of anarchy is a well-studied concept in algorithmic game theory for problems such as load balancing, routing, and network design. Upper bounds on diameter of equilibrium graphs translate to approximately equal upper bounds on the price of anarchy but not necessarily vice versa.

Notation

Formally, we define four games depending on the objective (sum or max) and the consent (unilateral or bilateral). In all versions, we have n players; call them $1, 2, \ldots, n$. The strategy of player i is specified by a subset s_i of $1, 2, \ldots, n \setminus i$, which corresponds to the set of neighbors to which player i forms a link. Together, let $s = \{s_1, s_2, \ldots, s_n\}$ denote the strategies of all players.

To define the cost of a strategy, we introduce an undirected graph G_s with vertex set $\{1, 2, \ldots, n\}$. In the unilateral game, G_s has an edge (i, j) if either $i \in s_j$ or $j \in s_i$. In the bilateral game, G_s has an edge (i, j) if both $i \in s_j$ and $j \in s_i$. Define $d_s(i, j)$ to be the distance (the number of edges in a shortest path) between vertices i and j in graph G_s. In the sum game, the cost incurred by player i is $c_i(s) = \alpha|s_i| + \sum_{j=1}^{n} d_s(i, j)$, and in the max game, the cost incurred by player i is $c_i(s) = \alpha|s_i| + \max_{j=1}^{n} d_s(i, j)$. In both cases, the total cost incurred by strategy s is $c(s) = \sum_{i=1}^{n} c_i(s)$. In the unilateral game, a (pure) Nash equilibrium is a strategy s such that $c_i(s) \leq c_i(s')$ for all strategies s' that differ from s in only one player i. The price of anarchy is then the maximum cost of a Nash equilibrium divided by the minimum cost of any strategy (called the social optimum).

In the bilateral game, Nash equilibria are not so interesting because the game requires coalition between two players to create an edge (in general). For example, if every player i chooses the empty strategy $s_i = \emptyset$, then we obtain a Nash equilibrium inducing an empty graph G_s, which has an infinite cost $c(s)$. To address this issue, Corbo and Parkes [6] use the notion of pairwise stability [13]: a strategy is pairwise stable if (1) for any edge (i, j) of G_s, both $c_i(s) \leq c_i(s')$ and $c_j(s) \leq c_j(s')$ where s' differs from s only in deleting edge (i, j) from G_s and (2) for any non-edge (i, j) of G_s, either $c_i(s) < c_i(s')$ or $c_j(s) < c_j(s')$ where s' differs from s only in adding edge (i, j) to $G_{s'}$. The price of anarchy is then the maximum cost of a pairwise-stable strategy divided by the social optimum (the minimum cost of any strategy).

Key Results

We start by the sum unilateral games. Fabrikant et al. [11] introduce these games and prove an upper bound of $O(\sqrt{\alpha})$ on the price of anarchy for all α. Albers et al. [1] prove that the price of anarchy is constant for $\alpha = O(\sqrt{n})$, as well as for the larger range $\alpha \geq 12n \lg(n)$. In addition, Albers et al. prove a general upper bound of $15\left(1 + \left(\min\{\frac{\alpha^2}{n}, \frac{n^2}{\alpha}\}\right)^{1/3}\right)$. The latter bound shows the first sublinear worst-case bound, $O(n^{1/3})$, for all α. Demaine et al. [10] prove the first $o(n^{\epsilon})$ upper bound on the price of anarchy for general α, namely, $2^{O(\sqrt{\log(n)})}$. They also prove that price of anarchy is constant for $\alpha = O(n^{1\epsilon})$ for any fixed $\epsilon > 0$, substantially reducing the range of α for which constant bounds have not been obtained. Demain et al. also prove that in the max unilateral games, the price of anarchy is at most 2 for $\alpha \geq n$, $O\left(\min\{4^{\sqrt{\log(n)}}, (n/\alpha)^{1/3}\}\right)$ for $2\sqrt{\log(n)} \leq \alpha \leq n$, and $O(n^{2/\alpha})$ for $\alpha < 2\sqrt{\log(n)}$. Alon et al. [2] consider a natural version of network creation games in which nodes only can switch their edges instead of drastically changing their strategies. In these simpler games, they achieve similar bounds on the price of anarchy. The advantage of their model is its simplicity in both agents strategies at each point and the fact that there is no α to be considered in their model.

The bilateral variation on the network creation game, considered by Corbo and Parkes [6], requires both nodes to agree before they can create a link between them. In the sum bilateral network creation game, Corbo and Parkes prove that the price of anarchy is $O(\min\{\sqrt{\alpha}, n/\sqrt{\alpha}\})$. Demaine et al. [10] prove that this upper bound is tight by showing a matching lower bound of $\Omega(\min\{\sqrt{\alpha}, n/\sqrt{\alpha}\})$. For the max bilateral case, Demaine et al. show that the price of anarchy is $\Theta(\frac{n}{1+\alpha})$ for $\alpha \leq n$ and at most 2 for $\alpha > n$.

Finding a polylogarithmic upper bound on price of anarchy for all values of α in these four network creation settings remains an open problem. In an effort to reduce the upper bounds, Demaine et al. [9] introduce the cooperative network creation games in which all agents can contribute in the construction of any edge even if they are not an endpoint of the edge. They prove that in the sum cooperative network creation game the price of anarchy is at most polylogarithmic in terms of the number of nodes. As a result, they exhibit the small-world phenomenon (polylogarithmic diameter) in the equilibrium graphs of these games. To reduce the price of anarchy even further, Demaine et al. [8] consider a special version of network creation games, and using some kind of an advertising campaign, they show that the price of anarchy is a constant number independent of the number of nodes.

Techniques

To keep this survey of results short, we just overview some of the nice combinatorial techniques in this area. Albers et al. [1] observe that any node u has the option of just connecting to another node v and exploits the BFS tree rooted at v. In an equilibrium graph G_s, this should not be a better strategy for u. Applying this trick and summing up all these inequalities for different choices of u, Albers et al. prove that for any Nash equilibrium s and any vertex v in G_s, the cost $c(s)$ is at most $2\alpha(n-1)+n\text{Dist}(v)+(n-1)^2$ where $\text{Dist}(v)=\sum_{v'\in V(G_s)} d_s(v,v')$.

Demaine et al. use this lemma to prove that price of anarchy is $O(D)$ where D is the diameter of the graph. To upper bound the diameter, they develop different techniques for different ranges of α. For instance, for $\alpha = O(n^{1-\epsilon})$, they prove that the neighborhood sizes around any node grows exponentially with a rate of $n/\alpha = \Omega(n^{\epsilon})$. Formally, they prove that when the radius of the neighborhood around a node is doubled, the number of nodes inside the neighborhood is multiplied by n/α until this radius becomes comparable with the diameter of the graph D. Clearly, it takes $O(1/\epsilon)$ rounds of doubling the neighborhood radius to cover all nodes which means that the diameter is at most exponentially growing in $1/\epsilon$ which is a constant for a fixed ϵ. For other ranges of α, more complicated bounds are needed.

Recommended Reading

1. Albers S, Eilts S, Even-Dar E, Mansour Y, Roditty L (2006) On nash equilibria for a network creation game. In: Proceedings of the seventeenth annual ACM-SIAM symposium on discrete algorithms (SODA 2006), Miami, 22–26 Jan 2006, pp 89–98. http://dl.acm.org/citation.cfm?id= 1109557.\penalty-\@M1109568
2. Alon N, Demaine ED, Hajiaghayi MT, Leighton T (2013) Basic network creation games. SIAM J Discret Math 27(2):656–668. doi:10.1137/090771478, http://dx.doi.org/10.1137/090771478
3. Anshelevich E, Dasgupta A, Tardos É, Wexler T (2003) Near-optimal network design with selfish agents. In: Proceedings of the 35th annual ACM symposium on theory of computing, San Diego, 9–11 June 2003, pp 511–520. doi:10.1145/780542.780617, http://doi.acm.org/10.1145/780542.780617
4. Anshelevich E, Dasgupta A, Kleinberg JM, Tardos É, Wexler T, Roughgarden T (2004) The price of stability for network design with fair cost allocation. In: Proceedings of the 45th symposium on foundations of computer science (FOCS 2004), Rome, 17–19 Oct 2004, pp 295–304. doi:10.1109/FOCS.2004.68, http://dx.doi.org/10.1109/FOCS.2004.68
5. Chun B, Fonseca R, Stoica I, Kubiatowicz J (2004) Characterizing selfishly constructed overlay routing networks. In: Proceedings IEEE INFOCOM 2004, the 23rd annual joint conference of the IEEE computer and communications societies, Hong Kong, 7–11 Mar 2004. http://www.ieee-infocom.org/2004/ Papers/28_4.PDF
6. Corbo J, Parkes DC (2005) The price of selfish behavior in bilateral network formation. In: Proceedings of the twenty-fourth annual ACM symposium on principles of distributed computing (PODC 2005), Las Vegas, 17–20 July 2005, pp 99–107. doi:10.1145/1073814.1073833, http://doi.acm.org/10.1145/1073814.1073833
7. Czumaj A, Vöcking B (2002) Tight bounds for worst-case equilibria. In: Proceedings of the thirteenth annual ACM-SIAM symposium on discrete algorithms, San Francisco, 6–8 Jan 2002, pp 413–420. http://dl.acm.org/citation.cfm?id=545381.545436
8. Demaine ED, Zadimoghaddam M (2012) Constant price of anarchy in network-creation games via public-service advertising. Internet Math 8(1–2):29–45. doi:10.1080/15427951.2012.625251, http://dx.doi.org/10.1080/15427951.2012.625251
9. Demaine ED, Hajiaghayi MT, Mahini H, Zadimoghaddam M (2009) The price of anarchy in cooperative network creation games. SIGecom Exch 8(2):2. doi:10.1145/1980522.1980524, http://doi.acm.org/10.1145/1980522.1980524
10. Demaine ED, Hajiaghayi MT, Mahini H, Zadimoghaddam M (2012) The price of anarchy in network creation games. ACM Trans Algorithms 8(2):13. doi:10.1145/2151171.2151176, http://doi.acm.org/10.1145/2151171.2151176
11. Fabrikant A, Luthra A, Maneva EN, Papadimitriou CH, Shenker S (2003) On a network creation game. In: Proceedings of the twenty-second ACM symposium on principles of distributed computing (PODC 2003), Boston, 13–16 July 2003, pp 347–351. doi:10.1145/872035.872088, http://doi.acm.org/10.1145/872035.872088
12. Jackson MO (2003) A survey of models of network formation: stability and efficiency. In: Demange G, Wooders M (eds) Group formation in economics: networks, clubs and coalitions. Cambridge University Press, Cambridge
13. Jackson M, Wolinsky A (1996) A strategic model of social and economic networks. J Econ Theory 71(1):44–74. http://EconPapers.repec.org/RePEc:eee:jetheo:v:71:y:1996:i:1:p:44-74
14. Koutsoupias E, Papadimitriou CH (1999) Worst-case equilibria. In: Proceedings of the 16th annual symposium on theoretical aspects of computer science (STACS 99), Trier, 4–6 Mar 1999, pp 404–413. doi:10.1007/3-540-49116-3_38, http://dx.doi.org/10.1007/3-540-49116-3_38
15. Papadimitriou CH (2001) Algorithms, games, and the internet. In: Proceedings on 33rd annual ACM symposium on theory of computing, Heraklion, 6–8 July 2001, pp 749–753. doi:10.1145/380752.380883, http://doi.acm.org/10.1145/380752.380883
16. Roughgarden T (2002) The price of anarchy is independent of the network topology. In: Proceed-

ings on 34th annual ACM symposium on theory of computing, Montréal, 19–21 May 2002, pp 428–437. doi:10.1145/509907.509971, http://doi.acm.org/10.1145/509907.509971
17. Roughgarden T (2002) Selfish routing. PhD thesis, Cornell University

Non-approximability of Bimatrix Nash Equilibria

Xi Chen[1,2] and Xiaotie Deng[3,4]
[1]Computer Science Department, Columbia University, New York, NY, USA
[2]Computer Science and Technology, Tsinghua University, Beijing, China
[3]AIMS Laboratory (Algorithms-Agents-Data on Internet, Market, and Social Networks), Department of Computer Science and Engineering, Shanghai Jiao Tong University, Shanghai, China
[4]Department of Computer Science, City University of Hong Kong, Hong Kong, China

Keywords

Approximate Nash equilibrium

Years and Authors of Summarized Original Work

2006; Chen, Deng, Teng

Problem Definition

In this entry, the following two problems are considered: (1) the problem of finding an approximate Nash equilibrium in a positively normalized bimatrix (or two-player) game; and (2) the smoothed complexity of finding an exact Nash equilibrium in a bimatrix game. It turns out that these two problems are strongly correlated [3].

Let $\mathcal{G} = (\mathbf{A}, \mathbf{B})$ be a bimatrix game, where $\mathbf{A} = (a_{i,j})$ and $\mathbf{B} = (b_{i,j})$ are both $n \times n$ matrices. Game $\mathcal{G}$ is said to be positively normalized, if $0 \leq a_{i,j}, b_{i,j} \leq 1$ for all $1 \leq i, j \leq n$.

Let $\mathbb{P}^n$ denote the set of all probability vectors in $\mathbb{R}^n$, i.e., non-negative vectors whose entries sum to 1. A Nash equilibrium [8] of $\mathcal{G} = (\mathbf{A}, \mathbf{B})$ is a pair of mixed strategies $(\mathbf{x}^* \in \mathbb{P}^n, \mathbf{y}^* \in \mathbb{P}^n)$ such that for all $\mathbf{x}, \mathbf{y} \in \mathbb{P}^n$,

$$(\mathbf{x}^*)^{\mathrm{T}}\mathbf{A}\mathbf{y}^* \geq \mathbf{x}^{\mathrm{T}}\mathbf{A}\mathbf{y}^* \text{ and } (\mathbf{x}^*)^{\mathrm{T}}\mathbf{B}\mathbf{y}^* \geq (\mathbf{x}^*)^{\mathrm{T}}\mathbf{B}\mathbf{y},$$

while an ϵ-approximate Nash equilibrium is a pair $(\mathbf{x}^* \in \mathbb{P}^n, \mathbf{y}^* \in \mathbb{P}^n)$ that satisfies

$$(\mathbf{x}^*)^{\mathrm{T}}\mathbf{A}\mathbf{y}^* \geq \mathbf{x}^{\mathrm{T}}\mathbf{A}\mathbf{y}^* - \epsilon \quad \text{and}$$
$$(\mathbf{x}^*)^{\mathrm{T}}\mathbf{B}\mathbf{y}^* \geq (\mathbf{x}^*)^{\mathrm{T}}\mathbf{B}\mathbf{y} - \epsilon, \quad \text{for all} \quad \mathbf{x}, \mathbf{y} \in \mathbb{P}^n.$$

In the smoothed analysis [11] of bimatrix games, a perturbation of magnitude $\sigma > 0$ is first applied to the input game: For a positively normalized $n \times n$ game $\mathcal{G} = (\overline{\mathbf{A}}, \overline{\mathbf{B}})$, let $\mathbf{A}$ and $\mathbf{B}$ be two matrices with

$$a_{i,j} = \overline{a}_{i,j} + r^A_{i,j} \quad \text{and} \quad b_{i,j} = \overline{b}_{i,j} + r^B_{i,j},$$
$$\forall 1 \leq i, j \leq n,$$

while $r^A_{i,j}$ and $r^B_{i,j}$ are chosen independently and uniformly from interval $[-\sigma, \sigma]$ or from Gaussian distribution with variance σ^2. These two kinds of perturbations are referred to as *σ-uniform* and *σ-Gaussian perturbations*, respectively. An algorithm for bimatrix games has *polynomial smoothed complexity* (under σ-uniform or σ-Gaussian perturbations) [11], if it finds a Nash equilibrium of game $(\mathbf{A}, \mathbf{B})$ in expected time poly $(n, 1/\sigma)$, for all $(\overline{\mathbf{A}}, \overline{\mathbf{B}})$.

Key Results

The complexity class **PPAD** [9] is defined in entry ▶ Complexity of Bimatrix Nash Equilibria. The following theorems are proved in [3].

Theorem 1 *For any constant $c > 0$, the problem of computing a $1/n^c$-approximate Nash equilibrium of a positively normalized $n \times n$ bimatrix game is* **PPAD**-*complete.*

Theorem 2 *The problem of computing a Nash equilibrium in a bimatrix game is not in smoothed polynomial time, under uniform or Gaussian perturbations, unless* **PPAD** $\subseteq$ **RP**.

Corollary 1 *The smoothed complexity of the Lemke-Howson algorithm is not polynomial, under uniform or Gaussian perturbations, unless* **PPAD** $\subseteq$ **RP**.

Applications

See entry ▶ Complexity of Bimatrix Nash Equilibria.

Open Problems

There remains a complexity gap on the approximation of Nash equilibria in bimatrix games: The result of [7] shows that, an ϵ-approximate Nash equilibrium can be computed in $n^{O(\log n/\epsilon^2)}$-time, while [3] show that no algorithm can find an ϵ-approximate Nash equilibrium in poly$(n, 1/\epsilon)$-time for ϵ of order $1/\text{poly}(n)$, unless **PPAD** is in **P**. However, the hardness result of [3] does not cover the case when ϵ is a constant between 0 and 1. Naturally, it is unlikely that the problem of finding an ϵ-approximate Nash equilibrium is **PPAD**-complete when ϵ is an absolute constant, for otherwise, all the search problems in **PPAD** would be solvable in $n^{O(\log n)}$-time, due to the result of [7]. An interesting open problem is that, for every constant $\epsilon > 0$, is there a polynomial-time algorithm for finding an ϵ-approximate Nash equilibrium? The following conjectures are proposed in [3]:

Conjecture 1 There is an $O(n^{k+\epsilon^{-c}})$-time algorithm for finding an ϵ-approximate Nash equilibrium in a bimatrix game, for some constants c and k.

Conjecture 2 There is an algorithm to find a Nash equilibrium in a bimatrix game with smoothed complexity $O(n^{k+\sigma^{-c}})$ under perturbations with magnitude σ, for some constants c and k.

It is also conjectured in [3] that Corollary 1 remains true without any complexity assumption on class **PPAD**. A positive answer would extend the result of [10] to the smoothed analysis framework.

Cross-References

▶ Complexity of Bimatrix Nash Equilibria
▶ General Equilibrium
▶ Leontief Economy Equilibrium

Recommended Reading

1. Chen X, Deng X (2005) 3-Nash is PPAD-complete. ECCC, TR05-134
2. Chen X, Deng X (2006) Settling the complexity of two-player Nash equilibrium. In: FOCS'06: proceedings of the 47th annual IEEE symposium on foundations of computer science, pp 261–272
3. Chen X, Deng X, Teng SH (2006) Computing Nash equilibria: approximation and smoothed complexity. In: FOCS'06: proceedings of the 47th annual IEEE symposium on foundations of computer science, pp 603–612
4. Daskalakis C, Goldberg PW, Papadimitriou CH (2006) The complexity of computing a Nash equilibrium. In: STOC'06: proceedings of the 38th ACM symposium on theory of computing, pp 71–78
5. Daskalakis C, Papadimitriou CH (2005) Three-player games are hard. ECCC, TR05-139
6. Goldberg PW, Papadimitriou CH (2006) Reducibility among equilibrium problems. In: STOC'06: proceedings of the 38th ACM symposium on theory of computing, pp 61–70
7. Lipton R, Markakis E, Mehta A (2003) Playing large games using simple strategies. In: Proceedings of the 4th ACM conference on electronic commerce, pp 36–41
8. Nash JF (1950) Equilibrium point in n-person games. Proc Natl Acad Sci USA 36(1):48–49
9. Papadimitriou CH (1994) On the complexity of the parity argument and other inefficient proofs of existence. J Comput Syst Sci 48:498–532
10. Savani R, von Stengel B (2004) Exponentially many steps for finding a Nash equilibrium in a bimatrix game. In: FOCS'04: proceedings of the 45th annual IEEE symposium on foundations of computer science, Rome, pp 258–267

11. Spielman DA, Teng SH (2006) Smoothed analysis of algorithms and heuristics: progress and open questions. In: Pardo LM, Pinkus A, Süli E, Todd MJ (eds) Foundations of computational mathematics, Cambridge University Press, Cambridge, pp 274–342

Non-shared Edges

Wing-Kai Hon
Department of Computer Science, National Tsing Hua University, Hsin Chu, Taiwan

Keywords

Robinson-Foulds distance; Robinson-Foulds metric

Years and Authors of Summarized Original Work

1985; Day

Problem Definition

Phylogenies are binary trees whose leaves are labeled with distinct leaf labels. This problem in this article is concerned with a well-known measurement, called *non-shared edge distance*, for comparing the dissimilarity between two phylogenies. Roughly speaking, the non-shared edge distance counts the number of edges that differentiate one phylogeny from the other.

Let e be an edge in a phylogeny T. Removing e from T splits T into two subtrees. The leaf labels are partitioned into two subsets according to the subtrees. The edge e is said to *induce* a partition of the set of leaf labels. Given two phylogenies T and T' having the same number of leaves with the same set of leaf labels, an edge e in T is *shared* if there exists some edge e' in T' such that the edges e and e' induce the same partition of the set of leaf labels in their corresponding tree. Otherwise, e is *non-shared*. Notice that T and T' have the same number of edges, so that the number of non-shared edges in T (with respect to T') is the same as the number of non-shared edges in T' (with respect to T). Such a number is called the *non-shared edge distance* between T and T'. Two problems are defined as follows:

Non-shared Edge Distance Problem

INPUT: Two phylogenies on the same set of leaf labels

OUTPUT: The non-shared edge distance between the two input phylogenies

All-Pairs Non-shared Edge Distance Problem

INPUT: A collection of phylogenies on the same set of leaf labels

OUTPUT: The non-shared edge distance between each pair of the input phylogenies

Extension

Phylogenies that are commonly used in practice have weights associated to the edges. The notion of non-shared edge can be easily extended for edge-weighted phylogenies. In this case, an edge e will induce a partition of the set of leaf labels as well as the multi-set of edge weights (here, edge weights are allowed to be non-distinct). Given two edge-weighted phylogenies R and R' having the same set of leaf labels and the same multi-set of edge weights, an edge e in R is *shared* if there exists some edge e' in R' such that the edges e and e' induce the same partition of the set of leaf labels and the multi-set of edge weights. Otherwise, e is *non-shared*. The *non-shared edge distance* between R and R' is similarly defined, giving the following problem:

General Non-shared Edge Distance Problem

INPUT: Two edge-weighted phylogenies on the same set of leaf labels and same multi-set of edge weights

OUTPUT: The non-shared edge distance between the two input phylogenies

Key Results

Day [3] proposed the first linear-time algorithm for the Non-shared Edge Distance Problem.

Theorem 1 *Let T and T' be two input phylogenies with the same set of leaf labels and n be the number of leaves in each phylogeny. The non-shared edge distance between T and T' can be computed in $O(n)$ time.*

Let Δ be a collection of k phylogenies on the same set of leaf labels and n be the number of leaves in each phylogeny. The All-Pairs Non-shared Edge Distance Problem can be solved by applying Theorem 1 on each pair of phylogenies, thus solving the problem in a total of $O(k^2 n)$ time. Pattengale and Moret [9] proposed a randomized result based on [7] to solve the problem approximately, whose running time is faster when $n \le k \le 2^n$.

Theorem 2 *Let ε be a parameter with $\varepsilon > 0$. Then, there exists a randomized algorithm such that with probability at least $1 - k^{-2}$, the non-shared edge distance between each pair of phylogenies in Δ can be approximated within a factor of $(1 + \varepsilon)$ from the actual distance; the running time of the algorithm is $O(k(n^2 + k \log k) / \varepsilon^2)$.*

For general phylogenies, let R and R' be two input phylogenies with the same set of leaf labels and the same multi-set of edge weights and n be the number of leaves in each phylogeny. The General Non-shared Edge Distance Problem can be solved easily in $O(n^2)$ time by applying Theorem 1 in a straightforward manner. The running time is improved by Hon et al. in [5].

Theorem 3 *The non-shared edge distance between R and R' can be computed in $O(n \log n)$ time.*

Applications

Phylogenies are commonly used by biologists to model the evolutionary relationship among species. Many reconstruction methods (such as maximum parsimony, maximum likelihood, compatibility, distance matrix) produce different phylogenies based on the same set of species, and it is interesting to compute the dissimilarities between them. Also, through the comparison, information about rare genetic events such as recombinations or gene conversions may be uncovered. The most common dissimilarity measure is the Robinson-Foulds metric [11], which is exactly the same as the non-shared edge distance.

Other dissimilarity measures, such as the nearest-neighbor interchange (NNI) distance and the subtree-transfer (STT) distance (see [2] for details), are also proposed in the literature. These measures are sometimes preferred by the biologists since they can be used to deduce the biological events that create the dissimilarity. Nevertheless, these measures are usually difficult to compute. In particular, computing the NNI distance and the STT distance is shown to be NP-hard by DasGupta et al. [1, 2]. Approximation algorithms are devised for these problems (NNI, [4, 8]; STT, [1, 6]). Interestingly, all these algorithms make use of the non-shared edge distance to bound their approximation ratios.

N

Recommended Reading

1. DasGupta B, He X, Jiang T, Li M, Tromp J (1999) On the linear-cost subtree-transfer distance between phylogenetic trees. Algorithmica 25(2–3):176–195
2. DasGupta B, He X, Jiang T, Li M, Tromp J, Zhang L (1997) On distances between phylogenetic trees. In: Proceedings of the eighth ACM-SIAM annual symposium on discrete algorithms (SODA), New Orleans. SIAM, pp 427–436
3. Day WHE (1985) Optimal algorithms for comparing trees with labeled leaves. J Classif 2:7–28
4. Hon WK, Lam TW (2001) Approximating the nearest neighbor intercharge distance for non-uniform-degree evolutionary trees. Int J Found Comp Sci 12(4):533–550
5. Hon WK, Kao MY, Lam TW, Sung WK, Yiu SM (2004) Non-shared edges and nearest neighbor interchanges revisited. Inf Process Lett 91(3):129–134
6. Hon WK, Lam TW, Yiu SM, Kao MY, Sung WK (2004) Subtree transfer distance for degree-D phylogenies. Int J Found Comp Sci 15(6):893–909
7. Johnson W, Lindenstrauss J (1984) Extensions of lipschitz mappings into a hilbert space. Contemp Math 26:189–206

8. Li M, Tromp J, Zhang L (1996) Some notes on the nearest neighbour interchange distance. J Theor Biol 26(182):463–467
9. Pattengale ND, Moret BME (2006) A sublinear-time randomized approximation scheme for the robinson-foulds metric. In: Proceedings of the tenth ACM annual international conference on research in computational molecular biology (RECOMB), Venice, pp 221–230
10. Robinson DF (1971) Comparison of labeled trees with valency three. J Comb Theor 11: 105–119
11. Robinson DF, Foulds LR (1981) Comparison of phylogenetic trees. Math Biosci 53:131–147

Nowhere Crownful Classes of Directed Graphs

Stephan Kreutzer
Chair for Logic and Semantics, Technical University, Berlin, Germany

Keywords

Algorithmic digraph structure theory; Algorithms for directed graphs; Digraph width measures; Dominating set problems; Nowhere dense classes of graphs; Sparse digraphs

Years and Authors of Summarized Original Work

2012; Kreutzer, Tazari

Problem Definition

Many common computational problems on directed graphs are computationally intractable; they are NP-complete and sometimes even harder. Examples include domination problems such as directed dominating set, Kernel, directed Steiner networks, directed disjoint paths, and many other problems.

For undirected graphs, there is an extensive structure theory available to help dealing with this computational intractability. In particular, there is a well-developed hierarchy of classes of undirected graphs and a rich set of algorithmic tools which allow to solve hard computational problems on these classes of graphs. Most notably in this context are classes of graphs of bounded tree width, planar graphs or graphs embeddable on any other fixed surface, classes excluding a fixed minor, and many other graph classes. This theory is closely related to parameterized complexity theory.

For directed graphs, to date, there is no comparable theory available. A directed version of tree width was introduced by Reed [9] and Johnson et al. [4]. Further proposals for "tree width"-like width measures for directed graphs have been made in the literature; see, e.g., references in [1]. Algorithmically, the main application is that on classes of bounded directed tree width, the directed k-disjoint paths problem can be solved in polynomial time for any fixed value of k.

Almost all of these proposals have in common that the class of acyclic digraphs (DAGs) have small width, i.e., acyclic digraphs are taken as particularly simple digraphs. While this is certainly useful for problems such as directed disjoint paths, problems such as directed dominating set remain NP-complete and fixed-parameter intractable on acyclic directed graphs.

What is needed, therefore, are digraph parameters and structural classes of digraphs which separate acyclic digraphs into simple and hard instances. Nowhere crownful classes propose a solution to this problem based on the concept of excluded directed minors.

Key Results

While there is a well-defined concept of a minor for undirected graphs, there is as yet no commonly agreed concept of directed minors. A widely used, and very conservative, version of directed minor is a *butterfly minor* (see, e.g., [4]) in which a directed edge (u, v) is contractible if it is the only outgoing edge of u or the only incoming edge of v. In [5] a much more general concept of directed minors is used to give a classification of classes of digraphs in terms of shallow directed minors. For the sake of brevity, we introduce directed minors here only for digraphs called *crowns*, which is enough for defining nowhere crownful classes of digraphs.

An *out-branching* is a digraph H whose underlying undirected graph is a tree and in which there is a unique vertex r, the *root* of H, such that all edges are oriented away from the root, i.e., every vertex in H is reachable by a unique directed path from the root. An *in-branching* is the same as an out-branching but all edges are oriented towards the root.

Definition 1 A *crown* of order q, for $q > 0$, is the graph S_q with

- $V(S_q) := \{v_1, \ldots, v_q\} \dot\cup \{u_{i,j} : 1 \le i < j \le q\}$ and
- $E(S_q) := \{(u_{i,j}, v_i), (u_{i,j}, v_j) : 1 \le i < j \le q\}$.

Definition 2 Let H with $V(H) := \{v_1, \ldots, v_q\} \dot\cup \{u_{i,j} : 1 \le i < j \le q\}$ be a crown of order q, for some $q > 0$. A digraph G contains H as a *directed minor*, if for every v_i, $1 \le i \le q$, there is an in-branching $T_i \subseteq G$ and for every $u_{i,j}$, $1 \le i < j \le q$, there is an out-branching $S_{i,j} \subseteq G$ such that all subgraphs T_i, $S_{i,j}$ are pairwise vertex disjoint and for all $1 \le i < j \le q$, there are edges e_i, e_j from a vertex in $S_{i,j}$ to a vertex in T_i and T_j, respectively.

H is a *depth-r-minor* of G, or an *r-shallow minor* of G, denoted $H \preceq_r^d G$, for some $r \ge 0$, if all $S_{i,j}$ and all T_i are of height at most r.

Definition 3 A class $\mathcal{C}$ of directed graphs is *nowhere crownful* if for every $r \ge 0$ there exists a $q = q(r)$ so that $S_q \not\preceq_r^d G$ for all $G \in \mathcal{C}$. If the function taking each r to $q(r)$ as above is computable, then we call $\mathcal{C}$ *effectively nowhere crownful.*

Nowhere crownful classes of digraphs are very general. For instance, if $\mathcal{C}$ is a class of digraphs and $\tilde{\mathcal{C}}$ is the class of underlying undirected graphs (obtained from digraphs in $\mathcal{C}$ by ignoring edge direction), then if $\tilde{\mathcal{C}}$ has bounded genus, excludes a fixed minor, or is nowhere dense, then $\mathcal{C}$ is nowhere crownful. But there are nowhere crownful classes $\mathcal{C}$ of digraphs such that $\tilde{\mathcal{C}}$ does not have any of the properties above. On the other hand, the class of acyclic digraphs is not nowhere crownful as it contains every crown.

Nowhere crownful classes of digraphs can be characterized equivalently as follows. Let G be a digraph and $d \ge 0$. A set $U \subseteq V(G)$ is *d-scattered* if there is no $v \in V(G)$ and $u \ne u' \in U$ with $u, u' \in N_d^+(v)$. That is, no two elements of U can be reached from a single vertex v by paths of length at most d.

Definition 4 A class $\mathcal{C}$ of directed graphs is *uniformly quasi-wide* if there are functions $s : \mathbb{N} \to \mathbb{N}$ and $N : \mathbb{N} \times \mathbb{N} \to \mathbb{N}$ such that for every $G \in \mathcal{C}$ and all $d, m \in \mathbb{N}$ and $W \subseteq V(G)$ with $|W| > N(d, m)$, there is a set $S \subseteq V(G)$ with $|S| \le s(d)$ and a set $U \subseteq W$ with $|U| = m$ such that U is d-scattered in $G - S$. The functions s, N are called the *margin* of $\mathcal{C}$. If s and N are computable, then we call $\mathcal{C}$ effectively uniformly quasi-wide.

Theorem 1 *A class $\mathcal{C}$ of digraphs is nowhere crownful if, and only if, it is directed uniformly quasi-wide.*

Nowhere crownful classes of digraphs were defined as a directed analogue to the concept of *nowhere dense* classes of undirected graphs, for which a similar equivalence to uniformly quasi-wideness can be proved. See [2, 3, 6–8] for nowhere dense classes of graphs and algorithmic applications. As mentioned above, nowhere crownful classes properly generalise nowhere denseness to digraphs.

Applications

A *directed dominating set* in a digraph G is a set $X \subseteq V(G)$ such that every $u \in V(G) \setminus X$ is the out-neighbour of a vertex in X. A *distance-d directed dominating set* is a set $X \subseteq V(G)$ such that every vertex $v \in V(G)$ can be reached from a vertex in X by a directed path of length at most d. An important variant of the undirected dominating set problem is the *connected dominating set problem*, where we are asked to find a dominating set D of size k such that D induces a connected subgraph. There are various natural translations of this problem to the directed case: we can require the dominating set to induce a strongly connected subgraph or we can simply require it to induce an out-branching. The second variation, which we call *dominating out-branching*, still captures the idea that information can flow from the root to all vertices in the dominating set.

There is an easy reduction from the undirected dominating set problem to its directed counterpart proving that the directed dominating set problem is fixed-parameter intractable and NP-complete. In fact, the problem to decide whether an undirected graph G contains a dominating set of order k can be reduced to the question whether an acyclic digraph contains a directed dominating set of order k. The result of the reduction is a crown (plus one extra vertex). So, classes of digraphs where this problem and its extension to distance-d versions are to become tractable should exclude crowns. This observation was one of the motivations for defining and studying nowhere crownful classes of digraphs. Furthermore, if besides the directed dominating set we also want to solve its distance-d version, we need to exclude crown minors in some form.

However, excluding shallow-crowns is sufficient for these problems to become fixed-parameter tractable.

Theorem 2 *Let $\mathcal{C}$ be a class of directed graphs which is nowhere crownful. Then the directed (independent or unrestricted) dominating set problem, the dominating out-branching problem, as well as their distance-d versions are fixed-parameter tractable on $\mathcal{C}$.*

In the same way, several other similar problems can be shown to become tractable on nowhere crownful classes.

Open Problems

As mentioned before, nowhere crownful classes are modelled after nowhere dense classes of undirected graphs. For such classes, many other equivalent characterizations and powerful algorithmic applications are known. For instance, nowhere dense classes of graphs allow for very efficient sparse neighborhood covers (and this is again if, and only if) and can be defined by a game yielding bounded search tree techniques. Furthermore, there is a close connection between nowhere dense classes of graphs and generalized colouring numbers (see, e.g., [8]).

It is open in how far these characterizations and applications can be generalized to the digraph setting.

A particular open problem is the tractability of strongly connected Steiner networks and strongly connected dominating set on nowhere crownful classes of digraphs.

Nowhere crownful classes provide a way for dealing with domination-type problems. Directed tree width on the other hand provides a way of dealing with linkage problems such as disjoint paths. It is open how to bring the two concepts together.

Cross-References

▶ Efficient Dominating and Edge Dominating Sets for Graphs and Hypergraphs
▶ Exact Algorithms for Dominating Set

Recommended Reading

1. Ganian R, Hlinený P, Kneis J, Langer A, Obdrzálek J, Rossmanith P (2009) On digraph width measures in parameterized algorithmics. In: International workshop in parameterized and exact computation (IWPEC), Copenhagen, pp 185–197
2. Grohe M, Kreutzer S, Siebertz S (2013) Characterisations of nowhere dense graphs. In: Foundations of software technology and theoretical computer science (FSTTCS 2013), Guwahati, pp 21–40
3. Grohe M, Kreutzer S, Siebertz S (2014) Deciding first-order properties of nowhere dense graphs. In: 46th annual symposium on the theory of computing (STOC), New York
4. Johnson T, Robertson N, Seymour PD, Thomas R (2001) Directed tree-width. J Comb Theory Ser B 82(1):138–154
5. Kreutzer S, Tazari S (2012) Directed nowhere dense classes of graphs. In: Proceedings of the 23rd ACM-SIAM symposium on discrete algorithms (SODA), Kyoto, pp 1552–1562
6. Nešetřil J, Ossona de Mendez P (2008) Grad and classes with bounded expansion I–III. Eur J Comb 29. Series of 3 papers appearing in volumes (3) and (4)

7. Nešetřil J, Ossona de Mendez P (2010) First order properties of nowhere dense structures. J Symb Log 75(3):868–887
8. Nešetřil J, Ossona de Mendez P (2012) Sparsity. Vol. 28 of Algorithms and Combinatorics. Springer Heidelberg
9. Reed B (1999) Introducing directed tree-width. Electron Notes Discret Math 3:222–229

Nucleolus

Qizhi Fang
School of Mathematical Sciences, Ocean University of China, Qingdao, Shandong Province, China

Keywords

Kernel; Nucleon

Years and Authors of Summarized Original Work

2006; Deng, Fang, Sun

Problem Definition

Cooperative game theory considers how to distribute the total income generated by a set of participants in a joint project to individuals. The Nucleolus, trying to capture the intuition of minimizing dissatisfaction of players, is one of the most well-known solution concepts among various attempts to obtain a unique solution. In Deng, Fang, and Sun's work [3], they study the Nucleolus of flow games from the algorithmic point of view. It is shown that, for a flow game defined on a simple network (arc capacity being all equal), computing the Nucleolus can be done in polynomial time, and for flow games in general cases, both the computation and the recognition of the Nucleolus are $\mathcal{NP}$-hard.

A cooperative (profit) game (N, v) consists of a player set $N = \{1, 2, \cdots, n\}$ and a characteristic function $v : 2^N \to R$ with $v(\emptyset) = 0$, where the value $v(S)(S \subseteq N)$ is interpreted as the profit achieved by the collective action of players in S. Any vector $x \in R^n$ with $\sum_{i \in N} x_i = v(N)$ is an allocation. An allocation x is called an *imputation* if $x_i \geq v(\{i\})$ for all $i \in N$. Denote by $\mathcal{I}(v)$ the set of imputations of the game.

Given an allocation x, the *excess* of a coalition $S(S \subseteq N)$ at x is defined as

$$e(S, x) = x(S) - v(S),$$

where $x(S) = \sum_{i \in S} x_i$ for $S \subseteq N$. The value $e(S, x)$ can be interpreted as a measure of satisfaction of coalition S with the allocation x. The core of the game (N, v), denoted by $\mathcal{C}(v)$, is the set of allocations whose excesses are all nonnegative. For an allocation x of the game (N, v), let $\theta(x)$ denote the $(2^n - 2)$-dimensional vector whose components are the nontrivial excesses $e(S, x)$, $\emptyset \neq S \neq N$, arranged in a non-decreasing order. That is, $\theta_i(x) \leq \theta_j(x)$, for $1 \leq i < j \leq 2^n - 2$. Denote by $\succeq l$ the "lexicographically greater than" relationship between vectors of the same dimension.

Definition 1 The Nucleolus $\eta(v)$ of game (N, v) is the set of imputations that lexicographically maximize $\theta(x)$ over all imputations $x \in \mathcal{I}(v)$. That is,

$$\eta(v) = \{x \in \mathcal{I}(v) : \theta(x) \succeq_l \theta(y) \text{ for all } y \in \mathcal{I}(v)\}.$$

Even though, the Nucleolus may contain multiple points by the definition, it was proved by Schmeidler [14] that the Nucleolus of a game with nonempty imputation set contains exactly one element. Kopelowitz [12] proposed that the Nucleolus can be obtained by recursively solving sequential linear programs (SLP):

$$\text{LP}_k : \quad \begin{array}{ll} \max \varepsilon & \\ x(S) = v(S) + \varepsilon_r & \forall S \in \mathcal{J}_r \\ r = 0, 1, \cdots, k-1 & \\ x(S) \geq v(S) + \varepsilon & \forall S \in 2^N \setminus \bigcup_{r=0}^{k-1} \mathcal{J}_r \\ x \in \mathcal{I}(v). & \end{array}$$

N

Here, $\mathcal{J}_0 = \{\emptyset, N\}$ and $\varepsilon_0 = 0$ initially; the number ε_r is the optimum value of the rth program (LP_r) and $\mathcal{J}_r = \{S \in 2^N : x(S) = v(S) + \varepsilon_r$ for every $x \in X_r$, where $X_r = \{x \in \mathcal{I}(v) : (x, \varepsilon_r)$ is an optimal solution to $LP_r\}$. It can be shown that after at most $n - 1$ iterations, one arrives at a unique optimal solution (x^*, ε_k), where x^* is just the Nucleolus of the game. In addition, the set of optimal solutions X_1 to the first program LP_1 is called the least core of the game.

The definition of the Nucleolus entails comparisons between vectors of exponential length. And with linear programming approach, each linear programs in SLP may possess exponential size in the number of players. Clearly, both do not provide an efficient solution in general.

Flow games, first studied in Kailai and Zemel [9, 10], arise from the profit distribution problem related to the maximum flow in a network. Let $D = (V, E; \omega; s, t)$ be a directed flow network, where V is the vertex set, E is the arc set, $\omega : E \to R^+$ is the arc capacity function, and s and t are the source and the sink of the network, respectively. The network D is *simple* if $\omega(e) = 1$ for each $e \in E$, which is denoted briefly by $D = (V, E; s, t)$.

Definition 2 The flow game $\Gamma_f = (E, v)$ associated with network $D = (V, E; \omega; s, t)$ is defined by:

(i) The player set is E.
(ii) $\forall S \subseteq E$, $v(S)$ is the value of a maximum flow from s to t in the subnetwork of D consisting only of arcs belonging to S.

Problem 1 (Computing the Nucleolus)

INSTANCE: A flow network $D = (V, E; \omega; s, t)$.
QUESTION: Is there a polynomial time algorithm to compute the Nucleolus of the flow game associated with D?

Problem 2 (Recognizing the Nucleolus)

INSTANCE: A flow network $D = (V, E; \omega; s, t)$ and $y : E \to R^+$.
QUESTION: Is it true that y is the Nucleolus of the flow game associated with D?

Key Results

Theorem 1 *Let $D = (V, E; s, t)$ be a simple network and $\Gamma_f = (E, \nu)$ be the associated flow game. Then the Nucleolus $\eta(\nu)$ can be computed in polynomial time.*

By making use of duality technique in linear programming, Kalai and Zemel [10] gave a characterization on the core of a flow game. They further conjectured that their approach may serve as a practical basis for computing the Nucleolus. In fact, the proof of Theorem 1 in the work of Deng, Fang, and Sun [3] is just an elegant application of Kalai and Zemel's approach (especially the duality technique) and hence settling their conjecture.

Theorem 2 *Given a flow game $\Gamma_f = (E, \nu)$ defined on network $D = (V, E; \omega; s, t)$, computing the Nucleolus $\eta(\nu)$ is $\mathcal{NP}$-hard.*

Theorem 3 *Given a flow game $\Gamma_f = (E, \nu)$ defined on network $D = (V, E; \omega; s, t)$ and an imputation $y \in \mathcal{I}(\nu)$, checking whether y is the Nucleolus of Γ_f is $\mathcal{NP}$-hard.*

Although a flow game can be formulated as a linear production game [2], the size of reduction may in general be exponential in space, and consequently, their complexity results on the Nucleolus are independent. However, in the $\mathcal{NP}$-hardness proof of Theorems 2 and 3, the flow game constructed possesses a polynomial size formulation of linear production game [3]. Therefore, as a direct corollary, the same $\mathcal{NP}$-hardness conclusions for linear production games are obtained. That is, both computing and recognizing the Nucleolus of a linear production game are $\mathcal{NP}$-hard.

Applications

As an important solution concept in economics and game theory, the Nucleolus and related solution concepts have been applied to insurance policies, real estate and bankruptcy, etc. However, it is a challenging problem to decide what classes of cooperative games permit polynomial time computation of the Nucleolus.

The first polynomial time algorithm for Nucleolus in a special tree game was proposed by Megiddo [13], in advocation of efficient algorithms for cooperative game solutions. Subsequently, some efficient algorithms have been developed for computing the Nucleolus, such as for assignment games [15] and matching games [1, 11]. On the negative side, $\mathcal{NP}$-hardness result was obtained for minimum cost spanning tree games [5] and weighted voting games [4].

Granot, Granot, and Zhu [8] observed that most of the efficient algorithms for computing the Nucleolus are based on the fact that the information needed to completely characterize the Nucleolus is much less than that dictated by its definition. Therefore, they introduced the concept of a characterization set for the Nucleolus to embody the notion of "minimum" relevant information needed for the Nucleolus. Furthermore, based on the sequential linear programs (SLP), they established a general relationship between the size of a characterization set and the complexity of computing the Nucleolus. Following this approach, some known efficient algorithms for computing the Nucleolus are derived directly.

Another approach to computing the Nucleolus is taken by Faigle, Kern, and Kuipers [6], which is motivated by Schmeidler's observation that the Nucleolus of a game lies in the kernel [14]. In the case where the kernel of the game contains exactly one core vector and the minimum excess for any given allocation can be computed efficiently, their approach derives a polynomial time algorithm for the Nucleolus. However, their algorithm uses the ellipsoid method as a subroutine, implying that the efficiency of the algorithm is of a more theoretical kind.

Open Problems

The field of combinatorial optimization has much to offer for the study of cooperative games. It is usually the case that the values of subgroups can be obtained via a combinatorial optimization problem, where the game is called a combinatorial optimization game. This class of games leads to the applications of a variety of combinatorial optimization techniques in design and analysis of algorithms, as well as establishing complexity results. One of the most interesting result is the LP duality characterization of the core [2]. However, little work dealt with the Nucleolus by using the duality technique so far. Hence, the work of Deng, Fang, and Sun [3] on computing the Nucleolus may be of independent interest.

There are still many unsolved complexity questions concerning the Nucleolus. For the computation of the Nucleolus of matching games, Kern and Paulusma [11] proposed an efficient algorithm in unweighted case and conjectured that it is in general $\mathcal{NP}$-hard. Biro, Kern, and Paulusma [1] partly settled the conjecture by showing that in weighted case, when the matching game has a nonempty core, the Nucleolus can be computed in polynomial time. Since both the flow game and the matching game fall into the class of packing/covering games, it is interesting to know the complexity of computing the Nucleolus for other game models in this class, such as vertex covering games and minimum coloring games.

For cooperative games arising from $\mathcal{NP}$-hard combinatorial optimization problems, the computation of the Nucleolus may in general be a hard task. For example, in a traveling salesman game, nodes of the graph are the players and an extra node 0, and the value of a subgroup S of players is the length of a minimum Hamiltonian tour in the subgraph induced by $S \cup \{0\}$ [2]. It would not be surprising if one shows that both the computation and the recognition of the Nucleolus for this game model are $\mathcal{NP}$-hard. However, this is not known yet. The same questions are proposed for facility location games [7], though there have been efficient algorithms for some special cases.

Moreover, when the computation of the Nucleolus is difficult, it is also interesting to seek for meaningful approximation concepts of the Nucleolus, especially from the political and economic background.

Cross-References

▸ Complexity of Core

Recommended Reading

1. Biro P, Kern W, Paulusma D (2012) Computing solutions for matching games. Int J Game Theory 41:75–90
2. Deng X (1998) Combinatorial optimization and coalition games. In: Du D, Pardalos PM (eds) Handbook of combinatorial optimization, vol 2. Kluwer, Boston, pp 77–103
3. Deng X, Fang Q, Sun X (2006) Finding nucleolus of flow games. In: Proceedings of the 17th annual ACM-SIAM symposium on discrete algorithm (SODA 2006), Miami. Lecture Notes in Computer Science, vol 3111, pp 124–131
4. Elkind P, Goldberg LA, Goldberg PW, Wooldridge M (2009) On the computational complexity of weighted voting games. Ann Math Artif Intell 56:109–131
5. Faigle U, Kern W, Kuipers J (1998) Computing the nucleolus of min-cost spanning tree games is $\mathcal{NP}$-hard. Int J Game Theory 27:443–450
6. Faigle U, Kern W, Kuipers J (2001) On the computation of the nucleolus of a cooperative game. Int J Game Theory 30:79–98
7. Goemans MX, Skutella M (2004) Cooperative facility location games. J Algorithms 50:194–214
8. Granot D, Granot F, Zhu WR (1998) Characterization sets for the nucleolus. Int J Game Theory 27:359–374
9. Kalai E, Zemel E (1982) Totally balanced games and games of flow. Math Oper Res 7:476–478
10. Kalai E, Zemel E (1982) Generalized network problems yielding totally balanced games. Oper Res 30:998–1008
11. Kern W, Paulusma D (2003) Matching games: the least core and the nucleolus. Math Oper Res 28:294–308
12. Kopelowitz A (1967) Computation of the kernels of simple games and the nucleolus of n-person games. RM-31, Math. Dept., The Hebrew University of Jerusalem, Jerusalem
13. Megiddo N (1978) Computational complexity and the game theory approach to cost allocation for a tree. Math Oper Res 3:189–196
14. Schmeidler D (1969) The nucleolus of a characteristic function game. SIAM J Appl Math 17:1163–1170
15. Solymosi T, Raghavan TES (1994) An algorithm for finding the nucleolus of assignment games. Int J Game Theory 23:119–143

O

O(log log n)-Competitive Binary Search Tree

Chengwen Chris Wang and Daniel Sleator
Department of Computer Science, Carnegie Mellon University, Pittsburgh, PA, USA

Keywords

Competitive BST algorithms; Splay trees; Tango trees

Years and Authors of Summarized Original Work

2004; Demaine, Harmon, Iacono, Patrascu

Problem Definition

Here is a precise definition of BST algorithms and their costs. This model is implied by most BST papers and developed in detail by Wilber [22]. A static set of n keys is stored in the nodes of a binary tree. The keys are from a totally ordered universe, and they are stored in symmetric order. Each node has a pointer to its left child, to its right child, and to its parent. Also, each node may keep $o(\log\ n)$ bits of additional information but no additional pointers.

A BST algorithm is required to process a sequence of m accesses (without insertions or deletions), $S = s_1, s_2, s_3, s_4 \ldots s_m$. The ith access starts from the root and follows pointers until s_i is reached. The algorithm can update the fields in any node or rotate any edges that it touches along the way. The cost of the algorithm to execute an access sequence is defined to be the number of nodes touched plus the number of rotations. A BST algorithm is *on-line* if it processes access s_i without making use of anything after s_i in the access sequence.

Let A be any online BST algorithm, and define $A(S)$ to be the cost to algorithm A of processing sequence S and $\text{OPT}(S, T_0)$ to be the minimum possible (online or off-line) cost to process the sequence S, starting from an initial tree T_0. The algorithm A is T-competitive if for all possible sequences S, $A(S) \leq T^*\ \text{OPT}(S, T_0) + O(m+n)$.

Since the number of rotations needed to change any binary tree of n keys into another one (with the same n keys) is at most $2n - 6$ [4,5,12,13,15], it follows that $\text{OPT}(S, T_0)$ differs from $\text{OPT}(S, T_0')$ by at most $2n - 6$. Thus, if $m > n$, then the initial tree can only affect the constant factor.

Key Results

The *interleave bound* is a lower bound on $\text{OPT}(S, T_0)$ that depends only on S. Consider any binary search tree P of all the elements in T_0. For each node y in P, define the *left side*

M.-Y. Kao (ed.), *Encyclopedia of Algorithms*,
DOI 10.1007/978-1-4939-2864-4

of y to include all nodes in y's left subtree and y. And define the *right side* of y to include all nodes in y's right subtree. For each node y, label each access s_i in S by whether it is in the left or right side of y, ignoring all accesses not in y's subtree. Denote the number of times the label changes for y as $\text{IB}(S, y)$. The *interleave* bound $\text{IB}(S)$ is $\sum_y \text{IB}(S, y)$.

Theorem 1 (Interleave Lower Bound [6, 22]) *$\text{IB}(S)/2 - n$ is a lower bound on $\text{OPT}(S, T_0)$.*

Demaine et al. observe that it is impossible to use this lower bound to improve the competitive ratio beyond $\Theta(\log\log n)$.

Theorem 2 (Tango is $O(\log\log n)$-competitive BST [6]) *The running time of Tango BST on a sequence S of m accesses is $O(((\text{OPT}(S, T_0)) + n)^*(1 + \log\log n))$.*

Applications

Binary search tree (BST) is one of the oldest data structures in the history of computer science. It is frequently used to maintain an ordered set of data. In the last 40 years, many specialized binary search trees have been designed for specific applications. Almost every one of them supports access, insertion, and deletion in worst-case $O(\log n)$ time on average for random sequences of access. This matches the best theoretically possible worst-case bound. For most of these data structures, a random sequence of m accesses will use $\Theta(m \log n)$ time.

While it is impossible to have better asymptotic performance for a random sequence of m accesses, many of the real-world access sequences are not random. For instance, if the set of accesses are randomly drawn from a *small* subset of k element, it's possible to answer all the accesses in $O(m \log k)$ time. A notable binary search tree is splay tree. It is proved to perform well for many access patterns [2, 3, 8, 14, 16–18]. As a result, Sleator and Tarjan [14] conjectured that splay tree is $O(1)$-competitive to the optimal off-line BST. After more than 20 years, the conjecture remains an open problem.

Over the years, several restricted types of optimality have been proved. Many of these restrictions and usage patterns are based on real-world applications. If each access is drawn independently at random from a fixed distribution, D, Knuth [11] constructed a BST based on D that is expected to run in optimal time up to a constant factor. Sleator and Tarjan [14] achieve the same bound without knowing D ahead of time. Other types includes key-independent optimality [10] and BST with free rotations [1].

In 2004, Demaine et al. suggested searching for alternative BST algorithms that have small but nonconstant competitive factors [6]. They proposed *Tango*, the first data structure proved to achieve a nontrivial competitive factor of $O(\log\log n)$. This is a major step toward developing a $O(1)$-competitive BST, and this line of research could potentially replace a large number of specialized BSTs.

Extensions and Promising Research Directions

Following this paper, several new $O(\log\log n)$-competitive BSTs have emerged [9, 21]. A notable example is multi-splay trees [21]. It generalizes the interleave bound to include insertions and deletions. Multi-splay trees also have many theorems analogous to splay trees [20, 21], such as the access lemma and the working set theorem. Wang [21] conjectured that multi-splay trees is $O(1)$-competitive, but it remains an open problem.

Returning to the original motivation for this research, the problem of finding an $o(\log\log n)$-competitive online BST remains open. Several attempts have been made to improve the lower bound [6, 7, 22], but none of them have led to a lower competitive ratio. Even in the off-line model, the problem of finding an $O(1)$-competitive BST is difficult. The best known off-line constant competitive algorithm uses dynamic programming and requires exponential time.

In 2009 Demaine et al. [23] described a *geometric view* of BST algorithms. This is an equivalent model of BST algorithms, but sufficiently different that it has allowed progress to be made in a number of directions. First of all it has simplified and unified BST lower bounds. It has also allowed progress to be made toward proving a different algorithm to be O(1) competitive. The algorithm is called *GreedyFuture* and was proposed as an off-line algorithm in 1988 by Joan Lucas [24]. After each access, the algorithm restructures the access path according to the future accesses. Specifically if the next access is on this path, then that node is made the new root and the left and right sides are built in a similar fashion recursively. If the next access is to a subtree of that path, then the path node to the left of that subtree is made the root, and the path node to the right of it is made the right child of the root, and the process again continues recursively. A remarkable result of the geometric view is that it shows how the GreedyFuture algorithm can actually be implemented as an online algorithm. At the moment, this seems to be the most likely candidate to be proven to be O(1) competitive.

Cross-References

▸ B-trees

Recommended Reading

Based on Wilber [22]'s lower bound, Tango [6] is the first $O(\log\log n)$-competitive binary search tree. Using many of the ideas in Tango and Link-cut Trees [14, 19], Multi-Splay Trees [21] generalize the competitive framework to include insertion and deletion. The recommended readings are *Self-adjusting binary search trees* by Sleator and Tarjan, *Lower bounds for accessing binary search trees with rotations* by Wilber, *Dynamic Optimality – Almost* by Demaine, et al., and *$O(\log\log n)$ dynamic competitive binary search tree by Wang, et al.*

Recommended Reading

1. Blum A, Chawla S, Kalai A (2003) Static optimality and dynamic search-optimality in lists and trees. Algorithmica 36:249–260
2. Cole R (2000) On the dynamic finger conjecture for splay trees II: the proof. SIAM J Comput 30(1):44–85
3. Cole R, Mishra B, Schmidt J, Siegel A (2000) On the dynamic finger conjecture for splay trees I: splay sorting log n-block sequences. SIAM J Comput 30(1):1–43
4. Crane CA (1972) Linear lists and priority queues as balanced binary trees. Technical report STAN-CS-72-259, Computer Science Department, Stanford University
5. Culik II K, Wood D (1982) A note on some tree similarity measures. Inf Process Lett 15(1): 39–42
6. Demaine ED, Harmon D, Iacono J, Patrascu M (2007) Dynamic optimality-almost. SIAM J Comput 37(1):240–251
7. Derryberry J, Sleator DD, Wang CC (2005) A lower bound framework for binary search trees with rotations. Technical report CMU-CS-05-187, Carnegie Mellon University
8. Elmasry A (2004) On the sequential access theorem and deque conjecture for splay trees. Theor Comput Sci 314(3):459–466
9. Georgakopoulos GF (2005) How to splay for log log n-competitiveness. In: Proceedings of the 4th international workshop on experimental and efficient algorithms (WEA), Santorini Island, pp 570–579
10. Iacono J (2005) Key-independent optimality. Algorithmica 42(1):3–10
11. Knuth DE (1971) Optimum binary search trees. Acta Inf 1:14–25
12. Luccio F, Pagli L (1989) On the upper bound on the rotation distance of binary trees. Inf Process Lett 31(2):57–60
13. Mäkinen E (1988) On the rotation distance of binary trees. Inf Process Lett 26(5):271–272
14. Sleator DD, Tarjan RE (1985) Self-adjusting binary search trees. J ACM 32(3):652–686
15. Sleator DD, Tarjan RE, Thurston WP (1986) Rotation distance, triangulations, and hyperbolic geometry. In: Proceedings 18th ACM symposium on theory of computing (STOC), Berkeley, pp 122–135
16. Sundar R (1989) Twists, turns, cascades, deque conjecture, and scanning theorem. In: Proceedings 30th IEEE symposium on foundations of computer science (FOCS), pp 555–559
17. Sundar R (1992) On the deque conjecture for the splay algorithm. Combinatorica 12(1):95–124
18. Tarjan R (1985) Sequential access in play trees takes linear time. Combinatorica 5(4):367–378
19. Tarjan RE (1983) Data structures and network algorithms. In: CBMS-NSF regional conference series in applied mathematics, vol. 44. SIAM, Philadelphia
20. Wang CC (2006) Multi-splay trees. Ph.D. thesis, Carnegie Mellon University
21. Wang CC, Derryberry J, Sleator DD (2006) $O(\log\log n)$-competitive dynamic binary search trees. In: Proceedings of the 17th annual ACM-SIAM symposium on discrete algorithms (SODA), Miami, pp 374–383

22. Wilber R (1989) Lower bounds for accessing binary search trees with rotations. SIAM J Comput 18(1):56–67
23. Demaine ED, Harmon D, Iacono J, Kane D, Patrascu, M (2009) The Geometry of binary search trees. In: Proceedings of the 20th annual ACM-SIAM symposium on discrete algorithms (SODA), New York, pp 496–505
24. Lucas J (1988) Canonical forms for competitive binary search tree algorithms. Technical report DCS-TR-250, Rutgers University

Oblivious Routing

Nikhil Bansal
Eindhoven University of Technology, Eindhoven, The Netherlands

Keywords

Fixed path routing

Years and Authors of Summarized Original Work

2002; Räcke

Problem Definition

Consider a communication network, for example, the network of cities across the country connected by communication links. There are several sender-receiver pairs on this network that wish to communicate by sending traffic across the network. The problem deals with routing all the traffic across the network such that no link in the network is overly congested. That is, no link in the network should carry too much traffic relative to its capacity. The obliviousness refers to the requirement that the routes in the network must be designed without the knowledge of the actual traffic demands that arise in the network, i.e., the route for every sender-receiver pair stays fixed irrespective of how much traffic any pair chooses to send. Designing a good oblivious routing strategy is useful since it ensures that the network is robust to changes in the traffic pattern.

Notations

Let $G = (V, E)$ be an undirected graph with nonnegative capacities $c(e)$ on edges $e \in E$. Suppose there are k source-destination pairs (s_i, t_i) for $i = 1, \dots, k$, and let d_i denote the amount of flow (or demand) that pair i wishes to send from s_i to t_i. Given a routing of these flows on G, the congestion of an edge e is defined as $u(e)/c(e)$, the ratio of the total flow crossing edge e divided by its capacity. The congestion of the overall routing is defined as the maximum congestion over all edges. The congestion minimization problem is to find the routing that minimizes the maximum congestion. Observe that specifying a flow from s_i to t_i is equivalent to finding a probability distribution (not necessarily unique) on a collection of paths from s_i to t_i.

The congestion minimization problem can be studied in many settings. In the offline setting, the instance of the flow problem is provided in advance, and the goal is to find the optimum routing. In the online setting, the demands arrive in an arbitrary adversarial order, and a flow must be specified for a demand immediately upon arrival; this flow is fixed forever and cannot be rerouted later when new demands arrive. Several distributed approaches have also been studied where each pair routes its flow in a distributed manner based on some global information such as the current congestion on the edges.

In this note, the oblivious setting is considered. Here, a routing scheme is specified for each pair of vertices in advance without any knowledge of which demands will actually arrive. Note that an algorithm in the oblivious setting is severely restricted. In particular, if d_i units of demand arrive for pair (s_i, t_i), the algorithm must necessarily route this demand according to the prespecified paths irrespective of the other demands or any other information such as congestion of other edges. Thus, given a network graph G, the oblivious flows need to be computed just once. After this is done, the job of the routing algorithm is trivial; whenever a demand arrives, it simply routes it along the precomputed path. An oblivious routing scheme is called $c$competitive if for any collection of demands D, the maximum congestion of the oblivious routing is no more than c times the congestion of the optimum offline solution for D. Given this stringent requirement on the quality of oblivious routing, it is not a priori clear that any reasonable oblivious routing scheme should exist at all.

Key Results

Oblivious routing was first studied in the context of permutation routing where the demand pairs form a permutation and have unit value each. It was shown that any oblivious routing that specifies a single path (instead of a flow) between every two vertices must necessarily perform badly. This was first shown by Borodin and Hopcroft [6] for hypercubes, and the argument was later extended to general graphs by Kaklamanis, Krizanc, and Tsantilas [10], who showed the following.

Theorem 1 ([6, 10]) *For every graph G of size n and maximum degree d and every oblivious routing strategy using only a single path for every source-destination pair, there is a permutation that causes an overlap of at least $(n/d)^{1/2}$ paths at some node. Thus, if each edge in G has unit capacity, the edge congestion is at least $(n/d)^{1/2}/d$.*

Since there exists constant degree graphs such as the butterfly graphs that can route any permutation with logarithmic congestion, this implies that such oblivious routing schemes must necessarily perform poorly on certain graphs.

Fortunately, the situation is substantially better if the single path requirement is relaxed and a probability distribution on paths (equivalently a flow) is allowed between each pair of vertices. In a seminal paper, Valiant and Brebner [13] gave the first oblivious permutation routing scheme with low congestion on the hypercube. It is instructive to consider their scheme. Consider an hypercube with $N = 2^n$ vertices. Represent vertex i by the binary expansion of i. For any two vertices s and t, there is a canonical path (of length at most $n = \log N$) from s to t obtained by starting from s and flipping the bits of s in left to right order to match with that of t. Consider routing scheme that for a pair s and t, it first chooses some node p uniformly at random, routes the flow from s to p along the canonical path, and then routes it again from p to t along the canonical path (or equivalently it sends $1/N$ units of flow from s to each intermediate vertex p and then routes it to t). A relatively simple analysis shows that

Theorem 2 ([13]) *The above oblivious routing scheme achieves a congestion of $O(1)$ for hypercubes.*

Subsequently, oblivious routing schemes were proposed for few other special classes of networks. However, the problem of designing oblivious routing schemes for general graphs remained open until recently, when in a breakthrough result Räcke showed the following.

Theorem 3 ([11]) *For any undirected capacitated graph $G = (V, E)$, there exist an oblivious routing scheme with congestion $O(\log^3 n)$ where n is the number of vertices in G.*

The key to Räcke's theorem is a hierarchical decomposition procedure of the underlying graph (described in further detail below). This hierarchical decomposition is a fundamental combinatorial result about the cut structure of graphs and has found several other applications, some of which are mentioned in section "Applications." Räcke's proof of Theorem 3 only showed the existence of a good hierarchical decomposition and did not give an efficient polynomial time algorithm to find it. In subsequent work, Harrelson, Hildrum, and Rao [9] gave a polynomial time procedure to find the decomposition and improved the competitive ratio of the oblivious routing to $O(\log^2 n \log\log n)$.

Theorem 4 ([9]) *There exists an $O(\log^2 n \log\log n)$-competitive oblivious routing scheme for general graphs, and moreover, it can be found in polynomial time.*

Recently, Räcke [12] has given a tight O(log n)-competitive oblivious routing scheme together with an efficient algorithm to find it. His algorithm is based on an elegant connection to probabilistic embedding of arbitrary metrics into tree metrics.

Interestingly, Azar et al. [4] show that the problem of finding the optimum oblivious routing for a graph can be formulated as a linear program. They consider a formulation with exponentially many constraints, one for each possible demand matrix that has optimum congestion 1 that enforces that the oblivious routing should have low congestion for this demand matrix. Azar et al. [4] give a separation oracle for this problem, and hence, it can be solved using the ellipsoid method. A more practical polynomial size linear program was given later by Applegate and Cohen [2]. Bansal et al. [5] considered a more general variant referred to as the online oblivious routing that can also be used to find an optimum oblivious routing. However, note that without Räcke's result, it would not be clear whether these optimum routings were any good. Moreover, these techniques do not give a hierarchical decomposition and hence may be less desirable in certain contexts. On the other hand, they may be more useful sometimes since they produce an optimum routing (while [9] implies an $O(\log^2 n \log\log n)$-competitive routing for any graph, the best oblivious routing could have a much better guarantee for a specific graph).

Oblivious routing has also been studied for directed graphs; however, the situation is much worse here. Azar et al. [4] show that there exist directed graphs where any oblivious routing is $\Omega(\sqrt{n})$ competitive. Some positive results are also known. Hajiaghayi et al. [7] show a substantially improved guarantee of $O(\log^2 n)$ for directed graphs in the random demands model. Here, each source-sink pair has a distribution (which is known by the algorithm) from which it chooses its demand independently. A relaxation of oblivious routing known as semi-oblivious routing has also been studied recently [8].

Techniques

This section describes the high-level idea of Räcke's result. For a subset $S \subset V$, let $\mathrm{cap}(S)$ denote the total capacity of the edges that cross the cut $(S, V\setminus S)$, and let $\mathrm{dem}(S)$ denote the total demand that must be routed across the cut $(S, V\setminus S)$. Observe that $q = \max_{S\subset V} \mathrm{dem}(S)/\mathrm{cap}(S)$ is a lower bound on the congestion of any solution. On the other hand, the key result [3, 13] relating multicommodity flows and cuts implies that there is a routing such that the maximum congestion is at most $O(q \log k)$ where k is the number of distinct source sink pairs. However, note that this by itself does not suffice to obtain good oblivious routings, since a pair (s_i, t_i) can have different routing for different demand sets. The main idea of Räcke was to impose a treelike structure for routing on the graph to achieve obliviousness. This is formalized by a hierarchical decomposition described below.

Consider a hierarchical decomposition of the graph $G = (V, E)$ as follows. Starting from the set $S = V$, the sets are partitioned successively until each set becomes singleton vertex. This hierarchical decomposition can be viewed naturally as a tree T, where the root corresponds to the set V and leaves corresponds to the singleton sets $\{v\}$. Let S_i denote the subset of V corresponding to node i in T. For an edge (i, j) in the tree where i is the child of j, assign it a capacity equal to $\mathrm{cap}(S_i)$ (note that this is the capacity from S_i to the rest of G and not just capacity between S_i and S_j in G). The tree T is used to simulate routing in G and vice versa. Given a demand from u to v in G, consider the corresponding (unique) route among leaves corresponding to $\{u\}$ and $\{v\}$ in T. For any set of demands, it is easily seen that the congestion in T is no more than the congestion in G. Conversely, Räcke showed that there also exists a tree T where the routes in T can be mapped back to flows in G, such that for any set of demands, the congestion in G is at most $O(\log^3 n)$ times that in T. In this mapping, a flow along the (i, j) in the tree T corresponds to a suitably constructed flow between sets S_i and S_j in G. Since route between any two vertices in T is unique, this gives an oblivious routing in G.

Räcke uses very clever ideas to show the existence of such a hierarchical decomposition. Describing the construction is beyond the scope of this note, but it is instructive to understand the properties that must be satisfied by such a decomposition. First, the tree T should capture the bottlenecks in G, i.e., if there is a set of demands that produces high congestion in G, then it should also produce a high congestion in T. A natural approach to construct T would be to start with V, split V along a bottleneck (formally, along a cut with low sparsity), and recurse. However, this approach is too simple to work. As discussed below, T must also satisfy two other natural conditions, known as the *bandwidth* property and the *weight* property which are motivated as follows. Consider a node i connected to its parent j in T. Then, i needs to route dem(S_i) flow out of S_i, and it incurs congestion dem(S_i)/cap(S_i) in T. However, when T is mapped back to G, all the flow going out of S_i must pass via S_j. To ensure that the edges from S_i to S_j are not overloaded, it must be the case that the capacity from S_i to S_j is not too small compared to the capacity from S_i to the rest of the graph $V \backslash S_i$. This is referred to as the bandwidth property. Räcke guarantees that this is ratio is always $\Omega(1/\log n)$ for every S_i and S_j corresponding to edges (i, j) in the tree. The weight property is motivated as follows. Consider a node j in T with children $i_1, \ldots, i_p$; then the weight property essentially requires that the sets $S_{i1}, \ldots, S_{ip}$ should be well connected among themselves even when restricted to the subgraph S_j. To see why this is needed, consider any communication between, say, nodes i_1 and i_2 in T. It takes the route i_1 to j to i_2, and hence, in G,S_{i1} cannot use edges that lie outside S_j to communicate with S_{i2}. Räcke shows that these conditions suffice and that a decomposition can be obtained that satisfies them.

The factor $O(\log {}^3 n)$ in Räcke's guarantee arises from three sources. The first logarithmic factor is due to the flow-cut gap [3, 13]. The second is due to the logarithmic height of the tree, and the third is due to the loss of a logarithmic factor in the bandwidth and weight properties.

Applications

The problem has widespread applications to routing in networks. In practice, it is often required that the routes must be a single path (instead of flows). This can often be achieved by randomized rounding techniques (sometimes under an assumption that the demands to capacity ratios be not too large). The flow formulation provides a much cleaner framework for studying the problems above.

Interestingly, the hierarchical decomposition also found widespread uses in other seemingly unrelated areas such as obtaining good preconditioners for solving systems of linear equations, finding edge-disjoint paths and multicommodity flow problems, online network optimization, speeding up the running time of graph algorithms, and so on.

Cross-References

▶ Routing
▶ Separators in Graphs
▶ Sparsest Cut

Recommended Reading

1. Alon N, Awerbuch B, Azar Y, Buchbinder N, Naor J (2004) A general approach to online network optimization problems. In: Symposium on discrete algorithms, pp 570–579
2. Applegate D, Cohen E (2003) Making intra-domain routing robust to changing and uncertain traffic demands: understanding fundamental tradeoffs. In: SIGCOMM, pp 313–324
3. Aumann Y, Rabani Y (1998) An O(log k) approximate min-cut max-flow theorem and approximation algorithm. SIAM J Comput 27(1):291–301
4. Azar Y, Cohen E, Fiat A, Kaplan H, Räcke H (2003) Optimal oblivious routing in polynomial time. In: Proceedings of the 35th ACM symposium on the theory of computing, pp 383–388
5. Bansal N, Blum A, Chawla S, Meyerson A (2003) Online oblivious routing. In: Symposium on parallelism in algorithms and architectures, pp 44–49
6. Borodin A, Hopcroft J (1985) Routing, merging and sorting on parallel models of computation. J Comput Syst Sci 10(1):130–145
7. Hajiaghayi M, Kim JH, Leighton T, Räcke H (2005) Oblivious routing in directed graphs with random demands. In: Symposium on theory of computing, pp 193–201
8. Hajiaghayi M, Kleinberg R, Leighton T (2007) Semi-oblivious routing: lower bounds. In: Proceedings of

the 18th ACM-SIAM symposium on discrete algorithms, pp 929–938
9. Harrelson C, Hildrum K, Rao S (2003) A polynomial-time tree decomposition to minimize congestion. In: Proceedings of the 15th annual ACM symposium on parallel algorithms and architectures, pp 34–43
10. Kaklamanis C, Krizanc D, Tsantilas T (1991) Tight bounds for oblivious routing in the hypercube. In: Proceedings of the 3rd annual ACM symposium on parallel algorithms and architectures, pp 31–36
11. Räcke H (2002) Minimizing congestion in general networks. In: Proceedings of the 43rd annual symposium on the foundations of computer science, pp 43–52
12. Räcke H (2008) Optimal hierarchical decompositions for congestion minimization networks. In: Symposium on theory of computing, pp 255–264
13. Valiant L, Brebner GJ (1981) Universal schemes for parallel communication. In: Proceedings of the 13th ACM symposium on theory of computing, pp 263–277

Oblivious Subspace Embeddings

Jelani Nelson
Harvard John A. Paulson School of Engineering and Applied Sciences, Cambridge, MA, USA

Keywords

Dimensionality reduction; Oblivious subspace embeddings; Randomized linear algebra

Years and Authors of Summarized Original Work

2006; Sarlós

Problem Definition

This entry surveys some of the applications of "oblivious subspace embeddings," introduced by Sarlós in [19], to problems in linear algebra.

Definition 1 ([19]) Given $0 < \varepsilon < 1/2$ and a d-dimensional subspace $E \subseteq \mathbb{R}^n$, we say an $m \times n$ matrix Π is an *ε-subspace embedding for E* if

$$\forall x \in E\ (1-\varepsilon)\|x\|_2^2 \le \|\Pi x\|_2^2 \le (1+\varepsilon)\|x\|_2^2.$$

The goal is to have m small so that Π provides dimensionality reduction for E.

Given $0 < \varepsilon, \delta < 1/2$, and integers $1 \le d \le n$, an *$(\varepsilon, \delta, d, n)$-oblivious subspace embedding (OSE)* is a distribution $\mathcal{D}$ over $m \times n$ matrices such that for every d-dimensional linear subspace $E \subseteq \mathbb{R}^N$ of dimension d,

$$\mathop{\mathbb{P}}_{\Pi \sim \mathcal{D}} (\Pi \text{ is an } \varepsilon\text{-subspace embedding for } E) > 1-\delta.$$

Sometimes we omit a subset of the variables $\varepsilon, \delta, d, n$ if they are understood from context.

In the definition of an OSE, note that we can write $E = \{Ux : x \in \mathbb{R}^d\}$ where $U \in \mathbb{R}^{n\times d}$ and $U^T U = I$. That is, the columns of U form an orthonormal basis for E. Therefore, we would like that $\|\Pi U x\|_2^2 \approx \|Ux\|_2^2 = x^T U^T U x = \|x\|_2^2$ for all $x \in \mathbb{R}^d$. Letting $\|\cdot\|$ denote operator norm and noting that $\|A\| = \sup_x |x^T A x|$ for any real symmetric A, we see that being an OSE as above is equivalent to the following holding for all $U \in \mathbb{R}^{n\times d}$ with orthonormal columns:

$$\mathop{\mathbb{P}}_{\Pi \sim \mathcal{D}} \left(\|(\Pi U)^T (\Pi U) - I\| > \varepsilon \right) < \delta. \quad (1)$$

Sarlós introduced OSEs [19] to provide faster approximate algorithms for least squares regression and low-rank approximation. In these problems, the input is a tall and skinny matrix $A \in \mathbb{R}^{n\times d}$ $(n \gg d)$. For regression, we also are given $b \in \mathbb{R}^n$. The goal is to solve some computational problem given A, and naturally the running time depends on both n and d. The basic idea is to instead run the computation on ΠA for Π sampled from an OSE and (1) prove that the quality of solution found is near optimal if ε is small and (2) enjoy faster computation time to find a solution since the dimensionality of the problem is reduced (ΠA is $m \times d$, whereas A is $n \times d$, $m \ll n$).

Key Results

As mentioned above, Sarlós showed how to use OSEs to speed up least squares regression and low-rank approximation. Below we first discuss constructions of OSEs, and then we elaborate on applications.

Constructing OSEs

One OSE is to pick the entries of $\Pi \in \mathbb{R}^{m \times n}$ i.i.d. from a Gaussian distribution with mean zero and variance $1/m$, where $m = \Theta((d + \log(1/\delta))/\varepsilon^2)$. In fact it suffices to pick any "Johnson-Lindenstrauss transform," i.e., a Π which preserves the Euclidean norms of a certain set of $2^{O(d)}$ vectors up to $1 + \varepsilon$ (see [7]). This setting of m is optimal for any OSE [17]. The downside of such constructions is multiplying ΠA then takes time $\Theta(nmd)$, which is in fact worse than the time to solve the problems considered here.

Sarlós remedied this by picking Π from the "Fast Johnson-Lindenstrauss" distribution [1], which improved this time to $O(nd \log n) + \text{poly}(d)/\varepsilon^2$. A related construction, the "Subsampled Randomized Hadamard Transform" (SRHT), with improved bounds for OSEs was analyzed in [14, 21] using matrix concentration inequalities. In this construction, one chooses $\Pi = \sqrt{n/m} \cdot SHD$ where $D \in \mathbb{R}^{n \times n}$ is diagonal with random signs on the diagonal, H is any bounded orthonormal system that supports matrix-vector multiplication in $O(n \log n)$ time (i.e., H should be orthogonal with $\max_{i,j} |H_{i,j}| = O(1/\sqrt{n})$), and S is a sampling matrix with m rows. That is, the rows of S are independent, and each row of S has exactly a single 1 in a uniformly random location and zeroes elsewhere. The works [14, 21] showed one can take $m = O(d \log(d/\delta)/\varepsilon^2)$. Note we can multiply ΠA in time $O(nd \log n + m)$ by individually multiplying Π by each column of A.

Subsequently, Clarkson and Woodruff [7] showed that the Thorup-Zhang sketch [20] provides an OSE with small m. In particular, consider a random Π with independent columns where in each column there is a single nonzero entry placed in a uniformly random location. The value of this nonzero is uniform in $\{-1, 1\}$. Note that with this construction, one can multiply ΠA in time $O(\text{nnz}(A))$, where $\text{nnz}(\cdot)$ counts nonzero entries. They showed this distribution is an OSE for $m = O(d^2 \log^6(d/\varepsilon)/(\varepsilon^2\delta))$. This bound was improved independently in [15, 16] to $m = O(d^2/(\varepsilon^2\delta))$ via the moment method (see also an observation of Nguyễn [12, Remark 6.4]). Note also a valid OSE is the product of the SRHT with this construction, yielding m as for the SRHT and with multiplication time $O(\text{nnz}(A) + d^3 \log(d/(\varepsilon\delta))/\varepsilon^2)$ to apply to A.

Nelson and Nguyễn [16] analyzed the "Sparse Johnson-Lindenstrauss Transform" (SJLT) of [12] in the context of OSEs. In particular, they showed one can choose an OSE with $m = O(d \log^6(d/\delta)/\varepsilon^2)$ and $s = O(\log^3(d/\delta)/\varepsilon)$ nonzero entries per column. See also [5]. Note ΠA can be computed in time $O(s \cdot \text{nnz}(A))$. One could also choose $m = O(d^{1+\gamma}/\varepsilon^2), s = O(1/\varepsilon)$ for any fixed constant $0 < \gamma < 1$, in which case $\delta = 1/d^c$ for any desired constant $c > 0$. A conjecture of [16] is that $m = O((d + \log(1/\delta))/\varepsilon^2), s = O(\log(d/\delta)/\varepsilon)$ suffices.

Applying OSEs

Least Squares Regression

The input is $A \in \mathbb{R}^{n \times d}$, $b \in \mathbb{R}^n$. The goal is to compute

$$x^* = \text{argmin}_{x \in \mathbb{R}^d} \|Ax - b\|_2.$$

By optimality of x^*, Ax^* must be the projection of b onto the column span of A. Write the singular value decomposition (SVD) $A = U\Sigma V^T$, where $U \in \mathbb{R}^{n \times r}, V \in \mathbb{R}^{d \times r}$ have orthonormal columns, and r is the rank of A. Also, $\Sigma \in \mathbb{R}^{r \times r}$ is diagonal with strictly positive entries on the diagonal (the "singular values" of A). Then the column spans of U and of A are identical, and so $Ax^* = UU^T b$ is the desired projection. Thus, we can choose $x^* = V\Sigma^{-1}U^T b$. Alternatively one can write $x^* = (A^T A)^+ A^T b$, where the pseudoinverse of a matrix B with SVD LDW^T is $B^+ = WD^{-1}L^T$.

Simply computing $A^T A$ in the formula for x^* naively takes $O(nd^2)$ time (or $O(nd^{\omega-1})$ if using fast matrix multiplication). Note the following observation.

Observation 1 *Let E be the subspace spanned by b and the columns of A. Let Π be an ε-subspace embedding for E, and write $\tilde{x} = argmin_x \|\Pi Ax - \Pi b\|_2$. Then*

$$\|A\tilde{x} - b\|_2 \leq \sqrt{\frac{1+\varepsilon}{1-\varepsilon}} \cdot \|Ax^* - b\|_2$$

Proof By optimality of $\tilde{x}$, $\|\Pi A\tilde{x} - \Pi b\|_2 \leq \|\Pi Ax^* - \Pi b\|_2$. Furthermore, since $A\tilde{x} - b, Ax^* - b \in E$, we have $(1-\varepsilon)\|A\tilde{x} - b\|_2^2 \leq \|\Pi A\tilde{x} - \Pi b\|_2^2$ and $\|\Pi Ax^* - \Pi b\|_2^2 \leq (1+\varepsilon)\|Ax^* - b\|_2^2$. The claim follows.

The above observation informs us that minimizing $\|\Pi Ax - \Pi b\|_2$ for a subspace embedding Π is sufficient to obtain a high-quality solution to the original problem. Using an OSE with m rows, one can compute $\tilde{x}$ in $O(md^2)$ time (or faster using fast matrix multiplication).

Sarlós also provided another method of using OSEs for least squares regression. First we provide a definition.

Definition 2 We call a distribution $\mathcal{D}$ over $\mathbb{R}^{m\times n}$ an (ε, δ)-*AMM*$_F$ distribution if, for every pair of matrices A, B each with n rows,

$$\mathop{\mathbb{P}}_{\Pi\sim\mathcal{D}}(\|(\Pi A)^T(\Pi B) - A^T B\| > \varepsilon\|A\|_F\|B\|_F) < \delta$$

where $\|A\|_F = (\sum_{i,j} A_{i,j}^2)^{1/2}$ is the Frobenius norm of A. ("AMM" here stands for "approximate matrix multiplication.")

The work [10] was the first to propose using AMM$_F$ and (non-oblivious) subspace embeddings in the context of low-rank approximation. Sarlós showed that any Johnson-Lindenstrauss (JL) distribution also provides AMM$_F$, but with δ increased by some factor involving the dimensions of A, B. This factor was removed for random sign matrices in [8] and later for a fairly general class of JL distributions in [12, Theorem 6.2]. We state the relevant definition and theorem for this general class.

Definition 3 ([12]) We say a distribution $\mathcal{D}$ over $\mathbb{R}^{m\times n}$ has (ε, δ, p)-*JL moments* if for any $x \in \mathbb{R}^n$ of unit Euclidean norm,

$$\mathop{\mathbb{E}}_{\Pi\sim\mathcal{D}} \left|\|\Pi x\|_2^2 - 1\right|^p < \varepsilon^p \delta.$$

Theorem 1 ([12]) *Given $\varepsilon, \delta \in (0, 1/2)$, let $\mathcal{D}$ be any distribution with the (ε, δ, p)-JL moment property for some $p \geq 2$. Then $\mathcal{D}$ is a $(3\varepsilon, \delta)$-AMM$_F$ distribution.*

For constant δ, for example, it thus follows from [20] that the Thorup-Zhang sketch with $m = O(1/\varepsilon^2)$ provides AMM$_F$. Sarlós then proved the following.

Theorem 2 *Suppose $\tilde{x} = argmin \|\Pi Ax - \Pi b\|_2$ where the distribution Π is drawn from is (1) an $(O(1), \delta)$-OSE for d-dimensional subspaces (with a distortion parameter independent of ε), and (2) a $(\sqrt{\varepsilon/d}, \delta)$-AMM$_F$ distribution. Then with probability $1 - 2\delta$,*

$$\|A\tilde{x} - b\|_2 \leq (1 + O(\varepsilon))\|Ax^* - b\|_2.$$

The above theorem combined with [16] allows, for example, picking Π as the SJLT with $m = O(d^{1+\gamma} + d/\varepsilon)$, $s = O(1)$ for any constant $0 < \gamma < 1$ to achieve $(1+\varepsilon)$-multiplicative error for least squares regression.

Low-Rank Approximation

The input is $A \in \mathbb{R}^{n\times d}$ and positive integer k, and the goal is to compute

$$A_k = \mathop{argmin}_{B:\mathrm{rank}(B)\leq k} \|A - B\|_F.$$

Given the SVD $A = U\Sigma V^T$, the Schmidt approximation theorem (later rediscovered as the Eckart-Young theorem) yields that $A_k = U\Sigma_k V^T$, where Σ_k retains only the k largest elements of Σ and zeroes out the rest. Up to terms logarithmic in dimension and precision, the SVD can be computed in time $nd^{\omega-1}$ [9] where ω is the exponent of square matrix multiplication.

Thus, low-rank approximation can be solved in the same time bound.

A scheme based on OSEs and AMM_F was given by Sarlós. For matrices B, S, let $\mathrm{Proj}_{S,k}(B)$ denote the best rank-k approximation to B in the column span of S. Equivalently, it is the best rank-k approximation to the matrix formed by projecting each column of B to the column span of S. Sarlós' theorem is as follows.

Theorem 3 ([19]) *Let Π be drawn from a distribution which is (1) an $(O(1), \delta)$-OSE for k-dimensional subspaces and (2) a $(\sqrt{\varepsilon/k}, \delta)$-$\mathrm{AMM}_F$ distribution. Then with probability $1-2\delta$,*

$$\|A - \mathrm{Proj}_{A\Pi^T,k}(A)\|_F \le (1 + O(\varepsilon))\|A - A_k\|_F.$$

The above theorem has led to low-rank approximation algorithms with $(1+\varepsilon)$-multiplicative error running in time $O(\mathrm{nnz}(A)) + \tilde{O}(nk^2/\mathrm{poly}(\varepsilon))$ [7, 15, 16] or even $O(\mathrm{nnz}(A)) + \tilde{O}(nk^{\omega-1}/\mathrm{poly}(\varepsilon))$ [16]. Here $\tilde{O}(\cdot)$ hides logarithmic factors. These algorithms, in these running times, can output a decomposition $L \in \mathbb{R}^{n\times k}, D \in \mathbb{R}^{k\times k}, W \in \mathbb{R}^{d\times k}$ with D diagonal and L, W having orthonormal columns such that

$$\|A - LDW^T\|_F \le (1+\varepsilon)\|A - A_k\|_F.$$

Other Applications

OSEs have also found other applications, e.g., to approximating leverage scores [11], distributed principle component analysis [13], k-means clustering [6], canonical correlation analysis [3], support vector machines [18], ℓ_p regression [22], ridge regression [14], CUR matrix factorization [4], and streaming approximation of eigenvalues [2]. The reader may investigate these references for more details.

Recommended Reading

1. Ailon N, Chazelle B (2009) The fast Johnson–Lindenstrauss transform and approximate nearest neighbors. SIAM J Comput 39(1):302–322
2. Andoni A, Nguyễn HL (2013) Eigenvalues of a matrix in the streaming model. In: Proceedings of the 24th annual ACM-SIAM symposium on discrete algorithms (SODA), New Orleans, pp 1729–1737
3. Avron H, Boutsidis C, Toledo S, Zouzias A (2013) Efficient dimensionality reduction for canonical correlation analysis. In: Proceedings of the 30th international conference on machine learning (ICML), Atlanta, pp 347–355
4. Boutsidis C, Woodruff DP (2014) Optimal CUR matrix decompositions. In: Proceedings of the 46th ACM symposium on theory of computing (STOC), New York, pp 353–362
5. Bourgain J, Nelson J (2013) Toward a unified theory of sparse dimensionality reduction in Euclidean space. CoRR abs/1311.2542
6. Christos Boutsidis, Anastasios Zouzias, Mahoney MW, Drineas P (2011) Stochastic dimensionality reduction for k-means clustering. CoRR abs/1110.2897
7. Clarkson KL, Woodruff DP (2013) Low rank approximation and regression in input sparsity time. In: Proceedings of the 45th ACM symposium on theory of computing (STOC), Palo Alto, pp 81–90
8. Clarkson KL, Woodruff DP (2009) Numerical linear algebra in the streaming model. In: Proceedings of the 41st ACM symposium on theory of computing (STOC), Bethesda, pp 205–214
9. Demmel J, Dumitriu I, Holtz O (2007) Fast linear algebra is stable. Numer Math 108(1):59–91
10. Drineas P, Mahoney MW, Muthukrishnan S (2006) Subspace sampling and relative-error matrix approximation: column-based methods. In: Proceedings of the 10th international workshop on randomization and computation (RANDOM), Barcelona, pp 316–326
11. Drineas P, Magdon-Ismail M, Mahoney MW, Woodruff DP (2012) Fast approximation of matrix coherence and statistical leverage. J Mach Learn Res 13:3475–3506
12. Kane DM, Nelson J (2014) Sparser Johnson-Lindenstrauss transforms. J ACM 61(1):4
13. Kannan R, Vempala S,Woodruff DP (2014) Principal component analysis and higher correlations for distributed data. In: Proceedings of the 27th annual conference on learning theory (COLT), Barcelona, pp 1040–1057
14. Lu Y, Dhillon P, Foster D, Ungar L (2013) Faster ridge regression via the subsampled randomized Hadamard transform. In: Proceedings of the 26th annual conference on advances in neural information processing systems (NIPS), Lake Tahoe
15. Mahoney MW, Meng X (2013) Low-distortion subspace embeddings in input-sparsity time and applications to robust linear regression. In: Proceedings of the 45th ACM symposium on theory of computing (STOC), Palo Alto, pp 91–100
16. Nelson J, Nguyễn HL (2013) OSNAP: faster numerical linear algebra algorithms via sparser subspace embeddings. In: Proceedings of the 54th annual IEEE symposium on foundations of computer science (FOCS), Berkeley, pp 117–126

17. Nelson J, Nguyễn HL (2014) Lower bounds for oblivious subspace embeddings. In: Proceedings of the 41st international colloquium on automata, languages, and programming, Copenhagen, pp 883–894
18. Paul S, Boutsidis C, Magdon-Ismail M, Drineas P (2013) Random projections for support vector machines. In: Proceedings of the sixteenth international conference on artificial intelligence and statistics (AISTATS), Scottsdale, pp 498–506
19. Sarlós T (2006) Improved approximation algorithms for large matrices via random projections. In: Proceedings of the 47th annual IEEE symposium on foundations of computer science (FOCS), Berkeley, pp 143–152
20. Thorup M, Zhang Y (2012) Tabulation-based 5-independent hashing with applications to linear probing and second moment estimation. SIAM J Comput 41(2):293–331
21. Tropp JA (2011) Improved analysis of the subsampled randomized Hadamard transform. Adv Adapt Data Anal 3(1–2, special issue, "Sparse Representations of Data and Images,"):115–126
22. Woodruff DP, Zhang Q (2013) Subspace embeddings and ℓ_p-regression using exponential random variables. In: Proceedings of the 26th annual conference on learning theory (COLT), Princeton, pp 546–567

Obstacle Avoidance Algorithms in Wireless Sensor Networks

Sotiris Nikoletseas[1,2] and Olivier Powell[3]
[1]Computer Engineering and Informatics Department, University of Patras, Patras, Greece
[2]Computer Technology Institute and Press "Diophantus", Patras, Greece
[3]Informatics Department, University of Geneva, Geneva, Switzerland

Keywords

Greedy geographic routing; Routing holes

Years and Authors of Summarized Original Work

2007; Powell, Nikoletseas

Problem Definition

Wireless sensor networks are composed of many small devices called sensor nodes with sensing, computing and radio frequency communication capabilities. Sensor nodes are typically deployed in an ad hoc manner and use their sensors to collect environmental data. The emerging network collectively processes, aggregates and propagates data to regions of interest, e.g., from a region where an event is being detected to a base station or a mobile user. This entry is concerned with the data propagation duty of the sensor network in the presence of obstacles.

For different reasons, including energy conservation and limited transmission range of sensor nodes, information propagation is achieved via multi-hop message transmission, as opposed to single-hop long range transmission. As a consequence, message routing becomes necessary. Routing algorithms are usually situated at the network layer of the protocol stack where the most important component is the (dynamic) communication graph.

Definition 1 (Communication graph) A wireless sensor network is viewed as a graph $G = (V, E)$ where vertexes correspond to sensor nodes and edges represent wireless links between nodes.

Wireless sensor networks have stringent constraints that make classical routing algorithms inefficient, unreliable or even incorrect. Therefore, the specific requirements of wireless sensor networks have to be addressed [2] and geographic routing offers the possibility to design particularly well adapted algorithms.

Geographic Routing

A geographic routing algorithm takes advantage of the fact that sensor nodes are location aware, i.e., they know their position in a coordinate system following the use of a localization protocol [7]. Although likely to introduce a significant overhead, the use of a localization protocol is also likely to be inevitable in many applications where environmental data collected by the sensors would be useless if not related to some geographical information. For those applications, node location awareness can be assumed to be available for routing purposes at no additional cost.

The Power of Simple Geographic Routing

The early "most forward within range" (MFR) or greedy geographic routing algorithms [14] route messages by maximizing, at each hop, the progress on a projected line towards the destination or, alternatively, minimizing the remaining distance to the message's destination. Both of these greedy heuristics are referred to as *greedy forwarding* (GF). Greedy forwarding is a very appealing routing technique for wireless sensor networks. Among explanations for the attractiveness of GF are the following. (1) GF, as is almost imperatively required, is *fully distributed.* (2) It is *lightweight* in the sense that it induces no topology control overhead. (3) It is *all-to-all* (as opposed to all-to-one). (4) Making no assumptions on the structure of the communication graph, which can be directed, undirected, stable or dynamic (e.g., nodes may be mobile or wireless links may appear and disappear, for example following environmental fluctuation or as a consequence of lower protocol stack layers such as sleep/awake schemes for energy saving purposes), it is *robust.* (5) It is *on-demand*: no routing table or gradient has to be built prior to message propagation. (6) *Efficiency* is featured as messages find short paths to their destination in terms of hop count. (7) It is very *simple* and thus *easy to implement.* (8) It is *memory efficient* in the sense that (8a) the only information stored in the message header is the message's destination and that (8b) it is "ecologically sound" because no "polluting" information is stored on the sensor nodes visited by messages.

Problem Statement

Although very appealing, GF suffers from a major flaw: when a message reaches a local minimum where no further progress towards the destination is possible the routing algorithm fails. There are two major reasons for the occurrence of local minimums: routing holes [1] and obstacles.

Definition 2 The so called *routing holes* are low density regions of the network where no sensor nodes are available for next-hop forwarding.

Even in uniform-randomly deployed networks, routing holes appear as the manifestation of statistical variance of node density. Although increasing as network density diminishes, routing holes have a severe impact on the performance of GF even for very high density networks [12].

Definition 3 A *transmission blocking obstacle* is a region of the network where no sensors are deployed and through which radio signals do not propagate.

Clearly, large obstacles lying between a message and its destination tend to make GF fail.

The problem reported in this entry is to find a geographic routing algorithm that *maintains the advantages of greedy forwarding* listed in section "Geographic Routing" such as simplicity, light weight, robustness and efficiency while *overcoming its weaknesses*: the inability to escape local minimum nodes created by routing holes and large transmission blocking obstacles such as those seen in Fig. 1.

Problem 1 (Escaping routing holes) The first problem is to route messages out of the many routing holes which are statistically doomed to occur even in dense networks.

Problem 2 (Contouring obstacles) The second problem is to design a protocol capable of routing messages around large transmission blocking obstacles.

Problem 1 can be considered a simplified instance of Problem 2. Lightweight solutions to problem 1 have been previously proposed, usually using limited backtracking [6] or controlled flooding combined with a GF heuristic [4, 13]. However, as shown in [5] where an integrated model for obstacles is proposed and where different algorithms are compared with respect to their obstacle avoidance capability, those solutions do not

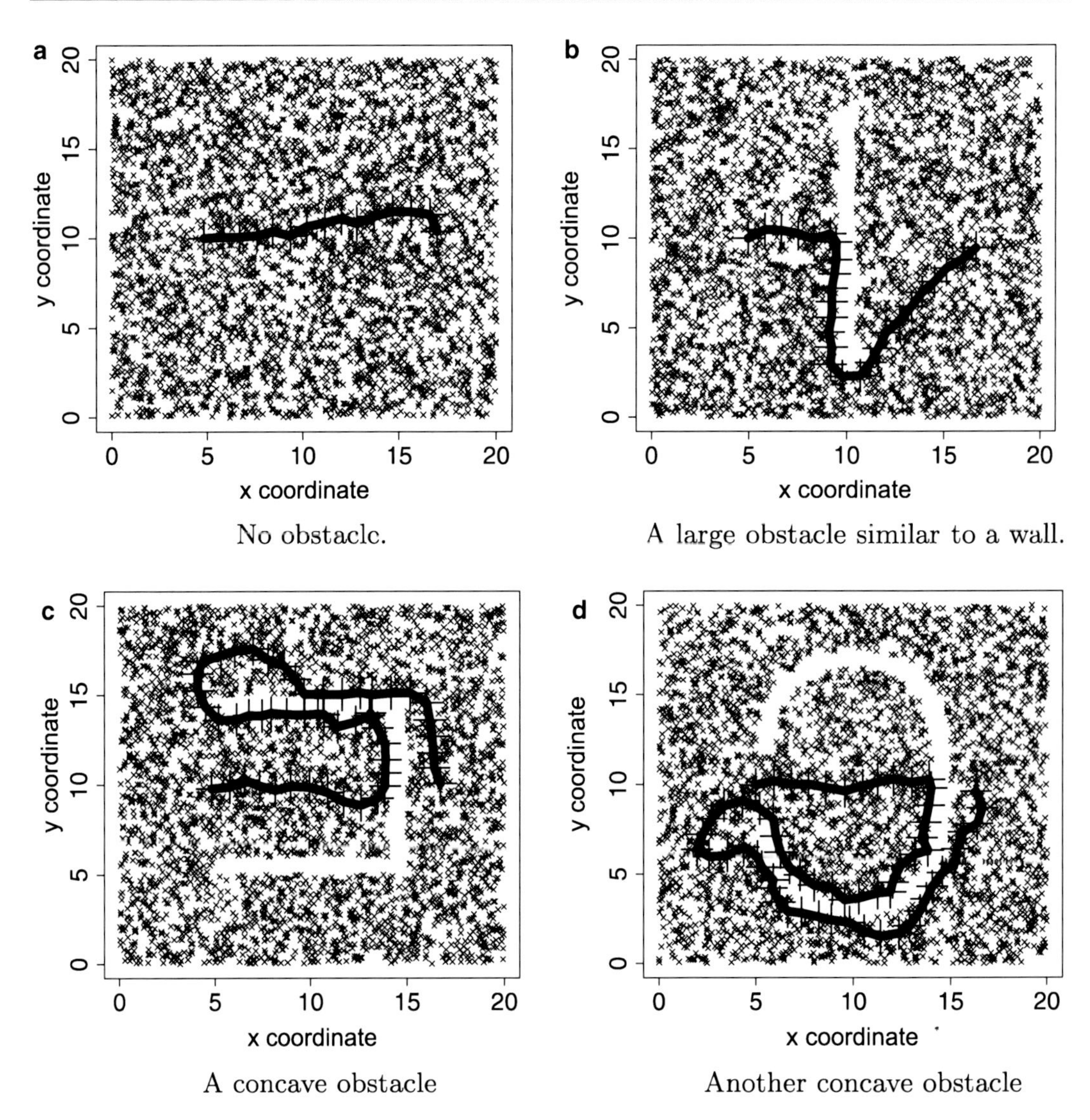

Obstacle Avoidance Algorithms in Wireless Sensor Networks, Fig. 1 Typical path followed by GRIC to bypass certain obstacles

satisfactorily solve Problem 2 in the sense that only small and simple obstacles are efficiently bypassed.

Key Results

In [12] a new **g**eographic **r**out**i**ng around obsta**c**les (GRIC) algorithm was proposed to address the problems described in the previous section.

Basic Idea of the Algorithm

In GF, the strategy is to always propagate the message to the neighbor that maximizes progress *towards the destination*. Similarly, GRIC also maximizes progress in a chosen direction. However, this direction is not necessarily the message's destination but an *ideal direction of progress* which has to be computed according to one of two possible strategies: the *inertia mode* or the *rescue mode* described below. Finally, it was found that performance is better in the presence of slightly unstable networks, cf. Result 4, and thus in the case where the communication graph is very stable it is recommended to use a randomized version of GRIC where nodes about to take a routing decision randomly mark as either passive or active each outbound wireless link of the communication graph. Only active wireless links can be used for message propagation, and link status is re-evaluated each time a new routing decision is taken. Marking links as active with a probability of $p = 0.95$ was found to be a good choice for practical purposes [12].

Inertia Mode

The idea of the inertia mode is that a message should have a strong incentive to go towards its destination but this incentive should be moderated by one to follow the straight ahead direction of current motion "*... like a celestial body in a planet system ...*" [12]. The inertia mode aims at making messages follow closely the perimeter of routing holes and obstacles in order to eventually bypass them and ensure final routing to the destination. To implement the inertia mode, a single additional assumption is made: sensor nodes should be aware of the position of the node from which they receive a message. As an example, this could be done by piggy-backing this 1-hop away routing path history in the message header. Knowing its own position p, the message's destination and the 1-hop away previous position of the message a sensor node can compute the vectors v_{cur} and v_{dst} starting at position p and pointing in the direction of current motion and the direction to the message's destination respectively. The inertia mode defines the ideal direction of progress, v_{idl}, as a vector starting at point p and lying "somewhere in between" v_{cur} and v_{dst}. More precisely, let α be the only angle in $[-\pi, \pi[$ such that v_{dst} is obtained by applying a rotation of angle α to v_{cur}, then v_{idl} is the vector obtained by applying a rotation of angle α' to v_{cur}, where $\alpha' = \text{sign}(\alpha) \cdot \min\left\{\frac{\pi}{6}, |\alpha|\right\}$. Finally, the message is greedily forwarded to the neighbor node maximizing progress in the computed ideal direction of progress v_{idl}.

Rescue Mode

In order to improve overall performance and to bypass complex obstacles, the rescue mode imitates the right-hand rule (RHR) which is a well known wall follower technique to find one's way out of a maze. A high-level description of the RHR component of GRIC is given below while details will be found in [12]. In GRIC, the RHR makes use of a virtual compass and a flag. The virtual compass assigns to v_{cur} a cardinal point value, treating the message's destination as the north. Considering the angle α defined in the previous section, the compass returns a string x-y with x equal to north or south if $|\alpha|$ is smaller or greater than $\frac{\pi}{2}$ respectively, while y is equal to west or east if α is negative or positive respectively. The first time the compass returns a south value, the flag is raised and tagged with the (x, y) value of the compass. Raising the flag means that the message is being routed around an obstacle using the RHR rule if the compass indicates south-west. In the case where the compass indicates south-east, a symmetric case not discussed here for brevity is applied using the left-hand rule (LHR) instead of the RHR. Once the flag is raised, it stays up with its tag unchanged until the compass indicates north, meaning that the obstacle has been bypassed. In fact, a small optimization can be made by lowering the flag only if the compass points to the north-west (in the case of the RHR) and not if it points north-east, but cf. [12] for details. According to the RHR the obstacle's perimeter should be followed closely and kept on the right side of the message's current direction. If ever the compass and the flag's tag disagree, i.e., if the flag is tagged with south-west and the compass returns south-east, it is assumed that the message is turning left too much, that it risks going away from the obstacle and that the RHR is at risk of being violated (a symmetric case applies for the LHR). When this is so, GRIC responds by calling the rescue mode which changes the default way of computing v_{idl}: in rescue mode the message is forced to turn right (or left if the LHR is applied), by defining v_{idl} as the vector obtained by applying to v_{cur} a rotation of angle α'' (instead of α' in inertia mode) where $\alpha'' = -sign(\alpha)(2\pi - |\alpha|)/6$.

Main Findings

The performance of GRIC was evaluated through simulations. The main parameters were the presence (or absence) of different shapes of large communication blocking obstacles and the network density which ranged from very low to very high and controls the average degree of the communication graph and the occurrence of routing holes. The main performance metrics were the success rate, i.e., the percentage of messages routed to destination, and the path length. The main findings are that GRIC efficiently, i.e., using short paths, bypasses routing holes and obstacles but that in the presence of hard obstacles, the performance decreases with network density. In Fig. 1. typical routing paths found by GRIC for different obstacle shapes are illustrated, cf. [12] for details on the simulation environment.

Result 1 *In the absence of obstacles, routing holes are bypassed for every network density: The success rate is close to 100 % as long as the source and the destination are connected. Also, routing is efficient in the sense that path lengths are very short.*

Result 2 *Some convex obstacles such as the one in* Fig. 1b *are bypassed with almost 100% success rate and using short paths, even for low densities. When the density gets very low performance diminishes: If the density gets below the critical level guaranteeing the communication graph to be connected with high probability, then the success probability diminishes quickly and successful routings use longer routing paths.*

Result 3 *Some large concave obstacles such as those in* Fig. 1c *and d are efficiently bypassed. However, when facing such obstacles performance becomes more sensitive to network density. The success rate drops and routing paths become longer when the density gets below a certain level depending on the exact obstacle shape.*

Result 4 (Robustness) *Similarly to GF, GRIC is robust to link instability. Furthermore, it was observed that limited link instability has a significantly positive impact on performances. This can be understood as the fact that messages are less likely to enter endless routing loops in a "hot" system than in a "cold" system.*

Applications

Replacement for Greedy Forwarding

Because it makes no compromise with the advantages of GF except the fact that it may be somehow more complicated to implement and because it overcomes GF's main limitations, GRIC can probably replace GF for most routing scenarios including but not exclusively wireless sensor networks. As an example opportunistic-routing strategies [11] could be applied to GRIC rather than to GF.

Wireless Sensor Networks with Large Obstacles

GRIC successfully bypasses large communication blocking obstacles. However, it does so efficiently only if the network density is high enough. This suggests that the obstacle avoidance feature of GRIC may be more useful for dense wireless networks than for sparse networks. Wireless sensor networks are an example of networks which are usually considered to be dense.

Dynamic Networks

There exist some powerful alternatives to GRIC such as the celebrated guaranteed delivery protocols GFG [3], GPSR [8] or GOAFR [10]. Those protocols rely on a planarization phase such as the lazy cross-link detection protocol (CLDP) [9]. LCR implies significant topology maintenance overhead which would be amortized over time if the network is stable enough. On the contrary, if the network is highly dynamic the necessity for frequent updates could make this topology maintenance overhead prohibitive. GRIC may thus be a preferable choice for dynamic networks where the communication graph is not a stable structure.

Open Problems

(1) Hard concave obstacles such as the one in Fig. 1d are still a challenge for lightweight protocol since in this configuration GRIC's performance is strongly dependent on network density. (2) Low to very low densities are challenging when combined with large obstacles, even when they are "simple" convex obstacles like the one in Fig. 1b. (3) The problem reported in this entry in the case of 3-dimensional networks is open. Inertia may be of some help, however the virtual compass and the right-hand rule seem quite strongly dependant on the 2-dimensional plane. (4) GRIC is not loop free. A mechanism to detect loops or excessively long routing paths would be quite important for practical purposes. (5) The understanding of GRIC could be improved. Analytical results are lacking and new metrics could be considered such as network lifetime, energy consumption or traffic congestion.

Cross-References

▸ Probabilistic Data Forwarding in Wireless Sensor Networks

Recommended Reading

1. Ahmed N, Kanhere SS, Jha S (2005) The holes problem in wireless sensor networks: a survey. SIGMOBILE Mob Comput Commun Rev 9:4–18
2. Al-Karaki JN, Kamal AE (2004) Routing techniques in wireless sensor networks: a survey. Wirel Commun IEEE 11:6–28
3. Bose P, Morin P, Stojmenovic I, Urrutia J (1999) Routing with guaranteed delivery in ad hoc wireless networks. In: Discrete algorithms and methods for mobile computing and communications
4. Chatzigiannakis I, Dimitriou T, Nikoletseas S, Spirakis P (2004) A probabilistic forwarding protocol for efficient data propagation in sensor networks. In: European wireless conference on mobility and wireless systems beyond 3G (EW2004), Barcelona, pp 344–350
5. Chatzigiannakis I, Mylonas G, Nikoletseas S (2006) Modeling and evaluation of the effect of obstacles on the performance of wireless sensor networks. In: 39th ACM/IEEE simulation symposium (ANSS), Los Alamitos. IEEE Computer Society, pp 50–60
6. Chatzigiannakis I, Nikoletseas S, Spirakis P (2005) Smart dust protocols for local detection and propagation. J Mob Netw (MONET) 10:621–635
7. Karl H, Willig A (2005) Protocols and architectures for wireless sensor networks. Wiley, West Sussex
8. Karp B, Kung HT (2000) GPSR: greedy perimeter stateless routing for wireless networks. In: Mobile computing and networking. ACM, New York
9. Kim YJ, Govindan R, Karp B, Shenker S (2006) Lazy cross-link removal for geographic routing. In: Embedded networked sensor systems. ACM, New York
10. Kuhn F, Wattenhofer R, Zhang Y, Zollinger A (2003) Geometric ad-hoc routing: of theory and practice. In: Principles of distributed computing. ACM, New York
11. Lee S, Bhattacharjee B, Banerjee S (2005) Efficient geographic routing in multihop wireless networks. In: MobiHoc'05: proceedings of the 6th ACM international symposium on mobile ad hoc networking and computing. ACM, New York, pp 230–241
12. Powell O, Nikolesteas S (2007) Simple and efficient geographic routing around obstacles for wireless sensor networks. In: WEA 6th workshop on experimental algorithms, Rome. Springer, Berlin
13. Stojmenovic I, Lin X (2001) Loop-free hybrid single-path/flooding routing algorithms with guaranteed delivery for wireless networks. IEEE Trans Parallel Distrib Syst 12:1023–1032
14. Takagi H, Kleinrock L (1984) Optimal transmission ranges for randomly distributed packet radio terminals. IEEE Trans Commun [legacy, pre-1988] 32:46–257

O

Online Interval Coloring

Leah Epstein
Department of Mathematics, University of Haifa, Haifa, Israel

Keywords

Interval graphs; Online algorithms; Vertex coloring

Years and Authors of Summarized Original Work

1981; Kierstead, Trotter

Problem Definition

Online interval coloring is a graph coloring problem. In such problems the vertices of a graph are presented one by one. Each vertex is presented in turn, along with a list of its edges in the graph, which are incident to previously presented vertices. The goal is to assign colors (which without loss of generality are assumed to be nonnegative integers) to the vertices, so that two vertices which share an edge receive different colors and the total number of colors used (or alternatively, the largest index of any color that is used) is minimized. The smallest number of colors, for which the graph still admits a valid coloring, is called the chromatic number of the graph.

The interval coloring problem is defined as follows. Intervals on the real line are presented one by one, and the online algorithm must assign each interval a color before the next interval arrives, so that no two intersecting intervals receive the same color. The goal is again to minimize the number of colors used to color any interval. The last problem is equivalent to coloring of interval graphs. These are graphs which have a representation (or realization) where each interval represents a vertex and two vertices share an edge if and only if they intersect. It is assumed that the interval graph arrives online together with its realization.

Given an interval graph, denote the size of the largest cardinality clique (complete subgraph) in it by ω. Interval graphs have the special property that in a realization, the set of vertices in a clique have a common point in which they all intersect.

Before discussing the online problem, some properties of interval graphs need to be stated. There exists a simple offline algorithm which produces an optimal coloring of interval graphs. An algorithm applies First Fit, if each time it needs to assign a color to an interval, it assigns a smallest index color which still produces a valid coloring. The optimal algorithm simply considers intervals sorted from left to right by their left end points and applies First Fit. Note that the resulting coloring never uses more than ω colors. Indeed, interval graphs are perfect. A graph G is perfect if any induced subgraph of G, G' (including G), can be colored using $\omega(G')$ colors, where $\omega(G')$ is the size of the largest cardinality clique in G'. (For any graph, ω is a clear lower bound on its chromatic number.)

However, once intervals arrive in an arbitrary order, it is impossible to design an optimal coloring. Consider a simple example where the two intervals $[1, 3]$ and $[6, 8]$ are introduced. If they are colored using two distinct colors, this is already suboptimal, since using the same color for both of them is possible. However, if the sequence of intervals is augmented with $[2, 5]$ and $[4, 7]$, these two new intervals cannot receive the color of the previous intervals or the same color for both new intervals. Thus, three colors are used, even though a valid coloring using two colors can be designed. Note that even if it is known in advance that the input can be colored using exactly two colors, not knowing whether the additional intervals are as defined above, or alternatively, a single interval $[2, 7]$ arrives instead, leads to the usage of three colors instead of only two.

Online coloring is typically hard, which already applies to some simple graph classes such as trees. This is due to the lower bound of $\Omega(\log n)$ (where n is the number of vertices), given by Gyárfás and Lehel [9] on the competitive ratio of online coloring of trees. There are very few classes for which constant bounds are known. One such class is line graphs, for which Bar-Noy, Motwani, and Naor [3] showed that First Fit is 2-competitive (specifically it uses at most $2 \cdot \text{OPT} - 1$ colors, where OPT is the number of colors in an optimal coloring), and this is the best possible bound. This result was later generalized to $k \cdot \text{OPT} - k + 1$ for $(k + 1)$-claw-free graphs by [8] (note that line graphs are 3-claw-free).

Key Results

The paper of Kierstead and Trotter [11] provides a solution to the online interval coloring problem.

They show that the best possible competitive ratio is 3 which is achieved by an algorithm they design. More accurately, the following theorem is proved in the paper.

Theorem 1 *Given an interval graph which is introduced online and presented via its realization, any online algorithm uses at least $3\omega - 2$ colors to color the graph, and there exists an algorithm which achieves this bound.*

The algorithm does not need to know ω in advance. Moreover, even though the algorithm is deterministic, it was shown in [13] that the lower bound of 3 on the competitive ratio of online algorithms for interval coloring holds for randomized algorithms as well. Thus, [11] gives a complete solution for the problem.

The main idea of the algorithm is creation of "levels." At the time of arrival of an interval, it is classified into a level as follows. Denote by A_k the union of sets of intervals which currently belong to all levels $1, \ldots, k$. Intervals are classified so that the largest cardinality clique in A_k is of size k. Thus, A_1 is simply a set of non-intersecting intervals. On arrival of an interval, the algorithm finds the smallest k such that the new interval can join level k, without violating the rule above. It can be shown that each level can be colored using two colors by an offline algorithm. Since the algorithm defined here is online, such a coloring cannot be found in general (see example above). However it is shown in [11] that at most three colors are required for each such level, and a coloring using three colors can be found by applying First Fit on each level (with disjoint sets of colors). Moreover, the first level can always be colored using a single color, and ω is equal exactly to the number of levels. Thus a total number of colors, which is at most $3(\omega - 1) + 1 = 3\omega - 2$, is used.

Applications

In this section, both real-world applications of the problem and applications of the methods of Kierstead and Trotter [11] to related problems are discussed.

Many applications arise in various communication networks. The need for connectivity all over the world is rapidly increasing. On the other hand, networks are still composed of very expensive parts. Thus application of optimization algorithms is required in order to save costs.

Consider a network with a line topology that consists of links. Each connection request is for a path between two nodes in the network. The set of requests assigned to a channel must consist of disjoint paths. The goal is to minimize the number of channels (colors) used. A connection request from a to b corresponds to an interval $[a, b]$, and the goal is to minimize the number of required channels to serve all requests.

Another network-related application is that if the requests have constant duration c and all requests have to be served as fast as possible. In this case the colors correspond to time slots, and the total number of colors corresponds to the schedule length. The problem can be described as a scheduling problem as well, and it is clearly of theoretical interest being a natural online graph coloring problem. Two later studies are of possible interest here, both due to their relevance to the original problem and for the usage of related methods.

The applications in networks stated above raise a generalized problem studied in the recent years. In these applications, it is assumed that once a connection request between two points is satisfied, the channel is blocked at least for the duration of this request. An interesting question that was raised by Adamy and Erlebach [1] is the following. Assume that a request consists not only of a requested interval but also from a bandwidth requirement. That is, a customer of a communication channel specifies exactly how much of the channel is needed. Thus, in some cases it is possible to have overlapping requests sharing the same channel. It is required that at every point, the sum of all bandwidth requirements of requests sharing a color cannot exceed the value 1, which is the capacity of the channel. This problem is called *online interval coloring with bandwidth*. In the paper [1], a (large) constant competitive algorithm was designed for the problem. The original interval coloring problem is a special case of this problem where all bandwidth requests are 1. Note that this problem is a generalization of *bin packing* as well, since bin packing is the special case of the problem where all requests have a common point. Azar et al. [2] designed an algorithm of competitive ratio of at most 10 for this problem. This was done by partitioning the requests into four classes based on their bandwidth requirements and coloring each such class separately. The class of requests with bandwidth in $(\frac{1}{2}, 1]$ was colored using the basic algorithm of [11], since no two such requests colored with one color can overlap. The two other classes, which are $(0, \frac{1}{4}]$ and $(\frac{1}{4}, \frac{1}{2}]$, were colored using adaptations of the algorithm of [11]. Epstein and Levy [6,7] designed improved lower bounds on the competitive ratio, showing that online interval coloring with bandwidth is harder than online interval coloring.

Another problem related to coloring is the *max coloring* problem [5, 14, 15]. In this problem each interval is given a nonnegative weight. Given a coloring, the weight of a color is the maximum weight of any vertex of this color. The goal is to minimize the sum of weights of the used colors. Note that if all weights are 1, max coloring reduces to the graph coloring problem. Several papers [5, 15], starting with that of Pemmaraju, Raman, and Varadarajan [15], designed algorithms for max coloring that are based on the algorithm of [11] (sometimes as a black box).

Open Problems

Since the paper [11] provided a nice and clean solution to the online interval coloring problem, it does not directly raise open problems. Yet, one related problem is of interest to researchers over the last 30 years, which is the performance of First Fit on this problem. It was shown by Kierstead [10] that First Fit uses at most 40ω colors, thus implying that First Fit has a constant competitive ratio. The quest after the exact competitive ratio was never completed. The best current published results are an upper bound of 10ω by [15] and a lower bound of 4.4ω by Chrobak and Slusarek [4]. See [16] for recent developments. It particular, it is mentioned that a lower bound of $4.99999\omega - C$ (for a fixed $C > 0$) was proved by Kierstead and Trotter in 2004, later improved to a lower bound of $(5-\varepsilon)\omega$ for any $\varepsilon > 0$, implying a lower bound of 5 on the competitive ratio of First Fit [12]. It is interesting to note that for online interval coloring with bandwidth, First Fit has an unbounded competitive ratio [1].

Cross-References

▶ Graph Coloring

Recommended Reading

1. Adamy U, Erlebach T (2003) Online coloring of intervals with bandwidth. In: Proceedings of the first international workshop on approximation and online algorithms (WAOA2003), Budapest, Hungary, pp 1–12
2. Azar Y, Fiat A, Levy M, Narayanaswamy NS (2006) An improved algorithm for online coloring of intervals with bandwidth. Theor Comput Sci 363(1):18–27
3. Bar-Noy A, Motwani R, Naor J (1992) The greedy algorithm is optimal for on-line edge coloring. Inf Process Lett 44(5):251–253
4. Chrobak M, Ślusarek M (1988) On some packing problems relating to dynamical storage allocation. RAIRO J Inf Theory Appl 22:487–499
5. Epstein L, Levin A (2012) On the max coloring problem. Theor Comput Sci 462(1):23–38
6. Epstein L, Levy M (2005) Online interval coloring and variants. In: Proceedings of the 32nd international colloquium on automata, languages and programming (ICALP2005), Lisbon, Portugal, pp 602–613
7. Epstein L, Levy M (2008) Online interval coloring with packing constraints. Theor Comput Sci 407(1–3):203–212
8. Epstein L, Levin A, Woeginger GJ (2011) Graph coloring with rejection. J Comput Syst Sci 77(2):439–447
9. Gyárfás A, Lehel J (1991) Effective on-line coloring of P_5-free graphs. Combinatorica 11(2):181–184
10. Kierstead HA (1988) The linearity of first-fit coloring of interval graphs. SIAM J Discret Math 1(4):526–530
11. Kierstead HA, Trotter WT (1981) An extremal problem in recursive combinatorics. Congr Numer 33:143–153
12. Kierstead HA, Smith DA, Trotter WT (2013) Manuscript. https://math.la.asu.edu/~halk/Publications/biwall6.pdf
13. Leonardi S, Vitaletti A (1998) Randomized lower bounds for online path coloring. In: Proceedings of

the second international workshop on randomization and approximation techniques in computer science (RANDOM'98), Barcelona, Spain, pp 232–247
14. Nonner T (2011) Clique clustering yields a PTAS for max-coloring interval graphs. In: Proceedings of the 38th international colloquium on automata, languages and programming (ICALP2011), Zurich, Switzerland, pp 183–194
15. Pemmaraju S, Raman R, Varadarajan KS (2011) Max-coloring and online coloring with bandwidths on interval graphs. ACM Trans Algorithms 7(3):35
16. Trotter WT (2014) Current research problems: first fit colorings of interval graphs. http://people.math.gatech.edu/~trotter/rprob.html

Online Learning and Optimization

Yaoliang Yu
Machine Learning Department, Carnegie Mellon University, Pittsburgh, PA, USA

Keywords

Convex programming; Hannan consistency; Online learning; Regret; Subgradient descent

Years and Authors of Summarized Original Work

2003; Zinkevich

Problem Definition

Suppose we are going to invest in a stock market. Our neighbor, for mysterious reasons, happens to know how the market evolves. But he cannot change his portfolio (proportions of holding stocks) once committed (to avoid being caught by regulators, say). On the other hand, we, the normal investor, do not have any inside information but can sell and buy at will. If we and our prescient neighbor invest the same amount of money, is there a (computationally feasible) way for us to perform comparably well to our neighbor, without knowing his investing strategy? Surprisingly (as contrary to our real-life experience perhaps), the answer is yes, and we will see it through the lens of online learning. Disclaimer: The reader is at his own risk if he decides to practice the beautiful theoretical results we describe below.

The online learning problem is best described as a multi-round two-person game between the "learner" and the "environment," following the protocol:

The Online Learning Protocol

For $t = 1, \ldots, T$
Learner predicts $x_t \in D$;
Environment responds with a cost function $f_t : D \to \mathbb{R}$;
Learner suffers an immediate cost $f_t(x_t)$;
Learner learns some information of f_t.

Through the multi-round interactions with the environment, the learner tries to learn the behavior of the environment so as to minimize its cumulative cost in the time horizon $t \in [1, T]$, where we could allow the game to continue indefinitely, i.e., $T = \infty$.

The online learning framework is particularly relevant in real applications where (1) *sequential* decisions are needed, (2) *average* good performance is desired, and (3) the process is too *complicated* to be modeled statistically. In our stock example above, the learner will be us (normal stock holder), and the environment will be the market. Each day we submit our portfolio x_t, carefully constructed based on the past information and perhaps also mingled with some randomness (coin tosses for luck). The market responds with rises and falls of the stock prices, represented as the cost function f_t. We suffer the loss $f_t(x_t)$ and learn something about the market (e.g., f_t), and the life moves on to the next day. (If it feels more comfortable, one can negate f and call it *gain*. We shall not do this, because "a true warrior faces her bleak life bravely.") Our adventure ends at day T, which is prefixed. (For $T = \infty$, the adventure never ends.) Of course, the goal is to earn *on average* as much money as possible; it is OK if we lose occasionally. Also, for an average person (us), it is perhaps too complicated to have a clear idea what is exactly going on in that stock market. As mentioned, we would like to compete against our "prescient neighbor." This is formalized as the regret below.

To evaluate the performance of the learner, the following notion of *regret* plays a central role:

$$\mathsf{R}_T(x) := \sum_{t=1}^{T} \Big(f_t(x_t) - f_t(x)\Big), \ \forall x \in C \subseteq D. \tag{1}$$

Intuitively, the learner compares itself with the baseline (e.g., the "prescient" neighbor) that *constantly* predicts $x \in C$ in each round. We are interested in bounding the learner's regret with respect to the "best" competitor in the set C (although our notation drops the dependence on C):

$$\mathsf{R}_T := \sup_{x \in C} \mathsf{E}(\mathsf{R}_T(x)), \tag{2}$$

where the expectation $\mathsf{E}(\cdot)$ is taken with respect to any internal randomization the learner or the environment might use. The learner is said to be (Hannan) consistent if

$$\frac{\mathsf{R}_T}{T} \to 0, \text{ as } T \to \infty, \text{ i.e., } \mathsf{R}_T = o(T).$$

In other words, the learner performs, on average, as well as the best constant competitor in the long run.

We adopted the notion of regret *not* because we believe a constant (unchanging) predictor is the best strategy for our problem. Instead, the regret should be interpreted as a bare minimum requirement: If there does exist a constant predictor that performs reasonably well on our task, it would be unacceptable if our algorithm is not even on par with it. More often than not, we would like to do *better* than any constant predictor, but this can be highly nontrivial (either computationally or statistically).

We have allowed the learner to operate on a larger set D than its "competitors" (which are restricted to C). Of course this buys the learner some advantage, which sometimes is necessary for consistency, particularly when C is a nonconvex set. For instance, consider the game where the sets $C = D = \{0, 1\}$ and the cost functions

$$f_t(x) = \begin{cases} 1, & \text{if } x = x_t \\ 0, & \text{otherwise} \end{cases}. \tag{3}$$

Recall that in our online learning protocol, we have no control on how the environment reacts. In the very worst case, the environment may appear to be completely "hostile." For instance, the cost function f_t in (3) is thus defined to make the learner always suffer unit cost in each round. On the other hand, the best constant competitor in C suffers cost at most $T/2$ in T rounds. Hence, $\frac{\mathsf{R}_T}{T} \geq \frac{1}{2}$ for all T, meaning that any learner that follows our protocol cannot be consistent. The lesson is, of course, that we cannot compete under a very adversarial environment. However, if we allow the learner to *randomize* its decisions and correspondingly pay *expected* cost, then it is again possible to devise consistent learners for this game [8], provided that the environment is oblivious, i.e., it does not adapt to the learner's randomization, thus constraining its "hostility." Intuitively, randomization and averaging *smooth* out the possible worst-case (but oblivious) reactions of the environment. This is also equivalent to allowing the learner to operate on $D = [0, 1]$, the convex hull of $C = \{0, 1\}$. Indeed, for binary x we can interpret the cost function in (3) as $f_t(x) = |x - y_t|$, where in the worse case the environment could happen to choose $y_t = 1 - x_t$ from the set C. In the randomized setting, the learner first picks $x \in D$, the convex hull of C, and then chooses 1 with probability x and 0 otherwise. Provided that the environment still chooses (however adversarial) $y_t \in C$, the *expected* cost the learner suffers is again $f_t(x) = |x - y_t|$, but this time extended to the convex domain D. The claim that there exists a consistent learner under this randomized setting follows from Theorem 2 below. Intuitively, now the learner sits in the middle ($x = 1/2$) and leans toward the better constant predictor fast enough.

The previous example shows that consistency may not always be achievable. Consequently, the interesting questions in online learning include (but are not limited to) the following:

- Identifying settings under which consistency can be achieved
- Determining the correct order of the regret tending to infinity
- Devising computationally efficient and order optimal learners

These questions heavily depend on what the learner can learn in each round. For instance, in the full information setting, the learner observes the entire cost function f_t; in the bandit setting, the learner only observes its incurred cost $f_t(x_t)$, while in the partial monitoring setting, the learner only observes some quantity related to its cost. The geometry of the decision set D and the competitor set C, as well as the structural property (such as convexity, smoothness, etc.) of the cost functions, also play a significant role. In the next section, we will consider a special case where a particularly simple algorithm known as online gradient descent suffices to achieve the optimal regret. For more complete and thorough discussions, please refer to the excellent book [3] and surveys [2, 8].

Online Convex Programming

We further simplify our online learning protocol as follows:

Online Convex Programming (on the real line)

- $D \subseteq \mathbb{R}$ is a closed convex set, with $r = \max_{x,y \in D} |x - y| < \infty$;
- $\forall t \leq T$, f_t is convex and differentiable on some open set containing D;
- The gradient is uniformly bounded: $\sup_{x \in D, t \leq T} |\nabla f_t(x)| \leq M < \infty$;
- The learner gets to observe $\nabla f_t(x_t)$ in round t.

The third condition is satisfied if each f_t is M-Lipschitz continuous, i.e.,

$$\forall x, y \in D,\ |f_t(x) - f_t(y)| \leq M \cdot |x - y|, \tag{4}$$

while the last condition is certainly met if the cost function f_t is revealed to the learner in each round. Under this setting, Zinkevich [9] first analyzed the online learner that simply follows the (projected) gradient update:

$$\forall t \geq 1,\ x_{t+1} = \mathsf{P}_D(x_t - \eta_t \nabla f_t(x_t)), \tag{5}$$

where $\eta_t \geq 0$ is a small step size that we determine later and

$$\mathsf{P}_D(x) = \operatorname*{argmin}_{y \in D} |x - y|, \tag{6}$$

is the (Euclidean) projection of x onto the closed set D, i.e., the closest point in D to x. The projection is needed since the learner's prediction x_{t+1} is restricted to the decision set D.

Before we analyze the regret of the above online gradient algorithm, let us first observe that

$$\begin{aligned} \mathsf{R}_T(x) &= \sum_{t=1}^{T} \Big(f_t(x_t) - f_t(x) \Big) \\ &\leq \sum_{t=1}^{T} \Big(\nabla f_t(x_t) \cdot (x_t - x) \Big) \\ &\leq M \sum_{t=1}^{T} |x_t - x|, \end{aligned} \tag{7}$$

where the first inequality follows from the convexity of f_t. Interestingly, the right-hand side is the worst-case regret for the special case where each f_t is a linear function, say, $w_t x_t$ for some $|w_t| \leq M$. In other words, we could have restricted the game to linear cost functions, instead of the seemingly more general convex functions.

The regret of an online learner can be bounded by analyzing its progress with respect to some *potential* function. Here we choose the familiar quadratic potential. Note that for any $x \in C \subseteq D$, clearly $\mathsf{P}_D(x) = x$; hence,

$$|x_{t+1} - x|^2 = |\mathsf{P}_D(x_t - \eta_t \nabla f_t(x_t)) - \mathsf{P}_D(x)|^2$$
$$\le |x_t - \eta_t \nabla f_t(x_t) - x|^2$$
$$\le |x_t - x|^2 - 2\eta_t \nabla f_t(x_t) \cdot (x_t - x) + \eta_t^2 M^2$$
$$\le |x_t - x|^2 - 2\eta_t (f_t(x_t) - f_t(x)) + \eta_t^2 M^2, \tag{8}$$

where the first inequality follows from the 1-Lipschitz continuity of the projection $\mathsf{P}_D(\cdot)$ and the last inequality is due to the convexity of f_t. Dividing (8) by $2\eta_t$, summing the indices from $t = 1$ to $t = T$, and rearranging, we have

$$\mathsf{R}_T(x) = \sum_{t=1}^{T} \left(f_t(x_t) - f_t(x) \right)$$
$$\le \sum_{t=1}^{T} \frac{1}{2\eta_t} (|x_t - x|^2 - |x_{t+1} - x|^2) + M^2 \sum_{t=1}^{T} \frac{\eta_t}{2} \tag{9}$$
$$\le \frac{1}{2\eta_1} |x_1 - x|^2 + \sum_{t=2}^{T} \left(\frac{1}{2\eta_t} - \frac{1}{2\eta_{t-1}} \right) |x_t - x|^2 + M^2 \sum_{t=1}^{T} \frac{\eta_t}{2}. \tag{10}$$

Setting the step size η_t properly leads to our key results, summarized in the next section.

Key Results

If the horizon T is finite and known in advance, then we can use a constant step size $\eta_t \equiv \eta$. Optimizing with respect to $\eta \ge 0$ from (10) yields

Theorem 1 (e.g., [8,9]) *Let $\eta_t \equiv \eta = \frac{c}{M\sqrt{T}}$ for some constant $c > 0$; then the online gradient learner achieves sublinear regret*

$$\mathsf{R}_T(x) \le M\sqrt{T} \frac{c^2 + |x - x_1|^2}{2c}$$
$$\le M\sqrt{T} \frac{c^2 + r^2}{2c} \tag{11}$$

for the online convex programming problem.

If the horizon is not know in advance, inspired by the step size in Theorem 1, we can try setting $\eta_t = \frac{c}{M\sqrt{t}}$. Note that η_t is decreasing with respect to t. Continuing from (10):

$$\mathsf{R}_T(x) \le r^2 \left(\frac{1}{2\eta_1} + \sum_{t=2}^{T} \left(\frac{1}{2\eta_t} - \frac{1}{2\eta_{t-1}} \right) \right) + cM \sum_{t=1}^{T} \frac{1}{2\sqrt{t}}.$$

Using integration, $\sum_{t=1}^{T} \frac{1}{2\sqrt{t}} \le \int_0^T \frac{1}{2\sqrt{t}} \mathrm{d}t \le \sqrt{T}$. Thus, we have proved

Theorem 2 (Zinkevich [9]) *Let $\eta_t = \frac{c}{M\sqrt{t}}$ for some constant $c > 0$; then the online gradient learner achieves sublinear regret (simultaneously for all T)*

$$\mathsf{R}_T(x) \le M\sqrt{T} \frac{2c^2 + r^2}{2c} \tag{12}$$

for the online convex programming problem.

Comparing to Theorem 1, we only lose a constant 2 in Theorem 2, but the result now holds simultaneously for all T – a property sometimes called *anytime*. Theorems 1 and 2 not only imply the consistency of the online gradient learner but also demonstrate that $\mathsf{R}_T = O(\sqrt{T})$, since the right-hand sides of (11) and (12) are independent of the competitor x. In fact, this rate is optimal, i.e., there exists an instantiation where no learner (efficient or not) can do better; see, e.g., [6]. Thanks to the convexity assumption on f_t (and the decision set D), the online gradient algorithm can be efficiently implemented if the gradient ∇f_t and the projection $\mathsf{P}_D(\cdot)$ can be efficiently computed.

Doubling Trick When the horizon T is not known in advance, we can also use the doubling trick, which divides the time into exponentially increasing phases

$$\bigcup_{i=1}^{\lceil \log_2(T+1) \rceil} \{2^{i-1}, \ldots, 2^i - 1\},$$

and on the ith phase, we use the constant step size $\eta_i = O(1/\sqrt{2^{i-1}})$ suggested in Theorem 1. The overall regret is bounded by

$$\sum_{i=1}^{\lceil \log_2(T+1) \rceil} O(\sqrt{2^{i-1}}) = \frac{\sqrt{2}}{\sqrt{2}-1} O(\sqrt{T+1}).$$

So asymptotically we only lose a factor of $\frac{\sqrt{2}}{\sqrt{2}-1} \approx 3.41$.

Other Rates It is possible to tighten the regret rate if the cost functions are more "regular." Intuitively, this means the environment is more constrained hence can only be less adversarial. Indeed, if $f_t - \frac{\sigma}{2}|\cdot|^2$ is convex, namely, f_t is σ-strongly convex, Hazan et al. [6] showed that the online gradient learner equipped with a smaller step size $\eta_t \propto \frac{1}{\sigma t}$ suffers only logarithmic regret $O(\log(T))$ – an exponential improvement compared to Theorem 2. Just like the time horizon, it is possible to achieve the same logarithmic regret without knowing the parameter σ; see [1]. Similarly, if f_t is so-called exponentially concave, a similar logarithmic regret can be achieved using a second-order Newton-type learner [6].

Extension to High Dimensions The above analysis easily extends to high dimensions. In fact, Theorems 1 and 2 hold in any abstract Hilbert space, with virtually the same proof (provided that we replace the absolute value with the Hilbert norm). The cost functions f_t need not be differentiable either; picking an arbitrary subgradient in the subdifferential $\partial f_t(x_t)$ would suffice.

Extension to Composite Functions The regret can be extended to include a penalty function g as follows:

$$\mathsf{R}_T(x) = \sum_{t=1}^{T} (f_t(x_t) + g(x_t) - f_t(x) - g(x)). \quad (13)$$

Our previous definition in (1) corresponds to the setting where $g(x) = 0$ iff $x \in D$ (otherwise the regret is set to ∞). We could simply treat $f_t + g$ as a whole and apply the online gradient algorithm without any modification. A different approach, resulting in a similar regret bound, upgrades the projection to the proximity operator (of g):

$$\mathsf{P}_g^\eta(x) = \operatorname*{argmin}_y \frac{1}{2\eta}|x-y|^2 + g(y), \quad (14)$$

where $\eta > 0$ is the step size to be chosen appropriately. The latter approach is not only more general but also leads to more *structured* intermediate predictions [4]. For instance, if $g(x) = \sum_i |x_i|$ is the ℓ_1 norm, then $[\mathsf{P}_g^\eta(x)]_i = \operatorname{sign}(x_i) \cdot \max\{|x_i| - \eta, 0\}$, which would be exactly zero if $|x_i|$ is small and η is large. In contrast, if we apply online gradient descent directly to $f_t + g$, we would almost never get sparse intermediate predictions.

Without Projections The online gradient learner is computationally efficient only when the projection $\mathsf{P}_D(\cdot)$ in (6) (or more generally the proximity operator in (14)) can be efficiently implemented. In some applications, this is unfortunately not the case. Instead, Hazan and Kale [5] proposed a different learner that bypasses the projection step. Basically, the learner iteratively finds the vertexes of the decision set D and then takes suitable convex combinations of them to make progress.

Connection to Stochastic Optimization The regret bound in Theorem 2 is closely related to some results in stochastic optimization, for the following problem [7]:

$$\inf_{x \in D} f(x), \text{ where } f(x) := \mathsf{E}_\xi(F(x,\xi)), \quad (15)$$

and ξ is some random variable. The stochastic (sub)gradient method is a popular iterative algorithm for optimizing (15). In each iteration, it randomly draws an independent sample ξ_t and follows the projected (sub)gradient update:

$$x_{t+1} = \mathsf{P}_D(x_t - \eta_t \nabla_x F(x_t, \xi_t)),$$

for some small step size $\eta_t \geq 0$. The similarity to the online gradient learner is apparent once we identify $f_t(x) := F(x, \xi_t)$. Thus, the regret bound in Theorem 2 implies

$$\begin{aligned} \mathrm{O}\left(\frac{1}{\sqrt{T}}\right) &= \sup_{x\in D} \frac{1}{T}\mathsf{E}\Big[\sum_{t=1}^{T}\left(f_t(x_t) - f_t(x)\right)\Big] \\ &= \sup_{x\in D} \frac{1}{T}\mathsf{E}\Big[\sum_{t=1}^{T}\left(F(x_t,\xi_t) - F(x,\xi_t)\right)\Big] \\ &= \mathsf{E}\left(\frac{1}{T}\sum_{t=1}^{T} f(x_t)\right) - \inf_{x\in D} f(x) \\ &\geq \mathsf{E}\left(f\left(\frac{1}{T}\sum_{t=1}^{T} x_t\right)\right) - \inf_{x\in D} f(x), \end{aligned}$$

provided that the random sample ξ_t is independent of x_t and $F(\cdot, \xi)$ is convex for (almost) every realization of ξ. In other words, the ergodic mean $\frac{1}{T}\sum_{t=1}^{T} x_t$ approaches, in expectation, the infimum in (15) at the rate $\mathrm{O}(1/\sqrt{T})$.

Cross-References

▶ Multi-armed Bandit Problem

Recommended Reading

1. Bartlett PL, Hazan E, Rakhlin A (2007) Adaptive online gradient descent. In: Platt JC, Koller D, Singer Y, Roweis ST (eds) Advances in neural information processing systems 20 (NIPS). Curran Associates, Inc., Vancouver, pp 257–269
2. Bubeck S, Cesa-Bianchi N (2012) Regret analysis of stochastic and nonstochastic multi-armed bandit problems. Found Trends Mach Learn 5(1):1–122
3. Cesa-Bianchi N, Lugosi G (2006) Prediction, learning, and games. Cambridge University Press, New York
4. Duchi JC, Shalev-Shwartz S, Singer Y, Tewari A (2010) Composite objective mirror descent. In: Kalai AT, Mohri M (eds) The 23rd conference on learning theory (COLT). Haifa, pp 14–26
5. Hazan E, Kale S (2012) Projection-free online learning. In: Langford J, Pineau J (eds) The 29th international conference on machine learning (ICML), Edinburgh. Omnipress, pp 521–528
6. Hazan E, Agarwal A, Kale S (2007) Logarithmic regret algorithms for online convex optimization. Mach Learn 69:169–192
7. Nemirovski A, Juditsky A, Lan G, Shapiro A (2009) Robust stochastic approximation approach to stochastic programming. SIAM J Optim 19(4):1574–1609
8. Shalev-Shwartz S (2011) Online learning and online convex optimization. Found Trends Mach Learn 4(2):107–194
9. Zinkevich M (2003) Online convex programming and generalized infinitesimal gradient approach. In: Fawcett T, Mishra N (eds) The 20th international conference on machine learning (ICML), Washington. AAAI Press, pp 928–936

Online List Update

Shahin Kamali
David R. Cheriton School of Computer Science, University of Waterloo, Waterloo, ON, Canada

Keywords

Competitive analysis; Data compression; Online computation; Self-adjusting lists

Years and Authors of Summarized Original Work

1985; Sleator, Tarjan

Problem Definition

List update is one of the classic problems in the context of online computation. The main motivation for the study of the problem is self-adjusting lists. Consider a linear list which represents a dictionary abstract data type. There are three elementary operations in the dictionary, namely, insertion, deletion, and lookup (search). To perform these operations on an item x, an algorithm needs to search for x, i.e., examine the list items, one by one, to find x. For the case of an insertion, all items should be sequentially checked to ensure that the inserted item is not already in the list. A deletion also requires finding the item that is being deleted. In this manner, all operations can be translated into a sequence of lookups or *accesses* to the items in the list. To access an item at index i, an algorithm examines

i items and therefore incurs an access cost of i. Immediately after the access, the algorithm can move the accessed item to any position closer to the front of the list at no extra cost; this is called a *free exchange*. It is also possible to exchange any two consecutive items at a cost of 1 through a *paid exchange*. The objective is to organize the list, using free and paid exchanges, so that the total cost (for accesses and paid exchanges) is minimized.

The list update is naturally an online problem, i.e., at the time of accessing an item, it is not clear what items will be requested in the future. An online algorithm has to take its decision without any knowledge about the forthcoming requests. For example, Move-To-Front (MTF) is a well-known list update algorithm which moves an accessed item to front using a free exchange. In taking its decision, MTF does not rely on any information about future requests. Among other classic list update algorithms, we might mention Transpose (TRANS) and Frequency Count (FC). After accessing an item, TRANS moves it one step closer to the front, i.e., it exchanges the position of x with its preceding item. FC maintains the list in a way that more frequent items appear closer to the front. In doing so, it maintains a counter for each item x which indicates the number of previous requests to x.

Competitive analysis is the standard method for the study and classification of list update algorithms. An algorithm is said to be c-competitive if the cost of serving any request sequence never exceeds c times the optimal cost of an *offline* algorithm OPT which knows the entire sequence in advance. More precisely, an algorithm $\mathcal{A}$ is c-competitive if $\mathcal{A}(\sigma) \leq c\,\text{OPT}(\sigma) + b$ for any sequence σ. Here, $\mathcal{A}(\sigma)$ and $\text{OPT}(\sigma)$ respectively denote the costs of $\mathcal{A}$ and OPT for serving σ, and c and b are constants.

Key Results

List update algorithms were initially studied in regard to their typical behavior on sequences that follow probability distributions. The average cost ratio of an algorithm $\mathcal{A}$ is the ratio between the expected cost of $\mathcal{A}$ for a random sequence and the cost of an optimal offline algorithm which arranges items in nonincreasing order by probability. Under this setting, the ratio achieved by FC is 1 [14], while that of MTF is $\pi/2$ [7]. Moreover, there are distributions in which TIMESTAMP has a better ratio than MTF [14]. These results indicate that FC and TIMESTAMP are better than MTF. However, in practice, MTF has an advantage over the other algorithms. This is partially because the input sequences do not necessarily follow a fixed probability distribution.

In their seminal paper, Sleator and Tarjan proved that MTF is 2-competitive, while TRANS and FC do not achieve a constant competitive ratio [15]. At the same time, for sufficiently long lists, no algorithm can achieve a competitive ratio better than 2. For a while, MTF was the only algorithm with optimal competitive ratio until Albers introduced the TIMESTAMP algorithm [1]. After accessing an item x, TIMESTAMP inserts x in front of the first item y that is before x in the list and is requested at most once since the last request for x. If there is no such item y, or if this is the first access to x, TIMESTAMP does not reorganize the list. TIMESTAMP is also 2-competitive [1].

Randomized Algorithms

No randomized list update algorithm can be better than 2-competitive against adaptive adversaries. However, there are randomized algorithms with better competitive ratios against oblivious adversaries. Reingold et al. introduced a randomized algorithm called BIT which assigns a bit to each item x and initially sets it, uniformly and randomly, to be 0 or 1. At the time of an access to an element x, the bit of x is complemented, and if it becomes 1, the algorithm moves x to the front. BIT has a competitive ratio of 1.75 [13]. Albers et al. proposed a hybrid algorithm, called COMB, which randomly selects between TIMESTAMP and BIT strategies [4]. Upon a request to an item, the algorithm applies BIT strategy with probability 0.8 and TIMESTAMP with probability 0.2. COMB has a competitive ratio of 1.6 [4] which is the best among existing algorithms. Teia proved that no randomized algorithm can be better than 1.5-competitive [16].

Locality of Reference

Real-life sequences usually exhibit locality of reference which implies that the currently requested item is more likely to be requested again. One model of locality is *concave analysis* in which the sequences are consistent with a concave function f so that the number of distinct requests in any window of size τ is at most $f(\tau)$. MTF is the unique optimal solution under *bijective analysis* for sequences that have locality of reference with respect to concave analysis [6]. Bijective analysis is an alternative to competitive analysis that directly compares two algorithms based on their worst-case and average-case behavior (see [6] for details).

Inspired by the concave analysis, Dorrigiv et al. [8] defined the *nonlocality* of a sequence σ of length n, denoted by $\hat{\lambda}(\sigma)$, as $\sum_{i=1}^{n} d_i$ in which d_i is the number of distinct items requested since the last request to the ith item in σ. For the first request to an item, d_i is equal to the length l of the list. For any sequence σ, the cost of any online algorithm is at least $\hat{\lambda}(\sigma)$, while MTF has the same cost of $\hat{\lambda}(\sigma)$. The cost of TIMESTAMP is at least $2\hat{\lambda}(\sigma)$, and TRANS and FC both have a cost of at least $l/2 \times \hat{\lambda}(\sigma)$ [8]. These results imply an advantage for MTF when sequences have high locality.

Albers and Lauer defined an alternative locality model which assigns a value $\lambda \in [0, 1]$ for each sequence [2]. The larger values for λ imply a higher locality. Using this notion of locality, the competitive ratio of MTF is at most $\frac{2}{1+\lambda}$, i.e., for sequences with high locality, MTF is 1-competitive. The ratio of TIMESTAMP does not improve on request sequences satisfying λ locality, i.e., it remains 2-competitive. The same holds for algorithm COMB, i.e., it remains 1.6-competitive. However, for the algorithm BIT the competitive ratio improves to $\min\{1.75, \frac{2+\lambda}{1+\lambda}\}$.

Applications

As mentioned earlier, the basic application of the list update is in maintaining self-adjusting lists. Martínez and Roura [11], and also Munro [12], observed that a complete rearrangement of items which precede an item at position i is proportional to i rather than i^2. For example, accessing the item at the end of a list and reversing the list can be done in linear time, while under the standard model, it has a quadratic cost. In the MRM model, after an access to an item at index i, the preceding items can be arranged free of charge. It is known that any online algorithm has a competitive ratio of $\Omega(l/\lg l)$ for a list of length l under the MRM model [11, 12], i.e., under this practical setting of the problem, no online algorithm can be competitive (see [9] for details).

List update is widely used for compression purposes. Consider each character of a text as an item in the list and the text as the input sequence. A compression algorithm writes an arbitrary initial configuration in the compressed file, as well as the access costs of a list update algorithm $\mathcal{A}$ for serving each character. In the decompression phase, the algorithm starts from the same initial configuration and follows the steps of the algorithm by reading the access cost written in the compressed file. In order to enhance the performance of the compression schemes, the Burrows-Wheeler Transform (BWT) can be applied to the input string to increase the locality of the input. The bzip2 compression program which applies MTF after BWT transform outperforms the widely used gzip program by more than 5 % on the standard Canterbury corpus.

To theoretically study the list update problem in the context of compression, it is better to assume the cost of accessing an item at index i is $\Theta(\log i)$ rather than $\Theta(i)$. This is because, when an item is accessed in the ith position, the value of i is written as a binary code rather than unary. Sleator and Tarjan show that MTF is 2-competitive if the access cost is a convex function. On the other hand, some algorithms are competitive when the access cost is linear and noncompetitive when the access cost is $\Theta(\log i)$. However, it is not the case for MTF and it has been shown that MTF is 2-competitive when the access cost is logarithmic [10]. In other words,

MTF is useful for compression, even for the sequences (files) generated by an adversary.

Open Problems

The competitive ratio of the best randomized algorithm lies in the range [1.5, 1.6]. Closing this gap is an important direction for future research. Almost all existing algorithms have the *projective property* which informally means that the relative position of any two items in the lists maintained by these algorithms only depend on the requests to these items. Projective algorithms are analyzed under the partial cost model where the cost of accessing an item at index i is $i - 1$. It is known that no online algorithm with the projective property can achieve a competitive ratio better than 1.6 under the partial cost model [5]. Hence, to introduce an algorithm with competitive ratio better than 1.6 of BIT, one needs to deviate from the projective property.

Reingold et al. introduced another model, called *d-paid exchange model*, in which the cost of paid exchanges is scaled up by a value $d \geq 1$, while free exchanges are not allowed [13]. Under this model, no deterministic algorithm can be better than 3-competitive [13], while the best existing algorithm is 4.56-competitive (reported in [3]). For the particular case of $d = 1$, the best lower and upper bound are, respectively, 3 and 4 (MTF is 4-competitive). Closing these gaps is another direction for future research.

Cross-References

- ► Alternative Performance Measures in Online Algorithms
- ► Burrows-Wheeler Transform
- ► Online Paging and Caching

Recommended Reading

1. Albers S (1998) Improved randomized on-line algorithms for the list update problem. SIAM J Comput 27:682–693
2. Albers S, Lauer S (2008) On list update with locality of reference. In: Proceedings of the 35th international colloquium on automata, languages, and programming (ICALP), Reykjavik. Lecture notes in computer science, vol 5125. Springer, pp 96–107
3. Albers S, Westbrook J (1996) Self-organizing data structures. In: Online algorithms: the state of the art, Dagstuhl. Lecture notes in computer science, vol 1442. Springer, pp 13–51
4. Albers S, von Stengel B, Werchner R (1995) A combined BIT and TIMESTAMP algorithm for the list update problem. Inf Process Lett 56(3):135–139
5. Ambühl C, Gärtner B, von Stengel B (2013) Optimal lower bounds for projective list update algorithms. ACM Trans Algorithms 9(4):31
6. Angelopoulos S, Dorrigiv R, López-Ortiz A (2008) List update with locality of reference. In: Proceedings of the 8th Latin American theoretical informatics symposium (LATIN), Búzios. Lecture notes in computer science, vol 4957. Springer, pp 399–410
7. Chung FRK, Hajela DJ, Seymour PD (1988) Self-organizing sequential search and Hilbert's inequality. J Comput Syst Sci 36(2):148–157
8. Dorrigiv R, Ehmsen MR, López-Ortiz A (2009) Parameterized analysis of paging and list update algorithms. In: Proceedings of the 7th international workshop on approximation and online algorithms (WAOA), Copenhagen. Lecture notes in computer science, vol 5893. Springer, pp 104–115
9. Kamali S, López-Ortiz A (2013) A survey of algorithms and models for list update. In: Space-efficient data structures, streams, and algorithms, Waterloo. Lecture notes in computer science, vol 8066. Springer, pp 251–266
10. Kamali S, López-Ortiz A (2014) Better compression through better list update algorithms. In: Proceedings of the 23rd data compression conference (DCC), Snowbird, pp 372–381
11. Martínez C, Roura S (2000) On the competitiveness of the move-to-front rule. Theor Comput Sci 242(1–2):313–325
12. Munro JI (2000) On the competitiveness of linear search. In: Proceedings of the 8th European symposium on algorithms (ESA), Saarbrücken. Lecture notes in computer science, vol 1879. Springer, pp 338–345
13. Reingold N, Westbrook J, Sleator DD (1994) Randomized competitive algorithms for the list update problem. Algorithmica 11:15–32
14. Rivest R (1976) On self-organizing sequential search heuristics. Commun ACM 19:63–67
15. Sleator D, Tarjan RE (1985) Amortized efficiency of list update and paging rules. Commun ACM 28:202–208
16. Teia B (1993) A lower bound for randomized list update algorithms. Inf Process Lett 47:5–9

Online Load Balancing of Temporary Tasks

Leah Epstein
Department of Mathematics, University of Haifa, Haifa, Israel

Keywords

Online scheduling of temporary tasks

Years and Authors of Summarized Original Work

1994; Azar, Broder, Karlin
1997; Azar, Kalyanasundaram, Plotkin, Pruhs, Waarts

Problem Definition

Load balancing of temporary tasks is an online problem. In this problem, arriving tasks (or jobs) are to be assigned to processors, which are also called machines. In this entry, deterministic online load balancing of temporary tasks with unknown duration is discussed. The input sequence consists of departures and arrivals of tasks. If the sequence consists of arrivals only, the tasks are called permanent. Events happen one by one, so that the next event appears after the algorithm completes dealing with the previous event.

Clearly, the problem with temporary tasks is different from the problem with permanent tasks. One such difference is that for permanent tasks, the maximum load is always achieved in the end of the sequence. For temporary tasks, this is not always the case. Moreover, the maximum load may be achieved at different times for different algorithms.

In the most general model, there are m machines $1, \ldots, m$. The information of an arriving job j is a vector p_j of length m, where p_j^i is the load or size of job j if it is assigned to machine i. As stated above, each job is to be assigned to a machine before the next arrival or departure. The load of a machine i at time t is denoted by L_i^t and is the sum of the loads (on machine i) of jobs which are assigned to machine i that arrived by time t and did not depart by this time. The goal is to minimize the maximum load of any machine over all times t. This machine model is known as *unrelated machines* (see [3] for a study of the load-balancing problem of permanent tasks on unrelated machines). Many more specific models were defined. In the sequel, a few such models are described.

For an algorithm $\mathcal{A}$, denote its cost by $\mathcal{A}$ as well. The cost of an optimal offline algorithm that knows the complete sequence of events in advance is denoted by OPT. Load balancing is studied in terms of the (absolute) competitive ratio. The competitive ratio of $\mathcal{A}$ is the infimum $\mathcal{R}$ such that for any input, $\mathcal{A} \leq \mathcal{R} \cdot \text{OPT}$. If the competitive ratio of an online algorithm is at most $\mathcal{C}$, it is also called $\mathcal{C}$-competitive.

Uniformly related machines [3, 12] are machines with speeds associated with them; thus, machine i has speed s_i, and the information that a job j needs to provide upon its arrival is just its size, or the load that it incurs on a unit speed machine, which is denoted by p_j. Then, let $p_j^i = p_j/s_i$. If all speeds are equal, this results in *identical machines* [13].

Restricted assignment [8] is a model where each job may be run only on a subset of the machines. A job j is associated with running time, which is the time to run it on any of its permitted machines M_j. Thus, if $i \in M_j$, then $p_j^i = p_j$, and otherwise, $p_j^i = \infty$.

Key Results

The known results in all four models are surveyed below.

Identical Machines

Interestingly, the well-known algorithm of Graham [13], List Scheduling, which is defined for identical machines, is valid for temporary tasks as well as permanent tasks. This algorithm greedily assigns a new job to the least loaded machine.

The competitive ratio of this algorithm is $2-1/m$, which is best possible (see [5]). Note that the competitive ratio is the same as for permanent tasks, but for permanent tasks, it is possible to achieve a competitive ratio which does not tend to 2 for large m, see, e.g., [11].

Uniformly Related Machines

The situation for uniformly related machines is not very different. In this case, the algorithms of Aspnes et al. [3] and of Berman et al. [12] cannot be applied as they are, and some modifications are required. The algorithm of Azar et al. [7] has competitive ratios of at most 20, and it is based on the general method introduced in [3]. The algorithm of [3] keeps a guess value λ, which is an estimation of the cost of an optimal offline algorithm OPT. An invariant that must be kept is $\lambda \leq 2OPT$. At each step, a procedure is applied for some value of λ (which can be initialized as the load of the first job on the fastest machine). The procedure for a given value of λ is applied until it fails, and some job cannot be assigned while satisfying all conditions. The procedure is designed so that if it fails, then it must be the case that $OPT > \lambda$, the value of λ is doubled, and the procedure is reinvoked for the new value, ignoring all assignments that were done for small values of λ. This method is called doubling and results in an algorithm with a competitive ratio which is at most four times the competitive ratio achieved by the procedure. The procedure for a given λ acts as follows. Let c be a target competitive ratio for the procedure. The machines are sorted according to speed. Each job is assigned to the first machine in the sorted order such that the job is assignable to it. A job j arriving at time t is assignable to machine i if $p_j/s_i \leq \lambda$ and $L_i^{t-1} + p_j/s_i \leq c\lambda$. It is shown in [7] that $c = 5$ allows the algorithm to succeed in the assignment of all jobs (i.e., to have at least one assignable machine for each job) as long as $OPT \leq \lambda$. Note that the constant c for permanent tasks used in [3] is 2. As for lower bounds, it is shown in [7] that the competitive ratio $\mathcal{R}$ of any algorithm satisfies $\mathcal{R} \geq 3 - o(1)$. The upper bound has been improved to $6 + 2\sqrt{5} \approx 10.47$ by Bar-Noy et al. [9].

Restricted Assignment

As for restricted assignment, temporary tasks make this model much more difficult than permanent tasks. The competitive ratio $O(\log m)$ which is achieved by a simple greedy algorithm (see [8]) does not hold in this case. In fact, the competitive ratio of this algorithm becomes $\Omega(m^{\frac{2}{3}})$ [4]. Moreover, in the same paper, a lower bound of $\Omega\sqrt{m}$ on the competitive ratio of any algorithm was shown. The construction was quite involved; however, Ma and Plotkin [14] gave a simplified construction which yields the same result.

The construction of [14] selects a value p, which is the largest integer that satisfies $p + p^2 \leq m$. Clearly, $p = \Theta\left(\sqrt{m}\right)$. The lower bound uses two sets of machines, p machines which are called "the small group" and p^2 machines which are called "the large group." The construction consists of p^2 phases, each of which consists of p jobs and is dedicated to one machine in the large group. In phase i, job k of this phase can run either on the k-th machine of the small group or the i-th machine of the large group. After this arrival, only one of these p jobs does not depart. An optimal offline algorithm assigns all jobs in each phase to the small group except for the one job that will not depart. Thus, when the construction is completed, it has one job on each machine of the large group. The maximum load ever achieved by OPT is 1. However, the algorithm does not know at each phase which job will not depart. If no job is assigned to the small group in phase i, then the load of machine i becomes p. Otherwise, a job that the algorithm assigns to the small group is chosen as the one that will not depart. In this way, after p phases, a total load of p^2 is accumulated on the small group, which means that at least one machine there has load p. This completes the construction.

An alternative algorithm called ROBIN HOOD was designed in [7]. This algorithm keeps a lower bound on OPT, which is the maximum between the following two functions. The first one is the maximum average machine load over time. The second is the maximum job size that has ever arrived. Denote this lower bound at time t (after

O

t events have happened) by B^t. A machine i is called rich at time t if $L_i^t \geq \sqrt{m}B^t$. Otherwise, it is called poor. The windfall time of a rich machine i at time t is the time t' such that i is poor at time $t' - 1$ and rich at times $t', \ldots, t$, i.e., the last time that machine i became rich. Clearly, machines can become poor due to an update of B^t or departure of jobs. A machine can become rich due to arrival of jobs that are assigned to it.

The algorithm assigns a job j to a poor machine in $M(j)$ if such a machine exists. Otherwise, j is assigned to the machine in $M(j)$ with the most recent windfall time. The analysis makes use of the fact that at most $\sqrt{m}$ machines can be rich simultaneously.

Note that for small values of m ($m \leq 5$), the competitive ratio of the greedy algorithm is still best possible, as shown in [1]. In this paper, it was shown that these bounds are $(m+3)/2$ for $m = 3, 4, 5$. It is not difficult to see that for $m = 2$, the best bound is 2.

Unrelated Machines

The most extreme difference occurs for unrelated machines. Unlike the case of permanent tasks, where an upper bound of $O(\log m)$ can be achieved [3], it was shown in [2] that any algorithm has a competitive ratio of $\Omega(m/\log m)$. Note that a trivial algorithm, which assigns each job to the machine where it has a minimum load, has a competitive ratio of at most m [3].

Applications

In [10], a hierarchical model was studied. This is a special case of restricted assignment where for each job j, $M(j)$ is a prefix of the machines. They showed that even for temporary tasks, an algorithm of constant competitive ratio exists for this model.

In [6], which studied resource augmentation in load balancing, temporary tasks were considered as well. Resource augmentation is a type of analysis where the online algorithm is compared to an optimal offline algorithm which has less machines.

Open Problems

Small gaps still remain for both uniformly related machines and for unrelated machines. For unrelated machines, it could be interesting to find if there exists an algorithm of competitive ratio $o(m)$ or whether the simple algorithm stated above has optimal competitive ratio (up to a multiplicative factor).

Cross-References

▶ List Scheduling

Recommended Reading

1. Armon A, Azar Y, Epstein L, Regev O (2003) On-line restricted assignment of temporary tasks with unknown durations. Inf Process Lett 85(2):67–72
2. Armon A, Azar Y, Epstein L, Regev O (2003) Temporary tasks assignment resolved. Algorithmica 36(3):295–314
3. Aspnes J, Azar Y, Fiat A, Plotkin S, Waarts O (1997) On-line load balancing with applications to machine scheduling and virtual circuit routing. J ACM 44:486–504
4. Azar Y, Broder AZ, Karlin AR (1994) On-line load balancing. Theor Comput Sci 130:73–84
5. Azar Y, Epstein L (2004) On-line load balancing of temporary tasks on identical machines. SIAM J Discret Math 18(2):347–352
6. Azar Y, Epstein L, van Stee R (2000) Resource augmentation in load balancing. J Sched 3(5):249–258
7. Azar Y, Kalyanasundaram B, Plotkin S, Pruhs K, Waarts O (1997) On-line load balancing of temporary tasks. J Algorithms 22(1):93–110
8. Azar Y, Naor J, Rom R (1995) The competitiveness of on-line assignments. J Algorithms 18:221–237
9. Bar-Noy A, Freund A, Naor J (2000) New algorithms for related machines with temporary jobs. J Sched 3(5):259–272
10. Bar-Noy A, Freund A, Naor J (2001) On-line load balancing in a hierarchical server topology. SIAM J Comput 31:527–549
11. Bartal Y, Fiat A, Karloff H, Vohra R (1995) New algorithms for an ancient scheduling problem. J Comput Syst Sci 51(3):359–366
12. Berman P, Charikar M, Karpinski M (2000) On-line load balancing for related machines. J Algorithms 35:108–121
13. Graham RL (1966) Bounds for certain multiprocessing anomalies. Bell Syst Tech J 45:1563–1581

14. Ma Y, Plotkin S (1997) Improved lower bounds for load balancing of tasks with unknown duration. Inf Process Lett 62:31–34

Online Node-Weighted Problems

Debmalya Panigrahi
Department of Computer Science, Duke University, Durham, NC, USA

Keywords

Network design; Node-weighted graphs; Online algorithms; Primal dual algorithms

Years and Authors of Summarized Original Work

2011; Naor, Panigrahi, Singh
2013; Hajiaghayi, Liaghat, Panigrahi
2014; Hajiaghayi, Liaghat, Panigrahi

Problem Definition

We are given an undirected graph $G = (V, E)$ offline, where node v has a given weight w_v. Initially, the output graph $H \subseteq G$ is the empty graph. In the generic online Steiner network design problem, each online step has a connectivity request C_i and the online algorithm must augment the output graph H to meet the new request. We will consider the following problems in this domain:

- *Steiner tree.* Each connectivity request C_i comprises a new vertex $t_i \in V$ (called a *terminal*) that must be connected in H to all previous terminals. (The first terminal t_0 is often called the *root* and the constraint C_i can then be restated as connecting terminal t_i to the root.)
- *Steiner forest.* Each connectivity request C_i comprises a new vertex pair (s_i, t_i) (called a *terminal pair*) that must be connected in H.
- *Group Steiner tree.* Each connectivity request C_i comprises a new set (group) of vertices $T_i \subseteq V$ (called a *terminal group*). The first terminal group T_0 is a single vertex r called the *root*. At least one vertex in each terminal group must be connected in H to the root.
- *Group Steiner forest.* Each connectivity request C_i comprises a new pair of sets (groups) of vertices (S_i, T_i) (called a *terminal group pair*). For each terminal group pair, at least one vertex in S_i must be connected in H to at least one vertex in T_i.
- *Prize-collecting Steiner tree (resp., Prize-collecting Steiner forest).* Each connectivity request comprises a new terminal t_i (resp., a new terminal pair (s_i, t_i)) and a penalty $\pi_i > 0$; the algorithm must either *pay the penalty* π_i or augment graph H to connect terminal t_i to the root (resp., augment graph H to connect the terminal pair (s_i, t_i)).

In the (group) Steiner tree and (group) Steiner forest problems, the objective is to minimize the total weight (i.e., sum of weights of vertices) of graph H. In the prize-collecting versions of these problems, the objective is to minimize the sum of the total weight of H and the sum of penalties paid by the algorithm.

Key Results

The following theorem was obtained by Naor et al. [7] for the online node-weighted Steiner tree problem.

Theorem 1 *There is a randomized online algorithm for the node-weighted Steiner tree problem that has a competitive ratio of $O(\log^2 k \log n)$ and runs in polynomial time.*

This was the first result to obtain a polylogarithmic competitive ratio for the online node-weighted Steiner tree problem. The competitive ratio for this problem was later improved to $O(\log k \log n)$ (see [5]), which is tight up to constants.

The lower bound follows from the observation that the online set cover problem is a special case of the online node-weighted Steiner tree problem. For the online set cover problem, a lower bound of $\Omega\left(\frac{\log m \log n}{\log\log m+\log\log n}\right)$ for deterministic algorithms was obtained by Alon et al. [1], where m is the number of sets and n is the number of elements. This was later improved and extended to a lower bound of $\Omega(\log m \log n)$ for randomized algorithms by Korman [6]. An online set cover instance can be encoded as an online node-weighted Steiner tree instance where the terminals are the elements and the nonterminals are the sets. This encoding yields a lower bound of $\Omega(\log k \log n)$ for the online node-weighted Steiner tree problem and its generalizations discussed below.

In addition to the Steiner tree problem, Naor et al. [7] also considered the online node-weighted Steiner forest problem and the online node-weighted group Steiner tree problem. In fact, they obtained the following theorem for the online node-weighted group Steiner forest problem which generalizes both these problems.

Theorem 2 *There is a randomized online algorithm for the node-weighted group Steiner forest problem that has a competitive ratio polylogarithmic in n and k and runs in quasi-polynomial time.*

For edge-weighted graphs, the same competitive ratio was obtained with a polynomial-time algorithm.

Subsequent to this work, Hajiaghayi et al. [4] investigated the online node-weighted Steiner forest problem and obtained the first polynomial-time algorithm with a polylogarithmic competitive ratio.

Theorem 3 *There is a randomized online algorithm for the node-weighted Steiner forest problem that has a competitive ratio of $O(\log^2 k \log n)$ and runs in polynomial time.*

The competitive ratio is tight up to a logarithmic factor owing to the online set cover lower bound described above. For graphs with an excluded minor (such as planar graphs), they gave an improved competitive ratio of $O(\log n)$ for this problem, which is tight up to constants. Moreover, the result can be extended to all $\{0, 1\}$-proper functions which were introduced by Goemans and Williamson [3] to capture a broad range of connectivity problems and extended later to node-weighted graphs by Demaine et al. [2].

For the prize-collecting variants of the online node-weighted Steiner tree and Steiner forest problems, Hajiaghayi et al. [5] gave the first algorithms with a polylogarithmic competitive ratio by showing that these problems can be reduced to the fractional versions of their non prize-collecting variants while losing only a logarithmic factor in the competitive ratio. This led to the following results.

Theorem 4 *There is a randomized online algorithm for the prize-collecting node-weighted Steiner tree problem that has a competitive ratio of $O(\log k \log^2 n)$. For the node-weighted prize-collecting Steiner forest problem, there is a randomized online algorithm that has a competitive ratio of $O(\log^2 k \log^2 n)$. Both these algorithms run in polynomial time.*

Corresponding results for edge-weighted graphs were previously known [8].

Applications

Online node-weighted Steiner problems have broad applications in designing communication networks where the clientele grows over time.

Open Problems

Suppose we are given a node-weighted undirected graph $G = (V, E)$. In the online edge-connectivity (resp., vertex connectivity) version of the survivable network design problem (SNDP), the online connectivity requirement C_i comprises a pair of terminals (s_i, t_i) and an integer *requirement* $r_i > 0$. The online algorithm must augment the output graph H so that there are r_i edge-disjoint (resp., node-disjoint) paths between s_i and t_i in H. The objective is to minimize the total weight of H.

An interesting open problem is to obtain an algorithm with competitive ratio $O(r_{\max}^{\alpha} \log^{\beta} n)$ for any constants α, β for the online node-weighted SNDP problem with either edge or vertex connectivity requirements, where $r_{\max} = \max_i r_i$.

Experimental Results

No experimental results are known.

Cross-References

- ▸ Generalized Steiner Network
- ▸ Steiner Forest
- ▸ Steiner Trees

Recommended Reading

1. Alon N, Awerbuch B, Azar Y, Buchbinder N, Naor J (2009) The online set cover problem. SIAM J Comput 39(2):361–370
2. Demaine ED, Hajiaghayi MT, Klein PN (2009) Node-weighted steiner tree and group steiner tree in planar graphs. In: ICALP (1), Rhodes, pp 328–340
3. Goemans MX, Williamson DP (1995) A general approximation technique for constrained forest problems. SIAM J Comput 24(2):296–317
4. Hajiaghayi MT, Liaghat V, Panigrahi D (2013) Online node-weighted steiner forest and extensions via disk paintings. In: FOCS, Berkeley, pp 558–567
5. Hajiaghayi MT, Liaghat V, Panigrahi D (2014) Near-optimal online algorithms for prize-collecting steiner problems. In: ICALP (1), Copenhagen, pp 576–587
6. Korman S (2005) On the use of randomization in the online set cover problem. M.S. thesis, Weizmann Institute of Science
7. Naor J, Panigrahi D, Singh M (2011) Online node-weighted steiner tree and related problems. In: FOCS, Palm Springs, pp 210–219
8. Qian J, Williamson DP (2011) An $O(\log n)$-competitive algorithm for online constrained forest problems. In: ICALP (1), Zurich, pp 37–48

Online Paging and Caching

Neal E. Young
Department of Computer Science and Engineering, University of California, Riverside, CA, USA

Keywords

Caching; Competitive analysis; Competitive ratio; k-server problem; Least-recently-used; Online algorithms; Paging

Years and Authors of Summarized Original Work

1985–2013; multiple authors

Synonyms

Caching; File caching; Paging; Weighted caching; Weighted paging

Problem Definition

A *file-caching* problem instance specifies a cache size k (a positive integer) and a sequence of requests to files, each with a *size* (a positive integer) and a *retrieval cost* (a nonnegative number). The goal is to maintain the cache to satisfy the requests while minimizing the retrieval cost. Specifically, for each request, if the file is not in the cache, one must retrieve it into the cache (paying the retrieval cost) and remove other files to bring the total size of files in the cache to k or less. *Weighted caching* or *weighted paging* is the special case when each file size is 1. *Paging* is the special case when each file size and each retrieval cost is 1 (then the retrieval cost is the number of *cache misses*, and the *fault rate* is the average retrieval cost per request).

An algorithm is *online* if its response to each request is independent of later requests. In practice this is generally necessary. Standard worst-case analysis is not meaningful for online algorithms – any algorithm will have some input sequence that forces a retrieval for every request. Yet worst-case analysis can be done meaningfully as follows. An algorithm is *$c(h,k)$-competitive* if on *any* sequence σ the total (expected) retrieval cost incurred by the algorithm using a cache of size k is at most $c(h,k)$ times the *minimum* cost to handle σ with a cache of size h (plus a constant independent of σ). Then the algorithm has *competitive ratio* $c(h,k)$. The study of competitive ratios is called *competitive analysis*. (In the larger context of approximation algorithms for combinatorial optimization, this ratio is commonly called the *approximation ratio*.)

Algorithms. Here are definitions of a number of caching algorithms; first is LANDLORD. LANDLORD gives each file "credit" (equal to its cost) when the file is requested and not in cache. When necessary, LANDLORD reduces all cached file's credits proportionally to file size, then evicts files as they run out of credit.

File-caching algorithm LANDLORD
Maintain real value credit[f] with each file f (credit[f] $= 0$ if f is not in the cache).
When a file g is requested:
1. **if** g is not in the cache:
2. **until** the cache has room for g:
3. **for each** cached file f: decrease credit[f] by $\Delta \cdot$ size[f],
4. where $\Delta = \min_{f \in \text{cache}}$ credit[f]/size[f].
5. Evict from the cache any subset of the zero-credit files f.
6. Retrieve g into the cache; set credit[g] $\leftarrow$ cost(g).
7. **else** Reset credit[g] anywhere between its current value and cost(g).

For weighted caching, file sizes equal 1. GREEDY DUAL is LANDLORD for this special case. BALANCE is the further special case obtained by leaving credit unchanged in line 7.

For paging, files sizes and costs equal 1. FLUSH-WHEN-FULL is obtained by evicting *all* zero-credit files in line 5; FIRST-IN-FIRST-OUT is obtained by leaving credits unchanged in line 7 and evicting the file that entered the cache earliest in line 5; LEAST-RECENTLY-USED is obtained by raising credits to 1 in line 7 and evicting the least-recently requested file in line 5. The MARKING algorithm is obtained by raising credits to 1 in line 7 and evicting a *random* zero-credit file in line 5. (LANDLORD generalizes to arbitrary covering problems with submodular costs as described in [10].)

Key Results

This entry focuses on competitive analysis of paging and caching strategies as defined above. Competitive analysis has been applied to many problems other than paging and caching, and much is known about other methods of analysis (mainly empirical or average case) of paging and caching strategies, but these are outside scope of this entry.

Paging

In a seminal paper, Sleator and Tarjan showed that LEAST-RECENTLY-USED, FIRST-IN-FIRST-OUT, and FLUSH-WHEN-FULL are $\frac{k}{k-h+1}$-competitive [13]. Sleator and Tarjan also showed that this competitive ratio is the best possible for any deterministic online algorithm. Fiat et al. showed that the MARKING algorithm is $2H_k$-competitive and that no randomized online algorithm is better than H_k-competitive [6]. Here $H_k = 1 + 1/2 + \cdots + 1/k \approx 0.58 + \ln k$. McGeoch and Sleator gave an optimal H_k-competitive randomized online paging algorithm [12].

Weighted Caching

For weighted caching, Chrobak et al. showed that the deterministic online BALANCE algorithm is k-competitive [4]. Young showed that GREEDY DUAL is $\frac{k}{k-h+1}$-competitive and that GREEDY DUAL is a primal-dual algorithm – it generates a solution to the linear-programming dual which proves the near-optimality of the primal solution [14]. Bansal et al., resolving a long-standing open problem, used the primal-dual framework to give an $O(\log k)$-competitive randomized algorithm for weighted caching [2].

File Caching

When each cost equals 1 (the goal is to minimize the *number* of retrievals), or when each file's cost equals the file's size (the goal is to minimize the total number of *bytes* retrieved), Irani gave $O(\log^2 k)$-competitive randomized online algorithms [7].

For general file caching, Irani and Cao showed that a restriction of LANDLORD is k-competitive [3]. Independently, Young showed that LANDLORD is $\frac{k}{k-h+1}$-competitive [15].

Other Theoretical Models

Practical performance can be better than the worst case studied in competitive analysis. Refinements of the model have been proposed to increase realism. Borodin et al. [1], to model locality of reference, proposed the *access-graph* model (see also [8, 9]). Koutsoupias and Papadimitriou proposed the *comparative ratio* (for comparing classes of online algorithms directly) and the *diffuse-adversary model* (where the adversary chooses requests probabilistically subject to restrictions) [11]. Young showed that any $\frac{k}{k-h+1}$-competitive algorithm is also *loosely* $O(1)$-competitive: for any fixed $\varepsilon, \delta > 0$, on any sequence, for all but a δ-fraction of cache sizes k, the algorithm either is $O(1)$-competitive or pays at most ε times the sum of the retrieval costs [15].

Analyses of Deterministic Algorithms

Here is a competitive analysis of GREEDY DUAL for weighted caching.

Theorem 1 GREEDY DUAL *is* $\frac{k}{k-h+1}$-*competitive for weighted caching.*

Proof Here is an amortized analysis (in the spirit of Sleator and Tarjan, Chrobak et al., and Young; see [14] for a different primal-dual analysis). Define potential

$$\Phi = (h-1) \cdot \sum_{f \in \text{GD}} \text{credit}[f] + k \cdot \sum_{f \in \text{OPT}} \big(\text{cost}(f) - \text{credit}[f]\big),$$

where GD and OPT denote the current caches of GREEDY DUAL and OPT (the optimal off-line algorithm that manages the cache to minimize the total retrieval cost), respectively. After each request, GREEDY DUAL and OPT take (some subset of) the following steps in order.

OPT evicts a file f: Since credit$[f] \le$ cost(f), Φ cannot increase.

OPT retrieves requested file g: OPT pays cost(g); Φ increases by at most k cost(g).

GREEDY DUAL decreases credit$[f]$ **for all $f \in$ GD:** The cache is full and the requested file is in OPT but not yet in GD. So $|\text{GD}| = k$ and $|\text{OPT} \cap \text{GD}| \le h-1$. Thus, the total decrease in Φ is $\Delta[(h-1)|\text{GD}| - k\,|\text{OPT} \cap \text{GD}|] \ge \Delta[(h-1)k - k(h-1)] = 0$.

GREEDY DUAL evicts a file f: Since credit$[f] = 0$, Φ is unchanged.

GREEDY DUAL retrieves requested file g and sets credit$[g]$ **to** cost(g)**:** GREEDY DUAL pays c = cost(g). Since g was not in GD but is in OPT, credit$[g] = 0$ and Φ decreases by $-(h-1)c + k\,c = (k-h+1)c$.

GREEDY DUAL resets credit$[g]$ **between its current value and** cost(g)**:** Since $g \in$ OPT and credit$[g]$ only increases, Φ decreases.

So, with each request: (1) when OPT retrieves a file of cost c, Φ increases by at most kc; (2) at no other time does Φ increase; and (3) when GREEDY DUAL retrieves a file of cost c, Φ decreases by at least $(k-h+1)c$. Since initially $\Phi = 0$ and finally $\Phi \ge 0$, it follows that GREEDY DUAL's total cost times $k-h+1$ is at most OPT's cost times k.

Extension to File Caching

Although the proof above easily extends to LANDLORD, it is more informative to analyze LANDLORD via a *general reduction* from file caching to weighted caching:

Corollary 1 LANDLORD *is* $\frac{k}{k-h+1}$-*competitive for file caching.*

Proof Let W be any deterministic c-competitive weighted-caching algorithm. Define file-caching algorithm F_W as follows. Given request sequence σ, F_W simulates W on weighted-caching sequence σ' as follows. For each file f, break f

into size(f) "pieces" $\{f_i\}$ each of size 1 and cost cost(f)/size(f). When f is requested, give a batch $(f_1, f_2, \ldots, f_s)^{N+1}$ of requests for pieces to W. Take N large enough so W has all pieces $\{f_i\}$ cached after the first sN requests of the batch.

Assume that W *respects equivalence*: after each batch, for every file f, all or none of f's pieces are in W's cache. After each batch, make F_W update its cache correspondingly to $\{f : f_i \in \text{cache}(W)\}$. F_W's retrieval cost for σ is at most W's retrieval cost for σ', which is at most c OPT(σ'), which is at most c OPT(σ). Thus, F_W is c-competitive for file caching.

Now, observe that GREEDY DUAL can be made to respect equivalence. When GREEDY DUAL processes a batch of requests $(f_1, f_2, \ldots, f_s)^{N+1}$ resulting in retrievals, for the last s requests, make GREEDY DUAL set credit$[f_i]$ = cost(f_i) = cost$(f)/s$ in line 7. In general, restrict GREEDY DUAL to raise credits of equivalent pieces f_i equally in line 7. After each batch the credits on equivalent pieces f_i will be the same. When GREEDY DUAL evicts a piece f_i, make GREEDY DUAL evict all other equivalent pieces f_j (all will have zero credit).

With these restrictions, GREEDY DUAL respects equivalence. Finally, taking W to be GREEDY DUAL above, F_W is LANDLORD.

Analysis of the Randomized MARKING Algorithm.

Here is a competitive analysis of the MARKING algorithm

Theorem 2 *The* MARKING *algorithm is* $2H_k$*-competitive for paging.*

Proof Given a paging request sequence σ, partition σ into contiguous *phases* as follows. Each phase starts with the request after the end of the previous phase and continues as long as possible subject to the constraint that it should contain requests to at most k distinct pages. (Each phase starts when the algorithm runs out of zero-credit files and reduces all credits to zero.)

Say a request in the phase is *new* if the item requested was not requested in the previous phase. Let m_i denote the number of new requests in the ith phase. During phases $i-1$ and i, $k + m_i$ distinct files are requested. OPT has at most k of these in cache at the start of the $i-1$st phase, so it will retrieve at least m_i of them before the end of the ith phase. So OPT's total cost is at least $\max\{\sum_i m_{2i}, \sum_i m_{2i+1}\} \geq \sum_i m_i/2$.

Say a non-new request is *redundant* if it is to a file with credit 1 and nonredundant otherwise. Each new request costs the MARKING algorithm 1. The jth nonredundant request costs the MARKING algorithm at most $m_i/(k-j+1)$ in expectation because, of the $k-j+1$ files that if requested would be nonredundant, at most m_i are not in the cache (and each is equally likely to be in the cache). Thus, in expectation MARKING pays at most $m_i + \sum_{j=1}^{k-m_i} m_i/(k-j+1) \leq m_i H_k$ for the phase and at most $H_k \sum_i m_i$ total.

Applications

Variants of GREEDY DUAL and LANDLORD have been incorporated into file-caching software such as Squid [5].

Open Problems

None to report.

Experimental Results

For a study of competitive ratios on practical inputs, see, for example, [3, 5, 14].

Cross-References

- ▶ Algorithm DC-TREE for k-Servers on Trees
- ▶ Alternative Performance Measures in Online Algorithms
- ▶ Online List Update
- ▶ Price of Anarchy
- ▶ Work-Function Algorithm for k-Servers

Recommended Reading

1. Borodin A, Irani S, Raghavan P, Schieber B (1995) Competitive paging with locality of reference. J Comput Syst Sci 50(2):244–258. Elsevier

2. Buchbinder N, Naor J (2009) Online primal-dual algorithms for covering and packing. Math Oper Res 34(2):270–286. INFORMS
3. Cao P, Irani S (1997) Cost-aware WWW proxy caching algorithms. In: USENIX symposium on internet technologies and systems, Monterey, vol 12(97), pp 193–206
4. Chrobak M, Karloff H, Payne T, Vishwanathan S (1991) New results on server problems. SIAM J Discret Math 4(2):172–181
5. Dilley J, Arlitt M, Perret S (1999) Enhancement and validation of Squid's cache replacement policy. Technical report HPL-1999-69, Hewlett-Packard Laboratories, also in 4th International Web Caching Workshop
6. Fiat A, Karp RM, Luby M, McGeoch LA, Sleator DD, Young NE (1991) Competitive paging algorithms. J Algorithms 12:685–699
7. Irani S (2002) Page replacement with multi-size pages and applications to web caching. Algorithmica 33(3):384–409
8. Irani S, Karlin AR, Phillips S (1996) Strongly competitive algorithms for paging with locality of reference. SIAM J Comput 25(3):477–497. SIAM
9. Karlin AR, Phillips SJ, Raghavan P (2000) Markov paging. SIAM J Comput 30(3):906–922
10. Koufogiannakis C, Young NE (2013) Greedy Δ-approximation algorithm for covering with arbitrary constraints and submodular cost. Algorithmica 66(1):113–152
11. Koutsoupias E, Papadimitriou C (2000) Beyond competitive analysis. SIAM J Comput 30(1):300–317
12. McGeoch L, Sleator D (1991) A strongly competitive randomized paging algorithm. Algorithmica 6(6):816–825
13. Sleator D, Tarjan RE (1985) Amortized efficiency of list update and paging rules. Commun ACM 28:202–208
14. Young NE (1994) The k-server dual and loose competitiveness for paging. Algorithmica 11:525–541
15. Young NE (2002) On-line file caching. Algorithmica 33(3):371–383

Online Preemptive Scheduling on Parallel Machines

Jiří Sgall
Computer Science Institute, Charles University, Prague, Czech Republic

Keywords

Makespan; Online scheduling; Preemption

Years and Authors of Summarized Original Work

2009; Ebenlendr, Jawor, Sgall
2010; Ebenlendr
2011; Ebenlendr, Sgall

Problem Definition

We consider an online version of the classical problem of preemptive scheduling on uniformly related machines.

We are given m machines with *speeds* $s_1 \geq s_2 \geq \cdots \geq s_m$ and a sequence of jobs, each described by its *processing time* (length). The actual time needed to process a job with length p on a machine with speed s is p/s. In the *preemptive version*, each job may be divided into several pieces, which can be assigned to different machines in disjoint time slots. (A job may be scheduled in several time slots on the same machine, and there may be times when a partially processed job is not running at all.) The objective is to find a schedule of all jobs in which the *maximal completion time (makespan)* is minimized.

In the *online problem*, jobs arrive one by one and the algorithm needs to assign each incoming job to some time slots on some machines, without any knowledge of the jobs that arrive later. This problem, also known as list scheduling, was first studied by Graham [8] for identical machines (i.e., $s_1 = \cdots = s_m = 1$), without preemption. In the preemptive version, upon the arrival of a job, its complete assignment at all times must be given and the algorithm is not allowed to change this assignment later. In other words, the online nature of the problem is in the order of the input sequence, and it is not related to possible preemptions and the time in the schedule.

Key Results

The main result is an optimal online algorithm RatioStretch for preemptive scheduling on uniformly related machines [4]. RatioStretch achieves the best possible competitive ratio not only in the general case but also for any number of machines and any particular combination of machine speeds. Although RatioStretch is deterministic, its competitive ratio matches the best competitive ratio of any randomized algorithm. This proves that randomization does not help for this variant of preemptive scheduling.

For any fixed set of speeds, the competitive ratio of the algorithm RatioStretch can be computed by solving a linear program. However, its worst-case value over all speed combinations is not known. Nevertheless, using the fact that there exists an e-competitive randomized algorithm [5], it is possible to conclude that RatioStretch also achieves the ratio of at most $e \approx 2.718$. The best lower bound shows that RatioStretch (and thus any algorithm) is not better than 2.112-competitive, by providing an explicit numerical instance on 200 machines [3].

Key Techniques

The idea of the algorithm RatioStretch is fairly natural. Suppose that the algorithm is given a ratio R which we are trying to achieve. For each arriving job, RatioStretch computes the optimal makespan for jobs that have arrived so far and runs the incoming job as slow as possible so that it finishes at R times the computed optimal makespan. There are many ways of creating such a schedule given the flexibility of preemptions. RatioStretch chooses a particular one based on the notion of a *virtual machine* from [5]. Given a schedule, the ith virtual machine at each time corresponds to the ith fastest real machine that is idle. (In particular, before the first job, the virtual machines are the real machines.) This assignment of the real machines to the virtual machines can vary at different times in the schedule. Due to preemption, a virtual machine can be thought of and used as a single machine with changing speed. The key idea of RatioStretch is to schedule each job on two adjacent virtual machines.

If RatioStretch fails on some input for a given R, it is possible to use the lower bound technique from [7] and show that there is no R-competitive algorithm. This implies that if the algorithm knows the optimal competitive ratio R, it never fails and thus it is R-competitive.

It remains to find the optimal competitive ratio R. Since the lower bound technique from [7] results in a linear condition, one can show that R can be computed by a linear program for each combination of speeds.

Semi-online Scheduling

The algorithm RatioStretch can be extended to *semi-online* scenarios [6]. This term encompasses situations where some partial information about the input is given to the scheduler in advance. Already Graham [9] studied a semi-online variant of scheduling on identical machines: he proved that if the jobs are presented in non-increasing order of their processing times, the approximation ratio of list scheduling decreases from 2 to 4/3. Since then numerous semi-online models of scheduling have been studied; typical examples include (sequences of) jobs with decreasing processing times, jobs with bounded processing times, sequences with known total processing time of jobs, and so on. Most of these models can be viewed as online algorithms on a restricted set of input sequences.

RatioStretch can be generalized so that it is optimal for any chosen semi-online restriction. This means not only the cases listed above – the restriction can be given as an arbitrary set of sequences that are allowed as inputs. Again, for any semi-online restriction, RatioStretch achieves the best possible approximation ratio for any number of machines and any particular combination of machine speeds; it is deterministic, but its approximation ratio matches the best possible approximation ratio of any randomized algorithm. This result also provides a clear separation between the design of the algorithm and the analysis of the optimal approximation ratio. While the algorithm is always the same, the analysis of the optimal ratio depends on the studied restrictions.

For typical semi-online restrictions, the optimal ratio can again be computed by linear programs (with machine speeds as parameters). Then we can study the relations between the optimal approximation ratios for different semi-online restrictions and give some bounds for a large number of machines by analysis of these linear programs. One interesting result is that the overall ratio with known sum of processing times is the same as in the purely online case – even though for a small fixed number of machines, knowing the sum provides a significant advantage.

Some basic restrictions form an inclusion chain: the inputs where the first job has the maximal processing time (which is equivalent to known maximal processing time) include the inputs with non-increasing processing times, which in turn include the inputs with all jobs of equal processing time. The restriction to non-increasing processing times gives the same approximation ratio as when all jobs have equal processing times, even for any particular combination of speeds. The overall approximation ratio of these two equivalent problems is at most 1.52. For known maximal processing time of a job, there exists a computer-generated hard instance with approximation ratio 1.88 with 120 machines. Thus, restricting the jobs to be non-increasing helps the algorithm much more than just knowing the maximal processing time of a job. This is very different from identical machines, where knowing the maximal processing time is equally powerful as knowing that all the jobs are equal; see [10].

Small Number of Machines

For two, three, and sometimes four machines, it is possible to give an exact formula for the competitive ratio for any speed combination [2,3]. This is a fairly routine task which can be simplified (but not completely automated) using standard mathematical software. Once the solution is known, verification amounts to checking the given primal and dual solutions for the linear program.

Open Problems

The main remaining open problem is to develop techniques for determining or bounding the overall competitive ratio of the optimal algorithm RatioStretch. In particular, it would be interesting to obtain a tight bound in the online case.

It is also open if similar techniques can be used for the non-preemptive problem. In this case, the currently best algorithms were obtained by a doubling approach. This means that a competitive algorithm is designed for the case when the optimum is approximately known in advance, and then, without this knowledge, it is used in phases with geometrically increasing guesses of the optimum. Such an approach probably cannot lead to an optimal algorithm for this type of scheduling problems. The best lower and upper bounds for non-preemptive scheduling on uniformly related machines are 2.438 and 5.828 for deterministic algorithms (see [1]) and 2 and 4.311 for randomized algorithms (see [1,7]). Thus, it is still open whether randomized algorithms are better than deterministic.

Cross-References

▸ Approximation Schemes for Makespan Minimization
▸ List Scheduling
▸ Online Load Balancing of Temporary Tasks

O

Recommended Reading

1. Berman P, Charikar M, Karpinski M (2000) On-line load balancing for related machines. J. Algorithms 35:108–121
2. Ebenlendr T (2010) Semi-online preemptive scheduling: study of special cases. In: Proceedings of 8th international conference on parallel processing and applied mathematics (PPAM 2009), part II, Wroclaw. Lecture notes in computer science, vol 6068. Springer, pp 11–20
3. Ebenlendr T (2011) Combinatorial algorithms for online problems: semi-online scheduling on related machines. Ph.D. thesis, Charles University, Prague
4. Ebenlendr T, Jawor W, Sgall J (2009) Preemptive online scheduling: optimal algorithms for all speeds. Algorithmica 53:504–522
5. Ebenlendr T, Sgall J (2004) Optimal and online preemptive scheduling on uniformly related machines. In: Proceedings of 21st symposium on theoretical aspects of computer science (STACS), Montpellier. Lecture notes in computer science, vol 2996. Springer, pp 199–210
6. Ebenlendr T, Sgall J (2011) Semi-online preemptive scheduling: one algorithm for all variants. Theory Comput Syst 48:577–613

7. Epstein L, Sgall J (2000) A lower bound for on-line scheduling on uniformly related machines. Oper Res Lett 26:17–22
8. Graham RL (1966) Bounds for certain multiprocessing anomalies. Bell Syst Tech J 45: 1563–1581
9. Graham RL (1969) Bounds on multiprocessing timing anomalies. SIAM J Appl Math 17: 263–269
10. Seiden S, Sgall J, Woeginger GJ (2000) Semi-online scheduling with decreasing job sizes. Oper Res Lett 27:215–221

Optimal Crowdsourcing Contests

Balasubramanian Sivan
Microsoft Research, Redmond, WA, USA

Keywords

All-pay auctions; Bayesian Nash equilibrium; Contests

Years and Authors of Summarized Original Work

2012; Chawla, Hartline, Sivan

Problem Definition

With the ever-increasing reach of the Internet, crowdsourcing contests have become an increasingly convenient alternative for completing tasks, compared to traditional hire-and-pay methods. There are several websites dedicated to providing users a platform for creating their own crowdsourcing contests. For instance, Taskcn.com allows users to post tasks, collect submissions from registered users, and provide a monetary reward to the best submission. The reach of crowdsourcing is far beyond tedious/labor-intensive tasks. Netflix, for instance, issued a million-dollar contest for developing a collaborative filtering algorithm to predict user ratings for films, instead of hiring an in-house research team to develop this. The Indian Government used a crowdsourcing contest to pick a new symbol for its rupee currency.

The Questions

In designing a crowdsourcing contest, a principal, with a preallocated sum of monetary reward in hand, seeks to identify the format of the contest that optimizes the quality of the best submission. For instance, Topcoder.com issues 2/3 of the reward to the best submission and 1/3 of the reward to the second-best submission. Is this the format best suited to optimize the best submission? Or should the entire award be given to the winner? More generally, should the precise division of rewards be even announced prior to the contest, or should they be announced only as a function of the quality of the submissions received? In a different direction, crowdsourcing contests make several people to expend efforts in producing submissions, but often only the best submission is put to use. How much effort is getting "burnt" in this process compared to conventional hire-and-pay?

The Model

Formally, let there be n contestants, and let the monetary reward be normalized to \$1. Contestants enter their submissions which are ranked according to their qualities. Agent i's submission quality p_i is a function of their skill v_i and their effort e_i, given by $p_i = v_i \cdot e_i$. The skill v_i can be interpreted as the rate at which agent i can do useful work. The contest designer can observe only the submission qualities p_i's and not the skills v_i's. However, the distribution F from which the v_i's are drawn (independently) is common knowledge to all contestants and the contest designer. Every contestant's goal is to maximize their utility, namely, their reward minus the effort they expended. If x_i is the probability that agent i gets the reward, then their utility is given by $x_i - e_i = x_i - \frac{p_i}{v_i}$.

We model crowdsourcing contests as all-pay auctions, following the contest architecture literature [4, 6]. In an all-pay auction with n bidders, a seller auctions a good that bidder i values at v_i. The value v_i is private to bidder i, but the distribution F from which v_i's are independently drawn is common knowledge to the seller and the bidders. The seller solicits sealed bids from the agents, and all bidders agree to pay their bids regardless of which bidder gets the good (corresponding to all contestants losing their effort irrespective of which contestant wins the contest). Which agent gets the good depends on the allocation rule of the auction. Given the rules of the auction, each bidder aims to maximize his utility. If x_i is the probability that agent i receives the good, then agent i maximizes his utility of $v_i x_i - p_i$. Note that this utility is precisely v_i times the utility of a contestant in the crowdsourcing contest defined in the previous paragraph. From agent i's perspective, v_i is just a constant. Thus, the incentives in the contest and the all-pay auction are identical. Thus, designing a contest to maximize the quality of the best submission, namely, maximize $\max_i p_i$, is the same as designing an all-pay auction to maximize the maximum payment. Thus, we have an all-pay auction design problem where the objective is not the traditional one of maximizing revenue $(\sum_i p_i)$ but requires maximizing $\max_i p_i$.

We assume that the space of possible valuations V is an interval and the density $f(\cdot)$ of the value distribution is nonzero everywhere in this interval.

Bayesian Nash Equilibrium

In an all-pay auction it is not strategic for an agent to bid his true value v: the probability he wins the good is at most 1 (in which case he gets a value v), where he is sure to lose his bid. Thus, agents submit bids smaller than their true value. An agent's bidding function $b_i(\cdot)$ maps their true value to bids. A profile of bidding functions $(b_1(\cdot), \ldots, b_n(\cdot))$ is a Bayesian Nash equilibrium (BNE) if the bidding functions are mutual best responses, i.e., if values are drawn from F and other agents bid according to their bidding functions, agent i weakly prefers following his own bidding strategy $b_i(\cdot)$ over submitting any other bid. For a given outcome $x_i(\mathbf{v})$, let $x_i(v_i) = \mathbf{E}_{\mathbf{v}_{-i}}[x_i(\mathbf{v})]$, and let $p_i(v_i) = \mathbf{E}_{\mathbf{v}_{-i}}[p_i(\mathbf{v})]$.

We will appeal to the following result from [2] that shows that for most of the all-pay auctions that we discuss, there exists a unique Bayesian Nash equilibrium, and it is also symmetric. That is, in the unique BNE, all agents have the same bidding function.

Theorem 1 ([2]) *In the all-pay auction parameterized by a reserve price and a nonincreasing sequence of rewards $a_1, \ldots, a_n$, where the agents whose bids meet the reserve are assigned to the rewards in decreasing order of bids, a symmetric BNE exists and is the unique equilibrium.*

Key Results

We now present the key results concerning the design of optimal crowdsourcing contests (from [3]).

Rank-Based-Reward Contests

Consider the class of contests that predetermine the division of rewards into fractions $a_1, \ldots, a_n$, s.t. $\sum_i a_i = 1$, $a_k \geq a_{k+1}$, and $a_i \geq 0$. That is, agents are ordered by submission qualities and the ith best submission receives a_i fraction of the reward. In this notation, Topcoder's contest will be $a_1 = 2/3$, $a_2 = 1/3$, and $a_k = 0$ for $k > 2$. The first key result is that if the goal is to maximize the maximum payment, the optimal all-pay auction is to award the good completely to the highest bidder. In contest language, the optimal contest is a winner-takes-all contest. Note that by Theorem 1 this contest format has a unique BNE.

Theorem 2 *When the contestant skills are distributed i.i.d., the optimal rank-based-reward contest is a winner-takes-all contest.*

Optimal Symmetric Contest

Is there an even better contest in the larger space of contests? Suppose we allow contests that announce rewards as a function of agents' submission qualities, what is the optimal contest? We focus on the class of symmetric contests and optimize over their symmetric equilibria. For a large class of distributions, including distributions that satisfy the monotone hazard rate property (e.g., uniform, normal, exponential), the optimal auction will turn out to have a unique equilibrium that is also symmetric.

Theorem 3 *When the contestant skills are distributed i.i.d. from a distribution that satisfies the monotone hazard rate condition, the optimal* *symmetric contest is highest-submission-wins contest subject to a minimum submission quality.*

Proof (Sketch) We prove this result through an argument that mirrors Myerson's revenue optimal auction argument [8]. Recall that in auction theory terms, we have to prove that the optimal auction is a highest-bidder-wins auction subject to a minimum bid reserve. Writing out the expression for the expected maximum payment and using the characterization of BNE payments in terms of allocation, we realize that the expected maximum payment is just the expected virtual welfare. That is, let $\phi(\cdot)$ be a distribution-dependent transformation that is applied to each agent's value v_i to obtain $\phi(v_i) = v_i F(v_i)^{n-1} - \frac{1-F(v_i)^{n-1}}{nf(v_i)}$. The expected virtual welfare of an outcome is just $\mathbf{E}_{\mathbf{v}}\left[\sum_i \phi(v_i)x_i(\mathbf{v})\right]$. If this is the quantity to maximize, it is immediate that the optimal outcome is to allocate completely to the agent with the highest virtual value subject to the highest virtual value being nonnegative. If the virtual value transformation were a strictly increasing function (whenever it is positive), the bidder with the highest value, and hence also the highest bid because of our focus on symmetric equilibria, will also be the bidder with the highest virtual value. Thus the highest-bidder-wins auction subject to a minimum bid reserve will implement the desired outcome. Now, for the distribution-dependent transformation $\phi(\cdot)$ to be strictly increasing, it is enough for the distribution to satisfy the monotone hazard rate condition. Finally, this contest has a unique BNE from Theorem 1.

Theorem 4 *For any setting with i.i.d. values, the optimal symmetric contest is defined by a minimum submission quality and a subset of submission qualities called forbidden qualities that has the following format: the contest solicits submissions and rounds them down to the nearest non-forbidden quality; it then distributes the reward equally among the highest submissions subject to the submissions being above the minimum submission quality.*

Proof (Sketch) Continuing with the proof of Theorem 3, if the virtual value transformation were not increasing, allocating to the highest virtual value is no more a BNE outcome. In this case, the transformation ϕ is "ironed" to obtain a nondecreasing function $\bar{\phi}(\cdot)$, such that the expected maximum payment is equal to the expected ironed virtual surplus. To optimize this quantity, the outcome should be to allocate completely to the agent with the highest ironed virtual value subject to it being nonnegative. In case of a tie, all agents with the highest ironed virtual value get equal allocations. Such an allocation will result in a discontinuous allocation function and hence a discontinuous payment function. That is, some payments are forbidden. Correspondingly to ensure that some bids are forbidden, the auction explicitly says that bids in certain regions will be rounded down so that no rational agent will bid inside that interval. This explains the format of the optimal contest specified in the theorem.

Utilization Ratio of Crowdsourcing

In a crowdsourcing contest, which is like an all-pay auction, every agent's submission is collected, but only the best submission is used. In contrast, in conventional contracting, which is like first- or second-price auctions, only the winner makes any submission at all, and thus there is no underutilization. One way of measuring the amount of work that actually gets utilized in crowdsourcing as opposed to getting "burnt" is to study the ratio of the maximum payment and the sum of all payments in an all-pay auction. It turns out that the utilization ratio in a large class of contests is at least a $1/2$.

Theorem 5 *In any highest-submission-wins contest with a minimum submission quality, the quality of the best submission is at least half of the sum total of the qualities of all the submissions.*

Related Work

Other objectives that have been studied in contest design include maximizing the sum of submission qualities instead of the maximum submission quality [5–7] and maximizing the sum of submission qualities less the normalized reward [1]. The rank-based-reward result in

Theorem 2 is quite robust and continues to hold in many of these other models as well. Moldovanu and Sela [7] study multi-round contests and show that there are situations where it is better to split contestants into two divisions and to have a final among the divisional winners. DiPalantino and Vojnovic [4] study crowdsourcing websites as a matching market. Yang et al. [9] and DiPalantino and Vojnovic [4] study contestant behavior from contest website Taskcn.com and observe that experienced contestants strategize well.

Open Problems

Multi-round Contests

The optimality result discussed here is restricted to single-round contests. If one were allowed to do a tournament-style multi-round contest, what is the optimal contest in this large class of contests? How significant is the difference in objective value when one is allowed to organize more than one round of contest? How does the objective value grow with the number of rounds?

Cross-References

► Algorithmic Mechanism Design
► Competitive Auction
► Generalized Vickrey Auction

Recommended Reading

1. Archak N, Sundararajan A (2009) Optimal design of crowdsourcing contests. In: International conference on information systems (ICIS), Phoenix
2. Chawla S, Hartline JD (2013) Auctions with unique equilibria. In: Proceedings of the fourteenth ACM conference on electronic commerce (EC '13), New York. ACM, pp 181–196
3. Chawla S, Hartline JD, Sivan B (2012) Optimal crowdsourcing contests. In: SODA, Kyoto, pp 856–868
4. DiPalantino D, Vojnovic M (2009) Crowdsourcing and all-pay auctions. In: Proceedings of the 10th ACM conference on electronic commerce (EC '09), Stanford, pp 119–128
5. Minor D (2011) Increasing effort through rewarding the best less. Mansucript. http://www.kellogg.northwestern.edu/faculty/minor/Papers/Increasing%20Effort%20%28with%20Figures%29.pdf
6. Moldovanu B, Sela A (2001) The optimal allocation of prizes in contests. Am Econ Rev 91(3):542–558
7. Moldovanu B, Sela A (2006) Contest architecture. J Econ Theory 126(1):70–97
8. Myerson R (1981) Optimal auction design. Math Oper Res 6:58–73
9. Yang J, Adamic LA, Ackerman MS (2008) Crowdsourcing and knowledge sharing: strategic user behavior on taskcn. In: Proceedings of the 9th ACM conference on electronic commerce (EC '08), Chicago, pp 246–255

Optimal Probabilistic Synchronous Byzantine Agreement

Juan Garay
Bell Laboratories, Murray Hill, NJ, USA

Keywords

Byzantine generals problem; Distributed consensus

Years and Authors of Summarized Original Work

1988; Feldman, Micali

Problem Definition

The Byzantine agreement problem (BA) is concerned with multiple processors (parties, "players") all starting with some initial value, agreeing on a common value, despite the possible disruptive or even malicious behavior of some them. BA is a fundamental problem in fault-tolerant distributed computing and secure multi-party computation.

The problem was introduced by Pease, Shostak and Lamport in [17], who showed that the number of faulty processors must be less than a third of the total number of processors for the problem to have a solution. They also presented a protocol matching this bound, which

O

requires a number of communication rounds proportional to the number of faulty processors – exactly $t + 1$, where t is the number of faulty processors. Fischer and Lynch [10] later showed that this number of rounds is necessary in the worst-case run of any deterministic BA protocol. Furthermore, the above assumes that communication takes place in synchronous rounds. Fischer, Lynch and Patterson [11] proved that no completely asynchronous BA protocol can tolerate even a single processor with the simplest form of misbehavior – namely, ceasing to function at an arbitrary point during the execution of the protocol ("crashing").

To circumvent the above-mentioned lower bound on the number of communication rounds and impossibility result, respectively, researchers beginning with Ben-Or [1] and Rabin [18], and followed by many others (e.g., [3, 5]) explored the use of randomization. In particular, Rabin showed that linearly resilient BA protocols in expected *constant* rounds were possible, provided that all the parties have access to a "common coin" (i.e., a common source of randomness) – essentially, the value of the coin can be adopted by the non-faulty processors in case disagreement at any given round is detected, a process that is repeated multiple times. This line of research culminated in the unconditional (or information-theoretic) setting with the work of Feldman and Micali [9], who showed an efficient (i.e., polynomial-time) probabilistic BA protocol tolerating the maximal number of faulty processors (Karlin and Yao, Probabilistic lower bounds for the byzantine generals problem, unpublished manuscript showed that the maximum number of faulty processors for probabilistic BA is also $t < \frac{n}{3}$, where n is the total number of processors.) that runs in expected constant number of rounds. The main achievement of the Feldman–Micali work is to show how to obtain a shared random coin with constant success probability in the presence of the maximum allowed number of misbehaving parties "from scratch".

Randomization has also been applied to BA protocols in the computational (or cryptographic) setting and for weaker failure models. See [6] for an early survey on the subject.

Notations

Consider a set $\mathcal{P} = \{P_1, P_2, \cdots, P_n\}$ of processors (probabilistic polynomial-time Turing machines) out of which t, $t < n$ may not follow the protocol, and even collude and behave in arbitrary ways. These processors are called *faulty*; it is useful to model the faulty processors as being coordinated by an adversary, sometimes called a *t-adversary*.

For $1 \le i \le n$, let b_i, $b_i \in \{0, 1\}$ denote party P_i's initial value. The work of Feldman and Micali considers the problem of designing a probabilistic BA protocol in the model defined below.

System Model

The processors are assumed to be connected by point-to-point private channels. Such a network is assumed to be synchronous, i.e., the processors have access to a global clock, and thus the computation of all processors can proceed in a lock-step fashion. It is customary to divide the computation of a synchronous network into *rounds*. In each round, processors send messages, receive messages, and perform some local computation.

The t-adversary is computationally unbounded, *adaptive* (i.e., it chooses which processors to corrupt on the fly), and decides on the messages the faulty processors send in a round depending on the messages sent by the non-faulty processors in all previous rounds, including the current round (this is called a *rushing* adversary).

Given the model above, the goal is to solve the problem stated in the ▶ Byzantine Agreement; that is, for every set of inputs and any behavior of the faulty processors, to have the non-faulty processors output a common value, subject to the additional condition that if they all start the computation with the same initial value, then that should be the output value. The difference with respect to the other entry is that, thanks to randomization, BA protocols here run in expected constant rounds.

Problem 1 (BA)
INPUT: Each processor P_i, $1 \leq i \leq n$, has bit b_i.
OUTPUT: Eventually, each processor P_i, $1 \leq i \leq n$, outputs bit d_i satisfying the following two conditions:

- Agreement: *For any two non-faulty processors* P_i *and* P_j, $d_i = d_j$.
- Validity: *If* $b_i = b_j = b$ *for all non-faulty processors* P_i *and* P_j, *then* $d_i = b$ *for all non-faulty processors* P_i.

In the above definition input and output values are from $\{0, 1\}$. This is without loss of generality, since there is a simple two-round transformation that reduces a multi-valued agreement problem to the binary problem [19].

Key Results

Theorem 1 *Let* $t < \frac{n}{3}$. *Then there exists a polynomial-time BA protocol running in expected constant number of rounds.*

The number of rounds of the Feldman–Micali BA protocol is expected constant, but there is no bound in the worst case; that is, for every r, the probability that the protocol proceeds for more than r rounds is very small, yet greater than 0 – in fact, equal to $2^{-O(r)}$. Further, the non-faulty processors may not terminate in the same round. (Indeed, it was shown by Dwork and Moses [7] that at least $t + 1$ rounds are necessary for simultaneous termination. In [13], Goldreich and Petrank combine "the best of both worlds" by showing a BA protocol running in expected constant number of rounds which always terminates within $t + O(\log t)$ rounds.)

The Feldman–Micali BA protocol assumes synchronous rounds. As mentioned above, one of the motivations for the use of randomization was to overcome the impossibility result due to Fischer, Lynch and Paterson [11] of BA in asynchronous networks, where there is no global clock, and the adversary is also allowed to schedule the arrival time of a message sent to a non-faulty processor (of course, faulty processors may not send any message(s)). In [8], Feldman mentions that the Feldman–Micali BA protocol can be modified to work on asynchronous networks, at the expense of tolerating $t < \frac{n}{4}$ faults. In [4], Canetti and Rabin present a probabilistic asynchronous BA protocol for $t < \frac{n}{3}$ that differs from the Feldman–Micali approach in that it is a Las Vegas protocol – i.e., it has non-terminating runs, but when it terminates, it does so in constant expected rounds.

Applications

There exists a one-to-one correspondence, possibility- and impossibility-wise between BA in the unconditional setting as defined above and a formulation of the problem called the "Byzantine generals" by Lamport, Shostak and Pease [15], where there is a distinguished source among the parties sending a value, call it b_s, and the rest of the parties having to agree on it. The Agreement condition remains unchanged; the Validity condition becomes

- VALIDITY: If the source is non-faulty, then $d_i = b_s$ for all non-faulty processors P_i.

A protocol for this version of the problem realizes a functionality called a "broadcast channel" on a network with only point-to-point connectivity. Such a tool is very useful in the context of cryptographic protocols and secure multi-party computation [12]. Probabilistic BA is particularly relevant here, since it provides a constant-round implementation of the functionality. In this respect, without any optimizations, the reported actual number of expected rounds of the Feldman–Micali BA protocol is at most 57. Recently, Katz and Koo [14] presented a probabilistic BA protocol with an expected number of rounds at most 23.

BA has many other applications. Refer to the ▸ Byzantine Agreement, as well as to, e.g., [16] for further discussion of other application areas.

Cross-References

▶ Asynchronous Consensus Impossibility
▶ Atomic Broadcast
▶ Byzantine Agreement
▶ Randomized Energy Balance Algorithms in Sensor Networks

Recommended Reading

1. Ben-Or M (1983) Another advantage of free choice: completely asynchronous agreement protocols. In: Proceedings of the 22nd annual ACM symposium on the principles of distributed computing, pp 27–30
2. Ben-Or M, El-Yaniv R (2003) Optimally-resilient interactive consistency in constant time. Distrib Comput 16(4):249–262
3. Bracha G (1987) An O(log n) expected rounds randomized Byzantine generals protocol. J Assoc Comput Mach 34(4):910–920
4. Canetti R, Rabin T (1993) Fast asynchronous Byzantine agreement with optimal resilience. In: Proceedings of the 25th annual ACM symposium on the theory of computing, San Diego, 16–18 May 1993, pp 42–51
5. Chor B, Coan B (1985) A simple and efficient randomized Byzantine agreement algorithm. IEEE Trans Softw Eng SE-11(6):531–539
6. Chor B, Dwork C (1989) Randomization in Byzantine agreement. Adv Comput Res 5:443–497
7. Dwork C, Moses Y (1990) Knowledge and common knowledge in a Byzantine environment: crash failures. Inf Comput 88(2):156–186, Preliminary version in TARK'86
8. Feldman P (1988) Optimal algorithms for Byzantine agreement. PhD thesis, MIT
9. Feldman P, Micali S (1997) An optimal probabilistic protocol for synchronous Byzantine agreement. SIAM J Comput 26(4):873–933, Preliminary version in STOC'88
10. Fischer MJ, Lynch NA (1982) A lower bound for the time to assure interactive consistency. Inf Process Lett 14(4):183–186
11. Fischer MJ, Lynch NA, Paterson MS (1985) Impossibility of distributed consensus with one faulty processor. J ACM 32(2):374–382
12. Goldreich O (2001/2004) Foundations of cryptography, vols 1 and 2. Cambridge University Press, Cambridge
13. Goldreich O, Petrank E (1990) The best of both worlds: guaranteeing termination in fast randomized agreement protocols. Inf Process Lett 36(1):45–49
14. Katz J, Koo C (2006) On expected constant-round protocols for Byzantine agreement. In: Proceedings of advances in cryptology–CRYPTO 2006, Santa Barbara. Springer, Berlin/Heidelberg/New York, pp 445–462
15. Lamport L, Shostak R, Pease M (1982) The Byzantine generals problem. ACM Trans Program Lang Syst 4(3):382–401
16. Lynch N (1996) Distributed algorithms. Morgan Kaufmann, San Francisco
17. Pease M, Shostak R, Lamport L (1980) Reaching agreement in the presence of faults. J ACM 27(2):228–234
18. Rabin M (1983) Randomized Byzantine generals. In: Proceedings of the 24th annual IEEE symposium on foundations of computer science, pp 403–409
19. Turpin R, Coan BA (1984) Extending binary Byzantine agreement to multivalued Byzantine agreement. Inf Process Lett 18(2):73–76

Optimal Stable Marriage

Robert W. Irving
School of Computing Science, University of Glasgow, Glasgow, UK

Keywords

Optimal stable matching

Years and Authors of Summarized Original Work

1987; Irving, Leather, Gusfield

Problem Definition

The classical *stable marriage problem* (SM), first studied by Gale and Shapley [5], is introduced in ▶ Stable Marriage. An instance of SM comprises a set $\mathcal{M} = \{m_1, \ldots, m_n\}$ of n men and a set $\mathcal{W} = \{w_1, \ldots, w_n\}$ of n women and for each person a *preference list*, which is a total order over the members of the opposite sex. A man's (respectively woman's) preference list indicates his (respectively her) strict order of preference over the women (respectively men). A matching M is a set of n man-woman pairs in which each person appears exactly once. If the pair (m, w) is in the matching M, then m and w are *partners* in M, denoted by $w = M(m)$ and $m = M(w)$. Matching M is stable if there is no man m and woman w such that m prefers w to $M(m)$ and w prefers m to $M(w)$.

The key result established in [5] is that at least one stable matching exists for every instance of SM. In general, there may be many stable matchings, so the question arises as to what is an appropriate definition for the "best" stable matching and how such a matching may be found.

Gale and Shapley described an algorithm to find a stable matching for a given instance of SM. This algorithm may be applied either from the men's side or from the women's side. In the former case, it finds the so-called *man-optimal* stable matching, in which each man has the best partner, and each woman the worst partner, that is possible in any stable matching. In the latter case, the *woman-optimal* stable matching is found, in which these properties are interchanged. For some instances of SM, the man-optimal and woman-optimal stable matchings coincide, in which case this is the unique stable matching. In general, however, there may be many other stable matchings between these two extremes. Knuth [13] was first to show that the number of stable matchings can grow exponentially with n.

Because of the imbalance inherent, in general, in the man-optimal and woman-optimal solutions, several other notions of optimality in SM have been proposed.

A stable matching M is *egalitarian* if the sum

$$\sum_i r(m_i, M(m_i)) + \sum_j r(w_j, M(w_j))$$

is minimized over all stable matchings, where $r(m, w)$ represents the rank, or position, of w in m's preference list and similarly for $r(w, m)$. An egalitarian stable matching incorporates an optimality criterion that does not overtly favor the members of one sex – though it is easy to construct instances having many stable matchings in which the unique egalitarian stable matching is in fact the man (or woman) optimal.

A stable matching M is *minimum regret* if the value $\max(r(p, M(p)))$ is minimized over all stable matchings, where the maximum is taken over all persons p. A minimum-regret stable matching involves an optimality criterion based on the least happy member of the society, but again, minimum regret can coincide with man optimal or woman optimal in some cases, even when there are many stable matchings.

A stable matching is *rank maximal* (or *lexicographically maximal*) if, among all stable matchings, the largest number of people have their first choice partner and, subject to that, the largest number have their second choice partner and so on.

A stable matching M is *sex equal* if the difference

$$\left| \sum_i r(m_i, M(m_i)) - \sum_j r(w_j, M(w_j)) \right|$$

is minimized over all stable matchings. This definition is an explicit attempt to ensure that one sex is treated no more favorably than the other, subject to the overriding criterion of stability.

In the *weighted* stable marriage problem (WSM), each person has, as before, a strictly ordered preference list, but the entries in this list have associated costs or weights – $wt(m, w)$ represents the weight associated with woman w in the preference list of man m and likewise for $wt(w, m)$. It is assumed that the weights are strictly increasing along each preference list.

A stable matching M in an instance of WSM is *optimal* if

$$\sum_i wt(m_i, M(m_i)) + \sum_j wt(w_j, M(w_j))$$

is minimized over all stable matchings.

A stable matching M in an instance of WSM is *balanced* if

$$\max \left(\sum_i wt(m_i, M(m_i)), \sum_j wt(w_j, M(w_j)) \right)$$

is minimized over all stable matchings.

These same forms of optimality may be defined in the more general context of the stable marriage problem with incomplete preference lists (SMI); see ▶ Stable Marriage for a formal definition of this problem.

Again as described in ▶ Stable Marriage, the *stable roommates* problem (SR) is a non-bipartite generalization of SM, also introduced by Gale and Shapley [5]. In contrast to SM, an instance of SR may or may not admit a stable matching. Irving [9] gave the first polynomial-time algorithm to determine whether an SR instance admits a stable matching and if so to find one such matching.

There is no concept of man or woman optimal in the SR context, and nor is there any analogue of sex-equal or balanced matching. However, the other forms of optimality introduced above can be defined also for instances of SR and WSR (weighted stable roommates).

A comprehensive treatment of many aspects of the stable marriage problem, as of 1989, appears in the monograph of Gusfield and Irving [8], and a more recent detailed exposition is given by Manlove [14].

Key Results

The key to providing efficient algorithms for the various kinds of optimal stable matching is an understanding of the algebraic structure underlying an SM instance and the discovery of methods to exploit this structure. Knuth [13] attributes to Conway the observation that the set of stable matchings for an SM instance forms a distributive lattice under a natural dominance relation. Irving and Leather [10] characterized this lattice in terms of the so-called rotations – essentially minimal differences between lattice elements – which can be efficiently computed directly from the preference lists. The rotations form a natural partial order, the *rotation poset*, and there is a one-to-one correspondence between the stable matchings and the closed subsets of the rotation poset.

Building on these structural results, Gusfield [6] gave a $O(n^2)$ algorithm to find a Minimum-regret stable matching, improving an earlier $O(n^4)$ algorithm described by Knuth [13] and attributed to Selkow. Irving et al. [11] showed how the application of network flow methods to the rotation poset yields efficient algorithms for egalitarian and rank-maximal stable matchings as well as for an optimal stable matching in WSM. These algorithms have complexities $O(n^4)$, $O(n^5 \log n \log n)$ and $O(n^4 \log n)$, respectively. Subsequently, by using an interpretation of a stable marriage instance as an instance of 2-SAT and with the aid of a faster network flow algorithm exploiting the special structure of networks representing SM instances, Feder [3, 4] reduced these complexities to $O(n^3)$, $O(n^{3.5})$, and $O(\min(n, \sqrt{K})n^2 \log(K/n^2 + 2))$, respectively, where K is the weight of an optimal solution.

By way of contrast, and perhaps surprisingly, the problems of finding a sex-equal stable matching for an instance of SM and of finding a balanced stable matching for an instance of WSM have been shown to be NP-hard [2, 12].

The following theorem summarizes the current state of knowledge regarding the various flavors of optimal stable matching in SM and WSM.

Theorem 1 *For an instance of SM:*

1. *A minimum-regret stable matching can be found in $O(n^2)$ time.*
2. *An egalitarian stable matching can be found in $O(n^3)$ time.*
3. *A rank-maximal stable matching can be found in $O(n^{3.5})$ time.*
4. *The problem of finding a sex-equal stable matching is NP-hard.*

For an instance of WSM:

1. *An optimal stable matching can be found in $O(\min(n, \sqrt{K})n^2 \log(K/n^2+2))$ time, where K is the weight of an optimal solution.*

2. *The problem of finding a balanced stable matching is NP-hard, but can be approximated within a factor of 2 in $O(n^2)$ time.*

Among related problems that can also be solved efficiently by exploitation of the rotation structure of an instance of SM are the following [6]:

- All stable pairs, i.e., pairs that belong to at least one stable matching, can be found in $O(n^2)$ time.
- All stable matchings can be enumerated in $O(n^2 + kn)$ time, where k is the number of such matchings.

Results analogous to those of Theorem 1 are known for the more general SMI problem. In the case of the stable roommates problem (SR), some of these problems appear to be harder, as summarized in the following theorem.

Theorem 2 *For an instance of SR:*

1. *A minimum-regret stable matching can be found in $O(n^2)$ time [7].*
2. *The problem of finding an egalitarian stable matching is NP-hard. It can be approximated in polynomial time within a factor of α if and only if minimum vertex cover can be approximated within α [1, 2].*

For an instance of WSR (weighted stable roommates):

1. *The problem of finding an optimal stable matching is NP-hard, but can be approximated within a factor of 2 in $O(n^2)$ time [3].*

Applications

The best known and most important applications of stable matching algorithms are in centralized matching schemes in the medical and educational domains. ▸ Hospitals/Residents Problem includes a summary of some of these applications.

Open Problems

There remains the possibility of improving the complexity bounds for some of the optimization problems discussed and for finding better polynomial-time approximation algorithms for the various NP-hard problems.

Cross-References

▸ Hospitals/Residents Problem
▸ Ranked Matching
▸ Stable Marriage and Discrete Convex Analysis
▸ Stable Marriage with Ties and Incomplete Lists
▸ Stable Partition Problem

Recommended Reading

1. Feder T (1989) A new fixed point approach for stable networks and stable marriages. In: Proceedings of 21st ACM symposium on theory of computing, Seattle, May 1989. ACM, New York, pp 513–522
2. Feder T (1991) Stable networks and product graphs. Ph.D. thesis, Stanford University
3. Feder T (1992) A new fixed point approach for stable networks and stable marriages. J Comput Syst Sci 45:233–284
4. Feder T (1994) Network flow and 2-satisfiability. Algorithmica 11:291–319
5. Gale D, Shapley LS (1962) College admissions and the stability of marriage. Am Math Mon 69:9–15
6. Gusfield D (1987) Three fast algorithms for four problems in stable marriage. SIAM J Comput 16(1):111–128
7. Gusfield D (1988) The structure of the stable roommate problem: efficient representation and enumeration of all stable assignments. SIAM J Comput 17(4):742–769
8. Gusfield D, Irving RW (1989) The stable marriage problem: structure and algorithms. MIT, Cambridge
9. Irving RW (1985) An efficient algorithm for the stable roommates problem. J Algorithms 6:577–595
10. Irving RW, Leather P (1986) The complexity of counting stable marriages. SIAM J Comput 15(3):655–667
11. Irving RW, Leather P, Gusfield D (1987) An efficient algorithm for the "optimal stable" marriage. J ACM 34(3):532–543
12. Kato A (1993) Complexity of the sex-equal stable marriage problem. Jpn J Ind Appl Math 10:1–19
13. Knuth DE (1976) Mariages stables. Les Presses de L'Université de Montréal, Montréal
14. Manlove DF (2013) Algorithmics of matching under preferences. World Scientific, Singapore

Optimal Triangulation

Tiow-Seng Tan
School of Computing, National University of Singapore, Singapore, Singapore

Keywords

Edge-flip; Edge-insertion; Max-min height; Max-min length; Minimum weight; Min-max angle; Min-max eccentricity; Min-max length; Min-max slope

Years and Authors of Summarized Original Work

1972; Lawson
1992; Edelsbrunner, Tan, Waupotitsch
1993; Bern, Edelsbrunner, Eppstein, Mitchell, Tan
1993; Edelsbrunner, Tan

Problem Definition

Let S be a set of n *points* or *vertices* in $\mathbb{R}^2$. An *edge* is a closed line segment connecting two points. Let E be a collection of edges determined by vertices of S. The graph $\mathcal{G} = (S, E)$ is a *plane geometric graph* if (i) no edge contains a vertex other than its endpoints, that is, $ab \cap S = \{a, b\}$ for every edge $ab \in E$, and (ii) no two edges *cross*, that is, $ab \cap cd \in \{a, b\}$ for every two edges $ab \neq cd$ in E. A *triangulation* of S is a plane geometric graph $\mathcal{T} = (S, E)$ with E being maximal. Here maximality means that edges in E bound the convex hull of S, i.e., the smallest convex set in $\mathbb{R}^2$ that contains S, and subdivide its interior into disjoint faces bounded by triangles.

A plane geometric graph $\mathcal{G} = (S, E)$ can be augmented with an edge set E' so that it is a triangulation $\mathcal{T} = (S, E \cup E')$, referred to as a triangulation of $\mathcal{G}$. In this case, E is the set of *constraining edges* if it is not empty. Some triangulations of $\mathcal{G}$ are classified as *optimal triangulations* according to various shape criteria. Many of these criteria are defined as *max-min*, short for maximizing the minimum, or *min-max*, short for minimizing the maximum. The first quantifier is over all possible triangulations of $\mathcal{G}$ while the second one is over all measures (e.g., angles) μ of triangles of a triangulation. For example, in the case of a min-max μ criterion, we define the *measure* of a triangulation $\mathcal{A}$ as $\mu(\mathcal{A}) = \max\{\mu(t) : t \text{ is a triangle of } \mathcal{A}\}$. If $\mathcal{A}$ and $\mathcal{B}$ are two triangulations of $\mathcal{G}$, then $\mathcal{B}$ is called an *improvement* of $\mathcal{A}$ if either $\mu(\mathcal{B}) < \mu(\mathcal{A})$ or $\mu(\mathcal{B}) = \mu(\mathcal{A})$ and the set of triangles t of $\mathcal{B}$ with $\mu(t) = \mu(\mathcal{B})$ is a proper subset of that of $\mathcal{A}$. Triangulation $\mathcal{A}$ is *optimal* for μ, i.e., a min-max μ triangulation of $\mathcal{G}$, if there exists no improvement of $\mathcal{A}$. Hence, the computational problem addressed here is to find a specific optimal triangulation for a given plane geometric graph $\mathcal{G}$.

Key Results

There are a few algorithmic paradigms or approaches to solve the optimal triangulation problems in $\mathbb{R}^2$.

The Edge-Flip Approach

The most notable one is the edge-flip approach [11] to solve the max-min angle triangulation problem of a point set S. Given a triangulation $\mathcal{A}$ of $\mathcal{G} = (S, \emptyset)$, edge-flip is a local optimization method that operates on two adjacent triangles whose union forms a convex polygon. It replaces (or flips) the edge bd shared by triangles abd and cdb with the edge ac when the smallest angle of these triangles is smaller than that of abc and acd. In effect, an edge-flip replaces two existing triangles with two new triangles to (possibly) obtain an improvement of $\mathcal{A}$. By repeating the edge-flip until no such an edge bd exists, the algorithm produces a specific max-min angle triangulation of S, known as the Delaunay triangulation, in

$O(n^2)$ time. Besides being a max-min angle triangulation [3], Delaunay triangulation is also the min-max circumscribed circle and the min-max smallest enclosing circle [2] triangulation. Note that other approaches exist to compute the Delaunay triangulation more efficiently in $\Theta(n \log n)$ time [3].

The Edge-Insertion Approach

The edge-insertion approach is considered as an extension of the edge-flip approach, to replace one or more edges in each operation. The basic idea is to iteratively improve a current triangulation $\mathcal{A}$ by an *edge-insertion* step which adds an appropriate, new edge say qs to $\mathcal{A}$, deleting edges in $\mathcal{A}$ that cross qs and re-triangulating the resulting polygons to the left and the right of qs. In other words, the method starts by constructing an arbitrary triangulation $\mathcal{A}$ of $\mathcal{G}$ and then subsequently applies the edge-insertion steps until no further improvement exists. Same as in the case of edge-flip, this does not work for all measures μ as some may lead to suboptimal solutions. The approach is known to be applicable if the conditions of the so-called *Cake-Cutting Lemma*, which guarantees the existence of an improvement, are fulfilled; see [1,5] for details. The next theorem summarizes the results obtained by the edge-insertion approach.

Theorem 1 *For a plane geometric graph* $\mathcal{G} = (S, E)$ *of* $n = |S|$ *vertices:*

1. *A min-max angle triangulation of* $\mathcal{G}$ *can be computed in time* $O(n^2 \log n)$ *and storage* $O(n)$*.*
2. *A max-min height triangulation of* $\mathcal{G}$ *can be computed in time* $O(n^2 \log n)$ *and storage* $O(n)$*.*
3. *A min-max eccentricity triangulation of* $\mathcal{G}$ *can be computed in time* $O(n^3)$ *and storage* $O(n^2)$*.*
4. *A min-max slope triangulation of* $\mathcal{G}$ *can be computed in time* $O(n^3)$ *and storage* $O(n^2)$*.*

Let us go through those measures mentioned in the theorem. The *height* of a triangle is the minimum distance from a vertex to the opposite edge. The *eccentricity* of a triangle is the infimum over all distances between the center of the circumcircle of the triangle and points in its closure. To define the slope of a triangle, the triangulation is given as a 2D projection of a 2.5D piecewise-linear surface where each vertex of S has a third coordinate, and the slope of each triangle is its slope in $\mathbb{R}^3$.

The Subgraph Approach

The subgraph approach constructs a desired optimal triangulation by first computing a substructure of the optimal triangulation and then completes the computation by solving the smaller problems defined by the substructure. This approach works when (i) the substructure can be computed efficiently and (ii) the substructure subdivides the problem into smaller problems such as polygons that can be solved efficiently. For instance, the approach has successfully solved the min-max length triangulation problem using a substructure called relative neighborhood graph [4]. Here the length of a triangle is the length of its longest edge.

Theorem 2 *A min-max length triangulation of a set of n points in* $\mathbb{R}^2$ *can be constructed in* $O(n^2)$ *time and storage.*

Note that the theorem is formulated with reference to a set of n points instead of the general plane geometric graph. In fact, this theorem is valid for the latter provided the minimization condition is defined over all edges (of triangles) including those constraining edges. In both cases, the correctness of the theorem follows from the fact that every point set S in $\mathbb{R}^2$ has a min-max length triangulation $\mathrm{mlt}(S)$ such that $\mathrm{rng}(S) \cup \mathrm{ch}(S) \subseteq \mathrm{mlt}(S)$ where $\mathrm{rng}(S)$ is the relative neighborhood graph of S and $\mathrm{ch}(S)$ is the set of edges bounding the convex hull of S. Since $\mathrm{rng}(S)$ and $\mathrm{ch}(S)$ can each be computed in $O(n \log n)$ time, and $\mathrm{rng}(S) \cup \mathrm{ch}(S)$ is a connected graph of S, the min-max length triangulation problem can be solved by first constructing $\mathrm{rng}(S) \cup \mathrm{ch}(S)$ and then computing an optimal triangulation within each polygon defined by edges of $\mathrm{rng}(S) \cup \mathrm{ch}(S)$. The latter is solvable in

$O(n^2)$ time. Besides Euclidean metric, Theorem 2 can be extended to general normed metrics as stated in the next theorem.

Theorem 3 *Let S be a set of n points in $\mathbb{R}^2$ equipped with a normed metric. Given the relative neighborhood graph, a min-max length triangulation of S can be constructed in time $O(n^2)$.*

Examples of normed metrics are the ℓ_p-metrics, for $p = 1, 2, 3, \ldots$, and the so-called A-metric used in VLSI applications. Note that the relative neighborhood graph under the ℓ_p-metrics can be computed in $O(n \log n)$ time. As for the other normed metrics, a relative neighborhood graph can be constructed in time $O(n^3)$ with a trivial approach to test all $\binom{n}{2}$ edges, each in time $O(n)$.

We note that min-max length is currently the only nontrivial length criterion known to be solvable in polynomial time. The max-min length triangulation problem for an input point set is shown to be NP-complete, while the same problem for a convex polygon is known to be solvable in linear time [6]. Another related problem is to find the minimum weight triangulation of $\mathcal{G}$, where the weight of a triangulation is the sum of length of its edges. This problem is proven to be NP-hard [12].

Applications

Triangulation is a prominent meshing method that decomposes a domain into a collection of triangles. Such decomposition is used in many areas of engineering and scientific applications such as physics simulation, visualization, approximation theory, numerical analysis, computer-aided geometric design, etc. It is desirable to obtain an optimal triangulation, often with respect to the angle, edge length, aspect ratio, etc., as the quality of the subsequent computation depends on the shapes of the triangles. Two popular techniques that greatly depend on such optimal triangulations are finite element analysis and surface interpolation; see, for example, the survey in [7].

Open Problems

There are a few other interesting measures one can define over a triangulation, such as area, aspect ratio, and vertex degree. The min-max area and max-min area triangulation problems for a point set are still open, though the special case of a convex polygon can be solved in polynomial time [10]. The problem to triangulate a plane geometric graph with degree at most seven is known to be NP-complete [8], and the min-max degree problem for an arbitrary biconnected plane geometric graph is also NP-complete [9]. Its general problem without any constraining edges is still open.

URLs to Code and Data Sets

A version of the edge-insertion approach was implemented by Roman Waupotitsch. It is known to be available at: ftp://ftp.ncsa.uiuc.edu/SGI/MinMaxer/

Cross-References

▶ Minimum Weight Triangulation

Recommended Reading

1. Bern M, Edelsbrunner H, Eppstein D, Mitchell S, Tan TS (1993) Edge insertion for optimal triangulations. Discret Comput Geom 10(1):47–65
2. D'Azevedo E, Simpson R (1989) On optimal interpolation triangle incidences. SIAM J Sci Stat Comput 10(6):1063–1075
3. Edelsbrunner H (2000) Triangulations and meshes in computational geometry. Acta Numer 2000 9:133–213
4. Edelsbrunner H, Tan TS (1993) A quadratic time algorithm for the minmax length triangulation. SIAM J Comput 22(3):527–551
5. Edelsbrunner H, Tan TS, Waupotitsch R (1992) An $O(n^2 \log n)$ time algorithm for the minmax angle triangulation. SIAM J Sci Stat Comput 13(4):994–1008
6. Fekete SP (2012) The complexity of maxmin length triangulation. CoRR abs/1208.0202

7. Ho-Le K (1988) Finite element mesh generation methods: a review and classification. Comput Aided Des 20(1):27–38
8. Jansen K (1993) One strike against the min-max degree triangulation problem. Comput Geom 3(2):107–120
9. Kant G, Bodlaender HL (1997) Triangulating planar graphs while minimizing the maximum degree. Inf Comput 135(1):1–14
10. Keil JM, Vassilev TS (2006) Algorithms for optimal area triangulations of a convex polygon. Comput Geom Theory Appl 35(3):173–187
11. Lawson CL (1972) Transforming triangulations. Discret Math 3(4):365–372
12. Mulzer W, Rote G (2008) Minimum-weight triangulation is NP-hard. J ACM 55(2):11:1–11:29

Optimal Two-Level Boolean Minimization

Robert P. Dick
Department of Electrical Engineering and Computer Science, University of Michigan, Ann Arbor, MI, USA

Keywords

Logic minimization; Quine–McCluskey algorithm; Tabular method

Years and Authors of Summarized Original Work

1952; Quine
1955; Quine
1956; McCluskey

Problem Definition

Find a minimal sum-of-products expression for a Boolean function. Consider a Boolean algebra with elements *False* and *True*. A Boolean function $f(y_1, y_2, \ldots, y_n)$ of n Boolean input variables specifies, for each combination of input variable values, the function's value. It is possible to represent the same function with various expressions. For example, the first and last expressions in Table 1 correspond to the same function. Assuming access to complemented input variables, straightforward implementations of these expressions would require two *AND* gates and an *OR* gate for $\left(\overline{a} \wedge \overline{b}\right) \vee (\overline{a} \wedge b)$ and only a wire for $\overline{a}$. Although the implementation efficiency depends on target technology, in general terser expressions enable greater efficiency. Boolean minimization is the task of deriving the tersest expression for a function. Elegant and optimal algorithms exist for solving the variant of this problem in which the expression is limited to two levels, i.e., a layer of *AND* gates followed by a single *OR* gate or a layer of *OR* gates followed by a single *AND* gate.

Key Results

This survey will start by introducing the Karnaugh Map visualization technique, which will be used to assist in the subsequent explanation of the Quine–McCluskey algorithm for two-level Boolean minimization. This algorithm is optimal for its constrained problem variant. It is one of the fundamental algorithms in the field of computer-aided design and forms the basis or inspiration for many solutions to more general variants of the Boolean minimization problem.

Karnaugh Maps

Karnaugh Maps [4] provide a method of visualizing adjacency in Boolean space. A Karnaugh Map is a projection of an n-dimensional hypercube onto a two-dimensional surface such that adjacent points in the hypercube remain adjacent in the two-dimensional projection. Figure 1 illustrates Karnaugh Maps of 1, 2, 3, and 4 variables: a, b, c, and d.

A *literal* is a single appearance of a complemented or uncomplemented input variable in a Boolean expression. A product term or *implicant* is the Boolean product, or *AND*, of one or more literals. Every implicant corresponds to the repeated balanced bisection of Boolean space, or of the corresponding Karnaugh Map, i.e., an

Optimal Two-Level Boolean Minimization, Table 1 Equivalent representations with different implementation complexities

Expression	Meaning in English	Boolean Logic Identity
$\bar{a} \wedge \bar{b} \vee \bar{a} \wedge b$	not a and not b or not a and b	Distributivity
$\bar{a} \wedge (\bar{b} \vee b)$	Not a and either not b or b	Complements
$\bar{a} \wedge True$	Not a and *True*	Boundeness
$\bar{a}$	not a	

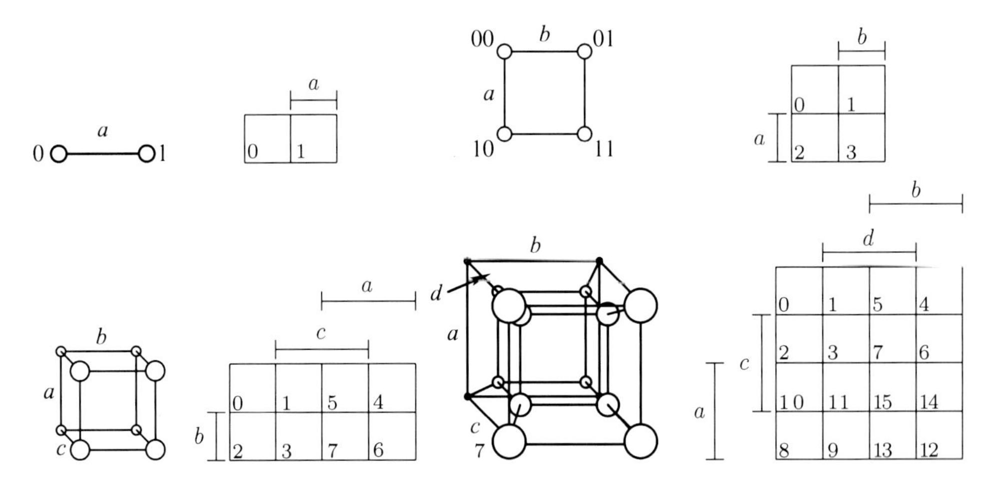

Optimal Two-Level Boolean Minimization, Fig. 1 Boolean function spaces from one to four dimensions and their corresponding Karnaugh Maps

implicant is a rectangle in a Karnaugh Map with width m and height n where $m = 2^j$ and $n = 2^k$ for arbitrary nonnegative integers j and k, e.g., the ovals in Fig. 2(ii–v). An *elementary implicant* is an implicant in which, for each variable of the corresponding function, the variable or its complement appears, e.g., the circles in Fig. 2(ii). Implicant A *covers* implicant B if every elementary implicant in B is also in A.

Prime implicants are implicants that are not covered by any other implicants, e.g., the ovals and circle in Fig. 2(iv). It is unnecessary to consider anything but prime implicants when seeking a minimal function representation because, if non-prime implicants could be used to cover some set of elementary implicants, there is guaranteed to exist a prime implicant that covers those elementary implicants and contains fewer literals. One can draw the largest implicants covering each elementary implicant and covering no positions for which the function is *False*, thereby using Karnaugh Maps to identify prime implicants. One can then manually seek a compact subset of prime implicants covering all elementary implicants in the function.

This Karnaugh Map-based approach is effective for functions with few inputs, i.e., those with low dimensionality. However, representing and manipulating Karnaugh Maps for functions of many variables is challenging. Moreover, the Karnaugh Map method provides no clear set of rules to follow when selecting a minimal subset of prime implicants to implement a function.

The Quine–McCluskey Algorithm

The Quine–McCluskey algorithm provides a formal, optimal way of solving the two-level Boolean minimization problem. W. V. Quine laid the essential theoretical groundwork for optimal two-level logic minimization [7, 8]. However, E. J. McCluskey first proposed a precise algorithm to fully automate the process [6]. Both are built upon the ideas of M. Karnaugh [4].

The Quine–McCluskey method has two phases: (1) produce all prime implicants and (2) select a minimal subset of prime implicants

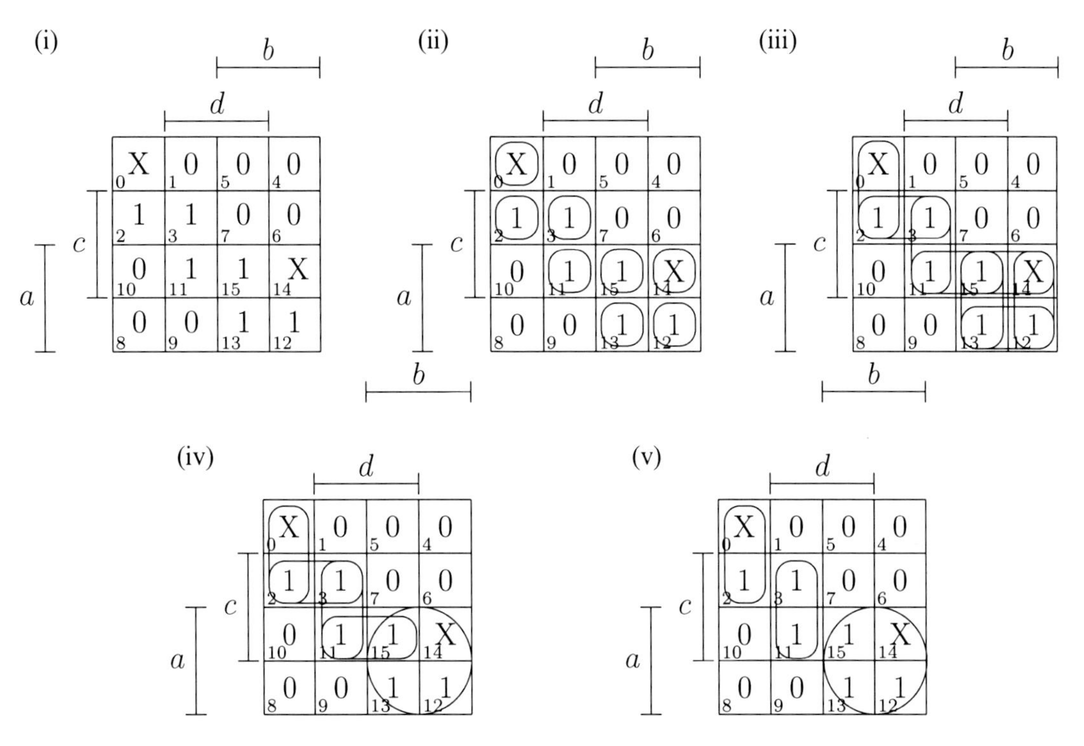

Optimal Two-Level Boolean Minimization, Fig. 2 (**i**) Karnaugh Map of function $f(a,b,c,d)$, (**ii**) elementary implicants, (**iii**) second-order implicants, (**iv**) prime implicants, and (**v**) a minimal cover

covering the function. In the first phase, the elementary implicants of a function are iteratively combined to produce implicants with fewer literals. Eventually, all prime implicants are thus produced. In the second phase, a minimal subset of prime implicants covering the on-set elementary implicants is selected using unate covering [5].

The Quine–McCluskey method may be illustrated using an example. Consider the function indicated by the Karnaugh Map in Fig. 2(i) and the truth table in Table 2. For each combination of Boolean input variable values, the function $f(a,b,c,d)$ is required to output a 0 (*False*), a 1 (*True*), or has no requirements. The lack of requirements is indicated with an X, or don't-care symbol.

Expanding implicants as much as possible will ultimately produce the prime implicants. To do this, combine on-set and don't-care elementary implicants using the reduction theorem ($\overline{a}b \vee ab = b$) shown in Table 1. The elementary implicants are circled in Fig. 2(ii) and listed in the second column of Table 3. In this figure, 0s indicate complemented variables, and 1s indicate uncomplemented variables, e.g., 0010 corresponds to $\overline{a}\overline{b}c\overline{d}$. It is necessary to determine all possible combinations of implicants. It is impossible to combine nonadjacent implicants, i.e., those that differ in more than one variable. Therefore, it is not necessary to consider combining any pair of implicants with a number of uncomplemented variables differing by any value other than 1. This

Optimal Two-Level Boolean Minimization, Table 2 Truth table of function $f(a,b,c,d)$

Elementary implicant (a,b,c,d)	Function value (a,b,c,d)	Elementary implicant	Function value
0000	X	1000	0
0001	0	1001	0
0010	1	1010	0
0011	1	1011	1
0100	0	1100	1
0101	0	1101	1
0110	0	1110	X
0111	0	1111	0

Optimal Two-Level Boolean Minimization, Table 3 Identifying prime implicants

Number of ones	Elementary Implicant (a, b, c, d)	Second-order Implicant	Third-order Implicant
0	0000 ✓	00X0	
1	0010 ✓	001X	
2	0011 ✓	X011	
	1100 ✓	110X ✓	11XX
		11X0 ✓	
3	1011 ✓	1X11	
	1101 ✓	11X1 ✓	
	1110 ✓	111X ✓	
4	1111 ✓		

Optimal Two-Level Boolean Minimization, Table 4 Solving unate covering problem to select minimal cover

Requirements (elementary implicants)	Resources (prime implicants)				
	00X0	001X	X011	1X11	11XX
0010	✓	✓			
0011		✓	✓		
1011			✓	✓	
1100					✓
1101					✓
1111				✓	✓

fact can be exploited by organizing the implicants based on the number of ones they contain, as indicated by the first column in Table 3. All possible combinations of implicants in adjacent subsets are considered. For example, consider combining 0010 with 0011, which results in 001X or $\overline{a}\overline{b}c$, and also consider combining 0010 with 1100, which is impossible due to differences in more than one variable. Whenever an implicant is successfully merged, it is marked. These marked implicants are clearly not prime implicants because the implicants they produced cover them and contain fewer literals. Note that marked implicants should still be used for subsequent combinations. The merged implicants in the third column of Table 3 correspond to those depicted in Fig. 2(iii).

After all combinations of elementary implicants have been considered, and successful combinations listed in the third column, this process is repeated on the second-order merged implicants in the third column, producing the implicants in the fourth column. Implicants that contain don't-care marks in different locations may not be combined. This process is repeated until a column yielding no combinations is arrived at. The unmarked implicants in Table 3 are the prime implicants, which correspond to the implicants depicted in Fig. 2(iv).

After a function's prime implicants have been identified, it is necessary to select a minimal subset that covers the function. The problem can be formulated as unate covering. As shown in Table 4, label each column of a table with a prime implicant; these are resources that may be used to fulfill the requirements of the function. Label each row with an elementary implicant from the on-set; these rows correspond to requirements. Do not add rows for don't cares. Don't cares impose no requirements, although they were useful in simplifying prime implicants. Mark each row–column intersection for which the elementary implicant corresponding to the row is covered by the prime implicant corresponding to the column. If a column is selected, all the rows for which the column contains marks are *covered*, i.e., those requirements are satisfied. The goal is to cover all rows with a minimal-cost subset of columns. McCluskey defined minimal cost as having a minimal number of prime implicants, with ties broken by selecting the prime implicants containing the fewest literals. The most appropriate cost function depends on the implementation technology. One can also use a similar formulation with other cost functions, e.g., minimize the total number of literals by labeling each column with a cost corresponding to the number of literals in the corresponding prime implicant.

One can use a number of heuristics to accelerate solution of the unate covering problem, e.g., neglect rows that have a superset of the marks of any other row, for they will be implicitly covered and neglect columns that have a subset of the marks of any other column if their costs are as high, for the other column is at least as useful. One can easily select columns as long as there exists a row with only one mark because the marked column is required for a valid solution. However, there exist problem instances in which each row contains multiple two marks. In the worst case, the best existing algorithms are required to make tentative decisions, determine the consequences, and then backtrack and evaluate alternative decisions.

The unate covering problem appears in many applications. It is $\mathcal{NP}$-complete [5], even for the instances arising during two-level minimization [9]. Its use in the Quine–McCluskey method predates its categorization as an $\mathcal{NP}$-complete problem by 16 years. A detailed treatment of this problem would go well beyond the scope of this entry. However, Gimpel [3] as well as Coudert and Madre [2] provide good starting points for further reading.

Some families of logic functions have optimal two-level representations that grow in size exponentially in the number of inputs, but have more compact multilevel implementations. These families are frequently encountered in arithmetic, e.g., a function indicating whether the number of on inputs is odd. Efficient implementation of such functions requires manual design or multilevel minimization [1].

Applications

Digital computers are composed of precisely two things: (1) implementations of Boolean logic functions and (2) memory elements. The Quine–McCluskey method is used to permit efficient implementation of Boolean logic functions in a wide range of digital logic devices, including computers. The Quine–McCluskey method served as a starting point or inspiration for most currently used logic minimization algorithms. Its direct use is contradicted when functions are not amenable to efficient two-level implementation, e.g., many arithmetic functions.

Cross-References

▶ Greedy Set-Cover Algorithms

Recommended Reading

1. Brayton RK, Hachtel GD, Sangiovanni-Vincentelli AL (1990) Multilevel logic synthesis. Proc IEEE 78(2):264–300
2. Coudert O, Madre JC (1995) New ideas for solving covering problems. In: Proceedings of the design automation conference, San Francisco, pp 641–646
3. Gimpel JF (1965) A reduction technique for prime implicant tables. IEEE Trans Electron Comput 14(4):535–541
4. Karnaugh M (1953) The map method for synthesis of combinational logic circuits. Trans AIEE Commun Electron 72:593–599
5. Karp RM (1972) Reducibility among combinatorial problems. In: Miller RE, Thatcher JW (eds) Complexity of computer computations. Plenum Press, New York, pp 85–103
6. McCluskey EJ (1956) Minimization of Boolean functions. Bell Syst Tech J 35(6):1417–1444
7. Quine WV (1952) The problem of simplyfying truth functions. Am Math Mon 59(8):521–531
8. Quine WV (1955) A way to simplify truth functions. Am Math Mon 62(9):627–631
9. Umans C, Villa T, Sangiovanni-Vincentelli AL (2006) Complexity of two-level logic minimization. IEEE Trans Comput-Aided Des Integr Circuits Syst 25(7):1230–1246

Orienteering Problems

Nitish Korula
Google Research, New York, NY, USA

Keywords

k-Stroll; k-TSP; k-MST; Orienteering; Routing with deadlines or time windows; Vehicle routing

Years and Authors of Summarized Original Work

2004; Bansal, Blum, Chawla, Meyerson
2007; Blum, Chawla, Karger, Lane, Meyerson, Minkoff
2011; Nagarajan, Ravi
2012; Chekuri, Korula, Pál

Problem Definition

The Orienteering problem and its variants are in the large class of vehicle routing problems, also containing the traveling salesperson problem (TSP), in which the goal is to find a short route that visits several potential destinations. Typically, the input is represented by a graph $G(V, E)$ with an associated length function $\ell: E \to R^+$, where each destination is a vertex $v \in V$, and an edge $e = (u, v)$ has length $\ell(e)$ representing the distance between u and v or the time it takes to travel between them. Unlike TSP, where the goal is to find a short tour visiting all vertices, Orienteering and its variants typically involve finding short walks that visit many vertices; having to choose the set of vertices to visit adds additional complexity to the problem.

In Orienteering, we are given a bound on the maximum length of the walk (also referred to as a budget), and the goal is to visit as many vertices as possible. A closely related problem is k-Stroll; here, we are given an integer k, and the goal is to find a walk that is as short as possible, subject to visiting at least k vertices. For both these problems, the walks are allowed to traverse an edge multiple times; the length of a walk W is $\sum_{e \in W} \ell(e)$. Hence, w.l.o.g., one can assume that the input graph is complete (by working with its metric completion) or, equivalently, that the input is represented by a metric.

We focus mainly on the "point-to-point" versions of these problems, in which the start and end vertices of the walk are also specified; here, the goal is to find a short walk from the specified start vertex to the specified end vertex that visits many other vertices. One can also consider the variant in which only the start vertex is specified (and the algorithm can choose where to end the walk) or the one in which neither the start nor the end vertex is specified. These are referred to as the "rooted" and "unrooted" variants, respectively. We define the problems formally below.

Problem 1 (Orienteering)

INPUT: Graph $G(V, E)$, with an associated length function $\ell: E \to R^+$, start and end vertices $s, t \in V$, and a budget/length bound L.

OUTPUT: An s-t walk of total length at most L.

OBJECTIVE: Maximize the number of distinct vertices in the walk.

Problem 2 (k-Stroll)

INPUT: Graph $G(V, E)$, with an associated length function $\ell: E \to R^+$, start and end vertices $s, t \in V$, and an integer k.

OUTPUT: An s-t walk containing at least k distinct vertices.

OBJECTIVE: Minimize the total length of the walk.

Orienteering and k-Stroll are "dual" problems. They are equivalent in terms of exact solvability; a polynomial-time optimal algorithm for one can be used to obtain a polynomial-time optimal algorithm for the other. However, this is not true from the standpoint of approximability; an α-approximation for one does not immediately imply an α-approximation for the other.

Orienteering with time windows (Orient-TW) is a generalization of Orienteering in which each vertex v has an associated time interval or window $[R(v), D(v)]$, and a vertex is considered "visited" (i.e., is counted toward the objective) only if the total length of the walk from the start vertex up to v is in the range $[R(v), D(v)]$. (For intuition, if the length of an edge is interpreted as the time taken to traverse it, then the vertex is counted if the time at which it is visited falls within its time window.) $R(v)$ and $D(v)$ are referred to as the release time and deadline of vertex v, respectively. A special case of this problem (sometimes called orienteering with deadlines or even deadline-TSP) is when $R(v) = 0$ for all vertices v.

Problem 3 (Orienteering with Time Windows)

INPUT: Graph $G(V, E)$, with an associated length function $\ell: E \to R^+$, start and end vertices $s, t \in V$, a budget/length bound L, and a time interval $[R(v), D(v)]$ for each vertex $v \in V$.

OUTPUT: An s-t walk of total length at most L.

OBJECTIVE: Maximize the number of distinct vertices in the walk that are visited during their time intervals. A vertex is counted as visited only if the walk visits v at a time $t \in [R(v), D(v)]$; we assume it takes ℓ units of time to cross an edge of length ℓ.

See [9] for an overview and applications of many vehicle routing problems related to Orienteering and its variants.

Key Results

Orienteering is both NP-hard and APX-hard [3]; the same applies for k-Stroll, as a generalization of TSP. Therefore, we focus primarily on approximability.

Undirected Graphs

Arkin et al. [1] gave a 2-approximation for rooted Orienteering in the Euclidean plane. Chen and Har-Peled [7] improved this to a PTAS for the plane and higher-dimensional Euclidean metrics.

For general undirected graphs (i.e., symmetric metrics), the first constant-factor approximation for *rooted* Orienteering was given by Blum et al. [3]. They obtained many of the key insights used in subsequent papers, reducing Orienteering to k-Stroll. Blum et al. [3] showed that an α-approximation to k-Stroll gives a $1 + \lceil \frac{3\alpha}{2} - \frac{1}{2} \rceil$-approximation for rooted Orienteering. This was improved by Bansal et al. [2] to $\lceil \frac{3\alpha}{2} - \frac{1}{2} \rceil$ even for the harder point-to-point variant. Since there is a $2 + \varepsilon$-approximation for k-Stroll due to [4], this gives a 3-approximation for Orienteering. Chekuri et al. [6] reduced Orienteering to a *bicriteria* version for k-Stroll; this gives the current best $2+\varepsilon$-approximation for Orienteering, matching the ratio for k-Stroll.

A key challenge in Orienteering is the hard constraint on the total length of the walk L. In particular, if the shortest path from the source s to the destination t has length close to L, then even a small detour from the shortest path to visit a cluster of many vertices might result in reaching t after the deadline L. Roughly speaking, [3] shows that an optimum walk can be broken down into segments which are "monotonic," meaning that they visit vertices in increasing order of their (shortest path) distance from s, and segments which are "non-monotonic," in which case the length of the segment is at least 3 times the shortest path between its endpoints. To find a good walk, one can use a dynamic program to "enumerate" all relevant segmentations of the optimal path; for each monotonic segment, one can find the optimum sub-path using dynamic programming. For the non-monotonic segments, one can "skip" the reward from some of them, which saves considerable distance since the detours taken by the path in such segments are large. This saving allows one to take a little extra distance to collect reward in the remaining non-monotonic segments (using an approximation algorithm for k-Stroll) while still keeping the total length of the walk at most L.

Directed Graphs

Orienteering is more challenging in directed graphs, or asymmetric metrics. The first poly-logarithmic approximation algorithms were due independently to [6, 8]; the former gave an $O(\log^2 n/\log\log n)$-approximation using an LP-based approach, while the latter gave an $O(\log^2 \text{OPT})$-approximation using combinatorial techniques. The ratio of [6] is better when OPT, the number of vertices visited by an optimal walk, is much less than n and is slightly worse otherwise; on the other hand, the LP of [8] is based on the well-known Held-Karp relaxation for asymmetric TSP, and a conjectured improved upper bound on the integrality gap of this relaxation would immediately give an improved approximation ratio for directed Orienteering. Both these papers also obtain poly-logarithmic approximations for the closely related problem-Directed k-TSP, which is the special case of k-Stroll when $s = t$.

Chekuri and Pál [5] gave a *quasi-polynomial-time* $O(\log n)$-approximation for directed Orienteering and several generalizations, including Orient-TW. This algorithm is based on repeatedly "guessing" the midpoint of sub-paths, and hence it does not appear easy to obtain a polynomial-time equivalent.

Orienteering with Time Windows

Orient-TW in undirected graphs, with arbitrary distinct release times and deadlines for each vertex, was first studied by [2], which gave an $O(\log n)$-approximation for the case with only deadlines (i.e., where $R(v) = 0$ for all v) and $O(\log^2 n)$ for the general problem. Chekuri et al. [6] later gave an $O(\max\{\log \text{OPT}, \log R\})$-approximation for Orient-TW, where R denotes the ratio between the length of the longest and shortest time windows; when all time windows are polynomially bounded, this is an $O(\log n)$-approximation. (For directed graphs, via [6, 8], we lose additional poly-logarithmic factors).

The general approach taken by these papers, following [2], is to use combinatorial techniques to reduce the given instance of the problem to a collection of subproblems in which all vertices have identical or disjoint time windows. For a set of vertices with identical time windows, these windows can be effectively ignored, yielding an instance of Orienteering; walks for different sets with disjoint time windows can be combined using dynamic programming. Thus, Orient-TW can be reduced to Orienteering with the loss of logarithmic factors in the approximation ratio.

Open Problems

There are several natural open problems related to Orienteering.

1. Is there a PTAS for Orienteering in undirected planar graphs? In recent years, PTASes have been obtained for many related problems, including TSP, STEINER TREE, and their prize-collecting versions, but extending these techniques to Orienteering and k-Stroll (or even the easier k-MST problem) seems challenging.
2. Can one obtain an $O(\log n)$ or even $O(1)$-approximation for directed Orienteering? The quasi-polynomial-time approximation of [5] provides some evidence that it may be possible. Can one obtain *any* poly-logarithmic approximation for directed k-Stroll? Currently, only bicriteria approximations are known.
3. Is there an $O(\log n)$-approximation for Orient-TW?

Recommended Reading

1. Arkin E, Mitchell J, Narasimhan G (1998) Resource-constrained geometric network optimization. In: Symposium on computational geometry, Minneapolis, pp 307–316
2. Bansal N, Blum A, Chawla S, Meyerson A (2004) Approximation algorithms for deadline-TSP and vehicle routing with time-windows. In: Proceedings of the 36th annual ACM symposium on theory of computing, Chicago. ACM, New York, pp 166–174
3. Blum A, Chawla S, Karger D, Lane T, Meyerson A, Minkoff M (2007) Approximation algorithms for orienteering and discounted-reward TSP. SIAM J Comput 37(2):653–670
4. Chaudhuri K, Godfrey B, Rao S, Talwar K (2003) Paths, trees, and minimum latency tours. In: 44th annual symposium on foundations of computer science, Cambridge. IEEE Computer Society, pp 36–45
5. Chekuri C, Pál M (2005) A recursive greedy algorithm for walks in directed graphs. In: Proceedings of the 46th annual symposium on foundations of computer science, Pittsburgh. IEEE Computer Society, pp 245–253
6. Chekuri C, Korula N, Pál M (2012) Improved algorithms for orienteering and related problems. ACM Trans Algorithms (TALG) 8(3):23
7. Chen K, Har-Peled S (2008) The orienteering problem in the plane revisited. SIAM J Comput 38(1):385–397, preliminary version in Proceedings of the ACM SoCG, Sedona, 2006, pp 247–254
8. Nagarajan V, Ravi R (2011) The directed orienteering problem. Algorithmica 60(4):1017–1030
9. Toth P, Vigo D (eds) (2001) The vehicle routing problem. SIAM monographs on discrete mathematics and applications. Society for Industrial and Applied Mathematics, Philadelphia

Orthogonal Range Searching on Discrete Grids

Yakov Nekrich
David R. Cheriton School of Computer Science, University of Waterloo, Waterloo, ON, Canada

Keywords

Orthogonal range searching; Word RAM model

Years and Authors of Summarized Original Work

1988; Chazelle
2000; Alstrup, Brodal, Rauhe
2004; JaJá, Mortensen, Shi
2007; Pătraşcu
2009; Karpinski, Nekrich
2011; Chan, Larsen, Pătraşcu
2013; Chan

Problem Definition

Let S be a set of n d-dimensional points. In the orthogonal range searching problem we keep S in a data structure, so that for an arbitrary query rectangle $Q = [a_1, b_1] \times \cdots \times [a_d, b_d]$ information about points in $Q \cap S$ can be found. Range searching is a fundamental computational geometry problem with numerous applications in data bases, text indexing, string processing, and network analysis. In computational geometry it is frequently assumed that point coordinates are real and the data structure works in the real RAM model. In a vast majority of practical situations we can, however, make a stronger assumption that point coordinates are discrete values. This scenario is captured by the word RAM model of computation: all coordinates are integers that fit into a machine word and standard operations on words can be performed in constant time. In this case, we say that points are on an integer grid. If point coordinates are also bounded by a parameter U, we say that points are on a grid of size U (also called a $U \times U$ grid if points are two-dimensional).

The discrete grid assumption leads to improved results for many range searching problems. In this entry we consider several such problems. We remark that a problem on an integer grid can be always reduced to the same problem on a grid of size n using the technique called reduction to rank space [1, 8]. This reduction takes $O(n)$ additional space and increases the query time by an additive term $\text{pred}(n, U)$, where $\text{pred}(n, U) = O(\min(\log\log U, \sqrt{\log n / \log\log n}))$ is the time needed to answer a predecessor query [2, 21]. Unless specified otherwise, we will assume that all points are on a grid of size n. Throughout this entry, the space usage of described data structures is measured in words; each word consists of $w \geq \log n$ bits and can hold a coordinate of any point from the input set.

Key Results

Orthogonal Range Reporting

Two Dimensions

The problem is to keep a set of points S in a data structure, so that all points in $Q \cap S$ for any query rectangle Q can be reported. Overmars [17] was the first to consider this problem in the discrete grid scenario. His data structure needs $O(n \log n)$ words of space and supports two-dimensional queries on $U \times U$ grid in $O(\log\log U + k)$ time, where k is the number of reported points. Alstrup et al. [1] improved the space usage and described a data structure that uses $O(n \log^{\varepsilon} n)$ space and supports queries in $O(\log\log n + k)$ time. Henceforth, ε denotes an arbitrarily small positive constant. There are also data structures that use less space but need more than constant time per reported point. Chan et al. [6] describe an $O(n)$-space data structure that answers two-dimensional queries in $O((k + 1)\log^{\varepsilon} n)$ time and an $O(n \log\log n)$-space data structure that answers queries in $O((k+1)\log\log n)$ time. Any data structure that uses $n \log^{O(1)} n$ space needs $\Omega(\log\log n + k)$ time to answer two-dimensional range reporting queries. This follows from the reduction of two-dimensional range reporting problem to the predecessor problem [12] and the lower bound for the predecessor problem [20].

In the special case of three-sided range reporting queries, the query range is bounded on three sides. Thus, a three-sided query range is a product of a closed interval and a half-open interval. Three-sided range reporting queries can be answered in $O(k + 1)$ time using a linear space data structure [1]. The restriction on point coordinates can be relaxed for one dimension in the case of three-sided queries: there is a linear-space data structure for points on $n \times \mathbb{N}$ grid that answers queries of the form $[a,b] \times (-\infty, d]$ or $[a,b] \times [c, +\infty)$ in $O(k+1)$ time. Symmetrically, there is also a linear-space data structure for points on $\mathbb{N} \times n$ grid that answers queries $[a, +\infty) \times [c,d]$ or $(-\infty, b] \times [c,d]$ in $O(k+1)$ time. We list the best currently known results for two-dimensional orthogonal range reporting in rows 1–4 of Table 1.

Three Dimensions

Three-dimensional queries can be answered in optimal $O(\log\log n + k)$ time and $O(n \log^{1+\varepsilon} n)$ space [6]. There is also an $O(n \log n (\log\log n)^3)$-space data structure that answers queries in $O((\log\log n)^2 + k \log\log n)$ time; this result is achieved by combining the approach of [11] with the point location data structure of [4]. Alternatively, there is a data structure that uses $O(n \log n)$ space and answers queries in $O(\log^{1+\varepsilon} n + k \log^{\varepsilon} n)$ time [7]. Better space bounds can be achieved for some special cases. A three-dimensional query that is a product of three closed intervals, $Q = [a,b] \times [c,d] \times [e,f]$, is called a $(2,2,2)$-sided query; this is the most general case of three-dimensional queries. We say that a three-dimensional orthogonal range reporting query Q is a $(2,2,1)$-sided query if it is a product of two closed intervals and one half-open interval: $Q = [a,b] \times [c,d] \times [e, +\infty)$. There exists a data structure that uses $O(n \log^{\varepsilon} n)$ space and answers $(2,2,1)$-sided queries in $O(\log\log n + k)$ time [6]. Since $(2,2,1)$-sided queries are not easier to answer than two-dimensional queries, this query time is optimal. There is also a data structure that uses $O(n(\log\log n)^3)$ space and answers $(2,2,1)$-sided queries in $O((\log\log n)^2 + k \log\log n)$ time; this result is obtained by combining the approach of Karpinski and Nekrich [11] with the point location data structure of Chan [4]. Finally, we can also answer $(2,2,1)$-sided queries in $O(\log^{1+\varepsilon} n + k \log^{\varepsilon} n)$ time using a linear-space data structure [7]. A $(1,1,1)$-sided query (also known as dominance query) Q is a product of three half-open intervals: $Q = (-\infty, a] \times (-\infty, b] \times (-\infty, c]$. Chan [4] describes a linear space data structure that answers $(1,1,1)$-sided queries in optimal $O(\log\log n + k)$ time. See Table 2.

Orthogonal Range Searching on Discrete Grids, Table 1 Two-dimensional orthogonal range reporting. Four-sided queries denote general two-dimensional queries

Ref.	Space	Query time	Grid	Remarks
[1]	$O(n)$	$O(k+1)$	$n \times \mathbb{N}$	3-sided
[1]	$O(n \log^{\varepsilon} n)$	$O(\log\log n + k)$	$n \times n$	4-sided
[6]	$O(n \log\log n)$	$O((k+1) \log\log n)$	$n \times n$	4-sided
[6]	$O(n)$	$O((k+1) \log^{\varepsilon} n)$	$n \times n$	4-sided
[16]	$O(n \log^{\varepsilon} n)$	$O(\log\log n + k')$	$n \times n$	Sorted 4-sided
[22]	$O(n \log\log n)$	$O((k'+1) \log\log n)$	$n \times n$	Sorted 4-sided
[16]	$O(n)$	$O((k'+1) \log^{\varepsilon} n)$	$n \times n$	Sorted 4-sided

Orthogonal Range Searching on Discrete Grids, Table 2 Three-dimensional orthogonal range reporting

Ref.	Space	Query time	Query type
[6]	$O(n \log^{1+\varepsilon} n)$	$O(\log\log n + k)$	$(2,2,2)$-sided
[11]+[4]	$O(n \log n (\log\log n)^3)$	$O((\log\log n)^2 + k \log\log n)$	$(2,2,2)$-sided
[7]	$O(n \log n)$	$O(\log^{1+\varepsilon} n + k \log^{\varepsilon} n)$	$(2,2,2)$-sided
[6]	$O(n \log^{\varepsilon} n)$	$O(\log\log n + k)$	$(2,2,1)$-sided
[11]+[4]	$O(n(\log\log n)^3)$	$O((\log\log n)^2 + k \log\log n)$	$(2,2,1)$-sided
[7]	$O(n)$	$O(\log^{1+\varepsilon} n + k \log^{\varepsilon} n)$	$(2,2,1)$-sided
[4]	$O(n)$	$O(\log\log n + k)$	$(1,1,1)$-sided

Multi-dimensional Queries

Using range trees [3], we can extend the three-dimensional data structures so that d-dimensional queries for any integer constant d can be answered. The query time and space usage grow by $O(\log^{d-3} n)$. We can also increase the arity of range trees, so that each internal node has $O(\log^{\varepsilon/(2(d-3))} n)$ children. In this case, the query time grows by $O((\log n/\log\log n)^{d-3})$ and the space usage grows by $O(\log^{d-3+\varepsilon} n)$. Thus, there is a data structure that uses $O(n \log^{2+\varepsilon} n)$ space and answers four-dimensional range reporting queries in $O(\log n + k)$ time [6]. This result almost matches the lower bound of Pătraşcu [19] stating that any $n \log^{O(1)} n$-space data structure answers four-dimensional orthogonal range reporting queries in $\Omega(\log n/\log\log n)$ time. The best currently known d-dimensional data structure needs $O((\log n/\log\log n)^{d-3} \log\log n + k)$ time and uses $O(n \log^{d-2+\varepsilon} n)$ space [6].

Emptiness Queries and One-Reporting Queries

An orthogonal range emptiness query Q asks whether Q contains any points of the input set S. A one-reporting query Q asks for an arbitrary point p in $Q \cap S$ if $Q \cap S \neq \emptyset$; if $Q \cap S = \emptyset$, a special $NULL$ value is returned. We can employ all data structures described above for answering emptiness and one-reporting queries. Any previously described data structure that answers range reporting queries in time $O(q(n) + k \cdot q'(n))$ can be used to answer emptiness and one-reporting queries (for the same dimension and the same query type) in $O(q(n))$ time and $O(q(n) + q'(n))$ time, respectively.

Two-Dimensional Range Successor and Sorted Range Reporting Queries

An orthogonal range successor (also known as range next value) query $Q = [a,b] \times [c,d]$ asks for the leftmost point in $S \cap Q$. In [16] the authors considered the following generalization of range successor queries: for a query range $Q = [a,b] \times [c,d]$ report all points in $Q \cap S$ in left-to-right order. Sorted range reporting queries can also be answered in online modus: a query can be terminated when the k' leftmost points in $Q \cap S$ are reported for any $k' \leq |Q \cap S|$ and k' can be specified at query time. Nekrich and Navarro [16] describe data structures that use $O(n)$ and $O(n \log^{\varepsilon} n)$ space and answer sorted range reporting queries in $O((k'+1)\log^{\varepsilon} n)$ and $O(\log\log n + k')$ time, respectively. The data structure of [22] needs $O(n \log\log n)$ time and supports queries in $O((k'+1)\log\log n)$ time. See Table 1. Sorted range reporting queries for $k' = 1$ are equivalent to range successor queries. Thus, data structures for sorted range reporting match the complexity of the best currently known structures for standard two-dimensional point reporting (respectively, the data structures for range successor queries match data structures for one-reporting in two dimensions).

Orthogonal Range Counting

The problem is to keep a set of points S in a data structure so that for any query rectangle Q, the number of points in $Q \cap S$ can be computed. The data structure of [10] uses $O(n(\log n/\log\log n)^{d-2})$ space and answers d-dimensional dominance counting queries (i.e., counts points in a range that is a product of d half-open intervals) in $O((\log n/\log\log n)^{d-1})$ time. We can count points in any d-dimensional rectangle by answering $O(2^d)$ d-dimensional dominance queries. Hence, d-dimensional range counting queries can also be answered in $O((\log n/\log\log n)^{d-1})$ time and $O(n(\log n/\log\log n)^{d-2})$ space for any constant d. Dynamic data structures that use $O(n)$ space, answer two-dimensional range counting queries in $O((\log n/\log\log n)^2)$ time, and support updates in poly-logarithmic time are described in [14] and [9]. Query time of the static data structure is optimal for $d = 2$ dimensions: by the lower bound of [18], any two-dimensional data structure that uses $n \log^{O(1)} n$ space needs $\Omega(\log n/\log\log n)$ time to answer queries.

We can, however, reduce the query cost if k is small, where $k = |Q \cap S|$ is the number of points in a query range. Chan and Wilkinson [5] describe a data structure that uses $O(n \log\log n)$ words of space and answers two-dimensional range counting queries in optimal $O(\log \log n + \log_w k)$ time, where w is the size of the machine word. Nekrich [15] describes a data structure that uses $O(n)$ words and counts the number of points in a two-dimensional three-sided range in optimal $O(\log_w k)$ time. There are also data structures that provide an approximate answer for two- and three-dimensional range counting queries [5, 13, 15]. These data structures return a value $\overline{k}$ such that $(1 - \delta)k \leq \overline{k} \leq (1 + \delta)k$ for an arbitrary query Q and some fixed constant $\delta > 0$. We can answer two-dimensional approximate counting queries in $O(\log \log n)$ time using an $O(n \log \log n)$-space data structure [5]; we can answer three-sided approximate counting queries in $O(1)$ time using an $O(n)$-space data structure [15]. An approximate answer to a three-dimensional dominance counting query can be obtained in $O((\log \log n)^3)$ time using an $O(n)$-space data structure; we can estimate the number of points in any three-dimensional range within the same time using an $O(n \log^3 n)$-space data structure [13]. If we plug the point location data structure of [4] into the data structure of [13], then the query time of approximate three dimensional queries is reduced to $O((\log \log n)^2)$.

Open Problems

In spite of extensive research and significant achievements, many important questions are still open. The best currently known data structure that supports two-dimensional reporting queries in optimal time needs $O(n \log^{\varepsilon} n)$ space [1]. Existence of a data structure that uses $o(n \log^{\varepsilon} n)$ space for any $\varepsilon > 0$ and achieves optimal query time is an interesting open question. Another important problem is improving the space complexity of d-dimensional range reporting for $d > 2$ dimensions and query time of d-dimensional range reporting for $d > 4$ dimensions.

Cross-References

▶ Predecessor Search

Recommended Reading

1. Alstrup S, Brodal G. S., Rauhe T (2000) New data structures for orthogonal range searching. In: Proceedings of the 41st annual symposium on foundations of computer science (FOCS 2000), Redondo Beach, pp 198–207
2. Beame P, Fich F. E. (2002) Optimal bounds for the predecessor problem and related problems. J Comput Syst Sci 65(1):38–72
3. Bentley J. L. (1980) Multi- dimensional divide-and-conquer. Commun ACM 23(4):214–229
4. Chan T. M. (2013) Persistent predecessor search and orthogonal point location on the word RAM. ACM Trans Algorithms 9(3):22
5. Chan T. M., Wilkinson B. T. (2013) Adaptive and approximate orthogonal range counting. In: Proceedings of the 24th annual ACM-SIAM symposium on discrete algorithms (SODA 2013), New Orleans, pp 241–251
6. Chan T. M., Larsen K. G., Pătraşcu M (2011) Orthogonal range searching on the RAM, revisited. In: Proceedings of the 27th SoCG, Paris, pp 1–10
7. Chazelle B (1988) A functional approach to data structures and its use in multi-dimensional searching. SIAM J Comput 17(3):427–462
8. Gabow H. N., Bentley J. L., Tarjan RE (1984) Scaling and related techniques for geometry problems. In: Proceedings of the 16th annual ACM symposium on theory of computing (STOC 84), Washington, DC, pp 135–143
9. He M, Munro J. I. (2014) Space-efficient data structures for dynamic orthogonal range counting. Comput Geom 47(2):268–281
10. JáJá J, Mortensen C. W., Shi Q (2004) Space-efficient and fast algorithms for multi-dimensional dominance reporting and counting. In: Proceedings of the 15th international symposium on algorithms and computation (ISAAC 2004), HongKong, pp 558–568
11. Karpinski M, Nekrich Y (2009) Space-efficient multi-dimensional range reporting. In: Proceedings of the 15th annual international conference on computing and combinatorics (COCOON 2009), Niagara Falls, pp 215–224
12. Miltersen P. B., Nisan N, Safra S, Wigderson A (1998) On data structures and asymmetric communication complexity. J Comput Syst Sci 57(1):37–49
13. Nekrich Y (2009) Data structures for approximate orthogonal range counting. In: Proceedings of the 20th international symposium on algorithms and computation (ISAAC 2009), Honolulu, pp 183–192
14. Nekrich Y (2009) Orthogonal range searching in linear and almost-linear space. Comput Geom 42(4):342–351
15. Nekrich Y (2014) Efficient range searching for categorical and plain data. ACM Trans Database Syst 39(1):9
16. Nekrich Y, Navarro G (2012) Sorted range reporting. In: Proceedings of the 13th Scandinavian symposium

and workshops on algorithm theory (SWAT 2012), Helsinki, pp 271–282
17. Overmars M. H. (1988) Efficient data structures for range searching on a grid. J Algorithms 9(2):254–275
18. Pătraşcu M (2007) Lower bounds for 2-dimensional range counting. In: Proceedings of the 39th annual ACM symposium on theory of computing (STOC 2007), San Diego, pp 40–46
19. Pătraşcu M (2010) Towards polynomial lower bounds for dynamic problems. In: Proceedings of the 42nd ACM symposium on theory of computing (STOC 2010), Cambridge, pp 603–610
20. Pătraşcu M, Thorup M (2006) Time-space trade-offs for predecessor search. In: Proceedings of the 38th annual ACM symposium on theory of computing (STOC 2006), Seattle, pp 232–240
21. van Emde Boas P, Kaas R, Zijlstra E (1977) Design and implementation of an efficient priority queue. Math Syst Theory 10:99–127
22. Zhou G (2013) Sorted range reporting revisited. ArXiv e-prints 12044509 1308.3326

P

P2P

Dahlia Malkhi
Microsoft, Silicon Valley Campus,
Mountain View, CA, USA

Keywords

CDN; Content delivery network; DHT; Distributed hash table; File sharing; Peer to peer; Overlay; Overlay network; Resource sharing

Years and Authors of Summarized Original Work

2001; Stoica, Morris, Karger, Kaashoek, Balakrishnan

Problem Definition

This problem is concerned with efficiently designing a serverless infrastructure for a federation of hosts to store, index and locate information, and for efficient data dissemination among the hosts. The key services of peer-to-peer (P2P) overlay networks are:

1. A keyed lookup protocol locates information at the server(s) that hold it.
2. Data store, update and retrieve operations maintain a distributed persistent data repository.
3. Broadcast and multicast support information dissemination to multiple recipients.

Because of their symmetric, serverless nature, these networks are termed *P2P* networks. Below, we often refer to hosts participating in the network as *peers*.

The most influential mechanism in this area is *consistent hashing*, pioneered in a paper by Karger et al. [21]. The idea is roughly the following. Frequently, a good way of arranging a lookup directory is a hash table, giving a fast $O(1)$-complexity data access. In order to scale and provide highly available lookup services, we partition the hash table and assign different chunks to different servers. So, for example, if the hash table has entries 1 through n, and there are k participating servers, we can have each server select a virtual identifier from 1 to n at random. Server i will then be responsible for key values that are closer to i than to any other server identifier. With a good randomization of the hash keys, we can have a more or less balanced distribution of information between our k servers. In expectation, each server will be responsible for (n/k) keys. Furthermore, the departure/arrival of a server perturbs only one or two other servers with adjacent virtual identifiers.

M.-Y. Kao (ed.), *Encyclopedia of Algorithms*,
DOI 10.1007/978-1-4939-2864-4

A network of servers that implement consistent hashing is called a *distributed hash table* (DHT). Many current-generation resource sharing networks, and virtually all academic research projects in the area, are built around a DHT idea.

The challenge in maintaining DHTs is two-fold:

Overlay routing Given a hash key *i*, and starting from any node *r* in the network, the problem is to find the server *s* whose key range contains *i*. The key name *i* bears no relation to any real network address, such as the IP address of a node, and therefore we cannot use the underlying IP infrastructure to locate *s*. An overlay routing network links the nodes, and provides them with a routing protocol, such that *r* can route toward *s* using the routing target *i*.

Dynamic maintenance DHTs must work in a highly dynamic environment in which the size of the network is not known a priori, and where there are no permanent servers for maintaining either the hash function or the overlay network (all servers are assumed to be ephemeral). This is especially acute in P2P settings, where the servers are transient users who may come and go as they wish. Hence, there must be a decentralized protocol, executed by joining and leaving peers, that incrementally maintains the structure of the system. Additionally, a joining peer should be able to correctly execute this protocol while initially only having knowledge of a single, arbitrary participating network node.

One of the first overlay network projects was **Chord** [35], after which this encyclopedia entry is named (2001; Stoica, Morris, Karger, Kaashoek, Balakrishnan). More details about Chord are given below.

Key Results

The P2P area is very dynamic and rapidly evolving. The current entry provides a mere snapshot, covering dominant and characteristic strategies, but not offering an exhaustive survey.

Unstructured Overlays

Many of the currently deployed widespread resource-sharing networks have little or no particular overlay structure. More specifically, early systems such as Gnutella version 0.4 had no overlay structure at all, and allowed every node to connect to other nodes arbitrarily. This resulted in severe load and congestion problems.

Two-tier networks were introduced to reduce communication overhead and solve the scalability issues that early networks like Gnutella version 0.4 had. Two-tier networks consist of one tier of relatively stable and powerful nodes, called servers (superpeers, ultrapeers), and a larger tier of clients that search the network though servers. Most current networks, including Edonkey/Emule, KaZaa, and Gnutella, are built using two tiers. Servers provide directory store and search facilities. Searching is either limited to servers to which clients directly connect (eDonkey/eMule) or done by limited-depth flooding among the servers (Gnutella). The two-tier design considerably enhances the scalability and reliability of P2P networks. Nevertheless, the connections among servers and between clients/servers is done in a completely ad hoc manner. Thus, these networks provide no guarantee for the success of searches, nor a bound on their costs.

Structured Overlays Without Locality Awareness

Chord

The Chord system was built at MIT and is currently being developed under FNSF's **IRIS** project (http://project-iris.net/). Several aspects of the Chord [35] design have influenced subsequent systems. We briefly explain the core structure of Chord here. Nodes have binary identifiers, assigned uniformly at random. Nodes are arranged in a linked ring according to their virtual identifiers. In addition, each node has shortcut links to other nodes along the ring, link *i* to a node 2^i away in the virtual identifier space.

In this way, one can move gradually to the target by decreasing the distance by half at every step. Routing takes on average $\log n$ hops to reach any target, in a network containing n nodes. Each node maintains approximately $\log n$ links, providing the ability to route to geometrically increasing distances.

Constant Per-Node State

Several overlay network algorithms were developed with the goal of pushing the amount of network state kept by each node in the overlay to a minimum. We refer to the state kept by a node as its *degree*, as it mostly reflects the number of connections to other nodes. **Viceroy** [23] was the first to demonstrate a dynamic network in which each node stores only five links to other network nodes, and routes to any other node in a logarithmic number of hops, $\log n$ for a network of n nodes. Viceroy provided a dynamic emulation of a butterfly network (see [11] for a textbook exposition of interconnect networks like butterfly). Later, several emulations of De Bruijn networks emerged, including the generic one of Abraham et al. (AAABMP) [1], the **distance halving** network [26], **D2B** [13], and **Koorde** [20]. Constant-degree overlay networks are too fragile for practical purposes, and may easily degrade in performance or even partition in the face of failures. A study of overlay networks under churn demonstrated these points [18]. Indeed, to the best of our knowledge, none of these constant-degree networks were built. Their main contribution, and the main reason for mentioning these works here, is to know that it is possible in principle to bring the per-node state to a bare, small constant.

Content Addressable Network

The Content Addressable Network (CAN) [31] developed at ICSI builds the network as virtual d-dimensional space, giving every node a d-dimensional identifier. The routing topology resembles a d-dimensional torus. Routing is done by following the Euclidean coordinates in every dimension, yielding a $dn^{1/d}$ hop routing strategy. The parameter d can be tuned by the network administrator. Note that for $d = \log n$, CAN's features are the same as in Chord, namely, logarithmic degree and logarithmic routing hop count.

Overlay Routing Inspired by "Small-World" Networks

The **Symphony** [24] algorithm emulates routing in a small world. Nodes have k links to nodes whose virtual identifiers are chosen at random according to a routable small-world distribution [22]. With k links, Symphony is expected to find a target in $\log^2 / k$ hops.

Overlay Networks Supporting Range Queries

One of the deficiencies of DHTs is that they support only exact key lookup; hence, they do not address well the need to locate a range of keys, or to have a fuzzy search, e.g., search for any key that matches some prefix. **SkipGraphs** [4] and the **SkipNet** [19] scheme from Microsoft (project Herald) independently developed a similar DHT based on a randomized skip list [28] that supports range queries over a distributed network. The approach in both of these networks is to link objects into a double-linked list, sorted by object names, over which "shortcut" pointers are built. Pointers from each object skip to a geometric sequence of distances in the sorted list, i.e., the first pointer jumps two items away, the second four items, and so on, up to pointer $\log n - 1$, which jumps over half of the list. Logarithmic, load-balanced lookup is achieved in this scheme in the same manner as in Chord. Because the identifier space is sorted by object names, rather than hash identifiers, ranges of objects can be scanned efficiently simply by routing to the lowest value in the range; the remaining range nodes reside contiguously along the ring.

By prefixing organization names to object names, SkipNet achieves contiguity of nodes belonging to a single organization along the ring, and the ability to map objects on nodes in their local organizations. In this way SkipNet achieves resource proximity and isolation the only system besides RP [33] to have this feature.

Whereas the SkipGraphs work focuses on randomized load-balancing strategies and proofs, the SkipNet system considers issues of dynamic

P2P, Table 1 Comparison of various measures of lookup schemes with no locality awareness

Overlay lookup scheme	Topology resemblance	Hops	Degree
Chord	Hypercube	$\log n$	$\log n$
Viceroy	Butterfly	$\log n$	5
AAABMP, Distance-halving, Koorde, D2B	De Bruijn	$\log n$	4
Symphony	Small world	$\log^2 n/k$	k
SkipGraphs/SkipNet	Skip list	$\log n$	$\log n$
CAN	Torus	$dn^{1/d}$	d

maintenance, variable base sizes, and adopts the locality-awareness strategy of Pastry [33], which is described below.

Summary of Non-Locality-Aware Networks

Each of the networks mentioned above is distinct in one or more of the following properties: The (intuitive) emulated **topology**; the expected number of **hops** required to reach a target; and the per-node **degree**. Table 1 summarizes these properties.

Locality Awareness

The problem with the approaches listed above is that they ignore the proximity of nodes in the underlying networks, and allow hopping back and forth across large physical distances in search of content. Recent studies of scalable content exchange networks [17] have indicated that up to 80 % of Internet searches could be satisfied by local hosts within one's own organization. Therefore, even one far hop might be too costly. The next systems we encounter consider proximity relations among nodes in order to obtain locality awareness, i.e., that lookup costs are proportional to the actual distance of interacting parties.

Growth-Bounded Networks

Several locality-aware lookup networks were built around a bit-fixing protocol that borrows from the seminal work of Plaxton et al. [27] (PRR). The *growth bounded* network model for which this scheme is aimed views the network as a metric space, and assumes that the densities of nodes in different parts of the network are not terribly different. The PRR [27] lookup scheme uses prefix routing, similar to Chord. It differs from Chord in that a link for flipping the ith identifier bit connects with any node whose length-i prefix matches the next hop. In this way, the scheme favors the closest one in the network. This strategy builds *geometric routing*, whose characteristic is that the routing steps toward a target increase geometrically in distance. This is achieved by having large flexibility in the choice of links for each prefix at the beginning of a route, and narrowing it down as the route progresses. The result is an overlay routing scheme that finds any target with a cost that is proportional to the shortest-distance route.

The systems that adopt the PRR algorithm are **Pastry** [33], **Tapestry** [36], and **Bamboo** [32]. A very close variant is **Kademlia** [25], in which links are symmetric. It is worth mentioning that the LAND scheme [2] improves PRR in providing a nearly optimal guaranteed locality guarantee; however, LAND has not been deployed.

Applications

Caching

The Coral network [14] from NYU, built on top of DSHT [15], has been operational since around 2004. It provides free content delivery services on top of the PlanetLab-distributed test bed [9], similar to the commercial services offered by the Akamai network. People use it to provide multiple, fast access points to content they wish to publish on the Web.

Coral optimizes access locality and download rate using locality-aware lookup provided by DSHT. Within Coral, DSHT is utilized to support locality-aware object location in two applications. First, Coral contains a collection of HTTP proxies that serve as content providers; DSHT is used by clients for locating a close-by proxy. Second, proxy servers themselves use DSHT to locate a near-by copy of content requested by the client, thus making use of copies of the content that are stored in the network, rather than going to the source of the content.

Multicast

Several works deploy an event notification or publish–subscribe service over an existing routing overlay by building reverse-routing multicast paths from a single "target" to all "sources." For example, multicast systems built in this way include the Bayeux network [38], which is built over Tapestry [36], and SCRIBE [5], which is built over Pastry. In order to publish a file, the source advertises using flooding a tuple which contains the semantic name of a multicast session and a unique ID. This tuple is hashed to obtain a node identifier which becomes the session root node. Each node can join this multicast session by sending a message to the root. Nodes along the way maintain membership information, so that a multicast tree is formed in the reverse direction. The file content (and any updates) is flooded down the tree. Narada [8] is built with the same general architecture, but differs in its choice of links, and the maintenance of data.

Routing Infrastructure

A DHT can serve well to store routing and (potentially dynamic) location information of virtual host names. This idea has been utilized in a number of projects. A naming system for the Internet called CoDoNS [30] was built at Cornell University over the BeeHive overlay [29]. CoDoNS provides a safety net and is a possible replacement for the Domain Name System, the current service for looking up host names. Support for virtual IPv6 network addresses is provided in [37] by mapping names to their up-to-date, reachable IPv4 address. The Internet Indirection Infrastructure [34] built at the University of California, Berkeley provides support for virtual Internet host addresses that allows mobility.

Collaborative Content Delivery

Recent advances provide collaborative content delivery solutions that address both load balance and resilience via striping. The content is split into pieces (quite possibly with some redundancy through error-correcting codes). The source pushes the pieces of the file to an initial group of nodes, each of which becomes a source of a distribution tree for its piece, and pushes it to all other nodes. These works demonstrate clearly the advantages of data striping, i.e., of simultaneously exchanging stripes of data, over a tree-based dissemination of the full content.

SplitStream [6] employs the Pastry routing overlay in order to construct multiple trees, such that each participating node is an inner node in only one tree. It then supports parallel download of stripes within all trees. SplitStream [6] strives to obtain load balancing between multicast nodes. It achieves that by splitting the published content into several parts, called stripes, and publishing each part separately. Each stripe is published using a tree-based multicast. The workload is divided between the participating nodes by sending each stripe using a different multicast tree. Load balance is achieved by carefully choosing the multicast trees so that each node serves as an interior node in at most one tree. This reduces the number of "free riders" who only receive data.

A very popular file-distribution network is the BitTorrent system [10]. Nodes in BitTorrent are divided into *seed* nodes and *clients*. Seed nodes contain the desired content in full (either by being original providers, or by having completed a recent download of the content). Client nodes connect with a seed node or several seed nodes, as well as a *tracker* node, whose goal is to keep track of currently downloading clients. Each client selects a group (currently, of size about 20) of other downloading clients, and exchanges chunks of data obtained from the seed(s) with them. BitTorrent employs several intricate strategies for selecting which chunks to request from what other clients, in order to obtain fair load sharing of the content distribution and, at the same time, achieve fast download.

BitTorrent currently does not contain P2P-searching facilities. It relies on central sites known as "trackers" to locate content, and to coordinate the BitTorrent download process. Recent announcements by Bram Cohen (the creator of BitTorrent) and creators of other BitTorrent clients state that new protocols based on BitTorrent will be available soon, in which the role of trackers is eliminated, and searching and coordination is done in a completely P2P manner.

Experience with BitTorrent and similar systems indicates that the main problem with this approach is that towards the end of a download, many peers may be missing the same rare chunks, and the download slows down. Fairly sophisticated approaches were published in an attempt to overcome this issue.

Recently, a number of works at Microsoft Research have demonstrated the benefits of network coding in efficient multicast, e.g., [7] and Avalanche [16]. We do not cover these techniques in detail here, but only briefly state the principal ideas that underlie them.

The basic approach in network coding is to re-encode all the chunks belonging to the file, so that each one that is shared is actually a linear combination of all the pieces. The blocks are then distributed with a description of the content. Once a node obtains these re-encoded chunks, it can generate new combinations from the ones it has, and can send those out to other peers. The main benefit is that peers can make use of any new piece, instead of having to wait for specific chunks that are missing. This means no one peer can become a bottleneck, since no piece is more important than any other. Once a peer collects sufficiently many such chunks, it may use them to reconstruct the whole file.

It is worth noting that in unstructured settings, it was recently shown that network coding offers no advantage [12].

Cross-References

▸ Geometric Spanners
▸ Routing
▸ Sparse Graph Spanners

Recommended Reading

1. Abraham I, Awerbuch B, Azar Y, Bartal Y, Malkhi D, Pavlov E (2003) A generic scheme for building overlay networks in adversarial scenarios. In: Proceedings of the international parallel and distributed processing symposium (IPDPS 2003)
2. Abraham I, Malkhi D, Dobzinski O (2004) LAND: stretch $(1 + \varepsilon)$ locality aware networks for DHTs. In: Proceedings of the ACM-SIAM symposium on discrete algorithms (SODA'04)
3. Abraham I, Badola A, Bickson D, Malkhi D, Maloo S, Ron S (2005) Practical locality-awareness for large scale information sharing. In: The 4th annual international workshop on peer-to-peer systems (IPTPS'05)
4. Aspnes J, Shah G (2003) Skip graphs. In: Fourteenth annual ACM-SIAM symposium on discrete algorithms, Baltimore, pp 384–393
5. Castro M, Druschel P, Rowstron A (2002) Scribe: a large-scale and decentralised application-level multicast infrastructure. IEEE J Sel Areas Commun (JSAC) 20(8):1489–1499, Spec Issue Netw Support Multicast Commun, ISSN:0733-8716
6. Castro M, Druschel P, Kermarrec A-M, Nandi A, Rowstron A, Singh A (2003) Splitstream: high-bandwidth multicast in a cooperative environment. In: SOSP'03
7. Chou P, Wu Y, Jain K (2004) Network coding for the internet. In: IEEE communication theory workshop
8. Chu Y, Rao SG, Zhang H (2000) A case for end system multicast. In: Proceedings of ACM SIGMETRICS, Santa Clara, pp 1–12
9. Chun B, Culler D, Roscoe T, Bavier A, Peterson L, Wawrzoniak M, Bowman M (2003) Planetlab: an overlay testbed for broadcoverage services. ACM SIGCOMM Comput Commun Rev 33:3–12
10. Cohen B (2003) Incentives build robustness in bittorrent. In: Proceedings of P2P economics workshop
11. Cormen TH, Leiserson CE, Rivest RL (1990) Introduction to algorithms. MIT
12. Fernandess Y, Malkhi D (2006) On collaborative content distribution using multi-message gossip. In: Twentieth IEEE international parallel and distributed processing symposium (IPDPS'06), Greece
13. Fraigniaud P, Gauron P (2003) The content-addressable network D2B. Technical report 1349, LRI, University Paris-Sud
14. Freedman MJ, Freudenthal E, Mazières D (2004) Democratizing content publication with coral. In: Proceedings of the 1st USENIX/ACM symposium on networked systems design and implementation (NSDI'04)
15. Freedman MJ, Mazières D (2003) Sloppy hashing and self-organizing clusters. In: Proceedings of the 2nd international workshop on peer-to-peer systems (IPTPS'03)
16. Gkantsidis C, Rodriguez P (2005) Network coding for large scale content distribution. In: IEEE/INFOCOM

17. Gummadi KP, Dunn RJ, Saroiu S, Gribble SD, Levy HM, Zahorjan J (2003) Measurement, modeling, and analysis of a peer-to-peer file-sharing workload. In: Proceedings of the nineteenth ACM symposium on Operating systems principles, ACM, pp 314–329
18. Gummadi K, Gummadi R, Gribble S, Ratnasamy S, Shenker S, Stoica I (2003) The impact of DHT routing geometry on resilience and proximity. In: Proceedings of the 2003 conference on applications, technologies, architectures, and protocols for computer communications, ACM, pp 381–394
19. Harvey NJA, Jones MB, Saroiu S, Theimer M, Wolman A (2003) Skipnet: a scalable overlay network with practical locality properties. In: Proceedings of fourth USENIX symposium on internet technologies and systems (USITS'03)
20. Kaashoek F, Karger DR (2003) Koorde: a simple degree-optimal hash table. In: 2nd international workshop on peer-to-peer systems (IPTPS'03)
21. Karger D, Lehman E, Leighton FT, Levine M, Lewin D, Panigrahy R (1997) Consistent hashing and random trees: distributed caching protocols for relieving hot spots on the world wide web. In: Proceedings of the 29th annual ACM symposium on theory of computing (STOC'97), pp 654–663
22. Kleinberg J (2000) The small-world phenomenon: an algorithmic perspective. In: Proceedings of the 32nd ACM symposium on theory of computing (STOC'00), pp 163–170
23. Malkhi D, Naor M, Ratajczak D (2002) Viceroy: a scalable and dynamic emulation of the butterfly. In: Proceedings of the 21st ACM symposium on principles of distributed computing (PODC'02), pp 183–192
24. Manku GS, Bawa M, Raghavan P (2003) Symphony: distributed hashing in a small world. In: Proceedings of the 4th USENIX symposium on internet technologies and systems (USITS'03), pp 127–140
25. Maymounkov P, Mazières D (2002) Kademlia: a peer-to-peer information system based on the XOR-metric. In: Proceedings of the 1st international workshop on peer-to-peer systems (IPTPS'02), pp 53–65
26. Naor M, Wieder U (2003) Novel architectures for p2p applications: the continuous-discrete approach. In: The fifteenth annual ACM symposium on parallelism in algorithms and architectures (SPAA'03)
27. Plaxton C, Rajaraman R, Richa A (1997) Accessing nearby copies of replicated objects in a distributed environment. In: Proceedings of the ninth annual ACM symposium on parallel algorithms and architectures (SPAA'97), pp 311–320
28. Pugh W (1989) Skip lists: a probabilistic alternative to balanced trees. In: Workshop on algorithms and data structures, pp 437–449
29. Ramasubramanian V, Sirer EG (2004) Beehive: O(1) lookup performance for power-law query distributions in peer-to-peer overlays. In: Proceedings of networked system design and implementation (NSDI)
30. Ramasubramanian V, Sirer EG (2004) The design and implementation of a next generation name service for the internet. In: Proceedings of SIGCOMM
31. Ratnasamy S, Francis P, Handley M, Karp R, Shenker S (2001) A scalable content-addressable network. In: Proceedings of the ACM SIGCOMM 2001 Technical Conference
32. Rhea S, Geels D, Roscoe T, Kubiatowicz J (2003) Handling churn in a dht. Technical report, UCB//CSD-03-1299. The University of California, Berkeley
33. Rowstron A, Druschel P (2001) Pastry: scalable, distributed object location and routing for large-scale peer-to-peer systems. In: IFIP/ACM international conference on distributed systems platforms (Middleware), pp 329–350
34. Stoica I, Adkins D, Zhuang S, Shenker S, Surana S (2002) Internet indirection infrastructure. In: Proceedings of ACM SIGCOMM, pp 73–88
35. Stoica I, Morris R, Karger D, Kaashoek MF, Balakrishnan H (2001) Chord: a scalable peer-to-peer lookup service for internet applications. In: Proceedings of the SIGCOMM
36. Zhao BY, Huang L, Stribling J, Rhea SC, Joseph AD, Kubiatowicz J (2003) Tapestry: a resilient global-scale overlay for service deployment. IEEE J Sel Areas Commun
37. Zhou L, van Renesse R, Marsh M (2002) Implementing IPv6 as a peer-to-peer overlay network. In: Proceedings of the 21st IEEE symposium on reliable distributed systems (SRDS'02), p 347
38. Zhuang SQ, Zhao BY, Joseph AD, Katz RH, Kubiatowicz J (2001) Bayeux: an architecture for scalable and fault-tolerant widearea data dissemination. In: Proceedings of the eleventh international workshop on network and operating system support for digital audio and video (NOSSDAV'01)

P

PAC Learning

Joel Ratsaby
Department of Electrical and Electronics Engineering, Ariel University of Samaria, Ariel, Israel

Keywords

Computational learning theory; Probably approximately correct learning

Years and Authors of Summarized Original Work

1984; Valiant

Problem Definition

Valiant's work defines a model for representing the general problem of learning a Boolean concept from examples. The motivation comes from classical fields of artificial intelligence [2], pattern classification [8], and machine learning [13]. Classically, these fields have employed numerous heuristics for representing knowledge and defining criteria by which computer algorithms can learn. The pioneering work of [16, 17] provided the leap from heuristic-based approaches to a rigorous statistical theory of pattern recognition (see also [1, 7, 14]). Their main contribution was the introduction of probabilistic upper bounds on the generalization error which hold uniformly over a whole class of concepts. Valiant's main contribution is in formalizing this probabilistic theory into a general model for computational inference. This model which is known as the *Probably Approximately Correct* (PAC) model of learnability is concerned with computational complexity of learning. In his formulation, learning is depicted as an interaction between a teacher and a learner with two main procedures, one which provides randomly drawn examples x of the concept c that is being learned and the second acts as an oracle which provides the correct classification label $c(x)$. Based on a finite number of such examples drawn identically and independently according to *any* fixed probability distribution, the aim of the learner is to infer an approximation of c which is correct with high confidence. Using the terminology of [12] suppose X denotes the space of instances, i.e., objects which a learner can obtain as training examples. A *concept* over X is a Boolean mapping from X to $\{0,1\}$. Let $\mathbb{P}$ be any fixed probability distribution over X and c a fixed *target* concept to be learned. For any hypothesis concept h over X define by $L(h) = \mathbb{P}(c(x) \neq h(x))$ the *error* of h, i.e., the probability that h disagrees with c on a test instance x which is drawn according to $\mathbb{P}$. Then according to Valiant, an algorithm $\mathbb{A}$ for learning c is one which runs in time t and with a sample of size m where both t and m are polynomials with respect to some parameters (to be specified below) and produces a hypothesis concept h such that with high confidence $L(h)$ is small.

Key Results

The main result of Valiant's work is a formal definition of what constitutes a *learnable* problem. Formally, this is stated as follows: let $\mathcal{H}$ be a class of concepts over X. Then $\mathcal{H}$ is *learnable* if there exists an algorithm $\mathbb{A}$ with the following property: for every possible target concept $c \in \mathcal{H}$, for every probability distribution $\mathbb{P}$ on X (this is sometimes referred to as the "distribution-independence" assumption), for all values of a confidence parameter $0 < \delta < 1/2$ and an approximation accuracy parameter $0 < \epsilon < 1/2$, if $\mathbb{A}$ receives as input the value of δ, ϵ and a sample $S = \{(x_i, c(x_i))\}_{i=1}{}^m$ of cardinality m (which may depend on ε and δ) which consists of examples x_i that are randomly drawn according to $\mathbb{P}$ and labeled by an oracle as $c(x_i)$, then with probability $1 - \delta$, $\mathbb{A}$ outputs a hypothesis concept $h \in \mathcal{H}$ such that the error $L(h) \leq \epsilon$. That ϵ can be arbitrarily close to zero follows from what is known as the "noise-free" assumption, i.e., that the labels comprise the true value of the target concept. If $\mathbb{A}$ runs in time t and if t and m are polynomial in $1/\epsilon$ and $1/\delta$ (and possibly other relevant parameters, such as n if the space of instance X is $\{0,1\}^n$ or $\mathbb{R}^n$), then $\mathcal{H}$ is *efficiently* PAC learnable.

Valiant has shown that the following classes are all efficiently PAC learnable: the class of conjunctive normal form expressions with a bounded number of literals in each clause, the class of monotone disjunctive normal form expressions (here the learner requires in addition to S also an oracle that can answer membership queries, i.e., provide the true label $c(x)$ for an x in question), and the class of arbitrary expressions in which each variable occurs just once (using more powerful oracles). The work following Valiant's paper (see [11] for references) has shown that the classes of k-DNF, k-CNF, and k-decision lists are efficiently PAC learnable for each fixed k. Under suitable complexity-theoretic hardness assumptions, the class of concepts in the form of a disjunction of two conjunctions is not PAC learnable and neither is the class of existential conjunctive concepts on structural instance spaces with two objects. Linear threshold concepts (perceptrons) are PAC learnable on both Boolean and real-valued instance spaces, but the class of concepts in the form of a conjunction of two linear threshold concepts is not PAC learnable. The same holds for disjunctions and linear thresholds of linear thresholds (i.e., multilayer perceptrons with two hidden units). If the weights are restricted to 1 and 0 (but the threshold is arbitrary), then linear threshold concepts on Boolean instance spaces are not PAC learnable.

It should be noted that the notion of PAC learnability discussed throughout this entry is sometimes referred to as "proper" PAC learnability because of the requirement that, when learning a concept class $\mathcal{H}$, the learning algorithm must output a hypothesis that also belongs to $\mathcal{H}$. Several of the negative results mentioned above can be circumvented in a model of "improper" PAC learning, where the learning algorithm is allowed to output hypotheses from a broader class of functions than $\mathcal{H}$. This is sometimes referred to as agnostic PAC learnability (see [15], Ch. 3, [12] and the proceedings of the COLT conferences for many results of this type).

Applications

Valiant's paper is a milestone in the history of the area known as *Computational Learning Theory* (see proceedings of COLT conferences). The PAC model has been criticized in that the distribution-independence assumption and the notion of target concepts with noise-free training data are unrealistic in practice, e.g., in machine learning and AI. There has thus been much work on learning models that relax several of the assumptions in Valiant's PAC model. These include models which allow noisy labels or remove the assumptions on the independence of training examples, relax the assumption on the probability distribution to be fixed, allow the bounds to be distribution dependent, permit the training sample to be picked by the learner and labeled by the oracle instead of the random sample or chosen by a helpful teacher, allow learning regression, and use generalized loss functions. For references, see Sec. 2.6 of [1] and Ch. 3 of [15]. An important follow-up of Valiant's model was the work of [6] who unified his model with the uniform convergence results of [16]. They showed the important dependence between the notion of learnability and certain combinatorial properties of concept classes, one of which is known as the Vapnik-Chervonenkis (VC) dimension (see Sec. 3.4 of [1] for history on the VC-dimension).

Cross-References

▸ Attribute-Efficient Learning
▸ Hardness of Proper Learning
▸ Learning Constant-Depth Circuits
▸ Learning DNF Formulas
▸ Learning with Malicious Noise
▸ Learning with the Aid of an Oracle

Recommended Reading

For a recommended collection of works on the PAC model and its extensions see [5–7].

1. Anthony M, Bartlett PL (1999) Neural network learning: theoretical foundations. Cambridge University Press, Cambridge, U.K
2. Barr A, Feigenbaum EA, Cohen PR (eds) (1981) The handbook of artificial intelligence, vol 1. William Kaufmann, Los Altos
3. Barr A, Feigenbaum EA, Cohen PR (eds) (1982) The handbook of artificial intelligence, vol 2. William Kaufmann, Los Altos
4. Barr A, Feigenbaum EA, Cohen PR (eds) (1989) The handbook of artificial intelligence, vol 4. Addison-Wesley. Reading
5. Cohen PR, Feigenbaum EA, Cohen PR (eds) (1982) The handbook of artificial intelligence, volu 3. HeurisTech Press/William Kaufmann, Stanford/Los Altos

6. Blumer A, Ehrenfeucht A, Haussler D, Warmuth M (1989) Learnability and the Vapnik-Chervonenkis dimension. J ACM 36(4):929–965
7. Devroye L, Gyorfi L, Lugosi G (1996) A probabilistic theory of pattern recognition. Springer, New York
8. Duda RO, Hart PE, Stork DG (2000) Pattern classification. Wiley, New York
9. Haussler D (1990) Applying Valiant's learning framework to AI concept learning problems. In: Michalski R, Kodratoff Y (eds) Machine learning: an artificial intelligence approach, vol III. Morgan Kaufmann, Los Altos, pp 641–669
10. Haussler D (1992) Decision theoretic generalizations of the PAC model for neural net and other learning applications. Inf Comput 100(1):78–150
11. Haussler D (1996) Probably approximately correct learning and decision-theoretic generalizations. In: Smolensky P, Mozer M, Rumelhart D (eds) Mathematical perspectives on neural networks. L. Erlbaum Associates, Mahwah, pp 651–718
12. Kearns MJ, Vazirani UV (1997) An introduction to computational learning theory. MIT, Cambridge
13. Mitchell T (1997) Machine learning. McGraw Hill, New York
14. Pearl J (1979) Capacity and error-estimates for boolean classifiers with limited complexity. IEEE Trans Pattern Recognit Mach Intell PAMI-1(4): 350–356
15. Shalev-Shwartz S, Ben-David S (2014) Understanding machine learning; From theory to algorithms. Cambridge University Press, New York
16. Vapnik VN (1982) Estimations of dependences based on statistical data. Springer, New York
17. Vapnik VN, Chervonenkis AY (1971) On the uniform convergence of relative frequencies of events to their probabilities. Theory Probab Appl 16:264–280

Packet Routing

Lenore J. Cowen
Department of Computer Science, Tufts University, Medford, MA, USA

Keywords

Job shop scheduling; Store-and-forward routing

Years and Authors of Summarized Original Work

1988; Leighton, Maggs, Rao

Problem Definition

A collection of packets need to be routed from a set of specified sources to a set of specified destinations in an arbitrary network. Leighton, Maggs and Rao [5] looked at a model where this task is divided into two separate tasks: the first is the *path selection* task, where for each specified packet i with source s_i and packet destination t_i, a simple (meaning edges don't repeat) path P_i through the network from s_i to t_i is pre-selected. Packets traverse the network in a *store and forward* manner: each time a packet is forwarded it travels along the next link in the pre-selected path. It is assumed that only one packet can cross each individual link at each given global (synchronous) timestep. Thus, when there is contention for a link, packets awaiting traversal are stored in the local link's queue (special source and sink queues of unbounded size are also defined that store packets at their origins and destinations). Thus, the second task, and the focus of the Leighton, Maggs and Rao result (henceforth called the LMR result) is the *scheduling* task: a determination, when a link's queue is not empty, of which packet gets to traverse the link in the next timestep (where it is assumed to immediately join the link queue for its next hop). The goal is to schedule the packets so that the *maximum* time that it takes any packet to reach its destination is minimized.

There are two parameters of the network together with the pre-selected paths that are clearly relevant. One is the *congestion* c, defined as the maximum number of paths that all use the same link. The other is the *dilation* d, which is simply the length of the longest path that any packet traverses in the network. Clearly each of c and d is a lower-bound on the length of any schedule that routes all the packets to their destinations. It is easy to see that a schedule of length at most cd always exists. In fact, any schedule that never lets a link go idle if there is a packet that can use that link at that timestep is guaranteed to terminate in cd steps, because each packet traverses at most d links, and at any link can be delayed by at most $c - 1$ other packets.

Key Results

The surprising and beautiful result of LMR is as follows:

Theorem ([5]) *For any network G with a pre-specified set of paths P with congestion c and dilation d, there exists a schedule of length* $O(c + d)$*, where the queue sizes at each edge are always bounded by a constant.*

The original proof of the LMR paper is non-constructive. That is, it uses the Local Lemma [3] to prove the existence of such a schedule, but does not give a way to find it. In his book [10], Scheideler showed that in fact, a $O(c + d)$ schedule exists with edge queue sizes bounded by 2 (and gave a simpler proof of the original LMR result). A subsequent paper of Leighton, Maggs and Richa in 1999 [6] provides a constructive version of the original LMR paper as follows:

Theorem ([6]) *For any network G with a pre-specified set of paths P with congestion c and dilation d, there exists a schedule of length* $O(c + d)$*. Furthermore, such a schedule can be found in* $O(p \log^{1+\epsilon} p \log^{*}(c + d))$ *time for any* $\epsilon > 0$*, where p is the sum of the lengths of the paths taken by the packets and ε is incorporated into the constant hidden by the big-O in the schedule length.*

The algorithm in the paper is a randomized one, though the authors claim that it can be derandomized using the method of conditional probabilities. However, even though the algorithm of Leighton, Maggs and Richa is constructive, it is still an offline algorithm: namely, it requires full knowledge of all packets in the network and the precise paths that each will traverse in order to construct the schedule. The original LMR paper also gave a simple randomized *online* algorithm, that, by assigning delays to packets independently and uniformly at random from an appropriate interval, results in a schedule which is much better than greedy schedules, though not as good as the offline constructions.

Theorem ([5]) *There is a simple randomized online algorithm for producing, with high probability, a schedule of length* $O(c + d \log(Nd))$ *using queues of size* $O(\log(Nd))$*, where c is the congestion, d is the dilation, and N is the number of packets.*

In the special case where it is assumed that all packets follow shortest paths in the network, Meyer, auf der Heide and Vöcking produced a simple randomized online algorithm that produces, with high probability, a schedule of length $O(c + d + \log Nd)$ steps, but queues can be as large as $O(c)$ [7]. For arbitrary paths, the LMR online result was ultimately improved to $O(c + d + \log^{1+\epsilon} N)$ steps, for any $\epsilon > 0$ with high probability, in a series of two papers by Rabani and Tardos [9], and Rabani and Ostrovsky [8]. Online protocols have also been studied in a setting where additional packets are dynamically injected into the network in adversarial settings, see [10] for a survey.

The discussion is briefly returned to the first task, namely to pre-construct the set of paths. Clearly, the goal is to find, for a particular set of packets with pre-specified sources and destinations, a set of paths that minimizes $c + d$. Srinivasan and Teo [12] designed an off-line algorithm that produces a set of paths whose $c + d$ is provably within a constant factor of optimal. Together with the offline LMR result, that gives a constant-factor approximation problem for the offline store-and-forward packet routing problem. Note that the approach of trying to minimize $c + d$ rather than c alone seems crucial; producing schedules within a constant factor of optimal congestion c is hard, and in fact has been shown to be related to the integrality gap for multicommodity flow [1, 2].

Applications

Network Emulations

Typically, a guest network G is emulated by a host network H by embedding G into H. Nodes of G are mapped to nodes of H, while edges of G are mapped to paths in H. If P is the set of

e paths (each corresponding to an edge in the guest network G), the congestion and dilation can be defined analogously as in the main result for the set of paths P, namely c denotes the maximum number of paths that use any one edge of H, and d is the length of the longest path in P. In addition, the *load* l is defined to be the maximum number of nodes in G that are mapped to a single node of H. Once G is embedded in H, H can emulate G as follows: Each node of H emulates the local computations performed by the l (or fewer) nodes mapped to it in $O(l)$ time. Then for each packet sent along an edge of G, H sends a packet along the corresponding path in the embedding; using the offline LMR result this takes $O(c + d)$ steps. Thus, H can emulate each step of G in $O(c + d + l)$ steps.

Job Shop Scheduling

Consider a scheduling problem with jobs $j_1, \dots j_r$ and machines $m_1, \dots, m_s$ for which each job must be performed on a specified sequence of machines (in a specified order). Assume each job spends unit time on each machine, and that no machine has to work on any job more than once (In the language of job-shop scheduling, this is the *non-preemptive, acyclic, job-shop scheduling problem*, with unit jobs). There is a mapping of sequences of machines to paths and jobs to packets so that this becomes an encoding of the main packet routing problem, where if c is now to be the maximum number of jobs that have to be run on any one machine, and d to be the maximum number of different machines that work on any single job, there becomes $O(c)$ congestion and $O(d)$ dilation for the corresponding packet-routing instance. Then the offline LMR result shows that there is a schedule that completes all jobs in $O(c + d)$ steps, where in addition, each job waits at most a constant number of steps in between consecutive machines (and the queue of jobs waiting for any particular machine will always be bounded by a constant). Similar techniques to those developed in the LMR paper have subsequently been applied to more general instances of Job-Shop Scheduling; see [4, 11].

Open Problems

The main open problem is whether there is a randomized *online* packet scheduling that matches the offline LMR bound of $O(c + d)$. The bound of [8] is close, but still grows logarithmically with the total number of packets.

For job shop scheduling, it is unknown whether the constant-factor approximation algorithm for the non-preemptive acyclic job-shop scheduling problem with unit length jobs implied by LMR can be improved to a PTAS. It is also unknown whether there is a constant-factor approximation in the case of arbitrary-length jobs.

Recommended Reading

1. Andrews M, Zhang L (2005) Hardness of the undirected congestion minimization problem. In: Proceedings of the 37th annual ACM symposium on theory of computing, pp 284–293
2. Chuzhoy J, Naor J (2004) New hardness results for congestion minimization and machine scheduling. In: Proceedings of the 36th annual ACM symposium on theory of computing. ACM, New York, pp 28–34
3. Erdös P, Lovász L (1975) Problems and results on 3-chromatic hypergraphs and some related questions. Colloq Math Soc János Bolyai 10:609–627
4. Goldberg LA, Patterson M, Srinivasan A, Sweedick E (2001) Better approximation guarantees for job-shop scheduling. SIAM J Discret Math 14(1):67–92
5. Leighton FT, Maggs BM, Rao SB (1994) Packet routing and jobs-hop scheduling in O(congestion+dilation) steps. Combinatorica 14(2):167–180
6. Leighton FT, Maggs BM, Richa AW (1999) Fast algorithms for finding O(congestion + dilation) packet routing schedules. Combinatorica 19(3):375–401
7. Meyer auf der Heide F, Vöcking B (1999) Shortest-path routing in arbitrary networks. J Algorithms 31(1):105–131
8. Ostrovsky R, Rabani Y (1997) Universal O (congestion + dilation + $\log^{1+\varepsilon} N$) local control packet switching algorithm. In: Proceedings of the twenty-ninth ACM symposium on theory of computing, pp 644–653
9. Rabani Y, Tardos E (1996) Distributed packet switching in arbitrary networks. In: The 28th ACM symposium on theory of computing, pp 366–376
10. Scheideler C (1998) Universal routing strategies for interconnection networks, vol 1390, Lecture Notes in Computer Science. Springer

11. Shmoys DB, Stein C, Wein J (1994) Improved approximation algorithms for shop scheduling problems. SIAM J Comput 23(3):617–632
12. Srinivasan A, Teo CP (2000) A constant-factor approximation algorithm for packet routing and balancing local vs. global criteria. SIAM J Comput 30(6):2051–2068

Packet Switching in Multi-queue Switches

Markus Schmidt
Institute for Computer Science, University of Freiburg, Freiburg, Germany

Keywords

Online packet buffering; Online packet routing

Years and Authors of Summarized Original Work

2004; Azar, Richter, Albers, Schmidt

Problem Definition

A multi-queue network switch serves m incoming queues by transmitting data packets arriving at m input ports through one single output port. In each time step, an arbitrary number of packets may arrive at the input ports, but only one packet can be passed through the common output port. Each packet is marked with a value indicating its priority in the Quality of Service (QoS) network. Since each queue has bounded capacity B and the rate of arriving packets can be much higher than the transmission rate, packets can be lost due to insufficient queue space. The goal is to maximize the throughput which is defined as the total value of transmitted packets. The problem comprises two dependent questions: buffer management, namely which packets to admit into the queues, and scheduling, i.e., which (FIFO) queue to use for transmission in each time step.

Two scenarios are distinguished: (a) unit packet value (All packets have the same value.), (b) arbitrary packet values.

The problem is considered as an online problem, i.e., at time step t, only the packet arrivals until t are known, but nothing about future packet arrivals. The online switch performance in QoS based networks is studied by using competitive analysis in which the throughput of the online algorithm is compared to the throughput of an optimal offline algorithm knowing the whole arrival sequence in advance.

If not stated otherwise, the admission control is assumed to allows preemption, i.e., packets once enqueued need not necessarily be transmitted, but can be discarded.

Problem 1 (Unit Value Problem) All packets have value 1. Since all packets are thus equally important, the admission control policies simplify: All arriving packets are to be enqueued; in the case of buffer overflow, it does not matter which packets are stored in the queue and which packets are discarded.

Problem 2 (General problem) Each packet has its individual value where usually a range $[1, \alpha]$ is given for all packets. A special case consists in the two value model where the values are restricted to $\{1, \alpha\}$.

Key Results

Unit Value Packets

Deterministic Algorithms

Theorem 1 ([1]) *For any buffer size B, the competitive ratio of each deterministic online algorithm is not smaller than* $(e_B + \frac{2}{B})/(e_B - 1 + \frac{1}{B}) \geq \frac{e}{e-1} \approx 1.58$ *where* $e_B = ((B+1)/B)^B$.

Theorem 2 ([4]) *Every work-conserving online algorithm is 2-competitive.*

Theorem 3 ([1]) *For any buffer size B, the competitive ratio of any greedy algorithm, which* *always serves a longest queue (LQF), is at least* $2 - \frac{1}{B}$ *if* $m \gg B$.

Algorithm: *SGR* (Semi-Greedy) In each time step, the algorithm executes the first rule that applies to the current buffer configuration.

1. If there is a queue buffering more than $\lfloor B/2 \rfloor$ packets, serve the queue currently having the maximum load.
2. If there is a queue the hitherto maximum load of which is less than B, serve among these queues the one currently having the maximum load.
3. Serve the queue currently having the maximum load.

Ties are broken by choosing the queue with the smallest index. The hitherto maximum load is reset to 0 for all queues whenever all queues are unpopulated in *SGR*'s configuration.

Theorem 4 ([1]) *If B is even, then SGR is* $\frac{17}{9} \approx 1.89$*-competitive. If B is odd, then SGR is* $(\frac{17}{9} + \frac{\delta_B}{9})$ *competitive where* $\delta_B = \frac{2}{B+1}$.

Theorem 5 ([3]) *Algorithm* $E^{M^{\hat{EP}'}}$ *(not stated in detail due to space limitation), which is based on a water level algorithm and uses a fractional matching in an online constructed graph, achieves a competitiveness of* $e/(e-1)(1 + (\lfloor H_m + 1 \rfloor)/B)$, *where* H_m *denotes the* m^{th} *harmonic number. Thus,* $E^{M^{\hat{EP}'}}$ *is asymptotically* $\frac{e}{e-1}$*-competitive for* $B \gg \log m$.

Randomized Algorithms

Theorem 6 ([1]) *The competitive ratio of each randomized online algorithm is at least* $\varrho = 1.4659$ *for any buffer size B (* $\varrho = 1 + \frac{1}{\alpha+1}$ *where* α *is the unique positive root of* $e^{\alpha} = \alpha + 2$*).*

Theorem 7 (Generalizing technique [9]) *If there is a randomized c-competitive algorithm A for* $B = 1$*, then there is a randomized c-competitive algorithm* $\tilde{A}$ *for all B.*

Algorithm: *RS* (Random Schedule)

1. The algorithm uses m auxiliary queues $Q_1, \ldots, Q_m$ of sizes $B_1, \ldots, B_m$ (different buffer sizes at the distinct ports are allowed), respectively. These queues contain real numbers from the range (0,1), where each number is labeled as either marked or unmarked. Initially, these queues are empty.
2. Packet arrival: If a new packet arrives at queue q_i, then the algorithm chooses uniformly at random a real number from the range (0,1) that is inserted into queue Q_i and labeled as unmarked. If queue Q_i was full when the packet arrived, the number at the head of the queue is deleted prior to the insertion of the new number.
3. Packet transmission: Check whether queues $Q_1, \ldots, Q_m$ contain any unmarked number. If there are unmarked numbers, let Q_i be the queue containing the largest unmarked number. Change the label of the largest number to "marked" and select queue q_i for transmission. Otherwise (no unmarked number), transmit a packet from any non-empty queue if such exists.

Theorem 8 ([4]) *Randomized algorithm RS is* $\frac{e}{e-1} \approx 1.58$*-competitive.*

Algorithm: *RP* (Random Permutation) Let $\mathcal{P}$ be the set of permutations of $\{1, \ldots, m\}$, denoted as m-tuples. Choose $\pi \in \mathcal{P}$ according to the uniform distribution and fix it. In each transmission step, choose among the populated queues that one whose index is most to the front in the m-tuple π.

Theorem 9 ([9]) *Randomized algorithm RP is* $\frac{3}{2}$*-competitive for* $B = 1$*. By Theorem 7, there is a randomized algorithm* $\tilde{RP}$ *that is* $\frac{3}{2}$*-competitive for arbitrary B.*

Arbitrary Value Packets

Definition 1 A switching algorithm *ALG* is called *comparison-based* if it bases its decisions on the relative order between packet values (by performing only comparisons), with no regard to the actual values.

Theorem 10 (Zero-one principle [5]) *Let ALG be a comparison-based switching algorithm (deterministic or randomized). ALG is c-competitive if and only if ALG achieves a c-competitiveness for all packet sequences whose values are restricted to $\{0, 1\}$ for every possible way of breaking ties between equal values.*

Algorithm: *GR* (Greedy) Enqueue a new packet if

- the queue is not full
- or a packet with the smallest value in the queue has a lower value then the new packet. In this case, a smallest value packet is discarded and the new packet in enqueued.

Algorithm: *TLH* (Transmit Largest Head)

1. Buffer management: Use algorithm *GR* independently in all m incoming queues.
2. Scheduling: At each time step, transmit the packet with the largest value among all packets at the head of the queues.

Theorem 11 ([5]) *Algorithm TLH is 3-competitive.*

Algorithm: *TL* (Transmit Largest)

1. Buffer management: Use algorithm *GR* independently in all m incoming queues.
2. Scheduling: At each time step, transmit the packet with the largest value among all packets stored in the queues.

Algorithm: GS^A (Generic Switch)

1. Buffer management: Apply buffer management policy A to all m incoming queues.
2. Scheduling: Run a simulation of algorithm *TL* (in the preemptive relaxed model) with the online input sequence σ. Adopt all scheduling decisions of *TL*, i.e., at each time step, transmit the packet at the head of the queue used by *TL* simulation.

Theorem 12 (General reduction [4]) *Let GS^A denote the algorithm obtained by running algorithm GS with the event-driven single-queue buffer management policy A (preemptive or non-preemptive) and let c_A be the competitive ratio of A. The competitive ratio of GS^A satisfies $c_{GS^A} \leq 2 \cdot c_A$.*

Applications

The unit value scenario models most current networks, e.g., IP networks which only support a "best effort" service in which all packet streams are treated equally, whereas the scenario with arbitrary packet values integrates full QoS capabilities.

The general reduction technique allows to restrict oneself to investigate single-queue buffer problems. It can be applied to a 1.75-competitive algorithm named *PG* by Bansal et al. [7], which achieves the best ratio known today, and yields an algorithm GS^{PG} that is 3.5-competitive for multi-queue buffers (3.5 is still higher than 3 which is the competitive ratio of TLH). In the 2-value preemptive model, Lotker and Patt-Shamir [8] presented a *mark&flush* algorithm *mf* that is 1.30-competitive for single queue buffers and that the general reduction technique transforms into a 2.60-competitive algorithm GS^{mf} for multi-queue buffers.

For the general non-preemptive model, Andelman et al. [2] presented a policy for a single queue called *Exponential-Interval Round-Robin* (*EIRR*), which is $(e\lceil\ln\alpha\rceil)$-competitive, and showed also a lower bound of $\Theta(\log\alpha)$. In the multi-queue buffer case, the general reduction technique provides a non-preemptive $(e\lceil\ln\alpha\rceil)$-competitive algorithm.

Open Problems

It is known from Theorem 3 that the competitive ratio of any greedy algorithm in the unit value model is at least 2 if $m \gg B$. Which is the tight upper bound for greedy algorithms in the opposite case $B \gg m$?

The proof of the lower bound $e/(e-1)$ in Theorem 1 uses $m \gg B$ whereas Theorem 5 achieves $e/(e-1)$ as an upper bound for $B \gg \log m$. In [4], a lower bound of 1.366 is shown, independent of B and m. Which is the optimal competitive ratio for arbitrary B and m?

Due to the general reduction technique in Theorem 7, the competitive ratio for multi-queue buffer algorithms can be improved if better competitiveness results for single queue buffer algorithms are achieved. Currently, $\frac{\sqrt{13}+5}{6} \approx 1.43$ [2] and 1.75 [7] are the best known lower and upper bounds, respectively. How to reduce this gap?

Cross-References

▶ Online Paging and Caching
▶ Packet Switching in Single Buffer

Recommended Reading

1. Albers S, Schmidt M (2005) On the performance of greedy algorithms in packet buffering. SIAM J Comput 35:278–304
2. Andelman N, Mansour Y, Zhu A (2003) Competitive queueing policies for QoS switches. In: Proceedings of the 14th ACM-SIAM symposium on discrete algorithms (SODA), pp 761–770
3. Azar Y, Litichevskey M (2004) Maximizing throughput in multiqueue switches. In: Proceedings of the 12th annual European symposium on algorithms (ESA), pp 53–64
4. Azar Y, Richter Y (2003) Management of multi-queue switches in QoS networks. In: Proceedings of the 35th ACM symposium on theory of computing (STOC), pp 82–89
5. Azar Y, Richter Y (2004) The zero-one principle for switching networks. In: Proceedings of the 36th ACM symposium on theory of computing (STOC), pp 64–71
6. Azar Y, Richter Y (2004) An improved algorithm for CIOQ switches. In: Proceedings of the 12th annual European symposium on algorithms (ESA). LNCS, vol 3221, pp 65–76
7. Bansal N, Fleischer L, Kimbrel T, Mahdian M, Schieber B, Sviridenko M (2004) Further improvements in competitive guarantees for QoS buffering. In: Proceedings of the 31st international colloquium on automata, languages, and programming (ICALP), pp 64–71
8. Lotker Z, Patt-Shamir B (2003) Nearly optimal FIFO buffer management for two packet classes. Comput Netw 42(4):481–492
9. Schmidt M (2005) Packet buffering: randomization beats deterministic algorithms. In: Proceedings of the 22nd annual symposium on theoretical aspects of computer science (STACS). LNCS, vol 3404, pp 293–304

Packet Switching in Single Buffer

Rob van Stee
University of Leicester, Leicester, UK

Keywords

Buffering

Years and Authors of Summarized Original Work

2003; Bansal, Fleischer, Kimbrel, Mahdian, Schieber, Sviridenko

Problem Definition

In this entry, consider a quality-of-service (QoS) buffering system that is able to hold B packets. Time is slotted. At the beginning of a time step, a set of packets (possibly empty) arrives, and at the end of the time step, a single packet may leave the buffer to be transmitted. Since the buffer has a bounded size, at some point packets may need to be dropped. The buffer management algorithm has to decide at each step which of the packets to drop and which packets to transmit, subject to the buffer capacity constraint. The value of a packet p is denoted by $v(p)$. The system obtains the value of the packets it sends and gains no value otherwise. The aim of the buffer management algorithm is to maximize the total value of transmitted packets.

In the FIFO model, the packet transmitted at time t is always the first (oldest) packet in the buffer.

In the *nonpreemptive* model, packets accepted to the queue will be transmitted eventually and cannot be dropped. In this model, the best competitive ratio achievable is $\Theta(\log \alpha)$, where α is the ratio of the maximum value of a packet to the minimum [1,2].

In the *preemptive* model, packets can also be dropped at some later time before they are served. The rest of this entry focuses on this model. Mansour, Patt-Shamir, and Lapid [11] were the first to study preemptive queuing policies for a single FIFO buffer, proving that the natural greedy algorithm (see definition in Fig. 1) maintains a competitive ratio of at most 4. This bound was improved to the tight value of 2 by Kesselman, Lotker, Mansour, Patt-Shamir, Schieber, and Sviridenko [8]. An alternative proof of the 2-competitiveness, due to Kimbrel [9], is presented in Epstein and van Stee's survey on buffer management [5].

The greedy algorithm is not optimal since it never preempts a packet until the buffer is full and this might be too late. The first algorithm with a competitive ratio strictly below 2 was presented by Kesselman, Mansour, and van Stee [7]. This algorithm uses a parameter β and introduces an extra rule for processing arrivals that is executed before rules (1) and (2) of the greedy algorithm. This rule is formulated in Fig. 2.

It is shown in [7] that by taking $\beta = 15$, the algorithm preemptive greedy (PG) has a competitive ratio of 1.983. The analysis is rather complicated and is done by assigning the value of packets served by the offline algorithm to packets served by PG.

A lower bound of 5/4 for this problem was shown in [11]. This was improved to $\sqrt{2}$ in [2] and then to 1.419 in [7].

Key Results

A modification of PG was presented by Bansal, Fleischer, Kimbrel, Mahdian, Schieber, and Sviridenko [3]. It changes rule 0 to rule 0′ (Fig. 3).

Thus, the modification compared to PG is that this algorithm finds a "locally optimal" packet to evict. We will denote modified preemptive greedy by MPG.

Packet Switching in Single Buffer, Fig. 1 The natural greedy algorithm

The Greedy Algorithm.
When a packet of value $\nu(p)$ arrives:
1. Accept p if there is free space in the buffer.
2. Otherwise, reject (drop or preempt) the packet p' that has minimal value among p and the packets in the buffer. If $p' \neq p$, accept p.

0. Preempt (drop) the first packet p' in the FIFO order such that $\nu(p') \leq \nu(p)/\beta$, if any (*p preempts p′*).

Packet Switching in Single Buffer, Fig. 2 Extra rule for the preemptive greedy algorithm

Packet Switching in Single Buffer, Fig. 3 Modified preemptive greedy

0'. Find the first (i.e., closest to the front of the buffer) packet p' such that p' has value less than $\nu(p)/\beta$ and not more than the value of the packet after p' in the buffer (if any). If such a packet exists, drop it (*p preempts p′*).

Theorem 1 ([3]) *For $\beta = 4$, MPG has a competitive ratio of 1.75.*

The proof begins by showing that in order to analyze the performance of MPG, it is sufficient to consider only input instances in which the value of each packet is either 0 or β^i for some $i \geq 0$, but ties are allowed to be broken by the adversary.

The authors then define an *interval structure* for input instances. An interval I is said to be of type i if at every step $t \in I$, MPG outputs a packet of value at least β^i, and I is a maximal interval with this property.

$\mathcal{I}_i$ is the collection of maximal intervals of type i, and $\mathcal{I}$ is the union of all $\mathcal{I}_i$'s. This is a multiset, since an interval of type i can also be contained in an interval of one or more types $j < i$.

This induces an interval structure which is a sequence of ordered rooted trees in a natural way: The root of each tree is an interval in $\mathcal{I}_0$, and the children of each interval $I \in \mathcal{I}_i$ are all the maximal intervals of type $i + 1$ which are contained in I. These children are ordered from left to right based on time, as are the trees themselves. The intervals of type i (and the vertices that represent them) are distinguished by whether or not an eviction of a packet of value at least β^i occurred during the interval.

To complete the proof, the authors show that for every interval structure $\mathcal{T}$, the competitive ratio of MPG on instances with interval structure $\mathcal{T}$ can be bounded by the solution of a linear program indexed by $\mathcal{T}$. Finally, it is shown that for every $\mathcal{T}$ and every $\beta \geq 4$, the solution of this program is at most $2 - 1/\beta$.

Applications

In recent years, there has been a lot of interest in quality-of-service networks. In regular IP networks, packets are indistinguishable, and in case of overload, any packet may be dropped. In a commercial environment, it is much more preferable to allow better service to higher-paying customers or customers with critical requirements. The idea of quality-of-service guarantees is that packets are marked with values which indicate their importance.

This naturally leads to decision problems at network switches when many packets arrive and overload occurs. The algorithm presented in this entry can be used to maximize network performance in a network which supports quality of service.

Open Problems

Despite substantial advances in improving the upper bound for this problem, a fairly large gap remains. Sgall (quoted in Jawor [6]) showed that the performance of PG is as good as that of MPG. Englert and Westermann [4] showed that PG has a competitive ratio of at most $\sqrt{3} \approx 1.732$ and at least $1 + 1/2\sqrt{2} \approx 1.707$. Thus, to improve further, a different algorithm will be needed.

The authors also note that Lotker and Patt-Shamir [10] studied the special case of two packet values and derived a 1.3-competitive algorithm, which closely matches the corresponding lower bound of 1.28 from Mansour et al. [11]. An open problem is to close the remaining small gap.

Cross-References

▸ Single and Multiple Buffer Processing

Recommended Reading

1. Aiello W, Mansour Y, Rajagopolan S, Rosen A (2000) Competitive queue policies for differentiated services. In: Proceedings of the IEEE INFOCOM, Tel-Aviv. IEEE, pp 431–440
2. Andelman N, Mansour Y, Zhu A (2003) Competitive queueing policies in QoS switches. In: Proceedings of the 14th symposium on discrete algorithms (SODA), Baltimore. ACM/SIAM, San Francisco, pp 761–770
3. Bansal N, Fleischer L, Kimbrel T, Mahdian M, Schieber B, Sviridenko M (2004) Further improvements in competitive guarantees for QoS buffering. In: Proceedings of the 31st international colloquium on automata, languages, and programming (ICALP),

Turku. Lecture notes in computer science, vol 3142. Springer, Berlin, pp 196–207
4. Englert M, Westermann M (2006) Lower and upper bounds on FIFO buffer management in qos switches. In: Azar Y, Erlebach T (eds) Proceedings of the 14th annual European symposium on algorithms – ESA 2006, Zurich. Lecture notes in computer science, vol 4168. Springer, Berlin, pp 352–363
5. Epstein L, van Stee R (2004) Buffer management problems. SIGACT News 35(3):58–66
6. Jawor W (2005) Three dozen papers on online algorithms. SIGACT News 36(1):71–85
7. Kesselman A, Mansour Y, van Stee R (2003) Improved competitive guarantees for QoS buffering. In: Di Battista G, Zwick U (eds) Proceedings of the eleventh annual European symposium on algorithms – ESA 2003, Budapest. Lecture notes in computer science, vol 2380. Springer, Berlin, pp 361–373
8. Kesselman A, Lotker Z, Mansour Y, Patt-Shamir B, Schieber B, Sviridenko M (2004) Buffer overflow management in QoS switches. SIAM J Comput 33(3):563–583
9. Kimbrel T (2004) A simple proof of the 2-competitiveness of the greedy FIFO buffering algorithm. Technical report RC23272, IBM Research
10. Lotker Z, Patt-Shamir B (2002) Nearly optimal FIFO buffer management for DiffServ. In: Proceedings of the 21st ACM symposium on principles of distributed computing (PODC 2002), Monterey. ACM, New York, pp 134–142
11. Mansour Y, Patt-Shamir B, Lapid O (2000) Optimal smoothing schedules for real-time streams. In: Proceedings of the 19th symposium on principles of distributed computing (PODC), Portland. ACM, New York, pp 21–29

PageRank Algorithm

Monika Henzinger
University of Vienna, Vienna, Austria

Keywords

Hyperlink analysis; Web information retrieval

Years and Authors of Summarized Original Work

1998; Brin, Page

Problem Definition

Given a user query, current web search services retrieve all web pages that contain the query terms resulting in a huge number of web pages for the majority of searches. Thus it is crucial to reorder or *rank* the resulting documents with the goal of placing the most relevant documents first. Frequently, ranking uses two types of information: (1) query-specific information and (2) query-independent information. The query-specific part tries to measure how relevant the document is to the query. Since it depends to a large part on the content of the page, it is mostly under the control of the page's author. The query-independent information tries to estimate the quality of the page in general. To achieve an objective measure of page quality, it is important that the query-independent information incorporates a measure that is not controlled by the author. Thus the problem is to find a measure of page quality that: (a) cannot be easily manipulated by the web page's author and (b) works well for *all* web pages. This is challenging as web pages are extremely heterogeneous.

Key Results

The hyperlink structure of the web is a good source for basing such a measure as it is hard for one author or a small set of authors to influence the whole structure, even though they can manipulate a subset of the web pages. Brin and Page showed that a relatively simple analysis of the hyperlink structure of the web can be used to produce a quality measure for web documents that leads to large improvements in search quality. The measure is called the *PageRank* measure.

Linear Algebra-Based Definition

Let n be the total number of web pages. The PageRank vector is an $n$dimensional vector with one dimension for each web page. Let d be a small constant, like 1/8, let $deg(p)$ denote the number of hyperlinks in the body text of page p, and let $PR(p)$ denote the PageRank value of page p. Assume first that every page contains at least one hyperlink. In such a collection of web pages, the PageRank vector is computed by solving a system of linear equations that contains for each page p the equation

$$PR(p) = d/n + (1-d) * \sum_{q \text{ has hyperlink to } p} PR(q)/deg(q)$$

In matrix notation, the PageRank vector is the Eigenvector with 1Norm one of the matrix A with $d/n + (1-d)/deg(q)$ for entry A_{qp} if q has a hyperlink to p and d/n otherwise.

If web pages without hyperlinks are allowed in this linear system, then they might become "PageRank sinks", i.e., they would "receive" PageRank from the pages pointing to them, but would not "give out" their PageRank, potentially resulting in an "unusually high" PageRank value for themselves. Brin and Page proposed two ways to deal with web pages without out-links, namely either to recursively delete them until no such web pages exist anymore in the collection or to add a hyperlink from each such page to *every* other page.

Random Surfer Model

Let the *web graph* $G = (V, E)$ be a directed graph such that each node corresponds to a web page and every hyperlink corresponds to a directed edge from the referencing node to the referenced node. The PageRank can also be interpreted as the following random walk in the web graph. The random walk starts at a random node in the graph. Assume in step k it visits page q. Then it flips a biased coin, and with probability d or if q has no out-edges, it selects a random node out of V and visits it in step $k+1$. Otherwise it selects a random out-edge of the current node and visits it in step $k+1$. (Note that this corresponds to adding a directed edge from every page without hyperlink to *every* node in the graph.) Under certain conditions (which do not necessarily hold on the web) the stationary distribution of this random walk corresponds to the PageRank vector. See [1, 4] for details.

Brin and Page also suggested computing the PageRank vector approximately using the power method, i.e., by setting all initial values to $1/n$ and then repeatedly using the PageRank vector of the previous iteration to compute the PageRank vector of the current iteration using the above linear equations. After a hundred iterations, barely any values change and the computation is stopped.

Applications

The PageRank measure is used as one of the factors by Google in its ranking of search results. The PageRank computation can be applied to other domains as well. Two examples are reputation management in P2P networks and learning word dependencies in natural language processing. In relational databases PageRank was used to weigh database tuples in order to improve keyword searching when a user does not know the schema. Finally, in rank aggregation PageRank can be used to find a permutation that minimally violates a set of given orderings. See [1] for more details.

Variations of PageRank were studied as well. Personalizing the PageRank computation such that the values reflect the interest of a user has received a lot of attention. See [3] for a survey on this topic. It can also be modified to be used for detecting web search spam, i.e., web pages that try to manipulate web search results [1].

Recommended Reading

1. Berkhin P (2005) A survey on PageRank computing. Internet Math 2(1):73–120

2. Brin S, Page L (1998) The anatomy of a large-scale hypertextual Web search engine. In: Proceedings of the 7th international conference on World Wide Web. Elsevier, Amsterdam, Brisbane, Australia, pp 107–117
3. Haveliwala T, Kamvar S, Jeh G (2003) An analytical comparison of approaches to personalizing PageRank. Technical report, Stanford University, Stanford
4. Langville AN, Meyer CD (2004) Deeper inside PageRank. Internet Math 1(3):335–380
5. Page L, Brin S, Motwani R, Winograd T (1998) The PageRank citation ranking: bringing order to the Web. Technical report, Stanford University, Stanford

Parallel Algorithms for Two Processors Precedence Constraint Scheduling

Maria Serna
Department of Language and System Information, Technical University of Catalonia, Barcelona, Spain

Keywords

Optimal scheduling for two processors

Years and Authors of Summarized Original Work

2003; Jung, Serna, Spirakis

Problem Definition

In the general form of *multiprocessor precedence scheduling problems* a set of n tasks to be executed on m processors is given. Each task requires exactly one unit of execution time and can run on any processor. A directed acyclic graph specifies the precedence constraints where an edge from task x to task y means task x must be completed before task y begins. A solution to the problem is a schedule of shortest length indicating when each task is started. The work of Jung, Serna, and Spirakis provides a parallel algorithm (on a PRAM machine) that solves the above problem for the particular case that $m = 2$, that is where there are two parallel processors.

The *two processor precedence constraint scheduling problem* is defined by a directed acyclic graph (dag) $G = (V, E)$. The vertices of the graph represent unit time tasks, and the edges specify precedence constraints among the tasks. If there is an edge from node x to node y then x is an *immediate predecessor* of y. *Predecessor* is the transitive closure of the relation immediate predecessor, and *successor* is its symmetric counterpart. A *two processor schedule* is an assignment of the tasks to time units $1, \ldots, t$ so that each task is assigned exactly one time unit, at most two tasks are assigned to the same time unit, and if x is a predecessor of y then x is assigned to a lower time unit than y. The length of the schedule is t. A schedule having minimum length is an *optimal* schedule. Thus the problem is the following:

Name	Two processor precedence constraint scheduling
Input	A directed acyclic graph
Output	A minimum length schedule preserving the precedence constraints.

Preliminaries

The algorithm assume that tasks are partitioned into levels as follows:

(i) Every task will be assigned to only one level
(ii) Tasks having no successors will be assigned to level 1 and
(iii) For each level i, all tasks which are immediate predecessors of tasks in level i will be assigned to level $i + 1$.

Clearly topological sort will accomplish the above partition, and this can be done by an NC algorithm that uses $O(n^3)$ processors and $O(\log n)$ time, see [3]. Thus, from now on, it is assumed that a level partition is given as part of the input. For sake of convenience two special tasks, t_0 and t^* are added, in such a way that the original graph could be taught as the graph formed by all tasks that are successors of t_0 and predecessors of t^*. Thus t_0 is a predecessor of all tasks in the system (actually an immediate predecessor of tasks in level the highest level $L(G)$) and t^* is a successor of all tasks in the system (an immediate successor of level 1 tasks).

Notice that if two tasks are at the same level they can be paired. But when x and y are at different levels, they can be paired only when neither of them is a predecessor of the other. Let $L(G)$ denote the number of levels in a given precedence graph G. A *level schedule* schedules tasks level by level. More precisely, suppose levels $L(G), \ldots, i+1$ have already been scheduled and there are k unscheduled tasks remaining on level i. When k is even, those tasks with are paired with each other. When k is odd, $k-1$ of the tasks are paired with each other, while the remaining task may (but not necessarily) be paired with a task from a lower level.

Given a level schedule level *i jumps to level i'* $(i' < i)$ if the last time step containing a task from level i also contains a task from level i'. If the last task from level i is scheduled with an empty slot, it is said that level *i jumps to level* 0. The *jump sequence* of a level schedule is the list of levels jumped to. A *lexicographically first jump schedule* is a level schedule whose jump sequence is lexicographically greater than any other jump sequence resulting from a level schedule.

Given a graph G a *level partition* of G is a partition of the nodes in G into two sets in such a way that levels $0, \ldots, k$ are contained in one set (the upper part) denoted by U, and levels $k+1, \ldots, L$ in the other (the lower part) denoted by L. Given a graph G and a level i, the *i-partition* of G (or the partition at level i) is formed by the graphs U_i and L_i defined as U_i contains all nodes x such that level$(x) < i$ and L_i contains all nodes x with level$(x) > i$. Note that each i-partition determines two different level partitions depending on whether level i nodes are assigned to the upper or the lower part. A task $x \in U_i$ is called *free* with respect to a partition at level i if x has no predecessors in L_i.

Auxiliary Problems

The algorithm for the two processors precedence constraint scheduling problem uses as a building block an algorithm for solving a matching problem in a particular graph class.

A *full convex bipartite graph* G is a triple (V, W, E), where $V = \{v_1, \ldots, v_k\}$ and $W = \{w_1, \ldots, w_{k'}\}$ are disjoint sets of vertices. Furthermore the edge set E satisfies the following property: If $(v_i, w_j) \in E$ then $(v_q, w_j) \in E$ for all $q \geq i$. Thus, from now on it is assumed that the graph is connected.

A set $F \subseteq E$ is a *matching* in the graph $G = (V, W, E)$ iff no two edges in F have a common endpoint. A *maximal matching* is a matching that cannot be extended by the addition of any edge in G. A *lexicographically first maximal matching* is a maximal matching whose sorted list of edges is lexicographically first among all maximal matchings in G.

Key Results

When the number of processors m is arbitrary the problem is known to be NP-complete [8]. For any $m \geq 3$, the complexity is open [6]. Here the case of interest has been $m = 2$. For two processors a number of efficient algorithms has been given. For sequential algorithms see [2, 4, 5] among others. The first deterministic parallel algorithm was given by Helmbold and Mayr [7], thus establishing membership in the class NC. Previously [9] gave a randomized NC algorithm for the problem. Jung, Serna and Spirakis present a new parallel algorithm for the two processors scheduling problem that takes time $O(\log^2 n)$ and uses $O(n^3)$ processors on a CREW PRAM. The algorithm improves the number of processors of the algorithm given in [7] from $O(n^7 L(G)^2)$, where $L(G)$ is the number of levels in the precedence graph, to $O(n^3)$. Both algorithms compute a level schedule that has a lexicographically first jump sequence.

To match jumps with tasks it is used a solution to the problem of computing the lexicographically first matching for a special type of convex bipartite graphs, here called *full convex bipartite graphs*. A geometric interpretation of this problem leads to the discovery of an efficient parallel algorithm to solve it.

Theorem 1 *The lexicographically first maximal matching of full convex bipartite graphs can be computed in time $O(\log n)$ on a CREW PRAM with $O(n^3/\log n)$ processors, where n is the number of nodes.*

The previous algorithm is used to solve efficiently in parallel two related problems.

Theorem 2 *Given a precedence graph G, there is a PRAM parallel algorithm that computes all levels that jump to level 0 in the graph L_i and all tasks in level $i-1$ that can be scheduled together with a task in level i, for $i = 1, \ldots, L(G)$, using $O(n^3)$ processors and $O(\log^2 n)$ time.*

Theorem 3 *Given a level partition of a graph G together with the levels in the lower part in which one task remains to be matched with some other task in the upper part of the graph. There is a PRAM parallel algorithm that computes the corresponding tasks in time $O(\log n)$ using $n^3/\log n$ processors.*

With those building blocks the algorithm for two processor precedence constraint scheduling starts by doing some preprocessing and after that an adequate decomposition that insure that at each recursive call a number of problems of half size are solved in parallel. This recursive schema is the following:

Algorithm Schedule

0. Preprocessing
1. Find a level i such that $|U_i| \leq n/2$ and $|L_i| \leq n/2$.
2. Match levels that jump to free tasks in level i.
3. Match levels that jump to free tasks in U_i.
4. If level i (or $i+1$) remain unmatched try to match it with a non free task.
5. Delete all tasks used to match jumps.
6. Apply (1)–(5) in parallel to L_i and the modified U_i.

Algorithm **Schedule** stops whenever the corresponding graph has only one level.

The correction an complexity bounds for algorithm **Schedule** follows from the previous results, leading to:

Theorem 4 *There is an NC algorithm which finds an optimal two processors schedule for any precedence graph in time $O(\log^2 n)$ using $O(n^3)$ processors.*

Applications

A fundamental problem in many applications is to devise a proper schedule to satisfy a set of constrains. Assigning people to jobs, meetings to rooms, or courses to final exam periods are all different examples of scheduling problems. A key and critical algorithm in parallel processing is the one mapping tasks to processors. In the performance of such an algorithm relies many properties of the system, like load balancing, total execution time, etc. Scheduling problems differ widely in the nature of the constraints that must be satisfied, the type of processors, and the type of schedule desired.

The focus on precedence-constrained scheduling problems for directed acyclic graphs has a most direct practical application in problems arising in parallel processing. In particular in systems where computations are decomposed, prior to scheduling into approximately equal sized tasks and the corresponding partial ordering among them is computed. These constraints must define a directed acyclic graph, acyclic because a cycle in the precedence constraints represents a Catch situation that can never be resolved.

Open Problems

The parallel deterministic algorithm for the two processors scheduling problem presented here improves the number of processors of the Helmbold and Mayr algorithm for the problem [7]. However, the complexity bounds are far from optimal: recall that the sequential algorithm given in [5] uses time $O(e + n\alpha(n))$, where e is the number of edges in the precedence graph and $\alpha(n)$ is an inverse Ackermann's function. Such an optimal algorithm might have a quite different approach, in which the levelling algorithm is not used.

Interestingly enough computing the lexicographically first matching for full convex bipartite graphs is in NC, in contraposition with the results given in [1] which show that many problems defined through a lexicographically first procedure in the plane are P-complete. It is an interesting problem to show whether all these problems fall in NC when they are convex.

Cross-References

- ▶ Approximation Schemes for Makespan Minimization
- ▶ List Scheduling
- ▶ Maximum Matching
- ▶ Shortest Elapsed Time First Scheduling
- ▶ Stochastic Scheduling
- ▶ Voltage Scheduling

Recommended Reading

1. Attallah M, Callahan P, Goodrich M (1993) P-complete geometric problems. Int J Comput Geom Appl 3(4):443–462
2. Coffman EG, Graham RL (1972) Optimal scheduling for two processors systems. Acta Informatica 1:200–213
3. Dekel E, Nassimi D, Sahni S (1981) Parallel matrix and graph algorithms. SIAM J Comput 10:657–675
4. Fujii M, Kasami T, Ninomiya K (1969) Optimal sequencing of two equivalent processors. SIAM J Comput 17:784–789
5. Gabow HN (1982) An almost linear time algorithm for two processors scheduling. J ACM 29(3):766–780
6. Garey MR, Johnson DS (1979) Computers and intractability: a guide to the theory of NP completeness. Freeman, San Francisco
7. Helmbold D, Mayr E (1987) Two processor scheduling is in NC. SIAM J Comput 16(4):747–756
8. Ullman JD (1975) NP-complete scheduling problems. J Comput Syst Sci 10:384–393
9. Vazirani U, Vazirani V (1989) Two-processor scheduling problem is in random NC. SIAM J Comput 18(4):1140–1148

Parallel Connectivity and Minimum Spanning Trees

Tak-Wah Lam
Department of Computer Science, University of Hong Kong, Hong Kong, China

Keywords

EREW PRAM algorithms for finding connected components and minimum spanning trees

Years and Authors of Summarized Original Work

1995; Ka Wong Chong, Lam
2001; Ka Wong Chong, Han, Lam

Problem Definition

Given a weighted undirected graph G with n vertices and m edges, compute a minimum spanning tree (or spanning forest) of G on a parallel random access machine (PRAM) without concurrent write capability.

A minimum spanning tree of a graph is a spanning tree with the smallest possible sum of edge weights. The parallel random access machine (PRAM) is an abstract model for designing parallel algorithms and understanding the power of parallelism. In this model, processors (each being a random access machine) work in a synchronous manner and communicate through a shared memory. PARM can be further classified according to whether it is allowed for more than one processor to read and write into the same shared memory location simultaneously. The strongest model is CRCW (concurrent-read, concurrent-write) PRAM, and the weakest is EREW (exclusive-read, exclusive-write) PRAM. For an introduction to PRAM algorithms, one can refer to Karp and Ramachandran [8] and JáJá [5].

The input graph G is assumed to be given in the form of adjacency lists. Furthermore, isolated (degree-0) vertices are removed, and hence it is assumed that $m \geq n$.

Key Results

The MST problem is related to the connected component (CC) problem, which is to find the connected components of an undirected graph. Sequential algorithms for solving the CC problem and the MST problem in $O(m)$ time and $O(m \log n)$ time, respectively, were known a few decades ago. A number of more efficient MST algorithms have since been published, the most recent of which is Pettie and Ramachandran's algorithm [9], which is provably optimal.

In the parallel context, both problems are often solved in a similar way. With respect to CRCW PRAM, the two problems can be solved using $O(\log n)$ time and $n + m$ processors (see, e.g., Cole and Vishkin [3]). Using randomization, $(n + m)/\log n$ processors are sufficient to solve these problems in $O(\log n)$ expected time [2, 10].

For EREW PRAM, $O(\log^2 n)$ time algorithms for the CC and MST problems were developed in the early 1980s. For a while, it was believed that the exclusive write models (including both concurrent read and exclusive read) could not overcome the $O(\log^2 n)$ time bound [8]. The first breakthrough was due to Johnson and Metaxas [6]; they devised $O(\log^{1.5} n)$ time algorithms for the CC problem and the MST problem. These results were soon improved to $O(\log n \log\log n)$ time by Chong and Lam. If randomization is allowed, the time complexity can be improved to $O(\log n)$ expected time and optimal work [7, 10, 11]. Finally, Chong, Han, and Lam [1] obtained an algorithm for MST (and CC) using $O(\log n)$ time and $n + m$ processors. This algorithm does not need randomization. Notice that $\Theta(\log n)$ is optimal since these graphs' problems are at least as hard as computing the OR of n bits, and Cook et al. [4] have proven that the latter requires $\Omega(\log n)$ time on exclusive-write PRAM no matter how many processors are used.

Below is a sketch of some ideas for computing a minimum spanning tree in parallel without using concurrent write.

Without loss of generality, assume that the edge weights are all distinct. Thus, G has a unique minimum spanning tree, which is denoted by T_G^*. Let B be a subset of edges in G which contains no cycle. B induces a set of trees $F = \{T_1, T_2, \cdots, T_l\}$ in a natural sense – two vertices in G are in the same tree if they are connected by an edge of B. B is said to be a λ-forest if each tree $T \in F$ has at least λ vertices. For example, if B is the empty set, then B is a 1forest; a spanning tree is an n-forest.

Suppose that B is a λ-forest and its edges are all found in T_G^*. Then B can be augmented to give a 2λ-forest using a greedy approach: Let F' be an arbitrary subset of F including all trees $T \in F$ with fewer than 2λ vertices. For every tree in F', pick its minimum external edge (i.e., the smallest-weight edge connecting to a vertex outside the tree). Denote B' as this set of edges. It can be proven that B' consists of edges in T_G^* only and $B \cup B'$ is a 2λ-forest. The above idea allows us to find T_G^* in $\lfloor \log n \rfloor$ stages as follows:

1. $B \leftarrow \phi$
2. **For** $i = 1$ to $\lfloor \log n \rfloor$ **do** /* Stage i */
   1. Let F be the set of trees induced by B on G. Let F' be an arbitrary subset of F such that F' includes all trees $T \in F$ with fewer than 2^i vertices.
   2. $B_i \leftarrow \{e \mid e$ is the minimum external edge of $T \in F'\}$; $B \leftarrow B \cup B_i$
3. **return** B

Different strategies for choosing the set F' in Step 1(a) may lead to different B_i's. Nevertheless, $B[1, i]$ is always a subset of T_G^* and induces a 2^i -forest. In particular, $B[1, \lfloor \log n \rfloor]$ induces exactly one tree, which is exactly T_G^*. Using standard parallel algorithmic techniques, each stage can be implemented in $O(\log n)$ time on EREW PRAM using a linear number of processors (see, e.g., [5],). Therefore, T_G^* can be found in $O(\log^2 n)$ time. In fact, most parallel algorithms for finding MST are based on a similar approach. These parallel algorithms are "sequential" in the sense that the computation of B_i starts only after B_{i-1} is available.

The $O(\log n)$-time EREW algorithm in [1] is based on some structural properties related to MST and can compute the B_i's in a more parallel fashion. In this algorithm, there are $\lfloor \log n \rfloor$ concurrent threads (a thread is simply a group of processors). For $1 \leq i \leq \lfloor \log n \rfloor$, Thread i aims at computing B_i, and it actually starts long before Thread $i-1$ has computed B_{i-1}, and it receives the output of Threads 1 to $i-1$ (i.e., $B_1, \cdots, B_{i-1}$) incrementally. More specifically, the algorithm runs in $\lfloor \log n \rfloor$ supersteps, where each superstep lasts $O(1)$ time. Thread i delivers B_i at the end of the ith superstep. The computation of Thread i is divided into $\lfloor \log i \rfloor$ phases. Let us first consider a simple case when i is a power of 2. Phase 1 of Thread i starts at the $(i/2+1)$th superstep, i.e., when $B_1, \cdots, B_{i/2}$ are available. Phase 1 takes no more than $i/4$ supersteps, ending at the $(i/2+i/4)$th superstep. Phase 2 starts at the $(i/2+i/4+1)$th superstep (i.e., when $B_{i/2+1}, \cdots, B_{i/2+i/4}$ are available) and uses $i/8$ supersteps. Each subsequent phase uses half as many supersteps as the preceding phase. The last phase (Phase $\log i$) starts and ends within the ith superstep; note that B_{i-1} is available after $(i-1)$th superstep.

Applications

Finding connected components or MST is a key step in several parallel algorithms for other graph problems. For example, the Chong-Han-Lam algorithm implies an $O(\log n)$ time algorithm for finding ear decomposition and biconnectivity without using concurrent write.

Cross-References

- ▶ Graph Connectivity
- ▶ Randomized Parallel Approximations to Max Flow

Recommended Reading

1. Chong KW, Han Y, Lam TW (2001) Concurrent threads and optical parallel minimum spanning trees algorithm. J ACM 48(2):297–323
2. Cole R, Klein PN, Tarjan RE (1996) Finding minimum spanning forests in logarithmic time and linear work using random sampling. In: Proceedings of the 8th annual ACM symposium on parallel architectures and algorithms, Padua, pp 243–250
3. Cole R, Vishkin U (1986) Approximate and exact parallel scheduling with applications to list, tree, and graph problems. In: Proceedings of the 27th annual IEEE symposium on foundations of computer science, Toronto, pp 478–491
4. Cook SA, Dwork C, Reischuk R (1986) Upper and lower time bounds for parallel random access machines without simultaneous writes. SIAM J Comput 15(1):87–97
5. JáJá J (1992) An introduction to parallel algorithms. Addison-Wesley, Boston
6. Johnson DB, Metaxas P (1991) Connected components in $O(\lg^{3/2} |V|)$ parallel time for the CREW PRAM. In: Proceedings of the 32nd annual IEEE symposium on foundations of computer science, San Juan, pp 688–697
7. Karger DR (1995) Random sampling in graph optimization problems. Ph.D. thesis, Department of computer science, Stanford University
8. Karp RM, Ramachandran V (1990) Parallel algorithms for shared-memory machines. In: Van Leeuwen Ed J (ed) Handbook of theoretical computer science, vol A, pp 869–941. MIT Press, Massachusetts
9. Pettie S, Ramachandran V (2002) An optimal minimum spanning tree algorithm. J ACM 49(1):16–34
10. Pettie S, Ramachandran V (2002) A randomized time-work optimal parallel algorithm for finding a minimum spanning forest. SIAM J Comput 31(6):1879–1895
11. Poon CK, Ramachandran V (2003) A randomized linear-work EREW PRAM algorithm to find a minimum spanning forest. Algorithmica 35(3):257–268

Parameterization in Computational Social Choice

Piotr Faliszewski[1] and Rolf Niedermeier[2,3]
[1]AGH University of Science and Technology, Krakow, Poland
[2]Department of Mathematics and Computer Science, University of Jena, Jena, Germany
[3]Institut für Softwaretechnik und Theoretische Informatik, Technische Universität Berlin, Berlin, Germany

Keywords

Bribery in elections; Control in elections; NP-hard problems; Parameterized and multivariate complexity analysis; Voting problems; Winner determination

Years and Authors of Summarized Original Work

1989; Bartholdi, Tovey, Trick
2009; Betzler, Uhlmann
2011; Liu, Feng, Zhu, Luan
2012; Dorn, Schlotter

Problem Definition

Computational Social Choice is an interdisciplinary research area involving Economics, Political Science, and Social Science on the one side and Mathematics and Computer Science (including artificial intelligence and multi-agent systems) on the other side. Concrete questions addressed in this area include the following three: How efficiently can one determine the winner of an election, given a number of votes with preferences over a number of alternatives? Is it possible to obtain a desirable outcome of an election by executing a number of campaigning actions? (Formally, such problems are often modeled as bribery.) Can the chair of an election influence the result of an election by modifying the set of available alternatives (e.g., by encouraging some alternatives (candidates) to participate in the run)?

The main objective of parameterized complexity is to analyze computationally hard problems with respect to multiple input parameters and to classify these problems according to their inherent difficulty when "viewed through different parameterizations." The complexity of a problem typically depends on the values of a multitude of input parameters, so this approach allows to classify NP-hard problems on a much finer scale than in classical complexity theory (where the complexity of a problem is only measured relative to the input size). In particular, a parameterized problem consisting of an input instance I and a parameter k is called *fixed-parameter tractable* if it can be solved in $f(k) \cdot |I|^{O(1)}$ time for a computable function f which typically grows at least exponentially in k. This still means, however, that the problem is efficiently solvable for small parameter values.

Parameterized complexity analysis, which still adheres to a worst-case complexity scenario, has been successfully applied in many areas. Computational Social Choice, although being a more recent application area [5], is among the most natural ones. To make this more precise, we next discuss in more detail three prominent voting problems that were already briefly mentioned in the introductory part: winner determination, campaign management/bribery, and control.

An election $E := (C, V)$ consists of a set C of m alternatives $c_1, c_2, \ldots, c_m$ and a list V of n voters $v_1, v_2, \ldots, v_n$. Each voter v has a linear order $\succ_v$ over the set C which we call a *preference order*. For example, if $C = \{c_1, c_2, c_3\}$, then the preference order $c_1 \succ_v c_2 \succ_v c_3$ of voter v indicates that v likes c_1 most (the 1st position), then c_2, and c_3 least (the 3rd position). For any two distinct alternatives c and c', we write $c \succ_v c'$ if voter v prefers c over c'. For an election $E = (C, V)$, an alternative $c \in C$ is a *Condorcet winner* if any other alternative $c' \in C \setminus \{c\}$ satisfies

$$|\{v \in V \mid c \succ_v c'\}| > |\{v \in V \mid c' \succ_v c\}|.$$

It is important to note that voting problems typically come along with many natural parameterizations, the two most obvious ones being the number of candidates (which is typically small in political elections, say) or the number of voters (which is often small in applications concerning multi-agent systems or decision making by committees). A further, standard type of parameter refers to the size (or the value) of the solution that we seek.

We now give an example of a winner determination problem, focusing on the voting rule due to Dodgson (also known as the writer Lewis Carroll). In Dodgson's system, the score of a candidate is the smallest number of swaps of adjacent candidates needed to ensure that this candidate is a Condorcet winner. In the DODGSON SCORE problem we ask about the score of a given candidate in an election:

Input: An election $E := (C, V)$, a distinguished alternative $c \in C$, and a nonnegative integer k.

Question: Can one make c the Condorcet winner by swapping a total number of at most k pairs of neighboring alternatives (i.e., k "bubble sort operations") in the voters' preference orders?

Our campaign management/bribery example is based on the t-Approval voting rule. In t-Approval, every voter can assign one point to each of t most preferred alternatives, and the alternatives with maximum total number of points win. Notably, 1-Approval simply is the frequently used Plurality voting rule. Now, the NP-hard problem SWAP BRIBERY FOR t-APPROVAL reads as follows.

Input: An election $E := (C, V)$, a distinguished alternative $c \in C$, a cost function assigning a nonnegative integer cost to every swap-of-consecutive-candidates operation, and a nonnegative integer β called the budget.

Question: Can one make c a winner by swap operations of total cost at most β?

Intuitively, each swap operation means a campaigning effort that convinces the voter that one candidate is better than the other and comes at a given cost (measured in time or money or some other way).

Finally, our control example focuses on control by adding alternatives in an election based on the plurality voting role. Note that other control actions include, for example, deleting alternatives or adding/deleting voters (assuming a powerful and corrupted chair of an election). The goal of the chair can either be to ensure someone's victory (constructive control) or preclude someone from winning (destructive control); we focus on the former. The NP-hard problem PLURALITY CONSTRUCTIVE CONTROL BY ADDING ALTERNATIVES reads as follows:

Input: An election $E := (C, V)$, a distinguished alternative $c \in C$, a set of "spoiler alternatives" for possible addition, and a nonnegative integer k.

Question: Can one make c a winner by adding at most k spoiler alternatives?

Note that we assume that every voter has a clear linear order over all alternatives (including the spoiler alternatives) and that this is known by the manipulating election chair.

Key Results

We again start with our winner determination example. Bartholdi, Tovey, and Trick [1] were the first to provide an "ILP-based" fixed-parameter tractability result in the context of Computational Social Choice (actually the result was stated implicitly). They developed an integer linear program (ILP) to solve the NP-hard DODGSON SCORE problem and gave a running time bound based on a famous result of Lenstra, concerning the exact solvability of integer linear programs with "few" variables. Without having explicitly stated this in their publication, this implies fixed-parameter tractability for DODGSON SCORE with respect to the parameter number m of alternatives.

Bartholdi et al. [1]'s integer linear program for DODGSON SCORE reads as follows. It computes the Dodgson score of an alternative c.

$$\min \sum_{i,j} j \cdot x_{i,j} \text{ subject to}$$

$$\forall i \in \tilde{V} : \sum_{j} x_{i,j} = N_i,$$

$$\forall y \in C : \sum_{i,j} e_{i,j,y} \cdot x_{i,j} \geq d_y,$$

$$x_{i,j} \geq 0.$$

Here, $\tilde{V}$ denotes the set of preference order types (i.e., the set of *different* preference orders in the given election), N_i denotes the number of voters of type i, $x_{i,j}$ denotes the number of voters with preference order of type i for which alternative c will be moved upward by j positions,

$e_{i,j,y}$ is 1 if the result of moving alternative c by j positions upward in a preference order of type i is that c gains an additional voter support against alternative y, and 0 otherwise. Furthermore, d_y is the deficit of c with respect to alternative y, that is, the minimum number of voter supports that c must gain against y to defeat y in a pairwise comparison. If c already defeats y, then $d_y = 0$. Altogether, the integer linear program contains at most $m \cdot m!$ variables $x_{i,j}$, where m denotes the number of alternatives. Thus, the number of variables in the described integer linear program is upper-bounded by a function in parameter m, yielding fixed-parameter tractability due to Lenstra's result.

We remark that beyond the above parameterization by the number m of alternatives, it is also known that DODGSON SCORE is fixed-parameter tractable with respect to the parameter total number of swaps [4] (this is an example of a parameter that measures the solution value).

Now we briefly discuss some parameterized complexity results for SWAP BRIBERY FOR t-APPROVAL due to Dorn and Schlotter [8]. The SWAP BRIBERY problem was introduced by Elkind et al. [9], who have shown that the problem is NP-complete for a variety of voting rules, including t-Approval (for $t \geq 2$). Dorn and Schlotter provided a detailed discussion of the complexity of the problem for t-Approval, and, in particular, they have shown that the problem is fixed-parameter tractable when parameterized by the number of voters. On the contrary, if we take t to be the parameter (i.e., the problem no longer considers a single, fixed voting rule but a whole family of them), then the problem is W[1]-hard (and, unless unlikely complexity class collapses occur, the problem is not fixed-parameter tractable).

The complexity of control problems was first studied by Bartholdi, Tovey, and Trick [2], who – in particular – have shown that PLURALITY CONSTRUCTIVE CONTROL BY ADDING ALTERNATIVES (PCCAC) is NP-complete. Control problems have received relatively little attention from the parameterized complexity perspective. For example, among other issues, Betzler and Uhlmann [3] considered parameterization by the number of voters (for Copeland voting rule, which we do not discuss here). Liu et al. [10] considered the parameterization by the solution size (i.e., the number of candidates to be added) and, among other results, obtained W[2]-hardness for PCCAC (in essence, this already follows from the proof of Bartholdi, Tovey, and Trick).

Open Problems

1. We sketched the ILP-based fixed-parameter tractability result for DODGSON SCORE. A key question for this and many other problems shown to be fixed-parameter tractable using Lenstra's result is whether the (impractical) ILP formulation can be replaced by a direct combinatorial algorithm (still providing fixed-parameter tractability); we point to a recent survey [6] for a broader exposition on that. Concerning DODGSON SCORE, it is also interesting to settle its parameterized complexity with respect to the number of votes [4].
2. One of the most intriguing questions regarding the complexity of SWAP BRIBERY is whether the problem (for some given voting rule) is fixed-parameter tractable when parameterized by the number of candidates (or, if not, then if at least there is a fixed-parameter tractable approximation scheme; interestingly, for SHIFT BRIBERY, a significantly simpler variant of the problem, such an approximation scheme indeed exists [7]). Dorn and Schlotter [8] showed that this problem is fixed-parameter tractable, but their proof only applies (in essence) to the case where each swap has the same cost.
3. While there is quite a number of NP-completeness results regarding control problems and various voting rules, there are relatively few parameterized results. Can one

turn some of these NP-completeness results into fixed-parameter tractability results for some natural parameters?

A recent survey article [6] contains several more research challenges concerning the parameterized complexity of problems from Computational Social Choice.

Recommended Reading

1. Bartholdi JJ III, Tovey CA, Trick MA (1989) Voting schemes for which it can be difficult to tell who won the election. Soc Choice Welf 6(2): 157–165
2. Bartholdi JJ III, Tovey CA, Trick MA (1992) How hard is it to control an election? Math Comput Model 16(8/9):27–40
3. Betzler N, Uhlmann J (2009) Parameterized complexity of candidate control in elections and related digraph problems. Theor Comput Sci 410(52): 43–53
4. Betzler N, Guo J, Niedermeier R (2010) Parameterized computational complexity of Dodgson and Young elections. Inf Comput 208(2): 165–177
5. Betzler N, Bredereck R, Chen J, Niedermeier R (2012) Studies in computational aspects of voting – a parameterized complexity perspective. In: Bodlaender HL, Downey R, Fomin FV, Marx D (eds) The multivariate algorithmic revolution and beyond. LNCS, vol 7370. Springer, Berlin/New York, pp 318–363
6. Bredereck R, Chen J, Faliszewski P, Guo J, Niedermeier R, Woeginger GJ (2014) Parameterized algorithmics for computational social choice: nine research challenges. Tsinghua Sci Technol 19(4):358–373
7. Bredereck R, Chen J, Nichterlein A, Faliszewski P, Niedermeier R (2014) Prices matter for the parameterized complexity of shift bribery. In: Proceedings of the 28th Conference on Artificial Intelligence (AAAI'14), Quebéc City, pp 1398–1404
8. Dorn B, Schlotter I (2012) Multivariate complexity analysis of swap bribery. Algorithmica 64(1):126–151
9. Elkind E, Faliszewski P, Slinko A (2009) Swap bribery. In: Proceedings of the 2nd International Symposium on Algorithmic Game Theory (SAGT '09). LNCS, vol 5814. Springer, Berlin/Heidelberg, pp 299–310
10. Liu H, Feng H, Zhu D, Luan J (2009) Parameterized computational complexity of control problems in voting systems. Theor Comput Sci 410(27–29):2746–2753

Parameterized Algorithms for Drawing Graphs

Henning Fernau
Fachbereich 4, Abteilung Informatik wissenschaften, University of Trier, Trier, Germany
Institute for Computer Science, University of Trier, Trier, Germany

Years and Authors of Summarized Original Work

2004; Dujmovic, Whitesides

Problem Definition

ONE-SIDED CROSSING MINIMIZATION (OSCM) can be viewed as a specific form of drawing a bipartite graph $G = (V_1, V_2, E)$, where all vertices from partition V_i are assigned to the same line (also called layer) L_i in the plane, with L_1 and L_2 being parallel. The vertex assignment to L_1 is fixed, while that to L_2 is free and should be chosen in a way to minimize the number of crossings between etdes drawn as straight-line segments.

Notations

A graph G is described by its vertex set V and its edge set E, i.e., $G=(V, E)$, with $E \subseteq V \times V$. The (open) *neighborhood* of a vertex v, denoted $N(v)$, collects all vertices that are adjacent to v. $N[v] = N(v) \cup \{v\}$ denotes the *closed neighborhood* of v. $\deg(v) = |N(v)|$ is the *degree* of v. For a vertex set S, $N(S) = \bigcup_{v \in S} N(v)$, and $N[S] = N(S) \cup S$. $G[S]$ denotes the graph induced by vertex set S, i.e., $G[S] = (S, E \cap (S \times S))$. A graph $G = (V, E)$ with vertex set V and edge set $E \subseteq V \times V$ is *bipartite* if there is a partition of V into two sets V_1 and V_2 such that $V = V_1 \cup V_2$, $V_1 \cap V_2 = \emptyset$, and $E \subseteq V_1 \times V_2$. For clarity, $G = (V_1, V_2, E)$ is written in this case.

A *two-layer drawing* of a bipartite graph $G = (V_1, V_2, E)$ can be described by two linear orders $<_1$ on V_1 and $<_2$ on V_2. This drawing can be realized as follows: the vertices of V_1 are placed on a line L_1 (also called *layer*) in the order induced by $<_1$ and the vertices of V_2 are placed on a second layer L_2 (parallel to the first one) in the order induced by $<_2$; then, draw a straight-line segment for each edge $e = (u_1, u_2)$ in E connecting the points that represent u_1 and u_2, respectively. A *crossing* is a pair of edges $e = (u_1, u_2)$ and $f = (v_1, v_2)$ that cross in the realization of a two-layer drawing $(G, <_1, <_2)$. It is well-known that two edges cross if and only if $u_1 <_1 v_1$ and $v_2 <_2 u_2$; in other word, this notion is a purely combinatorical object, independent of the concrete realization of the two-layer drawing. $\mathrm{cr}(G, <_1, <_2)$ denotes the number of crossings in the described two-layer drawing. In the graph drawing context, it is of course desirable to draw graphs with few crossings. In its simplest (yet probably most important) form, the vertex order in one layer is fixed, and the aim is to minimize crossings by choosing an order of the second layer. Formally, this means:

Problem 1 (*k*–OSCM)

INPUT: A simple n-vertex bipartite graph $G = (V_1, V_2, E)$ and a linear order $<_1$ on V_1, a nonnegative integer k (the parameter).

OUTPUT: If possible, a linear order $<_2$ on V_2 such that $\mathrm{cr}(G, <_1, <_2) \leq k$. If no such order exists, the algorithm should tell so.

Given an instance $G = (V_1, V_2, E)$ and $<_1$ of OSCM and two vertices $u, v \in V_2$,

$$c_{uv} = \mathrm{cr}(G[N[\{u, v\}]], <_1 \cap (N(\{u, v\}) \times N(\{u, v\})), \{(u, v)\}) .$$

Hence, the closed neighborhoods of u and v are considered when assuming the ordering $u <_2 v$.

Consider the following as a running example:

Example 1 In Fig. 1, a concrete drawing of a bipartite graph is shown. Is this drawing optimal with respect to the number of crossings, assuming the ordering of the upper layer being fixed? At some points, more than two edges cross; in that case, a number is shown to count the crossings. All crossings are emphasized by a surrounding box.

Let us now compute the *crossing number matrix* (c_{uv}) for this graph.

c_{uv}	a	b	c	d	e
a	–	4	5	0	1
b	1	–	1	0	0
c	3	3	–	0	1
d	3	2	3	–	1
e	2	3	2	0	–

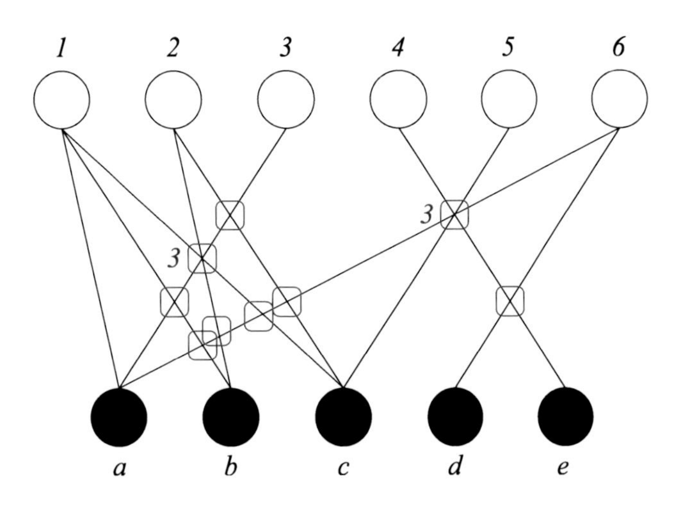

Parameterized Algorithms for Drawing Graphs, Fig. 1 The running example for OSCM

The number of crossings in the given drawing can be hence computed as

$$c_{ab} + c_{ac} + c_{ad} + c_{ae} + c_{bc} + c_{bd} + c_{be} + c_{cd} + c_{ce} + c_{de} = 13.$$

Key Results

Exact exponential-time algorithms are mostly interesting when dealing with problems for which no polynomial-time algorithm is expected to exist.

Theorem 1 ([6]) *The decision problem corresponding to k-OSCM is $\mathcal{NP}$-complete.*

In the following, to state the results, let $G = (V_1, V_2, E)$ be an instance of OSCM, where the ordering $<_1$ of V_1 is fixed.

It can be checked in polynomial time if an order of V_2 exists that avoids any crossings. This observation can be based on either of the following graph-theoretic characterizations:

Theorem 2 ([3]) $\mathrm{cr}(G, <_1, <_2) = 0$ *if and only if G is acyclic and, for every path (x, a, y) of G with $x, y \in V_1$, it holds: for all $u \in V_1$ with $x <_1 u <_1 y$, the only edge incident to u (if any) is (u, a).*

The previously introduced notion is crucial due to the following facts:

Lemma 3 $\sum_{u,v \in V_2, u <_2 v} c_{uv} = \mathrm{cr}(G, <_1, <_2)$.

Theorem 4 ([9]) *If k is the minimum number of edge crossings in an OSCM instance $(G = (V_1, V_2, E), <_1)$, then*

$$\sum_{u,v \in V_2, u \neq v} \min\{c_{uv}, c_{vu}\} \leq k < 1.4664 \sum_{u,v \in V_2, u \neq v} \min\{c_{uv}, c_{vu}\} .$$

In fact, Nagamochi also presented an approximation algorithm with a factor smaller than 1.4664.

Furthermore, for any $u \in V_2$ with $\deg(u) > 0$, let l_u be the leftmost neighbor of u on L_1, and r_u be the rightmost neighbor of u. Two vertices $u, v \in V_2$ are called *unsuited* if there exists some $x \in N(u)$ with $l_v <_1 x <_1 r_v$, or there exists some $x \in N(v)$ with $l_u <_1 x <_1 r_u$. Otherwise, they are called *suited*. Observe that, for $\{u, v\}$ suited, $c_{uv} \cdot c_{vu} = 0$. Dujmović and Whitesides have shown:

Lemma 5 ([5]) *In any optimal ordering $<_2$ of the vertices of V_2, $u <_2 v$ is found if $r_u \leq_1 l_v$.*

This means that all suited pairs appear in their *natural ordering*.

This already allows us to formulate a first parameterized algorithm for OSCM, which is a simple search tree algorithm. In the course of this algorithm, a suitable ordering $<_2$ on V_2 is gradually constructed; when settling the ordering between u and v on V_2, $u <_2 v$ or $v <_2 u$ is *committed.* A *generalized instance* of OSCM therefore contains, besides the bipartite graph $G = (V_1, V_2, E)$, a partial ordering $<_2$ on V_2. A vertex $v \in V_2$ is *fully committed* if, for all $u \in V_2 \setminus \{u, v\}$, $\{u, v\}$ is committed.

Lemma 5 allows us to state the following rule:

RR1: For every pair of vertices $\{u, v\}$ from V_2 with $c_{uv} = 0$, commit $u <_2 v$. In the example, d would be fully committed by applying RR1, since the d-column in the crossing number matrix is all zeros; hence, ignore d in what follows.

Algorithm 1 is a simple search tree algorithm for OSCM that repeatedly uses Rule RR1.

Lemma 6 *OSCM can be solved in time $\mathcal{O}^*(2^k)$.*

Proof Before any branching can take place, the graph instance is reduced, so that every pair of vertices $\{u, v\}$ from V_2 which is not committed satisfies $\min\{c_{uv}, c_{vu}\} \geq 1$. Therefore, each recursive branch reduces the parameter by at least one. □

It is possible to improve on this very simple search tree algorithm. A first observation is that it is not necessary to branch at $\{x, y\} \subset V_2$ with $c_{xy} = c_{yx}$. This means two modifications to Algorithm 1:

Algorithm 1 A search tree algorithm solving OSCM, called OSCM-ST-simple

Require: a bipartite graph $G = (V_1, V_2, E)$, an integer k, a linear ordering $<_1$ on V_1, a partial ordering $<_2$ on V_2
Ensure: YES iff the given OSCM instance has a solution

```
    repeat
       Exhaustively apply the reduction rules, adjusting <_2 and k accordingly.
       Determine the vertices whose order is settled by transitivity and adjust <_2 and k accordingly.
    until there are no more changes to <_2 and to k
 5: if k < 0 or <_2 contains both (x, y) and (y, x) then
       return NO.
    else if ∃{x, y} ⊆ V_2 : neither x <_2 y nor y <_2 x is settled then
       if OSCM-ST-simple(G, k − 1, <_1, <_2 ∪ {(x, y)}) then
          return YES
10:    else
          return OSCM-ST-simple(G, k − 1, <_1, <_2 ∪ {(y, x)})
       end if
    else
       return YES
15: end if
```

- Line 5 should exclude $c_{xy} = c_{yx}$.
- Line 12 should arbitrary commit some $\{x, y\} \subset V_2$ that are not yet committed and recurse; only if all $\{x, y\} \subset V_2$ are committed, YES is to be returned.

These modifications immediately yield an $\mathcal{O}^*(1.6182^k)$ algorithm for OSCM. This is also the core of the algorithm proposed by Dujmović and Whitesides [5]. There, more details are discussed, as, for example:

- How to efficiently calculate all the crossing numbers c_{xy} in a preprocessing phase.
- How to integrate branch and cut elements in the algorithm that are surely helpful from a practical perspective.
- How to generalize the algorithm for instances that allow integer weights on the edges (multiple edges).

Further improvements are possible if one gives a deeper analysis of local patterns $\{x, y\} \in V_2$ such that $c_{xy}c_{yx} \leq 2$. This way, it has been shown:

Theorem 7 ([4]) *OSCM can be solved in time* $\mathcal{O}^*(1.4656^k)$.

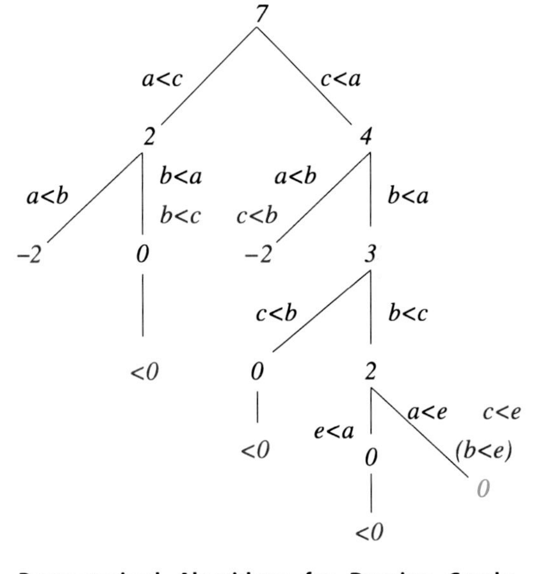

Parameterized Algorithms for Drawing Graphs, Fig. 2 A search tree example for OSCM

A possible run of the improved search tree algorithm is displayed in Fig. 2, with the (optimal) outcome shown in Fig. 3.

Variants and Related Problems have been discussed in the literature.

1. Change the goal of the optimization: minimize the number of edges involved in cross-

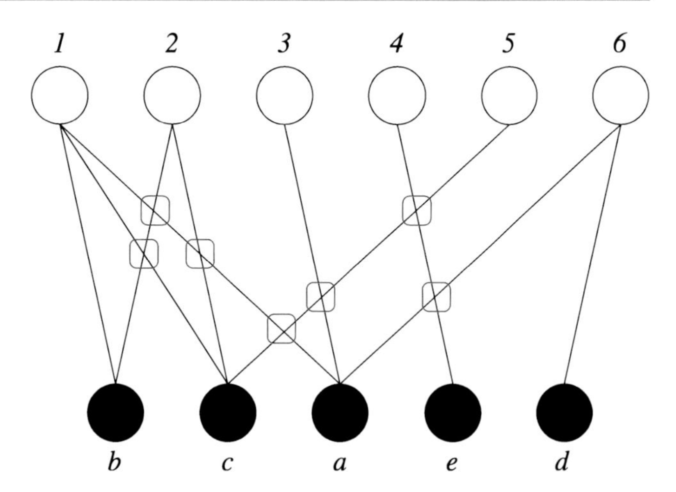

Parameterized Algorithms for Drawing Graphs, Fig. 3 An optimal solution to the example instance

ings (ONE-LAYER PLANARIZATION (OLP)). As observed in [7, 10], Theorem 2 almost immediately leads to an $\mathcal{O}^*(3^k)$ algorithm for OLP that was subsequently improved down to $\mathcal{O}^*(2^k)$ in [10].

2. One could allow more degrees of freedom by considering two (or more) layer assignments at the same time. For both the crossing minimization and the planarization variants, parameterized algorithms are reported in [3, 7, 10].
3. One can consider other additional constraints on the drawings or the admissible orderings; in [8], parameterized algorithms for two-layer assignment problems are discussed where the admissible orderings are restricted by binary trees.

Applications

Besides seeing the question of drawing bipartite graphs as an interesting problem in itself, e.g., for nice drawings of relational diagrams, this question quite naturally shows up in the so-called Sugiyama approach to hierarchical graph drawing, see [12]. This very popular approach tackles the problem of laying out a hierarchical graph in three phases: (1) cycle removal (2) assignment of verticesto layers, (3) assignment of vertices to layers. The last phase is usually performed in a sweep-line fashion, intermediately solving many instances of OSCM. The third variant in the discussion above has important applications in computational biology.

Open Problems

As with all exponential-time algorithms, it is always a challenge to further improve on the running times of the algorithms or to prove lower bounds on those running times under reasonable complexity theoretic assumptions. Let us notice that the tacit assumptions underlying the approach by parameterized algorithmics are well met in this application scenario: e.g., one would not accept drawings with many crossings anyways (if such a situation is encountered in practice, one would switch to another way of representing the information); so, one can safely assume that the parameter is indeed small.

This is also true for other $\mathcal{NP}$-hard subproblems that relate to the Sugiyama approach. However, no easy solutions should be expected. For example, the DIRECTED FEEDBACK ARC SET PROBLEM [1] that is equivalent to the first phase is not known to admit a nice parameterized algorithm, see [2].

Experimental Results

Suderman [10] reports on experiments with nearly all problem variants discussed above, also see [11] for a better accessible presentation of some of the experimental results.

URL to Code

Suderman presents several JAVA applets related to the problems discussed in this article, see http://cgm.cs.mcgill.ca/~msuder/.

Cross-References

Other parameterized search tree algorithms are explained in the contribution ▶ Vertex Cover Search Trees by Chen, Kanj, and Jia.

Recommended Reading

1. Chen J, Liu Y, Lu S, O'Sullivan B, Razgon I (2008) A fixed-parameter algorithm for the directed feedback vertex set problem. In: 40th ACM symposium on theory of computing STOC 2008, Victoria, 17–20 May 2008
2. Downey RG, Fellows MR (1999) Parameterized complexity. Springer, Berlin
3. Dujmovic V, Fellows MR, Hallett M, Kitching M, Liotta G, McCartin C, Nishimura N, Ragde P, Rosamond FA, Suderman M, Whitesides S, Wood DR (2006) A fixed-parameter approach to 2-layer planarization. Algorithmica 45: 159–182
4. Dujmovic V, Fernau H, Kaufmann M (2004) Fixed parameter algorithms for one-sided crossing minimization revisited. In: Liotta G (ed) Graph drawing, 11th international symposium GD 2003. LNCS, vol 2912. Springer, Berlin, pp 332–344, A journal version has been accepted to J Discret Algorithms see doi:10.1016/j.jda.2006.12.008
5. Dujmovic V, Whitesides S (2004) An efficient fixed parameter tractable algorithm for 1-sided crossing minimization. Algorithmica 40:15–32
6. Eades P, Wormald NC (1994) Edge crossings in drawings of bipartite graphs. Algorithmica 11: 379–403
7. Fernau H (2005) Two-layer planarization: improving on parameterized algorithmics. J Graph Algorithms Appl 9:205–238
8. Fernau H, Kaufmann M, Poths M (2005) Comparing trees via crossing minimization. In: Ramanujam R, Sen S (eds) Foundations of software technology and theoretical computer science FSTTCS 2005. LNCS, vol 3821. Springer, Berlin, pp 457–469
9. Nagamochi H (2005) An improved bound on the one-sided minimum crossing number in two-layered drawings. Discret Comput Geom 33:569–591
10. Suderman M (2005) Layered graph drawing. PhD thesis, McGill University, Montréal
11. Suderman M, Whitesides S (2005) Experiments with the fixed-parameter approach for two-layer planarization. J Graph Algorithms Appl 9(1): 149–163
12. Sugiyama K, Tagawa S, Toda M (1981) Methods for visual understanding of hierarchical system structures. IEEE Trans Syst Man Cybern 11(2): 109–125

Parameterized Pattern Matching

Moshe Lewenstein
Department of Computer Science, Bar-Ilan University, Ramat-Gan, Israel

Years and Authors of Summarized Work

1994; Amir, Farach, Muthukrishnan
1995; Kosaraju
1996; Baker

Problem Definition

Parameterized strings, or *p-strings*, are strings that contain both ordinary symbols from an alphabet Σ and parameter symbols from an alphabet Π. Two equal-length p-strings s and s' are a parameterized match, or *p-match*, if one p-string can be transformed into the other by applying a one-to-one function that renames the parameter

P

symbols. The following example of a p-match is one with both ordinary and parameter symbols. The ordinary symbols are in lowercase and the parameter symbols are in uppercase:

$$s = A\ b\ A\ b\ C\ A\ d\ b\ A\ C\ d\ d$$
$$s' = D\ b\ D\ b\ E\ D\ d\ b\ D\ E\ d\ d$$

In some of the problems to be considered, it will be sufficient to solve for p-strings in which all symbols are parameter symbols, as this is the more difficult part of the problem. In other words, the case in which $\Sigma = \emptyset$. In this case, the definition can be reformulated so that s and s' are a p-match if there exists a bijection $\pi : \Pi_s \rightarrow \Pi'_s$, such that $\pi(s) = s'$, where $\pi(s)$ is the renaming of each character of s via π.

The following problems will be considered. *Parameterized matching* – given a parameterized pattern p of length m and parameterized text t, find all locations i of a parameterized text t for which p p-matches $t_i \ldots t_{i+m-1}$, where $m = |p|$. The same problem is also considered in two dimensions. *Approximate parameterized matching* – find all substrings of a parameterized text t that are approximate parameterized matches of a parameterized pattern p (to be fully defined later).

Key Results

Baker [4] introduced parameterized matching in the framework of her seminal work on discovering duplicate code within large programs for the sake of code minimization. An example of two code fragments that p-match taken from the X Windows system can be found in [4].

Parameterized Suffix Trees

In [4] and in the follow-up journal versions [6,7], a novel method was presented for parameterized matching by constructing *parameterized suffix trees*. The advantage of the parameterized suffix tree is that it supports indexing, i.e., one can preprocess a text and subsequently answer parameterized queries p in $O(|p|)$ time. In order to achieve parameterized suffix trees, it is necessary to introduce the concept of a *predecessor string*. A *predecessor string* of a string s has at each location i the distance between i and the location containing the previous appearance of the symbol. The first appearance of each symbol is replaced with a 0. For example, the predecessor string of *aabbaba* is 0,1,0,1,3,2,2. A simple and well-known fact is that:

Observation 1 ([7]) s and s' p-match if and only if they have the same predecessor string.

Notice that this implies transitivity of parameterized matching, since if s and s' p-match and s' and s'' p-match, then, by the observation, s and s' have the same predecessor string and, likewise, s' and s'' have the same predecessor string. This implies that s and s'' have the same predecessor string and hence, by the observation, p-match.

Moreover, one may also observe that if r is a prefix of s, then the predecessor string of r, by definition, is exactly the $|r|$-length prefix of the predecessor string of s. Hence, similar to regular pattern matching, a parameterized pattern p p-matches at location i of t if and only if the $|p|$-length predecessor string of p is equal to the $|p|$-length prefix of the predecessor string of the suffix $t_i \ldots t_n$. Combining these observations, it is natural to do as follows: create a (parameterized suffix) tree with a leaf for each suffix where the path from the root to the leaf corresponding to a given suffix will have its predecessor string labeling the path. Branching in the parameterized suffix tree, as with suffix trees, occurs according to the labels of the predecessor strings. See [4, 6, 7] for an example.

Baker's method essentially mimics the McCreight suffix tree construction [18]. However, while the suffix tree and the parameterized suffix tree are very similar, there is a slight hitch. A strong component of the suffix tree construction is the suffix link. This is used for the construction and, sometimes, for later pattern searches. The suffix link is based on the *distinct right context* property, which does not hold for the parameterized suffix tree. In fact, the node that is pointed to by the suffix link may not even exist. The main parts of [6, 7] are dedicated to circumventing this problem.

In [7] Baker added the notion of "bad" suffix links, which point to the vertex just above, i.e., closer to the root than the desired place, and of updating them with a lazy evaluation when they are used. The algorithm runs in time $O(n|\Pi|\log|\Sigma|)$. In [6] (which is chronologically later than [7] despite being the first to appear) Baker changed the definition of "bad" suffix links to point to just below the desired place. This turns out to have nice properties, and one can use more sophisticated data structures to improve the construction time to $O(n(|\Pi| + \log|\Sigma|))$.

Kosaraju [16] made a careful analysis of Baker's properties utilized in the algorithm of [6] which suffer from the $|\Pi|$ factor. He pointed out two sources for this large factor. He handled these two issues by using a concatenable queue and maintaining it in a lazy manner. This is sufficient to reduce the $|\Pi|$ factor to a $\log|\Pi|$ factor, yielding an algorithm of time $O(n(\log|\Pi| + \log|\Sigma|))$.

Obviously if the alphabet or symbol set is large, the construction time may be $O(n \log n)$. Cole and Hariharan [9] showed how to construct the parameterized suffix trees in randomized $O(n)$ time for alphabets and parameters taken from a polynomially sized range, e.g., $[1, \ldots, n^c]$. They did this by adding additional nodes to the tree in a back-propagation manner which is reminiscent of fractional cascading. They showed that this adds only $O(n)$ nodes and allows the updating of the missing suffix links. However, this causes other problems and one may find the details of how this is handled in their paper.

More Methods for Parameterized Matching

Obviously the parameterized suffix tree efficiently solves the parameterized matching problem. Nevertheless, a couple of other results on parameterized matching are worth mentioning.

First, in [6] it was shown how to construct the parameterized suffix tree for the pattern and then to run the parameterized text through it, giving an algorithm with $O(m)$ space instead of $O(n)$.

Amir et al. [2] presented a simple method to solve the parameterized matching problem by mimicking the algorithm of Knuth, Morris, and Pratt. Their algorithm works in $O(n * \min(\log|\Pi|, m))$ time independent of the alphabet size ($|\Sigma|$). Moreover, they proved that the log factor cannot be avoided for large symbol sets.

In [5] parameterized matching was solved with a Boyer-Moore type algorithm. In [10] the problem was solved with a Shift-Or type algorithm. Both handle the average case efficiently. In [10] emphasis was also put on the case of multiple parameterized matching, which was previously solved in [14] with an Aho-Corasick automaton-style algorithm.

Two-Dimensional Parameterized Matching

Two-dimensional parameterized matching arises in applications of image searching; see [13] for more details. Two-dimensional parameterized matching is the natural extension of parameterized matching where one seeks pmatches of a two-dimensional parameterized pattern p within a two-dimensional parameterized text t. It must be pointed out that classical methods for two-dimensional pattern matching, such as the L-suffix tree method, fail for parameterized matching. This is because known methods tend to cut the text and pattern into pieces to avoid going out of boundaries of the pattern. This is fine because each pattern piece can be individually evaluated (checked for equality) to a text piece. However, in parameterized matching, there is a strong dependency between the pieces.

In [1] an innovative solution for the problem was given based on a collection of linearizations of the pattern and text with the property to be currently described. Consider a linearization. Two elements with the same character, say "a," in the pattern are defined to be neighbors if there is no other "a" between them in this linearization. Now take all the "a"s of the pattern and create a graph G_a with "a"s as the nodes and edges between the two if they are neighbors in some linearization. We say that two "a"s are chained if there is a path from one to the other in G_a. Applying one-dimensional parameterized matching on these linearizations ensures that any two elements that are chained will be evaluated to map to the same text value (the parameterized property). A collection of linearizations has the *fully chained* property if every two locations in p with the same character are chained. It was shown in [1] that one can obtain a collection of log m linearizations that is fully chained and that does not exceed pattern boundary limits. Each such linearization is solved with a convolution-based pattern-matching algorithm. This takes $O(n^2 \log m)$ time for each linearization, where the text size is n^2. Hence, overall the time is $O(n^2 \log^2 m)$.

A different solution was proposed in [13], where it was shown that it is possible to solve the problem in $O(n^2 + m^{2.5}\text{polylog } m)$, where the text size is $O(n^2)$ and the pattern size is $O(m^2)$. Clearly, this is more efficient for large texts.

Approximate Parameterized Matching

Our last topic relates to parameterized matching in the presence of errors. Errors occur in various applications and it is natural to consider parameterized matching with the Hamming distance metric or the edit distance metric.

In [8] the parameterized matching problem was considered in conjunction with the edit distance. Here the definition of edit distance was slightly modified so that the edit operations are defined to be insertion, deletion, and parameterized replacements, i.e., the replacement of a substring with a string that p-matches it. An algorithm for finding the "parameterized edit distance" of two strings was devised whose efficiency is close to the efficiency of the algorithms for computing the classical edit distance.

However, it turns out that the operation of parameterized replacement relaxes the problem to an easier problem. The reason that the problem becomes easier is that two substrings that participate in two parameterized replacements are independent of each other (in the parameterized sense).

A more rigid, but more realistic, definition for the Hamming distance variant was given in [3]. For a pair of equal-length strings s and s' and a bijection π defined on the alphabet of s, the π-mismatch is the Hamming distance between the image under π of s and s'. The minimal π-mismatch over all bijections π is the approximate parameterized match. The problem considered in [3] is to find for each location i of a text t the approximate parameterized match of a pattern p with the substring beginning at location i. In [3] the problem was defined and linear-time algorithms were given for the case where the pattern is binary or the text is binary. However, this solution does not carry over to larger alphabets.

Unfortunately, under this definition, the methods for classical string matching with errors for the Hamming distance, also known as pattern matching with mismatches, seem to fail. Following is an outline of a classical method [17] for pattern matching with mismatches that uses suffix trees.

The pattern is compared separately with each suffix of the text, beginning at locations $1 \leq i \leq n - m + 1$. Using a suffix tree of the text and precomputed longest common ancestor information (which can be computed once in linear time [11]), one can find the longest common prefix of the pattern and the corresponding suffix (in constant time). There must be a mismatch immediately afterwards. The algorithm jumps over the mismatch and repeats the process, taking into consideration the offsets of the pattern and suffix.

When attempting to apply this technique to a parameterized suffix tree, it fails. To illustrate this, consider the first matching substring (up until the first error) and the next matching substring (after the error). Both of these substrings p-match the substring of the text that they are aligned with. However, it is possible that combined they do not form a p-match. See the example below. In the example *abab* p-matches *cdcd* followed by a mismatch and subsequently followed by *abaa* p-matching *efee*. However, different π's are required for the local p-matches. This example also emphasizes why the definition of [8] is a simplification. Specifically, each local p-matching substring is one replacement, i.e., *abab* with *cdcd* is one replacement and *abaa* with *efee* is one more replacement. However, the definition of [3] captures the globality of the parameterized matching, not allowing, in this case, *abab* to p-match to two different substrings.

$$p = a\ b\ a\ b\ a\ a\ b\ a\ a \ldots$$

$$t = \ldots c\ d\ c\ d\ d\ e\ f\ e\ e \ldots \cdot$$

In [12] the problem of *parameterized matching with k mismatches* was considered. The parameterized matching problem with k mismatches seeks all locations i in text t where the minimal π-mismatch between p to $t_i \ldots t_{i+m-1}$ has at most k mismatches. An $O(nk^{1.5} + mk \log m)$ time algorithm was presented in [12]. At the base of the algorithm, i.e., for the case where $|p| = |t| = m$, an $O(m + k^{1.5})$ algorithm is used based on maximum matching algorithms. Then the algorithm uses a doubling scheme to handle the growing distance between potential parameterized matches (with at most k mismatches). Also shown in [12] is a strong relationship between maximum matching algorithms in sparse graphs and parameterized matching with k errors.

The rigid, but more realistic, definition for the Hamming distance version given in [3] can be naturally extended to the edit distance. Lately, it was shown that this problem is nondeterministic polynomial-time complete [15].

Applications

Parameterized matching has applications in code duplication detection in programming languages, in homework plagiarism detection, and in image processing, among others [1,4].

Cross-References

- ▶ Approximate String Matching
- ▶ Multidimensional String Matching
- ▶ String Matching
- ▶ Suffix Tree Construction

Recommended Reading

1. Amir A, Aumann Y, Cole R, Lewenstein M, Porat E (2003) Function matching: algorithms, applications and a lower bound. In: Proceedings of the 30th international colloquium on automata, languages and programming (ICALP), Eindhoven, pp 929–942
2. Amir A, Farach M, Muthukrishnan S (1994) Alphabet dependence in parameterized matching. Inf Process Lett 49:111–115
3. Apostolico A, Erdös P, Lewenstein M (2007) Parameterized matching with mismatches. J Discret Algorithms 5(1):135–140
4. Baker BS (1993) A theory of parameterized pattern matching: algorithms and applications. In: Proceedings of the 25th annual ACM symposium on the theory of computation (STOC), San Diego, pp 71–80
5. Baker BS (1995) Parameterized pattern matching by Boyer-Moore-type algorithms. In: Proceedings of the 6th annual ACM-SIAM symposium on discrete algorithms (SODA), San Francisco, pp 541–550
6. Baker BS (1996) Parameterized pattern matching: algorithms and applications. J Comput Syst Sci 52(1):28–42
7. Baker BS (1997) Parameterized duplication in strings: algorithms and an application to software maintenance. SIAM J Comput 26(5):1343–1362
8. Baker BS (1999) Parameterized diff. In: Proceedings of the 10th annual ACM-SIAM symposium on discrete algorithms (SODA), Baltimore, pp 854–855
9. Cole R, Hariharan R (2000) Faster suffix tree construction with missing suffix links. In: Proceedings of the 32nd ACM symposium on theory of computing (STOC), Portland, pp 407–415
10. Fredriksson K, Mozgovoy M (2006) Efficient parameterized string matching. Inf Process Lett 100(3):91–96
11. Harel D, Tarjan RE (1984) Fast algorithms for finding nearest common ancestor. J Comput Syst Sci 13:338–355
12. Hazay C, Lewenstein M, Sokol D (2007) Approximate parameterized matching. ACM Trans Algorithms 3(3):29
13. Hazay C, Lewenstein M, Tsur D (2005) Two dimensional parameterized matching. In: Proceedings of the 16th symposium on combinatorial pattern matching (CPM), Jeju Island, pp 266–279
14. Idury RM, Schäffer AA (1996) Multiple matching of parameterized patterns. Theor Comput Sci 154(2):203–224
15. Keller O, Kopelowitz T, Lewenstein M. Parameterized LCS and edit distance are NP-complete. Manuscript
16. Kosaraju SR (1995) Faster algorithms for the construction of parameterized suffix trees. In: Proceedings of the 36th annual symposium on foundations of computer science (FOCS), Milwaukee, pp 631–637

17. Landau GM, Vishkin U (1988) Fast string matching with k differences. J Comput Syst Sci 37(1): 63–78
18. McCreight EM (1976) A space-economical suffix tree construction algorithm. J ACM 23: 262–272

Parameterized SAT

Stefan Szeider
Department of Computer Science, Durham University, Durham, UK

Keywords

Structural parameters for SAT

Years and Authors of Summarized Original Work

2003; Szeider

Problem Definition

Much research has been devoted to finding classes of propositional formulas in conjunctive normal form (CNF) for which the recognition problem as well as the propositional satisfiability problem (SAT) can be decided in polynomial time. Some of these classes form infinite chains $C_1 \subset C_2 \subset \cdots$ such that every CNF formula is contained in some C_k for k sufficiently large. Such classes are typically of the form $C_k = \{F \in \text{CNF}: \pi(F) \leq k\}$, where π is a computable mapping that assigns to CNF formulas F a non-negative integer $\pi(F)$; we call such a mapping a *satisfiability parameter*. Since SAT is an NP-complete problem (actually, the first problem shown to be NP-complete [1]), we must expect that, the larger k, the longer the worst-case running times of the polynomial-time algorithms that recognize and decide satisfiability of formulas in C_k. Whence there is a certain tradeoff between the generality of classes and the performance guarantee for the corresponding algorithms. Szeider [12] initiates a broad investigation of this tradeoff in the conceptional framework of *parameterized complexity* [2, 3, 6]. This investigation draws attention to satisfiability parameters π for which the following holds: recognition and satisfiability decision of formulas F with $\pi(F) \leq k$ can be carried out in *uniform polynomial time*, that is, by algorithms with running time bounded by a polynomial whose order is independent of k (albeit, possibly involving a constant factor that is exponential in k). If a satisfiability parameter π allows satisfiability decision in uniform polynomial time, we say that *SAT is fixed-parameter tractable with respect to* π.

Satisfiability Parameters Based on Graph Invariants

One can define satisfiability parameters by means of certain graphs associated with CNF formulas. The *primal graph* of a CNF formula is the graph whose vertices are the variables of the formula; two variables are joined by an edge if the variables occur together in a clause. The *incidence graph* of a CNF formula is the bipartite graph whose vertices are the variables and clauses of the formula; a variable and a clause are joined by an edge if the variable occurs in the clause.

Satisfiability Parameters Based on Backdoor Sets

The concept of backdoor sets [13] gives rise to several interesting satisfiability parameters. Let C be a class of CNF formulas. A set B of variables of a CNF formula F is a *strong* C-*backdoor set* if for every partial truth assignment $\tau: B \to \{\text{true}, \text{false}\}$, the restriction of F to τ belongs to C. Here, the restriction of F to τ is the CNF formula obtained from F by removing all clauses that contain a literal that is true under τ and by removing from the remaining clauses all literals that are false under τ.

Key Results

Theorem 1 (Gottlob, Scarcello, and Sideri [4]) *SAT is fixed-parameter tractable with respect to the treewidth of primal graphs.*

Several satisfiability parameters that generalize the treewidth of primal graphs, such as the *treewidth and clique-width of incidence graphs*, have been studied [5, 10, 12].

The *maximum deficiency* of a CNF formula F is the number of clauses remaining exposed by a maximum matching of the incidence graph of F.

Theorem 2 (Szeider [11]) *SAT is fixed-parameter tractable with respect to maximum deficiency.*

A CNF formula is *minimal unsatisfiable* if it is unsatisfiable but removing any of its clauses makes it satisfiable. Recognition of minimal unsatisfiable formulas is DP-complete [9].

Corollary 1 (Szeider [11]) *Recognition of minimal unsatisfiable CNF formulas is fixed-parameter tractable with respect to the difference between the number of clauses and the number of variables.*

Theorem 3 (Nishimura, Ragde, and Szeider [7]) *SAT is fixed-parameter tractable with respect to the size of strong Horn-backdoor sets and with respect to the size of strong 2CNF-backdoor sets.*

Applications

Satisfiability provides a powerful and general formalism for solving various important problems including hardware and software verification and planning. Instances stemming from applications usually contain a "hidden structure" (see, e.g., [13]). The satisfiability parameters considered above are designed to make this hidden structure explicit in the form of small values for the parameter. Thus, satisfiability parameters are a way to make the hidden structure accessible to an algorithm.

Open Problems

A new line of research is concerned with the identification of further satisfiability parameters that allow a fixed-parameter tractable SAT decision are more general than the known parameters and apply well to real-world problem instances.

Cross-References

▸ Maximum Matching
▸ Treewidth of Graphs

Recommended Reading

1. Cook SA (1971) The complexity of theorem-proving procedures. In: Proceedings of the 3rd annual symposium on theory of computing, Shaker Heights, pp 151–158
2. Downey RG, Fellows MR (1999) Parameterized complexity, Monographs in computer science. Springer, Berlin
3. Flum J, Grohe M (2006) Parameterized complexity theory. Texts in theoretical computer science. An EATCS series, vol XIV. Springer, Berlin
4. Gottlob G, Scarcello F, Sideri M (2002) Fixed-parameter complexity in AI and nonmonotonic reasoning. Artif Intellect 138:55–86
5. Gottlob G, Szeider S (2007) Fixed-parameter algorithms for artificial intelligence, constraint satisfaction, and database problems. Comput J Spec Issue Parameterized Complex Adv Access
6. Niedermeier R (2006) Invitation to fixed-parameter algorithms, Oxford lecture series in mathematics and its applications. Oxford University Press, Oxford
7. Nishimura N, Ragde P, Szeider S (2004) Detecting backdoor sets with respect to Horn and binary clauses. In: Informal proceedings of SAT 2004, 7th international conference on theory and applications of satisfiability testing, Vancouver, 10–13 May 2004, pp 96–103
8. Nishimura N, Ragde P, Szeider S (2007) Solving SAT using vertex covers. Acta Inf 44(7–8):509–523
9. Papadimitriou CH, Wolfe D (1988) The complexity of facets resolved. J Comput Syst Sci 37:2–13

P

10. Samer M, Szeider S (2007) Algorithms for propositional model counting. In: Proceedings of LPAR 2007, 14th international conference on logic for programming, artificial intelligence and reasoning, Yerevan, 15–19 Oct 2007. Lecture notes in computer science, vol 4790. Springer, Berlin, pp 484–498
11. Szeider S (2004) Minimal unsatisfiable formulas with bounded clause-variable difference are fixed-parameter tractable. J Comput Syst Sci 69:656–674
12. Szeider S (2004) On fixed-parameter tractable parameterizations of SAT. In: Giunchiglia E, Tacchella A (eds) Theory and applications of satisfiability, 6th international conference, SAT 2003, selected and revised papers. Lecture notes in computer science, vol 2919. Springer, Berlin, pp 188–202
13. Williams R, Gomes C, Selman B (2003) On the connections between backdoors, restarts, and heavy-tailedness in combinatorial search. In: Informal proceedings of SAT 2003 sixth international conference on theory and applications of satisfiability testing, 5–8 May 2003, S. Margherita Ligure – Portofino, pp 222–230

Parity Games

Tim A.C. Willemse and Maciej Gazda
Department of Mathematics and Computer Science, Eindhoven University of Technology, Eindhoven, The Netherlands

Keywords

Automata; Computer-aided verification; Determinacy; Infinite duration games; Perfect information; Two-player games

Years and Authors of Summarized Original Work

1991; Emerson, Jutla
1991; Mostowski

Problem Definition

A parity game is an infinite duration game, played by players *odd* and *even*, denoted by $\Box$ and $\Diamond$, respectively on a directed, finite graph. Throughout this note, we let $\bigcirc$ denote an arbitrary player and we write $\bar{\bigcirc}$ for $\bigcirc$'s opponent; i.e. $\bar{\Diamond} = \Box$ and $\bar{\Box} = \Diamond$.

Definition 1 (Parity game) A parity game is a tuple $(V, E, \Omega, (V_\Diamond, V_\Box))$, where

- V is a set of vertices, partitioned in a set $V_\Diamond$ of vertices owned by player $\Diamond$, and a set of vertices $V_\Box$ owned by player $\Box$,
- $E \subseteq V \times V$ is a total edge relation,
- $\Omega : V \to \mathbb{N}$ is a priority function that assigns priorities to vertices.

The graph (V, E) underlying a parity game is often referred to as the *arena*. Parity games are depicted as graphs with diamond-shaped vertices representing vertices owned by player $\Diamond$ and box-shaped vertices representing those owned by player $\Box$. The priorities associated with vertices are written inside vertices; see the game depicted in Fig. 1.

Imagine the following game, played on an arena. One starts by placing a token on a vertex. Then, players perpetually move this token according to a single simple rule: if the token is on some vertex $v \in V_\bigcirc$, player $\bigcirc$ gets to move the token to an adjacent vertex. The infinite sequence of vertices visited this way is referred to as a *play* and the *parity* of the *lowest* priority (associated with the vertices) that occurs infinitely often on the play defines its *winner*: player $\Diamond$ wins if and only if this priority is even.

It does not immediately follow from the above notion of winning that there is always a player that can win all plays that start in a given vertex. The most elementary of all problems concerning parity games is thus as follows.

Problem 1 Are parity games *determined*? That is, for a given parity game and some vertex in that game, is there always a way in which one of the players can play such that, regardless of how her opponent moves, the resulting plays are won by her?

The determinacy problem can be formalized by describing the choices players make when

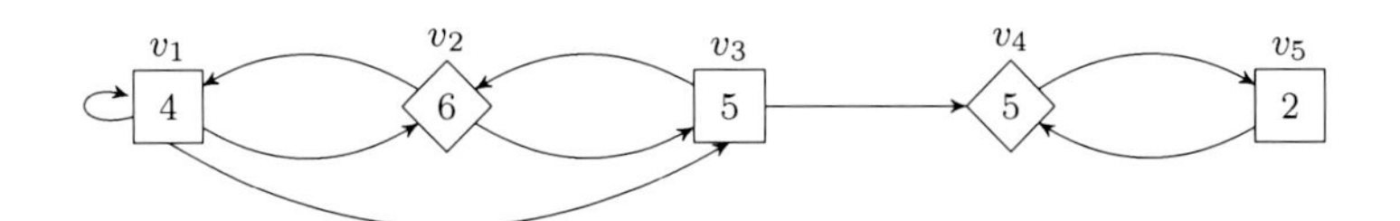

Parity Games, Fig. 1 Parity games depicted as graphs with priorities associated with vertices, written inside vertices

they engage in a play; that is, we can formally capture their *strategies*. More specifically, a strategy for player ○ determines for a vertex v belonging to that player which of its adjacent vertices the token will be moved to next once the game play moves the token to v. Of course, such a decision can be based on the history of the play so far. We write vE to denote the set $\{w \in V \mid (v, w) \in E\}$. A strategy for player ○ is thus described adequately by a partial function $\sigma : V^* V_\bigcirc \rightarrow V$ satisfying that for all sequences $v_1 \ldots v_n \in V^*$ for which σ is defined, we have $\sigma(v_1 \ldots v_n) \in v_n E$. For a given strategy σ, a play p *conforms* to σ if for all finite prefixes q of p, whenever σ is defined for q, also $q\sigma(q)$ is a prefix of p. We say that a strategy σ for player ○ is *winning* from a vertex v if and only if ○ is the winner of every play that starts in v and that conforms to σ. The problem of determinacy can then be rephrased as follows: *for given vertex v in a game, is there always a player with a winning strategy from v?*

A vertex is won by some player if that player has a winning strategy from that vertex. The vertices won by player ◇ and player □ are denoted $W_\Diamond$ and $W_\Box$, respectively; determining these sets is typically referred to as *solving* a game. This immediately leads to the second fundamental question:

Problem 2 Can we compute $W_\Diamond$ and $W_\Box$?

Solving the above computational problem may not necessarily involve computing the winning strategies for both players. Nonetheless, the winning strategies themselves have their own merit, if only that they serve as certificates in proving that the winning sets for both players are indeed just that. This leads to the third problem of interest:

Problem 3 Can we, for all vertices that are won by one of the players, compute winning strategies for that player?

Key Results

The first claim, stated below, positively answers the determinacy problem for parity games.

Theorem 1 *Parity games are determined: for every vertex either player ◇ or player □ has a winning strategy from that vertex.*

Determinacy of parity games already follows from a general result due to Martin [9] who showed that Borel games (which subsume parity games) are determined. The proof of the latter result employs strategies that require infinite memory. A deep result by Emerson and Jutla [1], and found independently by Mostowski [11], states that parity games are in fact *memoryless determined*: if a player has a winning strategy from a vertex v, she also has a memoryless strategy that is winning for her; that is, player ○ always has a strategy σ for which $\sigma(pv) = \sigma(p'v)$ for all sequences pv and $p'v$ for which it is defined and which can thus be represented by a partial function $\sigma' : V_\bigcirc \rightarrow V$.

A simpler and constructive proof of memoryless determinacy for parity games with a finite arena was subsequently proposed by McNaughton [10] and extended to games with an infinite arena by Zielonka [15]. From the memoryless determinacy result, it follows that the problem of deciding whether player ◇ has a winning strategy from a given vertex is in NP: essentially one can guess a memoryless strategy for player ◇ and check in polynomial time whether it is winning; the latter can be done efficiently by showing the absence of odd cycles using e.g. [8]. In a

P

similar manner, one can prove that the problem is in coNP, making the decision problem one of those interesting problems that are in NP ∩ coNP. In fact, in 1998, Jurdziński [5] showed that the problem is in UP ∩ coUP. The complexity class UP is a subclass of NP and is defined to contain all problems that can be recognized by a non-deterministic polynomial time Turing machine that for every input has at most one accepting computation. From the memoryless determinacy, one also immediately obtains a positive answer to the second question.

Theorem 2 *The problem of deciding whether $v \in W_\diamond$ is in NP ∩ coNP, and the sets $W_\diamond$ and $W_\square$ can be computed.*

McNaughton and Zielonka's constructive proofs can be converted into a recursive algorithm for computing $W_\diamond$ and $W_\square$ for games with a finite arena. This algorithm, which we will introduce shortly, can be modified in a straightforward manner to also produce winning (memoryless) strategies for both players, thus also answering the third question.

Theorem 3 *Winning strategies for players $\diamond$ and $\square$ can be computed.*

Algorithms for Parity Games

While the problem of computing the winning sets of a parity game is decidable, its exact complexity is still open, but over the years, the upper bound has been improved. We therefore summarize the various algorithms that have been invented for solving parity games so far, starting with a brief exposition of the recursive algorithm which is, as we described above, also interesting for its theoretical consequences. For the remainder of this section, fix a parity game $G = (V, E, \Omega, (V_\diamond, V_\square))$ with n vertices, m edges, and d different priorities.

The recursive algorithm effectively decomposes a game into subgames and solves these. An essential ingredient in this decomposition is the notion of an $\bigcirc$-attractor into a set of vertices U, denoting $Attr^0_\bigcirc(U)$, which is the least set A, satisfying $A \supseteq U$ and

Algorithm 1: The recursive algorithm

function SOLVE(G)
  Input: parity game $G = (V, E, \Omega, (V_\diamond, V_\square))$
  Output: winning partition $(W_\diamond, W_\square)$

  if $V = \emptyset$ **then** $(W_\diamond, W_\square) \leftarrow (\emptyset, \emptyset)$
  else
    $m \leftarrow \min\{\Omega(v) \mid v \in V\}$
    if $m \bmod 2 = 0$ **then** $p \leftarrow \diamond$ **else** $p \leftarrow \square$
end if
    $U \leftarrow \{v \in V \mid \Omega(v) = m\}$
    $A \leftarrow Attr^0_p(U)$
    $(W'_\diamond, W'_\square) \leftarrow$ SOLVE$(G \setminus A)$
    if $W'_{\bar{p}} = \emptyset$ **then** $(W_p, W_{\bar{p}}) \leftarrow (A \cup W'_p, \emptyset)$
    else
      $B \leftarrow Attr^0_{\bar{p}}(W'_{\bar{p}})$
      $(W'_\diamond, W'_\square) \leftarrow$ SOLVE$(G \setminus B)$
      $(W_p, W_{\bar{p}}) \leftarrow (W'_p, W'_{\bar{p}} \cup B)$
    end if
  end if
  return $(W_\diamond, W_\square)$
end function

- for all $v \in V_\bigcirc$, if $vE \cap A \neq \emptyset$, then also $v \in A$;
- for all $v \in V_{\bar{\bigcirc}}$, if $vE \subseteq A$, then also $v \in A$.

Intuitively, the $\bigcirc$-attractor into U contains U and exactly those vertices for which $\bigcirc$ can force play into U.

Writing $G \cap A$ for the parity game $(V \cap A, E \cap (A \times A), \Omega|_A, (V_\diamond \cap A, V_\square \cap A))$, where $\Omega|_A$ is the priority function Ω restricted to the vertices in A, and writing $G \setminus A$ for $G \cap (V \setminus A)$, the recursive algorithm is defined as in Algorithm 1.

The correctness of Algorithm 1 leans on the observation that lower priorities in the game dominate higher priorities and that revisiting these lower priorities is beneficial to the player with the "same parity" as this priority. The algorithm runs in polynomial space and its runtime complexity is $\mathcal{O}(m \cdot n^d)$.

The first improvement on this runtime that maintains a polynomial space complexity was achieved by an algorithm due to Jurdziński [6] in 2000. This algorithm, colloquially known as the *small progress measures* (SPM) algorithm, runs in time $\mathcal{O}(d \cdot m \cdot (\frac{n}{\lfloor \frac{d}{2} \rfloor})^{\lfloor \frac{d}{2} \rfloor})$. Jurdziński's algorithm builds on decorations of the game arena, called *parity progress measures* and *game parity progress measures*; the former exist for a parity game with only odd-owned vertices if and only if the game only has even cycles (and thus is won by player $\Diamond$); the latter extend parity progress measures to arbitrary games and are essentially witnesses for winning strategies for player $\Diamond$. The SPM algorithm computes game parity progress measures using a fixpoint iteration, ensuring the measures are the least measures that decrease along a play with each bad odd priority that is encountered and only increase when reaching beneficial even priorities.

The next improvement came in 2006 and was based on a modification of the recursive algorithm, resulting in a subexponential algorithm with running time $n^{\mathcal{O}(\sqrt{n})}$; see [7]. It relies on a notion called a dominion for a player: a set of vertices D that is won by that player by staying *within* D and without allowing her opponent to leave D. The main idea behind the algorithm is that it identifies small dominions (of size at most $\sqrt{2n}$) using a dedicated algorithm and removes them from the game prior to the recursive calls. This algorithm, in turn, inspired Schewe [13] to improve on the runtime complexity for games with a small number of priorities. Rather than using a brute-force method for searching and eliminating dominions, Schewe's algorithm utilizes a modified SPM algorithm while executing the standard recursive algorithm. As a result, this reduces the complexity of solving parity games to $\mathcal{O}(m \cdot (\frac{\kappa \cdot n}{d})^{\gamma(d)})$, where κ is a small constant and $\gamma(d) \approx \frac{d}{3}$.

In parallel to the abovementioned solutions, a different family of algorithms has been developed. They are based on the notion of *strategy improvement*, which has been known in the game theory since the 1960s. The first algorithm of this kind designed specifically for parity games is due to Vöge and Jurdziński [14]. In this approach, one requires devising an order on strategies $\prec_{\bigcirc}$ that satisfies two conditions. Firstly, the maximal strategy w.r.t $\prec_{\bigcirc}$ is winning for $\bigcirc$ on $W_{\bigcirc}$. Secondly, there has to be an (efficiently computable) improvement procedure which, for every strategy σ that is not $\prec_{\bigcirc}$-maximal, computes a better strategy $\sigma' \succ_{\bigcirc} \sigma$. Strategy improvement algorithms start with a certain initial strategy and perform a sequence of improvement steps, until the maximal strategy is reached. While a single iteration (improvement step) is typically efficient, so far no policy guaranteeing a polynomial number of iterations has been found. In fact, Friedmann [3, 4] has proved that the key strategy improvement algorithms have worst-case exponential running time.

Applications

Parity games underlie a number of problems in theoretical computer science. For instance, they served as a vehicle for elegantly proving the complementation lemma for automata on infinite trees, a crucial lemma in Rabin's proof of the decidability of a particular second-order mathematical theory. Parity games are also used in word and emptiness problems for a variety of (alternating) automata [1]. Moreover, they are closely related to other two-player, infinite duration games with perfect information such as mean payoff games, which have, among others, applications in scheduling.

The practical significance of parity games stems from the fact that they have proved to be of great value in computer-aided software and hardware verification and synthesis. Of particular importance is the result that parity games are polynomial-time equivalent to model checking for the modal μ-calculus (see, for instance, [2]), a modal logic that expressively subsumes most of the popular temporal logics used in computer-aided verification. We present this transformation to illustrate the tight connection between game theory and logic.

Parity Games for Model Checking

Say we are given a structure (S, A, R), where A is a set of atomic actions, (S, R) is a directed graph in which S is a set of states, and $R \subseteq S \times A \times S$ is a (for simplicity) total edge-labeled transition relation. The structure (S, A, R) is often referred to as a *Labeled Transition System* (LTS), and it serves the purpose of modeling the behavior of software or hardware. The modal μ-calculus allows for reasoning about such behaviors; the logic is defined through the following grammar:

$$f, g ::= true \mid false \mid X \mid f \wedge g \mid$$
$$f \vee g \mid [a]f \mid \langle a \rangle f \mid \nu X.f \mid \mu X.f$$

where $a \in A$ and X is a propositional variable, taken from a sufficiently large set of variables $\mathcal{X}$. We write σ to denote either μ or ν. For simplicity, we assume that a propositional variable X in f is bound at most once (by some σ) and occurrences of X are all within the scope of its binder. Expressions in the logic are interpreted in the context of an LTS (S, A, R) and a mapping $e{:}\mathcal{X} \to 2^S$, typically referred to as an environment, assigning sets of states to propositional variables. The modal operators $\langle\_\rangle\_$ and $[\_]\_$ allow for reasoning with the transition relation of an LTS; e.g.,, $\langle a \rangle f$ will hold in states that have *some* a-successor satisfying f, whereas $[a]f$ will hold in states for which *all* a-successors (if any) satisfy f. More formally, the meaning of a formula f is established by stating which states in the LTS satisfy it; this satisfaction relation, denoting $s, e \models f$, is defined inductively as follows:

$$
\begin{array}{lll}
s, e \models true & & \\
s, e \not\models false & & \\
s, e \models X & \text{iff} & s \in e(X) \\
s, e \models f \wedge g & \text{iff} & s, e \models f \text{ and } s, e \models g \\
s, e \models f \vee g & \text{iff} & s, e \models f \text{ or } s, e \models g \\
s, e \models [a]f & \text{iff} & \text{for all } (s, a, t) \in R \;\; t, e \models f \text{ holds} \\
s, e \models \langle a \rangle f & \text{iff} & \text{exists } (s, a, t) \in R \text{ such that } t, e \models f \text{ holds} \\
s, e \models \nu X.f & \text{iff} & s \in \bigcup\{S' \subseteq S \mid S' \subseteq F(S')\} \\
s, e \models \mu X.f & \text{iff} & s \in \bigcap\{S' \subseteq S \mid F(S') \subseteq S'\}
\end{array}
$$

where $F(T) = \{t \in S \mid t, e[X \mapsto T] \models f\}$ is a monotone operator in the complete lattice $(2^S, \subseteq)$, and where $e[X \mapsto T]$ is the environment e in which X is assigned set T. Perhaps due to its extreme expressive power, expressions in the modal μ-calculus are famously known for being hard to understand. Expressions using only one fixpoint are reasonably straightforward to interpret. For instance, the formula $\nu X.\langle a \rangle X$ holds for a state s with an infinite a-path: essentially, we are looking for the largest solution (its *fixpoint*) to the equation "$X = \langle a \rangle X$," or, more semantically, the largest set $T \subseteq S$ that can be assigned to X that satisfies the equation $T = \{s \in S \mid \exists t \in T : (s, a, t) \in R\}$. An expression such as $\mu Y.\langle a \rangle Y \vee \langle a \rangle true$ holds for a state s whenever there is a finite sequence of b-transitions leading to a state having an a-successor. Mixing fixpoints allows for expressing more complicated properties, unfortunately at the expense of readability, making formulas such as $\nu X.\mu Y.\langle a \rangle X \vee \langle b \rangle Y$ (expressing that, when it holds, there is an infinite sequence of a, b steps in which stretch of b-steps (if any) is of finite length), hard to understand.

The model checking problem is to decide, given some formula f, a state s in an LTS and some environment e, whether $s, e \models f$. This problem can be reduced to solving a parity game as follows: define a parity game $(V, E, \Omega, (V_\Diamond, V_\Box))$ in which $V = S \times \Phi(f)$, where $\Phi(f)$ is the set of all subformulas of f, and in which E, $V_\Diamond$, and $V_\Box$ are defined structurally as follows:

vertex	successor(s)	owner
$(s, true)$	$(s, true)$	$\Diamond$
$(s, false)$	$(s, false)$	$\Diamond$
(s, X) and $\sigma X.g \in \Phi(f)$	$(s, \sigma X.g)$	$\Diamond$
(s, X) and $\sigma X.g \notin \Phi(f)$	(s, X)	$\Diamond$
$(s, f \wedge g)$	(s, f) and (s, g)	$\Box$
$(s, f \vee g)$	(s, f) and (s, g)	$\Diamond$
$(s, [a]f)$	all (t, f) for $(s, a, t) \in R$	$\Box$
$(s, \langle a \rangle f)$	all (t, f) for $(s, a, t) \in R$	$\Diamond$
$(s, \nu X.f)$	(s, f)	$\Diamond$
$(s, \mu X.f)$	(s, f)	$\Diamond$

The priority function is assigned in such a way that it meets the following conditions:

- $\Omega((s, true)) = 0$ and $\Omega((s, false)) = 1$;
- if $\sigma X.g \notin \Phi(f)$, then $\Omega((s, X)) = 0$ if $s \in e(X)$ and $\Omega((s, X)) = 1$ otherwise;
- if $\sigma X.g \in \Phi(f)$, then
  - $\Omega((s, X))$ is *even* if $\sigma = \nu$ and *odd* otherwise, and
  - $\Omega((s, X)) \leq \Omega((t, Y))$ if Y depends on X (if X is *free* in g in $\sigma Y.g$);
- $\Omega((s, g))$ is maximal for other formulas $g \in \Phi(f)$.

There is an "optimal" assignment of priorities that does not assign values larger than the alternation depth of the μ-calculus formula f [12]. Intuitively, the alternation depth is a measure of the degree of semantic alternations between μ- and ν-operators. Using an optimal priority assignment yields better complexity bounds for the model checking problem. The theorem below establishes the connection between the model checking problem and the parity game solving problem.

Theorem 4 *$s, e \models f$ iff player $\Diamond$ wins (s, f) in the constructed parity game.*

On the one hand, through the above reduction, practical algorithmic progress in solving parity games directly impacts the performance and scalability of tooling for conducting the verification and synthesis. Abstracting from syntactic details, the more elementary parity games setting, on the other hand, permits studying the true complexity of model checking and, at the same time, provides a better understanding of the dynamics of the modal μ-calculus.

Open Problems

Parity games are among the few problems that are in NP ∩ co-NP for which no polynomial time algorithm has been found. The key open problem is thus whether there is a polynomial time algorithm for solving parity games.

Problem 4 Can $W_\Diamond$ and $W_\Box$ be computed in polynomial time?

Cross-References

▸ Symbolic Model Checking

Recommended Reading

1. Emerson E, Jutla C (1991) Tree automata, mu-calculus and determinacy. In: FOCS'91. IEEE Computer Society, Washington, DC, pp 368–377. 10.1109/SFCS.1991.185392
2. Emerson E, Jutla C, Sistla A (2001) On model checking for the μ-calculus and its fragments. Theor Comput Sci 258(1–2):491–522. 10.1016/S0304-3975(00)00034-7
3. Friedmann O (2011) An exponential lower bound for the latest deterministic strategy iteration algorithms. Log Methods Comput Sci 7(3)
4. Friedmann O (2013) A superpolynomial lower bound for strategy iteration based on snare memorization. Discret Appl Math 161(10–11):1317–1337
5. Jurdziński M (1998) Deciding the winner in parity games is in UP ∩ co-UP. IPL 68(3):119–124. 10.1016/S0020-0190(98)00150-1
6. Jurdziński M (2000) Small progress measures for solving parity games. In: STACS'00. LNCS, vol 1770. Springer, pp 290–301. 10.1007/3-540-46541-3_24
7. Jurdziński M, Paterson M, Zwick U (2006) A deterministic subexponential algorithm for solving parity games. In: SODA'06. ACM/SIAM, pp 117–123. 10.1145/1109557.1109571
8. King V, Kupferman O, Vardi MY (2001) On the complexity of parity word automata. In: FOSSACS. LNCS, vol 2030. Springer, pp 276–286. 10.1007/3-540-45315-6_18
9. Martin D (1975) Borel determinacy. Ann Math 102:363–371. 10.2307/1971035
10. McNaughton R (1993) Infinite games played on finite graphs. APAL 65(2):149–184. 10.1016/0168-0072(93)90036-D
11. Mostowski A (1991) Hierarchies of weak automata and weak monadic formulas. Theor Comput Sci 83(2):323–335. 10.1016/0304-3975(91)90283-8
12. Niwinski D (1997) Fixed point characterization of infinite behavior of finite-state systems. Theor Comput Sci 189(1–2):1–69. 10.1016/S0304-3975(97)00039-X
13. Schewe S (2007) Solving parity games in big steps. In: FSTTCS'07. LNCS, vol 4855. Springer, pp 449–460. 10.1007/978-3-540-77050-3
14. Vöge J, Jurdzinski M (2000) A discrete strategy improvement algorithm for solving parity games. In: Emerson EA, Sistla AP (eds) CAV. LNCS, vol 1855. Springer, Heidelberg, pp 202–215
15. Zielonka W (1998) Infinite games on finitely coloured graphs with applications to automata on infinite trees. TCS 200(1–2):135–183. 10.1016/S0304-3975(98)00009-7

P

Pattern Matching on Compressed Text

Masayuki Takeda[1] and Ayumi Shinohara[2]
[1]Department of Informatics, Kyushu University, Fukuoka, Japan
[2]Graduate School of Information Sciences, Tohoku University, Sendai, Japan

Keywords

Collage system; Compressed pattern matching; Straight-line program

Years and Authors of Summarized Original Work

2003; Kida, et al.
2011; Gawrychowski
2013; Gawrychowski

Problem Definition

Let **c** be a given compression function that maps strings A to their compressed representations $\mathbf{c}(A)$. The problem of *compressed pattern matching (CPM)* is defined as follows:

Problem 1 (Compressed Pattern Matching) Given a pattern string P and a compressed text string $\mathbf{c}(T)$, determine whether there is an occurrence of P in T, without decompressing T.

A CPM algorithm is said to be *optimal* if it runs in $O(|P| + |\mathbf{c}(T)|)$ time. The time/space complexity of the CPM problem can be a new criterion to evaluate compression schemes in addition to the traditional ones: the compression ratio and the time/space complexity of compression/decompression.

The CPM problem was first defined in the work of Amir and Benson [1], and many studies have been made over different compression formats. Kida et al. [9] introduced a useful CPM-oriented abstraction of compression formats, named *collage systems*. Outputs of various compression algorithms – not only dictionary-based compression algorithms but also grammar-based compression algorithms – can be regarded as collage systems, and hence algorithmic research working on collage systems is of great significance. They presented in the same paper a general Knuth-Morris-Pratt (KMP) algorithm on collage systems. A general Boyer-Moore (BM) algorithm on collage systems was also designed by almost the same authors [17].

Collage Systems

Let Σ be a fixed finite alphabet. A *collage system* on Σ is a pair $\langle \mathcal{D}, \mathcal{S} \rangle$ defined as follows:

- $\mathcal{D}$ is a sequence of assignments $X_1 = expr_1;\ X_2 = expr_2;\ \cdots;\ X_n = expr_n$, where, for each $k = 1, \ldots, n$, X_k is a variable and $expr_k$ is any of the form:
- $\mathcal{S}$ is a sequence $X_{i_1} \cdots X_{i_\ell}$ of variables defined in $\mathcal{D}$.

By *the j length prefix (resp. suffix) truncation*, we mean an operation on strings which takes a string w and returns the string obtained from w by removing its prefix (resp. suffix) of length j. The variables X_k represent the strings $val(X_k)$ obtained by evaluating their expressions. A collage system $\langle \mathcal{D}, \mathcal{S} \rangle$ represents the string obtained by concatenating the strings $val(X_{i_1}), \ldots, val(X_{i_\ell})$ represented by variables $X_{i_1}, \ldots, X_{i_\ell}$ of $\mathcal{S}$.

The *size* of $\mathcal{D}$ is the number n of assignments and denoted by $|\mathcal{D}|$. The *height* of $\mathcal{D}$, denoted by *height*($\mathcal{D}$), is defined to be the longest path length of the dependency graph of $\mathcal{D}$, namely, a directed acyclic graph such that (1) the vertices are the variables in $\mathcal{D}$ and (2) a directed edge from X_k to X_i exists if and only if $\mathcal{D}$ contains a non-primitive assignment $X_k = expr_k$ such that X_i appears in $expr_k$. The *length* of $\mathcal{S}$ is the number ℓ of variables in $\mathcal{S}$ and denoted by $|\mathcal{S}|$. The *size* of collage system $\langle \mathcal{D}, \mathcal{S} \rangle$ is defined to be $|\mathcal{D}| + |\mathcal{S}|$.

It should be noted that any collage system can be converted into an equivalent one with $|\mathcal{S}| = 1$, by adding a series of assignments with concatenation operations into $\mathcal{D}$. This may imply $\mathcal{S}$ is unnecessary. However, a variety of compression schemes can be captured naturally by separating $\mathcal{D}$ (defining *phrases*) from $\mathcal{S}$ (giving a factorization of text T into phrases). How to express outputs of existing compression algorithms is found in [9].

A collage system $\langle \mathcal{D}, \mathcal{S} \rangle$ is said to be *truncation-free* if $\mathcal{D}$ contains no truncation operation, *repetition-free* if $\mathcal{D}$ contains no repetition operation, and *regular* if it is truncation- and repetition-free. A regular collage system $\langle \mathcal{D}, \mathcal{S} \rangle$ is *simple* if $|val(Y)| = 1$ or $|val(Z)| = 1$ for every assignment $X = YZ$ of $\mathcal{D}$.

Outputs of grammar-based compression algorithms such as Re-Pair, Sequitur, and Byte Pair Encoding (BPE) fall into the class of regular collage systems, and outputs of LZ78/LZW fall into the class of simple collage systems. LZ77 factorization is an abstraction of LZ77 and its variants, which has two variations depending upon whether self-referencing is allowed. The LZ77 factorization Z of T with (resp. without) self-referencing can be transformed into a collage system (resp. a repetition-free collage system) of size $O(|Z| \cdot \log |Z|)$ generating T (see [5]).

It should be mentioned that the so-called straight-line programs (SLPs) are the regular collage systems with $|\mathcal{S}| = 1$.

Key Results

Theoretical Aspect

Amir et al. [2] presented two solutions to CPM for LZW with time complexities $O(|\mathbf{c}(T)| \log |P| + |P|)$ and $O(|\mathbf{c}(T)| + |P|^2)$, respectively. The latter was generalized by Kida et al. [9] via the unified framework of collage systems.

Theorem 1 (Kida et al. [9]) *CPM for collage systems can be solved in* $O\big((|\mathcal{D}| + |\mathcal{S}|) \cdot height(\mathcal{D}) + |P|^2\big)$ *time using* $O(|\mathcal{D}| + |P|^2)$ *space. The factor* $height(\mathcal{D})$ *is dropped for truncation-free collage systems.*

We briefly sketch the algorithm of [9]. It is originally intended to solve the *all-occurrence* version of the CPM problem and reports all locations of T at which P occurs with additional time linearly proportional to the number of pattern occurrences. The algorithm has two stages: First, it preprocesses $\mathcal{D}$ and P, and second it processes the variables of $\mathcal{S}$. In the second stage, it simulates the move of KMP automaton running on uncompressed text, by using two functions *Jump* and *Output*, both take as input a state q and a variable X. The former is used to substitute just one state transition for the consecutive state transitions of the KMP automaton for the string $val(X)$ for each variable X of $\mathcal{S}$, and the latter is used to report all pattern occurrences found during the state transitions. Let δ be the state-transition function of the KMP automaton. Then $Jump(q, X) = \delta(q, val(X))$, and $Output(q, X)$ is the set of lengths $|w|$ of nonempty prefixes w of $val(X)$ such that $\delta(q, w)$ is the final state. A naive two-dimensional array implementation of the two functions requires $\Omega(|\mathcal{D}| \cdot |P|)$ space, and the size of $Output(q, X)$ can be exponential in $|\mathcal{D}|$. The data structures of [9] use only $O(|\mathcal{D}| + |P|^2)$ space, are built in $O(|\mathcal{D}| \cdot height(\mathcal{D}) + |P|^2)$ time, and enable us to compute $Jump(q, X)$ in $O(1)$ time and enumerate the set $Output(q, X)$ in $O(height(\mathcal{D}) + \ell)$ time where $\ell = |Output(q, X)|$. The factor $height(\mathcal{D})$ is dropped for truncation-free collage systems.

By replacing $|\mathcal{D}| + |\mathcal{S}|$ with $|\mathbf{c}(T)|$, the above theorem means that the existence version of CPM can be solved in $O(|\mathbf{c}(T)| + |P|^2)$ time for *any* compression formats that fall into the class of truncation-free collage systems. $|P|^2$ is acceptable since it is often smaller than $|\mathbf{c}(T)|$ in

a	for $a \in \Sigma \cup \{\varepsilon\}$,	*(primitive assignment)*
$X_i X_j$	for $i, j < k$,	*(concatenation)*
$^{[j]}X_i$	for $i < k$ and a positive integer j,	*(j length prefix truncation)*
$X_i^{[j]}$	for $i < k$ and a positive integer j,	*(j length suffix truncation)*
$(X_i)^j$	for $i < k$ and a positive integer j.	*(j times repetition)*

practice. But removing the quadratic dependency on $|P|$ is of great interest in theory.

Consider a maximal variable sequence $\mathcal{S}[i..j]$ of $\mathcal{S}[1..\ell]$ such that the string $val(\mathcal{S}[k])$ is a factor of pattern P for any $k \in [i..j]$. Take the longest suffix h of $val(\mathcal{S}[i-1])$ and the longest prefix t of $val(\mathcal{S}[j+1])$ to obtain the sequence $h, val(\mathcal{S}[i]), \ldots, val(\mathcal{S}[j]), t$ of pattern factors. Any pattern occurrence must appear in such a pattern factor sequence, except for the case that some variable $X = \mathcal{S}[k]$ contains a pattern occurrence in its string $val(X)$. The CPM task can thus be reduced into a number of instances of pattern matching in a sequence of pattern factors. It should be noted that the task of pattern matching in a pattern factor sequence depends only on P, not depending on T nor its compression format. Gawrychowski [7] described an elaborate technique to perform this task in linear time. On the other hand, in the reduction task, we are faced with the need to solve the so-called factor concatenation problem [9]:

> Preprocess P to build a data structure that returns in constant time the vertex representing the factor xy, for any two (explicit or implicit) vertices of suffix tree of P representing factors x, y of P.

An $O(|P|^2)$ time preprocessing was presented in [9]. For LZW or, more generally, simple collage systems, it is rather straightforward to see that if the alphabet is of constant size, the preprocessing requires only $O(|P|)$ time since either x or y is of length 1, and the reduction task thus takes $O(|\mathcal{D}|+|\mathcal{S}|+|P|)$ time. CPM for simple collage systems can therefore be solved in optimal linear time for a constant alphabet. Gawrychowski [7] further described how to keep it linear even in the case of integer alphabet for LZW.

Theorem 2 (Gawrychowski [7]) *CPM for LZW can be solved in optimal linear time even for a polynomial size integer alphabet, assuming the word RAM model.*

For LZ77, one possible solution is to convert the input LZ77 factorization into a truncation-free collage system and then apply the CPM algorithm of [9]. We can convert an LZ77 factorization of T into a truncation-free collage system with an increase in size by a factor of $O(\log \frac{|T|}{|\mathbf{c}(T)|})$ in time linear to the output size (see [4]). The resulting algorithm thus has time complexity of $O(|\mathbf{c}(T)| \log \frac{|T|}{|\mathbf{c}(T)|} + |P|^2)$. Gawrychowski in [6] successfully removed the quadratic dependency on $|P|$ again.

Theorem 3 (Gawrychowski [6]) *CPM for LZ77 can be solved in* $O(|\mathbf{c}(T)| \log \frac{|T|}{|\mathbf{c}(T)|} + |P|)$ *time.*

Table 1 summarizes the best known solutions to CPM for several compression formats.

Practical Aspect

From a practical viewpoint, we have two goals. One is to perform the CPM task in less time compared with a decompression followed by an ordinary search (Goal 1), and the other is to perform it in less time compared with an ordinary search over uncompressed text (Goal 2). An optimal CPM algorithm theoretically achieves the two goals if $|\mathbf{c}(T)| = o(|T|)$. However, we often observe $|\mathbf{c}(T)| = \Theta(|T|)$ in practice, and hence reducing the constant factors hidden behind the O-notation of time complexity of CPM algorithms plays a crucial role in achieving the two goals, especially for Goal 2. For example, code words are limited to multiples of 8 bits to avoid bit manipulation.

Kida et al. [8] reported the first experimental results in this area, achieving Goal 1 for LZW. Navarro and Tarhio [14] presented BM-type algorithms for LZ78/LZW compression schemes and showed they are twice as fast as a decompression followed by a search using the best algorithms.

Data compression can be regarded as a preprocessing that allows a fast search in the context of Goal 2. An appropriate compression format would be chosen for this purpose. We note that in general, some occurrences of the encoded pattern can be false matches, and/or the pattern possibly occurs in several different forms within the encoded text. There are two lines of research work addressing Goal 2. One is to put a restriction on the compression scheme so that every pattern occurrence can be identified simply as a substring of the encoded text that is identical to the encoded

Pattern Matching on Compressed Text, Table 1 Best known solutions to CPM for several compression formats

Compression formats	Time complexity	Work
Run-length	$O(\lvert c(T)\rvert + \lvert P\rvert)$	Trivial
LZW	$O(\lvert c(T)\rvert + \lvert P\rvert)$	[7]
LZ77	$O(\lvert c(T)\rvert \log \frac{\lvert T\rvert}{\lvert c(T)\rvert} + \lvert P\rvert)$	[6]
Simple collage systems	$O((\lvert\mathcal{D}\rvert + \lvert\mathcal{S}\rvert) + \lvert P\rvert)$	[7]
Truncation-free collage systems	$O((\lvert\mathcal{D}\rvert + \lvert\mathcal{S}\rvert) + \lvert P\rvert^2)$	[9]
Collage systems	$O((\lvert\mathcal{D}\rvert + \lvert\mathcal{S}\rvert) \cdot height(\mathcal{D}) + \lvert P\rvert^2)$	[9]

pattern. The advantage is that *any* favored pattern matching algorithm can be used to search the encoded text for the encoded pattern. The works of Manber [10] and Rautio et al. [15] are along this line. The latter is based on a combination of the so-called stopper encoding and the Boyer-Moore-Horspool (BMH) algorithm and is regarded as the fastest combination that achieves Goal 2. The drawback of this line is, however, that the restriction considerably sacrifices the compression ratio (e.g., 60–70 % for typical English texts). In the case of natural language texts written in western languages such as English (having explicit word boundaries), there are some compression formats that enable us to achieve Goal 2 by using a modification of byte-oriented Huffman coding on words (see, e.g., [3]).

The other line is to suppress a false detection or detection omission by an algorithmic device, without putting such a restriction on the compression scheme. The work of Miyazaki et al. [13] for Huffman encoding and the works of Shibata et al. [16, 17] for BPE are along this line. While all of the works [10, 13, 15–17] mentioned here achieve Goal 2, the compression ratios are poor: BPE is the best among them. A BPE compressed text is a regular collage system with limitation $\lvert\mathcal{D}\rvert \leq 256$. Matsumoto [12] extended BPE to get a higher compression ratio by easing the limitation and using the byte-oriented Huffman coding for representing the variables occuring in $\mathcal{S}$. Their CPM algorithm runs fast to achieve Goal 2, but memory requirement increases as $\lvert\mathcal{D}\rvert$ grows. Maruyama [11] introduced a new compression scheme, called the *context-sensitive grammar transform*, of which compression ratio is a match for gzip and Re-Pair. The search speed of their CPM algorithm is almost twice faster than the KMP-type CPM algorithm of [16] on BPE and faster than [15] for short patterns.

Cross-References

- ▶ Grammar Compression
- ▶ Huffman Coding
- ▶ Lempel-Ziv Compression
- ▶ String Matching
- ▶ Suffix Trees and Arrays

Recommended Reading

1. Amir A, Benson G (1992) Efficient two-dimensional compressed matching. In: Proceedings of the data compression conference (DCC'92), Snowbird, p 279
2. Amir A, Benson G, Farach M (1996) Let sleeping files lie: pattern matching in Z-compressed files. J Comput Syst Sci 52(2):299–307
3. Brisaboa N, Fariña A, Navarro G, Paramá J (2007) Lightweight natural language text compression. Inf Retr 10:1–33
4. Charikar M, Lehman E, Liu D, Panigrahy R, Prabhakaran M, Rasala A, Sahai A, Shelat A (2002) Approximating the smallest grammar: Kolmogorov complexity in natural models. In: STOC, Montréal, pp 792–801
5. Gąsieniec L, Karpinski M, Plandowski W, Rytter W (1996) Efficient algorithms for Lempel-Ziv encoding. In: Proceedings of the 5th Scandinavian workshop on algorithm theory (SWAT'96), Reykjavík. Lecture notes in computer science, vol 1097, pp 392–403
6. Gawrychowski P (2011) Pattern matching in Lempel-Ziv compressed strings: fast, simple, and deterministic. In: ESA, Saarbrücken, pp 421–432
7. Gawrychowski P (2013) Optimal pattern matching in LZW compressed strings. ACM Trans Algorithms 9(3):25
8. Kida T, Takeda M, Shinohara A, Miyazaki M, Arikawa S (2000) Multiple pattern matching in LZW compressed text. J Discret Algorithms 1(1):133–158

9. Kida T, Matsumoto T, Shibata Y, Takeda M, Shinohara A, Arikawa S (2003) Collage systems: a unifying framework for compressed pattern matching. Theor Comput Sci 298(1):253–272
10. Manber U (1997) A text compression scheme that allows fast searching directly in the compressed file. ACM Trans Inf Syst 15(2):124–136
11. Maruyama S, Tanaka Y, Sakamoto H, Takeda M (2010) Context-sensitive grammar transform: compression and pattern matching. IEICE Trans Inf Syst 93-D(2):219–226
12. Matsumoto T, Hagio K, Takeda M (2009) A runtime efficient implementation of compressed pattern matching automata. Int J Found Comput Sci 20(4):717–733
13. Miyazaki M, Fukamachi S, Takeda M, Shinohara T (1998) Speeding up the pattern matching machine for compressed texts. Trans Inf Process Soc Jpn 39(9):2638–2648
14. Navarro G, Tarhio J (2005) LZgrep: A Boyer-Moore string matching tool for Ziv-Lempel compressed text. Softw Pract Exp 35(12):1107–1130
15. Rautio J, Tanninen J, Tarhio J (2002) String matching with stopper encoding and code splitting. In: Proceedings of the 13th annual symposium on combinatorial pattern matching (CPM'02), Fukuoka, pp 42–52
16. Shibata Y, Kida T, Fukamachi S, Takeda M, Shinohara A, Shinohara T, Arikawa S (2000) Speeding up pattern matching by text compression. In: Proceedings of the 4th Italian conference on algorithms and complexity (CIAC'00), Rome. Lecture notes in computer science, vol 1767. Springer, pp 306–315
17. Shibata Y, Matsumoto T, Takeda M, Shinohara A, Arikawa S (2000) A Boyer-Moore type algorithm for compressed pattern matching. In: Proceedings of the 11th annual symposium on combinatorial pattern matching (CPM'00), Montréal. Lecture notes in computer science, vol 1848. Springer, pp 181–194

Patterned Self-Assembly Tile Set Synthesis

Shinnosuke Seki
Department of Computer Science, Helsinki Institute for Information Technology (HIIT), Aalto University, Aalto, Finland

Keywords

Computer-assisted proof; Heuristic algorithms; Minimization; NP-hardness; Pattern self-assembly; Rectilinear tile assembly system (RTAS)

Years and Authors of Summarized Original Work

2008; Ma, Lombardi
2013; Seki
2014; Göös, Lampiäinen, Czeizler, Orponen
2014; Kari, Kopecki, Meunier, Patitz, Seki

Problem Definition

A *tile type* is a colored unit square each of whose four sides is provided with a glue. An *assembly* is a partial function from $\mathbb{Z}^2$ (2D-grid) to a tile type set T. A (rectangular) pattern P (of width w and height h) is a function from the rectangular domain $[w] \times [h]$ to a set of colors, where $[m] = \{1, \ldots, m\}$ for $m \in \mathbb{N}$. If at most k colors appear on P, we say P is *k-colored.* Tiles being colored, an assembly of domain $[w] \times [h]$ induces a unique pattern of width w and height h.

The rectilinear tile assembly system (RTAS) is a variant of Winfree's aTAM system [10]. Figure 1 illustrates how an RTAS self-assembles the binary counter pattern. An RTAS is a pair of a finite set of tile types and an L-shape seed, which is an assembly of domain $\{(0,0)\} \cup [w] \times \{0\} \cup \{0\} \times [h]$. Starting from the L-shape seed, it tiles the plain according to the following rule:

RTAS's tiling rule: A tile can attach at a position (x, y) if its west glue matches the east glue of the tile at $(x-1, y)$ and its south glue matches the north glue label of the tile at $(x, y-1)$.

A tile hence finds the sole attachable position (1, 1) on the L-shape seed. The attachment of tile there makes the positions (2, 1) and (1, 2) attachable. Tiles attach in this manner one after another and an assembly grows rectilinearly.

An RTAS is *directed* (a.k.a. *deterministic*) if no two tile types share the west and south glues, as the one in Fig. 1. A directed RTAS admits a unique assembly $\mathcal{A}_\square$ to which no tile can attach any more. Then we say that it *uniquely self-assembles the pattern of* $\mathcal{A}_\square$.

Patterned self-assembly tile set synthesis (PATS), proposed by Ma and Lombardi [7], aims

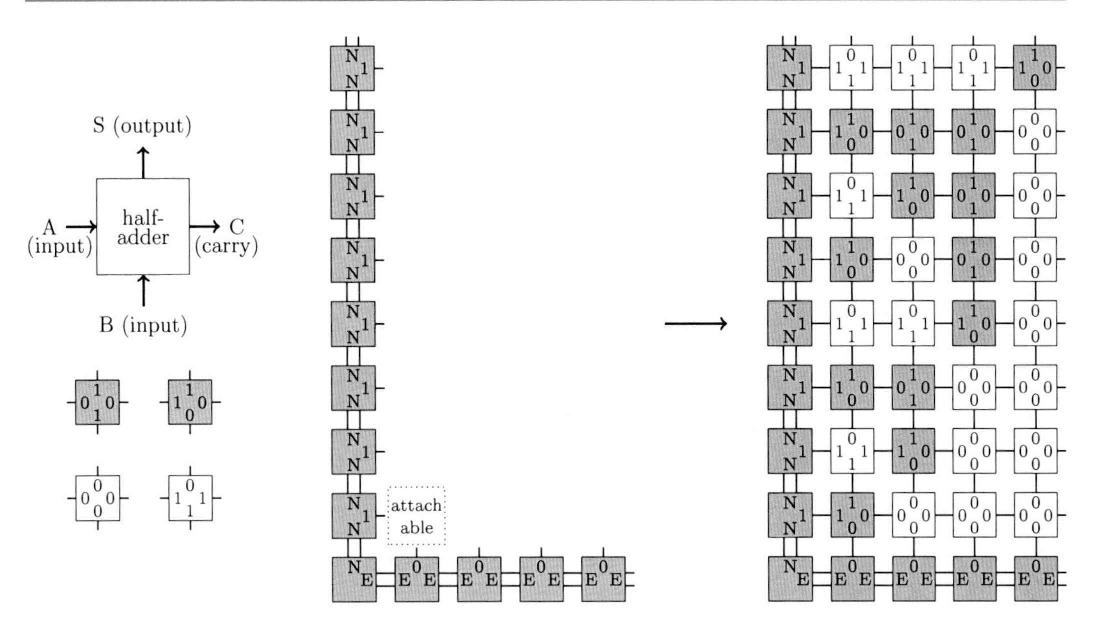

Patterned Self-Assembly Tile Set Synthesis, Fig. 1 (*Left*) Four tile types implement together the half-adder with two inputs A, B from the west and south, the output S to the north, and the carryout C to the east (*Right*) Copies of the "half-adder" tile types turn the L-shape seed into the binary counter pattern

at minimizing the RTAS that self-assembles a given pattern, where an RTAS is measured by the cardinality of its tile type set. Since the minimum RTAS is known to be directed [2], this problem is formulated as follows:

Patterned self-assembly tile set synthesis (PATS) [7]

GIVEN: A (rectangular) pattern P
FIND: A minimum directed RTAS that uniquely self-assembles P

For $k \geq 1$, the k-colored PATS (k-PATS) is a practical variant of PATS which takes only the k-colored patterns as input.

Key Results

PATS and k-PATS have been studied mainly in two research directions so far: its computational complexity and algorithms.

The **NP**-hardness of 2-PATS was claimed in [8], but what was proved **NP**-hard there was something different. Czeizler and Popa proved that PATS is **NP**-hard [1] (its concise proof is in [5]) by a polynomial-time reduction of 3SAT to the decision variant of PATS: given a pattern P and $n \in \mathbb{N}$, decide whether P can be uniquely self-assembled by a directed RTAS with n tile types. Variables and clauses of a given 3SAT instance ϕ are color-coded so that the number of colors in the reduced pattern is in proportion to the size of ϕ.

Potential of geometry, or more precisely, configuration of colors, as a medium of encoding a 3SAT encoding a 3SAT instance ϕ was suggested by Seki [9]. In the reduction, a set T_{eval} of 84 tile types is designed as a 3SAT-verifier, i.e., using tiles in the set, a directed RTAS evaluates ϕ to be true and assembles a pattern $P(\phi)$, starting from a seed encoding ϕ and some satisfying assignment. A subpattern GADGET of $P(\phi)$ endows $P(\phi)$ with the property that in order for a directed RTAS with a set T of at most 84 tile types to assemble $P(\phi)$, T must be isomorphic to T_{eval}. The pattern $P(\phi)$ is 60-colored. Hence, ϕ is satisfiable if and only if $(P(\phi), 84)$ is a yes-instance of 60-PATS. Johnsen, Kao, and Seki have refined the original design so that the number of colors decreased to 29 [3] and further to 11 [4].

In these proofs, quite a few colors are devoted just to make the property of GADGET manually checkable. Giving up the manual checkability of the GADGET property has yielded a computer-assisted proof of **NP**-hardness of 2-PATS by Kari et al. [6]. The proof was verified in two different environments. Note that in this proof, everything but the GADGET property is manually checkable.

The **NP**-hardness of 2-PATS makes it essentially indispensable for exact PATS-solvers to search for the exponential number of solution candidates. Göös et al. designed PATS-solvers: an exhaustive partition-search branch-and-bound (PS-BB) algorithm, its heuristic modification (PS-H), and an ASP-solver-based algorithm [2]. PS-BB is an exact algorithm for PATS, running in practical time just for small patterns, say 7×7, while PS-H works even for larger patterns in exchange for the loss of guarantee on the minimality of its output.

Let P be an input pattern of width w and height h. The colors on P induce a partition $\pi(c)$ of the P's domain $[w] \times [h]$. In principle, among all possible partitions, PS-BB searches for the one $\pi_{\min}$ of least cardinality satisfying:

- $\pi(c)$ is coarser than $\pi_{\min}$ in the sense that $\forall q \in \pi_{\min}, \exists p \in \pi(c), q \subseteq p$.
- One can associate each class p of $\pi_{\min}$ with a quadruple of glues such that a directed RTAS with the associated tile type set uniquely self-assembles P.

Note that the finest partition $\pi_{\max} = \{\{(x, y)\} \mid (x, y) \in [w] \times [h]\}$ satisfies these conditions (associated tile types "hardcode their position" in glues). The coarser-finer relation yields a tree of partitions whose root is $\pi_{\max}$. This is the search tree of PS-BB (and PS-H).

PS-BB employs branch-pruning by bounding function to save computational resources. PS-H more greedily optimizes the order in which the coarsenings of a partition are explored by preferring some search paths to the others. Random choice of the one among the preferred lets PS-H perform differently at each run. PS-H_n is a variant of PS-H which runs multiple independent searches in parallel for efficiency.

Open Problems

The lack of guarantee on the minimality of tile type sets output by the heuristic algorithms or on their running time motivates the design of a polynomial-time approximation algorithm for PATS. The ratio $14/13 \approx 1.077$ is known to be unachievable in polynomial-time, unless $\mathbf{P} = \mathbf{NP}$ [6].

A manually checkable proof for the **NP**-hardness of 2-PATS is of not practical but theoretical interest.

URLs to Code and Data Sets

The computer program for the computer-assisted proof is available online:
http://self-assembly.net/wiki/index.php?title=2PATS-tileset-search.

Cross-References

▸ Circuit Placement
▸ Combinatorial Optimization and Verification in Self-Assembly
▸ Experimental Implementation of Tile Assembly

Recommended Reading

1. Czeizler E, Popa A (2012) Synthesizing minimal tile sets for complex patterns in the framework of patterned DNA self-assembly. In: Proceedings of the DNA 18, Aarhus. LNCS, vol 7433. Springer, pp 58–72
2. Göös M, Lampiäinen T, Czeizler E, Orponen P (2014) Search methods for tile sets in patterned DNA self-assembly. J Comput Syst Sci 80:297–319
3. Johnsen A, Kao MY, Seki S (2013) Computing minimum tile sets to self-assemble color patterns. In: Proceedings of the ISAAC 2013, Hong Kong. LNCS, vol 8283. Springer, pp 699–710
4. Johnsen A, Kao MY, Seki S (2014, Submitted) A manually-checkable proof for the NP-hardness of 11-color pattern self-assembly tile set synthesis
5. Kari L, Kopecki S, Seki S (2013) 3-color bounded patterned self-assembly. In: Proceedings of the DNA 19, Tempe. LNCS, vol 8141. Springer, pp 105–117

6. Kari L, Kopecki S, Étienne Meunier P, Patitz MJ, Seki S (2014) Binary pattern tile set synthesis is NP-hard. arXiv: 1404.0967
7. Ma X, Lombardi F (2008) Synthesis of tile sets for DNA self-assembly. IEEE Trans Comput-Aided Des Integr Circuits Syst 27(5):963–967
8. Ma X, Lombardi F (2009) On the computational complexity of tile set synthesis for DNA self-assembly. IEEE Trans Circuits Syst II 56(1):31–35
9. Seki S (2013) Combinatorial optimization in pattern assembly (extended abstract). In: Proceedings of the UCNC 2013, Milan. LNCS, vol 7956. Springer, pp 220–231
10. Winfree E (1998) Algorithmic self-assembly of DNA. PhD thesis, California Institute of Technology

Peptide De Novo Sequencing with MS/MS

Bin Ma
David R. Cheriton School of Computer Science, University of Waterloo, Waterloo, ON, Canada
Department of Computer Science, University of Western Ontario, London, ON, Canada

Keywords

De novo sequencing; Peptide sequencing

Years and Authors of Summarized Original Work

2005; Ma, Zhang, Liang

Problem Definition

De novo sequencing arises from the identification of peptides by using tandem mass spectrometry (MS/MS). A peptide is a sequence of amino acids in biochemistry and can be regarded as a string over a finite alphabet from a computer scientist's point of view. Each letter in the alphabet represents a different kind of amino acid and is associated with a mass value. In the biochemical experiment, a tandem mass spectrometer is utilized to fragment many copies of the peptide into pieces and to measure the mass values (in fact, the mass to charge ratios) of the fragments simultaneously. This gives a tandem mass spectrum. Since different peptides normally produce different spectra, it is possible, and now a common practice, to deduce the amino acid sequence of the peptide from its spectrum. Often this deduction involves the searching in a database for a peptide that can possibly produce the spectrum. But in many cases such a database does not exist or is not complete, and the calculation has to be done without looking for a database. The latter approach is called de novo sequencing.

A general form of de novo sequencing problems is described in [2]. First, a score function $f(P, S)$ is defined to evaluate the pairing of a peptide P and a spectrum S. Then the de novo sequencing problem seeks for a peptide P such that $f(P, S)$ is maximized for a given spectrum S.

When the peptide is fragmented in the tandem mass spectrometer, many types of fragments can be generated. The most common fragments are the so called b-ions and y-ions. b-ions correspond to the prefixes of the peptide sequence, and y-ions the suffixes. Readers are referred to [8] for the biochemical details of the MS/MS experiments and the possible types of fragment ions. For clarity, in what follows only b-ions and y-ions are considered, and the de novo sequencing problem will be formulated as a pure computational problem.

A spectrum $S = \{(x_i, h_i)\}$ is a set of peaks, each has a mass value x_i and an intensity value h_i. A peptide $P = a_1 a_2 \ldots a_n$ is a string over a finite alphabet Σ. Each $a \in \Sigma$ is associated with a positive mass value $m(a)$. For any string $t = t_1 t_2 \ldots t_k$, denote $m(t) \sum_{i=1}^{k} m(t_i)$. The mass of a length-$i$ prefix (b-ion) of P is defined as $b_i = c_b + m(a_1 a_2 \ldots a_i)$. The mass of a length-$i$ suffix (y-ion) of P is defined as $y_i = c_y + m(a_{k-i+1} \ldots a_{k-1} a_k)$. Here c_b and c_y are two constants related to the nature of the MS/MS experiments. If the mass unit used for measuring each amino acid is dalton, then $c_b = 1$ and $c_y = 19$.

Let δ be a mass error tolerance that is associated with the mass spectrometer. For mass value m, the peaks *matched* by m is defined as $D(m) = \{(x_i, h_i) \in S \mid |x_i - m| \leq \delta\}$. The general idea of de novo sequencing is to maximize the number and intensities of the peaks matched by all b and y ions. Normally, δ is far less than the minimum mass of an amino acid. Therefore, for different i and j, $D(b_i) \cap D(b_j) = \emptyset$ and $D(y_i) \cap D(y_j) = \emptyset$. However, $D(b_i)$ and $D(y_j)$ may share common peaks. So, if not defined carefully, a peak may be counted twice in the score function. There are two different definitions of de novo sequencing problem, corresponding to two different ways of handling this situation.

Definition 1 (Anti-symmetric de novo sequencing)

Instance: A spectrum S, a mass value M, and an error tolerance δ.
Solution: A peptide P such that $m(P) = M$, and $D(b_i) \cap D(y_j) = \emptyset$ for any i, j.
Objective: Maximize $\sum_{k=1}^{n} \sum_{(x_i,h_i) \in D(b_k) \cup D(y_k)} h_i$.

This definition discards the peptides that give a pair of b_i and y_j with similar mass values, because this happens rather infrequently in practice. Another definition allows the peptides to have pairs of b_i and y_j with similar mass values. However, when a peak is matched by multiple ions, it is counted only once. More precisely, define the matched peaks by P as

$$D(P) = \bigcup_{i=1}^{n} (D(b_i) \cup D(y_i)).$$

Definition 2 (De novo sequencing)

Instance: A spectrum S, a mass value M, and an error tolerance δ.
Solution: A peptide P such that $m(P) = M$.

$$f(P, S) = \sum_{(x_i,h_i) \in D(P)} h_i.$$

Objective: Maximize

Key Results

Anti-symmetric de novo sequencing was studied in [1, 2]. These studies convert the spectrum into a spectrum graph. Each peak in the spectrum generates a few of nodes in the spectrum graph, corresponding to the different types of ions that may produce the peak. Each edge in the graph indicates that the mass difference of the two adjacent nodes is approximately the mass of an amino acid, and the edge is labeled with the amino acid. When at least one of each pair of b_i and y_{n-i} matches a peak in the spectrum, the de novo sequencing problem is reduced to the finding of the "anti-symmetric" longest path in the graph. A dynamic programming algorithm for such purpose was published in [1].

Theorem 1 ([1]) *The longest anti-symmetric path in a spectrum graph* $G = \langle V, E \rangle$ *can be found in* $O(|V|\,|E|)$ *time.*

Under Definition 2, de novo sequencing was studied in [6] and a polynomial time algorithm was provided. The algorithm is again a dynamic programming algorithm. For two mass values (m, m'), *the dynamic programming calculates an optimal pair of prefix* Aa *and suffix* $a'A'$, *such that*

1. $m(Aa) = m$ *and* $m(a'A') = m'$.
2. *Either* $c_b + m(A) < c_y + m(a'A') \leq c_b + m(Aa)$ *or* $c_y + m(A') \leq c_b + m(A) < c_y + m(a'A')$.

The calculation for (m, m') *is based on the optimal solutions of smaller mass values. Because of the second above requirement, it is proved in [6] that not all pairs of* (m, m') *are needed. This is used to speed up the algorithm. A carefully designed strategy can eventually output a prefix and a suffix so that their concatenation form the optimal solution of the de novo sequencing problem. More specifically, the following theorem holds.*

Theorem 2 ([5]) *The de novo sequencing problem has an algorithm that gives the optimal peptide in* $O(|\Sigma| \times \delta \times \max_{a \in \Sigma} m(a) \times M)$.

Because $|\Sigma|$, δ, $\max_{a\in\Sigma} m(a)$ are all constants, the algorithm in fact runs in linear time with a large coefficient.

Although the above algorithms are designed to maximize the total intensities of the matched peaks, they can be adapted to work on more sophisticated score functions. Some studies of other score functions can be found in [2–5]. Some of these score functions require new algorithms.

Applications

The algorithms have been implemented into software programs to assist the analyses of tandem mass spectrometry data. Software using the spectrum graph approach includes Sherenga [2]. The de novo sequencing algorithm under the second definition was implemented in PEAKS [5]. More complete lists of the de novo sequencing software and their comparisons can be found in [7, 9].

URL to Code

PEAKS free trial version is available at http://www.bioinfor.com/.

Recommended Reading

1. Chen T, Kao M-Y, Tepel M, Rush J, Church G (2001) A dynamic programming approach to de novo peptide sequencing via tandem mass spectrometry. J Comput Biol 8(3):325–337
2. Dančík V, Addona T, Clauser K, Vath J, Pevzner P (1999) De novo protein sequencing via tandem mass-spectrometry. J Comput Biol 6:327–341
3. Fischer B, Roth V, Roos F, Grossmann J, Baginsky S, Widmayer P, Gruissem W, Buhmann J (2005) NovoHMM: a hidden Markov model for de novo peptide sequencing. Anal Chem 77:7265–7273
4. Frank A, Pevzner P (2005) Pepnovo: De novo peptide sequencing via probabilistic network modeling. Anal Chem 77:964–973
5. Ma B, Zhang K, Lajoie G, Doherty-Kirby A, Hendrie C, Liang C, Li M (2003) PEAKS: powerful software for peptide de novo sequencing by tandem mass spectrometry. Rapid Commun Mass Spectrom 17(20):2337–2342
6. Ma B, Zhang K, Liang C (2005) An effective algorithm for the peptide de novo sequencing from MS/MS spectrum. J Comput Syst Sci 70:418–430
7. Pevtsov S, Fedulova I, Mirzaei H, Buck C, Zhang X (2006) Performance evaluation of existing de novo sequencing algorithms. J Proteome Res 5(11):3018–3028. ASAP Article 10.1021/pr060222h
8. Steen H, Mann M (2004) The ABC's (and XYZ's) of peptide sequencing. Nat Rev Mol Cell Biol 5(9):699–711
9. Xu C, Ma B (2006) Software for computational peptide identification from MS/MS. Drug Discov Today 11(13/14):595–600

Perceptron Algorithm

Shai Shalev-Shwartz
School of Computer Science and Engineering, The Hebrew University, Jerusalem, Israel

Keywords

Online learning; Single-layer neural network

Years and Authors of Summarized Original Work

1959; Rosenblatt

Problem Definition

The Perceptron algorithm [1, 13] is an iterative algorithm for learning classification functions. The Perceptron was mainly studied in the online learning model. As an online learning algorithm, the Perceptron observes instances in a sequence of trials. The observation at trial t is denoted by $\mathbf{x}_t$. After each observation, the Perceptron predicts a yes/no (+/−) outcome, denoted $\hat{y}_t$, which is calculated as follows:

$$\hat{y}_t = \text{sign}(\langle \mathbf{w}_t, \mathbf{x}_t \rangle),$$

where $\mathbf{w}_t$ is a weight vector which is learned by the Perceptron and $\langle \cdot, \cdot \rangle$ is the inner product operation. Once the Perceptron has made a prediction, it receives the correct outcome, denoted y_t, where $y_t \in \{+1, -1\}$. If the prediction

P

of the Perceptron was incorrect, it updates its weight vector, presumably improving the chance of making an accurate prediction on subsequent trials. The update rule of the Perceptron is

$$\mathbf{w}_{t+1} = \begin{cases} \mathbf{w}_t + y_t \mathbf{x}_t & \text{if } \hat{y}_t \neq y_t \\ \mathbf{w}_t & \text{otherwise} \end{cases}. \quad (1)$$

The quality of an online learning algorithm is measured by the number of prediction mistakes it makes along its run. Novikoff [12] and Block [2] have shown that whenever the Perceptron is presented with a sequence of linearly separable examples, it makes a bounded number of prediction mistakes which does not depend on the length of the sequence of examples. Formally, let $(\mathbf{x}_i, y_1), \ldots, (\mathbf{x}_T, y_T)$ be a sequence of instance-label pairs. Assume that there exists a unit vector $\mathbf{u}(\|\mathbf{u}\|_2 = 1)$ and a positive scalar $\gamma > 0$ such that for all t, $y_t(\mathbf{u} \cdot \mathbf{x}_t) \geq \gamma$. In words, $\mathbf{u}$ separates the instance space into two half-spaces such that positively labeled instances reside in one half-space, while the negatively labeled instances belong to the second half-space. Moreover, the distance of each instance to the separating hyperplane, $\{\mathbf{x} : \mathbf{u} \cdot \mathbf{x} = 0\}$, is at least γ. The scalar γ is often referred to as the margin attained by $\mathbf{u}$ on the sequence of examples. Novikoff and Block proved that the number of prediction mistakes the Perceptron makes on a sequence of linearly separable examples is at most $(R/\gamma)^2$, where $R = \max_t \|\mathbf{x}_t\|_2$ is the minimal radius of an origin-centered ball enclosing all the instances. In 1969, Minsky and Papert [11] underscored serious limitations of the Perceptron by showing that it is impossible to learn many classes of patterns using the Perceptron (e.g., XOR functions). This fact caused a significant decrease of interest in the Perceptron. The Perceptron has gained back its popularity after Freund and Schapire [9] proposed to use it in conjunction with kernels. The kernel-based Perceptron not only can handle non-separable datasets but can also be utilized for efficiently classifying nonvectorial instances such as trees and strings (see, e.g., [5]).

To implement the Perceptron in conjunction with kernels, one can utilize the fact that at each trial of the algorithm, the weight vector $\mathbf{w}_t$ can be written as a linear combination of the instances

$$\mathbf{w}_t = \sum_{x \in I_t} y_i \mathbf{x}_i,$$

where $I_t = \{i < t : \hat{y}_i \neq y_i\}$ is the set of indices of trials in which the Perceptron made a prediction mistake. Therefore, the prediction of the algorithm can be rewritten as

$$\hat{y}_t = \text{sign}\left(\sum_{i \in I_t} y_i \langle \mathbf{x}_i, \mathbf{x}_t \rangle\right),$$

and the update rule of the weight vector can be replaced with an update rule for the set of erroneous trials

$$I_{t+1} = \begin{cases} I_t \cup \{t\} & \text{if } \hat{y}_t \neq y_t \\ I_t & \text{otherwise} \end{cases}. \quad (2)$$

In the kernel-based Perceptron, the inner product $\langle \mathbf{x}_i, \mathbf{x}_t \rangle$ is replaced with a Mercer kernel

Perceptron Algorithm, Table 1 Correspondence between the standard Perceptron algorithm and the kernel-based Perceptron

Online Perceptron	Kernel-based online Perceptron
INITIALIZATION: $\mathbf{w}_1 = \mathbf{0}$	INITIALIZATION: $\mathbf{I}_1 = \{\cdot\}$
For $t = 1, 2, \ldots$	For $t = 1, 2, \ldots$
Receive an instance $\mathbf{x}_t$	Receive an instance $\mathbf{x}_t$
Predict an outcome $\hat{y}_t = \text{sign}(\langle \mathbf{x}_t, \mathbf{x}_t \rangle)$	Predict an outcome $\hat{y}_t = \text{sign}\left(\sum_{i \in I_t} K(\mathbf{x}_i, \mathbf{x}_t)\right)$
Receive correct outcome $y_t \in \{+1, -1\}$	Receive correct outcome $y_t \in \{+1, -1\}$
Update: $\mathbf{w}_{t+1} = \begin{cases} \mathbf{w}_t + y_t \mathbf{x}_t & \text{if } \hat{y}_t \neq y_t \\ \mathbf{w}_t & \text{otherwise} \end{cases}$	Update: $I_{t+1} = \begin{cases} I_t \cup \{t\} & \text{if } \hat{y}_t \neq y_t \\ I_t & \text{otherwise} \end{cases}$

function, $K(\mathbf{x}_i, \mathbf{x}_t)$, without any further changes to the algorithm (for a discussion on Mercer kernels, see, e.g., [15]). Intuitively, the kernel function $K(\mathbf{x}_i, \mathbf{x}_t)$ implements an inner product $\langle\phi(\mathbf{x}_i), \phi(\mathbf{x}_t)$ where ϕ is a nonlinear mapping from the original instance space into another (possibly high-dimensional) Hilbert space. Even if the original instances are not linearly separable, the images of the instances due to the nonlinear mapping ϕ can be linearly separable and thus the kernel-based Perceptron can handle non-separable datasets. Since the analysis of the Perceptron does not depend on the dimensionality of the instances, all of the formal results still hold when the algorithm is used in conjunction with kernel functions (Table 1).

Key Results

In the following a mistake bound for the Perceptron in the non-separable case (see, e.g., [10, 14]) is provided.

Theorem Assume that the Perceptron is presented with the sequence of examples $(\mathbf{x}_1, y_1), \ldots, (\mathbf{x}_T, y_T)$ and denote $R = \max_t \|\mathbf{x}_t\|_2$. Let $\mathbf{u}$ be a unit length weight vector ($\|\mathbf{u}\|_2 = 1$), let $\gamma > 0$ be a scalar, and denote

$$L = \sum_{t=1}^{T} \max\{0, 1 - y_t \langle \mathbf{u}/\gamma, \mathbf{x}_t \rangle\}.$$

Then, the number of prediction mistakes the Perceptron makes on the sequence of example is at most

$$L + \left(\frac{R}{\gamma}\right)^2 + \frac{R\sqrt{L}}{\gamma}.$$

Note that if there exists $\mathbf{u}$ and γ such that $y_t\langle \mathbf{u}, \mathbf{x}_t\rangle \geq \gamma$ for all t, then $L = 0$ and the above bound reduces to Novikoff's bound,

$$\left(\frac{R}{\gamma}\right)^2.$$

Note also that the bound does not depend on the dimensionality of the instances. Therefore, it holds for the kernel-based Perceptron as well with $R = \max_t K(\mathbf{x}_t, \mathbf{x}_t)$.

Applications

So far the Perceptron has been viewed in the prism of online learning. Freund and Schapire [9] proposed a simple conversion of the Perceptron algorithm to the batch learning setting. A batch learning algorithm receives as input a training set of examples $\{(\mathbf{x}_1, y_1), \ldots, (\mathbf{x}_T, y_T)\}$ sampled independently from an underlying joint distribution over the instance and label space. The algorithm is required to output a single classification function which performs well on unseen examples as long as the unseen examples are sampled from the same distribution as the training set. The conversion of the Perceptron to the batch setting proposed by Freund and Schapire is called the voted Perceptron algorithm. The idea is to simply run the online Perceptron on the training set of examples, thus producing a sequence of weight vectors $\mathbf{w}_1, \ldots, \mathbf{w}_T$. Then, the single classification function to be used for unseen examples is a majority vote over the predictions of the weight vectors. That is,

$$f(\mathbf{x}) = \begin{cases} +1 \text{ if} |\{t : \langle \mathbf{w}_t, \mathbf{x}\rangle > 0\}| > \\ \qquad |\{t : \langle \mathbf{w}_t, \mathbf{x}\rangle < 0\}| \\ -1 \text{ otherwise} \end{cases}$$

It was shown (see again [9]) that if the number of prediction mistakes the Perceptron makes on the training set is small, then $f(\mathbf{x})$ is likely to perform well on unseen examples as well.

Finally, it should be noted that the Perceptron algorithm was used for other purposes such as solving linear programming [3] and training support vector machines [14]. In addition, variants of the Perceptron were used for numerous additional problems such as online learning on a budget [4,8], multiclass categorization and ranking problems [6,7], and discriminative training for hidden Markov models [5].

Cross-References

▶ Support Vector Machines

Recommended Reading

1. Agmon S (1954) The relaxation method for linear inequalities. Can J Math 6(3):382–392
2. Block HD (1962) The perceptron: a model for brain functioning. Rev Mod Phys 34:123–135
3. Blum A, Dunagan JD (2002) Smoothed analysis of the perceptron algorithm for linear programming. In: Proceedings of the thirteenth annual symposium on discrete algorithms, San Francisco
4. Cesa-Bianchi N, Gentile C (2006) Tracking the best hyperplane with a simple budget perceptron. In: Proceedings of the nineteenth annual conference on computational learning theory, Pittsburgh
5. Collins M (2002) Discriminative training methods for hidden Markov models: theory and experiments with perceptron algorithms. In: Conference on empirical methods in natural language processing, Philadelphia
6. Crammer K, Dekel O, Keshet J, Shalev-Shwartz S, Singer Y (2006) Online passive aggressive algorithms. J Mach Learn Res 7:551–585
7. Crammer K, Singer Y (2002) A new family of online algorithms for category ranking. In: Proceedings of the 25th annual international ACM SIGIR conference on research and development in information retrieval, Tampere
8. Dekel O, Shalev-Shwartz S, Singer Y (2005) The Forgetron: a kernel-based perceptron on a fixed budget. Adv neural Inf Process Syst 18: 259–266
9. Freund Y, Schapire RE (1998) Large margin classification using the perceptron algorithm. In: Proceedings of the eleventh annual conference on computational learning theory, Madison
10. Gentile C (2002) The robustness of the p-norm algorithms. Mach Learn 53(3), 265–299
11. Minsky M, Papert S (1969) Perceptrons: an introduction to computational geometry. MIT, Cambridge
12. Novikoff ABJ (1962) On convergence proofs on perceptrons. In: Proceedings of the symposium on the mathematical theory of automata, New York, vol XII, pp 615–622
13. Rosenblatt F (1958) The perceptron: a probabilistic model for information storage and organization in the brain. Psychol Rev 65:386–407
14. Shalev-Shwartz S, Singer Y (2005) A new perspective on an old perceptron algorithm. In: Proceedings of the eighteenth annual conference on computational learning theory, Bertinoro, 264–278
15. Vapnik VN (1998) Statistical learning theory. Wiley, New York

Perfect Phylogeny (Bounded Number of States)

Jesper Jansson
Laboratory of Mathematical Bioinformatics, Institute for Chemical Research, Kyoto University, Gokasho, Uji, Kyoto, Japan

Keywords

Bounded number of states; Character state matrix; Phylogenetic reconstruction; Phylogenetic tree; Perfect phylogeny

Years and Authors of Summarized Original Work

1994; Agarwala, Fernández-Baca
1997; Kannan, Warnow

Problem Definition

Let $S = \{s_1, s_2, \ldots, s_n\}$ be a set of elements called *objects* and let $C = \{c_1, c_2, \ldots, c_m\}$ be a set of functions called *characters* such that each $c_j \in C$ is a function from S to the set $\{0, 1, \ldots, r_j - 1\}$ for some integer r_j. For every $c_j \in C$, the set $\{0, 1, \ldots, r_j - 1\}$ is called the set of *allowed states* of character c_j, and for any $s_i \in S$ and $c_j \in C$, it is said that the *state of s_i on c_j* is α, or that the *state of c_j for s_i* is α, where $\alpha = c_j(s_i)$. The *character state matrix* for S and C is the $(n \times m)$-matrix in which entry (i, j) for any $i \in \{1, 2, \ldots, n\}$ and $j \in \{1, 2, \ldots, m\}$ equals the state of s_i on c_j.

In this encyclopedia entry, a *phylogeny for S* is an unrooted tree whose leaves are bijectively labeled by S. For every $c_j \in C$ and $\alpha \in \{0, 1, \ldots, r_j - 1\}$, define the set $S_{c_j,\alpha}$ by $S_{c_j,\alpha} = \{s_i \in S : \text{the state of } s_i \text{ on } c_j \text{ is } \alpha\}$. A *perfect phylogeny for* (S, C) (if one exists) is a phylogeny T for S such that the following holds: for each $c_j \in C$ and pair of allowed states α, β of c_j with $\alpha \neq \beta$, the minimal subtree of T

that connects $S_{c_j,\alpha}$ and the minimal subtree of T that connects $S_{c_j,\beta}$ are vertex disjoint. See Fig. 1 for an example. *The Perfect Phylogeny Problem* (also called *the Character Compatibility Problem* in the literature [2, 9]) is the following:

Problem 1 (The Perfect Phylogeny Problem)

INPUT: An $(n \times m)$-character state matrix M for some S and C.

OUTPUT: A perfect phylogeny for (S, C), if one exists; otherwise, *null*.

Below, we define $r = \max_{j \in \{1,2,\ldots,m\}} r_j$ for the input character state matrix M.

Key Results

The following negative result was proved by Bodlaender, Fellows, and Warnow [2] and, independently, by Steel [14]:

Theorem 1 ([2, 14]) *The Perfect Phylogeny Problem is NP-hard.*

On the other hand, certain restrictions of the Perfect Phylogeny Problem can be solved efficiently. One such special case occurs if the number of allowed states of each character is limited. For this case, Agarwala and Fernández-Baca [1] designed a dynamic programming-based algorithm that builds perfect phylogenies on certain subsets of S called *c-clusters* (also referred to as *proper clusters* in [5, 10] and as *character subfamilies* in [6]) in a bottom-up fashion. Each c-cluster G has the property that: (1) G and $S \setminus G$ share at most one state of each character; and (2) for at least one character, G and $S \setminus G$ share no states. The number of c-clusters is at most $2^r m$, and the algorithm's total running time is $O(2^{3r}(nm^3 + m^4))$, i.e., exponential in r. Hence, the Perfect Phylogeny Problem is polynomial-time solvable when the number of allowed states of every character is upper-bounded by $O(\log(m + n))$. Subsequently, Kannan and Warnow [10] presented a modified algorithm with improved running time. They restructured the algorithm of [1] to eliminate one of the three nested loops that steps through all possible c-clusters and added a preprocessing step which speeds up the innermost loop. The resulting time complexity is given by:

Theorem 2 ([10]) *The algorithm of Kannan and Warnow in [10] solves the Perfect Phylogeny Problem in $O(2^{2r} nm^2)$ time.*

A perfect phylogeny T for (S, C) is called *minimal* if no tree which results by contracting an edge of T is a perfect phylogeny for (S, C). In [10], Kannan and Warnow also showed how to extend their algorithm to enumerate all minimal perfect phylogenies for (S, C) by constructing a directed acyclic graph that implicitly stores the set of all perfect phylogenies for (S, C).

Theorem 3 ([10]) *The extended algorithm of Kannan and Warnow in [10] enumerates the set of all minimal perfect phylogenies for (S, C) so*

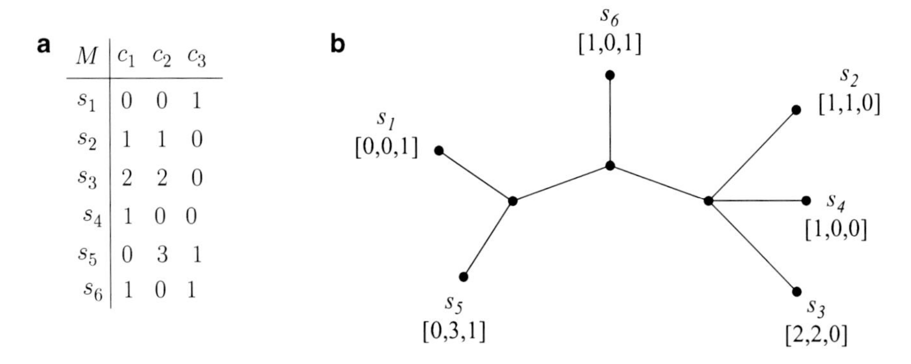

a

M	c_1	c_2	c_3
s_1	0	0	1
s_2	1	1	0
s_3	2	2	0
s_4	1	0	0
s_5	0	3	1
s_6	1	0	1

Perfect Phylogeny (Bounded Number of States), Fig. 1 (**a**) An example of a character state matrix M for $S = \{s_1, s_2, \ldots, s_6\}$ and $C = \{c_1, c_2, c_3\}$ with $r_1 = 3$, $r_2 = 4$, and $r_3 = 2$, i.e., $r = 4$. (**b**) A perfect phylogeny for (S, C). For convenience, the states of all three characters for each object in S are shown

Perfect Phylogeny (Bounded Number of States), Table 1 The running times of the fastest known algorithms for the Perfect Phylogeny Problem with a bounded number of states

r	Running time	Reference
2	$O(nm)$	[11] together with [7]
3	$\min\{O(nm^2), O(n^2m)\}$	[3, 10] together with [9]
4	$\min\{O(nm^2), O(n^2m)\}$	[10] together with [9]
≥ 5	$O(2^{2r}nm^2)$	[10]

that the maximum computation time between two consecutive outputs is $O(2^{2r}nm^2)$.

For small values of r, even faster algorithms are known. Refer to the table in Table 1 for a summary. If $r = 2$, then the problem can be solved in $O(nm)$ time by reducing it to the *Directed* Perfect Phylogeny Problem for Binary Characters (see, e.g., Encyclopedia entry ▶ Directed Perfect Phylogeny (Binary Characters) for a definition of this variant of the problem) using $O(nm)$ time [7, 11] and then applying Gusfield's $O(nm)$-time algorithm [7]. If $r = 3$ or $r = 4$, the problem is solvable in $O(n^2m)$ time by another algorithm by Kannan and Warnow [9], which is faster than the algorithm from Theorem 2 when $n < m$. Also note that for the case $r = 3$, there exists an older algorithm by Dress and Steel [3] whose time complexity coincides with that of the algorithm in Theorem 2.

For other special cases of the Perfect Phylogeny Problem that can be solved efficiently, see Encyclopedia entry ▶ Directed Perfect Phylogeny (Binary Characters) or the survey by Fernández-Baca [5].

Applications

Computational evolutionary biology relies on efficient methods for inferring, from some given data, a phylogenetic tree that accurately describes the evolutionary relationships among a set of objects (e.g., biological species, proteins, genes, etc.) assumed to have been produced by an evolutionary process. One of the most widely used techniques for reconstructing a phylogenetic tree is to represent the objects as vectors of character states and look for a tree that clusters objects which have a lot in common. The Perfect Phylogeny Problem can be regarded as the ideal special case of this approach in which the given data contains no errors, evolution is treelike, and each character state can emerge only once in the evolutionary history.

However, data obtained experimentally seldom admits a perfect phylogeny, so various optimization versions of the problem such as *maximum parsimony* and *maximum compatibility* are often considered in practice. These strategies generally lead to NP-complete problems, but there exist heuristics that work well for most inputs. See, e.g., [4, 5, 12] for a discussion. Nevertheless, algorithms for the Perfect Phylogeny Problem may be useful even when the data does not admit a perfect phylogeny, for example, if there exists a perfect phylogeny for $m - O(1)$ of the characters in C. In fact, in one crucial step of their proposed character-based methodology for determining the evolutionary history of a set of related natural languages, Warnow, Ringe, and Taylor [15] consider all subsets of C in decreasing order of cardinality, repeatedly applying the algorithm of [10] until a largest subset of C which admits a perfect phylogeny is found. The ideas behind the algorithms of [1] and [10] have also been utilized and extended by Fernández-Baca and Lagergren [6] in their algorithm for computing *near-perfect phylogenies* in which the constraints on the output have been relaxed in order to permit non-perfect phylogenies whose so-called penalty score is less than or equal to a prespecified parameter q; see [6] for details. (See also [13] for a fixed-parameter tractable algorithm for this problem variant when $r = 2$.)

The motivation for considering a bounded number of states is that characters based on directly observable traits are, by the way they are defined, naturally bounded by some small number (often 2). When biomolecular data is used to define characters, the number of allowed states is typically bounded by a constant, e.g., $r = 2$ for SNP markers, $r = 4$ for DNA or RNA sequences, or $r = 20$ for amino acid sequences. (see also Encyclopedia entry ▶ Directed Perfect Phylogeny (Binary Characters)). Moreover, characters with $r = 2$ can be useful in comparative linguistics [8].

Open Problems

An open problem is to determine whether the time complexity of the algorithm of Kannan and Warnow [10] can be improved. As noted in [5], it would be interesting to find out if the Perfect Phylogeny Problem is solvable in $O(2^{2r}nm)$ time for any r, or more generally, in $O(f(r) \cdot nm)$ time, where f is a function of r which does not depend on n or m, since this would match the fastest known algorithm for the special case $r = 2$ (see Table 1). Another open problem is to establish lower bounds on the computational complexity of the Perfect Phylogeny Problem with a bounded number of states.

Cross-References

▶ Directed Perfect Phylogeny (Binary Characters)

▶ Perfect Phylogeny Haplotyping

Acknowledgments JJ was funded by the Hakubi Project at Kyoto University and KAKENHI grant number 26330014.

Recommended Reading

1. Agarwala R, Fernández-Baca D (1994) A polynomial-time algorithm for the perfect phylogeny problem when the number of character states is fixed. SIAM J Comput 23(6):1216–1224
2. Bodlaender HL, Fellows MR, Warnow T (1992) Two strikes against perfect phylogeny. In: Proceedings of the 19th international colloquium on automata, languages and programming (ICALP 1992), Vienna. Lecture notes in computer science, vol 623. Springer, Berlin/Heidelberg, pp 273–283
3. Dress A, Steel M (1992) Convex tree realizations of partitions. Appl Math Lett 5(3):3–6
4. Felsenstein J (2004) Inferring phylogenies. Sinauer Associates, Sunderland
5. Fernández-Baca D (2001) The perfect phylogeny problem. In: Cheng X, Du DZ (eds) Steiner trees in industry. Kluwer Academic, Dordrecht, pp 203–234
6. Fernández-Baca D, Lagergren J (2003) A polynomial-time algorithm for near-perfect phylogeny. SIAM J Comput 32(5):1115–1127
7. Gusfield DM (1991) Efficient algorithms for inferring evolutionary trees. Networks 21:19–28
8. Kanj IA, Nakhleh L, Xia G (2006) Reconstructing evolution of natural languages: Complexity and parameterized algorithms. In: Proceedings of the 12th annual international computing and combinatorics conference (COCOON 2006), Taipei. Lecture notes in computer science, vol 4112. Springer, Berlin/Heidelberg, pp 299–308
9. Kannan S, Warnow T (1994) Inferring evolutionary history from DNA sequences. SIAM J Comput 23(4):713–737
10. Kannan S, Warnow T (1997) A fast algorithm for the computation and enumeration of perfect phylogenies. SIAM J Comput 26(6):1749–1763
11. McMorris FR (1977) On the compatibility of binary qualitative taxonomic characters. Bull Math Biol 39(2):133–138
12. Setubal JC, Meidanis J (1997) Introduction to Computational Molecular Biology. PWS Publishing Company, Boston
13. Sridhar S, Dhamdhere K, Blelloch GE, Halperin E, Ravi R, Schwartz R (2007) Algorithms for efficient near-perfect phylogenetic tree reconstruction in theory and practice. IEEE/ACM Trans Comput Biol Bioinform 4(4):561–571
14. Steel MA (1992) The complexity of reconstructing trees from qualitative characters and subtrees. J Classif 9(1):91–116
15. Warnow T, Ringe D, Taylor A (1996) Reconstructing the evolutionary history of natural langauges. In: Proceedings of the 7th annual ACM-SIAM symposium on discrete algorithms (SODA'96), Atlanta, pp 314–322

Perfect Phylogeny Haplotyping

Giuseppe Lancia
Department of Mathematics and Computer Science, University of Udine, Udine, Italy

Keywords

Alleles phasing

P

Years and Authors of Summarized Original Work

2005; Ding, Filkov, Gusfield

Problem Definition

In the context of the *perfect phylogeny haplotyping* (PPH) problem, each vector $h \in \{0,1\}^m$ is called a *haplotype*, while each vector $g \in \{0,1,2\}^m$ is called a *genotype*. Haplotypes are binary encodings of DNA sequences, while genotypes are ternary encodings of *pairs* of DNA sequences (one sequence for each of the two homologous copies of a certain chromosome).

Two haplotypes h' and h'' are said to *resolve* a genotype g if, at each position j: (i) if $g_j \in \{0,1\}$ then both $h'_j = g_j$ and $h''_j = g_j$; (ii) if $g_j = 2$ then either $h'_j = 0$ and $h''_j = 1$ or $h'_j = 1$ and $h''_j = 0$. If h' and h'' resolve g, we write $g = h' + h''$. An instance of the PPH problem consists of a set $G = \{g^1, g^2, \ldots, g^n\}$ of genotypes. A set H of haplotypes such that, for each $g \in G$, there are $h', h'' \in H$ with $g = h' + h''$, is called a *resolving set* for G.

A *perfect phylogeny* for a set H of haplotypes is a rooted tree T for which

- the set of leaves is H and the root is labeled by some binary vector r;
- each index $j \in \{1, \ldots, m\}$ labels exactly one edge of T;
- if an edge e is labeled by an index k, then, for each leaf h that can be reached from the root via a path through e, it is $h_k \neq r_k$.

Without loss of generality, it can be assumed that the vector labeling the root is $r = 0$. Within the PPH problem, T is meant to represent the evolution of the sequences at the leaves from a common ancestral sequence (the root). Each edge labeled with an index represents a point in time when a mutation happened at a specific site. This model of evolution is also known as *coalescent* [11]. It can be shown that a perfect phylogeny for H exists if and only if for all choices of four haplotypes $h^1, \ldots, h^4 \in H$ and two indices i, j,

$$\{h_i^a h_j^a, 1 \leq a \leq 4\} \neq \{00, 01, 10, 11\}.$$

Given the above definitions, the problem surveyed in this entry is the following:

Perfect Phylogeny Haplotyping Problem (PPH)

Given a set G of genotypes, find a resolving set H of haplotypes and a perfect phylogeny T for H, or determine that such a resolving set does not exist.

In a slightly different version of the above problem, one may require to find all perfect phylogenies for H instead of just one (in fact, all known algorithms for PPH do find all perfect phylogenies).

The perfect phylogeny problem was introduced by Gusfield [7], who also proposed a nearly linear-time $O(nm\,\alpha(nm))$-algorithm for its solution (where $\alpha()$ is the extremely slowly growing inverse Ackerman function). The algorithm resorted to a reduction to a complex procedure for the *graph realization* problem (Bixby and Wagner [2]), of very difficult understanding and implementation. Later approaches for PPH proposed much simpler, albeit slower, $O(nm^2)$-algorithms (Bafna et al. [1]; Eskin et al. [6]). However, a major question was left open: *does there exist a linear-time algorithm for PPH?* In [7], Gusfield conjectured that this should be the case. The 2005 algorithm by Ding, Filkov, and Gusfield [5] surveyed in this entry settles the above conjecture in the affirmative.

Key Results

The main idea of the algorithm is to find the maximal sub-graphs that are common to all PPH solutions. Let us call *P-class* a maximal subgraph of all PPH trees for G. The authors show that each P-class consists of two sub-trees which, in each PPH solution, can appear in either one of two possible ways (called *flips* of the P-class) with respect to any fixed P-class taken as a reference. Hence, if there are k P-classes, there are 2^{k-1} distinct PPH solutions.

The algorithm resorts to an original and effective data structure, called the *shadow tree*, which gives an implicit representation of all P-classes. The data structure is built incrementally, by processing one genotype at a time. The total cost for building and updating the shadow tree is linear in the input size (i.e., in nm). A detailed description of the shadow tree requires a rather large number of definitions, possibly accompanied by figures and examples. Here, we will introduce only its basic features, those that allow us to state the main theorems of [5].

The shadow tree is a particular type of directed rooted tree, which contains both *edges* and *links* (strictly speaking, the latter are just arcs, but they are called links to underline their specific use in the algorithm). The edges are of two types: *tree-edges* and *shadow-edges*, and are associated to the indices $\{1, \ldots, m\}$. For each index i, there is a tree-edge labeled t_i and a shadow-edge labeled s_i. Both edges and links are oriented, with their head closer to the root than their tail. Other than the root, each node of the shadow tree is the endpoint of exactly one tree-edge or one shadow-edge (while the root is the head of two "dummy" links). The links are used to connect certain tree- and shadow-edges. A link can be either *free* or *fixed*. The head of a free link can still be changed during the execution of the algorithm, but once a link is fixed, it cannot be changed any more.

Tree-edges, shadow-edges and *fixed* links are organized into *classes*, which are sub-graphs of the shadow tree. Each fixed link is contained in exactly one class, while each free link connects one class to another, called its *parent*. For each index i, if the tree-edge t_i is in a class X, then the shadow-edge s_i is in X as well, so that a class can be seen as a pair of "twin" sub-trees of the shadow tree. The free links point out from the root of the sub-trees (the *class roots*). Classes change during the running of the algorithm. Specifically, classes are *created* (containing a single tree- and shadow-edge) when a new genotype is processed; a class can be *merged* with its parent, by fixing a pair of free edges; finally, a class can be *flipped*, by switching the heads of the two free links that connect the class roots to the parent class.

A tree T is said to be "*contained in*" a shadow tree if T can be obtained by flipping some classes in the shadow tree, followed by contracting all links and shadow-edges. Let us call *contraction* of a class the sub-graph (consisting of a pair of sub-trees, made of tree-edges only) that is obtained from a class X of the shadow tree by contracting all fixed links and shadow-edges of X. The following are the main results obtained in [5]:

Proposition 1 *Every P-class can be obtained by contraction of a class of the final shadow tree produced by the algorithm. Conversely, every contraction of a class of the final shadow tree is a P-class.*

Theorem 1 *Every PPH solution is contained in the final shadow tree produced by the algorithm. Conversely, every tree contained in the final shadow tree is a distinct PPH solution.*

Theorem 2 *The total time required for building and updating the shadow tree is $O(nm)$.*

Applications

The PPH problem arises in the context of *Single Nucleotide Polymorphisms* (SNP's) analysis in human genomes. A SNP is the site of a single nucleotide which varies in a statistically significant way in a population. The determination of SNP locations and of common SNP patterns (haplotypes) are of uttermost importance. In fact, SNP analysis is used to understand the nature of several genetic diseases, and the international Haplotype Map Project is devoted to SNP study (Helmuth [9]).

The values that a SNP can take are called its *alleles*. Almost all SNPs are bi-allelic, i.e., out of the four nucleotides A, C, T, G, only two are observed at any SNP. Humans are *diploid* organisms, with DNA organized in pairs of chromosomes (of paternal and of maternal origin). The sequence of alleles on a chromosome copy is called a *haplotype*. Since SNPs are bi-allelic, haplotypes can be encoded as binary strings. For a given SNP, an individual can be either *homozygous*, if both parents contributed the same allele, or *heterozygous*, if the paternal and maternal alleles are different.

Haplotyping an individual consists of determining his two haplotypes. Haplotyping a population consists of haplotyping each individual of the population. While it is today economically infeasible to determine the haplotypes directly, there is a cheap experiment which can determine the (less informative and often ambiguous) *genotypes*.

A genotype of an individual contains the *conflated* information about the two haplotypes. For each SNP, the genotype specifies which are the two (possibly identical) alleles, but does not specify their origin (paternal or maternal). The ternary encoding that is used to represent a genotype g has the following meaning: at each SNP j, it is $g_j = 0$ (respectively, 1) if the individual is homozygous for the allele 0 (respectively, 1), and $g_j = 2$ if the individual is heterozygous. There may be many possible pairs of haplotypes that justify a particular genotype (there are 2^{k-1} pairs of haplotypes that can resolve a genotype with k heterozygous SNPs). Given a set of genotypes, in order to determine the correct resolving set out of the exponentially many possibilities, one imposes some "biologically meaningful" constraints that the solution must possess. The perfect phylogeny model (coalescent) requires that the resolving set must fit a particular type of evolutionary tree. That is, all haplotypes should descend from some ancestral haplotype, via mutations that happened (only once) at specific sites over time. The coalescent model is accurate especially for short haplotypes (for longer haplotypes there is also another type of evolutionary event, *recombination*, that should be taken into account).

The linear-time PPH algorithm is of significant practical value in two respects. First, there are instances of the problem where the number of SNPs considered is fairly large (genotypes can extend over several kilo-bases). For these long instances, the advantage of an $O(nm)$ algorithm with respect to the previous $O(nm^2)$ approach is evident. On the other hand, when genotypes are relatively short, the benefit of using the linear-time algorithm is not immediately evident (both algorithms run extremely quickly). Nevertheless, there are situations in which one has to solve a large set of haplotyping problems, where each single problem is defined over short genotypes. For instance, this is the case in which one examines all "small" subsets of SNPs in order to determine the subsets for which there is a PPH solution. In this type of application, the gain of efficiency with the use of the linear-time PPH algorithm is significant (Chung and Gusfield [4]; Wiuf [14]).

Open Problems

A linear-time algorithm is the best possible for PPH, and no open problems are listed in [5].

Experimental Results

The algorithm has been implemented in C and its performance has been compared with the previous fastest PPH algorithm, i.e., DPPH (Bafna et al. [1]). In the case of $m = 2000$ and $n = 1000$, the algorithm is about 250-times faster than DPPH, and is capable of solving an instance in an average time of 2 s, versus almost 8 min needed by DPPH (on a "standard" 2005 Personal Computer). The smaller instances (e.g., with $m = 50$ SNPs) are such that the superior performance of the algorithm is not as evident, with an average running time of 0.07 s versus 0.2 s. However, as already remarked, when the small instances are executed within a loop, the speed-up turns out to be again of two or more orders of magnitude.

Data Sets

The data sets used in [5] have been generated by the program *ms* (Hudson [12]), which is the widely used standard for instance generation reflecting the coalescent model of SNP sequence evolution. Real-life instances can be found at the HapMap web site http://www.hapmap.org.

URL to Code

http://wwwcsif.cs.ucdavis.edu/~gusfield/lpph/

Cross-References

- ▸ Directed Perfect Phylogeny (Binary Characters)
- ▸ Perfect Phylogeny (Bounded Number of States)

Recommended Reading

For surveys about computational haplotyping problems in general, see Bonizzoni et al. [3], Gusfield and Orzack [8], Halldorsson et al. [10], and Lancia [13].

1. Bafna V, Gusfield D, Lancia G, Yooseph S (2003) Haplotyping as perfect phylogeny: a direct approach. J Comput Biol 10(3–4):323–340
2. Bixby RE, Wagner DK (1988) An almost linear-time algorithm for graph realization. Math Oper Res 13:99–123
3. Bonizzoni P, Della Vedova G, Dondi R, Li J (2004) The haplotyping problem: an overview of computational models and solutions. J Comput Sci Technol 19(1):1–23
4. Chung RH, Gusfield D (2003) Empirical exploration of perfect phylogeny haplotyping and haplotypes. In: Proceedings of annual international conference on computing and combinatorics (COCOON). Lecture notes in computer science, vol 2697. Springer, Berlin, pp 5–9
5. Ding Z, Filkov V, Gusfield D (2005) A linear-time algorithm for the perfect phylogeny haplotyping problem. In: Proceedings of the annual international conference on computational molecular biology (RECOMB), New York. ACM, New York
6. Eskin E, Halperin E, Karp R (2003) Efficient reconstruction of haplotype structure via perfect phylogeny. J Bioinf Comput Biol 1(1):1–20
7. Gusfield D (2002) Haplotyping as perfect phylogeny: conceptual framework and efficient solutions. In: Myers G, Hannenhalli S, Istrail S, Pevzner P, Waterman M (eds) Proceedings of the annual international conference on computational molecular biology (RECOMB). ACM, New York, pp 166–175
8. Gusfield D, Orzack SH (2005) Haplotype inference. In: Aluru S (ed) Handbook of computational molecular biology. Champman and Hall/CRC, Boca Raton, pp 1–28
9. Helmuth L (2001) Genome research: map of the human genome 3.0. Science 293(5530): 583–585
10. Halldorsson BV, Bafna V, Edwards N, Lippert R, Yooseph S, Istrail S (2004) A survey of computational methods for determining haplotypes. In: Computational methods for SNP and haplotype inference: DIMACS/RECOMB satellite workshop. Lecture notes in computer science, vol 2983. Springer, Berlin, pp 26–47
11. Hudson R (1990) Gene genealogies and the coalescent process. Oxf Surv Evol Biol 7:1–44
12. Hudson R (2002) Generating samples under the wright-fisher neutral model of genetic variation. Bioinformatics 18(2):337–338
13. Lancia G (2008) The phasing of heterozygous traits: algorithms and complexity. Comput Math Appl 55(5):960–969
14. Wiuf C (2004) Inference on recombination and block structure using unphased data. Genetics 166(1): 537–545

P

Performance-Driven Clustering

Rajmohan Rajaraman
Department of Computer Science, Northeastern University, Boston, MA, USA

Keywords

Circuit clustering; Circuit partitioning

Years and Authors of Summarized Original Work

1993; Rajaraman, Wong

Problem Definition

Circuit partitioning consists of dividing the circuit into parts, each of which can be implemented as a separate component (e.g., a chip) that satisfies the design constraints. The work of Rajaraman and Wong [5] considers the problem of dividing a circuit into components, subject to area constraints, such that the maximum delay at the outputs is minimized.

A combinational circuit can be represented as a directed acyclic graph $G = (V, E)$, where V is the set of nodes and E is the set of directed edges. Each node represents a gate in the network and each edge (u, v) in E represents an interconnection between gates u and v in the network. The *fanin* of a node is the number of edges incident into it, and the *fanout* of a node is the number of edges incident out of it. A *primary input* (PI) is a node with fanin 0, while a *primary output* (PO) is a node with fanout 0. Each node has a *weight* and a *delay* associated with it.

Definition 1 A clustering of a network $G = (V, E)$ is a triple (H, ϕ, Σ), where

1. $H = (V', E')$ is a directed acyclic graph.
2. φ is a function mapping V' to V such that
   - For every edge $(u', v') \in E'$, $(\phi(u'), \phi(v')) \in E$.
   - For every node $v' \in V'$ and edge $(u, \phi(v')) \in E$, there exists a unique $u' \in V'$ such that $\phi(u') = u$ and $(u', v') \in E'$.
   - For every PO node $v \in V$, there exists a unique $v' \in V'$ such that $\phi(v') = v$.
3. Σ is a partition of V'.

Let $\Gamma = (H = (V', E'), \phi, \Sigma)$ be a clustering of G. For $v \in V, v' \in V'$, if $\phi(v') = v$, we call v' a *copy* of v. The set V' consists of all the copies of the nodes in V that appear in the clustering. A node v' is a PI (respectively, PO) in Γ if $\phi(v')$ is a PI (respectively, PO) in G. It follows from the definition of φ that H is logically equivalent to G.

The weights and delays on the individual nodes in G yield weights and delays of nodes in H' and a delay for the clustering Γ. The weight (respectively, delay) of a node v' in V' is the weight (respectively, delay) of $\phi(v)$. The weight of any cluster $C \in \Sigma$, denoted by $W(C)$, is the sum of the weights of the nodes in C. The delay of a clustering is given by the general delay model of Murgai et al. [3], which is as follows. The delay of an edge $(u', v') \in E'$ is D (which is a given parameter) if u' and v' belong to different elements of Σ and zero otherwise. The delay along a path in H' is simply the sum of the delays of the edges of the path. Finally, the delay of Γ is the delay of a maximum-delay path in H', among all the paths from a PI node to a PO node in H'.

Definition 2 Given a combinational network $G = (V, E)$ with weight function $w : V \rightarrow R^+$, weight capacity M, and a delay function $\delta : V \rightarrow R^+$, we say that a clustering $\Gamma = (H, \phi, \Sigma)$ is *feasible* if for every cluster $C \in \Sigma, W(C)$ is at most M. The *circuit clustering problem* is to compute a feasible clustering Γ of G such that the delay of Γ is minimum among all feasible clusterings of G.

An early work of Lawler et al. [2] presented a polynomial-time optimal algorithm for the circuit clustering problem in the special case where all the gate delays are zero (i.e., $\delta(v) = 0$ for all v).

Key Results

Rajaraman and Wong [5] presented an optimal polynomial-time algorithm for the circuit clustering problem under the general delay model.

Theorem 1 *There exists an algorithm that computes an optimal clustering for the circuit clustering problem in $O(n^2 \log n + nm)$ time, where n and m are the vertices and edges, respectively, of the given combinational network.*

This result can be extended to compute optimal clusterings under any monotone clustering constraint. A clustering constraint is monotone if any connected subset of nodes in a feasible cluster is also monotone [2].

Theorem 2 *The circuit clustering problem can be solved optimally under any monotone clustering constraint in time polynomial in the size of the circuit.*

Applications

Circuit partitioning/clustering is an important component of very large scale integration design. One application of the circuit clustering problem formulated above is to implement a circuit on multiple field programmable gate array chips. The work of Rajaraman and Wong focused on clustering combinational circuits to minimize delay under area constraints. Related studies have considered other important constraints, such as pin constraints [1] and a combination of area and pin constraints [6]. Further work has also included clustering sequential circuits (as opposed to combinational circuits) with the objective of minimizing the clock period [4].

Experimental Results

Rajaraman and Wong reported experimental results on five ISCAS (International Symposium on Circuits and Systems) circuits. The number of nodes in these circuits ranged from 196 to 913. They reported the maximum delay of the clusterings and running times of their algorithm on a Sun Sparc workstation.

Cross-References

▶ FPGA Technology Mapping

Recommended Reading

1. Cong J, Ding Y (1992) An optimal technology mapping algorithm for delay optimization in lookup-table based fpga design. In: Proceedings of IEEE international conference on computer-aided design, Santa Clara, pp 48–53
2. Lawler EL, Levitt KN, Turner J (1966) Module clustering to minimize delay in digital networks. IEEE Trans Comput C-18:47–57
3. Murgai R, Brayton RK, Sangiovanni-Vincentelli A (1991) On clustering for minimum delay/area. In: Proceedings of IEEE international conference on computer-aided design, Santa Clara, pp 6–9
4. Pan P, Karandikar AK, Liu CL (1998) Optimal clock period clustering for sequential circuits with retiming. IEEE Trans Comput-Aided Des Integr Circuits Syst 17:489–498
5. Rajaraman R, Wong DF (1995) Optimum clustering for delay minimization. IEEE Trans Comput-Aided Des Integr Circ Syst 14:1490–1495
6. Yang HH, Wong DF (1997) Circuit clustering for delay minimization under area and pinconstraints. IEEE Trans Comput-Aided Des Integr Circ Syst 16:976–986

Permutation Enumeration

Katsuhisa Yamanaka
Department of Electrical Engineering and Computer Science, Iwate University, Iwate, Japan

Keywords

Combinatorial Gray code; Enumeration; Permutation; Partition search; Reverse search

Years and Authors of Summarized Original Work

1962; Trotter
1963; Johnson
1977; Sedgewick
2008; Sekine, Yamanaka, Nakano

Problem Definition

Permutation Enumeration

Let S_n be the set of permutations of $[n] = \{1, 2, \ldots, n\}$. We write a permutation as a sequence of elements in $[n]$ such that each element appears exactly once. A *permutation enumeration* is to list all permutations in S_n. For example, there are 24 permutations of [4]: 1234, 1243, 1324, 1342, 1423, 1432, 2134, 2143, 2314, 2341, 2413, 2431, 3124, 3142, 3214, 3241, 3412, 3421, 4123, 4132, 4213, 4231, 4312, 4321. The enumeration of permutations is a basic and long-standing enumeration problem, and it was surveyed by Sedgewick [8].

Note that the above example lists all the permutations of [4], and we have listed them in lexicographic order, which is the most natural way to enumerate them. The purpose of this paper is to introduce representative methods for enumeration problems by showing how these methods are applied to permutation enumeration.

Efficiency of Enumeration Algorithms

The efficiency of an algorithm is measured by the time and space complexity for a given input size. However, many enumeration problems have an exponential number of outputs (solutions) for a given input. Hence, an enumeration algorithm may require exponential time in order to output the solutions. Typically, an enumeration algorithm is measured by the "delay time." We say that an enumeration algorithm has delay d if (1) it takes at most d time to output the first object, and (2) it takes at most d time between two consecutive outputs. See [1, 3, 4] for further details. Note that the delay time does not include the time required to output the objects, since this is typically ignored when estimating the time complexity of an enumeration algorithm. The space complexity of an enumeration algorithm is an estimate of the amount of working memory required by the algorithm (as in the usual sense).

Key Results

Enumeration by Partition Search

In the partition search enumeration method, objects are listed by repeatedly partitioning the set of objects. As an example, we will apply the partition search method to permutation enumeration. We will partition S_n by fixing the first element of a permutation. Denote by $S_n(i) \subseteq S_n$ the set of permutations in which i in $[n]$ is the first element. Then, S_n is partitioned into $S_n(1), S_n(2), \ldots, S_n(n)$. Hence, if we have the list of all permutations of $[n] \setminus \{i\}$ for each $i = 1, 2, \ldots, n$, then we can enumerate all permutations in S_n by appending i as the first element to every permutation. This recursive structure gives the algorithm shown as Algorithm 1. To begin, Algorithm 1 is called with the empty sequence and the empty set. The algorithm recursively fixes the first element, and we then obtain all permutations in S_n. Figure 1 illustrates a tree structure of recursive calls of the algorithm. The root corresponds to the empty sequence. Each vertex of the tree corresponds to the prefix of a permutation, and it is obtained by removing the last element of its child. The leaves correspond to the permutations in S_n. Algorithm 1 traverses the tree structure in a depth-first manner.

Algorithm 1: PARTITION-SEARCH(π, S)

```
1 π is the current subpermutation, and S is the set of
  elements in π;
2 if S = [n] then    /* π is a permutation
  in S_n */
3 |  Output π;
4 |  return;
5 foreach i ∈ [n] \ S do
6 |  PARTITION-SEARCH(π + i, S ∪ {i});
  |        /* The operation '+' is a
  |  concatenation */
```

We now estimate the running time of Algorithm 1. Let T be the running time for traversing an edge of the search tree. Since the depth of the search tree is n, the delay time of the algorithm is $O(nT)$ time in worst case.

Theorem 1 *One can enumerate all permutations by the partition search method. The running time of the algorithm is $O(nT)$, where T is the time for traversing an edge of the search tree. The required working space is $O(n)$.*

Enumeration by Gray Code

A *combinatorial Gray code* is a list of all of the objects in some class such that two consecutive objects in the list differ only by a small amount. Since the list contains all of the ob-

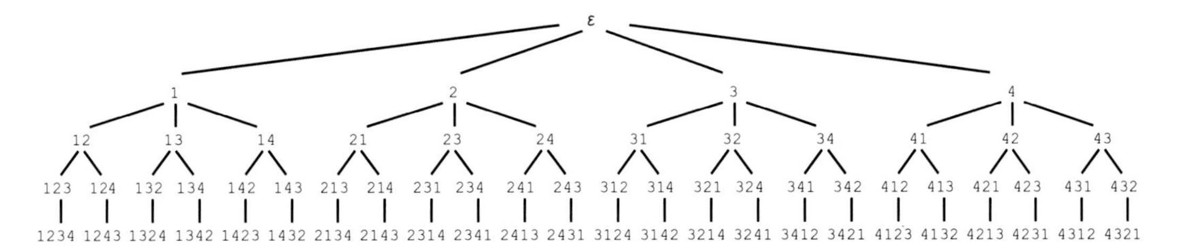

Permutation Enumeration, Fig. 1 Search tree of the partition search method

jects, an algorithm that generates a combinatorial Gray code can be regarded as an enumeration algorithm. There are combinatorial Gray codes for various combinatorial objects, and these have been surveyed [7]. For a permutation, a difference between two consecutive objects would be a swap of two adjacent elements, where we say that the i-th and the $(i + 1)$-th element in a permutation are *adjacent*. For permutations, the well-known Steinhaus-Johnson-Trotter algorithm (or Johnson-Trotter algorithm) [5,8,10,11] generates the combinatorial Gray code for the permutations of $[n]$. This algorithm is regarded as an enumeration of permutations.

Let $\pi = p_1 p_2 \dots p_{n-1}$ in S_{n-1}, and denote by $\pi(i) = p_1 p_2 \dots p_i n p_{i+1} \dots p_{n-1}$ for $i = 0, 1, \dots, n-1$ the permutation obtained from π by inserting n between p_i and p_{i+1}. Then, the list of $\pi(0), \pi(1), \dots, \pi(n-1)$ or $\pi(n-1), \pi(n-2), \dots, \pi(0)$ is a combinatorial Gray code for a subset of S_n. Such lists can be defined for all permutations in S_{n-1}, and the lists for all permutations in S_{n-1} contain all permutations in S_n. Assume that we have a combinatorial Gray code for S_{n-1}. Let π_i be the i-th permutation in the list. Then, we construct the list $\pi_i(0), \pi_i(1), \dots, \pi_i(n-1)$ if i is even, and $\pi_i(n-1), \pi_i(n-2), \dots, \pi_i(0)$ if i is odd. The obtained list is a combinatorial Gray code for S_n. Note that if i is even, then $\pi_{i+1}(n-1)$ is obtained from $\pi_i(n-1)$ by swapping two adjacent elements, where π_{i+1} is the $(i+1)$-th permutation in a combinatorial Gray code for S_{n-1}. Similarly, if i is odd, then $\pi_{i+1}(0)$ is obtained from $\pi_i(0)$ by swapping two adjacent elements. By recursively applying this idea, we can design a combinatorial Gray code for S_n.

Now we explain the details of Steinhaus-Johnson-Trotter algorithm. The algorithm first outputs the identity permutation $\iota = 12\dots n$. Let us consider the case of $\pi = p_1 p_2 \dots p_n$ in $S_n \setminus \{\iota\}$, and let us assume that π is generated from $\pi' = p'_1 p'_2 \dots p'_n$ by swapping two adjacent elements in π'. We construct the next permutation of π by swapping n and its left-adjacent or its right-adjacent element. More precisely, the rule of swapping is as follows.

1. n is the last element of π.
   - 1-1. n is the second-to-last element of π'.
     In this case, π is obtained from π' by swapping $p'_{n-1} = n$ and p'_n by Step 3. We swap two elements in $[n-1]$ by recursively applying this swapping algorithm to the subpermutation $p_1 p_2 \dots p_{n-1}$, which is obtained from π by removing the element n. Let π_s be the obtained subpermutation. Then, we append n to π_s as the last element. The obtained permutation is the next permutation.
   - 1-2. n is also the last element of π'.
     We construct the next permutation of π by swapping p_{n-1} and $p_n = n$.
2. n is the first element of π. Similar to Step 1, we construct the next permutation of π, as follows:
   - 2-1. n is the second element of π'.
     We recursively apply this swapping algorithm to the subpermutation $p_2 p_3 \dots p_n$. Let π_t be the obtained subpermutation. Then, we append n to π_t as the first element. The obtained permutation is the next permutation.

Permutation Enumeration, Table 1 List of S_4 in combinatorial Gray code ($n = 4$ is underlined)

1234	3124	2314
1243	3142	2341
1423	3412	2431
4123	4312	4231
4132	4321	4213
1432	3421	2413
1342	3241	2143
1324	3214	2134

2-2. n is also the first element of π'.

We construct the next permutation of π by swapping $p_1 = n$ and p_2.

3. Otherwise.

We swap $p_i = n$ and p_{i+1} if π is obtained from π' by swapping $p'_{i-1} = n$ and p'_i of π', and swap p_{i-1} and $p_i = n$ if π is obtained from π' by swapping p'_i and $p'_{i+1} = n$ of π'.

Table 1 shows the list of permutations in S_4 enumerated by the above swapping algorithm. Note that any permutation can be obtained by swapping two adjacent elements.

Pseudocodes for the above algorithm are shown in Algorithm 2, which is the main routine, and Algorithm 3, which is a subroutine which generates the next permutation of a given permutation. These pseudocodes assume that a direction vector $d = (d(1), d(2), \ldots, d(n))$ is stored in global memory. Each $d(i)$ for $i = 1, 2, \ldots, n$ represents the direction in which the element i in the current permutation goes to obtain the next permutation. More precisely, an instruction of "left" or "right" is stored in each $d(i)$. By using the direction vector, we know in which direction two adjacent elements were swapped without needing to check the current permutation and the preceding permutation.

Our implementations (Algorithms 2 and 3) are not efficient, but an efficient loopless algorithm was given by Sedgewick [8], as in the following theorem.

Theorem 2 ([8]) *After constructing the identity permutation in $O(n)$ time, one can enumerate all permutations in S_n in the order of a combinatorial Gray code with a constant time delay.*

It can be observed that the combinatorial Gray coder order defined by the algorithm represents a Hamiltonian path of the permutohedron. Figure 2 shows a permutohedron of S_4 and its Hamiltonian path corresponding to the combinatorial Gray code.

Enumeration by Reverse Search

Avis and Fukuda [2] proposed a reverse search enumeration method. The idea of the reverse search method is as follows: We first define a rooted tree structure for the objects such that each vertex corresponds to an object and each edge corresponds to a relation between two objects. Then, by traversing the tree structure, we enumerate all of the objects.

Now, we illustrate the reverse search method by applying it to permutation enumeration [9]. We define a rooted tree structure for S_n, as follows. To construct a tree structure, we define its root and the parent of each permutation

Algorithm 2: GRAY-CODE(n)

```
1  d(i) for i = 1,2,...,n represents in which
   direction the element i goes;
2  foreach i = 1,2,...,n do
     /* Initialization of direction
     vector d(i) */
3    d(i) ← left
4  π ← 12...n          /* Set the identity
   permutation to π. */;
5  foreach i = 1,2,...,n! do
6    Output π;
7    π ← SWAP(n,π);
```

Algorithm 3: SWAP($n, \pi = p_1 p_2 \ldots p_n$)

```
1  if n = 1 then return π
2  else if p_n = n and d(n) = right then
3    π' ← SWAP(n − 1, p_1 p_2 ... p_{n−1});
4    d(n) ← left;
5    return π' + n
6  else if p_1 = n and d(n) = left then
7    π' ← SWAP(n − 1, p_2 p_3 ... p_n);
8    d(n) ← right;
9    return n + π'
10 else return the permutation obtained from π by
   swapping n and its left-adjacent or its right-adjacent
   element depending on d(n)
```

Permutation Enumeration, Fig. 2 Permutohedron of S_4 and its Hamiltonian path

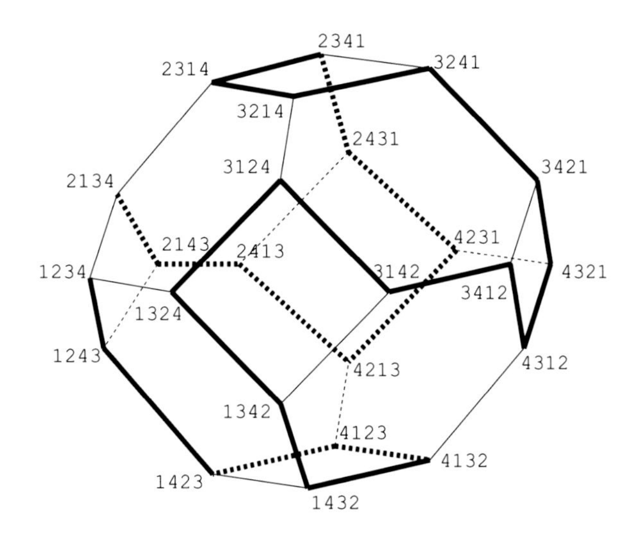

in S_n except the root. We define the identity permutation $\iota = 12\ldots n$ as the *root* of the tree structure. Then, we define the parent of a permutation by an adjacent swap of two elements. Intuitively, the parent is defined so that there is greater similarity between it and ι than there is between the child and ι. A formal definition is as follows. Let $\pi = p_1 p_2 \ldots p_n \in S_n \setminus \{\iota\}$ be a permutation, and let i be the minimum index such that $p_i > p_{i+1}$ holds. The *parent permutation* of π, denoted by $P(\pi)$, is the permutation obtained from π by swapping p_i and p_{i+1}. Then, we call π a *child permutation* of $P(\pi)$. Note that the parent permutation of π is uniquely defined. For example, $P(3421) = 3241$, $P(3241) = 2341$, $P(2341) = 2314$, $P(2314) = 2134$, and $P(2134) = 1234$ are obtained. By repeatedly finding the parent permutations, we have the sequence of permutations in S_n, which ends up with the identity permutation. By merging these sequences, we have the tree structure, called the *family tree* T_n of S_n. Figure 3 shows the family tree T_4.

We next design an algorithm that traverses the family tree by recursively generating all of the child permutations of any permutation. Intuitively, the operation to generate a child permutation is the reverse of finding the parent permutation.

We introduce some notation. Let $\pi = p_1 p_2 \ldots p_n$ be a permutation in S_n. Then, we denote by $\pi[i] = p_1 p_2 \ldots p_{i-1} p_{i+1} p_i p_{i+2} \ldots p_n$, the permutation obtained from π by swapping p_i and p_{i+1}. Note that $\pi[i]$ is a child permutation if and only if $\pi = P(\pi[i])$ holds. A key point to find i with $\pi = P(\pi[i])$ is to maintain the *reverse point* $r(\pi)$, which is the minimum index of π such that $p_{r(\pi)} > p_{r(\pi)+1}$. For convenience, we set $r(\iota) = n$ for the identity permutation in S_n. Note that the subpermutation $p_1 p_2 \ldots p_{r(\pi)}$ is the maximal increasing prefix of π. If we know $r(\pi)$ of π, all child permutations are generated, as follows. For each $i = 1, 2, \ldots, r(\pi) - 1$, $\pi[i]$ is a child permutation of π since $\pi = P(\pi[i])$ holds. If $p_{r(\pi)} < p_{r(\pi)+2}$ holds, then $\pi[r(\pi)+1]$ is a child permutation of π. Otherwise, $\pi[r(\pi) + 1]$ is not a child permutation. For each $i = r(\pi), r(\pi) + 2, r(\pi) + 3, \ldots, n - 1$, $\pi[i]$

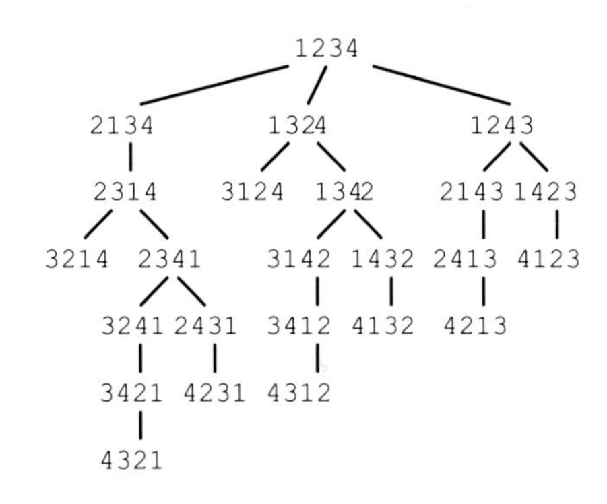

Permutation Enumeration, Fig. 3 The family tree T_4

Algorithm 4: REVERSE-SEARCH($\pi = p_1 p_2 \cdots p_n$)

1 Let $r(\pi)$ be a reverse point of π;
2 Output π;
3 **for** *each* $i = 1, 2, \ldots, r(\pi) - 1$ **do**
4 REVERSE-SEARCH($\pi[i]$)
5 **if** $r(\pi) \leq n - 2$ *and* $p_{r(\pi)} < p_{r(\pi)+2}$ **then** REVERSE-SEARCH($\pi[r(\pi) + 1]$)

is not a child permutation. Based on the above observation, we obtain the enumeration algorithm shown in Algorithm 4. To begin, Algorithm 4 is called with the identity permutation which is the root of the family tree.

By maintaining the reverse point of the current permutation in a traverse of the family tree, we can use a stack to generate each child permutation in O(1) time. To estimate the running time of the algorithm, note that the algorithm can traverse each edge of the family tree in O(1) time. However, the delay time of the algorithm is not bounded by O(1) time for the case that the next permutation is output after deep recursive calls without outputting any permutation. However, by applying the speed-up method proposed by Nakano and Uno [6], we have the following lemma.

Theorem 3 ([9]) *After constructing the root (the identity permutation) in $O(n)$ time, one can enumerate all the permutations in S_n by the reverse search method with a constant time delay. The required working space is $O(n)$.*

Recommended Reading

1. Arimura H, Uno T (2007) An efficient polynomial space and polynomial delay algorithm for enumeration of maximal motifs in a sequence. J Comb Optim 13:243–262
2. Avis D, Fukuda K (1996) Reverse search for enumeration. Discret Appl Math 65(1–3):21–46
3. Goldberg L (1992) Efficient algorithm for listing unlabeled graphs. J Algorithms 13:128–143
4. Goldberg L (1993) Polynomial space polynomial delay algorithms for listing families of graphs. In: Proceedings of the 25th annual ACM symposium on theory of computing, San Diego, pp 218–225
5. Johnson S (1963) Generation of permutations by adjacent transposition. Math Comput 17:282–285
6. Nakano S, Uno T (2004) Constant time generation of trees with specified diameter. In: Proceedings of the 30th workshop on graph-theoretic concepts in computer science (WG 2004), Bad Honnef. LNCS, vol 3353, pp 33–45
7. Savage C (1997) A survey of combinatorial gray codes. SIAM Rev 39(4):605–629
8. Sedgewick R (1977) Permutation generation methods. Comput Surv 9(2):137–164
9. Sekine K, Yamanaka K, Nakano S (2008) Enumeration of permutations. IEICE Trans Fundam Electron Commun Comput Sci J91-A(5):543–549 (in Japanese)
10. Steinhaus H (1964) One hundred problems in elementary mathematics. Basic Books, New York
11. Trotter H (1962) Perm (algorithm 115). Commun ACM 5(8):434–435

Phylogenetic Tree Construction from a Distance Matrix

Jesper Jansson
Laboratory of Mathematical Bioinformatics, Institute for Chemical Research, Kyoto University, Gokasho, Uji, Kyoto, Japan

Keywords

Additive; Dissimilarity matrix; Distance matrix; Phylogenetic tree; Phylogenetic reconstruction; Tree-realizable

Years and Authors of Summarized Original Work

1968; Boesch
1989; Hein
1989; Culberson, Rudnicki
2003; King, Zhang, Zhou

Problem Definition

Let n be a positive integer. A *distance matrix of order n* is a matrix D of size $(n \times n)$ which satisfies (1) $D_{i,j} > 0$ for all $i, j \in \{1, 2, \ldots, n\}$ with $i \neq j$; (2) $D_{i,j} = 0$ for all $i, j \in \{1, 2, \ldots, n\}$ with $i = j$; and (3) $D_{i,j} = D_{j,i}$ for all $i, j \in \{1, 2, \ldots, n\}$. In the literature, a distance matrix of order n is also called a *dissimilarity matrix of order* n.

Below, all trees are assumed to be unrooted and edge-weighted. For any tree $\mathcal{T}$, the *distance* between two nodes u and v in $\mathcal{T}$ is defined as the sum of the weights of all edges on the unique path in $\mathcal{T}$ between u and v and is denoted by $d^{\mathcal{T}}_{u,v}$. A tree $\mathcal{T}$ is said to *realize* a given distance matrix D of order n if and only if it holds that $\{1, 2, \ldots, n\}$ is a subset of the nodes of $\mathcal{T}$ and $d^{\mathcal{T}}_{i,j} = D_{i,j}$ for all $i, j \in \{1, 2, \ldots, n\}$. Finally, a distance matrix D is called *additive* or *tree-realizable* if and only if there exists a tree which realizes D. See Fig. 1 for an example.

Problem 1 (The Phylogenetic Tree from Distance Matrix Problem)
INPUT: A distance matrix D of order n
OUTPUT: A tree which realizes D and has the smallest possible number of nodes, if D is additive, otherwise *null*

In the time complexities listed below, the time needed to input all of D is not included. Instead, $O(1)$ is charged to the running time whenever an algorithm requests to know the value of any specified entry of D.

Key Results

Several authors have independently shown how to solve the Phylogenetic Tree from Distance Matrix Problem in $O(n^2)$ time. (See [5] for a short survey of older algorithms which do not run in $O(n^2)$ time.)

Theorem 1 ([2, 4, 5, 7, 14]) *There exists an algorithm which solves the Phylogenetic Tree from Distance Matrix Problem in $O(n^2)$ time.*

Although the various existing algorithms are different, it can be proved that:

Theorem 2 ([8, 14]) *For any given distance matrix, the solution to the Phylogenetic Tree from Distance Matrix Problem is unique.*

Furthermore, the algorithms referred to in Theorem 1 have optimal running time since any algorithm for the Phylogenetic Tree from Distance Matrix Problem must in the worst case query all $\Omega(n^2)$ entries of D to make sure that D is additive. However, if it is known in advance that the input distance matrix is additive, then the time complexity improves as follows.

Theorem 3 ([9, 12]) *There exists an algorithm which solves the Phylogenetic Tree from Distance Matrix Problem restricted to additive distance matrices in $O(kn \log_k n)$ time, where k is the maximum degree of the tree that realizes the input distance matrix.*

The algorithm of Hein [9] starts with a tree containing just two nodes and then successively inserts each node i into the tree by repeatedly choosing a pair of existing nodes and computing where on the path between them that i should be attached, until i's position has been determined. The same basic technique is used in the $O(n^2)$-time algorithm of Waterman et al. [14] referenced to by Theorem 1 above, but the algorithm of Hein selects paths which are more efficient at discriminating between the possible positions for i. According to [12], the running time of Hein's algorithm is $O(kn \log_k n)$.

A lower bound that implies the optimality of Theorem 3 is given by the next theorem.

Theorem 4 ([10]) *The Phylogenetic Tree from Distance Matrix Problem restricted to additive distance matrices requires $\Omega(kn \log_k n)$ queries to the distance matrix D, where k is the maximum degree of the tree that realizes D, even if restricted to trees in which all edge weights are equal to* 1.

Independently of [9], Culberson and Rudnicki [5] presented an algorithm for the Phylogenetic Tree from Distance Matrix Problem and claimed it to have $O(kn \log_k n)$ time complexity when restricted to additive distance

a

$$D = \begin{bmatrix} 0 & 18 & 23 & 25 & 6 \\ 18 & 0 & 15 & 17 & 12 \\ 23 & 15 & 0 & 16 & 17 \\ 25 & 17 & 16 & 0 & 19 \\ 6 & 12 & 17 & 19 & 0 \end{bmatrix}$$

b

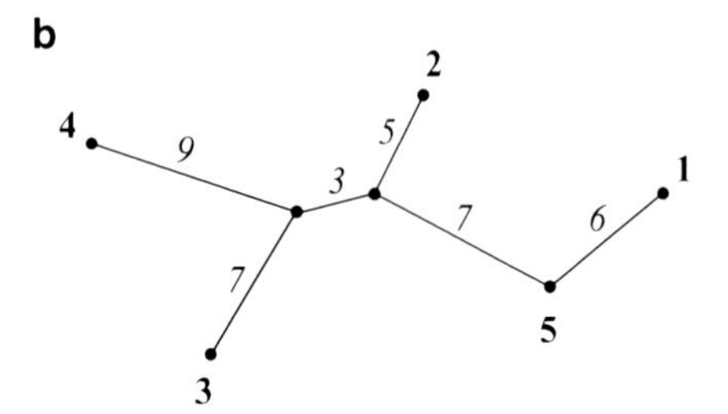

Phylogenetic Tree Construction from a Distance Matrix, Fig. 1 (**a**) An additive distance matrix D of order 5. (**b**) A tree $\mathcal{T}$ which realizes D. Here, $\{1, 2, \ldots, 5\}$ forms a subset of the nodes of $\mathcal{T}$

matrices and trees in which all edge weights are equal to 1. As pointed out by Reyzin and Srivastava [12], the algorithm actually runs in $\Theta(n^{3/2}\sqrt{k})$ time. See [12] for a counterexample to [5] and a correct analysis. On the positive side, the following special case is solvable in linear time by the Culberson-Rudnicki algorithm:

Theorem 5 ([5]) *There exists an* $O(n)$*-time algorithm which solves the Phylogenetic Tree from Distance Matrix Problem restricted to additive distance matrices for which the realizing tree contains two leaves only and has all edge weights equal to* 1.

Applications

The main application of the Phylogenetic Tree from Distance Matrix Problem is in the construction of a tree (a so-called *phylogenetic tree*) that represents evolutionary relationships among a set of studied objects (e.g., species or other taxa, populations, proteins, genes, etc.). Here, it is assumed that the objects are indeed related according to a treelike branching pattern caused by an evolutionary process and that their true pairwise evolutionary distances are proportional to the measured pairwise dissimilarities. See, e.g., [1, 6, 7, 14] for examples and many references as well as discussions on how to estimate pairwise dissimilarities based on biological data. Other applications of the Phylogenetic Tree from Distance Matrix Problem can be found in psychology, for example, to describe semantic memory organization [1], in comparative linguistics to infer the evolutionary history of a set of languages [11], or in the study of the filiation of manuscripts to trace how manuscript copies of a text (whose original version may have been lost) have evolved in order to identify discrepancies among them or to reconstruct the original text [1, 3, 13].

In general, real data seldom forms additive distance matrices [14]. Therefore, in practice, researchers consider optimization versions of the Phylogenetic Tree from Distance Matrix Problem which look for a tree that "almost" realizes D. Many alternative definitions of "almost" have been proposed, and numerous heuristics and approximation algorithms have been developed. A comprehensive description of some of the most popular methods for phylogenetic reconstruction from a non-additive distance matrix such as *Neighbor-joining* [16] as well as more background information can be found in, e.g., Chapter 11 of [6]. See also [1] and [15] and the references therein.

Cross-References

- ▶ Distance-Based Phylogeny Reconstruction (Fast-Converging)
- ▶ Distance-Based Phylogeny Reconstruction: Safety and Edge Radius

Acknowledgments JJ was funded by the Hakubi Project at Kyoto University and KAKENHI grant number 26330014.

Recommended Reading

1. Abdi H (1990) Additive-tree representations. In: Dress A, von Haeseler A (eds) Trees and hierarchical structures: proceedings of a conference held at Bielefeld, FRG, Oct 5–9th, 1987. Lecture Notes in Biomathematics, vol 84. Springer, Berlin/Heidelberg, pp 43–59
2. Batagelj V, Pisanski T, Simões-Pereira JMS (1990) An algorithm for tree-realizability of distance matrices. Int J Comput Math 34(3–4):171–176
3. Bennett CH, Li M, Ma B (2003) Chain letters and evolutionary histories. Sci Am 288(6):76–81
4. Boesch FT (1968) Properties of the distance matrix of a tree. Quart Appl Math 26:607–609
5. Culberson JC, Rudnicki P (1989) A fast algorithm for constructing trees from distance matrices. Inf Process Lett 30(4):215–220
6. Felsenstein J (2004) Inferring phylogenies. Sinauer Associates, Sunderland
7. Gusfield DM (1997) Algorithms on strings, trees, and sequences. Cambridge University Press, New York
8. Hakimi SL, Yau SS (1964) Distance matrix of a graph and its realizability. Quart Appl Math 22:305–317
9. Hein J (1989) An optimal algorithm to reconstruct trees from additive distance data. Bull Math Biol 51(5):597–603
10. King V, Zhang L, Zhou Y (2003) On the complexity of distance-based evolutionary tree construction. In: Proceedings of the 14th annual ACM-SIAM symposium on discrete algorithms (SODA 2003), Baltimore, pp 444–453
11. Nakhleh L, Warnow T, Ringe D, Evans SN (2005) A comparison of phylogenetic reconstruction methods on an Indo-European dataset. Trans Philol Soc 103(2):171–192
12. Reyzin L, Srivastava N (2007) On the longest path algorithm for reconstructing trees from distance matrices. Inf Process Lett 101(3):98–100
13. The Canterbury Tales Project. University of Birmingham, Brigham Young University, University of Münster, New York University, Virginia Tech, and Keio University. Website: http://www.petermwrobinson.me.uk/canterburytalesproject.com/
14. Waterman MS, Smith TF, Singh M, Beyer WA (1977) Additive evolutionary trees. J Theor Biol 64(2):199–213
15. Wu BY, Chao K-M, Tang CY (1999) Approximation and exact algorithms for constructing minimum ultrametric trees from distance matrices. J Combin Optim 3(2–3):199–211
16. Saitou N, Nei M (1987) The Neighbor-joining Method: A New Method for Reconstructing Phylogenetic Trees. Mol Biol Evol 4(4):406–425

Planar Directed k-VERTEX-DISJOINT PATHS Problem

Marcin Pilipczuk
Institute of Informatics, University of Bergen, Bergen, Norway
Institute of Informatics, University of Warsaw, Warsaw, Poland

Keywords

Directed graphs; Fixed-parameter tractability; Graph decomposition; Planar graphs; VERTEX-DISJOINT PATHS problem

Years and Authors of Summarized Original Work

1994; Schrijver
2013; Cygan, Marx, Pilipczuk, Pilipczuk

Problem Definition

In the classic VERTEX-DISJOINT PATHS problem, the input consists of an n-vertex graph G and k pairs of terminals $(s_i, t_i)_{i=1}^{k}$, and the question is whether there exist pairwise VERTEX-DISJOINT PATHS $P_1, P_2, \ldots, P_k$ such that for every $1 \leq i \leq k$, the path P_i starts in s_i and ends in t_i. In this entry we are interested in the complexity of this problem restricted to planar directed graphs.

Key Results

An algorithm for the VERTEX-DISJOINT PATHS problem in undirected graphs with running time $f(k)n^3$ for some function f is one of the key ingredients of the minor testing algorithm of Robertson and Seymour [8]. The approach can be summarized as follows: either the input graph has treewidth bounded by a function of k, in which case we can apply standard dynamic

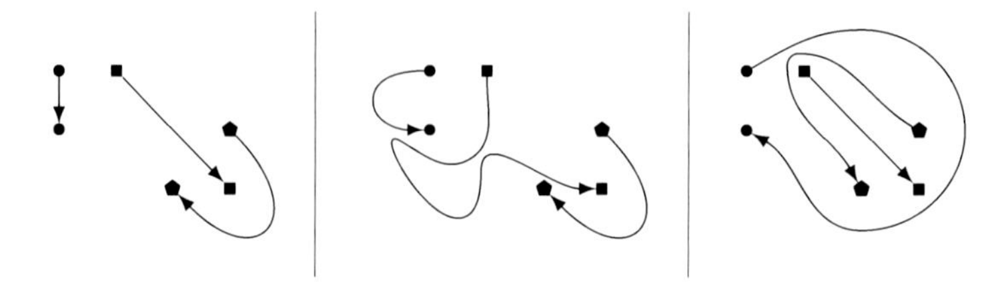

Planar Directed k-VERTEX-DISJOINT PATHS Problem, Fig. 1 Different homotopy classes of a solution: in the first two figures, the solutions are of the same class, whereas on the third figure the homotopy class is different

programming techniques, or, by the Excluded Grid Theorem, the input graph contains a large grid minor. In the second case, we may deduce that a middle vertex of the grid minor is irrelevant for the problem and can be discarded.

The original proof of the irrelevancy of a middle vertex of a grid minor by Robertson and Seymour [8] is not only highly involved but also leads to an extremely large dependency on k in the running time bound. A more recent algorithm by Kawarabayashi and Wollan [6] improves upon the original approach in both these aspects, but is still very complex.

As already observed by Robertson and Seymour in [7], a situation becomes dramatically simpler if we restrict ourselves to planar graphs. A short self-contained argument of irrelevancy of the middle vertex of a $c^k \times c^k$ grid minor, for some universal constant c, is due to Adler, Kolliopoulos, Krause, Lokshtanov, Saurabh, and Thilikos [1]. It is worth noting that the exponential dependency on k in the irrelevant vertex argument is necessary. The intuitive reason why planarity greatly helps in the VERTEX-DISJOINT PATHS problem is that, on the plane, the solution paths need to correspond to noncrossing curves and one path serves as a separator for other paths. This allows us to use a wide variety of topological arguments.

In directed graphs, the VERTEX-DISJOINT PATHS problem is already NP-hard for two paths ($k = 2$) [3], so we cannot hope for similar results. However, it turns out that in the directed case the planarity assumption is very useful, too. More than 20 years ago, Schrijver showed that the VERTEX-DISJOINT PATHS problem in n-vertex planar directed graphs can be solved in time $n^{\mathcal{O}(k)}$ [9]. Recently, Cygan, Marx, Pilipczuk, and Pilipczuk presented a fixed-parameter algorithm for this problem, running in time $2^{2^{\mathcal{O}(k^2)}} n^{\mathcal{O}(1)}$ [2].

Key Techniques for Planar Directed Graphs

The Schrijver's Algorithm

The approach of Schrijver [9] can be summarized as follows. The main observation is that there are $n^{\mathcal{O}(k)}$ homotopy types of the solution, where two different solutions are considered homotopical if the paths of one solution can be "shifted" (modified by a homotopy) to obtain the second solution, without crossing any face that contains a terminal (without loss of generality, we may assume that all terminals are of degree one, and the notion of a face containing a terminal is well defined). See also Fig. 1 for an illustration. A second ingredient of Schrijver's approach is a polynomial-time algorithm that essentially checks if there exists a solution in one homotopy class. (It should be noted that this statement is a significant simplification, as the Schrijver's algorithm operates on the notions of (co)homologies and in fact searches for a solution in a significant superset of one homotopy class, but that is sufficient for our needs.)

The Fixed-Parameter Algorithm

The first step in the fixed-parameter algorithm of [2] is an appropriate irrelevant vertex rule for the problem. Unfortunately, for directed graphs there is no Excluded Grid Theorem which as convenient as it is in the undirected case. Al-

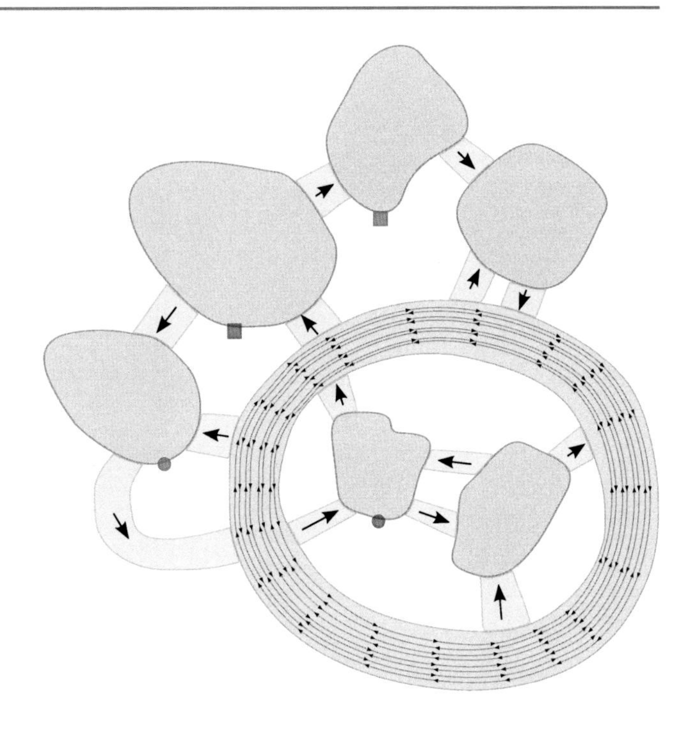

Planar Directed k-VERTEX-DISJOINT PATHS Problem, Fig. 2 An example of a decomposition with six disk components and a single ring component

though the conjecture of Johnson, Robertson, Seymour, and Thomas [4] about the connections between directed treewidth and directed grid minors has been recently proven by Kawarabayashi and Kreutzer for graph excluding a fixed minor [5], neither directed treewidth seems well suited for dynamic programming algorithm for the VERTEX-DISJOINT PATHS problem [10], nor it is clear whether a directed grid minor can be useful for an irrelevant vertex argument. In [2], it is proven that a family of sufficiently many concentric cycles with alternating direction, without any terminal enclosed by the outermost cycle, is sufficient to make an irrelevant vertex argument.

In the light of such an irrelevant vertex rule, the next question is: what is the structure of a graph without many concentric cycles with alternating directions? The answer provided in [2] can be informally stated as follows: such a graph can be decomposed into a bounded (in k) number of disk and ring components, connected by a bounded number of *bundles*, where every bundle is a set of edges in one direction that lie close to each other on the plane (see Fig. 2).

Unfortunately, it is not easy to make algorithmic use of such a decomposition; this should be contrasted with the undirected case, where bounded treewidth immediately yields efficient algorithms via standard dynamic programming approach. The approach taken in [2] is to make use of Schrijver's approach and use the decomposition to enumerate only $2^{2^{\mathcal{O}(k^2)}} n^{\mathcal{O}(1)}$ "reasonable" homotopy types of the solution.

Recall that the $n^{\mathcal{O}(k)}$ term in the time complexity of Schrijver's algorithm comes from the number of homotopy classes. It is quite easy to see that this bound cannot be improved: each of k solution paths can "wind" arbitrary number of times in some part of the graph, leading to different homotopy classes. This is also the case in the decomposition, as can be seen on Fig. 3. To deal with this issue, an involved technical argumentation is developed in [2] to show that for any such place in the graph as on Fig. 3, there is some "canonical" number of turns, and we may assume that the solution will take approximately this number of turns, up to an additive $f(k)$ factor, for some computable function f. This leads to an FPT bound on the number of "reasonable" homotopy classes to consider and, consequently, to an FPT running time bound.

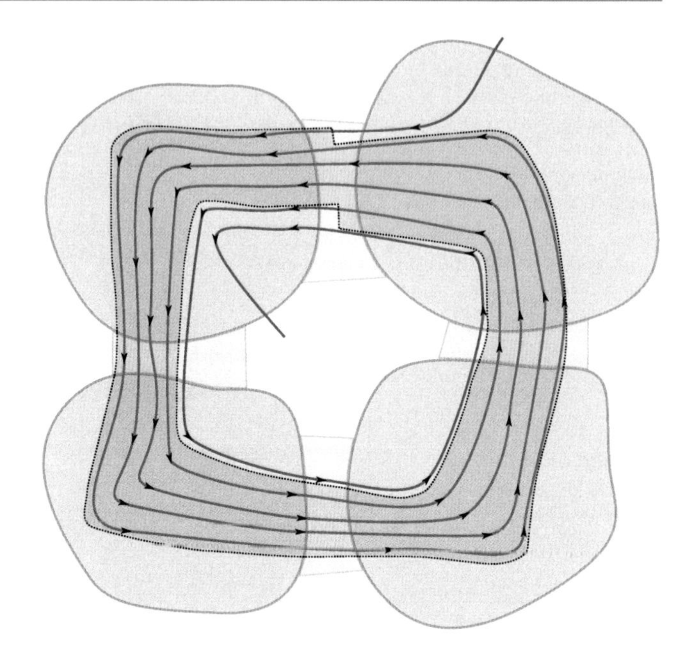

Planar Directed k-Vertex-Disjoint Paths Problem, Fig. 3 An example of a solution path winding many times between four disk components

Cross-References

▶ Bidimensionality

Recommended Reading

1. Adler I, Kolliopoulos SG, Krause PK, Lokshtanov D, Saurabh S, Thilikos DM (2011) Tight bounds for linkages in planar graphs. In: Aceto L, Henzinger M, Sgall J (eds) ICALP (1), Zurich. Lecture notes in computer science, vol 6755, pp 110–121. Springer
2. Cygan M, Marx D, Pilipczuk M, Pilipczuk M (2013) The planar directed k-Vertex-Disjoint Paths problem is fixed-parameter tractable. In: FOCS, Berkeley. IEEE Computer Society, pp 197–206
3. Fortune S, Hopcroft J, Wyllie J (1980) The directed subgraph homeomorphism problem. Theor Comput Sci 10(2):111–121
4. Johnson T, Robertson N, Seymour PD, Thomas R (2001) Directed tree-width. J Comb Theory Ser B 82(1):138–154
5. Kawarabayashi K, Kreutzer S (2014) An excluded grid theorem for digraphs with forbidden minors. In: Chekuri C (ed) SODA, Portland. SIAM, pp 72–81
6. Kawarabayashi K, Wollan P (2010) A shorter proof of the graph minor algorithm: the unique linkage theorem. In: Schulman LJ (ed) STOC, Cambridge. ACM, pp 687–694
7. Robertson N, Seymour PD (1988) Graph minors. VII. Disjoint paths on a surface. J Comb Theory Ser B 45(2):212–254
8. Robertson N, Seymour PD (1995) Graph minors. XIII. The disjoint paths problem. J Comb Theory Ser B 63(1):65–110
9. Schrijver A (1994) Finding k disjoint paths in a directed planar graph. SIAM J Comput 23(4):780–788
10. Slivkins A (2010) Parameterized tractability of edge-disjoint paths on directed acyclic graphs. SIAM J Discret Math 24(1):146–157

Planar Geometric Spanners

Joachim Gudmundsson[1,2], Giri Narasimhan[3,5], and Michiel Smid[4]

[1]DMiST, National ICT Australia Ltd, Alexandria, Australia

[2]School of Information Technologies, University of Sydney, Sydney, NSW, Australia

[3]Department of Computer Science, Florida International University, Miami, FL, USA

[4]School of Computer Science, Carleton University, Ottawa, ON, Canada

[5]School of Computing and Information Sciences, Florida International University, Miami, FL, USA

Keywords

Dilation; Detour; Geometric network

Years and Authors of Summarized Original Work

2005; Bose, Smid, Gudmundsson

Problem Definition

Let S be a set of n points in the plane and let G be an undirected graph with vertex set S, in which each edge (u, v) has a weight, which is equal to the Euclidean distance $|uv|$ between the points u and v. For any two points p and q in S, their shortest-path distance in G is denoted by $\delta_G(p, q)$. If $t \geq 1$ is a real number, then G is a *t-spanner* for S if $\delta_G(p, q) \leq t|pq|$ for any two points p and q in S. Thus, if t is close to 1, then the graph G contains close approximations to the $\binom{n}{2}$ Euclidean distances determined by the pairs of points in S. If, additionally, G consists of $O(n)$ edges, then this graph can be considered a sparse approximation to the complete graph on S. The smallest value of t for which G is a t-spanner is called the *stretch factor* (or *dilation*) of G. For a comprehensive overview of geometric spanners, see the book by Narasimhan and Smid [16].

Assume that each edge (u, v) of G is embedded as the straight-line segment between the points u and v. The graph G is said to be *plane* if its edges intersect only at their common vertices.

In this entry, the following two problems are considered:

Problem 1 Determine the smallest real number $t > 1$ for which the following is true: For every set S of n points in the plane, there exists a plane graph with vertex set S, which is a t-spanner for S. Moreover, design an efficient algorithm that constructs such a plane t-spanner.

Problem 2 Determine the smallest positive integer D for which the following is true: There exists a constant t, such that for every set S of n points in the plane, there exists a plane graph with vertex set S and maximum degree at most D, which is a t-spanner for S. Moreover, design an efficient algorithm that constructs such a plane t-spanner.

Key Results

Let S be a finite set of points in the plane that is in *general position*, i.e., no three points of S are on a line and no four points of S are on a circle. The *Delaunay triangulation* of S is the plane graph with vertex set S, in which (u, v) is an edge if and only if there exists a circle through u and v that does not contain any point of S in its interior. (Since S is in general position, this graph is a triangulation.) The Delaunay triangulation of a set of n points in the plane can be constructed in $O(n \log n)$ time. Dobkin, Friedman and Supowit [10] were the first to show that the stretch factor of the Delaunay triangulation is bounded by a constant: They proved that the Delaunay triangulation is a t-spanner for $t = \pi(1 + \sqrt{5})/2$. The currently best known upper bound on the stretch factor of this graph is due to Keil and Gutwin [12]:

Theorem 1 *Let S be a finite set of points in the plane. The Delaunay triangulation of S is a t-spanner for S, for $t = 4\pi\sqrt{3}/9$.*

A slightly stronger result was proved by Bose et al. [3]. They proved that for any two points p and q in S, the Delaunay triangulation contains a path between p and q, whose length is at most $(4\pi\sqrt{3}/9)|pq|$ and all edges on this path have length at most $|pq|$.

Levcopoulos and Lingas [14] generalized the result of Theorem 1: Assume that the Delaunay triangulation of the set S is given. Then, for any real number $r > 0$, a plane graph G with vertex set S can be constructed in $O(n)$ time, such that G is a t-spanner for S, where $t = (1+1/r)4\pi\sqrt{3}/9$, and the total length of all edges in G is at most $2r + 1$ times the weight of a minimum spanning tree of S.

The Delaunay triangulation can alternatively be defined to be the dual of the *Voronoi diagram* of the set S. By considering the Voronoi diagram for a metric other than the Euclidean metric, a corresponding Delaunay triangulation is obtained. Chew [7] has shown that the Delaunay triangulation based on the Manhattan-metric is a $\sqrt{10}$-spanner (in this spanner, path-lengths are measured in the Euclidean metric). The currently best result for Problem 1 is due to Chew [8]:

Theorem 2 *Let S be a finite set of points in the plane, and consider the Delaunay triangulation of S that is based on the convex distance function defined by an equilateral triangle. This plane graph is a 2-spanner for S (where path-lengths are measured in the Euclidean metric).*

Das and Joseph [9] have generalized the result of Theorem 1 in the following way (refer to Fig. 1). Let G be a plane graph with vertex set S and let α be a real number with $0 < \alpha < \pi/2$. For any edge e of G, let Δ_1 and Δ_2 be the two isosceles triangles with base e and base angle α. The edge e is said to satisfy the *α-diamond property*, if at least one of the triangles Δ_1 and Δ_2 does not contain any point of S in its interior. The plane graph G is said to satisfy the *α-diamond property*, if every edge e of G satisfies this property. For a real number $d \geq 1$, G satisfies the *d-good polygon property*, if for every face f of G, and for every two vertices p and q on the boundary of f, such that the line segment joining them is completely inside f, the shortest path between p and q along the boundary of f has length at most $d|pq|$. Das and Joseph [9] proved that any plane graph satisfying both the α-diamond property and the d-good polygon property is a t-spanner, for some real number t that depends only on α and d. A slight improvement on the value of t was obtained by Lee [13]:

Theorem 3 *Let $\alpha \in (0, \pi/2)$ and $d \geq 1$ be real numbers, and let G be a plane graph that satisfies the α-diamond property and the d-good polygon property. Then, G is a t-spanner for the vertex set of G, where*

$$t = \frac{8(\pi - \alpha)^2 d}{\alpha^2 \sin^2(\alpha/4)}.$$

To give some examples, it is not difficult to show that the Delaunay triangulation satisfies the α-diamond property with $\alpha = \pi/4$. Drysdale et al. [11] have shown that the minimum weight triangulation satisfies the α-diamond property with $\alpha = \pi/4.6$. Finally, Lee [13] has shown that the greedy triangulation satisfies the α-diamond property with $\alpha = \pi/6$. Of course, any triangulation satisfies the d-good polygon property with $d = 1$.

Now consider Problem 2, that is, the problem of constructing plane spanners whose maximum degree is small. The first result for this problem is due to Bose et al. [2]. They proved that the Delaunay triangulation of any finite point set contains a subgraph of maximum degree at most 27, which is a t-spanner (for some constant t). Li and Wang [15] improved this result, by showing that the Delaunay triangulation contains a t-spanner of maximum degree at most 23. Given the Delaunay triangulation, the subgraphs in [2, 15] can be constructed in $O(n)$ time. The currently best result for Problem 2 is by Bose et al. [6]:

Theorem 4 *Let S be a set of n points in the plane. The Delaunay triangulation of S contains a subgraph of maximum degree at most 17, which is a t-spanner for S, for some constant t. Given the Delaunay triangulation of S, this subgraph can be constructed in $O(n)$ time.*

In fact, the result in [6] is more general:

Theorem 5 *Let S be a set of n points in the plane, let $\alpha \in (0, \pi/2)$ be a real number, and let G be a triangulation of S that satisfies the α-diamond property. Then, G contains a subgraph of maximum degree at most $14 + \lceil 2\pi/\alpha \rceil$, which is a t-spanner for S, where t depends only on α. Given the triangulation G, this subgraph can be constructed in $O(n)$ time.*

Applications

Plane spanners have applications in on-line path-finding and routing problems that arise in, for example, geographic information systems and communication networks. In these application areas, the complete environment is not known, and routing has to be done based only on the source, the destination, and the neighborhood of the current position. Bose and Morin [4, 5] have shown that, in this model, good routing strategies

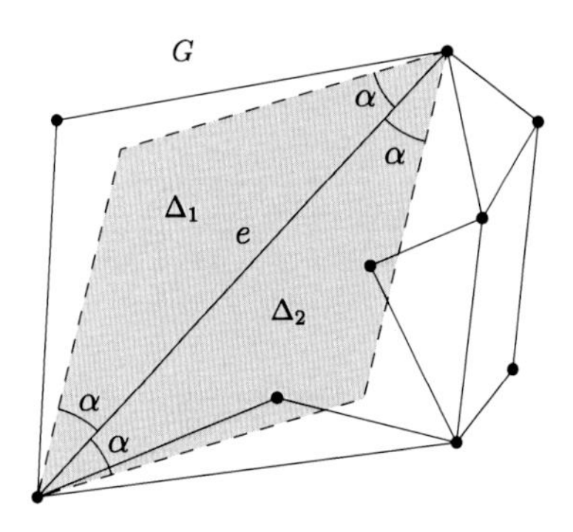

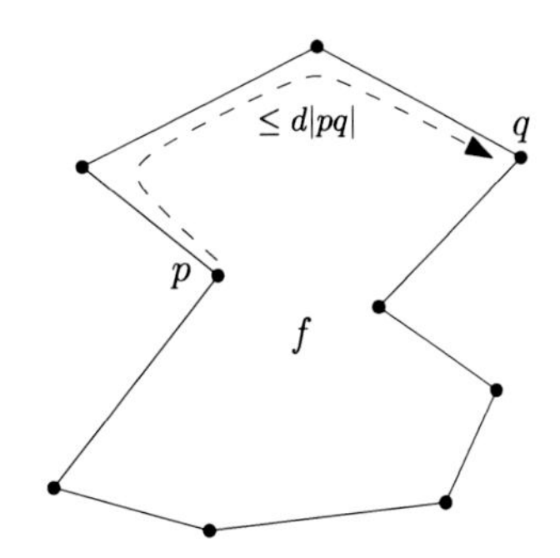

Planar Geometric Spanners, Fig. 1 On the left, the α-diamond property is illustrated. At least one of the triangles Δ_1 and Δ_2 does not contain any point of S in its interior. On the right, the d-good polygon property is illustrated. p and q are two vertices on the same face f which can see each other. At least one of the two paths between p and q along the boundary of f has length at most $d|pq|$

exist for plane graphs, such as the Delaunay triangulation and graphs that satisfy both the α-diamond property and the d-good polygon property. These strategies are competitive, in the sense that the paths computed have lengths that are within a constant factor of the Euclidean distance between the source and destination. Moreover, these routing strategies use only a limited amount of memory.

Open Problems

None of the results for Problems 1 and 2 that are mentioned in section "Key Results" seem to be optimal. The following problems are open:

1. Determine the smallest real number t, such that the Delaunay triangulation of any finite set of points in the plane is a t-spanner. It is widely believed that $t = \pi/2$. By Theorem 1, $t \leq 4\pi\sqrt{3}/9$.
2. Determine the smallest real number t, such that a plane t-spanner exists for any finite set of points in the plane. By Theorem 2, $t \leq 2$. By taking S to be the set of four vertices of a square, it follows that t must be at least $\sqrt{2}$.
3. Determine the smallest integer D, such that the Delaunay triangulation of any finite set of points in the plane contains a t-spanner (for some constant t) of maximum degree at most D. By Theorem 4, $D \leq 17$. It follows from results in Aronov et al. [1] that the value of D must be at least 3.
4. Determine the smallest integer D, such that a plane t-spanner (for some constant t) of maximum degree at most D exists for any finite set of points in the plane. By Theorem 4 and results in [1], $3 \leq D \leq 17$.

Cross-References

- ▶ Applications of Geometric Spanner Networks
- ▶ Dilation of Geometric Networks
- ▶ Geometric Spanners
- ▶ Sparse Graph Spanners

Recommended Reading

1. Aronov B, de Berg M, Cheong O, Gudmundsson J, Haverkort H, Vigneron A (2005) Sparse geometric graphs with small dilation. In: Proceedings of the 16th international symposium on algorithms and computation. Lecture notes in computer science, vol 3827. Springer, Berlin, pp 50–59
2. Bose P, Gudmundsson J, Smid M (2005) Constructing plane spanners of bounded degree and low weight. Algorithmica 42:249–264
3. Bose P, Maheshwari A, Narasimhan G, Smid M, Zeh N (2004) Approximating geometric bottleneck shortest paths. Comput Geom Theory Appl 29:233–249
4. Bose P, Morin P (2004a) Competitive online routing in geometric graphs. Theor Comput Sci 324:273–288
5. Bose P, Morin P (2004b) Online routing in triangulations. SIAM J Comput 33:937–951
6. Bose P, Smid M, Xu D (2006) Diamond triangulations contain spanners of bounded degree. In: Pro-

P

ceedings of the 17th international symposium on algorithms and computation. Lecture notes in computer science, vol 4288. Springer, Berlin, pp 173–182
7. Chew LP (1986) There is a planar graph almost as good as the complete graph. In: Proceedings of the 2nd ACM symposium on computational geometry, pp 169–177
8. Chew LP (1989) There are planar graphs almost as good as the complete graph. J Comput Syst Sci 39:205–219
9. Das G, Joseph D (1989) Which triangulations approximate the complete graph? In: Proceedings of the international symposium on optimal algorithms. Lecture notes in computer science, vol 401. Springer, Berlin, pp 168–192
10. Dobkin DP, Friedman SJ, Supowit KJ (1990) Delaunay graphs are almost as good as complete graphs. Discret Comput Geom 5:399–407
11. Drysdale RL, McElfresh S, Snoeyink JS (2001) On exclusion regions for optimal triangulations. Discret Appl Math 109:49–65
12. Keil JM, Gutwin CA (1992) Classes of graphs which approximate the complete Euclidean graph. Discret Comput Geom 7:13–28
13. Lee AW (2004) Diamonds are a plane graph's best friend. Master's thesis, School of Computer Science, Carleton University, Ottawa
14. Levcopoulos C, Lingas A (1992) There are planar graphs almost as good as the complete graphs and almost as cheap as minimums panning trees. Algorithmica 8:251–256
15. Li X-Y, Wang Y (2004) Efficient construction of low weighted bounded degree planar spanner. Int J Comput Geom Appl 14:69–84
16. Narasimhan G, Smid M (2007) Geometric spanner networks. Cambridge University Press, Cambridge

Planar Maximum Flow – Multiple-Source Multiple-Sink Maximum Flow in Directed Planar Graphs

Glencora Borradaile
Department of Computer Science, Brown University, Providence, RI, USA
School of Electrical Engineering and Computer Science, Oregon State University, Corvallis, OR, USA

Keywords

Maximum flow; Planar graphs

Years and Authors of Summarized Original Work

2011; Borradaile, Klein, Mozes, Nussbaum, Wulff-Nilsen

Problem Definition

Given a directed, planar graph $G = (V, E)$ with arc capacities $c : E \to \Re^+$, a subset S of source vertices, and a subset T of sink vertices, the goal is to find a maximum flow from the source vertices to the sink vertices:

$$\max \sum_{su:s\in S, su\in E} f_{su}$$

$$\text{s.t.} \sum_{uv:uv\in E} f_{uv} - \sum_{vw:vw\in E} f_{vw} = 0$$

$$\forall v \in V \setminus (S \cup T) \qquad (1)$$

$$0 \le f_e \le c_e \qquad \forall e \in E \qquad (2)$$

Key Results

In general (i.e., nonplanar) graphs, multiple sources and sinks can be reduced to the single-source, single-sink case by introducing an artificial source and sink and connecting them to all the sources and sinks, respectively, but this reduction does not preserve planarity. Using Orlin's algorithm for sparse graphs [21] leads to a running time of $O(n^2/\log n)$. For integer capacities less than U, one could instead use the algorithm of Goldberg and Rao [9], which leads to a running time of $O(n^{1.5} \log n \log U)$.

Maximum flow in planar graphs with multiple sources and sinks was first studied by Miller and Naor [19]. They gave a divide-and-conquer algorithm for the case where all the sinks and the sources are on the boundary of a single face. Plugging in the linear-time shortest-path algorithm of Henzinger et al. [12] yields a running time of $O(n \log n)$. Borradaile and Harutyunyan have given an iterative algorithm with the same running time [2]. Miller and Naor also gave an algorithm for the case where the sources and the sinks reside on the boundaries of k different faces. Using the $O(n \log n)$ time single-source, single-sink maximum flow algorithm of Borradaile and Klein [3] yields a running time of $O(k^2 n \log^2 n)$. Miller and Naor show that, when it is known how much of the commodity is produced/consumed at each source and each sink, finding a consistent routing of flow that respects arc capacities can be reduced to negative-length shortest paths [19], which can be solved in planar graphs in $O(n \log^2 n / \log \log n)$ time [20].

Near-Linear Time Algorithm

Borradaile et al. gave the first $O(n \text{poly} \log n)$ time algorithm for the multiple-source, multiple-sink maximum flow problem in directed planar graphs. The approach uses pseudoflows [10, 14] (flows which may violate the balance constraints (1) in a limited way) and a divide-and-conquer scheme influenced by that of Johnson and Venkatesan [15] and that of Miller and Naor [19], using the separators introduced by Miller: a (triangulated) planar graph G can be separated by a simple cycle C of $O(\sqrt{n})$ vertices [18].

In each of the two subgraphs, a more general problem is solved in which, after the two recursive calls have been executed, within each of the two subgraphs, there is no residual path from any source to any sink nor from any source to C or from C to any sink. Then, since C is a separator, there is no residual path from any source to any sink in G, but, however, the balance constraints (1) may not be satisfied for vertices in C. The flow is then balanced among the vertices in C by augmenting the flow so that there is no residual path in G from a vertex with positive inflow to a vertex with positive outflow. The resulting flow can then be turned into a maximum flow in linear time.

The core of the algorithm is this final balancing procedure which involves a series of $|C| - 1$ max-flow computations in G. Since $|C|$ is $O(\sqrt{n})$, the challenge is carrying out all these max-flow computations in near-linear time. The procedure uses a succinct representation to keep track of the changes to the pseudoflow without explicitly storing the changes. The representation relies on the relationship between circulations in G and shortest paths in the dual, and the computations make use of an adaptation of Fakcharoenphol and Rao's efficient implementation of Dijkstra's algorithm [7]. The resulting running time to balance the flow is $O(n \log^2 n)$ time for an overall running time of $O(n \log^3 n)$ time for the original multiple-source, multiple-sink maximum flow problem.

Applications

Multiple-source, multiple-sink min-cut arises in several computer vision problems including image segmentation (or binary labeling) [11]. For the case of more than two labels, there is a powerful and effective heuristic [5] using a very large-neighborhood [1] local search; the inner loop consists of solving the two-label case.

Maximum matching in a bipartite planar graph reduces to multiple-source, multiple-sink maximum flow. Multiple-source, multiple-sink maximum flow can also be used for finding *orthogonal drawings of planar graphs with a minimum number of bends* [6] and *uniformly monotone subdivisions of polygons* [23].

P

Recommended Reading

1. Ahuja R, Ergun O, Orlin J, Punnen A (2002) A survey of very large scale neighborhood search techniques. Discret Appl Math 23:75–102. doi:10.1016/S0166-218X(01)00338-9
2. Borradaile G, Harutyunyan A (2013) Boundary-to-boundary flows in planar graphs. In: Proceedings of IWOCA, Rouen, pp 67–80. doi:10.1007/978-3-642-45278-9_7
3. Borradaile G, Klein PN (2009) An $O(n \log n)$ algorithm for maximum st-flow in a directed planar graph. J ACM 56(2):9:1–9:30. doi:10.1145/1502793.1502798, http://doi.acm.org/10.1145/1502793.1502798
4. Boykov Y, Kolmogorov V (2004) An experimental comparison of min-cut/max-flow algorithms for en-

ergy minimization in vision. IEEE Trans Pattern Anal Mach Intell 26(9):1124–1137. doi:http://dx.doi.org/10.1109/TPAMI.2004.60
5. Boykov Y, Veksler O, Zabih R (2001) Efficient approximate energy minimization via graph cuts. IEEE Trans Pattern Anal Mach Intell 20(12):1222–1239. doi:10.1109/TPAMI.2003.1233908
6. Cornelsen S, Karrenbauer A (2012) Accelerated bend minimization. J Graph Algorithm Appl 16(3):635–650
7. Fakcharoenphol J, Rao S (2006) Planar graphs, negative weight edges, shortest paths, and near linear time. J Comput Syst Sci 72(5):868–889. doi:http://dx.doi.org/10.1016/j.jcss.2005.05.007
8. Geman S, Geman D (1984) Stochastic relaxation, Gibbs distributions, and the Bayesian relation of images. IEEE Trans Pattern Anal Mach Intell 6(6):721–742
9. Goldberg A, Rao S (1998) Beyond the flow decomposition barrier. J ACM 45(5):783–797. doi:10.1145/290179.290181
10. Goldberg A, Tarjan R (1988) A new approach to the maximum-flow problem. J ACM 35(4):921–940
11. Greig D, Porteous B, Seheult A (1989) Exact maximum a posteriori estimation for binary images. J R Stat Soc B 51(2):271–279
12. Henzinger MR, Klein PN, Rao S, Subramanian S (1997) Faster shortest-path algorithms for planar graphs. J Comput Syst Sci 55(1):3–23. doi:10.1145/195058.195092
13. Hochbaum DS (2001) An efficient algorithm for image segmentation, Markov random fields and related problems. J ACM 48(4):686–701
14. Hochbaum DS (2008) The pseudoflow algorithm: a new algorithm for the maximum-flow problem. Oper Res 56(4):992–1009
15. Johnson DB, Venkatesan SM (1983) Partition of planar flow networks. In: Proceeding of 24th FOCS, Tucson, pp 259–264. doi:10.1109/SFCS.1983.44
16. Kleinberg J, Tardos E (2002) Approximation algorithms for classification problems with pairwise relationships: metric labeling and markov random fields. J ACM 49(5):616–639. doi:10.1145/585265.585268, http://doi.acm.org/10.1145/585265.585268
17. Kohli P, Torr PHS (2007) Dynamic graph cuts for efficient inference in Markov random fields. IEEE Trans Pattern Anal Mach Intell 29(12):2079–2088
18. Miller GL (1986) Finding small simple cycle separators for 2-connected planar graphs. J Comput Syst Sci 32(3):265–279. doi:10.1016/0022-0000(86)90030-9
19. Miller GL, Naor J (1995) Flow in planar graphs with multiple sources and sinks. SIAM J Comput 24(5):1002–1017. doi:10.1137/S0097539789162997
20. Mozes S, Wulff-Nilsen C (2010) Shortest paths in planar graphs with real lengths in $O(n \log^2 n / \log \log n)$ time. In: Proceedings of 18th ESA, Liverpool, pp 206–217. doi:10.1007/978-3-642-15781-3_18
21. Orlin J (2013) Max flows in $O(nm)$ time, or better. In: Proceedings of STOC, Palo Alto, pp 765–774
22. Schmidt FR, Toeppe E, Cremers D (2009) Efficient planar graph cuts with applications in computer vision. In: Proceedings of CVPR, Miami, pp 351–356
23. Wei X, Joneja A, Mount DM (2012) Optimal uniformly monotone partitioning of polygons with holes. Comput Aided Des 44(12):1235–1252

Planar Maximum *s-t* Flow

Glencora Borradaile
Department of Computer Science, Brown University, Providence, RI, USA
School of Electrical Engineering and Computer Science, Oregon State University, Corvallis, OR, USA

Keywords

Maximum flow; Planar graphs

Years and Authors of Summarized Original Work

2006; Borradaile, Klein
2009; Borradaile, Klein
2010; Erickson

Problem Definition

Given a directed, planar graph $G = (V, E)$ with arc capacities $c : E \to \Re^+$, a source vertex s, and a sink vertex t, the goal is to find a flow assignment f_e for each arc $e \in E$ such that

$$\max \sum_{su:su \in E} f_{su}$$

$$\text{s.t.} \sum_{uv:uv \in E} f_{uv} - \sum_{vw:vw \in E} f_{vw} = 0 \quad \forall v \in V \setminus \{s, t\} \quad (1)$$

$$0 \le f_e \le c_e \qquad \forall e \in E \quad (2)$$

Key Results

In the paper proposing the maximum flow problem in general graphs, Ford and Fulkerson [5] gave a generic method for computing a maximum flow: the augmenting-path algorithm. The algorithm is iterative: find a path P from the source to the sink such that capacity constraint (2) is loose for each arc on P (*residual*); increase the flow on each arc in P by a constant chosen so that at least one of the capacity constraints become tight; update the capacities of each arc, making note that the reverse of these arcs now have *residual capacity*; and repeat until there is no path from the source to the sink along which the flow can be augmented. By augmenting the flow along a path, the balance constraints (1) are always satisfied.

st-Planar Graphs

Ford and Fulkerson further showed that, in the case of planar graphs when the source and the sink are on a common face (st-planar graphs), by selecting the augmenting paths to be as far to the left as possible in each iteration (viewing s on the bottom and t on the top), each arc is saturated at most once, resulting in at most $|E|$ iterations [5]. In 1979, Itai and Shiloach showed that each iteration of this algorithm could be implemented in $O(\log n)$ time using a priority queue and gave a simple example showing that any implementation of this algorithm is capable of sorting n numbers [11]. In 1991, Hassin demonstrated that such a maximum st-flow could be derived from shortest-path distances in the planar dual G^* of G where capacities in G are interpreted as lengths in G^* [7]. Faster algorithms for computing shortest paths in planar graphs culminated in a linear-time algorithm for this case of maximum st-flow in planar graphs with s and t on a common face [9].

Undirected Planar Graphs

For undirected planar graphs, Reif gave an algorithm for computing the maximum st-flow where s and t need not be on a common face, by way of several shortest-path computations in the dual [19]. The algorithm finds a shortest path P in G^* from a vertex adjacent to the face corresponding to s to a vertex adjacent to the face corresponding to t. Reif proves that C only crosses P once; by finding the minimum separating cycle C_v through each vertex v of P, we will surely find C: C is the minimum of the cycles C_v. These cycles can be found in time $\log n$ times the time for one shortest-path computation via divide and conquer over the length of P. Hassin and Johnson show that the corresponding maximum flow can be computed within this framework by computing shortest-path distances between the nested cycles C_v [8]. The shortest-path algorithms of Henzinger et al. [9] or Klein [15] can be used to reimplement these algorithms in $O(n \log n)$ time. Italiano et al. [12] further improved this running time to $O(n \log \log n)$ by using an r-division to break the graph into sufficiently small pieces through which shortest paths can be efficiently computed.

If the capacities are all units, the maximum st-flow can be computed in linear time [1].

Directed Planar Graphs

Maximum st-flow in directed graphs is more general since the problem of maximum st-flow in an undirected graph can be converted to a directed problem by introducing two oppositely oriented arcs of equal capacity for each edge. Johnson and Venkatesan gave a divide-and-conquer algorithm that finds a flow of input value v in $O(n^{1.5} \log n)$ time [13]. The algorithm divides the graph using balanced separators, finding a flow in each side of value v. However, the flow on the $O(\sqrt{n})$-boundary edges of each subproblem might not be feasible. Each boundary edge is made feasible via an st-planar flow computation. Miller and Naor showed that finding a directed st-flow of value v could be reduced to computing shortest-path distances in a graph with positive and negatives lengths [17]. Here, v units of flow are routed (perhaps violating the capacity constraints) along any s-to-t path P. For those arcs whose capacity are violated, we must route the excess flow through the rest of the graph. This is a feasible circulation problem and can be solved using shortest-path distances in the dual graph, where lengths may be negative (representing the negative or violated capacities). Using an $O(n \operatorname{poly}\log n)$-time algorithm for computing shortest paths in a planar graph with negative edge lengths [4, 16, 18] gives an $O(n \text{ poly} \log n \log C)$-time algorithm where C is the sum of the capacities.

If the capacities are all unit, the maximum st-flow can be computed in linear time [21].

Leftmost-Path Algorithm

Borradaile and Klein gave an augmenting-path, $O(n \log n)$-time algorithm for the maximum st-flow problem in directed planar graphs. The algorithm is a generalization of the algorithm for the st-planar case, augmenting flow repeatedly along the leftmost path from s to t. However, with s and t not on a common face, what leftmost is not clear. With the graph embedded such that t is on the external face and the clockwise cycles saturated, a leftmost path is well-defined and can be found with a left-first, depth-first search into t. Clockwise cycles can be initially saturated with a circulation defined by potentials on the faces given by shortest-path distances in the dual graph [14], and clockwise cycles remain saturated under leftmost augmentations. Borradaile and Klein, and Erickson improved the analysis [3] showed that under these conditions an arc and its reverse can be saturated at most once, resulting in at most $2n$ augmentations. Augmentations can be performed in $O(\log n)$ time using a dynamic tree data structure, resulting in an $O(n \log n)$ running time.

Applications

Maximum st-flow in directed planar graphs has applications to computer vision problems. Schmidt et al. [20] use it as a black box for image segmentation and Greig et al. [6] provide an example for smoothing noisy images.

Open Problems

Currently, maximum st-flow in undirected planar graphs can be computed more quickly than in directed. Can this gap be closed?

Experimental Results

Schmidt et al. [20] have implemented this algorithm and compared its performance on an image segmentation problem.

URLs to Code and Data Sets

Hoch and Wang have provided an open-source implementation of the algorithm [10]. Eisenstat has an implementation of the linear-time algorithm for unit-capacity graphs [2].

Cross-References

▶ Planar Maximum Flow – Multiple-Source Multiple-Sink Maximum Flow in Directed Planar Graphs

Recommended Reading

1. Coupry L (1997) A simple linear algorithm for the edge-disjoint (s,t)-paths problem in undirected planar graphs. Inf Process Lett 64:83–86
2. Eisenstat D (2013) Trickle: linear-time maximum flow in planar graphs with unit capacities (java). http://www.davideisenstat.com/trickle/
3. Erickson J (2010) Maximum flows and parametric shortest paths in planar graphs. In: 21st SODA, Austin, pp 794–804
4. Fakcharoenphol J, Rao S (2006) Planar graphs, negative weight edges, shortest paths, and near linear time. J Comput Syst Sci 72(5):868–889. doi:http://dx.doi.org/10.1016/j.jcss.2005.05.007
5. Ford C, Fulkerson D (1956) Maximal flow through a network. Can J Math 8:399–404
6. Greig D, Porteous B, Seheult A (1989) Exact maximum a posteriori estimation for binary images. J R Stat Soc B 51(2):271–279
7. Hassin R (1981) Maximum flow in (s,t) planar networks. Inf Process Lett 13(3):107
8. Hassin R, Johnson DB (1985) An $O(n \log^2 n)$ algorithm for maximum flow in undirected planar networks. SIAM J Comput 14:612–624. doi:http://locus.siam.org/SICOMP/volume-14/art_0214045.html
9. Henzinger MR, Klein PN, Rao S, Subramanian S (1997) Faster shortest-path algorithms for planar graphs. J Comput Syst Sci 55(1):3–23. doi:10.1145/195058.195092
10. Hoch J, Wang J (2012) Max flow in a directed planar graph. https://github.com/jrshoch/msmsmaxflow

11. Itai A, Shiloach Y (1979) Maximum flow in planar networks. SIAM J Comput 8:135–150
12. Italiano GF, Nussbaum Y, Sankowski P, Wulff-Nilsen C (2011) Improved algorithms for min cut and max flow in undirected planar graphs. In: 43rd STOC, San Jose, pp 313–322
13. Johnson DB, Venkatesan SM (1983) Partition of planar flow networks. In: 24th FOCS, Tucson, pp 259–264. doi:10.1109/SFCS.1983.44
14. Khuller S, Naor J, Klein P (1993) The lattice structure of flow in planar graphs. SIAM J Discret Math 6(3):477–490. doi:10.1137/0406038
15. Klein PN (2005) Multiple-source shortest paths in planar graphs. In: 16th SODA, Vancouver, pp 146–155. doi:10.1145/1070454
16. Klein PN, Mozes S, Weimann O (2010) Shortest paths in directed planar graphs with negative lengths: a linear-space $O(n \log^2 n)$-time algorithm. TALG 6(2):1–18. doi:http://doi.acm.org/10.1145/1721837.1721846
17. Miller GL, Naor J (1995) Flow in planar graphs with multiple sources and sinks. SIAM J Comput 24(5):1002–1017. doi:10.1137/S0097539789162997
18. Mozes S, Wulff-Nilsen C (2010) Shortest paths in planar graphs with real lengths in $O(n \log^2 n / \log \log n)$ time. In: 18th ESA, Liverpool, pp 206–217
19. Reif J (1983) Minimum s-t cut of a planar undirected network in $O(n \log^2 n)$ time. SIAM J Comput 12:71–81. doi:http://locus.siam.org/SICOMP/volume-12/art_0212005.html
20. Schmidt FR, Toeppe E, Cremers D (2009) Efficient planar graph cuts with applications in computer vision. In: CVPR, Miami, pp 351–356
21. Weihe K (1994) Edge-disjoint (s,t)-paths in undirected planar graphs in linear time. In: Proceedings of the European symposium on algorithms, Edinburgh. LNCS, vol 855, pp 130–137

Planarisation and Crossing Minimisation

Markus Chimani
Faculty of Mathematics/Computer Science, Theoretical Computer Science, Osnabrück University, Osnabrück, Germany

Keywords

Approximations; Exact algorithms; Graph theory; Heuristics; Insertion problems; Mathematical programming

Years and Authors of Summarized Original Work

2005; Gutwenger, Mutzel, Weiskircher
2008; Chimani, Mutzel, Bomze

Problem Definition

Given a graph $G = (V, E)$, the *crossing number* $cr(G)$ of G is the smallest number of edge crossings possible for any drawing of G into the plane. Since its introduction in the mid-1940s, the crossing number problem has proved to be notoriously difficult. Even some of the oldest and seemingly simplest questions remain unanswered, despite the large amount of research.

Already the problem definition is more ambiguous than it may seem. We will usually only consider drawings where vertices are mapped to distinct points in the plane and edges to continuous non-self-intersecting curves between their end vertices. Any non-vertex point may only be contained in at most two edge curves, in which these curves have to meet transversally (i.e., *cross*). There are several different and specialized related crossing number variants; see [16] for a comprehensive annotated list.

A *planarization* of a nonplanar graph G is a planar graph obtained from G by drawing G into the plane and replacing the crossings by dummy vertices of degree 4. Observe that in other literature, the term *planarization* is also sometimes used to denote a (large) planar spanning subgraph of G.

Key Results

The (decision version of the) crossing number problem is NP-complete, even when all vertices have degree at most three [13] or when the removal of a single edge would give a planar graph [3]. There is, however, a fixed parameter tractable (FPT) algorithm to test in linear time for a *constant* k (not part of the input), whether G allows a crossing number of at most k [15]. The dependency on k, however, is doubly exponential, and the algorithm is far from being applicable in practice. Most questions regarding the problem's approximability remain open; see below for details.

Planarization Algorithms

The practically strongest heuristic is the *planarization approach*, cf. Fig. 1: First, we seek a maximal planar subgraph (observe that finding a maximum planar subgraph is already NP-hard). In other words, we temporarily remove edges from G until it becomes planar. Then, we reinsert those edges with as few crossings as possible one after another. After each step, crossings are replaced by dummy vertices, such that we can consider a sequence of *edge insertion problems*: Given a planar graph H and an edge $e \notin E(H)$, let $H_e := H + e$. Since $cr(H_e)$ is still NP-hard to obtain, we ask for a crossing-minimum solution under the side-constraint that H is drawn planarly. An *embedding* is an equivalence class over planar drawings, based on the cyclic order of the edges around their incident vertices. For a fixed embedding of H, the insertion problem is trivially solvable via a breadth-first search in H's dual graph. However, the number of embeddings of H is exponential in general.

The seminal paper [12] shows that it is possible to find the best embedding *in linear time* using SPR-trees, or formally:

Theorem 1 *Let G be a planar graph and $v, w \in V(G)$ two vertices. We can find a planar embedding of G in $\mathcal{O}(|V(G)|)$ time, into which an edge $e = (v, w)$ can be inserted with the least possible number of edge crossings over all possible embeddings.*

To discuss this result, we first need to describe SPR-trees, which are used to decompose graphs into their triconnectivity structures. While their graph-theoretic foundation is based on Tutte [17], the data structure was first suggested by Di Battista and Tamassia [11] under the name *SPQR-tree*. Nowadays, we often drop the "Q" from the abbreviation, as the corresponding node type is not necessary, and use the following contemporary definition (see, e.g., [7]), illustrated in Fig. 2.

Definition 1 (SPR-tree) The *SPR-tree* $\mathcal{T}$ of a biconnected graph G is the unique smallest tree satisfying the following properties:

1. Each node ν in $\mathcal{T}$ holds a graph $S_\nu = (V_\nu, E_\nu)$, $V_\nu \subseteq V(G)$, called *skeleton*. Each edge of E_ν is either a *real* edge from $E(G)$ or a *virtual* edge $f = (u, v)$ where $\{u, v\}$ forms a 2-cut (a *split pair*) in G.
2. $\mathcal{T}$ has only three different node types with the following skeleton structures:
   S: S_ν is a simple cycle – it represents a *serial* component.
   P: S_ν consists of two vertices connected by multiple edges – a *parallel* component.
   R: S_ν is a simple triconnected graph.
3. For every edge (ν, μ) in $\mathcal{T}$, S_ν (S_μ) contains a specific virtual edge e_μ (e_ν) which "represents" S_μ (S_ν, respectively). Both edges e_μ and e_ν connect the same vertices.
4. The original graph G can be obtained by recursively applying the following operation: For the edge (ν, μ) in $\mathcal{T}$, let e_μ, e_ν be the virtual edges as in (3) connecting the same vertices u, v. A merged graph $(S_\nu \cup S_\mu) - e_\mu - e_\nu$ is obtained by gluing the skeletons together at u, v and removing e_μ, e_ν.

There are several essential properties of any SPR-tree: First, it has linear size and can be constructed in linear time. Second, a simple triconnected graph – the skeleton of an R-node – allows only a unique embedding and its mirror; the embedding of an S-node skeleton is unique, and the possible embeddings of a P-node are precisely all cyclic permutations of its edges. Moreover, each embedding of G can be precisely described via the subembeddings of the skeletons.

In order to obtain Theorem 1, it is shown that we only need to consider the unique shortest path P in G's SPR-tree from any node whose skeleton contains v to any node whose skeleton contains w. We will specify an embedding for each skeleton S_μ of the nodes $\mu \in P$; the embedding of all other skeletons is irrelevant. Moreover, the optimum embedding of each of the former skeletons can be chosen independent of the others: For each S_μ, we can specify a source s and a target t and ask for an embedding such that edge (s,t) can be inserted with the least possible number of edge crossings into planar S_μ: In the first (last) skeleton along P, the source (target) simply is v (w, respectively). In the other cases, the source (target) is the virtual edge corresponding to the predecessor (successor) along P. For simplicity, it may be helpful to consider a subdivision of such a virtual edge, such that each source and target can be represented by a vertex. If μ is an S-node, its skeleton has a unique embedding, (s,t) will require no crossings, and there is nothing to specify. If μ is a P-node, both the source and the target are virtual edges due to minimality of P; we pick any embedding where the two virtual edges appear consecutively, so that (s,t) again requires no crossings. Finally, if μ is an R-node, its skeleton allows only a unique embedding and its mirror and hence has a unique dual graph. We compute a shortest path between s and t via a simple breadth-first-search in the dual of S_μ as for the fixed embedding case.

Finally, traverse P and let μ, μ' be any two consecutive nodes. We want to establish that if the insertion path in S_μ enters its target virtual edge "from the left" (corresponding to some arbitrary predefined orientation of the virtual edges), it leaves the source virtual edge in $S_{\mu'}$ "to the right" or vice versa. If this is not already the case, it suffices to flip the embedding of $S_{\mu'}$ (we can ignore the case when μ' is an S-node). This establishes a suitable embedding of G.

This algorithmic breakthrough has allowed the planarization heuristic to perform extraordinarily well in practice, both in terms of running time and of solution quality. The algorithm and its proofs also give rise to several strong pre- and postprocessing routines. Furthermore, it is pivotal to several later results such as insertion of stars [6] or insertion-based approximation algorithms [2, 7, 8]. The strongest of the latter considers the problem of inserting several edges F simultaneously into a planar graph G [7]: it can be shown that this multi-edge-insertion problem approximates $cr(G + F)$ [8], but is unfortunately itself already NP-hard. However, this insertion problem can in turn also be approximated. This approximation chain gives the currently only practically relevant approximation algorithm. In fact, it arguably gives the best running time vs. solution-quality trade-off among all known algorithms in practice [5].

Exact Approaches

In certain cases, a heuristic or approximate solution is not good enough, e.g., when the result is to be used as a base case in some formal graph-theoretic proof. There are exact approaches based on integer linear programs. The currently strongest one – constituting the second central paper of this entry – is able to solve typical "real-world graph drawing" instances (i.e., relatively

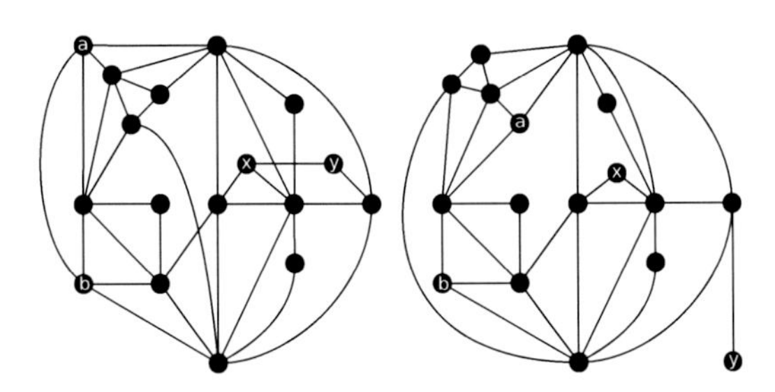

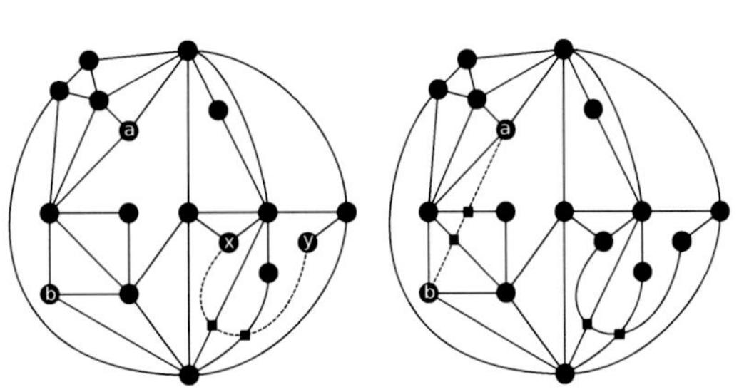

Planarisation and Crossing Minimisation, Fig. 1 The planarization heuristic: the first figure shows a graph drawn with the optimal number of three crossings. The planarization heuristic starts with a maximal (in the figure, in fact, maximum) planar subgraph (second figure) where the edges (x, y) and (a, b) are removed. Then, iteratively, we find optimal embeddings to reinsert these edges into the planarly drawn graph, simulating crossings by dummy vertices (shown as *squares*). Although the insertion problems are solved optimally, the result requires four crossings

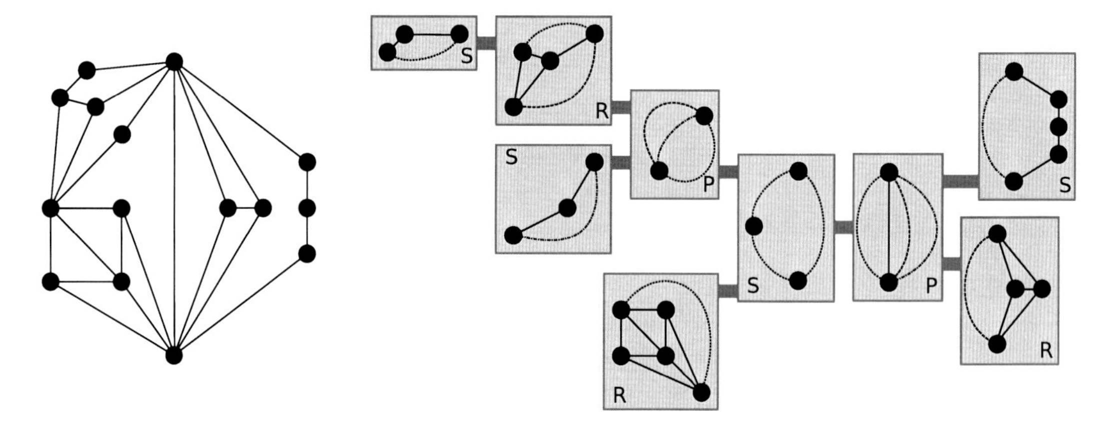

Planarisation and Crossing Minimisation, Fig. 2 A graph (to the *left*) and its decomposition into triconnectivity structures, arranged within an SPR-tree. *Thick dashed* or *dotted edges* are virtual

sparse graphs with up to 80–100 vertices) to optimality in a couple of minutes [9]. The idea is to introduce binary indicator variables $x_{e,f}$ for each edge pair $e, f \in E$, which are 1 if and only if the two edges cross in the optimum solution. Minimizing the sum of these variables gives the desired objective function.

It remains to ensure that the variables are set correctly. To this end, we introduce *Kuratowski constraints*. The famous Kuratowski theorem states that planar graphs are characterized by the absence of subdivisions of certain small subgraphs (namely, K_5 and $K_{3,3}$). In other words, for each such subgraph K, we can require that at least one edge pair $e, f \in E(K)$ crosses. Unfortunately, there may be an exponential number of such subgraphs in G, and we also have to take care of Kuratowski subgraphs that only arise because certain other edge pairs cross (establishing a dummy vertex). Even when solving all these challenges, a further crucial problem remains: Consider a (presumably optimum) 0/1-assignment to the x-variables, satisfying all Kuratowski constraints. It is still NP-complete to decide whether this solution is at all feasible! Consider an edge e that is crossed by edges f and g. Our x-variables establish these crossings, but we do not know the order of f and g along e. Deciding whether any feasible order exists is what makes the problem still hard. There are two methods to solve the problem: We can subdivide each edge e sufficiently often and allow at most one crossing per edge segment. Alternatively we can introduce additional variables $y_{e,f,g}$ to explicitly describe a linear ordering of all edges crossing edge e, for all edges e. Of course, we have to modify the Kuratowski constraints accordingly.

While the second modeling approach is practically more efficient, we like to showcase the central ideas only with a simplified version of the first model here. Let U be any upper bound on the crossing number of G, e.g., found via the planarization heuristic described above. Clearly, any optimum solution will have at most U crossings on any edge. Therefore, let H be the graph obtained from G by subdividing each edge into U *edge segments*. Now, we solve the problem on H instead of G, where we can require that each segment is crossed at most once:

$$\min \sum_{e,f \in E(H), e \neq f} x_{\{e,f\}}, \text{ subject to} \quad (1)$$

$$\sum_{f \in E(H) \setminus \{e\}} x_{\{e,f\}} \leq 1 \qquad \forall e \in E(H) \quad (2)$$

$$x_{\{e,f\}} \in \{0,1\} \quad \forall e, f \in E(H), e \neq f \quad (3)$$

This model minimizes the sum of the variables indicating a crossing between a pair of edge segments and allows at most one crossing per segment. It remains to discuss the Kuratowski constraints to establish that only graph-theoretically feasible solutions are allowed. Let $\mathcal{X}_s$ denote the set of all binary solution vectors satisfying (2), and consider any solution vector $\bar{x} \in \mathcal{X}_s$. Furthermore, let $R(\bar{x})$ be the set of edge pairs $\{e, f\}$ for which $\bar{x}_{\{e,f\}} = 1$. Starting with H, we can *realize* $\bar{x}$ by, for each $\{e, f\} \in R(\bar{x})$, subdividing e and f and identifying the two new vertices. This vertex may be called a *dummy vertex*, representing the crossing. We obtain a final graph $H[\bar{x}]$ and let $\mathcal{K}(\bar{x})$ denote the set of all Kuratowski subdivisions in $H[\bar{x}]$. Intuitively, for any subdivision $K \in \mathcal{K}(\bar{x})$, we need to require at least one crossing on K, if the crossings $R(\bar{x})$ exist. Formally, this establishes the final constraint class:

$$\sum_{e,f \in K, e \neq f} x_{\{e,f\}} \geq 1 - \sum_{\{e,f\} \in R(\bar{x})} (1 - x_{\{e,f\}})$$
$$\forall \bar{x} \in \mathcal{X}_s, K \in \mathcal{K}(\bar{x}) \qquad (4)$$

Independent on whether we use the just described subdivision-based model or the stronger one based on linear orderings, the obtained ILP models are much too large to solve directly. We require both special separation and column-generation routines, in order to produce the actually necessary constraints and variables on the fly. The approaches' practical applicability is furthermore only possible due to the strong heuristics described above (which often give optimum upper bounds early on), heavy preprocessing [4], and efficient planarity testing routines.

Open Problems

The area of crossing numbers is filled with interesting open questions. Let us pinpoint two of them:

The original question, as stated in 1944 by Pál Turán, was for the crossing number of complete bipartite graphs. We still do not know the answer for this graph class, nor do we for complete graphs. While we have upper bounds, which are conjectured to be optimal, we are stuck with partial proofs and positive results for small graphs.

We know that crossing number is APX-hard [1], i.e., there cannot be a polynomial approximation scheme. However, even for graphs with bounded maximum degree, the best approximation ratios are only slightly sublinear in $|V|$ [10] or dependent on parameters like the graph's genus [14] or the number of edges required to remove in order to become planar [7]. Does there exist a constant factor approximation to the crossing number problem? At least for graphs with bounded degree?

URLs to Code and Data Sets

The free (GPL) Open Graph Drawing Framework (OGDF) contains implementations of the strongest planarization heuristics and the exact algorithms: http://www.ogdf.net.

A web front-end to the exact crossing minimizer is freely available at http://crossings.uos.de.

Recommended Reading

1. Cabello S (2013) Hardness of approximation for crossing number. Discret Comput Geom 49(2):348–358
2. Cabello S, Mohar B (2011) Crossing number and weighted crossing number of near-planar graphs. Algorithmica 60(3):484–504
3. Cabello S, Mohar B (2013) Adding one edge to planar graphs makes crossing number and 1-planarity hard. SIAM J Comput 42(5):1803–1829
4. Chimani M, Gutwenger C (2009) Non-planar core reduction of graphs. Discret Math 309(7):1838–1855
5. Chimani M, Gutwenger C (2012) Advances in the planarization method: effective multiple edge insertions. J Graph Algorithms Appl 16(3):729–757
6. Chimani M, Gutwenger C, Mutzel P, Wolf C (2009) Inserting a vertex into a planar graph. In: SODA, New York, pp 375–383
7. Chimani M, Hlinený P (2011) A tighter insertion-based approximation of the crossing number. In: Aceto L, Henzinger M, Sgall J (eds) ICALP (1), Zurich. Lecture notes in computer science, vol 6755. Springer, pp 122–134
8. Chimani M, Hlinený P, Mutzel P (2012) Vertex insertion approximates the crossing number of apex graphs. Eur J Comb 33(3):326–335

9. Chimani M, Mutzel P, Bomze I (2008) A new approach to exact crossing minimization. In: Halperin D, Mehlhorn K (eds) ESA, Karlsruhe. Lecture notes in computer science, vol 5193. Springer, pp 284–296
10. Chuzhoy J (2011) An algorithm for the graph crossing number problem. In: Proceedings of the 43rd annual ACM symposium on theory of computing (STOC'11), San Jose. ACM, pp 303–312
11. Di Battista G, Tamassia R (1996) On-line planarity testing. SIAM J Comput 25:956–997
12. Gutwenger C, Mutzel P, Weiskircher R (2005) Inserting an edge into a planar graph. Algorithmica 41:289–308
13. Hliněný P (2006) Crossing number is hard for cubic graphs. J Comb Theory Ser B 96:455–471
14. Hliněný P, Chimani M (2010) Approximating the crossing number of graphs embeddable in any orientable surface. In: Proceedings of the twenty-first annual ACM-SIAM symposium on discrete algorithms (SODA '10), Austin. Society for Industrial and Applied Mathematics, pp 918–927
15. Kawarabayashi K-i, Reed B (2007) Computing crossing number in linear time. In: Proceedings of the thirty-ninth annual ACM symposium on theory of computing (STOC '07), San Diego. ACM, pp 382–390
16. Schaefer M (2013) The graph crossing number and its variants: a survey. Electron J Comb Dyn Sur #21
17. Tutte WT (1966) Connectivity in graphs. Mathematical expositions, vol 15. University of Toronto Press, Toronto

Planarity Testing

Glencora Borradaile
Department of Computer Science, Brown University, Providence, RI, USA
School of Electrical Engineering and Computer Science, Oregon State University, Corvallis, OR, USA

Keywords

Planar embedding; Planarity testing

Years and Authors of Summarized Original Work

1976; Booth, Lueker

Problem Definition

The problem is to determine whether or not the input graph G is planar. The definition pertinent to planarity-testing algorithms is: G is planar if there is an *embedding* of G into the plane (vertices of G are mapped to distinct points and edges of G are mapped to curves between their respective endpoints) such that edges do not cross. Algorithms that test the planarity of a graph can be modified to obtain such an embedding of the graph.

Key Results

Theorem 1 *There is an algorithm that given a graph G with n vertices, determines whether or not G is planar in $O(n)$ time.*

The first linear-time algorithm was obtained by Hopcroft and Tarjan [5] by analyzing an iterative version of a recursive algorithm suggested by Auslander and Parter [1] and corrected by Goldstein [4]. The algorithm is based on the observation that a connected graph is planar if and only if all its biconnected components are planar. The recursive algorithm works with each biconnected component in turn: find a separating cycle C and partition the edges of G not in C; define a component of the partition as consisting of edges connected by a path in G that does not use an edge of C; and, recursively consider each cyclic component of the partition. If each component of the partition is planar and the components can be combined with C to give a planar graph, then G is planar.

Another method for determining planarity was suggested by Lempel, Even, and Cederbaum [6]. The algorithm starts with embedding a single vertex and the edges adjacent to this vertex. It then considers a vertex adjacent to one of these edges. For correctness, the vertices must be considered in a particular order. This algorithm was first implemented in $O(n)$ time by Booth and Lueker [2] using an efficient implementation of the PQ-trees data structure. Simpler implementations of this algorithm have been given by Boyer and Myrvold [3] and Shih and Hsu [8].

Tutte gave an algebraic method for giving a *straight-line embedding* of a graph that, if the input graph is 3-connected and planar, is guaranteed to generate a planar embedding. The key idea is to fix the vertices of one face of the graph to be the corners of a convex polygon and then embed every other vertex as the geometric average of its neighbors.

Applications

Planarity testing has applications to computer-aided circuit design and VLSI layout by determining whether a given network can be realized in the plane.

URL to Code

LEDA has an efficient implementation of the Hopcroft and Tarjan planarity testing algorithm [7]: http://www.algorithmic-solutions.info/leda_guide/graph_algorithms/planar_kuratowski.html

Cross-References

► Fully Dynamic Planarity Testing

Recommended Reading

1. Auslander L, Parter SV (1961) On imbedding graphs in the plane. J Math Mech 10:517–523
2. Booth KS, Lueker GS (1976) Testing for the consecutive ones property, interval graphs, and graph planarity using PQ-tree algorithms. J Comput Syst Sci 13:335–379
3. Boyer J, Myrvold W (1999) Stop minding your P's and Q's: a simplified O(n) planar embedding algorithm. In: SODA'99: proceedings of the tenth annual ACM-SIAM symposium on discrete algorithms, Philadelphia. Society for Industrial and Applied Mathematics, pp 140–146
4. Goldstein AJ (1963) An efficient and constructive algorithm for testing whether a graph can be embedded in the plane. In: Graph and combinatorics conference
5. Hopcroft J, Tarjan R (1974) Efficient planarity testing. J ACM 21:549–568
6. Lempel A, Even S, Cederbaum I (1967) An algorithm for planarity testing of graphs. In: Rosentiehl P (ed) Theory of graphs: international symposium. Gordon and Breach, New York, pp 215–232
7. Mehlhorn K, Mutzel P, Näher S (1993) An implementation of the hopcroft and tarjan planarity test. Technical report, MPI-I-93-151, Saarbrücken
8. Shih W-K, Hsu W-L (1999) A newplanarity test. Theor Comput Sci 223:179–191

Point Location

Marcel Roeloffzen
Graduate School of Information Sciences, Tohoku University, Sendai, Japan

Keywords

Jump and walk; Nearest neighbor; Point location; Trapezoidal decomposition; Triangulation

Years and Authors of Summarized Original Work

1983; Kirkpatrick
1991; Seidel
2009; Haran, Halperin

Problem Definition

Point location is a well-studied problem in computational geometry with many applications in geometric information systems and computer-aided design. In general terms, the problem is to find which element of a given object contains a given query point. More precisely, we are given a subdivision S of a metric space, usually the Euclidean plane. The goal is then to preprocess S so that for a query point p we can determine which region of the subdivision contains p. There are many variants of the problem, with different constraints on the subdivision. The most common variant and the focus of this overview is planar point location, where S is a polygonal subdivision of the Euclidean plane. In fact, for several data structures, we assume that the input is a triangulation. (Note that a polygonal subdivision can be triangulated in linear time [1].) As for most query data structures, the efficiency of a point location structure is measured by its space requirement, the time needed to preprocess the input, and the query time. The efficiency of the algorithms and data structures described here will be expressed as a function of n, the number of vertices in the subdivision.

Key Results

Most solutions for the point location problem build upon one (or a combination) of three basic ideas: walking in a triangulation, a trapezoidal decomposition, or a hierarchical triangulation. The three methods provide a basic trade-off between space usage and query time and are described in more detail below.

Walking in a Triangulation

This approach is the simplest of the three and requires no additional storage or preprocessing if the input is provided in a suitable format in which it takes constant time to access the neighbors of a triangle. For each query, we start in some triangle or vertex of the triangulation and walk to the query point by traversing the triangulation, walking from the current triangle to a neighbor in each step. The triangle to visit next is determined by a walking strategy. Devillers et al. [3] describe three walking strategies that are described below and illustrated in Fig. 1.

- *Straight-line walk.* Starting at a vertex of the triangulation we walk along a straight line toward the query point.
- *Orthogonal walk.* Instead of walking along a straight line, we first walk parallel to the x-axis and then parallel to the y-axis.
- *Visibility walk.* For each triangle visited, we pick an edge of the triangle (at random or in some specific order) and test if its supporting line separates the query point from the triangle. If this is the case, then we continue our walk by crossing that edge into a new triangle.

The worst-case behavior of each of the walking strategies is very bad as there are triangulations where each walk visits $\Omega(n)$ triangles in expectation given a random starting point and query point. In fact, the visibility walk is not guaranteed to reach the query point if edges are picked in a fixed order and the randomized version may require exponential time [3]. However, this seems to require many long and thin triangles, which do not occur in most practical applications, where the subdivision is often closer to a Delaunay triangulation. Experiments on Delaunay triangulations of random point sets suggest that the methods are comparable in total query time, though they provide a trade-off between the number of triangles visited and the time spent per triangle.

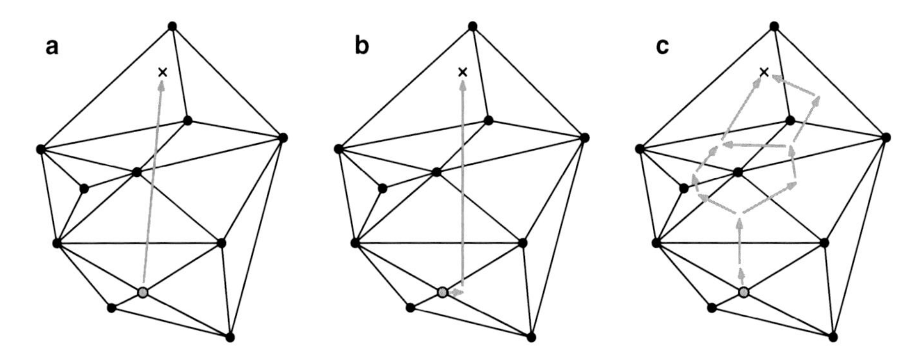

Point Location, Fig. 1 The paths traversed by (**a**) a straight-line walk, (**b**) an orthogonal walk, and (**c**) a visibility walk. Note that any path following the *arrows* is possible in the visibility walk

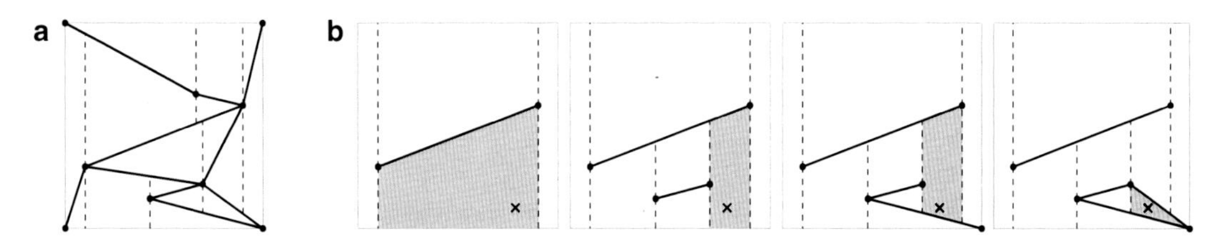

Point Location, Fig. 2 (**a**) Part of a trapezoidal decomposition with (**b**) a sequence of trapezoids visited by a query

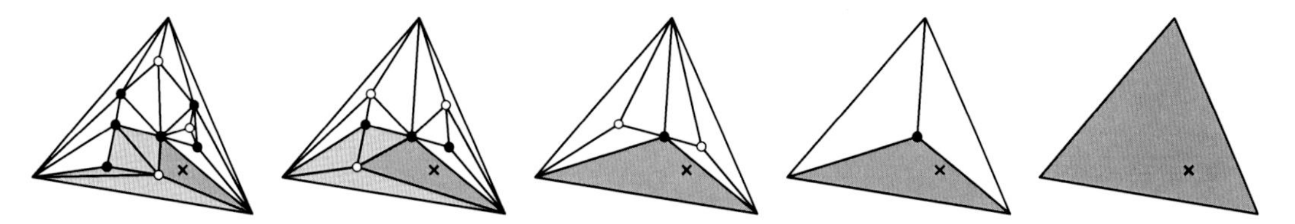

Point Location, Fig. 3 The triangulations of the hierarchy from left to right. Triangles containing the query point are marked with *dark gray* and other triangles inspected with *light gray*

That is, the straight-line walk visits the fewest triangles, but needs most time to determine which triangle to visit next, whereas the visibility walk visits the most triangles, but spends the least time per triangle.

Trapezoidal Decomposition

The first data structure for point location to achieve $O(\log n)$ query time for point was provided by Dobkin and Lipton [6]. The structure is created by cutting the subdivision into slices using vertical lines through each of its vertices. Point location can then be done using two binary searches, first on the slices and then within a slice. This creates $O(n^2)$ trapezoids, but Sarnak and Tarjan [11] show we don't have to explicitly store all of them and instead only need $O(n)$ space and $O(n \log n)$ time to store them implicitly in a search structure.

A subdivision into trapezoids can also be created in a more careful way so that only $O(n)$ trapezoids are created. For a polygonal subdivision, its *trapezoidal decomposition* is defined as the result of shooting vertical rays upward and downward from each vertex of the subdivision until they hit an edge of the subdivision – if they do not hit such an edge, they extend to infinity (see also Fig. 2). A trapezoidal decomposition can be constructed by incrementally adding the edges of the subdivision [10]. Each insertion of an edge can interrupt some rays and introduces two new ones, namely, the rays from the endpoints of the edge. Seidel [12] showed that the history of this process can be recorded in a directed acyclic graph, which can be used for point location. Using randomized incremental construction, that is, adding the edges in random order, the longest path in the graph has an expected length of $O(\log n)$. Building the structure itself takes $O(n \log n)$ expected time and $O(n)$ expected space. For a guaranteed query time of $O(\log n)$, we can find the longest path and reconstruct if needed resulting in $O(n \log n)$ expected preprocessing time and $O(\log n)$ guaranteed query time [13].

Hierarchical Triangulation

Point location can also be done using a so-called hierarchical triangulation. We start with a triangulation of the subdivision and in each level of the hierarchy, we remove a subset of the points and compute a triangulation of the remainder. Each triangle of this new triangulation then stores a list of triangles from the previous triangulation that intersect it as illustrated in Fig. 3. This process is repeated until only one triangle remains. (Here, we assume that the outer face of the input triangulation is a triangle and these vertices are never removed.) To query for a point p, we traverse this hierarchy of triangles starting at the top level consisting of only one triangle. At each level, we know the triangle T that contains p and find the triangle T' from the previous level that contains p by a linear search on the triangles that intersect T (see Fig. 3 from right to left).

The query time depends on the number of levels and the number of intersections each triangle has with the previous level. Kirkpatrick [9] showed that it is possible to find a constant size set of points, such that its removal creates triangles that each intersect a constant number of triangles from the previous level. He also shows that such a set of points can be found in $O(n)$ time. By picking and removing points from each level in this way, we create at most $O(\log n)$ levels and each triangle will intersect at most $O(1)$ triangles from the previous level. As a result, a point location query takes $O(\log n)$ time in total, while the structure requires $O(n)$ space.

Hybrid and Refined Approaches

The walking strategy requires no additional memory and is fast on small instances, but has very poor performance on larger instances, both in theory and in practice. To make the walking approach more feasible for larger inputs, several structures have been proposed that combine walking strategies with other methods to obtain fast query times with low overhead costs in terms of space and preprocessing time.

*Delaunay Hierarchy***.** Hierarchical triangulations can be combined with walking algorithms to reduce the number of levels in the hierarchy. Finding the correct triangle in the next level of the hierarchy is done using a walking algorithm as opposed to a linear search. Devillers [2,3] showed that this results in a very fast query time without requiring a lot of preprocessing time or space.

*Jump and Walk***.** In most walking strategies, the starting point is chosen at random. In the jump-and-walk approach, we use a set S of several starting points and start our walk from the one that is nearest to the query point. The set S can either be chosen at random for each new query [5] or picked more carefully and stored in a data structure for nearest-neighbor searches [4]. Depending on the size of S and the complexity of the search structure, query times between $O(\sqrt[3]{n})$ and $O(\log n)$ can be achieved.

Experimental Results

Several solutions have been implemented in CGAL, the Computational Geometry Algorithm Library. Haran and Halperin [7] compared several of the implementations from CGAL. Their results show that the various methods provide a trade-off between how much memory and preprocessing time is used and the resulting query time. Overall, they conclude that a jump-and-walk algorithm performs well, if the set of potential starting points is carefully chosen and stored in an efficient search structure. Recently, the CGAL variant of the trapezoidal decomposition approach has received a major overhaul [8]. Unlike some of the other variants, this implementation guarantees a $O(\log n)$ query time, while experiments show that the implementation is still competitive with other approaches.

Cross-References

- ► Triangulation Data Structures
- ► Voronoi Diagrams and Delaunay Triangulations

Recommended Reading

1. Chazelle B (1991) Triangulation a simple polygon in linear time. Discret Comput Geom 6(1):485–524
2. Devillers O (2002) The Delaunay hierarchy. Int J Found Comput Sci 13:163–181
3. Devillers O, Pion S, Teillaud M (2002) Walking in a triangulation. Int J Found Comput Sci 13:181–199
4. Devroye L, Lemaire C, Moreau JM (2004) Expected time analysis for Delaunay point location. Comput Geom Theory Appl 29:61–89
5. Devroye L, Mücke E, Zhu B (1998) A note on point location in Delaunay triangulations of random points. Algorithmica (Spec Issue Aver Case Anal Algorithms) 22(4):477–482
6. Dobkin D, Lipton RJ (1976) Multidimensional searching problems. SIAM J Comput 5(2):181–186

7. Haran I, Halperin D (2009) An experimental study of point location in planar arrangements in CGAL. J Exp Algorithm 13:Article 3
8. Hemmer M, Kleinbort M, Halperin D (2012) Improved implementation of point location in general two-dimensional subdivisions. In: Proceedings of the 20th European symposium on algorithms (ESA), Ljubljana, pp 611–623
9. Kirkpatrick D (1983) Optimal search in planer subdivisions. SIAM J Comput 12(1):28–35
10. Mulmuley K (1990) A fast planar partition algorithm, I. J Symb Comput 10(3):253–280
11. Sarnak N, Tarjan RE (1986) Planar point location using persistent search trees. Commun ACM 29(7):669–679
12. Seidel R (1991) A simple and fast incremental randomized algorithm for computing trapezoidal decompositions and for triangulating polygons. Comput Geom: Theory Appl 1(1):51–64
13. Seidel R, Adamy U (2000) On the exact worst case query complexity of planar point location. J Algorithms 37:189–217

Point Pattern Matching

Veli Mäkinen and Esko Ukkonen
Department of Computer Science, Helsinki Institute for Information Technology (HIIT), University of Helsinki, Helsinki, Finland

Keywords

Geometric alignment; Geometric matching; Largest common point set; Point set matching

Years and Authors of Summarized Original Work

2003; Ukkonen, Lemström, Mäkinen

Problem Definition

Let $\mathbb{R}$ denote the set of reals and $\mathbb{R}^d$ the d-dimensional real space. A finite subset of $\mathbb{R}^d$ is called a *point set*. The set of all point sets (subsets of $\mathbb{R}^d$) is denoted $\mathcal{P}(\mathbb{R}^d)$.

Point pattern matching problems ask for finding similarities between point sets under some transformations. In the basic set–up a *target* point set $T \subset \mathbb{R}^d$ and a *pattern* point set (*point pattern*) $P \subset \mathbb{R}^d$ are given, and the problem is to locate a subset I of T (if it exists) such that P *matches* I. Matching here means that P becomes exactly or approximately equal to I when a *transformation* from a given set $\mathcal{F}$ of transformations is applied on P.

Set $\mathcal{F}$ can be, for example, the set of all *translations* (a constant vector added to each point in P), or all compositions of translations and *rotations* (after a translation, each point is rotated with respect to a common origin; this preserves the distances and is also called a *rigid movement*), or all compositions of translations, rotations, and *scales* (after translating and rotating, distances to the common origin are multiplied by a constant).

The problem variant with exact matching, called the *Exact Point Pattern Matching (EPPM)* problem, requires that $f(P) = I$ for some $f \in \mathcal{F}$. In other words, the EPPM problem is to decide whether or not there is an allowed transformation f such that $f(P) \subset T$. For example, if $\mathcal{F}$ is the set of translations, the problem is simply to decide whether $P + t \subset T$ for some $t \in \mathbb{R}^d$.

Approximate matching is a better model of many situations that arise in practice. Then the quality of the matching between $f(P)$ and I is controlled using a *threshold parameter* $\varepsilon \geq 0$ and a *distance function* $\delta\colon (\mathcal{P}(\mathbb{R}^d), \mathcal{P}(\mathbb{R}^d)) \rightarrow \mathbb{R}$ for measuring distances between point sets. Given $\varepsilon \geq 0$, the *Approximate Point Pattern Matching (APPM)* problem is to determine whether there is a subset $I \subset T$ and a transformation $f \in \mathcal{F}$ such that $\delta(f(P), I) \leq \varepsilon$.

The choice of the distance function δ is another source of diversity in the problem statement. A variant requires that there is a *one-to-one* mapping between $f(P)$ and I, and each point p of $f(P)$ is ε-*close* to its one-to-one counterpart p^* in I, that is, $|p - p^*| \leq \varepsilon$. A commonly studied relaxed version uses matching under a *many-to-one* mapping: it is only required that each point of $f(P)$ has *some* point of I that is ε-*close*; this distance is also known as the *directed Hausdorff distance*. Still more variants come from the choice of the *norm* $|\cdot|$ to measure the distance between points.

Another form of approximation is obtained by allowing a minimum amount of unmatched points in *P*: The *Largest Common Point Set (LCP)* problem asks for the largest $I \subset T$ such that $I \subset f(P)$ for some $f \in \mathcal{F}$. In the *Largest Approximately Common Point Set (LACP)* problem each point $p^* \in I$ must occur *ε-close* to a point $p \in f(P)$.

Finally, a problem closely related to point pattern matching is to evaluate for point sets *A* and *B* their smallest distance $\min_{f \in \mathcal{F}} \delta(f(A), B)$ under transformations $\mathcal{F}$ or to test if this distance is $\leq \varepsilon$. This problem is called the *distance evaluation problem*.

Key Results

A folk theorem is a voting algorithm to solve EPPM under translations in $O(|P||T| \log(|T||P|))$ time: Collect all translations mapping each point of *P* to each point of *T*, sort the set, and report the translation getting most votes. If some translation gets $|P|$ votes, then a subset *I* such $f(P) = I$ is found. With some care in organizing the sorting, one can achieve $O(|P||T| \log |P|)$ time [13].

The voting algorithm also solves the LCP problem under translations. A faster algorithm specific to EPPM is as follows: Let $p_1, p_2, \cdots p_m$ and $t_1, t_2, \cdots t_n$ be the lists of pattern and target points, respectively, *lexicographicly ordered* according to their *d*-dimensional coordinate values. Consider the translation $f_{i_1} = t_{i_1} - p_1$, for any $1 \leq i_1 \leq n$. One can scan the target points in the lexicographic order to find a point t_{i_2} such that $p_2 + f_{i_1} = t_{i_2}$. If such is found, one can continue scanning from t_{i_2+1} on to find t_{i_3} such that $p_3 + f_{i_1} = t_{i_3}$. This process is continued until a translated point of *P* does not occur in *T* or until a translated occurrence of the entire *P* is found. Careful implementation of this idea leads to the following result showing that the time bound of the naive string matching algorithm is possible also for the exact point pattern matching under translations.

Theorem 1 (Ukkonen et al. 2003 [13]) *The EPPM problem under translations for point pattern P and target T can be solved in O(mn) time and O(n) space where* $m = |P| \leq |T| = n$.

Quadratic running times are probably the best one can achieve for PPM algorithms:

Theorem 2 (Clifford et al. 2006 [10]) *The LCP problem under translations is 3SUM-hard.*

This means that an $o(|P||T|)$ time algorithm for LCP would yield an $o(n^2)$ algorithm for the `3SUM` problem, where $|T| = n$ and $|P| = \Theta(n)$. The `3SUM` problem asks, given *n* numbers, whether there are three numbers *a*, *b*, and *c* among them such that $a + b + c = 0$; finding a sub-quadratic algorithm for `3SUM` would be a surprise [5]. For a more in-depth combinatorial characterization of the geometric properties of the EPPM problem, see [7].

For the distance evaluation problems there are plethora of results. An excellent survey of the key results until 1999 is by Alt and Guibas [2]. As an example, consider in the 2-dimensional case how one can decide in $O(n \log n)$ time whether there is a transformation *f* composed of translation, rotation and scale, such that $f(A) = B$, where $A, B \subset \mathbb{R}^2$ and $n = |A| = |B|$: The idea is to convert *A* and *B* into an invariant form such that one can easily check their congruence under the transformations. First, scale is taken into account by scaling *A* to have the same *diameter* as *B* (in $O(n \log n)$ time). If *A* and *B* are congruent, then they must have the same *centroids* (which can be computed $O(n)$ time). Consider rotating a line from the centroid and listing the angles and distances to other points in the order they are met during the rotation. Having done this (in $O(n \log n)$ time) on both *A* and *B*, the lists of angles and distances should be *cyclic shifts* of each other; the list L_A of *A* occurs as a substring in $L_B L_B$, where L_B is the list of *B*. This latter step can be done in $O(n)$ time using any linear time exact string matching algorithm. One obtains the following result.

Theorem 3 (Atkinson 1987 [4]) *It is possible to decide in* $O(n \log n)$ *time whether there is* *a transformation f composed of translation, rotation and scale, such that $f(A) = B$, where $A, B \subset \mathbb{R}^2$ and $|A| = |B| = n$.*

Approximate variant of the above problem is much harder. Denote by $f(A) =^{\varepsilon} B$ the *directed approximate congruence* of point sets A and B, meaning that there is a one-to-one mapping from $f(A)$ to B such that for each point in $f(A)$ its image in B is ε-close. The following result demonstrates the added difficulty.

Theorem 4 (Alt et al. 1988 [3]) *It is possible to decide in $O(n^6)$ time whether there is a translation f such that $f(A) =^{\varepsilon} B$, where $A, B \subset \mathbb{R}^2$ and $|A| = |B| = n$. The same algorithm solves the corresponding LACP problem for point pattern P and target T under the one-to-one matching condition in $O((mn)^3)$ time, where $m = |P| \leq |T| = n$.*

To get an idea of the techniques to achieve the $O((mn)^3)$ time algorithm for LACP, consider first the one-dimensional version, i.e., let $P, T \subset \mathbb{R}$. Observe, that if there is a translation f' such that $f'(P) =^{\varepsilon} T$, then there is a translation f such that $f(P) =^{\varepsilon} T$ *and* a point $p \in P$ that is mapped *exactly* at ε-distance of a point $t \in T$. This lets one concentrate on these $2mn$ *representative translations*. Consider these translations sorted from left to right. Denote the left-most translation by f. Create a bipartite graph, whose nodes are the points in P and in T on the different parties. There is an edge between $p \in P$ and $t \in T$ if and only if $f(p)$ is ε-close to t. Finding a maximum matching in this graph tells the size of the largest approximately common point set after applying the translation f. One can repeat this on each representative translation to find the overall largest common point set. When the representative translations are considered from left to right, the bipartite graph instances are such that one can compute the maximum matchings greedily at each translation in time $O(|P|)$ [6]. Hence, the algorithm solves the one-dimensional LACP problem under translations and one-to-one matching condition in time $O(m^2 n)$, where $m = |P| \leq |T| = n$.

In the two-dimensional case, the set of representative translations is more implicitly defined: In short, the mapping of each point $p \in P$ ε-close to each point $t \in T$, gives mn circles. The boundary of each such circle is partitioned into intervals such that the end points of these intervals can be chosen as representative translations. There are $O((mn)^2)$ such representative translations. As in the one-dimensional case, each representative translation defines a bipartite graph. Once the representative translations along a circle are processed e.g., counterclockwise, the bipartite graph changes only by one edge at a time. This allows an $O(mn)$ time update for the maximum matching at each representative translation yielding an overall $O((mn)^3)$ time algorithm [3].

More efficient algorithms for variants of this problem have been developed by Efrat, Itai, and Katz [11], as by-products of more efficient bipartite matching algorithms for points on a plane. Their main result is the following:

Theorem 5 (Efrat et al. 2001 [11]) *It is possible to decide in $O(n^5 \log n)$ time whether there is a translation f such that $f(A) =^{\varepsilon} B$, where $A, B \subset \mathbb{R}^2$ and $|A| = |B| = n$.*

The problem becomes somewhat easier when the one-to-one matching condition is relaxed; one-to-one condition seems to necessitate the use of bipartite matching in one form or another. Without the condition, one can match the points independently of each other. This gives many tools to preprocess and manipulate the point sets during the algorithm using dynamic geometric data structures. Such techniques are exploited e.g., in the following result.

Theorem 6 (Chew and Kedem 1992 [8]) *The LACP problem under translations and using directed Hausdorff distance and the L_1 norm, can be solved in $O(mn \log n)$ time, where $P, T \subset \mathbb{R}^2$ and $m = |P| \leq |T| = n$. The distance evaluation problem for directed Hausdorff distance can be solved in $O(n^2 \log^2 n)$ time.*

Most algorithms revisited here have relatively high running times. To obtain faster algorithms, it seems that randomization and approximation techniques are necessary. See [9] for a comprehensive summary of the main achievements in that line of development.

Finally, note that the linear transformations considered here are not always enough to model a real-world problem–even when approximate congruence is allowed. Sometimes the proper transformation between two point sets (or between their subsets) is non-linear, without an easily parametrizable representation. Unfortunately, the formulations trying to capture such non-uniformness have been proven NP-hard [1] or even NP-hard to approximate within any constant factor [12].

Applications

Point pattern matching is a fundamental problem that naturally arises in many application domains such as computer vision, pattern recognition, image retrieval, music information retrieval, bioinformatics, dendrochronology, and many others.

Cross-References

- ▶ Assignment Problem
- ▶ Maximum Cardinality Stable Matchings
- ▶ Multidimensional String Matching
- ▶ String Matching

Recommended Reading

1. Akutsu T, Kanaya K, Ohyama A, Fujiyama A (2003) Point matching under non-uniform distortions. Discret Appl Math 127:5–21
2. Alt H, Guibas L (1999) Discrete geometric shapes: matching, interpolation, and approximation. In: Sack JR, Urrutia J (eds) Handbook of computational geometry. Elsevier, North-Holland, pp 121–153
3. Alt H, Mehlhorn K, Wagener H, Welzl E (1988) Congruence, similarity and symmetries of geometric objects. Discret Comput Geom 3:237–256
4. Atkinson MD (1997) An optimal algorithm for geometric congruence. J Algorithm 8:159–172
5. Barequet G, Har-Peled S (2001) Polygon containment and translational min-hausdorff-distance between segment sets are 3SUM-hard. Int J Comput Geom Appl 11(4):465–474
6. Böcker S, Mäkinen V (2005) Maximum line-pair stabbing problem and its variations. In: Proceedings of the 21st European workshop on computational geometry (EWCG'05). Technische Universität Eindhoven, pp 183–186
7. Brass P, Pach J (2005) Problems and results on geometric patterns. In: Avis D et al (eds) Graph theory and combinatorial optimization. Springer, New York, pp 17–36
8. Chew LP, Kedem K (1992) Improvements on geometric pattern matching problems. In: Proceedings of the Scandinavian workshop algorithm theory (SWAT). LNCS, vol 621. Springer, Berlin, pp 318–325
9. Choi V, Goyal N (2006) An efficient approximation algorithm for point pattern matching under noise. In: Proceedings of the 7th Latin American symposium on theoretical informatics (LATIN 2006). LNCS, vol 3882. Springer, Berlin, pp 298–310
10. Clifford R, Christodoukalis M, Crawford T, Meredith D, Wiggins G (2006) A fast, randomised, maximum subset matching algorithm for document-level music retrieval. In: Proceedings of the international conference on music information retrieval (ISMIR 2006), University of Victoria
11. Efrat A, Itai A, Katz M (2001) Geometry helps in bottleneck matching and related problems. Algorithmica 31(1):1–28
12. Mäkinen V, Ukkonen E (2002) Local similarity based point-pattern matching. In: Proceedings of the 13th annual symposium on combinatorial pattern matching (CPM 2002). LNCS, vol 2373. Springer, Berlin, pp 115–132
13. Ukkonen E, Lemström K, Mäkinen V (2003) Sweepline the music! In: Klein R, Six HW, Wegner L (eds) Computer science in perspective, essays dedicated to Thomas Ottmann. LNCS, vol 2598. Springer, pp 330–342

Polygon Triangulation

Jan Vahrenhold
Department of Computer Science, Westfälische Wilhelms-Universität Münster, Münster, Germany

Keywords

Computational geometry; Partitioning; Trapezoidation; Triangulation

Years and Authors of Summarized Original Work

1991; Chazelle

Problem Definition

Definition 1 A *simple polygon* is a polygon whose interior is simply connected, i.e., it consists of a single connected component and does not contain holes.

Definition 2 A triangulation of a simple polygon P with N vertices is a partition of the polygon, considered as a full-dimensional subset of the plane, into $N - 2$ nonoverlapping triangles such that the set of vertices of these triangles is the set of vertices of P, such that no edge of a triangle lies outside of P, and such that no triangle edges intersect except in their common endpoints.

Key Results

In addition to the regularization-based approach by Garey et al. [7], three other $\mathcal{O}(N \log N)$-time algorithms are milestones on the way toward an optimal linear-time algorithm. In the first of these algorithms [2], Chazelle uses a linear-time "polygon-cutting" approach to partition a simple polygon by a suitably chosen diagonal; the resulting divide-and-conquer scheme yields an $\mathcal{O}(N \log N)$-time algorithm for simple polygons. Hertel and Mehlhorn [8] present an $\mathcal{O}(N \log N)$-time plane-sweep algorithm and refine its analysis to yield an $\mathcal{O}(N + R \log R)$-time upper bound, where R is the number of concave polygon angles. Chazelle and Incerpi [4] present a divide-and-conquer algorithm with $\mathcal{O}(N + S \log S)$ running time; here, S denotes the maximum number of times the boundary of the polygon changes from "spiraling" to "antispiraling."

Polygon Triangulation in $o(N \log N)$ Time

Algorithms with $o(N \log N)$ running time were developed by Tarjan and van Wyk [14] (using Jordan sorting and finger trees) and Kirkpatrick, Klawe, and Tarjan [10] (using efficient point-location structures). Both algorithms can be shown to run in $\mathcal{O}(N \log \log N)$ time; the algorithm by Kirkpatrick et al. can be made to run in $\mathcal{O}(N \log^* N)$ time if the polygon's vertices have bounded integer coordinates. Clarkson, Tarjan, and van Wyk [7] restate the algorithm by Tarjan and van Wyk in a randomized setting using random sampling and develop a Las Vegas algorithm with $\mathcal{O}(N \log^* N)$ expected running time. The same expected running time can be obtained by a considerably simpler randomized incremental construction presented by Seidel [7]; as an added benefit, this algorithm constructs an efficient data structure for vertical ray shooting among a set of line segments.

Polygon Triangulation via Trapezoidation

The key to an efficient polygon triangulation algorithm was that polygon triangulation is linear-time equivalent to polygon trapezoidation. Here, the task is to compute for each vertex v of a simple polygon P the point (if any) of the boundary of P that is visible from v when shooting horizontal rays (*chords*) from v toward $\pm\infty$ through the interior of P. The resulting structure is called the *visibility map* of P.

Theorem 1 ([6]) *Given the trapezoidal decomposition of a simple polygon P, a triangulation of P can be computed in linear time and vice versa.*

The proof builds upon the fact that a trapezoidation for a set of points in general position, i.e., a set in which no two points share the same y-coordinate, consists of trapezoids, which may degenerate into triangles that have exactly two polygon vertices on their boundary. A trapezoid T is said to be a *class-A*-trapezoid if these vertices lie on the same side of T (this includes the case of triangles); otherwise, it is said to be a *class-B*-trapezoid. Fournier and Montuno observed that a polygon can be partitioned into so-called unimonotone polygons by adding diagonals between the vertices of class-B-trapezoids

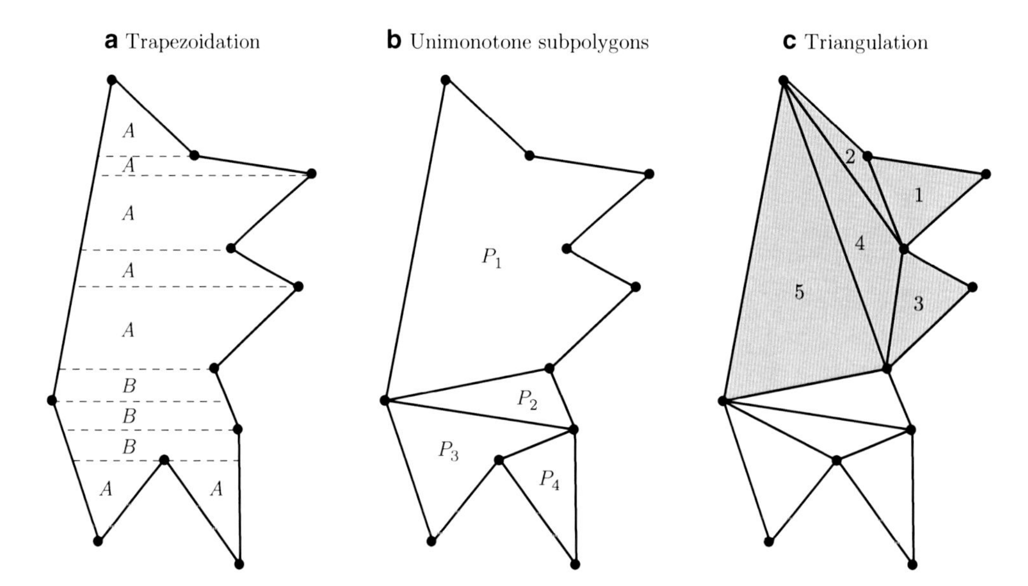

Polygon Triangulation, Fig. 1 Phases of Fournier and Montuno's algorithm [6]. By connecting the vertices of class-B-trapezoids in (**a**), the polygon is subdivided in unimonotone polygons as shown in (**b**). The numbers in (**c**) indicate the order in which the ear-cutting algorithms construct the triangles in the triangulation of the unimonotone subpolygon P_1 (shown in *gray*)

and that these unimonotone polygons can be triangulated independently in linear time – see Fig. 1.

Polygon Triangulation in Linear Time

The only deterministic linear-time algorithm for triangulating a simple polygon known so far is due to Chazelle [3]. Chazelle's algorithm uses a divide-and-conquer approach to compute the visibility map of a simple polygon. Since the divide-and-conquer approach subdivides the polygon's boundary into polygonal chains, there is no proper notion of the interior of the polygon through which the chords are supposed to pass. Instead, the polygon is embedded into the spherical plane on which the chords can "warp around infinity," and the visibility map is computed by always shooting rays in both directions.

Stated in terms of visibility maps, the algorithm's task now can be reduced to merging two visibility maps. To avoid linear-time merging steps which in turn would lead to $\Theta(N \log N)$ runtime, the algorithm proceeds as follows: in a first, bottom-up phase, the algorithm repeatedly merges the visibility maps of two subchains of the same length that share a common vertex. To ensure a sublinear running time, however, the algorithm does not compute the full visibility map, i.e., the map obtained by shooting rays from each vertex. Instead, for a subchain consisting of $2^\lambda + 1$ vertices, the algorithm maintains a $2^{\lceil \beta\lambda \rceil}$-*granular* visibility map.

Definition 3 A visibility map for a polygonal chain P is γ-granular for some $\gamma > 0$ if no part of the boundary of any region consists of more than γ consecutive edges of P and if no two adjacent regions can be merged without violating this property.

The consequence of this definition is that for polygons in general position, i.e., polygons for which no two vertices share the same y-coordinate, a γ-granular visibility map consists of $\mathcal{O}\left(\frac{N}{\gamma} + 1\right)$ regions, each of which is bounded by a constant number of chords and polygonal chains with a total complexity of $\mathcal{O}(\gamma)$. This enables a compact representation of each submap along with a uniform upper bound on the coarseness of the approximation of the visibility map.

For each subchain π considered throughout the bottom-up phase, the algorithm computes two "oracle" data structures, which are reused in the final phase of the algorithm. The first oracle, the so-called ray shooter, returns in $\mathcal{O}(f(\gamma))$ time for any point in the plane the first point of the γ-granular visibility map of π that is seen when shooting a horizontal ray in either direction. This oracle, whose construction is based upon Lipton and Tarjan's planar separator theorem [11], is used when merging the visibility map of π with the visibility map of a subchain π' that shares a common vertex p with π. Starting with p, the algorithm walks along π and π' and uses the respective ray shooter to update the visibility information for as many vertices as needed to guarantee the desired granularity. Due to Chazelle's polygon-cutting theorem [2], each region in either submap is closed under visibility. This implies that the "ray-shooter" oracle can be defined for each region separately, and only one oracle for a region in π needs to be queried for any vertex in π' (and vice versa) as long as the algorithm keeps track of the region the vertex currently under consideration lies in.

The second oracle, the so-called arc cutter, subdivides π in $\mathcal{O}(g(\gamma))$ time into $g(\gamma)$ subarcs each of which is given along with an $h(\gamma)$-granular visibility map. Using these two oracles, merging a γ_1-granular visibility map for a subchain consisting of N_1 vertices with a γ_2-granular visibility map for a subchain consisting of N_2 vertices, $\gamma_2 \geq \gamma_1$, can be done in

$$\mathcal{O}\left(\left(\frac{N_1}{\gamma_1} + \frac{N_2}{\gamma_2} + 1\right) f(\gamma_2)g(\gamma_2)(h(\gamma_2) + \log(N_1 + N_2))\right)$$

time. Chazelle proves that one can maintain these oracles such that $f(x) \in x^{0.74}$, $g(x) \in \mathcal{O}(\log x)$, and $h(x) \in \mathcal{O}(x^{0.20})$, with $\gamma_2 \in \mathcal{O}(N_2^{0.20})$; this eventually implies a sublinear merging step and hence a linear-time complexity of the bottom-up phase.

The final, top-down phase incrementally refines all regions of the visibility maps produced in the bottom-up phase. Using the "arc-cutter" oracle, each polygonal chain on the boundary of a region is subdivided into an appropriate number of subchains. As a result of the bottom-up phase and by carefully aligning the subchains considered in that phase with the results of the arc cutter, visibility maps, ray shooters, and arc cutters are available for each of these chains. The algorithm then uses the ray shooters to construct new chords and the arc cutters to further refine the visibility maps until the recursion bottoms out, and visibility maps of constant size can be refined by a brute-force algorithm. An inductive proof yields a linear runtime for the refinement of each region in the visibility map that was constructed in the bottom-up phase; hence, the overall running time is linear.

While Chazelle's algorithm uses only reasonably complicated data structures and subroutines, the analysis of both phases strongly suggests large constant factors hidden in the Big-Oh notation. In addition, the algorithm requires rather delicate implementation issues to be solved, in particular regarding the representation of the visibility maps, and thus it is not surprising that Chazelle mentioned developing a simpler, randomized algorithm with expected linear runtime as a major open problem.

Randomized Polygon Triangulation in Expected Linear Time

Over a decade after the publication of Chazelle's deterministic, linear-time algorithm, Amato, Goodrich, and Ramos [1] affirmatively answered Chazelle's question. Their algorithm follows Chazelle's two-phase approach and uses a bottom-up phase to preprocess helper data structures for so-called portal queries in the subchains' visibility maps. The top-down phase also subdivides the polygonal chain into smaller chains and refines the visibility maps. In contrast to Chazelle's algorithm, however, this refinement step is done on a random sample of the subchains only. As the sampling probability tends to one as the size of the subchain approaches $\mathcal{O}(1)$, both correctness and an expected linear runtime can be shown.

Applications

Being able to efficiently triangulate simple polygons has a variety of applications in computational geometry, computer graphics, and geographic information systems. For some of the problems considered, using a linear-time polygon triangulation algorithm is the key to obtaining an optimal algorithm. One such example among a variety of results is the optimal point-location scheme presented by Kirkpatrick [9] whose preprocessing time is linear assuming the availability of a linear-time triangulation algorithm. Several other applications are covered in O'Rourke's textbook on art gallery problems [12].

Open Problems

The main open problem is to devise an optimal deterministic algorithm that is reasonably efficient in practice.

Experimental Results

Due to its inherent complexity, Chazelle's algorithm has eluded a rigorous experimental evaluation so far. Preliminary results reported by Vahrenhold [15], however, seem to indicate that running even the first nontrivial stage of the bottom-up process takes significantly more time than running, e.g., Seidel's randomized algorithm [13] in full. Similarly, Amato et al. [1] conjecture that their randomized algorithm, despite its expected optimal runtime, is "not likely to be of practical value," either. Hence, for practical purposes, the simplicity of the deterministic algorithm by Hertel and Mehlhorn [8] and the randomized algorithm by Seidel [13] strongly advocates their use.

Cross-References

▸ Geometric Shortest Paths in the Plane
▸ Point Location

Recommended Reading

1. Amato NM, Goodrich MT, Ramos EA (2001) A randomized algorithm for triangulating a simple polygon in linear time. Discret Comput Geom 26:245–265
2. Chazelle BM (1982) A theorem on polygon cutting with applications. In: Proceedings of the twenty-third annual symposium on foundations of computer science, Chicago. IEEE, pp 339–349
3. Chazelle BM (1991) Triangulating a simple polygon in linear time. Discret Comput Geom 6:485–524
4. Chazelle BM, Incerpi JM (1984) Triangulation and shape-complexity. ACM Trans Graph 3(2):135–152
5. Clarkson KL, Tarjan RE, Van Wyk CJ (1989) A fast Las Vegas algorithm for triangulating a simple polygon. Discret Comput Geom 4:423–432
6. Fournier A, Montuno DY (1984) Triangulating simple polygons and equivalent problems. ACM Trans Graph 3(2):153–174
7. Garey MR, Johnson DS, Preparata FP, Tarjan RE (1978) Triangulating a simple polygon. Inf Process Lett 7(4):175–179
8. Hertel S, Mehlhorn K (1983) Fast triangulation of simple polygons. In: Karpinski M (ed) Proceedings of the 4th international conference on fundamentals of computation theory. Lecture notes in computer science, vol 158. Springer, Berlin, pp 207–218
9. Kirkpatrick DG (1983) Optimal search in planar subdivisions. SIAM J Comput 12(1):28–35
10. Kirkpatrick DG, Klawe MM, Tarjan RE (1992) Polygon triangulation in $O(n \log \log n)$ time with simple data structures. Discret Comput Geom 7:329–346
11. Lipton RJ, Tarjan RE (1979) A planar separator theorem for planar graphs. SIAM J Appl Math 36:177–189
12. O'Rourke J (1987) Art gallery theorems and algorithms. International series of monographs on computer science, vol 3. Oxford University Press, New York
13. Seidel R (1991) A simple and fast incremental randomized algorithm for computing trapezoidal decompositions and for triangulating polygons. Comput Geom Theory Appl 1(1):51–64
14. Tarjan RE, Van Wyk CJ (1988) An $O(n \log \log n)$-time algorithm for triangulating a simple polygon. SIAM J Comput 17(1):143–178; erratum in 17(5):1061 (1988)
15. Vahrenhold J (1996) Triangulierung eines einfachen Polygons in linearer Zeit. Master's thesis, Department of Computer Science, University of Münster (in German)

Position Auction

Aries Wei Sun
Department of Computer Science, City University of Hong Kong, Hong Kong, China

Keywords

Adword auction

Years and Authors of Summarized Original Work

2005; Varian

Problem Definition

This problem is concerned with the Nash equilibria of a game based on the ad auction used by Google and Yahoo. This research work [5] is motivated by the huge revenue that the adword auction derives every year. It defines two types of Nash equilibrium in the position auction game, applies economic analysis to the equilibria, and provides some empirical evidence that the Nash equilibria of the position auction describes the basic properties of the prices observed in Google's adword auction reasonably accurately. The problem being studied is closely related to the assignment game studied by [4, 1, 3]. And [2] has independently examined the problem and developed related results.

The Model and Its Notations

Consider the problem of assigning agents $a = 1, 2, \ldots, A$ to slots $s = 1, 2, \ldots, S$ where agent a's valuation for slot s is given by $u_{as} = v_a x_s$. The slots are numbered such that $x_1 > x_2 > \cdots > x_S$. It is assumed that $x_S = 0$ for all $s > S$ and the number of agents is greater than the number of slots. A higher position receives more clicks, so x_s can be interpreted as the click-through rate for slot s. The value $v_a > 0$ can be interpreted as the expected profit per click so $u_{as} = v_a x_s$ indicates the expected profit to advertiser a from appearing in slot s.

The slots are sold via an auction. Each agent bids an amount b_a, with the slot with the best click through rate being assigned to the agent with the highest bid, the second-best slot to the agent with the second highest bid, and so on. Renumbering the agents if necessary, let v_s be the value per click of the agent assigned to slot s. The price agent s faces is the bid of the agent immediately below him, so $p_t = b_{t+1}$. Hence the net profit that agent a can expect to make if he acquires slot s is $(v_a - p_s)\, x_s = (v_a - b_{s+1})\, x_s$.

Definitions

Definition 1 A Nash equilibrium set of prices *(NE)* satisfies

$$(v_s - p_s)\, x_s \geq (v_s - p_t)\, x_t, \text{for } t > s$$
$$(v_s - p_s)\, x_s \geq (v_s - p_{t-1})\, x_t, \text{for } t < s$$

where $p_t = b_{t+1}$.

Definition 2 A symmetric Nash equilibrium set of prices *(SNE)* satisfies

$$(v_s - p_s)\, x_s \geq (v_s - p_t)\, x_t \text{ for all } t \text{ and } s.$$

Equivalently,

$$v_s\,(x_s - x_t) \geq p_s x_s - p_t x_t \text{ for all } t \text{ and } s.$$

Key Results

Facts of NE and SNE

Fact 1 (Non-negative surplus) In an SNE, $v_s \geq p_s$.

Fact 2 (Monotone values) In an SNE, $v_{s-1} \geq v_s$, for all s.

Fact 3 (Monotone prices) In an SNE, $p_{s-1} x_{s-1} > p_s x_s$ and $p_{s-1} \geq p_s$ for all s. If $v_s > p_s$ then $p_{s-1} > p_s$.

Fact 4 ($NE \supset SNE$) If a set of prices is an SNE then it is an NE.

Fact 5 (One-step solution) If a set of bids satisfies the symmetric Nash equilibria inequalities for $s + 1$ and $s - 1$, then it satisfies these inequalities for all s.

Fact 6 The maximum revenue NE yields the same revenue as the upper recursive solution to the SNE.

A Sufficient and Necessary Condition of the Existence of a Pure Strategy Nash Equilibrium in the Position Auction Game

Theorem 1 *In the position auction described before, a pure strategy Nash equilibrium exists if and only if all the intervals*

$$\left[\frac{p_s x_s - p_{s+1} x_{s+1}}{x_s - x_{s+1}}, \frac{p_{s-1} x_{s-1} - p_s x_s}{x_{s-1} - x_s}\right], \quad \text{for } s = 2, 3, \ldots, S$$

are non-empty.

Applications

The model studied in this paper is a simple and elegant abstraction of the real adword auctions used by search engines such as Google and Yahoo. Different search engines have slightly different rules. For example, Yahoo ranks the advertisers according to their bids, while Google ranks the advertisers not only according to their bids but also according to the likelihood of their links being clicked.

However, similar analysis can be applied to real world situations, as the author has demonstrated above for the Google adword auction case.

Cross-References

▶ Adwords Pricing

Recommended Reading

1. Demange G, Gale D, Sotomayor M (1986) Multi-item auctions. J Polit Econ 94(4):863–872
2. Edelman B, Ostrovsky M, Schwartz M (2005) Internet advertising and the generalized second price auction. NBER Working Paper, 11765, Nov 2005
3. Roth A, Sotomayor M (1990) Two-sided matching. Cambridge University Press, Cambridge
4. Shapely L, Shubik M (1972) The assignment game I: the core. Int J Game Theory 1:111–130
5. Varian HR (2007) Position auctions. Int J Ind Organ 25(6):1163–1178

Power Grid Analysis

Sachin S. Sapatnekar
Department of Electrical and Computer Engineering, University of Minnesota, Minneapolis, MN, USA

Keywords

EDA; Electronic design automation; Noise; On-chip variations; Reliability; Signal integrity

Years and Authors of Summarized Original Work

2002; Zhao, Panda, Sapatnekar, Blaauw
2002; Kozhaya, Nassif, Najm
2005; Qian, Nassif, Sapatnekar

Problem Definition

The power grid of an integrated system is responsible for providing reliable supply and ground voltages to every circuit element in the system. Degradations in the supply voltage levels can result in parametric failures due to increased delays, whereby circuits no longer meet their specifications, as well as catastrophic failures due to incorrect gate switching. Further, power grids are susceptible to reliability faults due to catastrophic failure modes such as electromigration. Therefore, accurate power grid analysis is a vital step in integrated circuit design.

Power grids may be analyzed under DC waveforms that reflect the steady-state currents drawn by the circuit or under transient analysis that captures the response of the grid to specific time-varying current waveforms; inductive effects, particularly in the integrated circuit package; and decoupling capacitors that are deliberately placed in the circuit to temper the effect of large transients. For both DC and transient analysis, the problem can be abstracted as solving a set of linear algebraic equations of the following form:

$$G\mathbf{V} = \mathbf{E}, \quad (1)$$

where $G \in \Re^{N\times N}$ models the conductances in the system, $\mathbf{V} \in \Re^{N}$ is the vector of unknown node voltages, and $\mathbf{E} \in \Re^{N}$ is the right-hand side (RHS) vector, modeling the current loads. In case of DC analysis, a single such system must be solved, while in case of transient analysis, one such system is solved at each time step. For computational efficiency, a constant time step is often used during transient analysis of power grids in order to ensure that the G matrix, whose entries depend on the time step, remains unchanged through the simulation.

Given a power grid topology with $|E|$ resistors, these equations can be formulated using modified nodal analysis [2] in $O(|E|)$ time. Matrix G is sparse and diagonally dominant ($\sum_{i\neq j} |g_{ij}| \leq g_{ii}, \forall i$), and all off-diagonals of G are less than or equal to zero.

The task of power grid analysis is to determine all voltage levels in the system and verify that the maximum deviation from their ideal values is within a user-specified bound and to ensure that the current density in each wire is within a user-specified limit in order to assure resilience to electromigration failure.

Key Results

Mainstream methods for solving such systems include direct methods such as LU/Cholesky factorization and iterative methods. Due to the favorable structural properties of the power grid, notably sparsity and diagonal dominance, it is possible to solve these systems efficiently. However, the scale of the problem, where power grids may have hundreds of millions or billions of resistors, poses large memory and computation challenges even to the most efficient solvers. As a result, there has been considerable work on developing specialized solvers for power grids. Notable contributions in this direction are described below.

Hierarchy-Based Solvers

In real designs, the power grid is inherently hierarchical since it is created as a part of a hierarchical design process, where individual blocks with locally constructed power are first designed individually and then assembled at the chip level. This structure is exploited in [9] to build a hierarchical solution to the grid.

Based on inherent hierarchy, the power grid has k local partitions, corresponding to blocks, and a global partition that connects the power grids within these blocks. The *global grid* is then defined to include the set of nodes that lie in the global partition and the port nodes, while the grid in a local partition constitutes a *local grid*. The local grid is connected to the global grid through a set of nodes called *ports*, and due to the hierarchical structure, the number of port nodes is a small fraction of all nodes. The technique consists of the following steps:

Macromodeling Each of the k local grids may be modeled as a multi-port linear element represented by a macromodel of the type $\mathbf{I} = A \cdot \mathbf{V} + \mathbf{S}$, where $\mathbf{I}$ and $\mathbf{V}$ are the vectors of port voltages and A and $\mathbf{S}$ are, respectively, a constant matrix and a constant vector. Here, the A matrix is sparsified with bounded and minimal loss of accuracy using an integer knapsack scheme.

Solution of the global grid Once the macromodels for all the local grids are generated, the entire network is abstracted simply as the global grid, with the macromodel elements connected to it at the port nodes. This system is solved to determine the voltages at all ports.

Solution of the local grids Given the port voltages, the local grids are then each separately solved to provide the solution to the entire system.

P

Multigrid-Based Solvers

Multigrid-based approaches are an effective way of solving large systems of equations and have been customized to solve power grids [3]. The solution proceeds by creating a coarsened form of the network with a reduced number of nodes, which can be solved efficiently, and then by propagating the result of this solution to the full network. The technique consists of four steps:

Grid reduction, in which the large power grid is coarsened by selecting a subset of nodes that are to be maintained, while the other nodes are removed. The number of variables is therefore significantly reduced from from n to m.

Interpolation, in which an $n \times m$ interpolation operator matrix P is defined to map the original grid to the coarsened grid. This interpolation operator relates the voltages on the removed nodes to those on the coarsened grid, thereby allowing the solution of the coarsened grid to accurately reflect that of the original grid.

Solution of the coarsened grid, in which a solution is found for the voltages in the coarsened grid by solving the above linear equations.

Mapping the solution from the coarsened grid to the original grid by applying the interpolation operator concludes the process.

Random Walk-Based Solvers

The diagonal dominance of the power grid enables a special property that creates an exact mapping between the solution of the power grid equations and the use of random walks on a network. This idea has been used in [8] and further sped up in [6]. Unlike other approaches that require all (or most) nodes in a system to be solved together, random walk approaches allow for a single node to be solved alone. This is particularly useful during incremental analysis and optimization [1].

The family of random walk methods has been extended to solve entire systems, in a marriage with iterative linear solvers based on conjugate gradient methods. The intuition is that since random walks provide approximate solutions rapidly, they can be used to build effective preconditioners for iterative solvers [7].

Applications

Power grid analysis is a vital ingredient in the design of every integrated circuit, and there are several commercial offerings of design automation tools that analyze power grids. Aside from the problem of solving the linear system, the issue of determining the worst-case excitation is also a difficult problem. In spite of numerous efforts, automated tools for this purpose have been excessively pessimistic and therefore ineffective. It is generally accepted that user-specified patterns are the most effective way to provide input excitations, particularly in a design world where the power grid must be analyzed at multiple corners and multiple modes (corresponding to different supply voltages that could be applied to the circuit).

Experimental Results

Intelligent solutions for solving linear systems of equations have found extensive use in the analysis of power grids. Solvers that are used include direct solvers as well as iterative solvers based on methods such as preconditioned conjugate gradient-based solvers. Preconditioners based on methods such as support trees have been found to be useful, and random walk preconditioners have also been shown to outperform conventional methods. Due to the computational nature of this task, there has been active work on developing parallel and multithreaded power grid solvers. For example, the 2011 and 2012 editions of the Tau workshop have hosted contests on solving power grid problems [4, 5].

URLs to Code and Data Sets

A set of power grid benchmarks have been made available to the community at http://dropzone.tamu.edu/~pli/PGBench.

Recommended Reading

1. Boghrati B, Sapatnekar SS (2014) Incremental analysis of power grids using backward random walks. ACM Trans Des Autom Electron Syst 19(3), Article 31. doi:10.1145/2611763. http://doi.acm.org/10.1145/2611763
2. Ho CW, Ruehli AE, Brennan PA (1975) The modified nodal approach to network analysis. IEEE Trans Circuits Syst 22(6):505–509
3. Kozhaya J, Nassif SR, Najm FN (2002) A multigrid-like technique for power grid analysis. IEEE Trans Comput-Aided Des Integr Circuits Syst 21(10):1148–1160
4. Li Z, Balasubramanian R, Liu F, Nassif S (2011) 2011 TAU power grid simulation contest: benchmark suite and results. In: Proceedings of the IEEE/ACM international conference on computer-aided design, San Jose, pp 478–481
5. Li Z, Balasubramanian R, Liu F, Nassif S (2012) 2012 TAU power grid simulation contest: benchmark suite and results. In: Proceedings of the IEEE/ACM international conference on computer-aided design, San Jose, pp 643–646
6. Miyakawa T, Yamanaga K, Tsutsui H, Ochi H, Sato T (2011) Acceleration of random-walk-based linear circuit analysis using importance sampling. In: Proceedings of the Great Lakes symposium on VLSI, Lausanne, pp 211–216
7. Qian H, Sapatnekar SS (2008) Stochastic preconditioning for diagonally dominant matrices. SIAM J Sci Comput 30(3):1178–1204
8. Qian H, Nassif SR, Sapatnekar SS (2005) Power grid analysis using random walks. IEEE Trans Comput-Aided Des Integr Circuits Syst 24(8):1204–1224
9. Zhao M, Panda RV, Sapatnekar SS, Blaauw D (2002) Hierarchical analysis of power distribution networks. IEEE Trans Comput-Aided Des Integr Circuits Syst 21(2):159–168

Predecessor Search

Mihai Pătraşcu
Computer Science and Artificial Intelligence Laboratory (CSAIL), Massachusetts Institute of Technology (MIT), Cambridge, MA, USA

Keywords

IP lookup; Predecessor problem; Successor problem

Years and Authors of Summarized Original Work

2006; Pătraşcu, Thorup

Problem Definition

Consider an ordered universe U, and a set $T \subset U$ with $|T| = n$. The goal is to preprocess T, such that the following query can be answered efficiently: given $x \in U$, report the predecessor of x, i.e., $\max\{y \in T \mid y < x\}$. One can also consider the dynamic problem, where elements are inserted and deleted into T. Let t_q be the query time, and t_u the update time.

This is a fundamental search problem, with an impressive number of applications. Later, this entry discusses IP lookup (forwarding packets on the Internet), orthogonal range queries and persistent data structures as examples.

The problem was considered in many computational models. In fact, most models below were initially defined to study the predecessor problem.

Comparison model: The problem can be solved through binary search in $\Theta(\lg n)$ comparisons. There is a lot of work on adaptive bounds, which may be sublogarithmic. Such bounds may depend on the finger distance, the working set, entropy etc.

Binary search trees: Predecessor search is one of the fundamental motivations for binary search trees. In this restrictive model, one can hope for an instance optimal (competitive) algorithm. Attempts to achieve this are described in a separate entry. ($O(\log\log n)$-*competitive Binary Search Trees* (2004; Demaine, Harmon, Iacono, Pătraşcu))

Word RAM: Memory is organized as words of b bits, and can be accessed through indirection. Constant-time operations include the standard operations in a language such as C (addition, multiplication, shifts and bitwise operations).

It is standard to assume the universe is $U = \{1, \ldots, 2^{\ell}\}$, i.e., one deals with ℓ-bit integers.

The floating point representation was designed so that order is preserved when values are interpreted as integers, so any algorithm will also work for ℓ-bit floating point numbers.

The standard *transdichotomous* assumption is that $b = \ell$, so that an input integer is represented in a word. This implies $b \geq \lg n$.

Cell-probe model: This is a nonuniform model stronger than the word RAM, in which the operations are arbitrary functions on the memory words (cells) which have already been probed. Thus, t_q only counts the number of cell probes. This is an ideal model for lower bounds, since it does not depend on the operations implemented by a particular computer.

Communication games: Let Alice have the query x, and Bob have the set T. They are trying to find the predecessor of x through τ rounds of communication, where in each round Alice sends m_A bits, and Bob replies with m_B bits.

This can simulate the cell-probe model when $m_B = b$ and m_A is the logarithm of the memory size. Then $\tau \leq t_q$ and one can use communication complexity to obtain cell-probe lower bounds.

External memory: The unit of access is a page, containing B words of ℓ bits each. B-trees solve the problem with query and update time $O(\log_B n)$, and one can also achieve this oblivious to the value of B (See *Cache-oblivious B-tree* (2005; Bender, Demaine, Farach-Colton).). The cell-probe model with $b = B \cdot \ell$ is stronger than this model.

AC^0 **RAM:** This is a variant of the word RAM in which allowable operations are functions that have constant depth, unbounded fan-in circuits. This excludes multiplication from the standard set of operations.

RAMBO: this is a variant of the RAM with a nonstandard memory, where words of memory can overlap in their bits. In the static case this is essentially equivalent to a normal RAM. However, in the dynamic case updates can be faster due to the word overlap [5].

The worst-case logarithmic bound for comparison search is not particularly informative when efficiency really matters. In practice, B-trees and variants are standard when dealing with huge data sets. Solutions based on RAM tricks are essential when the data set is not too large, but a fast query time is crucial, such as in software solutions to IP lookup [7].

Key Results

Building on a long line of research, Pătraşcu and Thorup [15, 16] finally obtained matching upper and lower bounds for the static problem in the word RAM, cell-probe, external memory and communication game models.

Let S be the number of words of space available. (In external memory, this is equivalent to S/B pages.) Define $a = \lg S \cdot \ell/n$. Also define $\lg x = \lceil \log_2(x+2) \rceil$, so that $\lg x \geq 1$ even if $x \in [0, 1]$. Then the optimal search time is, up to constant factors:

$$\min \begin{cases} \log_b n = \Theta(\min\{\log_B n, \log_\ell n\}) \\ \lg \frac{\ell - \lg n}{a} \\ \dfrac{\lg \frac{\ell}{a}}{\lg\left(\frac{a}{\lg n} \cdot \lg \frac{\ell}{a}\right)} \\ \dfrac{\lg \frac{\ell}{a}}{\lg\left(\frac{\lg \frac{\ell}{a}}{\lg \frac{\lg n}{a}}\right)} \end{cases} \tag{1}$$

The bound is achieved by a deterministic query algorithm. For any space S, the data structure can be constructed in time $O(S)$ by a randomized algorithm, starting with the set T given in sorted order. Updates are supported in expected time $t_q + O(S/n)$. Thus, besides locating the element through one predecessor query, updates change a minimal fraction of the data structure.

Lower bounds hold in the powerful cell-probe model, and hold even for randomized algorithms. When $S \geq n^{1+\varepsilon}$, the optimal trade-off for communication games coincides to (1). Note that the case $S = n^{1+o(1)}$ essentially disappears in the reduction to communication complexity, because Alice's messages only depends on $\lg S$.

Thus, there is no asymptotic difference between $S = O(n)$ and, say, $S = O(n^2)$.

Upper Bounds

The following algorithmic techniques give the optimal result:

- *B-trees* give $O(\log_B n)$ query time with linear space.
- *Fusion trees*, by Fredman and Willard [10], achieve a query time of $O(\log_b n)$. The basis of this is a *fusion node*, a structure which can search among b^ε values in constant time. This is done by recognizing that only a few bits of each value are essential, and packing the relevant information about all values in a single word.
- *Van Emde Boas search* [18] can solve the problem in $O(\lg \ell)$ time by binary searching for the length of the longest common prefix between the query and a value in T. Beginning the search with a table lookup based on the first $\lg n$ bits, and ending when there is enough space to store all answers, the query time is reduced to $O(\lg((\ell - \lg n)/a))$.
- A technique by *Beame and Fich* [4] can perform a multiway search for the longest common prefix, by maintaining a careful balance between ℓ and n. This is relevant when the space is at least $n^{1+\varepsilon}$, and gives the third branch of (1). Pătraşcu and Thorup [15] show how related ideas can be implemented with smaller space, yielding the last branch of (1).

Observe that external memory only features in the optimal trade-off through the $O(\log_B n)$ term coming from B-trees. Thus, it is optimal to either use the standard, comparison-based B-trees, or use the best word RAM strategy which completely ignores external memory.

Lower Bounds

All lower bounds before [15] where shown in the communication game model. Ajtai [1] was the first to prove a superconstant lower bound. His results, with a correction by Miltersen [12], show that for polynomial space, there exists n as a function of ℓ making the query time $\Omega(\sqrt{\lg \ell})$, and likewise there exists ℓ a function of n making the query complexity $\Omega(\sqrt[3]{\lg n})$.

Miltersen et al. [13] revisited Ajtai's proof, extending it to randomized algorithms. More importantly, they captured the essence of the proof in an independent *round elimination lemma*, which is an important tool for proving lower bounds in asymmetric communication.

Beame and Fich [4] improved Ajtai's lower bounds to $\Omega(\lg \ell / \lg \lg \ell)$ and $\Omega(\sqrt{\lg n / \lg \lg n})$ respectively. Sen and Venkatesh [17] later gave an improved round elimination lemma, which can reprove these lower bounds, but also for randomized algorithms.

Finally, using the message compression lemma of [6] (an alternative to round elimination), Pătraşcu and Thorup [15] showed an optimal trade-off for communication games. This is also an optimal lower bound in the other models when $S \geq n^{1+\varepsilon}$, but not for smaller space.

More importantly, [15] developed the first tools for proving lower bounds exceeding communication complexity, when $S = n^{1+o(1)}$. This showed the first separation ever between a data structure or polynomial size, and one of near linear size. This is fundamentally impossible through a direct communication lower bound, since the reduction to communication games only depends on $\lg S$.

The full result of Pătraşcu and Thorup [15] it the trade-off (1). Initially, this was shown only for deterministic query algorithms, but eventually it was extended to a randomized lower bound as well [16]. Among the surprising consequences of this result was that the classic van Emde Boas search is optimal for near-linear space (and thus for dynamic data structures), whereas with quadratic space it can be beaten by the technique of Beame and Fich.

A key technical idea of [15] is to analyze many queries simultaneously. Then, one considers a communication game involving all queries, and proves a direct-sum version of the round elimination lemma. Arguably, the proof is even simpler than for the regular round elimination lemma. This is achieved by considering a stronger model for the inductive analysis, in which the algorithm is allowed to *reject* a large fraction of the queries before starting to make probes.

Bucketing

The rich recursive structure of the problem can not only be used for fast queries, but also to optimize the space and update time – of course, within the limits of (1). The idea is to place ranges of consecutive values in buckets, and include a single representative of each bucket in the predecessor structure. After performing a query on the predecessor structure (now with fewer elements), one need only search within the relevant bucket.

Because buckets of size $w^{O(1)}$ can be handled in constant time by fusion trees, it follows that factors of w in space are irrelevant. A more extreme application of the idea is given by *exponential trees* [3]. Here buckets have size $\Theta(n^{1-\gamma})$, where γ is a sufficiently small constant. Buckets are handled recursively in the same way, leading to $O(\lg\lg n)$ levels. If the initial query time is at least $t_q \geq \lg^{\varepsilon} n$, the query times at each level decrease geometrically, so overall time only grows by a constant factor. However, any polynomial space is reduced to linear, for an appropriate choice of γ. Also, the exponential tree can be updated in $O(t_q)$ time, even if the original data structure was static.

Applications

Perhaps the most important application of predecessor search is IP lookup. This is the problem solved by routers for each packet on the Internet, when deciding which subnetwork to forward the packet to. Thus, it is probably the most run algorithmic problem in the world. Formally, this is an *interval stabbing* query, which is equivalent to predecessor search in the static case [9]. As this is a problem where efficiency really matters, it is important to note that the fastest deployed software solutions [7] use integer search strategies (not comparison-based), as theoretical results would predict.

In addition, predecessor search is used pervasively in data structures, when reducing problems to *rank space*. Given a set T, one often wants to relabel it to the simpler $\{1, \ldots, n\}$ ("rank space"), while maintaining order relations. If one is presented with new values dynamically, the need for a predecessor query arises. Here are a couple of illustrative examples:

- In orthogonal range queries, one maintains a set of points in U^d, and queries for points in some rectangle $[a_1, b_1] \times \cdots \times [a_d, b_d]$. Though bounds typically grow exponentially with the dimension, the dependence on the universe can be factored out. At query time, one first runs $2d$ predecessor queries transforming the universe to $\{1, \ldots, n\}^d$.
- To make pointer data structures persistent [8], an outgoing link is replaced by a vector of pointers, each valid for some period of time. Deciding which link to follow (given the time being queried) is a predecessor problem.

Finally, it is interesting to note that the lower bounds for predecessor hold, by reductions, for all applications described above. To make these reductions possible, the lower bounds are in fact shown for the weaker *colored predecessor* problem. In this problem, the values in T are colored red or blue, and the query only needs to find the color of the predecessor.

Open Problems

It is known [2] how to implement fusion trees with AC^0 instructions, but not the other query strategies. What is the best query trade-off achievable on the AC^0 RAM? In particular, can van Emde Boas search be implemented with AC^0 instructions?

For the dynamic problem, can the update times be made deterministic? In particular, can van Emde Boas search be implemented with fast deterministic updates? This is a very appealing problem, with applications to deterministic dictionaries [14]. Also, can fusion nodes be updated deterministically in constant time? Atomic heaps [11] achieve this when searching only among $(\lg n)^{\varepsilon}$ elements, not b^{ε}.

Finally, does an update to the predecessor structure require a query? In other words, can $t_u = o(t_q)$ be obtained, while still maintaining efficient query times?

Cross-References

▶ Cache-Oblivious B-Tree
▶ O(log log n)-Competitive Binary Search Tree

Recommended Reading

1. Ajtai M (1988) A lower bound for finding predecessors in Yao's cell probemodel. Combinatorica 8(3):235–247
2. Andersson A, Miltersen PB, Thorup M (1999) Fusion trees can be implemented with AC^0 instructions only. Theor Comput Sci 215(1–2):337–344
3. Andersson A, Thorup M (2002) Dynamic ordered sets with exponential search trees. CoRR cs.DS/0210006. See also FOCS'96, STOC'00.
4. Beame P, Fich FE (2002) Optimal bounds for the predecessor problem and related problems. J Comput Syst Sci 65(1):38–72, See also STOC'99
5. Brodnik A, Carlsson S, Fredman ML, Karlsson J, Munro JI (2005) Worst case constant time priority queue. J Syst Softw 78(3):249–256, See also SODA'01
6. Chakrabarti A, Regev O (2004) An optimal randomised cell probe lower bound for approximate nearest neighbour searching. In: Proceedings of the 45th IEEE symposium on foundations of computer science (FOCS), pp 473–482
7. Degermark M, Brodnik A, Carlsson S, Pink S (1997) Small forwarding tables for fast routing lookups. In: Proceedings of the ACM SIGCOMM, pp 3–14
8. Driscoll JR, Sarnak N, Sleator DD, Tarjan RE (1989) Making data structures persistent. J Comput Syst Sci 38(1):86–124, See also STOC'86
9. Feldmann A, Muthukrishnan S (2000) Tradeoffs for packet classification. In: Proceedings of the IEEE INFOCOM, pp 1193–1202
10. Fredman ML, Willard DE (1993) Surpassing the information theoretic bound with fusion trees. J Comput Syst Sci 47(3):424–436, See also STOC'90
11. Fredman ML, Willard DE (1994) Trans-dichotomous algorithms for minimum spanning trees and shortest paths. J Comput Syst Sci 48(3):533–551, See also FOCS'90
12. Miltersen PB (1994) Lower bounds for union-split-find related problems on random access machines. In: 26th ACM symposium on theory of computing (STOC), pp 625–634
13. Miltersen PB, Nisan N, Safra S, Wigderson A (1998) On data structures and asymmetric communication complexity. J Comput Syst Sci 57(1):37–49, See also STOC'95
14. Pagh R (2000) A trade-off for worst-case efficient dictionaries. Nord J Comput 7:151–163, See also SWAT'00
15. Pătraşcu M, Thorup M (2006) Time-space trade-offs for predecessor search. In: Proceedings of the 38th ACM symposium on theory of computing (STOC), pp 232–240
16. Pătraşcu, M, Thorup M (2007) Randomization does not help searching predecessors. In: Proceedings of the 18th ACM/SIAM symposium on discrete algorithms (SODA)
17. Sen P, Venkatesh S (2003) Lower bounds for predecessor searching in the cell probe model. arXiv:cs.CC/0309033. See also ICALP'01, CCC'03
18. van Emde Boas P, Kaas R, Zijlstra E (1977) Design and implementation of an efficient priority queue. Math Syst Theory 10:99–127, Announced by van Emde Boas alone at FOCS'75

Predecessor Search, String Algorithms and Data Structures

Djamal Belazzougui
Department of Computer Science, Helsinki Institute for Information Technology (HIIT), University of Helsinki, Helsinki, Finland

Keywords

Cell probe model; Integer search; RAM model

Years and Authors of Summarized Original Work

1975; Van Emde Boas
1993; Fredman, Willard
2002; Beame, Fich
2006; Pătraşcu, Thorup
2007; Andersson, Thorup
2014; Pătraşcu, Thorup

Problem Definition

Given a set S of n keys from the set $[1 \ldots 2^\ell]$, the goal of *predecessor search* is to return, given a key $y \in [1 \ldots 2^\ell]$, the largest key $x \in S$ such that $x \leq y$. We have the following models:

Comparison model Balanced binary search trees [1, 6] can solve the problem in optimal $O(\log n)$ time in the comparison model, in which the key can only be manipulated through comparisons with each other.

External memory model In this model, it is assumed that the data is read and written into blocks of B elements (integers in our case) and the cost of a query or algorithm is the number of read or written blocks. In this model, B-trees [7] can solve the problem in $O(\log_B n)$ time and $O(n)$ space.

RAM model This is the main subject of study. The model assumes that all standard arithmetic and logic operations (including multiplication) on integers of length w take constant time, where w is the computer word size. It is assumed that $w \geq \ell \geq \log n$.

AC^0 RAM model This model is similar to the RAM model except that it only contains instructions that can be implemented with circuits of polynomial size, constant depth and unfounded fan-in. The only affected instruction is multiplication which cannot be implemented with such circuits. While this model has been often considered in the literature in the past, modern computers support multiplications very fast, and the bottleneck is usually the memory access.

Cell probe model This model is used to prove lower bounds. It also has an associated word size w, but the cost of a query or an update is just the number of accessed memory words (computations have zero cost).

Useful Concepts

The two main techniques used for predecessor search are *cardinality reduction* and *length reduction*. At every step of the query, one would wish to either reduce the cardinality of the searched set or reduce the length of the searched keys.

Balanced Binary Search Tree

Cardinality reduction is achieved through the use of *balanced binary search tree*. This allows doing a predecessor search in $O(\log n)$ time in the comparison model. The B-tree is a generalization of the balanced binary search trees, where every node can have B children instead of just two. A static balanced binary search tree allows one to divide the set of searched keys by 2 at every level. A static B-tree allows one to divide the set by B. In the dynamic case, the cardinality can be reduced by a factor less than 2 (less than B for a dynamic B-tree) at every level, but it is guaranteed that the cardinality goes to one after $\Theta(\log n)$ levels ($\Theta(\log_B n)$ for a B-tree).

Trie

A key concept is that of a *trie*. A predecessor search for a key in a trie takes $O(\ell)$ time. A trie built on a set S of n keys from $[1 \ldots 2^\ell]$ is a binary tree with ℓ levels numbered top-down. All the leaves of the trie are at level ℓ. Every edge in the tree has a label that is either 0 or 1. Let $x[i]$ denote the bit number i of the integer x, where $i \in [1 \ldots \ell]$ and the bits are numbered from the most to the least significant bit ($x[1]$ is the most significant bit). Denote by $x[i \ldots j]$ the binary string that consists in the concatenation of the bits $x[i], \ldots x[j]$. A node of the trie at depth $d \in [0, \ell]$ will be labeled by a binary string of length d, formed by concatenating the edge labels from the root to the node. There will be a node of depth d labeled by binary string p if and only if there exists at least an element $x \in S$ such that $x[1 \ldots d] = p$. A node at depth $d < \ell$ labeled by string p will have as children the nodes at depth $d + 1$ labeled by $p0$ and $p1$ (if they both exist, otherwise it will only have one child). The leaves of the trie are labeled by strings of S and the root is labeled by the empty string. A trie occupies $O(n\ell \log n)$ bits. In order to support predecessor searches, every internal node of the trie stores two elements of S. The two elements stored by a node labeled by binary string p are the largest element prefixed by p and the predecessor of $p0^{w-|p|}$. A predecessor search on a compacted trie for a key y is then done by traversing the trie top-down, at level $i \in [1 \dots \ell]$ following the edge labeled by character $y[i+1]$. The node at which the traversal stops is called the *locus* of y. The predecessor of y is then easily determined from its locus. If the locus is a leaf labeled by x, then necessarily $x = y$ and y is returned as the predecessor. Otherwise, the locus is some internal node and the predecessor is one of the two elements of S stored at that internal node. Suppose that the internal node is at level i and is labeled by string p. Then, if $x[i+1] = 1$, the predecessor is the largest element prefixed by p; otherwise, it is the predecessor of $p0^{w-p}$. In a *compacted trie* (*Patricia trie*), only leaves and internal nodes with degree 2 are kept, resulting in a tree of $2n-1$ nodes, n leaves, and $n-1$ internal nodes (see Fig. 1). The trie will then occupy only $O(n(\ell + \log n))$ bits. A predecessor search can be implemented on a compacted trie in a way similar to the non-compacted trie. The main difference is that the locus in a compacted trie is either a leaf or a location in the middle of a compacted edge.

Key Results

Van Emde Boas

The *Van Emde Boas tree* [17–19] is a compacted trie representation supporting predecessor search by doing a binary search on the trie levels. Since the number of levels is ℓ, the binary search takes time $O(\log \ell)$. The original structure used space $O(2^{\ell})$. Later, Willard [20] showed how to use hashing to reduce the space usage to $O(n)$ while keeping the search time bounded by $O(\log \ell)$.

Searching a Compacted Trie (Fusion Trees)

The fusion tree reduces the predecessor search problem to the search on small compacted tries, each encoded in one memory word. The idea of the fusion tree dates back to a paper by Ajtai et al. [2], where it was shown how to implement predecessor search using a compacted trie in which the compacted paths are omitted and only their lengths are stored (the trie on the right in Fig. 1). This allows encoding a trie in $O(n(\log \ell + \log n))$ bits only compared to the $O(n(\ell + \log n))$ bits for the ordinary compacted trie. Then, a predecessor search for a key y is done by first doing a top-down traversal of the compacted trie, by always pretending that a comparison between the compacted paths and the bits of searched key is successful (only bits that label the edges are compared to bits of the searched key). At the end, the search terminates at a leaf that points to an integer $x \in S$ that is one of the elements that share the longest prefix with the searched key. Then, the locus (and hence the predecessor) is determined by filling the traversed compacted paths with bits from x and comparing those bits with the corresponding bits of y.

The main observation is that a predecessor search can be supported in constant time in the cell probe model whenever $n \log \ell \leq w$, since it

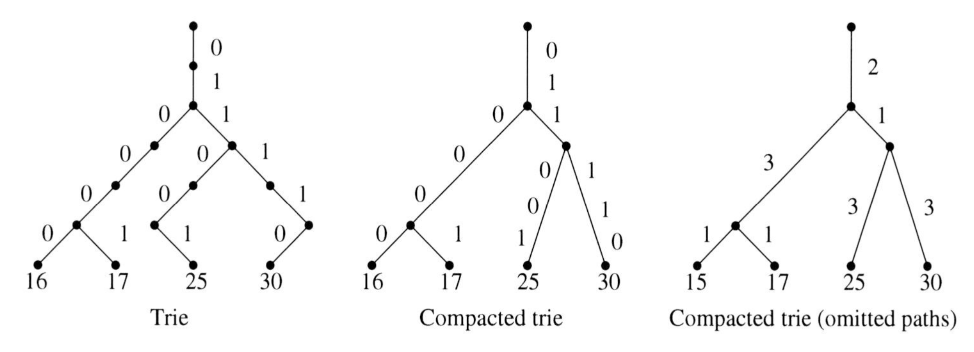

Predecessor Search, String Algorithms and Data Structures, Fig. 1 Three trie variants used for storing the integers 16, 17, 25, and 30. On the *left*, an ordinary trie; in the *middle*, a compacted trie; and on the *right*, a compacted trie in which the compacted paths are omitted and replaced by their lengths. Notice that the compacted trie has exactly $2 \cdot 4 - 1 = 7$ nodes involves reading a constant number of memory locations corresponding to the trie encoded in a constant number of words and to two elements of S (the traversal of the trie is charged zero cost). Later, Fredman and Willard [11] invented the fusion node, where they implemented a predecessor search in constant time in the RAM model, when $n \leq w^{1/c}$ for $c = 6$. This allows searching the predecessors among a set of $w^{1/c}$ keys for $c = 6$ in constant time. By implementing a B-tree with block size $B = w^{1/c}$, one can achieve query time $O(\log_B n) = O(\log n / \log w)$. Finally, Pătraşcu and Thorup [15] have shown how to implement the approach with constant time updates and queries on $B = w^{1/c}$ keys for $c = 4$. This allows searches and dynamic updates in deterministic $O(\log n / \log w)$ time.

Beame and Fich

Beame and Fich [8] use a more advanced search that combines cardinality and length reductions. As a building block, they use a data structure that recursively reduces a search over n keys of length ℓ to a search over a group of n' keys of length ℓ' such that either $\ell' = \ell$ but $n' \leq q$ or $\ell' = \ell / h$ for some parameters h and q. This reduction technique was taken from an algorithm used for integer sorting [5]. Their search time is $O\left(\frac{\log \ell}{\log \log \ell}\right)$ and the space is quadratic. Combining with the fusion tree, this gives query time $O\left(\sqrt{\frac{\log n}{\log \log n}}\right)$ with quadratic space. This is achieved by using the fusion tree when $\log \ell \geq \sqrt{\log n \log \log n}$ and the new data structure when $\log \ell < \sqrt{\log n \log \log n}$. They also prove a matching lower bound, by showing that one cannot achieve space polynomial in n with query time $o\left(\frac{\log \ell}{\log \log \ell}\right)$ for all values of n and w or query time $o\left(\sqrt{\frac{\log n}{\log \log n}}\right)$ for all values of ℓ and w. Later, Sen and Venkatesh show that their lower bound holds even if randomization is allowed [16].

Exponential Search Trees

The *exponential search tree* [4] allows one to transform any predecessor data structure with polynomial space and preprocessing time, and query time $q(n, \ell, w)$, into a data structure with $O(n)$ space and preprocessing time, and query time at most $O(\log \log n \cdot q(n, \ell, w))$. Given some predecessor data structure P that supports queries in time $q(n, \ell, w)$ and uses space and preprocessing time $S(n) = O(n^c)$, the exponential search tree is a recursive data structure built from a sorted set $x_1 < \cdots < x_n$ of n keys as follows:

1. The root has degree $d = n^{1/(c+1)}$.
2. The n keys are partitioned into d blocks of sizes $b = \frac{n}{d}$ each and the data structure P is built on the set which consists in the first element of each of the blocks $2 \ldots d$ (the elements $x_{b+1}, \ldots, x_{(d-1)b+1}$).
3. The root has d children, where every child is itself an exponential search tree built on the $b = n^{c/(c+1)}$ elements of a block.

The recursion stops whenever we have a tree of constant size, in which the predecessor search is trivially supported in constant time. The construction time $C(n)$ follows the equation

$$C(n) = S(n^{1/(c+1)}) + n^{1/(c+1)} \cdot C(n^{c/(c+1)})$$

which gives $C(n) = O(n^{c/(c+1)}) + n^{1/(c+1)} \cdot C(n^{c/(c+1)})$. By iteratively expanding the term $C(n^{c/(c+1)})$, we get $C(n) = O(n)$.

A query is done by traversing the $\log \log n$ levels of the tree and doing predecessor searches at the structure P of each traversed node. The query time follows the equation $Q(n, \ell, w) = q(n^{1/(c+1)}, \ell, w) + Q(n^{c/(c+1)}, \ell, w)$, which solves to $Q(n, \ell, w) = O(\log \log n \cdot q(n, \ell, w))$. In the same paper [4], Andersson and Thorup show how to insert or delete an element in worst-case constant time, once a pointer to its predecessor (or to itself in case of a deletion) has been determined.

Deterministic Dynamic Bounds

When applied to Beame and Fich's solution with time $O\left(\frac{\log \ell}{\log \log \ell}\right)$, the exponential search tree gives $O(n)$ space with time $O\left(\log \log n \cdot \frac{\log \ell}{\log \log \ell}\right)$, and when applied to the solution with $O\left(\sqrt{\frac{\log n}{\log\log n}}\right)$ time and quadratic space, it keeps the same time bound and achieves $O(n)$ space. By combining the two bounds with the bounds of the fusion tree [15], one gets the following time bounds which are the best known ones for dynamic deterministic predecessor search:

$$O\left(1+\min\left\{\begin{array}{c}\frac{\log n}{\log w}\\ \sqrt{\frac{\log n}{\log\log n}}\\ \log\log n\cdot\frac{\log \ell}{\log\log \ell}\end{array}\right\}\right) \quad (1)$$

The space is $O(n)$ for all the three branches. The bounds refer to the maximum of update and search times.

Optimal Static Bounds

Pătraşcu and Thorup [13] obtained optimal lower and upper bounds for the static case. They obtain optimal trade-offs between the time and the space usage. Define $a = \log\frac{S}{n} + \log w$, where S is the space usage. Then, the optimal time bound is

$$\Theta\left(1+\min\left\{\begin{array}{c}\frac{\log n}{\log w}\\ \log\frac{\ell-\log n}{\log w}\\ \frac{\log\frac{\ell}{a}}{\log\left(\frac{a}{\log n}\cdot\log\frac{\ell}{a}\right)}\\ \frac{\log\frac{\ell}{a}}{\log\left(\log\frac{\ell}{a}\Big/\log\frac{\log n}{a}\right)}\end{array}\right\}\right) \quad (2)$$

They later show that their lower bound holds even if randomization is allowed [14]. The first branch of the upper bound corresponds to the fusion tree. The second branch is obtained by first reducing the length from ℓ to $\ell - \log n$ bits, by dividing the universe into n buckets according to their most significant $\log n$ bits, and then by storing a separate Van Emde Boas structure for every bucket. The query time is then reduced from $O(\log(\ell - \log n))$ to $\log\frac{\ell-\log n}{\log w}$, by stopping the Van Emde Boas recursion when the key length gets to $\log w$ bits. Finally, the last two branches are implemented using a refinement of the technique by Beame and Fich.

Optimal Randomized Dynamic Bounds

When allowing randomization, optimal bounds can be achieved [15]. Again, the bounds are for the maximum of query and update times:

$$\Theta\left(1+\min\left\{\begin{array}{c}\frac{\log n}{\log w}\\ \log\left(\frac{\log(2^\ell-n)}{\log w}\right)\\ \frac{\log\frac{\ell}{\log w}}{\log(\log\frac{\ell}{\log w}/\log\frac{\log n}{\log w})}\end{array}\right\}\right) \quad (3)$$

The space usage is linear ($O(n)$) or almost linear $O(nw^{O(1)})$.

The first branch of the upper bound corresponds to the dynamic version of the fusion tree [15] and the third branch to a dynamic version of the fourth branch of the optimal static upper bound. The second branch is similar to the second branch of the optimal static upper bound, with the difference that the term $\ell - \log n$ is replaced by $\log(2^\ell - n)$.

The first and third branches are trivially optimal, since the bounds are the same as the static ones and any lower bound that applies to the latter also applies to the former. The authors show that the second branch is also optimal by proving a corresponding lower bound that is stronger than the static one.

Applications

Range queries are very important in databases. Answering to range queries is an obvious and natural application of predecessor search. A query (in one dimension) asks, given a range $[a, b] \subseteq [1 \ldots 2^\ell]$, to return every element x in the set $S \cap [a, b]$. This can obviously be solved by doing two predecessor queries for a and b and then reporting all elements between the two predecessors (excluding the predecessor of a). In the comparison model and the external memory model, this is the best one can hope for, and the optimal time bounds are $O(\log n + |S \cap [a, b]|)$ and $O(\log_B n +$

P

$|S \cap [a,b]|/B)$ respectively. Surprisingly, in the RAM model, there exists a linear-space static solution with $O(|S \cap [a,b]|)$ query time [3]. An important application of predecessor search is *IP forwarding problem*, which must be solved by every internet router. The router contains a database of subnetworks specified by their IP address prefixes, and each received packet has to be forwarded to the subnetwork with the longest matching prefix. The IP forwarding problem is an instance of the *longest common prefix problem*, which in the static case is equivalent to the predecessor problem.

The lower and upper bounds on predecessor search can be used to prove bounds for other problems through reductions. For example, predecessor search can be reduced to *two-dimensional range search*, allowing one to prove a lower bound of $\Omega(\log\log n)$ time for the two-dimensional range-emptiness problem on sets of n points on a grid of n rows by n columns. Optimal bounds can also be proved for rank queries on sequence representations through reduction to and from predecessor queries [9].

Open Problems

The deterministic complexity of the dynamic predecessor search is still open. Another open problem is whether updates can be supported faster than searches when the search time is optimal or near optimal (of course, one can always support constant update time when the query time is the trivial $\Theta(n)$). For the moment, this is not disallowed by any lower bound and has been achieved for the related *dynamic ranking problem* [10], in which a set $S \subset [1\ldots 2^\ell]$ is maintained under updates and a query asks, given an integer $y \in S$ to count the number of elements of S smaller than y.

Cross-References

- ▶ Monotone Minimal Perfect Hash Functions
- ▶ Orthogonal Range Searching on Discrete Grids
- ▶ Rank and Select Operations on Sequences

Recommended Reading

1. Georgy Adelson-Velsky G, Landis E M (1962) An algorithm for the organization of information. Proc USSR Acad Sci (in Russian) 146:263–266. English translation by Myron J Ricci in Sov Math Dokl 3:1259–1263 (1962)
2. Ajtai M, Fredman M, Komlós J (1984) Hash functions for priority queues. Inf Control 63(3):217–225
3. Alstrup S, Brodal GS, Rauhe T (2001) Optimal static range reporting in one dimension. In: Proceedings of the thirty-third annual ACM symposium on theory of computing, Heraklion. ACM, pp 476–482
4. Andersson A, Thorup M (2007) Dynamic ordered sets with exponential search trees. J ACM (JACM) 54(3):13
5. Andersson A, Hagerup T, Nilsson S, Raman R (1998) Sorting in linear time? J Comput Syst Sci 57:74–93
6. Bayer R (1972) Symmetric binary b-trees: data structure and maintenance algorithms. Acta Inform 1(4):290–306
7. Bayer R, McCreight EM (1972) Organization and maintenance of large ordered indexes. Acta Inform 1:173–189
8. Beame P, Fich FE (2002) Optimal bounds for the predecessor problem and related problems. J Comput Syst Sci 65(1):38–72
9. Belazzougui D, Navarro G (2012) New lower and upper bounds for representing sequences. In: Algorithms–ESA 2012, Ljubljana. Springer, pp 181–192
10. Chan TM, Pătraşcu M (2010) Counting inversions, offline orthogonal range counting, and related problems. In: Proceedings of the twenty-first annual ACM-SIAM symposium on discrete algorithms, Austin. Society for Industrial and Applied Mathematics, pp 161–173
11. Fredman ML, Willard DE (1993) Surpassing the information theoretic bound with fusion trees. J Comput Syst Sci 47(3):424–436
12. Mortensen CW, Pagh R, Pătraşcu M (2005) On dynamic range reporting in one dimension. In: Proceedings of the thirty-seventh annual ACM symposium on theory of computing, Baltimore. ACM, pp 104–111
13. Pătraşcu M, Thorup M (2006) Time-space trade-offs for predecessor search. In: Proceedings of the thirty-eighth annual ACM symposium on theory of computing, San Diego. ACM, pp 232–240
14. Pătraşcu M, Thorup M (2007) Randomization does not help searching predecessors. In: Proceedings of the eighteenth annual ACM-SIAM symposium on discrete algorithms, New Orleans. Society for Industrial and Applied Mathematics, pp 555–564
15. Pătraşcu M, Thorup M (2014) Dynamic integer sets with optimal rank, select, and predecessor search. In: Proceedings of of the 55th annual IEEE symposium on foundations of computer science, Philadelphia. arXiv preprint arxiv:1408.3045

16. Sen P, Venkatesh S (2008) Lower bounds for predecessor searching in the cell probe model. J Comput Syst Sci 74:364–385
17. van Emde Boas P (1975) Preserving order in a forest in less than logarithmic time. In: FOCS, Berkeley, pp 75–84
18. van Emde Boas P (1977) Preserving order in a forest in less than logarithmic time and linear space. Inf Process Lett 6(3):80–82
19. van Emde Boas P, Kaas R, Zijlstra E (1976) Design and implementation of an efficient priority queue. Math Syst Theory 10(1):99–127
20. Willard DE (1983) Log-logarithmic worst-case range queries are possible in space $\Theta(n)$. Inf Process Lett 17(2):81–84

Price of Anarchy

George Christodoulou
University of Liverpool, Liverpool, UK

Keywords

Congestion games; Network games; Price of anarchy

Synonyms

Coordination ratio

Years and Authors of Summarized Original Work

2005; Koutsoupias

Problem Definition

The *price of anarchy* captures the lack of coordination in systems where users are selfish and may have conflicted interests. It was first proposed by Koutsoupias and Papadimitriou in [8], where the term *coordination ratio* was used instead, but later Papadimitriou in [12] coined the term price of anarchy that finally prevailed in the literature.

Roughly, the price of anarchy is the system cost (e.g., makespan, average latency) of the worst-case Nash equilibrium over the optimal system cost that would be achieved if the players were forced to coordinate. Although it was originally defined in order to analyze a simple load-balancing game, it was soon applied to numerous variants and to more general games. The family of *(weighted) congestion games* [11, 13] is a nice abstract form to describe most of the alternative settings. (We focus our presentation on cost minimization problems in congestion games. We mention some utility maximization problems where price of anarchy analysis has been used in the Applications section.)

The price of anarchy may vary, depending on the

- Equilibrium solution concept *(e.g., pure, mixed, correlated equilibria)*
- Characteristics of the congestion game
  - Players Set *(e.g., atomic – non-atomic)*
  - Strategy Set *(e.g., symmetric asymmetric, parallel machines-network-general)*
  - Players' cost functions *(e.g., linear, polynomial)*
- Social cost *(e.g., maximum, sum, total latency)*

Notation

Let G be a (finite) game that is determined by the triple $(N, (\mathcal{S}_i)_{i \in N}, (c_i)_{i \in N})$. $N = \{1, \ldots, n\}$ is the set of the players that participate in the game. $\mathcal{S}_i$ is a *pure strategy set* for player i. An element $A_i \in \mathcal{S}_i$ is a *pure strategy* for player $i \in N$. A *pure strategy profile* $A = (A_1, \ldots, A_n)$ is a vector of pure strategies, one for each player. The set of all possible pure strategy profiles is denoted by $\mathcal{S} = \mathcal{S}_1 \times \cdots \times \mathcal{S}_n$. The *cost* of a player $i \in N$, for a pure strategy, is determined by a cost function $c_i : \mathcal{S} \mapsto \mathbb{R}$.

A pure strategy profile A is a *pure Nash equilibrium*, if none of the players $i \in N$ can benefit, by unilaterally deviating to another pure strategy $s_i \in \mathcal{S}_i$:

$$c_i(A) \le c_i(A_{-i}, s_i) \qquad \forall i \in N, \quad \forall s_i \in \mathcal{S}_i,$$

where (A_{-i}, s_i) is the simple strategy profile that results when just the player i deviates from strategy $A_i \in \mathcal{S}_i$ to strategy $s_i \in \mathcal{S}_i$.

Similarly, a *mixed strategy* p_i for a player $i \in N$ is a probability distribution over her pure strategy set $\mathcal{S}_i$. A mixed strategy profile p is the tuple $p = (p_1, \ldots p_n)$, where player i chooses mixed strategy p_i. The expected cost of a player $i \in N$ with respect to the p is

$$c_i(p) = \sum_{A \in \mathcal{S}} p(A) c_i(A),$$

where $p(A) = \prod_{i \in N} p_i(A_i)$ is the probability that pure strategy A occurs, with respect to $(p_i)_{i \in N}$. A mixed strategy profile p is a *Nash equilibrium*, if and only if

$$c_i(p) \leq c_i(p_{-i}, s_i) \qquad \forall i \in N, \quad \forall s_i \in \mathcal{S}_i$$

The *social cost* of a pure strategy profile A, denoted by $\mathrm{SC}(A)$, is the maximum cost of a player $\mathrm{MAX}(A) = \max_{i \in N} c_i(A)$ or the average cost of a player. For simplicity, the sum of the players' cost is considered (i.e., n times the average cost) $\mathrm{SUM}(A) = \sum_{i \in N} c_i(A)$. The same definitions extend naturally for the case of mixed strategies, but with expected costs in this case.

The (mixed) *price of anarchy* [8] for a game is the worst-case ratio, among all the (mixed) Nash equilibria, of the social cost over the optimal cost, $\mathrm{OPT} = \min_{P \in \mathcal{S}} \mathrm{SC}(P)$.

$$\mathrm{PA} = \max_{p \text{ is N.E.}} \frac{\mathrm{SC}(p)}{\mathrm{OPT}}$$

The price of anarchy for a class of games is the maximum (supremum) price of anarchy among all the games of this class.

Congestion Games Here, a general class of games is described that captures most of the games for which price of anarchy is studied in the literature. A *congestion game [11, 13]*, is defined by the tuple $(N, E, (\mathcal{S}_i)_{i \in N}, (f_e)_{e \in E})$, where $N = \{1, \ldots, n\}$ is a set of players, E is a set of *facilities*, $\mathcal{S}_i \subseteq 2^E$ is the pure strategy set for player i, a pure strategy $A_i \in \mathcal{S}_i$ is a subset of the facility set, and f_e is a *cost (or latency) function* (Unless otherwise stated, linear cost functions are only considered throughout this article. See [14] and references therein for results about more general cost functions, and for additional results see entries 00260, 00251, 00053.) with respect to the facility $e \in E$.

A pure strategy profile $A = (A_1, \ldots, A_n)$ is a vector of pure strategies, one for each player. The cost $c_i(A)$ of player i for the pure strategy profile A is given by

$$c_i(A) = \sum_{e \in A_i} f_e(n_e(A)),$$

where $n_e(A)$ is the number of the players that use facility e in A.

A congestion game is called *symmetric or single commodity*, if all the players have the same strategy set: $\mathcal{S}_i = \mathcal{C}$. The term *asymmetric or multi-commodity* is used to refer to all the games including the symmetric ones. A special class is the class of *network congestion games*. In these games, the facilities are edges of a (multi)graph $G(V, E)$. The pure strategy set for a player $i \in N$ is the simple paths set from a source $s_i \in V$ to a destination $t_i \in V$. In network symmetric congestion games, all the players have the same source and destination.

A natural generalization of congestion games are the *weighted congestion games*, where every player controls an amount of traffic w_i. The cost of each facility $e \in E$ depends on the total load of the facility. In this case, a well-studied social cost function is the weighted sum of players costs, or *total latency*.

In a congestion game with *splittable weights (divisible demands)*, every player $i \in N$, instead of fixing a single pure strategy, is allowed to distribute her demand among her pure strategy set.

Finally, in a *non-atomic congestion game*, there are k different player types $1 \ldots k$. Players are infinitesimal and for each player type i the continuum of the players is denoted by the interval $[0, n_i]$. In general, each player type contributes in a different way to the congestion on the facility $e \in E$, and this contribution is determined by a positive *rate of consumption* $r_{s,e}$ for a strategy $s \in S_i$ and a facility $e \in s$. Each player chooses a strategy that results in a *strategy distribution* $x = (x_s)_{s \in \mathcal{S}}$, with $\sum_{s \in S_i} x_s = n_i$.

Key Results

Maximum Social Cost

First, we review results on the price of anarchy w.r.t maximum social cost that was historically the first social cost considered in [8]. Formally, for a pure strategy profile A, the social cost is

$$\mathrm{SC}(A) = \mathrm{MAX}(A) = \max_{i \in N} c_i(A)$$

The definition naturally extends to mixed strategies.

Theorem 1 ([7–10]) *The price of anarchy for m identical machines is* $\Theta\left(\frac{\log m}{\log \log m}\right)$.

Theorem 2 ([7]) *The price of anarchy for m uniformly related machines with speeds $s_1 \geq s_2 \geq \cdots \geq s_m$ is*

$$\Theta\left(\min\left\{\frac{\log m}{\log\log\log m}, \frac{\log m}{\log\left(\frac{\log m}{\log(s_1/s_m)}\right)}\right\}\right).$$

Theorem 3 ([4]) *The price of anarchy for pure equilibria is $\Theta(\sqrt{n})$ for asymmetric but at most $5/2$ for symmetric congestion games.*

Average Social Cost: Total Latency

Here, we consider as social cost the (weighted) sum (total latency) of the players' cost for (weighted) congestion games, i.e.,

$$\mathrm{SC}(A) = \mathrm{SUM}(A) = \sum_{i \in N} c_i(A),$$

$$\mathrm{SC}(A) = \mathrm{C}(A) = \sum_{i \in N} w_i c_i(A).$$

The definition naturally extends for mixed strategies.

Theorem 4 ([2–4]) *The price of anarchy is $5/2$ for asymmetric and $\frac{5n-2}{2n+1}$ for symmetric congestion games.*

Theorem 5 ([2, 3]) *The price of anarchy for weighted congestion games is $1 + \phi \approx 2.618$.*

Theorem 6 ([6]) *The price of anarchy is at most $3/2$ for congestion games with splittable weights.*

Theorem 7 ([15, 16]) *The price of anarchy for non-atomic congestion games is $4/3$.*

Key Proof Technique: Smoothness Most of the above results on atomic congestion games have been generalized for polynomial latencies [1–3] and hold for various equilibrium concepts. Roughgarden's *smoothness* framework [14] distills the main ideas in these proofs and provides a general, canonical proof recipe to obtain price of anarchy bounds. He also shows how smoothness provides tight bounds for congestion games with general cost functions.

Applications

The efficiency of large-scale networks, in which selfish users interact, is highly affected due to the users' selfish behavior. The price of anarchy is a quantitative measure of the lack of coordination in such systems. It is a useful theoretical tool for the analysis and design of telecommunication and traffic networks, where selfish users compete on system's resources motivated by their atomic interests and are indifferent to the social welfare.

The price of anarchy has been also studied in utility maximization Problems; see, for example, Valid Utility Games [17]. Finally, a line of work shows that price of anarchy can be used to evaluate the performance of mechanisms; see, for example, [5] for an analysis of simultaneous Second-Price Auctions.

Cross-References

▸ Best Response Algorithms for Selfish Routing
▸ Computing Pure Equilibria in the Game of Parallel Links

▸ Price of Anarchy for Machines Models
▸ Selfish Unsplittable Flows: Algorithms for Pure Equilibria

Recommended Reading

1. Aland S, Dumrauf D, Gairing M, Monien B, Schoppmann F (2006) Exact price of anarchy for polynomial congestion games. In: 23rd annual symposium on theoretical aspects of computer science (STACS), Marseille, pp 218–229
2. Awerbuch B, Azar Y, Epstein A (2005) Large the price of routing unsplittable flow. In: Proceedings of the 37th annual ACM symposium on theory of computing (STOC), Baltimore, pp 57–66
3. Christodoulou G, Koutsoupias E (2005) On the price of anarchy and stability of correlated equilibria of linear congestion games. In: Algorithms – ESA 2005, 13th annual European symposium, Palma de Mallorca, pp 59–70
4. Christodoulou G, Koutsoupias E (2005) The price of anarchy of finite congestion games. In: Proceedings of the 37th annual ACM symposium on theory of computing (STOC), Baltimore, pp 67–73
5. Christodoulou G, Kovács A, Schapira M (2008) Bayesian combinatorial auctions. In: ICALP '08: proceedings of the 35th international colloquium on automata, languages and programming, part I, Reykjavik, pp 820–832
6. Cominetti R, Correa JR, Moses NES (2006) Network games with atomic players. In: Automata, languages and programming, 33rd international colloquium (ICALP), Venice, pp 525–536
7. Czumaj A, Vöcking B (2002) Tight bounds for worst-case equilibria. In: Proceedings of the thirteenth annual ACM-SIAM symposium on discrete algorithms (SODA), San Francisco, pp 413–420
8. Koutsoupias E, Papadimitriou CH (1999) Worst-case equilibria. In: Proceedings of the 16th annual symposium on theoretical aspects of computer science (STACS), Trier, pp 404–413
9. Koutsoupias E, Mavronicolas M, Spirakis PG (2003) Approximate equilibria and ball fusion. Theory Comput Syst 36:683–693
10. Mavronicolas M, Spirakis PG (2001) The price of selfish routing. In: Proceedings on 33rd annual ACM symposium on theory of computing (STOC), Hersonissos, pp 510–519
11. Monderer D, Shapley L (1996) Potential games. Games Econ Behav 14:124–143
12. Papadimitriou CH (2001) Algorithms, games, and the internet. In: Proceedings on 33rd annual ACM symposium on theory of computing (STOC), Hersonissos, pp 749–753
13. Rosenthal RW (1973) A class of games possessing pure-strategy Nash equilibria. Int J Game Theory 2:65–67
14. Roughgarden T (2012) Intrinsic robustness of the price of anarchy. Commun ACM 55(7):116–123
15. Roughgarden T, Tardos E (2002) How bad is selfish routing? J ACM 49:236–259
16. Roughgarden T, Tardos E (2004) Bounding the inefficiency of equilibria in nonatomic congestion games. Games Econ Behav 47:389–403
17. Vetta A (2002) Nash equilibria in competitive societies, with applications to facility location, traffic routing and auctions. In: Proceedings of the 43rd annual symposium on foundations of computer science (FOCS), Vancouver, pp 416–425

Price of Anarchy for Machines Models

Artur Czumaj[1] and Berthold Vöcking[2]
[1]Department of Computer Science, Centre for Discrete Mathematics and Its Applications, University of Warwick, Coventry, UK
[2]Department of Computer Science, RWTH Aachen University, Aachen, Germany

Keywords

Algorithmic game theory; Coordination ratio; Nash equilibria; Noncooperative networks; Price of anarchy; Selfish routing; Selfish strategies; Traffic routing

Years and Authors of Summarized Original Work

2002; Czumaj, Vöcking

Problem Definition

This entry considers a selfish routing model formally introduced by Koutsoupias and Papadimitriou [10], in which the goal is to route the traffic on parallel links with linear latency functions. One can describe this model as a scheduling problem with m independent machines with speeds $s_1, \ldots, s_m$ and n independent tasks with weights $w_1, \ldots, w_n$. The goal is to allocate the tasks to the machines to minimize the maximum load of the links in the system.

It is assumed that all tasks are assigned by noncooperative agents. The set of *pure strategies* for task i is the set $\{1,\ldots,m\}$, and a *mixed strategy* is a distribution on this set.

Given a combination $(j_1,\ldots,j_n) \in \{1,\ldots,m\}^n$ of pure strategies, one for each task, the *cost* for task i is $\sum_{j_k=j_i}\frac{w_k}{s_{j_i}}$, which is the time needed for machine j_i chosen by task i to complete all tasks allocated to that machine. Similarly, for a combination of pure strategies $(j_1,\ldots,j_n) \in \{1,\ldots,m\}^n$, the *load* of machine j is defined as $\sum_{j_k=j}\frac{w_k}{s_j}$.

Given n tasks of length $w_1,\ldots,w_n$ and m machines with the speeds $s_1,\ldots,s_m$, let opt denote the *social optimum*, that is, the minimum cost over all combinations of pure strategies:

$$\mathsf{opt} = \min_{(j_1,\ldots,j_n)\in\{1,\ldots,m\}^n} \max_{1\le j\le m} \sum_{i:j_i=j}\frac{w_i}{s_j}.$$

For example, if all machines have the same unit speed ($s_j = 1$ for every j, $1 \le j \le m$) and all tasks have the same unit weight ($w_i = 1$ for every i, $1 \le i \le n$), then the social optimum is $\lceil \frac{n}{m} \rceil$.

It is also easy to see that in any system

$$\mathsf{opt} \ge \frac{\max_i w_i}{\max_j s_j}.$$

It is known that computing the social optimum is $\mathcal{NP}$-hard even for identical speeds (see [10]).

For mixed strategies, let p_i^j denote the probability that an agent i sends the entire traffic w_i to a machine j. Let ℓ_j denote the *expected load* on a machine j, that is,

$$\ell_j = \frac{1}{s_j}\cdot\sum_{i=1}^{n} w_i\, p_i^j.$$

For a task i, the *expected cost of task i on machine j* is equal to

$$c_i^j = \frac{w_i}{s_j} + \sum_{t\ne i}\frac{w_t\, p_t^j}{s_j} = \ell_j + (1-p_i^j)\,\frac{w_i}{s_j}.$$

The expected cost c_i^j corresponds to the expected finish time of task i on machine j under the processor sharing scheduling policy. This is an appropriate cost model with respect to the underlying traffic routing application.

Definition 1 (Nash equilibrium) The probabilities $(p_i^j)_{1\le i\le n, 1\le j\le m}$ define a *Nash equilibrium* if and only if any task i will assign nonzero probabilities only to machines that minimize c_i^j, that is, $p_i^j > 0$ implies $c_i^j \le c_i^q$, for every q, $1 \le q \le m$.

As an example, in the system considered above in which all machines have the same unit speed and all weights are the same, the uniform probabilities $p_i^j = \frac{1}{m}$ for all $1 \le j \le m$ and $1 \le i \le n$ define a system in a Nash equilibrium.

The existence of a Nash equilibrium over mixed strategies for noncooperative games was shown by Nash [13]. In fact, the routing game considered here admits an equilibrium even if all players are restricted to pure strategies, what has been shown by Fotakis et al. [7].

Fix an arbitrary Nash equilibrium, that is, fix the probabilities $(p_i^j)_{1\le i\le n, 1\le j\le m}$ that define a Nash equilibrium. Consider the randomized allocation strategies in which each task i is allocated to a single machine chosen independently at random according to the probabilities p_i^j, that is, task i is allocated to machine j with probability p_i^j. Let C_j, $1 \le j \le m$, be the random variable indicating the *load of machine* j in our random experiment. Observe that C_j is the weighted sum of independent 0–1 random variables J_i^j, $\mathbf{Pr}[J_i^j = 1] = p_i^j$, such that

$$C_j = \frac{1}{s_j}\sum_{i=1}^{n} w_i \cdot J_i^j.$$

Let $\mathfrak{c}$ denote the *maximum expected load* over all machines, that is,

$$\mathfrak{c} = \max_{1\le j\le m} \ell_j.$$

Notice that $\mathbf{E}[C_j] = \ell_j$, and therefore, $\mathfrak{c} = \max_{1\le j\le m}\mathbf{E}[C_j]$.

P

Finally, let the *social cost* C be defined as the expected maximum load (instead of maximum expected load), that is,

$$\mathsf{C} = \mathbf{E}[\max_{1 \le j \le m} C_j].$$

Observe that $c \le C$ and possibly $c \ll C$. The goal is to estimate the *price of anarchy* (also called the *worst-case coordination ratio*) which is the worst-case ratio

$$R = \max \frac{\mathsf{C}}{\mathtt{opt}},$$

where the maximum is over all Nash equilibria.

Key Results

Early Work

The study of the price of anarchy has been initiated by Koutsoupias and Papadimitriou [10], who showed also some very basic results for this model. For example, they proved that for two identical machines, the price of anarchy is exactly $\frac{3}{2}$, and for two machines (with possibly different speeds), the price of anarchy is at least $\phi = \frac{1+\sqrt{5}}{2}$. Koutsoupias and Papadimitriou showed also that for m identical machines, the price of anarchy is $\Omega(\frac{\log m}{\log\log m})$ and it is at most $\mathcal{O}(\sqrt{m \ln m})$, and for m arbitrary machines, the price of anarchy is $\mathcal{O}(\sqrt{\frac{s_1}{s_m} \sum_{j=1}^{m} \frac{s_j}{s_m}} \sqrt{\log m})$, where $s_1 \ge s_2 \ge \cdots \ge s_m$ [10].

Koutsoupias and Papadimitriou [10] conjectured also that the price of anarchy for m *identical machines* is $\Theta(\frac{\log m}{\log\log m})$. In the quest to resolve this conjecture, Mavronicolas and Spirakis [12] considered the problem in the so-called fully mixed model, which is a special class of Nash equilibria in which all p_i^j are strictly positive. In this model, Mavronicolas and Spirakis [12] showed that for m identical machines in the fully mixed Nash equilibrium, the price of anarchy is $\Theta(\frac{\log m}{\log\log m})$. Similarly, they proved also that for m (not necessarily identical) machines and n identical weights in the fully mixed Nash equilibrium, if $m \le n$, then the price of anarchy is $\Theta(\frac{\log n}{\log\log n})$.

The motivation behind studying fully mixed equilibria is the so-called fully mixed Nash equilibrium conjecture stating that these equilibria maximize the price of anarchy because they maximize the randomization. The conjecture seems to be quite appealing as a fully mixed equilibrium can be computed in polynomial time, which led to numerous studies of this kind of equilibria with the hope to obtain efficient algorithms for computing or approximating the price of anarchy with respect to mixed equilibria. However, Fischer and Vöcking [6] disproved the fully mixed Nash equilibrium conjecture and showed that there is a mixed Nash equilibrium whose expected cost is larger than the expected cost of the fully mixed Nash equilibrium by a factor of $\Omega(\frac{\log m}{\log\log m})$. Furthermore, they presented polynomial time algorithms for approximating the price of anarchy for mixed equilibria on identical machines up to a constant factor.

Tight Bounds for the Price of Anarchy

Czumaj and Vöcking [4] entirely resolved the conjecture of Koutsoupias and Papadimitriou [10] and gave an exact description of the price of anarchy as a function of the number of machines and the ratio of the speed of the fastest machine over the speed of the slowest machine. (To simplify the notation, for any real $x \ge 0$, let $\log x$ denote $\log x = \max\{\log_2 x, 1\}$. Also, following standard convention, $\Gamma(N)$ is used to denote the *gamma (factorial) function*, which for any natural N is defined by $\Gamma(N+1) = N!$ and for an arbitrary real $x > 0$ is $\Gamma(x) = \int_0^{\infty} t^{x-1} e^{-t}\, dt$. For the inverse of the gamma function, $\Gamma^{(-1)}(N)$, it is known that $\Gamma^{(-1)}(N) = x$ such that $\lfloor x \rfloor! \le N - 1 \le \lceil x \rceil!$. It is well known that $\Gamma^{(-1)}(N) = \frac{\log N}{\log\log N}(1 + o(1))$.)

Theorem 1 ([4] Upper bound)

The price of anarchy for m machines is bounded from above by

$$\mathcal{O}\left(\min\left\{\frac{\log m}{\log\log\log m}, \frac{\log m}{\log\left(\frac{\log m}{\log(s_1/s_m)}\right)}\right\}\right),$$

where it is assumed that the speeds satisfy $s_1 \ge \cdots \ge s_m$.

In particular, the price of anarchy for m machines is $\mathcal{O}\left(\frac{\log m}{\log\log\log m}\right)$.

The theorem follows directly from the following two results [4]: that the maximum expected load $\mathfrak{c}$ satisfies $\mathfrak{c} = \mathtt{opt} \cdot \Gamma^{(-1)}(m) = \mathtt{opt} \cdot \mathcal{O}\left(\min\left\{\frac{\log m}{\log\log m}, \log\left(\frac{s_1}{s_m}\right)\right\}\right)$ and that the social cost C satisfies $\mathsf{C} = \mathtt{opt} \cdot \mathcal{O}\left(\frac{\log m}{\log\left(\frac{\mathtt{opt}\cdot\log m}{\mathfrak{c}}\right)} + 1\right)$.

If one applied these results to systems in which all agents follow only *pure strategies*, then since then $\ell_j = C_j$ for every j, it holds that $\mathsf{C} = \mathfrak{c}$. This leads to the following result.

Corollary 1 ([4]) *For pure strategies the price of anarchy for m machines is upper bounded by*

$$\mathcal{O}\left(\min\left\{\frac{\log m}{\log\log m}, \log\left(\frac{s_1}{s_m}\right)\right\}\right),$$

where it is assumed that the speeds satisfy $s_1 \geq \cdots \geq s_m$.

Theorem 3 below proves that this corollary gives an asymptotically tight bound for the price of anarchy for pure strategies.

By Theorem 1, in the special case when all *machines are identical*, the price of anarchy is $\mathcal{O}\left(\frac{\log m}{\log\log m}\right)$; this result has been also obtained independently by Koutsoupias et al. [11]. However, in this special case one can get a stronger bound that is tight up to an additive constant.

Theorem 2 ([4]) *For m identical machines the price of anarchy is at most*

$$\Gamma^{(-1)}(m) + \Theta(1) = \frac{\log m}{\log\log m} \cdot (1 + o(1)).$$

One can obtain a lower bound for the price of anarchy for m *identical machines* by considering the system in which $p_i^j = \frac{1}{m}$ for every i, j. Gonnet [9] proved that then the price of anarchy is $\Gamma^{(-1)}(m) - \frac{3}{2} + o(1)$, which implies that Theorem 2 is tight up to an additive constant.

The next theorem shows that the upper bound in Theorem 1 is asymptotically tight.

Theorem 3 ([4] Lower bound) *The price of anarchy for m machines is lower bounded by*

$$\Omega\left(\min\left\{\frac{\log m}{\log\log\log m}, \frac{\log m}{\log\left(\frac{\log m}{\log(s_1/s_m)}\right)}\right\}\right).$$

In particular, the price of anarchy for m machines is $\Omega\left(\frac{\log m}{\log\log\log m}\right)$.

In fact, it can be shown [4] (analogously to the upper bound) that for every positive integer m, positive real r, and $S \geq 1$, there exists a set of m machines with $\frac{s_1}{s_m} = S$ being in a Nash equilibrium and satisfying $\mathtt{opt} = r$, $\mathfrak{c} = \mathtt{opt} \cdot \Omega\left(\min\left\{\frac{\log m}{\log\log m}, \log\left(\frac{s_1}{s_m}\right)\right\}\right)$, and $C = \mathtt{opt} \cdot \Omega\left(\frac{\log m}{\log\left(\frac{\mathtt{opt}\cdot\log m}{c}\right)}\right)$.

Applications

The model discussed here has been extended in the literature in numerous ways, in particular in [1,5,8]; see also survey presentations in [3, 14].

Open Problems

An interesting attempt that adds an algorithmic or constructive element to the analysis of the price of anarchy is made in [2]. The idea behind "coordination mechanisms" is not to study the price of anarchy for a fixed system, but to design the system in such a way that the increase in cost or the loss in performance due to selfish behavior is as small as possible. This is a promising direction of research that might result in practical guidelines of how to build a distributed system that does not suffer from selfish behavior but might even exploit the selfishness of the agents.

Experimental Results

None is reported.

URLs to Code and Data Sets

None is reported.

Cross-References

- ▶ Computing Pure Equilibria in the Game of Parallel Links
- ▶ Price of Anarchy

Recommended Reading

1. Awerbuch B, Azar Y, Richter Y, Tsur D (2006) Tradeoffs in worst-case equilibria. Theor Comput Sci 361(2–3):200–209
2. Christodoulou G, Koutsoupias E, Nanavati A (2009) Coordination mechanisms. Theor Comput Sci 410(36):3327–3336
3. Czumaj A (2004) Selfish routing on the Internet. In: Leung J (ed) Handbook of scheduling: algorithms, models, and performance analysis. CRC, Boca Raton
4. Czumaj A, Vöcking B (2007) Tight bounds for worst-case equilibria. ACM Trans Algorithms 3(1):Article 4
5. Czumaj A, Krysta P, Vöcking B (2010) Selfish traffic allocation for server farms. SIAM J Comput 39(5):1957–1987
6. Fischer S, Vöcking B (2007) On the structure and complexity of worst-case equilibria. Theor Comput Sci 378(2):165–174
7. Fotakis D, Kontogiannis S, Koutsoupias E, Mavronicolas M, Spirakis P (2009) The structure and complexity of Nash equilibria for a selfish routing game. Theor Comput Sci 410(36):3305–3326
8. Gairing M, Lücking T, Mavronicolas M, Monien B (2006) The price of anarchy for polynomial social cost. Theor Comput Sci 369(1–3):116–135
9. Gonnet G (1981) Expected length of the longest probe sequence in hash code searching. J Assoc Comput Mach 28(2):289–304
10. Koutsoupias E, Papadimitriou CH (1999) Worst-case equilibria. In: Proceeding of the 16th annual symposium on theoretical aspects of computer science (STACS), Trier, pp 404–413
11. Koutsoupias E, Mavronicolas M, Spirakis P (2003) Approximate equilibria and ball fusion. Theory Comput Syst 36(6):683–693
12. Mavronicolas M, Spirakis P (2001) The price of selfish routing. In: Proceeding of the 33rd annual ACM symposium on theory of computing (STOC), Heraklion, pp 510–519
13. Nash JF Jr (1951) Non-cooperative games. Ann Math 54(2):286–295
14. Vöcking B (2007) Selfish load balancing. In: Nisan N, Roughgarden T, Tardos É, Vazirani V (eds) Algorithmic game theory. Cambridge University Press, Cambridge

Privacy Preserving Auction

Zhiyi Huang
Department of Computer Science, The University of Hong Kong, Hong Kong, Hong Kong

Keywords

Differential privacy; Mechanism design

Years and Authors of Summarized Original Work

2012; Huang, Kannan
2014; Hsu, Huang, Roth, Roughgarden, Wu

Problem Definition

Let there be n agents and a set of feasible outcomes Ω. For concreteness, readers may think of Ω as the set of allocations of m items to n agents. Each agent has a private value function $v_i : \Omega \mapsto [0, 1]$ over feasible outcomes. We focus on direct revelation mechanisms, which first let each agent i report a value function $\bar{v}_i$, then choose a feasible outcome $\omega \in \Omega$ and a payment p_i for each agent i according to the reported value functions. Let $\omega(\bar{v})$ and $p(\bar{v})$ denote the outcome and payment vector chosen by the mechanism. Note that both $\omega(\bar{v})$ and $p(\bar{v})$ may be random variables.

We hope to achieve the following three objectives:

Objective 1: Maximizing Social Welfare

The social welfare of a feasible outcome $\omega \in \Omega$ is the sum of the agents' values for the outcome, namely, $\sum_{i=1}^{n} v_i(\omega)$. We hope to approximately maximize the expected social welfare of the chosen outcome over the randomness of the mechanism, a widely considered objective in mechanism design.

Approximately maximizing social welfare given the true value functions is a well-studied algorithmic problem (e.g., [4, 16]). In our mechanism design setting, agents may choose not to report the true value if it fits their interests. So the mechanism has an additional challenge of motivating the agents to report truthfully.

Objective 2: Incentive Compatibility

We adopt the standard assumption that each agent i aims to maximize the expectation of his quasi-linear utility, which equals his value for the chosen outcome less his payment. A mechanism is *incentive compatible* if truth telling maximizes an agent's expected utility regardless of the reported values of other agents, that is, for any agent i, any true value v_i, reported value $\bar{v}_i$, and any reported values of other agents $\bar{v}_{-i}$, we have $\mathbb{E}[v_i(\omega(v_i, \bar{v}_{-i})) - p_i(v_i, \bar{v}_{-i})] \geq \mathbb{E}[v_i(\omega(\bar{v}_i, \bar{v}_{-i})) - p_i(\bar{v}_i, \bar{v}_{-i})]$. We also consider a relaxed notion called α*-incentive compatibility*, where an agent's expected utility of truth telling can be worst off by at most an α additive factor comparing to his utility of reporting any alternative value.

There is a vast literature on designing incentive compatible mechanisms with approximately optimal social welfare (see, e.g., [11] for a comprehensive survey). We remark the Vickrey-Clarke-Groves (VCG) mechanism [2, 5, 15], which chooses an outcome that maximizes the social welfare and uses payments to align the interests of the agents and the mechanism designer. When computational efficiency is not of concern, the VCG mechanism gives optimal social welfare and is incentive compatible for arbitrary problems. However, it does not achieve the next objective.

Objective 3: Protecting Agents' Privacy

Our last objective is to protect the agents' privacy by ensuring that the chosen outcome and payments do not reveal too much information about any individual agent' private value function. Agents may care about their privacy for both exogenous and endogenous reasons. On the one hand, privacy is a basic desideratum. On the other hand, violating an agent's privacy could explicitly hurt the agent's utility in the future, e.g., companies may post higher reserve prices based on an agent's past values if such information is revealed by previous mechanisms.

Definition 1 ([3]) A mechanism is (ϵ, δ)-differentially private if for any agent i, any value v_i, alternative value v_i', any values of other agents v_{-i}, and any subset of feasible outcome $S \subseteq \Omega$,

$$\Pr[\omega(v_i, v_{-i}) \in S] \leq e^{\epsilon} \cdot \Pr[\omega(v_i', v_{-i}) \in S] + \delta \ .$$

We remark that the payments may violate the agents' privacy as well. Nevertheless, we can make the prices privacy preserving without changing the agents' expected utilities by adding any scale of noise with expectation zero to the payments. (Having arbitrarily large variance in the payment and, thus, in the utility of an agent is an undesirable property. In some settings, it is possible to privately compute prices without having large variance. Readers are referred to Hsu et al. [6] for details, which we will omit due to space constraint.) So we focus on the privacy property of the outcome in the above definition.

We provide two informal interpretations of differential privacy (for sufficiently small δ). Information theoretically, a mechanism being (ϵ, δ)-differentially private implies that the outcome reveals at most $O(\epsilon^2)$ bits of information about any individual agent's private value. Game theoretically, it implies that truth telling may decrease an agent's future utility by at most a factor of $e^{-\epsilon} \approx 1 - \epsilon$.

For some mechanism design problems, such as auctioning m items to n agents, no (ϵ, δ)-differentially private mechanisms can approximately maximize social welfare. For such resource allocation problems, let ω_{-i} denote the allocation to all agents except agent i. We consider the following relaxed notion of privacy:

Definition 2 ([8]) A mechanism is (ϵ, δ)-jointly differentially private if for any agent i, any value v_i, alternative value v_i', any values of other agents v_{-i}, and any subset of feasible outcome $S \subseteq \Omega$,

$$\Pr[\omega_{-i}(v_i, v_{-i}) \in S] \leq e^{\epsilon} \cdot \Pr[\omega_{-i}(v_i', v_{-i}) \in S] + \delta \ .$$

In settings where each agent can see only his own allocation, a mechanism being (ϵ, δ)-jointly differentially private also implies that it reveals at most $O(\epsilon^2)$ bits of information of an agent's value and that truth telling decreases an agent's future utility by at most $e^{-\epsilon}$, even if the adversary colludes with all other agents.

Related Work

The problem we consider in this article falls into the growing literature on the interface of game theory and differential privacy (see, e.g., Pai and Roth [14] for a survey). McSherry and Talwar [10] proposed using differentially private mechanisms to design auctions by pointing out that differential privacy implies approximate incentive compatibility and resilience to collusion. They also proposed the exponential mechanism, which is an important building block in one of the results we will discuss. Nissim et al. [13] showed how to convert differentially private mechanisms into exactly incentive compatible mechanisms in some settings, but the final mechanisms no longer protect agents' privacy. Xiao [17] proposed mechanisms that are both incentive compatible and differentially private in some special cases. Unfortunately, it does not seem possible to extend the results of Nissim et al. [13] and Xiao [17] to more general problems. Finally, Xiao [17], Chen et al. [1], and Nissim et al. [12] considered modeling the agents' concern for privacy in the utility functions and introduced incentive compatible mechanisms for some special cases in this model. In sum, most previous techniques apply only to special cases. In this article, we summarize two recent techniques for designing privacy-preserving auctions for a large family of mechanism design problems.

Key Results

Almost all mechanism design problems can be classified into two families: social choice problems and resource allocation problems. In a social choice problem, the set of feasible outcome is independent of the number of agents n. In particular, the number of feasible outcome is independent of n. For example, leader elections and choosing a subset of public projects subject a budget constraint fall into this family. In a resource allocation problem, such as allocating m items to n agents, the set of feasible outcome depends on the number of agents. In particular, the number of feasible outcome grows exponentially with n. Below we discuss two techniques by Huang and Kannan [7] and Hsu et al. [6] for designing privacy-preserving auctions for these two families of problems, respectively.

Social Choice Problems

Huang and Kannan [7] proposed a technique for designing incentive compatible and ϵ-differentially private mechanisms for arbitrary mechanism design problems. For social choice problems, in particular, this technique also gives nearly optimal social welfare.

Theorem 1 ([7]) *For any mechanism design problem, there is an incentive compatible and ϵ-differentially private mechanism that gives at least* $\mathrm{OPT} - \frac{2}{\epsilon}(\ln|\Omega| + \ln\frac{1}{\beta})$ *social welfare with probability at least* $1 - \beta$.

This mechanism is based on the exponential mechanism by McSherry and Talwar [10], a general differentially private mechanism that can be applied to a large family of problems. The social welfare guarantee and ϵ-differential privacy in Theorem 1 follow directly from properties of the exponential mechanism. However, the exponential mechanism is not incentive compatible in general. Huang and Kannan [7] noticed that the exponential mechanism can be viewed as maximizing a linear combination of the social welfare and the Shannon entropy of the outcome distribution. Therefore, its allocation rule is equivalent to that of the VCG mechanism in a virtual market where the set of feasible outcomes are distributions over the original outcomes, the set of agents are the original n agents plus an additional agent whose value equals the entropy of the chosen distribution. As a result, using the payments in the virtual market along with the exponential mechanism achieves incentive compatibility.

In social choice problems, $\ln|\Omega|$ is a constant independent of n. So the loss in social welfare is a constant independent of n. On the other hand, the social welfare is the sum of n agents' values, each of which is between 0 and 1. Hence, in a large market with many agents, it is reasonable to expect the optimal social welfare (if not of scale $\Theta(n)$) to be much larger than the additive loss in Theorem 1 in practical instances.

In resource allocation problems, however, $|\Omega|$ grows exponentially in n and, thus, $\ln|\Omega|$ is of scale $\Omega(n)$. For instance, consider matching $m = n$ items to n agents. Then, $|\Omega| = n!$ and $\ln|\Omega| = \Omega(n \ln n)$. Even if the optimal social welfare is of scale $\Theta(n)$, we would need ϵ to be at least $\Omega(\ln n)$ to have nontrivial social welfare guarantee in Theorem 1. This means that the mechanism would reveal $\Omega(\ln^2 n)$ bits of information of an agent's private value, and truth telling may decrease an agent's future utility by a poly(n) factor. Further, this is not only a limitation of the current technique. Huang and Kannan [7] showed that no ϵ-differentially private can give nontrivial social welfare guarantee for $\epsilon = o(\ln n)$, even without incentive compatibility.

Resource Allocation Problems

Given the obstacles for applying differential privacy to resource allocation problems, Hsu et al. [6] looked into a relaxed notion of privacy, namely, joint differential privacy. In particular, they considered matching m items to n agents where each item has a supply of at least s copies and then generalized the results to combinatorial auctions with gross substitute value functions (e.g., [9]). Their first result is a jointly differentially private (yet not incentive compatible) mechanism with nearly optimal social welfare when the supply s is polylogarithmic in n and m. Their main technique is a noisy variant of the deferred-acceptance algorithm by Kelso and Crawford [9].

Theorem 2 ([6]) *For combinatorial auctions with gross substitute valuations, there is an ϵ-jointly differentially private algorithm that gives at least $OPT - \alpha n$ social welfare with probability at least $1 - \beta$ if $s = \Omega(\frac{1}{\epsilon\alpha^3} \text{polylog}(n, m, \frac{1}{\alpha}, \frac{1}{\beta}))$.*

They also showed a supply of $\omega(1)$ is needed for a jointly differentially private mechanism to achieve $o(n)$ additive loss in social welfare. More precisely, they showed:

Theorem 3 ([6]) *No jointly differentially private algorithm can compute matchings with social welfare at least $OPT - \alpha n$ if $s \leq O(\frac{1}{\sqrt{\alpha}})$.*

Their approach can also be used to design approximately incentive compatible and jointly differentially private mechanisms, but the supply needs to be polynomially large.

Theorem 4 (implicit in [6]) *For combinatorial auctions with gross substitute valuations, there is an α-incentive compatible and ϵ-jointly differentially private algorithm that gives at least $OPT - \alpha n$ social welfare with probability at least $1 - \beta$ if $s = \Omega(m)$, where the constant depends on ϵ, δ, α, and β.*

Open Problems

The results of Huang and Kannan [7] and Hsu et al. [6] provided a preliminary step towards designing auctions that protect agents' privacy. There are still many open problems in this topic, some of which we sketch below.

First, the techniques of Hsu et al. [6] fundamentally rely on properties of gross substitute value functions and, thus, cannot be extended to more general value functions. Further, there are many important families of value functions beyond gross substitute, e.g., sub-modular functions, sub-additive functions, etc. So it is natural to seek for techniques that work for more general families of value functions.

Problem 1 Are there jointly differentially private mechanisms that achieve nearly optimal social welfare for arbitrary value functions?

Even if we restrict our attention to gross substitute value functions or even to matching markets, Theorems 2 and 3 leave a large gap between the upper and lower bounds on the supply needed by jointly differentially private mechanisms to get nearly optimal social welfare. Closing this gap would advance our understanding on joint differential privacy.

Problem 2 What is the minimal supply needed so that a jointly differentially private mechanism can achieve nearly optimal social welfare in combinatorial auctions? In particular, is the logarithmic dependency on n and m in Theorem 2 necessary?

Finally, the current technique for achieving both approximate incentive compatibility and joint differential privacy requires a polynomially large supply of items, much larger than the supply needed for achieving joint differential privacy alone. Does approximate incentive compatibility make the problem fundamentally harder? Or is it just a limitation of the current technique?

Problem 3 What is the minimal supply needed so that an approximately incentive compatible and jointly differentially private mechanism can achieve nearly optimal social welfare in combinatorial auctions? In particular, is the polynomial dependency on m in Theorem 4 necessary?

Cross-References

▶ Mechanism Design and Differential Privacy

Recommended Reading

1. Chen Y, Chong S, Kash IA, Moran T, Vadhan S (2013) Truthful mechanisms for agents that value privacy. In: 14th conference on electronic commerce. ACM, New York, pp 215–232
2. Clarke EH (1971) Multipart pricing of public goods. Public Choice 11(1):17–33
3. Dwork C, McSherry F, Nissim K, Smith A (2006) Calibrating noise to sensitivity in private data analysis. In: Theory of cryptography. Springer, Berlin/Heidelberg, pp 265–284
4. Feige U (2009) On maximizing welfare when utility functions are subadditive. SIAM J Comput 39(1):122–142
5. Groves T (1973) Incentives in teams. Econom J Econom Soc 41:617–631
6. Hsu J, Huang Z, Roth A, Roughgarden T, Wu SZ (2014) Private matchings and allocations. In: 46th annual symposium on theory of computing (STOC). ACM, New York
7. Huang Z, Kannan S (2012) The exponential mechanism for social welfare: private, truthful, and nearly optimal. In: 53rd annual symposium on foundations of computer science (FOCS). IEEE, Washington, DC, pp 140–149
8. Kearns M, Pai M, Roth A, Ullman J (2014) Mechanism design in large games: incentives and privacy. In: 5th conference on innovations in theoretical computer science. ACM, New York, pp 403–410
9. Kelso AS, Crawford VP (1982) Job matching, coalition formation, and gross substitutes. Econom J Econom Soc 50:1483–1504
10. McSherry F, Talwar K (2007) Mechanism design via differential privacy. In: 48th annual symposium on foundations of computer science (FOCS). IEEE, pp 94–103
11. Nisan N, Roughgarden T, Tardos E, Vazirani VV (2007) Algorithmic game theory. Cambridge University Press, Cambridge/New York
12. Nissim K, Orlandi C, Smorodinsky R (2012) Privacy-aware mechanism design. In: 13th conference on electronic commerce. ACM, New York, pp 774–789
13. Nissim K, Smorodinsky R, Tennenholtz M (2012) Approximately optimal mechanism design via differential privacy. In: 3rd conference on innovations in theoretical computer science. ACM, New York, pp 203–213
14. Pai MM, Roth A (2013) Privacy and mechanism design. ACM SIGecom Exch 12(1):8–29
15. Vickrey W (1961) Counterspeculation, auctions, and competitive sealed tenders. J Financ 16(1):8–37
16. Vondrák J (2008) Optimal approximation for the submodular welfare problem in the value oracle model. In: 40th annual symposium on theory of computing. ACM, New York, pp 67–74
17. Xiao D (2013) Is privacy compatible with truthfulness? In: 4th conference on innovations in theoretical computer science. ACM, New York, pp 67–86

Private Spectral Analysis

Moritz Hardt
IBM Research – Almaden, San Jose, CA, USA

Keywords

Differential privacy; Power method; Singular value decomposition; Spectral analysis

Years and Authors of Summarized Original Work

2014; Dwork, Talwar, Thakurta, Zhang
2014; Hardt, Price

Problem Definition

Spectral analysis refers to a family of popular and effective methods that analyze an input matrix by exploiting information about its eigenvectors or singular vectors. Applications include principal component analysis, low-rank approximation, and spectral clustering. Many of these applications are commonly performed on data sets that feature sensitive information such as patient records in a medical study. In such cases privacy is a major concern. Differential privacy is a powerful general-purpose privacy definition. This entry explains how differential privacy may be applied to task of approximately computing the top singular vectors of a matrix.

Generally speaking, the input is a real-valued matrix $A \in \mathbb{R}^{m \times n}$ and a parameter $k \in \mathbb{N}$. We think of the input matrix as specifying n attributes for m individuals. The goal of the algorithm is to approximately compute the first $k \leq \min\{m, n\}$ singular vectors of A while achieving differential privacy. There are several notions of approximation as well as several variants of differential privacy that make sense in this context.

Approximation Guarantee

Let $A = U\Sigma V^\top$ denote the singular value decomposition of A with singular values $\sigma_1 \geq \sigma_2 \geq \cdots \geq \sigma_{\max\{m,n\}} \geq 0$. Further, let U_k and V_k represent the first k columns of U and V, respectively. In other words, U_k consists of the first k left singular vectors of A and V_k consists of the first k right singular vectors.

Principal Angle

Principal angles are a useful tool for comparing the distance between subspaces. The sine of the largest principal angle between subspaces X, Y of equal dimension represented by orthonormal matrices is defined as $\sin\theta(X, Y) = \left\|(I - XX^\top)Y\right\|_2$, where the norm refers to the spectral norm (or ℓ_2-operator norm).

A natural objective is to require that the algorithm $\mathcal{M}$ outputs an orthonormal matrix $X \in \mathbb{R}^{m \times k}$ so as to minimize $\left\|(I - XX^\top)V_k\right\|$. We call this the *principal angle objective*. The angle is of course 0 when $X = V_k$. We will also be interested in the case where the rank of X is larger than the rank of V_k. Note that our objective is still well defined.

Expressed Variance

Another natural objective is to output an orthonormal matrix $X \in \mathbb{R}^{m \times k}$ so as to maximize the variance captured by the subspace spanned by the columns of X. A convenient way to express this objective is to maximize the quantity $\|AX\|_F^2$, where the norm refers to the Frobenius norm. It is not difficult to show that this objective is maximized for $X = V_k$. Again, the objective is still well defined when the rank of X is larger than that of X.

Privacy Guarantee

Differential privacy requires the definition of a *neighborhood relation* on matrices, denoted $A \sim A'$. Pairs of matrices in this relation are called *neighboring*. Differential privacy requires that the algorithm maps neighboring databases to nearly indistinguishable output distributions.

Definition 1 Given a neighborhood relation $\sim$, we say that a randomized algorithm $\mathcal{M}$ satisfies (ϵ, δ)-differential privacy if for all neighboring matrices $A \sim A'$ and for every measurable set S in the output space of the algorithm, we have that

$$\Pr\{\mathcal{M}(A) \in S\} \leq \exp(\epsilon) \Pr\{\mathcal{M}(A') \in S\} + \delta. \tag{1}$$

Neighborhood Relations. Typically in differential privacy, the neighborhood relation is chosen to be the set of all pairs of matrices that differ in at most one row. Unfortunately, this definition is unattainable in the spectral setting as the privacy definition is sensitive to the scale of the row vector that's being changed. Indeed, if we replace a single row of the matrix by a vector Δu of norm Δ, then as we let Δ tend to infinity, the top right singular vector of the matrix will tend to the vector u (say, in angle).

To circumvent this problem, we will generally specify a norm bound in each neighbor relation. It is important to note that the strength of privacy definition then depends on the scaling of the matrix.

- We say that A, A' are *entry neighbors* if they differ by at most 1 in absolute value.
- We say that A, A' are *row neighbors* if they differ in at most one row by a vector whose Euclidean norm is bounded by 1.

All entry neighbors are of course also row neighbors so that the privacy definition based on row neighbors is stronger than that of entry neighbors. It is sometimes natural to scale the matrix such that either all entries have magnitude at most 1 or all rows have Euclidean norm at most 1. While this may strengthen the privacy guarantee, it also leads to a corresponding deterioration in the utility guarantee of the algorithm as the signal-to-noise ratio decreases. It is tempting to nonuniformly scale each row by a different factor. However, this can dramatically change the singular vector decomposition and does not in general lead to an easily interpretable guarantee.

Key Results

We describe two simple and effective methods that lead to nearly optimal approximation guarantees in various settings we introduced above. The first algorithm is based on the well-known power method. The other uses a simple Gaussian noise addition step (Fig. 1).

Noisy Power Method

For simplicity we describe the algorithm in the case where A is a symmetric $n \times n$ matrix. The algorithm extends straightforwardly to rectangular and asymmetric matrices as explained in [5]. We first state a general-purpose analysis of PPM.

Theorem 1 ([3]) *Let $k \leq p$. Then, the private power method satisfies (ϵ, δ)-differential privacy under the entry neighbor relation, and after $L = O(\frac{\sigma_k}{\sigma_k - \sigma_{k+1}} \log(n))$ iterations, we have with probability 9/10 that*

$$\|(I - X_L X_L^{\top})V_k\| \leq$$

$$O\Big(\frac{\sigma \max_{\ell=0}^{L} \|X_\ell\|_\infty \sqrt{n \log L}}{\sigma_k - \sigma_{k+1}} \cdot \frac{\sqrt{p}}{\sqrt{p} - \sqrt{k-1}}\Big).$$

When $p = k + \Omega(k)$, the trailing factor becomes a constant. If $p = k$, it creates a factor k overhead. In the worst case we can always bound $\|X_\ell\|_\infty$ by 1 since X_ℓ is an orthonormal basis. However, in principle, we could hope that a much better bound holds provided that the target basis V_k has small coordinates. Hardt and Roth [4, 5] suggested a way to accomplish a stronger bound by considering a notion of *coherence* of A, denoted as $\mu(A)$. The coherence parameter varies between 1 and n but is often sublinear in n. Intuitively, the coherence measures the correlation between the singular vectors of the matrix with the standard basis. Low coherence means that the singular vectors have small coordinates in the standard basis. Many results on matrix completion and robust PCA crucially rely on such an assumption though the exact notion is somewhat different here. Specifically, if $A = V \Sigma V^{\top}$ is a singular vector decomposition of A, we define $\mu(A) \stackrel{\text{def}}{=} n \max_{i,j \in [n} |V_{ij}|^2$.

Theorem 2 ([3]) *Under the assumptions of Theorem 3, we have the conclusion*

Input: Symmetric $A \in \mathbb{R}^{n \times n}$, L, p, privacy parameters $\varepsilon, \delta > 0$
1. Let X_0 be a random orthonormal basis and put $\sigma = \varepsilon^{-1}\sqrt{4pL\log(1/\delta)}$
2. For $\ell = 1$ to L:
   a) $Y_\ell \leftarrow AX_{\ell-1} + G_\ell$ where $G_\ell \sim \mathrm{N}(0, \|X_{\ell-1}\|_\infty^2 \sigma^2)^{n \times p}$.
   b) Compute the QR-factorization $Y_\ell = X_\ell R_\ell$

Output: Matrix X_L

Private Spectral Analysis, Fig. 1 Private power method. Here $\|X\|_\infty = \max_{ij} |X_{ij}|$

Input: Matrix $A \in \mathbb{R}^{m \times n}$, privacy parameters $\varepsilon, \delta > 0$, parameter $p \in \mathbb{N}$
1. Let E be a symmetric matrix where the upper triangle (including the diagonal) is sampled i.i.d. from $N(0, \sigma^2)$ where $\sigma = \sqrt{2\ln(1.25/\delta)}/\varepsilon$
2. $C \leftarrow A^\top A + E$

Output: Top p singular vectors $X \in \mathbb{R}^{n \times p}$ of $\widetilde{C}$

Private Spectral Analysis, Fig. 2 Gaussian mechanism

$$\|(I - X_L X_L^\top)V_k\| \leq O\left(\frac{\sigma\sqrt{\mu(A)\log n \log L}}{\sigma_k - \sigma_{k+1}} \cdot \frac{\sqrt{p}}{\sqrt{p} - \sqrt{k-1}}\right).$$

Gaussian Mechanism

The Gaussian mechanism first appeared in [1] and was recently revisited [2]. The algorithm simply computes the covariance matrix of the data set and adds suitably scaled (symmetric) Gaussian noise to the covariance matrix. The result is differentially private, and the top singular vectors of the perturbed covariance matrix serve as an approximation of the true singular vectors (Fig. 2).

Theorem 3 ([2]) *Let $k \leq p$. .Then, the Gaussian mechanism satisfies (ϵ, δ)-differential privacy under the row neighbor relation, and with probability $1 - o(1)$, we have $\|AX\|_F^2 \geq \|AV_k\|_F^2 - O\left(\sigma k\sqrt{n}\right)$. Moreover, with probability $1 - o(1)$,*

$$\|AX\|_F^2 \geq \|AV_k\|_F^2 - O\left(\frac{\sigma^2 pn}{\sigma_k^2 - \sigma_{p+1}^2}\right).$$

Applications

Principal Component Analysis

In principal component analysis, the goal is to compute the top k singular vectors of the $n \times n$ matrix $A^\top A$. Recall that we identified data points with row vectors in A. The singular vectors of $A^\top A$ are identical to the right singular vectors of A. Hence, both algorithms we previously discussed immediately solve this problem.

Low-Rank Approximation

In low-rank approximation, the goal is to output a matrix B of rank k such that $\|A - B\|_\nu$ is small, where $\nu \in \{2, F\}$. For either norm, the optimal solution is given by the truncated singular value decomposition $B = U_k \Sigma_k V_k^\top$. In the context of privacy-preserving spectral analysis, a good approximation $\tilde{V}_k$ to V_k typically leads to a good low-rank approximation by performing a privacy-preserving multiplication step $\tilde{\Sigma}_k \tilde{U}_k = A\tilde{V}_k + N$, where N is suitably chosen noise matrix. See, for example, [5].

Open Problems

1. Is it possible to obtain an expressed variance guarantee for the noisy power method? For instance, can we match the bounds achieved by Gaussian noise addition via the power method? The problem with the Gaussian mechanism is that it computes the matrix $A^\top A$ which is impractical when n is large but A may be sparse. In this case, the power method is computationally far more efficient.
2. Can we weaken the incoherence assumption in Theorem 2?
3. Theorem 3 depends on the separation between σ_k and σ_{k+1} even when $p > k$. Is it possible to replace the dependence on $\sigma_k - \sigma_{k+1}$ with a dependence on $\sigma_k - \sigma_{p+1}$?

Recommended Reading

1. Blum A, Dwork C, McSherry F, Nissim K (2005) Practical privacy: the SuLQ framework. In: Proceedings of the 24th PODS. ACM, pp 128–138
2. Dwork C, Talwar K, Thakurta A, Zhang L (2014) Analyze gauss: optimal bounds for privacy-preserving

P

principal component analysis. In: Proceedings of the 46th symposium on theory of computing (STOC). ACM, pp 11–20
3. Hardt M, Price E (2014) The noisy power method: A meta algorithm with applications. CoRR abs/1311.2495v2. http://arxiv.org/abs/1311.2495
4. Hardt M, Roth A (2012) Beating randomized response on incoherent matrices. In: Proceedings of the 44th symposium on theory of computing (STOC). ACM, pp 1255–1268
5. Hardt M, Roth A (2013) Beyond worst-case analysis in private singular vector computation. In: Proceedings of the 45th Symposium on Theory of Computing (STOC). ACM

Probabilistic Data Forwarding in Wireless Sensor Networks

Sotiris Nikoletseas
Computer Engineering and Informatics, University of Patras, Patras, Greece
Computer Technology Institute and Press "Diophantus", Patras, Greece

Keywords

Data propagation; Routing

Years and Authors of Summarized Original Work

2004; Chatzigiannakis, Dimitriou, Nikoletseas, Spirakis

Problem Definition

An important problem in wireless sensor networks is that of *local detection and propagation*, i.e., the local sensing of a crucial event and the energy and time efficient propagation of data reporting its realization to a control center (for a graphical presentation, see Fig. 1). This center (called the *"sink"*) could be some human authorities responsible of taking action upon the realization of the crucial event. More formally:

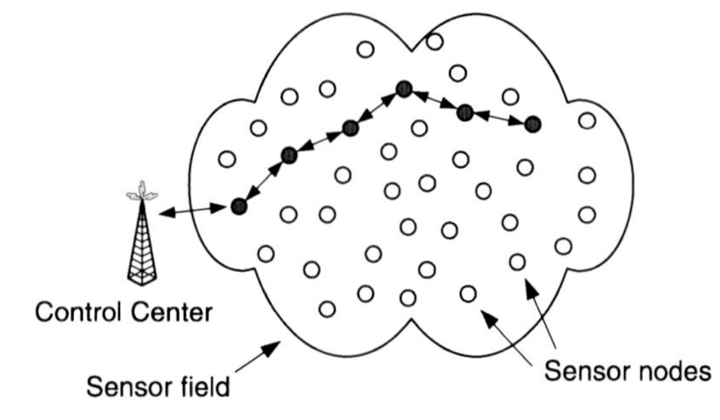

Probabilistic Data Forwarding in Wireless Sensor Networks, Fig. 1 A sensor network

Definition 1 Assume that a single sensor, E, senses the realization of a *local event* $\mathcal{E}$. Then the *propagation problem* is the following: "How can sensor P, via cooperation with the rest of the sensors in the network, efficiently propagate information reporting the realization of the event to the sink S?"

Note that this problem is in fact closely related to the more general problem of data propagation in sensor networks.

Wireless Sensor Networks

Recent dramatic developments in micro-electro-mechanical systems (MEMS), wireless communications and digital electronics have led to the development of small in size, low-power, low-cost sensor devices. Such extremely small (soon in the cubic millimetre scale) devices integrate sensing, data processing and wireless communication capabilities. Examining each such device individually might appear to have small utility, however the effective *distributed self-organization* of large numbers of such devices into an ad-hoc network may lead to the efficient accomplishment of large sensing tasks. Their wide range of applications is based on the use of various sensor types (i.e., thermal, visual, seismic, acoustic, radar, magnetic, etc.) to monitor a wide variety of conditions (e.g., temperature, object presence and movement, humidity, pressure, noise levels etc.). For a survey on wireless sensor networks see [1] and also [6, 9].

A Simple Model

Sensor networks are comprised of a vast number of ultra-small homogeneous sensors, which are called *"grain" particles*. Each grain particle is a fully-autonomous computing and communication device, characterized mainly by its available power supply (battery) and the energy cost of computation and transmission of data. Such particles (in the model here) do not move. Each particle is equipped with a set of monitors (sensors) for light, pressure, humidity, temperature etc. and has a *broadcast* (digital radio) *beacon mode*.

It is assumed that grain particles are *randomly deployed* in a given area of interest. Such a placement may occur e.g., when throwing sensors from an airplane over an area. A special case is considered, when the network being a lattice (or grid) deployment of sensors. This grid placement of grain particles is motivated by certain applications, where it is possible to have a pre-deployed sensor network, where sensors are put (possibly by a human or a robot) in a way that they form a *2-dimensional lattice*.

It is assumed that each particle has the following abilities: (i) It can estimate the direction of a received transmission (e.g., via the technology of direction-sensing antennae). (ii) It can estimate the distance from a nearby particle that did the transmission (e.g., via estimation of the attenuation of the received signal). (iii) It knows the direction towards the sink S. This can be implemented during a set-up phase, where the (powerful) sink broadcasts the information about itself to all particles. (iv) All particles have a common co-ordinates system. Notice that GPS information is not assumed. Also, there is no need to know the global structure of the network.

Key Results

The Basic Idea

For the above problem [3] proposes a protocol which tries to minimize energy consumption by *probabilistically favoring certain paths of local data transmissions towards the sink*. Thus this protocol is called PFR (Probabilistic Forwarding Protocol). Its basic idea is to *avoid flooding* by favoring (in a probabilistic manner) data propagation along sensors which lie "close" to the (optimal) transmission line, ES, that connects the sensor node detecting the event, E, and the sink, S. This is implemented by locally calculating the angle $\phi = (\widehat{EPS})$, whose corner point P is the sensor currently running the local protocol, having received a transmission from a nearby sensor, previously possessing the event information (see Fig. 2). If φ is equal or greater to a predetermined threshold, then p will transmit (and thus propagate the information further). Else, it decides whether to transmit with probability equal to $\frac{\phi}{\pi}$. Because of the probabilistic nature of data propagation decisions and to prevent the propagation process from early failing, the protocol initially uses (for a short time period which is evaluated) a flooding mechanism that leads to a sufficiently large *"front"* of sensors possessing the data under propagation. When such a "front" is created, probabilistic Forwarding is performed.

The PFR Protocol

The protocol evolves in two phases:

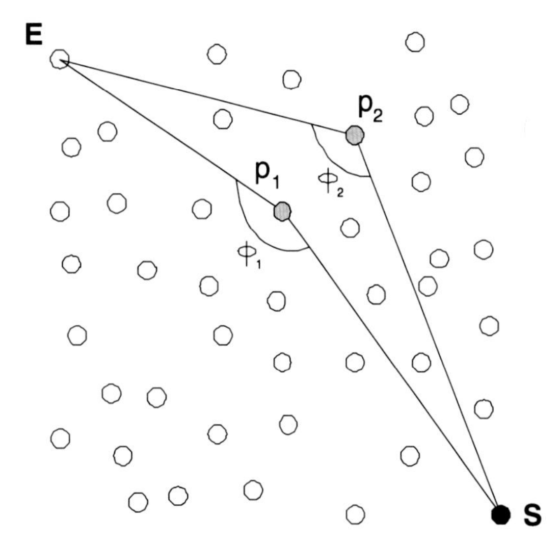

Probabilistic Data Forwarding in Wireless Sensor Networks, Fig. 2 Angle φ and proximity to the optimal line

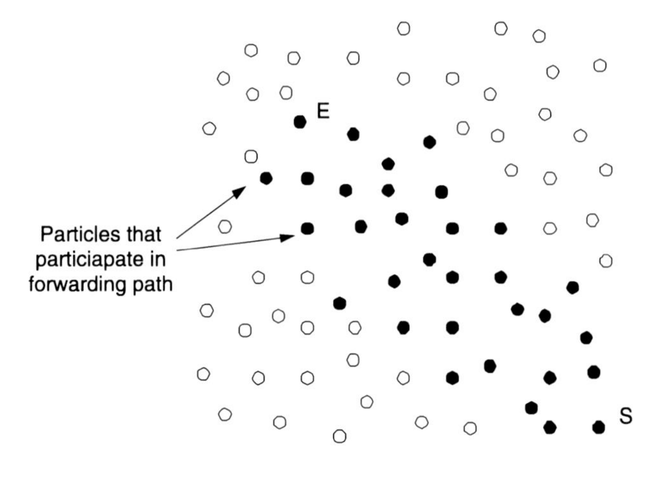

Probabilistic Data Forwarding in Wireless Sensor Networks, Fig. 3 Thin zone of particles

Phase 1: The "Front" Creation Phase

Initially the protocol builds (by using a limited, in terms of rounds, flooding) a sufficiently large "front" of particles, to guarantee the survivability of the data propagation process. During this phase, each particle having received the data to be propagated, deterministically forwards them towards the sink.

Phase 2: The Probabilistic Forwarding Phase

Each particle P possessing the information under propagation (called $info(\mathcal{E})$ hereafter), calculates an angle φ by calling the subprotocol "φ-calculation" (see description below) and broadcasts $info(\mathcal{E})$ to all its neighbors with probability $\mathbb{P}_{fwd}$ (or it does not propagate any data with probability $1 - \mathbb{P}_{fwd}$) as follows:

$$\mathbb{P}_{fwd} = \begin{cases} 1 & \text{if } \phi \geq \phi_{\text{threshold}} \\ \frac{\phi}{\pi} & \text{otherwise} \end{cases}$$

where φ is the $(\widehat{EPS})$ angle and $\phi_{\text{threshold}} = 134°$ (the selection reasons of this value are discussed in [3]).

If the density of particles is appropriately large, then for a line *ES* there is (with high probability) a sequence of points "closely surrounding *ES*" whose angles φ are larger than $\phi_{\text{threshold}}$ and so that successive points are within transmission range. All such points broadcast and thus essentially they follow the line *ES* (see Fig. 3).

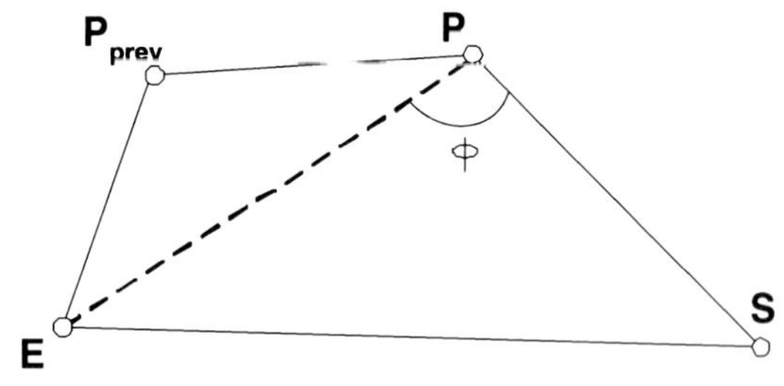

Probabilistic Data Forwarding in Wireless Sensor Networks, Fig. 4 Angle φ calculation example

The φ-calculation Subprotocol (see Fig. 4)

Let P_{prev} the particle that transmitted $info(E)$ to P.

1. When P_{prev} broadcasts $info(E)$, it also attaches the info $|EP_{\text{prev}}|$ and the direction $\overrightarrow{P_{\text{prev}}E}$.
2. P estimates the direction and length of line segment $P_{\text{prev}}P$, as described in the model.
3. P now computes angle $(\widehat{EP_{\text{prev}}P})$, and computes $|EP|$ and the direction of $\overrightarrow{PE}$ (this will be used in further transmission from P).
4. P also computes angle $(\widehat{P_{\text{prev}}PE})$ and by subtracting it from $(\widehat{P_{\text{prev}}PS})$ it finds φ.

Performance Properties of PFR

Any protocol Π solving the data propagation problem must satisfy the following three properties: **(a) Correctness.** Π must guarantee that data arrives to the position S, given that the whole network exists and is operational. **(b) Robustness.**

Π must guarantee that data arrives at enough points in a small interval around S, in cases where part of the network has become inoperative. **(c) Efficiency**. If Π activates k particles during its operation then Π should have a small ratio of the number of activated over the total number of particles $r = \frac{k}{N}$. Thus r is an energy efficiency measure of Π. It is shown that this is indeed the case for PFR.

Consider a partition of the network area into small squares of a fictitious grid G (see Fig. 5). When particle density is high enough, occupancy arguments guarantee that with very high probability (tending to 1) all squares get particles. All the analysis is conditioned on this event, call it F, of at least one particle in each square. Below only sketches of proofs are provided (full proofs can be found in [3]).

The Correctness of PFR

Consider any square Σ intersecting the ES line. By the occupancy argument above, there is w.h.p. a particle in this square. Clearly, the worst case is when the particle is located in one of the corners of Σ (since the two corners located most far away from the ES line have the smallest φ-angle among all positions in Σ). By geometric calculations, [3] proves that the angle φ of this particle is $\phi > 134°$. But the initial square (i.e., that containing E) always broadcasts and any intermediate intersecting square will be notified(by induction) and thus broadcast because of the argument above. Thus the sink will be reached if the whole network is operational:

Lemma 1 ([3]) *PFR succeeds with probability 1 given the event F.*

The Energy Efficiency of PFR

Consider a "lattice-shaped" network like the one in Fig. 5 (all results will hold for any random deployment "in the limit"). The analysis of the energy efficiency considers particles that are active but are as far as possible from ES. [3] estimates an upper bound on the number of particles in an $n \times n$ (i.e., $N = n \times n$) lattice. If k is this number then $r = \frac{k}{n^2}$ $(0 < r \leq 1)$ is the "energy efficiency ratio" of PFR. More specifically, in [3] the authors prove the (very satisfactory) result below. They consider the area around the ES line, whose particles participate in the propagation process. The number of active particles is thus, roughly speaking, captured by the size of this area, which in turn is equal to $|ES|$ times the maximum distance from $|ES|$. This maximum distance is clearly a random variable. To calculate the expectation and variance of this variable, the authors in [3] basically "upper bound" the stochastic process of the distance from ES by a random walk on the line, and subsequently "upper bound" this random walk by a well-known stochastic process (i.e., the "discouraged arrivals"

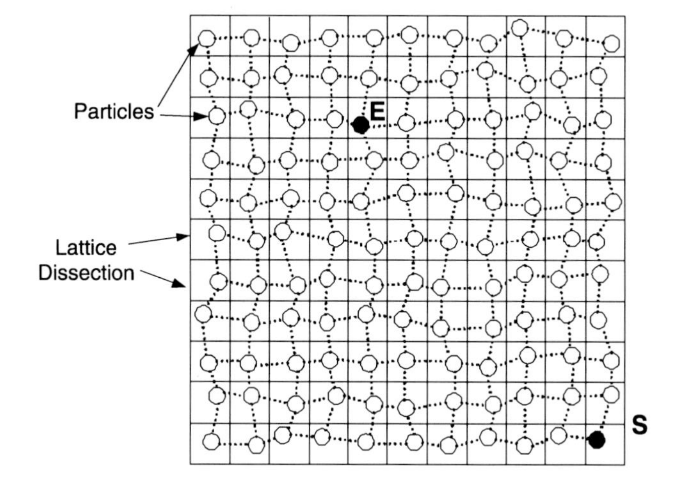

Probabilistic Data Forwarding in Wireless Sensor Networks, Fig. 5 A lattice dissection G

birth and death Markovian process. Thus they prove:

Theorem 2 ([3]) *The energy efficiency of the PFR protocol is* $\Theta\left(\left(\frac{n_0}{n}\right)^2\right)$ *where* $n_0 = |ES|$ *and* $n = \sqrt{N}$, *where N is the number of particles in the network. For* $n_0 = |ES| = o(n)$, *this is* $o(1)$.

The Robustness of PFR

Consider particles "very near" to the *ES* line. Clearly, such particles have large φ-angles (i.e., $\phi > 134°$). Thus, even in the case that some of these particles are not operating, the probability that none of those operating transmits (during phase 2) is very small. Thus:

Lemma 3 ([3]) PFR *manages to propagate the crucial data across lines parallel to ES, and of constant distance, with fixed nonzero probability (not depending on* n, $|ES|$*).*

Applications

Sensor networks can be used for continuous sensing, event detection, location sensing as well as micro-sensing. Hence, sensor networks have several important applications, including (a) security (like biological and chemical attack detection), (b) environmental applications (such as fire detection, flood detection, precision agriculture), (c) health applications (like telemonitoring of human physiological data) and (d) home applications (e.g., smart environments and home automation). Also, sensor networks can be combined with other wireless networks (like mobile) or fixed topology infrastructures (like the Internet) to provide transparent wireless extensions in global computing scenaria.

Open Problems

It would be interesting to come up with formal models for sensor networks, especially with respect to energy aspects; in this respect, [10] models energy dissipation using stochastic methods. Also, it is important to investigate fundamental trade-offs, such as those between energy and time. Furthermore, the presence of mobility and/or multiple sinks (highly motivated by applications) creates new challenges (see e.g., [2, 11]). Finally, heterogeneity aspects (e.g., having sensors of various types and/or combinations of sensor networks with other types of networks like p2p, mobile and the Internet) are very important; in this respect see e.g., [5, 13].

Experimental Results

An implementation of the PFR protocol along with a detailed comparative evaluation (using simulation) with greedy forwarding protocols can be found in [4]; with clustering protocols (like LEACH, [7]) in [12]; with tree maintenance approaches (like Directed Diffusion, [8]) in [5]. Several performance measures are evaluated, like the success rate, the latency and the energy dissipation. The simulations mainly suggest that PFR behaves best in sparse networks of high dynamics.

Cross-References

- ▸ Communication in Ad Hoc Mobile Networks Using Random Walks
- ▸ Obstacle Avoidance Algorithms in Wireless Sensor Networks
- ▸ Randomized Energy Balance Algorithms in Sensor Networks

Recommended Reading

1. Akyildiz IF, Su W, Sankarasubramaniam Y, Cayirci E (2002) Wireless sensor networks: a survey. J Comput Netw 38:393–422
2. Chatzigiannakis I, Kinalis A, Nikoletseas S (2006) Sink mobility protocols for data collection in wireless sensor networks. In: Proceedings of the 4th ACM/IEEE international workshop on mobility management and wireless access protocols (MobiWac). ACM, pp 52–59
3. Chatzigiannakis I, Dimitriou T, Nikoletseas S, Spirakis P (2004) A probabilistic algorithm for efficient

and robust data propagation in smart dust networks. In: Proceedings of the 5th European wireless conference on mobile and wireless systems (EW 2004), pp 344–350. Also in: Ad-Hoc Netw J 4(5):621–635 (2006)
4. Chatzigiannakis I, Dimitriou T, Mavronicolas M, Nikoletseas S, Spirakis P (2003) A comparative study of protocols for efficient data propagation in smart dust networks. In: Proceedings of the 9th European symposium on parallel processing (EuroPar), distinguished paper. Lecture notes in computer science, vol 2790. Springer, pp 1003–1016. Also in the Parall Process Lett (PPL) J 13(4):615–627 (2003)
5. Chatzigiannakis I, Kinalis A, Nikoletseas S (2005) An adaptive power conservation scheme for heterogeneous wireless sensors. In: Proceedings of the 17th annual ACM symposium on parallelism in algorithms and architectures (SPAA 2005). ACM, pp 96–105. Also in: Theory Comput Syst (TOCS) J 42(1):42–72 (2008)
6. Estrin D, Govindan R, Heidemann J, Kumar S (1999) Next century challenges: scalable coordination in sensor networks. In: Proceedings of the 5th ACM/IEEE international conference on mobile computing (MOBICOM)
7. Heinzelman WR, Chandrakasan A, Balakrishnan H (2000) Energy-efficient communication protocol for wireless microsensor networks. In: Proceedings of the 33rd Hawaii international conference on system sciences (HICSS)
8. Intanagonwiwat C, Govindan R, Estrin D (2000) Directed diffusion: a scalable and robust communication paradigm for sensor networks. In: Proceedings of the 6th ACM/IEEE international conference on mobile computing (MOBICOM)
9. Kahn JM, Katz RH, Pister KSJ (1999) Next century challenges: mobile networking for smart dust. In: Proceedings of the 5th ACM/IEEE international conference on mobile computing, pp 271–278
10. Leone P, Rolim J, Nikoletseas S (2005) An adaptive blind algorithm for energy balanced data propagation in wireless sensor networks. In: Proceedings of the IEEE international conference on distributed computing in sensor networks (DCOSS). Lecture notes in computer science (LNCS), vol 3267. Springer, pp 35–48
11. Luo J, Hubaux J-P (2005) Joint mobility and routing for lifetime elongation in wireless networks. In: Proceedings of the 24th INFOCOM
12. Nikoletseas S, Chatzigiannakis I, Antoniou A, Efthymiou C, Kinalis A, Mylonas G (2004) Energy efficient protocols for sensing multiple events in smart dust networks. In: Proceedings of the 37th annual ACM/IEEE simulation symposium (ANSS'04). IEEE Computer Society Press, pp 15–24
13. Triantafillou P, Ntarmos N, Nikoletseas S, Spirakis P (2003) NanoPeer networks and P2P worlds. In: Proceedings of the 3rd IEEE international conference on peer-to-peer computing (P2P 2003), pp 40–46

Probe Selection

Wen Xu[4], Weili Wu[1,2,4], and Ding-Zhu Du[3,4]
[1]College of Computer Science and Technology, Taiyuan University of Technology, Taiyuan, Shanxi Province, China
[2]Department of Computer Science, California State University, Los Angeles, CA, USA
[3]Computer Science, University of Minnesota, Minneapolis, MN, USA
[4]Department of Computer Science, The University of Texas at Dallas, Richardson, TX, USA

Keywords

Approximation algorithm; Group testing; Probe selection; Virus detection

Years and Authors of Summarized Original Work

2006; Du, Hwang
2007; Wang, Du, Jia, Deng, Wu, MacCallum

Problem Definition

The virus identification is an important research topic in molecular biology. One method is using probes. A probe is a short oligonucleotide of size 8–25, which plays a role of ID when identify a virus in a biological sample through hybridization. If each probe hybridizes to a unique virus, then identification of virus is straightforward. However, unique probes are very hard to be obtained, especially for virus subtypes which are closely related. Therefore, how to identify virus with the minimum number of nonunique probes becomes an interesting problem.

Given a biological sample and a set of possibly nonunique probes, how to select a minimum subset of probes to identify viruses in the biological sample. This problem is called the *nonunique probe selection.*

Key Results

Suppose the biological sample contains only one virus. The problem is to determine what is this virus. To do so, it is sufficient to select probes satisfying the condition that different viruses hybridize different subsets of probes. This condition enables us to find the virus easily from a test outcome.

In general, suppose the biological sample contains at most d viruses. Then selected probes should satisfy the condition that different sets of at most d viruses should hybridize different subsets of selected probes. Schilep, Torney, and Rahman [9] first pointed out that this is actually a nongroup testing group testing problem [3].

Consider each virus as an item and each probe as a pool consisting of all viruses hybridized by the probe. A nonadaptive group testing with n items and t pools can be represented, and $t \times n$ binary matrix with rows labeled by pools and columns labeled by items and cell (i, j) contains 1-entry if and only if the ith pool contains item j. This binary matrix is called the incidence matrix of the nonadaptive group testing. In theory of nonadaptive group testing, the above condition means that the incidence matrix is $\bar{d}$-separable. Actually, a binary matrix is $\bar{d}$-separable if all Boolean sums of at most d columns are distinct. Here, by Boolean sum, we mean the following: If each column is seen as a set of rows corresponding to 1-entries in the column, then the Boolean sum can be seen as a union of columns. The Boolean sum is a classic statement in the study of group testing. With a $\bar{d}$-separable matrix, the test outcome can identify up to d viruses in biological sample.

In nonadaptive group testing, each test is on a pool. Thus, each probe can also be seen as a test. The test outcome is *positive* if the probe is hybridized by some virus in a biological sample and *negative* otherwise. Test outcomes for all selected probes can be written as a column vector which is exactly the union of columns corresponding viruses contained in the biological sample, where 1-entry denotes a positive outcome and 0-entry denotes a negative outcome. Therefore, the definition of $\bar{d}$-separable matrix means that different sets of at most d viruses receive different test-outcome t-dimensional vectors.

The nonunique probe selection problem can also be formulated as follows:

> MIN-$\bar{d}$-SS (Minimum $\bar{d}$-Separable Submatrix). Given a binary matrix M, find the minimum of rows to form a $\bar{d}$-separable submatrix.

For any fixed d, MIN-$\bar{d}$-SS is NP-hard [3]. Moreover, from the test outcome obtained from $\bar{d}$-separable, it may take time $O(n^t)$ to find all existing viruses. This means that it is hard to decode the test outcome from a $\bar{d}$-separable matrix [3]. Therefore, Thai et al. [10] considered to use a d-disjunct matrix instead of $\bar{d}$-separable matrix. A binary matrix is *d-disjunct* if any union of d columns cannot contain the $(d + 1)$th column. Decoding test outcome from a d-disjunct matrix is very easy [3]. This introduces another minimization problem:

> MIN-d-DS (Minimum d-Disjunct Submatrix). Given a d-disjunct binary matrix M, find a minimum subset of rows to form d-disjunct submatrix.

Theoretically, there is another similar problem as follows:

> MIN-d-SS (Minimum d-Separable Submatrix). Given a d-separable binary matrix M, find a minimum subset of rows to form d-separable submatrix where a binary matrix is d-separable if all Boolean sums of exactly d columns are distinct.

For $d = 1$, MIN-d-SS is exactly the minimum test cover problem [5], also called the minimum test set problem [2] or the minimum test collection [6], which has a greedy approximation with performance $1 + 2 \ln n$ where n is the number of items [2]. This fact makes a suggestion that design greedy approximations for MIN-d-SS, MIN-$\bar{d}$-SS, and MIN-d-DS.

In fact, it is easy to construct greedy approximations with performance ratio $1 + 2d \ln n$ for MIN-d-SS, $1 + (d + 1) \ln n$ for MIN-d-DS, and $1 + 2d \ln(n + 1)$ for MIN-$\bar{d}$-SS. For example, let us study MIN-d-DS. Consider the collection S of all possible pairs (C, D) of one column

C and a submatrix D with d columns. Clearly $|S| < n^{d+1}$. A row is said to *cover* such a pair (C, D) if and only if at this row, the entry of column C is 1 and all entries of columns in D are 0. Now, MIN-d-DS is equivalent to the problem of finding the minimum number of rows covering all such pairs. This is a special case of the set cover problem. It is well known that there is a greedy algorithm for the set cover problem with performance ratio $1 + \ln |S| < 1 + (d + 1) \ln n$.

This greedy algorithm works well only for small d because its running time is $O(n^{d+1})$. When d is large, it runs too slow. Therefore, we must look for other smart ways. Schilep, Torney, and Rahman [9] proposed an algorithm which adds probe one by one until the incidence matrix with considered viruses forms a $\bar{d}$-separable matrix. This does not work for large d, neither. In fact, if d is not bounded, then testing whether a binary matrix is d-separable, or $\bar{d}$-separable, or d-disjunct is co-NP-complete [3]. There exist more methods [8] in the literature, which work well for small d. However, no efficient method has been found to produce good solutions for larger d.

<?pag ?>In some applications, the pool size cannot be too big due to the sensitivity of tests. For example, UNH suggested in ADS testing, each pool should not contain more than five blood samples. When the pool size is bounded, the problem becomes easier. For instance, let us consider the case that every pool has size at most 2 so that all pools of size 2 together with items form a graph G where pools are edges and item are vertices. Halldórsson et al. [6] and De Bontridder et al. [2] proved that in this case, MIN-1-SS is still APX-hard, which means that there is no polynomial-time approximation scheme for it unless NP=P. They also showed that MIN-1-SS in this case has a polynomial-time approximation with performance ratio $7/6 + \varepsilon$ for any fixed $\varepsilon > 0$.

A surprising result was showed by Wang et al. [11] that a subgraph H of G represents a d-disjunct matrix if and only if every vertex in H has degree at least $d + 1$, and hence, finding such an H with minimum number of edges is polynomial-time solvable. What about the case that all pools have size 3 Wang et al. proved that in this case MIN-d-DS is still NP-hard. However, there exist polynomial-time approximations with better performance.

Applications

In practice, we may select nonunique probes in the following steps [9]:

Step 1. Estimate an upper bound d for the number of viruses existing in a given biological sample. Collect a large set of nonunique probes to form a $\bar{d}$-separable matrix.

Step 2. From this large set of probes, find a subset of probes to identify up to d viruses by computing an approximation solution for MIN-d-DS or MIN-$\bar{d}$-SS.

Step 3. Decode the presence or absence of viruses in the given biological sample from test outcome.

Open Problems

When d is not fixed, MIN-d-DS belongs to Σ_2^p and is conjectured to be Σ_2^p-complete [3].

Recommended Reading

1. Berman P, Dasgupta B, Kao M-Y (2005) Tight approximability results for test set problems in bioinformatics. J Comput Syst Sci 71:145–162
2. De Bontridder KMJ, Halldórsson BV, Halldórsson MM, Hurkens CAJ, Lenstra JK, Ravi R, Stougie L (2003) Approximation algorithms for the test cover problem. Math Program 98:477–491
3. Du D-Z, Hwang FK (2006) Pooling designs and nonadaptive group testing. World Scientific, New Jersey
4. Du D-Z, Ko K-I (2000) Theory of computational complexity. Wiley, New York
5. Garey MR, Johnson DS (1979) Computers and intractability. W.H. Freeman, San Francisco
6. Halldórsson BV, Halldórsson MM, Ravi R (2001) On the approximability of the minimum test collection problem. Lect Notes Comput Sci 2161:158–169
7. Karp RM, Stougton R, Yeung KY (1999) Algorithms for choosing differential gene expression experiments. In: Proceedings of the third annual international conference on computational molecular biology, Lyon, pp 208–217

8. Klau G, Rahmann S, Schliep A, Vingron M, Reinert K (2004) Optimal robust non-unique probe selection using integer linear programming. Bioinformatics 20:I186–I193
9. Schliep A, Torney DC, Rahmann S (2003) Group testing with DNA chips: generating designs and decoding experiments. In: Proceedings of the 2nd IEEE computer society bioinformatics conference, Stanford
10. Thai M, Deng P, Wu W, Znati T (2006) Efficient algorithms for non-unique probes selection using d-disjunct matrix. Muscript
11. Wang F, Du H, Jia X, Deng P, Wu W, MacCallum D (2007) Non-unique probe selection and group testing. Theor Comput Sci 381:29–32<?pag ?>

Prophet Inequality and Online Auctions

Mohammad Taghi Hajiaghayi and Vahid Liaghat
Department of Computer Science, University of Maryland, College Park, MD, USA

Keywords

Auction design; Online algorithms; Online auctions; Prophet inequality

Years and Authors of Summarized Original Work

2007; Hajiaghayi, Kleinberg, Sandholm
2012; Kleinberg, Weinberg
2012; Alaei, Hajiaghayi, Liaghat
2015; Esfandiari, Hajiaghayi, Liaghat, Monemizadeh

Problem Definition

The topic of prophet inequality has been studied in optimal stopping theory since the 1970s [7, 9, 10] and more recently in computer science [1, 3, 6, 8]. In the prophet inequality setting, given (not necessary identical) independent distributions $D_1, \ldots, D_n$, a sequence of random variables $x_1, \ldots, x_n$ where x_i is drawn from D_i, a collection M of *feasible* subsets of $\{1, \ldots, n\}$, an onlooker has to choose from the succession of these values, where x_i is revealed to us at time step i. The onlooker starts with an empty set $S = \phi$. Upon the arrival of a value x_i, the onlooker can choose to either add x_i to the set S or discard it permanently. After the arrival of all values, the indices of values in S should form a feasible set in M. The revenue of the onlooker is the total value of variables in S. The onlooker's goal is to maximize his/her (expected) revenue compared to the hindsight expected revenue of a prophet who knows the drawn values in advance. The optimal offline solution (the prophet's revenue) is defined as $OPT = \mathrm{E}\left[\max_{I \in M} \sum_{i \in I} x_i\right]$. The *competitive ratio* of an algorithm for the onlooker is defined as the worst-case ratio of the expected revenue of the onlooker over OPT. This inequality ratio has been interpreted as meaning that a prophet with complete foresight has only a bounded advantage over an onlooker who observes the variables one by one, and this explains the name prophet inequality.

Different Variants

The *basic prophet inequality* discovered by Krengel, Sucheston, and Garling in the 1970s concerns the case in which the onlooker can only choose one value [9], i.e., M is the set of singletons. Decades later in 2007, Hajiaghayi, Kleinberg, and Sandholm [6] considered the k*-choice prophet secretary* variant in which sets with at most k elements are feasible. Later in 2012, Kleinberg and Weinberg [8] considered the more general *matroid prophet inequality*. In this variant the collection M contains the independent sets of a matroid.

Other prophet inequality settings (dependent D_i's, restricted prophets, etc.) have been considered in the literature as well. For an overview of these models, we refer the reader to [6, 8] and references therein.

Key Results

Krengel, Sucheston, and Garling [9] were first to consider basic prophet inequality. Using a very simple example, they showed no online algorithm can have a competitive ratio better than $\frac{1}{2}$: let

$q = \frac{1}{\epsilon}$. The first value, i.e., x_1 is always 1. The second value is either q with probability ϵ or 0 with probability $1 - \epsilon$. Observe that the expected revenue of any (randomized) online algorithm is at most $\max\left\{1, \epsilon\left(\frac{1}{\epsilon}\right)\right\} = 1$. However the prophet, i.e., the optimum offline solution, would choose x_2 if $x_2 = q$; otherwise he would choose the first value. Thus the optimum offline revenue is $(1-\epsilon) \times 1 + \epsilon\left(\frac{1}{\epsilon}\right) \approx 2$. We note that without considering stochastic assumptions, we cannot hope to get any constant competitive ratio.

An algorithm for the basic prophet inequality problem can be described by setting a threshold for every step: we stop at the first step that the arriving value is higher than the threshold of that step. The classical prophet inequality result [9] states that by choosing the same threshold $OPT/2$ for every step, one achieves the tight competitive ratio of $1/2$.

For the k-choice variant, Hajiaghayi et al. [6] show an algorithm with the competitive ratio $1 - O\left(\frac{\sqrt{\ln k}}{\sqrt{k}}\right)$. Later Alaei [1] improved this bound to $1 - \frac{1}{\sqrt{k+3}}$ using an involved randomized approach (gamma-conservative magician). Alaei, Hajiaghayi, and Liaghat simplified and generalized these results to the matching prophet inequality [2, 3]. Later they generalized their result to the *online stochastic generalized assignment problem* [4] (GAP) with slightly worse competitive ratio of $1 - \frac{1}{\sqrt{k}}$. In GAP, we have a set of items to be placed in a set of bins. The bins are known in advance, but the sequence of items arrives online; each item has a value and a size; upon arrival, an item can be placed in one of the bins or can be discarded permanently; the objective is to maximize the total value of the placement. Both value and size of an item may depend on the bin in which the item is placed; the size of an item is revealed only after it has been placed in a bin; distribution information is available about the value and size of each item in advance (not necessarily i.i.d.); however, items arrive in adversarial order (nonadaptive adversary). Alaei et al. [4] show an algorithm with the competitive ratio of $1 - \frac{1}{\sqrt{k}}$ where in this setting k is interpreted as the minimum number of items that can fill up the capacity of a bin.

Kleinberg and Weinberg [8] considered the matroid prophet inequality. They show an elegant algorithm that still achieves the competitive ratio of $1/2$. Generalizing their result still further, they show that under an intersection of p matroid constraints, the prophet's revenue exceeds the onlooker's by a factor of at most $O(p)$, and this factor is also tight. Kleinberg and Weinberg design the following algorithm for the matroid prophet inequality. The algorithm pretends that the online selection process is Phase 1 of a two-phase game; after each x_i has been revealed in Phase 1 and the algorithm has accepted some set A_1, Phase 2 begins. In Phase 2, a new weight will be sampled for every matroid element, independently of the Phase 1 weights, and the algorithm will play the role of the prophet on the Phase 2 weights, choosing the max-weight subset A_2 such that $A_1 \cup A_2$ is independent. However, the revenue for choosing an element in Phase 2 is only half of its value. When observing element i and deciding whether to select it, our algorithm can be interpreted as making the choice that would maximize its expected revenue if Phase 1 were to end immediately after making this decision and Phase 2 were to begin. Of course, Phase 2 is purely fictional: it never actually takes place, but it plays a key role in both the design and the analysis of the algorithm. The analysis of the algorithm is involved and relies on a careful analysis of the expected revenue at each step. For further intuitions about the analysis, we refer the reader to [8].

Applications

Beyond their interest as theorems about pure online algorithms or optimal stopping rules, prophet inequalities also have applications to mechanism design. Mechanism design has traditionally focused on the offline setting where all agents are present up front. However, many electronic commerce applications do not fit that model because the agents can arrive and depart dynamically. This is characteristic, for example, of online ticket auctions, search keyword auctions, Internet auctions, and scheduling computing jobs on a cloud. The online aspect is characteristic of some important traditional applications as well, such as the sale of a house, where the buyers arrive and depart dynamically.

The pioneer work of Hajiaghayi, Kleinberg, and Sandholm [6] initiated the research on the relationship between algorithmic mechanism design and prophet inequalities. They observed that algorithms used in the derivation of prophet inequalities, owing to their monotonicity properties, could be interpreted as (temporarily) truthful online auction mechanisms and that the prophet inequality in turn could be interpreted as the mechanism's approximation guarantee. Indeed, Bayesian optimal mechanism design problems provide a compelling application of prophet inequalities in economics. In such a Bayesian market, we have a set of n agents with private types sampled from (not necessary identical) known distributions. Upon receiving the reported types, a seller has to allocate resources and charge prices to the agents. The goal is to maximize the seller's revenue in equilibrium. Chawla et al. [5] pioneered the study of the approximability of a special class of such mechanisms, *sequential posted pricing* (SPM): the seller makes a sequence of take-it-or-leave-it offers to agents, offering an item for a specific price. They show although simple, SPMs approximate the optimal revenue in many different settings. Therefore prophet inequalities directly translate to approximation factors for the seller's revenue in these settings through standard machineries. Indeed one can analyze the so-called *virtual values* of winning bids introduced by Roger Myerson [11], to prove via prophet inequalities that the expected virtual value obtained by the SPM mechanism approximates an offline optimum that is with respect to the exact types. Chawla et al. [5] provide a type of prophet inequality in which one can choose the ordering of agents. As mentioned before, Kleinberg and Weinberg [8] later improved their result by giving an algorithm with the tight competitive ratio of 0.5 for an adversarial ordering.

Cross-Reference

▸ Algorithmic Mechanism Design

Recommended Reading

1. Alaei S (2011) Bayesian combinatorial auctions: expanding single buyer mechanisms to many buyers. In: FOCS, Palm Springs
2. Alaei S, Hajiaghayi MT, Liaghat V, Pei D, Saha B (2011) Adcell: ad allocation in cellular networks. In: ESA, Saarbrücken
3. Alaei S, Hajiaghayi M, Liaghat V (2012) Online prophet-inequality matching with applications to ad allocation. In: EC, Valencia, pp 18–35
4. Alaei S, Hajiaghayi M, Liaghat V (2013) The online stochastic generalized assignment problem. In: APPROX, Berkeley
5. Chawla S, Hartline JD, Malec DL, Sivan B (2010) Multi-parameter mechanism design and sequential posted pricing. In: STOC, Cambridge
6. Hajiaghayi MT, Kleinberg RD, Sandholm T (2007) Automated online mechanism design and prophet inequalities. In: AAAI, Vancouver
7. Kennedy DP (1987) Prophet-type inequalities for multi-choice optimal stopping. Stoch Proc Appl 24:77–88
8. Kleinberg R, Weinberg SM (2012) Matroid prophet inequalities. In: STOC, New York, pp 123–136
9. Krengel U, Sucheston L (1977) Semiamarts and finite values. Bull Am Math Soc 83:745–747
10. Krengel U, Sucheston L (1978) On semiamarts, amarts, and processes with finite value. In Kuelbs J (ed) Probability on banach spaces. M.L. Dekker, New York
11. Myerson RB (1981) Optimal auction design. Math Oper Res 6:58–73